Factoring and Special Case Products

$a^2 + 2ab + b^2 = (a + b)^2$

$a^2 - 2ab + b^2 = (a - b)^2$

$a^2 - b^2 = (a + b)(a - b)$

$a^3 + b^3 = (a + b)(a^2 - ab + b^2)$

$a^3 - b^3 = (a - b)(a^2 + ab + b^2)$

Complex Numbers

$i = \sqrt{-1}$ and $i^2 = -1$

For a real number $b > 0$, $\sqrt{-b} = i\sqrt{b}$

The complex numbers $a + bi$ and $a - bi$ are conjugates, and $(a + bi)(a - bi) = a^2 + b^2$.

Quadratic Formula

Given $ax^2 + bx + c = 0$, $a \neq 0$, the solutions are

$$x = \frac{-b \pm \sqrt{b^2 - 4ac}}{2a}$$

Absolute Value Equations/Inequalities

If $k \geq 0$, then

$|u| = k$ is equivalent to $u = k$ or $u = -k$.

$|u| = |w|$ is equivalent to $u = w$ or $u = -w$.

$|u| < k$ is equivalent to $-k < u < k$.

$|u| > k$ is equivalent to $u < -k$ or $u > k$.

Distance Formulas

The distance between two points a and b on a number line is given by $|a - b|$ or $|b - a|$.

The distance between (x_1, y_1) and (x_2, y_2) is

$$d = \sqrt{(x_2 - x_1)^2 + (y_2 - y_1)^2}$$

Midpoint Formula

The midpoint of the line segment between (x_1, y_1) and (x_2, y_2) is

$$M = \left(\frac{x_1 + x_2}{2}, \frac{y_1 + y_2}{2}\right)$$

Linear and Quadratic Functions

$f(x) = b$ constant function: slope 0 and y-intercept $(0, b)$

$f(x) = mx + b$ linear function: slope m and y-intercept $(0, b)$

$f(x) = ax^2 + bx + c$ $(a \neq 0)$ quadratic function: vertex $\left(\frac{-b}{2a}, f\left(\frac{-b}{2a}\right)\right)$

Slope and Average Rate of Change

Slope of a line through (x_1, y_1) and (x_2, y_2):

$$m = \frac{\Delta y}{\Delta x} = \frac{y_2 - y_1}{x_2 - x_1}$$

Average rate of change of $f(x)$ between (x_1, y_1) and (x_2, y_2): $\dfrac{f(x_2) - f(x_1)}{x_2 - x_1}$

Difference quotient: $\dfrac{f(x + h) - f(x)}{h}$

Circle

$(x - h)^2 + (y - k)^2 = r^2$ where $r > 0$

Parabola

Vertical Axis of Symmetry
$(x - h)^2 = 4p(y - k)$

Horizontal Axis of Symmetry
$(y - k)^2 = 4p(x - h)$

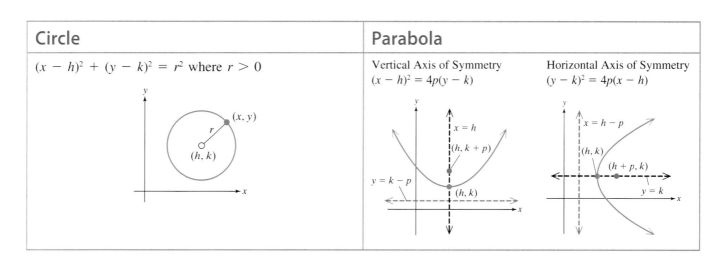

Special Binder-Ready Version

- This loose-leaf alternative will save you money

- Offers a flexible format

- Nonrefundable if shrink-wrap is removed

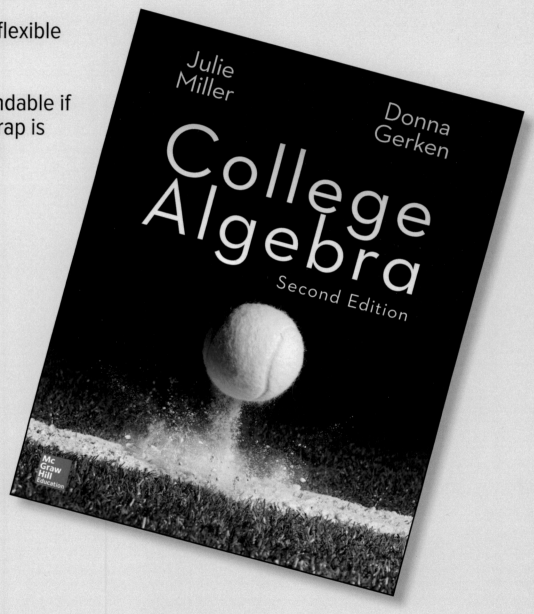

Julie Miller

Donna Gerken

College Algebra

Second Edition

Graphs of Basic Functions

Constant Function	Linear Function	Identity Function	Quadratic Function
$f(x) = b$	$f(x) = mx + b$	$f(x) = x$	$f(x) = x^2$

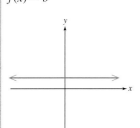 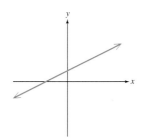 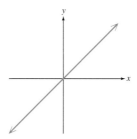 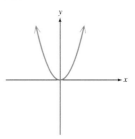

Cubic Function	Absolute Value Function	Square Root Function	Cube Root Function
$f(x) = x^3$	$f(x) = |x|$	$f(x) = \sqrt{x}$	$f(x) = \sqrt[3]{x}$

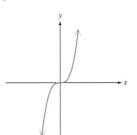 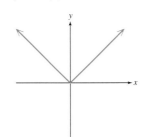 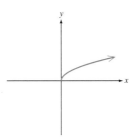 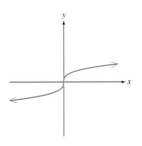

Reciprocal Function	Greatest Integer Function	Exponential Function	Logarithmic Function
$f(x) = \dfrac{1}{x}$	$f(x) = [\![x]\!]$	$f(x) = b^x$, where $b > 0$ and $b \neq 1$	$f(x) = \log_b x$, where $b > 0$ and $b \neq 1$
			$y = \log_b x \Leftrightarrow b^y = x$

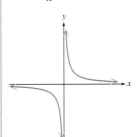 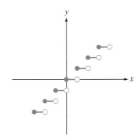 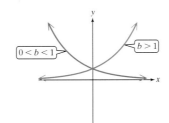 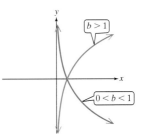

Ellipse

Major Axis Horizontal	Major Axis Vertical
$\dfrac{(x - h)^2}{a^2} + \dfrac{(y - k)^2}{b^2} = 1$	$\dfrac{(x - h)^2}{b^2} + \dfrac{(y - k)^2}{a^2} = 1$

$a > b$, and $c^2 = a^2 - b^2$, where $c > 0$.

Foci are c units from the center on the major axis.

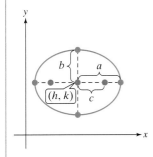 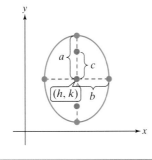

Hyperbola

Transverse Axis Horizontal	Transverse Axis Vertical
$\dfrac{(x - h)^2}{a^2} - \dfrac{(y - k)^2}{b^2} = 1$	$\dfrac{(y - k)^2}{a^2} - \dfrac{(x - h)^2}{b^2} = 1$

$c^2 = a^2 + b^2$, where $c > 0$.

Foci are c units from the center on the transverse axis.

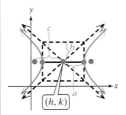 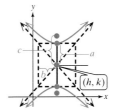

College Algebra

Second Edition

Julie Miller
Daytona State College

Donna Gerken
Miami-Dade College

Mc
Graw
Hill
Education

COLLEGE ALGEBRA, SECOND EDITION

10 BRP 22

ISBN 978–0–07–783634–4
MHID 0–07–783634–0

ISBN 978–1–259–57046–9 (Annotated Instructor's Edition)
MHID 1–259–57046–0

Senior Vice President, Products & Markets: *Kurt L. Strand*
Vice President, General Manager, Products & Markets: *Marty Lange*
Vice President, Content Design & Delivery: *Kimberly Meriwether David*
Managing Director: *Ryan Blankenship*
Senior Brand Manager: *Caroline Celano*
Director, Product Development: *Rose Koos*
Senior Product Developer: *Emily Windelborn*
Marketing Communications Manager: *Audrey Otto*
Marketing Specialist: *Cherie Harshman*
Director of Digital Content Development: *Robert Brieler*
Digital Product Analyst: *Michael Lemke*
Associate Digital Product Analyst: *Adam Fischer*
Director, Content Design & Delivery: *Linda Avenarius*
Program Manager: *Lora Neyens*
Content Project Manager: *Peggy J. Selle*
Assessment Content Project Manager: *Eric Dosmann*
Buyer: *Jennifer Pickel*
Design: *David W. Hash*
Content Licensing Specialist (Image): *Carrie Burger*
Cover Image, Photo Image: *Antonio Iacobelli/Getty Images*
Compositor: *Aptara®, Inc.*
Typeface: *10.5/12 pt. Times LT Std.*
Printer: *BR Printers*

Library of Congress Cataloging-in-Publication Data

Miller, Julie, 1962-
 College algebra / Julie Miller, Daytona State College, Donna Gerken, Miami Dade College. – Second edition.
   pages cm
 Includes index.
 ISBN 978–0–07–783634–4 (alk. paper) — ISBN 0–07–783634–0 (alk. paper) — ISBN 978–1–259–57046–9 (alk. paper) —
   ISBN 1–259–57046–0 (alk. paper) 1. Algebra. I. Gerken, Donna. II. Title.
 QA154.3.M54 2017
 512.9–dc23

                                                                                                    2015027054

About the Authors

Julie Miller is from Daytona State College where she has taught developmental and upper-level mathematics courses for 20 years. Prior to her work at DSC, she worked as a software engineer for General Electric in the area of flight and radar simulation. Julie earned a bachelor of science in applied mathematics from Union College in Schenectady, New York, and a master of science in mathematics from the University of Florida. In addition to this textbook, she has authored textbooks in developmental mathematics, trigonometry, and precalculus, as well as several short works of fiction and nonfiction for young readers.

"My father is a medical researcher, and I got hooked on math and science when I was young and would visit his laboratory. I remember doing simple calculations with him and using graph paper to plot data points for his experiments. He would then tell me what the peaks and features in the graph meant in the context of his experiment. I think that applications and hands-on experience made math come alive for me, and I'd like to see math come alive for my students."

Donna Gerken is a professor at Miami Dade College where she has taught developmental courses, honors classes, and upper-level mathematics classes for decades. Throughout her career she has been actively involved with many projects at Miami Dade including those on computer learning, curriculum design, and the use of technology in the classroom. Donna's bachelor of science in mathematics and master of science in mathematics are both from the University of Miami.

Letter from the Authors

For many students, college algebra serves as a gateway course to the higher levels of mathematics needed for a variety of careers. Our goal is to offer every student an opportunity for success in college algebra by bringing together a seamless integration of print and digital content delivery. The clear, concise writing style and pedagogical features of our textbook continue throughout the online content in ConnectMath, in our instructional videos, and in the adaptive reading and learning experience of SmartBook.

The main objectives of this college algebra textbook and our digital content are threefold:

- To provide students with a clear and logical presentation of fundamental concepts that will prepare them for continued study in mathematics.
- To help students develop logical thinking and problem-solving skills that will benefit them in all aspects of life.
- To motivate students by demonstrating the significance of mathematics in their lives through practical applications.

Julie Miller   julie.miller.math@gmail.com
donna gerken   donna.gerken.math@gmail.com

Dedications *To my parents Kent and Joanne Miller who have always taught me the value of education.* —Julie Miller

For all the students who keep apologizing for asking too many questions. Keep doing that, and I'll keep listening. —Donna Gerken

Table of Contents

CHAPTER 3   Polynomial and Rational Functions   285

CHAPTER 4   Exponential and Logarithmic Functions   401

CHAPTER 5   Systems of Equations and Inequalities   491

Key Features

Clear, Precise Writing

Because a diverse group of students take this course, Julie Miller has written this manuscript to use simple and accessible language. Through her friendly and engaging writing style, students are able to understand the material easily.

Exercise Sets

The exercises at the end of each section are graded, varied, and carefully organized to maximize student learning:

- **Prerequisite Review Exercises** begin the section-level exercises and ensure that students have the foundational skills to complete the homework sets successfully.
- **Concept Connections** prompt students to review the vocabulary and key concepts presented in the section.
- **Core Exercises** are presented next and are grouped by objective. These exercises are linked to examples in the text and direct students to similar problems whose solutions have been stepped-out in detail.
- **Mixed Exercises** do *not* refer to specific examples so that students can dip into their mathematical toolkit and decide on the best technique to use.
- **Write About It** exercises are designed to emphasize mathematical language by asking students to explain important concepts.
- **Technology Connections** require the use of a graphing utility and are found at the end of exercise sets. They can be easily skipped for those who do not encourage the use of calculators.
- **Expanding Your Skills Exercises** challenge and broaden students' understanding of the material.

Problem Recognition Exercises

Problem Recognition Exercises appear in strategic locations in each chapter of the text. These exercises provide students with an opportunity to synthesize multiple concepts and decide which problem-solving technique to apply to a given problem.

Examples

- The examples in the textbook are stepped-out in detail with thorough annotations at the right explaining each step.
- Following each example is a similar **Skill Practice** exercise to engage students by practicing what they have just learned.
- For the instructor, references to an even-numbered exercise are provided next to each example. These exercises are highlighted with blue circles in the exercise sets and mirror the related examples. With increased demands on faculty time, this has been a popular feature that helps faculty write their lectures and develop their presentation of material. If an instructor presents all of the highlighted exercises, then each objective of that section of text will be covered.

Modeling and Applications

One of the most important tools to motivate our students is to make the mathematics they learn meaningful in their lives. The textbook is filled with robust applications and numerous opportunities for mathematical modeling for those instructors looking to incorporate these features into their course.

Callouts

Throughout the text, popular tools are included to highlight important ideas. These consist of:

- **Tip** boxes that offer additional insight to a concept or procedure.
- **Avoiding Mistakes** boxes that fend off common mistakes.
- **Point of Interest** boxes that offer interesting and historical mathematical facts.
- **Instructor Notes** to assist with lecture preparation.

Graphing Calculator Coverage

Material is presented throughout the book illustrating how a graphing utility can be used to view a concept in a graphical manner. The goal of the calculator material is not to replace algebraic analysis, but rather, to enhance understanding with a visual approach. Graphing calculator examples are placed in self-contained boxes and may be skipped by instructors who choose not to implement the calculator. Similarly, the graphing calculator exercises are found at the end of the exercise sets and may also be easily skipped.

End-of-Chapter Materials

The textbook has the following end-of-chapter materials for students to review before test time:

- Brief summary with references to key concepts. A detailed summary is located at www.mhhe.com/millercollegealgebra.
- Chapter review exercises.
- Chapter test.
- Cumulative review exercises. These exercises cover concepts in the current chapter as well as all preceding chapters.

Updates to College Algebra:

- Two new sections, "Algebra for Calculus" and "Equations and Inequalities for Calculus," were added to Chapter R and Chapter 1. These additions provide STEM students an opportunity to connect current topics to what they'll learn in calculus.
- New "Prerequisite Review" exercises appear in every section. These allow students to ensure they have the necessary foundational skills to be successful in the section.
- Over 600 algorithmic homework exercises were added to Connect Math Hosted by ALEKS to ensure 90% textbook coverage.
- SmartBook content has been revised and enriched. For the first time, SmartBook is now available within Connect Math Hosted by ALEKS.
- 1200 new questions were added to the TestGen testbank.
- Graphing calculator screenshots have been updated to reflect the TI-84 Plus C.
- Section 2.7 for investigating increasing, decreasing, and constant behavior of a function now presents open intervals. This has also been updated in all of the digital materials accompanying the text.
- Sections R.1, R.2, 1.7 & 1.8 have been streamlined to provide greater clarity.
- New applications appear in Chapter 7 to provide students more real-world context for conic sections.
- Applications and real-world data have been updated, where appropriate, to ensure that content remains relevant and current.
- Wolfram Alpha Activities have been added to the Instructor's Resource Manual to allow students to explore college algebra in greater depth.

Supplement Package

Supplements for the Instructor

Author-Created Digital Media

Digital assets were created exclusively by the author team to ensure that the author voice is present and consistent throughout the supplement package.

- The coauthor, Donna Gerken, ensures that each algorithm in the online homework has a stepped-out solution that matches the textbook's writing style.
- Julie Miller created **video content** (lecture videos, exercise videos, graphing calculator videos, and Excel videos) to give students access to classroom-like instruction by the author.
- Julie Miller constructed over 50 **dynamic math animations** to accompany the college algebra text. The animations are diverse in scope and give students an interactive approach to conceptual learning. The animated content illustrates difficult concepts by leveraging the use of on-screen movement where static images in the text may fall short. They are organized in Connect hosted by ALEKS by chapter and section.

The *Instructor's Resource Manual* (IRM) is a printable electronic supplement put together by the author team. The IRM includes Guided Lecture Notes, Classroom Activities using Wolfram Alpha, and Group Activities.

- The Guided Lecture Notes are keyed to the objectives in each section of the text. The notes step through the material with a series of questions and exercises that can be used in conjunction with lecture.
- The Classroom Activities using Wolfram Alpha promote active learning in the classroom by using a powerful online resource.
- A Group Activity is available for each chapter of the book to promote classroom discussion and collaboration.

The *Instructor's Solution Manual* provides comprehensive, worked-out solutions to all exercises in the section exercises, review exercises, problem recognition exercises, chapter tests, and cumulative reviews. The steps shown in the solutions match the style and methodology of solved examples in the textbook.

TestGen is a computerized test bank utilizing algorithm-based testing software to create customized exams quickly. This user-friendly program enables instructors to search for questions by topic, format, or difficulty level; to edit existing questions, or to add new ones; and to scramble questions and answer keys for multiple versions of a single test. Hundreds of text-specific, open-ended, and multiple-choice questions are included in the question bank.

Annotated Instructor's Edition

- Answers to exercises appear adjacent to each exercise set, in a color used only for annotations.
- Instructors will find helpful notes within the margins to consider while teaching.
- References to even-numbered exercises appear in the margin next to each example for the instructor to use as Classroom Examples.

Power Points present key concepts and definitions with fully editable slides that follow the textbook. An instructor may project the slides in class or post to a website in an online course.

Supplements for the Student

Student Worksheets including guided lecture notes that step through the key objectives and Problem Recognition Exercise worksheets.

ALEKS® Prep for College Algebra

ALEKS Prep for College Algebra focuses on prerequisites and introductory material for College Algebra. These prep products can be used during the first 3 weeks of a course to prepare students for future success in the course and to increase retention and pass rates.

Connect Math® Hosted by ALEKS

Connect Math Hosted by ALEKS Corp. is an exciting, new assignment and assessment ehomework platform. Starting with an easily viewable, intuitive interface, students will be able to access key information, complete homework assignments, and utilize an integrated, media-rich eBook.

Smartbook® is the first and only adaptive reading experience available for the higher education market. Powered by the intelligent and adaptive LearnSmart engine, Smart-Book facilitates the reading process by identifying what content a student knows and doesn't know. As a student reads, the material continously adapts to ensure the student is focused on the content he or she needs the most to close specific knowledge gaps.

Detailed Chapter Summaries are available at www.mhhe.com/millercollegealgebra.

Efficient. Easy to Use. Effective. Engaging.
Gives Your Students the ALEKS Advantage

Are your students prepared for your course? Do they learn at different paces? Is there inconsistency between homework and test scores? ALEKS successfully addresses these core challenges and more.

With decades of scientific research behind its creation, ALEKS offers the most advanced adaptive learning technology that is proven to increase student success in math.

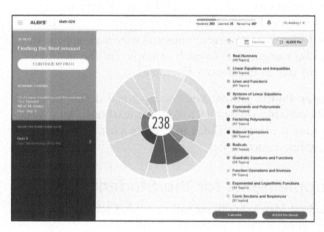

Student-Friendly Learning Experience

ALEKS is designed to meet the needs of today's students. A clean, modern, mobile-ready interface allows students to easily navigate their learning, track their progress and manage their assignments from anywhere.

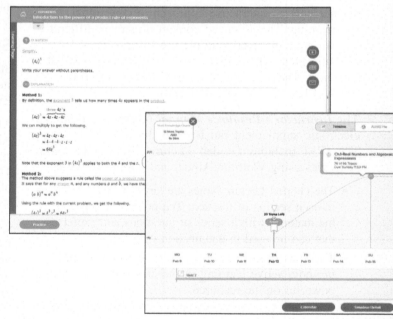

Optimized, Structured Learning

Using adaptive artificial intelligence, ALEKS identifies precisely what each student knows and doesn't know, and prescribes an individualized learning plan tailored to their unique strengths and weaknesses:

- Targets critical knowledge gaps
- Open-response environment
- Motivates student learning
- Presents only topics students are ready to learn
- Enhances learning with interactive resources
- Provides a structured learning path

> "I evaluated many different options, and ALEKS provided, by far, the best cycle of assessment and learning that allows for individualized instructional paths . . . no other program matches ALEKS."
> —**Professor Eliza Gallagher,** *Clemson University, SC*

Learn More: **Successinmath.com**

THE ALEKS Instructor Module includes intuitive customization and management features that help save you valuable time and effort. Easily manage your courses and track student progress, all through one simple interface.

- Create learning goals with due dates that align with your textbook/syllabi and to pace student learning
- Implementation services and training help make setup simple and timely
- 100% mobile-ready allows you to manage your classes from anywhere
- LMS integration offers single sign-on and gradebook sync capabilities

ALEKS Reporting provides detailed data on student progress, allowing you to quickly identify intervention needs and to better understand student learning behaviors. View data for individual students up to the multi-campus level.

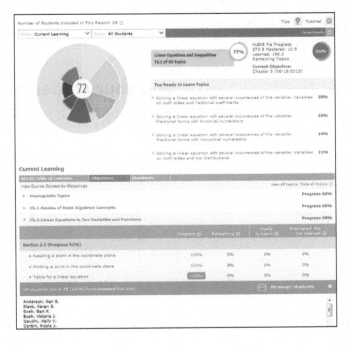

View student progress in a course area or at the individual topic level, including which topics students are struggling with the most.

Use this data to inform your teaching, group students based on similar knowledge levels, and shape a meaningful learning experience for your students.

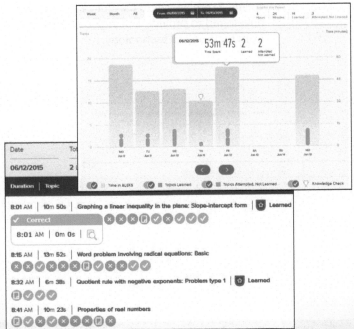

View a daily breakdown of how students learn in ALEKS, including the exact problems they attempt and their answers. This will help you to identify common mistakes and understand their behavior pattern, such as when they study and how frequently.

Learn More: Successinmath.com

Connect Math Hosted by ALEKS

Built By Today's Educators, For Today's Students

Fewer clicks means more time for you...

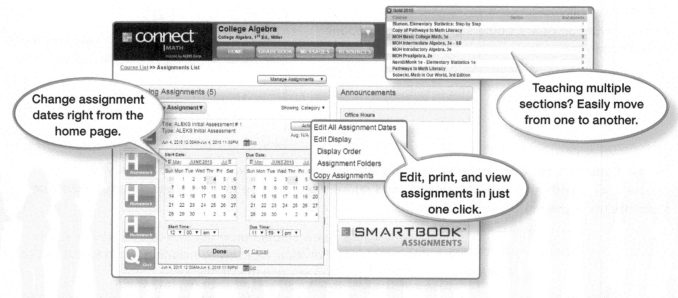

Change assignment dates right from the home page.

Teaching multiple sections? Easily move from one to another.

Edit, print, and view assignments in just one click.

...and your students.

Online Exercises were carefully selected and developed to provide a seamless transition from textbook to technology.

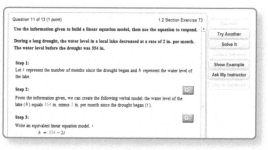

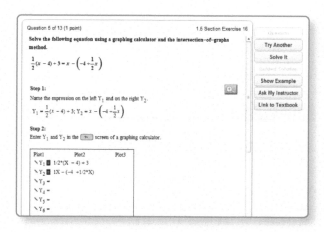

For consistency, the guided solutions match the style and voice of the original text as though the author is guiding the students through the problems.

Quality Content For Today's Online Learners

Why SmartBook? Because it's more than just words on a page.

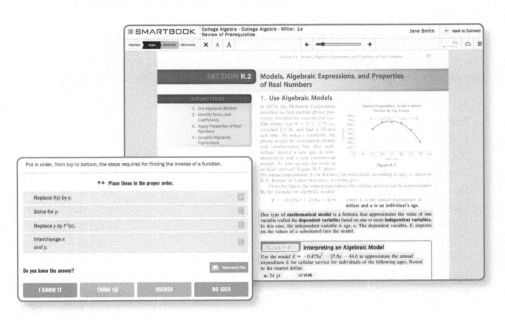

SMARTBOOK™ Students today thrive on efficiency, mobility, and motivation. And SmartBook delivers. SmartBook is the first adaptive reading experience designed to change the way students learn. Students and instructors can enjoy access to SmartBook anywhere, anytime (now available offline) with a new and improved mobile interface. If students still prefer holding a text as they study, they can **order a loose-leaf copy of their textbook for just $15.**

SmartBook breaks down the learning experience into four stages: *Preview, Read, Practice,* and *Recharge.* Each stage provides personalized guidance and just-in-time remediation to ensure students stay focused and learn as efficiently as possible. With LearnSmart technology, questions are designed to foster critical thinking and conceptual learning.

Reports are also available to both students and instructors that track progress and show each student's strengths and weaknesses. What does this mean for you? Teach a more informed classroom and provide more personalized guidance.

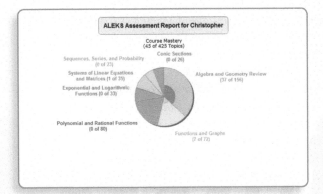

The ALEKS® Initial Assessment is an artificially intelligent (AI), diagnostic assessment that identifies precisely what a student knows. Instructors can then use this information to make more informed decisions on what topics to cover in more detail with the class.

ALEKS is a registered trademark of ALEKS Corporation.

successinmath.com

Hosted by **ALEKS Corp.**

Our Commitment to Market Development and Accuracy

Acknowledgments:

Paramount to the development of *College Algebra* was the invaluable feedback provided by the instructors from around the country who reviewed the manuscript or attended a market development event over the course of the several years the text was in development.

A Special Thanks to All of the Event Attendees Who Helped Shape *College Algebra*.

Focus groups and symposia were conducted with instructors from around the country to provide feedback to editors and the authors and ensure the direction of the text was meeting the needs of students and instructors.

Halina Adamska, *Broward College–Central*
Mary Beth Angeline, *West Virginia University*
Colleen Beaudoin, *University of Tampa*
Rachel Black, *Central New Mexico Community College*
Tony Bower, *Saint Phillips College*
Bowen Brawner, *Tarleton State University*
Denise Brown, *Collin College*
Wyatt Bryant, *Tarleton State University*
Christine Bush, *Palm Beach State College*
Michelle Carmel, *Broward College–North*
Lydia Casas, *Saint Phillips College*
Carlos Corona, *San Antonio College*
Deric Davenport, *Pikes Peak Community College*
Alan Dinwiddie, *Front Range Commmunity College–Fort Collins*
Marion Foster, *Houston Community College*
Charles Gabi, *Houston Community College*
Jason Geary, *Harper College*
Steve Gonzales, *Northwest Vista College*
Jeffrey Guild, *Broward-Central Campus*
Craig Hardesty, *Hillsborough Community College–Southshore*
Lori Hodges, *University of New Orleans*
Carolyn Horseman, *Polk College*
Kimber Kaushik, *Houston Community College*
Lynette Kenyon, *Collin College–Plano*
Daniel Kernler, *Elgin Community College*
Sharon Kobrin, *Broward-Central Campus*
Daniel Kopsas, *Ozarks Technical Community College*
Danny Lau, *University of North Georgia*
Andreas Lazari, *Valdosta State University*

Joyce Lee, *Polk College*
Wayne Lee, *Saint Philips College*
Domingo Litong, *Houston Community College*
Tammy Louie, *Portland Community College*
Susan May, *Meridian Community College*
Michael McClendon, *University of Central Oklahoma*
Jerry McCormack, *Tyler Junior College*
Mikal McDowell, *Cedar Valley College*
Meagan McNamee ,*Central Piedmont Community College*
Rebecca Muller, *Southeastern Louisiana University*
Denise Natasha, *Georgia Gwinnett College*
Lynne Nisbet, *St. Louis Community College*
Altay Ozgener, *State College of Florida–Manatee*
Denise Pendarvis, *Georgia Perimeter College*
Scott Peterson, *Oregon State University*
Davidson Pierre, *State College of Florida–Manatee*
Mihaela Poplicher, *University of Cincinnati*
Candace Rainer, *Meridian Community College*
Dee Dee Shaulis, *University of Colorado–Boulder*
Prem Singh, *Ohio University*
Rita Sowell, *Volunteer State Community College*
Pam Stogsdill, *Bossier Parish Community College*
Peter Surgent, *Community College of Baltimore*
Dustin Walsh, *Southeast Community College*
Kim Walters, *Mississippi State University*
Michael Warren, *Tarleton State University*
Jeff Weaver, *Baton Rouge Community College*
Carol Weideman, *Saint Petersburg College–Gibbs*
Benjamin Wescoatt, *Valdosta State University*
Sean Woodruff, *Saint Petersburg College–Tarpon Springs*

Manuscript Reviewers

Carol Abbott, *Ohio University*

Marylynne Abbott, *Ozarks Technical Community College*

Jay Abramson, *Arizona State University–Tempe*

Ryan Adams, *Northwest Florida State College*

Halina Adamska, *Broward College–Central*

Mark Ahrens, *Normandale Community College*

John Alford, *Sam Houston State University*

Kinnari Amin, *Georgia Perimeter College*

Patricia Anderson, *Arapahoe Community College*

Raji Ariyaratna, *Houston Community College*

Alvina Atkinson, *Georgia Gwinnett College*

Robin Ayers, *Western Kentucky University*

Mohamed Baghzali, *North Dakota State University*

Robert Banik, *Mississippi State University*

Terry Lee Barron, *Georgia Gwinnett College*

John Beatty, *Georgia Perimeter College*

Tim Bell, *San Jacint College*

Sergey Belyi, *Troy University*

Patricia Bezona, *Valdosta State University*

Nicholas Bianco, *Florida Gulf Coast University*

Susan Billimek, *Northwest Arkansas Community College*

Greg Bloxom, *Pensacola State College*

Kristina Bowers, *Tallahassee Community College*

Stephanie Branham, *University of Tampa*

James Brink, *Ohio University*

Annette Burden, *Youngstown State University*

James Carolan, *Wharton County Junior College*

Lydia Casas, *Saint Philips College*

Jason Cates, *Brookhaven College*

Tim Chappell, *MCC Penn Valley Community College*

Lars Christensen, *Texas Tech University*

Ivette Chuca, *El Paso Community College*

Carl Clark, *Indian River State College*

Beth Clickner, *Hillsborough Community College*

Thomas Cooper, *University of North Georgia*

Cindy Cummins, *Ozarks Technical Community College*

Nelson De La Rosa, *Miami Dade College*

Noemi DeHerrera, *Houston Community College*

Alan Dinwiddie, *Front Range Community College*

Christy Dittmar, *Austin Community College–Rio Grande*

Ginger Eaves, *Bossier Parish Community College*

Keith Erickson, *Georgia Gwinnett College*

Keith Erickson, *Georgia Gwinnett College*

Dihema Ferguson, *Georgia Perimeter College*

Elise Fischer, *Johnson County Community College*

Marion Foster, *Houston Community College*

Lana Fredrickson, *Front Range Community College*

John Fulk, *Georgia Perimeter College*

Darren Funk-Neubauer, *Colorado State University*

Valdez Gant, *North Lake College*

Kevin Gibbs, *University of Toledo*

Vijaya Gompam, *Troy University*

Jeff Gutliph, *Georgia Perimeter College*

Debbie Hanus, *Brookhaven College*

Craig Hardesty, *Hillsborough Community College*

Tom Hayes, *Montana State University*

Beata Hebda, *University of North Georgia*

Christy Hediger, *Lehigh Carbon Community College*

Jean Hindie, *Community College of Denver*

Gangadhar Hiremath, *University of North Carolina*

Linda Ho, *El Camino College*

Lori Hodges, *University of New Orleans*

Heidi Howard, *Florida State College*

Sharon Jackson, *Brookhaven College*

Erin Joseph, *Central New Mexico Community College*

Chandra Karnati, *University of North Georgia*

Susan Keith, *Georgia Perimeter College*

Lynette Kenyon, *Collin College Plano*

Raja Khoury, *Collin College–Plano*

Minsu Kim, *University of North Georgia*

Daniel Kopsas, *Ozarks Technical Community College*

Ramesh Krishnan, *South Plains College*

Bohdan Kunciw, *Salisbury University*

Weiling Landers, *Windward Community College*

Sungwook Lee, *University of Southern Mississippi*

Xuhui Li, *California State University–Long Beach*

Domingo Litong, *Houston Community College*

Rene Lumampao, *Austin Community College–Rio Grande*

Edmund MacPherson, *Tyler Junior College*

Anna Madrid-Larranaga, *Central New Mexico Community College*

Jason Malozzi, *Lehigh Carbon Community College*

Cynthia Martinez, *Temple College*

Shawna Masters, *Collin College–Plano*

Ramon Mata-Toledo, *James Madison University*

Janet Mayeux, *SE Louisiana University*

Roderick McBane, *Houston Community College*

Cynthia McGinnis, *Northwest Florida State College*

Christine McKenna, *University of Nevada Las Vegas*

Mary Merchant, *Cedar Valley College*

Chris Mizell, *Northwest Florida State College*

Monica Montalvo, *Collin College–Plano*

Malika T. Morris, *Houston Community College*

Dorothy Muhammad, *Houston Community College*

Linda Myers, *Harrisburg Community College*

Julie Nation, *Mississippi State University*

Mandri Obeyesekere, *Houston Community College*

Victor Obot, *Texas Southern University*

Charles Odion, *Houston Community College*

Donald Orr, *Miami Dade College*

Victor Pambuccian, *Arizona State University–West*

Stan Perrine, *Georgia Gwinnett College*

David Platt, *Front Range Community College*

Wendy Pogoda, *Hillsborough Community College*

John Polhill, *Bloomsburg University of Pennsylvania*

Jonathan Poritz, *Colorado State University*

Didi Quesada, *Miami Dade College*

Brooke Quinlan, *Hillsborough Community College*

Carolynn Reed, *Austin Community College–Rio Grande*
Denise Reid, *Valdosta State University*
Nancy Resseguie, *Arapahoe Community College*
Shelia Rivera, *University of West Georgia*
Ken Roblee, *Troy University*
Haazim Sabree, *Georgia Perimeter College*
Haazim Sabree, *Georgia Perimeter College*
Fatemeh Salehibakhsh, *Houston Community College*
Fary Sami, *Harford Community College*
Linda Schott, *Ozarks Technical Community College*
Mary Schuster, *University of Cinncinatti*
Lisa Shannon, *Meridian Community College*
Giorgi Shonia, *Ohio University*
Andrew Siefker, *Angelo State University*
Jennifer Siegel, *Broward College–Central*
Randell Simpson, *Temple College*

Premjit Singh, *Ohio University*
Sounny Slitine, *Saint Philips College*
David Slutzky, *University of North Georgia*
Mary Ann Sojda, *Montana State University*
Shannon Solis, *San Jacint College*
Scott Sorrell, *University of Louisiana Lafayette*
Malgorzata Surowiec, *Texas Tech University*
Vic Swaim, *SE Louisiana University*
Paula Talley, *Temple College*
Rae Tree, *Oklahoma State University*
Chris Turner, *Pensacola State College*
Phil Veer, *Johnson County Community College*
James Wan, *Long Beach City College*
Walter Wang, *Baruch College*
Emily Whaley, *Georgia Perimeter College*
Changyong Zhong, *Georgia State University*

Author Acknowledgments:

An editor once told us that publishing a book is like making a movie because there are so many people behind the scenes that make the final product a success. Words cannot begin to express our heartfelt thanks to all of you, but we'll do our best. First and foremost, we want to thank our editor Emily Windelborn who started with us on this project when it was just idea, and then lent her unwavering, day-to-day support through final publication. Without Emily, we'd still be on page 1. To Ryan Blankenship, Marty Lange, and Kurt Strand, we are forever grateful for the amazing opportunities you and McGraw-Hill have given us.

To our brand manager, Caroline Celano, as the pilot on this long journey you set the standard for leadership. Each day, you lead by example to unlock our full potential and inspire our best work. You're strong but not rude; kind, but not weak; humble, but not timid. To the marketing team Michelle Greco, Leigh Jacka, Simon Wong, Megan Farber, Sara Swangard, Ashley Swafford, Jill Gordon, and Alex Gay: Your artistry and creative ideas for our project are 90% inspiration, 90% innovation, and 100% perspiration. Thank you for making us shine.

Our heartfelt gratitude goes to the production manager Peggy Selle for steering the ship and keeping us all on task. To Laurie Janssen and David Hash, many thanks for a beautiful design, and to Carrie Burger for the beautiful photos and art. The book is gorgeous.

Special thanks to the digital team, Rob Brieler and Victor Pareja, for keeping the digital train on the tracks. Thank you for overseeing the enormous job of managing digital content and ensuring consistency of the author voice. Along these lines, we must express our utmost gratitude to digital authors Alina Coronel, Esmarie Kennedy, Tim Chappell, Stephen Toner, Michael Larkin, Lizette Foley, Meghan Clovis, and Lance Gooden for their diligence writing digital content. All of you are amazing. To Jason Wetherington and Mary Beth Headlee, thank you so much for your work on SmartBook and for the additional pairs of eyes on our manuscript. To Beth Clickner, many, many thanks for the beautiful and thorough PowerPoint presentations of our material. To our colleague and friend Kimberly Alacan, we're so grateful for your creativity in preparing the chapter openers and the group activities. To Nora Devlin, thank you for managing the solutions manual projects, lecture notes, and Internet Activities. No doubt, many instructors and students thank you as well.

To Patricia Steele, the best copy editor ever, thank you for mentoring us and for ensuring consistency throughout our work. To Elka Block, Jennifer Blue, and the team at DiacriTech, many thanks for doing multiple levels of accuracy checking. Your talents are absolutely amazing. Additional thanks to Julie Kennedy and Carey Lange for their tireless attention to detail proofreading pages.

Finally, to the dedicated people in the McGraw-Hill sales force, thank you so much for your continued confidence, encouragement, and support.

Most importantly, we want to give special thanks to all the students and instructors who use *College Algebra* in their classes.

—Julie Miller and Donna Gerken

Applications Index

Weather

Review of Prerequisites

Athletes know that in order to optimize their performance they need to pace themselves and be mindful of their target heart rate. For example, a 25-year-old with a maximum heart rate of 195 beats per minute should strive for a target heart rate zone of between 98 and 166 beats per minute. This correlates to between 50% and 85% of the individual's maximum heart rate (Source: American Heart Association, www.americanheart.org). The mathematics involved in finding maximum heart rate and an individual's target heart rate zone use a linear model relating age and resting heart rate. An introduction to modeling is presented here in Chapter R along with the standard order of operations used to carry out these calculations.

Chapter R reviews skills and concepts required for success in college algebra. Just as an athlete must first learn the basics of a sport and build endurance and speed, a student studying mathematics must focus on necessary basic skills to prepare for the challenge ahead. Preparation for algebra is comparable to an athlete preparing for a sporting event. Putting the time and effort into the basics here in Chapter R will be your foundation for success in later chapters.

SECTION R.1   Sets and the Real Number Line

1. Identify Subsets of the Set of Real Numbers

A hybrid vehicle gets 48 mpg in city driving and 52 mpg on the highway. The formula $A = \frac{1}{48}c + \frac{1}{52}h$ gives the amount of gas A (in gal) for c miles of city driving and h miles of highway driving. In the formula, A, c, and h are called **variables** and these represent values that are subject to change. The values $\frac{1}{48}$ and $\frac{1}{52}$ are called **constants** because their values do not change in the formula.

For a trip from Temple, Texas, to Dallas, Texas, a motorist travels 36 mi of city driving and 91 mi of highway driving. The amount of fuel used by this hybrid vehicle is given by

$$A = \frac{1}{48}(36) + \frac{1}{52}(91)$$
$$= 2.5 \text{ gal}$$

The numbers used in day-to-day life such as those used to determine fuel consumption come from the set of real numbers, denoted by $\mathbb{R}$. A **set** is a collection of items called **elements.** The braces { and } are used to enclose the elements of a set. For example, {gold, silver, bronze} represents the set of medals awarded to the top three finishers in an Olympic event. A set that contains no elements is called the **empty set** (or **null set**) and is denoted by { } or $\varnothing$.

When referring to individual elements of a set, the symbol $\in$ means "is an element of," and the symbol $\notin$ means "is not an element of." For example,

$5 \in \{1, 3, 5, 7\}$ is read as "5 is an element of the set of elements 1, 3, 5, and 7."

$6 \notin \{1, 3, 5, 7\}$ is read as "6 is *not* an element of the set of elements 1, 3, 5, and 7."

A set can be defined in several ways. Listing the elements in a set within braces is called the **roster method.** Using the roster method, the set of the even numbers between 0 and 10 is represented by {2, 4, 6, 8}. Another method to define this set is by using **set-builder notation.** This uses a description of the elements of the set. For example,

$$\{x \,|\, x \text{ is an even number between 0 and 10}\}$$

The set of   all x such that x is an even number between 0 and 10

In our study of college algebra, we will often refer to several important **subsets** (parts of) the set of real numbers (Table R-1).

Table R-1   Subsets of the Set of Real Numbers, $\mathbb{R}$

Set	Definition
Natural numbers, $\mathbb{N}$	$\{1, 2, 3, \ldots\}$
Whole numbers, $\mathbb{W}$	$\{0, 1, 2, 3, \ldots\}$
Integers, $\mathbb{Z}$	$\{\cdots, -3, -2, -1, 0, 1, 2, 3, \ldots\}$
Rational numbers, $\mathbb{Q}$	$\left\{ \dfrac{p}{q} \,\middle|\, p, q \in \mathbb{Z} \text{ and } q \neq 0 \right\}$ • Rational numbers can be expressed as a ratio of integers where the denominator is not zero. <u>Examples:</u> $-\frac{6}{11}$ (ratio of -6 and 11) and 9 (ratio of 9 and 1). • All terminating and repeating decimals are rational numbers. <u>Examples:</u> 0.71 (ratio of 71 and 100), $0.\overline{6} = 0.666\ldots$ (ratio of 2 and 3).
Irrational numbers, $\mathbb{H}$	Irrational numbers are real numbers that cannot be expressed as a ratio of integers. The decimal form of an irrational number is nonterminating and nonrepeating. <u>Examples:</u> π and $\sqrt{2}$

TIP Notice that the first five letters of the word *rational* spell *ratio.* This will help you remember that a rational number is a *ratio* of integers.

Real Numbers (R)

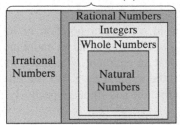

Figure R-1

TECHNOLOGY CONNECTIONS

Approximating Rational and Irrational Numbers

The number $-\frac{6}{11}$ is a rational number, and the number π is an irrational number. It is important to realize that for nonterminating decimals, a calculator or spreadsheet will only give approximate values, not exact values.

	A	B
1	Number	Decimal Approximation
2	-(6/11)	-0.545454545
3	Pi()	3.141592654

The relationships among the subsets of real numbers defined in Table R-1 are shown in Figure R-1. In particular, notice that together the elements of the set of rational numbers and the set of irrational numbers make up the set of real numbers.

EXAMPLE 1   Identifying Elements of a Set

Given $A = \left\{ \sqrt{3}, 0.\overline{83}, -\frac{19}{7}, 0.39, -16, 0, 11, 0.2020020002..., 0.444 \right\}$, determine which elements belong to the following sets.

Solution:

  a. $\mathbb{N}$ $11 \in \mathbb{N}$

  b. $\mathbb{W}$ $0, 11 \in \mathbb{W}$

  c. $\mathbb{Z}$ $-16, 0, 11 \in \mathbb{Z}$

  d. $\mathbb{Q}$ $0.\overline{83}, -\frac{19}{7}, 0.39, -16, 0, 11, 0.444 \in \mathbb{Q}$

  e. $\mathbb{H}$ $\sqrt{3}, 0.2020020002... \in \mathbb{H}$

  f. $\mathbb{R}$ $\sqrt{3}, 0.\overline{83}, -\frac{19}{7}, 0.39, -16, 0, 11, 0.2020020002..., 0.444 \in \mathbb{R}$

Skill Practice 1   Given set B, determine which elements belong to the following sets. $B = \left\{ -\frac{11}{7}, \sqrt{59}, 4.3, 0, 23, -13, \pi, 4.\overline{9} \right\}$

  a. $\mathbb{N}$ **b.** $\mathbb{W}$ **c.** $\mathbb{Z}$ **d.** $\mathbb{Q}$ **e.** $\mathbb{H}$ **f.** $\mathbb{R}$

2. Use Inequality Symbols and Interval Notation

All real numbers can be located on the real number line. We say that a is less than b (written symbolically as $a < b$) if a lies to the left of b on the number line. This is equivalent to saying that b is greater than a (written symbolically as $b > a$) because b lies to the right of a.

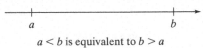

$a < b$ is equivalent to $b > a$

In Table R-2, we summarize other symbols used to compare two real numbers.

Answers

1. a. $23 \in \mathbb{N}$ **b.** $0, 23 \in \mathbb{W}$

  c. $0, 23, -13 \in \mathbb{Z}$

  d. $-\frac{11}{7}, 4.3, 0, 23, -13, 4.\overline{9} \in \mathbb{Q}$

  e. $\sqrt{59}, \pi \in \mathbb{H}$

  f. $-\frac{11}{7}, \sqrt{59}, 4.3, 0, 23, -13, \pi, 4.\overline{9} \in \mathbb{R}$

Table R-2   Summary of Inequality Symbols and Their Meanings

Inequality	Verbal Interpretation	Other Implied Meanings	Numerical Examples
$a < b$	a is less than b	b exceeds a b is greater than a	$5 < 7$
$a > b$	a is greater than b	a exceeds b b is less than a	$-3 > -6$
$a \leq b$	a is less than or equal to b	a is at most b a is no more than b	$4 \leq 5$ $5 \leq 5$
$a \geq b$	a is greater than or equal to b	a is no less than b a is at least b	$9 \geq 8$ $9 \geq 9$
$a = b$	a is equal to b		$-4.3 = -4.3$
$a \neq b$	a is not equal to b		$-6 \neq -7$
$a \approx b$	a is approximately equal to b		$-12.99 \approx -13$

Point of Interest

The infinity symbol ∞ is called a lemniscate from the Latin *lemniscus* meaning "ribbon." English mathematician John Wallis is credited with introducing the symbol in the seventeenth century. The symbols $-\infty$ and ∞ are not themselves real numbers, but instead refer to quantities without bound or end.

An interval on the real number line can be represented in set-builder notation or in interval notation. In Table R-3, observe that a parenthesis) or (indicates that an endpoint is not included in an interval. A bracket] or [indicates that an endpoint *is* included in the interval. The real number line extends infinitely far to the left and right. We use the symbols $-\infty$ and ∞ to denote the unbounded behavior to the left and right, respectively.

Table R-3   Summary of Interval Notation and Set-Builder Notation

Let a, b, and x represent real numbers.

Set-Builder Notation	Verbal Interpretation	Graph	Interval Notation
$\{x \mid x > a\}$	the set of real numbers greater than a		(a, ∞)
$\{x \mid x \geq a\}$	the set of real numbers greater than or equal to a		$[a, \infty)$
$\{x \mid x < b\}$	the set of real numbers less than b		$(-\infty, b)$
$\{x \mid x \leq b\}$	the set of real numbers less than or equal to b		$(-\infty, b]$
$\{x \mid a < x < b\}$	the set of real numbers between a and b		(a, b)
$\{x \mid a \leq x < b\}$	the set of real numbers greater than or equal to a and less than b		$[a, b)$
$\{x \mid a < x \leq b\}$	the set of real numbers greater than a and less than or equal to b		$(a, b]$
$\{x \mid a \leq x \leq b\}$	the set of real numbers between a and b, inclusive		$[a, b]$
$\{x \mid x \text{ is a real number}\}$ $\mathbb{R}$	the set of all real numbers		$(-\infty, \infty)$

TIP As an alternative to using parentheses and brackets to represent the endpoints of an interval, an open dot or closed dot may be used. For example, $\{x \mid a \leq x < b\}$ would be represented as follows.

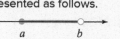

EXAMPLE 2   Expressing Sets in Interval Notation and Set-Builder Notation

Complete the table.

Graph	Interval Notation	Set-Builder Notation
$-5\ -4\ -3\ -2\ -1\quad 0\quad 1\quad 2\quad 3\quad 4\quad 5$		
	$\left(\frac{7}{2}, \infty\right)$	
		$\{y \mid -4 \leq y < 2.3\}$

Solution:

Graph	Interval Notation	Set-Builder Notation	Comments
-5 -4 -3 -2 -1 0 1 2 3 4 5	$(-\infty, 2]$	$\{x \mid x \leq 2\}$	The bracket at 2 indicates that 2 is included in the set.
-5 -4 -3 -2 -1 0 1 2 3 4 5	$\left(\frac{7}{2}, \infty\right)$	$\{x \mid x > \frac{7}{2}\}$	The parenthesis at $\frac{7}{2} = 3.5$ indicates that $\frac{7}{2}$ is *not* included in the set.
-5 -4 -3 -2 -1 0 1 2 3 4 5	$[-4, 2.3)$	$\{y \mid -4 \leq y < 2.3\}$	The set includes the real numbers between -4 and 2.3, including the endpoint -4.

Skill Practice 2

a. Write the set represented by the graph in interval notation and set-builder notation.

-5 -4 -3 -2 -1 0 1 2 3 4 5

b. Given the interval, $\left(-\infty, -\frac{4}{3}\right]$, graph the set and write the set-builder notation.

c. Given the set, $\{x \mid 1.6 < x \leq 5\}$, graph the set and write the interval notation.

3. Find the Union and Intersection of Sets

Two or more sets can be combined by the operations of union and intersection.

Union and Intersection of Sets

The **union** of sets A and B, denoted $A \cup B$, is the set of elements that belong to set A or to set B or to both sets A and B.

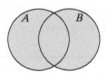

$A \cup B$
A union B
The elements in A or B or both

The **intersection** of sets A and B, denoted $A \cap B$, is the set of elements common to both set A and set B.

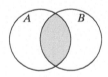

$A \cap B$
A intersection B
The elements common to A and B

In Examples 3 and 4, we practice finding the union and intersections of sets.

EXAMPLE 3   Finding the Union and Intersection of Sets

Find the union or intersection of sets as indicated.

$$A = \{-5, -3, -1, 1\} \qquad B = \{-5, 0, 5\} \qquad C = \{-4, -2, 0, 2, 4\}$$

a. $A \cap B$ **b.** $A \cup B$ **c.** $A \cap C$

Solution:

a. $A \cap B = \{-5\}$ The only element common to both A and B is -5.
$A = \{-5, -3, -1, 1\}, B = \{-5, 0, 5\}$

b. $A \cup B = \{-5, -3, -1, 0, 1, 5\}$ The union of A and B consists of all elements from A along with all elements from B.

c. $A \cap C = \{\ \}$ Sets A and C have no common elements.

Answers

2. a. Interval notation: $(-4, \infty)$;
Set-builder notation: $\{x \mid x > -4\}$

b.

-5 -4 -3 -2 -1 0 1 2 3 4 5 ;
Set-builder notation: $\{x \mid x \leq -\frac{4}{3}\}$

c.

-5 -4 -3 -2 -1 0 1 2 3 4 5 ;
Interval notation: $(1.6, 5]$

EXAMPLE 4   **Finding the Union and Intersection of Sets**

Find the union or intersection as indicated.

$$D = \{x \mid x < 4\} \qquad E = \{x \mid x \geq -2\} \qquad F = \{x \mid x \leq -3\}$$

   a. $D \cap E$ **b.** $D \cup E$ **c.** $D \cup F$ **d.** $E \cap F$

Solution:

Graph each individual set and take the union or intersection.

a.

   D:

   E:
      -2

   $D \cap E$:
      $-4 \ -3 \ -2 \ -1 \quad 0 \quad 1 \quad 2 \quad 3 \quad 4$

   The intersection is the region of overlap.

   Set notation: $D \cap E = \{x \mid -2 \leq x < 4\}$
   Interval notation: $[-2, 4)$

b.

   D:
      4

   E:
      -2

   $D \cup E$:
      $-4 \ -3 \ -2 \ -1 \quad 0 \quad 1 \quad 2 \quad 3 \quad 4$

   The union contains the elements from D along with those from E.

   Set notation: $D \cup E = \mathbb{R}$
   Interval notation: $(-\infty, \infty)$

c.

   D:
      4

   F:
      -3

   $D \cup F$:
      $-4 \ -3 \ -2 \ -1 \quad 0 \quad 1 \quad 2 \quad 3 \quad 4$

   Set F is contained within set D. The union is set D itself.

   Set notation: $D \cup F = \{x \mid x < 4\}$
   Interval notation: $(-\infty, 4)$

d.

   E:
      -2

   F:
      -3

   $E \cap F$:
      $-4 \ -3 \ -2 \ -1 \quad 0 \quad 1 \quad 2 \quad 3 \quad 4$

   There are no elements common to both sets E and F.

   Set notation: $E \cap F = \{ \ \}$

Answers
3. a. $\{0\}$ **b.** $\{-5, -4, -2, 0, 2, 4, 5\}$
   c. $\{-5, -4, -3, -2, -1, 0, 1, 2, 4\}$
4. a. $\{x \mid -1 \leq x < 2\}$; $[-1, 2)$
   b. $\mathbb{R}$; $(-\infty, \infty)$
   c. $\{x \mid x < -4\}$; $(-\infty, -4)$
   d. $\{ \ \}$

4. Evaluate Absolute Value Expressions

Every real number x has an opposite denoted by $-x$. For example, $-(4)$ is the opposite of 4 and simplifies to -4. Likewise, $-(-2.1)$ is the opposite of -2.1 and simplifies to 2.1.

The **absolute value** of a real number x, denoted by $|x|$, is the distance between x and zero on the number line. For example:

$|-5| = 5$ because -5 is 5 units from zero on the number line.

$|5| = 5$ because 5 is 5 units from zero on the number line.

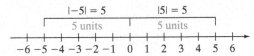

Notice that if a number is negative, its absolute value is the opposite of the number. If a number is positive, its absolute value is the number itself.

Definition of Absolute Value	
Let x be a real number. Then $|x| = \begin{cases} x \text{ if } x \geq 0 \\ -x \text{ if } x < 0 \end{cases}$	
Verbal Interpretation	**Numerical Example**
• If x is positive or zero, then $|x|$ is just x itself.	$|4| = 4$
	$|0| = 0$
• If x is negative, then $|x|$ is the opposite of x.	$|-4| = -(-4) = 4$

EXAMPLE 5   **Removing Absolute Value Symbols**

Use the definition of absolute value to rewrite each expression without absolute value bars.

a. $|\sqrt{3} - 3|$ **b.** $|3 - \sqrt{3}|$ **c.** $\dfrac{|x - 4|}{x - 4}$ for $x < 4$

Solution:

a. $|\sqrt{3} - 3| = -(\sqrt{3} - 3)$

$= -\sqrt{3} + 3$ or $3 - \sqrt{3}$

> The value $\sqrt{3} \approx 1.73 < 3$, which implies that $\sqrt{3} - 3 < 0$. Since the expression inside the absolute value bars is negative, take the opposite.

b. $|3 - \sqrt{3}| = 3 - \sqrt{3}$

> The value $\sqrt{3} \approx 1.73 < 3$, which implies that $3 - \sqrt{3} > 0$. Since the expression inside the absolute value bars is positive, the simplified form is the expression itself.

c. $\dfrac{|x - 4|}{x - 4}$ for $x < 4$

$= \dfrac{-(x - 4)}{x - 4}$

$= -1 \cdot \dfrac{x - 4}{x - 4}$

$= -1$

> The condition $x < 4$, implies that $x - 4 < 0$. Since the expression inside the absolute value bars is negative, take the opposite.

TIP: Calculator approximations can be used to show that $\sqrt{3} - 3 \approx -1.27$ is negative, and $3 - \sqrt{3} \approx 1.27$ is positive.

Skill Practice 5   Use the definition of absolute value to rewrite each expression without absolute value bars.

a. $|5 - \sqrt{7}|$ **b.** $|\sqrt{7} - 5|$ **c.** $\dfrac{x + 6}{|x + 6|}$ for $x > -6$

Answers

5. a. $5 - \sqrt{7}$
   b. $5 - \sqrt{7}$
   c. 1

5. Use Absolute Value to Represent Distance

Absolute value is also used to denote distance between two points on a number line.

> ### Distance Between Two Points on a Number Line
>
> The distance between two points a and b on a number line is given by
>
> $$|a - b| \qquad \text{or} \qquad |b - a|$$
>
> That is, the distance between two points on a number line is the absolute value of their difference.

> **EXAMPLE 6**   **Determining the Distance Between Two Points**
>
> Write an absolute value expression that represents the distance between 4 and -1 on the number line. Then simplify.
>
> **Solution:**
>
> $|4 - (-1)| = |5| = 5$ The distance between 4 and -1 is represented by
> $|-1 - 4| = |-5| = 5$ $|4 - (-1)|$ or by $|-1 - 4|$.
>
>

> **Skill Practice 6**   Write an absolute value expression that represents the distance between -9 and 2 on the number line. Then simplify.

6. Apply the Order of Operations

Repeated multiplication can be written by using exponential notation. For example, the product $5 \cdot 5 \cdot 5$ can be written as 5^3. In this case, 5 is called the base of the expression and 3 is the exponent (or power). The exponent indicates how many times the base is used as a factor.

> ### Definition of b^n
>
> Let b be a real number and let n represent a natural number. Then
>
> $$b^n = \underbrace{b \cdot b \cdot b \cdot \ldots \cdot b}$$
> $$b \text{ is used as a factor } n \text{ times}$$
>
> b^n is read as "b to the nth-power."
> b is the **base** and n is the **exponent** or **power.**

To find a square root of a nonnegative real number, we reverse the process to square a number. For example, a square root of 25 is a number that when squared equals 25. Both 5 and -5 are square roots of 25, because $5^2 = 25$ and $(-5)^2 = 25$. A radical sign $\sqrt{}$ is used to denote the principal square root of a number. The **principal square root** of a nonnegative real number is the square root that is greater than or equal to zero. Therefore, the principal square root of 25, denoted by $\sqrt{25}$, equals 5.

$$\sqrt{25} = 5 \text{ because } 5 \geq 0 \text{ and } 5^2 = 25$$

Answer
6. $|-9 - 2|$ or $|2 - (-9)|$;
   The distance is 11 units.

> **TIP** Note that the square of any real number is nonnegative. Therefore, there is no real number that is the square root of a negative number. For example,
>
> $\sqrt{-25}$ is not a real number because no real number when squared equals -25.
>
> *Note*: The value $\sqrt{-25}$ is an imaginary number and will be discussed in Section 1.3.

The symbol $\sqrt[3]{}$ represents the cube root of a number. For example:

$$\sqrt[3]{64} = 4 \text{ because } 4^3 = 64$$

Many expressions involve multiple operations. In such a case, it is important to follow the order of operations.

Order of Operations

Step 1 Simplify expressions within parentheses and other grouping symbols. These include absolute value bars, fraction bars, and radicals. If nested grouping symbols are present, start with the innermost symbols.

Step 2 Evaluate expressions involving exponents.

Step 3 Perform multiplication or division in the order in which they occur from left to right.

Step 4 Perform addition or subtraction in the order in which they occur from left to right.

EXAMPLE 7   **Simplifying a Numerical Expression**

Simplify. $7 - \left\{8 + 4\left[2 - \left(5 - \sqrt{64}\right)^2\right]\right\}$

Solution:

$7 - \left\{8 + 4\left[2 - \left(5 - \sqrt{64}\right)^2\right]\right\}$

$= 7 - \left\{8 + 4\left[2 - (5 - 8)^2\right]\right\}$ Simplify within inner parentheses, $\sqrt{64} = 8$.

$= 7 - \left\{8 + 4\left[2 - (-3)^2\right]\right\}$ Subtract within the inner parentheses.

$= 7 - [8 + 4(2 - 9)]$ Continue simplifying within the inner parentheses. Simplify $(-3)^2$ to get 9.

$= 7 - [8 + 4(-7)]$ Simplify $(2 - 9)$ to get (-7).

$= 7 - (8 - 28)$ Multiply before adding or subtracting.

$= 7 - (-20)$ Simplify within parentheses.

$= 27$

Skill Practice 7   Simplify. $50 - \left\{2 - \left[\sqrt{121} + 3(-1 - 3)^2\right]\right\}$

Answer

7. 107

When simplifying an expression, particular care must be taken with expressions involving division and zero.

Division Involving Zero

To investigate division involving zero, consider the expressions $\frac{5}{0}$, $\frac{0}{5}$, and $\frac{0}{0}$ and their related multiplicative forms.

1. **Division by zero is undefined.**
   <u>Example</u>: $\frac{5}{0} = n$ implies that $n \cdot 0 = 5$. No number, n, satisfies this requirement.
2. **Zero divided by any nonzero number is zero.**
   <u>Example</u>: $\frac{0}{5} = 0$ implies that $0 \cdot 5 = 0$ which is a true statement.
3. We say that $\frac{0}{0}$ is **indeterminate** (cannot be uniquely determined). This concept is investigated in detail in a first course in calculus.
   <u>Example</u>: $\frac{0}{0} = n$ implies that $n \cdot 0 = 0$. This is true for any number n. Therefore, the quotient cannot be uniquely determined.

---

7. Simplify Algebraic Expressions

An algebraic **term** is a product of factors that may include constants and variables. An algebraic **expression** is a single term or the sum of two or more terms. For example, the expression

$$-3xz^2 + \left(-\frac{4}{b}\right) + z\sqrt{x - y} + 5 \text{ has four terms.}$$

The terms $-3xz^2$, $-\frac{4}{b}$, and $z\sqrt{x - y}$ are **variable terms.** The term 5 is not subject to change and is called a **constant term.** The constant factor within each term is called the **coefficient** of the term. Writing the expressions as $-3xz^2 + \left(-4 \cdot \frac{1}{b}\right) + 1z\sqrt{x - y} + 5$, we identify the coefficients of the four terms as -3, -4, 1, and 5, respectively.

The properties of real numbers summarized in Table R-4 here and on page 11 are often helpful when working with algebraic expressions.

Table R-4   Properties of Real Numbers

Let a, b, and c represent real numbers or real-valued expressions.

Property	In Symbols and Words	Examples
Commutative property of addition	$a + b = b + a$ The order in which real numbers are added does not affect the sum.	<u>ex</u>: $4 + (-7) = -7 + 4$ <u>ex</u>: $6 + w = w + 6$
Commutative property of multiplication	$a \cdot b = b \cdot a$ The order in which real numbers are multiplied does not affect the product.	<u>ex</u>: $5 \cdot (-4) = -4 \cdot 5$ <u>ex</u>: $x \cdot 12 = 12x$
Associative property of addition	$(a + b) + c = a + (b + c)$ The order in which real numbers are grouped under addition does not affect the sum.	<u>ex</u>: $(3 + 5) + 2 = 3 + (5 + 2)$ <u>ex</u>: $-9 + (2 + t) = (-9 + 2) + t$ $= -7 + t$
Associative property of multiplication	$(a \cdot b) \cdot c = a \cdot (b \cdot c)$ The order in which real numbers are grouped under multiplication does not affect the product.	<u>ex</u>: $(6 \cdot 7) \cdot 3 = 6 \cdot (7 \cdot 3)$ <u>ex</u>: $8 \cdot \left(\frac{1}{8} \cdot y\right) = \left(8 \cdot \frac{1}{8}\right) \cdot y$ $= 1y$
Identity property of addition	$a + 0 = a$ and $0 + a = a$ The number 0 is called the **identity element of addition** because any number plus 0 is the number itself.	<u>ex</u>: $-5 + 0 = -5$ <u>ex</u>: $0 + \sqrt{z} = \sqrt{z}$
Identity property of multiplication	$a \cdot 1 = a$ and $1 \cdot a = a$ The number 1 is called the **identity element of multiplication** because any number times 1 is the number itself.	<u>ex</u>: $\sqrt{2} \cdot 1 = \sqrt{2}$ <u>ex</u>: $1 \cdot (2w + 3) = 2w + 3$

(Continued)

Property	In Symbols and Words	Examples
Inverse property of addition	$a + (-a) = 0$ and $(-a) + a = 0$ For any real number a, the value $-a$ is called the **additive inverse of a** (also called the **opposite of a**). The sum of any number and its additive inverse is the identity element for addition, 0.	ex: $4\pi + (-4\pi) = 0$ ex: $-ab^2 + ab^2 = 0$
Inverse property of multiplication	$a \cdot \frac{1}{a} = 1$ and $\frac{1}{a} \cdot a = 1$ where $a \neq 0$ For any nonzero real number a, the value $\frac{1}{a}$ is called the **multiplicative inverse of a** (also called the **reciprocal of a**). The product of any nonzero number and its multiplicative inverse is the identity element, 1.	ex: $-5 \cdot \left(-\frac{1}{5}\right) = 1$ ex: $\frac{1}{x^2} \cdot x^2 = 1$ for $x \neq 0$ *Note*: The number zero does not have a multiplicative inverse (reciprocal).
Distributive property of multiplication over addition	$a \cdot (b + c) = a \cdot b + a \cdot c$ The product of a number and a sum equals the sum of the products of the number and each term in the sum.	ex: $4 \cdot (5 + x) = 4 \cdot 5 + 4 \cdot x$ $\qquad = 20 + 4x$ ex: $2(x + \sqrt{3}) = 2x + 2\sqrt{3}$

The distributive property is used to "clear" parentheses when a factor outside parentheses is multiplied by multiple terms inside parentheses.

EXAMPLE 8   Applying the Distributive Property

Apply the distributive property.

a. $4(2x^2 - 5.1x + 3)$ **b.** $-(9y - \sqrt{2})$

Solution:

a. $4(2x^2 - 5.1x + 3)$

$= 4[2x^2 + (-5.1x) + 3]$ — Write subtraction in terms of addition.
$= 4 \cdot (2x^2) + 4 \cdot (-5.1x) + 4 \cdot (3)$ — Apply the distributive property.
$= 8x^2 - 20.4x + 12$ — Simplify.

b. $-(9y - \sqrt{2})$

$= -1[(9y + (-\sqrt{2})]$ — The negative sign in front of parentheses can be interpreted as -1 times the expression in parentheses.

$= -9y + \sqrt{2}$ — Apply the distributive property.

> **TIP** A negative factor preceding the parentheses will change the signs of the terms within parentheses.
> $-(+9y - \sqrt{2})$
> $= -9y + \sqrt{2}$

Skill Practice 8   Apply the distributive property.
a. $5(6y^3 - 4y^2 + 7)$ **b.** $-2(5t - \pi)$

Two terms in an expression are **like terms** if they have the same variables and the corresponding variables are raised to the same power. We can combine like terms by using the distributive property. For example,

$8x^2 + 6x^2 - x^2$
$= 8x^2 + 6x^2 - 1x^2$ — Note that the coefficient on the third term is -1.
$= (8 + 6 - 1)x^2$ — Apply the distributive property.
$= 13x^2$ — Simplify.

Answers
8. a. $30y^3 - 20y^2 + 35$
 b. $-10t + 2\pi$

> **TIP** Although the distributive property is used to add and subtract like terms, it is tedious to write each step. Adding or subtracting like terms can also be done by combining the coefficients and leaving the variable factor unchanged.
>
> $$8x^2 + 6x^2 - 1x^2 = 13x^2 \qquad \text{This method will be used throughout the text.}$$

In Example 9, we simplify an expression by applying the distributive property to "clear" parentheses and combine like terms.

EXAMPLE 9   Clearing Parentheses and Combining Like Terms

Simplify.

a. $5 - 2(4c - 8d) + 3(1 - d) + c$

b. $-3x^2 - \left[8 + \dfrac{1}{2}(2x^2 - 6) - 4x^2 \right]$

Solution:

> **TIP** After applying the distributive property, the original parentheses are removed. For this reason, we often call this process "*clearing parentheses*."

a. $5 - 2(4c - 8d) + 3(1 - d) + c$ Apply the distributive property to clear parentheses.

$= 5 - 8c + 16d + 3 - 3d + c$ Combine like terms.

$= 8 - 7c + 13d$

b. $-3x^2 - \left[8 + \dfrac{1}{2}(2x^2 - 6) - 4x^2 \right]$ Apply the distributive property.

$= -3x^2 - [8 + x^2 - 3 - 4x^2]$ Combine like terms inside brackets.

$= -3x^2 - [-3x^2 + 5]$ Apply the distributive property.

$= -3x^2 + 3x^2 - 5$ Combine like terms.

$= -5$

Skill Practice 9   Clear parentheses and combine like terms.

a. $12 - 3(5x - 2y) + 5(3 - x) - y$

b. $4w^3 - \left[3 - \dfrac{1}{4}(4 + 8w^3) - w^3 \right] + 2$

---

8. Write Algebraic Models

An important skill in mathematics and science is to develop mathematical models. Example 10 offers practice writing algebraic expressions based on verbal statements.

EXAMPLE 10   Writing an Algebraic Model

a. The maximum recommended heart rate M for adults is the difference of 220 and the person's age a. Write a model to represent an adult's maximum recommended heart rate in terms of age.

b. After eating at a restaurant, it is customary to leave a tip t for the server for at least 15% of the cost of the meal c. Write a model to represent the amount of the tip based on the cost of the meal.

Answers
9. a. $-20x + 5y + 27$ **b.** $7w^3$

Solution:

a. $M = 220 - a$ The word "difference" implies subtraction in the order given.

b. $t \geq 0.15c$ 15% of c implies multiplication.

Skill Practice 10

a. The sale price S on a lawn mower is the difference of the original price P and the amount of discount D. Write a model to represent the sale price.

b. The amount of simple interest I earned on a certain certificate of deposit is 4.65% of the amount of principal invested P. Write a model for the amount of interest.

Answers

10. a. $S = P - D$

   b. $I = 0.0465P$

SECTION R.1   Practice Exercises

Concept Connections

1. A(n) _____ is a collection of items called elements.

2. $\mathbb{Z} = \{..., -3, -2, -1, 0, 1, 2, 3, ...\}$ is called the set of _____.

3. $\mathbb{R}$ is the notation used to denote the set of _____ _____.

4. A(n) _____ number is a real number that cannot be expressed as a ratio of two integers.

5. Write an absolute value expression to represent the distance between a and b on the real number line.

6. The _____ properties of addition and multiplication indicate that the order in which two real numbers are added or multiplied does not affect the sum or product.

7. The associative property of addition indicates that $a + (b + c) =$ _____, and the associative property of multiplication indicates that $a(bc) =$ _____.

8. The statement $a(b + c) = ab + ac$ represents the _____ property of multiplication over addition.

Objective 1: Identify Subsets of the Set of Real Numbers

For Exercises 9–10, write an English sentence to represent the algebraic statement.

9. a. $3 \in \mathbb{N}$ **b.** $-3.1 \notin \mathbb{W}$

10. a. $\dfrac{2}{5} \in \mathbb{Q}$ **b.** $\pi \notin \mathbb{Q}$

For Exercises 11–12, determine whether the statement is true or false.

11. a. $-5 \in \mathbb{N}$ **b.** $-5 \in \mathbb{W}$

   c. $-5 \in \mathbb{Z}$ **d.** $-5 \in \mathbb{Q}$

12. a. $\dfrac{1}{3} \in \mathbb{N}$ **b.** $\dfrac{1}{3} \in \mathbb{W}$

   c. $\dfrac{1}{3} \in \mathbb{Z}$ **d.** $\dfrac{1}{3} \in \mathbb{Q}$

13. Refer to $A = \left\{\sqrt{5}, 0.\overline{3}, 0.33, -0.9, -12, \frac{11}{4}, 6, \frac{\pi}{6}\right\}$. Determine which elements belong to the given set. **(See Example 1)**

   a. $\mathbb{N}$ **b.** $\mathbb{W}$

   c. $\mathbb{Z}$ **d.** $\mathbb{Q}$

   e. $\mathbb{H}$ **f.** $\mathbb{R}$

14. Refer to $B = \left\{\frac{\pi}{2}, 0, -4, 0.\overline{48}, 1, -\sqrt{13}, 9.4\right\}$. Determine which elements belong to the given set.

   a. $\mathbb{N}$ **b.** $\mathbb{W}$

   c. $\mathbb{Z}$ **d.** $\mathbb{Q}$

   e. $\mathbb{H}$ **f.** $\mathbb{R}$

Objective 2: Use Inequality Symbols and Interval Notation

For Exercises 15–20, write each statement as an inequality.

15. a is at least 5.

16. b is at most -6.

17. $3c$ is no more than 9.

18. $8d$ is no less than 16.

19. The quantity $(m + 4)$ exceeds 70.

20. The quantity $(n - 7)$ is approximately equal to 4.

For Exercises 21–28, determine whether the statement is true or false.

21. $3.14 < \pi$

22. $-7 < -\sqrt{7}$

23. $6.7 \geq 6.7$

24. $-2.1 \leq -2.1$

25. $6.\overline{15} > 6.1\overline{5}$

26. $2.9\overline{3} > 2.\overline{93}$

27. $-\dfrac{9}{7} < -\dfrac{11}{8}$

28. $-\dfrac{5}{3} < -\dfrac{9}{5}$

For Exercises 29–34, write the interval notation and set-builder notation for each given graph. (See Example 2)

29.
-7

30.
2

31.
4.1

32.
-2.93

33.
$-6 \qquad 0$

34.
$2 \qquad 8$

For Exercises 35–40, graph the given set and write the corresponding interval notation. (See Example 2)

35. $\{x \,|\, x \leq 6\}$

36. $\{x \,|\, x < -4\}$

37. $\left\{x \,\middle|\, -\dfrac{7}{6} < x \leq \dfrac{1}{3}\right\}$

38. $\left\{x \,\middle|\, -\dfrac{4}{3} \leq x < \dfrac{7}{4}\right\}$

39. $\{x \,|\, 4 < x\}$

40. $\{x \,|\, -3 \leq x\}$

For Exercises 41–46, interval notation is given for several sets of real numbers. Graph the set and write the corresponding set-builder notation. (See Example 2)

41. $(-3, 7]$

42. $[-4, -1)$

43. $(-\infty, 6.7]$

44. $(-\infty, -3.2)$

45. $\left[-\dfrac{3}{5}, \infty\right)$

46. $\left(\dfrac{7}{8}, \infty\right)$

Objective 3: Find the Union and Intersection of Sets

For Exercises 47–50, refer to sets A, B, C, X, Y, and Z and find the union or intersection of sets as indicated. (See Example 3)

$A = \{0, 4, 8, 12\}, \qquad B = \{0, 3, 6, 9, 12\}, \qquad C = \{-2, 4, 8\}$

$X = \{1, 2, 3, 4, 5\}, \qquad Y = \{1, 2, 3\}, \qquad Z = \{6, 7, 8\}$

47. a. $A \cup B$ **b.** $A \cap B$ **c.** $A \cup C$

    d. $A \cap C$ **e.** $B \cup C$ **f.** $B \cap C$

48. a. $X \cup Z$ **b.** $Y \cup Z$ **c.** $Y \cap Z$

    d. $X \cup Y$ **e.** $X \cap Y$ **f.** $X \cap Z$

49. a. $A \cup X$ **b.** $A \cap Z$ **c.** $C \cap Y$

    d. $B \cap Y$ **e.** $C \cup Z$ **f.** $A \cup Z$

50. a. $A \cap X$ **b.** $B \cup Z$ **c.** $C \cup Y$

    d. $B \cap Z$ **e.** $C \cap Z$ **f.** $A \cup Y$

51. Refer to sets C, D, and F and find the union or intersection of sets as indicated. Write the answers in set notation. **(See Example 4)**

$C = \{x \mid x < 9\}, \quad D = \{x \mid x \geq -1\}, \quad F = \{x \mid x < -8\}$

a. $C \cup D$ **b.** $C \cap D$ **c.** $C \cup F$

d. $C \cap F$ **e.** $D \cup F$ **f.** $D \cap F$

52. Refer to sets M, N, and P and find the union or intersection of sets as indicated. Write the answers in set notation.

$M = \{y \mid y \geq -3\}, \quad N = \{y \mid y \geq 5\}, \quad P = \{y \mid y < 0\}$

a. $M \cup N$ **b.** $M \cap N$ **c.** $M \cup P$

d. $M \cap P$ **e.** $N \cup P$ **f.** $N \cap P$

For Exercises 53–56, find the union or intersection of the given intervals. Write the answers in interval notation.

53. a. $(-\infty, 4) \cup (-2, 1]$ **b.** $(-\infty, 4) \cap (-2, 1]$

54. a. $[0, 5) \cup [-1, \infty)$ **b.** $[0, 5) \cap [-1, \infty)$

55. a. $(-\infty, 5) \cup [3, \infty)$ **b.** $(-\infty, 5) \cap [3, \infty)$

56. a. $(-\infty, -1] \cup [-4, \infty)$ **b.** $(-\infty, -1] \cap [-4, \infty)$

Objective 4: Evaluate Absolute Value Expressions

For Exercises 57–68, simplify each expression by writing the expression without absolute value bars. (See Example 5)

57. $|-6|$ **58.** $|-4|$ **59.** $|0|$

60. $|1|$ **61.** $|\sqrt{17} - 5|$ **62.** $|\sqrt{6} - 6|$

63. a. $|\pi - 3|$ **64. a.** $|m - 11|$ for $m \geq 11$ **65. a.** $|x + 2|$ for $x \geq -2$

    b. $|3 - \pi|$ **b.** $|m - 11|$ for $m < 11$ **b.** $|x + 2|$ for $x < -2$

66. a. $|t + 6|$ for $t < -6$ **67. a.** $\dfrac{|z - 5|}{z - 5}$ for $z > 5$ **68. a.** $\dfrac{7 - x}{|7 - x|}$ for $x < 7$

    b. $|t + 6|$ for $t \geq -6$ **b.** $\dfrac{|z - 5|}{z - 5}$ for $z < 5$ **b.** $\dfrac{7 - x}{|7 - x|}$ for $x > 7$

Objective 5: Use Absolute Value to Represent Distance

For Exercises 69–74, write an absolute value expression to represent the distance between the two points on the number line. Then simplify without absolute value bars. (See Example 6)

69. 1 and 6   **70.** 2 and 9   **71.** 3 and -4

72. -8 and 2   **73.** 6 and 2π **74.** 3 and π

Objective 6: Apply the Order of Operations

For Exercises 75–88, simplify the expressions. (See Example 7)

75. a. 4^2 **b.** $(-4)^2$ **c.** -4^2 **d.** $\sqrt{4}$ **e.** $-\sqrt{4}$ **f.** $\sqrt{-4}$

76. a. 9^2 **b.** $(-9)^2$ **c.** -9^2 **d.** $\sqrt{9}$ **e.** $-\sqrt{9}$ **f.** $\sqrt{-9}$

77. a. $\sqrt[3]{8}$ **b.** $\sqrt[3]{-8}$ **c.** $-\sqrt[3]{8}$ **d.** $\sqrt{100}$ **e.** $\sqrt{-100}$ **f.** $-\sqrt{100}$

78. a. $\sqrt[3]{27}$ **b.** $\sqrt[3]{-27}$ **c.** $-\sqrt[3]{27}$ **d.** $\sqrt{49}$ **e.** $\sqrt{-49}$ **f.** $-\sqrt{49}$

79. $20 - 12(36 \div 3^2 \div 2)$ **80.** $200 - 2^2\left(6 \div \dfrac{1}{2} \cdot 4\right)$ **81.** $6 - \{-12 + 3[(1 - 6)^2 - 18]\}$

82. $-5 - \{4 - 6[(2 - 8)^2 - 31]\}$ **83.** $-4 \cdot \left(\dfrac{2}{5} - \dfrac{7}{10}\right)^2$ **84.** $6 \cdot \left[\left(\dfrac{1}{3}\right)^2 - \left(\dfrac{1}{2}\right)^2\right]$

85. $9 - (6 + ||3 - 7| - 8|) \div \sqrt{25}$ **86.** $8 - 2(4 + ||2 - 5| - 5|) \div \sqrt{9}$ **87.** $\dfrac{|11 - 13| - 4 \cdot 2}{\sqrt{12^2 + 5^2} - 3 - 10}$

88. $\dfrac{(4 - 9)^2 + 2^2 - 3^2}{|-7 + 4| + (-12) \div 4}$

Objective 7: Simplify Algebraic Expressions

For Exercises 89–92, apply the commutative property of addition.

89. $7 + x$

90. $9 + z$

91. $-3 + w$

92. $-11 + p$

For Exercises 93–96, apply the associative property of addition or multiplication. Then simplify if possible.

93. $(t + 3) + 9$

94. $(c + 4) + 5$

95. $\frac{1}{5}(5w)$

96. $-\frac{4}{9}\left(-\frac{9}{4}p\right)$

For Exercises 97–102, combine like terms.

97. $-14w^3 - 3w^3 + w^3$

98. $12t^5 - t^5 - 6t^5$

99. $3.9x^3y - 2.2xy^3 + 5.1x^3y - 4.7xy^3$

100. $0.004m^4n - 0.005m^3n^2 - 0.01m^4n + 0.007m^3n^2$

101. $\frac{1}{3}c^7d + \frac{1}{2}cd^7 - \frac{2}{5}c^7d - 2cd^7$

102. $\frac{1}{10}yz^4 - \frac{3}{4}y^4z + yz^4 + \frac{3}{2}y^4z$

For Exercises 103–108, apply the distributive property. (See Example 8)

103. $-(4x - \pi)$

104. $-\left(\frac{1}{2}k - \sqrt{7}\right)$

105. $-8(3x^2 + 2x - 1)$

106. $-6(-5y^2 - 3y + 4)$

107. $\frac{2}{3}(-6x^2y - 18yz^2 + 2z^3)$

108. $\frac{3}{4}(12p^5q - 8p^4q^2 - 6p^3q)$

For Exercises 109–116, simplify each expression. (See Example 9)

109. $2(4w + 8) + 7(2w - 4) + 12$

110. $3(2z - 4) + 8(z - 9) + 84$

111. $-(4u - 8v) - 3(7u - 2v) + 2v$

112. $-(10x - z) - 2(8x - 4z) - 3x$

113. $12 - 4[(8 - 2v) + 5(-3w - 4v)] - w$

114. $6 - 2[(9z + 6y) - 8(y - z)] - 11$

115. $2y^2 - \left[13 - \frac{2}{3}(6y^2 - 9) - 10\right] + 9$

116. $6 - \left[5t^2 - \frac{3}{4}(12 - 8t^2) + 5\right] + 11t^2$

Objective 8: Write Algebraic Models

117. Jake is 1 yr younger than Charlotte.

  a. Write a model for Jake's age J in terms of Charlotte's age C.

  b. Write a model for Charlotte's age C in terms of Jake's age J. (See Example 10)

118. For a recent NFL season, Aaron Rodgers had 9 more touchdown passes than Joe Flacco.

  a. Write a model for the number of touchdown passes R thrown by Rodgers in terms of the number thrown by Flacco F.

  b. Write a model for the number of touchdown passes F thrown by Flacco in terms of the number thrown by Rodgers R.

119. At the end of the summer, a store discounts an outdoor grill for at least 25% of the original price. If the original price is P, write a model for the amount of the discount D. (See Example 10)

120. When Ms. Celano has excellent service at a restaurant, the amount she leaves for a tip t is at least 20% of the cost of the meal c. Write a model representing the amount of the tip.

121. Suppose that an object is dropped from a height h. Its velocity v at impact with the ground is given by the square root of twice the product of the acceleration due to gravity g and the height h. Write a model to represent the velocity of the object at impact.

122. The height of a sunflower plant can be determined by the time t in weeks after the seed has germinated. Write a model to represent the height h if the height is given by the product of 8 and the square root of t.

123. A power company charges one household $0.12 per kilowatt-hour (kWh) and $14.89 in monthly taxes.

   a. Write a formula for the monthly charge C for this household if it uses k kilowatt-hours.

   b. Compute the monthly charge if the household uses 1200 kWh.

124. A utility company charges a base rate for water usage of $19.50 per month, plus $4.58 for every additional 1000 gal of water used over a 2000-gal base.

   a. Write a formula for the monthly charge C for a household that uses n thousand gallons over the 2000-gal base.

   b. Compute the cost for a family that uses a total of 6000 gal of water for a given month.

125. The cost C (in $) to rent an apartment is $640 per month, plus a $500 nonrefundable security deposit, plus a $200 nonrefundable deposit for each dog or cat.

   a. Write a formula for the total cost to rent an apartment for m months with n cats/dogs.

   b. Determine the cost to rent the apartment for 12 months, with 2 cats and 1 dog.

126. For a certain college, the cost C (in $) for taking classes the first semester is $105 per credit-hour, $35 for each lab, plus a one-time admissions fee of $40.

   a. Write a formula for the total cost to take n credit-hours and L labs the first semester.

   b. Determine the cost for the first semester if a student takes 12 credit-hours with 2 labs.

127. A hotel charges $159 per night plus an 11% nightly room tax.

   a. Write a formula to represent the total cost C for n nights in the hotel. (*Hint*: The total cost is the cost for n nights, plus the tax on the cost for n nights.)

   b. Determine the cost to stay in the hotel for four nights.

128. A hotel charges $149 per night plus a 16% nightly room tax. In addition, there is a one-time parking fee of $40.

   a. Write a formula to represent the total cost C for n nights in the hotel.

   b. Determine the cost to stay in the hotel for two nights.

Mixed Exercises

For Exercises 129–134, evaluate each expression for the given values of the variables.

129. $\dfrac{-b}{2a}$ for $a = -1, b = -6$

130. $\sqrt{b^2 - 4ac}$ for $a = 2, b = -6, c = 4$

131. $\sqrt{(x_2 - x_1)^2 + (y_2 - y_1)^2}$ for $x_1 = 2, x_2 = -1, y_1 = -4, y_2 = 1$

132. $\dfrac{y_2 - y_1}{x_2 - x_1}$ for $x_1 = -1.4, x_2 = 2, y_1 = 3.1, y_2 = -3.7$

133. $\dfrac{1}{3}\pi r^2 h$ for $\pi \approx \dfrac{22}{7}, r = 7, h = 6$

134. $\dfrac{4}{3}\pi r^3$ for $\pi \approx \dfrac{22}{7}, r = 3$

135. Under selected conditions, a sports car gets 15 mpg in city driving and 25 mpg for highway driving. The model $G = \frac{1}{15}c + \frac{1}{25}h$ represents the amount of gasoline used (in gal) for c miles driven in the city and h miles driven on the highway. Determine the amount of gas required to drive 240 mi in the city and 500 mi on the highway.

136. Under selected conditions, a sedan gets 22 mpg in city driving and 32 mpg for highway driving. The model $G = \frac{1}{22}c + \frac{1}{32}h$ represents the amount of gasoline used (in gal) for c miles driven in the city and h miles driven on the highway. Determine the amount of gas required to drive 220 mi in the city and 512 mi on the highway.

Write About It

137. When is a parenthesis used when writing interval notation?

138. When is a bracket used when writing interval notation?

139. Explain the difference between the commutative property of addition and the associative property of addition.

140. Explain why 0 has no multiplicative inverse.

Expanding Your Skills

141. If $n > 0$, then $n - |n| =$ _____.   **142.** If $n < 0$, then $n - |n| =$ _____.   **143.** If $n > 0$, then $n + |n| =$ _____.

144. If $n < 0$, then $n + |n| =$ _____.   **145.** If $n > 0$, then $-|n| =$ _____.   **146.** If $n < 0$, then $-|n| =$ _____.

For Exercises 147–150, determine the sign of the expression. Assume that a, b, and c are real numbers and $a < 0$, $b > 0$, and $c < 0$.

147. $\dfrac{ab^2}{c^3}$

148. $\dfrac{a^2c}{b^4}$

149. $\dfrac{b(a + c)^3}{a^2}$

150. $\dfrac{(a + b)^2(b + c)^4}{b}$

For Exercises 151–154, write the set as a single interval.

151. $(-\infty, 2) \cap (-3, 4] \cap [1, 3]$

152. $(-\infty, 5) \cap (-1, \infty) \cap [0, 3)$

153. $[(-\infty, -2) \cup (4, \infty)] \cap [-5, 3)$

154. $[(-\infty, 6) \cup (10, \infty)] \cap [8, 12)$

Technology Connections

For Exercises 155–158, use a calculator to approximate the expression to 2 decimal places.

155. $5000\left(1 + \dfrac{0.06}{12}\right)^{(12)(5)}$

156. $8500\left(1 + \dfrac{0.05}{4}\right)^{(4)(30)}$

157. $\dfrac{-3 + 5\sqrt{2}}{7}$

158. $\dfrac{6 - 3\sqrt{5}}{4}$

SECTION R.2   Integer Exponents and Scientific Notation

OBJECTIVES

1. Simplify Expressions with Zero and Negative Exponents
2. Apply Properties of Exponents
3. Apply Scientific Notation

1. Simplify Expressions with Zero and Negative Exponents

In Section R.1, we learned that exponents are used to represent repeated multiplication. Applications of exponents appear in many fields of study, including computer science. Computer engineers define a *bit* as a fundamental unit of information having just two possible values. These values are represented by either 0 or 1. A *byte* is usually taken as 8 bits. Computer programmers know that there are 2^n possible values for an n-bit variable. So 1 byte has $2^8 = 256$ possible values.

Three bytes are often used to represent color on a computer screen. The intensity of each of the colors red, green, and blue ranges from 0 to 255 (a total of 256 possible values each). So the number of colors that can be represented by this system is

$$2^8 \cdot 2^8 \cdot 2^8 = (256)(256)(256) = 16,777,216$$

There are over 16 million possible colors available using this system. For example, the color given by red 137, green 21, blue 131, is a deep pink. See Figure R-2.

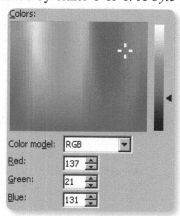

Figure R-2

The product $2^8 \cdot 2^8 \cdot 2^8$ can be visualized by expanding factors.

$$= \overbrace{(2 \cdot 2 \cdot 2 \cdot 2 \cdot 2 \cdot 2 \cdot 2 \cdot 2)}^{8 \text{ factors}} \overbrace{(2 \cdot 2 \cdot 2 \cdot 2 \cdot 2 \cdot 2 \cdot 2 \cdot 2)}^{8 \text{ factors}} \overbrace{(2 \cdot 2 \cdot 2 \cdot 2 \cdot 2 \cdot 2 \cdot 2 \cdot 2)}^{8 \text{ factors}}$$

$$= 2^{24} = 16{,}777{,}216$$

The same result can be obtained by adding exponents.

$$2^8 \cdot 2^8 \cdot 2^8 = 2^{8+8+8} = 2^{24} = 16{,}777{,}216$$

A similar approach can be used to simplify the product $b^m \cdot b^n$ for natural numbers m and n.

$$b^m \cdot b^n = \underbrace{(\overbrace{b \cdot b \cdot b \cdot \ldots \cdot b}^{m \text{ factors of } b})(\overbrace{b \cdot b \cdot b \cdot \ldots \cdot b}^{n \text{ factors of } b})}_{m + n \text{ factors of } b} = b^{m+n}$$

Add the exponents.

The base is unchanged.

We would like to extend this result to expressions where m and n are negative integers or zero. For example,

$$\underline{b^0} \cdot b^4 = b^{0+4} = b^4 \qquad \text{For this to be true, the value } b^0 \text{ must be 1.}$$
$$\underline{1} \cdot b^4 = b^4$$

Now consider an example involving a negative exponent.

$$\underline{b^{-4}} \cdot b^4 = b^{-4+4} = b^0 = 1 \qquad \text{For the product } b^{-4} \cdot b^4 \text{ to be equal to 1, it follows}$$
$$\underline{\frac{1}{b^4}} \cdot b^4 = 1 \qquad\qquad\qquad \text{that } b^{-4} \text{ must be the reciprocal of } b^4. \text{ That is, } b^{-4} = \frac{1}{b^4}.$$

These observations lead to two important definitions.

Definition of b^0 and b^{-n}	
If b is a nonzero real number and n is a positive integer, then	
$b^0 = 1$	<u>Examples:</u> $(5000)^0 = 1$ and $(-3x)^0 = 1$ for $x \neq 0$
$b^{-n} = \left(\dfrac{1}{b}\right)^n = \dfrac{1}{b^n}$	<u>Examples:</u> $4^{-1} = \left(\dfrac{1}{4}\right)^1 = \dfrac{1}{4}$ and $x^{-3} = \left(\dfrac{1}{x}\right)^3 = \dfrac{1}{x^3}$ for $x \neq 0$

The definitions given here have two important restrictions.

By definition, $b^0 = 1$ provided that $b \neq 0$. Therefore,

- The value of 0^0 is not defined here. The value of 0^0 is said to be indeterminate and is examined in calculus.

For a positive integer n, by definition, $b^{-n} = \dfrac{1}{b^n}$ provided that $b \neq 0$. Therefore,

- The value of 0^{-n} is not defined here.

All examples and exercises in the text will be given under the assumption that the variable expressions avoid these restrictions. For example, the expression x^0 will be stated with the implied restriction that $x \neq 0$.

EXAMPLE 1   **Simplifying Expressions with a Zero Exponent**

Simplify.

a. $6y^0$ **b.** $(6y)^0$ **c.** -5^0

Solution:

a. $6y^0 = 6 \cdot y^0 = 6 \cdot 1 = 6$ The base for the exponent of 0 is y, not $6y$.
                                                 The product of 6 and y^0 is $6 \cdot 1 = 6$.

b. $(6y)^0 = 1$ The base for the exponent of 0 is $6y$.
                                                 By definition, a nonzero base to the zero
                                                 power is 1.

c. $-5^0 = -1 \cdot 5^0 = -1 \cdot 1 = -1$ This is interpreted as the opposite of 5^0 or $-1 \cdot 5^0$.

Skill Practice 1   Simplify.

a. $4z^0$ **b.** $(4z)^0$ **c.** -4^0

EXAMPLE 2   **Simplifying Expressions with Negative Exponents**

Simplify.

a. 5^{-2} **b.** $\dfrac{1}{p^{-4}}$ **c.** $5x^{-8}y^2$

Solution:

a. $5^{-2} = \dfrac{1}{5^2}$ or $\dfrac{1}{25}$ Rewrite the expression using the reciprocal of 5 and change
                                                            the exponent to positive 2.

b. $\dfrac{1}{p^{-4}} = \dfrac{1}{\dfrac{1}{p^4}} = 1 \cdot \dfrac{p^4}{1} = p^4$ Rewrite p^{-4} as $\dfrac{1}{p^4}$. Then divide fractions.

c. $5x^{-8}y^2 = 5 \cdot x^{-8} \cdot y^2$ The exponent of -8 applies to x only, not to the
                                                     factors of 5 or y.

$= 5 \cdot \dfrac{1}{x^8} \cdot y^2 = \dfrac{5y^2}{x^8}$

Skill Practice 2   Simplify.

a. 2^{-3} **b.** $\dfrac{1}{n^{-5}}$ **c.** $-3a^4b^{-9}$

2. Apply Properties of Exponents

The property $b^m b^n = b^{m+n}$ is one of several important properties of exponents that
can be used to simplify an algebraic expression (Table R-5).

Answers

1. a. 4   **b.** 1   **c.** -1

2. a. $\dfrac{1}{2^3}$ or $\dfrac{1}{8}$ **b.** n^5

c. $\dfrac{-3a^4}{b^9}$

Table R-5 Properties of Exponents

Let a and b be real numbers and m and n be integers.*

Property	Example	Expanded Form
$b^m \cdot b^n = b^{m+n}$	$x^4 \cdot x^3 = x^{4+3} = x^7$	$x^4 \cdot x^3 = (x \cdot x \cdot x \cdot x)(x \cdot x \cdot x) = x^7$
$\dfrac{b^m}{b^n} = b^{m-n}$	$\dfrac{t^6}{t^2} = t^{6-2} = t^4$	$\dfrac{t^6}{t^2} = \dfrac{t \cdot t \cdot t \cdot t \cdot t \cdot t}{t \cdot t} = t^4$
$(b^m)^n = b^{m \cdot n}$	$(x^2)^3 = x^{2 \cdot 3} = x^6$	$(x^2)^3 = (x^2)(x^2)(x^2) = x^6$
$(ab)^m = a^m b^m$	$(4x)^3 = 4^3 x^3$	$(4x)^3 = (4x)(4x)(4x)$ $= (4 \cdot 4 \cdot 4)(x \cdot x \cdot x)$ $= 4^3 x^3$
$\left(\dfrac{a}{b}\right)^m = \dfrac{a^m}{b^m}$	$\left(\dfrac{2}{y}\right)^2 = \dfrac{2^2}{y^2}$	$\left(\dfrac{2}{y}\right)^2 = \left(\dfrac{2}{y}\right) \cdot \left(\dfrac{2}{y}\right) = \dfrac{2^2}{y^2}$

*The properties are stated under the assumption that the variables are restricted to avoid the expressions 0^0 and $\frac{1}{0}$. Expressions of the form 0^0 and $\frac{1}{0}$ are said to be indeterminate and are examined in calculus.

EXAMPLE 3 **Simplifying Expressions Containing Exponents**

Simplify. Write the answers with positive exponents only.

a. $13^0 + \left(\dfrac{1}{3}\right)^{-2} + \left(\dfrac{1}{9}\right)^{-1}$ **b.** $\left(\dfrac{14a^2 b^7}{2a^5 b}\right)^{-2}$ **c.** $(-3x)^{-4}(4x^{-2}y^3)^3$

Solution:

a. $13^0 + \left(\dfrac{1}{3}\right)^{-2} + \left(\dfrac{1}{9}\right)^{-1}$

$= 1 + 3^2 + 9^1$ By definition, $13^0 = 1$. Also simplify expressions with negative exponents.

$= 1 + 9 + 9$

$= 19$ Add.

b. $\left(\dfrac{14a^2 b^7}{2a^5 b}\right)^{-2} = (7a^{2-5} b^{7-1})^{-2}$ Simplify within parentheses first. To divide common bases, subtract the exponents.

$= (7a^{-3} b^6)^{-2}$

$= (7)^{-2}(a^{-3})^{-2}(b^6)^{-2}$ Raise each factor within parentheses to the -2 power.

$= 7^{-2} a^6 b^{-12}$

$= \dfrac{1}{7^2} \cdot a^6 \cdot \dfrac{1}{b^{12}}$ Simplify factors with negative exponents.

$= \dfrac{a^6}{49b^{12}}$ Simplify.

c. $(-3x)^{-4}(4x^{-2}y^3)^3$

$= (-3)^{-4} x^{-4} \cdot 4^3 (x^{-2})^3 (y^3)^3$ Raise each factor inside parentheses to the power outside parentheses.

$= (-3)^{-4} \cdot 4^3 \cdot x^{-4} \cdot x^{-6} \cdot y^9$ Regroup factors.

$= (-3)^{-4} \cdot 4^3 \cdot x^{-10} \cdot y^9$ To multiply the factors of x, add the exponents.

$= \dfrac{4^3 y^9}{(-3)^4 x^{10}}$ Simplify factors with negative exponents.

$= \dfrac{64 y^9}{81 x^{10}}$ Simplify.

Skill Practice 3   Simplify. Write the answers with positive exponents only.

a. $\left(\dfrac{2}{3}\right)^{-2} - 4^{-1} + 9^{0}$ **b.** $(3w)^{-3}(5wt^5)^2$ **c.** $\left(\dfrac{26c^5d^{-2}}{13cd^8}\right)^{-3}$

---

3. Apply Scientific Notation

In many applications of science, technology, and business we encounter very large or very small numbers. For example:

- eBay Inc. purchased the Internet communications company, Skype Technologies, for approximately \$2,600,000,000. (*Source:* www.ebay.com)
- The diameter of a capillary is measured as 0.000 005 m.
- The mean surface temperature of the planet Saturn is $-300°F$.

Very large and very small numbers are sometimes cumbersome to write because they contain numerous zeros. Furthermore, it is difficult to determine the location of the decimal point when performing calculations with such numbers. For these reasons, scientists will often write numbers using scientific notation.

Scientific Notation

A number expressed in the form $a \times 10^n$, where $1 \le |a| < 10$ and n is an integer is said to be in **scientific notation.**

Examples

Skype Purchase	Capillary Size	Saturn Temp.
\$2,600,000,000	0.000 005 m	$-300°F$
$= \$2.6 \times 1{,}000{,}000{,}000$	$= 5.0 \times 0.000001$ m	$= -3 \times 100°F$
$= \$2.6 \times 10^9$	$= 5.0 \times 10^{-6}$ m	$= -3 \times 10^2 \ °F$

To write a number in scientific notation, the number of positions that the decimal point must be moved determines the power of 10. Numbers 10 or greater require a positive exponent on 10. Numbers between 0 and 1 require a negative exponent on 10.

EXAMPLE 4   **Writing Numbers in Scientific Notation**

Write the numerical values in scientific notation.
   a. The size of the smallest visible object in an optical microscope is 0.0000002 m.
   b. One estimate for the number of stars in the Milky Way is 230 billion.
   c. The recommended daily intake of calcium is 1.2 g.

Solution:

a. 0.0000002 m $= 2.0 \times 10^{-7}$ m   The number 0.0000002 is between 0 and 1. Use a negative power of 10.

7 place positions

b. 230 billion $= 230{,}000{,}000{,}000$ First write 230 billion in standard form.

$230{,}000{,}000{,}000$ stars $= 2.3 \times 10^{11}$ stars   The number 230 billion is greater than 10. Use a positive power of 10.

11 place positions

c. 1.2 g $= 1.2 \times 10^0$ g   The decimal point is moved zero units, so the exponent on 10 is 0.

Answers
3. a. 3   **b.** $\dfrac{25t^{10}}{27w}$ **c.** $\dfrac{d^{30}}{8c^{12}}$

Skill Practice 4   Write the numerical values in scientific notation.

a. Salmonella bacteria are elongated bacteria and average 0.0000035 m in length.

b. The distance from Earth to Barnard's Star is 32,000,000,000,000 mi.

c. The average weight of a newborn baby is 7.5 lb.

EXAMPLE 5   **Writing Numbers in Standard Decimal Notation**

Write the numerical values in standard decimal notation.

a. The temperature at the core of the Sun is estimated to be 1.36×10^7 °C.

b. The thickness of a dollar bill is approximately 3.9×10^{-3} in.

Solution:

a. 1.36×10^7 °C = 13,600,000 °C

7 place positions

10^7 is greater than 10. Move the decimal point 7 places to the right. Insert zeros to the right as needed.

b. 3.9×10^{-3} in. = 0.0039 in.

3 place positions

10^{-3} is between 0 and 1. Move the decimal point 3 places to the left. Insert zeros to the left as needed.

Skill Practice 5   Write the numerical values in standard decimal notation.

a. Alaska is the largest state geographically with a land area of 5.86×10^5 mi².

b. A doctor orders 2.0×10^{-2} g of the drug atropine given by injection.

Example 6 demonstrates the process to multiply and divide numbers written in scientific notation.

EXAMPLE 6   **Performing Calculations with Scientific Notation**

a. A light-year is the distance that light travels in 1 yr. If light travels at a speed of 6.7×10^8 mph, how far will it travel in 1 yr (8.76×10^3 hr)?

b. California has a land area of 1.56×10^5 mi². If the population of California for a recent year was 3.9×10^7, determine the population density (number of people per square mile).

Solution:

a. Distance = (Rate)(Time)

$= (6.7 \times 10^8 \text{ mph})(8.76 \times 10^3 \text{ hr})$

$= (6.7)(8.76) \times (10^8)(10^3) \text{ mi}$

Regroup factors. Multiply and add the powers of 10.

$= 58.692 \times 10^{11} \text{ mi}$

The number 58.692 is not between 1 and 10. Rewrite this as 5.8692×10^1.

$= (5.8692 \times 10^1) \times 10^{11} \text{ mi}$

$= 5.8692 \times 10^{12} \text{ mi}$

One light-year is approximately 5.87 trillion miles.

b. $\dfrac{3.9 \times 10^7 \text{ people}}{1.56 \times 10^5 \text{ mi}^2} = \left(\dfrac{3.9}{1.56}\right) \times \left(\dfrac{10^7}{10^5}\right) \text{ people/mi}^2$

Population density is the number of people per square mile.

$= 2.5 \times 10^2 \text{ people/mi}^2$

At that time, California had a population density of 2.5×10^2 people/mi² or 250 people/mi².

Answers

4. a. 3.5×10^{-6} m
  b. 3.2×10^{13} mi
  c. 7.5×10^0 lb
5. a. 586,000 mi²
  b. 0.02 g

> ### Skill Practice 6
>
> **a.** A satellite travels 1.72×10^4 mph. How far does it travel in 24 hr $(2.4 \times 10^1$ hr)?
> **b.** The land area of Texas is 2.6×10^5 mi². If the population of Texas for a recent year was 2.5×10^7, determine the population density.

The calculations for Example 6 can also be performed on a calculator.

TECHNOLOGY CONNECTIONS

Using Scientific Notation on a Calculator

On many calculators, the EE key is used to enter the exponent for a number in scientific notation. The result on the screen shows the symbol E to indicate the exponent for the power of 10. For example, the number 5.8692E12 is read as 5.8692×10^{12}.

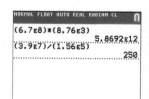

Answers
6. a. 4.128×10^5 mi
b. Approximately 9.6×10^1 people/mi² or 96 people/mi²

SECTION R.2   Practice Exercises

Concept Connections

1. For a nonzero real number b, the value of $b^0 = $ _____.

2. For a nonzero real number b, the value $b^{\boxed{}} = \dfrac{1}{b^n}$.

3. A number expressed in the form $a \times 10^n$, where $1 \le |a| < 10$ and n is an integer is said to be written in _____ notation.

4. From the properties of exponents, $b^m b^n = b^{\boxed{}}$.

5. If $b \ne 0$, then $\dfrac{b^m}{b^n} = b^{\boxed{}}$.

6. From the properties of exponents, $(b^m)^n = b^{\boxed{}}$.

Objective 1: Simplify Expressions with Zero and Negative Exponents

For Exercises 7–14, simplify each expression. (See Examples 1–2)

7. a. 8^0 **b.** -8^0 **c.** $8x^0$ **d.** $(8x)^0$

8. a. 7^0 **b.** -7^0 **c.** $7y^0$ **d.** $(7y)^0$

9. a. $\left(-\dfrac{2}{3}\right)^0$ **b.** $-\dfrac{2^0}{3}$ **c.** $-\dfrac{2}{3}p^0$ **d.** $\left(-\dfrac{2}{3}p\right)^0$

10. a. $\left(-\dfrac{3}{7}\right)^0$ **b.** $-\dfrac{3^0}{7}$ **c.** $-\dfrac{3}{7}w^0$ **d.** $\left(-\dfrac{3}{7}w\right)^0$

11. a. 8^{-2} **b.** $8x^{-2}$ **c.** $(8x)^{-2}$ **d.** -8^{-2}

12. a. 7^{-2} **b.** $7y^{-2}$ **c.** $(7y)^{-2}$ **d.** -7^{-2}

13. a. $\dfrac{1}{q^{-2}}$ **b.** q^{-2} **c.** $5p^3q^{-2}$ **d.** $5p^{-3}q^2$

14. a. $\dfrac{1}{t^{-4}}$ **b.** t^{-4} **c.** $11t^{-4}u^2$ **d.** $11t^4u^{-2}$

Objective 2: Apply Properties of Exponents

For Exercises 15–64, use the properties of exponents to simplify each expression. (See Example 3)

15. $2^5 \cdot 2^7$

16. $4^3 \cdot 4^8$

17. $x^7 \cdot x^6 \cdot x^{-2}$

18. $y^{-3} \cdot y^7 \cdot y^4$

19. $(-3c^2d^7)(4c^{-5}d)$

20. $(-7m^{-3}n^{-8})(3m^{-5}n)$

21. $\dfrac{y^{-3}y^6}{y^2}$

22. $\dfrac{z^{-8}z^{12}}{z^3}$

23. $\dfrac{6^5}{6^8}$

24. $\dfrac{7^2}{7^4}$

25. $\dfrac{18k^2p^9}{27k^5p^2}$

26. $\dfrac{10a^3b^{11}}{25a^7b^3}$

27. $\dfrac{2m^{-6}n^4}{6m^{-2}n^{-1}}$

28. $\dfrac{4p^{-7}q^5}{2p^{-2}q^{-2}}$

29. $(4^2)^3$

30. $(2^3)^5$

31. $(p^{-2})^7$

32. $(q^{-4})^2$

33. $(-2cd)^3$

34. $(-8mn)^2$

35. $\left(\dfrac{7a}{b}\right)^2$

36. $\left(\dfrac{pq}{4}\right)^4$

37. $(4x^2y^{-3})^2$

38. $(-3w^{-3}z^5)^2$

39. $\left(\dfrac{7k}{n^2}\right)^{-2}$

40. $\left(\dfrac{5w^5}{v}\right)^{-3}$

41. $\left(\dfrac{1}{8}\right)^{-2} - \left(\dfrac{1}{4}\right)^{-3} - \left(\dfrac{1}{2}\right)^0$

42. $\left(\dfrac{1}{9}\right)^{-2} - \left(\dfrac{1}{3}\right)^{-4} + \left(\dfrac{1}{27}\right)^0$

43. $\left(\dfrac{-16m^2n^7}{8m^5n^{-2}}\right)^{-2}$

44. $\left(\dfrac{-36a^6b^9}{9a^2b^{-4}}\right)^{-2}$

45. $\left(\dfrac{4x^3z^{-5}}{12y^{-2}}\right)^{-3}$

46. $\left(\dfrac{3z^2w^{-1}}{15p^{-4}}\right)^{-3}$

47. $(-2y)^{-3}(6y^{-2}z^8)^2$

48. $(-15z)^{-2}(5z^4w^{-6})^3$

49. $\left(\dfrac{1}{2} - \dfrac{1}{3} + \dfrac{1}{6}\right)^{-3}$

50. $\left(-\dfrac{1}{3} - \dfrac{1}{4} + \dfrac{1}{2}\right)^{-2}$

51. $\left(\dfrac{1}{2}\right)^{-3} - \left(\dfrac{1}{3}\right)^{-3} + \left(\dfrac{1}{6}\right)^{-3}$

52. $\left(-\dfrac{1}{3}\right)^{-2} - \left(\dfrac{1}{4}\right)^{-2} + \left(\dfrac{1}{2}\right)^{-2}$

53. $\left(\dfrac{1}{4x^3y^{-5}}\right)^{-2}\left(\dfrac{8}{x^{-13}y^{14}}\right)^{-1}$

54. $\left(\dfrac{1}{3a^2b^{-4}}\right)^{-3}\left(\dfrac{3}{a^{-5}b^4}\right)^{-1}$

55. $\left(\dfrac{(x^{-3})^{-4}x^{-2}}{x^{-6}}\right)^{-1}$

56. $\left(\dfrac{(y^2)^{-6}y^{-3}}{y^{-4}}\right)^{-1}$

57. $\dfrac{(4vw^{-3}x^2)^2}{(2v^2w^3x^{-2})^4} \cdot (-v^3w^2x^{-4})^{-5}$

58. $\dfrac{(14v^2w^{-3}x^{-2})^3}{(21v^{-5}w^3x^{-2})^{-1}} \cdot (-7v^4w^{-1}x^{-4})^{-4}$

59. $(3x + 5)^{14}(3x + 5)^{-2}$

60. $(2y - 7z)^{-4}(2y - 7z)^{13}$

61. $[(6v - 7)^{10}]^9$

62. $[(4x - 9)^5]^{11}$

63. $2^{-2} + 2^{-1} + 2^0 + 2^1 + 2^2$

64. $3^{-2} + 3^{-1} + 3^0 + 3^1 + 3^2$

Objective 3: Apply Scientific Notation

For Exercises 65–68, write the numbers in scientific notation. (See Example 4)

65. a. 350,000   **b.** 0.000035   **c.** 3.5

66. a. 2710   **b.** 0.00271   **c.** 2.71

67. a. 0.86   **b.** 8.6   **c.** 86

68. a. 0.792   **b.** 7.92   **c.** 79.2

69. The speed of light is approximately 29,980,000,000 cm/sec.

70. The mean distance between the Earth and the Sun is approximately 149,000,000 km.

71. The size of an HIV particle is approximately 0.00001 cm.

72. One picosecond is 0.000 000 000 001 sec.

73. For a test group of adult females between 18 and 20 yr old, the mean blood volume was 4.2 L.

74. The longest table tennis rally ever played lasted 8.25 hr.

For Exercises 75–78, write the number in standard decimal notation. (See Example 5)

75. a. 2.61×10^{-6} **b.** 2.61×10^{6} **c.** 2.61×10^{0}

76. a. 3.52×10^{-2} **b.** 3.52×10^{2} **c.** 3.52×10^{0}

77. a. 6.718×10^{-1} **b.** 6.718×10^{0} **c.** 6.718×10^{1}

78. a. 1.87×10^{-1} **b.** 1.87×10^{0} **c.** 1.87×10^{1}

79. A drop of water has approximately 1.67×10^{21} molecules of H_2O.

80. A computer with a 3-terabyte hard drive can store approximately 3.0×10^{12} bytes.

81. A typical red blood cell is 7.0×10^{-6} m.

82. The blue light used to read a laser disc has a wavelength of 4.7×10^{-7} m.

For Exercises 83–92, perform the indicated operation. Write the answer in scientific notation. (See Example 6)

83. $(2 \times 10^{-3})(4 \times 10^{8})$

84. $(3 \times 10^{4})(2 \times 10^{-1})$

85. $\dfrac{8.4 \times 10^{-6}}{2.1 \times 10^{-2}}$

86. $\dfrac{6.8 \times 10^{11}}{3.4 \times 10^{3}}$

87. $(6.2 \times 10^{11})(3 \times 10^{4})$

88. $(8.1 \times 10^{6})(2 \times 10^{5})$

89. $\dfrac{3.6 \times 10^{-14}}{5 \times 10^{5}}$

90. $\dfrac{3.68 \times 10^{-8}}{4 \times 10^{2}}$

91. $\dfrac{(6.2 \times 10^{5})(4.4 \times 10^{22})}{2.2 \times 10^{17}}$

92. $\dfrac{(3.8 \times 10^{4})(4.8 \times 10^{-2})}{2.5 \times 10^{-5}}$

93. For a recent year, the United States consumed about 1.0×10^{4} gal of petroleum per second. (*Source*: U.S. Energy Information Administration, www.eia.gov)

  a. How many seconds are in a year?

  b. How many gallons of petroleum did the United States use that year?

94. Geoff's average heart rate is 65 beats/min.

  a. How many minutes are in one day?

  b. How many times will Geoff's heart beat per day?

95. Jonas has a personal music player with 80 gigabytes of memory (80 gigabytes is approximately 8×10^{10} bytes). If each song requires an average of 4 megabytes of memory (approximately 4×10^{6} bytes), how many songs can Jonas store on the device?

96. Joelle has a personal web page with 60 gigabytes of memory (approximately 6×10^{10} bytes). She stores math videos on the site for her students to watch outside of class. If each video requires an average of 5 megabytes of memory (approximately 5×10^{6} bytes), how many videos can she store on her website?

97. A typical adult human has approximately 5 L of blood in the body. If 1 μL (1 microliter) contains 5×10^{6} red blood cells, how many red blood cells does a typical adult have? (*Hint*: 1 L = 10^{6} μL.)

98. The star Proxima Centauri is the closest star (other than the Sun) to the Earth. It is approximately 4.3 light-years away. If 1 light-year is approximately 5.9×10^{12} mi, how many miles is Proxima Centauri from the Earth?

Write About It

99. Explain the difference between the expressions $6x^{0}$ and $(6x)^{0}$.

100. Explain why scientific notation is used.

Expanding Your Skills

101. If $x < 0$ and m is an integer, can x^{m} be positive? If so, give an example.

102. If $x < 0$ and m is an integer, can x^{m} be negative? If so, give an example.

103. If x is a real number, can x^{-2} be negative? If so, give an example.

104. If $x > 10$ and m is an integer, can x^{m} be less than 1? If so, given an example.

For Exercises 105–106, refer to the formula $F = \dfrac{Gm_1m_2}{d^2}$. This gives the gravitational force F (in Newtons, N) between two masses m_1 and m_2 (each measured in kg) that are a distance of d meters apart. In the formula, $G = 6.6726 \times 10^{-11}$ N-m^2/kg^2.

105. Determine the gravitational force between the Earth (mass $= 5.98 \times 10^{24}$ kg) and Jupiter (mass $= 1.901 \times 10^{27}$ kg) if at one point in their orbits, the distance between them is 7.0×10^{11} m.

106. Determine the gravitational force between the Earth (mass $= 5.98 \times 10^{24}$ kg) and an 80-kg human standing at sea level. The mean radius of the Earth is approximately 6.371×10^{6} m.

For Exercises 107–110, without the assistance of a calculator, fill in the blank with the appropriate symbol $<$, $>$, or $=$.

107. a. 5^{15} ____ 5^{17} **b.** 5^{-15} ____ 5^{-17}

108. a. $\left(\dfrac{1}{5}\right)^{15}$ ____ $\left(\dfrac{1}{5}\right)^{17}$ **b.** $\left(\dfrac{1}{5}\right)^{-15}$ ____ $\left(\dfrac{1}{5}\right)^{-17}$

109. a. $(-1)^{86}$ ____ $(-1)^{87}$ **b.** $(1)^{86}$ ____ $(1)^{87}$

110. a. $(-1)^{0}$ ____ -1^{41} **b.** $(-1)^{42}$ ____ $(-1)^{0}$

For Exercises 111–122, simplify each expression. Assume that m and n are integers and that x and y are nonzero real numbers.

111. $x^m x^4$ **112.** $y^n y^7$ **113.** $x^{m+9} x^{m-2}$ **114.** $y^{n+9} y^{n-1}$

115. $\dfrac{x^m}{x^8}$ **116.** $\dfrac{y^n}{y^3}$ **117.** $\dfrac{x^{2m+7}}{x^{m+5}}$ **118.** $\dfrac{y^{3n+5}}{y^{2n-4}}$

119. $\left(x^{4m}\right)^{3n}$ **120.** $\left(y^{5m}\right)^{2n}$ **121.** $\dfrac{x^{4m-3} y^{5n+7}}{x^{m-7} y^{3n+2}}$ **122.** $\dfrac{x^{2n-4} y^{5n}}{x^{n+1} y^{3n-7}}$

SECTION R.3   Rational Exponents and Radicals

OBJECTIVES

1. Evaluate nth-Roots
2. Simplify Expressions of the Forms $a^{1/n}$ and $a^{m/n}$
3. Simplify Expressions with Rational Exponents
4. Simplify Radicals
5. Multiply Single-Term Radical Expressions
6. Add and Subtract Radicals

As scientists search for life beyond our solar system, they look for planets on which liquid water can exist. This means that for a planet with atmospheric pressure similar to Earth, the temperature of the planet must be greater than 0°C (the temperature at which water turns to ice) but less than 100°C (the temperature at which water turns to steam).

The following model approximates the surface temperature T_p (in °C) of an Earth-like planet based on its distance d (in km) from its primary star, the radius r (in km) of the star, and the temperature T_s (in °C) of the star.

$$T_p = 0.7(T_s + 273)\left(\frac{r}{d}\right)^{1/2} - 273$$

The expression on the right contains an exponent that is a rational number. In this section, we learn how to evaluate and simplify such expressions.

1. Evaluate nth-Roots

In Section R.2 we defined b^n, where b is a real number and n is an integer. In this section, we want to extend this definition to expressions in which the exponent, n, is a rational number.

First we need to understand the relationship between nth-powers and nth-roots. From Section R.1, we know that for $a \geq 0$, $\sqrt{a} = b$ if $b^2 = a$ and $b \geq 0$. Square roots are a special case of nth-roots.

> **Definition of an nth-Root**
>
> For a positive integer $n > 1$, the principal nth-root of a, denoted by $\sqrt[n]{a}$, is a number b such that
> $$\sqrt[n]{a} = b \text{ means that } b^n = a.$$
> If n is even, then we require that $a \geq 0$ and $b \geq 0$.

For the expression $\sqrt[n]{a}$, the symbol $\sqrt[n]{}$ is called a **radical sign,** the value a is called the **radicand,** and n is called the **index.**

EXAMPLE 1   Simplifying nth-Roots

Simplify.

  a. $\sqrt[5]{32}$ **b.** $\sqrt{\dfrac{49}{64}}$ **c.** $\sqrt[3]{-0.008}$ **d.** $\sqrt[4]{-1}$ **e.** $-\sqrt[4]{1}$

Solution:

  a. $\sqrt[5]{32} = 2$ $\sqrt[5]{32} = 2$ because $2^5 = 32$.

  b. $\sqrt{\dfrac{49}{64}} = \dfrac{7}{8}$ $\sqrt{\dfrac{49}{64}} = \dfrac{7}{8}$ because $\dfrac{7}{8} \geq 0$ and $\left(\dfrac{7}{8}\right)^2 = \dfrac{49}{64}$.

  c. $\sqrt[3]{-0.008} = -0.2$ $\sqrt[3]{-0.008} = -0.2$ because $(-0.2)^3 = -0.008$.

  d. $\sqrt[4]{-1}$ is not a real number $\sqrt[4]{-1}$ is not a real number because no real number when raised to the fourth power equals -1.

  e. $-\sqrt[4]{1} = -1 \cdot \sqrt[4]{1}$ $-\sqrt[4]{1}$ is interpreted as $-1\sqrt[4]{1}$. The factor of -1 is
  $\phantom{-\sqrt[4]{1}} = -1 \cdot 1$ outside the radical.
  $\phantom{-\sqrt[4]{1}} = -1$

Skill Practice 1   Simplify.

  a. $\sqrt[3]{-125}$ **b.** $\sqrt{\dfrac{144}{121}}$ **c.** $\sqrt[5]{0.00001}$ **d.** $\sqrt[6]{-64}$ **e.** $-\sqrt[6]{64}$

---

2. Simplify Expressions of the Forms $a^{1/n}$ and $a^{m/n}$

Next, we want to define an expression of the form a^n, where n is a rational number. Furthermore, we want a definition for which the properties of integer exponents can be extended to rational exponents. For example, we want

$$(25^{1/2})^2 = 25^{(1/2)\cdot 2} = 25^1 = 25$$
$$\uparrow$$

$25^{1/2}$ must be a square root of 25, because when squared, it equals 25.

Definition of $a^{1/n}$ and $a^{m/n}$

Let m and n be integers such that m/n is a rational number in lowest terms and $n > 1$. Then,

$$a^{1/n} = \sqrt[n]{a} \qquad \text{and} \qquad a^{m/n} = \sqrt[n]{a^m} = \left(\sqrt[n]{a}\right)^m$$

If n is even, we require that $a \geq 0$.

The definition of $a^{m/n}$ indicates that $a^{m/n}$ can be written as a radical whose index is the denominator of the rational exponent. The order in which the nth-root and exponent m are performed within the radical does not affect the outcome. For example:

Take the 4th $16^{3/4} = \left(\sqrt[4]{16}\right)^3$ or   Cube 16 $16^{3/4} = \sqrt[4]{16^3}$
root first: $\phantom{16^{3/4}} = (2)^3$ first: $\phantom{16^{3/4}} = \sqrt[4]{4096}$
                 $\phantom{16^{3/4}} = 8$ $\phantom{16^{3/4}} = 8$

EXAMPLE 2   **Simplifying Expressions of the Form $a^{1/n}$ and $a^{m/n}$**

Write the expressions using radical notation and simplify if possible.

   a. $25^{1/2}$ **b.** $\left(\dfrac{64}{27}\right)^{1/3}$ **c.** $(-81)^{1/4}$ **d.** $32^{3/5}$ **e.** $(-27)^{2/3}$

Solution:

  a. $25^{1/2} = \sqrt{25} = 5$

  b. $\left(\dfrac{64}{27}\right)^{1/3} = \sqrt[3]{\dfrac{64}{27}} = \dfrac{4}{3}$

  c. $(-81)^{1/4}$ is undefined because $\sqrt[4]{-81}$ is not a real number.

  d. $32^{3/5} = \sqrt[5]{32^3} = (\sqrt[5]{32})^3 = (2)^3 = 8$

  e. $(-27)^{2/3} = \sqrt[3]{(-27)^2} = (\sqrt[3]{-27})^2 = (-3)^2 = 9$

Skill Practice 2   Simplify the expressions if possible.

   a. $36^{1/2}$ **b.** $\left(\dfrac{1}{125}\right)^{1/3}$ **c.** $(-9)^{1/2}$ **d.** $(-1)^{4/3}$ **e.** $(16)^{3/4}$

---

3. Simplify Expressions with Rational Exponents

The properties of integer exponents learned in Section R.2 can be extended to expressions with rational exponents.

EXAMPLE 3   **Simplifying Expressions with Rational Exponents**

Simplify. Assume that all variables represent positive real numbers.

   a. $\dfrac{x^{4/7}x^{2/7}}{x^{1/7}}$ **b.** $\left(\dfrac{5c^{3/4}}{d^{1/2}}\right)^2\left(\dfrac{d^{5/3}}{2c^{1/2}}\right)^3$ **c.** $81^{-3/4}$

Solution:

  a.
$$\dfrac{x^{4/7}x^{2/7}}{x^{1/7}} = \dfrac{x^{4/7+2/7}}{x^{1/7}} = \dfrac{x^{6/7}}{x^{1/7}}$$
     Add the exponents in the numerator.

$$= x^{6/7-1/7} = x^{5/7}$$
     Subtract exponents.

  b. $\left(\dfrac{5c^{3/4}}{d^{1/2}}\right)^2\left(\dfrac{d^{5/3}}{2c^{1/2}}\right)^3 = \dfrac{5^2 c^{(3/4)\cdot 2}}{d^{(1/2)\cdot 2}} \cdot \dfrac{d^{(5/3)\cdot 3}}{2^3 c^{(1/2)\cdot 3}}$
     Raise each factor inside parentheses to the power outside parentheses.

$$= \dfrac{25c^{3/2}}{d} \cdot \dfrac{d^5}{8c^{3/2}}$$
     Multiply exponents.

$$= \dfrac{25d^4}{8}$$
     Simplify.

  c. $81^{-3/4} = \dfrac{1}{81^{3/4}}$
     Rewrite the expression to remove the negative exponent.

$$= \dfrac{1}{(\sqrt[4]{81})^3} = \dfrac{1}{(3)^3} = \dfrac{1}{27}$$
     Rewrite $81^{3/4}$ using radical notation.

Answers

2 a. 6   **b.** $\dfrac{1}{5}$

  c. Undefined (not a real number)

  d. 1   **e.** 8

Skill Practice 3   Simplify. Assume that all variables represent positive real numbers.

a. $\dfrac{c^{3/4}c^{7/4}}{c^{1/4}}$

b. $\left(\dfrac{4a^{2/3}}{b^{1/6}}\right)^3\left(\dfrac{b^{1/4}}{3a^{3/4}}\right)^2$

c. $(-125)^{-2/3}$

4. Simplify Radicals

In Example 3, we simplified several expressions with rational exponents. Next, we want to simplify radical expressions. First consider expressions of the form $\sqrt[n]{a^n}$. The value of $\sqrt[n]{a^n}$ is not necessarily a. Since $\sqrt[n]{a}$ represents the principal nth-root of a, then $\sqrt[n]{a}$ must be nonnegative for even values of n. For example:

$$\sqrt{(5)^2} = 5 \quad \text{and} \quad \sqrt{(-5)^2} = |-5| = 5$$

> The absolute value is needed here to guarantee a nonnegative result.

$$\sqrt[4]{(2)^4} = 2 \quad \text{and} \quad \sqrt[4]{(-2)^4} = |-2| = 2$$

In Table R-6, we generalize this result and give three other important properties of radicals.

Table R-6   Properties of Radicals

Let $n > 1$ be an integer and a and b be real numbers. The following properties are true provided that the given radicals are real numbers.

Property	Examples	
1. If n is even, $\sqrt[n]{a^n} = |a|$.	$\sqrt{x^2} = |x|$	$\sqrt[4]{x^8} = |x^2| = x^2$
2. If n is odd, $\sqrt[n]{a^n} = a$.	$\sqrt[3]{x^3} = x$	$\sqrt[5]{(y+4)^5} = y + 4$
3. Product Property $\sqrt[n]{a} \cdot \sqrt[n]{b} = \sqrt[n]{ab}$	$\sqrt[3]{7} \cdot \sqrt[3]{x} = \sqrt[3]{7x}$	$\sqrt{75} = \sqrt{25} \cdot \sqrt{3} = 5\sqrt{3}$
4. Quotient Property $\dfrac{\sqrt[n]{a}}{\sqrt[n]{b}} = \sqrt[n]{\dfrac{a}{b}}$	$\dfrac{\sqrt{125}}{\sqrt{5}} = \sqrt{\dfrac{125}{5}} = \sqrt{25} = 5$	
5. Nested Radical Property $\sqrt[m]{\sqrt[n]{a}} = \sqrt[m \cdot n]{a}$	$\sqrt[4]{\sqrt[3]{x}} = (x^{1/3})^{1/4} = x^{(1/3)(1/4)} = x^{1/12} = \sqrt[12]{x}$	

Properties 1–5 follow from the definition of $a^{1/n}$ and the properties of exponents. We use these properties to simplify radical expressions and must address four specific criteria.

Simplified Form of a Radical

Suppose that the radicand of a radical is written as a product of prime factors. Then the radical is simplified if all of the following conditions are met.

1. The radicand has no factor other than 1 that is a perfect nth-power. This means that all exponents in the radicand must be less than the index.
2. No fractions may appear in the radicand.
3. No denominator of a fraction may contain a radical.
4. The exponents in the radicand may not all share a common factor with the index.

Answers

3. a. $c^{9/4}$ **b.** $\dfrac{64a^{1/2}}{9}$ **c.** $\dfrac{1}{25}$

In Example 4, we simplify expressions that fail condition 1. Also notice that the expressions in Example 4 are assumed to have positive radicands. This eliminates the need to insert absolute value bars around the simplified form of $\sqrt[n]{a^n}$.

EXAMPLE 4   **Simplifying Radicals Using the Product Property**

Simplify each expression. Assume that all variables represent positive real numbers.

a. $\sqrt[3]{c^5}$ **b.** $\sqrt{50}$ **c.** $\sqrt[4]{32x^9y^6}$

Solution:

a. $\sqrt[3]{c^5} = \sqrt[3]{c^3 \cdot c^2}$ Write the radicand as a product of a perfect cube and another factor.

$\qquad = \sqrt[3]{c^3} \cdot \sqrt[3]{c^2}$ Apply the product property of radicals.

$\qquad = c\sqrt[3]{c^2}$ Simplify $\sqrt[3]{c^3}$ as c.

b. $\sqrt{50} = \sqrt{5^2 \cdot 2}$ Factor the radicand. The radical is not simplified because the radicand has a perfect square.

$\qquad = \sqrt{5^2} \cdot \sqrt{2}$ Apply the product property of radicals.

$\qquad = 5\sqrt{2}$ Simplify.

c. $\sqrt[4]{32x^9y^6}$

$\qquad = \sqrt[4]{2^5x^9y^6}$ Factor the radicand.

$\qquad = \sqrt[4]{(2^4x^8y^4)(2xy^2)}$ Write the radicand as the product of a perfect 4th power and another factor.

$\qquad = \sqrt[4]{2^4x^8y^4} \cdot \sqrt[4]{2xy^2}$ Apply the product property of radicals.

$\qquad = 2x^2y\sqrt[4]{2xy^2}$ Simplify the first radical.

Skill Practice 4   Simplify each expression. Assume that all variables represent positive real numbers.

a. $\sqrt[4]{d^7}$ **b.** $\sqrt{45}$ **c.** $\sqrt[3]{54x^{13}y^8}$

In Example 5, we will use the quotient property of radicals to simplify expressions that fail conditions 2 and 3 for a simplified radical. Removing a radical from the denominator of a fraction is called **rationalizing the denominator.**

EXAMPLE 5   **Applying the Quotient Property of Radicals**

Simplify the expressions. Assume that x and y are nonzero real numbers.

a. $\sqrt{\dfrac{x^3}{9}}$ **b.** $\dfrac{\sqrt[3]{3x^7y}}{\sqrt[3]{81xy^4}}$

Solution:

a. $\sqrt{\dfrac{x^3}{9}} = \dfrac{\sqrt{x^3}}{\sqrt{9}}$ Apply the quotient property of radicals.

$\qquad = \dfrac{\sqrt{x^2 \cdot x}}{3}$ Write the radicand as a product of a perfect square and another factor.

$\qquad = \dfrac{\sqrt{x^2} \cdot \sqrt{x}}{3}$ Apply the product property of radicals.

$\qquad = \dfrac{x\sqrt{x}}{3}$ Simplify.

TIP In Example 5(b), the purpose of writing the quotient of two radicals as a single radical is to simplify the resulting fraction in the radicand.

b. $\dfrac{\sqrt[3]{3x^7y}}{\sqrt[3]{81xy^4}} = \sqrt[3]{\dfrac{3x^7y}{81xy^4}}$ Apply the quotient property of radicals to write the expression as a single radical.

$= \sqrt[3]{\dfrac{x^6}{27y^3}}$ The numerator and denominator share common factors. Simplify the fraction.

$= \dfrac{x^2}{3y}$ Simplify.

Skill Practice 5   Simplify the expressions. Assume that x and y are nonzero real numbers.

    a. $\sqrt{\dfrac{y^5}{49}}$ **b.** $\dfrac{\sqrt[3]{625c^2d^{10}}}{\sqrt[3]{5c^5d}}$

5. Multiply Single-Term Radical Expressions

In Example 6, we use the product property of radicals to multiply two radical expressions. We can simplify a product of radicals, provided the indices are the same.

EXAMPLE 6   Multiplying Single-Term Radicals

Multiply. Assume that x represents a positive real number.

    a. $\sqrt{6} \cdot \sqrt{10}$ **b.** $\left(2\sqrt[4]{x^3}\right)\left(5\sqrt[4]{x^7}\right)$

Solution:

    a. $\sqrt{6} \cdot \sqrt{10} = \sqrt{60}$ The radicals have the same index. Apply the product property of radicals.

TIP When multiplying radicals, we have the option of factoring the individual radicands before multiplying. For example:

$\sqrt{6} \cdot \sqrt{10}$
$= \sqrt{2 \cdot 3} \cdot \sqrt{2 \cdot 5}$
$= \sqrt{2^2 \cdot 3 \cdot 5}$

$= \sqrt{2^2 \cdot 3 \cdot 5}$ Factor the radicand.

$= \sqrt{(2^2) \cdot (3 \cdot 5)}$ Write the radicand as the product of a perfect square and another factor.

$= \sqrt{2^2} \cdot \sqrt{3 \cdot 5}$ Apply the product property of radicals.

$= 2\sqrt{15}$ Simplify.

    b. $\left(2\sqrt[4]{x^3}\right)\left(5\sqrt[4]{x^7}\right) = 2 \cdot 5\sqrt[4]{x^3 \cdot x^7}$ Regroup factors, and apply the product property of radicals.

$= 10\sqrt[4]{x^{10}}$ Simplify the radicand.

Avoiding Mistakes

The product property of radicals can be applied only if the radicals have the same index.

$= 10\sqrt[4]{x^8 \cdot x^2}$ Write the radicand as the product of a perfect 4th power and another factor.

$= 10\sqrt[4]{x^8} \cdot \sqrt[4]{x^2}$ Apply the product property of radicals.

$= 10x^2\sqrt[4]{x^2}$ The expression $\sqrt[4]{x^2}$ fails condition 4 for a simplified radical. (The exponents in the radicand cannot all share a common factor with the index.)

$= 10x^2\sqrt[2]{x^1}$

$= 10x^2\sqrt{x}$ So, $\sqrt[4]{x^2} = x^{2/4} = x^{1/2} = \sqrt{x}$.

Answers

5. **a.** $\dfrac{y^2\sqrt{y}}{7}$ **b.** $\dfrac{5d^3}{c}$

6. **a.** $3\sqrt{35}$ **b.** $12y^2\sqrt[3]{y^2}$

Skill Practice 6   Multiply. Assume that y represents a positive real number.

    a. $\sqrt{15} \cdot \sqrt{21}$ **b.** $\left(3\sqrt[6]{y^5}\right)\left(4\sqrt[6]{y^{11}}\right)$

6. Add and Subtract Radicals

We can use the distributive property to add or subtract radical expressions. However, the radicals must be like radicals. This means that the radicands must be the same and the indices must be the same. For example:

$3\sqrt{2x}$ and $-5\sqrt{2x}$ are like radicals.

$3\sqrt{2x}$ and $-5\sqrt[3]{2x}$ are not like radicals because the indices are different.

$3\sqrt{2x}$ and $-5\sqrt{2y}$ are not like radicals because the radicands are different.

EXAMPLE 7   **Adding and Subtracting Radicals**

Add or subtract as indicated. Assume that all variables represent positive real numbers.

a. $5\sqrt[3]{7t^2} - 2\sqrt[3]{7t^2} + \sqrt[3]{7t^2}$

b. $x\sqrt{98x^3y} + 5\sqrt{18x^5y}$

c. $3\sqrt{5x} + 2x\sqrt{5x}$

Solution:

a. $5\sqrt[3]{7t^2} - 2\sqrt[3]{7t^2} + 1\sqrt[3]{7t^2}$ The radicals are like radicals. They have the same radicand and same index.

$= (5 - 2 + 1)\sqrt[3]{7t^2}$ Apply the distributive property.

$= 4\sqrt[3]{7t^2}$ Simplify.

b. $x\sqrt{98x^3y} + 5\sqrt{18x^5y}$ Each radical can be simplified.

$x\sqrt{98x^3y} = x\sqrt{(7^2x^2)\cdot(2xy)} = 7x^2\sqrt{2xy}$

$5\sqrt{18x^5y} = 5\sqrt{(3^2x^4)(2xy)} = 15x^2\sqrt{2xy}$

$= 7x^2\sqrt{2xy} + 15x^2\sqrt{2xy}$ The terms are like terms.

$= (7 + 15)x^2\sqrt{2xy}$ Apply the distributive property.

$= 22x^2\sqrt{2xy}$ Simplify.

c. $3\sqrt{5x} + 2x\sqrt{5x}$ The radicals are like radicals.

$= (3 + 2x)\sqrt{5x}$ Apply the distributive property. The expression cannot be further simplified because the terms within parentheses are not like terms.

Skill Practice 7   Add or subtract as indicated. Assume that all variables represent positive real numbers.

a. $-4\sqrt[3]{5w} + 9\sqrt[3]{5w} - 11\sqrt[3]{5w}$ **b.** $\sqrt{75cd^4} + 6d\sqrt{27cd^2}$

c. $8\sqrt{7z} + 3z\sqrt{7z}$

Answers

7. a. $-6\sqrt[3]{5w}$ **b.** $23d^2\sqrt{3c}$

   c. $(8 + 3z)\sqrt{7z}$

SECTION R.3   Practice Exercises

Concept Connections

1. b is an nth-root of a if $b^{\square} = a$.

2. Given the expression $\sqrt[n]{a}$, the value a is called the _____ and n is called the _____.

3. The expression $a^{m/n}$ can be written in radical notation as _____, provided that $\sqrt[n]{a}$ is a real number.

4. The expression $a^{1/n}$ can be written in radical notation as _____, provided that $\sqrt[n]{a}$ is a real number.

5. If x represents any real number, then $\sqrt{x^2} =$ _____.

6. If x represents any real number, then $\sqrt[3]{x^3} =$ _____.

7. The product property of radicals indicates that $\sqrt[n]{a} \cdot \sqrt[n]{b} =$ _____ provided that $\sqrt[n]{a}$ and $\sqrt[n]{b}$ represent real numbers.

8. Removing a radical from the denominator of a fraction is called _____ the denominator.

Objective 1: Evaluate nth-Roots

For Exercises 9–20, simplify the expression. (See Example 1)

9. $\sqrt[4]{81}$

10. $\sqrt[3]{125}$

11. $\sqrt{\dfrac{4}{49}}$

12. $\sqrt{\dfrac{9}{121}}$

13. $\sqrt{0.09}$

14. $\sqrt{0.16}$

15. $\sqrt[4]{-81}$

16. $\sqrt[4]{-625}$

17. $-\sqrt[4]{81}$

18. $-\sqrt[4]{625}$

19. $\sqrt[3]{-\dfrac{1}{8}}$

20. $\sqrt[3]{-\dfrac{64}{125}}$

Objective 2: Simplify Expressions of the Forms $a^{1/n}$ and $a^{m/n}$

For Exercises 21–30, simplify each expression. (See Example 2)

21. a. $25^{1/2}$ **b.** $(-25)^{1/2}$ **c.** $-25^{1/2}$

22. a. $36^{1/2}$ **b.** $(-36)^{1/2}$ **c.** $-36^{1/2}$

23. a. $27^{1/3}$ **b.** $(-27)^{1/3}$ **c.** $-27^{1/3}$

24. a. $125^{1/3}$ **b.** $(-125)^{1/3}$ **c.** $-125^{1/3}$

25. a. $\left(\dfrac{121}{169}\right)^{1/2}$ **b.** $\left(\dfrac{121}{169}\right)^{-1/2}$

26. a. $\left(\dfrac{49}{144}\right)^{1/2}$ **b.** $\left(\dfrac{49}{144}\right)^{-1/2}$

27. a. $16^{3/4}$ **b.** $16^{-3/4}$ **c.** $-16^{3/4}$
   d. $-16^{-3/4}$ **e.** $(-16)^{3/4}$ **f.** $(-16)^{-3/4}$

28. a. $81^{3/4}$ **b.** $81^{-3/4}$ **c.** $-81^{3/4}$
   d. $-81^{-3/4}$ **e.** $(-81)^{3/4}$ **f.** $(-81)^{-3/4}$

29. a. $64^{2/3}$ **b.** $64^{-2/3}$ **c.** $-64^{2/3}$
   d. $-64^{-2/3}$ **e.** $(-64)^{2/3}$ **f.** $(-64)^{-2/3}$

30. a. $8^{2/3}$ **b.** $8^{-2/3}$ **c.** $-8^{2/3}$
   d. $-8^{-2/3}$ **e.** $(-8)^{2/3}$ **f.** $(-8)^{-2/3}$

For Exercises 31–32, write the expression using radical notation. Assume that all variables represent positive real numbers.

31. a. $y^{4/11}$ **b.** $6y^{4/11}$ **c.** $(6y)^{4/11}$

32. a. $z^{3/10}$ **b.** $8z^{3/10}$ **c.** $(8z)^{3/10}$

For Exercises 33–40, write the expression using rational exponents. Assume that all variables represent positive real numbers.

33. $\sqrt[5]{a^3}$

34. $\sqrt[7]{z^4}$

35. $\sqrt{6x}$

36. $\sqrt{11t}$

37. $6\sqrt{x}$

38. $11\sqrt{t}$

39. $\sqrt[5]{a^5 + b^5}$

40. $\sqrt[3]{m^3 + n^3}$

Objective 3: Simplify Expressions with Rational Exponents

For Exercises 41–50, simplify each expression. Assume that all variable expressions represent positive real numbers. (See Example 3)

41. $\dfrac{a^{2/3}a^{5/3}}{a^{1/3}}$

42. $\dfrac{y^{7/5}y^{4/5}}{y^{1/5}}$

43. $\dfrac{3w^{-2/3}}{y^{-1/3}}$

44. $\dfrac{8d^{-5/7}}{c^{-3/4}}$

45. $(16x^{-8}y^{1/5})^{3/4}$

46. $(125a^6b^{-7/5})^{1/3}$

47. $\left(\dfrac{2m^{2/3}}{n^{3/4}}\right)^{12}\left(\dfrac{n^{1/5}}{2m^{1/2}}\right)^{10}$

48. $\left(\dfrac{3x^{1/2}}{y^{3/8}}\right)^{4}\left(\dfrac{y^{1/2}}{3x^{4/3}}\right)^{3}$

49. $\left(\dfrac{m^2}{m+n}\right)^{-1}\left(\dfrac{m^2}{m+n}\right)^{1/2}$

50. $\left(\dfrac{c^2}{c-d}\right)^{-2}\left(\dfrac{c^2}{c-d}\right)^{3/2}$

Objective 4: Simplify Radicals

51. a. For what values of t will the statement be true? $\sqrt{t^2} = t$

   b. For what value of t will the statement be true? $\sqrt{t^2} = |t|$

52. a. For what values of c will the statement be true? $\sqrt[4]{(c+8)^4} = c + 8$

   b. For what value of c will the statement be true? $\sqrt[4]{(c+8)^4} = |c + 8|$

For Exercises 53–60, simplify each expression.

53. $\sqrt{y^2}$

54. $\sqrt[4]{y^4}$

55. $\sqrt[3]{y^3}$

56. $\sqrt[5]{y^5}$

57. $\sqrt[4]{(2x-5)^4}$

58. $\sqrt{(3z+2)^2}$

59. $\sqrt{w^{12}}$

60. $\sqrt[4]{c^{32}}$

For Exercises 61–76, simplify each expression. Assume that all variable expressions represent positive real numbers. (See Examples 4–5)

61. a. $\sqrt{c^7}$
   b. $\sqrt[3]{c^7}$
   c. $\sqrt[4]{c^7}$
   d. $\sqrt[9]{c^7}$

62. a. $\sqrt{d^{11}}$
   b. $\sqrt[3]{d^{11}}$
   c. $\sqrt[4]{d^{11}}$
   d. $\sqrt[12]{d^{11}}$

63. a. $\sqrt{24}$
   b. $\sqrt[3]{24}$

64. a. $\sqrt{54}$
   b. $\sqrt[3]{54}$

65. $\sqrt[3]{250x^2y^6z^{11}}$

66. $\sqrt[3]{40ab^{13}c^{17}}$

67. $\sqrt[4]{96p^{14}q^7}$

68. $\sqrt[4]{243m^{19}n^{10}}$

69. $\sqrt{84(y-2)^3}$

70. $\sqrt{18(w-6)^3}$

71. $\sqrt{\dfrac{p^7}{36}}$

72. $\sqrt{\dfrac{q^{11}}{4}}$

73. $4\sqrt[3]{\dfrac{w^3z^5}{8}}$

74. $8\sqrt[3]{\dfrac{c^6d^7}{64}}$

75. $\dfrac{\sqrt[3]{5x^5y}}{\sqrt[3]{625x^2y^4}}$

76. $\dfrac{\sqrt[3]{2m^2n^7}}{\sqrt[3]{16m^{14}n^4}}$

5. Multiply Single-Term Radical Expressions

For Exercises 77–84, simplify each expression. Assume that all variables represent positive real numbers. (See Example 6)

77. $\sqrt{10} \cdot \sqrt{14}$

78. $\sqrt{6} \cdot \sqrt{21}$

79. $\sqrt[3]{xy^2} \cdot \sqrt[3]{x^2y}$

80. $\sqrt[4]{a^3b} \cdot \sqrt[4]{ab^3}$

81. $(3\sqrt[4]{a^3})(-5\sqrt[4]{a^3})$

82. $(7\sqrt[6]{t^5})(-2\sqrt[6]{t^5})$

83. $\left(-\dfrac{1}{2}\sqrt[3]{6a^2b^2c}\right)\left(\dfrac{4}{3}\sqrt[3]{4a^2c^2}\right)$

84. $\left(-\dfrac{3}{4}\sqrt[3]{9m^2n^5p}\right)\left(\dfrac{1}{6}\sqrt[3]{6m^2np^4}\right)$

6. Add and Subtract Radicals

For Exercises 85–94, add or subtract as indicated. Assume that all variables represent positive real numbers. (See Example 7)

85. $3\sqrt[3]{2y^2} - 9\sqrt[3]{2y^2} + \sqrt[3]{2y^2}$

86. $8\sqrt[4]{3z^3} - \sqrt[4]{3z^3} + 2\sqrt[4]{3z^3}$

87. $\frac{1}{5}\sqrt{50} - \frac{7}{3}\sqrt{18} + \frac{5}{6}\sqrt{72}$

88. $\frac{2}{5}\sqrt{75} - \frac{2}{3}\sqrt{27} - \frac{1}{2}\sqrt{12}$

89. $-3x\sqrt[3]{16xy^4} + xy\sqrt[3]{54xy} - 5\sqrt[3]{250x^4y^4}$

90. $8\sqrt[4]{32a^5b^6} - 5b\sqrt[4]{2a^5b^2} - ab\sqrt[4]{162ab^2}$

91. $12\sqrt{2y} + 5y\sqrt{2y}$

92. $-8\sqrt{3w} + 3w\sqrt{3w}$

93. $-\frac{1}{2}\sqrt{8z^3} + \frac{3}{7}\sqrt{98z}$

94. $\frac{2}{3}\sqrt{45c} + \frac{1}{2}\sqrt{20c^3}$

Mixed Exercises

For Exercises 95–96, use Heron's formula to determine the area A of a triangle with sides of length a, b, and c. Write each answer as a simplified radical. Heron's formula: $A = \sqrt{s(s-a)(s-b)(s-c)}$, where $s = \frac{1}{2}(a+b+c)$.

95.

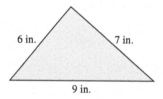

96.

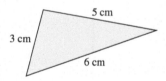

For Exercises 97–98, use the Pythagorean theorem to determine the length of the missing side. Write the answer as a simplified radical.

97.

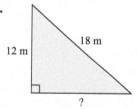

98.

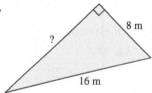

99. The size of a television is identified by the length of the diagonal. If Lynn's television is 48 in. across and 32 in. high, what size television does she have? Give the exact value and a decimal approximation to the nearest inch.

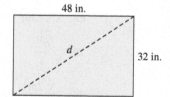

100. If the span of a roof is 36 ft and the rise is 12 ft, determine the length of the rafter R. Give the exact value and a decimal approximation to the nearest tenth of a foot.

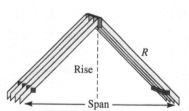

101. The slant length L for a right circular cone is given by $L = \sqrt{r^2 + h^2}$, where r and h are the radius and height of the cone. Find the slant length of a cone with radius 4 in. and height 10 in. Determine the exact value and a decimal approximation to the nearest tenth of an inch.

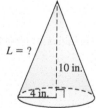

102. The lateral surface area A of a right circular cone is given by $A = \pi r\sqrt{r^2 + h^2}$, where r and h are the radius and height of the cone. Determine the exact value (in terms of π) of the lateral surface area of a cone with radius 6 m and height 4 m. Then give a decimal approximation to the nearest square meter.

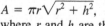

103. The depreciation rate for a car is given by $r = 1 - \left(\frac{S}{C}\right)^{1/n}$, where S is the value of the car after n years, and C is the initial cost. Determine the depreciation rate for a car that originally cost \$22,990 and was valued at \$11,500 after 4 yr. Round to the nearest tenth of a percent.

104. For a certain oven, the baking time t (in hr) for a turkey that weighs x pounds can be approximated by the model $t = 0.84x^{3/5}$. Determine the baking time for a 15-lb turkey. Round to 1 decimal place.

Write About It

105. Explain the similarity in simplifying the given expressions.

   a. $2x + 3x$

   b. $2\sqrt{x} + 3\sqrt{x}$

   c. $2\sqrt[3]{x} + 3\sqrt[3]{x}$

106. Explain why the given expressions cannot be simplified further.

   a. $2x + 3y$

   b. $2\sqrt{x} + 3\sqrt{y}$

   c. $2\sqrt[3]{x} + 3\sqrt{x}$

Expanding Your Skills

For Exercises 107–112, write each expression as a single radical for positive values of the variable. (*Hint*: Write the radicals as expressions with rational exponents and simplify. Then convert back to radical form.)

107. $\sqrt[5]{x^3 y^2} \cdot \sqrt[4]{x}$

108. $\sqrt[4]{a^2 b} \cdot \sqrt[3]{ab^2}$

109. $\sqrt[6]{m} \sqrt[3]{m^2}$

110. $\sqrt[5]{y} \sqrt[4]{y^3}$

111. $\sqrt{x \sqrt{x \sqrt{x}}}$

112. $\sqrt[3]{y \sqrt[3]{y \sqrt[3]{y}}}$

For Exercises 113–114, evaluate the expression without the use of a calculator.

113. $\sqrt{\dfrac{8.0 \times 10^{12}}{2.0 \times 10^4}}$

114. $\sqrt{\dfrac{1.44 \times 10^{16}}{9.0 \times 10^{10}}}$

For Exercises 115–116, simplify the expression.

115. $\sqrt{\sqrt[3]{6 + \sqrt[4]{16}} + \sqrt{\sqrt{25} + \sqrt{16}} + \sqrt{9}}$

116. $\sqrt{\sqrt{\sqrt[4]{11 + \sqrt[3]{125}} + \sqrt{\sqrt{81} + \sqrt[3]{1000}} + \sqrt{36}} + \sqrt{25}}$

The mean surface temperature T_p (in °C) of an Earth-like planet can be approximated based on its distance from its primary star d (in km), the radius of the star r (in km), and the temperature of the star T_s (in °C) by the following formula.

$$T_p = 0.7(T_s + 273)\left(\frac{r}{d}\right)^{1/2} - 273$$ For Exercises 117–118, use the model to find T_p.

117. The star Altair is relatively close to the Earth (16.8 light-years) and has a mean surface temperature of approximately 7700°C. Although not completely spherical in shape, Altair has a mean radius of approximately 1.26×10^6 km. If a planet with an atmosphere similar to that of the Earth is 4.3×10^8 km away from Altair, will the temperature on the surface of the planet be suitable for liquid water to exist? (Recall that under pressure similar to that at sea level on Earth, water freezes at 0°C and turns to steam at 100°C.)

118. Suppose the Sun has a mean surface temperature of 5700°C and a radius of approximately 7.0×10^5 km. If the Earth is a distance of 1.49×10^8 km from the Sun, approximate the mean surface temperature for the Earth.

Polynomials and Multiplication of Radicals

OBJECTIVES

1. Identify Key Elements of a Polynomial
2. Add and Subtract Polynomials
3. Multiply Polynomials
4. Identify and Simplify Special Case Products
5. Multiply Radical Expressions Involving Multiple Terms

1. Identify Key Elements of a Polynomial

The Environmental Protection Agency (EPA) is responsible for providing fuel economy data (gas mileage information) that is posted on the window stickers of new vehicles. Many variables contribute to fuel consumption including the speed of the vehicle. For example, for one midsize sedan tested, the gas mileage G (in miles per gallon, mpg) can be approximated by

$$G = -0.008x^2 + 0.748x + 13.5,$$ where x is the speed of the vehicle in mph and $15 \le x \le 75$ mph.

The expression on the right side of this equation is called a polynomial. A **polynomial** in the variable x is a finite sum of terms of the form ax^n. In each term, the coefficient, a, is a real number, and the exponent, n, is a whole number. The degree of ax^n is n.

The terms of the polynomial $-0.008x^2 + 0.748x + 13.5$ are written in **descending order** by degree. The term with highest degree is written first. This is called the **leading term,** and its coefficient is called the **leading coefficient.** The **degree of the polynomial** is the same as the degree of the leading term. Therefore, the polynomial $-0.008x^2 + 0.748x + 13.5$ is a degree 2 polynomial.

The preceding discussion is meant as an informal introduction to polynomials and associated key vocabulary. However, as your level of mathematical sophistication increases, you should strive to understand definitions written in a more concise mathematical language. Take a minute to read the formal definition of a polynomial.

> **TIP** The number 0 can be written in infinitely many ways: $0x$, $0x^2$, $0x^3$, and so on. For this reason, the degree of 0 is undefined.

> ### Definition of a Polynomial in *x*
>
> A **polynomial in the variable *x*** is an expression of the form:
>
> $$a_n x^n + a_{n-1} x^{n-1} + a_{n-2} x^{n-2} + \cdots + a_1 x + a_0$$
>
> The coefficients a_n, a_{n-1}, a_{n-2}, $\ldots$, a_0 are real numbers, where $a_n \ne 0$, and the exponents n, $n - 1$, $n - 2$, $\ldots$, 0 are whole numbers.
>
> The term $a_n x^n$ is called the **leading term,** the coefficient a_n is the **leading coefficient,** and the exponent n is the **degree of the polynomial.**

In the preceding definition, subscript notation a_n (read as "*a* sub *n*"), a_{n-1} (read as "*a* sub $n - 1$"), and so on, is used to denote the coefficients of the terms. Subscript notation is used rather than lettered variables such as a, b, c, and the like, when a large or undetermined number of terms is suggested.

A polynomial with one term is also called a **monomial.** A polynomial with two terms is called a **binomial,** and a polynomial with three terms is called a **trinomial.**

Some polynomials have more than one variable:

$$-4x^4 y^7 + x^2 y^5 + 5xy^4$$

This polynomial has three terms. The degree of each term is the sum of the exponents on the variable factors.

$-4x^4 y^7$	degree 11	(sum of $4 + 7$)
$x^2 y^5$	degree 7	(sum of $2 + 5$)
$5xy^4$	degree 5	(sum of $1 + 4$)

2. Add and Subtract Polynomials

To add or subtract polynomials, combine like terms. This is demonstrated in Example 1.

EXAMPLE 1   **Adding and Subtracting Polynomials**

Add or subtract as indicated, and simplify.

a. $(-4w^3 - 5w^2 + 6w + 3) + (8w^2 - 4w + 2)$
b. $(6.1a^2b + 2.9ab - 4.5b^2) - (2.6a^2b - 4.1ab + 2.1b^2)$

Solution:

a. $(-4w^3 - 5w^2 + 6w + 3) + (8w^2 - 4w + 2)$

$= -4w^3 - 5w^2 + 8w^2 + 6w - 4w + 3 + 2$ Group like terms.

$= -4w^3 + 3w^2 + 2w + 5$ Combine like terms.

b. $(6.1a^2b + 2.9ab - 4.5b^2) - (2.6a^2b - 4.1ab + 2.1b^2)$

$= (6.1a^2b + 2.9ab - 4.5b^2) + (-2.6a^2b + 4.1ab - 2.1b^2)$

$= 6.1a^2b - 2.6a^2b + 2.9ab + 4.1ab - 4.5b^2 - 2.1b^2$ Group like terms.

$= 3.5a^2b + 7ab - 6.6b^2$ Combine like terms.

Skill Practice 1   Add or subtract as indicated, and simplify.

a. $(7t^5 - 3t^2 - 2t) + (2t^5 + 3t^2 - 5t - 4)$
b. $(0.08x^3y - 0.02x^2y - 0.1xy) - (0.05x^3y - 0.07x^2y + 0.02xy)$

TIP Addition and subtraction of polynomials can also be done by aligning like terms in columns. The difference of polynomials from Example 1(b) is shown here.

$$6.1a^2b + 2.9ab - 4.5b^2$$
$$\underline{-(2.6a^2b - 4.1ab + 2.1b^2)} \longrightarrow$$
Add the opposite.

$$6.1a^2b + 2.9ab - 4.5b^2$$
$$\underline{+ (-2.6a^2b + 4.1ab - 2.1b^2)}$$
$$3.5a^2b + \; 7ab - 6.6b^2$$

---

3. Multiply Polynomials

- To multiply two monomials, use the commutative and associative properties of multiplication to regroup like factors. Then apply the properties of exponents to simplify.

$$(-4x^2y^5)\left(\frac{1}{2}x^3y\right) = -4 \cdot \frac{1}{2}x^2x^3y^5y = -2x^5y^6$$

- To multiply a polynomial by a monomial, apply the distributive property.

$$-3x^3(4x^2 - 2x + 6) = -3x^3(4x^2) + (-3x^3)(-2x) + (-3x^3)(6)$$
$$= -12x^5 + 6x^4 - 18x^3$$

- To multiply two polynomials with two or more terms, we also use the distributive property. Ultimately, each term in the first polynomial must be multiplied by each term in the second. Then simplify and combine like terms.

$$(4w + 7)(2w - 3) = 4w(2w) + 4w(-3) + 7(2w) + 7(-3)$$
$$= 8w^2 - 12w + 14w - 21$$
$$= 8w^2 + 2w - 21$$

Answers
1. a. $9t^5 - 7t - 4$
 b. $0.03x^3y + 0.05x^2y - 0.12xy$

EXAMPLE 2 **Multiplying Polynomials**

Multiply and simplify.

$$(4x + 2)\left(x^2 - 6x + \frac{1}{2}\right)$$

Solution:

$$(4x + 2)\left(x^2 - 6x + \frac{1}{2}\right)$$

Multiply each term in the first polynomial by each term in the second.

$$= 4x(x^2) + 4x(-6x) + 4x\left(\frac{1}{2}\right) + 2(x^2) + 2(-6x) + 2\left(\frac{1}{2}\right)$$

$$= 4x^3 - 24x^2 + 2x + 2x^2 - 12x + 1$$

Simplify.

$$= 4x^3 - 22x^2 - 10x + 1$$

Combine like terms.

TIP Multiplication of polynomials can also be performed vertically.

$$\begin{array}{r} x^2 - 6x + \frac{1}{2} \\ \times \quad\quad 4x + 2 \\ \hline 2x^2 - 12x + 1 \\ 4x^3 - 24x^2 + 2x \quad\quad\quad \\ \hline 4x^3 - 22x^2 - 10x + 1 \end{array}$$

Skill Practice 2 Multiply and simplify.

$$(3y - 6)\left(y^2 + 4y + \frac{2}{3}\right)$$

4. Identify and Simplify Special Case Products

Two expressions of the form $a - b$ and $a + b$ are called **conjugates.** The product of two conjugates results in a **difference of squares.**

TIP The product of conjugates equals the square of the first term from the binomials, minus the square of the second term from the binomials.

$$(a - b)(a + b) = a^2 + ab - ab - b^2$$
$$= a^2 - b^2 \quad \text{(difference of squares)}$$

An expression of the form $(a + b)^2$ or $(a - b)^2$ is called a **square of a binomial.** In expanded form, the product is a **perfect square trinomial.**

TIP The square of a binomial results in the square of the first term in the binomial, plus twice the product of terms in the binomial, plus the square of the second term in the binomial.

$$(a + b)^2 = (a + b)(a + b) = a^2 + ab + ab + b^2$$
$$= a^2 + 2ab + b^2 \quad \text{(perfect square trinomial)}$$

$$(a - b)^2 = (a - b)(a - b) = a^2 - ab - ab + b^2$$
$$= a^2 - 2ab + b^2 \quad \text{(perfect square trinomial)}$$

The patterns associated with the product of conjugates and the square of a binomial are important to understand and memorize. These will be used again in many more applications of algebra, including factoring in the next section.

Special Case Products

Product of Conjugates: $(a + b)(a - b) = a^2 - b^2$ (difference of squares)

Square of a Binomial: $(a + b)^2 = a^2 + 2ab + b^2$ (perfect square trinomial)

$(a - b)^2 = a^2 - 2ab + b^2$ (perfect square trinomial)

Answer

2. $3y^3 + 6y^2 - 22y - 4$

EXAMPLE 3   **Multiplying Conjugates**

Multiply and simplify.

a. $(2x + 5)(2x - 5)$ **b.** $\left(\dfrac{1}{3}c^2 - \dfrac{1}{2}d\right)\left(\dfrac{1}{3}c^2 + \dfrac{1}{2}d\right)$

Solution:

a. $(2x + 5)(2x - 5)$ This is a product of conjugates.

$\quad = (2x)^2 - (5)^2$ The product is a difference of squares.

$\quad = 4x^2 - 25$ Simplify.

b. $\left(\dfrac{1}{3}c^2 - \dfrac{1}{2}d\right)\left(\dfrac{1}{3}c^2 + \dfrac{1}{2}d\right)$ This is a product of conjugates.

$\quad = \left(\dfrac{1}{3}c^2\right)^2 - \left(\dfrac{1}{2}d\right)^2$ The product is a difference of squares.

$\quad = \dfrac{1}{9}c^4 - \dfrac{1}{4}d^2$ Simplify.

Skill Practice 3   Multiply and simplify.

a. $(3y - 7)(3y + 7)$ **b.** $\left(\dfrac{2}{5}t - \dfrac{1}{4}w^2\right)\left(\dfrac{2}{5}t + \dfrac{1}{4}w^2\right)$

EXAMPLE 4   **Squaring Binomials**

Square the binomials.

a. $(3x - 7)^2$ **b.** $(5t^2 + 2v^2)^2$

Solution:

a. $(3x - 7)^2$ This is the square of a binomial, $(a - b)^2$, where $a = 3x$ and $b = 7$.

$\quad = (3x)^2 - 2(3x)(7) + (7)^2$ The product is $a^2 - 2ab + b^2$.

$\quad = 9x^2 - 42x + 49$ Simplify.

b. $(5t^2 + 2v^2)^2$ This is the square of a binomial, $(a + b)^2$, where $a = 5t^2$ and $b = 2v^2$.

$\quad = (5t^2)^2 + 2(5t^2)(2v^2) + (2v^2)^2$ The product is $a^2 + 2ab + b^2$.

$\quad = 25t^4 + 20t^2v^2 + 4v^4$ Simplify.

Skill Practice 4   Square the binomials.

a. $(8z - 2)^2$ **b.** $(3c^2 + 4d^3)^2$

Answers

3. a. $9y^2 - 49$ **b.** $\dfrac{4}{25}t^2 - \dfrac{1}{16}w^4$

4. a. $64z^2 - 32z + 4$

   b. $9c^4 + 24c^2d^3 + 16d^6$

In Example 5, we apply operations on polynomials to geometric formulas.

EXAMPLE 5   **Applying Operations on Polynomials to Geometry**

a. Write a polynomial that represents the area of the rectangle.

b. Write a polynomial that represents the volume of the cube.

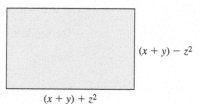

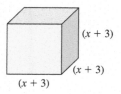

Solution:

a. $A = lw$ The area of a rectangle is length times width.

$$A = \overbrace{[(x + y) + z^2]}^{l}\overbrace{[(x + y) - z^2]}^{w}$$ Substitute $[(x + y) + z^2]$ for l, and $[(x + y) - z^2]$ for w.

$$= (x + y)^2 - (z^2)^2$$ This is a product of conjugates. The result is a difference of squares.

$$= x^2 + 2xy + y^2 - z^4$$ Square the binomial $(x + y)^2$ as $x^2 + 2xy + y^2$.

b. $V = s^3$ The volume of a cube is the product of the lengths of the sides.

$$V = (x + 3)^3$$ Substitute $(x + 3)$ for s.

$$= (x + 3)^2(x + 3)$$ This product can be written as $(x + 3)(x + 3)(x + 3)$ or as $(x + 3)^2(x + 3)$.

$$= (x^2 + 6x + 9)(x + 3)$$

$$= x^2(x) + x^2(3) + 6x(x) + 6x(3) + 9(x) + 9(3)$$

$$= x^3 + 3x^2 + 6x^2 + 18x + 9x + 27$$ Simplify.

$$= x^3 + 9x^2 + 27x + 27$$ Combine like terms.

Skill Practice 5

a. Write a polynomial that represents the area of a rectangle with length $(x + 3) + y$ and width $(x + 3) - y$.

b. Write a polynomial that represents the volume of a cube with sides of length $x + 2$.

---

5. Multiply Radical Expressions Involving Multiple Terms

The process used to multiply polynomials can be extended to algebraic expressions that are not polynomials. In Example 6 we multiply expressions containing two or more radical terms.

Answers

5. a. $x^2 + 6x + 9 - y^2$

  b. $x^3 + 6x^2 + 12x + 8$

> **EXAMPLE 6**   **Multiplying Radical Expressions**

Multiply and simplify.

   a. $3\sqrt{5}(2\sqrt{5} + 4\sqrt{2} + 1)$ **b.** $(3\sqrt{x} + 5)(2\sqrt{x} - 7)$

Solution:

a. $3\sqrt{5}(2\sqrt{5} + 4\sqrt{2} + 1)$ Apply the distributive
$= (3\sqrt{5})(2\sqrt{5}) + (3\sqrt{5})(4\sqrt{2}) + (3\sqrt{5})(1)$ property.

$= 6\sqrt{25} + 12\sqrt{10} + 3\sqrt{5}$ Apply the product
                                                                         property of radicals.
$= 6(5) + 12\sqrt{10} + 3\sqrt{5}$

$= 30 + 12\sqrt{10} + 3\sqrt{5}$ Simplify.

b. $(3\sqrt{x} + 5)(2\sqrt{x} - 7)$ Apply the distributive
                                                                         property.

$= (3\sqrt{x})(2\sqrt{x}) + (3\sqrt{x})(-7) + (5)(2\sqrt{x}) + (5)(-7)$

$= 6\sqrt{x^2} - 21\sqrt{x} + 10\sqrt{x} - 35$

$= 6x - 11\sqrt{x} - 35$ Combine like terms.

> **Skill Practice 6**   Multiply and simplify.
>   **a.** $2\sqrt{11}(2\sqrt{11} + 4\sqrt{2} - 3)$ **b.** $(4\sqrt{t} + 10)(3\sqrt{t} - 8)$

In Example 7, we evaluate special case products for expressions with radicals.

> **EXAMPLE 7**   **Simplifying Special Case Products**
> **Involving Radicals**

Multiply and simplify.

   a. $(4\sqrt{5} + \sqrt{6})(4\sqrt{5} - \sqrt{6})$ **b.** $(3x + \sqrt{2})^2$

Solution:

a. $(4\sqrt{5} + \sqrt{6})(4\sqrt{5} - \sqrt{6})$ This is a product of conjugates: $(a + b)(a - b)$.

$= (4\sqrt{5})^2 - (\sqrt{6})^2$ The product is a difference of squares. Note that
                                                         $(\sqrt{5})^2 = 5$ and $(\sqrt{6})^2 = 6$.
$= 16 \cdot 5 - 6$

$= 80 - 6$

$= 74$

b. $(3x + \sqrt{2})^2$ This is the square of a binomial, $(a + b)^2$, where
                                                         $a = 3x$ and $b = \sqrt{2}$.

$= (3x)^2 + 2(3x)(\sqrt{2}) + (\sqrt{2})^2$ The product is $a^2 + 2ab + b^2$.

$= 9x^2 + 6\sqrt{2}x + 2$ Simplify.

> **Skill Practice 7**   Multiply and simplify.
>   **a.** $(3\sqrt{7} - \sqrt{5})(3\sqrt{7} + \sqrt{5})$ **b.** $(4y - \sqrt{3})^2$

Answers
6. a. $44 + 8\sqrt{22} - 6\sqrt{11}$
    b. $12t - 2\sqrt{t} - 80$
7. a. 58   **b.** $16y^2 - 8\sqrt{3}y + 3$

SECTION R.4   Practice Exercises

Concept Connections

1. A _____ in the variable x is a finite sum of terms of the form ax^n where a is a real number and n is a whole number.

2. The polynomial $5x^3 - 2x^2 + 4$ is written in _____ order by degree.

3. The _____ term of a polynomial is the term of highest degree.

4. The leading _____ of a polynomial is the numerical factor of the leading term.

5. A _____ is a polynomial that has two terms, and a _____ is a polynomial with three terms.

6. The expanded form of the square of a binomial is a trinomial called a _____ square trinomial.

7. The product of conjugates $(a + b) \cdot$ _____ results in a difference of squares $a^2 - b^2$.

8. The conjugate of $3 - \sqrt{x}$ is _____.

Objective 1: Identify Key Elements of a Polynomial

For Exercises 9–10, determine if the expression is a polynomial.

9. **a.** $4a^2 + 7b - 3$ **b.** $\frac{3}{4}x^2y$ **c.** $6x + \frac{7}{x} + 5$ **d.** $\sqrt{p^2 + 2p - 5}$

10. **a.** $3x^5 - 9x^2 + \frac{2}{x^3}$ **b.** $\sqrt{2}ab^4$ **c.** $3|y| + 2$ **d.** $-7x^3 - 4x^2 + 2x - 5$

For Exercises 11–14, write the polynomial in descending order. Then identify the leading coefficient and degree of the polynomial.

11. $7.2x^3 - 18x^7 - 4.1$ 12. $9.1y^5 + 4.6y^2 - 1.7y^8$

13. $\frac{1}{3}y - y^2$ 14. $\frac{4}{5}c^2 + c^5$

For Exercises 15–16, determine the degree of the polynomial.

15. $-8p^2qr^5 + 4pq^8r^2 + 5p^3q^3r$ 16. $-4.7abc^4 - 5.2a^2bc^5 + 2.6a^3c$

Objective 2: Add and Subtract Polynomials

For Exercises 17–22, add or subtract as indicated and simplify. (See Example 1)

17. $(-8p^7 - 4p^4 + 2p - 5) + (2p^7 + 6p^4 + p^2)$

18. $(-7w^5 + 3w^3 - 6) + (9w^5 - 5w^3 + 4w - 3)$

19. $(0.05c^3b + 0.02c^2b^2 - 0.09cb^3) - (-0.03c^3b + 0.08c^2b^2 - 0.1cb^3)$

20. $(0.004mn^5 - 0.001mn^4 + 0.05mn^3) - (0.003mn^5 + 0.007mn^4 - 0.07mn^3)$

21. Subtract $\left(\frac{1}{2}x^2 - \frac{3}{4}x - \sqrt{2}\right)$ from $\left(\frac{1}{4}x^2 + \frac{5}{8}x + 5\sqrt{2}\right)$.

22. Subtract $\left(\frac{3}{10}y^3 + \frac{7}{5}y + 5\sqrt{7}\right)$ from $\left(\frac{2}{5}y^3 - \frac{1}{10}y + \sqrt{7}\right)$.

Objective 3: Multiply Polynomials

For Exercises 23–32, multiply and simplify. (See Example 2)

23. $(-6a^5b)\left(\frac{1}{3}a^2b^2\right)$ 24. $(-10c^2d^5)\left(\frac{1}{2}cd^7\right)$ 25. $7m^2(2m^4 - 3m + 4)$

26. $8p^5(-2p^2 - 5p - 1)$ 27. $(2x - 5)(x + 4)$ 28. $(w + 7)(6w - 3)$

29. $(4u^2 - 5v^2)(2u^2 + 3v^2)$ 30. $(2z^3 + 5u^2)(7z^3 - u^2)$ 31. $(3y + 6)\left(\frac{1}{3}y^2 - 5y - 4\right)$

32. $(10v - 5)\left(\frac{1}{5}v^2 - 3v + 1\right)$

Objective 4: Identify and Simplify Special Case Products

33. Write the expanded form for $(a + b)^2$.

34. Write the expanded form for $(a + b)(a - b)$.

35. Develop a formula for the expansion of the cube of $a + b$. [*Hint*: Write the expression $(a + b)^3$ as $(a + b)^2(a + b)$ and multiply.]

36. Develop a formula for the expansion of the cube of $a - b$. [*Hint*: Write the expression $(a - b)^3$ as $(a - b)^2(a - b)$ and multiply.]

For Exercises 37–50, perform the indicated operations and simplify. (See Examples 3–4)

37. $(4x - 5)(4x + 5)$

38. $(3p - 2)(3p + 2)$

39. $(3w^2 - 7z)(3w^2 + 7z)$

40. $(9v^3 + 2u)(9v^3 - 2u)$

41. $\left(\dfrac{1}{5}c - \dfrac{2}{3}d^3\right)\left(\dfrac{1}{5}c + \dfrac{2}{3}d^3\right)$

42. $\left(\dfrac{1}{6}n - \dfrac{4}{5}p^4\right)\left(\dfrac{1}{6}n + \dfrac{4}{5}p^4\right)$

43. $(5m - 3)^2$

44. $(7v - 2)^2$

45. $(4t^2 + 3p^3)^2$

46. $(2a^2 + 11b^3)^2$

47. $(w + 4)^3$

48. $(p - 2)^3$

49. $[(u + v) - w][(u + v) + w]$

50. $[(c + d) - a][(c + d) + a]$

Mixed Exercises

For Exercises 51–58, write an expression that represents the perimeter, area, or volume as indicated, and simplify. (See Example 5)

51. Perimeter

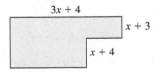

52. Perimeter

53. Area

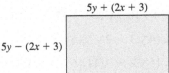

54. Area

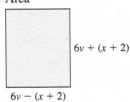

55. Volume

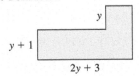

56. Volume

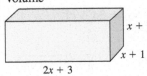

57. Area

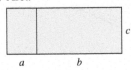

58. Area

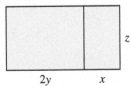

For Exercises 59–60, write an expression that represents the area of the shaded region and simplify the expression.

59.

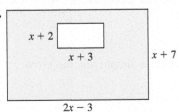

60.

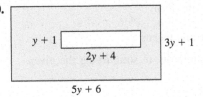

61. Suppose that x represents the smaller of two consecutive integers.

  a. Write a polynomial that represents the larger integer.

  b. Write a polynomial that represents the sum of the two integers. Then simplify.

  c. Write a polynomial that represents the product of the two integers. Then simplify.

  d. Write a polynomial that represents the sum of the squares of the two integers. Then simplify.

62. Suppose that x represents the larger of two consecutive odd integers.

  a. Write a polynomial that represents the smaller integer.

  b. Write a polynomial that represents the sum of the two integers. Then simplify.

  c. Write a polynomial that represents the product of the two integers. Then simplify.

  d. Write a polynomial that represents the difference of the squares of the two integers. Then simplify.

For Exercises 63–72, simplify each expression.

63. $(y + 7)^2 - 2(y - 3)^2$

64. $(x - 4)^2 - 6(x + 1)^2$

65. $(x^n + 3)(x^n - 7)$

66. $(y^n + 4)(y^n - 5)$

67. $(z^n + w^m)^2$

68. $(w^n - y^m)^2$

69. $(a^n - 5)(a^n + 5)$

70. $(b^n + 7)(b^n - 7)$

71. $(6x + 5)(6x - 5) - (6x + 5)^2$

72. $(2y - 7)(2y + 7) - (2y - 7)^2$

Objective 5: Multiply Radical Expressions Involving Multiple Terms

For Exercises 73–96, multiply and simplify. Assume that all variable expressions represent positive real numbers. (See Examples 6–7)

73. $5\sqrt{2}(2\sqrt{2} + 6\sqrt{3} - 4)$

74. $5\sqrt{7}(3\sqrt{7} - 2\sqrt{5} + 3)$

75. $3\sqrt{6}(5\sqrt{3} - 4\sqrt{2} - \sqrt{6})$

76. $4\sqrt{10}(7\sqrt{2} - 3\sqrt{5} + \sqrt{10})$

77. $(2\sqrt{y} - 3)(4\sqrt{y} + 5)$

78. $(5\sqrt{z} - 4)(3\sqrt{z} + 6)$

79. $(4\sqrt{3} - 2\sqrt{5})(6\sqrt{3} + 5\sqrt{5})$

80. $(3\sqrt{11} - 7\sqrt{2})(2\sqrt{11} + 9\sqrt{2})$

81. $(2\sqrt{3} + \sqrt{7})(2\sqrt{3} - \sqrt{7})$

82. $(5\sqrt{2} - \sqrt{11})(5\sqrt{2} + \sqrt{11})$

83. $(4x\sqrt{y} - 2y\sqrt{x})(4x\sqrt{y} + 2y\sqrt{x})$

84. $(5u\sqrt{v} - 6v\sqrt{u})(5u\sqrt{v} + 6v\sqrt{u})$

85. $(6z - \sqrt{5})^2$

86. $(2v - \sqrt{17})^2$

87. $(5a^2\sqrt{b} + 7b^2\sqrt{a})^2$

88. $(2c^3\sqrt{d} + 3d^3\sqrt{c})^2$

89. $(\sqrt{x + 1} - 5)(\sqrt{x + 1} + 5)$

90. $(\sqrt{y + 2} - 4)(\sqrt{y + 2} + 4)$

91. $(\sqrt{x + 1} - 5)^2$

92. $(\sqrt{y + 2} - 4)^2$

93. $(\sqrt{5 + 2\sqrt{x}})(\sqrt{5 - 2\sqrt{x}})$

94. $(\sqrt{7 + 3\sqrt{z}})(\sqrt{7 - 3\sqrt{z}})$

95. $(\sqrt{x + y} - \sqrt{x - y})^2$

96. $(\sqrt{a + 2} - \sqrt{a - 2})^2$

For Exercises 97–98, find the area of each triangle.

97.

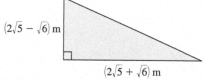

$(2\sqrt{5} - \sqrt{6})$ m

$(2\sqrt{5} + \sqrt{6})$ m

98.

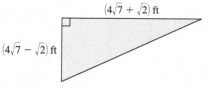

$(4\sqrt{7} + \sqrt{2})$ ft

$(4\sqrt{7} - \sqrt{2})$ ft

Write About It

99. A polynomial in the variable x is defined as an expression of the form
$a_n x^n + a_{n-1} x^{n-1} + a_{n-2} x^{n-2} + \cdots + a_1 x + a_0$.
Explain what this means.

100. Explain why $x + \sqrt{7}$ is a polynomial, but $\sqrt{x} + 7$ is not a polynomial.

101. Explain the similarity in simplifying the given expressions.

  a. $(3x + 2)(4x - 7)$

  b. $(3\sqrt{x} + \sqrt{2})(4\sqrt{x} - \sqrt{7})$

102. Explain the similarity in simplifying the given expressions.

  a. $(x + 3)(x - 3)$

  b. $(\sqrt{x} + \sqrt{3})(\sqrt{x} - \sqrt{3})$

Expanding Your Skills

For Exercises 103–106, determine if the statement is true or false.

103. The sum of two polynomials each of degree 5 will be degree 5.

104. The sum of two polynomials each of degree 5 will be less than or equal to degree 5.

105. The product of two polynomials each of degree 4 will be degree 8.

106. The product of two polynomials each of degree 4 will be less than degree 8.

PROBLEM RECOGNITION EXERCISES

Simplifying Algebraic Expressions

Many expressions in algebra look similar but the methods used to simplify the expressions may be different. For Exercises 1–14, simplify each expression. Assume that the variables are restricted so that each expression is defined.

1. a. $64^{1/2}$
   b. $64^{1/3}$
   c. $64^{2/3}$
   d. 64^{-1}
   e. $-64^{1/2}$
   f. $(-64)^{1/2}$
   g. $(-64)^{2/3}$
   h. $(-64)^{-2/3}$

2. a. $(5ab^3)^2$
   b. $(5a + b^3)^2$
   c. $(5ab^3)^{-2}$
   d. $(5a + b^3)^{-2}$

3. a. $(2x^4 y)^2$
   b. $(2x^4 - y)^2$
   c. $(2x^4 y)^{-2}$
   d. $(2x^4 - y)^{-2}$

4. a. $x^5 x^3$
   b. $\dfrac{x^5}{x^3}$
   c. $x^{-5} x^3$
   d. $(x^5)^{-3}$

5. a. $\sqrt{x^8}$
   b. $\sqrt[3]{x^8}$
   c. $\sqrt[5]{x^8}$
   d. $\sqrt[9]{x^8}$

6. a. $(3a + 4b^2) - (2a - b^2)$
   b. $(3a + 4b^2)(2a - b^2)$

7. a. $(a^2 - b^2) + (a^2 + b^2)$
   b. $(a^2 - b^2)(a^2 + b^2)$
   c. $(a - b)^2 - (a + b)^2$
   d. $(a - b)^2 (a + b)^2$

8. a. $|a - b|$ for $a < b$
   b. $|a - b|$ for $a > b$

9. a. $|x + 2|$ for $x > -2$
   b. $|x + 2|$ for $x < -2$

10. a. $(\sqrt{x})^2 + (\sqrt{y})^2$
   b. $\sqrt{x^2 + y^2}$
   c. $(\sqrt{x} + \sqrt{y})^2$
   d. $(\sqrt{x} + \sqrt{y})(\sqrt{x} - \sqrt{y})$

11. a. $\sqrt[3]{2x} \cdot \sqrt[3]{2x}$
   b. $\sqrt[3]{2x} + \sqrt[3]{2x}$

12. a. $\sqrt[4]{y} \cdot \sqrt[3]{y}$
   b. $\sqrt[4]{y} + \sqrt[3]{y}$

13. a. $36 \div 12 \div 6 \div 2$
   b. $36 \div (12 \div 6) \div 2$
   c. $36 \div 12 \div (6 \div 2)$
   d. $(36 \div 12) \div (6 \div 2)$

14. a. $\sqrt{6^2 + 8^2}$
   b. $\sqrt{6^2} + \sqrt{8^2}$

SECTION R.5   Factoring

OBJECTIVES

1. **Factor Out the Greatest Common Factor**
2. **Factor by Grouping**
3. **Factor Quadratic Trinomials**
4. **Factor Binomials**
5. **Apply a General Strategy to Factor Polynomials**
6. **Factor Expressions Containing Negative and Rational Exponents**

In Section R.4 we learned how to multiply polynomials. In this section, we reverse this process. The goal is to decompose a polynomial into a product of factors. This process is called factoring. Factoring is important in a variety of applications. In particular, factoring is often used to solve equations. For example, the two equations that follow are equivalent (that is, they have the same solution set).

$$x^2 - 19x + 88 = 0 \qquad (x - 8)(x - 11) = 0$$

Both equations ask, for what values of x will the left side equal zero? In the first equation it is difficult to determine the correct values of x. However, in the second equation with the left side factored, we can see by inspection that $x = 8$ and $x = 11$.

1. Factor Out the Greatest Common Factor

There are many techniques used to factor a polynomial, but the first step is always to factor out the greatest common factor.

The **greatest common factor (GCF)** of a polynomial is the expression of highest degree that divides evenly into each term of the polynomial. For example, the GCF of the following polynomial is $6x^3$.

x^3 is the greatest power of x common to all three terms.

$$12x^5 + 18x^4 - 24x^3 \qquad \text{The GCF is } 6x^3.$$

6 is the greatest integer that divides evenly into 12, 18, and -24.

To factor out the greatest common factor, we use the distributive property.

EXAMPLE 1   **Factoring Out the Greatest Common Factor**

Factor out the greatest common factor.

  a. $12x^5 + 18x^4 - 24x^3$ **b.** $3y(2y - 5) + (2y - 5)$

Solution:

  a. $12x^5 + 18x^4 - 24x^3$ The GCF is $6x^3$.

    $= 6x^3(2x^2) + 6x^3(3x) + 6x^3(-4)$ Write each term as a product of the GCF and another factor.

    $= 6x^3(2x^2 + 3x - 4)$ Apply the distributive property.

Check: $6x^3(2x^2 + 3x - 4)$
$= 12x^5 + 18x^4 - 24x^3$ ✓

> **Avoiding Mistakes**
>
> In Example 1(b), there is an understood factor of 1 in the second term: $1(2y - 5)$.
> Do not forget to include this in the factored form.

  b. $3y(2y - 5) + (2y - 5)$

    $= 3y(2y - 5) + 1(2y - 5)$ The GCF is the binomial $(2y - 5)$.

    $= (2y - 5)(3y + 1)$ Apply the distributive property.

Skill Practice 1   Factor out the greatest common factor.

  a. $9z^6 - 27z^4 + 12z^2$ **b.** $5w(w - 3) - (w - 3)$

Sometimes it is preferable to factor out a negative factor from a polynomial. For example, consider the polynomial, $-4x^2 - 8x + 12$. If we factor out -4, then the leading coefficient of the remaining polynomial will be positive.

EXAMPLE 2   **Factoring Out a Negative Factor**

Factor out -4 from the polynomial. $-4x^2 - 8x + 12$

Solution:

$-4x^2 - 8x + 12$

$= (-4)(x^2) + (-4)(2x) + (-4)(-3)$ Write each term as a product of -4 and another factor.

$= -4(x^2 + 2x - 3)$ Apply the distributive property.

> **TIP** When a negative factor is factored out of a polynomial, the remaining terms in parentheses will have signs opposite to those in the original polynomial.

Skill Practice 2   Factor out -6 from the polynomial. $-6y^4 + 24y^2 + 6$

Answers

1. a. $3z^2(3z^4 - 9z^2 + 4)$
  b. $(w - 3)(5w - 1)$
2. $-6(y^4 - 4y^2 - 1)$

2. Factor by Grouping

To factor a polynomial containing four terms, we often try factoring by grouping. This is demonstrated in Example 3.

EXAMPLE 3   **Factoring by Grouping**

Factor by grouping.

   a. $2ax - 6ay + 5x - 15y$ **b.** $m^2 - 3n + 3m - mn$

Solution:

a. $2ax - 6ay \mathrel{\vert} + 5x - 15y$ Consider the first pair of terms and second pair of terms separately.

$\quad = 2a(x - 3y) + 5(x - 3y)$ Factor out the GCF from the first two terms and from the second two terms.

$\quad = (x - 3y)(2a + 5)$ Factor out the common binomial factor $(x - 3y)$ from each term.

$\quad$ Check: $(x - 3y)(2a + 5)$

$\qquad = 2ax + 5x - 6ay - 15y$ ✓

b. $m^2 - 3n \mathrel{\vert} + 3m - mn$

$\quad = 1(m^2 - 3n) + m(3 - n)$ After factoring out the GCF from each pair of terms, the resulting binomial factors do not match. Therefore, try rearranging terms.

$\qquad\qquad$ These do not match.

$\quad m^2 + 3m \mathrel{\vert} - mn - 3n \xleftarrow{\text{rearrange}}$

$\quad = m(m + 3) - n(m + 3)$ Factor out the GCF from each pair of terms.

$\quad = (m + 3)(m - n)$ Factor out the common binomial factor.

> **Avoiding Mistakes**
>
> After factoring out the GCF from each pair of terms, the binomial factors must match to be able to complete the process of factoring by grouping.

> **TIP** Sometimes a polynomial has more than one factored form. For example, the polynomial $m^2 - 3n + 3m - mn$ can be written as $(m + 3)(m - n)$ or as $(-m - 3)(n - m)$ or as $-1(m + 3)(n - m)$. You can verify by multiplying factors.

Skill Practice 3   Factor by grouping.

   a. $7cd - c^2 + 14d - 2c$ **b.** $xy + 8x - 2x^2 - 4y$

---

3. Factor Quadratic Trinomials

Next we want to factor quadratic trinomials. These are trinomials of the form $ax^2 + bx + c$, where the coefficients a, b, and c are integers and $a \neq 0$. To understand the basis to factor a trinomial, first consider the product of two binomials.

product of $2x$ and x product of 3 and 2

$$(2x + 3)(x + 2) = 2x^2 + \underline{4x + 3x} + 6$$

sum of products of inner terms and outer terms

To factor a trinomial, this process is reversed.

Answers

3. a. $(7d - c)(c + 2)$
   b. $(y - 2x)(x - 4)$

EXAMPLE 4 **Factoring a Quadratic Trinomial with Leading Coefficient 1**

Factor. $x^2 - 8x + 12$

Solution:

factors of x^2

$x^2 - 8x + 12 = (\square x + \square)(\square x + \square)$

First fill in the blanks so that the product of first terms in the binomials is x^2. In this case, we have $1x$ and $1x$.

factors of 12

$x^2 - 8x + 12 = (1x + \square)(1x + \square)$

Fill in the remaining blanks with numbers whose product is 12 and whose sum is the middle term coefficient, -8. The numbers are -2 and -6.

$= (x - 2)(x - 6)$

<u>Check</u>: $(x - 2)(x - 6) = x^2 - 6x - 2x + 12$
$= x^2 - 8x + 12$ ✓

Skill Practice 4 Factor. $y^2 - 9y + 14$

The trinomial in Example 4 has a leading coefficient of 1. This simplified the factorization process. For any trinomial with a leading coefficient of 1 $(x^2 + bx + c)$, the constant terms in the binomial factors have a product of c and a sum equal to the middle term coefficient, b.

To factor a trinomial of the form $ax^2 + bx + c$, where $a \neq 1$, all possible factors of ax^2 must be tested with all factors of c until the correct middle term is found. This is demonstrated in Example 5.

EXAMPLE 5 **Factoring a Quadratic Trinomial with Leading Coefficient $\neq$1**

Factor. $27y + 10y^2 + 5$

Solution:

$10y^2 + 27y + 5$

First write the polynomial in the form $ay^2 + by + c$.

factors of $10y^2$

$10y^2 + 27y + 5 = (\square y + \square)(\square y + \square)$

factors of 5

Fill in the blanks so that the product of first terms in the binomials is $10y^2$. We have $1y$ and $10y$ or $2y$ and $5y$.

$(1y + \square)(10y + \square)$ or $(2y + \square)(5y + \square)$

Fill in the remaining blanks with factors of 5. Since the middle term of the trinomial is positive, consider only positive factors of 5: $(1)(5)$ and $(5)(1)$.

$(y + 1)(10y + 5) = 10y^2 + 15y + 5$ Incorrect. Wrong middle term.
$(y + 5)(10y + 1) = 10y^2 + 51y + 5$ Incorrect. Wrong middle term.
$(2y + 1)(5y + 5) = 10y^2 + 15y + 5$ Incorrect. Wrong middle term.
$(2y + 5)(5y + 1) = 10y^2 + 27y + 5$ Correct! ✓

The trinomial $10y^2 + 27y + 5$ factors as $(2y + 5)(5y + 1)$.

Skill Practice 5 Factor. $6x^2 + 23x + 7$

Answers
4. $(y - 7)(y - 2)$
5. $(2x + 7)(3x + 1)$

The technique shown in Examples 4 and 5 is called the trial-and-error method to factor a trinomial. The process is somewhat tedious, but we can often eliminate some of the possible binomials to test. For instance, the trinomial $10y^2 + 27y + 5$ has no common factor other than 1. Therefore, the terms within the binomials in the factored form must not have a common factor other than 1. From Example 5,

$(y + 1)(10y + 5)$ is not correct because $10y$ and 5 share a common factor of 5.

$(2y + 1)(5y + 5)$ is not correct because $5y$ and 5 share a common factor of 5.

Eliminating these possibilities leaves us with only two pairs of binomials to test.

EXAMPLE 6 **Factoring a Quadratic Trinomial**

Factor. $10x^3 + 105x^2y - 55xy^2$

Solution:

$10x^3 + 105x^2y - 55xy^2$

$= 5x(2x^2 + 21xy - 11y^2)$ Factor out the GCF, $5x$.

factors of $2x^2$

$= 5x(2x^2 + 21xy - 11y^2) = 5x(\square x + \square y)(\square x + \square y)$ Fill in the blanks so that the product of first terms in the binomials is $2x^2$. We have $2x$

factors of $-11y^2$ and x.

$= 5x(2x^2 + 21xy - 11y^2) = 5x(2x + \square)(x + \square)$ Fill in the remaining blanks with factors of $-11y^2$.

$(2x + 11y)(x - 1y) = 2x^2 + 9xy - 11y^2$ Incorrect.

$(2x + y)(x - 11y) = 2x^2 - 21xy - 11y^2$ Incorrect.

$(2x - 11y)(x + y) = 2x^2 - 9xy - 11y^2$ Incorrect.

$(2x - y)(x + 11y) = 2x^2 + 21xy - 11y^2$ Correct! ✓

The trinomial $10x^3 + 105x^2y - 55xy^2$ factors as $5x(2x - y)(x + 11y)$.

> **TIP** The order in which the product of factors is written does not affect the product. Therefore, $5x(2x - y)(x + 11y)$ equals $5x(x + 11y)(2x - y)$. This is guaranteed by the commutative property of multiplication.

Skill Practice 6 Factor. $30m^3 - 200m^2n - 70mn^2$

The process to factor a trinomial can be challenging if the coefficients are large. In fact, we will learn other techniques that will come in handy in such cases. One technique we show here is to recognize a perfect square trinomial and factor it as the square of a binomial. From Section R.4, we have the following formulas.

Factored Form of a Perfect Square Trinomial

$a^2 + 2ab + b^2 = (a + b)^2$ $a^2 - 2ab + b^2 = (a - b)^2$	A perfect square trinomial factors as the square of a binomial.

To identify a perfect square trinomial:	$4x^2 + 12x + 9 = (2x + 3)^2$
• First check whether the first and third terms are perfect squares.	perfect squares
• If so, label their principal square roots as a and b.	$a = 2x$ and $b = 3$
• Then determine if the absolute value of the middle term is $2ab$.	$\|12x\| \overset{?}{=} 2(2x)(3)$ ✓

EXAMPLE 7   Factoring a Perfect Square Trinomial

Factor.

a. $4x^2 - 28x + 49$ **b.** $81c^4 + 90c^2d + 25d^2$

Solution:

a. $4x^2 - 28x + 49$

$= (2x)^2 - 2(2x)(7) + (7)^2$ The trinomial fits the pattern $a^2 - 2ab + b^2$, where $a = 2x$ and $b = 7$.

$= (2x - 7)^2$ Factor as $a^2 - 2ab + b^2 = (a - b)^2$.

b. $81c^4 + 90c^2d + 25d^2$

$= (9c^2)^2 + 2(9c^2)(5d) + (5d)^2$ The trinomial fits the pattern $a^2 + 2ab + b^2$, where $a = 9c^2$ and $b = 5d$.

$= (9c^2 + 5d)^2$ Factor as $a^2 + 2ab + b^2 = (a + b)^2$.

Skill Practice 7   Factor.

a. $64y^2 + 16y + 1$ **b.** $9m^4 - 30m^2n + 25n^2$

---

4. Factor Binomials

Recall that the product of conjugates results in a difference of squares. Therefore, the factored form of a difference of squares is a product of conjugates.

Factored Form of a Difference of Squares

$$a^2 - b^2 = (a + b)(a - b)$$

Note: If a and b share no common factors other than 1, then a sum of squares $a^2 + b^2$ is not factorable over the set of real numbers.

EXAMPLE 8   Factoring a Difference of Squares

Factor completely.

a. $p^2 - 100$ **b.** $32y^4 - 162$

Solution:

a. $p^2 - 100$

$= (p)^2 - (10)^2$ The binomial fits the pattern $a^2 - b^2$, where $a = p$ and $b = 10$.

$= (p + 10)(p - 10)$ Factor as $a^2 - b^2 = (a + b)(a - b)$.

b. $32y^4 - 162$

$= 2(16y^4 - 81)$ Factor out the GCF, 2.

$= 2[(4y^2)^2 - (9)^2]$ The binomial fits the pattern $a^2 - b^2$, where $a = 4y^2$ and $b = 9$.

$= 2(4y^2 + 9)(4y^2 - 9)$ Factor as $a^2 - b^2 = (a + b)(a - b)$.

$= 2(4y^2 + 9)[(2y)^2 - (3)^2]$ The factor $(4y^2 - 9)$ is also a difference of squares, with $a = 2y$ and $b = 3$.

$= 2(4y^2 + 9)(2y + 3)(2y - 3)$ The polynomial is now factored completely.

Answers

7. a. $(8y + 1)^2$ **b.** $(3m^2 - 5n)^2$

> **Skill Practice 8** Factor completely.
>   **a.** $t^2 - 121$ **b.** $625z^5 - z$

A binomial can also be factored if it fits the pattern of a difference of cubes or a sum of cubes.

TIP The factored form of a sum or difference of cubes is a binomial times a trinomial. To help remember the pattern for the signs, remember **SOAP**: **S**ame sign, **O**pposite signs, **A**lways **P**ositive.

Factored Form of a Sum and Difference of Cubes

Sum of cubes: $a^3 + b^3 = (a + b)(a^2 - ab + b^2)$
Difference of cubes: $a^3 - b^3 = (a - b)(a^2 + ab + b^2)$

Same sign   Always Positive
Opposite signs

EXAMPLE 9   **Factoring a Sum and Difference of Cubes**

Factor completely.
  a. $x^3 + 125$ **b.** $8m^6 - 27n^3$

Solution:

a. $x^3 + 125$
$= (x)^3 + (5)^3$ — The binomial fits the pattern of a sum of cubes $a^3 + b^3$, where $a = x$ and $b = 5$.
$= (x + 5)[(x)^2 - (x)(5) + (5)^2]$ — Factor as $a^3 + b^3 = (a + b)(a^2 - ab + b^2)$.
$= (x + 5)(x^2 - 5x + 25)$ — Simplify.

b. $8m^6 - 27n^3$
$= (2m^2)^3 - (3n)^3$ — The binomial fits the pattern of a difference of cubes $a^3 - b^3$, where $a = 2m^2$ and $b = 3n$.
$= (2m^2 - 3n)[(2m^2)^2 + (2m^2)(3n) + (3n)^2]$ — Factor as $a^3 - b^3$ $= (a - b)(a^2 + ab + b^2)$.
$= (2m^2 - 3n)(4m^4 + 6m^2n + 9n^2)$ — Simplify.

> **Skill Practice 9** Factor completely.
>   **a.** $u^3 + 27$ **b.** $64v^3 - 125z^6$

5. Apply a General Strategy to Factor Polynomials

Factoring polynomials is a strategy game. The process requires that we identify the technique or techniques that best apply to a given polynomial. To do this we can follow the guidelines given in Table R-7.

Answers
8. a. $(t + 11)(t - 11)$
   b. $z(25z^2 + 1)(5z + 1)(5z - 1)$
9. a. $(u + 3)(u^2 - 3u + 9)$
   b. $(4v - 5z^2)(16v^2 + 20vz^2 + 25z^4)$

Table R-7 Factoring Strategy

First Step	Number of Terms	Technique
Factor out the GCF	4 or more terms	Try factoring by grouping.
	3 terms	If possible write the trinomial in the form $ax^2 + bx + c$. • If the trinomial is a perfect square trinomial, factor as $$a^2 + 2ab + b^2 = (a + b)^2$$ $$a^2 - 2ab + b^2 = (a - b)^2$$ • Otherwise try factoring by the trial-and-error method.
	2 terms	• If the binomial is a difference of squares, factor as $$a^2 - b^2 = (a + b)(a - b)$$ • If the binomial is a sum of cubes, factor as $$a^3 + b^3 = (a + b)(a^2 - ab + b^2)$$ • If the binomial is a difference of cubes, factor as $$a^3 - b^3 = (a - b)(a^2 + ab + b^2)$$ *Note*: A sum of squares, $a^2 + b^2$, cannot be factored over the real numbers.

EXAMPLE 10 Factoring a 4-Term Polynomial

Factor completely. $x^5 - 9x^3 + 8x^2 - 72$

Solution:

$x^5 - 9x^3 + 8x^2 - 72$ — Factor by grouping two terms with two terms.

$= x^3(x^2 - 9) + 8(x^2 - 9)$ — Factor out the GCF from the first two terms. Factor out the GCF from the second two terms.

$= (x^2 - 9)(x^3 + 8)$ — Factor the first binomial as a difference of squares and the second binomial as a sum of cubes.

$= (x + 3)(x - 3)(x + 2)(x^2 - 2x + 4)$

Skill Practice 10 Factor completely. $x^5 - x^3 - 64x^2 + 64$

EXAMPLE 11 Factoring by Grouping 1 Term with 3 Terms

Factor completely. $25 - x^2 - 4xy - 4y^2$

Solution:

This polynomial has 4 terms. However, the standard grouping method does not work even if we try rearranging terms.

$25 - \underbrace{x^2 - 4xy - 4y^2}_{\text{perfect square trinomial}}$ — We might try grouping 1 term with 3 terms. The reason is that after factoring out -1 from the last 3 terms we have a perfect square trinomial.

$= 25 - (x^2 + 4xy + 4y^2)$

$= 25 - (x + 2y)^2$ — This results in a difference of squares $a^2 - b^2$ where $a = 5$ and $b = (x + 2y)$.

$= (5)^2 - (x + 2y)^2$

$= [5 + (x + 2y)][5 - (x + 2y)]$ — Factor as $a^2 - b^2 = (a + b)(a - b)$.

$= (5 + x + 2y)(5 - x - 2y)$ — Simplify.

Answer

10. $(x + 1)(x - 1)(x - 4)(x^2 + 4x + 16)$

Skill Practice 11 Factor completely. $36 - m^2 - 6mn - 9n^2$

In Example 11, it may be helpful to write the expression $25 - (x + 2y)^2$ by using a convenient substitution. If we let $u = x + 2y$, then the expression becomes $25 - u^2$. This is easier to recognize as a difference of squares.

$$25 - u^2 = (5 + u)(5 - u)$$
$$= [5 + (x + 2y)][5 - (x + 2y)] \qquad \text{Back substitute.}$$
$$= (5 + x + 2y)(5 - x - 2y)$$

In Example 12, we practice making an appropriate substitution to convert a cumbersome expression into one that is more easily recognizable and factorable.

EXAMPLE 12 Factoring a Trinomial by Using Substitution

Factor completely. $(x^2 - 5)^2 + 2(x^2 - 5) - 24$

Solution:

$(x^2 - 5)^2 + 2(x^2 - 5) - 24$	Notice that the first two terms in the polynomial share a common factor of $(x^2 - 5)$. Suppose that we let u represent this expression. That is, $u = x^2 - 5$.
$= u^2 + 2u - 24$	
$= (u + 6)(u - 4)$	Factor.
$= [(x^2 - 5) + 6][(x^2 - 5) - 4]$	Back substitute. Replace u by $(x^2 - 5)$.
$= (x^2 + 1)(x^2 - 9)$	Simplify.
$= (x^2 + 1)(x + 3)(x - 3)$	Factor $x^2 - 9$ as a difference of squares.

Skill Practice 12 Factor completely. $(x^2 - 2)^2 + 5(x^2 - 2) - 14$

Some polynomials cannot be factored with the techniques learned thus far. For example, no combination of binomial factors with integer coefficients will produce the trinomial $3x^2 + 9x + 5$.

$$(3x + 5)(x + 1) = 3x^2 + 3x + 5x + 5 \qquad \text{Incorrect. Wrong middle term.}$$
$$= 3x^2 + 8x + 5$$
$$(3x + 1)(x + 5) = 3x^2 + 15x + x + 5 \qquad \text{Incorrect. Wrong middle term.}$$
$$= 3x^2 + 16x + 5$$

In such a case we say that the polynomial is a **prime polynomial.** Its only factors are 1 and itself.

6. Factor Expressions Containing Negative and Rational Exponents

We now revisit the process to factor out the greatest common factor. In some applications, it is necessary to factor out a variable factor with a negative integer exponent or a rational exponent. Before we demonstrate this in Example 13, take a minute to review a similar example with positive integer exponents.

$$2x^6 + 5x^5 + 7x^4 \qquad \text{The GCF is } x^4. \text{ This is } x \text{ raised to the } smallest$$
$$exponent \text{ to which it appears in any term.}$$

$$= x^4(2x^2 + 5x^1 + 7x^0) \qquad \text{The powers on the factors of } x \text{ within parentheses are}$$
$$\text{found by subtracting 4 from the original exponents.}$$

6 − 4 │ 5 − 4 │ 4 − 4

Answers
11. $(6 + m + 3n)(6 - m - 3n)$
12. $(x^2 + 5)(x + 2)(x - 2)$

EXAMPLE 13 Factoring Out Negative and Rational Exponents

Factor. Write the answers with no negative exponents.

a. $2x^{-6} + 5x^{-5} + 7x^{-4}$ **b.** $x(2x + 5)^{-1/2} + (2x + 5)^{1/2}$

Solution:

a. $2x^{-6} + 5x^{-5} + 7x^{-4}$ The smallest exponent on x is -6.
Factor out x^{-6}.

$\boxed{-6 - (-6)}\ \boxed{-5 - (-6)}\ \boxed{-4 - (-6)}$

$= x^{-6}(2x^0 + 5x^1 + 7x^2)$ The powers on the factors of x within parentheses are found by subtracting -6 from the original exponents.

$= x^{-6}(2 + 5x + 7x^2)$

$= \dfrac{7x^2 + 5x + 2}{x^6}$ Simplify the negative exponent.

$b^{-n} = \dfrac{1}{b^n}$

b. $x(2x + 5)^{-1/2} + (2x + 5)^{1/2}$ The smallest exponent on $(2x + 5)$ is $-\frac{1}{2}$.

$\boxed{-1/2 - (-1/2)}\ \boxed{1/2 - (-1/2)}$ Factor out $(2x + 5)^{-1/2}$.

$= (2x + 5)^{-1/2}\,[x(2x + 5)^0 + (2x + 5)^1]$ The powers on the factors of $(2x + 5)$ within parentheses are found by subtracting $-\frac{1}{2}$ from the original exponents.

$= (2x + 5)^{-1/2}\,[x + (2x + 5)]$ The expression $(2x + 5)^0 = 1$, for $2x + 5 \neq 0$.

$= (2x + 5)^{-1/2}\,(3x + 5)$ Simplify.

$= \dfrac{3x + 5}{(2x + 5)^{1/2}}$ Simplify the negative exponent.

Answers

13. a. $\dfrac{2a^2 - 3a + 11}{a^5}$

b. $\dfrac{5x + 3}{(4x + 3)^{3/4}}$

Skill Practice 13 Factor completely and write the answer with no negative exponents.

a. $11a^{-5} - 3a^{-4} + 2a^{-3}$ **b.** $x(4x + 3)^{-3/4} + (4x + 3)^{1/4}$

SECTION R.5   Practice Exercises

Concept Connections

1. The binomial $a^3 + b^3$ is called a sum of _____ and factors as _____.

2. The binomial $a^3 - b^3$ is called a _____ of cubes and factors as _____.

3. The trinomial $a^2 + 2ab + b^2$ is a _____ square trinomial. Its factored form is _____.

4. The binomial $a^2 - b^2$ is called a difference of _____ and factors as _____.

Objective 1: Factor Out the Greatest Common Factor

For Exercises 5–12, factor out the greatest common factor. (See Example 1)

5. $15c^5 - 30c^4 + 5c^3$ 6. $12m^4 + 15m^3 + 3m^2$ 7. $21a^2b^5 - 14a^3b^4 + 35a^4b$

8. $36p^4q^7 + 18p^3q^5 - 27p^2q^6$ 9. $5z(x - 6y) + 7(x - 6y)$ 10. $4t(u + 8v) - 3(u + 8v)$

11. $10k^2(3k^2 + 7) - 5k(3k^2 + 7)$ 12. $8j^3(4j + 9) + 4j^2(4j + 9)$

For Exercises 13–16, factor out the indicated common factor. (See Example 2)

13. a. Factor out 3. $-6x^2 + 12x + 9$
b. Factor out -3. $-6x^2 + 12x + 9$

14. a. Factor out 5. $-15y^2 - 10y + 25$
b. Factor out -5. $-15y^2 - 10y + 25$

15. Factor out $-4x^2y$. $-12x^3y^2 - 8x^4y^3 + 4x^2y$

16. Factor out $-7a^3b$. $-14a^4b^3 + 21a^3b^2 - 7a^3b$

Objective 2: Factor by Grouping

For Exercises 17–22, factor by grouping. (See Example 3)

17. $8ax + 18a + 20x + 45$

18. $6ty + 9y + 14t + 21$

19. $12x^3 - 9x^2 - 40x + 30$

20. $30z^3 - 35z^2 - 24z + 28$

21. $cd - 8d + 4c - 2d^2$

22. $7t - 6v^2 + tv - 42v$

Objective 3: Factor Quadratic Trinomials

For Exercises 23–36, factor the trinomials. (See Examples 4–7)

23. $p^2 + 2p - 63$

24. $w^2 + 5w - 66$

25. $2t^3 - 28t^2 + 80t$

26. $5u^4 - 40u^3 + 35u^2$

27. $25z + 6z^2 + 14$

28. $8 + 15m^2 + 26m$

29. $7y^3z - 40y^2z^2 - 12yz^3$

30. $11a^3b + 18a^2b^2 - 8ab^3$

31. $t^2 - 18t + 81$

32. $p^2 + 8p + 16$

33. $50x^3 + 160x^2y + 128xy^2$

34. $48y^3 - 72y^2z + 27yz^2$

35. $4c^4 - 20c^2d^3 + 25d^6$

36. $9m^4 + 42m^2n^4 + 49n^8$

Objective 4: Factor Binomials

For Exercises 37–48, factor the binomials. (See Examples 8–9)

37. $9w^2 - 64$

38. $16t^2 - 49$

39. $200u^4 - 18v^6$

40. $75m^6 - 27n^4$

41. $625p^4 - 16$

42. $81z^4 - 1$

43. $y^3 + 64$

44. $u^3 + 343$

45. $c^4 - 27c$

46. $d^4 - 8d$

47. $8a^6 - 125b^9$

48. $27m^{12} - 64n^9$

Objective 5: Apply a General Strategy to Factor Polynomials

For Exercises 49–76, factor completely. (See Examples 10–12)

49. $30x^4 + 70x^3 - 120x^2 - 280x$

50. $4y^4 - 10y^3 - 36y^2 + 90y$

51. $a^2 - y^2 + 10y - 25$

52. $c^2 - z^2 + 8z - 16$

53. $30x^3y + 125x^2y + 120xy$

54. $60t^4v + 78t^3v - 180t^2v$

55. $(x^2 - 2)^2 - 3(x^2 - 2) - 28$

56. $(y^2 + 2)^2 + 5(y^2 + 2) - 24$

57. $(x^3 + 12)^2 - 16$

58. $(y^3 + 34)^2 - 49$

59. $(x + y)^2 - z^2$

60. $(a + 5)^2 - y^2$

61. $(x + y)^3 + z^3$

62. $(a + 5)^3 - b^3$

63. $9m^2 + 42m(3n + 1) + 49(3n + 1)^2$

64. $4x^2 + 36x(7y - 1) + 81(7y - 1)^2$

65. $(c - 3)^2 - (2c - 5)^2$

66. $(d + 6)^2 - (4d - 3)^2$

67. $p^{11} - 64p^8 - p^3 + 64$

68. $t^7 + 27t^4 - t^3 - 27$

69. $m^6 + 26m^3 - 27$

70. $n^6 - 7n^3 - 8$

71. $16x^6z + 38x^3z - 54z$

72. $24y^7 + 21y^4 - 3y$

73. $x^2 - y^2 - x + y$

74. $a^2 - b^2 - a - b$

75. $a^2 + ac - 2c^2 - c + a$

76. $x^2 + 2xy - 3y^2 - y + x$

Objective 6: Factor Expressions Containing Negative and Rational Exponents

For Exercises 77–88, factor completely. Write the answers with positive exponents only. (See Example 13)

77. $2x^{-4} - 7x^{-3} + x^{-2}$

78. $5t^{-7} + 2t^{-6} - t^{-5}$

79. $y^{-2} - y^{-3} - 12y^{-4}$

80. $w^{-3} + 10w^{-4} + 9w^{-5}$

81. $2c^{7/4} + 4c^{3/4}$

82. $10y^{9/5} - 15y^{4/5}$

83. $\dfrac{8}{3}x^{1/3} + \dfrac{5}{3}x^{-2/3}$

84. $\dfrac{15}{2}x^{1/2} - \dfrac{3}{2}x^{-1/2}$

85. $5x(3x + 1)^{2/3} + (3x + 1)^{5/3}$

86. $7t(4t + 1)^{3/4} + (4t + 1)^{7/4}$

87. $x(3x + 2)^{-2/3} + (3x + 2)^{1/3}$

88. $x(5x - 8)^{-4/5} + (5x - 8)^{1/5}$

Mixed Exercises

For Exercises 89–90, the shaded and unshaded areas are squares. Write an expression for the shaded area A, and factor the expression.

89.

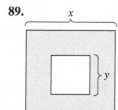

90.

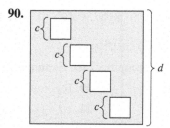

For Exercises 91–94, write the expression in factored form.

91. $2\pi r^2 + 2\pi rh$ Surface area of a right circular cylinder

92. $P + Prt$ Principal and simple interest

93. $\frac{4}{3}\pi R^3 - \frac{4}{3}\pi r^3$ Volume between two concentric spheres

94. $\pi R^2 - \pi r^2$ Area between two concentric circles

95. Simplify the expression $21^2 - 19^2$ by first factoring the expression. Do not use a calculator.

96. Simplify the expression $17^2 - 13^2$ by first factoring the expression. Do not use a calculator.

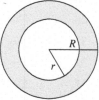

Write About It

97. Explain how to construct a perfect square trinomial.

98. Why is the sum of squares $a^2 + b^2$ not factorable over the real numbers?

99. Explain the similarity in the process to factor out the GCF in the following two expressions.

$5x^4 + 4x^3$ and $5x^{-4} + 4x^{-3}$

100. Explain the similarity in the process to factor out the GCF in the following two expressions.

$6x^5 + 5x^2$ and $6x^{5/3} + 5x^{2/3}$

Expanding Your Skills

We say that the expression $x^2 - 4$ is factorable over the integers as $(x + 2)(x - 2)$. Notice that the constant terms in the binomials are integers. The expression $x^2 - 3$ can be factored over the irrational numbers as $x^2 - 3 = (x + \sqrt{3})(x - \sqrt{3})$. For Exercises 101–106, factor each expression over the irrational numbers.

101. $x^2 - 5$

102. $y^2 - 11$

103. $z^4 - 36$

104. $w^4 - 49$

105. $x^2 - 2\sqrt{5}x + 5$

106. $c^2 - 2\sqrt{3}c + 3$

107. Consider the following binomials and their factored forms. (The factored forms can be verified by multiplying the expressions on the right.)

$x^2 - 1 = (x - 1)(x + 1)$

$x^3 - 1 = (x - 1)(x^2 + x + 1)$

$x^4 - 1 = (x - 1)(x^3 + x^2 + x + 1)$

a. Use the pattern to factor the expression $x^5 - 1$.

b. Use the pattern to write a generic formula for $x^n - 1$ where n is a positive integer.

108. For a positive integer n, the expression $a^n - b^n$ can be factored as

$a^n - b^n = (a - b)(a^{n-1} + a^{n-2}b + a^{n-3}b^2 + \cdots + ab^{n-2} + b^{n-1})$.

a. Use this formula to factor $a^5 - b^5$.

b. Check the result to part (a) by multiplication.

Consider the trinomial $ax^2 + bx + c$ with integer coefficients a, b, and c. The trinomial can be factored as the product of two binomials with integer coefficients if $b^2 - 4ac$ is a perfect square. For Exercises 109–114, determine whether the trinomial can be factored as a product of two binomials with integer coefficients.

109. $36p^2 - 33p - 12$

110. $24w^2 - 25w + 8$

111. $8x^2 + 2x - 15$

112. $6x^2 - 7x - 20$

113. $18y^2 + 45y - 48$

114. $54z^2 - 39z - 60$

SECTION R.6 Rational Expressions and More Operations on Radicals

1. Determine Restricted Values for a Rational Expression

Suppose that an object that is originally 35°C is placed in a freezer. The temperature T (in °C) of the object t hours after being placed in the freezer can be approximated by the model

$$T = \frac{350}{t^2 + 3t + 10}$$

For example, 2 hr after being placed in the freezer the temperature of the object is

$$T = \frac{350}{(2)^2 + 3(2) + 10} = 17.5°C$$

The expression $\dfrac{350}{t^2 + 3t + 10}$ is called a rational expression. A **rational expression** is a ratio of two polynomials. Since a rational expression may have a variable in the denominator, we must be careful to exclude values of the variable that make the denominator zero.

EXAMPLE 1   **Determining Restricted Values for a Rational Expression**

Determine the restrictions on the variable for each rational expression.

a. $\dfrac{x - 3}{x + 2}$ **b.** $\dfrac{x}{x^2 - 49}$ **c.** $\dfrac{4}{5x^2y}$

Solution:

a. $\dfrac{x - 3}{x + 2}$ $\boxed{x \neq -2}$

Division by zero is undefined.
For this expression $x \neq -2$. If -2 were substituted for x, the denominator would be zero. $-2 + 2 = 0$

b. $\dfrac{x}{x^2 - 49} = \dfrac{x}{(x + 7)(x - 7)}$ $\boxed{x \neq -7}$ $\boxed{x \neq 7}$

For this expression, $x \neq -7$ and $x \neq 7$. If x were -7, then $-7 + 7 = 0$. If x were 7, then $7 - 7 = 0$. In either case, the denominator would be zero.

c. $\dfrac{4}{5x^2y}$ $\boxed{x \neq 0}$ $\boxed{y \neq 0}$

For this expression, $x \neq 0$ and $y \neq 0$. If either x or y were zero, then the product would be zero.

Skill Practice 1   Determine the restrictions on the variable.

a. $\dfrac{x + 4}{x - 3}$ **b.** $\dfrac{5}{c^2 - 16}$ **c.** $\dfrac{6}{7ab^3}$

---

2. Simplify Rational Expressions

A rational expression is simplified (in lowest terms) if the only factors shared by the numerator and denominator are 1 or -1. To simplify a rational expression, factor the numerator and denominator, then apply the property of equivalent algebraic fractions.

Answers
1. **a.** $x \neq 3$ **b.** $c \neq -4$ and $c \neq 4$
   c. $a \neq 0$ and $b \neq 0$

Equivalent Algebraic Fractions

If a, b, and c represent real-valued expressions, then

$$\frac{ac}{bc} = \frac{a}{b} \quad \text{for } b \neq 0 \text{ and } c \neq 0 \qquad \text{Example: } \frac{10x}{15x} = \frac{2 \cdot \overset{1}{\cancel{5 \cdot x}}}{3 \cdot \underset{1}{\cancel{5 \cdot x}}} = \frac{2}{3}$$

EXAMPLE 2 Simplifying Rational Expressions

Simplify.

a. $\dfrac{x^2 - 16}{x^2 - x - 12}$ **b.** $\dfrac{8 + 2\sqrt{7}}{4}$

Solution:

a. $\dfrac{x^2 - 16}{x^2 - x - 12} = \dfrac{(x+4)(x-4)}{(x+3)(x-4)}$ Factor the numerator and denominator. We have the restrictions that $x \neq -3$, $x \neq 4$.

$= \dfrac{(x+4)\overset{1}{\cancel{(x-4)}}}{(x+3)\underset{1}{\cancel{(x-4)}}}$ Divide out common factors that form a ratio of $\frac{1}{1}$.

$= \dfrac{x+4}{x+3}$ for $x \neq -3$, $x \neq 4$ The same restrictions for the original expression also apply to the simplified expression.

b. $\dfrac{8 + 2\sqrt{7}}{4} = \dfrac{2(4 + \sqrt{7})}{2 \cdot 2} = \dfrac{\overset{1}{\cancel{2}}(4 + \sqrt{7})}{\underset{1}{\cancel{2}} \cdot 2} = \dfrac{4 + \sqrt{7}}{2}$

Skill Practice 2 Simplify.

a. $\dfrac{x^2 - 8x}{x^2 - 7x - 8}$ **b.** $\dfrac{3 + 9\sqrt{5}}{6}$

In Example 2(a), the expressions $\dfrac{x^2 - 16}{x^2 - x - 12}$ and $\dfrac{x+4}{x+3}$ are equal for all values of x for which *both* expressions are defined. This excludes the values $x = -3$ and $x = 4$.

The property of equivalent algebraic fractions tells us that we can divide out common factors that form a ratio of 1. We can also divide out factors that form a ratio of -1. For example:

TIP The expressions $4 - x$ and $x - 4$ are opposite polynomials (the signs of their terms are opposites). The ratio of two opposite factors is -1.

$$\frac{4-x}{x-4} \xrightarrow{\substack{\text{Factor out } -1 \text{ from} \\ \text{the numerator.}}} \frac{-1(-4+x)}{x-4} = \frac{-1(x-4)}{x-4} = -1$$

Numerator and denominator are opposite polynomials. Their ratio is -1.

EXAMPLE 3 Simplifying Rational Expressions

Simplify. $\dfrac{14 - 2x}{x^2 - 7x}$

Answers

2. a. $\dfrac{x}{x+1}$; $x \neq 8$, $x \neq -1$

  b. $\dfrac{1 + 3\sqrt{5}}{2}$

Solution:

$$\frac{14 - 2x}{x^2 - 7x} = \frac{2(7 - x)}{x(x - 7)}$$

Factor the numerator and denominator. We have the restrictions that $x \neq 0$, $x \neq 7$.

$$= \frac{2(7 - x)^{(-1)}}{x(x - 7)}$$

The factors $(7 - x)$ and $(x - 7)$ are opposites. Their ratio is -1.

Divide out opposite factors that form a ratio of -1.

$$= -\frac{2}{x} \quad \text{for } x \neq 0, x \neq 7$$

The same restrictions for the original expression also apply to the simplified expression.

Skill Practice 3   Simplify. $\dfrac{30 - 10m}{m^2 - 9}$

3. Multiply and Divide Rational Expressions

To multiply and divide rational expressions, use the following properties.

> **Multiplication and Division of Algebraic Fractions**
>
> Let a, b, c, and d be real-valued expressions.
>
> $$\frac{a}{b} \cdot \frac{c}{d} = \frac{ac}{bd} \quad \text{for } b \neq 0, d \neq 0$$
>
> $$\frac{a}{b} \div \frac{c}{d} = \frac{a}{b} \cdot \frac{d}{c} = \frac{ad}{bc} \quad \text{for } b \neq 0, c \neq 0, d \neq 0$$

To multiply or divide rational expressions, factor the numerator and denominator of each fraction completely. Then apply the multiplication and division properties for algebraic fractions.

Restriction Agreement

Operations on rational expressions are valid for all values of the variable for which the rational expressions are defined. In Examples 4–10 and in the exercises, we will perform operations on rational expressions without explicitly stating the restrictions. Instead, the restrictions on the variables will be implied.

EXAMPLE 4   **Multiplying Rational Expressions**

Multiply. $\dfrac{2xy}{x^2y + 3xy} \cdot \dfrac{x^2 + 6x + 9}{4x + 12}$

Solution:

$$\frac{2xy}{x^2y + 3xy} \cdot \frac{x^2 + 6x + 9}{4x + 12} = \frac{2xy}{xy(x + 3)} \cdot \frac{(x + 3)^2}{4(x + 3)}$$

$$= \frac{2 \, xy \cdot (x + 3)(x + 3)}{xy(x + 3) \cdot 2 \cdot 2(x + 3)}$$

$$= \frac{1}{2}$$

Answers

3. $-\dfrac{10}{m + 3}; m \neq 3, m \neq -3$

4. $\dfrac{1}{3}$

Skill Practice 4   Multiply. $\dfrac{7x - 7y}{x^2 - 2xy + y^2} \cdot \dfrac{x^2 - xy}{21x}$

EXAMPLE 5   Dividing Rational Expressions

Divide. $\dfrac{x^3 - 8}{4 - x^2} \div \dfrac{3x^2 + 6x + 12}{x^2 - x - 6}$

Solution:

$\dfrac{x^3 - 8}{4 - x^2} \div \dfrac{3x^2 + 6x + 12}{x^2 - x - 6}$

$= \dfrac{x^3 - 8}{4 - x^2} \cdot \dfrac{x^2 - x - 6}{3x^2 + 6x + 12}$ Multiply the first fraction by the reciprocal of the second fraction.

$= \dfrac{(x - 2)(x^2 + 2x + 4)}{(2 - x)(2 + x)} \cdot \dfrac{(x - 3)(x + 2)}{3(x^2 + 2x + 4)}$ Factor. Note that $x^3 - 8$ is a difference of cubes.

$= \dfrac{\overset{(-1)}{\cancel{(x - 2)}}\,\overset{1}{\cancel{(x^2 + 2x + 4)}} \cdot (x - 3)\overset{1}{\cancel{(x + 2)}}}{\cancel{(2 - x)}\cancel{(2 + x)} \cdot 3\cancel{(x^2 + 2x + 4)}}$ Note that $(x - 2)$ and $(2 - x)$ are opposite polynomials, and their ratio is -1.

$= -\dfrac{x - 3}{3}$

> **Avoiding Mistakes**
>
> For the expression $-\dfrac{x - 3}{3}$ do not be tempted to "divide out" the 3 in the numerator with the 3 in the denominator. The 3's are *terms*, not factors. Only common *factors* can be divided out.

Skill Practice 5   Divide. $\dfrac{y^3 - 27}{9 - y^2} \div \dfrac{5y^2 + 15y + 45}{y^2 + 8y + 15}$

4. Add and Subtract Rational Expressions

Recall that fractions can be added or subtracted if they have a common denominator.

> **Addition and Subtraction of Algebraic Fractions**
>
> Let a, b, c, and d be real-valued expressions.
>
> $\dfrac{a}{b} + \dfrac{c}{b} = \dfrac{a + c}{b}$ and $\dfrac{a}{b} - \dfrac{c}{b} = \dfrac{a - c}{b}$ for $b \neq 0$

If two rational expressions have different denominators, then it is necessary to convert the expressions to equivalent expressions with the same denominator. We do this by applying the property of equivalent fractions.

> **Property of Equivalent Fractions**
>
> Let a, b, and c be real-valued expressions.
>
> $\dfrac{a}{b} = \dfrac{ac}{bc}$, where $b \neq 0$ and $c \neq 0$ <u>Example:</u> $\dfrac{5}{x} = \dfrac{5 \cdot xy}{x \cdot xy} = \dfrac{5xy}{x^2 y}$

When adding or subtracting numerical fractions or rational expressions, it is customary to use the least common denominator (LCD) of the original expressions and to follow these guidelines.

Answer

5. $-\dfrac{y + 5}{5}$

> ### Adding and Subtracting Rational Expressions
>
> **Step 1** Factor the denominators and determine the LCD of all expressions. The LCD is the product of unique prime factors where each factor is raised to the greatest power to which it appears in any denominator.
>
> **Step 2** Write each expression as an equivalent expression with the LCD as its denominator.
>
> **Step 3** Add or subtract the numerators as indicated and write the result over the LCD.
>
> **Step 4** Simplify if possible.

EXAMPLE 6   Adding Rational Expressions

Add the rational expressions and simplify the result. $\dfrac{7}{4a} + \dfrac{11}{10a^2}$

Solution:

$\dfrac{7}{4a} + \dfrac{11}{10a^2} = \dfrac{7}{2^2 a} + \dfrac{11}{2 \cdot 5a^2}$

Step 1: Factor the denominators. The LCD is $2^2 \cdot 5a^2$ or $20a^2$.

$= \dfrac{7 \cdot (5a)}{2^2 a \cdot (5a)} + \dfrac{11 \cdot (2)}{2 \cdot 5a^2 \cdot (2)}$

Step 2: Multiply numerator and denominator of each expression by the factors missing from the denominators.

$= \dfrac{35a}{20a^2} + \dfrac{22}{20a^2}$

Step 3: Add the numerators and write the result over the common denominator.

$= \dfrac{35a + 22}{20a^2}$

Step 4: The expression is already simplified.

Skill Practice 6   Add the rational expressions and simplify the result.

$$\dfrac{8}{9y^2} + \dfrac{1}{15y}$$

EXAMPLE 7   Subtracting Rational Expressions

Subtract the rational expressions and simplify the result.

$$\dfrac{3x + 5}{x^2 + 4x + 3} - \dfrac{x - 5}{x^2 + 2x - 3}$$

Solution:

$\dfrac{3x + 5}{x^2 + 4x + 3} - \dfrac{x - 5}{x^2 + 2x - 3}$

$= \dfrac{3x + 5}{(x + 3)(x + 1)} - \dfrac{x - 5}{(x + 3)(x - 1)}$

Factor the denominators. The LCD is $(x + 3)(x + 1)(x - 1)$.

$= \dfrac{(3x + 5)(x - 1)}{(x + 3)(x + 1)(x - 1)} - \dfrac{(x - 5)(x + 1)}{(x + 3)(x - 1)(x + 1)}$

Answer

6. $\dfrac{40 + 3y}{45y^2}$

Avoiding Mistakes

It is very important to use parentheses around the second trinomial in the numerator. This will ensure that all terms that follow will be subtracted.

$$= \frac{(3x + 5)(x - 1) - (x - 5)(x + 1)}{(x + 3)(x + 1)(x - 1)}$$

Subtract the numerators and write the result over the common denominator.

$$= \frac{3x^2 + 2x - 5 - (x^2 - 4x - 5)}{(x + 3)(x + 1)(x - 1)}$$

$$= \frac{3x^2 + 2x - 5 - x^2 + 4x + 5}{(x + 3)(x + 1)(x - 1)}$$

$$= \frac{2x^2 + 6x}{(x + 3)(x + 1)(x - 1)}$$

$$= \frac{2x(x + 3)}{(x + 3)(x + 1)(x - 1)}$$

Factor the numerator and denominator and simplify.

$$= \frac{2x}{(x + 1)(x - 1)}$$

Skill Practice 7   Subtract the rational expressions and simplify the result.

$$\frac{t}{t^2 + 5t + 6} - \frac{2}{t^2 + 3t + 2}$$

5. Simplify Complex Fractions

A **complex fraction** (also called a **compound fraction**) is an expression that contains one or more fractions in the numerator or denominator. We present two methods to simplify a complex fraction. Method I is an application of the order of operations.

Simplifying a Complex Fraction: Order of Operations (Method I)

Step 1   Add or subtract the fractions in the numerator to form a single fraction. Add or subtract the fractions in the denominator to form a single fraction.

Step 2   Divide the resulting expressions.

Step 3   Simplify if possible.

EXAMPLE 8   Simplifying a Complex Fraction (Method I)

Simplify. $\dfrac{\dfrac{x}{4} - \dfrac{4}{x}}{\dfrac{1}{4} + \dfrac{1}{x}}$

Answer

7. $\dfrac{t - 3}{(t + 1)(t + 3)}$

Solution:

$$\frac{\dfrac{x}{4} - \dfrac{4}{x}}{\dfrac{1}{4} + \dfrac{1}{x}} = \frac{\dfrac{x \cdot x}{4 \cdot x} - \dfrac{4 \cdot 4}{x \cdot 4}}{\dfrac{1 \cdot x}{4 \cdot x} + \dfrac{1 \cdot 4}{x \cdot 4}} = \frac{\dfrac{x^2 - 16}{4x}}{\dfrac{x + 4}{4x}}$$

$$= \frac{x^2 - 16}{4x} \cdot \frac{4x}{x + 4}$$

$$= \frac{(x - 4)(x + 4)}{4x} \cdot \frac{\overset{1}{4x}}{x + 4}$$

$$= x - 4$$

Step 1: Subtract the fractions in the numerator. Add the fractions in the denominator.

Step 2: Multiply the rational expression from the numerator by the reciprocal of the expression from the denominator.

Step 3: Simplify by factoring and dividing out common factors.

Skill Practice 8   Simplify. $\dfrac{\dfrac{1}{7} + \dfrac{1}{y}}{\dfrac{y}{7} - \dfrac{7}{y}}$

In Example 9 we demonstrate another method (Method II) to simplify a complex fraction.

Simplifying a Complex Fraction: Multiply by the LCD (Method II)

Step 1   Multiply the numerator and denominator of the complex fraction by the LCD of all individual fractions.

Step 2   Apply the distributive property and simplify the numerator and denominator.

Step 3   Simplify the resulting expression if possible.

EXAMPLE 9   **Simplifying a Complex Fraction (Method II)**

Simplify. $\dfrac{d^{-2} - c^{-2}}{d^{-1} - c^{-1}}$

Solution:

$$\frac{d^{-2} - c^{-2}}{d^{-1} - c^{-1}} = \frac{\dfrac{1}{d^2} - \dfrac{1}{c^2}}{\dfrac{1}{d} - \dfrac{1}{c}}$$

First write the expression with positive exponents.

$$= \frac{c^2 d^2 \cdot \left(\dfrac{1}{d^2} - \dfrac{1}{c^2} \right)}{c^2 d^2 \cdot \left(\dfrac{1}{d} - \dfrac{1}{c} \right)}$$

Step 1: Multiply numerator and denominator by the LCD of all four individual fractions: $c^2 d^2$.

$$= \frac{\dfrac{c^2d^2}{1} \cdot \dfrac{1}{d^2} - \dfrac{c^2d^2}{1} \cdot \dfrac{1}{c^2}}{\dfrac{c^2d^2}{1} \cdot \dfrac{1}{d} - \dfrac{c^2d^2}{1} \cdot \dfrac{1}{c}}$$

Step 2: Apply the distributive property.

$$= \frac{c^2 - d^2}{c^2d - cd^2}$$

$$= \frac{(c - d)(c + d)}{cd(c - d)} = \frac{c + d}{cd}$$

Step 3: Simplify by factoring and dividing out common factors.

Skill Practice 9 Simplify. $\dfrac{4 - 6x^{-1}}{2x^{-1} - 3x^{-2}}$

EXAMPLE 10   **Simplifying a Complex Fraction (Method II)**

Simplify. $\dfrac{\dfrac{2}{1 + h} - 2}{h}$

Solution:

> **TIP** The expression given in Example 10 is a pattern we see in a first semester calculus course.

$$\frac{\dfrac{2}{1 + h} - 2}{h} = \frac{\dfrac{2}{1 + h} - \dfrac{2}{1}}{h} = \frac{(1 + h) \cdot \left(\dfrac{2}{1 + h} - \dfrac{2}{1} \right)}{(1 + h) \cdot (h)}$$

Step 1: Multiply numerator and denominator by the LCD, which is $(1 + h)$.

$$= \frac{\dfrac{(1 + h)}{1} \cdot \left(\dfrac{2}{1 + h} \right) - \dfrac{(1 + h)}{1} \cdot \left(\dfrac{2}{1} \right)}{(1 + h) \cdot (h)}$$

Step 2: Apply the distributive property.

$$= \frac{2 - 2(1 + h)}{h(1 + h)}$$

Step 3: Simplify.

$$= \frac{2 - 2 - 2h}{h(1 + h)} = \frac{-2h}{h(1 + h)} = \frac{-2}{1 + h} \quad \text{or} \quad -\frac{2}{1 + h}$$

Skill Practice 10 Simplify. $\dfrac{\dfrac{5}{1 + h} - 5}{h}$

---

6. Rationalize the Denominator of a Radical Expression

The same principle that applies to simplifying rational expressions also applies to simplifying algebraic fractions. For example, $\frac{5}{\sqrt{x}}$ is an algebraic fraction, but not a rational expression because the denominator is not a polynomial.

From Section R.3, we outlined the criteria for a radical expression to be simplified. Conditions 2 and 3 are stated here.

- No fraction may appear in the radicand.
- No denominator of a fraction may contain a radical.

In Example 11, we use the property of equivalent fractions to remove a radical from the denominator of a fraction. This is called **rationalizing the denominator.**

EXAMPLE 11   Rationalizing the Denominator

Simplify. Assume that x is a positive real number.

a. $\dfrac{5}{\sqrt{x}}$ **b.** $\dfrac{4}{\sqrt{7} - \sqrt{5}}$

Solution:

a. $\dfrac{5}{\sqrt{x}} = \dfrac{5 \cdot \sqrt{x}}{\sqrt{x} \cdot \sqrt{x}}$ Multiply numerator and denominator by $\sqrt{x}$ so that the radicand in the denominator is a perfect square.

$= \dfrac{5\sqrt{x}}{\sqrt{x^2}}$ Apply the product property of radicals.

$= \dfrac{5\sqrt{x}}{x}$ Simplify the radical in the denominator.

b. $\dfrac{4}{\sqrt{7} - \sqrt{5}} = \dfrac{4 \cdot \left(\sqrt{7} + \sqrt{5}\right)}{\left(\sqrt{7} - \sqrt{5}\right) \cdot \left(\sqrt{7} + \sqrt{5}\right)}$ Multiply numerator and denominator by the conjugate of the denominator.

$= \dfrac{4\left(\sqrt{7} + \sqrt{5}\right)}{\left(\sqrt{7}\right)^2 - \left(\sqrt{5}\right)^2}$ Recall that $(a - b)(a + b) = a^2 - b^2$.

$= \dfrac{4\left(\sqrt{7} + \sqrt{5}\right)}{7 - 5}$ Simplify the radicals in the denominator.

$= \dfrac{\overset{2}{4}\left(\sqrt{7} + \sqrt{5}\right)}{2}$ Simplify the fraction.

$= 2\left(\sqrt{7} + \sqrt{5}\right)$ or $2\sqrt{7} + 2\sqrt{5}$

TIP Keep the numerator in factored form until the denominator is simplified completely. By so doing, it will be easier to identify common factors in the numerator and denominator.

Skill Practice 11 Simplify.

a. $\dfrac{\sqrt{5}}{\sqrt{7}}$ **b.** $\dfrac{12}{\sqrt{13} - \sqrt{10}}$

In Example 11(b), we multiplied the numerator and denominator by the conjugate of the denominator. The rationale is that product $(a + b)(a - b)$ results in a difference of squares $a^2 - b^2$. If either a or b has a square root factor, then the product will simplify to an expression without square roots.

Answers

11. a. $\dfrac{\sqrt{35}}{7}$

   b. $4\left(\sqrt{13} + \sqrt{10}\right)$ or $4\sqrt{13} + 4\sqrt{10}$

SECTION R.6   Practice Exercises

Concept Connections

1. A _____ expression is a ratio of two polynomials.

2. The restricted values of the variable for a rational expression are those that make the denominator equal to _____.

3. The expression $\dfrac{5(x + 2)}{(x + 2)(x - 1)}$ equals $\dfrac{5}{x - 1}$ provided that $x \neq$ _____ and $x \neq$ _____.

4. The ratio of a polynomial and its opposite equals _____.

5. A _____ fraction is an expression that contains one or more fractions in the numerator or denominator.

6. The process to remove a radical from the denominator of a fraction is called _____ the denominator.

Objective 1: Determine Restricted Values for a Rational Expression

For Exercises 7–14, determine the restrictions on the variable. (See Example 1)

7. $\dfrac{x - 4}{x + 7}$

8. $\dfrac{y - 1}{y + 10}$

9. $\dfrac{a}{a^2 - 81}$

10. $\dfrac{t}{t^2 - 16}$

11. $\dfrac{a}{a^2 + 81}$

12. $\dfrac{t}{t^2 + 16}$

13. $\dfrac{6c}{7a^3b^2}$

14. $\dfrac{11z}{8x^5y}$

Objective 2: Simplify Rational Expressions

15. Determine which expressions are equal to $-\dfrac{5}{x - 3}$.

   a. $\dfrac{-5}{x - 3}$ **b.** $\dfrac{5}{3 - x}$

   c. $-\dfrac{5}{3 - x}$ **d.** $\dfrac{-5}{3 - x}$

16. Determine which expressions are equal to $\dfrac{-2}{a + b}$.

   a. $\dfrac{-2}{a - b}$ **b.** $-\dfrac{2}{a + b}$

   c. $\dfrac{2}{-a - b}$ **d.** $\dfrac{2}{a - b}$

For Exercises 17–26, simplify the expression and state the restrictions on the variable. (See Examples 2–3)

17. $\dfrac{x^2 - 9}{x^2 - 4x - 21}$

18. $\dfrac{y^2 - 64}{y^2 - 7y - 8}$

19. $-\dfrac{12a^2bc}{3ab^5}$

20. $-\dfrac{15tu^5v}{3t^3u}$

21. $\dfrac{10 - 5\sqrt{6}}{15}$

22. $\dfrac{12 + 4\sqrt{3}}{8}$

23. $\dfrac{2y^2 - 16y}{64 - y^2}$

24. $\dfrac{81 - t^2}{7t^2 - 63t}$

25. $\dfrac{4b - 4a}{ax - xb - 2a + 2b}$

26. $\dfrac{2z - 2y}{xy - xz + 3y - 3z}$

Objective 3: Multiply and Divide Rational Expressions

For Exercises 27–34, multiply or divide as indicated. The restrictions on the variables are implied. (See Examples 4–5)

27. $\dfrac{3a^5b^7}{a - 5b} \cdot \dfrac{2a - 10b}{12a^4b^{10}}$

28. $\dfrac{8x - 3y}{x^3y^4} \cdot \dfrac{6xy^8}{24x - 9y}$

29. $\dfrac{c^2 - d^2}{cd^{11}} \div \dfrac{8c^2 + 4cd - 4d^2}{8c^4d^{10}}$

30. $\dfrac{m^{11}n^2}{m^2 - n^2} \div \dfrac{18m^9n^5}{9m^2 + 6mn - 15n^2}$

31. $\dfrac{2a^2b - ab^2}{8b^2 + ab} \cdot \dfrac{a^2 + 16ab + 64b^2}{2a^2 + 15ab - 8b^2}$

32. $\dfrac{2c^2 - 2cd}{3c^2d + 2c^3} \cdot \dfrac{4c^2 + 12cd + 9d^2}{2c^2 + cd - 3d^2}$

33. $\dfrac{x^3 - 64}{16x - x^3} \div \dfrac{2x^2 + 8x + 32}{x^2 + 2x - 8}$

34. $\dfrac{3y^2 + 21y + 147}{25y - y^3} \div \dfrac{y^3 - 343}{y^2 - 12y + 35}$

Objective 4: Add and Subtract Rational Expressions

For Exercises 35–40, identify the least common denominator for each pair of expressions.

35. $\dfrac{7}{6x^5yz^4}$ and $\dfrac{3}{20xy^2z^3}$

36. $\dfrac{12}{35b^4cd^3}$ and $\dfrac{8}{25b^2c^3d}$

37. $\dfrac{2t + 1}{(3t + 4)^3(t - 2)}$ and $\dfrac{4}{t(3t + 4)^2(t - 2)}$

38. $\dfrac{5y - 7}{y(2y - 5)(y + 6)^4}$ and $\dfrac{6}{(2y - 5)^3(y + 6)^2}$

39. $\dfrac{x + 3}{x^2 + 20x + 100}$ and $\dfrac{3}{2x^2 + 20x}$

40. $\dfrac{z - 4}{4z^2 - 20z + 25}$ and $\dfrac{5}{12z^2 - 30z}$

For Exercises 41–56, add or subtract as indicated. (See Examples 6–7)

41. $\dfrac{m^2}{m + 3} + \dfrac{6m + 9}{m + 3}$

42. $\dfrac{n^2}{n + 5} + \dfrac{7n + 10}{n + 5}$

43. $\dfrac{2}{9c} + \dfrac{7}{15c^3}$

44. $\dfrac{6}{25x} + \dfrac{7}{10x^4}$

45. $\dfrac{9}{2x^2y^4} - \dfrac{11}{xy^5}$

46. $\dfrac{-2}{3m^3n} - \dfrac{5}{m^2n^4}$

47. $\dfrac{2}{x + 3} - \dfrac{7}{x}$

48. $\dfrac{4}{m - 2} - \dfrac{3}{m}$

49. $\dfrac{1}{x^2 + xy} - \dfrac{2}{x^2 - y^2}$

50. $\dfrac{4}{4a^2 - b^2} - \dfrac{1}{2a^2 - ab}$

51. $\dfrac{5}{y} + \dfrac{2}{y + 1} - \dfrac{6}{y^2}$

52. $\dfrac{5}{t^2} + \dfrac{4}{t + 2} - \dfrac{3}{t}$

53. $\dfrac{3w}{w - 4} + \dfrac{2w + 4}{4 - w}$

54. $\dfrac{2x - 1}{x - 7} + \dfrac{x + 6}{7 - x}$

55. $\dfrac{4}{x^2 + 6x + 5} - \dfrac{3}{x^2 + 7x + 10}$

56. $\dfrac{3}{x^2 - 4x - 5} - \dfrac{2}{x^2 - 6x + 5}$

Objective 5: Simplify Complex Fractions

For Exercises 57–68, simplify the complex fraction. (See Examples 8–10)

57. $\dfrac{\dfrac{1}{27x} + \dfrac{1}{9}}{\dfrac{1}{3} + \dfrac{1}{9x}}$

58. $\dfrac{\dfrac{1}{8x} + \dfrac{1}{4}}{\dfrac{1}{2} + \dfrac{1}{4x}}$

59. $\dfrac{\dfrac{x}{6} - \dfrac{5x + 14}{6x}}{\dfrac{1}{6} - \dfrac{7}{6x}}$

60. $\dfrac{\dfrac{x}{3} - \dfrac{2x + 3}{3x}}{\dfrac{1}{3} + \dfrac{1}{3x}}$

61. $\dfrac{2a^{-1} - b^{-1}}{4a^{-2} - b^{-2}}$

62. $\dfrac{3u^{-1} - v^{-1}}{9u^{-2} - v^{-2}}$

63. $\dfrac{\dfrac{3}{1 + h} - 3}{h}$

64. $\dfrac{\dfrac{4}{1 + h} - 4}{h}$

65. $\dfrac{\dfrac{7}{x + h} - \dfrac{7}{x}}{h}$

66. $\dfrac{\dfrac{8}{x + h} - \dfrac{8}{x}}{h}$

67. $\dfrac{\dfrac{3}{x - 1} - \dfrac{1}{x + 1}}{\dfrac{6}{x^2 - 1}}$

68. $\dfrac{\dfrac{1}{x + 1}}{\dfrac{-5}{x^2 - 3x - 4} + \dfrac{1}{x - 4}}$

Objective 6: Rationalize the Denominator of a Radical Expression

For Exercises 69–84, simplify the expression. Assume that the variable expressions represent positive real numbers. (See Example 11)

69. $\dfrac{4}{\sqrt{y}}$

70. $\dfrac{7}{\sqrt{z}}$

71. $\dfrac{4}{\sqrt[3]{y}}$

72. $\dfrac{7}{\sqrt[4]{z}}$

73. $\dfrac{\sqrt{12}}{\sqrt{x + 1}}$

74. $\dfrac{\sqrt{50}}{\sqrt{x - 2}}$

75. $\dfrac{8}{\sqrt{15} - \sqrt{11}}$

76. $\dfrac{12}{\sqrt{6} - \sqrt{2}}$

77. $\dfrac{x-5}{\sqrt{x}+\sqrt{5}}$

78. $\dfrac{y-3}{\sqrt{y}+\sqrt{3}}$

79. $\dfrac{2\sqrt{10}+3\sqrt{5}}{4\sqrt{10}+2\sqrt{5}}$

80. $\dfrac{3\sqrt{3}+\sqrt{6}}{5\sqrt{3}-2\sqrt{6}}$

81. $\dfrac{7}{\sqrt{3x}}+\dfrac{\sqrt{3x}}{x}$

82. $\dfrac{4}{\sqrt{11y}}+\dfrac{\sqrt{11y}}{y}$

83. $\dfrac{5}{w\sqrt{7}}-\dfrac{\sqrt{7}}{w}$

84. $\dfrac{13}{t\sqrt{2}}-\dfrac{\sqrt{2}}{t}$

Mixed Exercises

85. The average round trip speed S (in mph) of a vehicle traveling a distance of d miles each way is given by

$$S = \dfrac{2d}{\dfrac{d}{r_1}+\dfrac{d}{r_2}}.$$

In this formula, r_1 is the average speed going one way, and r_2 is the average speed on the return trip.

a. Simplify the complex fraction.

b. If a plane flies 400 mph from Orlando to Albuquerque and 460 mph on the way back, compute the average speed of the round trip. Round to 1 decimal place.

86. The formula $R = \dfrac{1}{\dfrac{1}{R_1}+\dfrac{1}{R_2}}$ gives the total electrical

resistance R (in ohms, Ω) when two resistors of resistance R_1 and R_2 are connected in parallel.

a. Simplify the complex fraction.

b. Find the total resistance when $R_1 = 12\ \Omega$ and $R_2 = 20\ \Omega$.

87. The concentration C (in ng/mL) of a drug in the bloodstream t hours after ingestion is modeled by

$$C = \dfrac{600t}{t^3+125}.$$

a. Determine the concentration at 1 hr, 12 hr, 24 hr, and 48 hr. Round to 1 decimal place.

b. What appears to be the limiting concentration for large values of t?

88. An object that is originally 35°C is placed in a freezer. The temperature T (in °C) of the object can be

approximated by the model $T = \dfrac{350}{t^2+3t+10}$,

where t is the time in hours after the object is placed in the freezer.

a. Determine the temperature at 2 hr, 4 hr, 12 hr, and 24 hr. Round to 1 decimal place.

b. What appears to be the limiting temperature for large values of t?

For Exercises 89–92, write a simplified expression for the perimeter or area as indicated.

89. Perimeter

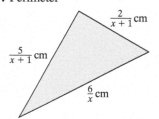

90. Perimeter

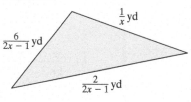

91. Area

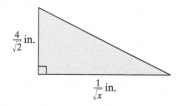

92. Area

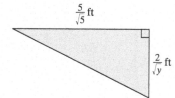

For Exercises 93–108, simplify the expression.

93. $\dfrac{2x^3y}{x^2y+3xy}\cdot\dfrac{x^2+6x+9}{2x+6}\div 5xy^4$

94. $\dfrac{2y^2+20y+50}{12-4y}\cdot\dfrac{y-3}{y^2+12y+35}\div(3y+15)$

95. $\left(\dfrac{4}{2t+1}-\dfrac{t}{2t^2+17t+8}\right)(t+8)$

96. $\left(\dfrac{2m}{6m+3}-\dfrac{1}{m+4}\right)(2m+1)$

97. $\dfrac{n-2}{n-4} + \dfrac{2n^2 - 15n + 12}{n^2 - 16} - \dfrac{2n-5}{n+4}$

98. $\dfrac{c^2 + 13c + 18}{c^2 - 9} + \dfrac{c+1}{c+3} - \dfrac{c+8}{c-3}$

99. $\dfrac{1 - a^{-1} - 6a^{-2}}{1 - 4a^{-1} + 3a^{-2}}$

100. $\dfrac{1 + t^{-1} - 12t^{-2}}{1 - 4t^{-1} + 3t^{-2}}$

101. $\dfrac{34}{2\sqrt{5} - \sqrt{3}}$

102. $\dfrac{13}{2\sqrt{7} + \sqrt{2}}$

103. $\dfrac{8 - \sqrt{48}}{6}$

104. $\dfrac{10 - \sqrt{50}}{15}$

105. $\dfrac{45 + 9x - 5x^2 - x^3}{x^3 - 3x^2 - 25x + 75}$

106. $\dfrac{98 - 49x - 2x^2 + x^3}{7x^2 - x^3 - 28 + 4x}$

107. $\dfrac{\dfrac{t+6}{1 + \dfrac{2}{t}} - t - 4}{}$

108. $\dfrac{\dfrac{m-4}{1 - \dfrac{2}{m}} - m + 2}{}$

For Exercises 109–116, write the expression as a single term, factored completely. Do not rationalize the denominator.

109. $1 - x^{-2} - 2x^{-3}$

110. $1 - 8x^{-5} + 30x^{-7}$

111. $\dfrac{3}{2\sqrt{3x}} + \sqrt{3x}$

112. $\dfrac{2}{\sqrt{x}} + \sqrt{x}$

113. $\dfrac{\sqrt{x^2 + 1} + \dfrac{x^2}{\sqrt{x^2 + 1}}}{x^2 + 1}$

114. $\dfrac{\dfrac{x^2}{\sqrt{x^2 + 9}} - \sqrt{x^2 + 9}}{x^2}$

115. $2\sqrt{4x^2 + 9} + \dfrac{8x^2}{\sqrt{4x^2 + 9}}$

116. $3\sqrt{9x^2 + 1} + \dfrac{27x}{\sqrt{9x^2 + 1}}$

Write About It

117. Explain why the expression $\dfrac{x}{x - y}$ is not defined for $x = y$.

118. Is the statement $\dfrac{3(x - 4)}{(x + 2)(x - 4)} = \dfrac{3}{x + 2}$ true for all values of x? Explain why or why not.

Expanding Your Skills

119. The numbers 1, 2, 4, 5, 10, and 20 are natural numbers that are factors of 20. There are other factors of 20 within the set of rational numbers and the set of irrational numbers. For example:

    a. Show that $\dfrac{14}{3}$ and $\dfrac{30}{7}$ are factors of 20 over the set of rational numbers.

    b. Show that $\left(5 - \sqrt{5}\right)$ and $\left(5 + \sqrt{5}\right)$ are factors of 20 over the set of irrational numbers.

120. a. Show that $\dfrac{15}{2}$ and $\dfrac{4}{5}$ are factors of 6 over the set of rational numbers.

    b. Show that $\left(3 - \sqrt{3}\right)$ and $\left(3 + \sqrt{3}\right)$ are factors of 6 over the set of irrational numbers.

For Exercises 121–128, simplify the expression.

121. $\dfrac{w^{3n+1} - w^{3n}z}{w^{n+2} - w^n z^2}$

122. $\dfrac{x^{2n+1} - x^{2n}y}{x^{n+3} - x^n y^3}$

123. $\sqrt{\dfrac{x - y}{x + y}}$

124. $\sqrt{\dfrac{m - 3}{m + 3}}$

125. $\dfrac{\sqrt{5}}{\sqrt[3]{2}}$

126. $\dfrac{\sqrt{7}}{\sqrt[4]{3}}$

127. $\dfrac{a - b}{\sqrt[3]{a} - \sqrt[3]{b}}$ (*Hint*: Factor the numerator as a difference of cubes over the set of irrational numbers.)

128. $\dfrac{x + y}{\sqrt[3]{x} + \sqrt[3]{y}}$

For Exercises 129–130, rationalize the numerator by multiplying numerator and denominator by the conjugate of the *numerator*.

129. $\dfrac{\sqrt{4+h}-2}{h}$

130. $\dfrac{\sqrt{x+h}-\sqrt{x}}{h}$

ALGEBRA FOR CALCULUS

For Exercises 1–2, write an inequality using an absolute value that represents the given condition.

1. The distance between y and L is less than ε (epsilon).

2. The distance between x and c is less than δ (delta).

3. Simplify $\dfrac{x+8}{|x+8|}$

   a. for $x > -8$.   **b.** for $x < -8$.

4. Simplify $\dfrac{14-x}{|x-14|}$

   a. for $x > 14$.   **b.** for $x < 14$.

For Exercises 5–10,

   a. Simplify the expression.

   b. Substitute 0 for h in the simplified expression.

5. $\dfrac{2(x+h)^2 + 3(x+h) - (2x^2 + 3x)}{h}$

6. $\dfrac{3(x+h)^2 - 4(x+h) - (3x^2 - 4x)}{h}$

7. $\dfrac{\dfrac{1}{(x+h)-2} - \dfrac{1}{x-2}}{h}$

8. $\dfrac{\dfrac{1}{2(x+h)+5} - \dfrac{1}{2x+5}}{h}$

9. $\dfrac{(x+h)^3 - x^3}{h}$

10. $\dfrac{(x+h)^4 - x^4}{h}$

For Exercises 11–12,

   a. Rationalize the numerator of the expression and simplify.

   b. Substitute 0 for h in the simplified expression.

11. $\dfrac{\sqrt{x+h}+1 - (\sqrt{x}+1)}{h}$

12. $\dfrac{\sqrt{2(x+h)} - \sqrt{2x}}{h}$

For Exercises 13–22, factor completely and write the answer with no negative exponents. Do not rationalize the denominator.

13. $\dfrac{3}{2}x^{1/2} + \dfrac{5}{2}x^{3/2}$

14. $\dfrac{7}{6}x^{1/6} - \dfrac{1}{6}x^{-5/6}$

15. $4(3x+1)^3(3)(x^2+2)^3 + (3x+1)^4(3)(x^2+2)^2(2x)$

16. $3(-2x+3)^2(-2)(4x^2-5)^2 + (-2x+3)^3(2)(4x^2-5)(8x)$

17. $\dfrac{6(t-1)^5(2t+5)^6 - 6(2t+5)^5(2)(t-1)^6}{[(2t+5)^6]^2}$

18. $\dfrac{6x^5(x^2+4)^3 - 3(x^2+4)^2(2x)x^6}{[(x^2+4)^3]^2}$

19. $(x^2+4)^{1/2} + x \cdot \dfrac{1}{2}(x^2+4)^{-1/2}(2x)$

20. $(2-x^2)^{1/2} + x \cdot \dfrac{1}{2}(2-x^2)^{-1/2}(-2x)$

21. $(x^2+1)^{-1/2} + x\left(-\dfrac{1}{2}\right)(x^2+1)^{-3/2}(2x)$

22. $6(3x-1)^{1/3} + 6x\left(-\dfrac{1}{3}\right)(3x-1)^{-2/3}(3)$

For Exercises 23–28, write the answer as a single term and simplify. It is not necessary to rationalize the denominator.

23. $\dfrac{2\sqrt{x+4} - \dfrac{x}{\sqrt{x+4}}}{\left(\sqrt{x+4}\right)^2}$

24. $\dfrac{2x\sqrt{16-x^2} + \dfrac{x^3}{\sqrt{16-x^2}}}{\left(\sqrt{16-x^2}\right)^2}$

25. $(x-4)^{1/3} + \dfrac{x}{3(x-4)^{2/3}}$

26. $(x+5)^{1/4} + \dfrac{x}{4(x+5)^{3/4}}$

27. $\dfrac{1}{2}\left(\dfrac{2x}{x+1}\right)^{-1/2}\left[\dfrac{(x+1)(2) - 2x(1)}{(x+1)^2}\right]$

28. $\dfrac{1}{3}\left(\dfrac{3x}{x^2+1}\right)^{-2/3}\left[\dfrac{(x^2+1)\cdot 3 - 3x(2x)}{(x^2+1)^2}\right]$

CHAPTER R   KEY CONCEPTS

SECTION R.1 Sets and the Real Number Line	Reference
Natural Numbers: $\mathbb{N} = \{1, 2, 3, \ldots\}$ **Whole Numbers:** $\mathbb{W} = \{0, 1, 2, 3, \ldots\}$ **Integers:** $\mathbb{Z} = \{\ldots, -3, -2, -1, 0, 1, 2, 3, \ldots\}$ **Rational Numbers:** $\mathbb{Q} = \{\frac{p}{q} \mid p, q \in \mathbb{Z} \text{ and } q \neq 0\}$ **Irrational Numbers:** $\mathbb{H}$ is the set of real numbers that cannot be expressed as a ratio of integers.	p. 2
$A \cup B$ is the **union** of A and B. This is the set of elements that belong to set A or set B or to both sets A and B. $A \cap B$ is the **intersection** of A and B. This is the set of elements common to both A and B. $A \cup B$ $\qquad$ $A \cap B$	p. 5
For a real number x, the **absolute value of x** is $\lvert x \rvert = \begin{cases} x & \text{if } x \geq 0 \\ -x & \text{if } x < 0 \end{cases}$	p. 7
The distance between two points a and b on a number line is given by $\lvert a - b \rvert$ or $\lvert b - a \rvert$.	p. 8
• Commutative property of addition $\qquad a + b = b + a$ • Commutative property of multiplication $\qquad a \cdot b = b \cdot a$ • Associative property of addition $\qquad (a + b) + c = a + (b + c)$ • Associative property of multiplication $\qquad (a \cdot b) \cdot c = a \cdot (b \cdot c)$ • Identity property of addition $\qquad a + 0 = a$ and $0 + a = a$ • Identity property of multiplication $\qquad a \cdot 1 = a$ and $1 \cdot a = a$ • Inverse property of addition $\qquad a + (-a) = 0$ and $(-a) + a = 0$ • Inverse property of multiplication $\qquad a \cdot \frac{1}{a} = 1$ and $\frac{1}{a} \cdot a = 1$ where $a \neq 0$ • Distributive property of multiplication over addition $\quad a(b + c) = ab + ac$	pp. 10–11

SECTION R.2 Integer Exponents and Scientific Notation	Reference
Properties of exponents and key definitions: $b^m \cdot b^n = b^{m+n}$ $\qquad$ $(b^m)^n = b^{m \cdot n}$ $\qquad$ $\left(\dfrac{a}{b}\right)^m = \dfrac{a^m}{b^m}$ $\qquad$ $b^{-n} = \dfrac{1}{b^n}$ $\dfrac{b^m}{b^n} = b^{m-n}$ $\qquad$ $(ab)^m = a^m b^m$ $\qquad$ $b^0 = 1$	p. 19
A number expressed in the form $a \times 10^n$, where $1 \leq \lvert a \rvert < 10$ and n is an integer is said to be in **scientific notation.**	p. 22

SECTION R.3 Rational Exponents and Radicals	Reference
b is an nth-root of a if $b^n = a$. $\sqrt[n]{a}$ represents the principal nth-root of a.	p. 27
If $n > 1$ is an integer and $\sqrt[n]{a}$ is a real number, then • $a^{1/n} = \sqrt[n]{a}$　　　　　• $a^{m/n} = \left(\sqrt[n]{a}\right)^m$　and　$a^{m/n} = \sqrt[n]{a^m}$	p. 28
Let $n > 1$ be an integer and a be a real number. • If n is *even* then $\sqrt[n]{a^n} = \lvert a \rvert$.　　• If n is *odd* then $\sqrt[n]{a^n} = a$. Product property of radicals: $\sqrt[n]{a} \cdot \sqrt[n]{b} = \sqrt[n]{ab}$. Quotient property of radicals: $\dfrac{\sqrt[n]{a}}{\sqrt[n]{b}} = \sqrt[n]{\dfrac{a}{b}}$. Property of nested radicals: $\sqrt[m]{\sqrt[n]{a}} = \sqrt[m \cdot n]{a}$.	p. 30

SECTION R.4 Polynomials and Multiplication of Radicals	Reference
Special Case Products: $(a + b)(a - b) = a^2 - b^2$ $(a + b)^2 = a^2 + 2ab + b^2$ $(a - b)^2 = a^2 - 2ab + b^2$	p. 40

SECTION R.5 Factoring	Reference
General factoring strategy: 1. Factor out the GCF. 2. Identify the number of terms in the polynomial. 　4 terms: Factor by grouping either 2 terms with 2 terms or 3 terms with 1 term. 　3 terms: If the trinomial is a perfect square trinomial, factor as the square of a binomial. 　　　$a^2 + 2ab + b^2 = (a + b)^2$ 　　　$a^2 - 2ab + b^2 = (a - b)^2$ 　　　Otherwise, factor by the trial-and-error method. 　2 terms: Determine whether the binomial fits one of the following patterns. 　　　$a^2 - b^2 = (a + b)(a - b)$ 　　　$a^3 + b^3 = (a + b)(a^2 - ab + b^2)$ 　　　$a^3 - b^3 = (a - b)(a^2 + ab + b^2)$	p. 54

SECTION R.6 Rational Expressions and More Operations on Radicals	Reference
A **rational expression** is a ratio of two polynomials. Values of the variable that make the denominator equal to zero are called restricted values of the variable.	p. 59
To simplify a rational expression, use the property of equivalent algebraic fractions. $$\dfrac{ac}{bc} = \dfrac{a}{b} \quad \text{for } b \neq 0, c \neq 0$$	p. 60
To multiply or divide rational expressions, $$\dfrac{a}{b} \cdot \dfrac{c}{d} = \dfrac{ac}{bd} \quad \text{for } b \neq 0, d \neq 0 \quad \text{and} \quad \dfrac{a}{b} \div \dfrac{c}{d} = \dfrac{a}{b} \cdot \dfrac{d}{c} = \dfrac{ad}{bc} \quad \text{for } b \neq 0, c \neq 0, d \neq 0$$	p. 61
To add or subtract rational expressions, write each fraction as an equivalent fraction with a common denominator. Then apply the following properties. $$\dfrac{a}{b} + \dfrac{c}{b} = \dfrac{a + c}{b} \quad \text{and} \quad \dfrac{a}{b} - \dfrac{c}{b} = \dfrac{a - c}{b} \quad \text{for } b \neq 0$$	p. 62
Removing a radical from the denominator of a fraction is called **rationalizing the denominator**.	p. 67

CHAPTER R   Review Exercises

SECTION R.1

1. Given $A = \left\{ \sqrt{6}, 0, -8, 1.\overline{45}, \sqrt{9}, -\dfrac{2}{3}, 3\pi \right\}$, determine which elements belong to the given set.

   a. $\mathbb{N}$ **b.** $\mathbb{W}$ **c.** $\mathbb{Z}$

   d. $\mathbb{Q}$ **e.** $\mathbb{H}$ **f.** $\mathbb{R}$

2. Write the statement as an inequality. "$6y$ is no more than 8."

3. Complete the table.

Graph	Interval Notation	Set-Builder Notation
a. $\quad$ $\underset{-3}{[}\;\;\;\;\;\;\underset{7}{)}$		
b.	$(2.1, \infty)$	
c.		$\{x \mid 4 \geq x\}$

4. Given $A = \{10, 11, 12, 13\}$, $B = \{10, 12, 14, 16\}$, and $C = \{7, 8, 9, 10, 11\}$, find

   a. $A \cup B$ **b.** $A \cap B$ **c.** $A \cup C$

   d. $A \cap C$ **e.** $B \cup C$ **f.** $B \cap C$

5. Given $X = \{x \mid x < 7\}$, $Y = \{x \mid x \geq -2\}$, and $Z = \{x \mid x < -3\}$, write the union or intersection in set-builder notation.

   a. $X \cup Y$ **b.** $X \cap Y$ **c.** $X \cup Z$

   d. $X \cap Z$ **e.** $Y \cup Z$ **f.** $Y \cap Z$

6. Simplify without absolute value bars.

   a. $|w - 4|$ for $w < 4$

   b. $|w - 4|$ for $w \geq 4$

7. a. Write an absolute value expression to represent the distance between 2 and $\sqrt{5}$ on the number line.

   b. Simplify the expression from part (a) without absolute value bars.

For Exercises 8–12, simplify the expression.

8. a. 16^2 **b.** $(-16)^2$ **c.** -16^2

   d. $\sqrt{16}$ **e.** $-\sqrt{16}$ **f.** $\sqrt{-16}$

9. $\sqrt{\dfrac{25}{4}}$

10. $72 \cdot \left[\left(\dfrac{2}{3}\right)^2 - \left(\dfrac{1}{2}\right)^3 \right]$

11. $\dfrac{32 - |-11 + 3|}{36 \div 2 \div 3 - 2}$

12. $-5 - 3\left[8 - 2\sqrt{4^2 + (2 - 5)^2} \right]$

13. Jesse makes \$150 more per week than Ethan.

   a. Write a model for Jesse's salary J in terms of Ethan's salary E.

   b. Write a model for Ethan's salary E in terms of Jesse's salary J.

14. The width of a rectangle is 8 ft less than twice the length. Write a model for the width W in terms of the length L.

15. For a single-story home with a low-pitch roof and asphalt shingles, a roofing company charges \$3.60 per square foot to tear off the old shingles and install new shingles. If the plywood underneath is damaged, the roofer charges \$50 per sheet replaced. The cost for high-impact skylights is \$250 each.

   a. Write a formula for the cost C (in \$) for a new roof, based on the square footage s, the number of plywood sheets replaced p, and the number of skylights n.

   b. Compute the cost to replace a 2100-ft² low-pitch roof that requires four sheets of new plywood and two high-impact skylights.

16. Given $13x^2z^4 - \dfrac{2}{z} + \sqrt{5x + z}$, list the terms of the expression and the coefficients of each term.

For Exercises 17–18, simplify the expression.

17. $15.2c^2d - 11.1cd + 8.7c^2d - 5.4cd$

18. $8 - \left[4x^2 - \dfrac{1}{2}(6 - 4x^2) + 3 \right] + 13x^2$

For Exercises 19–27, identify the property that makes the given statement true. Choose from

   a. Commutative property of addition
   b. Commutative property of multiplication
   c. Associative property of addition
   d. Associative property of multiplication
   e. Identity property of addition
   f. Identity property of multiplication
   g. Inverse property of addition
   h. Inverse property of multiplication
   i. Distributive property of multiplication over addition

19. $5(ab) = (5a)b$ **20.** $\dfrac{1}{5} \cdot 5 = 1$

21. $p + (q + r) = (p + q) + r$

22. $p(q + r) = pq + pr$

23. $-\pi + 0 = -\pi$

24. $-\pi + \pi = 0$

25. $x + (y + z) = (y + z) + x$

26. $1 \cdot x = x$

27. $(ab)c = c(ab)$

28. Evaluate $\sqrt{(x_2 - x_1)^2 + (y_2 - y_1)^2}$ for $x_1 = 5$, $x_2 = -2$, $y_1 = -6$, $y_2 = 3$.

SECTION R.2

For Exercises 29–35, simplify completely. Write the answers with positive exponents only.

29. a. 9^0 **b.** -9^0 **c.** $9x^0$ **d.** $(9x)^0$

30. a. $\dfrac{1}{m^{-5}}$ **b.** m^{-5} **c.** $8m^{-9}n^2$ **d.** $8m^9n^{-2}$

31. $p^{-8} \cdot p^{12} \cdot p^{-1}$

32. $\dfrac{m^{-4}m^{10}}{m^6}$

33. $(-12a^{-3}b^4)^2$

34. $\left(\dfrac{-81x^8y^5}{9x^6y^8}\right)^{-2}$

35. $\left(\dfrac{1}{2u^5v^{-2}}\right)^{-3}\left(\dfrac{4}{u^{-3}v^2}\right)^{-1}$

36. Write the numbers in scientific notation.

   a. 4920   **b.** 0.00492   **c.** 4.92

37. Write the numbers in standard decimal notation.

   a. 9.8×10^{-1} **b.** 9.8×10^0 **c.** 9.8×10^1

For Exercises 38–39, perform the indicated operations.

38. $(9.2 \times 10^4)(3.0 \times 10^5)$

39. $\dfrac{(8.6 \times 10^{-3})(4.1 \times 10^8)}{2.0 \times 10^{-6}}$

40. A healthy adult female will have approximately 5 million red blood cells per 1 μL of blood. If a woman donates 1 pint of blood to a blood bank, approximately how many red blood cells will be present? (*Hint:* 1 μL $= 10^{-6}$ L and 1 pint ≈ 0.47 L.)

SECTION R.3

41. Write the expressions using radical notation.

   a. $x^{2/7}$ **b.** $9x^{2/7}$ **c.** $(9x)^{2/7}$

42. Write the expressions using rational exponents.

   a. $12\sqrt{w}$ **b.** $\sqrt{12w}$

For Exercises 43–47, simplify the expression. Write exponential expressions with positive exponents only. Assume that all variables represent positive real numbers.

43. a. $-\sqrt[4]{256}$ **b.** $\sqrt[4]{-256}$

44. $\sqrt[3]{-\dfrac{8}{125}}$

45. a. $10{,}000^{3/4}$ **b.** $10{,}000^{-3/4}$

   c. $-10{,}000^{3/4}$ **d.** $-10{,}000^{-3/4}$

   e. $(-10{,}000)^{3/4}$ **f.** $(-10{,}000)^{-3/4}$

46. $\dfrac{p^{7/3}p^{-2/3}}{p^{2/3}}$

47. $(9m^{-4} n^{2/3})^{1/2}$

48. Simplify. Assume that t represents any real number.

   a. $\sqrt{(3t - 4)^2}$ **b.** $\sqrt[3]{(3t - 4)^3}$

   c. $\sqrt[4]{(3t - 4)^4}$

For Exercises 49–55, simplify the radical expressions. Assume that all variables represent positive real numbers.

49. $\sqrt[3]{54xy^{12}z^{14}}$

50. $\sqrt[4]{32b^3c^{15}d^8}$

51. $\sqrt{\dfrac{p^{13}}{9}}$

52. $\dfrac{\sqrt[3]{3xy^5}}{\sqrt[3]{81x^7y^2}}$

53. $\sqrt{10} \cdot \sqrt{35}$

54. $(-5\sqrt[5]{a^3})(6\sqrt[5]{a^4})$

55. $\sqrt[4]{cd^2} \cdot \sqrt[3]{c^2d}$

For Exercises 56–58, add or subtract as indicated. Assume that all variables represent positive real numbers.

56. $\dfrac{1}{5}\sqrt{125} + \dfrac{3}{2}\sqrt{20} - \dfrac{1}{4}\sqrt{80}$

57. $-2c\sqrt[3]{54c^2d^3} + 5cd\sqrt[3]{2c^2} - 10d\sqrt[3]{250c^5}$

58. $3\sqrt{5t} + 8t\sqrt{5t}$

SECTION R.4

59. Determine if the expression is a polynomial.

   a. $5a^3 - 4a^2 + 6a - 3$

   b. $\dfrac{7}{a^3} - \dfrac{4}{a} + 6$

   c. $\sqrt{6x} + 5$

60. Write the polynomial in descending order. Identify the leading coefficient and degree of the polynomial.

$$-y^5 + 7.61y^9 + 2.5y^{11}$$

61. Determine the degree of the polynomial.

$$-4ac^2d^3 + 5ac^3d^4 - a^2cd^2$$

For Exercises 62–71, perform the indicated operations.

62. $(4x^3 - 3x^2 + 7) + (-3x^3 - 6x^2 + 2x)$

63. $(-7.2a^2b^3 + 4.1ab^2 - 3.9b) - (0.8a^2b^3 - 3.2ab^2 - b)$

64. $(-8uv^3)\left(\dfrac{1}{4}u^2v\right)$

65. $(5w^3 + 6y^2)(2w^3 - y^2)$

66. $(4p - 6)\left(\dfrac{1}{2}p^2 - p + 4\right)$

67. $(9t - 4)(9t + 4)$

68. $\left(\dfrac{1}{3}m - \dfrac{1}{4}n^3\right)\left(\dfrac{1}{3}m + \dfrac{1}{4}n^3\right)$

69. $(5k - 3)^2$

70. $(6c^2 + 5d^3)^2$

71. $[(2v - 1) + w][(2v - 1) - w]$

For Exercises 72–76, multiply the radicals and simplify. Assume that all variable expressions represent positive real numbers.

72. $4\sqrt{3}(2\sqrt{3} - 5\sqrt{5} + \sqrt{7})$

73. $(6\sqrt{5} - 2\sqrt{3})(2\sqrt{5} + 5\sqrt{3})$

74. $(7\sqrt{2} - 2\sqrt{11})(7\sqrt{2} + 2\sqrt{11})$

75. $(2c^2\sqrt{d} - 5d^2\sqrt{c})^2$

76. $(\sqrt{x + 2} + 4)^2$

77. Write and simplify an expression that represents the volume.

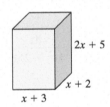

$2x + 5$

$x + 2$

$x + 3$

78. Determine the area of the triangle.

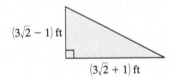

$(3\sqrt{2} - 1)$ ft

$(3\sqrt{2} + 1)$ ft

SECTION R.5

For Exercises 79–95, factor completely and write the answer with no negative exponents.

79. $80m^4n^8 - 48m^5n^3 - 16m^2n$

80. $11p^2(2p + 1) - 22p(2p + 1)$

81. $15ac - 14b - 10a + 21bc$

82. $-t + 12t^2 - 6$

83. $8x^3 - 40x^2y + 50xy^2$

84. $256a^4 - 625$

85. $3k^4 - 81k$

86. $(c + 2)^3 + d^3$

87. $25n^2 - m^2 - 12m - 36$

88. $(x^2 - 7)^2 + 8(x^2 - 7) + 15$

89. $(2p - 5)^2 - (4p + 1)^2$

90. $m^6 + 9m^3 + 8$

91. $x^4 + 6x^2y + 9y^2 - x^2 - 3y$

92. $7w^{-8} + 5w^{-7} + w^{-6}$

93. $12x^{7/2} - 4x^{5/2}$

94. $x(2x + 5)^{-3/4} + (2x + 5)^{1/4}$

95. $2x(1 - x^2)^{1/2} + x^2 \cdot \dfrac{1}{2}(1 - x^2)^{-1/2}(-2x)$

SECTION R.6

96. Determine the restrictions on the variable.

    a. $\dfrac{w - 2}{w^2 - 4}$ **b.** $\dfrac{w - 2}{w^2 + 4}$

For Exercises 97–98, simplify and state the restricted values on the variable.

97. $-\dfrac{24a^2c^5d^2}{16c^6d^7}$

98. $\dfrac{m^2 - 16}{m^2 - m - 12}$

For Exercises 99–107, perform the indicated operations.

99. $\dfrac{2x - 5y}{x^3y} \cdot \dfrac{x^5y^7}{6x - 15y}$

100. $\dfrac{4ac}{a^2 + 4ac} \cdot \dfrac{a^2 + 8ac + 16c^2}{8a + 32c}$

101. $\dfrac{5x^2 + 25x + 125}{16x - x^3} \div \dfrac{x^3 - 125}{x^2 - 9x + 20}$

102. $\dfrac{7}{20x} + \dfrac{2}{15x^4}$

103. $\dfrac{6}{9x^2 - y^2} - \dfrac{1}{3x^2 - xy}$

104. $\dfrac{4}{x^2} + \dfrac{3}{x + 3} - \dfrac{2}{x}$

105. $\dfrac{x - 1}{x - 6} + \dfrac{5}{6 - x}$

106. $\dfrac{\dfrac{1}{16x} + \dfrac{1}{8}}{\dfrac{1}{4} + \dfrac{1}{8x}}$

107. $\dfrac{5m^{-1} - n^{-1}}{25m^{-2} - n^{-2}}$

For Exercises 108–112, simplify the expression. Assume that all variable expressions represent positive real numbers.

108. $\dfrac{5}{\sqrt{k}}$

109. $\dfrac{5}{\sqrt[4]{k}}$

110. $\dfrac{15}{\sqrt{10} - \sqrt{7}}$

111. $\dfrac{x - 4}{\sqrt{x} + 2}$

112. $\dfrac{3}{\sqrt{5y}} + \dfrac{\sqrt{5y}}{y}$

CHAPTER R   Test

1. Refer to $B = \left\{ 0, \dfrac{\pi}{6}, \sqrt{-25}, 8, -\dfrac{5}{7}, 2.1, -0.\overline{4}, -3 \right\}$.

     Determine which elements belong to the given sets.

     a. $\mathbb{N}$ **b.** $\mathbb{W}$ **c.** $\mathbb{Z}$

     d. $\mathbb{Q}$ **e.** $\mathbb{H}$ **f.** $\mathbb{R}$

2. Max is twice as old as Jonas. If M represents Max's age and J represents Jonas's age, write an expression representing

     a. Max's age in terms of Jonas's age.

     b. Jonas's age in terms of Max's age.

3. Given $A = \{x \mid x < 2\}$, $B = \{x \mid x \geq 0\}$, and $C = (x \mid x < -1)$, write the union or intersection in set notation.

     a. $A \cup B$ **b.** $A \cap B$

     c. $A \cup C$ **d.** $A \cap C$

     e. $B \cup C$ **f.** $B \cap C$

4. Simplify without absolute value bars. $|5 - t|$ for $t > 5$.

5. a. Find the distance between $\sqrt{2}$ and 2 on the number line.

     b. Simplify without absolute value bars.

6. a. Determine the restrictions on y. $\dfrac{2y - 22}{y^2 - 9y - 22}$

     b. Simplify the expression.

For Exercises 7–26, perform the indicated operations and simplify the expression. Assume that all radical expressions represent real numbers.

7. $-\{6 + 4[9x - 2(3x - 5)] + 3\}$

8. $\dfrac{1}{3}a^2b - \dfrac{1}{2}ab^2 - \dfrac{5}{6}a^2b + 4ab^2$

9. $\dfrac{7^2 - 4[8 - (-2)] - 4^2}{-2|-4 + 6|}$

10. $(2a^3b^{-4})^2(4a^{-2}b^7)^{-1}$

11. $\left(\dfrac{1}{2}\right)^0 + \left(\dfrac{1}{3}\right)^{-3} + \left(\dfrac{1}{4}\right)^{-1}$

12. $\left(\dfrac{t^{2/5} \cdot t^{7/5}}{t^3}\right)^{10}$

13. $\sqrt[3]{80k^{15}m^2n^7}$

14. $\dfrac{3}{4}\sqrt[4]{8p^2q^3} \cdot \sqrt[4]{2p^3q}$

15. $\sqrt{125ab^3} - 3b\sqrt{20ab}$

16. $(12n^2 - 4)\left(\dfrac{1}{2}n^2 - 3n + 5\right)$

17. $[(3a + b) - c][(3a + b) + c]$

18. $\left(\dfrac{1}{4}\sqrt{z} - p^2\right)\left(\dfrac{1}{4}\sqrt{z} + p^2\right)$

19. $(5\sqrt{x} + \sqrt{2})(3\sqrt{x} - 2\sqrt{2})$

20. $(\sqrt{x} - 6z)^2$

21. $\dfrac{3x^2}{x^3 + 14x^2 + 49x} \div \dfrac{8x - 4}{4x^2 + 26x - 14}$

22. $\dfrac{x^2}{x - 5} + \dfrac{10x - 25}{5 - x}$

23. $\dfrac{-12}{y^3 + 4y^2} + \dfrac{1}{y} - \dfrac{3}{y^2 + 4y}$

24. $\dfrac{\dfrac{x}{4} - \dfrac{9}{4x}}{\dfrac{1}{4} - \dfrac{3}{4x}}$

25. $\dfrac{6}{\sqrt{13} + \sqrt{10}}$

26. $\dfrac{7}{t\sqrt{2}} - \dfrac{\sqrt{2}}{t}$

For Exercises 27–34, factor completely. Write the answer with positive exponents only.

27. $30x^3 + 2x^2 - 4x$

28. $xy + 5ay + 10ac + 2xc$

29. $x^5 + 2x^4 - 81x - 162$

30. $c^2 - 4a^2 - 44a - 121$

31. $27u^3 - v^6$

32. $4w^{-6} + 2w^{-5} + 7w^{-4}$

33. $y(2y - 1)^{-3/4} + (2y - 1)^{1/4}$

34. $2x(9 - x^2)^{1/2} + x^2 \cdot \dfrac{1}{2}(9 - x^2)^{-1/2}(-2x)$

35. A rectangle and square are shown. Write an expression in terms of x that represents the area of the shaded region.

36. Write the numbers in the given statement in scientific notation. "China uses 45,000,000,000 pairs of disposable chopsticks per year. This equates to approximately 1.66 million cubic meters of timber."

37. Write the number in the given statement in standard decimal notation. "The Ebola virus is approximately 8×10^{-7} m in length."

38. Perform the indicated operations.

$$\frac{(8.4 \times 10^{11})(6.0 \times 10^{-3})}{(4.2 \times 10^{-5})}$$

39. Write the interval notation representing $\{x \mid 2.7 < x\}$.

40. Write the set-builder notation representing the interval $[-3, 5)$.

Equations and Inequalities

According to legend, while taking a leisurely stroll through his orchard one day, the great astronomer and mathematician Sir Isaac Newton watched an apple fall from a tree. He most likely realized that the apple had no velocity when it was released, and yet it picked up speed as it moved toward the ground. As the legend is told, this event inspired Newton to begin his study on gravity. The resulting ideas incorporate gravity into the acceleration of the apple and may be generalized to any object launched into the air with an initial velocity.

In Chapter 1, we study equations and inequalities that will help us analyze many natural phenomena as well as applications in day-to-day life. The techniques we develop will enable us to solve equations that can be of assistance in making good financial decisions, buying the correct amount of materials for home projects, and even figuring the speed at which Sir Isaac Newton's apple might have hit the ground.

OBJECTIVES

1. Solve Linear Equations in One Variable
2. Identify Conditional Equations, Identities, and Contradictions
3. Solve Rational Equations
4. Solve Literal Equations for a Specified Variable

1. Solve Linear Equations in One Variable

Whether to lease an automobile, to buy an automobile, or to use public transportation is an important financial decision that affects a family's monthly budget. As part of an informed decision, it is useful to create a mathematical model of the cost for each option.

Suppose that a couple has an option to buy a used car for $8800 and that they expect it to last 3 yr without major repair. Another option is to lease a new automobile for an initial down payment of $2500 followed by 36 monthly payments of $225. After how many months will the cost to lease the new car equal the cost to buy the used car? The answer can be found by solving a type of equation called a linear equation in one variable (see Example 3).

Definition of a Linear Equation in One Variable

A **linear equation in one variable** is an equation that can be written in the form $ax + b = 0$, where a and b are real numbers, $a \neq 0$, and x is the variable.

A linear equation in one variable is also called a **first-degree equation** because the degree of the variable term must be exactly one.

Linear equation in one variable	Not a linear equation in one variable	
$5x + 35 = 0$	$5x^2 + 35 = 0$	(not first degree)
$\dfrac{x}{4} - 5 = 0$	$\dfrac{4}{x} - 5 = 0$	(not first degree because $\frac{4}{x}$ is $4x^{-1}$)
$3x + 4 = 7$	$3x + 4y = 7$	(contains two variables)
$0.7x - 0.8 = 0.1$	$0.7x - 0.8 - 0.1$	(This is an expression, not an equation.)

A **solution** to an equation is a value of the variable that makes the equation a true statement. The set of all solutions to an equation is called the **solution set** of the equation. **Equivalent equations** have the same solution set. To solve a linear equation in one variable, we form simpler, equivalent equations until we obtain an equation whose solution is obvious. The properties used to produce equivalent equations include the addition and multiplication properties of equality.

TIP Given $x + 4 = 9$, we can add the opposite of 4 to both sides or subtract 4 from both sides. Likewise, given $4x = 24$, we can multiply both sides by $\frac{1}{4}$ or divide both sides by 4.

Properties of Equality

Let a, b, and c be real-valued expressions.

Property Name	Statement of Property	Example
Addition property of equality	$a = b$ is equivalent to $a + c = b + c$.	$x + 4 = 9$ $x + 4 + (-4) = 9 + (-4)$ $x = 5$
Multiplication property of equality	$a = b$ is equivalent to $ac = bc$ $(c \neq 0)$.	$4x = 24$ $\dfrac{1}{4} \cdot 4x = \dfrac{1}{4} \cdot 24$ $x = 6$

To solve a linear equation in one variable, isolate the variable by following these guidelines.

Solving a Linear Equation in One Variable

Step 1   Simplify both sides of the equation.
- Use the distributive property to clear parentheses.
- Combine like terms.
- Consider clearing fractions or decimals by multiplying both sides of the equation by the least common denominator (LCD) of all terms.

Step 2   Use the addition property of equality to collect the variable terms on one side of the equation and the constant terms on the other side.

Step 3   Use the multiplication property of equality to make the coefficient of the variable term equal to 1.

Step 4   Check the potential solution in the original equation.

Step 5   Write the solution set.

EXAMPLE 1   **Solving a Linear Equation**

Solve. $-3(w - 4) + 5 = 10 - (w + 1)$

Solution:

$$-3(w - 4) + 5 = 10 - (w + 1)$$

$-3w + 12 + 5 = 10 - w - 1$	Apply the distributive property.
$-3w + 17 = 9 - w$	Combine like terms.
$-3w + w + 17 = 9 - w + w$	Add w to both sides of the equation.
$-2w + 17 = 9$	Combine like terms.
$-2w + 17 - 17 = 9 - 17$	Subtract 17 from both sides.
$-2w = -8$	Combine like terms.
$\dfrac{-2w}{-2} = \dfrac{-8}{-2}$	Divide both sides by -2 to obtain a coefficient of 1 for w.
$w = 4$	Check: $-3(w - 4) + 5 = 10 - (w + 1)$
	$-3[(4) - 4] + 5 \overset{?}{=} 10 - [(4) + 1]$
	$-3(0) + 5 \overset{?}{=} 10 - (5)$

The solution set is $\{4\}$. $5 \overset{?}{=} 5 \checkmark$ true

If a linear equation contains fractions, it is often helpful to clear the equation of fractions. This is done by multiplying both sides of the equation by the least common denominator (LCD) of all terms in the equation.

EXAMPLE 2   Solving a Linear Equation by Clearing Fractions

Solve. $\dfrac{m - 2}{5} - \dfrac{m - 4}{2} = \dfrac{m + 5}{15} + 2$

Solution:

TIP To find the least common multiple of 5, 2, and 15, first factor each number into prime factors.

$5 = 5 \cdot 1$
$2 = 2 \cdot 1$
$15 = 3 \cdot 5$
$\text{LCD} = 2 \cdot 3 \cdot 5 = 30$

$\dfrac{m - 2}{5} - \dfrac{m - 4}{2} = \dfrac{m + 5}{15} + 2$

The least common denominator is 30.

$30\left(\dfrac{m - 2}{5} - \dfrac{m - 4}{2}\right) = 30\left(\dfrac{m + 5}{15} + 2\right)$

Clear fractions by multiplying both sides by the LCD, 30.

$\dfrac{30^6}{1}\left(\dfrac{m - 2}{5_1}\right) - \dfrac{30^{15}}{1}\left(\dfrac{m - 4}{2_1}\right) = \dfrac{30^2}{1}\left(\dfrac{m + 5}{15_1}\right) + \dfrac{30}{1}\left(\dfrac{2}{1}\right)$

Apply the distributive property.

$6(m - 2) - 15(m - 4) = 2(m + 5) + 60$

$6m - 12 - 15m + 60 = 2m + 10 + 60$

Apply the distributive property.

$-9m + 48 = 2m + 70$

Combine like terms.

$-9m - 2m + 48 - 48 = 2m - 2m + 70 - 48$

Subtract $2m$ and 48 from both sides.

$-11m = 22$

Combine like terms.

$\dfrac{-11m}{-11} = \dfrac{22}{-11}$

Divide both sides by -11 to obtain a coefficient of 1 for m.

$m = -2$

The value -2 checks in the original equation.

The solution set is $\{-2\}$.

Skill Practice 2   Solve. $\dfrac{y + 5}{2} - \dfrac{y - 2}{4} = \dfrac{y + 7}{3} + 1$

In Example 3, we use a linear equation to solve the application given at the beginning of the section.

EXAMPLE 3   Using a Linear Equation in an Application

A couple must decide whether to buy a used car for $8800 or lease a new car for an initial down payment of $2500 followed by 36 monthly payments of $225.

a. Write a model for the cost C (in $) to lease the car for t months.
b. After how many months will the cost to lease the new car be equal to the cost to buy the new car?

Answers
1. $\{11\}$
2. $\{-4\}$

Solution:

a. The cost to lease the new car for t months includes the monthly payments plus the fixed down payment.

$$C = 225t + 2500$$

b.
$$C = 225t + 2500$$
$$8800 = 225t + 2500 \qquad \text{Substitute 8800 for the cost } C.$$
$$8800 - 2500 = 225t + 2500 - 2500 \qquad \text{Subtract 2500 from both sides.}$$
$$6300 = 225t$$
$$\frac{6300}{225} = \frac{225t}{225} \qquad \text{Divide both sides by 225.}$$
$$28 = t$$

In 28 months, the cost to lease a new car will be equal to the cost to buy a used car.

> **Skill Practice 3**   To rent a storage unit, a customer must pay a fixed deposit of \$150 plus \$52.50 in rent each month.
>
> **a.** Write a model for the cost C (in \$) to rent the unit for t months.
>
> **b.** If Winston has \$1200 budgeted for storage, for how many months can he rent the unit?

2. Identify Conditional Equations, Identities, and Contradictions

The linear equations examined thus far have all had exactly one solution. These equations are examples of conditional equations. A **conditional equation** is true for some values of the variable and false for other values.

An equation that is true for all values of the variable for which the expressions in an equation are defined is called an **identity**. An equation that is false for all values of the variable is called a **contradiction**. Example 4 presents each of these three types of equations.

> **EXAMPLE 4**   **Identifying Conditional Equations, Contradictions, and Identities**

Identify each equation as a conditional equation, a contradiction, or an identity. Then give the solution set.

a. $3(2x - 1) = 2(3x - 2)$ **b.** $3(2x - 1) = 2(3x - 2) + 1$

c. $3(2x - 1) = 5x - 4$

Solution:

a. $3(2x - 1) = 2(3x - 2)$
$$6x - 3 = 6x - 4 \qquad \text{Apply the distributive property.}$$
$$-3 = -4 \qquad \text{Subtract } 6x \text{ from both sides. Contradiction}$$

This equation is a contradiction.
The solution set is the empty set { }.

> **TIP** No real number substituted for x will make $-3 = -4$.

Answers
3. a. $C = 150 + 52.50t$
b. 20 months

b. $3(2x - 1) = 2(3x - 2) + 1$

$6x - 3 = 6x - 4 + 1$ Apply the distributive property.

$6x - 3 = 6x - 3$ Combine like terms.

$0 = 0$ Subtract $6x$ from both sides. Add 3 to both sides.

This equation is an identity.
The solution set is the set of all real numbers, $\mathbb{R}$.

c. $3(2x - 1) = 5x - 4$ Apply the distributive property.

$6x - 3 = 5x - 4$ Subtract $5x$ from both sides. Add 3 to both sides.

$x = -1$

This is a conditional equation.
The solution set is $\{-1\}$.   Conditional equation. The statement is true only under the condition that $x = -1$.

Skill Practice 4   Identify each equation as a conditional equation, a contradiction, or an identity. Then give the solution set.

a. $4x + 1 - x = 6x - 2$ **b.** $2(-5x - 1) = 2x - 12x + 6$

c. $2(3x - 1) = 6(x + 1) - 8$

3. Solve Rational Equations

One of the powerful features of mathematics is that methods used to solve one type of equation can sometimes be adapted to solve other types of equations. Two equations are shown here. The equation on the left is a linear equation. The equation on the right is a rational equation. A **rational equation** is an equation in which each term contains a rational expression. All linear equations are rational equations, but not all rational equations are linear. For example:

Linear Equation with Constants in the Denominator

$$\frac{x}{2} = \frac{2x}{3} - 1$$

Rational Equation with Variables in the Denominator

$$\frac{12}{x} = \frac{6}{2x} + 3$$

The linear equation can be solved by first multiplying both sides of the equation by the least common denominator of all the fractions. This is the same strategy used in Example 5 to solve a rational equation with a variable in the denominator. However, when a variable appears in the denominator of a fraction, we must restrict the values of the variable to avoid division by zero.

EXAMPLE 5   **Solving a Rational Equation**

Solve the equation and check the solution. $\frac{12}{x} = \frac{6}{2x} + 3$

Solution:

$\frac{12}{x} = \frac{6}{2x} + 3$ Restrict x so that $x \neq 0$.

$2x\left(\frac{12}{x}\right) = 2x\left(\frac{6}{2x} + \frac{3}{1}\right)$ Clear fractions by multiplying both sides by the LCD, $2x$. Since $x \neq 0$, this will produce an equivalent equation.

Answers
4. a. Conditional equation; $\{1\}$
   b. Contradiction; $\{\ \}$
   c. Identity; $\mathbb{R}$

$$\frac{2x}{1}\left(\frac{12}{x}\right) = \frac{2x}{1}\left(\frac{6}{2x}\right) + \frac{2x}{1}\left(\frac{3}{1}\right)$$ Apply the distributive property.

$$24 = 6 + 6x$$ Simplify.

$$18 = 6x$$ Subtract 6 from both sides.

$$3 = x$$

Check: $\dfrac{12}{x} = \dfrac{6}{2x} + \dfrac{3}{1}$

$$\frac{12}{(3)} \overset{?}{=} \frac{6}{2(3)} + \frac{3}{1}$$

The solution set is $\{3\}$. $4 \overset{?}{=} 1 + 3 \checkmark$ true

---

Skill Practice 5   Solve the equation and check the solution. $\dfrac{15}{y} = \dfrac{21}{3y} + 2$

In Example 6, we demonstrate the importance of determining restricted values of the variable in an equation and checking the potential solutions. You will see that for some equations, a potential solution does not check.

EXAMPLE 6   **Solving a Rational Equation**

Solve the equation and check the solution. $\dfrac{x}{x-4} = \dfrac{4}{x-4} - \dfrac{4}{5}$

Solution:

$$\frac{x}{x-4} = \frac{4}{x-4} - \frac{4}{5}$$ Restrict x so that $x \neq 4$.

$$5(x-4)\left(\frac{x}{x-4}\right) = 5(x-4)\left(\frac{4}{x-4} - \frac{4}{5}\right)$$ Clear fractions by multiplying both sides by the LCD $5(x-4)$.

$$\frac{5(x-4)}{1}\left(\frac{x}{x-4}\right) = \frac{5(x-4)}{1}\left(\frac{4}{x-4}\right) + \frac{5(x-4)}{1}\left(-\frac{4}{5}\right)$$ Apply the distributive property.

$$5x = 20 - 4(x-4)$$ Simplify.

$$5x = 20 - 4x + 16$$ Apply the distributive property.
   Combine like terms.

$$9x = 36$$

$x \neq 4$ This is a restricted value of x. Substituting 4 for x in the original equation results in division by 0.

Check: $\dfrac{x}{x-4} = \dfrac{4}{x-4} - \dfrac{4}{5}$

$$\frac{(4)}{(4)-4} \overset{?}{=} \frac{4}{(4)-4} - \frac{4}{5}$$

$$\underbrace{}_{\text{undefined}} \quad \underbrace{}_{\text{undefined}}$$

The solution set is $\{\ \}$.

Skill Practice 6   Solve the equation and check the solution.

$$\frac{y}{y+5} = \frac{-5}{y+5} + \frac{5}{4}$$

EXAMPLE 7   Solving a Rational Equation

Solve the equation and check the solution. $\dfrac{6}{y^2 + 8y + 15} - \dfrac{2}{y + 3} = \dfrac{-4}{y + 5}$

Solution:

$$\frac{6}{y^2 + 8y + 15} - \frac{2}{y + 3} = \frac{-4}{y + 5}$$

$$\frac{6}{(y + 3)(y + 5)} - \frac{2}{y + 3} = \frac{-4}{y + 5}$$

Restrict y so that $y \neq -3$, $y \neq -5$.

Clear fractions by multiplying both sides by the LCD $(y + 3)(y + 5)$.

$$(y + 3)(y + 5)\left(\frac{6}{(y + 3)(y + 5)} - \frac{2}{y + 3}\right) = (y + 3)(y + 5)\left(\frac{-4}{y + 5}\right)$$

$$\frac{(y + 3)(y + 5)}{1}\left(\frac{6}{(y + 3)(y + 5)}\right) - \frac{(y + 3)(y + 5)}{1}\left(\frac{2}{y + 3}\right)$$

$$= \frac{(y + 3)(y + 5)}{1}\left(\frac{-4}{y + 5}\right)$$

$6 - 2(y + 5) = -4(y + 3)$ Apply the distributive property.

$6 - 2y - 10 = -4y - 12$

$-2y - 4 = -4y - 12$ Combine like terms.

$2y = -8$

$y = -4$

The value -4 is *not* a restricted value.

Check: $\dfrac{6}{y^2 + 8y + 15} - \dfrac{2}{y + 3} = \dfrac{-4}{y + 5}$

$\dfrac{6}{(-4)^2 + 8(-4) + 15} - \dfrac{2}{(-4) + 3} \overset{?}{=} \dfrac{-4}{(-4) + 5}$

The solution set is $\{-4\}$.

$-6 + 2 \overset{?}{=} -4 \checkmark$ true

Skill Practice 7   Solve the equation.

$$\frac{11}{x^2 + 5x + 4} - \frac{3}{x + 4} = \frac{1}{x + 1}$$

4. Solve Literal Equations for a Specified Variable

Sometimes an equation contains multiple variables. For example, $d = rt$ relates the distance that an object travels to the rate of travel and time of travel. Such an equation is called a literal equation (an equation with many letters). We often want to manipulate a literal equation to solve for a specified variable. In such a case, we use the same techniques as we would with an equation containing one variable.

Answer

7. $\{1\}$

EXAMPLE 8 Solving an Equation for a Specified Variable

Solve for the indicated variable.

a. $d = rt$ for t **b.** $3x + 2y = 6$ for y

c. $A = \frac{1}{2}h(B + b)$ for B

Solution:

a. $d = rt$ for t

$$\frac{d}{r} = \frac{rt}{r}$$

$$\frac{d}{r} = t \quad \text{or} \quad t = \frac{d}{r}$$

The relationship between r and t is multiplication. Therefore, perform the inverse operation. Divide both sides by r.

b. $3x + 2y = 6$ for y

$$2y = -3x + 6$$

$$\frac{2y}{2} = \frac{-3x + 6}{2}$$

$$y = \frac{-3x + 6}{2} \quad \text{or} \quad y = -\frac{3}{2}x + 3$$

Subtract $3x$ from both sides to isolate the y term on one side of the equation.

Divide both sides by 2 to isolate y.

c. $A = \frac{1}{2}h(B + b)$ for B

> **TIP** Alternatively, after clearing fractions we can solve for B by first clearing parentheses.
>
> $$2A = h(B + b)$$
> $$2A = hB + hb$$
> $$2A - hb = hB$$
> $$\frac{2A - hb}{h} = B$$

$$2(A) = 2\left[\frac{1}{2}h(B + b)\right]$$

$$2A = h(B + b)$$

$$\frac{2A}{h} = \frac{h(B + b)}{h}$$

$$\frac{2A}{h} = B + b$$

$$\frac{2A}{h} - b = B \quad \text{or} \quad B = \frac{2A}{h} - b$$

First note that letters in algebra are case sensitive. The letters b and B represent different variables.

Multiply by 2 to clear fractions.

Divide by h.

Subtract b from both sides to isolate B.

Skill Practice 8 Solve for the indicated variable.

a. $I = Prt$ for t **b.** $4x + 3y = 12$ for y

c. $A = \frac{1}{2}h(B + b)$ for b

Answers

8. a. $t = \frac{I}{Pr}$

b. $y = \frac{-4x + 12}{3}$ or

    $y = -\frac{4}{3}x + 4$

c. $b = \frac{2A}{h} - B$

In Example 9, multiple occurrences of the variable x appear within the equation. Factoring is required to combine x terms so that we can isolate x.

EXAMPLE 9   **Solving an Equation for a Specified Variable**

Solve the equation for x. $ax + by = cx + d$

Solution:

$ax + by = cx + d$ Subtract cx from both sides to combine the x terms on one side.

$ax - cx = d - by$ Subtract by from both sides to combine the non-x terms on the other side.

$x(a - c) = d - by$ Factor out x as the GCF on the left side of the equation.

$\dfrac{x(a - c)}{(a - c)} = \dfrac{d - by}{(a - c)}$ Divide by $(a - c)$.

$x = \dfrac{d - by}{a - c}$

Answer

9. $x = \dfrac{w + z}{3 - a}$ or $x = -\dfrac{w + z}{a - 3}$

Skill Practice 9   Solve the equation for x. $3x - w = ax + z$

TIP In Example 9, the answer can be expressed in different forms. For example, if we had isolated x on the right side of the equation, the solution for x would be

$ax + by = cx + d$

$by - d = cx - ax$

$\dfrac{by - d}{(c - a)} = \dfrac{x(c - a)}{(c - a)}$

$x = \dfrac{by - d}{c - a}$

→ To show that $\dfrac{by - d}{c - a} = \dfrac{d - by}{a - c}$ multiply either expression by $\dfrac{-1}{-1}$.

$\dfrac{(by - d)}{(c - a)} \cdot \dfrac{(-1)}{(-1)} = \dfrac{-by + d}{-c + a} = \dfrac{d - by}{a - c}$

SECTION 1.1   **Practice Exercises**

Prerequisite Review

R.1. Clear parentheses and combine like terms.

$-3(-6 + 2w) - 9w + 2(w - 1)$

R.2. Simplify. $(x + 9)^2$

R.3. Factor completely. $2p^2 + p - 15$

For Exercises R.4–R.6, find the least common denominator (LCD).

R.4. $\dfrac{y}{6x}, \dfrac{y^2}{21}$

R.5. $\dfrac{n}{(5n - 2)(n - 2)}, \dfrac{7}{(5n - 2)(n + 7)}$

R.6. $\dfrac{1}{a - 5}, \dfrac{a - 6}{a^2 - 7a + 10}$

Concept Connections

1. An equation that can be written in the form $ax + b = 0$, where a and b are real numbers and $a \neq 0$, is called a _____ equation in one variable.

2. A linear equation is also called a _____ -degree equation because the degree of the variable is 1.

3. A _____ to an equation is the value of the variable that makes the equation a true statement.

4. A _____ equation is one that is true for some values of the variable and false for others.

5. An _____ is an equation that is true for all values of the variable for which the expressions in the equation are defined.

6. A _____ is an equation that is false for all values of the variable.

7. A _____ equation is an equation in which each term contains a rational expression.

8. If an equation has no solution, then the solution set is the _____ set and is denoted by _____ .

Objective 1: Solve Linear Equations in One Variable

For Exercises 9–10, determine if the equation is linear or nonlinear. If the equation is linear, find the solution set.

9. a. $-2x = 8$
      b. $\dfrac{-2}{x} = 8$
      c. $-\dfrac{1}{2}x = 8$
   d. $-2|x| = 8$
      e. $x - 2 = 8$

10. a. $12 = 4x$
       b. $12 = \dfrac{4}{x}$
       c. $12 = \dfrac{1}{4}x$
    d. $12 = 4\sqrt{x}$
       e. $12 = 4 + x$

For Exercises 11–30, solve the equation. (See Examples 1–2)

11. $-6x - 4 = 20$

12. $-8y + 6 = 22$

13. $4 = 7 - 3(4t + 1)$

14. $11 = 7 - 2(5p - 2)$

15. $-6(v - 2) + 3 = 9 - (v + 4)$

16. $-5(u - 4) + 2 = 11 - (u - 3)$

17. $2.3 = 4.5x + 30.2$

18. $9.4 = 3.5p - 0.4$

19. $0.05y + 0.02(6000 - y) = 270$

20. $0.06x + 0.04(10{,}000 - x) = 520$

21. $2(5x - 6) = 4[x - 3(x - 10)]$

22. $4(y - 3) = 3[y + 2(y - 2)]$

23. $\dfrac{1}{4}x - \dfrac{3}{2} = 2$

24. $\dfrac{1}{6}x - \dfrac{5}{3} = 1$

25. $\dfrac{1}{2}w - \dfrac{3}{4} = \dfrac{2}{3}w + 2$

26. $\dfrac{2}{5}p - \dfrac{3}{10} = \dfrac{7}{15}p - 1$

27. $\dfrac{y - 1}{5} + \dfrac{y}{4} = \dfrac{y + 3}{2} + 1$

28. $\dfrac{x - 6}{3} + \dfrac{x}{7} = \dfrac{x + 1}{3} + 2$

29. $\dfrac{n + 3}{4} - \dfrac{n - 2}{5} = \dfrac{n + 1}{10} - 1$

30. $\dfrac{t - 2}{3} - \dfrac{t + 7}{5} = \dfrac{t - 4}{10} + 2$

31. In the mid-nineteenth century, explorers used the boiling point of water to estimate altitude. The boiling temperature of water T (in °F) can be approximated by the model $T = -1.83a + 212$, where a is the altitude in thousands of feet.

   a. Determine the temperature at which water boils at an altitude of 4000 ft. Round to the nearest degree.

   b. Two campers hiking in Colorado boil water for tea. If the water boils at 193°F, approximate the altitude of the campers. Give the result to the nearest hundred feet.

32. For a recent year, the cost C (in $) for tuition and fees for x credit-hours at a public college was given by $C = 167.95x + 94$.

   a. Determine the cost to take 9 credit-hours.

   b. If Jenna spent $2445.30 for her classes, how many credit-hours did she take?

33. The annual per capita spending S (in $) for prescription drugs can be modeled by $S = 14.2t + 149$, where t is the number of years since 2004. If this model continues, in what year would the average spending for prescription drugs equal $362 per person? (*Source:* U.S. Centers for Medicare and Medicaid Services, www.census.gov)

34. The annual per capita spending S (in $) for home health care can be modeled by $S = 18t + 232$, where t is the number of years since 2000. If this model continues, in what year would the average spending for home health care equal $628 per person? (*Source:* U.S. Centers for Medicare and Medicaid Services, www.census.gov)

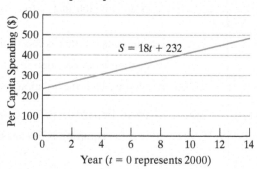

35. A motorist drives on State Road 417 to and from work each day and pays $3.50 in tolls one-way.

   a. Write a model for the cost for tolls C (in $) for x working days.

   b. The department of transportation has a prepaid toll program that discounts tolls for high-volume use. The motorist can buy a pass for $105 per month. How many working days are required for the motorist to save money by buying the pass? (**See Example 3**)

36. A subway ride is $2.25 per ride.

   a. Write a model for the cost C (in $) for x rides on the subway.

   b. A commuter can purchase an unlimited-ride MetroCard for $89 per month. How many rides are required for a commuter to save money by buying the MetroCard?

37. Helene considers two jobs. One pays $45,000/yr with an anticipated yearly raise of $2250. A second job pays $48,000/yr with yearly raises averaging $2000.

   a. Write a model representing the salary S_1 (in $) for the first job in x years.

   b. Write a model representing the salary S_2 (in $) for the second job in x years.

   c. In how many years will the salary from the first job equal the salary from the second?

38. Tasha considers two sales jobs for different pharmaceutical companies. One pays a base salary of $25,000 with a 16% commission on sales. The other pays $30,000 with a 15% commission on sales.

   a. Write a model representing the salary S_1 (in $) for the first job based on x dollars in sales.

   b. Write a model representing the salary S_2 (in $) for the second job based on x dollars in sales.

   c. For how much in sales will the two jobs result in equal salaries?

Objective 2: Identify Conditional Equations, Identities, and Contradictions

For Exercises 39–44, identify the equation as a conditional equation, a contradiction, or an identity. Then give the solution set. (See Example 4)

39. $2x - 3 = 4(x - 1) - 1 - 2x$

40. $4(3 - 5n) + 1 = -4n - 8 - 16n$

41. $-(6 - 2w) = 4(w + 1) - 2w - 10$

42. $-5 + 3x = 3(x - 1) - 2$

43. $\dfrac{1}{2}x + 3 = \dfrac{1}{4}x + 1$

44. $\dfrac{2}{3}y - 5 = \dfrac{1}{6}y - 4$

Objective 3: Solve Rational Equations

For Exercises 45–48, determine the restrictions on x.

45. $\dfrac{3}{x - 5} + \dfrac{2}{x + 4} = \dfrac{5}{7}$

46. $\dfrac{2}{x + 1} - \dfrac{5}{x - 7} = \dfrac{2}{3}$

47. $\dfrac{5}{2x - 3} - \dfrac{3}{5x} = \dfrac{1}{3 - x}$

48. $\dfrac{1}{2x} - \dfrac{1}{6 - x} = \dfrac{2}{4x - 5}$

For Exercises 49–66, solve the equation. (See Examples 5–7)

49. $\dfrac{1}{2} - \dfrac{7}{2y} = \dfrac{5}{y}$

50. $\dfrac{1}{3} - \dfrac{4}{3t} = \dfrac{7}{t}$

51. $\dfrac{w+3}{4w} + 1 = \dfrac{w-5}{w}$

52. $\dfrac{x+2}{6x} + 1 = \dfrac{x-7}{x}$

53. $\dfrac{c}{c-3} = \dfrac{3}{c-3} - \dfrac{3}{4}$

54. $\dfrac{7}{d-7} - \dfrac{7}{8} = \dfrac{d}{d-7}$

55. $\dfrac{1}{t-1} = \dfrac{3}{t^2-1}$

56. $\dfrac{1}{w+2} = \dfrac{5}{w^2-4}$

57. $\dfrac{2}{x-5} - \dfrac{1}{x+5} = \dfrac{11}{x^2-25}$

58. $\dfrac{2}{c+3} - \dfrac{1}{c-3} = \dfrac{10}{c^2-9}$

59. $\dfrac{-14}{x^2-x-12} - \dfrac{1}{x-4} = \dfrac{2}{x+3}$

60. $\dfrac{2}{x^2+5x+6} - \dfrac{2}{x+2} = \dfrac{1}{x+3}$

61. $\dfrac{5}{x^2-x-2} - \dfrac{2}{x^2-4} = \dfrac{4}{x^2+3x+2}$

62. $\dfrac{4}{x^2-2x-8} - \dfrac{1}{x^2-16} = \dfrac{2}{x^2+6x+8}$

63. $\dfrac{5}{m-2} = \dfrac{3m}{m^2+2m-8} - \dfrac{2}{m+4}$

64. $\dfrac{10}{n-6} = \dfrac{15n}{n^2-2n-24} - \dfrac{6}{n+4}$

65. $\dfrac{5x}{3x^2-5x-2} - \dfrac{1}{3x+1} = \dfrac{3}{2-x}$

66. $\dfrac{3x}{2x^2+x-3} - \dfrac{2}{2x+3} = \dfrac{4}{1-x}$

Objective 4: Solve Literal Equations for a Specified Variable

For Exercises 67–88, solve for the specified variable. (See Examples 8–9)

67. $A = lw$ for l

68. $E = IR$ for R

69. $P = a + b + c$ for c

70. $W = K - T$ for K

71. $\Delta s = s_2 - s_1$ for s_1

72. $\Delta t = t_f - t_i$ for t_i

73. $7x + 2y = 8$ for y

74. $3x + 5y = 15$ for y

75. $5x - 4y = 2$ for y

76. $7x - 2y = 5$ for y

77. $\dfrac{1}{2}x + \dfrac{1}{3}y = 1$ for y

78. $\dfrac{1}{4}x - \dfrac{2}{3}y = 2$ for y

79. $S = \dfrac{n}{2}(a + d)$ for d

80. $S = \dfrac{n}{2}[2a + (n-1)d]$ for a

81. $V = \dfrac{1}{3}\pi r^2 h$ for h

82. $V = \dfrac{1}{3}Bh$ for B

83. $6 = 4x + tx$ for x

84. $8 = 3x + kx$ for x

85. $6x + ay = bx + 5$ for x

86. $3x + 2y = cx + d$ for x

87. $A = P + Prt$ for P

88. $C = A + Ar$ for A

Mixed Exercises

For Exercises 89–102, solve the equation.

89. $\dfrac{5}{2n+1} = \dfrac{-2}{3n-4}$

90. $\dfrac{4}{5z-3} = \dfrac{-2}{4z+7}$

91. $5 - 2\{3 - [5v + 3(v - 7)]\} = 8v + 6(3 - 4v) - 61$

92. $6 - \{4 - 2[8u - 2(u - 3)]\} = -4u + 3(2 - u) + 8$

93. $(x - 7)(x + 2) = x^2 + 4x + 13$

94. $(m + 3)(2m - 5) = 2m^2 + 4m - 3$

95. $\dfrac{3}{c^2-4c} - \dfrac{9}{2c^2+3c} = \dfrac{2}{2c^2-5c-12}$

96. $\dfrac{4}{d^2-d} - \dfrac{5}{2d^2+5d} = \dfrac{2}{2d^2+3d-5}$

97. $\dfrac{1}{3}x + \dfrac{1}{2} = \dfrac{1}{2}(x + 1) - \dfrac{1}{6}x$

98. $\dfrac{1}{2}x + \dfrac{2}{5} = \dfrac{2}{5}(x + 1) + \dfrac{1}{10}x$

99. $(t + 2)^2 = (t - 4)^2$

100. $(y - 3)^2 = (y + 1)^2$

101. $\dfrac{3}{3a+4} = \dfrac{5}{5a-1}$

102. $\dfrac{8}{8x-3} = \dfrac{2}{2x+5}$

103. Suppose that 40 deer are introduced in a protected wilderness area. The population of the herd P can be approximated by $P = \frac{40 + 20x}{1 + 0.05x}$, where x is the time in years since introducing the deer. Determine the time required for the deer population to reach 200.

104. Starting from rest, an automobile's velocity v (in ft/sec) is given by $v = \frac{180t}{2t + 10}$, where t is the time in seconds after the car begins forward motion. Determine the time required for the car to reach a speed of 60 ft/sec ($\approx$ 41 mph).

105. Brianna's SUV gets 22 mpg in the city and 30 mpg on the highway. The amount of gas she uses A (in gal) is given by $A = \frac{1}{22}c + \frac{1}{30}h$, where c is the number of city miles driven and h is the number of highway miles driven. If Brianna drove 165 mi on the highway and used 7 gal of gas, how many city miles did she drive?

106. Dexter's truck gets 32 mpg on the highway and 24 mpg in the city. The amount of gas he uses A (in gal) is given by $A = \frac{1}{24}c + \frac{1}{32}h$, where c is the number of city miles driven and h is the number of highway miles driven. If Dexter drove 60 mi in the city and used 9 gal of gas, how many highway miles did he drive?

Write About It

107. Explain why the value 5 is not a solution to the equation $\frac{x}{x - 5} + \frac{1}{5} = \frac{5}{x - 5}$.

108. Explain why the value 2 is not the only solution to the equation $2x + 4 = 2(x - 3) + 10$.

109. Explain why $\frac{3}{x} + 12 = 0$ is not a linear equation in one variable.

110. Explain why $2\sqrt{x} + 6 = 0$ is not a linear equation in one variable.

111. Explain why the equation $x + 1 = x + 2$ has no solution.

112. Explain the difference in the process to clear fractions between the two equations.

$$\frac{x}{3} + \frac{1}{2} = 1 \quad \text{and} \quad \frac{3}{x} + \frac{1}{2} = 1$$

Expanding Your Skills

For Exercises 113–116, find the value of a so that the equation has the given solution set.

113. $ax + 6 = 4x + 14$ $\{4\}$

114. $ax - 3 = 2x + 9$ $\{3\}$

115. $a(2x - 5) + 6 = 5x + 7$ $\{16\}$

116. $a(2x + 4) + 12x = 3(2 - x)$ $\{34\}$

SECTION 1.2   Applications with Linear and Rational Equations

OBJECTIVES

1. Solve Applications Involving Simple Interest
2. Solve Applications Involving Mixtures
3. Solve Applications Involving Uniform Motion
4. Solve Applications Involving Rate of Work Done
5. Solve Applications Involving Proportions

1. Solve Applications Involving Simple Interest

In Examples 1–5, we use linear and rational equations to model physical situations to solve applications. While there is no magic formula to apply to all word problems, we do offer the following guidelines to help you organize the given information and to form a useful model.

> **Problem-Solving Strategy**
>
> 1. Read the problem carefully. Determine what the problem is asking for, and assign variables to the unknown quantities.
> 2. Make an appropriate figure or table if applicable. Label the given information and variables in the figure or table.
> 3. Write an equation that represents the verbal model. The equation may be a known formula or one that you create that is unique to the problem.
> 4. Solve the equation from step 3.
> 5. Interpret the solution to the equation and check that it is reasonable in the context of the problem.

Simple interest I for a loan or an investment is based on the principal P (amount of money invested or borrowed), the annual interest rate r, and the time of the loan t in years. The relationship among the variables is given by $I = Prt$.

For example, if $5000 is invested at 4% simple interest for 18 months (1.5 yr), then the amount of simple interest earned is

$$I = Prt$$
$$I = (\$5000)(0.04)(1.5)$$
$$= \$300$$

The formula for simple interest is used in Example 1.

EXAMPLE 1 Solving an Application Involving Simple Interest

Kent invested a total of $8000. He invested part of the money for 2 yr in a stock fund that earned the equivalent of 6.5% simple interest. He put the remaining money in an 18-month certificate of deposit (CD) that earned 2.5% simple interest. If the total interest from both investments was $855, determine the amount invested in each account.

Solution:

We can assign a variable to *either* the amount invested in the stock fund or the amount invested in the CD.

Let x represent the principal invested in the stock fund.
Then, $(8000 - x)$ is the remaining amount in the CD.

The interest from each account is computed from the formula $I = Prt$. Consider organizing this information in a table.

	Stock Fund (6.5% yield)	CD (2.5% yield)	Total
Principal ($)	x	$8000 - x$	$8000
Interest ($)	$x(0.065)(2)$	$(8000 - x)(0.025)(1.5)$	$855

> **Avoiding Mistakes**
>
> The CD was invested for 18 months. Be sure to convert to years.
>
> 18 months = 1.5 yr

To build an equation, note that

$$\left(\begin{array}{c}\text{Interest from}\\\text{stock fund}\end{array}\right) + \left(\begin{array}{c}\text{Interest}\\\text{from CD}\end{array}\right) = \left(\begin{array}{c}\text{Total}\\\text{interest}\end{array}\right)$$

$x(0.065)(2) + (8000 - x)(0.025)(1.5) = 855$	Second row of table
$0.13x + 0.0375(8000 - x) = 855$	Simplify.
$0.13x + 300 - 0.0375x = 855$	Apply the distributive property.
$0.0925x + 300 = 855$	Combine like terms.
$0.0925x = 555$	Subtract 300 from both sides.
$x = 6000$	

> **Avoiding Mistakes**
>
> Check that the answer is reasonable.
>
> Amount of interest:
> ($6000)(0.065)(2) = $780
> ($2000)(0.025)(1.5) = $75
> Total: $780 + $75 = $855 ✓

The amount invested in the stock fund is x: $6000.
The amount invested in the CD is $8000 - x = \$8000 - \$6000 = \$2000$.

Skill Practice 1 Franz borrowed a total of $10,000. Part of the money was borrowed from a lending institution that charged 5.5% simple interest. The rest of the money was borrowed from a friend to whom Franz paid 2.5% simple interest. Franz paid his friend back after 9 months (0.75 yr) and paid the lending institution after 2 yr. If the total amount Franz paid in interest was $735, how much did he borrow from each source?

Answer

1. Franz borrowed $4000 from his friend and $6000 from the lending institution.

2. Solve Applications Involving Mixtures

In Example 1, we "mixed" money between two different investments. We had to find the correct distribution of principal between two accounts to produce the given amount of interest. Example 2 presents a similar type of application that involves mixing different concentrations of a bleach solution to produce a third mixture of a given concentration.

For example, household bleach contains 6% sodium hypochlorite (active ingredient). This means that the remaining 94% of liquid is some other mixing agent such as water. Therefore, given 200 cL of household bleach, 6% would be pure sodium hypochlorite, and 94% would be some other mixing agent.

$$\text{Pure sodium hypochlorite} = (0.06)(200 \text{ cL}) = 12 \text{ cL}$$
$$\text{Other mixing agent} = (0.94)(200 \text{ cL}) = 188 \text{ cL}$$

To find the amount of pure sodium hypochlorite, we multiplied the concentration rate by the amount of solution.

EXAMPLE 2   Solving an Application Involving Mixtures

Household bleach contains 6% sodium hypochlorite. How much household bleach should be combined with 70 L of a weaker 1% hypochlorite solution to form a solution that is 2.5% sodium hypochlorite?

Solution:

Let x represent the amount of 6% sodium hypochlorite solution (in liters).
70 L is the amount of 1% sodium hypochlorite solution.
Therefore, $x + 70$ is the amount of the resulting mixture (2.5% solution).

The amount of pure sodium hypochlorite in each mixture is found by multiplying the concentration rate by the amount of solution.

	6% Solution	1% Solution	2.5% Solution
Amount of solution (L)	x	70	$x + 70$
Pure sodium hypochlorite (L)	$0.06x$	$0.01(70)$	$0.025(x + 70)$

To build an equation, note that

$$\begin{pmatrix} \text{Amount of sodium} \\ \text{hypochlorite in} \\ 6\% \text{ solution} \end{pmatrix} + \begin{pmatrix} \text{Amount of sodium} \\ \text{hypochlorite in} \\ 1\% \text{ solution} \end{pmatrix} = \begin{pmatrix} \text{Amount of sodium} \\ \text{hypochlorite in} \\ 2.5\% \text{ solution} \end{pmatrix}$$

$$0.06x + 0.01(70) = 0.025(x + 70) \qquad \text{Second row in the table}$$
$$0.06x + 0.7 = 0.025x + 1.75 \qquad \text{Solve the equation.}$$
$$0.035x = 1.05$$
$$x = 30$$

The amount of household bleach (6% sodium hypochlorite solution) needed is 30 L.

> **Avoiding Mistakes**
>
> Check that the answer is reasonable. The total amount of the resulting solution is 30 L + 70 L, which is 100 L.
>
> **Amount of sodium hypochlorite:**
> $0.06(30 \text{ L}) = 1.8 \text{ L}$
> $0.01(70 \text{ L}) = 0.7 \text{ L}$
> $\overline{0.025(100 \text{ L}) = 2.5 \text{ L}}$ ✓

Skill Practice 2   How much 4% acid solution should be mixed with 200 mL of a 12% acid solution to make a 9% acid solution?

Answer

2. 120 mL of the 4% acid solution should be used.

3. Solve Applications Involving Uniform Motion

Example 3 involves uniform motion. Recall that the distance that an object travels is given by

$$d = rt \quad \text{Distance} = (\text{Rate})(\text{Time})$$

EXAMPLE 3   **Solving an Application Involving Uniform Motion**

Donna participated in a 41-mi biathlon that included running and bicycling. She spent 1 hr 45 min on the bike and 45 min running. If her average speed on the bicycle was 12 mph faster than her average speed running, find her average speed running and her average speed riding.

Solution:

There are two unknowns: Donna's average speed on the bike and her average speed running.

  Let x represent Donna's average speed running.

  Then $x + 12$ represents her speed on the bicycle.

The remaining information can be organized in a table.

	Distance (mi)	Rate (mph)	Time (hr)
Run	$0.75x$	x	0.75
Bike	$1.75(x + 12)$	$x + 12$	1.75

The expressions in this column are found by $d = rt$.

Note that consistency in the units of measurement is important. The speed is given in miles per *hour*. Therefore, we want the time to be in hours.

1 hr 45 min = 1.75 hr
45 min = 0.75 hr

$$\left(\begin{array}{c} \text{Total} \\ \text{distance} \end{array} \right) = \left(\begin{array}{c} \text{Distance} \\ \text{running} \end{array} \right) + \left(\begin{array}{c} \text{Distance} \\ \text{riding} \end{array} \right)$$

$$41 = 0.75x + 1.75(x + 12)$$
$$41 = 0.75x + 1.75x + 21$$
$$20 = 2.5x$$
$$8 = x$$

To build an equation, note that the total distance equals the sum of the distance running and the distance riding.

Solve the equation.

Donna's speed running is 8 mph.
Her speed on the bicycle is $8 + 12 = 20$ mph.

Interpret the solution in the context of the problem.

Avoiding Mistakes

Check that the answer is reasonable by verifying that the total distance traveled is 41 mi.

Distance running:
(8 mph)(0.75 hr) = 6 mi

Distance riding:
(20 mph)(1.75 hr) = 35 mi

Total: 6 mi + 35 mi = 41 mi

Skill Practice 3   Rene drove from Miami to Orlando, a total distance of 240 mi. He drove for 1 hr in city traffic and for 3 hr on the highway. If his average speed on the highway was 20 mph faster than his speed in the city, determine his average speed driving in the city and his average speed driving on the highway.

Answer

3. Rene drove 45 mph in the city and 65 mph on the highway.

Point of Interest

The relationship $d = rt$ is a familiar formula indicating that distance equals rate times time, or equivalently that $t = \frac{d}{r}$. However, suppose that a spaceship travels to a distant planet and then returns to Earth. Einstein's theory of special relativity indicates that $t = \frac{d}{r}$ only represents the trip's duration for an observer on Earth. For a person on the spaceship, the time will be shorter by a factor of $\sqrt{1 - \frac{r^2}{c^2}}$, where r is the speed of the spaceship and c is the speed of light.

For example, suppose that a spaceship travels to a planet 10 light-years away (a light-year is the distance that light travels in 1 yr) and then returns. The round trip is 20 light-years. Further suppose that the spaceship travels at half the speed of light, that is, $r = 0.5c$.

To an observer on Earth, the elapsed time of travel (in yr) is

$$t_E = \frac{d}{r} = \frac{20}{0.5} = 40 \text{ yr}$$

To an observer on the spaceship, the elapsed time (in yr) is

$$t_S = \frac{d}{r}\sqrt{1 - \frac{r^2}{c^2}} = \frac{20}{0.5}\sqrt{1 - \frac{(0.5c)^2}{c^2}} \approx 34.6 \text{ yr}$$

The unit of measurement in each case is years. Thus, the observer on Earth perceives the time of travel to be 40 yr, whereas to an observer on the spaceship the time of travel is only 34.6 yr.

4. Solve Applications Involving Rate of Work Done

EXAMPLE 4   Solving an Application Involving "Work" Rates

At a mail-order company, Derrick can process 100 orders in 4 hr. Miguel can process 100 orders in 3 hr.

a. How long would it take them to process 100 orders if they work together?

b. How long would it take them to process 1400 orders if they work together?

Solution:

a. Let t represent the amount of time required to process 100 orders working together.

One method to approach this problem is to add the rates of speed at which each person works.

$$\left(\begin{array}{c}\text{Derrick's}\\\text{speed}\end{array}\right) + \left(\begin{array}{c}\text{Miguel's}\\\text{speed}\end{array}\right) = \left(\begin{array}{c}\text{Speed working}\\\text{together}\end{array}\right)$$

$$\frac{1 \text{ job}}{4 \text{ hr}} + \frac{1 \text{ job}}{3 \text{ hr}} = \frac{1 \text{ job}}{t \text{ hr}} \qquad 1 \text{ job} = 100 \text{ orders.}$$

$$\frac{1}{4} + \frac{1}{3} = \frac{1}{t}$$

$$12t\left(\frac{1}{4} + \frac{1}{3}\right) = 12t\left(\frac{1}{t}\right) \qquad \text{Multiply both sides by the LCD, } 12t.$$

$$3t + 4t = 12 \qquad \text{Apply the distributive properly.}$$

$$7t = 12$$

$$t = \frac{12}{7} \quad \text{or} \quad 1\frac{5}{7}$$

Derrick and Miguel can process 100 orders in $1\frac{5}{7}$ hr working together.

b. The time required to process 1400 orders is 14 times as long as the time to process 100 orders. $\left(\frac{12}{7} \text{ hr}\right)(14) = 24$ hr.

Skill Practice 4   Sheldon and Penny were awarded a contract to paint 16 offices in the new math building at a university. Once all the preparation work is complete, Sheldon can paint an office in 30 min and Penny can paint an office in 45 min.

   a. How long would it take them to paint one office working together?

   b. How long would it take them to paint all 16 offices?

5. Solve Applications Involving Proportions

An equation that equates two ratios or rates is called a **proportion.** Symbolically, we define a proportion as an equation of the form

$$\frac{a}{b} = \frac{c}{d}, \text{ where } b \neq 0 \text{ and } d \neq 0$$

The method of clearing fractions can be used to solve proportions.

EXAMPLE 5   **Solving an Application Involving a Proportion**

In a jury pool, there are 8 more men than women. If the ratio of men to women is 8 to 7, determine the number of men and women in the pool.

Solution:

Let x represent the number of women.   Label the variables.
Then $x + 8$ represents the number of men.

$$\text{number of men} \longrightarrow \frac{x + 8}{x} = \frac{8}{7} \longleftarrow \text{men}$$
$$\text{number of women} \longrightarrow \qquad\qquad \longleftarrow \text{women}$$
Set up a proportion.

$$7x\left(\frac{x + 8}{x}\right) = 7x\left(\frac{8}{7}\right)$$ Multiply by the LCD, $7x$.

$$7(x + 8) = x(8)$$

$$7x + 56 = 8x$$ Apply the distributive property.

$$56 = x$$

The number of women is 56. The number of men is $56 + 8 = 64$.

Answers

4. a. It would take 18 min to paint one office working together.
  b. It would take 288 min (4 hr 48 min) to paint 16 offices working together.
5. There were 52 Republicans and 48 Democrats.

Skill Practice 5   For the 104th Congress, there were 4 more Republicans than Democrats in the U.S. Senate. This resulted in a ratio of 13 Republicans to 12 Democrats. How many senators were Republican and how many were Democrat?

SECTION 1.2	Practice Exercises

Prerequisite Review

R.1. At a recent motorcycle rally, the number of men exceeded the number of women by 254. If x represents the number of men, write an expression for the number of women.

R.2. Rebecca downloaded five times as many songs as Nigel. If x represents the number of songs downloaded by Nigel, write an expression for the number of songs downloaded by Rebecca.

R.3. Sidney made $29 less than eight times Casey's weekly salary. If x represents Casey's weekly salary, write an expression for Sidney's weekly salary.

R.4. David scored 30 points more than six times the number of points Rich scored in a video game. If x represents the number of points scored by Rich, write an expression representing the number of points scored by David.

R.5. Write a proportion for $8.25 per hour is proportional to $82.50 per 10 hours.

Concept Connections

1. If $6000 is borrowed at 7.5% simple interest for 2 yr, then the amount of interest is _____.

2. Suppose that 8% of a solution is fertilizer by volume and the remaining 92% is water. How much fertilizer is in a 2-L bucket of solution?

3. If $d = rt$, then $t = \dfrac{\square}{\square}$

4. If $d = rt$, then $r = \dfrac{\square}{\square}$

5. The formula for the perimeter P of a rectangle with length l and width w is given by _____.

6. The sum of the measures of the angles inscribed inside a triangle is _____.

Objective 1: Solve Applications Involving Simple Interest

7. Rocco borrowed a total of $5000 from two student loans. One loan charged 3% simple interest and the other charged 2.5% simple interest, both payable after graduation. If the interest he owed after 1 yr was $132.50, determine the amount of principal for each loan. (**See Example 1**)

8. Laura borrowed a total of $22,000 from two different banks to start a business. One bank charged the equivalent of 4% simple interest, and the other charged 5.5% interest. If the total interest after 1 yr was $910, determine the amount borrowed from each bank.

9. Fernando invested money in a 3-yr CD (certificate of deposit) that returned the equivalent of 4.4% simple interest. He invested $2000 less in an 18-month CD that had a 3% return. If the total amount of interest from these investments was $706.50, determine how much was invested in each CD.

10. Ebony bought a 5-yr Treasury note that paid the equivalent of 2.8% simple interest. She invested $5000 more in a 10-yr bond earning 3.6% than she did in the Treasury note. If the total amount of interest from these investments was $5300, determine the amount of principal for each investment.

Objective 2: Solve Applications Involving Mixtures

11. Ethanol fuel mixtures have "E" numbers that indicate the percentage of ethanol in the mixture by volume. For example, E10 is a mixture of 10% ethanol and 90% gasoline. How much E5 should be mixed with 5000 gal of E10 to make an E9 mixture? (**See Example 2**)

12. A nurse mixes 60 cc of a 50% saline solution with a 10% saline solution to produce a 25% saline solution. How much of the 10% solution should he use?

13. The density and strength of concrete are determined by the ratio of cement and aggregate (aggregate is sand, gravel, or crushed stone). Suppose that a contractor has 480 ft^3 of a dry concrete mixture that is 70% sand by volume. How much pure sand must be added to form a new mixture that is 75% sand by volume?

14. Antifreeze is a compound added to water to reduce the freezing point of a mixture. In extreme cold (less than $-35°F$), one car manufacturer recommends that a mixture of 65% antifreeze be used. How much 50% antifreeze solution should be drained from a 4-gal tank and replaced with pure antifreeze to produce a 65% antifreeze mixture?

Objective 3: Solve Applications Involving Uniform Motion

15. Two passengers leave the airport at Kansas City, Missouri. One flies to Los Angeles, California, in 3.4 hr and the other flies in the opposite direction to New York City in 2.4 hr. With prevailing westerly winds, the speed of the plane to New York City is 60 mph faster than the speed of the plane to Los Angeles. If the total distance traveled by both planes is 2464 mi, determine the average speed of each plane. (**See Example 3**)

16. Two planes leave from Atlanta, Georgia. One makes a 5.2-hr flight to Seattle, Washington, and the other makes a 2.5-hr flight to Boston, Massachusetts. The plane to Boston averages 44 mph slower than the plane to Seattle. If the total distance traveled by both planes is 3124 mi, determine the average speed of each plane.

17. Darren drives to school in rush hour traffic and averages 32 mph. He returns home in mid-afternoon when there is less traffic and averages 48 mph. What is the distance between his home and school if the total traveling time is 1 hr 15 min?

18. Peggy competes in a biathlon by running and bicycling around a large loop through a city. She runs the loop one time and bicycles the loop five times. She can run 8 mph and she can ride 16 mph. If the total time it takes her to complete the race is 1 hr 45 min, determine the distance of the loop.

Objective 4: Solve Applications Involving Rate of Work Done

19. Joel can run around a $\frac{1}{4}$-mi track in 66 sec, and Jason can run around the track in 60 sec. If the runners start at the same point on the track and run in opposite directions, how long will it take the runners to cover $\frac{1}{4}$ mi? (**See Example 4**)

20. Marta can vacuum the house in 40 min. It takes her daughter 1 hr to vacuum the house. How long would it take them if they worked together?

21. One pump can fill a pool in 10 hr. Working with a second slower pump, the two pumps together can fill the pool in 6 hr. How fast can the second pump fill the pool by itself?

22. Brad and Angelina can mow their yard together with two lawn mowers in 30 min. When Brad works alone, it takes him 50 min. How long would it take Angelina to mow the lawn by herself?

Objective 5: Solve Applications Involving Proportions

23. At a construction site, cement, sand, and gravel are mixed to make concrete. The ratio of cement to sand to gravel is 1 to 2.4 to 3.6. If a 150-lb bag of sand is used, how much cement and gravel must be used? (**See Example 5**)

24. The property tax on a $180,000 house is $1296. At this rate, what is the property tax on a house that is $240,000?

25. In addition to measuring a person's individual HDL and LDL cholesterol levels, doctors also compute the ratio of total cholesterol to HDL cholesterol. Doctors recommend that the ratio of total cholesterol to HDL cholesterol be kept under 4. Suppose that the ratio of a patient's total cholesterol to HDL is 3.4 and her HDL is 60 mg/dL. Determine the patient's LDL level and total cholesterol. (Assume that total cholesterol is the sum of the LDL and HDL levels.)

26. For a recent Congress, there were 10 more Democrats than Republicans in the U.S. Senate. This resulted in a ratio of 11 Democrats to 9 Republicans. How many senators were Democrat and how many were Republican?

27. When studying wildlife populations, biologists sometimes use a technique called "mark-recapture." For example, a researcher captured and tagged 30 deer in a wildlife management area. Several months later, the researcher observed a new sample of 80 deer and determined that 5 were tagged. What is the total number of deer in the population?

28. To estimate the number of bass in a lake, a biologist catches and tags 24 bass. Several weeks later, the biologist catches a new sample of 40 bass and finds that 4 are tagged. How many bass are in the lake?

Mixed Exercises

29. Seismographs can record two types of wave energy (P waves and S waves) that travel through the Earth after an earthquake. Traveling through granite, P waves travel approximately 5 km/sec and S waves travel approximately 3 km/sec. If a geologist working at a seismic station measures a time difference of 40 sec between an earthquake's P waves and S waves, how far from the epicenter of the earthquake is the station?

30. Suppose that a shallow earthquake occurs in which the P waves travel 8 km/sec and the S waves travel 4.8 km/sec. If a seismologist measures a time difference of 20 sec between the arrival of the P waves and the S waves, how far is the seismologist from the epicenter of the earthquake?

31. Suppose that a merchant buys a patio set from the wholesaler for $180. At what price should the merchant mark the patio set so that it may be offered at a discount of 25% but still give the merchant a 40% profit on his $180 investment?

32. Suppose that a bookstore buys a textbook from the publisher for $80. At what price should the bookstore mark the textbook so that it may be offered at a discount of 10% but still give the bookstore a 35% profit on the $80 investment?

33. Henri needs to have a toilet repaired in his house. The cost of the new plumbing fixtures is $110 and labor is $60/hr.

    a. Write a model that represents the cost of the repair C (in $) in terms of the number of hours of labor x.

    b. After how many hours of labor would the cost of the repair job equal the cost of a new toilet of $350?

34. After a hurricane, repairs to a roof will cost $2400 for materials and $80/hr in labor.

    a. Write a model that represents the cost of the repair C (in $) in terms of the number of hours of labor x.

    b. If an estimate for a new roof is $5520, after how many hours of labor would the cost to repair the roof equal the cost of a new roof?

35. A student 5 ft tall measures the length of the shadow of the Washington Monument to be 444 ft. At the same time, her shadow is 4 ft. Approximate the height of the Washington Monument.

36. A 6-ft man is standing 40 ft from a light post. If the man's shadow is 20 ft, determine the height of the light post.

37. A vertical pole is placed in the ground at a campsite outside Salt Lake City, Utah. One winter day, $\frac{1}{8}$ of the pole is in the ground, $\frac{2}{3}$ of the pole is covered in snow, and 1.5 ft is above the snow. How long is the pole, and how deep is the snow?

38. The formula to convert temperature in Fahrenheit F to temperature in Celsius C is given by $C = \frac{5}{9}(F - 32)$. Determine the temperature at which the Celsius and Fahrenheit temperature readings are the same.

39. A tank contains 40 L of a mixture of plant fertilizer and water in which 20% of the mixture is fertilizer. How much of the mixture should be drained and replaced by an equal amount of water to dilute the mixture to 15% fertilizer?

40. How much water must be evaporated from 200 mL of a 5% salt solution to produce a 25% salt solution?

41. The perimeter of a rectangular lot of land is 440 ft. This includes an easement of x feet of uniform width inside the lot on which no building can be done. If the buildable area is 128 ft by 60 ft, determine the width of the easement.

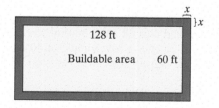

42. The Arthur Ashe Stadium tennis court is center court to the U.S. Open tennis tournament. The dimensions of the court are 78 ft by 36 ft, with a uniform border of x feet around the outside for additional play area. If the perimeter of the entire play area is 396 ft, determine the value of x.

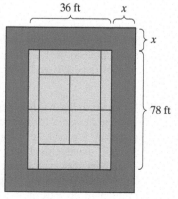

43. A contractor must tile a rectangular kitchen that is 4 ft longer than it is wide, and the perimeter of the kitchen is 48 ft.

    a. Find the dimensions of the kitchen.

    b. How many square feet of tile should be ordered if the contractor adds an additional 10% to account for waste?

    c. Determine the total cost if the tile costs $12/ft^2 and sales tax is 6%.

44. Max and Molly plan to put down all-weather carpeting on their porch. The length of the porch is 2 ft longer than twice the width, and the perimeter is 64 ft.

    a. Find the dimensions of the porch.

    b. How many square feet of carpeting should they buy if they add an additional 10% for waste?

    c. Determine the total cost if the carpeting costs $5.85/ft^2 and sales tax is 7.5%.

45. Aliyah earned an $8000 bonus from her sales job for exceeding her sales goals. After paying taxes at a 28% rate, she invested the remaining money in two stocks. One stock returned the equivalent of 11% simple interest after 1 yr, and the other returned 5% at the end of 1 yr. If her investments returned $453.60 (excluding commissions) how much did she invest in each stock?

46. Caitlin invested money in two mutual funds—a stock fund and a balanced fund. She invested twice as much in the stock fund as in the balanced fund. At the end of 1 yr, the stock fund earned the equivalent of 17% simple interest and the balanced fund earned 3.5%. If her total gain was $1125, determine how much she invested in each fund.

Proportions are used in geometry with similar triangles. If two triangles are similar, then the lengths of the corresponding sides are proportional.

For the similar triangles shown, $\dfrac{a}{x} = \dfrac{b}{y} = \dfrac{c}{z}$.

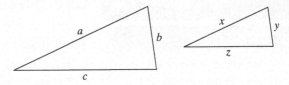

For Exercises 47–48, the triangles are similar with the corresponding sides oriented as shown. Solve for x and y.

47.

48.

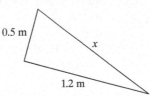

Write About It

49. Is it possible for the measures of the angles in a triangle to be represented by three consecutive odd integers? Explain.

50. Bob wants to change a $100 bill into an equal number of $20 bills, $10 bills, and $5 bills. Is this possible? Explain.

Expanding Your Skills

51. One number is 16 more than another number. The quotient of the larger number and smaller number is 3 and the remainder is 2. Find the numbers.

52. One number is 25 more than another number. The quotient of the larger number and the smaller number is 4 and the remainder is 1. Find the numbers.

53. The sum of the digits of a two-digit number is 14. If the digits are reversed, the new number is 18 more than the original number. Determine the original number.

54. The sum of the digits of a two-digit number is 9. If the digits are reversed, the new number is 45 less than the original number. Determine the original number.

Consider a seesaw with two children of masses m_1 and m_2 on either side. Suppose that the position of the fulcrum (pivot point) is labeled as the origin, $x = 0$. Further suppose that the position of each child relative to the origin is x_1 and x_2, respectively. The seesaw will be in equilibrium if $m_1x_1 + m_2x_2 = 0$. Use this equation for Exercises 55–58.

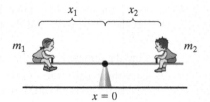

55. Find x_2 so that the system of masses is in equilibrium.

$m_1 = 30$ kg, $x_1 = -1.2$ m and $m_2 = 20$ kg, $x_2 = ?$

56. Find x_1 so that the system of masses is in equilibrium.

$m_1 = 64$ kg, $x_1 = ?$ and $m_2 = 80$ kg, $x_2 = 2$ m

57. Find the missing mass so that the system is in equilibrium. (*Hint:* Recall that positions to the left of 0 on the number line are negative.)

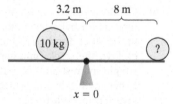

58. Find the missing mass so that the system is in equilibrium.

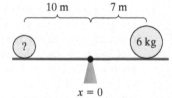

SECTION 1.3 | Complex Numbers

OBJECTIVES

1. **Simplify Imaginary Numbers**
2. **Write Complex Numbers in the Form** $a + bi$
3. **Perform Operations on Complex Numbers**

1. Simplify Imaginary Numbers

In our study of algebra thus far, we have worked exclusively with real numbers. However, as we encounter new types of equations, we need to look outside the set of real numbers to find solutions. For example, the equation $x^2 = 1$ has two solutions: 1 and -1. But what about the equation $x^2 = -1$? There is no real number x for which $x^2 = -1$. For this reason, mathematicians defined a new number i such that $i^2 = -1$. The number i is called an *imaginary number* and is used to represent $\sqrt{-1}$. Furthermore, the square root of any negative real number is an imaginary number that can be expressed in terms of i.

> **The Imaginary Number i**
>
> - $i = \sqrt{-1}$ and $i^2 = -1$
> - If b is a positive real number, then $\sqrt{-b} = i\sqrt{b}$.

> **EXAMPLE 1** Writing Imaginary Numbers in Terms of i
>
> Write each expression in terms of i.
>
> **a.** $\sqrt{-25}$ **b.** $\sqrt{-12}$ **c.** $\sqrt{-13}$
>
> **Solution:**
>
> **a.** $\sqrt{-25} = i\sqrt{25} = 5i$
>
> **b.** $\sqrt{-12} = i\sqrt{12} = i \cdot 2\sqrt{3}$ The value $i \cdot 2\sqrt{3}$ can be written as $2i\sqrt{3}$ or as
> $\quad\quad\quad = 2i\sqrt{3}$ or $2\sqrt{3}i$ $2\sqrt{3}i$. Note, however, that the factor i is written *outside* the radical.
>
> **c.** $\sqrt{-13} = i\sqrt{13}$ or $\sqrt{13}i$

> **Skill Practice 1** Write each expression in terms of i.
>
> **a.** $\sqrt{-81}$ **b.** $\sqrt{-50}$ **c.** $\sqrt{-11}$

Answers

1. a. $9i$ **b.** $5i\sqrt{2}$ **c.** $i\sqrt{11}$

In Example 2, we multiply and divide the square roots of negative real numbers. However, note that the multiplication and division properties of radicals can be used only if the radicals represent real-valued expressions.

$$\sqrt{a} \cdot \sqrt{b} = \sqrt{ab} \qquad \text{provided that the roots represent real numbers.}$$

$$\frac{\sqrt{a}}{\sqrt{b}} = \sqrt{\frac{a}{b}} \qquad \text{provided that the roots represent real numbers.}$$

For this reason, in Example 2 it is important to write the radical expressions in terms of i first, before applying the multiplication or division property of radicals.

EXAMPLE 2   **Simplifying Expressions in Terms of i**

Multiply or divide as indicated.

a. $\sqrt{-9} \cdot \sqrt{-25}$ **b.** $\sqrt{-15} \cdot \sqrt{-3}$ **c.** $\dfrac{\sqrt{-50}}{\sqrt{-2}}$

Solution:

a. $\sqrt{-9} \cdot \sqrt{-25} = i\sqrt{9} \cdot i\sqrt{25}$ — Write each radical in terms of i first, *before* multiplying.

$= 3i \cdot 5i$ — Simplify the radicals.

$= 15i^2$ — Multiply.

$= 15(-1)$ — By definition, $i^2 = -1$.

$= -15$

b. $\sqrt{-15} \cdot \sqrt{-3} = i\sqrt{15} \cdot i\sqrt{3}$ — Write each radical in terms of i first, *before* multiplying.

$= i^2\sqrt{45}$ — Apply the multiplication property of radicals.

$= (-1)\sqrt{3^2 \cdot 5}$ — Simplify $i^2 = -1$.

$= -3\sqrt{5}$

c. $\dfrac{\sqrt{-50}}{\sqrt{-2}} = \dfrac{i\sqrt{50}}{i\sqrt{2}}$ — Write each radical in terms of i first, *before* dividing.

$= \dfrac{\overset{1}{i}\sqrt{50}}{i\sqrt{2}}$ — Simplify the ratio of common factors to 1.

$= \sqrt{\dfrac{50}{2}}$ — Apply the division property of radicals.

$= \sqrt{25} = 5$ — Simplify.

Skill Practice 2   Multiply or divide as indicated.

a. $\sqrt{-16} \cdot \sqrt{-49}$ **b.** $\sqrt{-10}\sqrt{-2}$ **c.** $\dfrac{\sqrt{-48}}{\sqrt{-3}}$

2. Write Complex Numbers in the Form $a + bi$

We now define a new set of numbers that includes the real numbers and imaginary numbers. This is called the set of complex numbers.

Answers

2. a. -28 **b.** $-2\sqrt{5}$ **c.** 4

Complex Numbers

Given real numbers a and b, a number written in the form $a + bi$ is called a **complex number.** The value a is called the **real part** of the complex number and the value b is called the **imaginary part.**

$$\boxed{\text{Real part: 5}} \quad \boxed{\text{Imaginary part: } -7}$$

$$5 - 7i = 5 + (-7)i$$

Notes	Examples
• If $b = 0$, then $a + bi$ equals the real number a. This tells us that all real numbers are complex numbers.	$4 + 0i$ is generally written as the real number 4.
• If $a = 0$ and $b \neq 0$, then $a + bi$ equals bi, which we say is **pure imaginary.**	The number $0 + 8i$ is a pure imaginary number and is generally written as simply $8i$.

A complex number written in the form $a + bi$ is said to be in **standard form.** That being said, we sometimes write $a - bi$ in place of $a + (-b)i$. Furthermore, a number such as $5 + \sqrt{3}i$ is sometimes written as $5 + i\sqrt{3}$ to emphasize that the factor of i is not under the radical. In Example 3, we practice writing complex numbers in standard form.

EXAMPLE 3   **Writing Complex Numbers in Standard Form**

Simplify each expression and write the result in the form $a + bi$.

a. $3 - \sqrt{-100}$ **b.** $\dfrac{2 + 7i}{5}$ **c.** $\dfrac{-6 + \sqrt{-18}}{9}$

Solution:

a. $3 - \sqrt{-100} = 3 - 10i$ Simplify the expression.
$$\phantom{3 - \sqrt{-100}} = 3 + (-10)i$$ Although $3 + (-10)i$ is written in standard form, $3 - 10i$ is also acceptable.

b. $\dfrac{2 + 7i}{5} = \dfrac{2}{5} + \dfrac{7}{5}i$ Write the fraction as two separate terms.

c. $\dfrac{-6 + \sqrt{-18}}{9} = \dfrac{-6 + 3i\sqrt{2}}{9}$ Simplify the radical.
$$\sqrt{-18} = i\sqrt{18} = i\sqrt{3^2 \cdot 2} = 3i\sqrt{2}$$

$$\phantom{\dfrac{-6 + \sqrt{-18}}{9}} = \dfrac{-6}{9} + \dfrac{3i\sqrt{2}}{9}$$ Write the fraction as two separate terms.

$$\phantom{\dfrac{-6 + \sqrt{-18}}{9}} = -\dfrac{2}{3} + \dfrac{\sqrt{2}}{3}i$$ Simplify each fraction and write the result in the form $a + bi$.

Skill Practice 3 Simplify each expression and write the result in the form $a + bi$.

a. $4 + \sqrt{-49}$ **b.** $\dfrac{3 - 8i}{7}$ **c.** $\dfrac{10 + \sqrt{-75}}{20}$

Answers

3. a. $4 + 7i$ **b.** $\dfrac{3}{7} + \left(-\dfrac{8}{7}\right)i$

c. $\dfrac{1}{2} + \dfrac{\sqrt{3}}{4}i$

3. Perform Operations on Complex Numbers

By definition, $i^2 = -1$, but what about other powers of i? Consider the following pattern.

TIP Notice that even powers of i simplify to 1 or -1.
- If the exponent is a multiple of 4, then the expression equals 1.
- If the exponent is even but *not* a multiple of 4, then the expression equals -1.

$$i^1 = i$$
$$i^2 = -1$$
$$i^3 = i^2 \cdot i = (-1)i = -i$$
$$i^4 = i^2 \cdot i^2 = (-1)(-1) = 1$$
$$i^5 = i^4 \cdot i = (1)i = i$$
$$i^6 = i^4 \cdot i^2 = (1)(-1) = -1$$
$$i^7 = i^4 \cdot i^2 \cdot i = (1)(-1)i = -i$$
$$i^8 = i^4 \cdot i^4 = (1)(1) = 1$$

$$i^1 = i$$
$$i^2 = -1$$
$$i^3 = -i$$
$$i^4 = 1$$
Pattern: $i, -1, -i, 1$

$$i^5 = i$$
$$i^6 = -1$$
$$i^7 = -i$$
$$i^8 = 1$$
Pattern repeats: $i, -1, -i, 1, \ldots$

Notice that the fourth powers of i (i^4, i^8, i^{12}, ...) equal the real number 1. For other powers of i, we can write the expression as a product of a fourth power of i and a factor of i, i^2, or i^3, which equals i, -1, or $-i$, respectively.

EXAMPLE 4   **Simplifying Powers of *i***

Simplify.

a. i^{48} **b.** i^{23} **c.** i^{50} **d.** i^{-19}

Solution:

a. $i^{48} = (i^4)^{12} = (1)^{12} = 1$ The exponent 48 is a multiple of 4. Thus, i^{48} is equal to 1.

b. $i^{23} = i^{20} \cdot i^3$ Write i^{23} as a product of the largest fourth power of i and a remaining factor.
$$= (1) \cdot i^3 = -i$$

c. $i^{50} = i^{48} \cdot i^2$
$$= (1)(-1) = -1$$

d. $i^{-19} = i^{-20} \cdot i^1$
$$= (i^4)^{-5} \cdot i = (1) \cdot i = i$$

TIP To simplify i^n, divide the exponent, n, by 4. The remainder is the exponent of the remaining factor of i once the fourth power of i has been extracted.

Example: $i^{50} = i^{48} \cdot i^2 = (1)i^2$

$$\begin{array}{r} 12 \\ 4\overline{)50} \\ 48 \\ \hline 2 \end{array}$$

So $i^{50} = (1) \cdot i^2 = -1$

Skill Practice 4   Simplify.

a. i^{13} **b.** i^{103} **c.** i^{64} **d.** i^{-30}

To add or subtract complex numbers, add or subtract their real parts, and add or subtract their imaginary parts. That is,

$$(a + bi) + (c + di) = (a + c) + (b + d)i$$
$$(a + bi) - (c + di) = (a - c) + (b - d)i$$

EXAMPLE 5   **Adding and Subtracting Complex Numbers**

Add or subtract as indicated. Write the answer in the form $a + bi$.

a. $(-2 - 4i) + (5 + 2i) - (3 - 6i)$

b. Subtract $\left(\dfrac{1}{2} + \dfrac{2}{3}i\right)$ from $\left(\dfrac{3}{4} + \dfrac{9}{5}i\right)$.

Answers
4. a. i **b.** $-i$ **c.** 1   **d.** -1

Solution:

a. $(-2 - 4i) + (5 + 2i) - (3 - 6i)$ Combine the real parts and combine the
$= (-2 + 5 - 3) + [-4 + 2 - (-6)]i$ imaginary parts.
$= 0 + 4i$ Write the result in the form $a + bi$.

b. Subtract $\left(\dfrac{1}{2} + \dfrac{2}{3}i\right)$ from $\left(\dfrac{3}{4} + \dfrac{9}{5}i\right)$.

$\left(\dfrac{3}{4} + \dfrac{9}{5}i\right) - \left(\dfrac{1}{2} + \dfrac{2}{3}i\right)$ The statement "subtract x from y" is
equivalent to $y - x$. The order is
important.

$= \left(\dfrac{3}{4} - \dfrac{1}{2}\right) + \left(\dfrac{9}{5} - \dfrac{2}{3}\right)i$ Subtract the real parts. Subtract the
imaginary parts.

$= \left(\dfrac{3}{4} - \dfrac{2}{4}\right) + \left(\dfrac{27}{15} - \dfrac{10}{15}\right)i$ Write using common denominators.

$= \dfrac{1}{4} + \dfrac{17}{15}i$ Write the result in the form $a + bi$.

> **Skill Practice 5**   Add or subtract as indicated. Write the answer in the form $a + bi$.
>
> **a.** $(8 - 3i) - (2 + 4i) + (5 + 7i)$ **b.** Subtract $\left(\dfrac{1}{10} + \dfrac{1}{3}i\right)$ from $\left(\dfrac{3}{5} + \dfrac{5}{6}i\right)$

In Examples 6 and 7, we multiply complex numbers using a process similar to multiplying polynomials.

> ### EXAMPLE 6   Multiplying Complex Numbers
>
> Multiply. Write the results in the form $a + bi$.
>
> **a.** $-\dfrac{1}{2}i(4 + 6i)$ **b.** $(-2 + 6i)(4 - 3i)$

Solution:

a. $-\dfrac{1}{2}i(4 + 6i) = -2i - 3i^2$ Apply the distributive property.

$= -2i - 3(-1)$ Recall that $i^2 = -1$.
$= 3 - 2i$ or $3 + (-2)i$ Write the result in the form $a + bi$.

b. $(-2 + 6i)(4 - 3i)$ Apply the distributive property.

$= -2(4) + (-2)(-3i) + 6i(4) + 6i(-3i)$
$= -8 + 6i + 24i - 18i^2$
$= -8 + 30i - 18(-1)$ Recall that $i^2 = -1$.
$= -8 + 30i + 18$
$= 10 + 30i$ Write the result in the form $a + bi$.

> **Skill Practice 6**   Multiply. Write the result in the form $a + bi$.
>
> **a.** $-\dfrac{1}{3}i(9 - 15i)$ **b.** $(-5 + 4i)(3 - i)$

Answers

5. a. $11 + 0i$ **b.** $\dfrac{1}{2} + \dfrac{1}{2}i$

6. a. $-5 + (-3)i$ **b.** $-11 + 17i$

In Example 7, we make use of the special case products:

$$(a \pm b)^2 = a^2 \pm 2ab + b^2 \quad \text{and} \quad (a + b)(a - b) = a^2 - b^2$$

EXAMPLE 7   **Evaluating Special Products with Complex Numbers**

Multiply. Write the results in the form $a + bi$.

a. $(3 + 4i)^2$ **b.** $(5 + 2i)(5 - 2i)$

Solution:

a.
$$\begin{aligned}
(3 + 4i)^2 &= (3)^2 + 2(3)(4i) + (4i)^2 & \text{Apply the property}\\
&= 9 + 24i + 16i^2 & (a + b)^2 = a^2 + 2ab + b^2.\\
&= 9 + 24i + 16(-1)\\
&= 9 + 24i - 16\\
&= -7 + 24i & \text{Write the result in the form } a + bi.
\end{aligned}$$

b.
$$\begin{aligned}
(5 + 2i)(5 - 2i) &= (5)^2 - (2i)^2 & \text{Apply the property}\\
&= 25 - 4i^2 & (a + b)(a - b) = a^2 - b^2.\\
&= 25 - 4(-1)\\
&= 25 + 4\\
&= 29 \quad \text{or} \quad 29 + 0i & \text{Write the result in the form } a + bi.
\end{aligned}$$

Skill Practice 7   Multiply. Write the result in the form $a + bi$.

a. $(4 - 7i)^2$ **b.** $(10 - 3i)(10 + 3i)$

In Section R.4 we noted that the expressions $(a + b)$ and $(a - b)$ are conjugates. Similarly, the expressions $(a + bi)$ and $(a - bi)$ are called **complex conjugates.** Furthermore, as illustrated in Example 7(b), the product of complex conjugates is a real number.

$$\begin{aligned}
(a + bi)(a - bi) &= (a)^2 - (bi)^2\\
&= a^2 - b^2 i^2\\
&= a^2 - b^2(-1)\\
&= a^2 + b^2
\end{aligned}$$

Product of Complex Conjugates

If a and b are real numbers, then $(a + bi)(a - bi) = a^2 + b^2$.

Number	Standard Form	Conjugate	Product
$3 + 7i$	$3 + 7i$	$3 - 7i$	$(3 + 7i)(3 - 7i) = (3)^2 + (7)^2 = 58$
$\sqrt{-5}$	$0 + \sqrt{5}i$	$0 - \sqrt{5}i$	$(0 + \sqrt{5}i)(0 - \sqrt{5}i) = (0)^2 + (\sqrt{5})^2 = 5$

In Example 8, we demonstrate division of complex numbers such as $\frac{8 + 2i}{3 - 5i}$. The goal is to make the denominator a real number so that the quotient can be written in standard form $a + bi$. This can be accomplished by multiplying the denominator by its complex conjugate. Of course, this means that we must also multiply the numerator by the same quantity.

Answers
7. a. $-33 + (-56)i$ **b.** 109

EXAMPLE 8 Dividing Complex Numbers

Divide. Write the result in the form $a + bi$.

a. $\dfrac{8 + 2i}{3 - 5i}$ **b.** $\left(2 + \sqrt{3}i\right)^{-1}$ **c.** $\dfrac{-2}{5i}$

Solution:

a. $\dfrac{8 + 2i}{3 - 5i} = \dfrac{(8 + 2i) \cdot (3 + 5i)}{(3 - 5i) \cdot (3 + 5i)}$ Multiply numerator and denominator by the conjugate of the denominator.

$= \dfrac{24 + 40i + 6i + 10i^2}{(3)^2 + (5)^2}$ Apply the distributive property in the numerator.
Multiply conjugates in the denominator.

$= \dfrac{24 + 46i + 10(-1)}{9 + 25}$ Replace i^2 by -1.

$= \dfrac{14 + 46i}{34}$

$= \dfrac{14}{34} + \dfrac{46}{34}i = \dfrac{7}{17} + \dfrac{23}{17}i$ Write the result in the form $a + bi$.

TIP In Example 8(b) we left the answer as $\frac{2}{7} - \frac{\sqrt{3}}{7}i$ rather than as $\frac{2}{7} + \left(-\frac{\sqrt{3}}{7}\right)i$ because the expression written using addition is more cumbersome. Both answers are acceptable.

b. $\left(2 + \sqrt{3}i\right)^{-1} = \dfrac{1 \cdot \left(2 - \sqrt{3}i\right)}{\left(2 + \sqrt{3}i\right) \cdot \left(2 - \sqrt{3}i\right)}$ Multiply numerator and denominator by the conjugate of the denominator.

$= \dfrac{2 - \sqrt{3}i}{(2)^2 + \left(\sqrt{3}\right)^2}$

$= \dfrac{2 - \sqrt{3}i}{4 + 3} = \dfrac{2}{7} - \dfrac{\sqrt{3}}{7}i$ Simplify.

c. $\dfrac{-2}{5i} = \dfrac{-2 \cdot i}{5i \cdot i}$ In this example, it is sufficient to multiply numerator and denominator by i (rather than by the conjugate $-5i$) to produce a real number in the denominator.

$= \dfrac{-2i}{5i^2} = \dfrac{-2i}{5(-1)} = \dfrac{-2i}{-5} = \dfrac{2}{5}i$

$= 0 + \dfrac{2}{5}i$ Write the result in the form $a + bi$.

Skill Practice 8 Divide. Write the result in the form $a + bi$.

a. $\dfrac{5 + 6i}{2 - 7i}$ **b.** $\left(5 + \sqrt{7}i\right)^{-1}$ **c.** $\dfrac{-7}{10i}$

TECHNOLOGY CONNECTIONS

Operations on Complex Numbers

Most graphing calculators and some scientific calculators can perform operations on complex numbers. A graphing calculator may have two different modes: one for operations over the set of real numbers and one for operations over the set of complex numbers. Choose the "$a + bi$" mode on your calculator. Then evaluate the expressions.

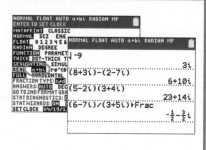

Answers

8. a. $-\dfrac{32}{53} + \dfrac{47}{53}i$ **b.** $\dfrac{5}{32} - \dfrac{\sqrt{7}}{32}i$

c. $0 + \dfrac{7}{10}i$

SECTION 1.3 | Practice Exercises

Prerequisite Review

R.1. Simplify the radical. $\sqrt{28}$

R.2. Simplify the radical. $4\sqrt{150}$

R.3. Simplify the radical. $\dfrac{7\sqrt{48}}{16}$

R.4. Multiply. $(p + \sqrt{3})(p - \sqrt{3})$

R.5. Rationalize the denominator. $\dfrac{9}{\sqrt{18}}$

R.6. Rationalize the denominator. $\dfrac{-33}{\sqrt{5} - 4}$

Concept Connections

1. The imaginary number i is defined so that $i = \sqrt{-1}$ and $i^2 =$ _____.

2. For a positive real number b, the value $\sqrt{-b} =$ _____.

3. Given a complex number $a + bi$, the value of a is called the _____ part and the value of b is called the _____ part.

4. Given a complex number $a + bi$, the expression $a - bi$ is called the complex _____.

Objective 1: Simplify Imaginary Numbers

For Exercises 5–22, write each expression in terms of i and simplify. (See Examples 1–2)

5. $\sqrt{-121}$

6. $\sqrt{-100}$

7. $\sqrt{-98}$

8. $\sqrt{-63}$

9. $\sqrt{-19}$

10. $\sqrt{-23}$

11. $-\sqrt{-16}$

12. $-\sqrt{-25}$

13. $\sqrt{-4}\sqrt{-9}$

14. $\sqrt{-1}\sqrt{-36}$

15. $\sqrt{-10}\sqrt{-5}$

16. $\sqrt{-6}\sqrt{-15}$

17. $\sqrt{-6}\sqrt{-14}$

18. $\sqrt{-10}\sqrt{-15}$

19. $\dfrac{\sqrt{-98}}{\sqrt{-2}}$

20. $\dfrac{\sqrt{-45}}{\sqrt{-5}}$

21. $\dfrac{\sqrt{-63}}{\sqrt{7}}$

22. $\dfrac{\sqrt{-80}}{\sqrt{5}}$

Objective 2: Write Complex Numbers in the Form $a + bi$

For Exercises 23–28, determine the real and imaginary parts of the complex number.

23. $3 - 7i$

24. $2 - 4i$

25. $19i$

26. $40i$

27. $-\dfrac{1}{4}$

28. $-\dfrac{4}{7}$

For Exercises 29–40, simplify each expression and write the result in standard form, $a + bi$. (See Example 3)

29. $4\sqrt{-4}$

30. $2\sqrt{-144}$

31. $2 + \sqrt{-12}$

32. $6 - \sqrt{-24}$

33. $\dfrac{8 + 3i}{14}$

34. $\dfrac{4 + 5i}{6}$

35. $\dfrac{-4 - 6i}{-2}$

36. $\dfrac{9 - 15i}{-3}$

37. $\dfrac{-18 + \sqrt{-48}}{4}$

38. $\dfrac{-20 + \sqrt{-50}}{-10}$

39. $\dfrac{14 - \sqrt{-98}}{-7}$

40. $\dfrac{-10 + \sqrt{-125}}{5}$

Objective 3: Perform Operations on Complex Numbers

For Exercises 41–44, simplify the powers of i. (See Example 4)

41. a. i^{20} **b.** i^{29} **c.** i^{50} **d.** i^{-41}

42. a. i^{32} **b.** i^{47} **c.** i^{66} **d.** i^{-27}

43. a. i^{37} **b.** i^{-37} **c.** i^{82} **d.** i^{-82}

44. a. i^{103} **b.** i^{-103} **c.** i^{52} **d.** i^{-52}

For Exercises 45–68, perform the indicated operations. Write the answers in standard form, $a + bi$. (See Examples 5–7)

45. $(2 - 7i) + (8 - 3i)$ **46.** $(6 - 10i) + (8 + 4i)$ **47.** $(15 + 21i) - (18 - 40i)$

48. $(250 + 100i) - (80 + 25i)$ **49.** $\left(\dfrac{1}{2} + \dfrac{2}{3}i\right) - \left(\dfrac{5}{6} + \dfrac{1}{12}i\right)$ **50.** $\left(\dfrac{3}{5} - \dfrac{1}{8}i\right) - \left(\dfrac{7}{10} + \dfrac{1}{6}i\right)$

51. $(2.3 + 4i) - (8.1 - 2.7i) + (4.6 - 6.7i)$ **52.** $(0.05 - 0.03i) + (-0.12 + 0.08i) - (0.07 + 0.05i)$

53. $-\dfrac{1}{8}(16 + 24i)$ **54.** $-\dfrac{1}{6}(60 - 30i)$ **55.** $2i(5 + i)$

56. $4i(6 + 5i)$ **57.** $\sqrt{-3}(\sqrt{11} - \sqrt{-7})$ **58.** $\sqrt{-2}(\sqrt{13} + \sqrt{-5})$

59. $(3 - 6i)(10 + i)$ **60.** $(2 - 5i)(8 + 2i)$ **61.** $(3 - 7i)^2$

62. $(10 - 3i)^2$ **63.** $(3 - \sqrt{-5})(4 + \sqrt{-5})$ **64.** $(2 + \sqrt{-7})(10 + \sqrt{-7})$

65. $4(6 + 2i) - 5i(3 - 7i)$ **66.** $-3(8 - 3i) - 6i(2 + i)$ **67.** $(2 - i)^2 + (2 + i)^2$

68. $(3 - 2i)^2 + (3 + 2i)^2$

For Exercises 69–72, for each given number, (a) identify the complex conjugate and (b) determine the product of the number and its conjugate.

69. $3 - 6i$ **70.** $4 - 5i$ **71.** $8i$ **72.** $9i$

For Exercises 73–88, perform the indicated operations. Write the answers in standard form, $a + bi$. (See Examples 7–8)

73. $(10 - 4i)(10 + 4i)$ **74.** $(3 - 9i)(3 + 9i)$ **75.** $(7i)(-7i)$

76. $(-5i)(5i)$ **77.** $(\sqrt{2} + \sqrt{3}i)(\sqrt{2} - \sqrt{3}i)$ **78.** $(\sqrt{5} + \sqrt{7}i)(\sqrt{5} - \sqrt{7}i)$

79. $\dfrac{6 + 2i}{3 - i}$ **80.** $\dfrac{5 + i}{4 - i}$ **81.** $\dfrac{8 - 5i}{13 + 2i}$

82. $\dfrac{10 - 3i}{11 + 4i}$ **83.** $(6 + \sqrt{5}i)^{-1}$ **84.** $(4 - \sqrt{3}i)^{-1}$

85. $\dfrac{5}{13i}$ **86.** $\dfrac{6}{7i}$ **87.** $\dfrac{-1}{\sqrt{-3}}$ **88.** $\dfrac{-2}{\sqrt{-11}}$

Mixed Exercises

For Exercises 89–92, evaluate $\sqrt{b^2 - 4ac}$ for the given values of a, b, and c, and simplify.

89. $a = 2$, $b = 4$, and $c = 6$ **90.** $a = 5$, $b = -5$, and $c = 10$

91. $a = 2$, $b = -6$, and $c = 5$ **92.** $a = 2$, $b = 4$, and $c = 4$

For Exercises 93–96, verify by substitution that the given values of x are solutions to the given equation.

93. $x^2 + 25 = 0$
  a. $x = 5i$
  b. $x = -5i$

94. $x^2 + 49 = 0$
  a. $x = 7i$
  b. $x = -7i$

95. $x^2 - 4x + 7 = 0$
  a. $x = 2 + i\sqrt{3}$
  b. $x = 2 - i\sqrt{3}$

96. $x^2 - 6x + 11 = 0$
  a. $x = 3 + i\sqrt{2}$
  b. $x = 3 - i\sqrt{2}$

97. Prove that $(a + bi)(c + di) = (ac - bd) + (ad + bc)i$.

98. Prove that $(a + bi)^2 = (a^2 - b^2) + (2ab)i$.

Write About It

99. Explain the flaw in the following logic.
$$\sqrt{-9} \cdot \sqrt{-4} = \sqrt{(-9)(-4)} = \sqrt{36} = 6$$

100. Discuss the difference between the products $(a + b)(a - b)$ and $(a + bi)(a - bi)$.

101. Give an example of a complex number that is its own conjugate.

102. Give an example of two complex numbers that are not real numbers, but whose product is a real number.

Expanding Your Skills

The variable z is often used to denote a complex number and $\bar{z}$ is used to denote its conjugate. If $z = a + bi$, simplify the expressions in Exercises 103–104.

103. $z \cdot \bar{z}$

104. $z^2 - \bar{z}^2$

For Exercises 105–110, factor the expressions over the set of complex numbers. For assistance, consider these examples.

- In Chapter R we saw that some expressions factor over the set of integers. For example: $x^2 - 4 = (x + 2)(x - 2)$.
- Some expressions factor over the set of irrational numbers. For example: $x^2 - 5 = (x + \sqrt{5})(x - \sqrt{5})$.
- To factor an expression such as $x^2 + 4$, we need to factor over the set of complex numbers. For example, verify that $x^2 + 4 = (x + 2i)(x - 2i)$.

105. a. $x^2 - 9$
b. $x^2 + 9$

106. a. $x^2 - 100$
b. $x^2 + 100$

107. a. $x^2 - 64$
b. $x^2 + 64$

108. a. $x^2 - 25$
b. $x^2 + 25$

109. a. $x^2 - 3$
b. $x^2 + 3$

110. a. $x^2 - 11$
b. $x^2 + 11$

Technology Connections

For Exercises 111–114, use a calculator to perform the indicated operations.

111. a. $\sqrt{-16}$
b. $(4 - 5i) - (2 + 3i)$
c. $(12 - 15i)(-2 + 9i)$

112. a. $\sqrt{-169}$
b. $(-11 - 2i) + (-4 + 9i)$
c. $(8 + 12i)(-3 - 7i)$

113. a. $(4 - 9i)^2$
b. $\dfrac{7}{2i}$
c. $\dfrac{14 + 8i}{3 - i}$

114. a. $(11 + 4i)^2$
b. $\dfrac{11}{10i}$
c. $\dfrac{5 + 7i}{6 + 8i}$

SECTION 1.4   Quadratic Equations

OBJECTIVES

1. Solve Quadratic Equations by Using the Zero Product Property
2. Solve Quadratic Equations by Using the Square Root Property
3. Complete the Square
4. Solve Quadratic Equations by Using the Quadratic Formula
5. Use the Discriminant
6. Solve an Equation for a Specified Variable

1. Solve Quadratic Equations by Using the Zero Product Property

A linear equation in one variable is an equation of the form $ax + b = 0$, where $a \neq 0$. A linear equation is also called a first-degree equation. We now turn our attention to a quadratic equation. This is identified as a second-degree equation.

> **Definition of a Quadratic Equation**
>
> Let a, b, and c represent real numbers, where $a \neq 0$. A **quadratic equation** in the variable x is an equation of the form
> $$ax^2 + bx + c = 0.$$

To solve a quadratic equation, we make use of the zero product property.

> **Zero Product Property**
>
> If $mn = 0$, then $m = 0$ or $n = 0$.

To solve a quadratic equation using the zero product property, set one side of the equation equal to zero and factor the other side.

EXAMPLE 1 Applying the Zero Product Property

Solve the equations.   **a.** $x^2 - 8x = 0$ **b.** $2x(2x - 7) = -12$

Solution:

a.

$$x^2 - 8x = 0 \qquad \text{One side of the equation is already zero.}$$
$$x(x - 8) = 0 \qquad \text{Factor the other side.}$$
$$x = 0 \text{ or } x - 8 = 0 \qquad \text{Set each factor equal to zero.}$$
$$x = 8 \qquad \underline{\text{Check:}} \quad x^2 - 8x = 0$$

$$(0)^2 - 8(0) \stackrel{?}{=} 0 \qquad (8)^2 - 8(8) \stackrel{?}{=} 0$$

The solution set is $\{0, 8\}$. $\qquad\qquad 0 - 0 = 0 \checkmark \qquad 64 - 64 = 0 \checkmark$

b.

$$2x(2x - 7) = -12$$
$$4x^2 - 14x = -12 \qquad \text{Apply the distributive property.}$$
$$4x^2 - 14x + 12 = 0 \qquad \text{Set one side to zero.}$$
$$2(2x^2 - 7x + 6) = 0 \qquad \text{Factor.}$$
$$2(x - 2)(2x - 3) = 0$$
$$2 = 0 \text{ or } x - 2 = 0 \text{ or } 2x - 3 = 0 \qquad \text{Set each factor equal to zero.}$$

$\uparrow$
contradiction $\qquad\qquad x = 2 \qquad\qquad x = \dfrac{3}{2} \qquad$ Both solutions check.

The solution set is $\left\{2, \dfrac{3}{2}\right\}$.

> **TIP** After applying the zero product property in Example 1(b), we have three equations. The first equation does not contain the variable x. It is a contradiction and does not yield a solution for x.

Skill Practice 1 Solve the equations.
a. $y^2 + 3y = 0$ **b.** $10(3x^2 - 13x) = -40$

---

2. Solve Quadratic Equations by Using the Square Root Property

The zero product property can be used to solve equations of the form $x^2 = k$.

$$x^2 = k$$
$$x^2 - k = 0$$
$$\left(x - \sqrt{k}\right)\left(x + \sqrt{k}\right) = 0$$
$$x - \sqrt{k} = 0 \quad \text{or} \quad x + \sqrt{k} = 0$$
$$x = \sqrt{k} \qquad\qquad x = -\sqrt{k}$$

The solutions can also be written as $x = \pm\sqrt{k}$. This is read as "x equals plus or minus the square root of k."

The solution set is $\left\{\pm\sqrt{k}\right\}$.

This result is formalized as the square root property.

Square Root Property

If $x^2 = k$, then $x = \pm\sqrt{k}$.

The solution set is $\left\{\sqrt{k}, -\sqrt{k}\right\}$ or more concisely $\left\{\pm\sqrt{k}\right\}$.

To apply the square root property to solve a quadratic equation, first isolate the square term on one side and the constant term on the other side.

Answers

1. a. $\{-3, 0\}$ **b.** $\left\{\dfrac{1}{3}, 4\right\}$

EXAMPLE 2   **Applying the Square Root Property**

Solve the equations by using the square root property.

a. $x^2 = 64$ **b.** $2y^2 + 36 = 0$ **c.** $(w + 3)^2 = 8$

Solution:

a. $x^2 = 64$

$x = \pm\sqrt{64}$ Apply the square root property.

$x = \pm 8$ Both solutions check in the original equation.

The solution set is $\{\pm 8\}$.

b. $2y^2 + 36 = 0$

$2y^2 = -36$ Isolate the square term.

$y^2 = -18$ Write the equation in the form $y^2 = k$.

$y = \pm\sqrt{-18}$ Apply the square root property.

$y = \pm 3i\sqrt{2}$ Simplify the radical.

$\sqrt{-18} = i\sqrt{18} = i\sqrt{3^2 \cdot 2} = 3i\sqrt{2}$

The solution set is $\{\pm 3i\sqrt{2}\}$.   Both solutions check in the original equation.

c. $(w + 3)^2 = 8$

$w + 3 = \pm\sqrt{8}$ Apply the square root property.

$w = -3 \pm\sqrt{8}$ Subtract 3 from both sides to isolate w.

$w = -3 \pm 2\sqrt{2}$ Simplify the radical. $\sqrt{8} = \sqrt{2^3} = 2\sqrt{2}$

The solution set is $\{-3 \pm 2\sqrt{2}\}$.   Both solutions check.

> **TIP** The solutions to the equation in Example 2(b) are written concisely as $\pm 3i\sqrt{2}$. Do not forget that this actually represents two solutions:
>
> $y = 3i\sqrt{2}$ and
> $y = -3i\sqrt{2}$

Skill Practice 2   Solve the equations by using the square root property.

a. $a^2 = 49$ **b.** $2c^2 + 80 = 0$ **c.** $(t + 4)^2 = 24$

Point of Interest

Unfortunate names? In the long history of mathematics, number systems have been expanded to accommodate meaningful solutions to equations. But the negative connotation of their names may suggest a reluctance by early mathematicians to accept these new concepts. Negative numbers for example are not unpleasant or disagreeable. Irrational numbers are not illogical or absurd, and imaginary numbers are not "fake." Instead these sets of numbers are necessary to render solutions to such equations as

$$2x + 10 = 0 \qquad x^2 - 5 = 0 \qquad x^2 + 4 = 0$$

3. Complete the Square

In Example 2(c), the left side of the equation is the square of a binomial and the right side is a constant. We can manipulate a quadratic equation $ax^2 + bx + c = 0$ ($a \neq 0$) to write it as the square of a binomial equal to a constant. First look at the relationship between a perfect square trinomial and its factored form.

Perfect Square Trinomial		**Factored Form**
$x^2 + 10x + 25$	$\longrightarrow$	$(x + 5)^2$
$t^2 - 6t + 9$	$\longrightarrow$	$(t - 3)^2$
$p^2 - 14p + 49$	$\longrightarrow$	$(p - 7)^2$

Answers

2. a. $\{\pm 7\}$ **b.** $\{\pm 2i\sqrt{10}\}$
   c. $\{-4 \pm 2\sqrt{6}\}$

For a perfect square trinomial with a leading coefficient of 1, the constant term is the square of one-half the linear term coefficient. For example:

$$x^2 + 10x + 25$$

$$\left[\tfrac{1}{2}(10)\right]^2$$

In general, an expression of the form $x^2 + bx + n$ is a perfect square trinomial if $n = \left(\tfrac{1}{2}b\right)^2$. The process to create a perfect square trinomial is called **completing the square.**

EXAMPLE 3   **Completing the Square**

Determine the value of n that makes the polynomial a perfect square trinomial. Then factor the result.

Solution:

Expression	Value of n	Complete the square
a. $x^2 + 18x + n$	$\left[\tfrac{1}{2}(18)\right]^2 = (9)^2 = 81$	$x^2 + 18x + 81 = (x + 9)^2$
b. $x^2 - 13x + n$	$\left[\tfrac{1}{2}(-13)\right]^2 = \left(-\tfrac{13}{2}\right)^2 = \tfrac{169}{4}$	$x^2 - 13x + \tfrac{169}{4} = \left(x - \tfrac{13}{2}\right)^2$
c. $x^2 + \tfrac{4}{7}x + n$	$\left[\tfrac{1}{2}\left(\tfrac{4}{7}\right)\right]^2 = \left(\tfrac{2}{7}\right)^2 = \tfrac{4}{49}$	$x^2 + \tfrac{4}{7}x + \tfrac{4}{49} = \left(x + \tfrac{2}{7}\right)^2$

TIP When factoring a perfect square trinomial, the constant term in the binomial will always be one-half the x term coefficient.

$$x^2 + 18x + 81$$
$$= (x + 9)^2$$

Note: $9 = \tfrac{1}{2}(18)$

Skill Practice 3   Determine the value of n that makes the polynomial a perfect square trinomial. Then factor the result.

a. $x^2 + 12x + n$ **b.** $x^2 + 5x + n$ **c.** $x^2 - \tfrac{2}{3}x + n$

We can solve a quadratic equation $ax^2 + bx + c = 0$ $(a \neq 0)$ by completing the square and then applying the square root property.

Solving a Quadratic Equation $ax^2 + bx + c = 0$ by Completing the Square and Applying the Square Root Property

Step 1 Divide both sides by a to make the leading coefficient 1.
Step 2 Isolate the variable terms on one side of the equation.
Step 3 Complete the square.
   • Add the square of one-half the linear term coefficient to both sides.
   • Factor the resulting perfect square trinomial.
Step 4 Apply the square root property and solve for x.

Answers

3. a. $n = 36; (x + 6)^2$
   b. $n = \dfrac{25}{4}; \left(x + \dfrac{5}{2}\right)^2$
   c. $n = \dfrac{1}{9}; \left(x - \dfrac{1}{3}\right)^2$

> **EXAMPLE 4**   **Completing the Square and Solving a Quadratic Equation**

Solve the equation by completing the square and applying the square root property. $x^2 - 3 = -10x$

Solution:

$x^2 - 3 = -10x$	Write the equation in the form $ax^2 + bx + c = 0$.
$x^2 + 10x - 3 = 0$	**Step 1:** Notice that the leading coefficient is already 1.
$x^2 + 10x + \underline{} = 3 + \underline{}$	**Step 2:** Add 3 to both sides to isolate the variable terms.
$x^2 + 10x + 25 = 3 + 25$	**Step 3:** Add $\left[\frac{1}{2}(10)\right]^2 = [5]^2 = 25$ to both sides.
$(x + 5)^2 = 28$	Factor.
$x + 5 = \pm\sqrt{28}$	**Step 4:** Apply the square root property and solve for x.
$x = -5 \pm 2\sqrt{7}$	Both solutions check in the original equation.
$\{-5 \pm 2\sqrt{7}\}$	Write the solution set.

> **Skill Practice 4**   Solve the equation by completing the square and applying the square root property. $x^2 - 2 = 8x$

In Example 5, we encounter a quadratic equation in which the leading coefficient is not 1. The first step is to divide both sides by the leading coefficient.

> **EXAMPLE 5**   **Completing the Square and Solving a Quadratic Equation**

Solve the equation by completing the square and applying the square root property. $-2x^2 - 3x - 5 = 0$

Solution:

$-2x^2 - 3x - 5 = 0$	The equation is in the form $ax^2 + bx + c = 0$.
$\dfrac{-2x^2}{-2} - \dfrac{3x}{-2} - \dfrac{5}{-2} = \dfrac{0}{-2}$	**Step 1:** Divide by the leading coefficient, -2.
$x^2 + \dfrac{3}{2}x + \dfrac{5}{2} = 0$	The new leading coefficient is 1.
$x^2 + \dfrac{3}{2}x + \underline{} = -\dfrac{5}{2} + \underline{}$	**Step 2:** Subtract $\frac{5}{2}$ from both sides to isolate the variable terms.
$x^2 + \dfrac{3}{2}x + \dfrac{9}{16} = -\dfrac{5}{2} + \dfrac{9}{16}$	**Step 3:** Add $\left[\frac{1}{2}\left(\frac{3}{2}\right)\right]^2 = \left[\frac{3}{4}\right]^2 = \frac{9}{16}$ to both sides.
$\left(x + \dfrac{3}{4}\right)^2 = -\dfrac{40}{16} + \dfrac{9}{16}$	Factor.
$\left(x + \dfrac{3}{4}\right)^2 = -\dfrac{31}{16}$	

Answer
4. $\{4 \pm 3\sqrt{2}\}$

$$x + \frac{3}{4} = \pm\sqrt{-\frac{31}{16}}$$

$$x = -\frac{3}{4} \pm \frac{\sqrt{31}}{4}i$$

$$\left\{-\frac{3}{4} \pm \frac{\sqrt{31}}{4}i\right\}$$

Step 4: Apply the square property and solve for x.

$$\sqrt{-\frac{31}{16}} = i\sqrt{\frac{31}{16}} = i\frac{\sqrt{31}}{\sqrt{16}} = \frac{\sqrt{31}}{4}i$$

The solutions both check in the original equation.

Skill Practice 5   Solve the equation by completing the square and applying the square root property. $-3x^2 + 5x - 7 = 0$

---

4. Solve Quadratic Equations by Using the Quadratic Formula

If we solve a general quadratic equation $ax^2 + bx + c = 0$ ($a \neq 0$) by completing the square and using the square root property, the result is a formula that gives the solutions for x in terms of a, b, and c.

$$ax^2 + bx + c = 0$$

Begin with a quadratic equation in standard form with $a > 0$.

$$\frac{ax^2}{a} + \frac{bx}{a} + \frac{c}{a} = \frac{0}{a}$$

Divide by the leading coefficient.

$$x^2 + \frac{b}{a}x + \frac{c}{a} = 0$$

$$x^2 + \frac{b}{a}x = -\frac{c}{a}$$

Isolate the terms containing x.

$$x^2 + \frac{b}{a}x + \left(\frac{1}{2} \cdot \frac{b}{a}\right)^2 = \left(\frac{1}{2} \cdot \frac{b}{a}\right)^2 - \frac{c}{a}$$

Add the square of $\frac{1}{2}$ the linear term coefficient to both sides of the equation.

$$\left(x + \frac{b}{2a}\right)^2 = \frac{b^2}{4a^2} - \frac{c}{a}$$

Factor the left side as a perfect square.

$$\left(x + \frac{b}{2a}\right)^2 = \frac{b^2 - 4ac}{4a^2}$$

Combine fractions on the right side by finding a common denominator.

$$x + \frac{b}{2a} = \pm\sqrt{\frac{b^2 - 4ac}{4a^2}}$$

Apply the square root property.

$$x + \frac{b}{2a} = \pm\frac{\sqrt{b^2 - 4ac}}{2a}$$

Simplify the denominator.

$$x = -\frac{b}{2a} \pm \frac{\sqrt{b^2 - 4ac}}{2a}$$

Subtract $\frac{b}{2a}$ from both sides.

$$= \frac{-b \pm \sqrt{b^2 - 4ac}}{2a}$$

Combine fractions.

The result is called the quadratic formula.

The Quadratic Formula

For a quadratic equation of the form $ax^2 + bx + c = 0$ ($a \neq 0$), the solutions are

$$x = \frac{-b \pm \sqrt{b^2 - 4ac}}{2a}$$

Answer

5. $\left\{\dfrac{5}{6} \pm \dfrac{\sqrt{59}}{6}i\right\}$

EXAMPLE 6   **Using the Quadratic Formula**

Solve the equation by applying the quadratic formula. $x(x - 6) = 3$

Solution:

$$x(x - 6) = 3$$
$$x^2 - 6x - 3 = 0$$

Write the equation in the form $ax^2 + bx + c = 0$.

$$a = 1, b = -6, c = -3$$

Identify the values of a, b, and c.

$$x = \frac{-(-6) \pm \sqrt{(-6)^2 - 4(1)(-3)}}{2(1)}$$

Apply the quadratic formula.
$$x = \frac{-b \pm \sqrt{b^2 - 4ac}}{2a}$$

$$= \frac{6 \pm \sqrt{48}}{2}$$

Simplify.

$$= \frac{6 \pm 4\sqrt{3}}{2}$$

Simplify the radical.
$$\sqrt{48} = \sqrt{2^4 \cdot 3} = 2^2\sqrt{3} = 4\sqrt{3}$$

$$= \frac{2(3 \pm 2\sqrt{3})}{2}$$

Factor the numerator.

$$= 3 \pm 2\sqrt{3}$$

Simplify the fraction.

The solutions both check in the original equation.

The solution set is $\{3 \pm 2\sqrt{3}\}$.

Skill Practice 6   Solve the equation by applying the quadratic formula.
$x(x - 8) = 3$

If a quadratic equation has fractional or decimal coefficients, we have the option of clearing fractions or decimals to create integer coefficients. This makes the application of the quadratic formula easier, as demonstrated in Example 7.

EXAMPLE 7   **Using the Quadratic Formula**

Solve the equation by applying the quadratic formula. $\dfrac{3}{10}x^2 - \dfrac{2}{5}x + \dfrac{7}{10} = 0$

Solution:

$$\frac{3}{10}x^2 - \frac{2}{5}x + \frac{7}{10} = 0$$

The equation is in the form $ax^2 + bx + c = 0$.

$$10\left(\frac{3}{10}x^2 - \frac{2}{5}x + \frac{7}{10}\right) = 10(0)$$

Multiply by 10 to clear fractions.

$$3x^2 - 4x + 7 = 0$$

$$a = 3, b = -4, c = 7$$

Identify the values of a, b, and c.

$$x = \frac{-(-4) \pm \sqrt{(-4)^2 - 4(3)(7)}}{2(3)}$$

Apply the quadratic formula.
$$x = \frac{-b \pm \sqrt{b^2 - 4ac}}{2a}$$

$$= \frac{4 \pm \sqrt{-68}}{6}$$

Simplify.

$$= \frac{4 \pm 2i\sqrt{17}}{6}$$

Simplify the radical.

Answer

6. $\{4 \pm \sqrt{19}\}$

$$= \frac{2(2 \pm i\sqrt{17})}{2 \cdot 3}$$

Factor the numerator and denominator.

$$= \frac{2 \pm i\sqrt{17}}{3}$$

Simplify the fraction.

$$= \frac{2}{3} \pm \frac{\sqrt{17}}{3}i$$

Write the solutions in standard form, $a + bi$.

The solution set is $\left\{ \frac{2}{3} \pm \frac{\sqrt{17}}{3}i \right\}$.

The solutions both check in the original equation.

Skill Practice 7 Solve the equation by applying the quadratic formula.

$$\frac{5}{12}x^2 - \frac{1}{2}x + \frac{1}{4} = 0$$

Three methods have been presented to solve a quadratic equation. We offer these guidelines to choose an appropriate and efficient method to solve a given quadratic equation.

Methods to Solve a Quadratic Equation

Method/Notes	**Examples**
Apply the Zero Product Property • Set one side of the equation equal to zero and factor the other side. Then apply the zero product property.	Solve. $\quad x^2 - x = 12$ $x^2 - x - 12 = 0$ $(x - 4)(x + 3) = 0$ $x = 4$ or $x = -3$
Complete the Square and Apply the Square Root Property • Good choice if the equation is in the form $x^2 = k$. • Good choice if the equation is in the form $ax^2 + bx + c = 0$, where $a = 1$ and b is an even real number.	Solve. $\quad c^2 = -6$ $c = \pm\sqrt{-6}$ $c = \pm i\sqrt{6}$ Solve. $\quad x^2 + 6x + 2 = 0$ $x^2 + 6x + 9 = -2 + 9$ $(x + 3)^2 = 7$ $x + 3 = \pm\sqrt{7}$ $x = -3 \pm \sqrt{7}$
Apply the Quadratic Formula • Applies in all situations. • Consider clearing fractions or decimals if the coefficients are not integer values.	Solve. $\quad 0.2x^2 + 0.5x + 0.1 = 0$ $10(0.2x^2 + 0.5x + 0.1) = 10(0)$ $2x^2 + 5x + 1 = 0$ $x = \dfrac{-(5) \pm \sqrt{(5)^2 - 4(2)(1)}}{2(2)}$ $x = \dfrac{-5 \pm \sqrt{17}}{4}$

5. Use the Discriminant

The solutions to a quadratic equation are given by $x = \dfrac{-b \pm \sqrt{b^2 - 4ac}}{2a}$. The radicand, $b^2 - 4ac$, is called the *discriminant*. The value of the discriminant tells us the number and type of solutions to the equation. We examine three different cases.

Using the Discriminant to Determine the Number and Type of Solutions to a Quadratic Equation

Given a quadratic equation $ax^2 + bx + c = 0$ $(a \neq 0)$, the quantity $b^2 - 4ac$ is called the **discriminant.**

Discriminant $b^2 - 4ac$	Number and Type of Solutions	Examples	Result of Quadratic Formula
$b^2 - 4ac < 0$	2 nonreal solutions	$2x^2 - 3x + 5 = 0$ $b^2 - 4ac = (-3)^2 - 4(2)(5)$ $\quad = -31$	$x = \dfrac{3 \pm \sqrt{-31}}{4}$
$b^2 - 4ac = 0$	1 real solution	$x^2 + 6x + 9 = 0$ $b^2 - 4ac = (6)^2 - 4(1)(9)$ $\quad = 0$	$x = \dfrac{-6 \pm \sqrt{0}}{2} = -3$
$b^2 - 4ac > 0$	2 real solutions	$2x^2 + 7x - 1 = 0$ $b^2 - 4ac = (7)^2 - 4(2)(-1)$ $\quad = 57$	$x = \dfrac{-7 \pm \sqrt{57}}{4}$

EXAMPLE 8   **Using the Discriminant**

Use the discriminant to determine the number and type of solutions for each equation.

a. $5x^2 - 3x + 1 = 0$ **b.** $2x^2 = 3 - 6x$ **c.** $4x^2 + 12x = -9$

Solution:

Equation	$b^2 - 4ac$	Solution Type and Number
a. $5x^2 - 3x + 1 = 0$	$(-3)^2 - 4(5)(1)$ $= -11$	Because $-11 < 0$, there are two nonreal solutions.
b. $2x^2 = 3 - 6x$ $\quad 2x^2 + 6x - 3 = 0$	$(6)^2 - 4(2)(-3)$ $= 60$	Because $60 > 0$, there are two real solutions.
c. $4x^2 + 12x = -9$ $\quad 4x^2 + 12x + 9 = 0$	$(12)^2 - 4(4)(9)$ $= 0$	Because the discriminant is 0, there is one real solution.

Skill Practice 8   Use the discriminant to determine the number and type of solutions for each equation.

a. $2x^2 - 4x + 5 = 0$ **b.** $25x^2 = 10x - 1$ **c.** $x^2 + 10x = -9$

Answers

8. a. Discriminant: -24
       (2 nonreal solutions)
   b. Discriminant: 0
       (1 real solution)
   c. Discriminant: 64
       (2 real solutions)

6. Solve an Equation for a Specified Variable

In Examples 9 and 10, we manipulate literal equations to solve for a specified variable.

EXAMPLE 9   Solving an Equation for a Specified Variable

Solve for r. $V = \dfrac{1}{3}\pi r^2 h$ $(r > 0)$

Solution:

TIP The equation $V = \frac{1}{3}\pi r^2 h$ is linear in the variables V and h, and quadratic in the variable r.

$$V = \frac{1}{3}\pi r^2 h$$

This equation is quadratic in the variable r. The strategy in this example is to isolate r^2 and then apply the square root property.

$$3(V) = 3\left(\frac{1}{3}\pi r^2 h\right)$$

Multiply both sides by 3 to clear fractions.

$$3V = \pi r^2 h$$

$$\frac{3V}{\pi h} = \frac{\pi r^2 h}{\pi h}$$

Divide both sides by πh to isolate r^2.

TIP The formula $V = \frac{1}{3}\pi r^2 h$ gives the volume of a right circular cone with radius r. Therefore, $r > 0$.

$$\frac{3V}{\pi h} = r^2$$

$$r = \sqrt{\frac{3V}{\pi h}} \text{ or } r = \frac{\sqrt{3V\pi h}}{\pi h}$$

Apply the square root property. Since $r > 0$, we take the positive square root only.

Skill Practice 9   Solve for v. $E = \dfrac{1}{2}mv^2$ $(v > 0)$

EXAMPLE 10   Solving an Equation for a Specified Variable

Solve for t. $mt^2 + nt = z$

Solution:

This equation is quadratic in the variable t. The strategy is to write the polynomial in descending order by powers of t. Then since there are two t terms with different exponents, we cannot isolate t directly. Instead we apply the quadratic formula.

$$mt^2 + nt = z$$
$$mt^2 + nt - z = 0$$

Write the polynomial in descending order by t.

Avoiding Mistakes

In the equation $mt^2 + nt - z = 0$, t is the variable, and m, n, and z are the coefficients.

$$a = m, b = n, c = -z$$

Identify the coefficients of each term.

$$t = \frac{-(n) \pm \sqrt{(n)^2 - 4(m)(-z)}}{2m}$$

Apply the quadratic formula.

$$t = \frac{-n \pm \sqrt{n^2 + 4mz}}{2m}$$

Simplify.

Skill Practice 10   Solve for p. $cp^2 - dp = k$

Answers

9. $v = \sqrt{\dfrac{2E}{m}}$ or $v = \dfrac{\sqrt{2Em}}{m}$

10. $p = \dfrac{d \pm \sqrt{d^2 + 4ck}}{2c}$

SECTION 1.4   Practice Exercises

Prerequisite Review

R.1. a. Multiply the binomials. $(x - 4)(x - 4)$

    b. Factor $x^2 - 8x + 16$.

For Exercises R.2–R.4, factor completely.

R.2. $5x^2 + 17xy - 12y^2$ **R.3.** $4p^3 - 48p^2q + 144pq^2$ **R.4.** $16x^2 + 40x + 25$

R.5. Simplify the expression. $\dfrac{12 - 4\sqrt{5}}{8}$

Concept Connections

1. A _____ equation is a second-degree equation of the form $ax^2 + bx + c = 0$ where $a \neq 0$.

2. A _____ equation is a first-degree equation of the form $ax + b = 0$ where $a \neq 0$.

3. The square root property indicates that if $x^2 = k$, then $x = $ _____.

4. The value of n that would make the trinomial $x^2 + 20x + n$ a perfect square trinomial is _____.

5. Given $ax^2 + bx + c = 0$ $(a \neq 0)$, write the quadratic formula.

6. For a quadratic equation $ax^2 + bx + c = 0$, the discriminant is given by the expression _____.

Objective 1: Solve Quadratic Equations by Using the Zero Product Property

For Exercises 7–18, solve by applying the zero product property. (See Example 1)

7. $(x - 3)(x + 7) = 0$ 8. $(t + 4)(t - 1) = 0$ 9. $n^2 + 5n = 24$

10. $y^2 = 18 - 7y$ 11. $8t(t + 3) = 2t - 5$ 12. $6m(m + 4) = m - 15$

13. $40p^2 - 90 = 0$ 14. $32n^2 - 162 = 0$ 15. $3x^2 = 12x$

16. $z^2 = 25z$ 17. $(m + 4)(m - 5) = -8$ 18. $(n + 2)(n - 4) = 27$

Objective 2: Solve Quadratic Equations by Using the Square Root Property

For Exercises 19–30, solve by using the square root property. (See Example 2)

19. $x^2 = 81$ 20. $w^2 = 121$ 21. $5y^2 - 35 = 0$

22. $6v^2 - 30 = 0$ 23. $4u^2 + 64 = 0$ 24. $8p^2 + 72 = 0$

25. $(k + 2)^2 = 28$ 26. $3(z + 11)^2 - 10 = 110$ 27. $2(w - 5)^2 + 5 = 23$

28. $(c - 3)^2 = 49$ 29. $\left(t - \dfrac{1}{2}\right)^2 = -\dfrac{17}{4}$ 30. $\left(a - \dfrac{1}{3}\right)^2 = -\dfrac{47}{9}$

Objective 3: Complete the Square

For Exercises 31–38, determine the value of n that makes the polynomial a perfect square trinomial. Then factor as the square of a binomial. (See Example 3)

31. $x^2 + 14x + n$ 32. $y^2 + 22y + n$ 33. $p^2 - 26p + n$

34. $u^2 - 4u + n$ 35. $w^2 - 3w + n$ 36. $v^2 - 11v + n$

37. $m^2 + \dfrac{2}{9}m + n$ 38. $k^2 + \dfrac{2}{5}k + n$

For Exercises 39–50, solve by completing the square and applying the square root property. (See Examples 4–5)

39. $y^2 + 22y - 4 = 0$

40. $x^2 + 14x - 3 = 0$

41. $t^2 - 8t = -24$

42. $p^2 - 24p = -156$

43. $4z^2 + 24z = -160$

44. $2m^2 + 20m = -70$

45. $2x(x - 3) = 4 + x$

46. $5c(c - 2) = 6 + 3c$

47. $-4y^2 - 12y + 5 = 0$

48. $-2x^2 - 14x + 5 = 0$

49. $3x^2 + 5x - 6 = 0$

50. $4x^2 + 3x - 8 = 0$

Objective 4: Solve Quadratic Equations by Using the Quadratic Formula

For Exercises 51–54, identify values of a, b, and c that could be used with the quadratic formula to solve the equation.

51. $x^2 = 7x - 4$

52. $x^2 = 3(x - 2)$

53. $5x^2 + 3x = 0$

54. $2x^2 - 18 = 0$

For Exercises 55–70, solve by using the quadratic formula. (See Examples 6–7)

55. $x^2 - 3x - 7 = 0$

56. $x^2 - 5x - 9 = 0$

57. $y^2 = -4y - 6$

58. $z^2 = -8z - 19$

59. $t(t - 6) = -10$

60. $m(m + 10) = -34$

61. $-7c + 3 = -5c^2$

62. $-5d + 2 = -6d^2$

63. $(6x + 5)(x - 3) = -2x(7x + 5) + x - 12$

64. $(5c + 7)(2c - 3) = -2c(c + 15) - 35$

65. $9x^2 + 49 = 0$

66. $121x^2 + 4 = 0$

67. $\frac{1}{2}x^2 - \frac{2}{7} = \frac{5}{14}x$

68. $\frac{1}{3}x^2 - \frac{7}{6} = \frac{3}{2}x$

69. $0.4y^2 = 2y - 2.5$

70. $0.09n^2 = 0.42n - 0.49$

Mixed Exercises

For Exercises 71–78, determine if the equation is linear, quadratic, or neither. If the equation is linear or quadratic, find the solution set.

71. $2y + 4 = 0$

72. $3z - 9 = 0$

73. $2y^2 + 4y = 0$

74. $3z^2 - 9z = 0$

75. $5x(x + 6) = 5x^2 + 27x + 3$

76. $3x(x - 4) = 3x^2 - 11x + 4$

77. $2x^2(x + 7) = x^2 + 3x + 1$

78. $-x(x^2 - 5) + 4 = x^2 + 5$

For Exercises 79–96, solve the equation by using any method.

79. $(3x - 4)^2 = 0$

80. $(2x + 1)^2 = 0$

81. $m^2 + 4m = -2$

82. $n^2 + 8n = -3$

83. $\frac{x^2 - 4x}{6} - \frac{5x}{3} = 1$

84. $\frac{m^2 + 2m}{7} - \frac{9m}{14} = \frac{3}{2}$

85. $2(x + 4) + x^2 = x(x + 2) + 8$

86. $3(y - 5) + y^2 = y(y + 3) - 15$

87. $\frac{3}{5}x^2 - \frac{1}{10}x = \frac{1}{2}$

88. $\frac{1}{12}x^2 - \frac{11}{24}x = -\frac{1}{2}$

89. $x^2 - 5x = 5x(x - 1) - 4x^2 + 1$

90. $p^2 - 4p = 4p(p - 1) - 3p^2 + 2$

91. $(2y + 7)(y + 1) = 2y^2 - 11$

92. $(3z - 8)(z + 2) = 3z^2 + 10$

93. $7d^2 + 5 = 0$

94. $11t^2 + 3 = 0$

95. $x^2 - \sqrt{5} = 0$

96. $y^2 - \sqrt{11} = 0$

Objective 5: Use the Discriminant

For Exercises 97–104, (a) evaluate the discriminant and (b) determine the number and type of solutions to each equation. (See Example 8)

97. $3x^2 - 4x + 6 = 0$

98. $5x^2 - 2x + 4 = 0$

99. $-2w^2 + 8w = 3$

100. $-6d^2 + 9d = 2$

101. $3x(x - 4) = x - 4$

102. $2x(x - 2) = x + 3$

103. $-1.4m + 0.1 = -4.9m^2$

104. $3.6n + 0.4 = -8.1n^2$

Objective 6: Solve an Equation for a Specified Variable

For Exercises 105–118, solve for the indicated variable. (See Examples 9–10)

105. $A = \pi r^2$ for $r > 0$

106. $V = \pi r^2 h$ for $r > 0$

107. $s = \dfrac{1}{2}gt^2$ for $t > 0$

108. $c = \dfrac{d^2 t}{2}$ for $d > 0$

109. $a^2 + b^2 = c^2$ for $a > 0$

110. $a^2 + b^2 + c^2 = d^2$ for $c > 0$

111. $L = c^2 I^2 Rt$ for $I > 0$

112. $I = cN^2 r^2 s$ for $N > 0$

113. $kw^2 - cw = r$ for w

114. $dy^2 + my = p$ for y

115. $s = v_0 t + \dfrac{1}{2}at^2$ for t

116. $S = 2\pi rh + \pi r^2 h$ for r

117. $LI^2 + RI + \dfrac{1}{C} = 0$ for I

118. $A = \pi r^2 + \pi rs$ for r

Write About It

119. Explain why the zero product property cannot be applied directly to solve the equation $(2x - 3)(x + 1) = 6$.

120. Given a quadratic equation, what is the discriminant and what information does it provide about the given quadratic equation?

Expanding Your Skills

For Exercises 121–122, solve for the indicated variable.

121. $x^2 - xy - 2y^2 = 0$ for x

122. $3a^2 + 2ab - b^2 = 0$ for a

For Exercises 123–132, write an equation with integer coefficients and the variable x that has the given solution set.
[*Hint*: Apply the zero product property in reverse. For example, to build an equation whose solution set is $\left\{2, -\dfrac{5}{2}\right\}$ we have $(x - 2)(2x + 5) = 0$, or simply $2x^2 + x - 10 = 0$.]

123. $\{4, -2\}$

124. $\{7, -1\}$

125. $\left\{\dfrac{2}{3}, \dfrac{1}{4}\right\}$

126. $\left\{\dfrac{3}{5}, \dfrac{1}{7}\right\}$

127. $\{\sqrt{5}, -\sqrt{5}\}$

128. $\{\sqrt{2}, -\sqrt{2}\}$

129. $\{2i, -2i\}$

130. $\{9i, -9i\}$

131. $\{1 \pm 2i\}$

132. $\{2 \pm 9i\}$

The solutions to the equation $ax^2 + bx + c = 0$ $(a \neq 0)$ are $x_1 = \dfrac{-b + \sqrt{b^2 - 4ac}}{2a}$ and $x_2 = \dfrac{-b - \sqrt{b^2 - 4ac}}{2a}$.
For Exercises 133–134, prove the given statements.

133. Prove that $x_1 + x_2 = -\dfrac{b}{a}$.

134. Prove that $x_1 x_2 = \dfrac{c}{a}$.

PROBLEM RECOGNITION EXERCISES

Simplifying Expressions Versus Solving Equations

For Exercises 1–8, identify the statement as an expression or as an equation. Then simplify the expression or solve the equation.

1. a. $(2x - 5)(3x + 1)$
  b. $(2x - 5)(3x + 1) = 0$

2. a. $\dfrac{5}{x - 3} - \dfrac{1}{x + 7} - \dfrac{2}{x^2 + 4x - 21}$
  b. $\dfrac{5}{x - 3} - \dfrac{1}{x + 7} = \dfrac{2}{x^2 + 4x - 21}$

3. a. $(2x - 3)^2 = 8$
  b. $(2x - 3)^2 - 8$

4. a. $5 - \{6 + 3[2 - 5(y - 2)] + 1\} = 7$
  b. $5 - \{6 + 3[2 - 5(y - 2)] + 1\}$

5. a. $x^2 - 11x + 28 = 0$
  b. $x^2 - 11x - 28 = 0$

6. a. $3x(x + 9) = 20 - x$
  b. $3(x + 9) = 20 - x$

7. a. $\dfrac{35}{x} + 12 + x = 0$
  b. $\dfrac{35}{x} + 12 + x$

8. a. $\dfrac{x}{x - 2} + \dfrac{2}{3} = \dfrac{2}{x - 2}$
  b. $\dfrac{x}{x - 2} + \dfrac{2}{3} - \dfrac{2}{x - 2}$

SECTION 1.5   Applications of Quadratic Equations

1. Solve Applications Involving Quadratic Equations and Geometry

In this section, we solve applications that involve quadratic equations. Examples 1 and 2 involve applications with geometric figures.

> **EXAMPLE 1   Solving an Application Involving Volume**
>
> A trough at the end of a gutter spout is meant to direct water away from a house. The homeowner makes the trough from a rectangular piece of aluminum that is 20 in. long and 12 in. wide. He makes a fold along the two long sides a distance of x inches from the edge. If he wants the trough to hold 360 in.^{3} of water, how far from the edge should he make the fold?

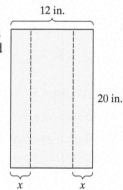

Solution:

Let x represent the distance between the edge of the sheet and the fold.

Information is given about the volume of the trough. When the fold is made, the trough will be in the shape of a rectangular solid with the ends missing. The volume is given by the product of length, width, and height.

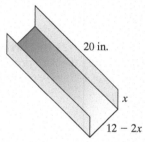

TIP In Example 1, we choose to move the terms to the left side of the equation so that the leading coefficient is positive. This makes the polynomial easier to factor.

$$V = lwh$$
$$360 = (20)(12 - 2x)(x)$$ The length is 20 in., the width is $12 - 2x$, and the height is x.

$$360 = 240x - 40x^2$$ Apply the distributive property.
$$40x^2 - 240x + 360 = 0$$ Set one side equal to zero.
$$40(x^2 - 6x + 9) = 0$$ Factor.
$$40(x - 3)^2 = 0$$
$$40 \neq 0 \quad \text{or} \quad x - 3 = 0$$ Apply the zero product property. Set each factor equal to zero.

$$x = 3$$ The first equation is a contradiction. The only solution is 3.

The sheet of aluminum should be folded 3 in. from the edges.

> **Skill Practice 1**   A box is to be formed by taking a sheet of cardboard and cutting away four 2-in. by 2-in. squares from each corner. Then the sides are turned up to form a box that holds 56 in.^{3}. If the length of the original piece of cardboard is 3 in. more than the width, find the dimensions of the original sheet of cardboard.

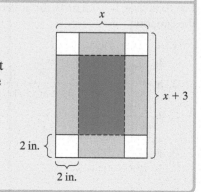

Answer

1. The sheet of cardboard is 8 in. by 11 in.

In Example 2, we use the Pythagorean theorem and a quadratic equation to find the lengths of the sides of a right triangle.

> ### EXAMPLE 2   Solving an Application Involving the Pythagorean Theorem
>
> A window is in the shape of a rectangle with an adjacent right triangle above (see figure). The length of one leg of the right triangle is 2 ft less than the length of the hypotenuse. The length of the other leg is 1 ft less than the length of the hypotenuse. Find the lengths of the sides.

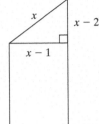

Solution:

Let x represent the length of the hypotenuse.
$x - 1$ represents the length of the longer leg.
$x - 2$ represents the length of the shorter leg.

Use the Pythagorean theorem to relate the lengths of the sides.

> ### Avoiding Mistakes
>
> In Example 2, be sure to square the binomials correctly. Recall that $(a - b)^2 = a^2 - 2ab + b^2$.
>
> Therefore,
> $(x - 1)^2 = x^2 - 2x + 1$
> $(x - 2)^2 = x^2 - 4x + 4$

$a^2 + b^2 = c^2$	Pythagorean theorem
$(x - 1)^2 + (x - 2)^2 = (x)^2$	Substitute $x - 1$, $x - 2$, and x for the lengths of the sides.
$x^2 - 2x + 1 + x^2 - 4x + 4 = x^2$	
$x^2 - 6x + 5 = 0$	Set one side of the equation equal to zero.
$(x - 1)(x - 5) = 0$	Factor.
$x - 1 = 0$ or $x - 5 = 0$	Apply the zero product property.
$\cancel{x = 1}$ or $x = 5$	Reject $x = 1$ because if x were 1, the lengths of the legs would be 0 ft and -1 ft, which is impossible.

The hypotenuse is 5 ft.
The length of the longer leg is given by $x - 1$:   5 ft $-$ 1 ft $=$ 4 ft.
The length of the shorter leg is given by $x - 2$:   5 ft $-$ 2 ft $=$ 3 ft.

> **Skill Practice 2**   A sail on a sailboat is in the shape of two adjacent right triangles. The hypotenuse of the lower triangle is 10 ft, and one leg is 2 ft shorter than the other leg. Find the lengths of the legs of the lower triangle.
>
>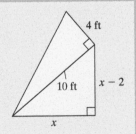

2. Solve Applications Involving Quadratic Models

In the study of physical science, a common model used to represent the vertical position of an object moving vertically under the influence of gravity is given in Table 1-1.

Answer

2. The longer leg is 8 ft and the shorter leg is 6 ft.

> **Table 1-1   Vertical Position of an Object**
>
> Suppose that an object has an initial vertical position of s_0 and initial velocity v_0 straight upward. The vertical position s of the object is given by
>
> $$s = -\frac{1}{2}gt^2 + v_0 t + s_0, \text{ where}$$
>
> | | |
> |---|---|
> | g | is the acceleration due to gravity. On Earth, $g = 32$ ft/sec^2 or $g = 9.8$ m/sec^2. |
> | t | is the time of travel. |
> | v_0 | is the initial velocity. |
> | s_0 | is the initial vertical position. |
> | s | is the vertical position of the object at time t. |

TIP The value of g is chosen to be consistent with the units for position and velocity. In this case, the initial height is given in ft. The initial velocity is given in ft/sec. Therefore, we choose g in ft/sec^2 rather than m/sec^2.

For example, suppose that a child tosses a ball straight upward from a height of 1.5 ft, with an initial velocity of 48 ft/sec.

The initial height is $s_0 = 1.5$ ft.
The initial velocity is $v_0 = 48$ ft/sec.
The acceleration due to gravity is $g = 32$ ft/sec^2.

The vertical position of the ball (in feet) is given by

$$s = -\frac{1}{2}gt^2 + v_0 t + s_0$$

$$s = -\frac{1}{2}(32)t^2 + (48)t + (1.5)$$

$$= -16t^2 + 48t + 1.5$$

EXAMPLE 3   Analyzing an Object Moving Vertically

A toy rocket is shot straight upward from a launch pad of 1 m above ground level with an initial velocity of 24 m/sec.

a. Write a model to express the height of the rocket s (in meters) above ground level.

b. Find the time(s) at which the rocket is at a height of 20 m. Round to 1 decimal place.

c. Find the time(s) at which the rocket is at a height of 40 m.

Solution:

a. $s = -\dfrac{1}{2}gt^2 + v_0 t + s_0$

$s = -\dfrac{1}{2}(9.8)t^2 + (24)t + (1)$

$= -4.9t^2 + 24t + 1$

In this example,
$s_0 = 1$ m
$v_0 = 24$ m/sec
$g = 9.8$ m/sec^2

TIP Choose $g = 9.8$ m/sec^2 because the height is given in meters and velocity is given in meters per second.

b. $20 = -4.9t^2 + 24t + 1$ Substitute 20 for s.

$4.9t^2 - 24t + 19 = 0$ Set one side equal to zero.

$t = \dfrac{-(-24) \pm \sqrt{(-24)^2 - 4(4.9)(19)}}{2(4.9)}$ Apply the quadratic formula.

$$t = \frac{24 \pm \sqrt{203.6}}{9.8}$$

$$t = \frac{24 + \sqrt{203.6}}{9.8} \approx 3.9$$

$$t = \frac{24 - \sqrt{203.6}}{9.8} \approx 1.0$$

The rocket will be at a height of 20 m at 1 sec and 3.9 sec after launch.

c. $40 = -4.9t^2 + 24t + 1$ Substitute 40 for s.

$4.9t^2 - 24t + 39 = 0$

$$t = \frac{-(-24) \pm \sqrt{(-24)^2 - 4(4.9)(39)}}{2(4.9)}$$ Apply the quadratic formula.

$$t = \frac{24 \pm \sqrt{-188.4}}{9.8}$$ The solutions are not real numbers.

There is no real number t for which the height of the rocket will be 40 m. The rocket will not reach a height of 40 m.

Skill Practice 3 A fireworks mortar is launched straight upward from a pool deck 2 m off the ground at an initial velocity of 40 m/sec.

a. Write a model to express the height of the mortar s (in meters) above ground level.

b. Find the time(s) at which the mortar is at a height of 60 m. Round to 1 decimal place.

c. Find the time(s) at which the rocket is at a height of 100 m.

Answers

3. a. $s = -4.9t^2 + 40t + 2$

  b. The mortar will be at a height of 60 m at 1.9 sec and 6.3 sec after launch.

  c. The mortar will not reach a height of 100 m.

SECTION 1.5   Practice Exercises

Prerequisite Review

R.1. Find the area of a rectangle that measures 28 ft by $9\frac{2}{7}$ ft.

R.2. Find the volume of a rectangular solid that measures 6 in. by 8 ft by 14 ft.

R.3. Find the length of the unknown side.

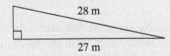

R.4. Find the area of the triangle.

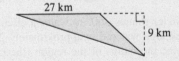

Concept Connections

1. Write a formula for the area of a triangle of base b and height h.

2. Write a formula for the area of a circle of radius r.

3. Write a formula for the volume of a rectangular solid of length l, width w, and height h.

4. Write the Pythagorean theorem for a right triangle with the lengths of the legs given by a and b and the length of the hypotenuse given by c.

Objective 1: Solve Applications Involving Quadratic Equations and Geometry

For Exercises 5–12, refer to the geometry formulas in the inside back cover.

  a. Write an equation in terms of x that represents the given relationship.

  b. Solve the equation to find the dimensions of the given shape.

5. The length of a rectangle is 3 yd more than twice the width x. The area is 629 yd^2.

6. The width of a rectangle is 2 m less than one-quarter of the length x. The area is 252 m^2.

7. The height of a triangle is 2 ft less than the base x. The area is 40 ft^2.

8. The height of a triangle is 4 yd longer than the base x. The area is 70 yd^2.

9. The width of a rectangular box is 8 in. The height is one-fifth the length x. The volume is 640 in.^{3}.

10. The height of a rectangular box is 4 ft. The length is 1 ft longer than twice the width x. The volume is 312 ft^3.

11. The length of the longer leg of a right triangle is 2 ft longer than the length of the shorter leg x. The hypotenuse is 2 ft shorter than twice the length of the shorter leg.

12. The longer leg of a right triangle is 7 cm longer than the length of the shorter leg x. The hypotenuse is 17 cm.

13. A rectangular garden covers 40 yd^2. The length is 2 yd longer than the width. Find the length and width. Round to the nearest tenth of a yard.

14. A rectangular piece of carpet covers 200 yd^2. The width is 9 yd less than the length. Find the length and width. Round to the nearest tenth of a yard.

15. The height of a triangular truss is 8 ft less than the base. The amount of drywall needed to cover the triangular area is 86 ft^2. Find the base and height of the triangle to the nearest tenth of a foot.

16. The base of a triangular piece of fabric is 6 in. more than the height. The area is 600 in.^{2}. Find the base and height of the triangle to the nearest tenth of an inch.

17. a. Write an equation representing the fact that the product of two consecutive even integers is 120.

    b. Solve the equation from part (a) to find the two integers.

18. a. Write an equation representing the fact that the product of two consecutive odd integers is 35.

    b. Solve the equation from part (a) to find the two integers.

19. a. Write an equation representing the fact that the sum of the squares of two consecutive integers is 113.

    b. Solve the equation from part (a) to find the two integers.

20. a. Write an equation representing the fact that the sum of the squares of two consecutive integers is 181.

    b. Solve the equation from part (a) to find the two integers.

21. On moving day, Guyton needs to rent a truck. The length of the cargo space is 12 ft, and the height is 1 ft less than the width. The brochure indicates that the truck can hold 504 ft^3. What are the dimensions of the cargo space? Assume that the cargo space is in the shape of a rectangular solid. (**See Example 1**)

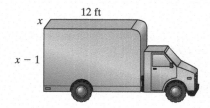

22. Lorene plans to make several open-topped boxes in which to carry plants. She makes the boxes from rectangular sheets of cardboard from which she cuts out 6-in. squares from each corner. The length of the original piece of cardboard is 12 in. more than the width. If the volume of the box is 1728 in.^{3}, determine the dimensions of the original piece of cardboard.

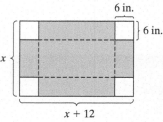

23. A sprinkler rotates 360° to water a circular region. If the total area watered is approximately 2000 yd^2, determine the radius of the region (the radius is length of the stream of water). Round the answer to the nearest yard.

24. An earthquake could be felt over a 46,000-mi^2 area. Up to how many miles from the epicenter could the earthquake be felt? Round to the nearest mile.

25. A patio is configured from a rectangle with two right triangles of equal size attached at the two ends. The length of the rectangle is 20 ft. The base of the right triangle is 3 ft less than the height of the triangle. If the total area of the patio is 348 ft^2, determine the base and height of the triangular portions.

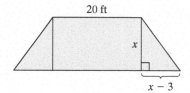

26. The front face of a house is in the shape of a rectangle with a Queen post roof truss above. The length of the rectangular region is 3 times the height of the truss. The height of the rectangle is 2 ft more than the height of the truss. If the total area of the front face of the house is 336 ft^2, determine the length and width of the rectangular region.

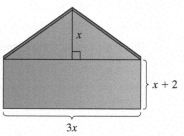

27. A baseball diamond is in the shape of a square with 90-ft sides. How far is it from home plate to second base? Give the exact value and give an approximation to the nearest tenth of a foot.

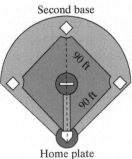

Second base

90 ft

90 ft

Home plate

28. The figure shown is a cube with 6-in. sides. Find the exact length of the diagonal through the interior of the cube d by following these steps.

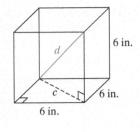

6 in.

d

c

6 in.

6 in.

a. Apply the Pythagorean theorem using the sides on the base of the cube to find the length of diagonal c.

b. Apply the Pythagorean theorem using c and the height of the cube as the legs of the right triangle through the interior of the cube.

29. The sail on a sailboat is in the shape of two adjacent right triangles. In the lower triangle, the shorter leg is 2 ft less than the longer leg. The hypotenuse is 2 ft more than the longer leg. **(See Example 2)**

4 ft

$x + 2$

$x - 2$

x

a. Find the lengths of the sides of the lower triangle.

b. Find the total sail area.

30. A portion of a roof truss is given in the figure. The triangle on the left is configured such that the longer leg is 7 ft longer than the shorter leg, and the hypotenuse is 1 ft more than twice the shorter leg.

$2x + 1$

$x + 7$

9 ft

x

a. Find the lengths of the sides of the triangle on the left.

b. Find the lengths of the sides of the triangle on the right.

31. The display area on a cell phone has a 3.5-in. diagonal.

a. If the aspect ratio of length to width is 1.5 to 1, determine the length and width of the display area. Round the values to the nearest hundredth of an inch.

b. If the phone has 326 pixels per inch, approximate the dimensions in pixels.

32. The display area on a computer has a 15-in. diagonal. If the aspect ratio of length to width is 1.6 to 1, determine the length and width of the display area. Round the values to the nearest hundredth of an inch.

Objective 2: Solve Applications Involving Quadratic Models

33. In a round-robin tennis tournament, each player plays every other player exactly one time. The number of matches N is given by $N = \frac{1}{2}n(n - 1)$, where n is the number of players in the tournament. If 28 matches were played, how many players were in the tournament?

34. The sum of the first n natural numbers, $S = 1 + 2 + 3 + \cdots + n$, is given by $S = \frac{1}{2}n(n + 1)$. If the sum of the first n natural numbers is 171, determine the value of n.

35. The population P of a culture of *Pseudomonas aeruginosa* bacteria is given by $P = -1718t^2 + 82{,}000t + 10{,}000$, where t is the time in hours since the culture was started. Determine the time(s) at which the population was 600,000. Round to the nearest hour.

36. The gas mileage for a certain vehicle can be approximated by $m = -0.04x^2 + 3.6x - 49$, where x is the speed of the vehicle in mph. Determine the speed(s) at which the car gets 30 mpg. Round to the nearest mph.

37. The distance d (in ft) required to stop a car that was traveling at speed v (in mph) before the brakes were applied depends on the amount of friction between the tires and the road and the driver's reaction time. After an accident, a legal team hired an engineering firm to collect data for the stretch of road where the accident occurred. Based on the data, the stopping distance is given by $d = 0.05v^2 + 2.2v$.

a. Determine the distance required to stop a car going 50 mph.

b. Up to what speed (to the nearest mph) could a motorist be traveling and still have adequate stopping distance to avoid hitting a deer 330 ft away?

38. Leptin is a hormone that has a central role in fat metabolism. One study published in the *New England Journal of Medicine* measured serum leptin concentrations versus the percentage of body fat for 275 individuals. The concentration of leptin c (in ng/mL) is approximated by $c = 219x^2 - 26.7x + 1.64$, where x is percentage of body fat.

   a. Determine the concentration of leptin in an individual with 22% body fat ($x = 0.22$). Round to 1 decimal place.

   b. If an individual has 3 ng/mL of leptin, determine the percentage of body fat. Round to the nearest whole percent.

   (*Source*: "Serum Immunoreactive-Leptin Concentrations in Normal-Weight and Obese Humans," *New England Journal of Medicine*, Feb., 1996)

For Exercises 39–42, use the model $s = -\frac{1}{2}gt^2 + v_0t + s_0$ with $g = 32$ ft/sec^2 or $g = 9.8$ m/sec^2. (See Example 3)

39. NBA basketball legend Michael Jordan had a 48-in. vertical leap. Suppose that Michael jumped from ground level with an initial velocity of 16 ft/sec.

   a. Write a model to express Michael's height (in ft) above ground level t seconds after leaving the ground.

   b. Use the model from part (a) to determine how long it would take Michael to reach his maximum height of 48 in. (4 ft).

40. At the time of this printing, the highest vertical leap on record is 60 in., held by Kadour Ziani. For this record-setting jump, Kadour left the ground with an initial velocity of $8\sqrt{5}$ ft/sec.

   a. Write a model to express Kadour's height (in ft) above ground level t seconds after leaving the ground.

   b. Use the model from part (a) to determine how long it would take Kadour to reach his maximum height of 60 in. (5 ft). Round to the nearest hundredth of a second.

41. A bad punter on a football team kicks a football approximately straight upward with an initial velocity of 75 ft/sec.

   a. If the ball leaves his foot from a height of 4 ft, write an equation for the vertical height s (in ft) of the ball t seconds after being kicked.

   b. Find the time(s) at which the ball is at a height of 80 ft. Round to 1 decimal place.

42. In a classic *Seinfeld* episode, Jerry tosses a loaf of bread (a marble rye) straight upward to his friend George who is leaning out of a third-story window.

   a. If the loaf of bread leaves Jerry's hand at a height of 1 m with an initial velocity of 18 m/sec, write an equation for the vertical position of the bread s (in meters) t seconds after release.

   b. How long will it take the bread to reach George if he catches the bread on the way up at a height of 16 m? Round to the nearest tenth of a second.

Expanding Your Skills

43. A **golden rectangle** is a rectangle in which the ratio of its length to its width is equal to the ratio of the sum of its length and width to its length: $\frac{L}{W} = \frac{L + W}{L}$ (values of L and W that meet this condition are said to be in the **golden ratio**).

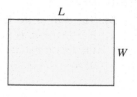

   a. Suppose that a golden rectangle has a width of 1 unit. Solve the equation to find the exact value for the length. Then give a decimal approximation to 2 decimal places.

   b. To create a golden rectangle with a width of 9 ft, what should be the length? Round to 1 decimal place.

44. An artist has been commissioned to make a stained glass window in the shape of a regular octagon. The octagon must fit inside an 18-in. square space. Determine the length of each side of the octagon. Round to the nearest hundredth of an inch.

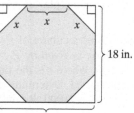

45. A farmer has 160 yd of fencing material and wants to enclose three rectangular pens. Suppose that x represents the length of each pen and y represents the width as shown in the figure.

   a. Assuming that the farmer uses all 160 yd of fencing, write an expression for y in terms of x.

   b. Write an expression in terms of x for the area of one individual pen.

   c. If the farmer wants to design the structure so that each pen encloses 250 yd^2, determine the dimensions of each pen.

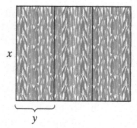

46. At noon, a ship leaves a harbor and sails south at 10 knots. Two hours later, a second ship leaves the harbor and sails east at 15 knots. When will the ships be 100 nautical miles apart? Round to the nearest minute.

SECTION 1.6 **More Equations and Applications**

1. Solve Polynomial Equations

In this section, we expand our repertoire of equations that we can recognize and solve. First, we use the zero product property to solve polynomial equations. The goal is to set one side of the equation equal to zero and factor the other side into linear or quadratic factors.

EXAMPLE 1   **Solving a Polynomial Equation**

Solve the equation. $4x^3 + 12x^2 - 9x - 27 = 0$

Solution:

$$4x^3 + 12x^2 - 9x - 27 = 0$$ This is a polynomial equation with one side already equal to zero.

$$4x^2(x + 3) - 9(x + 3) = 0$$ Factor by grouping.
$$(x + 3)(4x^2 - 9) = 0$$
$$(x + 3)(2x - 3)(2x + 3) = 0$$
$$x + 3 = 0 \quad \text{or} \quad 2x - 3 = 0 \quad \text{or} \quad 2x + 3 = 0$$ Apply the zero product property. Set each factor equal to zero.
$$x = -3 \quad \text{or} \quad x = \frac{3}{2} \quad \text{or} \quad x = -\frac{3}{2}$$

The solution set is $\left\{-3, \dfrac{3}{2}, -\dfrac{3}{2}\right\}$.   The solutions all check in the original equation.

Skill Practice 1   Solve the equation. $25y^3 + 100y^2 - y - 4 = 0$

EXAMPLE 2   **Solving a Polynomial Equation**

Solve the equation. $2x^5 = 16x^2$

Solution:

$$2x^5 = 16x^2$$
$$2x^5 - 16x^2 = 0$$ Set one side equal to zero.
$$2x^2(x^3 - 8) = 0$$ Factor out the GCF.
$$2x^2(x - 2)(x^2 + 2x + 4) = 0$$ Factor as a difference of cubes.
$$2x^2 = 0 \quad \text{or} \quad x - 2 = 0 \quad \text{or} \quad x^2 + 2x + 4 = 0$$
$$x = 0 \quad \text{or} \quad x = 2 \quad \text{or}$$
$$x = \frac{-(2) \pm \sqrt{(2)^2 - 4(1)(4)}}{2(1)}$$
$$= \frac{-2 \pm \sqrt{-12}}{2}$$
$$= \frac{-2 \pm 2i\sqrt{3}}{2}$$
$$= -1 \pm i\sqrt{3}$$

The solution set is $\{0, 2, -1 \pm i\sqrt{3}\}$.   The solutions all check in the original equation.

> **TIP** In Example 2, the third equation is a nonfactorable quadratic equation. Use the quadratic formula.

Skill Practice 2   Solve the equation. $w^4 = 27w$

Answers

1. $\left\{-4, \dfrac{1}{5}, -\dfrac{1}{5}\right\}$

2. $\left\{0, 3, \dfrac{-3 \pm 3i\sqrt{3}}{2}\right\}$

2. Solve Rational Equations

In Section 1.1, we solved rational equations by multiplying both sides of the equation by the LCD to clear fractions. In Example 3, we review this process. After clearing fractions in each term, the resulting equation is quadratic.

EXAMPLE 3 **Solving a Rational Equation**

Solve. $\dfrac{2x}{x-4} - \dfrac{3}{x+2} = \dfrac{x^2+14}{x^2-2x-8}$

Solution:

$$\dfrac{2x}{x-4} - \dfrac{3}{x+2} = \dfrac{x^2+14}{x^2-2x-8}$$

$$\dfrac{2x}{x-4} - \dfrac{3}{x+2} = \dfrac{x^2+14}{(x-4)(x+2)}$$

Factor the denominators.
The variable x has the restrictions that $x \neq 4$ and $x \neq -2$.

$$(x-4)(x+2)\left(\dfrac{2x}{x-4} - \dfrac{3}{x+2}\right) = (x-4)(x+2)\left[\dfrac{x^2+14}{(x-4)(x+2)}\right]$$

Multiply both sides by the LCD to clear fractions.

$$2x(x+2) - 3(x-4) = x^2 + 14$$ The resulting equation is quadratic.
$$2x^2 + 4x - 3x + 12 = x^2 + 14$$
$$x^2 + x - 2 = 0$$
$$(x+2)(x-1) = 0$$

Apply the zero product property.

$$\cancel{x = -2} \quad \text{or} \quad x = 1$$

The value -2 is not a solution because it is a restricted value. It does not check.

Check: $x = -2$

$$\dfrac{2(-2)}{(-2)-4} - \underbrace{\dfrac{3}{(-2)+2}}_{\text{undefined}} \overset{?}{=} \underbrace{\dfrac{(-2)^2+14}{(-2)^2-2(-2)-8}}_{\text{undefined}}$$

Check: $x = 1$

$$\dfrac{2(1)}{(1)-4} - \dfrac{3}{(1)+2} \overset{?}{=} \dfrac{(1)^2+14}{(1)^2-2(1)-8}$$
$$\dfrac{2}{-3} - 1 \overset{?}{=} \dfrac{15}{-9}$$
$$-\dfrac{5}{3} \overset{?}{=} -\dfrac{5}{3} \quad \checkmark \text{ true}$$

The solution set is $\{1\}$. The value -2 does not check.

Skill Practice 3 Solve. $\dfrac{3x}{x-5} = \dfrac{2}{x+1} + \dfrac{2x^2+40}{x^2-4x-5}$

In Example 4, we solve a uniform motion application that can be modeled by a rational equation.

EXAMPLE 4 **Solving an Application Involving Uniform Motion**

Trent takes his boat 6 mi downstream with a 1.5-mph current. The return trip against the current takes 1 hr longer. Find the speed of the boat in still water (in the absence of current).

Answer

3. $\{-6\}$; The value 5 does not check.

Solution:

Let b represent the speed of the boat in still water.

	Distance (mi)	Rate (mph)	Time (hr)
With current	6	$b + 1.5$	$\dfrac{6}{b + 1.5}$
Against current	6	$b - 1.5$	$\dfrac{6}{b - 1.5}$

Assign a variable to the unknown quantity.

Organize the given information in a figure or chart.

$$\left(\begin{array}{c}\text{Time of trip}\\ \text{against current}\end{array}\right) - \left(\begin{array}{c}\text{Time of trip}\\ \text{with current}\end{array}\right) = 1 \longleftarrow$$

The return trip against the current takes longer. The difference in time between the return trip and the original trip is 1 hr.

$$\frac{6}{b - 1.5} - \frac{6}{b + 1.5} = 1$$

The restrictions on b are $b \neq 1.5$ and $b \neq -1.5$.

$$(b - 1.5)(b + 1.5)\left(\frac{6}{b - 1.5} - \frac{6}{b + 1.5}\right) = (b - 1.5)(b + 1.5)(1)$$

$$6(b + 1.5) - 6(b - 1.5) = (b - 1.5)(b + 1.5)$$

Apply the distributive property.

$$6b + 9 - 6b + 9 = b^2 - 2.25$$
$$20.25 = b^2$$
$$b = \pm\sqrt{20.25}$$

Apply the square root property.

$$b = 4.5$$

Reject the negative solution because b represents the speed of the boat.

The speed of the boat in still water is 4.5 mph.

Skill Practice 4   A fishing boat can travel 60 km with a 2.5-km/hr current in 2 hr less time than it can travel 60 km against the current. Determine the speed of the fishing boat in still water.

---

3. Solve Absolute Value Equations

We now turn our attention to equations involving absolute value expressions. For example, given $|x| = 5$, the solution set is $\{-5, 5\}$. The equation can also be written as $|x - 0| = 5$, implying that the solutions are the values of x that are 5 units from 0 on the number line (Figure 1-1).

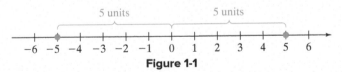

Figure 1-1

Given a nonnegative real number k, the generic absolute value equation $|u| = k$ can be solved directly from the definition of absolute value. Recall that

$$|u| = \begin{cases} u \text{ if } u \geq 0 \\ -u \text{ if } u < 0 \end{cases}$$

Thus, $|u| = k$ means that $u = k$ or $-u = k$. Solving for u, we have $u = k$ or $u = -k$.

This and three other properties summarized in Table 1-2 follow directly from the definition of absolute value.

Table 1-2   Absolute Value Equations
Let k represent a positive real number.
1. $|u| = k$ is equivalent to $u = k$ or $u = -k$.
2. $|u| = 0$ is equivalent to $u = 0$.
3. $|u| = -k$ has no solution.
4. $|u| = |w|$ is equivalent to $u = w$ or $u = -w$.

Answer

4. The boat travels 12.5 km/hr in still water.

To solve an absolute value equation, first isolate the absolute value. Then solve the equation by rewriting the equation in its equivalent form given in Table 1-2.

EXAMPLE 5   Solving Absolute Value Equations

Solve the equations.   **a.** $2|3 - 2t| = 6$ **b.** $2 = |7w - 3| + 8$

Solution:

a.

$2|3 - 2t| = 6$	Divide by 2 to isolate the absolute value.
$|3 - 2t| = 3$	The equation is in the form $|u| = k$, where $u = 3 - 2t$ and $k > 0$.

$$3 - 2t = 3 \quad \text{or} \quad 3 - 2t = -3 \quad \text{Rewrite the equation in the form } u = k \text{ or } u = -k.$$
$$-2t = 0 \quad \text{or} \quad -2t = -6$$
$$t = 0 \quad \text{or} \quad t = 3$$

Check: $t = 0$ Check: $t = 3$

$2|3 - 2(0)| \overset{?}{=} 6$ $2|3 - 2(3)| \overset{?}{=} 6$

The solution set is $\{3, 0\}$.

$2|3| = 6$ ✓ $2|3 - 6| \overset{?}{=} 6$

$2|-3| = 6$ ✓

b.

$2 = |7w - 3| + 8$	Subtract 8 from both sides to isolate the absolute value.
$-6 = |7w - 3|$	The equation is in the form $|u| = k$, where $u = 7w - 3$ and $k < 0$. By definition, an absolute value cannot be negative. There is no solution.

The solution set is $\{\ \}$.

Skill Practice 5   Solve the equations.

a. $5|2 - 4t| = 50$ **b.** $5 = |6c - 7| + 9$

In Example 6, we solve equations involving two absolute values by writing $|u| = |w|$ in the equivalent form $u = w$ or $u = -w$.

EXAMPLE 6   Solving Equations with Two Absolute Values

Solve the equations.   **a.** $|2x - 5| = |x + 1|$ **b.** $|6 - x| = |x - 6|$

Solution:

a. $|2x - 5| = |x + 1|$ The equation is in the form $|u| = |w|$.

$2x - 5 = x + 1 \quad \text{or} \quad 2x - 5 = -(x + 1)$ Rewrite the equation in the form $u = w$ or $u = -w$.

$x - 5 = 1 \qquad \text{or} \quad 2x - 5 = -x - 1$ Solve each individual equation.

$x = 6 \qquad \text{or} \qquad 3x = 4$

$x = \dfrac{4}{3}$

Both solutions check in the original equation.

The solution set is $\left\{ 6, \dfrac{4}{3} \right\}$.

b. $|6 - x| = |x - 6|$ The equation is in the form $|u| = |w|$.

$6 - x = x - 6 \quad \text{or} \quad 6 - x = -(x - 6)$ Rewrite the equation in the form $u = w$ or $u = -w$.

$-2x = -12 \quad \text{or} \quad 6 - x = -x + 6$ Solve each individual equation.

$x = 6 \qquad \text{or} \qquad 6 = 6 \text{ (identity)}$

The solution to this equation is all real numbers (including 6).

The solution set is $\mathbb{R}$.

Skill Practice 6   Solve the equations.

a. $|3x - 4| = |2x + 1|$ **b.** $|4 + x| = |-4 - x|$

4. Solve Radical Equations and Equations with Rational Exponents

An equation with one or more radicals containing a variable (such as $\sqrt[3]{x} = 5$) is called a **radical equation.** We can eliminate the radical by raising both sides of the equation to a power equal to the index of the radical.

$$\sqrt[3]{x} = 5$$
$$(\sqrt[3]{x})^3 = (5)^3 \qquad \text{The index is 3. Therefore, raise both sides}$$
$$x = 125 \qquad \text{to the third power.}$$

By raising each side of a radical equation to a power equal to the index, a new equation is produced. However, some (or all) of the solutions to the new equation may *not* be solutions to the original equation. These are called **extraneous solutions.** For this reason, it is necessary to check all potential solutions in the original equation. For example, consider the equation $\sqrt{x} = -10$. By inspection, this equation has no solution because the principal square root of x must be nonnegative. However, if we square both sides of the equation, it appears as though a solution exists:

Square both $\sqrt{x} = -10$ Solution set: { }
sides. $(\sqrt{x})^2 = (-10)^2$
          $x = 100$ The value 100 does not check in the original equation. Therefore, 100 is an extraneous solution.

Solving a Radical Equation

Step 1 Isolate the radical. If an equation has more than one radical, choose one of the radicals to isolate.

Step 2 Raise each side of the equation to a power equal to the index of the radical.

Step 3 Solve the resulting equation. If the equation still has a radical, repeat steps 1 and 2.

*Step 4 Check the potential solutions in the original equation and write the solution set.

*In solving radical equations, extraneous solutions potentially arise when both sides of the equation are raised to an even power. Therefore, an equation with only odd-indexed roots will not have extraneous solutions. However, it is still recommended that all potential solutions be checked.

EXAMPLE 7   **Solving a Radical Equation**

Solve. $\sqrt{x + 10} - 4 = x$

Solution:

$$\sqrt{x + 10} - 4 = x$$
$$\sqrt{x + 10} = x + 4 \qquad \text{Isolate the radical.}$$
$$(\sqrt{x + 10})^2 = (x + 4)^2 \qquad \text{The index is 2. Therefore, raise both sides to the second power.}$$
$$x + 10 = x^2 + 8x + 16 \qquad \text{The resulting equation is quadratic.}$$
$$0 = x^2 + 7x + 6 \qquad \text{Set one side equal to zero.}$$
$$0 = (x + 6)(x + 1) \qquad \text{Factor.}$$
$$x = -6 \quad \text{or} \quad x = -1 \qquad \text{Apply the zero product rule.}$$

Avoiding Mistakes

When raising both sides of an equation to a power, be sure to enclose both sides of the equation in parentheses.

Answers

6. a. $\left\{ \dfrac{3}{5}, 5 \right\}$ **b.** $\mathbb{R}$

Both sides of the equation were raised to an even power. Therefore, it is necessary to check the potential solutions.

$$\text{Check: } x = -6$$
$$\sqrt{x + 10} - 4 = x$$
$$\sqrt{(-6) + 10} - 4 \overset{?}{=} (-6)$$
$$\sqrt{4} - 4 \overset{?}{=} -6$$
$$2 - 4 \overset{?}{=} -6$$
$$-2 \overset{?}{=} -6 \text{ false}$$

$$\text{Check: } x = -1$$
$$\sqrt{x + 10} - 4 = x$$
$$\sqrt{(-1) + 10} - 4 \overset{?}{=} (-1)$$
$$\sqrt{9} - 4 \overset{?}{=} -1$$
$$3 - 4 \overset{?}{=} -1$$
$$-1 \overset{?}{=} -1 \checkmark \text{ true}$$

The solution set is $\{-1\}$. The value -6 does not check.

> **Skill Practice 7** Solve the equation. $\sqrt{t + 7} = t - 5$

In Example 8, we solve the equation $\sqrt{m - 1} - \sqrt{3m + 1} = -2$. The first step is to isolate one of the radicals. However, the presence of the constant term, -2, makes it impossible to isolate both radicals simultaneously. As a result, it is necessary to square both sides of the equation twice.

EXAMPLE 8 **Solving an Equation Containing Two Radicals**

Solve. $\sqrt{m - 1} - \sqrt{3m + 1} = -2$

Solution:

$$\sqrt{m - 1} - \sqrt{3m + 1} = -2$$
$$\sqrt{m - 1} = \sqrt{3m + 1} - 2 \qquad \text{Isolate one of the radicals.}$$
$$\left(\sqrt{m - 1}\right)^2 = \left(\sqrt{3m + 1} - 2\right)^2 \qquad \text{The index is 2. Therefore, raise both sides to the}$$
$$m - 1 = 3m + 1 - 4\sqrt{3m + 1} + 4 \qquad \text{second power.}$$
$$m - 1 = 3m + 5 - 4\sqrt{3m + 1}$$
$$4\sqrt{3m + 1} = 2m + 6 \qquad \text{Isolate the remaining radical.}$$
$$2\sqrt{3m + 1} = m + 3 \qquad \text{Divide both sides by 2 to simplify.}$$

$$\left(2\sqrt{3m + 1}\right)^2 = (m + 3)^2 \qquad \text{The resulting equation has another}$$
$$4(3m + 1) = m^2 + 6m + 9 \qquad \text{radical. Isolate the radical, and square both sides again.}$$
$$12m + 4 = m^2 + 6m + 9 \qquad \text{The resulting equation is quadratic.}$$
$$0 = m^2 - 6m + 5$$
$$0 = (m - 5)(m - 1) \qquad \text{Apply the zero product property.}$$
$$m = 5 \quad \text{or} \quad m = 1 \qquad \text{Both sides of the equation were raised to}$$
an even power. Check both potential solutions.

Avoiding Mistakes

Exercise caution when squaring the two-term expression on the right.
$$\left(\sqrt{3m + 1} - 2\right)^2$$
$$= \left(\sqrt{3m + 1}\right)^2$$
$$\quad - 2\left(\sqrt{3m + 1}\right)(2) + (2)^2$$
$$= 3m + 1$$
$$\quad - 4\sqrt{3m + 1} + 4$$

$$\text{Check: } m = 5$$
$$\sqrt{m - 1} - \sqrt{3m + 1} = -2$$
$$\sqrt{(5) - 1} - \sqrt{3(5) + 1} \overset{?}{=} -2$$
$$\sqrt{4} - \sqrt{16} \overset{?}{=} -2$$
$$2 - 4 \overset{?}{=} -2 \checkmark \text{ true}$$

$$\text{Check: } m = 1$$
$$\sqrt{m - 1} - \sqrt{3m + 1} = -2$$
$$\sqrt{(1) - 1} - \sqrt{3(1) + 1} \overset{?}{=} -2$$
$$\sqrt{0} - \sqrt{4} \overset{?}{=} -2$$
$$0 - 2 \overset{?}{=} -2 \checkmark \text{ true}$$

Both solutions check. The solution set is $\{1, 5\}$.

Answers

7. $\{9\}$; The value 2 does not check.
8. $\{5\}$; The value 0 does not check.

> **Skill Practice 8** Solve. $1 + \sqrt{n + 4} = \sqrt{3n + 1}$

In Example 9, we solve equations containing rational exponents.

> **EXAMPLE 9**   Solving an Equation Containing Rational Exponents
>
> Solve the equation. $2(x + 1)^{2/3} = 8$
>
> **Solution:**
>
> $2(x + 1)^{2/3} = 8$
> $2\sqrt[3]{(x + 1)^2} = 8$ Write the expression on the left in radical notation.
> $\sqrt[3]{(x + 1)^2} = 4$ Divide by 2 to isolate the radical.
> $\left[\sqrt[3]{(x + 1)^2}\right]^3 = (4)^3$ Raise both sides to a power equal to the index of the radical.
> $(x + 1)^2 = 64$
> $x + 1 = \pm\sqrt{64}$ Apply the square root property.
> $x = -1 \pm 8$
> $x = -1 - 8 = -9$ or $x = -1 + 8 = 7$
>
> Check: $x = -9$ Check: $x = 7$
> $(-9 + 1)^{2/3} \stackrel{?}{=} 4$ $(7 + 1)^{2/3} \stackrel{?}{=} 4$
> $(-8)^{2/3} = 4 \checkmark$ $(8)^{2/3} = 4 \checkmark$
>
> The solution set is $\{-9, 7\}$.
>
> **Skill Practice 9**   Solve the equation. $2(x - 4)^{3/4} = 54$

5. Solve Equations in Quadratic Form

In Section 1.4, we learned to solve quadratic equations by applying the quadratic formula or by completing the square and applying the square root property. This is particularly important because many other equations are quadratic in form. That is, with a simple substitution, these equations can be expressed as quadratic equations in a new variable. For example:

Equation in Quadratic Form   **New Equation**

$\left(2 + \dfrac{3}{x}\right)^2 - \left(2 + \dfrac{3}{x}\right) - 12 = 0$ $\xrightarrow{\text{Let } u = 2 + \frac{3}{x}}$ $u^2 - u - 12 = 0$

$2w^{2/3} - 3w^{1/3} - 20 = 0$ $\xrightarrow{\text{Let } u = w^{1/3}}$ $2u^2 - 3u - 20 = 0$

The equations on the right are quadratic and easily solved. Then using back substitution, we can solve for the original variable.

TIP For an equation written in quadratic form, notice that the expression for u is taken to be the variable expression from the middle term.

> **EXAMPLE 10**   Solving an Equation in Quadratic Form
>
> Solve. $(2x^2 - 3)^2 + 36(2x^2 - 3) + 35 = 0$
>
> **Solution:**
>
> $(2x^2 - 3)^2 + 36(2x^2 - 3) + 35 = 0$ The equation is in quadratic form.
> $u^2 + 36u + 35 = 0$ Let $u = 2x^2 - 3$.
> $(u + 35)(u + 1) = 0$ Set one side equal to zero and factor the other side.

Answer
9. $\{85\}$

$$u = -35 \quad \text{or} \quad u = -1$$

Apply the zero product property.

$$2x^2 - 3 = -35 \quad \text{or} \quad 2x^2 - 3 = -1$$

Back substitute. Replace u by $2x^2 - 3$.

$$2x^2 = -32 \quad \text{or} \quad 2x^2 = 2$$
$$x^2 = -16 \quad \text{or} \quad x^2 = 1$$

Isolate the square term.

$$x = \pm 4i \quad \text{or} \quad x = \pm 1$$

Apply the square root property.

The solution set is $\{\pm 4i, \pm 1\}$.

The solutions all check in the original equation.

Skill Practice 10 Solve. $(x^2 - 6)^2 + 33(x^2 - 6) + 62 = 0$

EXAMPLE 11 **Solving an Equation in Quadratic Form**

Solve. $2w^{2/3} = 3w^{1/3} + 20$

Solution:

$$2w^{2/3} = 3w^{1/3} + 20$$

Set one side equal to zero, and write the expression on the left in descending order.

$$2w^{2/3} - 3w^{1/3} - 20 = 0$$

$$2(w^{1/3})^2 - 3(w^{1/3}) - 20 = 0$$

The equation is in quadratic form.

$$2u^2 - 3u - 20 = 0$$

Let $u = w^{1/3}$.

$$(2u + 5)(u - 4) = 0$$

Factor.

$$u = -\frac{5}{2} \quad \text{or} \quad u = 4$$

Apply the zero product property.

$$w^{1/3} = -\frac{5}{2} \quad \text{or} \quad w^{1/3} = 4$$

Back substitute. Replace u by $w^{1/3}$.

$$(w^{1/3})^3 = \left(-\frac{5}{2}\right)^3 \quad \text{or} \quad (w^{1/3})^3 = (4)^3$$

Cube both sides.

$$w = -\frac{125}{8} \quad \text{or} \quad w = 64$$

Both solutions check in the original equation.

The solution set is $\left\{-\dfrac{125}{8}, 64\right\}$.

TIP Consider the equation from Example 11:

$$2w^{2/3} - 3w^{1/3} - 20 = 0$$

As an alternative to using substitution, the expression on the left can be factored directly.

$$(2w^{1/3} + 5)(w^{1/3} - 4) = 0$$

Applying the zero product property results in the same solutions.

Answers

10. $\{\pm 5i, \pm 2\}$ **11.** $\left\{-125, \dfrac{27}{8}\right\}$

Skill Practice 11 Solve. $2t^{2/3} = 15 - 7t^{1/3}$

SECTION 1.6   Practice Exercises

Prerequisite Review

R.1. Factor $9x^3 - 27x^2 - 16x + 48$.

R.2. Determine the restrictions on the variable.

$$\frac{a - 6}{a + 3}$$

R.3. Determine the restrictions on the variable.

$$\frac{a}{a^2 + 81}$$

R.4. Identify the least common denominator for the pair of expressions.

$$\frac{y + 3}{4y^2 + 12y + 9} \quad \text{and} \quad \frac{-8}{10y^2 + 15y}$$

R.5. Simplify the expression.

$$49^{-3/2}$$

R.6. Convert the expression to radical notation.

$$c^{2/7}$$

Concept Connections

1. An _____ value equation is an equation of the form $|u| = k$. If k is a positive real number then the solution set is _____.

2. If u and w represent real-valued expressions, then the equation $|u| = |w|$ can be written in an equivalent form without absolute value bars as _____.

3. The equation $m^{2/3} + 10m^{1/3} + 9 = 0$ is said to be in _____ form, because making the substitution $u =$ _____ results in a new equation that is quadratic.

4. Consider the equation $(4x^2 + 1)^2 + 4(4x^2 + 1) + 4 = 0$. If the substitution $u =$ _____ is made, then the equation becomes $u^2 + 4u + 4 = 0$.

Objective 1: Solve Polynomial Equations

For Exercises 5–20, solve the equation. (See Examples 1–2)

5. $-3x(2x - 1)(x + 6)^2 = 0$

6. $5y(3 - y)(4y + 1)^2 = 0$

7. $4(w^2 - 7)(w^2 + 4) = 0$

8. $-2(t^2 + 1)(t^2 - 5) = 0$

9. $75y^3 + 100y^2 - 3y - 4 = 0$

10. $98t^3 - 49t^2 - 8t + 4 = 0$

11. $x^3 + 7x^2 = 4(x + 7)$

12. $2m^3 + 3m^2 = 9(2m + 3)$

13. $2x^4 - 32 = 0$

14. $5m^4 - 5 = 0$

15. $2x^4 = -128x$

16. $10x^5 = -1250x^2$

17. $3n^2(n^2 + 3) = 20 - 2n^2$

18. $2y^2(y^2 - 2) = 18 + y^2$

19. $x^3 - 8 = x - 2$

20. $x^3 - 64 = x - 4$

Objective 2: Solve Rational Equations

For Exercises 21–28, solve the equation. (See Example 3)

21. $\dfrac{3x}{x + 2} - \dfrac{5}{x - 4} = \dfrac{2x^2 - 14x}{x^2 - 2x - 8}$

22. $\dfrac{4c}{c - 5} - \dfrac{1}{c + 1} = \dfrac{3c^2 + 3}{c^2 - 4c - 5}$

23. $\dfrac{m}{2m + 1} + 1 = \dfrac{2}{m - 3}$

24. $\dfrac{n}{n - 3} + 2 = \dfrac{3}{2n - 1}$

25. $2 - \dfrac{3}{y} = \dfrac{5}{y^2}$

26. $7 + \dfrac{20}{z} = \dfrac{3}{z^2}$

27. $\dfrac{18}{m^2 - 3m} + 2 = \dfrac{6}{m - 3}$

28. $\dfrac{48}{m^2 - 4m} + 3 = \dfrac{12}{m - 4}$

29. Jesse takes a 3-day kayak trip and travels 72 km south from Everglades City to a camp area in Everglades National Park. The trip to the camp area with a 2-km/hr current takes 9 hr less time than the return trip against the current. Find the speed that Jesse travels in still water. (**See Example 4**)

30. A plane travels 800 mi from Dallas, Texas, to Atlanta, Georgia, with a prevailing west wind of 40 mph. The return trip against the wind takes $\frac{1}{2}$ hr longer. Find the average speed of the plane in still air.

31. Jean runs 6 mi and then rides 24 mi on her bicycle in a biathlon. She rides 8 mph faster than she runs. If the total time for her to complete the race is 2.25 hr, determine her average speed running and her average speed riding her bicycle.

32. Barbara drives between Miami, Florida, and West Palm Beach, Florida. She drives 50 mi in clear weather and then encounters a thunderstorm for the last 15 mi. She drives 20 mph slower through the thunderstorm than she does in clear weather. If the total time for the trip is 1.5 hr, determine her average speed in nice weather and her average speed driving in the thunderstorm.

Objective 3: Solve Absolute Value Equations

For Exercises 33–54, solve the equations. (See Examples 5–6)

33. **a.** $|p| = 6$
    b. $|p| = 0$
    c. $|p| = -6$

34. **a.** $|w| = 2$
    b. $|w| = 0$
    c. $|w| = -2$

35. **a.** $|x - 3| = 4$
    b. $|x - 3| = 0$
    c. $|x - 3| = -7$

36. **a.** $|m + 1| = 5$
    b. $|m + 1| = 0$
    c. $|m + 1| = -1$

37. $2|3x - 4| + 7 = 9$

38. $4|2t + 7| + 2 = 22$

39. $-3 = -|c - 7| + 1$

40. $-4 = -|z + 8| - 3$

41. $2 = 8 + |11y + 4|$

42. $6 = 7 + |9z - 3|$

43. $\left| 4 - \frac{1}{2}w \right| - \frac{1}{3} = \frac{1}{2}$

44. $\left| 2 - \frac{1}{3}p \right| - \frac{7}{6} = \frac{1}{2}$

45. $|3y + 5| = |y + 1|$

46. $|2a - 3| = |a + 2|$

47. $|4 - x| = |2x + 1|$

48. $|3 - 2x| = |x + 5|$

49. $\left| \frac{1}{4}w \right| = |4w|$

50. $|3z| = \left| \frac{1}{3}z \right|$

51. $|x + 4| = |x - 7|$

52. $|k - 3| = |k + 3|$

53. $|2p - 1| = |1 - 2p|$

54. $|4d - 3| = |3 - 4d|$

Objective 4: Solve Radical Equations and Equations with Rational Exponents

For Exercises 55–80, solve the equation. (See Examples 7–9)

55. $\sqrt{2x - 4} = 6$

56. $\sqrt{3x + 1} = 11$

57. $1 = 3 + \sqrt{2x + 7}$

58. $6 = 9 + \sqrt{5 - 3x}$

59. $\sqrt{7x + 8} = x + 2$

60. $\sqrt{9x + 19} = x + 3$

61. $\sqrt{m + 18} + 2 = m$

62. $\sqrt{2n + 29} + 3 = n$

63. $-4\sqrt[3]{2x - 5} + 6 = 10$

64. $-3\sqrt[5]{4x - 1} + 2 = 8$

65. $\sqrt[4]{5y - 3} - \sqrt[4]{2y + 1} = 0$

66. $\sqrt[6]{y + 7} - \sqrt[6]{4y + 5} = 0$

67. $\sqrt{8 - p} - \sqrt{p + 5} = 1$

68. $\sqrt{d + 4} - \sqrt{6 + 2d} = -1$

69. $3 - \sqrt{y + 3} = \sqrt{2 - y}$

70. $\sqrt{k - 2} = \sqrt{2k + 3} - 2$

71. $2(x + 5)^{2/3} = 18$

72. $3(x - 6)^{2/3} = 48$

73. $(3x + 1)^{3/2} + 2 = 66$

74. $(2x - 1)^{3/2} - 3 = 122$

75. $m^{3/4} = 5$

76. $n^{5/6} = 3$

77. $2p^{4/5} = \frac{1}{8}$

78. $5t^{2/3} = \frac{1}{5}$

79. $(2v + 7)^{1/3} - (v - 3)^{1/3} = 0$

80. $(5u - 6)^{1/5} - (3u + 1)^{1/5} = 0$

Objective 5: Solve Equations in Quadratic Form

For Exercises 81–106, make an appropriate substitution and solve the equation. (See Examples 10–11)

81. $(2x + 5)^2 - 7(2x + 5) - 30 = 0$

82. $(3x - 7)^2 - 6(3x - 7) - 16 = 0$

83. $(x^2 + 2x)^2 - 18(x^2 + 2x) = -45$

84. $(x^2 + 3x)^2 - 14(x^2 + 3x) = -40$

85. $(x^2 + 2)^2 + (x^2 + 2) - 42 = 0$

86. $(y^2 - 3)^2 - 9(y^2 - 3) - 52 = 0$

87. $-\frac{2}{a^2} + \frac{4}{a} + 1 = 0$

88. $-\frac{4}{x^2} - \frac{4}{x} + 1 = 0$

89. $\frac{2}{(n + 2)^2} - \frac{3}{n + 2} = 5$

90. $\frac{3}{(m - 3)^2} - \frac{7}{m - 3} = -4$

91. $\left(m - \frac{10}{m} \right)^2 - 6\left(m - \frac{10}{m} \right) - 27 = 0$

92. $\left(x + \frac{6}{x} \right)^2 - 12\left(x + \frac{6}{x} \right) + 35 = 0$

93. $\left(2 + \frac{3}{t} \right)^2 - \left(2 + \frac{3}{t} \right) = 12$

94. $\left(\frac{5}{y} + 3 \right)^2 + 6\left(\frac{5}{y} + 3 \right) = -8$

95. $5c^{2/5} - 11c^{1/5} + 2 = 0$

96. $3d^{2/3} - d^{1/3} - 4 = 0$

97. $y^{1/2} - y^{1/4} - 6 = 0$

98. $n^{1/2} + 6n^{1/4} - 16 = 0$

99. $9y^{-4} - 10y^{-2} + 1 = 0$

100. $100x^{-4} - 29x^{-2} + 1 = 0$

101. $4t - 25\sqrt{t} = 0$

102. $9m - 16\sqrt{m} = 0$

103. $x^2(x^2 + 5) = 7$

104. $x^2(x^2 - 2) = x^2 + 13$

105. $30k^{-2} - 23k^{-1} + 2 = 0$

106. $3q^{-2} + 16q^{-1} + 5 = 0$

Mixed Exercises

For Exercises 107–118, solve for the indicated variable.

107. $\dfrac{1}{f} = \dfrac{1}{p} + \dfrac{1}{q}$ for p

108. $\dfrac{1}{R} = \dfrac{1}{R_1} + \dfrac{1}{R_2} + \dfrac{1}{R_3}$ for R_3

109. $E = kT^4$ for $T > 0$

110. $V = \dfrac{4}{3}\pi r^3$ for $r > 0$

111. $a = \dfrac{kF}{m}$ for m

112. $V = \dfrac{k}{P}$ for P

113. $16 + \sqrt{x^2 - y^2} = z$ for x

114. $4 + \sqrt{x^2 + y^2} = z$ for y

115. $\dfrac{P_1 V_1}{T_1} = \dfrac{P_2 V_2}{T_2}$ for T_1

116. $\dfrac{t_1}{s_1 v_1} = \dfrac{t_2}{s_2 v_2}$ for v_2

117. $T = 2\pi\sqrt{\dfrac{L}{g}}$ for g

118. $t = \sqrt{\dfrac{2s}{g}}$ for s

For Exercises 119–120, solve the equation in two ways.

a. Solve as a radical equation by first isolating the radical.

b. Solve by writing the equation in quadratic form and using an appropriate substitution.

119. $y + 4\sqrt{y} = 21$

120. $w - 3\sqrt{w} = 10$

For Exercises 121–122, solve the equation.

121. $\sqrt{x + \sqrt{x + 2}} = 3$

122. $\sqrt{1 + \sqrt{x + \sqrt{x + 1}}} = 2$

123. The equation $r = \sqrt[3]{\dfrac{3V}{4\pi}}$ gives the radius r of a sphere of volume V. If the radius of a sphere is 6 in., find the exact volume.

124. The distance d (in miles) that an observer can see on a clear day is approximated by $d = \dfrac{49}{40}\sqrt{h}$, where h is the height of the observer in feet. If Rita can see 24.5 mi, how far above ground is her eye level?

125. The percentage of drug released in the bloodstream t hours after being administered is affected by numerous variables including drug solubility and filler ingredients. For a particular drug and dosage, the percentage of drug released P is given by $P = 48t^{1/5}$ $(0 \le t \le 35)$. For example, the value $P = 50$ represents 50% of the drug released.

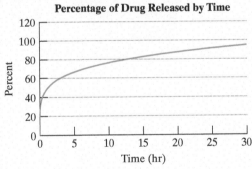

Percentage of Drug Released by Time

a. Determine the percentage of drug released after 2 hr. Round to the nearest percent.

b. How many hours is required for 75% of the drug to be released? Round to the nearest tenth of an hour.

126. A tomato plant is purchased at a garden supply store. The initial height of the plant is 25.4 in. The height of the plant h (in inches) is approximated by $h = 16(t + 4)^{1/3}$, where t is the time in days after planting.

a. Determine the height of the plant 14 days after planting. Round to the nearest inch.

b. How long after the plant is planted will it take for the height to reach 5 ft? Round to the nearest day.

127. If an object is dropped from a height of h meters, the velocity v (in m/sec) at impact is given by $v = \sqrt{2gh}$, where $g = 9.8$ m/sec^2 is the acceleration due to gravity.

a. Determine the impact velocity for an object dropped from a height of 10 m.

b. Determine the height required for an object to have an impact velocity of 26.8 m/sec (≈ 60 mph). Round to the nearest tenth of a meter.

128. The yearly depreciation rate for a certain vehicle is modeled by $r = 1 - \left(\dfrac{V}{C}\right)^{1/n}$, where V is the value of the car after n years, and C is the original cost.

a. Determine the depreciation rate for a car that originally cost $18,000 and is worth $12,000 after 3 yr. Round to the nearest tenth of a percent.

b. Determine the original cost of a truck that has a yearly depreciation rate of 15% and is worth $11,000 after 5 yr. Round to the nearest $100.

For Exercises 129–130,

a. Write an absolute value equation to represent each statement.

b. Solve the equation.

129. The distance between a number x and 4 on the number line is 6.

130. The distance between a number x and 3 on the number line is 8.

Write About It

131. Explain how to determine if an equation is in quadratic form.

132. Why must the potential solutions to a radical equation be checked in the original equation?

Expanding Your Skills

133. Joan and Henry both work for a mail-order company preparing packages for shipping. It takes Henry approximately 1 hr longer to fill 100 orders than Joan. If they work together, it takes 3 hr to fill 100 orders. Find the amount of time required for each individual to fill 100 orders working alone. Round to the nearest tenth of an hour.

134. Antonio and Jeremy work for a plumbing company that was recently awarded a contract to install the plumbing fixtures in a new office complex. There are 12 bathrooms in the building. It takes Jeremy 4 hr longer than Antonio to complete one bathroom. If they work together it takes 8 hr to complete a bathroom. Find the rate at which each individual can complete a bathroom working alone.

135. Pam is in a canoe on a lake 400 ft from the closest point on a straight shoreline. Her house is 800 ft up the road along the shoreline. She can row 2.5 ft/sec and she can walk 5 ft/sec. If the total time it takes for her to get home is 5 min (300 sec), determine the point along the shoreline at which she landed her canoe.

136. Martha is in a boat in the ocean 48 mi from point A, the closest point along a straight shoreline. She needs to dock the boat at a marina x miles farther up the coast, and then drive along the coast to point B, 96 mi from point A. Her boat travels 20 mph, and she drives 60 mph. If the total trip took 4 hr, determine the distance x along the shoreline.

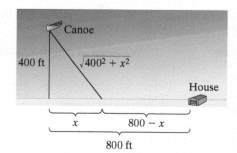

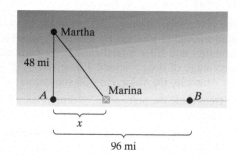

SECTION 1.7   Linear, Compound, and Absolute Value Inequalities

1. Solve Linear Inequalities in One Variable

Emily wants to earn an "A" in her College Algebra course and knows that the average of her tests and assignments must be at least 90. She has five test grades of 96, 84, 80, 98, and 88. She also has a score of 100 for online homework, and this carries the same weight as a test grade. She still needs to take the final exam and the final is weighted as two test grades. To determine the scores on the final exam that would result in an average of 90 or more, Emily would solve the following inequality (see Example 9):

$$\frac{96 + 84 + 80 + 98 + 88 + 100 + 2x}{8} \geq 90,$$ where x is Emily's score on the final.

A linear equation in one variable is an equation that can be written as $ax + b = 0$, where a and b are real numbers and $a \neq 0$. A **linear inequality** is any relationship of

TIP For a review of set-builder notation and interval notation, see Section R.1.

the form $ax + b < 0$, $ax + b \leq 0$, $ax + b > 0$, or $ax + b \geq 0$. The solution set to a linear equation consists of a single element that can be represented by a point on the number line. The solution set to a linear inequality contains an infinite number of elements and can be expressed in set-builder notation or in interval notation.

Equation/ Inequality	Solution Set	Graph
$x + 4 = 0$	$\{-4\}$	
$x + 4 \geq 0$	$\{x \mid x \geq -4\}$ or $[-4, \infty)$	
$x + 4 < 0$	$\{x \mid x < -4\}$ or $(-\infty, -4)$	

To solve a linear inequality in one variable, we use the following properties of inequality.

Properties of Inequality

Let a, b, and c represent real numbers.

1. If $x < a$, then $a > x$.
2. If $a < b$ and $b < c$, then $a < c$.
3. If $a < b$ and $c < d$, then $a + c < b + d$.
4. If $a < b$, then $a + c < b + c$ and $a - c < b - c$.
5. If c is *positive* and $a < b$, then $ac < bc$ and $\dfrac{a}{c} < \dfrac{b}{c}$.
6. If c is *negative* and $a < b$, then $ac > bc$ and $\dfrac{a}{c} > \dfrac{b}{c}$.

These statements are also true expressed with the symbols $>$, $\leq$, and $\geq$.

Property 6 indicates that if both sides of an inequality are multiplied or divided by a negative number, then the direction of the inequality sign must be reversed.

EXAMPLE 1   **Solving a Linear Inequality**

Solve the inequality. Graph the solution set and write the solution set in set-builder notation and in interval notation.

$$-6x + 4 < 34$$

Solution:

$$-6x + 4 < 34$$
$$-6x + 4 - 4 < 34 - 4 \qquad \text{Subtract } 4 \text{ from both sides.}$$
$$-6x < 30$$
$$\frac{-6x}{-6} > \frac{30}{-6} \qquad \text{Divide both sides by } -6. \text{ Reverse the inequality sign.}$$
$$x > -5$$

The solution set is $\{x \mid x > -5\}$.
Interval notation: $(-5, \infty)$

Answer

1.

$[t \mid t \leq -6]; (-\infty, -6]$

Skill Practice 1   Solve the inequality. Graph the solution set and write the solution set in set-builder notation and in interval notation. $-5t - 6 \geq 24$

> **TIP** In Example 1, the solution set to the inequality $-6x + 4 < 34$ is $\{x \mid x > -5\}$. This means that all numbers greater than -5 make the inequality a true statement. You can check by taking an arbitrary test point from the interval $(-5, \infty)$. For example, the value $x = -4$ makes the original inequality true.
>
>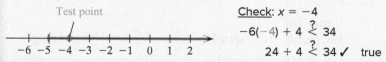
>
> Test point
>
> Check: $x = -4$
>
> $-6(-4) + 4 \overset{?}{<} 34$
>
> $24 + 4 \overset{?}{<} 34 \checkmark$ true

EXAMPLE 2   Solving a Linear Inequality Containing Fractions

Solve the inequality. Graph the solution set and write the solution set in set-builder notation and in interval notation.

$$\frac{x + 1}{3} - \frac{2x - 4}{6} \le -\frac{x}{2}$$

Solution:

$$\frac{x + 1}{3} - \frac{2x - 4}{6} \le -\frac{x}{2}$$

$$6\left(\frac{x + 1}{3} - \frac{2x - 4}{6}\right) \le 6\left(-\frac{x}{2}\right) \qquad \text{Multiply both sides by the LCD of } 6 \text{ to clear fractions.}$$

$$2(x + 1) - (2x - 4) \le -3x$$

$$2x + 2 - 2x + 4 \le -3x \qquad \text{Apply the distributive property.}$$

$$6 \le -3x$$

$$\frac{6}{-3} \ge \frac{-3x}{-3} \qquad \text{Divide both sides by } -3. \text{ Since } -3 \text{ is a negative number, reverse the inequality sign.}$$

$$-2 \ge x \quad \text{or} \quad x \le -2$$

The solution set is $\{x \mid x \le -2\}$.

Interval notation: $(-\infty, -2]$

> **Skill Practice 2**   Solve the inequality. Graph the solution set and write the solution set in set-builder notation and in interval notation.
>
> $$\frac{m - 4}{2} - \frac{3m + 4}{10} > -\frac{3m}{5}$$

2. Solve Compound Linear Inequalities

In Examples 3–5, we solve **compound inequalities.** These are statements with two or more inequalities joined by the word "and" or the word "or." For example, suppose that x represents the glucose level measured from a fasting blood sugar test.

- The normal glucose range is given by $x \ge 70$ mg/dL and $x \le 100$ mg/dL.
- An abnormal glucose level is given by $x < 70$ mg/dL or $x > 100$ mg/dL.

To find the solution sets for compound inequalities follow these guidelines.

Answer

2.

$\{m \mid m > 3\}$; $(3, \infty)$

> **Solving a Compound Inequality**
>
> **Step 1** To solve a compound inequality, first solve the individual inequalities.
>
> **Step 2** • If two inequalities are joined by the word "and," the solutions are the values of the variable that simultaneously satisfy each inequality. That is, we take the *intersection* of the individual solution sets.
>
> • If two inequalities are joined by the word "or," the solutions are the values of the variable that satisfy either inequality. Therefore, we take the *union* of the individual solution sets.

EXAMPLE 3   **Solving a Compound Inequality "Or"**

Solve. $x - 2 \leq 5$ or $\frac{1}{2}x > 6$

Solution:

$x - 2 \leq 5$ or $\frac{1}{2}x > 6$ First solve the individual inequalities. Then take the *union* of the individual solution sets.

$x \leq 7$ or $x > 12$

$x \leq 7$

$x > 12$

$x \leq 7 \text{ or } x > 12$

The solution set is $\{x \mid x \leq 7 \text{ or } x > 12\}$.

Interval notation: $(-\infty, 7] \cup (12, \infty)$

Skill Practice 3   Solve. $\frac{1}{4}y < -1$ or $3 + y \geq 5$

In Example 4, we solve a compound inequality in which the individual inequalities are joined by the word "and." In this case, we take the intersection of the individual solution sets.

EXAMPLE 4   **Solving a Compound Inequality "And"**

Solve. $-\frac{1}{4}t < 2$ and $0.52t \geq 1.3$

Solution:

$-\frac{1}{4}t < 2$ and $0.52t \geq 1.3$ First solve the individual inequalities. Multiply both sides of the first inequality by -4 (reverse the inequality sign). Divide the second inequality by 0.52.

$t > -8$ and $t \geq 2.5$ Take the *intersection* of the individual solution sets. The intervals overlap for values of t greater than or equal to 2.5.

Answer

3. $\{y \mid y < -4$ or $y \geq 2\}$;
$(-\infty, -4) \cup [2, \infty)$

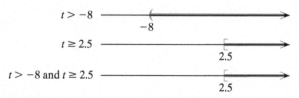

The solution set is $\{t|t \geq 2.5\}$. Interval notation: $[2.5, \infty)$

Skill Practice 4   Solve. $0.36w \leq 0.54$ and $-\dfrac{1}{2}w > 3$

Sometimes a compound inequality joined by the word "and" is written as a three-part inequality. For example:

$5 < -2x + 7$ and $-2x + 7 \leq 11$ In this example, two simultaneous conditions are imposed on the quantity $-2x + 7$.

$$5 < -2x + 7 \leq 11$$

To solve a three-part inequality, the goal is to isolate x in the middle region. This is demonstrated in Example 5.

EXAMPLE 5   **Solving a Three-Part Compound Inequality**

Solve. $5 < -2x + 7 \leq 11$

Solution:

$$5 < -2x + 7 \leq 11$$

$$5 - 7 < -2x + 7 - 7 \leq 11 - 7 \qquad \text{Subtract } 7 \text{ from all three parts of the inequality.}$$

$$-2 < -2x \leq 4$$

$$\frac{-2}{-2} > \frac{-2x}{-2} \geq \frac{4}{-2} \qquad \text{Divide all three parts by } -2.$$

$$1 > x \geq -2 \text{ or equivalently } -2 \leq x < 1.$$

The solution set is $\{x|-2 \leq x < 1\}$.
Interval notation: $[-2, 1)$

Skill Practice 5   Solve. $-16 \leq -3y - 4 < 2$

---

3. Solve Absolute Value Inequalities

TIP The solution set to $|x| < 3$ is the set of real numbers within 3 units of zero on the number line.

The solution set to $|x| > 3$ is the set of real numbers more than 3 units from zero on the number line.

We now investigate the solutions to absolute value inequalities. For example:

Inequality	Graph	Solution Set
$|x| < 3$		$\{x|-3 < x < 3\}$
$|x| > 3$		$\{x|x < -3 \text{ or } x > 3\}$

We can generalize these observations with the following properties involving absolute value inequalities.

Answers

4. $\{w|w < -6\}; (-\infty, -6)$
5. $\{y|-2 < y \leq 4\}; (-2, 4]$

Properties Involving Absolute Value Inequalities

For a real number $k > 0$,

> **1.** $|u| < k$ is equivalent to $-k < u < k$.   (1)
> **2.** $|u| > k$ is equivalent to $u < -k$ or $u > k$.   (2)

Note: The statements also hold true for the inequality symbols $\leq$ and $\geq$, respectively.

Properties (1) and (2) follow directly from the definition: $|u| = \begin{cases} u \text{ if } u \geq 0 \\ -u \text{ if } u < 0 \end{cases}$

By definition, $|u| < k$ is equivalent to

$$u < k \quad \text{and} \quad -u < k$$
$$u < k \quad \text{and} \quad u > -k$$
$$-k < u < k \qquad (1)$$

By definition, $|u| > k$ is equivalent to

$$u > k \quad \text{or} \quad -u > k$$
$$u > k \quad \text{or} \quad u < -k$$
$$u < -k \quad \text{or} \quad u > k \qquad (2)$$

EXAMPLE 6   **Solving an Absolute Value Inequality**

Solve the inequality and write the solution set in interval notation.

$$2|6 - m| - 3 < 7$$

Solution:

$2|6 - m| - 3 < 7$	First isolate the absolute value. Add 3 and divide by 2.
$|6 - m| < 5$	The inequality is in the form $|u| < k$, where $u = 6 - m$.
$-5 < 6 - m < 5$	Write the equivalent compound inequality $-k < u < k$.
$-11 < -m < -1$	Subtract 6 from all three parts.
$\dfrac{-11}{-1} > \dfrac{-m}{-1} > \dfrac{-1}{-1}$	Divide by -1 and reverse the inequality signs.

$11 > m > 1$ or equivalently $1 < m < 11$.

The solution set is $\{m \mid 1 < m < 11\}$.
Interval notation: $(1, 11)$

Skill Practice 6   Solve the inequality and write the solution set in interval notation. $3|5 - x| + 2 \leq 14$

EXAMPLE 7   **Solving an Absolute Value Inequality**

Solve the inequality and write the solution set in interval notation.

$$-4 \geq -2|3x + 1|$$

Solution:

$$-4 \geq -2|3x + 1|$$ First isolate the absolute value.

$$\frac{-4}{-2} \leq \frac{-2|3x + 1|}{-2}$$ Divide both sides by -2 and reverse the inequality sign.

$$2 \leq |3x + 1|$$ Write the absolute value on the left. Notice that the direction of the inequality sign is also changed. The inequality is now in the form $|u| \geq k$, where $u = 3x + 1$.

$$|3x + 1| \geq 2$$

$$3x + 1 \leq -2 \quad \text{or} \quad 3x + 1 \geq 2$$ Write the equivalent form $u \leq -k$ or $u \geq k$.

$$x \leq -1 \quad \text{or} \quad x \geq \frac{1}{3}$$ Take the union of the solution sets of the individual inequalities.

The solution set is $\left\{ x \mid x \leq -1 \quad \text{or} \quad x \geq \frac{1}{3} \right\}$.

Interval notation: $(-\infty, -1] \cup \left[\dfrac{1}{3}, \infty \right)$

Skill Practice 7   Solve the inequality and write the solution set in interval notation. $-18 > -3|2y - 4|$

An absolute value equation such as $|7x - 3| = -6$ has no solution because an absolute value cannot be equal to a negative number. We must also exercise caution when an absolute value is compared to a negative number or zero within an inequality. This is demonstrated in Example 8.

EXAMPLE 8   **Solving Absolute Value Inequalities with Special Case Solution Sets**

Solve the inequality and write the solution set in interval notation where appropriate.

a. $|x + 2| < -4$ **b.** $|x + 2| \geq -4$
c. $|x - 5| \leq 0$ **d.** $|x - 5| > 0$

Solution:

a. $|x + 2| < -4$ By definition an absolute value is greater than or equal to zero. Therefore, the absolute value of an expression cannot be less than zero or any negative number. This inequality has no solution.

The solution set is $\{ \ \}$.

b. $|x + 2| \geq -4$ An absolute value of any real number is greater than or equal to zero. Therefore, it is also greater than every negative number. This inequality is true for all real numbers, x.

The solution set is $\mathbb{R}$.
Interval notation: $(-\infty, \infty)$

c. $|x - 5| \leq 0$ The absolute value of x minus 5 cannot be less than zero, but it can be *equal* to zero.

$$|x - 5| = 0$$
$$x - 5 = 0$$
$$x = 5$$

The solution set is $\{5\}$.

Answer
7. $(-\infty, -1) \cup (5, \infty)$

d. $|x - 5| > 0$

$|x - 5| > 0$ for all values of x except 5. When $x = 5$, we have $|5 - 5| = 0$, and this is not greater than zero. The solution set is all real numbers excluding 5.

The solution set is $\{x | x < 5 \text{ or } x > 5\}$.
Interval notation: $(-\infty, 5) \cup (5, \infty)$

5

Skill Practice 8   Solve.

a. $|x - 3| < -2$ **b.** $|x - 3| > -2$
c. $|x + 1| \leq 0$ **d.** $|x + 1| > 0$

---

4. Solve Applications of Inequalities

In Example 9, we use a linear inequality to solve an application.

EXAMPLE 9   Using a Linear Inequality in an Application of Grades

Emily has test scores of 96, 84, 80, 98, and 88. Her score for online homework is 100 and is weighted as one test grade. Emily still needs to take the final exam, which counts as two test grades. What score does she need on the final exam to have an average of at least 90? (This is the minimum average to earn an "A" in the class.)

Solution:

Let x represent the grade needed on the final exam.

$$\left(\begin{array}{c}\text{Average of}\\\text{all scores}\end{array}\right) \geq 90$$

To earn an "A," Emily's average must be at least 90.

$$\frac{96 + 84 + 80 + 98 + 88 + 100 + 2x}{8} \geq 90$$

Take the sum of all grades. Divide by a total of eight grades.

$$\frac{546 + 2x}{8} \geq 90$$

$$8\left(\frac{546 + 2x}{8}\right) \geq 8(90)$$

Multiply by 8 to clear fractions.

$$546 + 2x \geq 720$$

$$2x \geq 174$$

Subtract 546 from both sides.

$$x \geq 87$$

Divide by 2.

Emily must earn a score of at least 87 to earn an "A" in the class.

Skill Practice 9   For a recent year, the monthly snowfall (in inches) for Chicago, Illinois, for November, December, January, and February was 2, 8.4, 11.2, and 7.9, respectively. How much snow would be necessary in March for Chicago to exceed its monthly average snowfall of 7.28 in. for these five months?

EXAMPLE 10   Applying an Absolute Value Inequality

Suppose that a machine is calibrated to dispense 8 fl oz of orange juice into a plastic bottle, with a measurement error of no more than 0.05 fl oz. Let x represent the actual amount of orange juice poured into the bottle.

Answers
8. a. $\{ \}$ **b.** $\mathbb{R}; (-\infty, \infty)$
c. $\{-1\}$ **d.** $(-\infty, -1) \cup (-1, \infty)$
9. Chicago would need more than 6.9 in. of snow in March.

a. Write an absolute value inequality that represents an interval in which to estimate x.

b. Solve the inequality and interpret the answer.

Solution:

a. The measurement error is ± 0.05 fl oz. This means that the value of x can deviate from 8 fl oz by as much as ± 0.05 fl oz.

$|x - 8| \le 0.05$ The distance between x and 8 is no more than 0.05 unit.

b. $|x - 8| \le 0.05$

$-0.05 \le x - 8 \le 0.05$ The amount of orange juice in the bottle is between

$7.95 \le x \le 8.05$ 7.95 fl oz and 8.05 fl oz, inclusive.

Skill Practice 10   A board is to be cut to a length of 24 in. The measurement error is no more than 0.02 in. Let x represent the actual length of the board.

a. Write an absolute value inequality that represents an interval in which to estimate x.

b. Solve the inequality from part (a) and interpret the meaning.

Answers

10. a. $|x - 24| \le 0.02$
   b. $23.98 \le x \le 24.02$; The actual length of the board is between 23.98 in. and 24.02 in., inclusive.

SECTION 1.7   Practice Exercises

Prerequisite Review

For Exercises R.1–R.3, write the set in set-builder notation (in terms of x) and in interval notation.

R.1. $\xleftarrow{\quad}$ at -1

R.2. $\xleftarrow{\quad}$ at 5

R.3. $\xleftarrow{\quad}$ at -3, 4

For Exercises R.4–R.6, solve the equation.

R.4. $2m + 60 = 84$

R.5. $\dfrac{4}{5} - \dfrac{z + 5}{20} = \dfrac{5z + 3}{4}$

R.6. $2[3 - (m + 6)] - 18 = 5(5 + m)$

Concept Connections

1. The multiplication and division properties of inequality indicate that if both sides of an inequality are multiplied or divided by a negative real number, the direction of the _____ sign must be reversed.

2. If a compound inequality consists of two inequalities joined by the word "and," the solution set is the _____ of the solution sets of the individual inequalities.

3. The compound inequality $a < x$ and $x < b$ can be written as the three-part inequality _____.

4. If a compound inequality consists of two inequalities joined by the word "or," the solution set is the _____ of the solution sets of the individual inequalities.

5. If k is a positive real number, then the inequality $|x| < k$ is equivalent to _____ $< x <$ _____.

6. If k is a positive real number, then the inequality $|x| > k$ is equivalent to $x <$ _____ or x _____ k.

7. If k is a positive real number, then the solution set to the inequality $|x| > -k$ is _____.

8. If k is a positive real number, then the solution set to the inequality $|x| < -k$ is _____.

Objective 1: Solve Linear Inequalities in One Variable

For Exercises 9–26, solve the inequality. Graph the solution set, and write the solution set in set-builder notation and interval notation. (See Examples 1–2)

9. $-2x - 5 > 17$

10. $-8t + 1 < 17$

11. $-3 \leq -\dfrac{4}{3}w + 1$

12. $8 \geq -\dfrac{5}{2}y - 2$

13. $-1.2 + 0.6a \leq 0.4a + 0.5$

14. $-0.7 + 0.3x \leq 0.9x - 0.4$

15. $-5 > 6(c - 4) + 7$

16. $-14 < 3(m - 7) + 7$

17. $\dfrac{4 + x}{2} - \dfrac{x - 3}{5} < -\dfrac{x}{10}$

18. $\dfrac{y + 3}{4} - \dfrac{3y + 1}{6} > -\dfrac{1}{12}$

19. $\dfrac{1}{3}(x + 4) - \dfrac{5}{6}(x - 3) \geq \dfrac{1}{2}x + 1$

20. $\dfrac{1}{2}(t - 6) - \dfrac{4}{3}(t + 2) \geq -\dfrac{3}{4}t - 2$

21. $5(7 - x) + 2x < 6x - 2 - 9x$

22. $2(3x + 1) - 4x > 2(x + 8) - 5$

23. $5 - 3[2 - 4(x - 2)] \geq 6\{2 - [4 - (x - 3)]\}$

24. $8 - [6 - 10(x - 1)] \geq 2\{1 - 3[2 - (x + 4)]\}$

25. $4 - 3k > -2(k + 3) - k$

26. $2x - 9 < 6(x - 1) - 4x$

Objective 2: Solve Compound Linear Inequalities

For Exercises 27–34, solve the compound inequality. Graph the solution set, and write the solution set in interval notation. (See Examples 3–4)

27. a. $x < 4$ and $x \geq -2$
  b. $x < 4$ or $x \geq -2$

28. a. $y \leq -2$ and $y > -5$
  b. $y \leq -2$ or $y > -5$

29. a. $m + 1 \leq 6$ or $\dfrac{1}{3}m < -2$
  b. $m + 1 \leq 6$ and $\dfrac{1}{3}m < -2$

30. a. $n - 6 > 1$ or $\dfrac{3}{4}n \geq 6$
  b. $n - 6 > 1$ and $\dfrac{3}{4}n \geq 6$

31. a. $-\dfrac{2}{3}y > -12$ and $2.08 \geq 0.65y$
  b. $-\dfrac{2}{3}y > -12$ or $2.08 \geq 0.65y$

32. a. $-\dfrac{4}{5}m < 8$ and $0.85 \leq 0.34m$
  b. $-\dfrac{4}{5}m < 8$ or $0.85 \leq 0.34m$

33. a. $3(x - 2) + 2 \leq x - 8$ or $4(x + 1) + 2 > -2x + 4$
  b. $3(x - 2) + 2 \leq x - 8$ and $4(x + 1) + 2 > -2x + 4$

34. a. $5(t - 4) + 2 > 3(t + 1) - 3$ or $2t - 6 > 3(t - 4) - 2$
  b. $5(t - 4) + 2 > 3(t + 1) - 3$ and $2t - 6 > 3(t - 4) - 2$

35. Write $-2.8 < y \leq 15$ as two separate inequalities joined by "and."

36. Write $-\frac{1}{2} \leq z < 2.4$ as two separate inequalities joined by "and."

For Exercises 37–42, solve the compound inequality. Graph the solution set, and write the solution set in interval notation. (See Example 5)

37. $-3 < -2x + 1 \leq 9$

38. $-6 \leq -3x + 9 < 0$

39. $1 \leq \dfrac{5x - 4}{2} < 3$

40. $-2 \leq \dfrac{4x - 1}{3} \leq 5$

41. $-2 \leq \dfrac{-2x + 1}{-3} \leq 4$

42. $-4 < \dfrac{-5x - 2}{-2} < 4$

Objective 3: Solve Absolute Value Inequalities

For Exercises 43–46, solve the equation or inequality. Write the solution set to each inequality in interval notation.

43. a. $|x| = 7$
  b. $|x| < 7$
  c. $|x| > 7$

44. a. $|y| = 8$
  b. $|y| < 8$
  c. $|y| > 8$

45. a. $|a + 9| + 2 = 6$
  b. $|a + 9| + 2 \leq 6$
  c. $|a + 9| + 2 \geq 6$

46. a. $|b + 1| - 4 = 1$
  b. $|b + 1| - 4 \leq 1$
  c. $|b + 1| - 4 \geq 1$

For Exercises 47–60, solve the inequality, and write the solution set in interval notation if possible. (See Examples 6–7)

47. $3|4 - x| - 2 < 16$

48. $2|7 - y| + 1 < 17$

49. $2|x + 3| - 4 \geq 6$

50. $5|x + 1| - 9 \geq -4$

51. $|4w - 5| + 6 \leq 2$

52. $|2x + 7| + 5 < 1$

53. $|5 - p| + 13 > 6$

54. $|12 - 7x| + 5 \geq 4$

55. $-11 \leq 5 - |2p + 4|$

56. $-18 \leq 6 - |3z + 3|$

57. $10 < |-5c - 4| + 2$

58. $15 < |-2d - 3| + 6$

59. $\left| \dfrac{y + 3}{6} \right| < 2$

60. $\left| \dfrac{m - 4}{2} \right| < 14$

For Exercises 61–68, write the solution set. (See Example 8)

61. a. $|x| = -9$
  b. $|x| < -9$
  c. $|x| > -9$

62. a. $|y| = -2$
  b. $|y| < -2$
  c. $|y| > -2$

63. a. $18 = 4 - |y + 7|$
  b. $18 \leq 4 - |y + 7|$
  c. $18 \geq 4 - |y + 7|$

64. a. $15 = 2 - |p - 3|$
  b. $15 \leq 2 - |p - 3|$
  c. $15 \geq 2 - |p - 3|$

65. a. $|z| = 0$
  b. $|z| < 0$
  c. $|z| \leq 0$
  d. $|z| > 0$
  e. $|z| \geq 0$

66. a. $|2w| = 0$
  b. $|2w| < 0$
  c. $|2w| \leq 0$
  d. $|2w| > 0$
  e. $|2w| \geq 0$

67. a. $|k + 4| = 0$
  b. $|k + 4| < 0$
  c. $|k + 4| \leq 0$
  d. $|k + 4| > 0$
  e. $|k + 4| \geq 0$

68. a. $|c - 3| = 0$
  b. $|c - 3| < 0$
  c. $|c - 3| \leq 0$
  d. $|c - 3| > 0$
  e. $|c - 3| \geq 0$

Objective 4: Solve Applications of Inequalities

For Exercises 69–72, write a three-part inequality to represent the given statement.

69. The normal range for the hemoglobin level x for an adult female is greater than or equal to 12.0 g/dL and less than or equal to 15.2 g/dL.

70. A tennis player must play in the "open" division of a tennis tournament if the player's age a is over 18 yr and under 25 yr.

71. The distance d that Zina hits a 9-iron is at least 90 yd, but no more than 110 yd.

72. A small plane's average speed s is at least 220 mph but not more than 410 mph.

73. Marilee wants to earn an "A" in a class and needs an overall average of at least 92. Her test grades are 88, 92, 100, and 80. The average of her quizzes is 90 and counts as one test grade. The final exam counts as 2.5 test grades. What scores on the final exam would result in Marilee's overall average of 92 or greater? **(See Example 9)**

74. A 10-yr-old competes in gymnastics. For several competitions she received the following "All-Around" scores: 36, 36.9, 37.1, and 37.4. Her coach recommends that gymnasts whose "All-Around" scores average at least 37 move up to the next level. What "All-Around" scores in the next competition would result in the child being eligible to move up?

75. Rita earns scores of 78, 82, 90, 80, and 75 on her five chapter tests for a certain class and a grade of 85 on the class project. The overall average for the course is computed as follows: the average of the five chapter tests makes up 60% of the course grade; the project accounts for 10% of the grade; and the final exam accounts for 30%. What scores can Rita earn on the final exam to earn a "B" in the course if the cut-off for a "B" is an overall score greater than or equal to 80, but less than 90? Assume that 100 is the highest score that can be earned on the final exam and that only whole-number scores are given.

76. Trent earns scores of 66, 84, and 72 on three chapter tests for a certain class. His homework grade is 60 and his grade for a class project is 85. The overall average for the course is computed as follows: the average of the three chapter tests makes up 50% of the course grade; homework accounts for 20% of the grade; the project accounts for 10%; and the final exam accounts for 20%. What scores can Trent earn on the final exam to pass the course if he needs a "C" or better? A "C" or better requires an overall score of 70 or better, and 100 is the highest score that can be earned on the final exam. Assume that only whole-number scores are given.

77. A car travels 50 mph and passes a truck traveling 40 mph. How long will it take the car to be more than 16 mi ahead?

78. A work-study job in the library pays $10.75/hr and a job in the tutoring center pays $16.25/hr. How long would it take for a tutor to make over $500 more than a student working in the library? Round to the nearest hour.

79. A rectangular garden is to be constructed so that the width is 100 ft. What are the possible values for the length of the garden if at most 800 ft of fencing is to be used?

80. The lengths of the sides of a triangle are given by three consecutive integers greater than 1. What are the possible values for the shortest side if the perimeter is not to exceed 24 ft?

81. For a certain bowling league, a beginning bowler computes her handicap by taking 90% of the difference between 220 and her average score in league play. Determine the average scores that would produce a handicap of 72 or less. Also assume that a negative handicap is not possible in this league.

82. Body temperature is usually maintained between 36.5°C and 37.5°C, inclusive. Determine the corresponding range of temperatures in Fahrenheit. Use the relationship between degrees Celsius C and degrees Fahrenheit F: $C = \dfrac{5}{9}(F - 32)$.

83. Donovan has offers for two sales jobs. Job A pays a base salary of $25,000 plus a 10% commission on sales. Job B pays a base salary of $30,000 plus 8% commission on sales.

   a. How much would Donovan have to sell for the salary from Job A to exceed the salary from Job B?

   b. If Donovan routinely sells more than $500,000 in merchandise, which job would result in a higher salary?

84. Nancy wants to vacation in Austin, Texas. Hotel A charges $179 per night with a 14% nightly room tax and free parking. Hotel B charges $169 per night with an 18% nightly room tax plus a one-time $40 parking fee. After how many nights will Hotel B be less expensive?

For Exercises 85–88,

a. Write an absolute value inequality to represent each statement.

b. Solve the inequality. Write the solution set in interval notation.

   85. The variation between the measured value v and 16 oz is less than 0.01 oz.

   86. The variation between the measured value t and 60 min is less than 0.2 min.

   87. The value of x differs from 4 by more than 1 unit.

   88. The value of y differs from 10 by more than 2 units.

   89. A refrigerator manufacturer recommends that the temperature t (in °F) inside a refrigerator be 36.5°F. If the thermostat has a margin of error of no more than 1.5°F,

   a. Write an absolute value inequality that represents an interval in which to estimate t. **(See Example 10)**

   b. Solve the inequality and interpret the answer.

90. A box of cereal is labeled to contain 16 oz. A consumer group takes a sample of 50 boxes and measures the contents of each box. The individual content of each box differs slightly from 16 oz, but by no more than 0.5 oz.

   a. If x represents the exact weight of the contents of a box of cereal, write an absolute value inequality that represents an interval in which to estimate x.

   b. Solve the inequality and interpret the answer.

91. The results of a political poll indicate that the leading candidate will receive 51% of the votes with a margin of error of no more than 3%. Let x represent the true percentage of votes received by this candidate.

   a. Write an absolute value inequality that represents an interval in which to estimate x.

   b. Solve the inequality and interpret the answer.

92. A police officer uses a radar detector to determine that a motorist is traveling 34 mph in a 25-mph school zone. The driver goes to court and argues that the radar detector is not accurate. The manufacturer claims that the radar detector is calibrated to be in error by no more than 3 mph.

   a. If x represents the motorist's actual speed, write an inequality that represents an interval in which to estimate x.

   b. Solve the inequality and interpret the answer. Should the motorist receive a ticket?

Mixed Exercises

For Exercises 93–98, determine the set of values for x for which the radical expression would produce a real number. For example, the expression $\sqrt{x - 1}$ is a real number if $x - 1 \geq 0$ or equivalently, $x \geq 1$.

93. a. $\sqrt{x - 2}$ **b.** $\sqrt{2 - x}$ **94. a.** $\sqrt{x - 6}$ **b.** $\sqrt{6 - x}$

95. a. $\sqrt{x + 4}$ **b.** $\sqrt[3]{x + 4}$ **96. a.** $\sqrt{x + 7}$ **b.** $\sqrt[3]{x + 7}$

97. a. $\sqrt{2x - 9}$ **b.** $\sqrt[4]{2x - 9}$ **98. a.** $\sqrt{3x - 7}$ **b.** $\sqrt[3]{3x - 7}$

For Exercises 99–102, answer true or false given that $a > 0$, $b < 0$, $c > 0$, and $d < 0$.

99. $cd > a$ **100.** $ab < c$ **101.** If $a > c$, then $ad < cd$.   **102.** If $a < c$, then $ab < bc$.

For Exercises 103–106, write an absolute value inequality whose solution set is shown in the graph.

103.
−3   7

104.
2   6

105.
4   10

106.
−1   11

Write About It

107. How is the process to solve a linear inequality different from the process to solve a linear equation?

108. Explain why $8 < x < 2$ has no solution.

109. Explain the difference between the solution sets for the following inequalities:

$$|x - 3| \leq 0 \quad \text{and} \quad |x - 3| > 0$$

110. Explain why $x^2 = 4$ is equivalent to the equation $|x| = 2$.

Expanding Your Skills

For Exercises 111–116, solve the inequality and write the solution set in interval notation.

111. $|x| + x < 11$ (*Hint:* Use the definition of $|x|$ to consider two cases.)
Case 1: $x + x < 11$ if $x \geq 0$.
Case 2: $-x + x < 11$ if $x < 0$.

112. $|x| - x > 10$

113. $1 < |x| < 9$

114. $2 < |y| < 11$

115. $5 \leq |2x + 1| \leq 7$

116. $7 \leq |3x - 5| \leq 13$

117. Solve the inequality for p: $|p - \hat{p}| < z\sqrt{\dfrac{\hat{p}\hat{q}}{n}}$. (Do not rationalize the denominator.)

118. Solve the inequality for μ: $|\mu - \bar{x}| < \dfrac{z\sigma}{\sqrt{n}}$. (Do not rationalize the denominator.)

PROBLEM RECOGNITION EXERCISES

Recognizing and Solving Equations and Inequalities

For Exercises 1–20,

   a. Identify the type of equation or inequality (some may fit more than one category).
   b. Solve the equation or inequality. Write the solution sets to the inequalities in interval notation if possible.

- Linear equation or inequality
- Quadratic equation
- Rational equation
- Absolute value equation or inequality

- Radical equation
- Equation in quadratic form
- Polynomial equation (degree > 2)
- Compound inequality

1. $(x^2 - 5)^2 - 5(x^2 - 5) + 4 = 0$

2. $2 \leq |3t - 1| - 6$

3. $\sqrt[3]{2y - 5} - 4 = -1$

4. $-9|3z - 7| + 1 = 4$

5. $\dfrac{2}{w - 3} + \dfrac{5}{w + 1} = 1$

6. $48x^3 + 80x^2 - 3x - 5 = 0$

7. $-2(m + 2) < -m + 5$ and $6 \geq m + 3$

8. $6 \leq -2c + 8$ or $\dfrac{1}{3}c - 2 < 2$

9. $(2p + 1)(p + 5) = 2p + 40$

10. $2x(x - 4) + 7 = 2x^2 - 3[x + 5 - (2 + x)]$

11. $\dfrac{a - 4}{2} - \dfrac{3a + 1}{4} \leq -\dfrac{a}{8}$

12. $3x^2 + 11 = 4$

13. $-1 \leq \dfrac{6 - x}{-5} \leq 7$

14. $5 = \sqrt{5 + 2n} + \sqrt{2 + n}$

15. $|4x - 5| = |3x - 2|$

16. $\dfrac{1}{d} - \dfrac{1}{2d - 1} + \dfrac{2d}{2d - 1} = 0$

17. $-|x + 4| + 8 > 3$

18. $y - 4\sqrt{y} - 12 = 0$

19. $c^{2/3} = 16$

20. $2|z - 14| + 8 > 4$

EQUATIONS AND INEQUALITIES FOR CALCULUS

In Calculus you will see the symbol y'. For Exercises 1–4, treat y' as a variable and solve the equation for y'.

1. $\dfrac{2x}{25} + \dfrac{2y}{9}y' = 0$

2. $2xy^3 + 3x^2y^2y' - y' = 1$

3. $3y^2y' + 6xy + 3x^2y' = 2y^2 + 4xyy'$

4. $3(x + y)^2 + 3(x + y)^2 y' - 3y^2y' = 3x^2$

For Exercises 5–7, simplify the expression. Do not rationalize the denominator.

5. $2x\sqrt{2x - 3} + x^2\left(\dfrac{1}{2}\right)\dfrac{1}{\sqrt{2x - 3}}(2)$

6. $\dfrac{2x(2x - 7)^{1/2} - x^2\left(\dfrac{1}{2}\right)(2x - 7)^{-1/2}(2)}{\left[(2x - 7)^{1/2}\right]^2}$

7. $\dfrac{(1)(x^2 - 9)^{1/2} - x\left(\dfrac{1}{2}\right)(x^2 - 9)^{-1/2}(2x)}{\left[(x^2 - 9)^{1/2}\right]^2}$

For Exercises 8–10,

  a. Simplify the expression. Do not rationalize the denominator.

  b. Find the values of x for which the expression equals zero.

  c. Find the values of x for which the denominator is zero.

8. $\dfrac{4x(4x - 5) - 2x^2(4)}{(4x - 5)^2}$

9. $\dfrac{-6x(6x + 1) - (-3x^2)(6)}{(6x + 1)^2}$

10. $\sqrt{4 - x^2} - x\left(\dfrac{1}{2}\right)\dfrac{1}{\sqrt{4 - x^2}}(2x)$

Some applications of calculus use a mathematical structure called a power series. To find the interval of convergence of a power series, it is often necessary to solve an absolute value inequality. For Exercises 11–12, solve the absolute value inequality to find the interval of convergence.

11. $\left|\dfrac{x + 1}{2}\right| < 1$

12. $\left|-\dfrac{x}{2}\right| < 1$

13. A 6-ft man walks away from a lamppost. At the instant the man is 14 ft away from the lamppost, the shadow is 10 ft. Find the height of the lamppost.

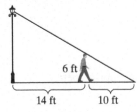

14. A water trough has a cross section in the shape of an equilateral triangle with sides of length 1 m. The length is 3 m. Determine the volume of water when the water level is $\frac{1}{2}$ m.

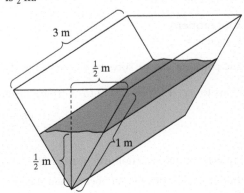

15. A contractor builds a swimming pool with cross section in the shape of a trapezoid. The deep end is 8 ft deep and the shallow end is 3 ft deep. The length of the pool is 50 ft and the width is 20 ft. As the pool is being filled, find the volume of water when the depth is 4 ft.

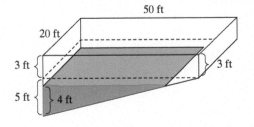

CHAPTER 1   KEY CONCEPTS

SECTION 1.1 Linear Equations and Rational Equations	Reference
A **linear equation in one variable** is an equation that can be written in the form $ax + b = 0$, where a and b are real numbers and $a \neq 0$.	p. 82
A **conditional equation** is an equation that is true for some values of the variable but false for others. A **contradiction** is false for all values of the variable. An **identity** is true for all values of the variable for which the expressions in the equation are defined.	p. 85
Solve a rational equation by multiplying both sides of the equation by the LCD of all fractions in the equation.	p. 86

SECTION 1.2 Applications with Linear and Rational Equations	Reference
Equations in algebra can be used to organize information from a physical situation.	p. 94

SECTION 1.3 Complex Numbers	Reference
$i = \sqrt{-1}$ and $i^2 = -1$. For a real number $b > 0$, $\sqrt{-b} = i\sqrt{b}$.	p. 104
To add or subtract complex numbers, combine the real parts, and combine the imaginary parts.	p. 106
Multiply complex numbers by using the distributive property.	p. 108
The product of complex conjugates: $(a + bi)(a - bi) = a^2 + b^2$	p. 109
Divide complex numbers by multiplying the numerator and denominator by the conjugate of the denominator.	p. 110

SECTION 1.4 Quadratic Equations	Reference
Let a, b, and c represent real numbers. A **quadratic equation** in the variable x is an equation of the form $ax^2 + bx + c = 0$, where $a \neq 0$.	p. 113
Zero product property: If $mn = 0$, then $m = 0$ or $n = 0$.	p. 113
Square root property: If $x^2 = k$, then $x = \pm\sqrt{k}$.	p. 114
A quadratic equation can be solved by completing the square and applying the square root property.	p. 116
The solutions to $ax^2 + bx + c = 0$ $(a \neq 0)$ are given by the quadratic formula.$$x = \frac{-b \pm \sqrt{b^2 - 4ac}}{2a}$$	p. 118
The discriminant to the equation $ax^2 + bx + c = 0$ $(a \neq 0)$ is given by $b^2 - 4ac$. The discriminant indicates the number of and type of solutions to the equation. • If $b^2 - 4ac < 0$, the equation has 2 nonreal solutions. • If $b^2 - 4ac = 0$, the equation has 1 real solution. • If $b^2 - 4ac > 0$, the equation has 2 real solutions.	p. 121

SECTION 1.5 Applications of Quadratic Equations	Reference
Quadratic equations are used to model applications with the Pythagorean theorem, volume, area, and objects moving vertically under the influence of gravity.	p. 126
The vertical position s of an object moving vertically under the influence of gravity is approximated by $s = -\frac{1}{2}gt^2 + v_0 t + s_0$, where • g is the acceleration due to gravity (at sea level on Earth: $g = 32$ ft/sec^2 or 9.8 m/sec^2). • t is the time after the start of the experiment. • v_0 is the initial velocity. • s_0 is the initial position (height). • s is the position of the object at time t.	p. 128

SECTION 1.6 More Equations and Applications	Reference
Polynomial equations: A polynomial equation with one side equal to zero and the other factored as a product of linear or quadratic factors can be solved by applying the zero product property.	p. 133
Absolute value equations: To solve an absolute value equation, isolate the absolute value. Then use one of the following properties for a positive real number k. 1. $\lvert u \rvert = k$ is equivalent to $u = k$ or $u = -k$. 2. $\lvert u \rvert = k$ is equivalent to $u = 0$. 3. $\lvert u \rvert = k$ has no solution. 4. $\lvert u \rvert = \lvert w \rvert$ is equivalent to $u = w$ or $u = -w$.	p. 135
Solving radical equations: 1. Isolate the radical. If an equation has more than one radical, choose one of the radicals to isolate. 2. Raise each side of the equation to a power equal to the index of the radical. 3. Solve the resulting equation. If the equation still has a radical, repeat steps 1 and 2. 4. Check the potential solutions in the original equation and write the solution set.	p. 137
Substitution can be used to solve equations that are in quadratic form.	p. 139

SECTION 1.7 Linear, Compound, and Absolute Value Inequalities	Reference
An inequality that can be written in one of the following forms is a **linear inequality in one variable.** $ax + b < 0$, $ax + b \leq 0$, $ax + b > 0$, or $ax + b \geq 0$	p. 144
Solving compound inequalities: • If two inequalities are joined by the word "and," take the *intersection* of the individual solution sets. • The inequality $a < x < b$ is equivalent to $a < x$ and $x < b$. • If two inequalities are joined by the word "or," take the *union* of the individual solution sets.	p. 147
Solving absolute value inequalities: For a positive real number k, 1. $\lvert u \rvert < k$ is equivalent to $-k < u < k$. 2. $\lvert u \rvert > k$ is equivalent to $u < -k$ or $u > k$.	p. 149

Expanded Chapter Summary available at www.mhhe.com/millercollegealgebra.

CHAPTER 1 Review Exercises

SECTION 1.1

1. Determine the restrictions on x for the equation

$$\frac{3}{x^2 - 4} + \frac{4}{2x - 7} = \frac{2}{3}$$

For Exercises 2–9, solve the equation.

2. $-8(t - 4) + 7 = 4[t - 3(1 - t)] + 6$

3. $\frac{4}{5}x - \frac{2}{3} = \frac{7}{10}x - 2$

4. $\frac{m + 2}{3} - \frac{m - 4}{4} = \frac{m + 1}{6} - 1$

5. $x - 5 + 2(x - 4) = 3(x + 1) - 5$

6. $0.2x + 1.6 = x - 0.8(x - 2)$

7. $(y - 4)^2 = (y + 3)^2$

8. $\frac{x + 3}{5x} + 2 = \frac{x - 4}{x}$

9. $\frac{1}{m - 1} = \frac{5m}{m^2 + 3m - 4} - \frac{3}{m + 4}$

For Exercises 10–12, solve for the indicated variable.

10. $4x - 3y = 6$ for y

11. $t_a = \frac{t_1 + t_2}{2}$ for t_2

12. $4x + 6y = ax + c$ for x

13. Dexter's hybrid car gets 41 mpg in the city and 36 mpg on the highway. The amount of gas he uses A (in gal) is given by $A = \frac{1}{41}c + \frac{1}{36}h$, where c is the number of city miles driven and h is the number of highway miles driven. If Dexter drove 288 mi on the highway and used 11 gal of gas, how many city miles did he drive?

SECTION 1.2

14. Shawna invested a total of $12,000 in two mutual funds: an international fund and a real estate fund. After 1 yr, the international fund earned the equivalent of 8.2% simple interest and the real estate fund returned 1.5%. If the total earnings at the end of the year was $749.50, determine the amount invested in each fund.

15. Cassandra bought a 10-yr Treasury note that paid the equivalent of 3.5% simple interest. She invested $4000 more in a 15-yr bond earning 4.1% than she did in the Treasury note. If the total amount of interest from these investments is $10,180, determine the amount of principal for each investment. Assume that each investment was held to maturity.

16. A chemist mixed 100 cc of a 60% acid solution with a 20% acid solution to produce a 25% acid solution. How much of the 20% solution did he use?

17. Suppose that 250 ft³ of dry concrete mixture is 50% sand by volume. How much pure sand must be added to form a new mixture that is 70% sand by volume?

18. When Kevin commuted to work one morning, his average speed was 45 mph. He averaged only 30 mph for the return trip because of an accident on the highway. If the total time for the round trip was 50 min ($\frac{5}{6}$ hr), determine the distance between his place of work and his home.

19. Two boats leave a marina at the same time. One travels south and the other travels north. The southbound boat travels 6 mph faster than the northbound boat. After 3 hr, the distance between the boats is 66 mi. Determine the speed of each boat.

20. Monique plans to join a gym so that she can use weights and participate in fitness classes. Gym A costs $300/yr plus $4 for each fitness class. Gym B costs $360/yr plus $2 for each class.

    a. Write a model representing the cost C_A (in $) for Gym A if Monique attends x fitness classes.

    b. Write a model representing the cost C_B (in $) for Gym B if Monique attends x fitness classes.

    c. For how many fitness classes will the cost for the two gyms be the same?

21. At a dance studio, each social dance has a $5 door fee.

    a. Write a model for the cost C (in $) for door fees to attend x social dances.

    b. A VIP membership at the studio costs $80 for 3 months, but the door fee is waived for all social dances. How many social dances would be required for a patron to save money by buying the VIP membership?

22. Petra and Dawn are typesetters for a publishing company. Petra can typeset 50 pages in 4 days and Dawn can typeset 50 pages in 3.5 days. How long would it take them to typeset a 150-page manuscript if they work together?

23. One pump can drain a pond in 22 hr. Working with a second pump, the two pumps together can drain the pond in 10 hr. How fast can the second pump drain the pond by itself?

24. On a police force, there are 60 more male officers than female officers. If the ratio of male to female officers is 10:7, determine the number of male and female officers on the force.

25. To estimate the number of turtles in a large pond, a biologist catches and tags 12 turtles. Several weeks later the biologist catches a new sample of 36 turtles from the pond and finds that 3 are tagged. How many turtles are in the pond?

SECTION 1.3

For Exercises 26–28, write each expression in terms of i and simplify.

26. $-\sqrt{-169}$ **27.** $\sqrt{-12}$ **28.** $\sqrt{-16} \cdot \sqrt{-4}$

29. Identify the real and imaginary parts of the complex number.

  a. $3 - 7i$ **b.** $2i$

30. Simplify the powers of i.

  a. i^{35} **b.** i^{56} **c.** i^{62}

  d. i^{17} **e.** i^{-5}

For Exercises 31–40, perform the indicated operations.

31. $\left(\frac{2}{3} + \frac{3}{5}i\right) - \left(\frac{1}{6} + \frac{2}{5}i\right)$ **32.** $3i(7 + 2i)$

33. $\sqrt{-5}(\sqrt{11} + \sqrt{-3})$ **34.** $(4 - 7i)(5 + i)$

35. $(4 - 6i)^2$ **36.** $(2 + \sqrt{-2})(4 + \sqrt{-2})$

37. $(8 - 3i)(8 + 3i)$ **38.** $\frac{4 + 3i}{3 - i}$

39. $(6 - \sqrt{5}i)^{-1}$ **40.** $\frac{7}{4i}$

SECTION 1.4

For Exercises 41–46, solve the equation.

41. $3y^2 - 4y = 8 - 6y$ **42.** $(2v + 3)^2 - 1 = 6$

43. $10t^2 + 1210 = 0$ **44.** $2d(d - 3) = 1 + 4d$

45. $x^2 - 5 = (x + 2)(x - 4)$ **46.** $\frac{1}{5}x^2 - \frac{2}{3} = \frac{7}{15}x$

For Exercises 47–48, determine the value of n that makes the polynomial a perfect square trinomial. Then factor as the square of a binomial.

47. $x^2 + 18x + n$ **48.** $x^2 + \frac{2}{7}x + n$

For Exercises 49–50, solve the equation by using three methods.

a. Factoring and applying the zero product property.

b. Completing the square and applying the square root property.

c. Applying the quadratic formula.

  49. $x^2 - 10x = -9$ **50.** $2x^2 - 3x - 5 = 0$

For Exercises 51–52, answer true or false.

51. The equation $(x - 4)^2 = 25$ is equivalent to $x - 4 = 5$.

52. The equation $(x + 1)^2 = 9$ is equivalent to $x + 1 = \pm 3$.

For Exercises 53–55, (a) evaluate the discriminant and (b) determine the number and type of solutions to each equation.

53. $4x^2 - 20x + 25 = 0$

54. $-2y^2 = 5y - 1$

55. $5t(t + 1) = 4t - 11$

For Exercises 56–58, solve for the indicated variable.

56. $H = kI^2Rt$ for $I > 0$

57. $(x - h)^2 + (y - k)^2 = r^2$ for y

58. $s = a_0t^2 + v_0t + s_0$ for t

SECTION 1.5

59. The area of a triangular plot of land is 52 yd^2. The base is 5 yd longer than the height. Find the base and height.

60. At a textile factory, rectangular pieces of cloth are cut to make tablecloths in which the width is 2 ft less than the length. However, the machine cuts the cloth with an additional 0.5 ft added to both the length and width to account for the hems on all sides. If the piece of cloth is 19.25 ft^2, determine the dimensions of the cloth.

61. A television has a 50-in. diagonal screen. If the aspect ratio (length to width) is 1.6 to 1, determine the length and width of the screen. Round to 1 decimal place.

62. A tablet computer has a 7-in. diagonal screen. The length is 2.7 in. more than the width. Find the length and width of the screen. Round to 1 decimal place.

63. The stopping distance d (in ft) for a car on a certain road is given by $d = 0.048v^2 + 2.2v$, where v is the speed of the car in mph the instant before the brakes were applied.

  a. If the car was traveling 50 mph before the brakes were applied, find the stopping distance.

  b. If the stopping distance is 390 ft, how fast was the car traveling before the brakes were applied? Round to the nearest mile per hour.

64. A fireworks mortar is shot straight upward with an initial velocity of 200 ft/sec from a platform 2 ft off the ground.

  a. Use the formula $s = -\frac{1}{2}gt^2 + v_0t + s_0$ to write a model for the height of the mortar s (in ft) at a time t seconds after launch. Assume that $g = 32$ ft/sec^2.

  b. How long will it take the mortar (on the way up) to clear a tree line that is 80 ft high? Round to the nearest tenth of a second.

SECTION 1.6

For Exercises 65–82, solve the equation.

65. $4x^3 - 6x^2 - 20x + 30 = 0$

66. $3x^2(x^2 + 2) = 20 - x^2$

67. $\sqrt{k + 7} - \sqrt{3 - k} = 2$

68. $\frac{n}{3n + 2} + 1 = \frac{4}{n - 2}$

69. $11v^{-2} + 23v^{-1} + 2 = 0$

70. $\sqrt[3]{4 - x} - \sqrt[3]{2x + 1} = 0$

71. $-2\sqrt{3m + 4} - 3 = 5$

72. $\sqrt{51 - 14x} + 4 = x - 2$

73. $(x - 11)^{2/3} = 9$

74. $(2x + 1)^{3/4} = 27$

75. $-2|3y - 10| + 4 = -6$

76. $|6 - w| + 7 = 2$

77. $|p - 4| = |2p - 3|$

78. $10w^{2/3} = \dfrac{1}{10}$

79. $6d^{2/3} - 7d^{1/3} - 3 = 0$

80. $(2u^2 - 1)^2 - 10(2u^2 - 1) + 9 = 0$

81. $2\left(\dfrac{4}{w} + 1\right)^2 - 10\left(\dfrac{4}{w} + 1\right) = 0$

82. $\dfrac{4v}{5v - 25} - \dfrac{10}{v - 5} = v + \dfrac{4}{5}$

For Exercises 83–85, solve for the indicated variable.

83. $m = \dfrac{1}{2}\sqrt{2a^2 + 2b^2 - 2c^2}$ for $a > 0$

84. $\dfrac{1}{a} = \dfrac{1}{b} + \dfrac{1}{c}$ for b

85. $\dfrac{a_1 t_1}{v_1} = \dfrac{a_2 t_2}{v_2}$ for v_2

SECTION 1.7

For Exercises 86–89, solve the inequality. Graph the solution set and write the solution set in set-builder notation and interval notation.

86. $-4 \le -\dfrac{2}{3}p + 14$

87. $-0.6 + 0.2x < 0.8x - 1.8$

88. $\dfrac{2 + y}{3} - \dfrac{y - 1}{4} < \dfrac{y}{6}$

89. $9 - [5 - 4(t - 1)] \ge 3\{2 - [5 - (t + 2)]\}$

For Exercises 90–91, determine the set of x values for which the radical expression would produce a real number.

90. a. $\sqrt{x - 12}$ **b.** $\sqrt{12 - x}$

91. a. $\sqrt{5x + 7}$ **b.** $\sqrt[3]{5x + 7}$

For Exercises 92–95, solve the compound inequality. Graph the solution set and write the solution set in interval notation.

92. a. $t + 2 \le 8$ or $\dfrac{1}{3}t < -4$

     b. $t + 2 \le 8$ and $\dfrac{1}{3}t < -4$

93. a. $-2(x - 1) + 4 < x + 3$ or $5(x + 2) - 3 \le 4x + 1$

     b. $-2(x - 1) + 4 < x + 3$ and $5(x + 2) - 3 \le 4x + 1$

94. $-11 \le -4x - 1 \le 7$ **95.** $0 < \dfrac{-3x + 9}{-4} < 6$

96. Write an inequality to represent the following statement. A pilot is instructed to keep a plane at an altitude a of over 29,000 ft, but not to exceed 31,000 ft.

97. The months of June, July, August, and September are the wettest months in Miami, Florida, averaging 7.83 in./month. If Miami gets 8.54 in. in June, 5.79 in. in July, and 8.63 in. in August, how much rain is needed in September to exceed the monthly average for these 4 months?

98. A homeowner wants to resod her 2000-ft^2 lawn. Sod varies in price from \$0.10/ft^2 to \$0.30/ft^2 depending on the type of grass. The cost of labor for her gardener to put down the sod is \$400. If the homeowner has budgeted \$850 for the project, determine the price range per square foot of sod that she can afford.

For Exercises 99–105, solve the equation or inequality. Write the solution set to the inequalities in interval notation if possible.

99. a. $|w + 2| + 1 = 6$ **100. a.** $3 = |7x + 1| + 4$

     b. $|w + 2| + 1 < 6$ **b.** $3 < |7x + 1| + 4$

     c. $|w + 2| + 1 \ge 6$ **c.** $3 \ge |7x + 1| + 4$

101. a. $|y + 5| - 3 = -3$ **102. a.** $|x - 1| = |3x + 5|$

     b. $|y + 5| - 3 < -3$ **b.** $|x - 1| = |x + 5|$

     c. $|y + 5| - 3 \le -3$ **c.** $|x - 1| = |1 - x|$

     d. $|y + 5| - 3 > -3$

     e. $|y + 5| - 3 \ge -3$

103. $4|x + 2| - 10 \ge -6$ **104.** $|0.5x - 8| < 0.01$

105. $-9 \le 4 - |2k - 1|$

For Exercises 106–107, (a) write an absolute value inequality that represents the given statement and (b) solve the inequality.

106. The distance between x and 3 on a number line is no more than 0.5.

107. The distance between t and -2 on a number line exceeds 0.01.

CHAPTER 1 Test

1. Write the expression in terms of i and simplify.
$\sqrt{-25} \cdot \sqrt{-4}$

2. Simplify the powers of i.

    a. i^{89} **b.** i^{46} **c.** i^{35} **d.** i^{120} **e.** i^{-11}

For Exercises 3–5, perform the indicated operations. Write the answers in the form $a + bi$.

3. $(4 - 7i)(6 + 2i)$ **4.** $(3 - 5i)^2$ **5.** $\dfrac{4 + 3i}{2 - 5i}$

For Exercises 6–8, (a) evaluate the discriminant and (b) determine the number and type of solutions to the equation.

6. $2x^2 - 4x + 7 = 0$

7. $x^2 + 25 = 10x$

8. $3x(x + 4) = 2x - 2$

For Exercises 9–24, solve the equation.

9. $3y + 2[5(y - 4) - 2] = 5y + 6(7 + y) - 3$

10. $\dfrac{2 + t}{6} - \dfrac{3t - 1}{4} = 1 - \dfrac{2t - 5}{3}$

11. $0.4(w + 1) + 0.8 = 0.1w + 0.3(4 + w)$

12. $\dfrac{-11}{2x^2 + x - 15} - \dfrac{2}{2x - 5} = \dfrac{1}{x + 3}$

13. $(3x - 4)^2 - 2 = 11$

14. $y^2 + 10y = 4$

15. $6t(2t + 1) = 5 - 5t$

16. $\dfrac{3x^2}{4} - x = -\dfrac{1}{2}$

17. $12y^3 + 24y^2 = 3y + 6$

18. $(2y - 3)^{1/3} - (4y + 5)^{1/3} = 0$

19. $\sqrt{2d} = 1 - \sqrt{d + 7}$

20. $\dfrac{c}{c + 6} - 4 = \dfrac{72}{c^2 - 36}$

21. $w^{4/5} - 11 = 0$

22. $\left(5 - \dfrac{2}{k}\right)^2 - 6\left(5 - \dfrac{2}{k}\right) - 27 = 0$

23. $-2 = |x - 3| - 6$

24. $|2v + 5| = |2v - 1|$

For Exercises 25–28, solve for the indicated variable.

25. $aP - 4 = Pt + 2$ for P

26. $\sqrt{a^2 - b^2} = c$ for $b > 0$

27. $-16t^2 + v_0 t + 2 = 0$ for t

28. $a^2 + b^2 + c^2 = 49$ for $c > 0$

For Exercises 29–31, solve the compound inequality. Write the answer in interval notation if possible.

29. $-2 \le \dfrac{4 - x}{3} \le 6$

30. $-\dfrac{4}{3}y < -24$ or $y + 7 \le 2y - 3$

31. $3(x - 5) + 1 \le 4(x + 2) + 6$ and $0.3x - 1.6 > 0.2$

For Exercises 32–35, solve the equations and inequalities. Write the answers to the inequalities in interval notation if possible.

32. $2 < -1 + |4w - 3|$ **33.** $-|8 - v| \ge -6$

34. a. $|7x + 4| + 11 = 2$ **b.** $|7x + 4| + 11 < 2$

    c. $|7x + 4| + 11 > 2$

35. a. $|x - 13| + 4 = 4$ **b.** $|x - 13| + 4 < 4$

    c. $|x - 13| + 4 \le 4$ **d.** $|x - 13| + 4 > 4$

    e. $|x - 13| + 4 \ge 4$

36. How much of an 80% antifreeze solution should be mixed with 2 gal of a 50% antifreeze solution to make a 60% antifreeze solution?

37. Two passengers leave the airport at Denver, Colorado. One takes a 2.3-hr flight to Seattle, Washington, and the other takes a 3.3-hr flight to New York City. The plane flying to New York flies 60 mph faster than the plane flying to Seattle. If the total distance traveled by both planes is 2662 mi, determine the average speed of each plane.

38. Kelly has an aboveground swimming pool. Using water from one hose requires 3 hr to fill the pool. If a second hose is turned on, the pool can be filled in 1.2 hr. How long would it take the second hose to fill the pool if it worked alone?

39. Total cholesterol is made up of LDL and HDL cholesterol. Suppose that the ratio of a patient's total cholesterol to HDL cholesterol is 3.8 and her HDL is 70 mg/dL. Determine the patient's LDL cholesterol level and total cholesterol.

40. A garden area is configured in the shape of a rectangle with two right triangles of equal size attached at the ends. The length of the rectangle is 18 ft. The height of the right triangles is 7 ft longer than the base. If the total area of the garden is 276 ft^2, determine the base and height of the triangular portions.

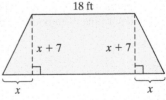

41. A varsity soccer player kicks a soccer ball approximately straight upward with an initial velocity of 60 ft/sec. The ball leaves the player's foot at a height of 2 ft.

 a. Use the formlua
 $s = -\frac{1}{2}gt^2 + v_0t + s_0$
 to write a model representing the height of the ball s (in ft), t seconds after being kicked. Assume that the acceleration due to gravity is $g = 32$ ft/sec^2.

 b. Determine the times at which the ball is 52 ft in the air.

42. A golfer plays 5 rounds of golf with the following scores: 92, 88, 85, 90, and 89. What score would he need on his sixth round to have an average below 88?

43. The equation $r = \sqrt{\dfrac{3V}{\pi h}}$ gives the radius r of a right circular cone of volume V and height h. If the radius is 9 in. for a right circular cone of volume 54π in.^{3}, determine the height of the cone.

CHAPTER 1   Cumulative Review Exercises

For Exercises 1–7, perform the indicated operations and simplify the expression.

1. $[(5x + 3)^2 - (5x - 3)^2]^2$

2. $(4\sqrt{3} + 2\sqrt{2})(4\sqrt{3} - 2\sqrt{2})$

3. $\dfrac{3x^2 - x - 4}{4x^2 - 8x - 12} \div \dfrac{3x - 4}{6x^2 - 54}$

4. $\dfrac{6}{x + 2} - \dfrac{5}{x - 2} + \dfrac{x}{x^2 - 4}$

5. $\dfrac{\dfrac{1}{5x} - \dfrac{3}{5}}{\dfrac{2}{x} + \dfrac{1}{5}}$

6. $\dfrac{2}{\sqrt{7} + \sqrt{3}}$

7. $\sqrt[3]{81y^5z^2w^{12}}$

8. a. Write an absolute value expression that represents the distance between 4π and 11 on the number line.

 b. Simplify the expression from part (a) without absolute value bars.

9. Factor. $4x^3 - 32y^6$

10. Divide and write the answer in the form $a + bi$.
 $\dfrac{3 - 7i}{2 + 5i}$

11. Stephan borrowed a total of $8000 in two student loans. One loan charged 4% simple interest and the other charged 5% simple interest, both payable after graduation. If the interest he owed after 1 yr was $380, determine the amount of principal for each loan.

For Exercises 12–17, solve the equation.

12. $(4x - 1)^2 + 3 = 6$

13. $2x(x - 4) = 2x + 5$

14. $2\left(\dfrac{x}{3} + 1\right)^2 + 5\left(\dfrac{x}{3} + 1\right) - 12 = 0$

15. $\sqrt{x + 4} - 2 = x$

16. $-|5 - x| + 6 = 4$

17. $x - 9 = \dfrac{72}{x - 8}$

18. Given $A = \{x \mid x < 11\}$, $B = \{x \mid x \geq 4\}$, and $C = \{x \mid x < 2\}$, find

 a. $A \cup B$ **b.** $A \cap B$ **c.** $A \cup C$

 d. $A \cap C$ **e.** $B \cup C$ **f.** $B \cap C$

For Exercises 19–20, solve the inequality and write the solution set in interval notation.

19. $|2x - 11| + 1 \leq 12$

20. $-\dfrac{3}{5}y < 15$

Functions and Relations

Chapter Outline

Each year the IRS (Internal Revenue Service) publishes tax rates that tell us how much federal income tax we need to pay based on our taxable income. For example, for a recent year, a single person with taxable income of more than $36,250 but not more than $87,850 pays $4991.25 plus 25% of the amount over $36,250 in federal income tax. However, finding taxable income is not always trivial. There are numerous variables that come into play. The IRS takes into account exemptions, deductions, and tax credits among other things.

In Chapter 2, we will look at mathematical relationships involving two or more variables, including the relationship between taxable income and federal income tax. To fully appreciate the connection among several variables, we will investigate their relationships algebraically, numerically, and graphically.

Schedule X—If your filling status is Single

If your taxable income is			
over—	but not over—	The tax is	of the amount over—
$0	$8925	$0 + 10%	$0
$8925	$36,250	$892.50 + 15%	$8925
$36,250	$87,850	$4991.25 + 25%	$36,250
$87,850	$183,250	$17,891.25 + 28%	$87,850
$183,250	$398,350	$44,603.25 + 33%	$183,250
$398,350	$400,000	$115,586.25 + 35%	$398,350
$400,000	———	$116,163.75 + 39.6%	$400,000

Source: Internal Revenue Service, www.irs.gov

SECTION 2.1   The Rectangular Coordinate System and Graphing Utilities

OBJECTIVES

1. Plot Points on a Rectangular Coordinate System
2. Use the Distance and Midpoint Formulas
3. Graph Equations by Plotting Points
4. Identify *x*- and *y*-Intercepts
5. Graph Equations Using a Graphing Utility

Websites, newspapers, sporting events, and the workplace all utilize graphs and tables to present data. Therefore, it is important to learn how to create and interpret meaningful graphs. Understanding how points are located relative to a fixed origin is important for many graphing applications. For example, computer game developers use a rectangular coordinate system to define the locations of objects moving around the screen.

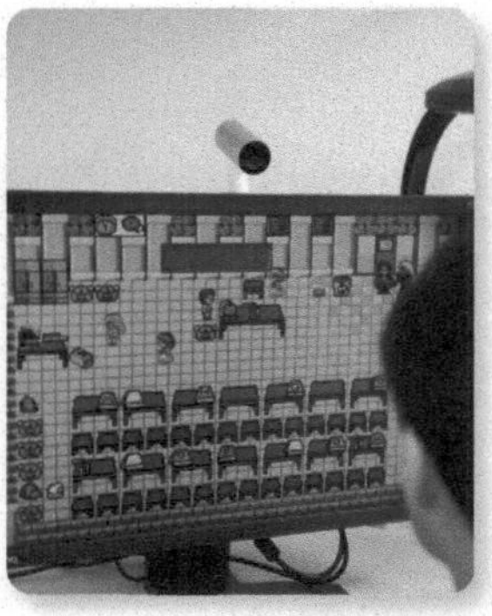

1. Plot Points on a Rectangular Coordinate System

Mathematician René Descartes (pronounced "day cart") (1597–1650) was the first to identify points in a plane by a pair of coordinates. He did this by intersecting two perpendicular number lines with the point of intersection called the **origin.** These lines form a **rectangular coordinate system** (also known in his honor as the **Cartesian coordinate system**) or simply a **coordinate plane.** The horizontal line is called the ***x*-axis** and the vertical line is called the ***y*-axis.** The *x*- and *y*-axes divide the plane into four **quadrants.** The quadrants are labeled counterclockwise as I, II, III, and IV (Figure 2-1).

Every point in the plane can be uniquely identified by using an ordered pair (x, y) to specify its coordinates with respect to the origin. In an ordered pair, the first coordinate is called the ***x*-coordinate,** and the second is called the ***y*-coordinate.** The origin is identified as $(0, 0)$. In Figure 2-2, six points have been graphed. The point $(-3, 5)$, for example, is placed 3 units in the negative *x* direction (to the left) and 5 units in the positive *y* direction (upward).

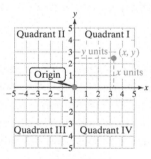

Figure 2-1

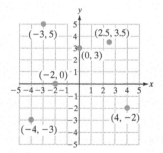

Figure 2-2

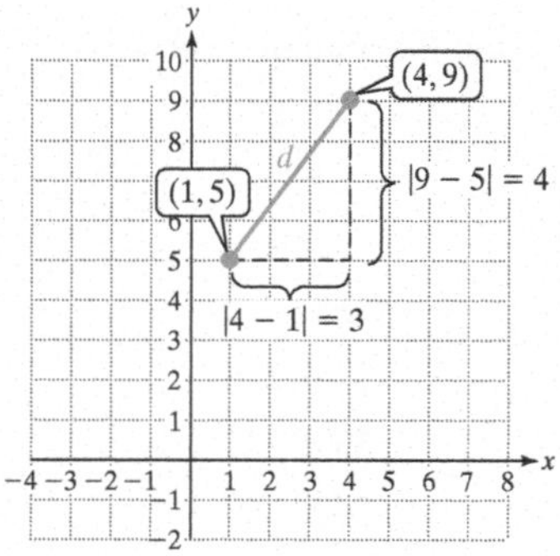

2. Use the Distance and Midpoint Formulas

Recall that the distance between two points *A* and *B* on a number line can be represented by $|A - B|$ or $|B - A|$. Now we want to find the distance between two points in a coordinate plane. For example, consider the points $(1, 5)$ and $(4, 9)$. The distance *d* between the points is labeled in Figure 2-3. The dashed horizontal and vertical line segments form a right triangle with hypotenuse *d*.

The horizontal distance between the points is $|4 - 1| = 3$.
The vertical distance between the points is $|9 - 5| = 4$.

Figure 2-3

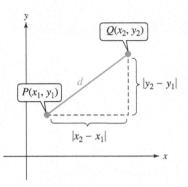

Figure 2-4

Applying the Pythagorean theorem, we have

$$d^2 = (3)^2 + (4)^2$$
$$d = \sqrt{(3)^2 + (4)^2} = \sqrt{25} = 5$$

Since d is a distance, reject the negative square root.

The distance between the points is 5 units.

We can make this process generic by labeling the points $P(x_1, y_1)$ and $Q(x_2, y_2)$. See Figure 2-4.

- The horizontal leg of the right triangle is $|x_2 - x_1|$ or equivalently $|x_1 - x_2|$.
- The vertical leg of the right triangle is $|y_2 - y_1|$ or equivalently $|y_1 - y_2|$.

Applying the Pythagorean theorem, we have

$$d^2 = (x_2 - x_1)^2 + (y_2 - y_1)^2$$
$$d = \sqrt{(x_2 - x_1)^2 + (y_2 - y_1)^2}$$

We can drop the absolute value bars because $|a|^2 = (a)^2$ for all real numbers a. Likewise $|x_2 - x_1|^2 = (x_2 - x_1)^2$ and $|y_2 - y_1|^2 = (y_2 - y_1)^2$.

> **TIP** Since
> $(x_2 - x_1)^2 = (x_1 - x_2)^2$ and
> $(y_2 - y_1)^2 = (y_1 - y_2)^2$,
>
> the distance formula can also be expressed as
> $d = \sqrt{(x_1 - x_2)^2 + (y_1 - y_2)^2}$.

Distance Formula

The distance between points (x_1, y_1) and (x_2, y_2) is given by

$$d = \sqrt{(x_2 - x_1)^2 + (y_2 - y_1)^2}$$

EXAMPLE 1   **Finding the Distance Between Two Points**

Find the distance between the points $(-5, 1)$ and $(7, -3)$. Give the exact distance and an approximation to 2 decimal places.

Solution:

$$(-5, 1) \quad \text{and} \quad (7, -3)$$
$$(x_1, y_1) \quad \text{and} \quad (x_2, y_2)$$

Label the points. Note that the choice for (x_1, y_1) and (x_2, y_2) will not affect the outcome.

$$d = \sqrt{[7 - (-5)]^2 + (-3 - 1)^2}$$

Apply the distance formula.
$d = \sqrt{(x_2 - x_1)^2 + (y_2 - y_1)^2}$

$$= \sqrt{(12)^2 + (-4)^2}$$

Simplify the radical.

$$= \sqrt{160}$$
$$= 4\sqrt{10} \approx 12.65$$

The exact distance is $4\sqrt{10}$ units. This is approximately 12.65 units.

Skill Practice 1   Find the distance between the points $(-1, 4)$ and $(3, -6)$. Give the exact distance and an approximation to 2 decimal places.

> **Avoiding Mistakes**
>
> A statement of the form "if p, then q" is called a **conditional statement.** Its **converse** is the statement "if q, then p." The converse of a statement is not necessarily true. However, in the case of the Pythagorean theorem, the converse is a true statement.

The Pythagorean theorem tells us that if a right triangle has legs of lengths a and b and hypotenuse of length c, then $a^2 + b^2 = c^2$. The following related statement is also true: If $a^2 + b^2 = c^2$, then a triangle with sides of lengths a, b, and c is a right triangle. We use this important concept in Example 2.

Answer

1. $2\sqrt{29}$ units ≈ 10.77 units

EXAMPLE 2 Determining if Three Points Form the Vertices of a Right Triangle

Determine if the points $M(-2, -3)$, $P(4, 1)$, and $Q(-1, 7)$ form the vertices of a right triangle.

Solution:

Determine the distance between each pair of points.

$d(M, P) = \sqrt{[4 - (-2)]^2 + [1 - (-3)]^2} = \sqrt{52}$

$d(P, Q) = \sqrt{(-1 - 4)^2 + (7 - 1)^2} = \sqrt{61}$

$d(M, Q) = \sqrt{[-1 - (-2)]^2 + [7 - (-3)]^2} = \sqrt{101}$

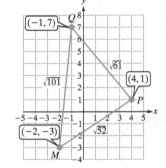

The line segment $\overline{MQ}$ is the longest and would potentially be the hypotenuse, c. Label the shorter sides as a and b.

Check the condition that $a^2 + b^2 = c^2$.

$(\sqrt{52})^2 + (\sqrt{61})^2 \stackrel{?}{=} (\sqrt{101})^2$

$52 + 61 \neq 101$

The points M, P, and Q do not form the vertices of a right triangle.

> **TIP** We denote the distance between points P and Q as
>
> $d(P, Q)$ or PQ.
>
> The second notation is the length of the line segment with endpoints P and Q.

Skill Practice 2 Determine if the points $X(-6, -4)$, $Y(2, -2)$, and $Z(0, 5)$ form the vertices of a right triangle.

Now suppose that we want to find the midpoint of the line segment between the distinct points (x_1, y_1) and (x_2, y_2). The **midpoint** of a line segment is the point equidistant (the same distance) from the endpoints (Figure 2-5).

The x-coordinate of the midpoint is the average of the x-coordinates from the endpoints. Likewise, the y-coordinate of the midpoint is the average of the y-coordinates from the endpoints.

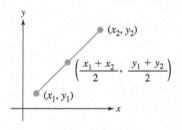

Figure 2-5

Midpoint Formula

The midpoint of the line segment with endpoints (x_1, y_1) and (x_2, y_2) is

$$M = \left(\frac{x_1 + x_2}{2}, \frac{y_1 + y_2}{2}\right)$$

average of x-coordinates   average of y-coordinates

> **Avoiding Mistakes**
>
> The midpoint of a line segment is an ordered pair (with two coordinates), not a single number.

EXAMPLE 3 Finding the Midpoint of a Line Segment

Find the midpoint of the line segment with endpoints $(4.2, -4)$ and $(-2.8, 3)$.

Solution:

$(4.2, -4)$ and $(-2.8, 3)$

(x_1, y_1) and (x_2, y_2) Label the points.

$M = \left(\dfrac{4.2 + (-2.8)}{2}, \dfrac{-4 + 3}{2}\right)$ Apply the midpoint formula.

$= \left(0.7, -\dfrac{1}{2}\right)$ or $(0.7, -0.5)$ Simplify.

Answer

2. No

> **Skill Practice 3**   Find the midpoint of the line segment with endpoints $(-1.5, -9)$ and $(-8.7, 4)$.

3. Graph Equations by Plotting Points

The relationship between two variables can often be expressed as a graph or expressed algebraically as an equation. For example, suppose that two variables, x and y, are related such that y is 2 more than x. An equation to represent this relationship is $y = x + 2$. A **solution to an equation** in the variables x and y is an ordered pair (x, y) that when substituted into the equation makes the equation a true statement.

For example, the following ordered pairs are solutions to the equation $y = x + 2$.

Solution	$y = x + 2$
$(0, 2)$	$2 = 0 + 2$ ✓
$(-4, -2)$	$-2 = -4 + 2$ ✓
$(2, 4)$	$4 = 2 + 2$ ✓

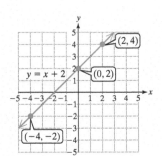

Figure 2-6

The set of all solutions to an equation is called the **solution set of the equation.** The graph of all solutions to an equation is called the **graph of the equation.** The graph of $y = x + 2$ is shown in Figure 2-6.

One of the goals of this text is to identify families of equations and the characteristics of their graphs. As we proceed through the text, we will develop tools to graph equations efficiently. For now, we present the point-plotting method to graph the solution set of an equation. In Example 4, we start by selecting several values of x and using the equation to calculate the corresponding values of y. Then we plot the points to form a general outline of the curve.

> **EXAMPLE 4**   **Graphing an Equation by Plotting Points**
>
> Graph the equation by plotting points. $y - |x| = -1$
>
> **Solution:**
>
> $\quad y - |x| = -1$ Solve for y in terms of x.
>
> $\qquad\ \ y = |x| - 1$ Arbitrarily select negative and positive values for x such as -3, -2, -1, 0, 1, 2, and 3. Then use the equation to calculate the corresponding y values.

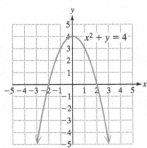

x	y	$y = \lvert x \rvert - 1$	Ordered pair
-3	2	$y = \lvert -3 \rvert - 1 = 2$	$(-3, 2)$
-2	1	$y = \lvert -2 \rvert - 1 = 1$	$(-2, 1)$
-1	0	$y = \lvert -1 \rvert - 1 = 0$	$(-1, 0)$
0	-1	$y = \lvert 0 \rvert - 1 = -1$	$(0, -1)$
1	0	$y = \lvert 1 \rvert - 1 = 0$	$(1, 0)$
2	1	$y = \lvert 2 \rvert - 1 = 1$	$(2, 1)$
3	2	$y = \lvert 3 \rvert - 1 = 2$	$(3, 2)$

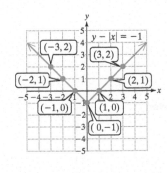

> **Skill Practice 4**   Graph the equation by plotting points. $x^2 + y = 4$

The graph of an equation in the variables x and y represents a relationship between a real number x and a corresponding real number y. Therefore, the values of x must be chosen so that when substituted into the equation, they produce a real number for y. Sometimes the values of x must be restricted to produce real numbers for y. This is demonstrated in Example 5.

EXAMPLE 5   Graphing an Equation by Plotting Points

Graph the equation by plotting points. $y^2 - 1 = x$

Solution:

$$y^2 - 1 = x \qquad \text{Solve for } y \text{ in terms of } x.$$
$$y^2 = x + 1$$
$$y = \pm\sqrt{x + 1} \qquad \text{Apply the square root property.}$$

Choose $x \geq -1$ so that the radicand is nonnegative.

> **TIP** In Example 5, we choose several convenient values of x such as -1, 0, 3, and 8 so that the radicand will be a perfect square.

x	y	$y = \pm\sqrt{x + 1}$	Ordered pairs
-1	0	$y = \pm\sqrt{(-1) + 1} = 0$	$(-1, 0)$
0	± 1	$y = \pm\sqrt{(0) + 1} = \pm 1$	$(0, 1), (0, -1)$
1	$\pm\sqrt{2}$	$y = \pm\sqrt{(1) + 1} = \pm\sqrt{2}$ ≈ 1.4	$(1, \sqrt{2}), (1, -\sqrt{2})$
3	± 2	$y = \pm\sqrt{(3) + 1} = \pm 2$	$(3, 2), (3, -2)$
8	± 3	$y = \pm\sqrt{(8) + 1} = \pm 3$	$(8, 3), (8, -3)$

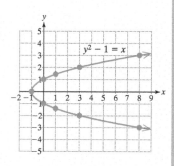

Skill Practice 5   Graph the equation by plotting points. $x + y^2 = 2$

4. Identify x- and y-Intercepts

When analyzing graphs, we want to examine their most important features. Two key features are the x- and y-intercepts of a graph. These are the points where a graph intersects the x- and y-axes.

Any point on the x-axis has a y-coordinate of zero. Therefore, an **x-intercept** is a point $(a, 0)$ where a graph intersects the x-axis (Figure 2-7). Any point on the y-axis has an x-coordinate of zero. Therefore, a **y-intercept** is a point $(0, b)$ where a graph intersects the y-axis (Figure 2-7).

Figure 2-7

> **TIP** In some applications, we may refer to an x-intercept as the x-coordinate of a point of intersection that a graph makes with the x-axis. For example, if an x-intercept is $(-4, 0)$, then the x-intercept may be stated simply as -4 (the y-coordinate is understood to be zero). Similarly, we may refer to a y-intercept as the y-coordinate of a point of intersection that a graph makes with the y-axis. For example, if a y-intercept is $(0, 2)$, then it may be stated simply as 2.

Answer
5.

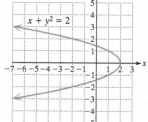

To find the x- and y-intercepts from an equation in x and y, follow these steps.

Determining x- and y-Intercepts from an Equation

Given an equation in x and y,

- Find the x-intercept(s) by substituting 0 for y in the equation and solving for x.
- Find the y-intercept(s) by substituting 0 for x in the equation and solving for y.

| EXAMPLE 6 | Finding *x*- and *y*-Intercepts |

Given the equation $y = |x| - 1$,

a. Find the *x*-intercept(s).   **b.** Find the *y*-intercept(s).

Solution:

a.
$$y = |x| - 1$$
$$0 = |x| - 1 \quad \text{To find the } x\text{-intercept(s), substitute } 0 \text{ for } y \text{ and solve for } x.$$
$$|x| = 1 \quad \text{Isolate the absolute value.}$$
$$x = 1 \quad \text{or} \quad x = -1 \quad \text{Recall that for } k > 0, |x| = k \text{ is equivalent to } x = k \text{ or } x = -k.$$

The *x*-intercepts are $(1, 0)$ and $(-1, 0)$.

b.
$$y = |x| - 1$$
$$= |0| - 1 \quad \text{To find the } y\text{-intercept(s), substitute } 0 \text{ for } x \text{ and solve for } y.$$
$$= -1$$

The *y*-intercept is $(0, -1)$.

The intercepts $(1, 0)$, $(-1, 0)$, and $(0, -1)$ are consistent with the graph of the equation $y = |x| - 1$ found in Example 4 (Figure 2-8).

Skill Practice 6   Given the equation $y = x^2 - 4$,

a. Find the *x*-intercept(s).   **b.** Find the *y*-intercept(s).

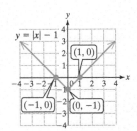

Figure 2-8

TIP Sometimes when solving for an *x*- or *y*-intercept, we encounter an equation with an imaginary solution. In such a case, the graph has no *x*- or *y*-intercept.

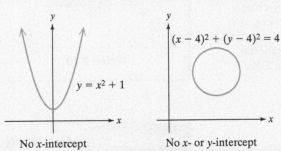

No *x*-intercept   No *x*- or *y*-intercept

5. Graph Equations Using a Graphing Utility

Graphing by the point-plotting method should only be considered a beginning strategy for creating the graphs of equations in two variables. We will quickly enhance this method with other techniques that are less cumbersome and use more analysis and strategy.

One weakness of the point-plotting method is that it may be slow to execute by pencil and paper. Also, the selected points must fairly represent the shape of the graph. Otherwise the sketch will be inaccurate. Graphing utilities can help with both of these weaknesses. They can graph many points quickly, and the more points that are plotted, the greater the likelihood that we see the key features of the graph. Graphing utilities include graphing calculators, spreadsheets, specialty graphing programs, and apps on phones.

Figures 2-9 and 2-10 show a table and a graph for $y = x^2 - 3$.

Answers
6. a. $(2, 0)$ and $(-2, 0)$
   b. $(0, -4)$

TECHNOLOGY CONNECTIONS

Using the Table Feature and Graphing an Equation

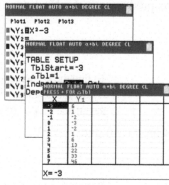

Figure 2-9

In Figure 2-9, we first enter the equation into the graphing editor. Notice that the calculator expects the equation represented with the y variable isolated.

To set up a table, enter the starting value for x, in this case, -3. Then set the increment by which to increase x, in this case 1. The x-increment is entered as ΔTbl (read "delta table"). Using the "Auto" setting means that the table of values for X and Y_1 will be automatically generated.

The table shows eleven x-y pairs but more can be accessed by using the up and down arrow keys on the keypad.

The graph in Figure 2-10 is shown between x and y values from -10 to 10. The tick marks on the axes are 1 unit apart. The viewing window with these parameters is denoted $[-10, 10, 1]$ by $[-10, 10, 1]$.

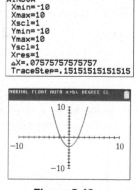

Figure 2-10

> **TIP** The Greek letter Δ ("delta") written before a variable represents an increment of change in that variable. In this context, it represents the change from one value of x to the next.

> **TIP** The calculator plots a large number of points and then connects the points. So instead of graphing a single smooth curve, it graphs a series of short line segments. This may give the graph a jagged look (Figure 2-10).

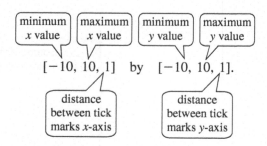

$$[-10, 10, 1] \quad \text{by} \quad [-10, 10, 1].$$

EXAMPLE 7   Graphing Equations Using a Graphing Utility

Use a graphing utility to graph $y = |x| - 15$ and $y = -x^2 + 12$ on the viewing window defined by $[-20, 20, 2]$ by $[-15, 15, 3]$.

Solution:

Enter the equations using the Y= editor.

Use the WINDOW editor to change the viewing window parameters. The variables Xmin, Xmax, and Xscl relate to $[-20, 20, 2]$. The variables Ymin, Ymax, and Yscl relate to $[-15, 15, 3]$.

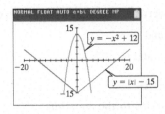

Select the **GRAPH** feature. Notice that the graphs of both equations appear. This provides us with a tool for visually examining two different models at the same time.

Answer

7.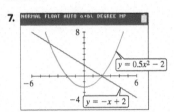

> **Skill Practice 7**   Use a graphing utility to graph $y = -x + 2$ and $y = 0.5x^2 - 2$ on the viewing window $[-6, 6, 1]$ by $[-4, 8, 1]$.

SECTION 2.1   Practice Exercises

Prerequisite Review

R.1. Simplify the radical. $\sqrt{48}$

R.2. Given a right triangle with a leg of length 7 km and hypotenuse of length 25 km, find the length of the unknown leg.

R.3. Solve for y. $ax + by = c$

R.4. Evaluate $x^2 + 4x + 5$ for $x = -5$

Concept Connections

1. In a rectangular coordinate system, the point where the x- and y-axes meet is called the _____.

2. The x- and y-axes divide the coordinate plane into four regions called _____.

3. The distance between two distinct points (x_1, y_1) and (x_2, y_2) is given by the formula _____.

4. The midpoint of the line segment with endpoints (x_1, y_1) and (x_2, y_2) is given by the formula _____.

5. A _____ to an equation in the variables x and y is an ordered pair (x, y) that makes the equation a true statement.

6. An x-intercept of a graph has a y-coordinate of _____.

7. A y-intercept of a graph has an x-coordinate of _____.

8. Given an equation in the variables x and y, find the y-intercept by substituting _____ for x and solving for _____.

Objective 1: Plot Points on a Rectangular Coordinate System

For Exercises 9–10, plot the points on a rectangular coordinate system.

9. $A(-3, -4)$ $B\left(\dfrac{5}{3}, \dfrac{7}{4}\right)$ $C(-1.2, 3.8)$ $D(\pi, -5)$ $E(0, 4.5)$ $F\left(\sqrt{5}, 0\right)$

10. $A(-2, -5)$ $B\left(\dfrac{9}{2}, \dfrac{7}{3}\right)$ $C(-3.6, 2.1)$ $D(5, -\pi)$ $E(3.4, 0)$ $F\left(0, \sqrt{3}\right)$

Objective 2: Use the Distance and Midpoint Formulas

For Exercises 11–18,

a. Find the exact distance between the points. **(See Example 1)**

b. Find the midpoint of the line segment whose endpoints are the given points. **(See Example 3)**

11. $(-2, 7)$ and $(-4, 11)$ **12.** $(-1, -3)$ and $(3, -7)$ **13.** $(-7, -4)$ and $(2, 5)$

14. $(3, 6)$ and $(-4, -1)$ **15.** $(2.2, -2.4)$ and $(5.2, -6.4)$ **16.** $(37.1, -24.7)$ and $(31.1, -32.7)$

17. $\left(\sqrt{5}, -\sqrt{2}\right)$ and $\left(4\sqrt{5}, -7\sqrt{2}\right)$ **18.** $\left(\sqrt{7}, -3\sqrt{5}\right)$ and $\left(2\sqrt{7}, \sqrt{5}\right)$

For Exercises 19–22, determine if the given points form the vertices of a right triangle. (See Example 2)

19. $(1, 3)$, $(3, 1)$, and $(0, -2)$ **20.** $(1, 2)$, $(3, 0)$, and $(-3, -2)$

21. $(-2, 4)$, $(5, 0)$, and $(-5, 1)$ **22.** $(-6, 2)$, $(3, 1)$, and $(1, -2)$

Objective 3: Graph Equations by Plotting Points

For Exercises 23–24, determine if the given points are solutions to the equation.

23. $x^2 + y = 1$

     a. $(-2, -3)$ **b.** $(4, -17)$ **c.** $\left(\dfrac{1}{2}, \dfrac{3}{4}\right)$

24. $|x - 3| - y = 4$

     a. $(1, -2)$ **b.** $(-2, -3)$ **c.** $\left(\dfrac{1}{10}, -\dfrac{11}{10}\right)$

For Exercises 25–30, identify the set of values x for which y will be a real number.

25. $y = \dfrac{2}{x - 3}$ **26.** $y = \dfrac{2}{x + 7}$ **27.** $y = \sqrt{x - 10}$

28. $y = \sqrt{x + 11}$ **29.** $y = \sqrt{1.5 - x}$ **30.** $y = \sqrt{2.2 - x}$

For Exercises 31–44, graph the equations by plotting points. (See Examples 4–5)

31. $y = x$ **32.** $y = x^2$ **33.** $y = \sqrt{x}$

34. $y = |x|$ **35.** $y = x^3$ **36.** $y = \dfrac{1}{x}$

37. $y - |x| = 2$ **38.** $|x| + y = 3$ **39.** $y^2 - x - 2 = 0$

40. $y^2 - x + 1 = 0$ **41.** $x = |y| + 1$ **42.** $x = |y| - 3$

43. $y = |x + 1|$ **44.** $y = |x - 2|$

Objective 4: Identify x- and y-Intercepts

For Exercises 45–50, estimate the x- and y-intercepts from the graph.

45.

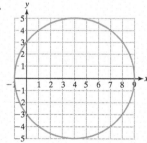

46.

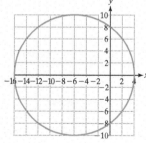

47.

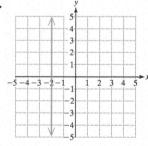

48.

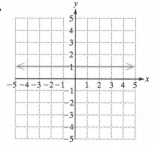

49.

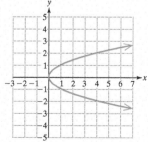

50.

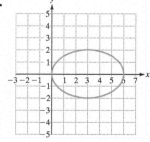

For Exercises 51–62, find the x- and y-intercepts. (See Example 6)

51. $-2x + 4y = 12$

52. $-3x - 5y = 60$

53. $x^2 + y = 9$

54. $x^2 = -y + 16$

55. $y = |x - 5| - 2$

56. $y = |x + 4| - 3$

57. $x = y^2 - 1$

58. $x = y^2 - 4$

59. $|x| = |y|$

60. $x = |5y|$

61. $\dfrac{(x - 3)^2}{4} + \dfrac{(y - 4)^2}{9} = 1$

62. $\dfrac{(x + 6)^2}{16} + \dfrac{(y + 3)^2}{4} = 1$

Mixed Exercises

63. A map of a wilderness area is drawn with the origin placed at the parking area. Two fire observation platforms are located at points A and B. If a fire is located at point C, determine the distance to the fire from each observation platform.

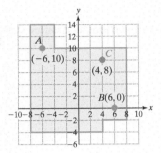

64. A map of a state park is drawn so that the origin is placed at the visitor center. The distance between grid lines is 1 mi. Suppose that two hikers are located at points A and B.

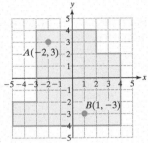

    a. Determine the distance between the hikers.

    b. If the hikers want to meet for lunch, determine the location of the midpoint between the hikers.

The position of an object in a video game is represented by an ordered pair. The coordinates of the ordered pair give the number of pixels horizontally and vertically from the origin. Use this scenario for Exercises 65–66.

65. a. Suppose that player A is located at (36, 315) and player B is located at (410, 53). How far apart are the players? Round to the nearest pixel.

    b. If the two players move directly toward each other at the same speed, where will they meet?

    c. If player A moves three times faster than player B, where will they meet? Round to the nearest pixel.

66. Suppose that a player is located at point $A(460, 420)$ and must move in a direct line to point $B(80, 210)$ and then in a direct line to point $C(120, 60)$ to pick up prizes before a 5-sec timer runs out. If the player moves at 120 pixels per second, will the player have enough time to pick up both prizes? Explain.

67. Verify that the points $A(0, 0)$, $B(x, 0)$, and $C\left(\dfrac{1}{2}x, \dfrac{\sqrt{3}}{2}x\right)$ make up the vertices of an equilateral triangle.

68. Verify that the points $A(0, 0)$, $B(x, 0)$, and $C(0, x)$ make up the vertices of an isosceles right triangle (an isosceles triangle has two sides of equal length).

For Exercises 69–70, assume that the units shown in the grid are in feet.

a. Determine the exact length and width of the rectangle shown.

b. Determine the perimeter and area.

69.

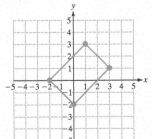

70.

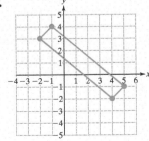

For Exercises 71–72, the endpoints of a diameter of a circle are shown. Find the center and radius of the circle.

71.

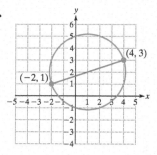

72.

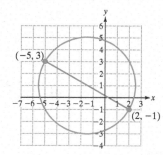

For Exercises 73–74, an isosceles triangle is shown. Find the area of the triangle. Assume that the units shown in the grid are in meters.

73.

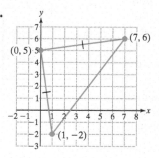

74.

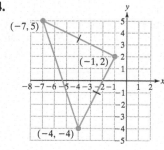

For Exercises 75–78, determine if points A, B, and C are collinear. Three points are collinear if they all fall on the same line. There are several ways that we can determine if three points, A, B, and C are collinear. One method is to determine if the sum of the lengths of the line segments $\overline{AB}$ and $\overline{BC}$ equals the length of $\overline{AC}$.

75. (2, 2), (4, 3), and (8, 5)

76. (2, 1.5), (4, 2), and (8, 3)

77. (−2, 8), (1, 2), and (4, −3)

78. (−1, 5), (0, 3), and (5, −13)

Write About It

79. Suppose that d represents the distance between two points (x_1, y_1) and (x_2, y_2). Explain how the distance formula is developed from the Pythagorean theorem.

80. Explain how you might remember the midpoint formula to find the midpoint of the line segment between (x_1, y_1) and (x_2, y_2).

81. Explain how to find the x- and y-intercepts from an equation in the variables x and y.

82. Given an equation in the variables x and y, what does the graph of the equation represent?

Expanding Your Skills

A point in three-dimensional space can be represented in a three-dimensional coordinate system. In such a case, a z-axis is taken perpendicular to both the x- and y-axes. A point P is assigned an ordered triple $P(x, y, z)$ relative to a fixed origin where the three axes meet. For Exercises 83–86, determine the distance between the two given points in space. Use the distance formula $d = \sqrt{(x_2 - x_1)^2 + (y_2 - y_1)^2 + (z_2 - z_1)^2}$.

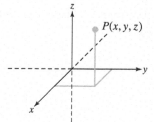

83. (5, −3, 2) and (4, 6, −1)

84. (6, −4, −1) and (2, 3, 1)

85. (3, 7, −2) and (0, −5, 1)

86. (9, −5, −3) and (2, 0, 1)

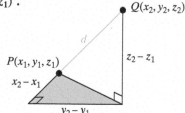

Objective 5: Graph Equations Using a Graphing Utility (Technology Connections)

87. What is meant by a viewing window on a graphing device?

88. Which of the viewing windows would show both the x- and y-intercepts of the graph of $780x - 42y = 5460$?

    a. $[-20, 20, 2]$ by $[-40, 40, 10]$

    b. $[-10, 10, 1]$ by $[-10, 10, 1]$

    c. $[-10, 10, 1]$ by $[-10, 150, 10]$

    d. $[-10, 10, 1]$ by $[-150, 10, 10]$

For Exercises 89–92, graph the equation with a graphing utility on the given viewing window. (See Example 7)

89. $y = 2x - 5$ on $[-10, 10, 1]$ by $[-10, 10, 1]$

90. $y = -4x + 1$ on $[-10, 10, 1]$ by $[-10, 10, 1]$

91. $y = 1400x^2 - 1200x$ on $[-5, 5, 1]$ by $[-1000, 2000, 500]$

92. $y = -800x^2 + 600x$ on $[-5, 5, 1]$ by $[-1000, 500, 200]$

For Exercises 93–94, graph the equations on the standard viewing window. (See Example 7)

93. a. $y = x^3$

    b. $y = |x| - 9$

94. a. $y = \sqrt{x + 4}$

    b. $y = |x - 2|$

SECTION 2.2   Circles

OBJECTIVES

1. **Write an Equation of a Circle in Standard Form**
2. **Write the General Form of an Equation of a Circle**

1. Write an Equation of a Circle in Standard Form

In addition to graphing equations by plotting points, we will learn to recognize specific categories of equations and the characteristics of their graphs. We begin by presenting the definition of a circle.

> **Definition of a Circle**
>
> A **circle** is the set of all points in a plane that are equidistant from a fixed point called the **center**. The fixed distance from any point on the circle to the center is called the **radius**.

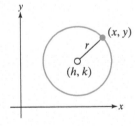

Figure 2-11

The radius of a circle is often denoted by r, where $r > 0$. It is also important to note that the center is not actually part of the graph of a circle. It will be drawn in the text as an open dot for reference only.

    Suppose that a circle is centered at the point (h, k) and has radius r (Figure 2-11). The distance formula can be used to derive an equation of the circle. Let (x, y) be an arbitrary point on the circle. Then by definition the distance between (h, k) and (x, y) must be r.

Apply the distance formula. $\sqrt{(x_2 - x_1)^2 + (y_2 - y_1)^2} = d$

$$\sqrt{(x - h)^2 + (y - k)^2} = r \qquad \text{Distance between } (h, k) \text{ and } (x, y)$$

$$(x - h)^2 + (y - k)^2 = r^2 \qquad \text{Squaring both sides of the equation results in the standard form of an equation of a circle.}$$

Standard Form of an Equation of a Circle

Given a circle centered at (h, k) with radius r, the **standard form** of an equation of the circle (also called **center-radius form**) is given by

$$(x - h)^2 + (y - k)^2 = r^2 \quad \text{where } r > 0.$$

Examples	Standard form	Center	Radius
$(x - 4)^2 + (y + 3)^2 = 25$	$(x - 4)^2 + [y - (-3)]^2 = (5)^2$	$(4, -3)$	5
$x^2 + (y - \frac{1}{2})^2 = 12$	$(x - 0)^2 + (y - \frac{1}{2})^2 = (\sqrt{12})^2$	$(0, \frac{1}{2})$	$2\sqrt{3}$
$x^2 + y^2 = 7$	$(x - 0)^2 + (y - 0)^2 = (\sqrt{7})^2$	$(0, 0)$	$\sqrt{7}$

In Example 1, we write an equation of a circle in standard form.

EXAMPLE 1 Writing an Equation of a Circle in Standard Form

a. Write the standard form of an equation of the circle with center $(-4, 6)$ and radius 2.

b. Graph the circle.

Solution:

a. $(h, k) = (-4, 6)$ and $r = 2$ Label the center (h, k) and the radius r.

$[x - (-4)]^2 + (y - 6)^2 = (2)^2$ Standard form: $(x - h)^2 + (y - k)^2 = r^2$

$(x + 4)^2 + (y - 6)^2 = 4$ Simplify.

b.

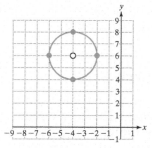

To graph the circle, first locate the center and draw a small open dot. Then plot points r units to the left, right, above, and below the center.

Draw the circle through the points.

Skill Practice 1

a. Write an equation of the circle with center $(3, -1)$ and radius 4.

b. Graph the circle.

EXAMPLE 2 Writing an Equation of a Circle in Standard Form

Write the standard form of an equation of the circle with endpoints of a diameter $(-1, 0)$ and $(3, 4)$.

Solution:

A sketch of this scenario is given in Figure 2-12. Notice that the midpoint of the diameter is the center of the circle.

$(-1, 0)$ and $(3, 4)$

(x_1, y_1) and (x_2, y_2) Label the points.

The center is $\left(\dfrac{-1 + 3}{2}, \dfrac{0 + 4}{2} \right) = (1, 2)$.

Answers

1. a. $(x - 3)^2 + (y + 1)^2 = 16$

   b.

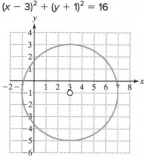

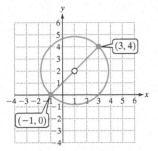

Figure 2-12

The radius of the circle is the distance between either endpoint of the diameter and the center. Using the endpoint $(-1, 0)$ as (x_1, y_1) and the center $(1, 2)$ as (x_2, y_2), apply the distance formula.

$$d = \sqrt{(x_2 - x_1)^2 + (y_2 - y_1)^2}$$
$$r = \sqrt{[1 - (-1)]^2 + (2 - 0)^2} = \sqrt{(2)^2 + (2)^2} = \sqrt{8}$$

An equation of the circle is: $(x - h)^2 + (y - k)^2 = r^2$.

$$(x - 1)^2 + (y - 2)^2 = \left(\sqrt{8}\right)^2$$
$$(x - 1)^2 + (y - 2)^2 = 8 \quad \text{(Standard form)}$$

> **Skill Practice 2**   Write the standard form of an equation of the circle with endpoints of a diameter $(-3, 3)$ and $(-1, -1)$.

2. Write the General Form of an Equation of a Circle

In Example 2 we have the equation $(x - 1)^2 + (y - 2)^2 = 8$. If we expand the binomials and combine like terms, we can write the equation in *general form*.

$$(x - 1)^2 + (y - 2)^2 = 8 \qquad \text{Standard form (center-radius form)}$$
$$x^2 - 2x + 1 + y^2 - 4y + 4 = 8 \qquad \text{Expand the binomials}$$
$$x^2 + y^2 - 2x - 4y - 3 = 0 \qquad \text{General form}$$

> **General Form of an Equation of a Circle**
>
> An equation of a circle written in the form $x^2 + y^2 + Ax + By + C = 0$ is called the **general form** of an equation of a circle.

By completing the square we can write an equation of a circle given in general form as an equation in standard form. The purpose of writing an equation of a circle in standard form is to identify the radius and center. This is demonstrated in Example 3.

> **EXAMPLE 3**   **Writing an Equation of a Circle in Standard Form**
>
> Write the equation of the circle in standard form. Then identify the center and radius.
>
> $$x^2 + y^2 + 10x - 6y + 25 = 0$$
>
> **Solution:**
>
> $$x^2 + y^2 + 10x - 6y + 25 = 0$$
> $$(x^2 + 10x \qquad) + (y^2 - 6y \qquad) = -25 \qquad \text{Group the } x \text{ terms. Group the } y \text{ terms. Move the constant term to the right.}$$
>
> $$(x^2 + 10x + 25) + (y^2 - 6y + 9) \qquad \text{Complete the squares.}$$
> $$= -25 + 25 + 9 \qquad \textit{Note: } \left[\tfrac{1}{2}(10)\right]^2 = 25, \left[\tfrac{1}{2}(-6)\right]^2 = 9$$
> $$(x + 5)^2 + (y - 3)^2 = 9 \qquad \text{Factor.}$$
>
> The center is $(-5, 3)$, and the radius is $\sqrt{9} = 3$. See Figure 2-13.

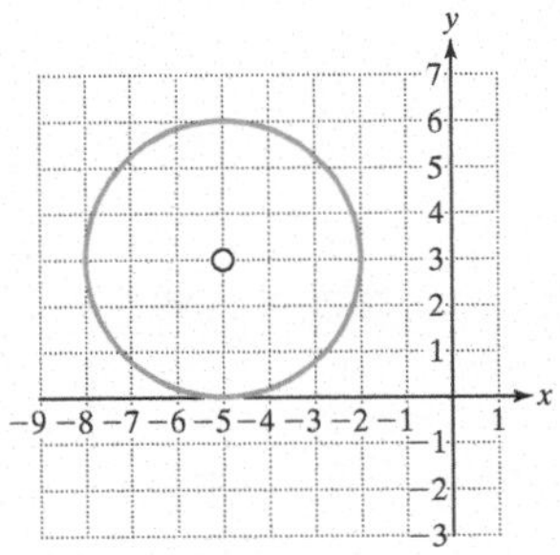

Figure 2-13

> **Skill Practice 3**   Write the equation of the circle in standard form. Then identify the center and radius. $x^2 + y^2 - 8x + 2y - 8 = 0$

Answers

2. $(x + 2)^2 + (y - 1)^2 = 5$

3. $(x - 4)^2 + (y + 1)^2 = 25$; Center: $(4, -1)$; Radius: 5

Not all equations of the form $x^2 + y^2 + Ax + By + C = 0$ represent the graph of a circle. Completing the square results in an equation of the form $(x - h)^2 + (y - k)^2 = c$, where c is a constant. In the case where $c > 0$, the graph of the equation is a circle with radius $r = \sqrt{c}$. However, if $c = 0$, or if $c < 0$, the graph will be a single point or nonexistent. These are called **degenerate cases**.

- If $c > 0$, then the graph will be a circle with radius $r = \sqrt{c}$.
- If $c = 0$, then the graph will be a single point, (h, k). The solution set is $\{(h, k)\}$.
- If $c < 0$, then the solution set is the empty set $\{\ \}$.

EXAMPLE 4 Determining if an Equation Represents the Graph of a Circle

Write the equation in the form $(x - h)^2 + (y - k)^2 = r^2$, and identify the solution set.

$$x^2 + y^2 - 14y + 49 = 0$$

Solution:

$$x^2 + y^2 - 14y + 49 = 0$$
$$x^2 + (y^2 - 14y\ \ ) = -49$$
Group the y terms and complete the square. Note that the x^2 term is already a perfect square: $(x - 0)^2$.

$$x^2 + (y^2 - 14y + 49) = -49 + 49$$
Complete the square: $\left[\frac{1}{2}(-14)\right]^2 = 49$.

$$x^2 + (y - 7)^2 = 0$$
Factor.

Since $r^2 = 0$, the solution set is $\{(0, 7)\}$. The sum of two squares will equal zero only if each individual term is zero. Therefore, $x = 0$ and $y = 7$.

Skill Practice 4 Write the equation in the form $(x - h)^2 + (y - k)^2 = r^2$, and identify the solution set. $x^2 + y^2 + 2x + 5 = 0$

TECHNOLOGY CONNECTIONS

Setting a Square Viewing Window and Graphing a Circle

A graphing calculator expects an equation with the y variable isolated. Therefore, to graph an equation of a circle such as $(x + 5)^2 + (y - 3)^2 = 9$, from Example 3, we first solve for y.

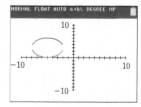

$$(x + 5)^2 + (y - 3)^2 = 9$$
$$(y - 3)^2 = 9 - (x + 5)^2$$
$$y - 3 = \pm\sqrt{9 - (x + 5)^2}$$
$$y = 3 \pm \sqrt{9 - (x + 5)^2}$$

Notice that the graph looks more oval-shaped than circular. This is because the calculator has a rectangular screen. If the scaling is the same on the x- and y-axes, the graph will appear elongated horizontally. To eliminate this distortion, use a ZSquare option, located in the Zoom menu.

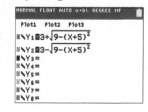

Also notice that the calculator display may not show the upper and lower semicircles connecting. The viewing window between $x = -16.1$ and $x = 16.1$ is divided by the number of pixels displayed horizontally to get the values of x used to graph the equation. These may not include x values at the leftmost and rightmost points on the circle. That is, the calculator may graph points *close* to $(-8, 3)$ and $(-2, 3)$ but not exactly at $(-8, 3)$ and $(-2, 3)$. Therefore, the upper and lower semicircles may not "hook up."

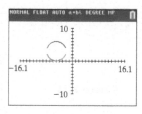

Answer

4. $(x + 1)^2 + y^2 = -4$; The solution set is $\{\ \}$.

SECTION 2.2   Practice Exercises

Prerequisite Review

For Exercises R.1–R.2, find the value of n so that the expression is a perfect square trinomial. Then factor the trinomial.

R.1. $c^2 - 8c + n$

R.2. $x^2 + \dfrac{2}{7}x + n$

R.3. Find the distance between $(2, 3)$ and $(-3, -2)$. Express your answer in simplified radical form.

R.4. Multiply by using the special case products. Simplify. $(x - 2)^2$

Concept Connections

1. A _____ is the set of all points in a plane equidistant from a fixed point called the _____.

2. The distance from the center of a circle to any point on the circle is called the _____ and is often denoted by r.

3. The standard form of an equation of a circle with center (h, k) and radius r is given by _____.

4. An equation of a circle written in the form $x^2 + y^2 + Ax + By + C = 0$ is called the _____ form of an equation of a circle.

Objective 1: Write an Equation of a Circle in Standard Form

5. Is the point $(2, 7)$ on the circle defined by $(x - 2)^2 + (y - 7)^2 = 4$?

6. Is the point $(3, 5)$ on the circle defined by $(x - 3)^2 + (y - 5)^2 = 36$?

7. Is the point $(-4, 7)$ on the circle defined by $(x + 1)^2 + (y - 3)^2 = 25$?

8. Is the point $(2, -7)$ on the circle defined by $(x + 6)^2 + (y + 1)^2 = 100$?

For Exercises 9–16, determine the center and radius of the circle.

9. $(x - 4)^2 + (y + 2)^2 = 81$

10. $(x + 3)^2 + (y - 1)^2 = 16$

11. $x^2 + (y - 2.5)^2 = 6.25$

12. $(x - 1.5)^2 + y^2 = 2.25$

13. $x^2 + y^2 = 20$

14. $x^2 + y^2 = 28$

15. $\left(x - \dfrac{3}{2}\right)^2 + \left(y + \dfrac{3}{4}\right)^2 = \dfrac{81}{49}$

16. $\left(x + \dfrac{1}{7}\right)^2 + \left(y - \dfrac{3}{5}\right)^2 = \dfrac{25}{9}$

For Exercises 17–32, information about a circle is given.

a. Write an equation of the circle in standard form.

b. Graph the circle. (See Examples 1–2)

17. Center: $(-2, 5)$; Radius: 1

18. Center: $(-3, 2)$; Radius: 4

19. Center: $(-4, 1)$; Radius: 3

20. Center: $(6, -2)$; Radius: 6

21. Center: $(-4, -3)$; Radius: $\sqrt{11}$

22. Center: $(-5, -2)$; Radius: $\sqrt{21}$

23. Center: $(0, 0)$; Radius: 2.6

24. Center: $(0, 0)$; Radius: 4.2

25. The endpoints of a diameter are $(-2, 4)$ and $(6, -2)$.

26. The endpoints of a diameter are $(7, 3)$ and $(5, -1)$.

27. The center is $(-2, -1)$ and a point on the circle is $(6, 5)$.

28. The center is $(3, 1)$ and a point on the circle is $(6, 5)$.

29. The center is $(4, 6)$ and the circle is tangent to the y-axis. (Informally, a line is tangent to a circle if it touches the circle in exactly one point.)

30. The center is $(-2, -4)$ and the circle is tangent to the x-axis.

31. The center is in Quadrant IV, the radius is 5, and the circle is tangent to both the x- and y-axes.

32. The center is in Quadrant II, the radius is 3, and the circle is tangent to both the x- and y-axes.

33. Write an equation that represents the set of points that are 5 units from $(8, -11)$.

34. Write an equation that represents the set of points that are 9 units from $(-4, 16)$.

35. Write an equation of the circle that is tangent to both axes with radius $\sqrt{7}$ and center in Quadrant I.

36. Write an equation of the circle that is tangent to both axes with radius $\sqrt{11}$ and center in Quadrant III.

Objective 2: Write the General Form of an Equation of a Circle

37. Determine the solution set for the equation $(x + 1)^2 + (y - 5)^2 = 0$.

38. Determine the solution set for the equation $(x - 3)^2 + (y + 12)^2 = 0$.

39. Determine the solution set for the equation $(x - 17)^2 + (y + 1)^2 = -9$.

40. Determine the solution set for the equation $(x + 15)^2 + (y - 3)^2 = -25$.

For Exercises 41–54, write the equation in the form $(x - h)^2 + (y - k)^2 = c$. Then if the equation represents a circle, identify the center and radius. If the equation represents a degenerate case, give the solution set. (See Examples 3–4)

41. $x^2 + y^2 + 6x - 2y + 6 = 0$

42. $x^2 + y^2 + 12x - 14y + 84 = 0$

43. $x^2 + y^2 - 22x + 6y + 129 = 0$

44. $x^2 + y^2 - 10x + 4y - 20 = 0$

45. $x^2 + y^2 - 20y - 4 = 0$

46. $x^2 + y^2 + 22x - 4 = 0$

47. $10x^2 + 10y^2 - 80x + 200y + 920 = 0$
(*Hint*: Divide by 10 to make the x^2 and y^2 term coefficients equal to 1.)

48. $2x^2 + 2y^2 - 32x + 12y + 90 = 0$

49. $x^2 + y^2 - 4x - 18y + 89 = 0$

50. $x^2 + y^2 - 10x - 22y + 155 = 0$

51. $4x^2 + 4y^2 - 20y + 25 = 0$

52. $4x^2 + 4y^2 - 12x + 9 = 0$

53. $x^2 + y^2 - x - \dfrac{3}{2}y - \dfrac{3}{4} = 0$

54. $x^2 + y^2 - \dfrac{2}{3}x - \dfrac{5}{3}y - \dfrac{5}{9} = 0$

Mixed Exercises

55. A cell tower is a site where antennas, transmitters, and receivers are placed to create a cellular network. Suppose that a cell tower is located at a point $A(4, 6)$ on a map and its range is 1.5 mi. Write an equation that represents the boundary of the area that can receive a signal from the tower. Assume that all distances are in miles.

56. A radar transmitter on a ship has a range of 20 nautical miles. If the ship is located at a point $(-32, 40)$ on a map, write an equation for the boundary of the area within the range of the ship's radar. Assume that all distances on the map are represented in nautical miles.

57. Suppose that three geological study areas are set up on a map at points $A(-4, 12)$, $B(11, 3)$, and $C(0, 1)$, where all units are in miles. Based on the speed of compression waves, scientists estimate the distances from the study areas to the epicenter of an earthquake to be 13 mi, 5 mi, and 10 mi, respectively. Graph three circles whose centers are located at the study areas and whose radii are the given distances to the earthquake. Then estimate the location of the earthquake.

58. Three fire observation towers are located at points $A(-6, -14)$, $B(14, 10)$, and $C(-3, 13)$ on a map where all units are in kilometers. A fire is located at distances of 17 km, 15 km, and 13 km, respectively, from the observation towers. Graph three circles whose centers are located at the observation towers and whose radii are the given distances to the fire. Then estimate the location of the fire.

Write About It

59. State the definition of a circle.

60. What are the advantages of writing an equation of a circle in standard form?

Expanding Your Skills

61. Find all values of y such that the distance between $(4, y)$ and $(-2, 6)$ is 10 units.

62. Find all values of x such that the distance between $(x, -1)$ and $(4, 2)$ is 5 units.

63. Find all points on the line $y = x$ that are 6 units from $(2, 4)$.

64. Find all points on the line $y = -x$ that are 4 units from $(-4, 6)$.

The general form of an equation of a circle is $(x - h)^2 + (y - k)^2 = r^2$. If we solve the equation for x we get equations of the form $x = h \pm \sqrt{r^2 - (y - k)^2}$. The equation $x = h + \sqrt{r^2 - (y - k)^2}$ represents the graph of the corresponding right-side semicircle, and the equation $x = h - \sqrt{r^2 - (y - k)^2}$ represents the graph of the left-side semicircle. Likewise, if we solve for y, we have $y = k \pm \sqrt{r^2 - (x - h)^2}$. These equations represent the top and bottom semicircles. For Exercises 65–68, graph the equations.

65. a. $y = \sqrt{16 - x^2}$
b. $y = -\sqrt{16 - x^2}$
c. $x = \sqrt{16 - y^2}$
d. $x = -\sqrt{16 - y^2}$

66. a. $y = \sqrt{9 - x^2}$
b. $y = -\sqrt{9 - x^2}$
c. $x = \sqrt{9 - y^2}$
d. $x = -\sqrt{9 - y^2}$

67. a. $x = -1 - \sqrt{9 - (y - 2)^2}$
b. $x = -1 + \sqrt{9 - (y - 2)^2}$
c. $y = 2 - \sqrt{9 - (x + 1)^2}$
d. $y = 2 + \sqrt{9 - (x + 1)^2}$

68. a. $x = 3 - \sqrt{4 - (y + 2)^2}$
b. $x = 3 + \sqrt{4 - (y + 2)^2}$
c. $y = -2 - \sqrt{4 - (x - 3)^2}$
d. $y = -2 + \sqrt{4 - (x - 3)^2}$

69. Find the shortest distance from the origin to a point on the circle defined by $x^2 + y^2 - 6x - 12y + 41 = 0$.

70. Find the shortest distance from the origin to a point on the circle defined by $x^2 + y^2 + 4x - 12y + 31 = 0$.

Technology Connections

For Exercises 71–74, use a graphing calculator to graph the circles on an appropriate square viewing window.

71. $x^2 + y^2 = 36$

72. $x^2 + y^2 = 49$

73. $(x - 18)^2 + (y + 20)^2 = 80$

74. $(x + 0.04)^2 + (y - 0.02)^2 = 0.01$

SECTION 2.3   Functions and Relations

OBJECTIVES

1. Determine Whether a Relation Is a Function
2. Apply Function Notation
3. Determine x- and y-Intercepts of a Function Defined by $y = f(x)$
4. Determine Domain and Range of a Function
5. Interpret a Function Graphically

1. Determine Whether a Relation Is a Function

In the physical world, many quantities that are subject to change are related to other variables. For example:

- The cost of mailing a package is related to the weight of a package.
- The minimum braking distance of a car depends on the speed of the car.
- The perimeter of a rectangle is a function of its length and width.
- The test score that a student earns is related to the number of hours of study.

In mathematics we can express the relationship between two values as a set of ordered pairs.

> **Definition of a Relation**
>
> A set of ordered pairs (x, y) is called a **relation** in x and y.
>
> - The set of x values in the ordered pairs is called the **domain** of the relation.
> - The set of y values in the ordered pairs is called the **range** of the relation.

EXAMPLE 1 Writing a Relation from Observed Data Points

Table 2-1 shows the score y that a student earned on an algebra test based on the number of hours x spent studying one week prior to the test.

Hours of Study, x	Test Score, y
8	92
3	58
11	98
5	72
8	86

Table 2-1

a. Write the set of ordered pairs that defines the relation given in Table 2-1.

b. Write the domain.

c. Write the range.

Avoiding Mistakes

Do not list the elements in a set more than once. The value 8 is listed in the domain one time only.

Solution:

a. Relation: $\{(8, 92), (3, 58), (11, 98), (5, 72), (8, 86)\}$

b. Domain: $\{8, 3, 11, 5\}$

c. Range: $\{92, 58, 98, 72, 86\}$

Skill Practice 1 For the table shown,

x	3	-2	5	1
y	-4	0	3	0

a. Write the set of ordered pairs that defines the relation.

b. Write the domain.

c. Write the range.

The data in Table 2-1 show two different test scores for 8 hr of study. That is, for $x = 8$, there are two different y values. In many applications, we prefer to work with relations that assign one and only one y value for a given value of x. Such a relation is called a function.

Definition of a Function

Given a relation in x and y, we say that **y is a function of x** if for each value of x in the domain, there is exactly one value of y in the range.

EXAMPLE 2 Determining if a Relation Is a Function

Determine if the relation defines y as a function of x.

a. $\{(3, 1), (2, 5), (-4, 2), (-1, 0), (3, -4)\}$

b. $\{(-1, 4), (2, 3), (3, 4), (-4, 5)\}$

Solution:

a.

same x values

$\{(3, 1), (2, 5), (-4, 2), (-1, 0), (3, -4)\}$

different y values

This relation is *not* a function.

When $x = 3$, there are two different y values: $y = 1$ and $y = -4$.

TIP A function may not have the same x value paired with different y values. However, it is acceptable for a function to have two or more x values paired with the same y value, as shown in Example 2(b).

b. $\{(-1, 4), (2, 3), (3, 4), (-4, 5)\}$

No two ordered pairs have the same x value but different y values.

This relation *is* a function.

Skill Practice 2 Determine if the relation defines y as a function of x.

a. $\{(8, 4), (3, -1), (5, 4)\}$ b. $\{(-3, 2), (9, 5), (1, 0), (-3, 1)\}$

Answers

1. a. $\{(3, -4), (-2, 0), (5, 3), (1, 0)\}$
   b. Domain: $\{3, -2, 5, 1\}$
   c. Range: $\{-4, 0, 3\}$
2. a. Yes   b. No

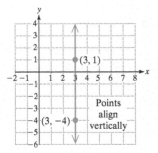

Figure 2-14

A relation that is not a function has at least one domain element x paired with more than one range element y. For example, the ordered pairs $(3, 1)$ and $(3, -4)$ do not make up a function. On a graph, these two points are aligned vertically. A vertical line drawn through one point also intersects the other point (Figure 2-14). This observation leads to the vertical line test.

Using the Vertical Line Test

Consider a relation defined by a set of points (x, y) graphed on a rectangular coordinate system. The graph defines y as a function of x if no vertical line intersects the graph in more than one point.

EXAMPLE 3   Applying the Vertical Line Test

The graphs of three relations are shown in blue. In each case, determine if the relation defines y as a function of x.

Solution:

a.

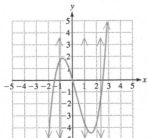

b.

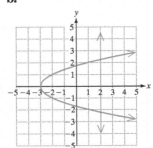

c.

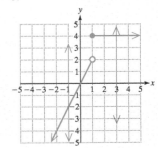

This is a function.
No vertical line intersects the graph in more than one point.

This is not a function.
There is at least one vertical line that intersects the graph in more than one point.

This is a function.
No vertical line intersects the graph in more than one point.

TIP In Example 3(c) there is only one y value assigned to $x = 1$. This is because the point $(1, 2)$ is *not* included in the graph of the function as denoted by the open dot.

Skill Practice 3   Determine if the given relation defines y as a function of x.

a.

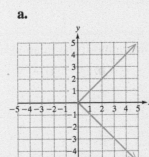

b.

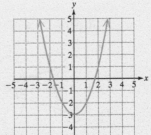

c.

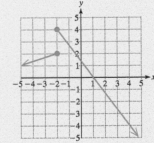

A relation can also be defined by a figure showing a "mapping" between x and y, or by an equation in x and y.

Answers
3. a. No   **b.** Yes   **c.** No

EXAMPLE 4   Determining if a Relation Is a Function

Determine if the relation defines y as a function of x.

a. x y **b.** $y^2 = x$ **c.** $(x - 2)^2 + (y + 1)^2 = 9$

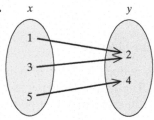

Solution:

a. This mapping defines the set of ordered pairs: $\{(1, 2), (3, 2), (5, 4)\}$.

This relation *is* a function.

No two ordered pairs have the same x value but different y values.

b. $y^2 = x$

$y = \pm\sqrt{x}$

Solve the equation for y.
For any $x > 0$, there are two corresponding y values.

x	y	Ordered pairs
0	0	$(0, 0)$
1	$1, -1$	$(1, 1), (1, -1)$
4	$2, -2$	$(4, 2), (4, -2)$
9	$3, -3$	$(9, 3), (9, -3)$

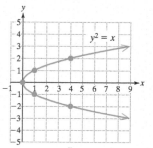

This relation is *not* a function.

c. $(x - 2)^2 + (y + 1)^2 = 9$

This equation represents the graph of a circle with center $(2, -1)$ and radius 3.

This relation is *not* a function because it fails the vertical line test.

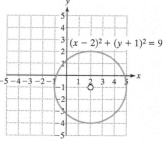

Skill Practice 4   Determine if the relation defines y as a function of x.

a. x y **b.** $|y + 1| = x$ **c.** $x^2 + y^2 = 25$

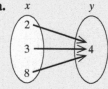

2. Apply Function Notation

A function may be defined by an equation with two variables. For example, the equation $y = x - 2$ defines y as a function of x. This is because for any real number x, the value of y is the unique number that is 2 less than x.

When a function is defined by an equation, we often use function notation. For example, the equation $y = x - 2$ may be written in function notation as

Answers
4. a. Yes   **b.** No   **c.** No

$$f(x) = x - 2 \text{ read as "}f \text{ of } x \text{ equals } x - 2.\text{"}$$

With function notation,

- f is the name of the function,
- x is an input variable from the domain,
- $f(x)$ is the function value (or y value) corresponding to x.

A function may be evaluated at different values of x by using substitution.

$$f(x) = x - 2$$
$$f(4) = (4) - 2 = 2 \qquad f(4) = 2 \text{ can be interpreted as } (4, 2).$$
$$f(1) = (1) - 2 = -1 \qquad f(1) = -1 \text{ can be interpreted as } (1, -1).$$

EXAMPLE 5   Evaluating a Function

Evaluate the function defined by $g(x) = 2x + 1$ for the given values of x.

a. $g(-2)$ **b.** $g(-1)$ **c.** $g(0)$ **d.** $g(1)$ **e.** $g(2)$

Solution:

a. $g(-2) = 2(-2) + 1$ Substitute -2 for x.
$\qquad = -3 \qquad\qquad g(-2) = -3$

b. $g(-1) = 2(-1) + 1$ Substitute -1 for x.
$\qquad = -1 \qquad\qquad g(-1) = -1$

c. $g(0) = 2(0) + 1$ Substitute 0 for x.
$\qquad = 1 \qquad\qquad g(0) = 1$

d. $g(1) = 2(1) + 1$ Substitute 1 for x.
$\qquad = 3 \qquad\qquad g(1) = 3$

e. $g(2) = 2(2) + 1$ Substitute 2 for x.
$\qquad = 5 \qquad\qquad g(2) = 5$

The function values represent the ordered pairs $(-2, -3)$, $(-1, -1)$, $(0, 1)$, $(1, 3)$, and $(2, 5)$. The line through the points represents all ordered pairs defined by this function. This is the graph of the function.

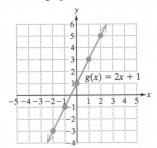

Skill Practice 5   Evaluate the function defined by $h(x) = 4x - 3$ for the given values of x.

a. $h(-3)$ **b.** $h(-1)$ **c.** $h(0)$ **d.** $h(1)$ **e.** $h(3)$

EXAMPLE 6   Evaluating a Function

Evaluate the function defined by $f(x) = 3x^2 + 2x$ for the given values of x.
a. $f(a)$ **b.** $f(x + h)$

Solution:

a. $f(a) = 3a^2 + 2a$ Substitute a for x.

b. $f(x + h) = 3(x + h)^2 + 2(x + h)$ Substitute $x + h$ for x.
$\qquad\qquad = 3(x^2 + 2xh + h^2) + 2x + 2h$ Simplify.
$\qquad\qquad\qquad$ Recall: $(a + b)^2 = a^2 + 2ab + b^2$
$\qquad\qquad = 3x^2 + 6xh + 3h^2 + 2x + 2h$

Skill Practice 6   Evaluate the function defined by $f(x) = -x^2 + 4x$ for the given values of x.

a. $f(t)$ **b.** $f(a + h)$

Answers
5. a. $h(-3) = -15$
   b. $h(-1) = -7$ c. $h(0) = -3$
   d. $h(1) = 1$ e. $h(3) = 9$
6. a. $f(t) = -t^2 + 4t$
   b. $f(a + h)$
      $= -a^2 - 2ah - h^2 + 4a + 4h$

3. Determine *x*- and *y*-Intercepts of a Function Defined by *y* = *f*(*x*)

Recall that to find an *x*-intercept(s) of the graph of an equation, we substitute 0 for *y* in the equation and solve for *x*. Using function notation, $y = f(x)$, this is equivalent to finding the real solutions of the equation $f(x) = 0$. To find the *y*-intercept, substitute 0 for *x* and solve the equation for *y*. Using function notation, this is equivalent to finding $f(0)$.

Finding Intercepts Using Function Notation

Given a function defined by $y = f(x)$,

- The *x*-intercepts are the real solutions to the equation $f(x) = 0$.
- The *y*-intercept is given by $f(0)$.

EXAMPLE 7   Finding the *x*- and *y*-Intercepts of a Function

Find the *x*- and *y*-intercepts of the function defined by $f(x) = x^2 - 4$.

Solution:

To find the *x*-intercept(s), solve the equation $f(x) = 0$.

$$f(x) = x^2 - 4$$
$$0 = x^2 - 4$$
$$x^2 = 4$$
$$x = \pm 2 \qquad \text{The } x\text{-intercepts are } (2, 0) \text{ and } (-2, 0).$$

To find the *y*-intercept, evaluate $f(0)$.

$$f(0) = (0)^2 - 4$$
$$= -4 \qquad \text{The } y\text{-intercept is } (0, -4).$$

Skill Practice 7   Find the *x*- and *y*-intercepts of the function defined by $f(x) = |x| - 5$.

4. Determine Domain and Range of a Function

Given a relation defining *y* as a function of *x*, the **domain** is the set of *x* values in the function, and the **range** is the set of *y* values in the function. In Example 8, we find the domain and range from the graph of a function.

EXAMPLE 8   Determining Domain and Range

Determine the domain and range for the functions shown.

a.   **b.** 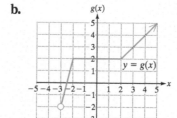   **c.**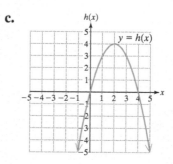

Answer

7. *x*-intercepts: (5, 0) and (−5, 0); *y*-intercept: (0, −5)

Solution:

a. The graph defines the set of ordered pairs:
$\{(-3, -4), (-1, 3), (0, 1), (2, 4), (4, 4)\}$

Domain: $\{-3, -1, 0, 2, 4\}$ The domain is the set of x values.

Range: $\{-4, 1, 3, 4\}$ The range is the set of y values.

b.

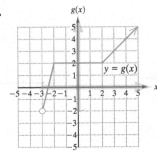

The domain is shown on the x-axis in green tint.
Domain: $\{x \mid x > -3\}$ or in interval notation: $(-3, \infty)$.

The range is shown on the y-axis in red tint.
Range: $\{y \mid y > -2\}$ or in interval notation: $(-2, \infty)$.

c.

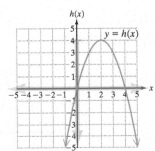

The graph extends infinitely far downward and infinitely far to the left and right. Therefore, the domain is the set of all real numbers, x.

The domain is shown on the x-axis in green tint.
Domain: $\mathbb{R}$ or in interval notation: $(-\infty, \infty)$.

The range is shown on the y-axis in red tint.
Range: $\{y \mid y \le 4\}$ or in interval notation: $(-\infty, 4]$.

Skill Practice 8   Determine the domain and range for the functions shown.

a.

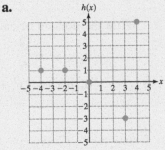

b.

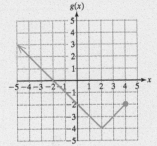

In some cases, a function may have restrictions on the domain. For example, consider the function defined by

$$f(x) = x^2 + 2 \quad \text{for} \quad x \ge 0$$

The restriction on x (that is, $x \ge 0$) is explicitly stated along with the definition of the function. If no such restriction is stated, then by default, the domain is all real numbers that when substituted into the function produce real numbers in the range.

Guidelines to Find Domain of a Function

To determine the implied domain of a function defined by $y = f(x)$,

- Exclude values of x that make the denominator of a fraction zero.
- Exclude values of x that make the radicand negative within an even-indexed root.

Answers
8. a. Domain: $\{-4, -2, 0, 3, 4\}$;
Range: $\{-3, 0, 1, 5\}$
b. Domain: $\{x \mid x \le 4\}$ or $(-\infty, 4]$;
Range: $\{y \mid y \ge -4\}$ or $[-4, \infty)$

EXAMPLE 9   Determining the Domain of a Function

Write the domain of each function in interval notation.

a. $f(x) = \dfrac{x + 3}{2x - 5}$ **b.** $g(x) = \dfrac{x}{x^2 + 4}$

c. $h(t) = \sqrt{2 - t}$ **d.** $m(a) = |4 + a|$

Solution:

a. $f(x) = \dfrac{x + 3}{2x - 5}$ The domain is all real numbers except those that make the denominator zero.

The variable x has the restriction that $2x - 5 \neq 0$. Therefore, $x \neq \frac{5}{2}$.

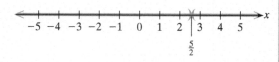

Domain: $\left(-\infty, \dfrac{5}{2}\right) \cup \left(\dfrac{5}{2}, \infty\right)$

b. $g(x) = \dfrac{x}{x^2 + 4}$ 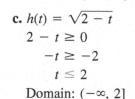 Denominator always positive (never zero)

The expression $x^2 \geq 0$ for all real numbers x. Therefore, $x^2 + 4 > 0$ for all real numbers x.

Domain: $(-\infty, \infty)$

c. $h(t) = \sqrt{2 - t}$ The domain is restricted to the real numbers that make the radicand greater than or equal to zero.

$2 - t \geq 0$

$-t \geq -2$ Divide by -1 and reverse the inequality sign.

$t \leq 2$

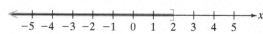

Domain: $(-\infty, 2]$

d. $m(a) = |4 + a|$ There are no fractions or radicals that would restrict the domain.

Domain: $(-\infty, \infty)$ The expression $|4 + a|$ is a real number for all real numbers a.

Skill Practice 9   Write the domain of each function in interval notation.

a. $f(x) = \dfrac{x - 2}{3x + 1}$ **b.** $g(x) = \dfrac{x^2}{5}$

c. $k(x) = \sqrt{x + 3}$ **d.** $p(x) = 2x^2 + 3x$

5. Interpret a Function Graphically

In Example 10, we will review the key concepts studied in this section by identifying characteristics of a function based on its graph.

EXAMPLE 10   Identifying Characteristics of a Function

Use the function f pictured to answer the questions.

a. Determine $f(2)$.

b. Determine $f(-5)$.

c. Find all x for which $f(x) = 0$.

d. Find all x for which $f(x) = 3$.

e. Determine the x-intercept(s).

f. Determine the y-intercept.

g. Determine the domain of f.

h. Determine the range of f.

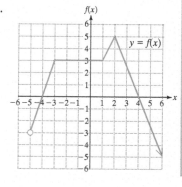

Answers

9. a. $\left(-\infty, -\dfrac{1}{3}\right) \cup \left(-\dfrac{1}{3}, \infty\right)$

b. $(-\infty, \infty)$

c. $[-3, \infty)$

d. $(-\infty, \infty)$

Solution:

a. $f(2) = 5$ $f(2) = 5$ because the function contains the point $(2, 5)$.

b. $f(-5)$ is not defined.   The point $(-5, -3)$ is not included in the function as indicated by the open dot.

c. $f(x) = 0$ for $x = -4$ and $x = 4$.   The points $(-4, 0)$ and $(4, 0)$ represent the points where $f(x) = 0$.

d. $f(x) = 3$ for all x on the interval $[-3, 1]$ and for $x = \frac{14}{5}$.

e. The x-intercepts are $(-4, 0)$ and $(4, 0)$.

f. The y-intercept is $(0, 3)$.

g. The domain is $(-5, \infty)$.

h. The range is $(-\infty, 5]$.

Skill Practice 10   Use the function f pictured to find:

a. $f(-2)$.

b. $f(4)$.

c. All x for which $f(x) = 3$.

d. All x for which $f(x) = 1$.

e. The x-intercept(s).

f. The y-intercept.

g. The domain of f.

h. The range of f.

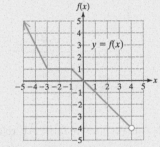

Answers

10. a. $f(-2) = 1$
   b. $f(4)$ is not defined.
   c. $x = -4$
   d. All x on the interval $[-3, -1]$
   e. $(0, 0)$
   f. $(0, 0)$
   g. $(-\infty, 4)$
   h. $(-4, \infty)$

SECTION 2.3   Practice Exercises

Prerequisite Review

R.1. Solve the equation using the square root property. $8x^2 - 40 = 0$

R.2. Solve. $4x^2 - 7x - 15 = 0$

R.3. Solve. Write the solution set in interval notation. $-3y - 9 \leq 15$

R.4. Solve. $|2n + 5| = 2$

R.5. Given $2x - 5y = 20$,
   a. Find the x-intercept.   **b.** Find the y-intercept.

Concept Connections

1. A set of ordered pairs (x, y) is called a _____ in x and y. The set of x values in the relation is called the _____ of the relation. The set of _____ values is called the range of the relation.

2. Given a function defined by $y = f(x)$, the statement $f(2) = 4$ is equivalent to what ordered pair?

3. Given a function defined by $y = f(x)$, to find the _____-intercept, evaluate $f(0)$.

4. Given a function defined by $y = f(x)$, to find the x-intercept(s), substitute 0 for _____ and solve for x.

5. Given $f(x) = \dfrac{x + 1}{x + 5}$, the domain is restricted so that $x \neq$ _____.

6. Given $g(x) = \sqrt{x - 5}$, the domain is restricted so that $x \geq$ _____.

7. Consider a relation that defines the height y of a tree for a given time t after it is planted. Does this relation define y as a function of t? Explain.

8. Consider a relation that defines a time y during the course of a year when the temperature T in Fort Collins, Colorado, is $70°$. Does this relation define y as a function of T? Explain.

Objective 1: Determine Whether a Relation Is a Function

For Exercises 9–12,

a. Write a set of ordered pairs (x, y) that defines the relation.

b. Write the domain of the relation.

c. Write the range of the relation.

d. Determine if the relation defines y as a function of x. (**See Examples 1–2**)

9.

Actor x	Number of Oscar Nominations y
Tom Hanks	5
Jack Nicholson	12
Sean Penn	5
Dustin Hoffman	7

10.

City x	Elevation at Airport (ft) y
Albany	285
Denver	5883
Miami	11
San Francisco	11

11.

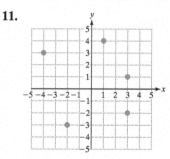

12.

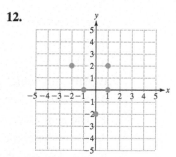

13. Answer true or false. All relations are functions.

14. Answer true or false. All functions are relations.

For Exercises 15–32, determine if the relation defines y as a function of x. (See Examples 3–4)

15.

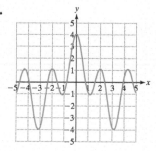

16.

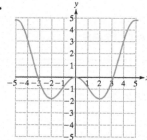

17.

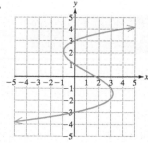

18.

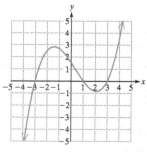

19.

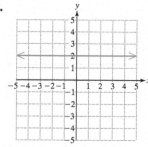

20.

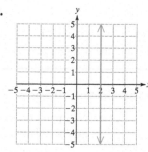

21.

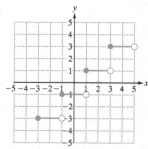

22.

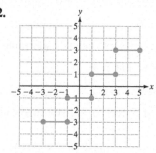

23.

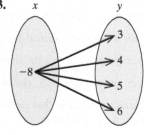

24.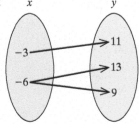

25.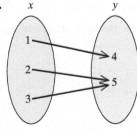

26.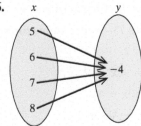

27. $(x + 1)^2 + (y + 5)^2 = 25$

28. $(x + 3)^2 + (y + 4)^2 = 1$

29. $y = x + 3$

30. $y = x - 4$

31. a. $y = x^2$
    b. $x = y^2$

32. a. $y = |x|$
    b. $x = |y|$

Objective 2: Apply Function Notation

33. The statement $f(4) = 1$ corresponds to what ordered pair?

34. The statement $g(7) = -5$ corresponds to what ordered pair?

For Exercises 35–56, evaluate the function for the given value of *x*. (See Examples 5–6)

$$f(x) = x^2 + 3x \qquad g(x) = \frac{1}{x} \qquad h(x) = 5 \qquad k(x) = \sqrt{x + 1}$$

35. a. $f(-2)$ **b.** $f(-1)$ **c.** $f(0)$ **d.** $f(1)$ **e.** $f(2)$

36. a. $g(-2)$ **b.** $g(-1)$ **c.** $g\left(-\frac{1}{2}\right)$ **d.** $g\left(\frac{1}{2}\right)$ **e.** $g(2)$

37. a. $h(-2)$ **b.** $h(-1)$ **c.** $h(0)$ **d.** $h(1)$ **e.** $h(2)$

38. a. $k(-2)$ **b.** $k(-1)$ **c.** $k(0)$ **d.** $k(1)$ **e.** $k(3)$

39. $g(3)$ **40.** $h(-7)$ **41.** $g\left(\frac{1}{3}\right)$

42. $h(7)$ **43.** $k(-5)$ **44.** $f(5)$

45. $k(8)$ **46.** $f(-5)$ **47.** $g(t)$

48. $f(a)$ **49.** $k(x + h)$ **50.** $h(x + h)$

51. $f(a + 4)$ **52.** $f(t - 3)$ **53.** $g(0)$

54. $k(-10)$ **55.** $f(x + h)$ **56.** $g(x + h)$

For Exercises 57–62, find and simplify $f(x + h)$. (See Example 6)

57. $f(x) = -4x^2 - 5x + 2$ **58.** $f(x) = -2x^2 + 6x - 3$ **59.** $f(x) = 7 - 3x^2$

60. $f(x) = 11 - 5x^2$ **61.** $f(x) = x^3 + 2x - 5$ **62.** $f(x) = x^3 - 4x + 2$

For Exercises 63–70, refer to the function $f = \{(2, 3), (9, 7), (3, 4), (-1, 6)\}$.

63. Determine $f(9)$.   **64.** Determine $f(-1)$.   **65.** Determine $f(3)$.

66. Determine $f(2)$.   **67.** For what value of x is $f(x) = 6$?   **68.** For what value of x is $f(x) = 7$?

69. For what value of x is $f(x) = 3$?   **70.** For what value of x is $f(x) = 4$?

71. Joe rides his bicycle an average of 18 mph. The distance Joe rides $d(t)$ (in mi) is given by $d(t) = 18t$, where t is the time in hours that he rides.
    a. Evaluate $d(2)$ and interpret the meaning.
    b. Determine the distance Joe travels in 40 min.

72. Frank needs to drive 250 mi from Daytona Beach to Miami. After having driven x miles, the distance remaining $r(x)$ (in mi) is given by $r(x) = 250 - x$.
    a. Evaluate $r(50)$ and interpret the meaning.
    b. Determine the distance remaining after 122 mi.

73. At a restaurant, if a party has eight or more people, the gratuity is automatically added to the bill. If x is the cost of the meal, then the total bill $C(x)$ with an 18% gratuity and a 6% sales tax is given by: $C(x) = x + 0.06x + 0.18x$. Evaluate $C(225)$ and interpret the meaning in the context of this problem.

74. A bookstore marks up the price of a book by 40% of the cost from the publisher. Therefore, the bookstore's price to the student, $P(x)$ (in \$) after a 7.5% sales tax, is given by $P(x) = 1.075(x + 0.40x)$, where x is the cost of the book from the publisher. Evaluate $P(60)$ and interpret the meaning in the context of this problem.

Objective 3: Determine *x*- and *y*-Intercepts of a Function Defined by $y = f(x)$

For Exercises 75–84, determine the *x*- and *y*-intercepts for the given function. (See Example 7)

75. $f(x) = 2x - 4$ **76.** $g(x) = 3x - 12$ **77.** $h(x) = |x| - 8$

78. $k(x) = -|x| + 2$ **79.** $p(x) = -x^2 + 12$ **80.** $q(x) = x^2 - 8$

81. $r(x) = |x - 8|$ **82.** $s(x) = |x + 3|$ **83.** $f(x) = \sqrt{x} - 2$

84. $g(x) = -\sqrt{x} + 3$

85. A student decides to finance a used car over a 5-yr (60-month) period. After making a down payment of $2000, the remaining cost of the car including tax and interest is $14,820. The amount owed $y = A(t)$ (in $) is given by $A(t) = 14{,}820 - 247t$, where t is the number of months after purchase and $0 \le t \le 60$. Determine the t-intercept and y-intercept and interpret their meanings in context.

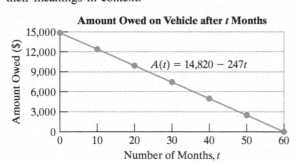

Amount Owed on Vehicle after *t* Months

$A(t) = 14{,}820 - 247t$

86. The amount spent on video games per person in the United States has been increasing since 2006. (*Source*: www.census.gov) The function defined by $f(x) = 9.4x + 35.7$ represents the amount spent $f(x)$ (in $) x years since 2006. Determine the y-intercept and interpret its meaning in context.

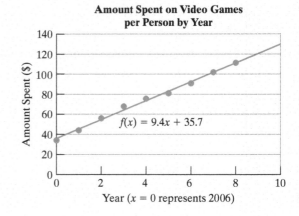

Amount Spent on Video Games per Person by Year

$f(x) = 9.4x + 35.7$

Year ($x = 0$ represents 2006)

Objective 4: Determine Domain and Range of a Function

For Exercises 87–96, determine the domain and range of the function. (See Example 8)

87.

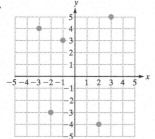

88.

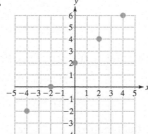

89.

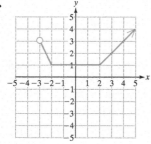

90.

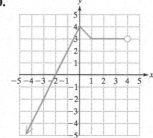

91.

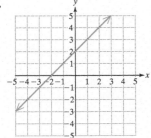

92.

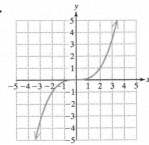

93.

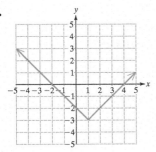

94.

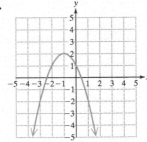

95.

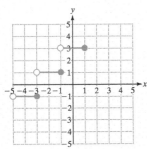

96.

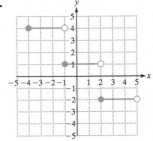

For Exercises 97–110, write the domain in interval notation. (See Example 9)

97. a. $f(x) = \dfrac{x-3}{x-4}$ **b.** $g(x) = \dfrac{x-3}{x^2-4}$ **c.** $h(x) = \dfrac{x-3}{x^2+4}$

98. a. $k(x) = \dfrac{x+6}{x-2}$ **b.** $j(x) = \dfrac{x+6}{x^2+2}$ **c.** $p(x) = \dfrac{x+6}{x^2-2}$

99. a. $a(x) = \sqrt{x+9}$ **b.** $b(x) = \sqrt{9-x}$ **c.** $c(x) = \dfrac{1}{\sqrt{x+9}}$

100. a. $y(t) = \sqrt{16-t}$ **b.** $w(t) = \sqrt{t-16}$ **c.** $z(t) = \dfrac{1}{\sqrt{16-t}}$

101. a. $f(t) = \sqrt[3]{t-5}$ **b.** $g(t) = \sqrt[3]{5-t}$ **c.** $h(t) = \dfrac{1}{\sqrt[3]{t-5}}$

102. a. $k(x) = \sqrt[5]{3+x}$ **b.** $m(x) = \sqrt[5]{x-3}$ **c.** $n(x) = \dfrac{1}{\sqrt[5]{x-3}}$

103. a. $f(x) = x^2 - 3x - 28$ **b.** $g(x) = \dfrac{x+2}{x^2-3x-28}$ **c.** $h(x) = \dfrac{x^2-3x-28}{x+2}$

104. a. $r(x) = x^2 - 4x - 12$ **b.** $s(x) = \dfrac{x^2-4x-12}{x+1}$ **c.** $t(x) = \dfrac{x+1}{x^2-4x-12}$

105. a. $w(x) = |x+1| + 4$ **b.** $y(x) = \dfrac{x}{|x+1|+4}$ **c.** $z(x) = \dfrac{x}{|x+1|-4}$

106. a. $f(a) = 8 - |a-2|$ **b.** $g(a) = \dfrac{5}{8-|a-2|}$ **c.** $h(a) = \dfrac{5}{8+|a-2|}$

107. a. $f(x) = \sqrt{x + 15}$　　　　　　**b.** $g(x) = \sqrt{x + 15} - 2$　　　　　　**c.** $k(x) = \dfrac{5}{\sqrt{x + 15} - 2}$

108. a. $f(c) = \sqrt{c + 20}$　　　　　　**b.** $g(c) = \sqrt{c + 20} - 1$　　　　　　**c.** $h(c) = \dfrac{-4}{\sqrt{c + 20} - 1}$

109. a. $p(x) = 2x + 1$　　　　　　**b.** $q(x) = 2x + 1; x \geq 0$　　　　　　**c.** $r(x) = 2x + 1; 0 \leq x < 7$

110. a. $m(x) = 3x - 7$　　　　　　**b.** $n(x) = 3x - 7; x < 0$　　　　　　**c.** $n(x) = 3x - 7; -2 < x < 2$

Objective 5: Interpret a Function Graphically

For Exercises 111–114, use the graph of $y = f(x)$ to answer the following. (See Example 10)

a. Determine $f(-2)$.　　　　　　**b.** Determine $f(3)$.　　　　　　**c.** Find all x for which $f(x) = -1$.

d. Find all x for which $f(x) = -4$.　　**e.** Determine the x-intercept(s).　　**f.** Determine the y-intercept.

g. Determine the domain of f.　　**h.** Determine the range of f.

111.

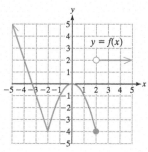

112.

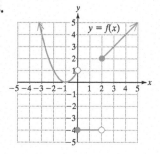

113.

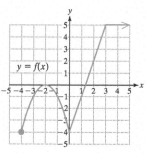

114.

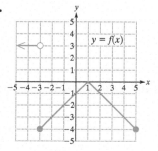

Mixed Exercises

For Exercises 115–122, write a function that represents the given statement.

115. Suppose that a phone card has 400 min. Write a relationship that represents the number of minutes remaining $r(x)$ as a function of the number of minutes already used x.

116. Suppose that a roll of wire has 200 ft. Write a relationship that represents the amount of wire remaining $w(x)$ as a function of the number of feet of wire x already used.

117. Given an equilateral triangle with sides of length x, write a relationship that represents the perimeter $P(x)$ as a function of x.

118. In an isosceles triangle, two angles are equal in measure. If the third angle is x degrees, write a relationship that represents the measure of one of the equal angles $A(x)$ as a function of x.

119. Two adjacent angles form a right angle. If the measure of one angle is x degrees, write a relationship representing the measure of the other angle $C(x)$ as a function of x.

120. Two adjacent angles form a straight angle (180°). If the measure of one angle is x degrees, write a relationship representing the measure of the other angle $S(x)$ as a function of x.

121. Write a relationship for a function whose $f(x)$ values are 2 less than three times the square of x.

122. Write a relationship for a function whose $f(x)$ values are 3 more than the principal square root of x.

Write About It

123. If two points align vertically then the points do not define y as a function of x. Explain why.

124. Given a function defined by $y = f(x)$, explain how to determine the x- and y-intercepts.

Expanding Your Skills

125. Given a square with sides of length s, diagonal of length d, perimeter P, and area A,

   a. Write P as a function of s.
   b. Write A as a function of s.
   c. Write A as a function of P.
   d. Write P as a function of A.
   e. Write d as a function of s.
   f. Write s as a function of d.
   g. Write P as a function of d.
   h. Write A as a function of d.

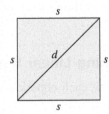

126. Given a circle with radius r, diameter d, circumference C, and area A,

   a. Write C as a function of r.
   b. Write A as a function of r.
   c. Write r as a function of d.
   d. Write d as a function of r.
   e. Write C as a function of d.
   f. Write A as a function of d.
   g. Write A as a function of C.
   h. Write C as a function of A.

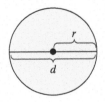

SECTION 2.4   Linear Equations in Two Variables and Linear Functions

OBJECTIVES

1. Graph Linear Equations in Two Variables
2. Determine the Slope of a Line
3. Apply the Slope-Intercept Form of a Line
4. Compute Average Rate of Change
5. Solve Equations and Inequalities Graphically

1. Graph Linear Equations in Two Variables

The median incomes for individuals for all levels of education have shown an increasing trend since 1990. However, the median income for individuals with a bachelor's degree is consistently greater than for individuals whose highest level of education is a high school degree or equivalent (Figure 2-15). (*Source*: U.S. Census Bureau, www.census.gov)

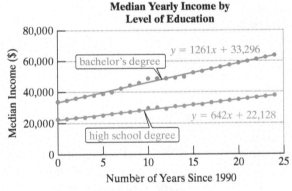

Median Yearly Income by Level of Education

Figure 2-15

The graph in Figure 2-15 is called a scatter plot. A **scatter plot** is a visual representation of a set of points. In this case, the x values represent the number of years since 1990, and the y values represent the median income in dollars. The line that models each set of data is called a **regression line** and is found by using techniques taught in a first course in statistics. The equations that represent the two lines are called linear equations in two variables.

> **TIP** For an equation in standard form, the value of A, B, and C are usually taken to be integers where A, B, and C share no common factors.

> **Linear Equation in Two Variables**
>
> Let A, B, and C represent real numbers such that A and B are not both zero. A **linear equation** in the variables x and y is an equation that can be written in the form:
>
> $Ax + By = C$ This is called the **standard form** of an equation of a line.
>
> *Note*: A linear equation $Ax + By = C$ has variables x and y each of first degree.

In Example 1, we demonstrate that the graph of a linear equation $Ax + By = C$ is a line. The line may be slanted, horizontal, or vertical depending on the coefficients A, B, and C.

> ### EXAMPLE 1   Graphing Linear Equations
>
> Graph the line represented by each equation.
>
> **a.** $2x + 3y = 6$ **b.** $x = -3$ **c.** $2y = 4$
>
> **Solution:**
>
> **a.** Solve the equation for y. Then substitute arbitrary values of x into the equation and solve for the corresponding values of y.

Avoiding Mistakes

The graph of a linear equation is a line. Therefore, a minimum of two points is needed to graph the line. A third point can be used to verify that the line is graphed correctly. The points must all line up.

$$2x + 3y = 6 \qquad \text{Solve the equation for } y.$$
$$3y = -2x + 6$$
$$y = -\frac{2}{3}x + 2$$

x	y
-3	4
0	2
3	0
6	-2

In the table we have selected convenient values of x that are multiples of 3.

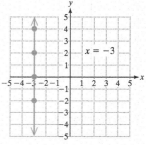

b. $x = -3$

The solutions to this equation must have an x-coordinate of -3. The y variable can be *any* real number.

x	y
-3	-2
-3	0
-3	2
-3	4

x must be -3. y can be any real number.

TIP The graph of a vertical line will have no y-intercept unless the line is the y-axis itself.

TIP The graph of a horizontal line will have no x-intercept unless the line is the x-axis itself.

c. $2y = 4$ Solve for y.
  $y = 2$

The solutions to this equation must have a y-coordinate of 2. The x variable can be *any* real number.

x	y
-2	2
0	2
2	2
4	2

x can be any real number. y must be 2.

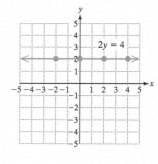

Answer

1. a.–c.

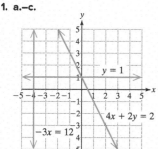

> **Skill Practice 1**   Graph the line represented by each equation.
>
> **a.** $4x + 2y = 2$ **b.** $y = 1$ **c.** $-3x = 12$

2. Determine the Slope of a Line

One of the important characteristics of a nonvertical line is that for every 1 unit of change in the horizontal variable, the vertical change is a constant m called the **slope** of the line. For example, consider the line representing the median income for individuals with a bachelor's degree, x years since the year 1990. The line in Figure 2-16 has a slope of $1261. This means that median income for individuals with a bachelor's degree increased on average by $1261 per year during this time period.

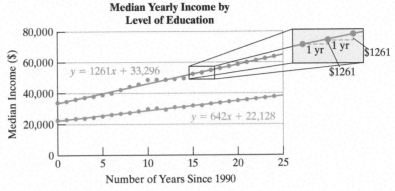

Figure 2-16

Consider any two distinct points (x_1, y_1) and (x_2, y_2) on a line (Figure 2-17). The slope m of the line through the points is the ratio between the change in the y values $(y_2 - y_1)$ and the change in the x values $(x_2 - x_1)$. In many applications in the sciences, the change in a variable is denoted by the Greek letter Δ (delta). Therefore, $(y_2 - y_1)$ can be represented by Δy and $(x_2 - x_1)$ can be represented by Δx.

Figure 2-17

Slope Formula

The **slope** of a line passing through the distinct points (x_1, y_1) and (x_2, y_2) is

$$m = \frac{\Delta y}{\Delta x} = \frac{y_2 - y_1}{x_2 - x_1} \text{ provided that } x_2 - x_1 \neq 0$$

change in y (rise)

change in x (run)

EXAMPLE 2   Finding the Slope of a Line Through Two Points

Find the slope of the line passing through the given points.

a. $(-3, -2)$ and $(2, 5)$ **b.** $\left(-\dfrac{5}{2}, 0\right)$ and $(1, -7)$

Solution:

a. $(-3, -2)$ and $(2, 5)$

$\quad (x_1, y_1) \quad$ and $(x_2, y_2) \qquad$ Label the points.

$$m = \frac{y_2 - y_1}{x_2 - x_1} = \frac{5 - (-2)}{2 - (-3)} = \frac{7}{5}$$

A line with a positive slope "*rises*" upward from left to right.

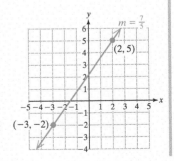

b. $\left(-\dfrac{5}{2}, 0\right)$ and $(1, -7)$

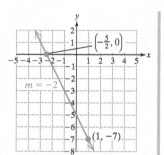

(x_1, y_1) and (x_2, y_2) Label the points.

$$m = \frac{y_2 - y_1}{x_2 - x_1} = \frac{-7 - 0}{1 - \left(-\frac{5}{2}\right)} = \frac{-7}{\frac{7}{2}} = -7 \cdot \frac{2}{7} = -2$$

A line with a negative slope "*falls*" downward from left to right.

Skill Practice 2   Find the slope of the line passing through the given points.

a. $(-4, 1)$ and $(2, -2)$ **b.** $\left(\dfrac{3}{4}, 2\right)$ and $(-3, 17)$

EXAMPLE 3   **Finding the Slope of Horizontal and Vertical Lines**

Find the slope of each line.

Solution:

a.

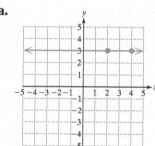

By inspection, we see that between any two points on the graph, the vertical change is zero, so the slope is zero.

To compute this numerically, select any two points on the line such as $(2, 3)$ and $(4, 3)$.

$$m = \frac{y_2 - y_1}{x_2 - x_1} = \frac{3 - 3}{4 - 2} = \frac{0}{2} = 0$$

b.

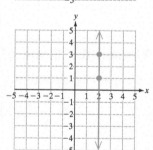

To find the slope, select any two points on the line such as $(2, 1)$ and $(2, 3)$.

$$m = \frac{y_2 - y_1}{x_2 - x_1} = \frac{3 - 1}{2 - 2} = \frac{2}{0} \quad \text{(undefined)}$$

By inspection, we see that between any two points on the line, the change in x is zero. This makes the slope undefined because the ratio representing the slope has a divisor of zero.

Skill Practice 3   Fill in the blank.

a. The slope of a vertical line is _____.
b. The slope of a horizontal line is _____.

From Example 1, we see that a linear equation may represent the graph of a slanted line, a horizontal line, or a vertical line. From Examples 2 and 3, we see that a line may have a positive slope, a negative slope, a zero slope, or an undefined slope.

Answers

2. a. $-\dfrac{1}{2}$ **b.** -4

3. a. Undefined   **b.** 0

Linear Equations and Slopes of Lines

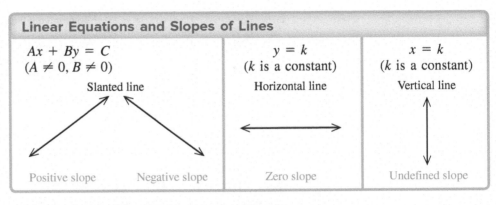

$Ax + By = C$ $(A \neq 0, B \neq 0)$	$y = k$ (k is a constant)	$x = k$ (k is a constant)
Slanted line	Horizontal line	Vertical line
Positive slope Negative slope	Zero slope	Undefined slope

3. Apply the Slope-Intercept Form of a Line

The slope formula can be used to develop the slope-intercept form of a line. Suppose that a line has a slope m and y-intercept $(0, b)$. Let (x, y) be any other point on the line. From the slope formula, we have:

$$\frac{y - b}{x - 0} = m \qquad \text{Slope formula}$$

$$y - b = mx \qquad \text{Multiply by } x.$$

$$y = mx + b \qquad \text{This is slope-intercept form. Slope}$$
intercept-form has the y variable isolated.

Avoiding Mistakes

An equation of a vertical line takes the form $x = k$, where k is a constant. Because there is no y variable and because the slope is undefined, an equation of a vertical line cannot be written in slope-intercept form.

Slope-Intercept Form of a Line

Given a line with slope m and y-intercept $(0, b)$, the **slope-intercept form** of the line is given by $y = mx + b$.

The slope-intercept form of a line is particularly useful because we can identify the slope and y-intercept by inspection. For example:

$$y = \frac{2}{3}x - 5 \qquad\qquad m = \frac{2}{3} \qquad y\text{-intercept: } (0, -5)$$

$$y = x + 4 \qquad\qquad m = 1 \qquad y\text{-intercept: } (0, 4)$$

$$y = 2x \quad (\text{or } y = 2x + 0) \qquad m = 2 \qquad y\text{-intercept: } (0, 0)$$

$$y = 6 \quad (\text{or } y = 0x + 6) \qquad m = 0 \qquad y\text{-intercept: } (0, 6)$$

If the slope and y-intercept of a line are known, we can graph the line. This is demonstrated in Example 4.

EXAMPLE 4   Using the Slope and y-Intercept to Graph a Line

Given $3x + 4y = 4$,

a. Write the equation in slope-intercept form.
b. Determine the slope and y-intercept.
c. Graph the line by using the slope and y-intercept.

Solution:

a. $3x + 4y = 4$

$$4y = -3x + 4 \qquad \text{To write an equation in slope-intercept}$$
form, isolate the y variable.

$$y = -\frac{3}{4}x + 1 \qquad \text{Slope-intercept form}$$

b. $m = -\dfrac{3}{4}$ and the y-intercept is $(0, 1)$.   The slope is the coefficient on x.
The constant term gives the y-intercept.

c.

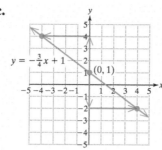

$y = -\frac{3}{4}x + 1$

(0, 1)

To graph the line, first plot the y-intercept $(0, 1)$.

Then begin at the y-intercept, and use the slope to find a second point on the line. In this case, the slope can be interpreted as the following two ratios:

$m = \dfrac{-3}{4}$ ◄——— Move down 3 units.
◄——— Move to the right 4 units.

$m = \dfrac{3}{-4}$ ◄——— Move up 3 units.
◄——— Move to the left 4 units.

Skill Practice 4   Given $2x + 4y = 8$,

  a. Write the equation in slope-intercept form.
  b. Determine the slope and y-intercept.
  c. Graph the line by using the slope and y-intercept.

Notice that the slope-intercept form of a line $y = mx + b$ has the y variable isolated and defines y in terms of x. Therefore, an equation written in slope-intercept form defines y as a function of x. In Example 4, $y = -\frac{3}{4}x + 1$ can be written using function notation as $f(x) = -\frac{3}{4}x + 1$.

Definition of Linear and Constant Functions

Let m and b represent real numbers where $m \neq 0$. Then,

- A function defined by $f(x) = mx + b$ is a **linear function.** The graph of a linear function is a slanted line.
- A function defined by $f(x) = b$ is a **constant function.** The graph of a constant function is a horizontal line.

The slope-intercept form of a line can be used as a tool to define a linear function given a point on the line and the slope.

EXAMPLE 5   **Writing an Equation of a Line Given a Point and the Slope**

Write an equation of the line that passes through the point $(2, -3)$ and has slope -4. Then write the linear equation using function notation, where $y = f(x)$.

Solution:

Given $m = -4$ and $(2, -3)$.   We need to find an equation of the form $y = mx + b$.

$y = mx + b$
$y = -4x + b$ The value of m is given as -4.
$-3 = -4(2) + b$ Substitute $x = 2$ and $y = -3$ from the given point $(2, -3)$.
$-3 = -8 + b$ Solve for b.
$b = 5$
$y = mx + b$
$y = -4x + 5$ Substitute $m = -4$ and $b = 5$ into the equation $y = mx + b$.
$f(x) = -4x + 5$ Write the relation using function notation.

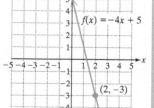

From the graph, we see that the graph of $f(x) = -4x + 5$ does indeed pass through the point $(2, -3)$ and has slope -4.

Answers

4. a. $y = -\dfrac{1}{2}x + 2$

  b. $m = -\dfrac{1}{2}$; y-intercept: $(0, 2)$

  c.

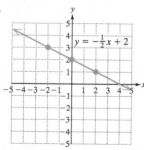

$y = -\frac{1}{2}x + 2$

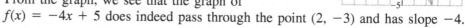

> **Skill Practice 5**   Write an equation of the line that passes through the point $(-1, -4)$ and has slope 3. Then write the equation using function notation.

4. Compute Average Rate of Change

The graphs of many functions are not linear. However, we often use linear approximations to analyze nonlinear functions on small intervals. For example, the graph in Figure 2-18 shows the blood alcohol concentration (BAC) for an individual over a period of 9 hr.

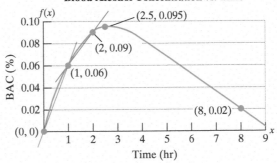

Figure 2-18

A line drawn through two points on a curve is called a **secant line.** In Figure 2-18, the average rate of change in BAC between two points on the graph is the slope of the secant line through the points. Notice that the slope of the secant line between $x = 0$ and $x = 1$ (shown in red) is greater than the slope of the secant line between $x = 1$ and $x = 2$ (shown in green). This means that the average increase in BAC is greater over the first hour than over the second hour.

Average Rate of Change of a Function

Suppose that the points (x_1, y_1) and (x_2, y_2) are points on the graph of a function f. Using function notation, these are the points $(x_1, f(x_1))$ and $(x_2, f(x_2))$.

If f is defined on the interval $[x_1, x_2]$, then the **average rate of change** of f on the interval $[x_1, x_2]$ is the slope of the secant line containing $(x_1, f(x_1))$ and $(x_2, f(x_2))$.

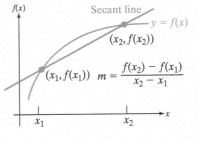

Average rate of change: $m = \dfrac{\Delta y}{\Delta x} = \dfrac{y_2 - y_1}{x_2 - x_1}$ or $m = \dfrac{f(x_2) - f(x_1)}{x_2 - x_1}$

EXAMPLE 6   **Computing Average Rate of Change**

Determine the average rate of change of blood alcohol level

a. from $x_1 = 0$ to $x_2 = 1$.

b. from $x_1 = 1$ to $x_2 = 2$.

c. Interpret the results from parts (a) and (b).

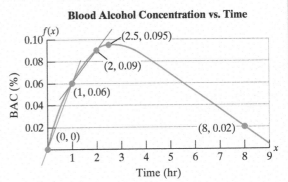

Answer

5. $y = 3x - 1$; $f(x) = 3x - 1$

Solution:

a. Average rate of change $= \dfrac{f(x_2) - f(x_1)}{x_2 - x_1} = \dfrac{f(1) - f(0)}{1 - 0} = \dfrac{0.06 - 0}{1}$

$$= 0.06$$

b. Average rate of change $= \dfrac{f(x_2) - f(x_1)}{x_2 - x_1} = \dfrac{f(2) - f(1)}{2 - 1} = \dfrac{0.09 - 0.06}{1}$

$$= 0.03$$

c. The blood alcohol concentration rose by an average of 0.06% per hour during the first hour.

The blood alcohol concentration rose by an average of 0.03% per hour during the second hour.

> **Skill Practice 6**   Refer to the graph in Example 6.
>
> **a.** Determine the average rate of change of blood alcohol level from $x_1 = 2.5$ to $x_2 = 8$. Round to 3 decimal places.
> **b.** Interpret the results from part (a).

EXAMPLE 7   **Computing Average Rate of Change**

Given the function defined by $f(x) = x^2 - 1$, determine the average rate of change from $x_1 = -2$ to $x_2 = 0$.

Solution:

Average rate of change $= \dfrac{f(x_2) - f(x_1)}{x_2 - x_1}$

$$= \dfrac{f(0) - f(-2)}{0 - (-2)} = \dfrac{-1 - 3}{2} = -2$$

The average rate of change is -2.

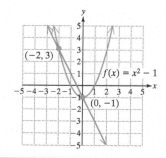

> **Skill Practice 7**   Given the function defined by $f(x) = x^3 + 2$, determine the average rate of change from $x_1 = -3$ to $x_2 = 0$.

---

5. Solve Equations and Inequalities Graphically

In many settings, the use of technology can provide a numerical and visual interpretation of an algebraic problem. For example, consider the equation $-x - 1 = x + 5$.

$$-x - 1 = x + 5$$
$$-6 = 2x$$
$$-3 = x \qquad \text{The solution set is } \{-3\}.$$

Now suppose that we create two functions from the left and right sides of the equation. We have $Y_1 = -x - 1$ and $Y_2 = x + 5$. Figure 2-19 shows that the graphs of the two lines intersect at $(-3, 2)$. The x-coordinate of the point of intersection is the solution to the equation $-x - 1 = x + 5$. That is, $Y_1 = Y_2$ when $x = -3$.

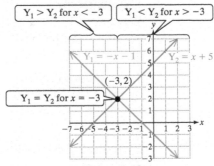

Figure 2-19

The graphs $Y_1 = -x - 1$ and $Y_2 = x + 5$ can also be used to find the solution sets to the related inequalities.

$-x - 1 < x + 5$ The solution set is the set of x values for which $Y_1 < Y_2$. This is the interval where the blue line is below the red line. The solution set is $(-3, \infty)$.

$-x - 1 > x + 5$ The solution set is the set of x values for which $Y_1 > Y_2$. This is the interval where the blue line is *above* the red line. The solution set is $(-\infty, -3)$.

EXAMPLE 8   **Solving Equations and Inequalities Graphically**

Solve the equations and inequalities graphically.

a. $2x - 3 = x - 1$ **b.** $2x - 3 < x - 1$ **c.** $2x - 3 > x - 1$

Solution:

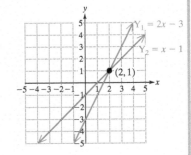

a. The left side of the equation is graphed as $Y_1 = 2x - 3$. The right side of the equation is graphed as $Y_2 = x - 1$. The point of intersection is $(2, 1)$. Therefore, $Y_1 = Y_2$ for $x = 2$. The solution set is $\{2\}$.

b. $Y_1 < Y_2$ to the *left* of $x = 2$. (That is, the blue line is below the red line for $x < 2$.) In interval notation the solution set is $(-\infty, 2)$.

c. $Y_1 > Y_2$ to the *right* of $x = 2$. (That is, the blue line is above the red line for $x > 2$.) In interval notation the solution set is $(2, \infty)$.

> **TIP** The solution set to the inequality $2x - 3 \leq x - 1$ includes equality, so the right endpoint would be included: $(-\infty, 2]$.
>
> The solution set to the inequality $2x - 3 \geq x - 1$ includes equality, so the left endpoint would be included: $[2, \infty)$.

Skill Practice 8   Use the graph to solve the equations and inequalities.

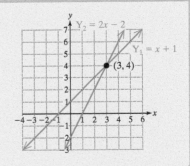

a. $x + 1 = 2x - 2$

b. $x + 1 \leq 2x - 2$

c. $x + 1 \geq 2x - 2$

Answers

8. a. $\{3\}$ **b.** $[3, \infty)$ **c.** $(-\infty, 3]$

TECHNOLOGY CONNECTIONS

Verifying Solutions to an Equation

We can verify the solutions to the equations and inequalities from Example 8 on a graphing calculator.

The solutions can be verified numerically by using the **Table** feature on the calculator. First enter $Y_1 = 2x - 3$ and $Y_2 = x - 1$.

Then display the table values for Y_1 and Y_2 for $x = 2$ and for x values less than and greater than 2.

Display the graphs of Y_1 and Y_2 and use the **Intersect** feature to determine the point of intersection.

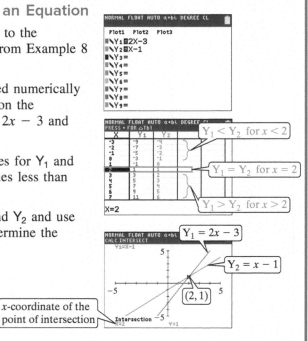

$Y_1 < Y_2$ for $x < 2$

$Y_1 = Y_2$ for $x = 2$

$Y_1 > Y_2$ for $x > 2$

$Y_1 = 2x - 3$

$Y_2 = x - 1$

x-coordinate of the point of intersection

In Example 9 we solve the equation $6x - 2(x + 2) - 5 = 0$. Notice that one side is zero. We can check the solution graphically by determining where the related function $Y_1 = 6x - 2(x + 2) - 5$ intersects the x-axis.

EXAMPLE 9   Solving Equations and Inequalities Graphically

a. Solve the equation $6x - 2(x + 2) - 5 = 0$ and verify the solution graphically on a graphing utility.

b. Use the graph to find the solution set to the inequality $6x - 2(x + 2) - 5 \leq 0$.

c. Use the graph to find the solution set to the inequality $6x - 2(x + 2) - 5 \geq 0$.

Solution:

a.
$$6x - 2(x + 2) - 5 = 0$$
$$6x - 2x - 4 - 5 = 0$$
$$4x - 9 = 0$$
$$x = \frac{9}{4}$$

The solution set is $\left\{ \dfrac{9}{4} \right\}$.

To verify the solution graphically enter the left side of the equation as $Y_1 = 6x - 2(x + 2) - 5$.

Using the **Zero** feature, we have $Y_1 = 0$ for $x = 2.25$. This is consistent with the solution $x = \frac{9}{4}$.

b. To solve $6x - 2(x + 2) - 5 \leq 0$ determine the values of x for which $Y_1 \leq 0$ (where the function is on or below the x-axis).

The solution set is $\left(-\infty, \dfrac{9}{4} \right]$.

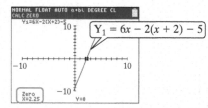

$Y_1 = 6x - 2(x + 2) - 5$

Answers

9. a. $\left\{\dfrac{5}{2}\right\}$

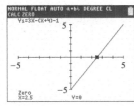

b. $\left(-\infty, \dfrac{5}{2}\right]$

c. $\left[\dfrac{5}{2}, \infty\right)$

c. To solve $6x - 2(x + 2) - 5 \geq 0$ determine the values of x for which $Y_1 \geq 0$ (where the function is on or above the x-axis).

The solution set is $\left[\dfrac{9}{4}, \infty\right)$.

> **Skill Practice 9**
>
> **a.** Solve the equation $3x - (x + 4) - 1 = 0$ and verify the solution graphically on a graphing utility.
> **b.** Use the graph to find the solution set to the inequality $3x - (x + 4) - 1 \leq 0$.
> **c.** Use the graph to find the solution set to the inequality $3x - (x + 4) - 1 \geq 0$.

SECTION 2.4   Practice Exercises

Prerequisite Review

R.1. Determine the x- and y-intercepts for $h(x) = 6x - 42$.

R.2. Solve $-7x - 8y = 1$ for y.

For Exercises R.3–R.4, solve the inequality. Write the solution set in interval notation.

R.3. $-4t + 5 < 13$

R.4. $6p - 2 \geq 5p + 8$

R.5. Given the function defined by $g(x) = -x^2 + 3x + 2$, find $g(-1)$.

Concept Connections

1. A _____ equation in the variables x and y can be written in the form $Ax + By = C$, where A and B are not both zero.

2. An equation of the form $x = k$ where k is a constant represents the graph of a _____ line.

3. An equation of the form $y = k$ where k is a constant represents the graph of a _____ line.

4. Write the formula for the slope of a line between the two distinct points (x_1, y_1) and (x_2, y_2).

5. The slope of a horizontal line is _____ and the slope of a vertical line is _____.

6. A function f is a linear function if $f(x) =$ _____, where m represents the slope and $(0, b)$ represents the y-intercept.

7. If f is defined on the interval $[x_1, x_2]$, then the average rate of change of f on the interval $[x_1, x_2]$ is given by the formula _____.

8. The graph of a constant function defined by $f(x) = b$ is a (horizontal/vertical) line.

Objective 1: Graph Linear Equations in Two Variables

For Exercises 9–20, graph the equation and identify the x- and y-intercepts. (See Example 1)

9. $-3x + 4y = 12$

10. $-2x + y = 4$

11. $2y = -5x + 2$

12. $3y = -4x + 6$

13. $x = -6$

14. $y = 4$

15. $5y + 1 = 11$

16. $3x - 2 = 4$

17. $0.02x + 0.05y = 0.1$

18. $0.03x + 0.07y = 0.21$

19. $2x = 3y$

20. $2x = -5y$

Objective 2: Determine the Slope of a Line

21. Find the average slope of the hill.

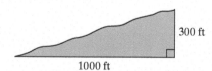

300 ft

1000 ft

22. Find the absolute value of the slope of the storm drainage pipe.

5 ft

80 ft

23. The road sign shown in the figure indicates the percent grade of a hill. This gives the slope of the road as the change in elevation per 100 horizontal feet. Given a 2.5% grade, write this as a slope in fractional form.

2.5% Grade

24. The pitch of a roof is defined as $\dfrac{\text{rafter rise}}{\text{rafter run}}$ and the fraction is typically written with a denominator of 12. Determine the pitch of the roof from point A to point C.

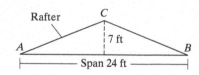

Rafter C

A 7 ft B

Span 24 ft

For Exercises 25–36, determine the slope of the line passing through the given points. (See Example 2)

25. $(4, -7)$ and $(2, -1)$

26. $(-3, -8)$ and $(4, 6)$

27. $(17, 9)$ and $(42, -6)$

28. $(-9, 4)$ and $(-1, -6)$

29. $(30, -52)$ and $(-22, -39)$

30. $(-100, -16)$ and $(84, 30)$

31. $(2.6, 4.1)$ and $(9.5, -3.7)$

32. $(8.5, 6.2)$ and $(-5.1, 7.9)$

33. $\left(\dfrac{3}{4}, 6\right)$ and $\left(\dfrac{5}{2}, 1\right)$

34. $\left(-3, \dfrac{2}{5}\right)$ and $\left(4, \dfrac{3}{10}\right)$

35. $(3\sqrt{6}, 2\sqrt{5})$ and $(\sqrt{6}, \sqrt{5})$

36. $(2\sqrt{11}, -3\sqrt{3})$ and $(\sqrt{11}, -5\sqrt{3})$

For Exercises 37–42, determine the slope of the line. (See Examples 2–3)

37.

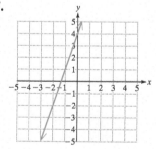

38.

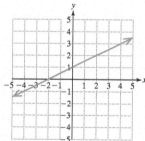

39.

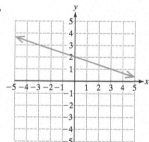

40.

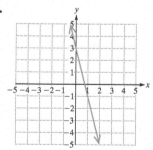

41.

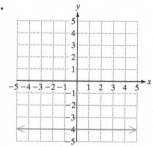

42.

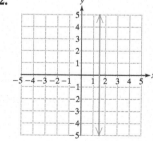

43. What is the slope of a line perpendicular to the *x*-axis?

44. What is the slope of a line parallel to the *x*-axis?

45. What is the slope of a line defined by $y = -7$?

46. What is the slope of a line defined by $x = 2$?

47. If the slope of a line is $\frac{4}{5}$, how much vertical change will be present for a horizontal change of 52 ft?

48. If the slope of a line is $\frac{5}{8}$, how much horizontal change will be present for a vertical change of 216 m?

49. Suppose that $y = P(t)$ represents the population of a city at time *t*. What does $\frac{\Delta P}{\Delta t}$ represent?

50. Suppose that $y = d(t)$ represents the distance that an object travels in time *t*. What does $\frac{\Delta d}{\Delta t}$ represent?

Objective 3: Apply the Slope-Intercept Form of a Line

For Exercises 51–62,

a. Write the equation in slope-intercept form if possible, and determine the slope and *y*-intercept.

b. Graph the equation using the slope and *y*-intercept. (**See Example 4**)

51. $2x - 4y = 8$

52. $3x - y = 6$

53. $3x = 2y - 4$

54. $5x = 3y - 6$

55. $3x = 4y$

56. $-2x = 3y$

57. $2y - 6 = 8$

58. $3y + 9 = 6$

59. $0.02x + 0.06y = 0.06$

60. $0.03x + 0.04y = 0.12$

61. $\dfrac{x}{4} + \dfrac{y}{7} = 1$

62. $\dfrac{x}{3} + \dfrac{y}{4} = 1$

For Exercises 63–64, determine if the function is linear, constant, or neither.

63. a. $f(x) = -\dfrac{3}{4}x$

b. $g(x) = -\dfrac{3}{4}x - 3$

c. $h(x) = -\dfrac{3}{4x}$

d. $k(x) = -\dfrac{3}{4}$

64. a. $m(x) = 5x + 1$

b. $n(x) = \dfrac{5}{x} + 1$

c. $p(x) = 5$

d. $q(x) = 5x$

For Exercises 65–74,

a. Use slope-intercept form to write an equation of the line that passes through the given point and has the given slope.

b. Write the equation using function notation where $y = f(x)$. (**See Example 5**)

65. $(0, 9)$; $m = \dfrac{1}{2}$

66. $(0, -4)$; $m = \dfrac{1}{3}$

67. $(1, -6)$; $m = -3$

68. $(2, -8)$; $m = -5$

69. $(-5, -3)$; $m = \dfrac{2}{3}$

70. $(-4, -2)$; $m = \dfrac{3}{2}$

71. $(2, 5)$; $m = 0$

72. $(-1, -3)$; $m = 0$

73. $(3.6, 5.1)$; $m = 1.2$

74. $(1.2, 2.8)$; $m = 2.4$

For Exercises 75–78,

a. Use slope-intercept form to write an equation of the line that passes through the two given points.

b. Then write the equation using function notation where $y = f(x)$.

75. $(4, 2)$ and $(0, -6)$

76. $(-8, 1)$ and $(0, -3)$

77. $(7, -3)$ and $(4, 1)$

78. $(2, -4)$ and $(-1, 3)$

Objective 4: Compute Average Rate of Change

For Exercises 79–80, find the slope of the secant line pictured in red. (See Example 6)

79.

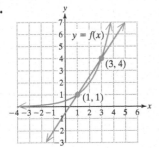

80.

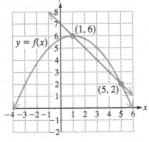

81. The function given by $y = f(x)$ shows the value of $5000 invested at 5% interest compounded continuously, x years after the money was originally invested.

   a. Find the average amount earned per year between the 5th year and 10th year.

   b. Find the average amount earned per year between the 20th year and the 25th year.

   c. Based on the answers from parts (a) and (b), does it appear that the rate at which annual income increases is increasing or decreasing with time?

Value of $5000 with Continuous Compounding at 5%

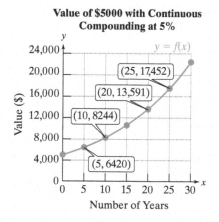

Number of Years

82. The function given by $y = f(x)$ shows the average monthly temperature (in °F) for Cedar Key. The value of x is the month number and $x = 1$ represents January.

   a. Find the average rate of change in temperature between months 3 and 5 (March and May).

   b. Find the average rate of change in temperature between months 9 and 11 (September and November).

   c. Comparing the results in parts (a) and (b), what does a positive rate of change mean in the context of this problem? What does a negative rate of change mean?

Average Monthly Temperature for Cedar Key, Florida

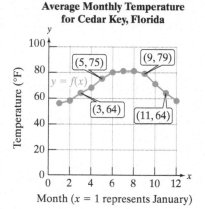

Month ($x = 1$ represents January)

83. The number $N(t)$ of new cases of a flu outbreak for a given city is given by $N(t) = 5000 \cdot 2^{-0.04t^2}$, where t is the number of months since the outbreak began.

   a. Find the average rate of change in the number of new flu cases between months 0 and 2, and interpret the result. Round to the nearest whole unit.

   b. Find the average rate of change in the number of new flu cases between months 4 and 6, and between months 10 and 12.

   c. Use a graphing utility to graph the function. Use the graph and the average rates of change found in parts (a) and (b) to discuss the pattern of the number of new flu cases.

84. The speed $v(L)$ (in m/sec) of an ocean wave in deep water is approximated by $v(L) = 1.2\sqrt{L}$, where L (in meters) is the wavelength of the wave. (The wavelength is the distance between two consecutive wave crests.)

   a. Find the average rate of change in speed between waves that are between 1 m and 4 m in length.

   b. Find the average rate of change in speed between waves that are between 4 m and 9 m in length.

   c. Use a graphing utility to graph the function. Using the graph and the results from parts (a) and (b), what does the difference in the rates of change mean?

For Exercises 85–90, determine the average rate of change of the function on the given interval. (See Example 7)

85. $f(x) = x^2 - 3$
   a. on $[0, 1]$
   b. on $[1, 3]$
   c. on $[-2, 0]$

86. $g(x) = 2x^2 + 2$
   a. on $[0, 1]$
   b. on $[1, 3]$
   c. on $[-2, 0]$

87. $h(x) = x^3$
   a. on $[-1, 0]$
   b. on $[0, 1]$
   c. on $[1, 2]$

88. $k(x) = x^3 - 2$
   a. on $[-1, 0]$
   b. on $[0, 1]$
   c. on $[1, 2]$

89. $m(x) = \sqrt{x}$
   a. $[0, 1]$
   b. $[1, 4]$
   c. $[4, 9]$

90. $n(x) = \sqrt{x - 1}$
   a. $[1, 2]$
   b. $[2, 5]$
   c. $[5, 10]$

Objective 5: Solve Equations and Inequalities Graphically

For Exercises 91–98, use the graph to solve the equation and inequalities. Write the solutions to the inequalities in interval notation. (See Examples 8–9)

91. a. $2x + 4 = -x + 1$
  b. $2x + 4 < -x + 1$
  c. $2x + 4 \geq -x + 1$

92. a. $4x - 2 = -3x + 5$
  b. $4x - 2 < -3x + 5$
  c. $4x - 2 \geq -3x + 5$

93. a. $-3x + 1 = -x - 3$
  b. $-3x + 1 > -x - 3$
  c. $-3x + 1 \leq -x - 3$

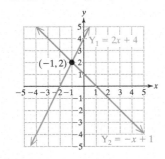
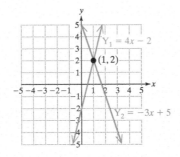
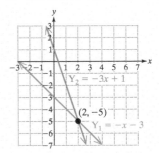

94. a. $-x - 2 = 2x - 5$
  b. $-x - 2 \leq 2x - 5$
  c. $-x - 2 > 2x - 5$

95. a. $-3(x + 2) + 1 = -x + 5$
  b. $-3(x + 2) + 1 \leq -x + 5$
  c. $-3(x + 2) + 1 \geq -x + 5$

96. a. $-4(x - 5) + 3x = -3x + 1$
  b. $-4(x - 5) + 3x \leq -3x + 1$
  c. $-4(x - 5) + 3x \geq -3x + 1$

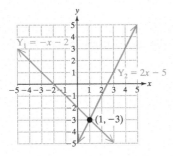
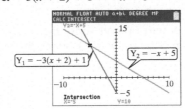
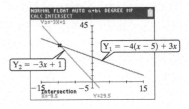

97. a. $4 - 2(x + 1) + 12 + x = 0$
  b. $4 - 2(x + 1) + 12 + x < 0$
  c. $4 - 2(x + 1) + 12 + x > 0$

98. a. $8 - 4(1 - x) - 7 - 2x = 0$
  b. $8 - 4(1 - x) - 7 - 2x < 0$
  c. $8 - 4(1 - x) - 7 - 2x > 0$

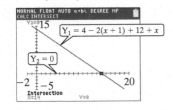
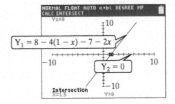

Write About It

99. Explain how you can determine from a linear equation $Ax + By = C$ (A and B not both zero) whether the line is slanted, horizontal, or vertical.

100. Explain how you can determine from a linear equation $Ax + By = C$ (A and B not both zero) whether the line passes through the origin.

101. What is the benefit of writing an equation of a line in slope-intercept form?

102. Explain how the average rate of change of a function f on the interval $[x_1, x_2]$ is related to slope.

Expanding Your Skills

103. Determine the area in the second quadrant enclosed by the equation $y = 2x + 4$ and the x- and y-axes.

104. Determine the area enclosed by the equations.
$$y = x + 6$$
$$y = -2x + 6$$
$$y = 0$$

105. Determine the area enclosed by the equations.
$$y = -\frac{1}{2}x - 2$$
$$y = \frac{1}{3}x - 2$$
$$y = 0$$

106. Determine the area enclosed by the equations.
$$y = \sqrt{4 - (x - 2)^2}$$
$$y = 0$$

107. Consider the standard form of a linear equation $Ax + By = C$ in the case where $B \neq 0$.

  a. Write the equation in slope-intercept form.

  b. Identify the slope in terms of the coefficients A and B.

  c. Identify the y-intercept in terms of the coefficients B and C.

108. Use the results from Exercise 107 to determine the slope and y-intercept for the graphs of the lines.

  a. $5x - 9y = 6$

  b. $0.052x - 0.013y = 0.39$

Technology Connections

For Exercises 109–112, solve the equation in part (a) and verify the solution on a graphing calculator. Then use the graph to find the solution set to the inequalities in parts (b) and (c). Write the solution sets to the inequalities in interval notation. (See Example 9)

109. a. $3.1 - 2.2(t + 1) = 6.3 + 1.4t$

  b. $3.1 - 2.2(t + 1) > 6.3 + 1.4t$

  c. $3.1 - 2.2(t + 1) < 6.3 + 1.4t$

110. a. $-11.2 - 4.6(c - 3) + 1.8c = 0.4(c + 2)$

  b. $-11.2 - 4.6(c - 3) + 1.8c > 0.4(c + 2)$

  c. $-11.2 - 4.6(c - 3) + 1.8c < 0.4(c + 2)$

111. a. $|2x - 3.8| - 4.6 = 7.2$

  b. $|2x - 3.8| - 4.6 \geq 7.2$

  c. $|2x - 3.8| - 4.6 \leq 7.2$

112. a. $|x - 1.7| + 4.95 = 11.15$

  b. $|x - 1.7| + 4.95 \geq 11.15$

  c. $|x - 1.7| + 4.95 \leq 11.15$

For Exercises 113–114, graph the lines in (a)–(c) on the standard viewing window. Compare the graphs. Are they exactly the same? If not, how are they different?

113. a. $y = 3x + 1$

  b. $y = 2.99x + 1$

  c. $y = 3.01x + 1$

114. a. $y = x + 3$

  b. $y = x + 2.99$

  c. $y = x + 3.01$

| SECTION 2.5 | Applications of Linear Equations and Modeling |

OBJECTIVES

1. Apply the Point-Slope Formula
2. Determine the Slopes of Parallel and Perpendicular Lines
3. Create Linear Functions to Model Data
4. Create Models Using Linear Regression

1. Apply the Point-Slope Formula

The slope formula can be used to develop the point-slope form of an equation of a line. Suppose that a line has a slope m and passes through a known point (x_1, y_1). Let (x, y) be any other point on the line. From the slope formula, we have

$$\frac{y - y_1}{x - x_1} = m \qquad \text{Slope formula}$$

$$\left(\frac{y - y_1}{x - x_1}\right)(x - x_1) = m(x - x_1) \qquad \text{Clear fractions.}$$

$$y - y_1 = m(x - x_1) \qquad \text{This is called the point slope formula for a line.}$$

The point-slope formula is useful to build an equation of a line given a point on the line and the slope of the line.

> ### Point-Slope Formula
>
> The **point-slope formula** for a line is given by $y - y_1 = m(x - x_1)$, where m is the slope of the line and (x_1, y_1) is a point on the line.

EXAMPLE 1   **Writing an Equation of a Line Given a Point on the Line and the Slope**

Use the point-slope formula to find an equation of the line passing through the point $(2, -3)$ and having slope -4. Write the answer in slope-intercept form.

Solution:

Label $(2, -3)$ as (x_1, y_1) and $m = -4$.

$y - y_1 = m(x - x_1)$	Apply the point-slope formula.
$y - (-3) = -4(x - 2)$	Substitute $x_1 = 2$, $y_1 = -3$, and $m = -4$.
$y + 3 = -4x + 8$	Simplify.
$y = -4x + 5$ (slope-intercept form)	

TIP The slope-intercept form of a line can also be used to write an equation of a line if a point on the line and the slope are known. See Example 5 in Section 2.4.

> **Skill Practice 1**   Use the point-slope formula to find an equation of the line passing through the point $(-5, 2)$ and having slope -3. Write the answer in slope-intercept form.

EXAMPLE 2   **Writing an Equation of a Line Given Two Points**

Use the point-slope formula to write an equation of the line passing through the points $(4, -6)$ and $(-1, 2)$. Write the answer in slope-intercept form.

Answer

1. $y = -3x - 13$

Solution:

To apply the point-slope formula, we first need to know the slope of the line.

$(4, -6)$ and $(-1, 2)$

(x_1, y_1) and (x_2, y_2)

Label the points. Either point can be labeled (x_1, y_1).

$m = \dfrac{y_2 - y_1}{x_2 - x_1} = \dfrac{2 - (-6)}{-1 - 4} = \dfrac{8}{-5} = -\dfrac{8}{5}$

Apply the slope formula.

$y - y_1 = m(x - x_1)$

Apply the point-slope formula.

$y - (-6) = -\dfrac{8}{5}(x - 4)$

Substitute $y_1 = -6$, $x_1 = 4$, and $m = -\dfrac{8}{5}$.

$y + 6 = -\dfrac{8}{5}x + \dfrac{32}{5}$

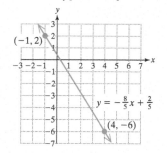

$y = -\dfrac{8}{5}x + \dfrac{32}{5} - 6$

$y = -\dfrac{8}{5}x + \dfrac{32}{5} - \dfrac{30}{5}$

$y = -\dfrac{8}{5}x + \dfrac{2}{5}$ (slope-intercept form)

To check, we see that the graph of the line passes through $(4, -6)$ and $(-1, 2)$ as expected.

> **TIP** In Example 2, the slope-intercept form of a line can also be used to find an equation of the line. Substitute $-\frac{8}{5}$ for m and $(4, -6)$ for (x, y).
>
> $y = mx + b$
> $-6 = -\frac{8}{5}(4) + b$
> $-6 = -\frac{32}{5} + b$
> $-6 + \frac{32}{5} = b$
> $\frac{2}{5} = b$
>
> Therefore, $y = mx + b$ is $y = -\frac{8}{5}x + \frac{2}{5}$.

Skill Practice 2 Write an equation of the line passing through the points $(2, -5)$ and $(7, -3)$.

---

2. Determine the Slopes of Parallel and Perpendicular Lines

Lines in the same plane that do not intersect are **parallel lines.** Nonvertical parallel lines have the same slope and different y-intercepts (Figure 2-20).

Lines that intersect at a right angle are **perpendicular lines.** If two nonvertical lines are perpendicular, then the slope of one line is the opposite of the reciprocal of the slope of the other line (Figure 2-21).

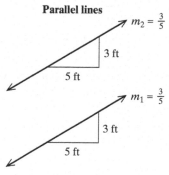

Figure 2-20

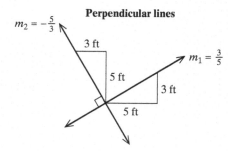

Figure 2-21

Slopes of Parallel and Perpendicular Lines

- If m_1 and m_2 represent the slopes of two nonvertical parallel lines, then $m_1 = m_2$.
- If m_1 and m_2 represent the slopes of two nonvertical perpendicular lines, then $m_1 = -\dfrac{1}{m_2}$ or equivalently $m_1 m_2 = -1$.

Answer

2. $y = \dfrac{2}{5}x - \dfrac{29}{5}$

In Examples 3 and 4, we use the point-slope formula to find an equation of a line through a specified point and parallel or perpendicular to another line.

EXAMPLE 3 **Writing an Equation of a Line Parallel to Another Line**

Write an equation of the line passing through the point $(-4, 1)$ and parallel to the line defined by $x + 4y = 3$. Write the answer in slope-intercept form and in standard form.

Solution:

$$x + 4y = 3$$
$$4y = -x + 3$$

The slope of the given line can be found from its slope-intercept form. Solve for y.

$$y = -\frac{1}{4}x + \frac{3}{4}$$

The slope of both lines is $-\frac{1}{4}$.

Apply the point-slope formula with $x_1 = -4$, $y_1 = 1$, and $m = -\frac{1}{4}$.

$$y - y_1 = m(x - x_1)$$

$$y - 1 = -\frac{1}{4}[x - (-4)]$$

$$y - 1 = -\frac{1}{4}(x + 4)$$

$$y - 1 = -\frac{1}{4}x - 1$$

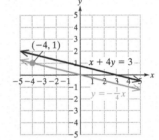

$$y = -\frac{1}{4}x \quad \text{(slope-intercept form)}$$

From the graph, we see that the line $y = -\frac{1}{4}x$ passes through the point $(-4, 1)$ and is parallel to the graph of $x + 4y = 3$.

$$4(y) = 4\left(-\frac{1}{4}x\right)$$

$$4y = -x$$

Clearing fractions, and collecting the x and y terms on one side of the equation gives us standard form.

$$x + 4y = 0 \quad \text{(standard form)}$$

Skill Practice 3 Write an equation of the line passing through the point $(-3, 2)$ and parallel to the line defined by $x + 3y = 6$. Write the answer in slope-intercept form and in standard form.

EXAMPLE 4 **Writing an Equation of a Line Perpendicular to Another Line**

Write an equation of the line passing through the point $(2, -3)$ and perpendicular to the line defined by $y = \frac{1}{2}x - 4$. Write the answer in slope-intercept form and in standard form.

Answer

3. $y = -\frac{1}{3}x + 1; \; x + 3y = 3$

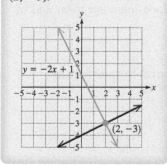

Solution:

From the slope-intercept form, $y = \frac{1}{2}x - 4$, the slope of given line is $\frac{1}{2}$.

$y - y_1 = m(x - x_1)$	The slope of a line perpendicular to the given line is -2.
$y - (-3) = -2(x - 2)$	Apply the point-slope formula with $x_1 = 2$, $y_1 = -3$, and $m = -2$.
$y + 3 = -2x + 4$	Simplify.

$y = -2x + 1$ (slope-intercept form)	Write the equation in slope-intercept form by solving for y.
$2x + y = 1$ (standard form)	Write the equation in standard form by collecting the x and y terms on one side of the equation.

Skill Practice 4   Write an equation of the line passing through the point $(-8, -4)$ and perpendicular to the line defined by $y = \frac{1}{6}x + 3$.

3. Create Linear Functions to Model Data

In many day-to-day applications, two variables are related linearly. By finding an equation of the line, we produce a model that relates the two variables. This is demonstrated in Example 5.

EXAMPLE 5   Using a Linear Function in an Application

A family plan for a cell phone has a monthly base price of $99 plus $12.99 for each additional family member added beyond the primary account holder.

a. Write a linear function to model the monthly cost $C(x)$ (in $) of a family plan for x additional family members added.

b. Evaluate $C(4)$ and interpret the meaning in the context of this problem.

Solution:

a. $C(x) = mx + b$ The base price $99 is the fixed cost with zero additional family members added. So the constant b is 99.

$C(x) = 12.99x + 99$ The rate of increase, $12.99 per additional family member, is the slope.

b. $C(4) = 12.99(4) + 99$ Substitute 4 for x.

$\quad\quad = 150.96$

The total monthly cost of the plan with 4 additional family members beyond the primary account holder is $150.96.

Skill Practice 5   A speeding ticket is $100 plus $5 for every 1 mph over the speed limit.

a. Write a linear function to model the cost $S(x)$ (in $) of a speeding ticket for a person caught driving x mph over the speed limit.

b. Evaluate $S(15)$ and interpret the meaning in the context of this problem.

Answers

4. $y = -6x - 52$; $6x + y = -52$
5. a. $S(x) = 5x + 100$
 b. $S(15) = 175$ means that a ticket costs $175 for a person caught speeding 15 mph over the speed limit.

Linear functions can sometimes be used to model the cost, revenue, and profit of producing and selling x items.

Linear Cost, Revenue, and Profit Functions

A **linear cost function** models the cost $C(x)$ to produce x items.

$C(x) = mx + b$ m is the variable cost per item. b is the fixed cost.	The fixed cost does not change relative to the number of items produced. For example, the cost to rent an office is a fixed cost. The variable cost per item is the rate at which cost increases for each additional unit produced. Variable costs include labor, material, and shipping.

A **linear revenue function** models revenue $R(x)$ for selling x items.

$R(x) = px$	The product px represents the price per item p times the number of items sold x.

A **linear profit function** models the profit for producing and selling x items.

$P(x) = R(x) - C(x)$	Subtract the cost to produce x items from the revenue brought in from selling x items.

EXAMPLE 6   Writing Linear Cost, Revenue, and Profit Functions

At a summer art show a vendor sells lemonade for $2.00 per cup. The cost to rent the booth is $120. Furthermore, the vendor knows that the lemons, sugar, and cups collectively cost $0.50 for each cup of lemonade produced.

  a. Write a linear cost function to produce x cups of lemonade.
  b. Write a linear revenue function for selling x cups of lemonade.
  c. Write a linear profit function for producing and selling x cups of lemonade.
  d. How much profit will the vendor make if 50 cups of lemonade are produced and sold?
  e. How much profit will be made for producing and selling 128 cups?
  f. Determine the break-even point.

Solution:

  a. $C(x) = 0.50x + 120$ — The fixed cost is $120 because it does not change relative to the number of cups of lemonade produced. The variable cost is $0.50 per lemonade.

  b. $R(x) = 2.00x$ — The price per cup of lemonade is $2.00. Therefore, the product $2.00x$ gives the amount of revenue for x cups of lemonade sold.

  c. $P(x) = R(x) - C(x)$ — Profit is defined as the difference of revenue and cost.
  $P(x) = 2.00x - (0.50x + 120)$
  $P(x) = 1.50x - 120$

  d. $P(50) = 1.50(50) - 120$ — Substitute 50 for x.
  $= -45$ — The vendor will lose $45.

e. $P(128) = 1.50(128) - 120$ Substitute 128 for x.

$= 72$ The vendor will make $72.

f. For what value of x will   The break-even point is defined as the point where
$R(x) = C(x)$ or $P(x) = 0$?   revenue equals cost. Alternatively, this can be stated
as the point where profit equals zero: $P(x) = 0$.

$$P(x) = 0$$
$$1.50x - 120 = 0$$ Solve for x.
$$1.50x = 120$$
$$x = 80$$

If the vendor produces and sells 80 cups of lemonade, the cost and revenue
will be equal, resulting in a profit of $0. This is the break-even point.

Skill Practice 6   Repeat Example 6 in the case where the vendor can cut
the cost to $0.40 per cup of lemonade, and sell lemonades for $1.50 per cup.

Figure 2-22 shows the graphs of the revenue and cost functions from Example 6.
Notice that R and C intersect at (80, 160). This means that if 80 cups of lemonade
are produced and sold, the revenue and cost are both $160. That is, $R(x) = C(x)$ and
the company breaks even. The graph of the profit function P is consistent with this
result. The value of $P(x)$ is 0 for 80 lemonades produced and sold (Figure 2-23).

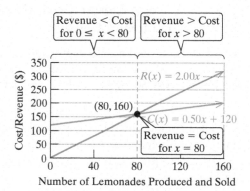

Figure 2-22

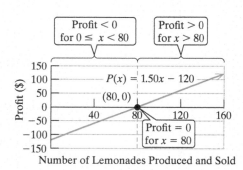

Figure 2-23

From Figures 2-22 and 2-23, we can draw the following conclusions.

- The company experiences a loss if fewer than 80 cups of lemonade are
produced and sold. That is, $R(x) < C(x)$, or equivalently $P(x) < 0$.
- The company experiences a profit if more than 80 cups of lemonade are
produced and sold. That is, $R(x) > C(x)$, or equivalently $P(x) > 0$.
- The company breaks even if exactly 80 cups of lemonade are produced and
sold. That is, $R(x) = C(x)$, or equivalently $P(x) = 0$.

Answers
6. a. $C(x) = 0.40x + 120$
   b. $R(x) = 1.50x$
   c. $P(x) = 1.10x - 120$
   d. $-$65
   e. $20.80
   f. Approximately 109 cups

EXAMPLE 7   **Writing a Linear Model to Relate Two Variables**

The data shown in the graph represent the age and systolic blood pressure for a sample of 12 randomly selected healthy adults.

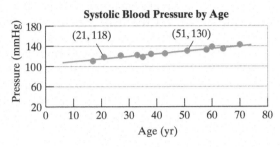

Systolic Blood Pressure by Age

a. Suppose that x represents the age of an adult (in yr), and y represents the systolic blood pressure (in mmHg). Use the points (21, 118) and (51, 130) to write a linear model relating y as a function of x.

b. Interpret the meaning of the slope in the context of this problem.

c. Use the model to estimate the systolic blood pressure for a 55-year-old. Round to the nearest whole unit.

Solution:

a. (21, 118) and (51, 130)

(x_1, y_1) and (x_2, y_2) Label the points.

$$m = \frac{130 - 118}{51 - 21} = 0.4$$ Determine the slope of the line.

$$y - y_1 = m(x - x_1)$$ Apply the point-slope formula.
$$y - 118 = 0.4(x - 21)$$
$$y = 0.4x + 109.6$$ The equation $y = 0.4x + 109.6$ relates an individual's age to an estimated systolic blood pressure for that age.

b. The slope is 0.4. This means that the average increase in systolic blood pressure for adults is 0.4 mmHg per year of age.

c. $y = 0.4x + 109.6$
   $y = 0.4(55) + 109.6$ Substitute 55 for x.
   $y = 131.6$

Based on the sample of data, the estimated systolic blood pressure for a 55-year-old is 132 mmHg.

Skill Practice 7   Suppose that y represents the average consumer spending on television services per year (in dollars), and that x represents the number of years since 2004.

a. Use the data points (2, 308) and (6, 408) to write a linear equation relating y to x.

b. Interpret the meaning of the slope in the context of this problem.

c. Interpret the meaning of the y-intercept in the context of this problem.

d. Use the model from part (a) to estimate the average consumer spending on television services for the year 2007.

Answers

7. a. $y = 25x + 258$
   b. The slope is 25 and means that consumer spending on television services rose $25 per year during this time period.
   c. (0, 258); The average consumer spending on television services for the year 2004 was $258.
   d. $333

4. Create Models Using Linear Regression

In Example 7, we used two given data points to determine a linear model for systolic blood pressure versus age. There are two drawbacks to this method. First, the equation is not necessarily unique. If we use two different data points, we may get a different equation. Second, it is generally preferable to write a model that is based on *all* the data points, rather than just two points. One such model is called the least-squares regression line.

The procedure to find the least-squares regression line is discussed in detail in a statistics course. Here we will give the basic premise and use a graphing utility to perform the calculations. Consider a set of data points (x_1, y_1), (x_2, y_2), (x_3, y_3), ... , (x_n, y_n). The **least-squares regression line,** $\hat{y} = mx + b$, is the unique line that minimizes the sum of the squared vertical deviations from the observed data points to the line (Figure 2-24).

On a calculator or spreadsheet, the equation $\hat{y} = mx + b$ may be denoted as $y = ax + b$ or as $y = b_0 + b_1 x$. In any event, the coefficient of x is the slope of the line, and the constant gives us the y-intercept. Although the exact keystrokes on different calculators and graphing utilities may vary, we will use the following guidelines to find the least-squares regression line.

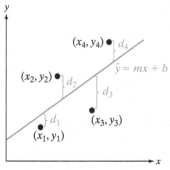

Figure 2-24

Creating a Linear Regression Model

1. Graph the data in a scatter plot.
2. Inspect the data visually to determine if the data suggest a linear trend.
3. Invoke the linear regression feature on a calculator, graphing utility, or spreadsheet.
4. Check the result by graphing the line with the data points to verify that the line passes through or near the data points.

EXAMPLE 8   Finding a Least-Squares Regression Line

The data given in the table represent the age and systolic blood pressure for a sample of 12 randomly selected healthy adults.

Age (yr)	17	21	27	33	35	38	43	51	58	60	64	70
Systolic blood pressure (mmHg)	110	118	121	122	118	124	125	130	132	138	134	142

a. Make a scatter plot of the data using age as the independent variable x and systolic pressure as the dependent variable y.

b. Based on the graph, does a linear model seem appropriate?

c. Determine the equation of the least-squares regression line.

d. Use the least-squares regression line to approximate the systolic blood pressure for a healthy 55-year-old. Round to the nearest whole unit.

Solution:

a. On a graphing calculator hit the STAT button and select EDIT to enter the x and y data into two lists (shown here as L1 and L2).

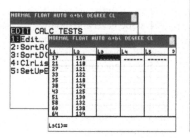

Select the STAT PLOT option and turn Plot1 to On. For the type of graph, select the scatter plot image.

Be sure that the window is set to accommodate x values between 17 and 70, and y values between 110 and 142, inclusive. Then hit the GRAPH key. The window settings shown here are [0, 80, 10] by [0, 200, 20].

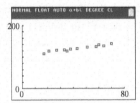

b. From the graph, the data appear to follow a linear trend.

c. Under the STAT menu, select CALC and then the LinReg(ax + b) option.

The command LinReg(ax + b) prompts the user to enter the list names (L_1 and L_2) containing the x and y data values. Then highlight Calculate and hit ENTER.

In the regression model $y = ax + b$, the values for the coefficients a and b are placed on the home screen.

Rounding the values of a and b gives us $y = 0.511x + 104$.

Enter the equation $Y_1 = 0.511x + 104$ into the equation editor and hit the GRAPH key. The graph of the regression line passes near or through the observed data points.

d. $y = 0.511x + 104$

$y = 0.511(55) + 104$ To approximate the systolic blood pressure for a 55-year-old,

$= 132.105$ substitute 55 for x.

The systolic blood pressure for a healthy 55-year-old would be approximately 132 mmHg.

> **TIP** The linear equation found in Example 7 was based on two data points. The least-squares regression line is based on all available data points. The estimate from each model for systolic blood pressure for a 55-year-old rounds to 132 mmHg.

> **TIP** In Example 8(d), the value of the function at $x = 55$ can also be found by selecting the CALC menu and selecting the VALUE function. Enter 55 for x and press the ENTER key.

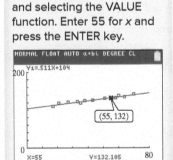

Skill Practice 8   The data given represent the class averages for individual students based on the number of absences from class.

Number of Absences (x)	3	7	1	11	2	14	2	5
Average in Class (y)	88	67	96	62	90	56	97	82

a. Find the equation of the least-squares regression line.

b. Use the model from part (a) to approximate the average for a student who misses 6 classes.

Answers

8. a. $y = -3.27x + 98.1$

b. The student's average would be approximately 78.5.

SECTION 2.5   Practice Exercises

Prerequisite Review

R.1. Use slope-intercept form to write an equation of the line that passes through $(3, -7)$ with slope -5.

R.2. Write the equation in slope-intercept form and determine the slope and y-intercept $3x - 5y = -15$.

R.3. Determine the slope of the line containing the points $(-4, -2)$ and $(-4, -7)$.

R.4. Determine the slope of the line containing the points $(3, -2)$ and $(5, -2)$.

R.5. The cost C (in dollars) to rent an apartment is \$850 per month, plus a \$450 nonrefundable security deposit, plus a \$250 deposit for each dog or cat. Write a formula for the total cost to rent an apartment for m months with n cats/dogs.

Concept Connections

1. Given a point (x_1, y_1) on a line with slope m, the point-slope formula is given by _____.

2. If two nonvertical lines have the same slope but different y-intercepts, then the lines are (parallel/perpendicular).

3. If m_1 and m_2 represent the slopes of two nonvertical perpendicular lines, then $m_1 m_2 =$ _____.

4. Suppose that $y = C(x)$ represents the cost to produce x items, and that $y = R(x)$ represents the revenue for selling x items. The profit $P(x)$ of producing and selling x items is defined by $P(x) =$ _____.

Objective 1: Apply the Point-Slope Formula

For Exercises 5–20, write an equation of the line having the given conditions. Write the answer in slope-intercept form (if possible). (See Examples 1–2)

5. Passes through $(-3, 5)$ and $m = -2$.

6. Passes through $(4, -6)$ and $m = 3$.

7. Passes through $(-1, 0)$ and $m = \dfrac{2}{3}$.

8. Passes through $(-4, 0)$ and $m = \dfrac{3}{5}$.

9. Passes through $(3.4, 2.6)$ and $m = 1.2$.

10. Passes through $(2.2, 4.1)$ and $m = 2.4$.

11. Passes through $(6, 2)$ and $(-3, 1)$.

12. Passes through $(-4, 8)$ and $(-7, -3)$.

13. Passes through $(0, 8)$ and $(5, 0)$.

14. Passes through $(0, -6)$ and $(11, 0)$.

15. Passes through $(2.3, 5.1)$ and $(1.9, 3.7)$.

16. Passes through $(1.6, 4.8)$ and $(0.8, 6)$.

17. Passes through $(3, -4)$ and $m = 0$.

18. Passes through $(-5, 1)$ and $m = 0$.

19. Passes through $\left(\dfrac{2}{3}, \dfrac{1}{5}\right)$ and the slope is undefined.

20. Passes through $\left(-\dfrac{4}{7}, \dfrac{3}{10}\right)$ and the slope is undefined.

21. Given a line defined by $x = 4$, what is the slope of the line?

22. Given a line defined by $y = -2$, what is the slope of the line?

Objective 2: Determine the Slopes of Parallel and Perpendicular Lines

For Exercises 23–28, the slope of a line is given.

a. Determine the slope of a line parallel to the given line, if possible.

b. Determine the slope of a line perpendicular to the given line, if possible.

23. $m = \dfrac{3}{11}$

24. $m = \dfrac{6}{7}$

25. $m = -6$

26. $m = -10$

27. $m = 1$

28. m is undefined

For Exercises 29–36, determine if the lines defined by the given equations are parallel, perpendicular, or neither.

29. $y = 2x - 3$
$y = -\dfrac{1}{2}x + 7$

30. $y = \dfrac{4}{3}x - 1$
$y = -\dfrac{3}{4}x + 5$

31. $8x - 5y = 3$
$2x = \dfrac{5}{4}y + 1$

32. $2x + 3y = 7$
$4x = -6y + 2$

33. $2x = 6$
$5 = y$

34. $3y = 5$
$x = 1$

35. $6x = 7y$
$\dfrac{7}{2}x - 3y = 0$

36. $5y = 2x$
$\dfrac{5}{2}x - y = 0$

For Exercises 37–44, write an equation of the line satisfying the given conditions. Write the answer in slope-intercept form (if possible) and in standard form ($Ax + By = C$) with no fractional coefficients. (See Examples 3–4)

37. Passes through $(2, 5)$ and is parallel to the line defined by $2x + y = 6$.

38. Passes through $(3, -1)$ and is parallel to the line defined by $-3x + y = 4$.

39. Passes through $(6, -4)$ and is perpendicular to the line defined by $x - 5y = 1$.

40. Passes through $(5, 4)$ and is perpendicular to the line defined by $x - 2y = 7$.

41. Passes through $(6, 8)$ and is parallel to the line defined by $3x = 7y + 5$.

42. Passes through $(7, -6)$ and is parallel to the line defined by $2x = 5y - 4$.

43. Passes through $(2.2, 6.4)$ and is perpendicular to the line defined by $2x = 4 - y$.

44. Passes through $(3.6, 1.2)$ and is perpendicular to the line defined by $4x = 9 - y$.

For Exercises 45–50, write an equation of the line that satisfies the given conditions.

45. Passes through $(8, 6)$ and is parallel to the x-axis.

46. Passes through $(-11, 13)$ and is parallel to the y-axis.

47. Passes through $\left(\dfrac{5}{11}, -\dfrac{3}{4}\right)$ and is perpendicular to the y-axis.

48. Passes through $\left(-\dfrac{7}{9}, \dfrac{7}{3}\right)$ and is perpendicular to the x-axis.

49. Passes through $(-61.5, 47.6)$ and is parallel to the line defined by $x = -12$.

50. Passes through $(-0.004, 0.009)$ and is parallel to the line defined by $y = 6$.

Objective 3: Create Linear Functions to Model Data

51. A sales person makes a base salary of $400 per week plus 12% commission on sales. **(See Example 5)**

   a. Write a linear function to model the sales person's weekly salary $S(x)$ for x dollars in sales.

   b. Evaluate $S(8000)$ and interpret the meaning in the context of this problem.

52. At a parking garage in a large city, the charge for parking consists of a flat fee of $2.00 plus $1.50/hr.

   a. Write a linear function to model the cost for parking $P(t)$ for t hours.

   b. Evaluate $P(1.6)$ and interpret the meaning in the context of this problem.

53. Millage rate is the amount per $1000 that is often used to calculate property tax. For example, a home with a $60,000 taxable value in a municipality with a 19 mil tax rate would require $(0.019)(\$60,000) = \$1140$ in property taxes. In one county, homeowners pay a flat tax of $172 plus a rate of 19 mil on the taxable value of a home.

   a. Write a linear function that represents the total property tax $T(x)$ for a home with a taxable value of x dollars.

   b. Evaluate $T(80,000)$ and interpret the meaning in the context of this problem.

54. The average water level in a retention pond is 6.8 ft. During a time of drought, the water level decreases at a rate of 3 in./day.

   a. Write a linear function W that represents the water level $W(t)$ (in ft) t days after a drought begins.

   b. Evaluate $W(20)$ and interpret the meaning in the context of this problem.

For Exercises 55–56, the fixed and variable costs to produce an item are given along with the price at which an item is sold. (See Example 6)

   a. Write a linear cost function that represents the cost $C(x)$ to produce x items.

   b. Write a linear revenue function that represents the revenue $R(x)$ for selling x items.

   c. Write a linear profit function that represents the profit $P(x)$ for producing and selling x items.

   d. Determine the break-even point.

55. Fixed cost: $2275
   Variable cost per item: $34.50
   Price at which the item is sold: $80.00

56. Fixed cost: $5625
   Variable cost per item: $0.40
   Price at which the item is sold: $1.30

57. The profit function P is shown for producing and selling x items. Determine the values of x for which

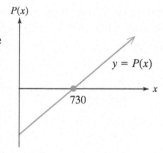

  a. $P(x) = 0$ (the company breaks even)

  b. $P(x) < 0$ (the company experiences a loss)

  c. $P(x) > 0$ (the company makes a profit)

58. The cost and revenue functions C and R are shown for producing and selling x items. Determine the values of x for which

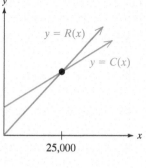

  a. $R(x) = C(x)$ (the company breaks even)

  b. $R(x) < C(x)$ (the company experiences a loss)

  c. $R(x) > C(x)$ (the company makes a profit)

59. A small business makes cookies and sells them at the farmer's market. The fixed monthly cost for use of a Health Department–approved kitchen and rental space at the farmer's market is $790. The cost of labor, taxes, and ingredients for the cookies amounts to $0.24 per cookie, and the cookies sell for $6.00 per dozen.
(See Example 6)

  a. Write a linear cost function representing the cost $C(x)$ to produce x dozen cookies per month.

  b. Write a linear revenue function representing the revenue $R(x)$ for selling x dozen cookies.

  c. Write a linear profit function representing the profit for producing and selling x dozen cookies in a month.

  d. Determine the number of cookies (in dozens) that must be produced and sold for a monthly profit.

  e. If 150 dozen cookies are sold in a given month, how much money will the business make or lose?

60. A lawn service company charges $60 for each lawn maintenance call. The fixed monthly cost of $680 includes telephone service and depreciation of equipment. The variable costs include labor, gasoline, and taxes and amount to $36 per lawn.

  a. Write a linear cost function representing the monthly cost $C(x)$ for x maintenance calls.

  b. Write a linear revenue function representing the monthly revenue $R(x)$ for x maintenance calls.

  c. Write a linear profit function representing the monthly profit $P(x)$ for x maintenance calls.

  d. Determine the number of lawn maintenance calls needed per month for the company to make money.

  e. If 42 maintenance calls are made for a given month, how much money will the lawn service make or lose?

61. The data in the graph show the wind speed y (in mph) for Hurricane Katrina versus the barometric pressure x (in millibars, mb). (*Source*: NOAA: www.noaa.gov)
(See Example 7)

  a. Use the points (950, 110) and (1000, 50) to write a linear model for these data.

  b. Interpret the meaning of the slope in the context of this problem.

  c. Use the model from part (a) to estimate the wind speed for a hurricane with a pressure of 900 mb.

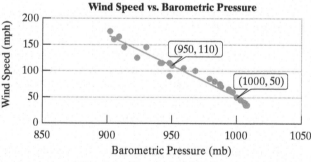

  d. The lowest barometric pressure ever recorded for an Atlantic hurricane was 882 mb for Hurricane Wilma in 2005. Would it be reasonable to use the model from part (a) to estimate the wind speed for a hurricane with a pressure of 800 mb?

62. Caroline adopted a puppy named Dodger from an animal shelter in Chicago. She recorded Dodger's weight during the first two months. The data in the graph show Dodger's weight y (in lb), x days after adoption.

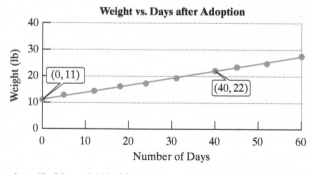

  a. Use the points (0, 11) and (40, 22) to write a linear model for these data.

  b. Interpret the meaning of the slope in context.

c. Interpret the meaning of the *y*-intercept in context.

d. If this linear trend continues during Dodger's growth period, how long will it take Dodger to reach 90% of his expected full-grown weight of 70 lb? Round to the nearest day.

e. Is the model from part (a) reasonable long term?

63. A pediatrician records the age *x* (in yr) and average height *y* (in inches) for girls between the ages of 2 and 10.

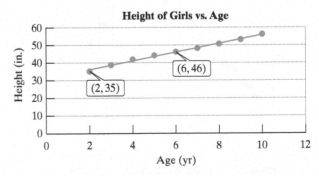

Height of Girls vs. Age

a. Use the points (2, 35) and (6, 46) to write a linear model for these data.

b. Interpret the meaning of the slope in context.

c. Use the model to forecast the average height of 11-yr-old girls.

d. If the height of a girl at age 11 is 90% of her full-grown adult height, use the result of part (c) to estimate the average height of adult women. Round to the nearest tenth of an inch.

64. The graph shows the number of students enrolled in public colleges for selected years (*Source*: U.S. National Center for Education Statistics, www.nces.ed.gov). The *x* variable represents the number of years since 1990 and the *y* variable represents the number of students (in millions).

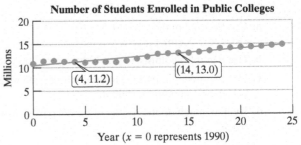

Number of Students Enrolled in Public Colleges

a. Use the points (4, 11.2) and (14, 13.0) to write a linear model for these data.

b. Interpret the meaning of the slope in the context of this problem.

c. Interpret the meaning of the *y*-intercept in the context of this problem.

d. In the event that the linear trend continues beyond the last observed data point, use the model in part (a) to predict the number of students enrolled in public colleges for the year 2020.

65. The table gives the number of calories and the amount of cholesterol for selected fast food hamburgers.

a. Graph the data in a scatter plot using the number of calories as the independent variable *x* and the amount of cholesterol as the dependent variable *y*.

b. Use the data points (480, 60) and (720, 90) to write a linear function that defines the amount of cholesterol $c(x)$ as a linear function of the number of calories *x*.

c. Interpret the meaning of the slope in the context of this problem.

d. Use the model from part (b) to predict the amount of cholesterol for a hamburger with 650 calories.

Hamburger Calories	Cholesterol (mg)
220	35
420	50
460	50
480	60
560	70
590	105
610	65
680	80
720	90

66. The table gives the average gestation period for selected animals and their corresponding average longevity.

a. Graph the data in a scatter plot using the number of days for gestation as the independent variable *x* and the longevity as the dependent variable *y*.

b. Use the data points (44, 8.5) and (620, 35) to write a linear function that defines longevity $L(x)$ as a linear function of the length of the gestation period *x*. Round the slope to 3 decimal places and the *y*-intercept to 2 decimal places.

c. Interpret the meaning of the slope in the context of this problem.

d. Use the model from part (b) to predict the longevity for an animal with an 80-day gestation period. Round to the nearest year.

Animal	Gestation Period (days)	Longevity (yr)
Rabbit	33	7.0
Squirrel	44	8.5
Fox	57	9.0
Cat	60	11.0
Dog	62	11.0
Lion	109	10.0
Pig	115	10.0
Goat	148	12.0
Horse	337	23.0
Elephant	620	35.0

Objective 4: Create Models Using Linear Regression

For Exercises 67–70, use the scatter plot to determine if a linear regression model appears to be appropriate.

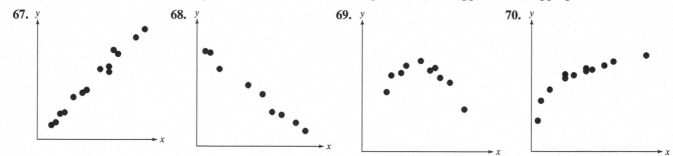

67. y   **68.** y   **69.** y   **70.** y

71. The graph in Exercise 61 shows the wind speed y (in mph) of a hurricane versus the barometric pressure x (in mb). The table gives a partial list of data from the graph. (**See Example 8**)

  a. Use the data in the table to find the least-squares regression line. Round the slope to 2 decimal places and the y-intercept to the nearest whole unit.

  b. Use a graphing utility to graph the regression line and the observed data.

  c. Use the model in part (a) to approximate the wind speed of a hurricane with a barometric pressure of 900 mb.

  d. By how much do the results of part (c) differ from the result of Exercise 61(c)?

Barometric Pressure (mb) (x)	Wind Speed (mph) (y)
1007	35
1003	45
1000	50
994	65
983	80
968	100
950	110
930	145
905	160

72. The graph in Exercise 62 shows the weight of Dodger, a puppy recently adopted from an animal shelter. The data in the table give Dodger's weight y (in lb), x days after adoption.

  a. Use the data in the table to find the least-squares regression line. Round the slope to 2 decimal places and the y-intercept to 1 decimal place.

  b. Use a graphing utility to graph the regression line and the observed data.

  c. Use the model in part (a) to approximate the time required for Dodger to reach 90% of his full-grown weight of 70 lb. Round to the nearest day.

  d. By how much do the results of part (c) differ from the result of Exercise 62(d)?

Number of Days (x)	Weight (lb) (y)
0	11.0
5	12.8
12	14.3
18	16.1
24	17.2
31	19.2
40	22.0
45	23.4
52	24.7
60	27.5

73. The graph in Exercise 63 shows the average height of girls based on their age. The data in the table give the average height y (in inches) for girls of age x (in yr).

  a. Use the data in the table to find the least-squares regression line. Round the slope to 2 decimal places and the y-intercept to 1 decimal place.

  b. Use a graphing utility to graph the regression line and the observed data.

  c. Use the model in part (a) to approximate the average height of 11-yr-old girls.

  d. If the height of a girl at age 11 is 90% of her full-grown adult height, use the result of part (c) to estimate the average height of adult women. Round to the nearest tenth of an inch.

  e. By how much do the results of part (d) differ from the result of Exercise 63(d)?

Age (yr) (x)	Height (in.) (y)
2	35.00
3	38.50
4	41.75
5	44.00
6	46.00
7	48.00
8	50.50
9	53.00
10	56.00

74. The graph in Exercise 64 shows the number of students y enrolled in public colleges for selected years x, where x is the number of years since 1990. The table gives a partial list of data from the graph.

Number of Years Since 1990 (x)	Enrollment (millions) (y)
0	10.8
4	11.2
8	11.1
12	12.8
16	13.2
20	14.2
24	14.8

a. Use the data in the table to find the least-squares regression line. Round the slope to 2 decimal places and the y-intercept to 1 decimal place.

b. Use a graphing utility to graph the regression line and the observed data.

c. Assuming that the linear trend continues, use the model from part (a) to predict the number of students enrolled in public colleges for the year 2020.

d. By how much do the results of part (c) differ from the result of Exercise 64(d)?

75. The data in Exercise 65 give the amount of cholesterol y for a hamburger with x calories.

a. Use these data to find the least-squares regression line. Round the slope to 3 decimal places and the y-intercept to 2 decimal places.

b. Use a graphing utility to graph the regression line and the observed data.

c. Use the regression line to predict the amount of cholesterol in a hamburger with 650 calories. Round to the nearest milligram.

76. The data in Exercise 66 give the average gestation period x (in days) for selected animals and their corresponding average longevity y (in yr).

a. Use these data to find the least-squares regression line. Round the slope to 3 decimal places and the y-intercept to 2 decimal places.

b. Use a graphing utility to graph the regression line and the observed data.

c. Use the regression line to predict the longevity for an animal with an 80-day gestation period. Round to the nearest year.

Mixed Exercises

77. Suppose that a line passes through the points $(4, -6)$ and $(2, -1)$. Where will it pass through the x-axis?

78. Suppose that a line passes through the point $(2, -5)$ and $(-4, 7)$. Where will it pass through the x-axis?

79. Write a rule for a linear function $y = f(x)$, given that $f(0) = 4$ and $f(3) = 11$.

80. Write a rule for a linear function $y = g(x)$, given that $g(0) = 7$ and $g(-2) = 4$.

81. Write a rule for a linear function $y = h(x)$, given that $h(1) = 6$ and $h(-3) = 2$.

82. Write a rule for a linear function $y = k(x)$, given that $k(-2) = 10$ and $k(5) = -18$.

Write About It

83. Explain how you can use slope to determine if two nonvertical lines are parallel or perpendicular.

84. State one application of using the point-slope formula.

85. Explain how cost and revenue are related to profit.

86. Explain how to determine the break-even point.

Expanding Your Skills

87. Find an equation of line L.

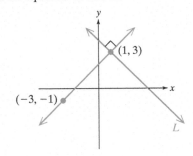

88. In geometry, it is known that the tangent line to a circle at a given point A on the circle is perpendicular to the radius drawn to point A. Suppose that line L is tangent to the given circle at the point $(4, 3)$. Write an equation representing line L.

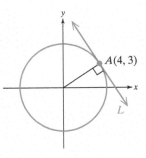

89. In calculus, we can show that the slope of the line drawn tangent to the curve $y = x^3 + 1$ at the point $(c, c^3 + 1)$ is given by $3c^2$. Find an equation of the line tangent to $y = x^3 + 1$ at the point $(-2, -7)$.

90. In calculus, we can show that the slope of the line drawn tangent to the curve $y = \frac{1}{x}$ at the point $\left(c, \frac{1}{c}\right)$ is given by $-\frac{1}{c^2}$. Find an equation of the line tangent to $y = \frac{1}{x}$ at the point $\left(2, \frac{1}{2}\right)$.

For Exercises 91–92, use the fact that a median of a triangle is a line segment drawn from a vertex of the triangle to the midpoint of the opposite side of the triangle.

91. Find an equation of the median of a triangle drawn from vertex $A(5, -2)$ to the side formed by $B(-2, 9)$ and $C(4, 7)$.

92. Find an equation of the median of a triangle drawn from vertex $A(6, -5)$ to the side formed by $B(-4, 1)$ and $C(12, 3)$.

PROBLEM RECOGNITION EXERCISES

Comparing Graphs of Equations

In Section 2.6, we will learn additional techniques to graph functions by recognizing characteristics of the functions. In many cases, we can also graph families of functions by relating them to one of several basic graphs. To prepare for the discussion in Section 2.6, use a graphing utility or plot points to graph the basic functions in Exercises 1–8.

1. $y = 1$

2. $y = x$

3. $y = x^2$

4. $y = x^3$

5. $y = \sqrt{x}$

6. $y = \sqrt[3]{x}$

7. $y = |x|$

8. $y = \dfrac{1}{x}$

For Exercises 9–18, graph the functions by plotting points or by using a graphing utility. Explain how the graphs are related.

9. a. $f(x) = x^2$
   b. $g(x) = x^2 + 2$
   c. $h(x) = x^2 - 4$

10. a. $f(x) = |x|$
   b. $g(x) = |x| + 2$
   c. $h(x) = |x| - 4$

11. a. $f(x) = \sqrt{x}$
   b. $g(x) = \sqrt{x - 2}$
   c. $h(x) = \sqrt{x + 4}$

12. a. $f(x) = x^2$
   b. $g(x) = (x - 2)^2$
   c. $h(x) = (x + 3)^2$

13. a. $f(x) = |x|$
   b. $g(x) = -|x|$

14. a. $f(x) = \sqrt{x}$
   b. $g(x) = -\sqrt{x}$

15. a. $f(x) = x^2$
   b. $g(x) = \dfrac{1}{2}x^2$
   c. $h(x) = 2x^2$

16. a. $f(x) = |x|$
   b. $g(x) = \dfrac{1}{3}|x|$
   c. $h(x) = 3|x|$

17. a. $f(x) = \sqrt{x}$
   b. $g(x) = \sqrt{-x}$

18. a. $f(x) = \sqrt[3]{x}$
   b. $g(x) = \sqrt[3]{-x}$

SECTION 2.6 **Transformations of Graphs**

OBJECTIVES

1. Recognize Basic Functions
2. Apply Vertical and Horizontal Translations (Shifts)
3. Apply Vertical and Horizontal Shrinking and Stretching
4. Apply Reflections Across the x- and y-Axes
5. Summarize Transformations of Graphs

1. Recognize Basic Functions

A function defined by $f(x) = mx + b$ is a linear function and its graph is a line in a rectangular coordinate system. In addition to linear functions, we will learn to identify other categories of functions and the shapes of their graphs (Table 2-2).

Table 2-2   Basic Functions and Their Graphs

1. Linear functions   Constant functions
   $f(x) = mx + b$ $f(x) = b$

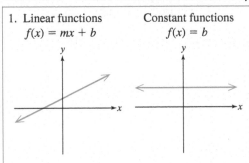

2. Identity function: $f(x) = x$

x	f(x)
-2	-2
-1	-1
0	0
1	1
2	2

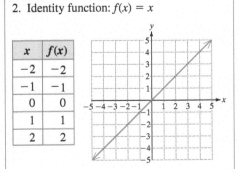

TIP The functions given in Table 2-2 were introduced in Section 2.1, Exercises 31–36, and in the Problem Recognition Exercises on page 228.

3. Quadratic function: $f(x) = x^2$
   (graph is a parabola)

x	f(x)
-2	4
-1	1
0	0
1	1
2	4

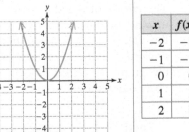

4. Cube function: $f(x) = x^3$

x	f(x)
-2	-8
-1	-1
0	0
1	1
2	8

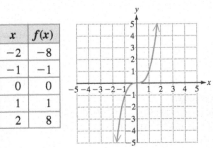

5. Square root function: $f(x) = \sqrt{x}$

x	f(x)
0	0
1	1
4	2
9	3
16	4

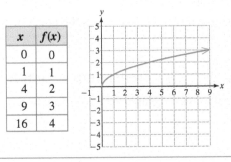

6. Cube root function: $f(x) = \sqrt[3]{x}$

x	f(x)
-8	-2
-1	-1
0	0
1	1
8	2

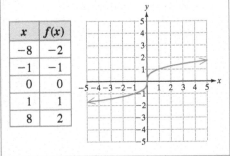

7. Absolute value function: $f(x) = |x|$

x	f(x)
-2	2
-1	1
0	0
1	1
2	2

8. Reciprocal function: $f(x) = \dfrac{1}{x}$

x	f(x)
-2	-½
-1	-1
-½	-2
½	2
1	1
2	½

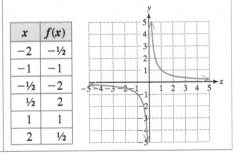

Notice that the graph of $f(x) = \frac{1}{x}$ gets close to (but never touches) the y-axis as x gets close to zero. Likewise, as x approaches ∞ and $-\infty$, the graph approaches the x-axis without touching the x-axis. The x- and y-axes are called **asymptotes** of f and will be studied in detail in Section 3.5.

2. Apply Vertical and Horizontal Translations (Shifts)

We will call the eight basic functions pictured in Table 2-2 "parent" functions. Other functions that share the characteristics of a parent function are grouped as a "family" of functions. For example, consider the functions defined by $g(x) = x^2 + 2$ and $h(x) = x^2 - 4$, pictured in Figure 2-25.

x	$f(x) = x^2$	$g(x) = x^2 + 2$	$h(x) = x^2 - 4$
-3	9	11	5
-2	4	6	0
-1	1	3	-3
0	0	2	-4
1	1	3	-3
2	4	6	0
3	9	11	5

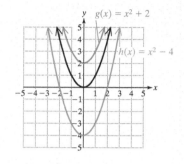

Figure 2-25

The graphs of g and h both resemble the graph of $f(x) = x^2$, but are shifted vertically upward or downward. The table of points reveals that for corresponding x values, the values of $g(x)$ are 2 more than the values of $f(x)$. Thus, the graph is shifted *upward* 2 units. Likewise, the values of $h(x)$ are 4 less than the values of $f(x)$ and the graph is shifted *downward* 4 units. Such shifts are called **translations.** These observations are consistent with the following rules.

TIP For each ordered pair (x, y) on the graph of $y = f(x)$, the corresponding point

- $(x, y + k)$ is on the graph of $y = f(x) + k$.
- $(x, y - k)$ is on the graph of $y = f(x) - k$.

Vertical Translations of Graphs

Consider a function defined by $y = f(x)$. Let k represent a positive real number.

- The graph of $y = f(x) + k$ is the graph of $y = f(x)$ shifted k units *upward.*
- The graph of $y = f(x) - k$ is the graph of $y = f(x)$ shifted k units *downward.*

EXAMPLE 1   Translating a Graph Vertically

Use translations to graph the given functions.

a. $g(x) = |x| - 3$ **b.** $h(x) = x^3 + 2$

Solution:

a.

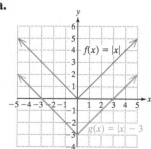

The parent function for $g(x) = |x| - 3$ is $f(x) = |x|$.

The graph of g (shown in blue) is the graph of f shifted *downward* 3 units. For example the point $(0, 0)$ on the graph of $f(x) = |x|$ corresponds to $(0, -3)$ on the graph of $g(x) = |x| - 3$.

b.

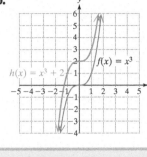

The parent function for $h(x) = x^3 + 2$ is $f(x) = x^3$.

The graph of h (shown in blue) is the graph of f shifted *upward* 2 units. For example:

The point $(0, 0)$ on the graph of $f(x) = x^3$ corresponds to $(0, 2)$ on the graph of $h(x) = x^3 + 2$.

The point $(1, 1)$ on the graph of $f(x) = x^3$ corresponds to $(1, 3)$ on the graph of $h(x) = x^3 + 2$.

Skill Practice 1   Use translations to graph the given functions.

   a. $g(x) = \sqrt{x} - 2$ **b.** $h(x) = \sqrt{x} + 3$

The graph of a function will be shifted to the right or left if a constant is added to or subtracted from the input variable x. In Example 2, we consider $g(x) = (x + 3)^2$.

EXAMPLE 2   **Translating a Graph Horizontally**

Graph the function defined by $g(x) = (x + 3)^2$.

Solution:

Because 3 is added to the x variable, we might expect the graph of $g(x) = (x + 3)^2$ to be the same as the graph of $f(x) = x^2$, but shifted in the x direction (horizontally). To determine whether the shift is to the left or right, we can locate the x-intercept of the graph of $g(x) = (x + 3)^2$. Substituting 0 for $g(x)$, we have:

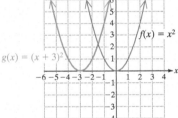

$$0 = (x + 3)^2$$
$$x = -3 \qquad \text{The } x\text{-intercept is } (-3, 0).$$

Therefore, the new x-intercept (and also the vertex of the parabola) is $(-3, 0)$. This means that the graph is shifted to the left.

Skill Practice 2   Graph the function defined by $g(x) = |x + 2|$.

Using similar logic as in Example 2, we can show that the graph of $h(x) = (x - 3)^2$ is the graph of $f(x) = x^2$ translated to the *right* 3 units. These observations are consistent with the following rules.

Horizontal Translations of Graphs

Consider a function defined by $y = f(x)$. Let h represent a positive real number.

• The graph of $y = f(x - h)$ is the graph of $y = f(x)$ shifted h units to the *right*.
• The graph of $y = f(x + h)$ is the graph of $y = f(x)$ shifted h units to the *left*.

TIP   Consider a positive real number h. To graph $y = f(x - h)$ or $y = f(x + h)$, shift the graph of $y = f(x)$ horizontally in the opposite direction of the sign within parentheses. The graph of $y = f(x - h)$ is a shift in the positive x direction. The graph of $y = f(x + h)$ is a shift in the negative x direction.

Answers

1. a.–b.

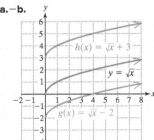

2.

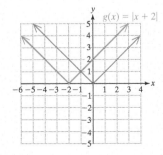

EXAMPLE 3   Translating a Function Horizontally and Vertically

Use translations to graph the function defined by $p(x) = \sqrt{x - 3} - 2$.

Solution:

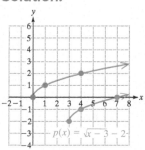

The parent function for $p(x) = \sqrt{x - 3} - 2$ is $f(x) = \sqrt{x}$.

The graph of p (shown in blue) is the graph of f shifted to the right 3 units and downward 2 units. We can plot several strategic points as an outline for the new curve.

- The point $(0, 0)$ on the graph of f corresponds to $(0 + 3, 0 - 2) = (3, -2)$ on the graph of p.
- The point $(1, 1)$ on the graph of f corresponds to $(1 + 3, 1 - 2) = (4, -1)$ on the graph of p.
- The point $(4, 2)$ on the graph of f corresponds to $(4 + 3, 2 - 2) = (7, 0)$ on the graph of p.

Skill Practice 3   Use translations to graph the function defined by $q(x) = \sqrt{x + 2} - 5$.

3. Apply Vertical and Horizontal Shrinking and Stretching

Horizontal and vertical translations of functions are called **rigid transformations** because the shape of the graph is not affected. We now look at **nonrigid transformations.** These operations cause a distortion of the graph (either an elongation or contraction in the horizontal or vertical direction). We begin by investigating the functions defined by $y = f(x)$ and $y = a \cdot f(x)$, where a is a positive real number.

EXAMPLE 4   Graphing a Function with a Vertical Stretch or Shrink

Graph the functions.

a. $f(x) = |x|$ **b.** $g(x) = 2|x|$ **c.** $h(x) = \dfrac{1}{2}|x|$

Solution:

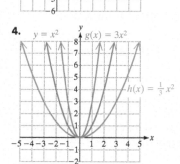
x	$f(x) = |x|$	$g(x) = 2|x|$	$h(x) = \frac{1}{2}|x|$
-3	3	6	$\frac{3}{2}$
-2	2	4	1
-1	1	2	$\frac{1}{2}$
0	0	0	0
1	1	2	$\frac{1}{2}$
2	2	4	1
3	3	6	$\frac{3}{2}$

double (from f to g column)

multiply by $\frac{1}{2}$

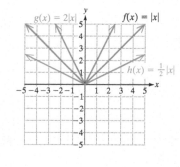

For a given value of x, the value of $g(x)$ is twice the value of $f(x)$. Therefore, the graph of g is elongated or stretched vertically by a factor of 2.

For a given value of x, the value of $h(x)$ is one-half that of $f(x)$. Therefore, the graph of h is shrunk vertically.

Skill Practice 4   Graph the functions.

a. $f(x) = x^2$ **b.** $g(x) = 3x^2$ **c.** $h(x) = \dfrac{1}{3}x^2$

Vertical Shrinking and Stretching of Graphs

Consider a function defined by $y = f(x)$. Let a represent a positive real number.

- If $a > 1$, then the graph of $y = af(x)$ is the graph of $y = f(x)$ stretched vertically by a factor of a.
- If $0 < a < 1$, then the graph of $y = af(x)$ is the graph of $y = f(x)$ shrunk vertically by a factor of a.

Note: For any point (x, y) on the graph of $y = f(x)$, the point (x, ay) is on the graph of $y = af(x)$.

A function may also be stretched or shrunk horizontally.

Horizontal Shrinking and Stretching of Graphs

Consider a function defined by $y = f(x)$. Let a represent a positive real number.

- If $a > 1$, then the graph of $y = f(ax)$ is the graph of $y = f(x)$ shrunk horizontally by a factor of $\frac{1}{a}$.
- If $0 < a < 1$, then the graph of $y = f(ax)$ is the graph of $y = f(x)$ stretched horizontally by a factor of $\frac{1}{a}$.

Note: For any point (x, y) on the graph of $y = f(x)$, the point $\left(\frac{x}{a}, y\right)$ is on the graph of $y = f(ax)$.

A point (x, y) on the graph of $y = f(x)$ corresponds to the point $\left(\frac{x}{a}, y\right)$ on the graph of $y = f(ax)$. Since the x-coordinate is multiplied by the *reciprocal* of a, values of a greater than 1 actually compress (shrink) the graph horizontally toward the y-axis. Values of a between 0 and 1 *stretch* the graph horizontally away from the y-axis. This is demonstrated in Example 5.

EXAMPLE 5   Graphing a Function with a Horizontal Shrink or Stretch

The graph of $y = f(x)$ is shown. Graph

a. $y = f(2x)$

b. $y = f\left(\dfrac{1}{2}x\right)$

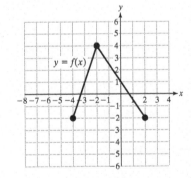

Solution:

a. $f(2x)$ is in the form $f(ax)$ with $a = 2 > 1$. The graph of $y = f(2x)$ is the graph of $y = f(x)$ horizontally compressed.

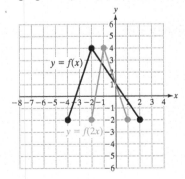

The graph of f has the following "strategic" points that define the shape of the function: $(-4, -2)$, $(-2, 4)$, and $(2, -2)$.

To graph $y = f(2x)$, divide each x value by 2.

$(-4, -2)$ becomes $\left(\frac{-4}{2}, -2\right) = (-2, -2)$.

$(-2, 4)$ becomes $\left(\frac{-2}{2}, 4\right) = (-1, 4)$.

$(2, -2)$ becomes $\left(\frac{2}{2}, -2\right) = (1, -2)$.

The graph of $y = f(2x)$ is shown in blue.

b. $f\left(\frac{1}{2}x\right)$ is in the form $f(ax)$ with $a = \frac{1}{2}$. The graph of $y = f\left(\frac{1}{2}x\right)$ is the graph of $y = f(x)$ stretched horizontally.

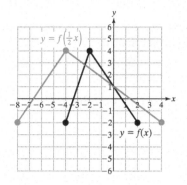

To graph $y = f\left(\frac{1}{2}x\right)$, divide each x value on the graph of $y = f(x)$ by $\frac{1}{2}$. For example:

$(-4, -2)$ becomes $\left(\dfrac{-4}{\frac{1}{2}}, -2\right) = (-8, -2)$.

$(-2, 4)$ becomes $\left(\dfrac{-2}{\frac{1}{2}}, 4\right) = (-4, 4)$.

$(2, -2)$ becomes $\left(\dfrac{2}{\frac{1}{2}}, -2\right) = (4, -2)$.

The graph of $y = f\left(\frac{1}{2}x\right)$ is shown in red.

Skill Practice 5 The graph of $y = f(x)$ is shown.

Graph. **a.** $y = f(2x)$ **b.** $y = f\left(\dfrac{1}{2}x\right)$

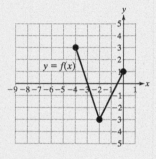

4. Apply Reflections Across the *x*- and *y*-Axes

The graphs of $f(x) = x^2$ (in black) and $g(x) = -x^2$ (in blue) are shown in Figure 2-26. Notice that a point (x, y) on the graph of f corresponds to the point $(x, -y)$ on the graph of g. Therefore, the graph of g is the graph of f reflected across the x-axis.

The graphs of $f(x) = \sqrt{x}$ (in black) and $g(x) = \sqrt{-x}$ (in blue) are shown in Figure 2-27. Notice that a point (x, y) on the graph of f corresponds to the point $(-x, y)$ on g. Therefore, the graph of g is the graph of f reflected across the y-axis.

Answers

5. a.

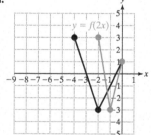

b.

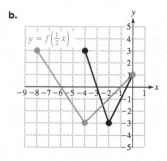

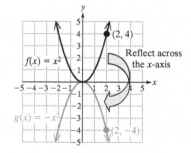

Figure 2-26

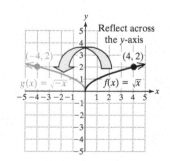

Figure 2-27

Reflections Across the *x*- and *y*-Axes

Consider a function defined by $y = f(x)$.

- The graph of $y = -f(x)$ is the graph of $y = f(x)$ reflected across the *x*-axis.
- The graph of $y = f(-x)$ is the graph of $y = f(x)$ reflected across the *y*-axis.

EXAMPLE 6   **Reflecting the Graph of a Function Across the *x*- and *y*-Axes**

The graph of $y = f(x)$ is given.

  a. Graph $y = -f(x)$.

  b. Graph $y = f(-x)$.

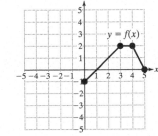

TIP For a given point (x, y), notice that $(-x, y)$ is on the opposite side of and equidistant to the *y*-axis. Likewise, $(x, -y)$ is on the opposite side of and equidistant from the *x*-axis as (x, y).

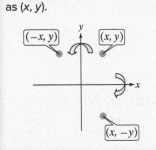

Solution:

  a. Reflect $y = f(x)$ across the *x*-axis.

  b. Reflect $y = f(x)$ across the *y*-axis.

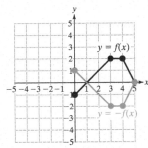
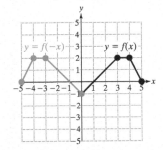

Skill Practice 6   The graph of $y = f(x)$ is given.

  a. Graph $y = -f(x)$.

  b. Graph $y = f(-x)$.

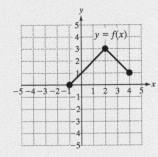

Answers

6. a.–b.

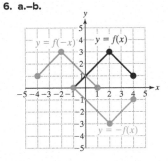

5. Summarize Transformations of Graphs

The operations of reflecting a graph of a function about an axis and shifting, stretching, and shrinking a graph are called **transformations**. Transformations give us tools to graph families of functions that are built from basic "parent" functions.

Transformations of Functions

Consider a function defined by $y = f(x)$. If h, k, and a represent positive real numbers, then the graphs of the following functions are related to $y = f(x)$ as follows.

Transformation	Effect on the Graph of f	Changes to Points on f
Vertical translation (shift) $y = f(x) + k$ $y = f(x) - k$	Shift upward k units Shift downward k units	Replace (x, y) by $(x, y + k)$. Replace (x, y) by $(x, y - k)$.
Horizontal translation (shift) $y = f(x - h)$ $y = f(x + h)$	Shift to the right h units Shift to the left h units	Replace (x, y) by $(x + h, y)$. Replace (x, y) by $(x - h, y)$.
Vertical stretch/shrink $y = a[f(x)]$	Vertical stretch (if $a > 1$) Vertical shrink (if $0 < a < 1$) Graph is stretched/shrunk vertically by a factor of a.	Replace (x, y) by (x, ay).
Horizontal stretch/shrink $y = f(a \cdot x)$	Horizontal shrink (if $a > 1$) Horizontal stretch (if $0 < a < 1$) Graph is shrunk/stretched horizontally by a factor of $\frac{1}{a}$.	Replace (x, y) by $\left(\frac{x}{a}, y\right)$.
Reflection $y = -f(x)$ $y = f(-x)$	Reflection across the x-axis Reflection across the y-axis	Replace (x, y) by $(x, -y)$. Replace (x, y) by $(-x, y)$.

When graphing a function requiring multiple transformations on the parent function, it is important to follow the correct sequence of steps.

Steps for Graphing Multiple Transformations of Functions

To graph a function requiring multiple transformations, use the following order.

1. Horizontal translation (shift)
2. Horizontal and vertical stretch and shrink
3. Reflections across the x- and y-axes
4. Vertical translation (shift)

EXAMPLE 7   **Using Transformations to Graph a Function**

Use transformations to graph the function defined by $n(x) = -\dfrac{1}{2}(x - 2)^2 + 3$.

Solution:

The graph of $n(x) = -\dfrac{1}{2}(x - 2)^2 + 3$ is the same as the graph of $f(x) = x^2$, with four transformations in the following order.

1. Shift the graph to the right 2 units.
2. Apply a vertical shrink (multiply the y values by $\frac{1}{2}$).
3. Reflect the graph over the x-axis.
4. Shift the graph upward 3 units.

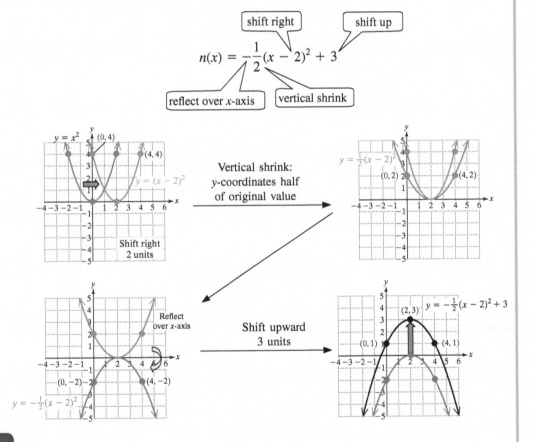

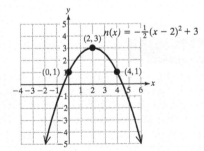

Avoiding Mistakes

As a means to check the graph of $y = n(x)$, substitute the x-coordinates of the strategic points $(0, 1)$, $(2, 3)$, and $(4, 1)$ into the function.

$n(0) = -\frac{1}{2}(0 - 2)^2 + 3 = 1$ ✓
$n(2) = -\frac{1}{2}(2 - 2)^2 + 3 = 3$ ✓
$n(4) = -\frac{1}{2}(4 - 2)^2 + 3 = 1$ ✓

> **Skill Practice 7** Use transformations to graph the function defined by $m(x) = -3|x - 2| - 4$.

EXAMPLE 8   Using Transformations to Graph a Function

Use transformations to graph the function defined by $v(x) = -\sqrt{-x + 2}$.

Solution:

The graph of $v(x) = -\sqrt{-x + 2}$ is the same as the graph of $f(x) = \sqrt{x}$, with three transformations in the following order.

1. Shift the graph to the left 2 units.
2. Reflect the graph across the y-axis.
3. Reflect the graph across the x-axis.
  (Note that the reflections in steps 2 and 3 can be applied in either order.)

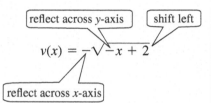

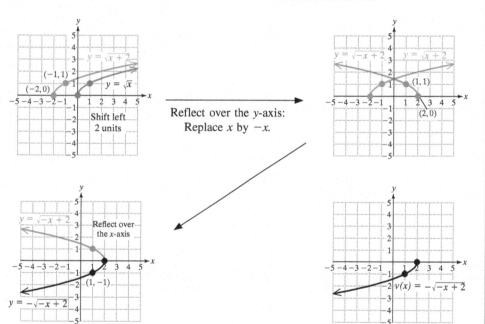

> **Skill Practice 8** Use transformations to graph the function defined by $r(x) = \sqrt[3]{-x + 1}$.

Avoiding Mistakes

Transformations involving a horizontal shrink, stretch, or reflection often introduce confusion when coupled with a horizontal shift. To further illustrate the rationale for the order of steps taken in Example 8, begin with the parent function $y = \sqrt{x}$. Performing a horizontal shift first means that we replace x by $x + 2$. This gives us $y = \sqrt{x + 2}$. Then to perform the reflection across the y-axis, we replace x by $-x$ to get $y = \sqrt{-x + 2}$. Performing these two transformations in the reverse order, would *not* result in the function we want. We would first have $y = \sqrt{-x}$, and then replacing x by $x + 2$ would give $y = \sqrt{-(x + 2)} = \sqrt{-x - 2}$ rather than $y = \sqrt{-x + 2}$.

Answers

7.

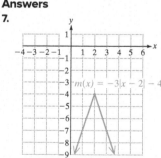

8.

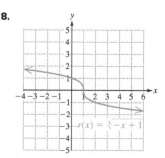

SECTION 2.6   Practice Exercises

Prerequisite Review

For Exercises R.1–R.3, graph each equation.

R.1. $y = -3x - 1$ **R.2.** $y = \dfrac{3}{5}x + 2$ **R.3.** $y = 1$

Concept Connections

1. Let c represent a positive real number. The graph of $y = f(x + c)$ is the graph of $y = f(x)$ shifted (up/down/left/right) c units.

2. Let c represent a positive real number. The graph of $y = f(x - c)$ is the graph of $y = f(x)$ shifted (up/down/left/right) c units.

3. Let c represent a positive real number. The graph of $y = f(x) - c$ is the graph of $y = f(x)$ shifted (up/down/left/right) c units.

4. The graph of $y = 3f(x)$ is the graph of $y = f(x)$ with a (choose one: vertical stretch, vertical shrink, horizontal stretch, horizontal shrink).

5. The graph of $y = f(3x)$ is the graph of $y = f(x)$ with a (choose one: vertical stretch, vertical shrink, horizontal stretch, horizontal shrink).

6. The graph of $y = f\left(\frac{1}{3}x\right)$ is the graph of $y = f(x)$ with a (choose one: vertical stretch, vertical shrink, horizontal stretch, horizontal shrink).

7. The graph of $y = \frac{1}{3}f(x)$ is the graph of $y = f(x)$ with a (choose one: vertical stretch, vertical shrink, horizontal stretch, horizontal shrink).

8. The graph of $y = -f(x)$ is the graph of $y = f(x)$ reflected across the _____ -axis.

Objective 1: Recognize Basic Functions

For Exercises 9–14, from memory match the equation with its graph.

9. $f(x) = \sqrt{x}$ 10. $f(x) = \sqrt[3]{x}$ 11. $f(x) = x^3$

12. $f(x) = x^2$ 13. $f(x) = |x|$ 14. $f(x) = \dfrac{1}{x}$

a. 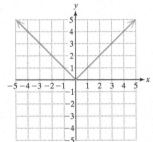   **b.** 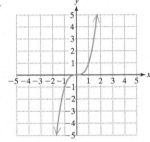   **c.**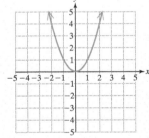

d. 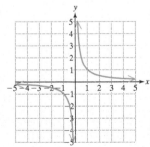   **e.** 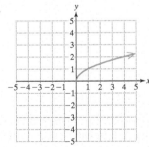   **f.**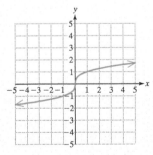

Objective 2: Apply Vertical and Horizontal Translations (Shifts)

For Exercises 15–26, use translations to graph the given functions. (See Examples 1–3)

15. $f(x) = |x| + 1$ 16. $g(x) = \sqrt{x} + 2$ 17. $k(x) = x^3 - 2$ 18. $h(x) = \dfrac{1}{x} - 2$

19. $g(x) = \sqrt{x + 5}$ 20. $m(x) = |x + 1|$ 21. $r(x) = (x - 4)^2$ 22. $t(x) = \sqrt[3]{x - 2}$

23. $a(x) = \sqrt{x + 1} - 3$ 24. $b(x) = |x - 2| + 4$ 25. $c(x) = \dfrac{1}{x - 3} + 1$ 26. $d(x) = \dfrac{1}{x + 4} - 1$

Objective 3: Apply Vertical and Horizontal Shrinking and Stretching

For Exercises 27–32, use transformations to graph the functions. (See Example 4)

27. $m(x) = 4\sqrt[3]{x}$ **28.** $n(x) = 3|x|$ **29.** $r(x) = \dfrac{1}{2}x^2$

30. $t(x) = \dfrac{1}{3}|x|$ **31.** $p(x) = |2x|$ **32.** $q(x) = \sqrt{2x}$

For Exercises 33–40, use the graphs of $y = f(x)$ and $y = g(x)$ to graph the given function. (See Example 5)

33. $y = \dfrac{1}{3}f(x)$ **34.** $y = \dfrac{1}{2}g(x)$

35. $y = 3f(x)$ **36.** $y = 2g(x)$

37. $y = f(3x)$ **38.** $y = g(2x)$

39. $y = f\left(\dfrac{1}{3}x\right)$ **40.** $y = g\left(\dfrac{1}{2}x\right)$

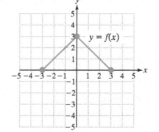 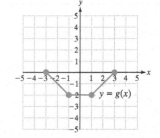

Objective 4: Apply Reflections Across the x- and y-Axes

For Exercises 41–46, graph the function by applying an appropriate reflection.

41. $f(x) = -\dfrac{1}{x}$ **42.** $g(x) = -\sqrt{x}$ **43.** $h(x) = -x^3$

44. $k(x) = -|x|$ **45.** $p(x) = (-x)^3$ **46.** $q(x) = \sqrt[3]{-x}$

For Exercises 47–50, use the graphs of $y = f(x)$ and $y = g(x)$ to graph the given function. (See Example 6)

47. $y = f(-x)$

48. $y = g(-x)$

49. $y = -f(x)$

50. $y = -g(x)$

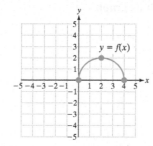 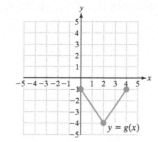

For Exercises 51–54, use the graphs of $y = f(x)$ and $y = g(x)$ to graph the given function. (See Example 6)

51. $y = f(-x)$

52. $y = g(-x)$

53. $y = -f(x)$

54. $y = -g(x)$

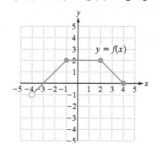 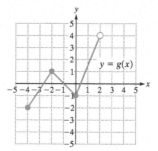

Objective 5: Summarize Transformations of Graphs

For Exercises 55–62, a function g is given. Identify the parent function from Table 2-2 on page 229. Then use the steps for graphing multiple transformations of functions on page 236 to list, in order, the transformations applied to the parent function to obtain the graph of g.

55. $g(x) = \dfrac{3}{1+x} - 2$ **56.** $g(x) = \dfrac{5}{x-4} + 1$ **57.** $g(x) = \dfrac{1}{3}(x - 2.1)^2 + 7.9$

58. $g(x) = \dfrac{1}{2}\sqrt{x + 4.3} - 8.4$ **59.** $g(x) = 2\sqrt{-2x + 5}$ **60.** $g(x) = 3\left|-\dfrac{1}{2}x - 4\right|$

61. $g(x) = -\sqrt{\dfrac{1}{3}x - 6}$ **62.** $g(x) = -|2x| + 8$

For Exercises 63–78, use transformations to graph the functions. (See Examples 7–8)

63. $v(x) = -(x + 2)^2 + 1$ **64.** $u(x) = -(x - 1)^2 - 2$ **65.** $f(x) = 2\sqrt{x + 3} - 1$ **66.** $g(x) = 2\sqrt{x - 1} + 3$

67. $p(x) = \dfrac{1}{2}|x - 1| - 2$ **68.** $q(x) = \dfrac{1}{3}|x + 2| - 1$ **69.** $r(x) = -\sqrt{-x} + 1$ **70.** $s(x) = -\sqrt{-x} - 2$

71. $f(x) = \sqrt{-x + 3}$ **72.** $g(x) = \sqrt{-x - 4}$ **73.** $n(x) = -\left|\dfrac{1}{2}x - 3\right|$ **74.** $m(x) = -\left|\dfrac{1}{3}x + 1\right|$

75. $f(x) = -\dfrac{1}{2}(x - 3)^2 + 8$ **76.** $g(x) = -\dfrac{1}{3}(x + 2)^2 + 3$ **77.** $p(x) = -4|x + 2| - 1$ **78.** $q(x) = -2|x - 1| + 4$

Mixed Exercises

For Exercises 79–86, the graph of $y = f(x)$ is given. Graph the indicated function.

79. Graph $y = -f(x - 1) + 2$.

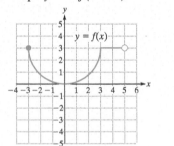

80. Graph $y = -f(x + 1) - 2$.

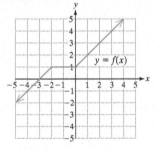

81. Graph $y = 2f(x - 2) - 3$.

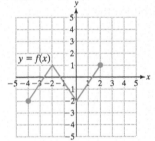

82. Graph $y = 2f(x + 2) - 4$.

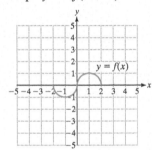

83. Graph $y = -3f(2x)$.

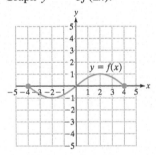

84. Graph $y = -\dfrac{1}{2}f\left(\dfrac{1}{2}x\right)$.

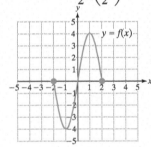

85. Graph $y = f(-x) - 2$.

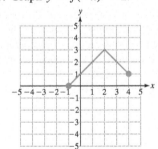

86. Graph $y = f(-x) + 3$.

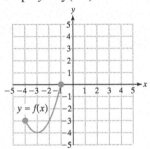

For Exercises 87–90, write a function based on the given parent function and transformations in the given order.

87. Parent function: $y = x^3$

   1. Shift 4.5 units to the left.
   2. Reflect across the y-axis.
   3. Shift upward 2.1 units.

88. Parent function $y = \sqrt[3]{x}$

   1. Shift 1 unit to the left.
   2. Stretch horizontally by a factor of 4.
   3. Reflect across the x-axis.

89. Parent function: $y = \dfrac{1}{x}$

   1. Stretch vertically by a factor of 2.
   2. Reflect across the x-axis.
   3. Shift downward 3 units.

90. Parent function: $y = |x|$

   1. Shift 3.7 units to the right.
   2. Shrink horizontally by a factor of $\dfrac{1}{3}$.
   3. Reflect across the y-axis.

Write About It

91. Explain why the graph of $g(x) = |2x|$ can be interpreted as a horizontal shrink of the graph of $f(x) = |x|$ or as a vertical stretch of the graph of $f(x) = |x|$.

92. Explain why the graph of $h(x) = \sqrt{\frac{1}{2}x}$ can be interpreted as a horizontal stretch of the graph of $f(x) = \sqrt{x}$ or as a vertical shrink of the graph of $f(x) = \sqrt{x}$.

93. Explain the difference between the graphs of $f(x) = |x - 2| - 3$ and $g(x) = |x - 3| - 2$.

94. Explain why $g(x) = \dfrac{1}{-x + 1}$ can be graphed by shifting the graph of $f(x) = \dfrac{1}{x}$ one unit to the left and reflecting across the y-axis, or by shifting the graph of f one unit to the right and reflecting across the x-axis.

Expanding Your Skills

For Exercises 95–100, use transformations on the basic functions presented in Table 2-2 to write a rule $y = f(x)$ that would produce the given graph.

95.

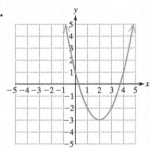

96.

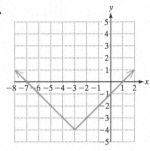

97.

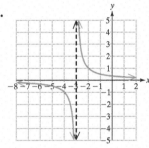

98.

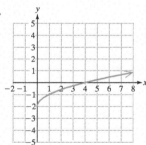

99.

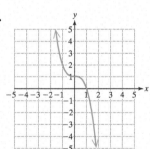

100.

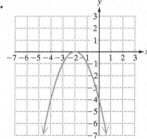

101. The graph shows the number of views y (in thousands) for a new online video, t days after it was posted. Use transformations on one of the parent functions from Table 2-2 on page 229 to model these data.

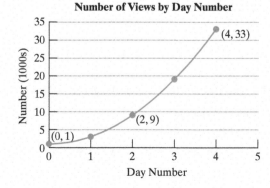

Number of Views by Day Number

102. The graph shows the cumulative number y of flu cases among passengers on a 25-day cruise, t days after the cruise began. Use transformations on one of the parent functions from Table 2-2 on page 229 to model these data.

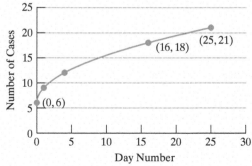

Cumulative Number of Flu Cases

Technology Connections

103. a. Graph the functions on the viewing window
$[-5, 5, 1]$ by $[-2, 8, 1]$.

$$y = x^2$$
$$y = x^4$$
$$y = x^6$$

c. Describe the general shape of the graph of $y = x^n$ where n is an even integer greater than 1.

b. Graph the functions on the viewing window
$[-4, 4, 1]$ by $[-10, 10, 1]$.

$$y = x^3$$
$$y = x^5$$
$$y = x^7$$

d. Describe the general shape of the graph of $y = x^n$ where n is an odd integer greater than 1.

SECTION 2.7 Analyzing Graphs of Functions and Piecewise-Defined Functions

OBJECTIVES

1. **Test for Symmetry**
2. **Identify Even and Odd Functions**
3. **Graph Piecewise-Defined Functions**
4. **Investigate Increasing, Decreasing, and Constant Behavior of a Function**
5. **Determine Relative Minima and Maxima of a Function**

1. Test for Symmetry

The photos in Figures 2-28 through 2-30 each show a type of symmetry.

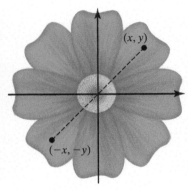

| Figure 2-28 | Figure 2-29 | Figure 2-30 |

The photo of the kingfisher (Figure 2-28) shows an image of the bird reflected in the water. Suppose that we superimpose the x-axis at the waterline. Every point (x, y) on the bird has a mirror image $(x, -y)$ below the x-axis. Therefore, this image is symmetric with respect to the x-axis.

A human face is symmetric with respect to a vertical line through the center (Figure 2-29). If we place the y-axis along this line, a point (x, y) on one side has a mirror image at $(-x, y)$. This image is symmetric with respect to the y-axis.

The flower shown in Figure 2-30 is symmetric with respect to the point at its center. Suppose that we place the origin at the center of the flower. Notice that a point (x, y) on the image has a corresponding point $(-x, -y)$ on the image. This image is symmetric with respect to the origin.

Given an equation in the variables x and y, use the following rules to determine if the graph is symmetric with respect to the x-axis, the y-axis, or the origin.

Tests for Symmetry

Consider an equation in the variables x and y.

- The graph of the equation is symmetric with respect to the y-axis if substituting $-x$ for x in the equation results in an equivalent equation.
- The graph of the equation is symmetric with respect to the x-axis if substituting $-y$ for y in the equation results in an equivalent equation.
- The graph of the equation is symmetric with respect to the origin if substituting $-x$ for x and $-y$ for y in the equation results in an equivalent equation.

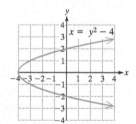

EXAMPLE 1   **Testing for Symmetry**

Determine whether the graph is symmetric with respect to the y-axis, x-axis, or origin.

   a. $y = |x|$ **b.** $x = y^2 - 4$

Solution:

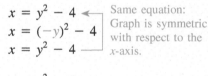

a. $y = |x|$ ← Same equation: Graph is symmetric with respect to the y-axis.
$y = |-x|$
$y = |x|$

Test for symmetry with respect to the y-axis.

Replace x by $-x$. Note that $|-x| = |x|$.

The resulting equation *is* equivalent to the original equation.

$y = |x|$ ←
$-y = |x|$ not the same
$y = -|x|$

Test for symmetry with respect to the x-axis.

Replace y by $-y$. The resulting equation is *not* equivalent to the original equation.

$y = |x|$ ←
$-y = |-x|$ not the same
$-y = |x|$
$y = -|x|$

Test for symmetry with respect to the origin.

Replace x by $-x$ and y by $-y$.

The resulting equation is *not* equivalent to the original equation.

The graph is symmetric with respect to the y-axis only.

b. $x = y^2 - 4$ ←
$-x = y^2 - 4$ not the same
$x = -y^2 + 4$

Test for symmetry with respect to the y-axis.

Replace x by $-x$. The resulting equation is *not* equivalent to the original equation.

$x = y^2 - 4$ ← Same equation: Graph is symmetric with respect to the x-axis.
$x = (-y)^2 - 4$
$x = y^2 - 4$

Test for symmetry with respect to the x-axis.

Replace y by $-y$. The resulting equation *is* equivalent to the original equation.

$x = y^2 - 4$ ←
$-x = (-y)^2 - 4$ not the same
$-x = y^2 - 4$
$x = -y^2 + 4$

Test for symmetry with respect to the origin.

Replace x by $-x$ and y by $-y$.

The resulting equation is *not* equivalent to the original equation.

The graph is symmetric with respect to the x-axis only (Figure 2-31).

Figure 2-31

Skill Practice 1   Determine whether the graph is symmetric with respect to the y-axis, x-axis, or origin.

   a. $y = x^2$ **b.** $|y| = x + 1$

EXAMPLE 2   **Testing for Symmetry**

Determine whether the graph is symmetric with respect to the y-axis, x-axis, or origin.

$$x^2 + y^2 = 9$$

Solution:

The graph of $x^2 + y^2 = 9$ is a circle with center at the origin and radius 3. By inspection, we can see that the graph is symmetric with respect to both axes and the origin.

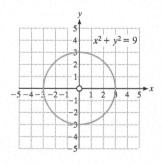

Answers
1. a. y-axis
   b. x-axis

Test for y-axis symmetry.
Replace x by $-x$.

$$\left.\begin{array}{l} x^2 + y^2 = 9 \\ (-x)^2 + y^2 = 9 \\ x^2 + y^2 = 9 \end{array}\right] \text{same}$$

Test for x-axis symmetry.
Replace y by $-y$.

$$\left.\begin{array}{l} x^2 + y^2 = 9 \\ x^2 + (-y)^2 = 9 \\ x^2 + y^2 = 9 \end{array}\right] \text{same}$$

Test for origin symmetry.
Replace x by $-x$ and y by $-y$.

$$\left.\begin{array}{l} x^2 + y^2 = 9 \\ (-x)^2 + (-y)^2 = 9 \\ x^2 + y^2 = 9 \end{array}\right] \text{same}$$

The graph is symmetric with respect to the y-axis, the x-axis, and the origin.

> **Skill Practice 2** Determine whether the graph is symmetric with respect to the y-axis, x-axis, or origin.
>
> $$\frac{x^2}{4} + \frac{y^2}{9} = 1$$

2. Identify Even and Odd Functions

A function may be symmetric with respect to the y-axis or to the origin. A function that is symmetric with respect to the y-axis is called an *even* function. A function that is symmetric with respect to the origin is called an *odd* function.

Even and Odd Functions

- A function f is an **even function** if $f(-x) = f(x)$ for all x in the domain of f. The graph of an even function is symmetric with respect to the y-axis.
- A function f is an **odd function** if $f(-x) = -f(x)$ for all x in the domain of f. The graph of an odd function is symmetric with respect to the origin.

EXAMPLE 3 **Identifying Even and Odd Functions**

By inspection determine if the function is even, odd, or neither.

Solution:

a.

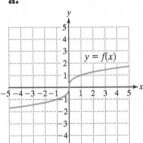

b.

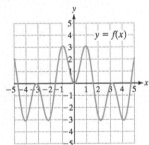

c.

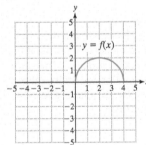

The function is symmetric with respect to the origin. Therefore, the function is an *odd* function.

The function is symmetric with respect to the y-axis. Therefore, the function is an *even* function.

The function is not symmetric with respect to either the y-axis or the origin. Therefore, the function is *neither* even nor odd.

Answer

2. y-axis, x-axis, and origin

Skill Practice 3 Determine if the function is even, odd, or neither.

a.   **b.**   **c.**

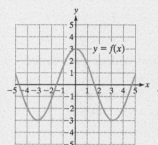   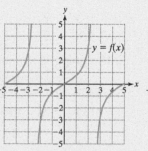   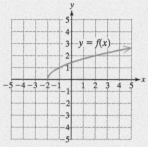

EXAMPLE 4   Identifying Even and Odd Functions

Determine if the function is even, odd, or neither.

a. $f(x) = -2x^4 + 5|x|$ **b.** $g(x) = 4x^3 - x$ **c.** $h(x) = 2x^2 + x$

Solution:

> **TIP** In Example 4(a), we suspect that f is an even function because each term is of the form x^{even} or $|x|$. In each case, replacing x by $-x$ results in an equivalent term.

a. $f(x) = -2x^4 + 5|x|$ Determine whether the function is even.
$f(-x) = -2(-x)^4 + 5|-x|$ same   Replace x by $-x$ to determine if $f(-x) = f(x)$.
$\quad\quad = -2x^4 + 5|x|$

Since $f(-x) = f(x)$, the function f is an even function.

There is no need to test whether f is an odd function because a function cannot be both even and odd unless all points are on the x-axis.

> **TIP** In Example 4(b), we suspect that g is an odd function because each term is of the form x^{odd}. In each case, replacing x by $-x$ results in the *opposite* of the original term.

b. $g(x) = 4x^3 - x$ Each term has x raised to an odd power. Therefore, replacing x by $-x$ will result in the *opposite* of the original term. Therefore, test whether g is an odd function. That is, test whether $g(-x) = -g(x)$.

Evaluate: $g(-x)$ Evaluate: $-g(x)$
$g(-x) = 4(-x)^3 - (-x)$ $-g(x) = -(4x^3 - x)$
$\quad\quad = -4x^3 + x$ ——— same ———→ $\quad\quad = -4x^3 + x$

Since $g(-x) = -g(x)$, the function g is an odd function.

> **TIP** In Example 4(c), $h(x)$ has a mixture of terms of the form x^{odd} and x^{even}. Therefore, we might suspect that the function is neither even nor odd.

c. $h(x) = 2x^2 + x$ Determine whether the function is even.
$h(-x) = 2(-x)^2 + (-x)$ not the same   Replace x by $-x$ to determine if $h(-x) = h(x)$.
$\quad\quad = 2x^2 - x$

Since $h(-x) \neq h(x)$, the function is not even.

Next, test whether h is an odd function. Test whether $h(-x) = -h(x)$.

Evaluate: $h(-x)$ Evaluate: $-h(x)$
$h(-x) = 2(-x)^2 + (-x)$ $-h(x) = -(2x^2 + x)$
$\quad\quad = 2x^2 - x$ ——— not the same ———→ $\quad\quad = -2x^2 - x$

Since $h(-x) \neq -h(x)$, the function is not an odd function. Therefore, h is neither even nor odd.

Answers
3. a. Even function
  b. Odd function
  c. Neither even nor odd
4. a. Odd function
  b. Even function
  c. Neither even nor odd

Skill Practice 4   Determine if the function is even, odd, or neither.
  a. $m(x) = -x^5 + x^3$ **b.** $n(x) = x^2 - |x| + 1$ **c.** $p(x) = 2|x| + x$

3. Graph Piecewise-Defined Functions

Suppose that a car is stopped for a red light. When the light turns green, the car undergoes a constant acceleration for 20 sec until it reaches a speed of 45 mph. It travels 45 mph for 1 min (60 sec), and then decelerates for 30 sec to stop at another red light. The graph of the car's speed y (in mph) versus the time x (in sec) after leaving the first red light is shown in Figure 2-32.

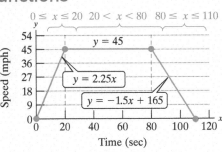

Figure 2-32

Notice that the graph can be segmented into three pieces. The first 20 sec is represented by a linear function with a positive slope, $y = 2.25x$. The next 60 sec is represented by the constant function $y = 45$. And the last 30 sec is represented by a linear function with a negative slope, $y = -1.5x + 165$.

To write a rule defining this function we use a **piecewise-defined function** in which we define each "piece" on a restricted domain.

$$f(x) = \begin{cases} 2.25x & \text{for } 0 \le x \le 20 \\ 45 & \text{for } 20 < x < 80 \\ -1.5x + 165 & \text{for } 80 \le x \le 110 \end{cases}$$

EXAMPLE 5   **Interpreting a Piecewise-Defined Function**

Evaluate the function for the given values of x.

$$f(x) = \begin{cases} -x - 1 & \text{for } x < -1 \\ -3 & \text{for } -1 \le x < 2 \\ \sqrt{x - 2} & \text{for } x \ge 2 \end{cases}$$

a. $f(-3)$ **b.** $f(-1)$
c. $f(2)$ **d.** $f(6)$

Solution:

a. $f(x) = -x - 1$ $x = -3$ is on the interval $x < -1$. Use the first rule
         $f(-3) = -(-3) - 1$ in the function: $f(x) = -x - 1$.
         $ = 2$

b. $f(x) = -3$ $x = -1$ is on the interval $-1 \le x < 2$. Use the second rule in
         $f(-1) = -3$ the function: $f(x) = -3$.

c. $f(x) = \sqrt{x - 2}$ $x = 2$ is on the interval $x \ge 2$. Use the third rule in the
         $f(2) = \sqrt{2 - 2}$ function: $f(x) = \sqrt{x - 2}$.
         $ = 0$

d. $f(x) = \sqrt{x - 2}$ $x = 6$ is on the interval $x \ge 2$. Use the third rule in the
         $f(6) = \sqrt{6 - 2}$ function: $f(x) = \sqrt{x - 2}$.
         $ = 2$

Skill Practice 5   Evaluate the function for the given values of x.

$$f(x) = \begin{cases} x + 7 & \text{for } x < -2 \\ x^2 & \text{for } -2 \le x < 1 \\ 3 & \text{for } x \ge 1 \end{cases}$$

a. $f(-3)$ **b.** $f(-2)$ **c.** $f(1)$ **d.** $f(4)$

TECHNOLOGY CONNECTIONS

Graphing a Piecewise-Defined Function

A graphing calculator can be used to graph a piecewise-defined function. The format to enter the function is as follows.

Y_1 = (first piece)/(first condition)
Y_2 = (second piece)/(second condition)
$\vdots$

Each condition in parentheses is an inequality and the calculator assigns it a value of 1 or 0 depending on whether the inequality is true or false. If an inequality is true, the function is divided by 1 on that interval and is "turned on." If an inequality is false, then the function is divided by 0. Since division by zero is undefined, the calculator does not graph the function on that interval, and the function is effectively "turned off."

Enter the function from Example 5 as shown. Note that the inequality symbols can be found in the TEST menu.

$$f(x) = \begin{cases} -x - 1 & \text{for } x < -1 \\ -3 & \text{for } -1 \le x < 2 \\ \sqrt{x - 2} & \text{for } x \ge 2 \end{cases}$$

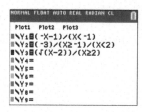 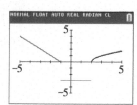

Notice that the individual "pieces" of the graph do not "hook-up." For this reason, it is also a good practice to put the calculator in DOT mode in the **MODE** menu.

In Examples 6 and 7, we graph piecewise-defined functions.

EXAMPLE 6   Graphing a Piecewise-Defined Function

Graph the function defined by $f(x) = \begin{cases} -3x & \text{for } x < 1 \\ -3 & \text{for } x \ge 1 \end{cases}$.

Solution:

- The first rule $f(x) = -3x$ defines a line with slope -3 and y-intercept $(0, 0)$. This line should be graphed only to the left of $x = 1$. The point $(1, -3)$ is graphed as an open dot, because the point is not part of the rule $f(x) = -3x$. See the blue portion of the graph in Figure 2-33.

- The second rule $f(x) = -3$ is a horizontal line for $x \ge 1$. The point $(1, -3)$ is a closed dot to show that it is part of the rule $f(x) = -3$. The closed dot from the red segment of the graph "overrides" the open dot from the blue segment. Taken together, the closed dot "plugs" the hole in the graph.

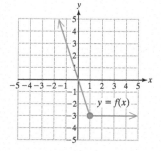

Figure 2-33

Skill Practice 6 Graph the function.

$$f(x) = \begin{cases} 2 & \text{for } x \le -1 \\ -2x & \text{for } x > -1 \end{cases}$$

> **TIP** The function in Example 6 has no "gaps," and therefore we say that the function is **continuous.** Informally, this means that we can draw the function without lifting our pencil from the page. The formal definition of a continuous function will be studied in calculus.

EXAMPLE 7   Graphing a Piecewise-Defined Function

Graph the function. $f(x) = \begin{cases} x + 3 & \text{for } x < -1 \\ x^2 & \text{for } -1 \le x < 2 \end{cases}$

Solution:

The first rule $f(x) = x + 3$ defines a line with slope 1 and y-intercept $(0, 3)$. This line should be graphed only for $x < -1$ (that is to the left of $x = -1$). The point $(-1, 2)$ is graphed as an open dot, because the point is not part of the function. See the red portion of the graph in Figure 2-34.

The second rule $f(x) = x^2$ is one of the basic functions learned in Section 2.6. It is a parabola with vertex at the origin. We sketch this function only for x values on the interval $-1 \le x < 2$. The point $(-1, 1)$ is a closed dot to show that it is part of the function. The point $(2, 4)$ is displayed as an open dot to indicate that it is not part of the function.

> **TIP** The function in Example 7 has a gap at $x = -1$, and therefore, we say that f is **discontinuous** at -1.

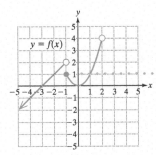

Figure 2-34

> **Avoiding Mistakes**
>
> Note that the function cannot have a closed dot at both $(-1, 1)$ and $(-1, 2)$ because it would not pass the vertical line test.

Answers

6.

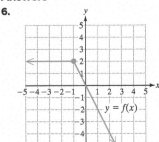

7.

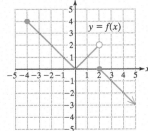

Skill Practice 7 Graph the function.

$$f(x) = \begin{cases} |x| & \text{for } -4 \le x < 2 \\ -x + 2 & \text{for } x \ge 2 \end{cases}$$

We now look at a special category of piecewise-defined functions called **step functions.** The graph of a step function is a series of discontinuous "steps." One important step function is called the **greatest integer function** or **floor function.** It is defined by

$$f(x) = [\![x]\!] \text{ where } [\![x]\!] \text{ is the greatest integer less than or equal to } x.$$

The operation $[\![x]\!]$ may also be denoted as **int**(x) or by **floor**(x). These alternative notations are often used in computer programming.

In Example 8, we graph the greatest integer function.

EXAMPLE 8 **Graphing the Greatest Integer Function**

Graph the function defined by $f(x) = [\![x]\!]$.

Solution:

x	$f(x) = [\![x]\!]$	
-1.7	-2	Evaluate f for several values of x.
-1	-1	Greatest integer less than or equal to -1.7 is -2.
-0.6	-1	Greatest integer less than or equal to -1 is -1.
0	0	Greatest integer less than or equal to -0.6 is -1.
0.4	0	Greatest integer less than or equal to 0 is 0.
1	1	Greatest integer less than or equal to 0.4 is 0.
1.8	1	Greatest integer less than or equal to 1 is 1.
2	2	Greatest integer less than or equal to 1.8 is 1.
2.5	2	Greatest integer less than or equal to 2 is 2.
		Greatest integer less than or equal to 2.5 is 2.

> **TIP** On many graphing calculators, the greatest integer function is denoted by int( ) and is found under the MATH menu followed by NUM.

From the table, we see a pattern and from the pattern, we form the graph.

If $-3 \le x < -2$, then $[\![x]\!] = -3$
If $-2 \le x < -1$, then $[\![x]\!] = -2$
If $-1 \le x < 0$, then $[\![x]\!] = -1$
If $0 \le x < 1$, then $[\![x]\!] = 0$
If $1 \le x < 2$, then $[\![x]\!] = 1$
If $2 \le x < 3$, then $[\![x]\!] = 2$
...

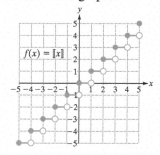

$f(x) = [\![x]\!]$

Skill Practice 8 Evaluate $f(x) = [\![x]\!]$ for the given values of x.
a. $f(1.7)$ **b.** $f(5.5)$ **c.** $f(-4)$ **d.** $f(-4.2)$

In Example 9, we use a piecewise-defined function to model an application.

EXAMPLE 9 **Using a Piecewise-Defined Function in an Application**

A salesperson makes a monthly salary of $3000 along with a 5% commission on sales over $20,000 for the month. Write a piecewise-defined function to represent the salesperson's monthly income $I(x)$ (in $) for x dollars in sales.

Solution:

Let x represent the amount in sales.
Then $x - 20,000$ represents the amount in sales over $20,000.

There are two scenarios for the salesperson's income.
Scenario 1: The salesperson sells $20,000 or less. In this case, the monthly income is a constant $3000. This is represented by

$$y = 3000 \quad \text{for } 0 \le x \le 20,000$$

Scenario 2: The salesperson sells over $20,000. In this case, the monthly income is $3000 plus 5% of sales over $20,000. This is represented by

$$y = 3000 + 0.05(x - 20,000) \quad \text{for } x > 20,000$$

Therefore, a piecewise-defined function for monthly income is

$$I(x) = \begin{cases} 3000 & \text{for } 0 \le x \le 20{,}000 \\ 3000 + 0.05(x - 20{,}000) & \text{for } x > 20{,}000 \end{cases}$$

Alternatively, we can simplify to get

$$I(x) = \begin{cases} 3000 & \text{for } 0 \le x \le 20{,}000 \\ 0.05x + 2000 & \text{for } x > 20{,}000 \end{cases}$$

A graph of $y = I(x)$ is shown in Figure 2-35. Notice that for $x = \$20{,}000$, both equations within the piecewise-defined function yield a monthly salary of $3000. Therefore, the two line segments in the graph meet at (20,000, 3000).

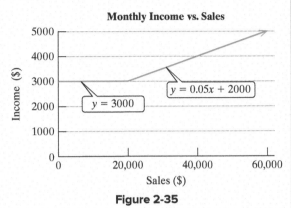

Monthly Income vs. Sales

$y = 3000$

$y = 0.05x + 2000$

Figure 2-35

> **Skill Practice 9**   A retail store buys T-shirts from the manufacturer. The cost is $7.99 per shirt for 1 to 100 shirts, inclusive. Then the price is decreased to $6.99 per shirt thereafter. Write a piecewise-defined function that expresses the cost $C(x)$ (in $) to buy x shirts.

4. Investigate Increasing, Decreasing, and Constant Behavior of a Function

The graph in Figure 2-36 approximates the altitude of an airplane, $f(t)$, at a time t minutes after takeoff.

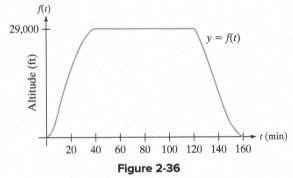

Figure 2-36

Notice that the plane's altitude increases up to the first 40 min of the flight. So we say that the function f is increasing on the interval (0, 40). The plane flies at a constant altitude for the next 1 hr 20 min, so we say that f is constant on the interval (40, 120). Finally, the plane's altitude decreases for the last 40 min, so we say that f is decreasing on the interval (120, 160).

Informally, a function is increasing on an interval in its domain if its graph rises from left to right. A function is decreasing on an interval in its domain if the graph "falls" from left to right. A function is constant on an interval in its domain if its graph is horizontal over the interval. These ideas are stated formally using mathematical notation.

Answer

9. $C(x) = \begin{cases} 7.99x & \text{for } 1 \le x \le 100 \\ 799 + 6.99(x - 100) & \text{for } x > 100 \end{cases}$

Intervals of Increasing, Decreasing, and Constant Behavior

Suppose that I is an interval contained within the domain of a function f.

- f is increasing on I if $f(x_1) < f(x_2)$ for all $x_1 < x_2$ on I.
- f is decreasing on I if $f(x_1) > f(x_2)$ for all $x_1 < x_2$ on I.
- f is constant on I if $f(x_1) = f(x_2)$ for all x_1 and x_2 on I.

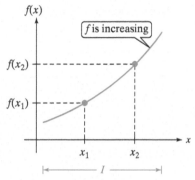

For all $x_1 < x_2$ on I, $f(x_1) < f(x_2)$

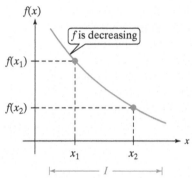

For all $x_1 < x_2$ on I, $f(x_1) > f(x_2)$

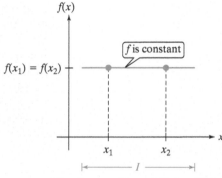

For all x_1 and x_2 on I, $f(x_1) = f(x_2)$

EXAMPLE 10   Determining the Intervals Over Which a Function Is Increasing, Decreasing, and Constant

Use interval notation to write the interval(s) over which f is

a. Increasing   **b.** Decreasing
c. Constant

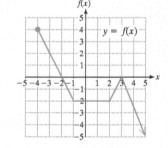

Solution:

a. f is increasing on the interval $(2, 3)$.
(Highlighted in red tint.)

b. f is decreasing on the interval $(-4, -1)$ and $(3, \infty)$.
(Highlighted in orange tint.)

c. f is constant on the interval $(-1, 2)$.
(Highlighted in green tint.)

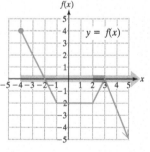

Skill Practice 10   Use interval notation to write the interval(s) over which f is

a. Increasing
b. Decreasing
c. Constant

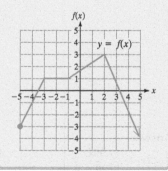

Answers
10. a. $(-5, -3) \cup (-1, 2)$
  b. $(2, \infty)$ **c.** $(-3, -1)$

5. Determine Relative Minima and Maxima of a Function

The intervals over which a function changes from increasing to decreasing behavior or vice versa tell us where to look for relative maximum values and relative minimum values of a function. Consider the function pictured in Figure 2-37.

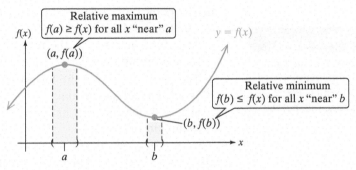

Figure 2-37

- The function has a relative maximum of $f(a)$. Informally, this means that $f(a)$ is the greatest function value relative to other points on the function nearby.
- The function has a relative minimum of $f(b)$. Informally, this means that $f(b)$ is the smallest function value relative to other points on the function nearby.

This is stated formally in the following definition.

> **TIP** The plural of maximum and minimum are **maxima** and **minima.**
>
> Note that relative maxima and minima are also called *local* maxima and minima.

Relative Minimum and Relative Maximum Values

- $f(a)$ is a **relative maximum** of f if there exists an open interval containing a such that $f(a) \geq f(x)$ for all x in the interval.
- $f(b)$ is a **relative minimum** of f if there exists an open interval containing b such that $f(b) \leq f(x)$ for all x in the interval.

Note: An open interval is an interval in which the endpoints are not included.

If an ordered pair $(a, f(a))$ corresponds to a relative minimum or relative maximum, we interpret the coordinates of the ordered pair as follows.

- The x-coordinate is the *location* of the relative maximum or minimum within the domain of the function.
- The y-coordinate is the *value* of the relative maximum or minimum. This tells us how "high" or "low" the graph is at that point.

EXAMPLE 11   Finding Relative Maxima and Minima

For the graph of $y = g(x)$ shown,

a. Determine the location and value of any relative maxima.

b. Determine the location and value of any relative minima.

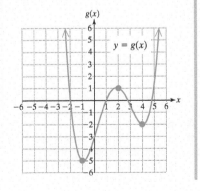

Solution:

a. The point (2, 1) is the highest point in a small interval surrounding $x = 2$. Therefore, at $x = 2$, the function has a relative maximum of 1.

b. The point $(-1, -5)$ is the lowest point in a small interval surrounding $x = -1$. Therefore, at $x = -1$, the function has a relative minimum of -5.

The point $(4, -2)$ is the lowest point in a small interval surrounding $x = 4$. Therefore, at $x = 4$, the function has a relative minimum of -2.

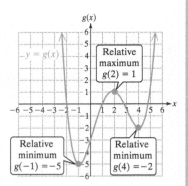

Avoiding Mistakes

Be sure to note that the value of a relative minimum or relative maximum is the y value of a function, not the x value.

Skill Practice 11   For the graph shown,

a. Determine the location and value of any relative maxima.

b. Determine the location and value of any relative minima.

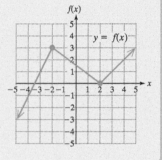

TECHNOLOGY CONNECTIONS

Determining Relative Maxima and Minima

Relative maxima and relative minima are often difficult to find analytically and require techniques from calculus. However, a graphing utility can be used to approximate the location and value of relative maxima and minima. To do so, we use the Minimum and Maximum features.

For example, enter the function defined by $Y_1 = x^3 - 4x^2 + 3x$. Then access the Maximum feature from the CALC menu.

The calculator asks for a left bound. This is a point slightly to the left of the relative maximum. Then hit ENTER.

The calculator asks for a right bound. This is a point slightly to the right of the relative maximum. Hit ENTER.

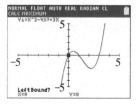

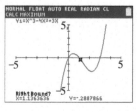

Answers

11. a. At $x = -2$, the function has a relative maximum of 3.
   b. At $x = 2$, the function has a relative minimum of 0.

The calculator asks for a guess. This is a point close to the relative maximum. Hit ENTER and the approximate coordinates of the relative maximum point are shown (0.45, 0.63).

To find the relative minimum, repeat these steps using the Minimum feature. The coordinates of the relative minimum point are approximately (2.22, −2.11).

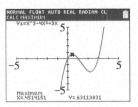

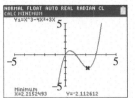

SECTION 2.7   Practice Exercises

Prerequisite Review

R.1. Given the function defined by $f(x) = 7x - 2$, find $f(-a)$.

For Exercises R.2–R.4, graph the set and express the set in interval notation.

R.2. $\{x \mid x < 8\}$ **R.3.** $\{x \mid -2.4 \le x < 5.8\}$ **R.4.** $\left\{x \mid x \ge -\dfrac{9}{2}\right\}$

Concept Connections

1. A graph of an equation is symmetric with respect to the _____-axis if replacing x by $-x$ results in an equivalent equation.

2. A graph of an equation is symmetric with respect to the _____-axis if replacing y by $-y$ results in an equivalent equation.

3. A graph of an equation is symmetric with respect to the _____ if replacing x by $-x$ and y by $-y$ results in an equivalent equation.

4. An even function is symmetric with respect to the _____.

5. An odd function is symmetric with respect to the _____.

6. The expression _____ represents the greatest integer, less than or equal to x.

Objective 1: Test for Symmetry

For Exercises 7–18, determine whether the graph of the equation is symmetric with respect to the x-axis, y-axis, origin, or none of these. (See Examples 1–2)

7. $y = x^2 + 3$ **8.** $y = -|x| - 4$ **9.** $x = -|y| - 4$ **10.** $x = y^2 + 3$

11. $x^2 + y^2 = 3$ **12.** $|x| + |y| = 4$ **13.** $y = |x| + 2x + 7$ **14.** $y = x^2 + 6x + 1$

15. $x^2 = 5 + y^2$ **16.** $y^4 = 2 + x^2$ **17.** $y = \dfrac{1}{2}x - 3$ **18.** $y = \dfrac{2}{5}x + 1$

Objective 2: Identify Even and Odd Functions

19. What type of symmetry does an even function have?

20. What type of symmetry does an odd function have?

For Exercises 21–26, use the graph to determine if the function is even, odd, or neither. (See Example 3)

21.

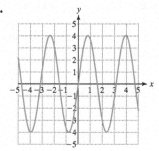

22.

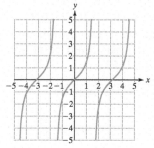

23.

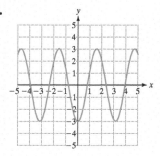

24.

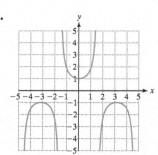

25.

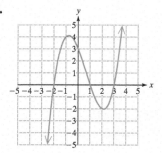

26.

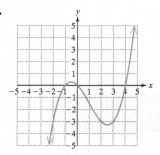

27. a. Given $f(x) = 4x^2 - 3|x|$, find $f(-x)$.
    b. Is $f(-x) = f(x)$?
    c. Is this function even, odd, or neither?

28. a. Given $g(x) = -x^8 + |3x|$, find $g(-x)$.
    b. Is $g(-x) = g(x)$?
    c. Is this function even, odd, or neither?

29. a. Given $h(x) = 4x^3 - 2x$, find $h(-x)$.
    b. Find $-h(x)$.
    c. Is $h(-x) = -h(x)$?
    d. Is this function even, odd, or neither?

30. a. Given $k(x) = -8x^5 - 6x^3$, find $k(-x)$.
    b. Find $-k(x)$.
    c. Is $k(-x) = -k(x)$?
    d. Is this function even, odd, or neither?

31. a. Given $m(x) = 4x^2 + 2x - 3$, find $m(-x)$.
    b. Find $-m(x)$.
    c. Is $m(-x) = m(x)$?
    d. Is $m(-x) = -m(x)$?
    e. Is this function even, odd, or neither?

32. a. Given $n(x) = 7|x| + 3x - 1$, find $n(-x)$.
    b. Find $-n(x)$.
    c. Is $n(-x) = n(x)$?
    d. Is $n(-x) = -n(x)$?
    e. Is this function even, odd, or neither?

For Exercises 33–46, determine if the function is even, odd, or neither. (See Example 4)

33. $f(x) = 3x^6 + 2x^2 + |x|$

34. $p(x) = -|x| + 12x^{10} + 5$

35. $k(x) = 13x^3 + 12x$

36. $m(x) = -4x^5 + 2x^3 + x$

37. $n(x) = \sqrt{16 - (x - 3)^2}$

38. $r(x) = \sqrt{81 - (x + 2)^2}$

39. $q(x) = \sqrt{16 + x^2}$

40. $z(x) = \sqrt{49 + x^2}$

41. $h(x) = 5x$

42. $g(x) = -x$

43. $f(x) = \dfrac{x^2}{3(x - 4)^2}$

44. $g(x) = \dfrac{x^3}{2(x - 1)^3}$

45. $v(x) = \dfrac{-x^5}{|x| + 2}$

46. $w(x) = \dfrac{-\sqrt[3]{x}}{x^2 + 1}$

Objective 3: Graph Piecewise-Defined Functions

For Exercises 47–50, evaluate the function for the given values of x. (See Example 5)

47. $f(x) = \begin{cases} -3x + 7 & \text{for } x < -1 \\ x^2 + 3 & \text{for } -1 \leq x < 4 \\ 5 & \text{for } x \geq 4 \end{cases}$

    a. $f(3)$ **b.** $f(-2)$ **c.** $f(-1)$
    d. $f(4)$ **e.** $f(5)$

48. $g(x) = \begin{cases} -2|x| - 3 & \text{for } x \leq -2 \\ 5x + 6 & \text{for } -2 < x < 3 \\ 4 & \text{for } x \geq 3 \end{cases}$

    a. $g(-3)$ **b.** $g(3)$ **c.** $g(-2)$
    d. $g(0)$ **e.** $g(4)$

49. $h(x) = \begin{cases} 2 & \text{for } -3 \le x < -2 \\ 1 & \text{for } -2 \le x < -1 \\ 0 & \text{for } -1 \le x < 0 \\ -1 & \text{for } 0 \le x < 1 \end{cases}$

   a. $h(-1.7)$ **b.** $h(-2.5)$ **c.** $h(0.05)$

   d. $h(-2)$ **e.** $h(0)$

50. $t(x) = \begin{cases} x & \text{for } 0 < x \le 1 \\ 2x & \text{for } 1 < x \le 2 \\ 3x & \text{for } 2 < x \le 3 \\ 4x & \text{for } 3 < x \le 4 \end{cases}$

   a. $t(1.99)$ **b.** $t(0.4)$ **c.** $t(3)$

   d. $t(1)$ **e.** $t(3.001)$

51. A sled accelerates down a hill and then slows down after it reaches a flat portion of ground. The speed of the sled $s(t)$ (in ft/sec) at a time t (in sec) after movement begins can be approximated by:

$$s(t) = \begin{cases} 1.5t & \text{for } 0 \le t \le 20 \\ \dfrac{30}{t-19} & \text{for } 20 < t \le 40 \end{cases}$$

Determine the speed of the sled after 10 sec, 20 sec, 30 sec, and 40 sec. Round to 1 decimal place if necessary.

52. A car starts from rest and accelerates to a speed of 60 mph in 12 sec. It travels 60 mph for 1 min and then decelerates for 20 sec until it comes to rest. The speed of the car $s(t)$ (in mph) at a time t (in sec) after the car begins motion can be modeled by:

$$s(t) = \begin{cases} \dfrac{5}{12}t^2 & \text{for } 0 \le t \le 12 \\ 60 & \text{for } 12 < t \le 72 \\ \dfrac{3}{20}(92-t)^2 & \text{for } 72 < t \le 92 \end{cases}$$

Determine the speed of the car 6 sec, 12 sec, 45 sec, and 80 sec after the car begins motion.

For Exercises 53–56, match the function with its graph.

53. $f(x) = x + 1$ for $x < 2$

54. $f(x) = x + 1$ for $-1 < x \le 2$

55. $f(x) = x + 1$ for $-1 \le x < 2$

56. $f(x) = x + 1$ for $x \ge 2$

a.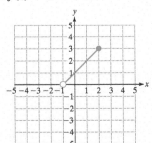

b.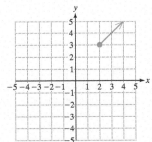

c.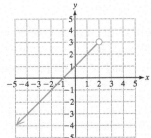

d.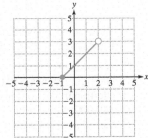

57. a. Graph $p(x) = x + 2$ for $x \le 0$.
    (See **Examples 6–7**)

   b. Graph $q(x) = -x^2$ for $x > 0$.

   c. Graph $r(x) = \begin{cases} x + 2 & \text{for } x \le 0 \\ -x^2 & \text{for } x > 0 \end{cases}$

58. a. Graph $f(x) = |x|$ for $x < 0$.

   b. Graph $g(x) = \sqrt{x}$ for $x \ge 0$.

   c. Graph $h(x) = \begin{cases} |x| & \text{for } x < 0 \\ \sqrt{x} & \text{for } x \ge 0 \end{cases}$

59. a. Graph $m(x) = \dfrac{1}{2}x - 2$ for $x \le -2$.

   b. Graph $n(x) = -x + 1$ for $x > -2$.

   c. Graph $t(x) = \begin{cases} \dfrac{1}{2}x - 2 & \text{for } x \le -2 \\ -x + 1 & \text{for } x > -2 \end{cases}$

60. a. Graph $a(x) = x$ for $x < 1$.

   b. Graph $b(x) = \sqrt{x - 1}$ for $x \ge 1$.

   c. Graph $c(x) = \begin{cases} x & \text{for } x < 1 \\ \sqrt{x - 1} & \text{for } x \ge 1 \end{cases}$

For Exercises 61–70, graph the function. (See Examples 6–7)

61. $f(x) = \begin{cases} |x| & \text{for } x < 2 \\ -x + 4 & \text{for } x \ge 2 \end{cases}$

62. $h(x) = \begin{cases} -2x & \text{for } x < 0 \\ \sqrt{x} & \text{for } x \ge 0 \end{cases}$

63. $g(x) = \begin{cases} x + 2 & \text{for } x < -1 \\ -x + 2 & \text{for } x \ge -1 \end{cases}$

64. $k(x) = \begin{cases} 3x & \text{for } x < 1 \\ -3x & \text{for } x \ge 1 \end{cases}$

65. $r(x) = \begin{cases} x^2 - 4 & \text{for } x \le 2 \\ 2x - 4 & \text{for } x > 2 \end{cases}$

66. $s(x) = \begin{cases} -x - 1 & \text{for } x \le -1 \\ \sqrt{x + 1} & \text{for } x > -1 \end{cases}$

67. $t(x) = \begin{cases} -3 & \text{for } -4 \le x < -2 \\ -1 & \text{for } -2 \le x < 0 \\ 1 & \text{for } 0 \le x < 2 \end{cases}$

68. $z(x) = \begin{cases} -1 & \text{for } -3 < x \le -1 \\ 1 & \text{for } -1 < x \le 1 \\ 3 & \text{for } 1 < x \le 3 \end{cases}$

69. $m(x) = \begin{cases} 3 & \text{for } -4 < x < -1 \\ -x & \text{for } -1 \le x < 3 \\ \sqrt{x-3} & \text{for } x \ge 3 \end{cases}$

70. $n(x) = \begin{cases} -4 & \text{for } -3 < x < -1 \\ x & \text{for } -1 \le x < 2 \\ -x^2 + 4 & \text{for } x \ge 2 \end{cases}$

71. a. Graph $f(x) = \begin{cases} -x & \text{for } x < 0 \\ x & \text{for } x \ge 0 \end{cases}$ **b.** To what basic function from Section 2.6 is the graph of f equivalent?

For Exercises 72–80, evaluate the step function defined by $f(x) = [\![x]\!]$ for the given values of x. (See Example 8)

72. $f(-3.7)$ **73.** $f(-4.2)$ **74.** $f(-0.5)$ **75.** $f(-0.09)$ **76.** $f(0.5)$

77. $f(0.09)$ **78.** $f(6)$ **79.** $f(-9)$ **80.** $f(-5)$

For Exercises 81–84, graph the function. (See Example 8)

81. $f(x) = [\![x + 3]\!]$ **82.** $g(x) = [\![x - 3]\!]$ **83.** $k(x) = \text{int}\left(\dfrac{1}{2}x\right)$ **84.** $h(x) = \text{int}(2x)$

85. For a recent year, the rate for first class postage was as follows. (**See Example 9**)

Weight not Over	Price
1 oz	$0.44
2 oz	$0.61
3 oz	$0.78
3.5 oz	$0.95

Write a piecewise-defined function to model the cost $C(x)$ to mail a letter first class if the letter is x ounces.

86. The water level in a retention pond started at 5 ft (60 in.) and decreased at a rate of 2 in./day during a 14-day drought. A tropical depression moved through at the beginning of the 15th day and produced rain at an average rate of 2.5 in./day for 5 days. Write a piecewise-defined function to model the water level $L(x)$ (in inches) as a function of the number of days x since the beginning of the drought.

87. A salesperson makes a base salary of $2000 per month. Once he reaches $40,000 in total sales, he earns an additional 5% commission on the amount in sales over $40,000. Write a piecewise-defined function to model the salesperson's total monthly salary $S(x)$ (in $) as a function of the amount in sales x.

88. A cell phone plan charges $49.95 per month plus $14.02 in taxes, plus $0.40 per minute for calls beyond the 600-min monthly limit. Write a piecewise-defined function to model the monthly cost $C(x)$ (in $) as a function of the number of minutes used x for the month.

Objective 4: Investigate Increasing, Decreasing, and Constant Behavior of a Function

For Exercises 89–96, use interval notation to write the intervals over which f is (a) increasing, (b) decreasing, and (c) constant. (See Example 10)

89.

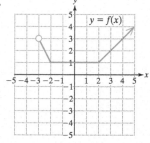

90.

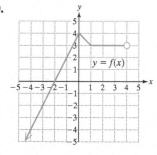

91.

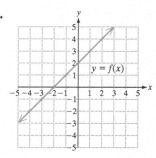

92.

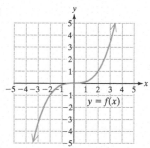

93.

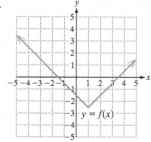

94.

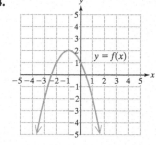

95.

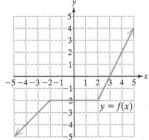

96.

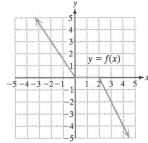

Objective 5: Determine Relative Minima and Maxima of a Function

For Exercises 97–102, identify the location and value of any relative maxima or minima of the function. (See Example 11)

97.

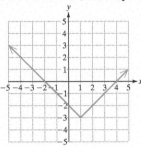

98.

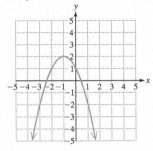

99.

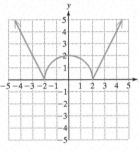

100.

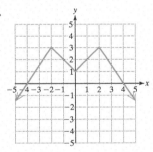

101.

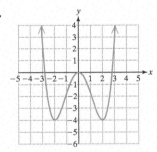

102.

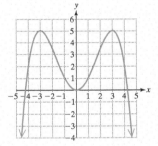

103. The graph shows the depth d (in ft) of a retention pond, t days after recording began.

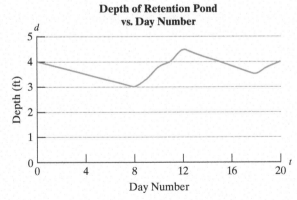

Depth of Retention Pond vs. Day Number

a. Over what interval(s) does the depth increase?

b. Over what interval(s) does the depth decrease?

c. Estimate the times and values of any relative maxima or minima on the interval (0, 20).

d. If rain is the only water that enters the pond, explain what the intervals of increasing and decreasing behavior mean in the context of this problem.

104. The graph shows the height h (in meters) of a roller coaster t seconds after the ride starts.

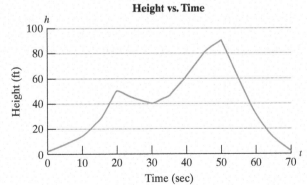

Height vs. Time

a. Over what interval(s) does the height increase?

b. Over what interval(s) does the height decrease?

c. Estimate the times and values of any relative maxima or minima on the interval (0, 70).

Mixed Exercises

For Exercises 105–110, produce a rule for the function whose graph is shown. (*Hint*: Consider using the basic functions learned in Section 2.6 and transformations of their graphs.)

105.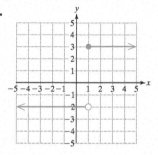

106.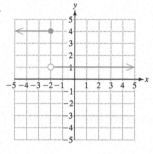

107.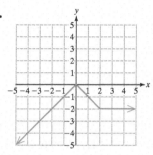

108.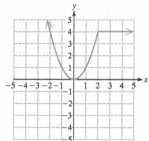

109.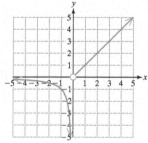

110.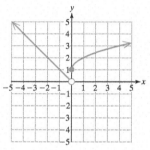

For Exercises 111–112,

a. Graph the function.

b. Write the domain in interval notation.

c. Write the range in interval notation.

d. Evaluate $f(-1)$, $f(1)$, and $f(2)$.

e. Find the value(s) of x for which $f(x) = 6$.

f. Find the value(s) of x for which $f(x) = -3$.

g. Use interval notation to write the intervals over which f is increasing, decreasing, or constant.

111. $f(x) = \begin{cases} -x^2 + 1 & \text{for } x \le 1 \\ 2x & \text{for } x > 1 \end{cases}$

112. $f(x) = \begin{cases} |x| & \text{for } x < 2 \\ -x & \text{for } x \ge 2 \end{cases}$

In computer programming, the greatest integer function is sometimes called the "floor" function. Programmers also make use of the "ceiling" function, which returns the smallest integer not less than x. For example: ceil(3.1) = 4. For Exercises 113–114, evaluate the floor and ceiling functions for the given value of x.

floor(x) is the greatest integer less than or equal to x.
ceil(x) is the smallest integer not less than x.

113. a. floor(2.8) **b.** floor(−3.1) **c.** floor(4) **114. a.** floor(5.5) **b.** floor(−0.1) **c.** floor(−2)

   d. ceil(2.8) **e.** ceil(−3.1) **f.** ceil(4)   **d.** ceil(5.5) **e.** ceil(−0.1) **f.** ceil(−2)

Write About It

115. From an equation in x and y, explain how to determine whether the graph of the equation is symmetric with respect to the x-axis, y-axis, or origin.

116. From the graph of a function, how can you determine if the function is even or odd?

117. Explain why the relation defined by

$$y = \begin{cases} 2x & \text{for } x \le 1 \\ 3 & \text{for } x \ge 1 \end{cases}$$

is not a function.

118. Explain why the function is discontinuous at $x = 1$.

$$f(x) = \begin{cases} 3x & \text{for } x < 1 \\ 3 & \text{for } x > 1 \end{cases}$$

119. Provide an informal explanation of a relative maximum.

120. Explain what it means for a function to be increasing on an interval.

Expanding Your Skills

121. Suppose that the average rate of change of a continuous function between any two points to the left of $x = a$ is negative, and the average rate of change of the function between any two points to the right of $x = a$ is positive. Does the function have a relative minimum or maximum at a?

122. Suppose that the average rate of change of a continuous function between any two points to the left of $x = a$ is positive, and the average rate of change of the function between any two points to the right of $x = a$ is negative. Does the function have a relative minimum or maximum at a?

A graph is *concave up* on a given interval if it "bends" upward. A graph is *concave down* on a given interval if it "bends" downward. For Exercises 123–126, determine whether the curve is (a) concave up or concave down and (b) increasing or decreasing.

123.    **124.**   **125.**   **126.**

127. For a recent year, the federal income tax owed by a taxpayer (single—no dependents) was based on the individual's taxable income. (*Source*: Internal Revenue Service, www.irs.gov)

If your taxable income is		The tax is	of the amount over—
over—	but not over—		
$0	$8925	$0 + 10%	$0
$8925	$36,250	$892.50 + 15%	$8925
$36,250	$87,850	$4991.25 + 25%	$36,250

Write a piecewise-defined function that expresses an individual's federal income tax $f(x)$ (in $) as a function of the individual's taxable income x (in $).

Technology Connections

For Exercises 128–131, use a graphing utility to graph the piecewise-defined function.

128. $f(x) = \begin{cases} 2.5x + 2 & \text{for } x \le 1 \\ x^2 - x - 1 & \text{for } x > 1 \end{cases}$

129. $g(x) = \begin{cases} -3.1x - 4 & \text{for } x < -2 \\ -x^3 + 4x - 1 & \text{for } x \ge -2 \end{cases}$

130. $k(x) = \begin{cases} -2.7x - 4.1 & \text{for } x \le -1 \\ -x^3 + 2x + 5 & \text{for } -1 < x < 2 \\ 1 & \text{for } x \ge 2 \end{cases}$

131. $z(x) = \begin{cases} 2.5x + 8 & \text{for } x < -2 \\ -2x^2 + x + 4 & \text{for } -2 \le x < 2 \\ -2 & \text{for } x \ge 2 \end{cases}$

For Exercises 132–135, use a graphing utility to

   a. Find the locations and values of the relative maxima and relative minima of the function on the standard viewing window. Round to 3 decimal places.

   b. Use interval notation to write the intervals over which f is increasing or decreasing.

132. $f(x) = -0.6x^2 + 2x + 3$

133. $f(x) = 0.4x^2 - 3x - 2.2$

134. $f(x) = 0.5x^3 + 2.1x^2 - 3x - 7$

135. $f(x) = -0.4x^3 - 1.1x^2 + 2x + 3$

SECTION 2.8   Algebra of Functions and Function Composition

1. Perform Operations on Functions

In Section 2.5, we learned that a profit function can be constructed from the difference of a revenue function and a cost function according to the following rule.

$$P(x) = R(x) - C(x)$$

As this example illustrates, the difference of two functions makes up a new function. New functions can also be formed from the sum, product, and quotient of two functions.

> **Sum, Difference, Product, and Quotient of Functions**
>
> Given functions f and g, the functions $f + g$, $f - g$, $f \cdot g$, and $\frac{f}{g}$ are defined by
>
> $$(f + g)(x) = f(x) + g(x)$$
> $$(f - g)(x) = f(x) - g(x)$$
> $$(f \cdot g)(x) = f(x) \cdot g(x)$$
> $$\left(\frac{f}{g}\right)(x) = \frac{f(x)}{g(x)} \text{ provided that } g(x) \neq 0$$
>
> The domains of the functions $f + g$, $f - g$, $f \cdot g$, and $\frac{f}{g}$ are all real numbers in the intersection of the domains of the individual functions f and g. For $\frac{f}{g}$, we further restrict the domain to exclude values of x for which $g(x) = 0$.

EXAMPLE 1   Adding Two Functions

Given $f(x) = \sqrt{25 - x^2}$ and $g(x) = 5$, find $(f + g)(x)$.

Solution:

By definition $(f + g)(x) = f(x) + g(x)$.
$$= \sqrt{25 - x^2} + 5$$

> **Skill Practice 1**   Given $m(x) = -|x|$ and $n(x) = 4$, find $(m + n)(x)$.

In Example 1, the graph of function f is a semicircle and the graph of function g is a horizontal line (Figure 2-38). Therefore, the graph of $y = (f + g)(x)$ is the graph of f with a vertical shift (shown in blue). Notice that each y value on $f + g$ is the sum of the y values from the individual functions f and g.

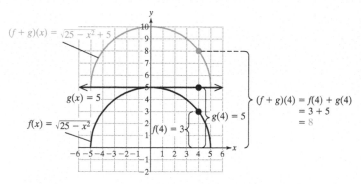

Figure 2-38

Answer

1. $(m + n)(x) = -|x| + 4$

In Example 2, we evaluate the difference, product, and quotient of functions for given values of x.

EXAMPLE 2 **Evaluating Functions for Given Values of x**

Given $m(x) = 4x$, $n(x) = |x - 3|$, and $p(x) = \dfrac{1}{x + 1}$, determine the function values if possible.

a. $(m - n)(-2)$ **b.** $(m \cdot p)(1)$ **c.** $\left(\dfrac{p}{n}\right)(3)$

Solution:

a. $(m - n)(-2) = m(-2) - n(-2)$ **b.** $(m \cdot p)(1) = m(1) \cdot p(1)$
$$= 4(-2) - |-2 - 3|$$
$$= -8 - 5$$
$$= -13$$

$$= 4(1) \cdot \dfrac{1}{1 + 1}$$
$$= 2$$

c. $\left(\dfrac{p}{n}\right)(3) = \dfrac{p(3)}{n(3)} = \dfrac{\dfrac{1}{3 + 1}}{|3 - 3|}$ The domain of $\dfrac{p}{n}$ excludes any values of x that make $n(x) = 0$. In this case, $x = 3$ is excluded from the domain.

$$= \dfrac{\frac{1}{4}}{0} \text{ (undefined)}$$

Skill Practice 2 Use the functions defined in Example 2 to find

a. $(n - m)(-6)$ **b.** $(n \cdot p)(0)$ **c.** $\left(\dfrac{p}{m}\right)(0)$

When combining two or more functions to create a new function, always be sure to determine the domain of the new function. Notice that in Example 2(c), the function $\dfrac{p}{n}$ is not defined for $x = -1$ or for $x = 3$.

$$\left(\dfrac{p}{n}\right)(x) = \dfrac{p(x)}{n(x)} = \dfrac{\dfrac{1}{x + 1}}{|x - 3|}$$
⟵ Denominator is zero for $x = -1$.
⟵ Denominator of the complex fraction is zero for $x = 3$.

EXAMPLE 3 **Combining Functions and Determining Domain**

Given $g(x) = 2x$, $h(x) = x^2 - 4x$, and $k(x) = \sqrt{x - 1}$,

a. Find $(g - h)(x)$ and write the domain of $g - h$ in interval notation.
b. Find $(g \cdot k)(x)$ and write the domain of $g \cdot k$ in interval notation.
c. Find $\left(\dfrac{k}{h}\right)(x)$ and write the domain of $\dfrac{k}{h}$ in interval notation.

Solution:

a. $(g - h)(x) = g(x) - h(x)$
$$= 2x - (x^2 - 4x)$$
$$= -x^2 + 6x$$

The domain of g is $(-\infty, \infty)$.
The domain of h is $(-\infty, \infty)$.
Therefore, the intersection of their domains is $(-\infty, \infty)$.

The domain is $(-\infty, \infty)$.

Answers
2. a. 33 **b.** 3 **c.** Undefined

b. $(g \cdot k)(x) = g(x) \cdot k(x)$

$\quad\quad\quad\quad\quad = 2x\sqrt{x - 1}$

The domain is $[1, \infty)$.

The domain of g is $(-\infty, \infty)$.
The domain of k is $[1, \infty)$.
Therefore, the intersection of their domains is $[1, \infty)$.

c. $\left(\dfrac{k}{h}\right)(x) = \dfrac{k(x)}{h(x)} = \dfrac{\sqrt{x - 1}}{x^2 - 4x}$

$\quad\quad\quad\quad\quad = \dfrac{\sqrt{x - 1}}{x(x - 4)}$

The domain of k is $[1, \infty)$.
The domain of h is $(-\infty, \infty)$.
The intersection of their domains is $[1, \infty)$.

However, we must also exclude values of x that make the denominator zero. In this case, exclude $x = 0$ and $x = 4$. The value $x = 0$ is already excluded because it is not on the interval $[1, \infty)$. Excluding $x = 4$, the domain of $\frac{k}{h}$ is $[1, 4) \cup (4, \infty)$.

The domain is $[1, 4) \cup (4, \infty)$.

Skill Practice 3   Given $m(x) = x + 3$, $n(x) = x^2 - 9$, and $p(x) = \sqrt{x + 1}$

a. Find $(n - m)(x)$ and write the domain of $n - m$ in interval notation.

b. Find $(m \cdot p)(x)$ and write the domain of $m \cdot p$ in interval notation.

c. Find $\left(\dfrac{p}{n}\right)(x)$ and write the domain of $\dfrac{p}{n}$ in interval notation.

---

2. Evaluate a Difference Quotient

In Section 2.4, we learned that if f is defined on an interval $[x_1, x_2]$, then the average rate of change of f between $(x_1, f(x_1))$ and $(x_2, f(x_2))$ is given by

$$m = \frac{f(x_2) - f(x_1)}{x_2 - x_1}.\quad \text{(Figure 2-39)}$$

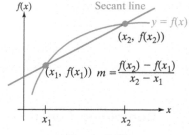

Figure 2-39

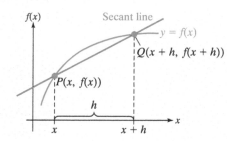

Figure 2-40

Now we look at a related idea. Let P be an arbitrary point $(x, f(x))$ on the function f. Let h be a positive real number and let Q be the point $(x + h, f(x + h))$. See Figure 2-40. The average rate of change between P and Q is the slope of the secant line and is given by:

$$m = \frac{f(x + h) - f(x)}{(x + h) - x}$$

$$= \frac{f(x + h) - f(x)}{h}\quad \text{(Difference quotient)}$$

Answers

3. a. $(n - m)(x) = x^2 - x - 12$;
Domain: $(-\infty, \infty)$

b. $(m \cdot p)(x) = (x + 3)\sqrt{x + 1}$;
Domain: $[-1, \infty)$

c. $\left(\dfrac{p}{n}\right)(x) = \dfrac{\sqrt{x + 1}}{x^2 - 9}$;
Domain: $[-1, 3) \cup (3, \infty)$

The expression on the right is called the **difference quotient** and is very important for the foundation of calculus. In Examples 4 and 5, we practice evaluating the difference quotient for two functions.

EXAMPLE 4   **Finding a Difference Quotient**

Given $f(x) = 3x - 5$,

  a. Find $f(x + h)$.

  b. Find the difference quotient, $\dfrac{f(x + h) - f(x)}{h}$.

Solution:

  a. $f(x + h) = 3(x + h) - 5$ Substitute $(x + h)$ for x.

  $= 3x + 3h - 5$

  b. $\dfrac{f(x + h) - f(x)}{h} = \dfrac{\overbrace{(3x + 3h - 5)}^{f(x+h)} - \overbrace{(3x - 5)}^{f(x)}}{h}$

  $= \dfrac{3x + 3h - 5 - 3x + 5}{h}$ Clear parentheses.

  $= \dfrac{3h}{h}$ Combine like terms.

  $= 3$ Simplify the fraction.

Skill Practice 4   Given $f(x) = 4x - 2$,

  a. Find $f(x + h)$.

  b. Find the difference quotient, $\dfrac{f(x + h) - f(x)}{h}$.

EXAMPLE 5   **Finding a Difference Quotient**

Given $f(x) = -2x^2 + 4x - 1$,

  a. Find $f(x + h)$.

  b. Find the difference quotient, $\dfrac{f(x + h) - f(x)}{h}$.

Solution:

  a. $f(x + h) = -2(x + h)^2 + 4(x + h) - 1$ Substitute $(x + h)$ for x.

  $= -2(x^2 + 2xh + h^2) + 4x + 4h - 1$

  $= -2x^2 - 4xh - 2h^2 + 4x + 4h - 1$

  b. $\dfrac{f(x + h) - f(x)}{h} = \dfrac{\overbrace{(-2x^2 - 4xh - 2h^2 + 4x + 4h - 1)}^{f(x+h)} - \overbrace{(-2x^2 + 4x - 1)}^{f(x)}}{h}$

  $= \dfrac{-2x^2 - 4xh - 2h^2 + 4x + 4h - 1 + 2x^2 - 4x + 1}{h}$ Clear parentheses.

  $= \dfrac{-4xh - 2h^2 + 4h}{h}$ Combine like terms.

  $= \dfrac{\overset{1}{\cancel{h}}(-4x - 2h + 4)}{\cancel{h}}$ Factor numerator and denominator, and simplify the fraction.

  $= -4x - 2h + 4$

Answers
4. a. $4x + 4h - 2$ **b.** 4

> **Skill Practice 5**   Given $f(x) = -x^2 - 5x + 2$,
>
> **a.** Find $f(x + h)$.
>
> **b.** Find the difference quotient, $\dfrac{f(x + h) - f(x)}{h}$.

3. Compose and Decompose Functions

The next operation on functions we present is called the composition of functions. Informally, this involves a substitution process in which the output from one function becomes the input to another function.

TIP It is important to note that the notation $(f \circ g)(x)$ represents the composition of functions, *not* multiplication of f, g, and x.

> **Composition of Functions**
>
> The **composition of f and g,** denoted $f \circ g$ is defined by $(f \circ g)(x) = f(g(x))$. The domain of $f \circ g$ is the set of real numbers x in the domain of g such that $g(x)$ is in the domain of f.

To visualize the composition of functions $(f \circ g)(x) = f(g(x))$, consider Figure 2-41.

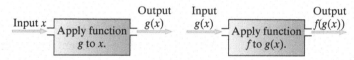

Figure 2-41

> **EXAMPLE 6   Composing Functions**
>
> Given $f(x) = x^2 + 2x$ and $g(x) = x - 4$, find
>
> **a.** $f(g(6))$ **b.** $g(f(-3))$ **c.** $(f \circ g)(0)$ **d.** $(g \circ f)(5)$
>
> **Solution:**
>
> **a.** $f(g(6)) = f(2)$ Evaluate $g(6)$ first. $g(6) = 6 - 4 = 2$.
>
> $\quad\quad\quad\quad = 8$ $f(2) = (2)^2 + 2(2) = 8$
>
> **b.** $g(f(-3)) = g(3)$ Evaluate $f(-3)$ first. $f(-3) = (-3)^2 + 2(-3) = 3$.
>
> $\quad\quad\quad\quad = -1$ $g(3) = 3 - 4 = -1$
>
> **c.** $(f \circ g)(0) = f(g(0))$ Evaluate $g(0)$ first. $g(0) = 0 - 4 = -4$.
>
> $\quad\quad\quad\quad = f(-4)$ $f(-4) = (-4)^2 + 2(-4) = 8$
>
> $\quad\quad\quad\quad = 8$
>
> **d.** $(g \circ f)(5) = g(f(5))$ Evaluate $f(5)$ first. $f(5) = (5)^2 + 2(5) = 35$.
>
> $\quad\quad\quad\quad = g(35)$ $g(35) = 35 - 4 = 31$
>
> $\quad\quad\quad\quad = 31$

TIP When composing functions, apply the order of operations. In Example 6(a), the value of $g(6)$ is found first.

> **Skill Practice 6**   Refer to functions f and g given in Example 6. Find
>
> **a.** $f(g(-4))$ **b.** $g(f(-5))$ **c.** $(f \circ g)(9)$ **d.** $(g \circ f)(10)$

Answers

5. a. $-x^2 - 2xh - h^2 - 5x - 5h + 2$
   b. $-2x - h - 5$

6. a. 48   **b.** 11   **c.** 35
   d. 116

In Example 7, we practice composing functions and identifying the domain of the composite function. This example also illustrates that function composition is not commutative. That is, $(f \circ g)(x) \neq (g \circ f)(x)$ for all functions f and g.

EXAMPLE 7   **Composing Functions and Determining Domain**

Given $f(x) = 2x - 6$ and $g(x) = \frac{1}{x + 4}$, write a rule for each function and write the domain in interval notation.

a. $(f \circ g)(x)$ **b.** $(g \circ f)(x)$

Solution:

a. $(f \circ g)(x) = f(g(x)) = 2(g(x)) - 6$

$= 2\left(\dfrac{1}{x + 4}\right) - 6$

$= \dfrac{2}{x + 4} - 6$ provided $x \neq -4$

Function g has the restriction that $x \neq -4$.

The domain of f is all real numbers. Therefore, no further restrictions need to be imposed.

The domain is $(-\infty, -4) \cup (-4, \infty)$.

b. $(g \circ f)(x) = g(f(x)) = \dfrac{1}{f(x) + 4}$

$= \dfrac{1}{(2x - 6) + 4}$ $f(x) \neq -4$

$= \dfrac{1}{2x - 2}$ provided $x \neq 1$

The domain is $(-\infty, 1) \cup (1, \infty)$.

The domain of f has no restrictions.

However, function g must not have an input value of -4. Therefore, we have the restriction $f(x) \neq -4$. Thus,

$2x - 6 = -4$
$2x = 2$
$x = 1$ must be excluded.

Skill Practice 7   Given $f(x) = 3x + 4$ and $g(x) = \frac{1}{x - 1}$, write a rule for each function and write the domain in interval notation.

a. $(f \circ g)(x)$ **b.** $(g \circ f)(x)$

EXAMPLE 8   **Composing Functions and Determining Domain**

Given $m(x) = \frac{1}{x - 5}$ and $p(x) = \sqrt{x - 2}$, find $(m \circ p)(x)$ and write the domain in interval notation.

Solution:

$(m \circ p)(x) = m(p(x)) = \dfrac{1}{p(x) - 5}$

$p(x) \neq 5$

$= \dfrac{1}{\sqrt{x - 2} - 5}$

$(m \circ p)(x) = \dfrac{1}{\sqrt{x - 2} - 5}$

The domain is $[2, 27) \cup (27, \infty)$.

First note that function p has the restriction that $x \geq 2$.

The input value for function m must not be 5. Therefore, $p(x) \neq 5$. We have

$\sqrt{x - 2} \neq 5$
$(\sqrt{x - 2})^2 \neq (5)^2$
$x - 2 \neq 25$
$x \neq 27$

Skill Practice 8   Given $f(x) = \sqrt{x - 1}$ and $g(x) = \frac{1}{x - 3}$, find $(g \circ f)(x)$ and write the domain of $g \circ f$ in interval notation.

Answers

7. a. $(f \circ g)(x) = \dfrac{3}{x - 1} + 4$;
   Domain: $(-\infty, 1) \cup (1, \infty)$

b. $(g \circ f)(x) = \dfrac{1}{3x + 3}$;
   Domain: $(-\infty, -1) \cup (-1, \infty)$

8. $(g \circ f)(x) = \dfrac{1}{\sqrt{x - 1} - 3}$;
   Domain: $[1, 10) \cup (10, \infty)$

EXAMPLE 9   Composing Functions and Determining Domain

Given $k(x) = \dfrac{x}{x-2}$ and $m(x) = \dfrac{6}{x^2-1}$, find $(k \circ m)(x)$ and write the domain in interval notation.

Solution:

$(k \circ m)(x) = k(m(x))$ Evaluate k at $m(x)$.

$= \dfrac{\left(\dfrac{6}{x^2-1}\right)}{\left(\dfrac{6}{x^2-1}\right) - 2}$ Substitute $\dfrac{6}{x^2-1}$ for x in $k(x)$.

                                             • m has the restriction that $x^2 - 1 \neq 0$. Therefore, $x \neq \pm 1$.

$= \dfrac{(x^2-1)}{(x^2-1)} \cdot \dfrac{\left(\dfrac{6}{x^2-1}\right)}{\left(\dfrac{6}{x^2-1} - 2\right)}$ Simplify the complex fraction by multiplying numerator and denominator by the LCD $x^2 - 1$.

$= \dfrac{6}{6 - 2(x^2-1)}$ Apply the distributive property.

$= \dfrac{6}{-2x^2 + 8}$ Simplify the denominator.

$= -\dfrac{6}{2x^2 - 8}$ Factor out -1 from the denominator.

$= -\dfrac{3}{x^2 - 4}$ Simplify the rational expression.

                                             • Note the added restriction that $x^2 - 4 \neq 0$ which means that $x \neq \pm 2$.

The domain is $(-\infty, -2) \cup (-2, -1) \cup (-1, 1) \cup (1, 2) \cup (2, \infty)$.

Skill Practice 9   Given $r(x) = \dfrac{x}{x+1}$ and $t(x) = \dfrac{5}{x^2-9}$, find $(r \circ t)(x)$ and write the domain in interval notation.

EXAMPLE 10   Applying Function Composition

At a popular website, the cost to download individual songs is \$1.49 per song. In addition, a first-time visitor to the website has a one-time coupon for \$1.00 off.

a. Write a function to represent the cost $C(x)$ (in \$) for a first-time visitor to purchase x songs.

b. The sales tax for online purchases depends on the location of the business and customer. If the sales tax rate on a purchase is 6%, write a function to represent the total cost $T(a)$ for a first-time visitor who buys a dollars in songs.

c. Find $(T \circ C)(x)$ and interpret the meaning in context.

d. Evaluate $(T \circ C)(10)$ and interpret the meaning in context.

Answer

9. $(r \circ t)(x) = \dfrac{5}{x^2-4}$;

   Domain: $(-\infty, -3) \cup (-3, -2) \cup$
   $(-2, 2) \cup (2, 3) \cup (3, \infty)$

Solution:

a. $C(x) = 1.49x - 1.00;\ x \geq 1$ The cost function is a linear function with \$1.49 as the variable rate per song.

b. $T(a) = a + 0.06a$ The total cost is the sum of the cost of the songs plus the sales tax.

    $= 1.06a$

c. $(T \circ C)(x) = T(C(x))$

$\qquad\qquad\quad = 1.06(C(x))$

$\qquad\qquad\quad = 1.06(1.49x - 1.00)$ $\qquad$ Substitute $1.49x - 1.00$ for $C(x)$.

$\qquad\qquad\quad = 1.5794x - 1.06$

$(T \circ C)(x) = 1.5794x - 1.06$ represents the total cost to buy x songs for a first-time visitor to the website.

d. $(T \circ C)(x) = 1.5794x - 1.06$

$\quad\ (T \circ C)(10) = 1.5794(10) - 1.06$

$\qquad\qquad\qquad = 14.734$

The total cost for a first-time visitor to buy 10 songs is $14.73.

Skill Practice 10   An artist shops online for tubes of watercolor paint. The cost is $16 for each 14-mL tube.

a. Write a function representing the cost $C(x)$ (in $) for x tubes of paint.

b. There is a 5.5% sales tax on the cost of merchandise and a fixed cost of $4.99 for shipping. Write a function representing the total cost $T(a)$ for a dollars spent in merchandise.

c. Find $(T \circ C)(x)$ and interpret the meaning in context.

d. Evaluate $(T \circ C)(18)$ and interpret the meaning in context.

TIP The decomposition of functions is not unique. For example, $h(x) = (x - 3)^2$ can also be written as $h(x) = f(g(x))$, where $g(x) = x^2 - 6x$ and $f(x) = x + 9$.

The composition of two functions creates a new function in which the output from one function becomes the input to the other. We can also reverse this process. That is, we can decompose a composite function into two or more simpler functions.

For example, consider the function h defined by $h(x) = (x - 3)^2$. To write h as a composition of two functions, we have $h(x) = (f \circ g)(x) = f(g(x))$. The function g is the "inside" function and f is the "outside" function. So one natural choice for g and f would be:

$\qquad g(x) = x - 3$ $\qquad$ Function g subtracts 3 from the input value.

$\qquad f(x) = x^2$ $\qquad$ Function f squares the result.

$\qquad h(x) = f(g(x)) = (g(x))^2 = (x - 3)^2$

EXAMPLE 11   **Decomposing Two Functions**

Given $h(x) = |2x^2 - 5|$, find two functions f and g such that $h(x) = (f \circ g)(x)$.

Solution:

We need to find two functions f and g such that $h(x) = (f \circ g)(x) = f(g(x))$. The function h first evaluates the expression $2x^2 - 5$, and then takes the absolute value. Therefore, it would be natural to take the absolute value of $g(x) = 2x^2 - 5$.

$\qquad$ We have: $g(x) = 2x^2 - 5$ and $f(x) = |x|$

$\qquad$ <u>Check:</u> $\qquad h(x) = (f \circ g)(x) = f(g(x)) = |g(x)|$

$\qquad\qquad\qquad\qquad\qquad\qquad\quad = |2x^2 - 5|$

Answers

10. a. $C(x) = 16x$

$\quad$ **b.** $T(a) = 1.055a + 4.99$

$\quad$ **c.** $(T \circ C)(x) = 16.88x + 4.99$ represents the total cost to buy x tubes of paint.

$\quad$ **d.** $(T \circ C)(18) = \$308.83$; The total cost to buy 18 tubes of paint is $308.83.

Skill Practice 11 Given $m(x) = \sqrt[3]{5x + 1}$, find two functions f and g such that $m(x) = (f \circ g)(x)$.

In Example 12, we have the graphs of two functions, and we apply function addition, subtraction, multiplication, and composition for selected values of x.

EXAMPLE 12 Estimating Function Values from a Graph

The graphs of f and g are shown. Evaluate the functions at the given values of x if possible.

a. $(f + g)(1)$
b. $(fg)(0)$
c. $(g - f)(-3)$
d. $(f \circ g)(3)$
e. $(g \circ f)(4)$
f. $f(g(1))$

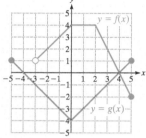

Solution:

a. $(f + g)(1) = f(1) + g(1)$
$= 4 + (-3)$
$= 1$

b. $(fg)(0) = f(0) \cdot g(0)$
$= (4)(-4)$
$= -16$

c. $(g - f)(-3) = g(-3) - f(-3)$ ⟵ $f(-3)$ is undefined.
$(g - f)(-3)$ is undefined.

d. $(f \circ g)(3) = f(g(3))$
$= f(-1)$
$= 3$

e. $(g \circ f)(4) = g(f(4))$
$= g(0)$
$= -4$

f. $f(g(1)) = f(-3)$ is undefined.

The open dot at $(-3, 1)$ indicates that -3 is not in the domain of f. The value $g(1) = -3$, but $f(-3)$ is undefined. Therefore, $f(g(1))$ is undefined.

Skill Practice 12 Refer to the functions f and g pictured in Example 12. Evaluate the functions at the given values of x if possible.

a. $(f - g)(-2)$
b. $\left(\dfrac{f}{g}\right)(3)$
c. $(gf)(5)$
d. $(g \circ f)(5)$
e. $(f \circ g)(5)$
f. $f(g(0))$

Answers
11. $g(x) = 5x + 1$ and $f(x) = \sqrt[3]{x}$
12. a. 4   **b.** −2   **c.** −2
      d. −2   **e.** 4   **f.** Undefined

SECTION 2.8   Practice Exercises

Prerequisite Review

For Exercises R.1–R.4, write the domain in interval notation.

R.1. $f(x) = \dfrac{x - 1}{x + 2}$

R.2. $r(x) = \sqrt{x + 3}$

R.3. $h(x) = \dfrac{4}{\sqrt{3 - x}}$

R.4. $p(x) = 2x^2 - 3x + 1$

R.5. Given $k(x) = x^2 - 2x + 3$, find $k(x + 3)$.

Concept Connections

1. The function $f + g$ is defined by $(f + g)(x) = $ _____ + _____.

2. The function $\dfrac{f}{g}$ is defined by $\left(\dfrac{f}{g}\right)(x) = $ _____ provided that _____ $\neq 0$.

3. Let h represent a positive real number. Given a function defined by $y = f(x)$, the difference quotient is given by _____.

4. The composition of f and g, denoted by $f \circ g$, is defined by $(f \circ g)(x) = $ _____.

Objective 1: Perform Operations on Functions

For Exercises 5–8, find $(f + g)(x)$ and identify the graph of $f + g$. (See Example 1)

5. $f(x) = |x|$ and $g(x) = 3$

6. $f(x) = |x|$ and $g(x) = -4$

7. $f(x) = x^2$ and $g(x) = -4$

8. $f(x) = x^2$ and $g(x) = 3$

a.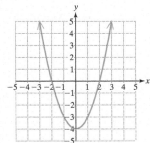

b.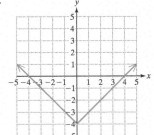

c.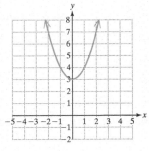

d.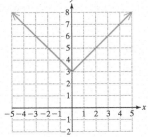

For Exercises 9–18, evaluate the functions for the given values of x. (See Example 2)

$$f(x) = -2x \qquad g(x) = |x + 4| \qquad h(x) = \dfrac{1}{x - 3}$$

9. $(f - g)(3)$

10. $(g - h)(2)$

11. $(f \cdot g)(-1)$

12. $(h \cdot g)(4)$

13. $(g + h)(0)$

14. $(f + h)(5)$

15. $\left(\dfrac{f}{g}\right)(8)$

16. $\left(\dfrac{h}{f}\right)(7)$

17. $\left(\dfrac{g}{f}\right)(0)$

18. $\left(\dfrac{h}{g}\right)(-4)$

For Exercises 19–26, refer to the functions r, p, and q. Find the indicated function and write the domain in interval notation. (See Example 3)

$$r(x) = -3x \qquad p(x) = x^2 + 3x \qquad q(x) = \sqrt{1 - x}$$

19. $(r - p)(x)$

20. $(p - r)(x)$

21. $(p \cdot q)(x)$

22. $(r \cdot q)(x)$

23. $\left(\dfrac{q}{p}\right)(x)$

24. $\left(\dfrac{q}{r}\right)(x)$

25. $\left(\dfrac{p}{q}\right)(x)$

26. $\left(\dfrac{r}{q}\right)(x)$

For Exercises 27–32, refer to functions s, t, and v. Find the indicated function and write the domain in interval notation. (See Example 3)

$$s(x) = \dfrac{x - 2}{x^2 - 9} \qquad t(x) = \dfrac{x - 3}{2 - x} \qquad v(x) = \sqrt{x + 3}$$

27. $(s \cdot t)(x)$

28. $\left(\dfrac{s}{t}\right)(x)$

29. $(s + t)(x)$

30. $(s - t)(x)$

31. $(s \cdot v)(x)$

32. $\left(\dfrac{v}{s}\right)(x)$

Objective 2: Evaluate a Difference Quotient

For Exercises 33–36, a function is given. (See Examples 4–5)

a. Find $f(x + h)$.

b. Find $\dfrac{f(x + h) - f(x)}{h}$.

33. $f(x) = 5x + 9$

34. $f(x) = 8x + 4$

35. $f(x) = x^2 + 4x$

36. $f(x) = x^2 - 3x$

For Exercises 37–44, find the difference quotient and simplify. (See Examples 4–5)

37. $f(x) = -2x + 5$ **38.** $f(x) = -3x + 8$ **39.** $f(x) = -5x^2 - 4x + 2$ **40.** $f(x) = -4x^2 - 2x + 6$

41. $f(x) = x^3 + 5$ **42.** $f(x) = x^3 - 2$ **43.** $f(x) = \dfrac{1}{x}$ **44.** $f(x) = \dfrac{1}{x + 2}$

45. Given $f(x) = 4\sqrt{x}$,
 a. Find the difference quotient (do not simplify).
 b. Evaluate the difference quotient for $x = 1$, and the following values of h: $h = 1$, $h = 0.1$, $h = 0.01$, and $h = 0.001$. Round to 4 decimal places.
 c. What value does the difference quotient seem to be approaching as h gets close to 0?

46. Given $f(x) = \dfrac{12}{x}$,
 a. Find the difference quotient (do not simplify).
 b. Evaluate the difference quotient for $x = 2$, and the following values of h: $h = 0.1$, $h = 0.01$, $h = 0.001$, and $h = 0.0001$. Round to 4 decimal places.
 c. What value does the difference quotient seem to be approaching as h gets close to 0?

Objective 3: Compose and Decompose Functions

For Exercises 47–62, refer to functions f, g, and h. Evaluate the functions for the given values of x. (See Example 6)

$$f(x) = x^3 - 4x \qquad g(x) = \sqrt{2x} \qquad h(x) = 2x + 3$$

47. $f(g(8))$ **48.** $h(g(2))$ **49.** $h(f(1))$ **50.** $g(f(3))$

51. $(f \circ g)(18)$ **52.** $(f \circ h)(-1)$ **53.** $(g \circ f)(5)$ **54.** $(h \circ f)(-2)$

55. $(h \circ f)(-3)$ **56.** $(h \circ g)(72)$ **57.** $(g \circ f)(1)$ **58.** $(g \circ f)(-4)$

59. $(f \circ f)(3)$ **60.** $(h \circ h)(-4)$ **61.** $(f \circ h \circ g)(2)$ **62.** $(f \circ h \circ g)(8)$

63. Given $f(x) = 2x + 4$ and $g(x) = x^2$,
 a. Find $(f \circ g)(x)$. **b.** Find $(g \circ f)(x)$.
 c. Is the operation of function composition commutative?

64. Given $k(x) = -3x + 1$ and $m(x) = \dfrac{1}{x}$,
 a. Find $(k \circ m)(x)$. **b.** Find $(m \circ k)(x)$.
 c. Is $(k \circ m)(x) = (m \circ k)(x)$?

For Exercises 65–76, refer to functions m, n, p, q, and r. Find the indicated function and write the domain in interval notation. (See Examples 7–9)

$$m(x) = \sqrt{x + 8} \qquad n(x) = x - 5 \qquad p(x) = x^2 - 9x \qquad q(x) = \dfrac{1}{x - 10} \qquad r(x) = |2x + 3|$$

65. $(n \circ p)(x)$ **66.** $(p \circ n)(x)$ **67.** $(m \circ n)(x)$ **68.** $(n \circ m)(x)$

69. $(q \circ n)(x)$ **70.** $(q \circ p)(x)$ **71.** $(q \circ r)(x)$ **72.** $(q \circ m)(x)$

73. $(n \circ r)(x)$ **74.** $(r \circ n)(x)$ **75.** $(q \circ q)(x)$ **76.** $(p \circ p)(x)$

For Exercises 77–80, find $(f \circ g)(x)$ and write the domain in interval notation. (See Example 9)

77. $f(x) = \dfrac{3}{x^2 - 16}, g(x) = \sqrt{2 - x}$

78. $f(x) = \dfrac{4}{x^2 - 9}, g(x) = \sqrt{3 - x}$

79. $f(x) = \dfrac{x}{x - 1}, g(x) = \dfrac{9}{x^2 - 16}$

80. $f(x) = \dfrac{x}{x + 4}, g(x) = \dfrac{3}{x^2 - 1}$

81. Given $f(x) = \dfrac{1}{x - 2}$, find $(f \circ f)(x)$ and write the domain in interval notation.

82. Given $g(x) = \sqrt{x - 3}$, find $(g \circ g)(x)$ and write the domain in interval notation.

For Exercises 83–86, find the indicated functions.

$$f(x) = 2x + 1 \qquad g(x) = x^2 \qquad h(x) = \sqrt[3]{x}$$

83. $(f \circ g \circ h)(x)$ **84.** $(g \circ f \circ h)(x)$

85. $(h \circ g \circ f)(x)$ **86.** $(g \circ h \circ f)(x)$

87. A law office orders business stationery. The cost is $21.95 per box. **(See Example 10)**

   a. Write a function that represents the cost $C(x)$ (in $) for x boxes of stationery.

   b. There is a 6% sales tax on the cost of merchandise and $10.99 for shipping. Write a function that represents the total cost $T(a)$ for a dollars spent in merchandise and shipping.

   c. Find $(T \circ C)(x)$.

   d. Find $(T \circ C)(4)$ and interpret its meaning in the context of this problem.

88. The cost to buy tickets online for a dance show is $60 per ticket.

   a. Write a function that represents the cost $C(x)$ (in $) for x tickets to the show.

   b. There is a sales tax of 5.5% and a processing fee of $8.00 for a group of tickets. Write a function that represents the total cost $T(a)$ for a dollars spent on tickets.

   c. Find $(T \circ C)(x)$.

   d. Find $(T \circ C)(6)$ and interpret its meaning in the context of this problem.

89. A bicycle wheel turns at a rate of 80 revolutions per minute (rpm).

   a. Write a function that represents the number of revolutions $r(t)$ in t minutes.

   b. For each revolution of the wheels, the bicycle travels 7.2 ft. Write a function that represents the distance traveled $d(r)$ (in ft) for r revolutions of the wheel.

   c. Find $(d \circ r)(t)$ and interpret the meaning in the context of this problem.

   d. Evaluate $(d \circ r)(30)$ and interpret the meaning in the context of this problem.

90. While on vacation in France, Sadie bought a box of almond croissants. Each croissant cost €2.40 (euros).

   a. Write a function that represents the cost $C(x)$ (in euros) for x croissants.

   b. At the time of the purchase, the exchange rate was $1 = €0.80. Write a function that represents the amount $D(C)$ (in $) for C euros spent.

   c. Find $(D \circ C)(x)$ and interpret the meaning in the context of this problem.

   d. Evaluate $(D \circ C)(12)$ and interpret the meaning in the context of this problem.

For Exercises 91–98, find two functions f and g such that $h(x) = (f \circ g)(x)$. (See Example 11)

91. $h(x) = (x + 7)^2$

92. $h(x) = (x - 8)^2$

93. $h(x) = \sqrt[3]{2x + 1}$

94. $h(x) = \sqrt[4]{9x - 5}$

95. $h(x) = |2x^2 - 3|$

96. $h(x) = |4 - x^2|$

97. $h(x) = \dfrac{5}{x + 4}$

98. $h(x) = \dfrac{11}{x - 3}$

Mixed Exercises

For Exercises 99–102, the graphs of two functions are shown. Evaluate the function at the given values of x, if possible. (See Example 12)

99. a. $(f + g)(0)$

   b. $(g - f)(2)$

   c. $(g \cdot f)(-1)$

   d. $\left(\dfrac{g}{f}\right)(1)$

   e. $(f \circ g)(4)$

   f. $(g \circ f)(0)$

   g. $g(f(4))$

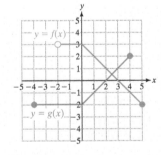

100. a. $(f + g)(0)$

   b. $(g - f)(1)$

   c. $(g \cdot f)(2)$

   d. $\left(\dfrac{f}{g}\right)(-3)$

   e. $(f \circ g)(3)$

   f. $(g \circ f)(0)$

   g. $g(f(-4))$

101. a. $(h + k)(-1)$

   b. $(h \cdot k)(4)$

   c. $\left(\dfrac{k}{h}\right)(-3)$

   d. $(k - h)(1)$

   e. $(k \circ h)(4)$

   f. $(h \circ k)(-2)$

   g. $h(k(3))$

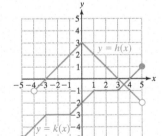

102. a. $(m + p)(1)$

   b. $(p - m)(-4)$

   c. $\left(\dfrac{m}{p}\right)(3)$

   d. $(m \cdot p)(3)$

   e. $(m \circ p)(0)$

   f. $(p \circ m)(0)$

   g. $p(m(-4))$

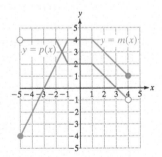

For Exercises 103–110, refer to the functions f and g and evaluate the functions for the given values of x.

$$f = \{(2, 4), (6, -1), (4, -2), (0, 3), (-1, 6)\} \quad \text{and} \quad g = \{(4, 3), (0, 6), (5, 7), (6, 0)\}$$

103. $(f + g)(4)$

104. $(g \cdot f)(0)$

105. $(g \circ f)(2)$

106. $(f \circ g)(0)$

107. $(g \circ g)(6)$

108. $(f \circ f)(-1)$

109. $(f \circ g)(5)$

110. $(g \circ f)(0)$

111. The diameter d of a sphere is twice the radius r. The volume of the sphere as a function of its radius is given by $V(r) = \dfrac{4}{3}\pi r^3$.

   a. Write the diameter d of the sphere as a function of the radius r.

   b. Write the radius r as a function of the diameter d.

   c. Find $(V \circ r)(d)$ and interpret its meaning.

112. Consider a right circular cone with given height h. The volume of the cone as a function of its radius r is given by $V(r) = \dfrac{1}{3}\pi r^2 h$. Consider a right circular cone with fixed height $h = 6$ in.

   a. Write the diameter d of the cone as a function of the radius r.

   b. Write the radius r as a function of the diameter d.

   c. Find $(V \circ r)(d)$ and interpret its meaning. Assume that $h = 6$ in.

113. An investment earns 4.5% interest paid at the end of 1 yr. If x is the amount of money initially invested, then $A(x) = 1.045x$ represents the amount of money in the account 1 yr later. Find $(A \circ A)(x)$ and interpret the result.

114. The population in a certain town has been decreasing at a rate of 2% per year. If x is the population at a certain fixed time, then $P(x) = 0.98x$ represents the population 1 yr later. Find $(P \circ P)(x)$ and interpret the result.

115. Suppose that a function H gives the high temperature $H(x)$ (in °F) for day x. Suppose that a function L gives the low temperature $L(x)$ (in °F) for day x. What does $\left(\dfrac{H + L}{2}\right)(x)$ represent?

116. For the given figure,

   a. What does $A_1(x) = \pi(x + 5)^2$ represent?

   b. What does $A_2(x) = \pi x^2$ represent?

   c. Find $(A_1 - A_2)(x)$ and interpret its meaning.

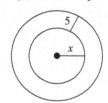

117. For the given figure,

   a. Write an expression $S_1(x)$ that represents the area of the rectangle.

   b. Write an expression $S_2(x)$ that represents the area of the semicircle.

   c. Find $(S_1 - S_2)(x)$ and interpret its meaning.

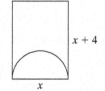

Write About It

118. Given functions f and g, explain how to determine the domain of $\left(\dfrac{f}{g}\right)(x)$.

119. Given functions f and g, explain how to determine the domain of $(f \circ g)(x)$.

120. Explain what the difference quotient represents.

Expanding Your Skills

121. Given $f(x) = \sqrt{x + 3}$,

   a. Find the difference quotient.

   b. Rationalize the numerator of the expression in part (a) and simplify.

   c. Evaluate the expression in part (b) for $h = 0$.

122. Given $f(x) = \sqrt{x - 4}$,

   a. Find the difference quotient.

   b. Rationalize the numerator of the expression in part (a) and simplify.

   c. Evaluate the expression in part (b) for $h = 0$.

123. A car traveling 60 mph (88 ft/sec) undergoes a constant deceleration until it comes to rest approximately 9.09 sec later. The distance $d(t)$ (in ft) that the car travels t seconds after the brakes are applied is given by $d(t) = -4.84t^2 + 88t$, where $0 \le t \le 9.09$. (**See Example 5**)

   a. Find the difference quotient $\dfrac{d(t + h) - d(t)}{h}$.

   Use the difference quotient to determine the average rate of speed on the following intervals for t.

   b. $[0, 2]$ (*Hint*: $t = 0$ and $h = 2$)

   c. $[2, 4]$ (*Hint*: $t = 2$ and $h = 2$)

   d. $[4, 6]$ (*Hint*: $t = 4$ and $h = 2$)

   e. $[6, 8]$ (*Hint*: $t = 6$ and $h = 2$)

124. A car accelerates from 0 to 60 mph (88 ft/sec) in 8.8 sec. The distance $d(t)$ (in ft) that the car travels t seconds after motion begins is given by $d(t) = 5t^2$, where $0 \le t \le 8.8$.

   a. Find the difference quotient $\dfrac{d(t + h) - d(t)}{h}$.

   Use the difference quotient to determine the average rate of speed on the following intervals for t.

   b. $[0, 2]$

   c. $[2, 4]$

   d. $[4, 6]$

   e. $[6, 8]$

125. If a is b plus eight, and c is the square of a, write c as a function of b.

126. If q is r minus seven, and s is the square root of q, write s as a function of r.

127. If x is twice y, and z is four less than x, write z as a function of y.

128. If m is one-third of n, and p is two less than m, write p as a function of n.

129. Given $f(x) = \sqrt[3]{4x^2 + 1}$, define functions m, n, h, and k such that $f(x) = (m \circ n \circ h \circ k)(x)$.

130. Given $f(x) = |-2x^3 - 4|$, define functions m, n, h, and k such that $f(x) = (m \circ n \circ h \circ k)(x)$.

CHAPTER 2   KEY CONCEPTS

SECTION 2.1 The Rectangular Coordinate System and Graphing Utilities	Reference
The **distance** between two points (x_1, y_1) and (x_2, y_2) in a rectangular coordinate system is given by $$d = \sqrt{(x_2 - x_1)^2 + (y_2 - y_1)^2}.$$	p. 167
The **midpoint** between the points is given by $M = \left(\dfrac{x_1 + x_2}{2}, \dfrac{y_1 + y_2}{2} \right)$.	p. 168
• To find an x-intercept $(a, 0)$ of the graph of an equation, substitute 0 for y and solve for x. • To find a y-intercept $(0, b)$ of the graph of an equation, substitute 0 for x and solve for y.	p. 170

SECTION 2.2 Circles	Reference
The **standard form** of an equation of a circle with radius r and center (h, k) is $(x - h)^2 + (y - k)^2 = r^2$.	p. 178
An equation of a circle written in the form $x^2 + y^2 + Ax + By + C = 0$ is called the **general form** of an equation of a circle.	p. 179

SECTION 2.3 Functions and Relations	Reference
A set of ordered pairs (x, y) is called a **relation** in x and y. The set of x values is the **domain** of the relation, and the set of y values is the **range** of the relation.	p. 183
Given a relation in x and y, we say that **y is a function of x** if for each value of x in the domain, there is exactly one value of y in the range.	p. 184
The vertical line test tells us that the graph of a relation defines y as a function of x if no vertical line intersects the graph in more than one point.	p. 185
Given a function defined by $y = f(x)$, • The x-intercepts are the real solutions to $f(x) = 0$. • The y-intercept is given by $f(0)$.	p. 188
Given $y = f(x)$, the domain of f is the set of real numbers x that when substituted into the function produce a real number. This excludes • Values of x that make the denominator zero. • Values of x that make a radicand negative within an even-indexed root.	p. 189

SECTION 2.4 Linear Equations in Two Variables and Linear Functions	Reference
Let A, B, and C represent real numbers where A and B are not both zero. A **linear equation** in the variables x and y is an equation that can be written as $Ax + By = C$.	p. 197
The slope of a line passing through the distinct points (x_1, y_1) and (x_2, y_2) is given by $m = \dfrac{\Delta y}{\Delta x} = \dfrac{y_2 - y_1}{x_2 - x_1}$.	p. 199
Given a line with slope m and y-intercept $(0, b)$, the **slope-intercept form** of the line is given by $y = mx + b$.	p. 207

If f is defined on the interval $[x_1, x_2]$, then the **average rate of change** of f on the interval $[x_1, x_2]$ is the slope of the secant line containing $(x_1, f(x_1))$ and $(x_2, f(x_2))$ and is given by $$m = \frac{f(x_2) - f(x_1)}{x_2 - x_1}$$	p. 203
The x-coordinates of the points of intersection between the graphs of $y = f(x)$ and $y = g(x)$ are the solutions to the equation $f(x) = g(x)$.	p. 204

SECTION 2.5 Applications of Linear Equations and Modeling	Reference
The **point-slope formula** for a line is given by $y - y_1 = m(x - x_1)$ where m is the slope of the line and (x_1, y_1) is a point on the line.	p. 213
• If m_1 and m_2 represent the slopes of two nonvertical parallel lines, then $m_1 = m_2$. • If m_1 and m_2 represent the slopes of two nonvertical perpendicular lines, then $m_1 = -\dfrac{1}{m_2}$ or equivalently $m_1 m_2 = -1$.	p. 214
In many-day-to-day applications, two variables are related linearly.	p. 219
• A linear model can be made from two data points that represent the general trend of the data. • Alternatively, the least-squares regression line is a model that utilizes *all* observed data points.	p. 220

SECTION 2.6 Transformations of Graphs	Reference
Consider a function defined by $y = f(x)$. Let h, k, and a represent positive real numbers. The graphs of the following functions are related to $y = f(x)$ as follows.	pp. 230–231
• Vertical translation (shift) $\quad y = f(x) + k \quad$ Shift upward $\quad y = f(x) - k \quad$ Shift downward • Horizontal translation (shift) $\quad y = f(x - h) \quad$ Shift to the right $\quad y = f(x + h) \quad$ Shift to the left	
• Vertical stretch/shrink $\quad y = af(x) \quad$ Vertical stretch (if $a > 1$) $\qquad\qquad\quad$ Vertical shrink (if $0 < a < 1$) • Horizontal stretch/shrink $\quad y = f(ax) \quad$ Horizontal shrink (if $a > 1$) $\qquad\qquad\quad$ Horizontal stretch (if $0 < a < 1$)	p. 233
• Reflection $\quad y = -f(x) \quad$ Reflection across the x-axis $\quad y = f(-x) \quad$ Reflection across the y-axis	p. 235
To graph a function requiring multiple transformations, use the following order. 1. Horizontal translation (shift) 2. Horizontal and vertical stretch and shrink 3. Reflections across the x- and y-axes 4. Vertical translation (shift)	p. 236

SECTION 2.7 Analyzing Graphs of Functions and Piecewise-Defined Functions	Reference
Consider the graph of an equation in x and y. • The graph of the equation is symmetric to the y-axis if substituting $-x$ for x results in an equivalent equation. • The graph of the equation is symmetric to the x-axis if substituting $-y$ for y results in an equivalent equation. • The graph of the equation is symmetric to the origin if substituting $-x$ for x and $-y$ for y results in an equivalent equation.	p. 243
• f is an even function if $f(-x) = f(x)$ for all x in the domain of f. • f is an odd function if $f(-x) = -f(x)$ for all x in the domain of f.	p. 245
To graph a piecewise-defined function, graph each individual function on its domain.	p. 248

The **greatest integer function,** denoted by $f(x) = [\![x]\!]$ or $f(x) = \text{int}(x)$ or $f(x) = \text{floor}(x)$ defines $f(x)$ as the greatest integer less than or equal to x.	p. 249
Suppose that I is an interval contained within the domain of a function f. • f is increasing on I if $f(x_1) < f(x_2)$ for all $x_1 < x_2$ on I. • f is decreasing on I if $f(x_1) > f(x_2)$ for all $x_1 < x_2$ on I. • f is constant on I if $f(x_1) = f(x_2)$ for all x_1 and x_2 on I.	p. 252
• $f(a)$ is a **relative maximum** of f if there exists an open interval containing a such that $f(a) \geq f(x)$ for all x in the interval. • $f(b)$ is a **relative minimum** of f if there exists an open interval containing b such that $f(b) \leq f(x)$ for all x in the interval.	p. 253

SECTION 2.8 Algebra of Functions and Function Composition	Reference
Given functions f and g, the functions $f + g, f - g, f \cdot g$, and $\frac{f}{g}$ are defined by $$(f + g)(x) = f(x) + g(x)$$ $$(f - g)(x) = f(x) - g(x)$$ $$(f \cdot g)(x) = f(x) \cdot g(x)$$ $$\left(\frac{f}{g}\right)(x) = \frac{f(x)}{g(x)} \text{ provided that } g(x) \neq 0$$	p. 262
The **difference quotient** represents the average rate of change of a function f between two points $(x, f(x))$ and $(x + h, f(x + h))$. $$\frac{f(x + h) - f(x)}{h} \quad \text{Difference quotient}$$	p. 264
The **composition of f and g,** denoted $f \circ g$ is defined by $(f \circ g)(x) = f(g(x))$. The domain of $f \circ g$ is the set of real numbers x in the domain of g such that $g(x)$ is in the domain of f.	p. 266

Expanded Chapter Summary available at http://www.mhhe.com/millercollegealgebra.

CHAPTER 2   Review Exercises

SECTION 2.1

For Exercises 1–2,

a. Find the exact distance between the points.

b. Find the midpoint of the line segment whose endpoints are the given points.

1. $(-1, 8)$ and $(4, -2)$

2. $\left(\sqrt{3}, -\sqrt{6}\right)$ and $\left(3\sqrt{3}, 4\sqrt{6}\right)$

3. Determine if the given ordered pair is a solution to the equation $4|x - 1| + y = 18$.

    a. $(-3, 2)$ **b.** $(5, -2)$

For Exercises 4–6, determine the x- and y-intercepts of the graph of the equation.

4. $-3y + 4x = 6$

5. $x = |y + 7| - 3$

6. $\dfrac{(x + 4)^2}{9} + \dfrac{y^2}{4} = 1$

7. Graph the equation by plotting points. $y = x^2 - 2x$

8. Find the length of the diagonal shown.

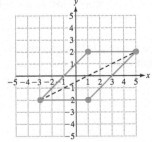

SECTION 2.2

For Exercises 9–10, determine the center and radius of the circle.

9. $(x - 4)^2 + (y + 3)^2 = 4$

10. $x^2 + \left(y - \dfrac{3}{2}\right)^2 = 17$

For Exercises 11–14, information about a circle is given.

a. Write an equation of the circle in standard form.

b. Graph the circle.

11. Center: $(-3, 1)$; Radius: $\sqrt{11}$

12. Center: $(0, 0)$; Radius: 3.2

13. Endpoints of a diameter $(7, 5)$ and $(1, -3)$

14. The center is in quadrant IV, the radius is 4, and the circle is tangent to both the x- and y-axes.

For Exercises 15–16, (a) write the equation of the circle in standard form and (b) identify the center and radius.

15. $x^2 + y^2 + 10x - 2y + 17 = 0$

16. $x^2 + y^2 - 8y + 3 = 0$

For Exercises 17–18, determine the solution set to the equation.

17. $(x + 3)^2 + (y - 5)^2 = 0$

18. $x^2 + y^2 + 6x - 4y + 15 = 0$

SECTION 2.3

19. The table lists four Olympic athletes and the number of Olympic medals won by the athlete.

Athlete (x)	Number of Medals (y)
Dara Torres (swimming)	12
Carl Lewis (track and field)	10
Bonnie Blair (speed skating)	6
Michael Phelps (swimming)	16

a. Write a set of ordered pairs (x, y) that defines the relation.

b. Write the domain of the relation.

c. Write the range of the relation.

d. Determine if the relation defines y as a function of x.

For Exercises 20–23, determine if the relation defines y as a function of x.

20.

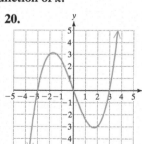

21.

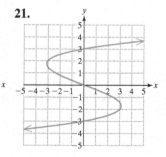

22. $x^2 + (y - 3)^2 = 4$ **23.** $x^2 + y - 3 = 4$

24. Evaluate $f(x) = -2x^2 + 4x$ for the values of x given.

   a. $f(0)$ **b.** $f(-1)$ **c.** $f(3)$

   d. $f(t)$ **e.** $f(x + 4)$

25. Given $f = \{(3, -1), (1, 5), (-2, 4), (0, 4)\}$,

   a. Determine $f(1)$.

   b. Determine $f(0)$.

   c. For what value(s) of x is $f(x) = -1$?

26. A department store marks up the price of a power drill by 32% of the price from the manufacturer. The price $P(x)$ (in \$) to a customer after a 6.5% sales tax is given by $P(x) = 1.065(x + 0.32x)$, where x is the cost of the drill from the manufacturer. Evaluate $P(189)$ and interpret the meaning in the context of this problem.

For Exercises 27–28, determine the x- and y-intercepts for the given function.

27. $p(x) = |x - 3| - 1$

28. $q(x) = -\sqrt{x} + 2$

For Exercises 29–30, determine the domain and range of the function.

29.

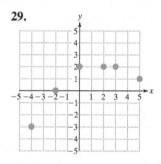

30.

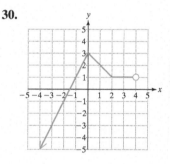

For Exercises 31–34, write the domain in interval notation.

31. $f(x) = \dfrac{x - 2}{x - 5}$ **32.** $g(x) = \dfrac{6}{|x| - 3}$

33. $m(x) = 2x^2 - 4x + 1$ **34.** $n(x) = \dfrac{10}{\sqrt{2 - x}}$

35. Use the graph of $y = f(x)$ to

   a. Determine $f(-2)$.

   b. Determine $f(3)$.

   c. Find all x for which $f(x) = -1$.

   d. Find all x for which $f(x) = -4$.

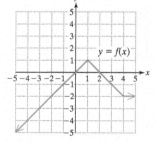

   e. Determine the x-intercept(s).

   f. Determine the y-intercept.

   g. Determine the domain of f.

   h. Determine the range of f.

36. Write a relationship for a function whose $f(x)$ value is 4 less than two times the square of x.

SECTION 2.4

For Exercises 37–40, graph the equation and determine the x- and y-intercepts.

37. $-2x + 4y = 8$
38. $-4x = 5y$
39. $y = 2$
40. $3x = 5$

For Exercises 41–43, determine the slope of the line passing through the given points.

41. $(4, -2)$ and $(-12, -4)$

42. $\left(-3, \dfrac{2}{3}\right)$ and $\left(1, -\dfrac{4}{3}\right)$

43. $(a, f(a))$ and $(b, f(b))$

44. What is the slope of a line parallel to the x-axis?

45. What is the slope of a line with equation $x = -2$?

46. What is the slope of a line perpendicular to a line with equation $y = 1$?

47. Suppose that $y = C(t)$ represents the average cost of a gallon of milk in the United States t years since 1980. What does $\dfrac{\Delta C}{\Delta t}$ represent?

48. Determine if the function is linear, constant, or neither.

a. $f(x) = -\dfrac{3}{2}x$ **b.** $g(x) = -\dfrac{3}{2x}$ **c.** $h(x) = -\dfrac{3}{2}$

For Exercises 49–50, use slope-intercept form to write an equation of the line that passes through the given point and has the given slope. Then write the equation using function notation where $y = f(x)$.

49. $(1, -5)$ and $m = -\dfrac{2}{3}$ **50.** $\left(2, \dfrac{1}{4}\right)$ and $m = 0$

51. Find the slope of the secant line pictured in red.

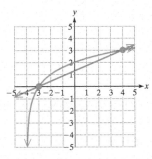

52. The function given by $y = f(x)$ shows the value of $8000 invested at 6% interest compounded continuously, x years after the money was originally invested.

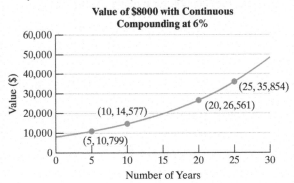

Value of $8000 with Continuous Compounding at 6%

a. Find the average amount earned per year between the 5th year and the 10th year.

b. Find the average amount earned per year between the 20th year and the 25th year.

c. Based on the answers from parts (a) and (b), does it appear that the rate at which annual income increases is increasing or decreasing with time?

53. Given $f(x) = -x^3 + 4$, determine the average rate of change of the function on the given intervals.

a. $[0, 2]$ **b.** $[2, 4]$

54. Use the graph to solve the equation and inequalities. Write the solutions to the inequalities in interval notation.

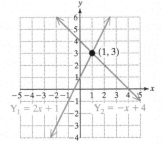

a. $2x + 1 = -x + 4$

b. $2x + 1 < -x + 4$

c. $2x + 1 \geq -x + 4$

SECTION 2.5

55. If the slope of a line is $\frac{2}{3}$,

a. Determine the slope of a line parallel to the given line.

b. Determine the slope of a line perpendicular to the given line.

56. Given a line L_1 defined by L_1: $2x - 4y = 3$, determine if the equations given in parts (a)–(c) represent a line parallel to L_1, perpendicular to L_1, or neither parallel nor perpendicular to L_1.

a. $12x + 6y = 6$ **b.** $3y = 1.5x - 5$

c. $4x + 8y = 8$

For Exercises 57–63, write an equation of the line having the given conditions. Write the answer in slope-intercept form if possible.

57. Passes through $(-2, -7)$ and $m = 3$.

58. Passes through $(0, 5)$ and $m = -\dfrac{2}{5}$.

59. Passes through $(1.1, 5.3)$ and $(-0.9, 7.1)$.

60. Passes through $(5, -7)$ and the slope is undefined.

61. Passes through $(2, -6)$ and is parallel to the line defined by $2x - y = 4$.

62. Passes through $(-2, 3)$ and is perpendicular to the line defined by $5y = 2x$.

63. The line is perpendicular to the y-axis and the y-intercept is $(0, 7)$.

64. A car has a 15-gal tank for gasoline and gets 30 mpg on a highway while driving 60 mph. Suppose that the driver starts a trip with a full tank of gas and travels 450 mi on the highway at an average speed of 60 mph.

a. Write a linear model representing the amount of gas $G(t)$ left in the tank t hours into a trip.

b. Evaluate $G(4.5)$ and interpret the meaning in the context of this problem.

65. A dance studio has fixed monthly costs of $1500 that include rent, utilities, insurance, and advertising. The studio charges $60 for each private lesson, but has a variable cost for each lesson of $35 to pay the instructor.

a. Write a linear cost function representing the cost to the studio $C(x)$ to hold x private lessons for a given month.

b. Write a linear revenue function representing the revenue $R(x)$ for holding x private lessons for the month.

c. Write a linear profit function representing the profit $P(x)$ for holding x private lessons for the month.

d. Determine the number of private lessons that must be held for the studio to make a profit.

e. If 82 private lessons are held during a given month, how much money will the studio make or lose?

66. The height y (in meters) of a volcano in the southeast Pacific Ocean is recorded in the table for selected years since 1960.

Number of Years Since 1960, x	Height (m) y
0	166
10	290
20	408
30	526
40	650
50	760
54	813

a. Graph the data in a scatter plot.

b. Use the points (0, 166) and (40, 650) to write a linear function that defines the height y of the volcano, x years since 1960.

c. Interpret the meaning of the slope in the context of this problem.

d. Use the model in part (b) to predict the height of the volcano in the year 2030 assuming that the linear trend continues.

67. Refer to the data given in Exercise 66.

a. Use a graphing utility to find the least-squares regression line. Round the slope to 1 decimal place and the y-intercept to the nearest whole unit.

b. Use a graphing utility to graph the regression line and the observed data.

c. In the event that the linear trend continues, use the model from part (a) to predict the height of the volcano in the year 2030.

SECTION 2.6

68. Write a function based on the given parent function and transformations in the given order.

Parent function: $y = x^2$

1. Shift 5 units to the left.

2. Reflect across the y-axis.

3. Shift downward 2 units.

For Exercises 69–78, use translations to graph the given functions.

69. $f(x) = |x| - 2$

70. $g(x) = \sqrt{x} + 1$

71. $h(x) = (x - 2)^2$

72. $k(x) = \sqrt[3]{x + 1}$

73. $r(x) = \sqrt{x - 3} + 1$

74. $s(x) = (x + 2)^2 - 3$

75. $t(x) = -2|x|$

76. $v(x) = -\dfrac{1}{2}|x|$

77. $m(x) = \sqrt{-x + 5}$

78. $n(x) = \sqrt{-x - 1}$

For Exercises 79–84, use the graph of $y = f(x)$ to graph the given function.

79. $y = f(2x)$

80. $y = f(\frac{1}{2}x)$

81. $y = -f(x + 1) - 3$

82. $y = -f(x - 4) - 1$

83. $y = 2f(x - 3) + 1$

84. $y = \frac{1}{2}f(x + 2) - 3$

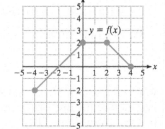

SECTION 2.7

For Exercises 85–88, determine if the graph of the equation is symmetric to the y-axis, x-axis, origin, or none of these.

85. $y = x^4 - 3$

86. $x = |y| + y^2$

87. $y = \dfrac{1}{3}x - 1$

88. $x^2 = y^2 + 1$

For Exercises 89–94, determine if the function is even, odd, or neither.

89. $f(x) = -4x^3 + x$

90. $g(x) = \sqrt[3]{x}$

91. $p(x) = \sqrt{4 - x^2}$

92. $q(x) = -|x|$

93. $k(x) = (x - 3)^2$

94. $m(x) = |x + 2|$

95. Evaluate the function for the given values of x.

$$f(x) = \begin{cases} -4x + 2 & \text{for } x < -1 \\ x^2 & \text{for } -1 \le x \le 2 \\ 5 & \text{for } x > 2 \end{cases}$$

a. $f(-4)$ **b.** $f(-1)$ **c.** $f(3)$ **d.** $f(2)$

For Exercises 96–98, graph the function.

96. $f(x) = \begin{cases} -4x - 3 & \text{for } x < 0 \\ x^2 & \text{for } x \ge 0 \end{cases}$

97. $g(x) = \begin{cases} |x| & \text{for } x \le 2 \\ 2 & \text{for } x > 2 \end{cases}$

98. $h(x) = \begin{cases} -3 & \text{for } x < -2 \\ 1 & \text{for } -2 \le x < 0 \\ \sqrt{x} & \text{for } x \ge 0 \end{cases}$

99. Evaluate $f(x) = [\![x - 1]\!]$ for the given values of x.

   a. $f(-1.5)$ **b.** $f(-2)$ **c.** $f(0.1)$ **d.** $f(6.3)$

For Exercises 100–101, use interval notation to write the interval(s) over which f is

   a. Increasing.   **b.** Decreasing.   **c.** Constant.

100.
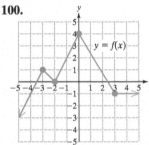

101.
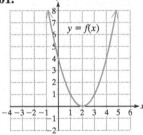

For Exercises 102–103, identify the location and value of any relative maxima or minima of the function.

102.
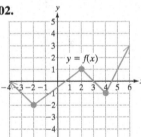

103.
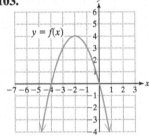

104. Write a rule for the graph of the function. Answers may vary.
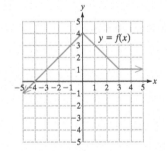

SECTION 2.8

For Exercises 105–109, evaluate the function for the given values of x.

$$f(x) = -3x \qquad g(x) = |x - 2| \qquad h(x) = \frac{1}{x + 1}$$

105. $(f - h)(2)$ **106.** $(g \cdot h)(3)$ **107.** $\left(\dfrac{g}{h}\right)(-5)$

108. $(f \circ g)(5)$ **109.** $(g \circ f)(5)$

110. Use the graphs of f and g to find the function values for the given values of x.

   a. $(f + g)(2)$

   b. $(g \cdot f)(-4)$

   c. $\left(\dfrac{g}{f}\right)(-3)$

   d. $f[g(-4)]$

   e. $(g \circ f)(-4)$

   f. $(g \circ f)(5)$

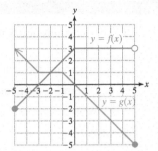

For Exercises 111–116, refer to the functions m, n, p, and q. Find the function and write the domain in interval notation.

$$m(x) = -4x \qquad n(x) = x^2 - 4x$$
$$p(x) = \sqrt{x - 2} \qquad q(x) = \frac{1}{x - 5}$$

111. $(n - m)(x)$ **112.** $\left(\dfrac{p}{n}\right)(x)$ **113.** $\left(\dfrac{n}{p}\right)(x)$

114. $(m \cdot p)(x)$ **115.** $(q \circ n)(x)$ **116.** $(q \circ p)(x)$

For Exercises 117–118, find the difference quotient, $\dfrac{f(x + h) - f(x)}{h}$.

117. $f(x) = -6x - 5$ **118.** $f(x) = 3x^2 - 4x + 9$

For Exercises 119–120, find two functions, f and g such that $h(x) = (f \circ g)(x)$.

119. $h(x) = (x - 4)^2$ **120.** $h(x) = \dfrac{12}{x + 5}$

121. A car traveling 60 mph on the highway gets 28 mpg.

   a. Write a function that represents the distance $d(t)$ (in miles) that the car travels in t hours.

   b. Write a function that represents the number of gallons of gasoline $n(d)$ used for d miles traveled.

   c. Find $(n \circ d)(t)$ and interpret the meaning in the context of this problem.

   d. Evaluate $(n \circ d)(7)$ and interpret the meaning in the context of this problem.

CHAPTER 2   Test

1. The endpoints of a diameter of a circle are $(-2, 3)$ and $(8, -5)$.

   a. Determine the center of the circle.

   b. Determine the radius of the circle.

   c. Write an equation of the circle in standard form.

2. Given $x = |y| - 4$,

   a. Determine the x- and y-intercepts of the graph of the equation.

   b. Does the equation define y as a function of x?

3. Given $x^2 + y^2 + 14x - 10y + 70 = 0$,

   a. Write the equation of the circle in standard form.

   b. Identify the center and radius.

For Exercises 4–5, determine if the relation defines y as a function of x.

4. x y

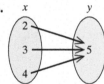

5.

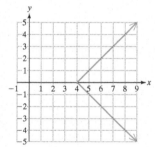

6. Given $f(x) = -2x^2 + 7x - 3$, find

   a. $f(-1)$.

   b. $f(x + h)$.

   c. The difference quotient: $\dfrac{f(x + h) - f(x)}{h}$.

   d. The x-intercepts of the graph of f.

   e. The y-intercept of the graph of f.

   f. The average rate of change of f on the interval $[1, 3]$.

7. Use the graph of $y = f(x)$ to estimate

   a. $f(0)$.

   b. $f(-4)$.

   c. The values of x for which $f(x) = 2$.

   d. The interval(s) over which f is increasing.

   e. The interval(s) over which f is decreasing.

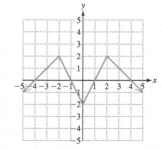

   f. Determine the location and value of any relative minima.

   g. Determine the location and value of any relative maxima.

h. The domain.

i. The range.

j. Whether f is even, odd, or neither.

For Exercises 8–9, write the domain in interval notation.

8. $f(w) = \dfrac{2w}{3w + 7}$

9. $f(c) = \sqrt{4 - c}$

10. Given $3x = -4y + 8$,

    a. Identify the slope.

    b. Identify the y-intercept.

    c. Graph the line.

    d. What is the slope of a line perpendicular to this line?

    e. What is the slope of a line parallel to this line?

11. Write an equation of the line passing through the point $(-2, 6)$ and perpendicular to the line defined by $x + 3y = 4$.

12. Use the graph to solve the equation and inequalities. Write the solutions to the inequalities in interval notation.

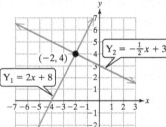

    a. $2x + 8 = -\dfrac{1}{2}x + 3$

    b. $2x + 8 < -\dfrac{1}{2}x + 3$

    c. $2x + 8 \geq -\dfrac{1}{2}x + 3$

For Exercises 13–16, graph the equation.

13. $x^2 + \left(y + \dfrac{5}{2}\right)^2 = 9$

14. $f(x) = 2|x + 3|$

15. $g(x) = -\sqrt{x + 4} + 3$

16. $h(x) = \begin{cases} -x + 3 & \text{for } x < 1 \\ \sqrt{x - 1} & \text{for } x \geq 1 \end{cases}$

17. Determine if the graph of the equation is symmetric to the y-axis, x-axis, origin, or none of these.
$$x^2 + |y| = 8$$

For Exercises 18–19, determine if the function is even, odd, or neither.

18. $f(x) = x^3 - x$ 19. $g(x) = x^4 + x^3 + x$

20. Evaluate the greatest integer function for the following values of x.

    a. 4.27 b. -4.27

For Exercises 21–26, refer to the functions f, g, and h defined here.

$$f(x) = x - 4 \qquad g(x) = \frac{1}{x - 3} \qquad h(x) = \sqrt{x - 5}$$

21. Evaluate $(f - h)(6)$.   **22.** Evaluate $(g \cdot h)(5)$.

23. Evaluate $(h \circ f)(1)$.

24. Find $(f \cdot g)(x)$ and state the domain in interval notation.

25. Find $\left(\dfrac{g}{f}\right)(x)$ and state the domain in interval notation.

26. Find $(g \circ h)(x)$ and state the domain in interval notation.

27. Write two functions f and g such that $h(x) = (f \circ g)(x)$.

$$h(x) = \sqrt[3]{x - 7}$$

28. For f and g pictured, estimate the following.

a. $(f + g)(3)$

b. $(f \cdot g)(0)$

c. $g(f(3))$

d. $(f \circ g)(2)$

e. The interval(s) over which f is increasing.

f. The interval(s) over which g is decreasing.

29. The number of people y that attend a weekly bingo game at an adult recreation center is given in the table for selected weeks, x.

Week Number, x	Number of attendees, y
1	8
3	21
6	30
9	40
12	46
15	56
18	68

a. Graph the data in a scatter plot.

b. Use the points $(1, 8)$ and $(9, 40)$ to write a linear function that defines the number of attendees as a function of week number.

c. Interpret the meaning of the slope in the context of this problem.

d. Use the model in part (b) to predict the number of attendees in week 24 assuming that the linear trend continues.

30. Refer to the data given in Exercise 29.

a. Use a graphing utility to find the least-squares regression line. Round the slope and y-intercept to 1 decimal place.

b. Use a graphing utility to graph the regression line and the observed data.

c. In the event that the linear trend continues, use the model from part (a) to predict the number of attendees in week 24.

CHAPTER 2 Cumulative Review Exercises

1. Use the graph of $y = f(x)$ to

a. Evaluate $f(2)$.

b. Find all x such that $f(x) = 0$.

c. Determine the domain of f.

d. Determine the range of f.

e. Determine the interval(s) over which f is increasing.

f. Determine the interval(s) over which f is decreasing.

g. Determine the intervals(s) over which f is constant.

h. Evaluate $(f \circ f)(-1)$.

2. Given the equation of the circle $x^2 + y^2 + 12x - 4y + 31 = 0$,

a. Write the equation in standard form.

b. Identify the center and radius.

For Exercises 3–7, refer to the functions f, g, and h defined here.

$$f(x) = -x^2 + 3x \qquad g(x) = \frac{1}{x} \qquad h(x) = \sqrt{x + 2}$$

3. Find $(g \circ f)(x)$ and write the domain in interval notation.

4. Find $(g \cdot h)(x)$ and write the domain in interval notation.

5. Find the difference quotient. $\dfrac{f(x + h) - f(x)}{h}$

6. Find the average rate of change of f over the interval $[0, 3]$.

7. Determine the x- and y-intercepts of f.

For Exercises 8–9, graph the function.

8. $f(x) = -\sqrt{x + 3}$

9. $g(x) = \begin{cases} -4 & \text{for } x < -2 \\ 1 & \text{for } -2 \le x < 0 \\ x^2 + 1 & \text{for } x \ge 0 \end{cases}$

10. Write an equation of the line passing through the points $(8, -3)$ and $(-2, 1)$. Write the final answer in slope-intercept form.

11. Write an absolute value expression that represents the distance between the points x and 7 on the number line.

12. Factor. $2x^3 - 128$

For Exercises 13–17, solve the equation or inequality. Write the solutions to the inequalities in interval notation.

13. $-3t(t - 1) = 2t + 6$

14. $7 = |4x - 2| + 5$

15. $x^{2/5} - 3x^{1/5} + 2 = 0$

16. $|3a + 1| - 2 \le 9$

17. $3 \le -2x + 1 < 7$

For Exercises 18–20, perform the indicated operations and simplify.

18. $\dfrac{6}{\sqrt{15} + \sqrt{11}}$

19. $3c\sqrt{8c^2 d^3} + c^2\sqrt{50d^3} - 2d\sqrt{2c^4 d}$

20. $\dfrac{2u^{-1} - w^{-1}}{4u^{-2} - w^{-2}}$

3

Polynomial and Rational Functions

Chapter Outline

Meteorology and the study of weather have a strong basis in mathematics. The factors impacting weather are not constant and change over time. For example, during the summer months, hot ocean temperatures in the Atlantic Ocean often produce breeding grounds for hurricanes off the coast of Africa or in the Caribbean. To predict the path of a hurricane, meteorologists collect data from satellites, weather stations around the world, and weather buoys in the ocean. Piecing together the data requires a variety of techniques of mathematical modeling using powerful computers. In the end, scientists combine a series of simple curves to approximate weather patterns that closely fit complicated models.

In this chapter, we study polynomial and rational functions. Both types of functions represent simple curves that can be used for modeling in a wide range of applications, including predictions for the path of a hurricane.

OBJECTIVES

1. Graph a Quadratic Function Written in Vertex Form
2. Write $f(x) = ax^2 + bx + c$ ($a \neq 0$) in Vertex Form
3. Find the Vertex of a Parabola by Using the Vertex Formula
4. Solve Applications Involving Quadratic Functions
5. Create Quadratic Models Using Regression

1. Graph a Quadratic Function Written in Vertex Form

In Chapter 2, we defined a function of the form $f(x) = mx + b$ ($m \neq 0$) as a linear function. The function defined by $f(x) = ax^2 + bx + c$ ($a \neq 0$) is called a *quadratic function*. Notice that a quadratic function has a leading term of second degree. We are already familiar with the graph of $f(x) = x^2$ (Figure 3-1). The graph is a parabola opening upward with vertex at the origin. Also note that the graph is symmetric with respect to the vertical line through the vertex called the **axis of symmetry**.

We can write $f(x) = ax^2 + bx + c$ ($a \neq 0$) in the form $f(x) = a(x - h)^2 + k$ by completing the square. Furthermore, from Section 2.6 we know that the graph of $f(x) = a(x - h)^2 + k$ is related to the graph of $y = x^2$ by a vertical shrink or stretch determined by a, a horizontal shift determined by h, and a vertical shift determined by k. Therefore, the graph of a quadratic function is a parabola with vertex at (h, k).

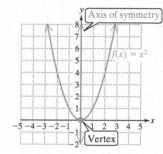

Figure 3-1

TIP A quadratic function is often used as a model for projectile motion. This is motion followed by an object influenced by an initial force and by the force of gravity.

Quadratic Function

A function defined by $f(x) = ax^2 + bx + c$ ($a \neq 0$) is called a **quadratic function.** By completing the square, $f(x)$ can be expressed in **vertex form** as $f(x) = a(x - h)^2 + k$.

- The graph of f is a parabola with vertex (h, k).
- If $a > 0$, the parabola opens upward, and the vertex is the minimum point. The minimum *value* of f is k.
- If $a < 0$, the parabola opens downward, and the vertex is the maximum point. The maximum *value* of f is k.
- The axis of symmetry is $x = h$. This is the vertical line that passes through the vertex.

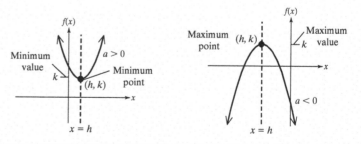

In Example 1, we analyze and graph a quadratic function by identifying the vertex, axis of symmetry, and x- and y-intercepts. From the graph, the minimum or maximum value of the function is readily apparent.

EXAMPLE 1   **Analyzing and Graphing a Quadratic Function**

Given $f(x) = -2(x - 1)^2 + 8$,

a. Determine whether the graph of the parabola opens upward or downward.

b. Identify the vertex.

c. Determine the x-intercept(s).

d. Determine the y-intercept.

e. Sketch the function.

f. Determine the axis of symmetry.

g. Determine the maximum or minimum value of f.

h. Write the domain and range in interval notation.

Solution:

a. $f(x) = -2(x - 1)^2 + 8$ The function is written as $f(x) = a(x - h)^2 + k$,
   The parabola opens downward.   where $a = -2$, $h = 1$, and $k = 8$. Since $a < 0$,
                                   the parabola opens downward.

b. The vertex is $(1, 8)$.   The vertex is (h, k), which is $(1, 8)$.

c. $f(x) = -2(x - 1)^2 + 8$
    $0 = -2(x - 1)^2 + 8$ To find the x-intercept(s), find all real solutions
    $-8 = -2(x - 1)^2$ to the equation $f(x) = 0$.
    $4 = (x - 1)^2$
    $\pm\sqrt{4} = x - 1$
    $1 \pm 2 = x$
    $x = 3$ or $x = -1$
   The x-intercepts are $(3, 0)$ and $(-1, 0)$.

d. $f(0) = -2(0 - 1)^2 + 8$ To find the y-intercept, evaluate $f(0)$.
       $= 6$
   The y-intercept is $(0, 6)$.

e. The graph of f is shown in Figure 3-2.

f. The axis of symmetry is the vertical line through the vertex: $x = 1$.

g. The maximum value is 8.

h. The domain is $(-\infty, \infty)$.
   The range is $(-\infty, 8]$.

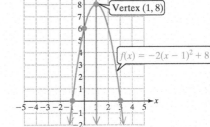

Figure 3-2

Skill Practice 1   Repeat Example 1 with $g(x) = (x + 2)^2 - 1$.

2. Write $f(x) = ax^2 + bx + c \ (a \neq 0)$ in Vertex Form

In Section 2.2, we learned how to complete the square to write an equation of a circle $x^2 + y^2 + Ax + By + C = 0$ in standard form $(x - h)^2 + (y - k)^2 = r^2$. We use the same process to write a quadratic function $f(x) = ax^2 + bx + c \ (a \neq 0)$ in vertex form $f(x) = a(x - h)^2 + k$. However, we will work on the right side of the equation only. This is demonstrated in Example 2.

Answers

1. a. Upward   **b.** $(-2, -1)$

   c. $(-3, 0)$ and $(-1, 0)$ **d.** $(0, 3)$

   e.

   $g(x) = (x + 2)^2 - 1$

   Vertex $(-2, -1)$

   f. $x = -2$

   g. The minimum value is -1.

   h. The domain is $(-\infty, \infty)$.
      The range is $[-1, \infty)$.

EXAMPLE 2   Writing a Quadratic Function in Vertex Form

Given $f(x) = 3x^2 + 12x + 5$,

   a. Write the function in vertex form: $f(x) = a(x - h)^2 + k$.
   b. Identify the vertex.
   c. Identify the x-intercept(s).
   d. Identify the y-intercept.
   e. Sketch the function.
   f. Determine the axis of symmetry.
   g. Determine the minimum or maximum value of f.
   h. Write the domain and range in interval notation.

Solution:

a. $f(x) = 3x^2 + 12x + 5$

$$= 3(x^2 + 4x \qquad) + 5$$

Factor out the leading coefficient of the x^2 term from the two terms containing x. The leading term within parentheses now has a coefficient of 1.

$$= 3(x^2 + 4x + 4 - 4) + 5$$

Complete the square within parentheses. Add and subtract $\left[\frac{1}{2}(4)\right]^2 = 4$ within parentheses.

$$= 3(x^2 + 4x + 4) + 3(-4) + 5$$

$$= 3(x + 2)^2 - 7 \text{ (vertex form)}$$

Remove -4 from within parentheses, along with a factor of 3.

b. The vertex is $(-2, -7)$.

c. $f(x) = 3x^2 + 12x + 5$

$$0 = 3x^2 + 12x + 5$$

To find the x-intercept(s), find the real solutions to the equation $f(x) = 0$.

$$x = \frac{-12 \pm \sqrt{(12)^2 - 4(3)(5)}}{2(3)}$$

The right side is not factorable. Apply the quadratic formula.

$$= \frac{-12 \pm \sqrt{84}}{6}$$

$$= \frac{-12 \pm 2\sqrt{21}}{6}$$

The x-intercepts are $\left(\dfrac{-6 + \sqrt{21}}{3}, 0\right)$ and $\left(\dfrac{-6 - \sqrt{21}}{3}, 0\right)$ or approximately $(-0.47, 0)$ and $(-3.53, 0)$.

$$= \frac{-6 \pm \sqrt{21}}{3} \begin{cases} x \approx -0.47 \\ x \approx -3.53 \end{cases}$$

d. $f(0) = 3(0)^2 + 12(0) + 5$

$$= 5$$

To find the y-intercept, evaluate $f(0)$. The y-intercept is $(0, 5)$.

e. The graph of f is shown in Figure 3-3.

f. The axis of symmetry is $x = -2$.

g. The minimum value is -7.

h. The domain is $(-\infty, \infty)$.

The range is $[-7, \infty)$.

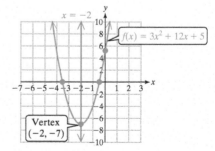

Figure 3-3

Answers

2. **a.** $f(x) = 3(x - 1)^2 - 2$ **b.** $(1, -2)$

   c. $\left(\dfrac{3 \pm \sqrt{6}}{3}, 0\right)$

   d. $(0, 1)$

   e.

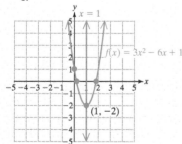

   f. $x = 1$
   g. The minimum value is -2.
   h. The domain is $(-\infty, \infty)$.
      The range is $[-2, \infty)$.

Skill Practice 2   Repeat Example 2 with $f(x) = 3x^2 - 6x + 1$.

3. Find the Vertex of a Parabola by Using the Vertex Formula

Completing the square and writing a quadratic function in the form $f(x) = a(x - h)^2 + k$ is one method to find the vertex of a parabola. Another method is to use the vertex formula. The vertex formula can be derived by completing the square on $f(x) = ax^2 + bx + c$.

$f(x) = ax^2 + bx + c \ (a \neq 0)$ Factor out a from the x terms, and complete the square within parentheses.

$= a\left(x^2 + \dfrac{b}{a}x + \dfrac{b^2}{4a^2} - \dfrac{b^2}{4a^2}\right) + c$ $\left[\dfrac{1}{2}\left(\dfrac{b}{a}\right)\right]^2 = \dfrac{b^2}{4a^2}$

$= a\left(x^2 + \dfrac{b}{a}x + \dfrac{b^2}{4a^2}\right) + a\left(-\dfrac{b^2}{4a^2}\right) + c$ Remove the term $-\dfrac{b^2}{4a^2}$ from within parentheses along with a factor of a.

$= a\left(x + \dfrac{b}{2a}\right)^2 - \dfrac{b^2}{4a} + c$ Factor the trinomial.

$= a\left(x + \dfrac{b}{2a}\right)^2 + \dfrac{4ac - b^2}{4a}$ Obtain a common denominator and add the terms outside parentheses.

$= a\left[x - \left(\dfrac{-b}{2a}\right)\right]^2 + \dfrac{4ac - b^2}{4a}$ $f(x)$ is now written in vertex form.

$f(x) = a(x - h)^2 \quad + \quad k$ $h = \dfrac{-b}{2a}$ and $k = \dfrac{4ac - b^2}{4a}$

The vertex is $\left(\dfrac{-b}{2a}, \dfrac{4ac - b^2}{4a}\right)$.

The y-coordinate of the vertex is given by $\dfrac{4ac - b^2}{4a}$ and is often hard to remember. Therefore, it is usually easier to evaluate the x-coordinate first from $\dfrac{-b}{2a}$, and then evaluate $f\left(\dfrac{-b}{2a}\right)$.

Vertex Formula to Find the Vertex of a Parabola

For $f(x) = ax^2 + bx + c \ (a \neq 0)$, the vertex is given by $\left(\dfrac{-b}{2a}, f\left(\dfrac{-b}{2a}\right)\right)$.

EXAMPLE 3   **Using the Vertex Formula**

Given $f(x) = -x^2 + 4x - 5$,

   a. State whether the graph of the parabola opens upward or downward.
   b. Determine the vertex of the parabola by using the vertex formula.
   c. Determine the x-intercept(s).
   d. Determine the y-intercept.
   e. Sketch the graph.
   f. Determine the axis of symmetry.
   g. Determine the minimum or maximum value of f.
   h. Write the domain and range in interval notation.

Solution:

a. $f(x) = -x^2 + 4x - 5$

The parabola opens downward.

> The function is written as $f(x) = ax^2 + bx + c$ where $a = -1$. Since $a < 0$, the parabola opens downward.

b. x-coordinate: $\dfrac{-b}{2a} = \dfrac{-(4)}{2(-1)} = 2$

y-coordinate: $f(2) = -(2)^2 + 4(2) - 5$
$\qquad\qquad\qquad = -1$

The vertex is $(2, -1)$.

c. Since the vertex of the parabola is below the x-axis and the parabola opens downward, the parabola cannot cross or touch the x-axis.

Therefore, there are no x-intercepts.

> Solving the equation $f(x) = 0$ to find the x-intercepts results in imaginary solutions:
>
> $0 = -x^2 + 4x - 5$
>
> $x = \dfrac{-(4) \pm \sqrt{(4)^2 - 4(-1)(-5)}}{2(-1)}$
>
> $x = 2 \pm i$

> **TIP** For more accuracy in the graph, plot one or two points near the vertex. Then use the symmetry of the curve to find additional points on the graph.
>
> For example, the points $(1, -2)$ and $(0, -5)$ are on the left branch of the parabola. The corresponding points to the right of the axis of symmetry are $(3, -2)$ and $(4, -5)$.

d. To find the y-intercept, evaluate $f(0)$.

$f(0) = -(0)^2 + 4(0) - 5 = -5$

The y-intercept is $(0, -5)$.

e. The graph of f is shown in Figure 3-4.

f. The axis of symmetry is $x = 2$.

g. The maximum value of f is -1.

h. The domain is $(-\infty, \infty)$.

The range is $(-\infty, -1]$.

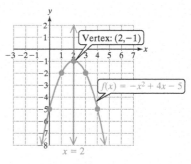

Vertex: $(2,-1)$

$f(x) = -x^2 + 4x - 5$

$x = 2$

Figure 3-4

> **Skill Practice 3**   Repeat Example 3 with $f(x) = -x^2 - 4x - 7$.

The x-intercepts of a quadratic function defined by $f(x) = ax^2 + bx + c$ are the real solutions to the equation $f(x) = 0$. The discriminant $b^2 - 4ac$ enables us to determine the number of real solutions to the equation and thus, the number of x-intercepts of the graph of the function.

Answers

3. a. Downward   **b.** $(-2, -3)$
   c. No x-intercepts   **d.** $(0, -7)$
   e.

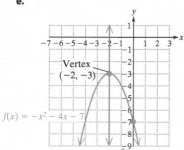

Vertex $(-2, -3)$

$f(x) = -x^2 - 4x - 7$

$x = -2$

   f. $x = -2$
   g. The maximum value is -3.
   h. The domain is $(-\infty, \infty)$.
      The range is $(-\infty, -3]$.

> **Using the Discriminant to Determine the Number of x-Intercepts**
>
> Given a quadratic function defined by $f(x) = ax^2 + bx + c$ ($a \neq 0$),
>
> - If $b^2 - 4ac = 0$, the graph of $y = f(x)$ has one x-intercept.
> - If $b^2 - 4ac > 0$, the graph of $y = f(x)$ has two x-intercepts.
> - If $b^2 - 4ac < 0$, the graph of $y = f(x)$ has no x-intercept.

From Example 2, the discriminant of $3x^2 + 12x + 5 = 0$ is $(12)^2 - 4(3)(5) = 84 > 0$. Therefore, the graph of $f(x) = 3x^2 + 12x + 5$ has two x-intercepts (Figure 3-3).

From Example 3, the discriminant of $-x^2 + 4x - 5 = 0$ is $(4)^2 - 4(-1)(-5) = -4 < 0$. Therefore, the graph of $f(x) = -x^2 + 4x - 5$ has no x-intercept (Figure 3-4).

4. Solve Applications Involving Quadratic Functions

Quadratic functions can be used in a variety of applications in which a variable is optimized. That is, the vertex of a parabola gives the maximum or minimum value of the dependent variable. We show three such applications in Examples 4–6.

EXAMPLE 4 **Using a Quadratic Function for Projectile Motion**

A stone is thrown from a 100-m cliff at an initial speed of 20 m/sec at an angle of 30° from the horizontal. The height of the stone can be modeled by $h(t) = -4.9t^2 + 10t + 100$, where $h(t)$ is the height in meters and t is the time in seconds after the stone is released.

a. Determine the time at which the stone will be at its maximum height. Round to 2 decimal places.

b. Determine the maximum height. Round to the nearest meter.

c. Determine the time at which the stone will hit the ground.

Solution:

a. The time at which the stone will be at its maximum height is the t-coordinate of the vertex.

$$t = \frac{-b}{2a} = \frac{-10}{2(-4.9)} \approx 1.02$$

The stone will be at its maximum height approximately 1.02 sec after release.

Given $h(t) = -4.9t^2 + 10t + 100$, the coefficients are $a = -4.9$, $b = 10$, and $c = 100$.

The vertex is given by
$$\left(\frac{-b}{2a}, h\!\left(\frac{-b}{2a}\right)\right).$$

b. The maximum height is the value of $h(t)$ at the vertex.
$$h(1.02) = -4.9(1.02)^2 + 10(1.02) + 100$$
$$\approx 105 \quad \text{The maximum height is 105 m.}$$

c. The stone will hit the ground when $h(t) = 0$.

$$h(t) = -4.9t^2 + 10t + 100$$
$$0 = -4.9t^2 + 10t + 100$$
$$t = \frac{-10 \pm \sqrt{(10)^2 - 4(-4.9)(100)}}{2(-4.9)}$$

$t \approx 5.65$ or $t \approx -3.61$ Reject the negative solution. The stone will hit the ground in approximately 5.65 sec.

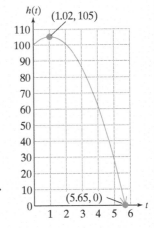

Skill Practice 4 A quarterback throws a football with an initial velocity of 72 ft/sec at an angle of 25°. The height of the ball can be modeled by $h(t) = -16t^2 + 30.4t + 5$, where $h(t)$ is the height (in ft) and t is the time in seconds after release.

a. Determine the time at which the ball will be at its maximum height.

b. Determine the maximum height of the ball.

c. Determine the amount of time required for the ball to reach the receiver's hands if the receiver catches the ball at a point 3 ft off the ground.

Answers
4. a. 0.95 sec **b.** 19.44 ft
  c. Approximately 1.96 sec

TECHNOLOGY CONNECTIONS

Compute Solutions to a Quadratic Equation

The syntax to compute the expressions from Example 4(c) is shown for a calculator in Classic mode and in Mathprint mode. In Classic mode, parentheses are required around the numerator and denominator of the fraction and around the radicand within the square root. In Mathprint mode, select the (ALPHA) key followed by F1 to access the fraction template.

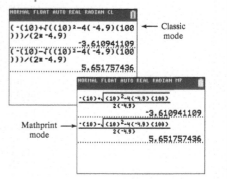

In Example 5, we present a type of application called an optimization problem. The goal is to maximize or minimize the value of the dependent variable by finding an optimal value of the independent variable.

EXAMPLE 5   Applying a Quadratic Function to Geometry

A parking area is to be constructed adjacent to a road. The developer has purchased 340 ft of fencing. Determine dimensions for the parking lot that would maximize the area. Then find the maximum area.

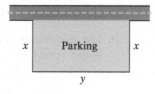

Solution:

Let x represent the width of the parking area.

Let y represent the length.

Let A represent the area.

Read the problem carefully, draw a representative diagram, and label the unknowns.

We need to find the values of x and y that maximize the area A of the rectangular region. The area is given by $A = $ (length)(width) $= yx = xy$.

To write the area as a function of one variable only, we need an equation that relates x and y. We know that the parking area is limited by a fixed amount of fencing. That is, the sum of the lengths of the three sides to be fenced can be at most 340 ft.

$2x + y = 340$

Solve for y.

The equation $2x + y = 340$ is called a **constraint equation.** This equation gives an implied restriction on x and y due to the limited amount of fencing.

$y = 340 - 2x$

Solve the constraint equation, $2x + y = 340$ for either x or y. In this case, we have solved for y.

$A = xy$

$A(x) = x(340 - 2x)$

Substitute $340 - 2x$ for y in the equation $A = xy$.

$\qquad = -2x^2 + 340x$

Function A is a quadratic function with a negative leading coefficient. The graph of the parabola opens downward, so the vertex is the maximum point on the function.

To check, verify that the value $A(85)$ is the same as the product of length and width, xy.

$A(85) = 14{,}450 \text{ ft}^2$

$xy = (85 \text{ ft})(170 \text{ ft})$

$\quad = 14{,}450 \text{ ft}^2 \checkmark$

x-coordinate of vertex:

$$x = \frac{-b}{2a} = \frac{-340}{2(-2)} = 85$$

$$y = 340 - 2(85) = 170$$

The x-coordinate of the vertex $\frac{-b}{2a}$ is the value of x that will maximize the area.

The second dimension of the parking lot can be determined from the constraint equation.

The values of x and y that would maximize the area are $x = 85$ ft and $y = 170$ ft.

$$A(85) = -2(85)^2 + 340(85) = 14{,}450$$

The value of the function at $x = 85$ gives the maximum area.

The maximum area is $14{,}450 \text{ ft}^2$.

Skill Practice 5 A farmer has 200 ft of fencing and wants to build three adjacent rectangular corrals. Determine the dimensions that should be used to maximize the area, and find the area of each individual corral.

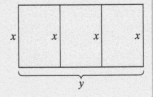

5. Create Quadratic Models Using Regression

In Section 2.5, we introduced linear regression. A regression line is a linear model based on all observed data points. In a similar fashion, we can create a quadratic function using regression. For example, suppose that a scientist growing bacteria measures the population of bacteria as a function of time. A scatter plot reveals that the data follow a curve that is approximately parabolic (Figure 3-5). In Example 6, we use a graphing calculator to find a quadratic function that models the population of the bacteria as a function of time.

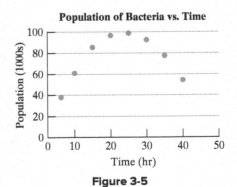

Figure 3-5

Creating a Quadratic Function Using Regression

The data in the table represent the population of bacteria $P(t)$ (in 1000s) versus the number of hours t since the culture was started.

a. Use regression to find a quadratic function to model the data. Round the coefficients to 3 decimal places.

b. Use the model to determine the time at which the population is the greatest. Round to the nearest hour.

c. What is the maximum population? Round to the nearest hundred.

Time (hr) t	Population (1000s) $P(t)$
5	37.7
10	60.9
15	85.3
20	96.3
25	98.6
30	92.4
35	77.5
40	54.1

Answer

5. The dimensions should be $x = 25$ ft and $y = 50$ ft. The area of each individual corral is $\frac{1250}{3} = 416.\overline{6} \text{ ft}^2$.

Solution:

a. From the graph in Figure 3-5, it appears that the data follow a parabolic curve. Therefore, a quadratic model would be reasonable.

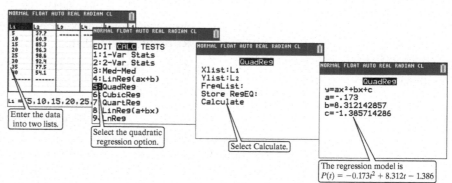

Enter the data into two lists.

Select the quadratic regression option.

Select Calculate.

The regression model is $P(t) = -0.173t^2 + 8.312t - 1.386$

b. From the graph, the time when the population is greatest is the t-coordinate of the vertex.

$$t = \frac{-b}{2a} = \frac{-(8.312)}{2(-0.173)} \approx 24$$

The population is greatest 24 hr after the culture is started.

c. The maximum population of the bacteria is the $P(t)$ value at the vertex.

$$P(24) = -0.173(24)^2 + 8.312(24) - 1.386$$

≈ 98.5 The maximum number of bacteria is approximately 98,500.

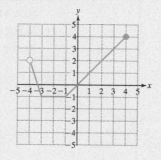

Skill Practice 6   The funding $f(t)$ (in \$ millions) for a drug rehabilitation center is given in the table for selected years t.

t	0	3	6	9	12	15
$f(t)$	3.5	2.2	2.1	3	4.9	8

a. Use regression to find a quadratic function to model the data.

b. During what year is the funding the least? Round to the nearest year.

c. What is the minimum yearly amount of funding received? Round to the nearest million.

Answers

6. a. $f(t) = 0.060t^2 - 0.593t + 3.486$
   b. Year 5   c. \$2 million

SECTION 3.1   Practice Exercises

Prerequisite Review

R.1. Solve the equation. $x^2 + 3x - 18 = 0$

R.2. a. Find the values of x for which $f(x) = 0$.

   b. Find $f(0)$.

$$f(x) = 3x^2 - 7x - 20$$

R.3. Solve the equation by completing the square and applying the square root property.

$$x^2 + 8x + 12 = 0$$

R.4. Find $g\left(-\frac{1}{2}\right)$ for $g(x) = -x^2 + 2x - 4$.

R.5. Write the domain and range in interval notation.

Concept Connections

1. A function defined by $f(x) = ax^2 + bx + c$ ($a \neq 0$) is called a _____ function.

2. The vertical line drawn through the vertex of a quadratic function is called the _____ of symmetry.

3. Given $f(x) = a(x - h)^2 + k$ ($a \neq 0$), the vertex of the parabola is the point _____.

4. Given $f(x) = a(x - h)^2 + k$, if $a < 0$, the parabola opens (upward/downward) and the (minimum/maximum) value is _____.

5. Given $f(x) = a(x - h)^2 + k$, if $a > 0$, the parabola opens (upward/downward) and the (minimum/maximum) value is _____.

6. The graph of $f(x) = a(x - h)^2 + k$, $a \neq 0$, is a parabola and the axis of symmetry is the line given by _____.

Objective 1: Graph a Quadratic Function Written in Vertex Form

For Exercises 7–14,

a. Determine whether the graph of the parabola opens upward or downward.

b. Identify the vertex.

c. Determine the x-intercept(s).

d. Determine the y-intercept.

e. Sketch the function.

f. Determine the axis of symmetry.

g. Determine the minimum or maximum value of the function.

h. Write the domain and range in interval notation.
   (**See Example 1**)

7. $f(x) = -(x - 4)^2 + 1$

8. $g(x) = -(x + 2)^2 + 4$

9. $h(x) = 2(x + 1)^2 - 8$

10. $k(x) = 2(x - 3)^2 - 2$

11. $m(x) = 3(x - 1)^2$

12. $n(x) = \dfrac{1}{2}(x + 2)^2$

13. $p(x) = -\dfrac{1}{5}(x + 4)^2 + 1$

14. $q(x) = -\dfrac{1}{3}(x - 1)^2 + 1$

Objective 2: Write $f(x) = ax^2 + bx + c$ $(a \neq 0)$ in Vertex Form

For Exercises 15–24,

a. Write the function in vertex form.

b. Identify the vertex.

c. Determine the x-intercept(s).

d. Determine the y-intercept.

e. Sketch the function.

f. Determine the axis of symmetry.

g. Determine the minimum or maximum value of the function.

h. Write the domain and range in interval notation.
   (**See Example 2**)

15. $f(x) = x^2 + 6x + 5$

16. $g(x) = x^2 + 8x + 7$

17. $p(x) = 3x^2 - 12x - 7$

18. $q(x) = 2x^2 - 4x - 3$

19. $c(x) = -2x^2 - 10x + 4$

20. $d(x) = -3x^2 - 9x + 8$

21. $h(x) = -2x^2 + 7x$

22. $k(x) = 3x^2 - 8x$

23. $p(x) = x^2 + 9x + 17$

24. $q(x) = x^2 + 11x + 26$

Objective 3: Find the Vertex of a Parabola by Using the Vertex Formula

For Exercises 25–32, find the vertex of the parabola by applying the vertex formula.

25. $f(x) = 3x^2 - 42x - 91$

26. $g(x) = 4x^2 - 64x + 107$

27. $k(a) = -\dfrac{1}{3}a^2 + 6a + 1$

28. $j(t) = -\dfrac{1}{4}t^2 + 10t - 5$

29. $f(c) = 4c^2 - 5$

30. $h(a) = 2a^2 + 14$

31. $P(x) = 1.2x^2 + 1.8x - 3.6$
    (Write the coordinates of the vertex as decimals.)

32. $Q(x) = 7.5x^2 - 2.25x + 4.75$
    (Write the coordinates of the vertex as decimals.)

For Exercises 33–42,

a. State whether the graph of the parabola opens upward or downward.

b. Identify the vertex.

c. Determine the x-intercept(s).

d. Determine the y-intercept.

e. Sketch the graph.

f. Determine the axis of symmetry.

g. Determine the minimum or maximum value of the function.

h. Write the domain and range in interval notation. **(See Example 3)**

33. $g(x) = -x^2 + 2x - 4$

34. $h(x) = -x^2 - 6x - 10$

35. $f(x) = 5x^2 - 15x + 3$

36. $k(x) = 2x^2 - 10x - 5$

37. $f(x) = 2x^2 + 3$

38. $g(x) = -x^2 - 1$

39. $f(x) = -2x^2 - 20x - 50$

40. $m(x) = 2x^2 - 8x + 8$

41. $n(x) = x^2 - x + 3$

42. $r(x) = x^2 - 5x + 7$

Objective 4: Solve Applications Involving Quadratic Functions

43. The monthly profit for a small company that makes long-sleeve T-shirts depends on the price per shirt. If the price is too high, sales will drop. If the price is too low, the revenue brought in may not cover the cost to produce the shirts. After months of data collection, the sales team determines that the monthly profit is approximated by $f(p) = -50p^2 + 1700p - 12,000$, where p is the price per shirt and $f(p)$ is the monthly profit based on that price. **(See Example 4)**

a. Find the price that generates the maximum profit.

b. Find the maximum profit.

c. Find the price(s) that would enable the company to break even.

44. The monthly profit for a company that makes decorative picture frames depends on the price per frame. The company determines that the profit is approximated by $f(p) = -80p^2 + 3440p - 36,000$, where p is the price per frame and $f(p)$ is the monthly profit based on that price.

a. Find the price that generates the maximum profit.

b. Find the maximum profit.

c. Find the price(s) that would enable the company to break even.

45. A long jumper leaves the ground at an angle of 20° above the horizontal, at a speed of 11 m/sec. The height of the jumper can be modeled by $h(x) = -0.046x^2 + 0.364x$, where h is the jumper's height in meters and x is the horizontal distance from the point of launch.

a. At what horizontal distance from the point of launch does the maximum height occur? Round to 2 decimal places.

b. What is the maximum height of the long jumper? Round to 2 decimal places.

c. What is the length of the jump? Round to 1 decimal place.

46. A firefighter holds a hose 3 m off the ground and directs a stream of water toward a burning building. The water leaves the hose at an initial speed of 16 m/sec at an angle of 30°. The height of the water can be approximated by $h(x) = -0.026x^2 + 0.577x + 3$, where $h(x)$ is the height of the water in meters at a point x meters horizontally from the firefighter to the building.

a. Determine the horizontal distance from the firefighter at which the maximum height of the water occurs. Round to 1 decimal place.

b. What is the maximum height of the water? Round to 1 decimal place.

c. The flow of water hits the house on the downward branch of the parabola at a height of 6 m. How far is the firefighter from the house? Round to the nearest meter.

47. The population $P(t)$ of a culture of the bacterium *Pseudomonas aeruginosa* is given by $P(t) = -1718t^2 + 82,000t + 10,000$, where t is the time in hours since the culture was started.

a. Determine the time at which the population is at a maximum. Round to the nearest hour.

b. Determine the maximum population. Round to the nearest thousand.

48. The gas mileage $m(x)$ (in mpg) for a certain vehicle can be approximated by $m(x) = -0.028x^2 + 2.688x - 35.012$, where x is the speed of the vehicle in mph.

a. Determine the speed at which the car gets its maximum gas mileage.

b. Determine the maximum gas mileage.

49. The sum of two positive numbers is 24. What two numbers will maximize the product? **(See Example 5)**

50. The sum of two positive numbers is 1. What two numbers will maximize the product?

51. The difference of two numbers is 10. What two numbers will minimize the product?

52. The difference of two numbers is 30. What two numbers will minimize the product?

53. Suppose that a family wants to fence in an area of their yard for a vegetable garden to keep out deer. One side is already fenced from the neighbor's property. (**See Example 5**)

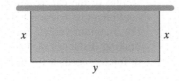

a. If the family has enough money to buy 160 ft of fencing, what dimensions would produce the maximum area for the garden?

b. What is the maximum area?

54. Two chicken coops are to be built adjacent to one another using 120 ft of fencing.

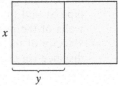

a. What dimensions should be used to maximize the area of an individual coop?

b. What is the maximum area of an individual coop?

55. A trough at the end of a gutter spout is meant to direct water away from a house. The homeowner makes the trough from a rectangular piece of aluminum that is 20 in. long and 12 in. wide. He makes a fold along the two long sides a distance of x inches from the edge.

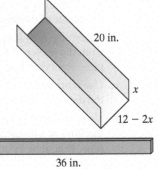

a. Write a function to represent the volume in terms of x.

b. What value of x will maximize the volume of water that can be carried by the gutter?

c. What is the maximum volume?

56. A rectangular frame of uniform depth for a shadow box is to be made from a 36-in. piece of wood.

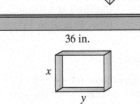

a. Write a function to represent the display area in terms of x.

b. What dimensions should be used to maximize the display area?

c. What is the maximum area?

Objective 5: Create Quadratic Models Using Regression

57. *Tetanus bacillus* bacteria are cultured to produce tetanus toxin used in an inactive form for the tetanus vaccine. The amount of toxin produced per batch increases with time and then decreases as the culture becomes unstable. The variable t is the time in hours after the culture has started, and $y(t)$ is the yield of toxin in grams. (**See Example 6**)

t	8	16	24	32	40	48
$y(t)$	0.60	1.12	1.60	1.78	1.90	2.00

t	56	64	72	80	88	96
$y(t)$	1.94	1.80	1.48	1.30	0.66	0.10

a. Use regression to find a quadratic function to model the data.

b. At what time is the yield the greatest? Round to the nearest hour.

c. What is the maximum yield? Round to the nearest gram.

58. Gas mileage is tested for a car under different driving conditions. At lower speeds, the car is driven in stop-and-go traffic. At higher speeds, the car must overcome more wind resistance. The variable x given in the table represents the speed (in mph) for a compact car, and $m(x)$ represents the gas mileage (in mpg).

x	25	30	35	40	45
$m(x)$	22.7	25.1	27.9	30.8	31.9

x	50	55	60	65
$m(x)$	30.9	28.4	24.2	21.9

a. Use regression to find a quadratic function to model the data.

b. At what speed is the gas mileage the greatest? Round to the nearest mile per hour.

c. What is the maximum gas mileage? Round to the nearest mile per gallon.

59. Fluid runs through a drainage pipe with a 10-cm radius and a length of 30 m (3000 cm). The velocity of the fluid gradually decreases from the center of the pipe toward the edges as a result of friction with the walls of the pipe. For the data shown, $v(x)$ is the velocity of the fluid (in cm/sec) and x represents the distance (in cm) from the center of the pipe toward the edge.

x	0	1	2	3	4
$v(x)$	195.6	195.2	194.2	193.0	191.5

x	5	6	7	8	9
$v(x)$	189.8	188.0	185.5	183.0	180.0

a. The pipe is 30 m long (3000 cm). Determine how long it will take fluid to run the length of the pipe through the center of the pipe. Round to 1 decimal place.

b. Determine how long it will take fluid at a point 9 cm from the center of the pipe to run the length of the pipe. Round to 1 decimal place.

c. Use regression to find a quadratic function to model the data.

d. Use the model from part (c) to predict the velocity of the fluid at a distance 5.5 cm from the center of the pipe. Round to 1 decimal place.

60. The braking distance required for a car to stop depends on numerous variables such as the speed of the car, the weight of the car, reaction time of the driver, and the coefficient of friction between the tires and the road. For a certain vehicle on one stretch of highway, the braking distances $d(s)$ (in ft) are given for several different speeds s (in mph).

s	30	35	40	45	50
$d(s)$	109	134	162	191	223

s	55	60	65	70	75
$d(s)$	256	291	328	368	409

a. Use regression to find a quadratic function to model the data.

b. Use the model from part (a) to predict the stopping distance for the car if it is traveling 62 mph before the brakes are applied. Round to the nearest foot.

c. Suppose that the car is traveling 53 mph before the brakes are applied. If a deer is standing in the road at a distance of 245 ft from the point where the brakes are applied, will the car hit the deer?

Mixed Exercises

For Exercises 61–64, given a quadratic function defined by $f(x) = ax^2 + bx + c$ $(a \neq 0)$, answer true or false. If an answer is false, explain why.

61. The graph of f can have two y-intercepts.

62. The graph of f can have two x-intercepts.

63. If $a < 0$, then the vertex of the parabola is the maximum point on the graph of f.

64. The axis of symmetry of the graph of f is the line defined by $y = c$.

For Exercises 65–70, determine the number of x-intercepts of the graph of $f(x) = ax^2 + bx + c$ $(a \neq 0)$, based on the discriminant of the related equation $f(x) = 0$. (*Hint*: Recall that the discriminant is $b^2 - 4ac$.)

65. $f(x) = 4x^2 + 12x + 9$

66. $f(x) = 25x^2 - 20x + 4$

67. $f(x) = -x^2 - 5x + 8$

68. $f(x) = -3x^2 + 4x + 9$

69. $f(x) = -3x^2 + 6x - 11$

70. $f(x) = -2x^2 + 5x - 10$

For Exercises 71–78, given a quadratic function defined by $f(x) = a(x - h)^2 + k$ $(a \neq 0)$, match the graph with the function based on the conditions given.

71. $a > 0$, $h < 0$, $k > 0$

72. $a > 0$, $h < 0$, $k < 0$

73. $a < 0$, $h < 0$, $k < 0$

74. $a < 0$, $h < 0$, $k > 0$

75. $a > 0$, axis of symmetry $x = 2$, $k < 0$

76. $a < 0$, axis of symmetry $x = 2$, $k > 0$

77. $a < 0$, $h = 2$, maximum value equals -2

78. $a > 0$, $h = 2$, minimum value equals 2

a.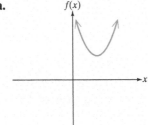

b.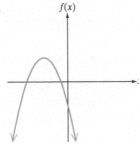

c.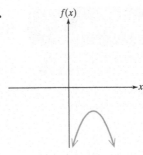

d.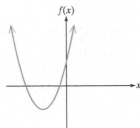

e.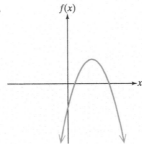

f.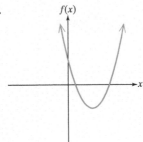

g.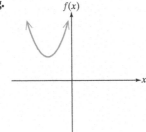

h.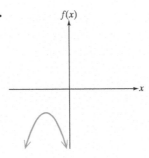

Write About It

79. Explain why a parabola opening upward has a minimum value but no maximum value. Use the graph of $f(x) = x^2$ to explain.

80. Explain why a quadratic function whose graph opens downward with vertex $(4, -3)$ has no x-intercept.

81. Explain why a quadratic function given by $f(x) = ax^2 + bx + c$ cannot have two y-intercepts.

82. Explain how to use the discriminant to determine the number of x-intercepts for the graph of $f(x) = ax^2 + bx + c$.

83. If a quadratic function given by $y = f(x)$ has x-intercepts of $(2, 0)$ and $(6, 0)$, explain why the vertex must be $(4, f(4))$.

84. Given an equation of a parabola in the form $y = a(x - h)^2 + k$, explain how to determine by inspection if the parabola has no x-intercepts.

Expanding Your Skills

For Exercises 85–88, define a quadratic function $y = f(x)$ that satisfies the given conditions.

85. Vertex $(2, -3)$ and passes through $(0, 5)$

86. Vertex $(-3, 1)$ and passes through $(0, -17)$

87. Axis of symmetry $x = 4$, maximum value 6, passes through $(1, 3)$

88. Axis of symmetry $x = -2$, minimum value 5, passes through $(2, 13)$

For Exercises 89–92, find the value of b or c that gives the function the given minimum or maximum value.

89. $f(x) = 2x^2 + 12x + c$; minimum value -9

90. $f(x) = 3x^2 + 12x + c$; minimum value -4

91. $f(x) = -x^2 + bx + 4$; maximum value 8

92. $f(x) = -x^2 + bx - 2$; maximum value 7

1. Determine the End Behavior of a Polynomial Function

A solar oven is to be made from an open box with reflective sides. Each box is made from a 30-in. by 24-in. rectangular sheet of aluminum with squares of length x (in inches) removed from each corner. Then the flaps are folded up to form an open box.

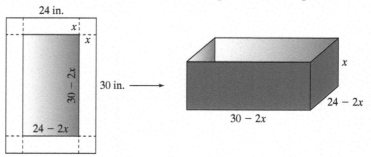

The volume $V(x)$ (in cubic inches) of the box is given by

$$V(x) = 4x^3 - 108x^2 + 720x, \text{ where } 0 < x < 12.$$

From the graph of $y = V(x)$ (Figure 3-6), the maximum volume appears to occur when squares of approximately 4 inches in length are cut from the corners of the sheet of aluminum. See Exercise 99.

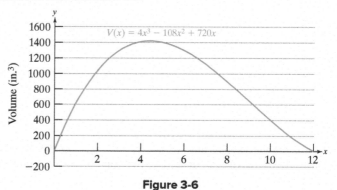

Figure 3-6

The function defined by $V(x) = 4x^3 - 108x^2 + 720x$ is an example of a polynomial function of degree 3.

Definition of a Polynomial Function

Let n be a whole number and $a_n, a_{n-1}, a_{n-2}, \ldots, a_1, a_0$ be real numbers, where $a_n \neq 0$. Then a function defined by

$$f(x) = a_n x^n + a_{n-1} x^{n-1} + a_{n-2} x^{n-2} + \cdots + a_1 x + a_0$$

is called a **polynomial function of degree n.**

The coefficients of each term of a polynomial function are real numbers, and the exponents on x must be whole numbers.

Polynomial Function	**Not a Polynomial Function**
$f(x) = 4x^5 - 3x^4 + 2x^2$	$f(x) = 4\sqrt{x} - \dfrac{3}{x} + (3 + 2i)x^2$

$\sqrt{x} = x^{1/2}$	$3/x = 3x^{-1}$	$(3 + 2i)$
Exponent not a whole number	Exponent not a whole number	Coefficient not a real number

TIP A third-degree polynomial function is referred to as a *cubic* polynomial function.

A fourth-degree polynomial function is referred to as a *quartic* polynomial function.

We have already studied several special cases of polynomial functions. For example:

$f(x) = 2$ constant function   (polynomial function, degree 0)

$g(x) = 3x + 1$ linear function   (polynomial function, degree 1)

$h(x) = 4x^2 + 7x - 1$ quadratic function   (polynomial function, degree 2)

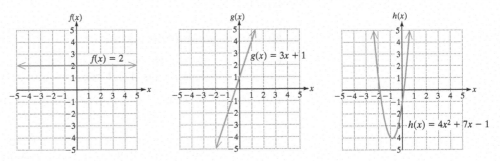

The domain of a polynomial function is all real numbers. Furthermore, the graph of a polynomial function is both continuous and smooth. Informally, a continuous function can be drawn without lifting the pencil from the paper. A smooth function has no sharp corners or points. For example, the first curve shown here could be a polynomial function, but the last three are not polynomial functions.

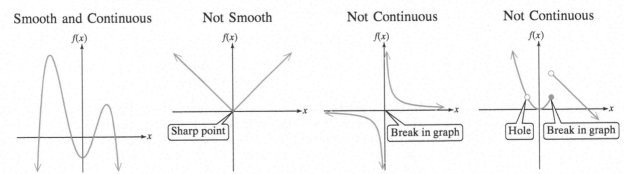

To begin our analysis of polynomial functions, we first consider the graphs of functions of the form $f(x) = ax^n$, where a is a real number and n is a positive integer. These fall into a category of functions called **power functions.** The graphs of three power functions with even degrees and positive coefficients are shown in Figure 3-7. The graphs of three power functions with odd degrees and positive coefficients are shown in Figure 3-8.

TIP For a positive integer n, the graph of the power function $y = x^n$ becomes "flatter" near the x-intercept for higher powers of n.

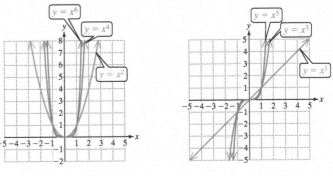

Figure 3-7   **Figure 3-8**

From Figure 3-7, notice that for even powers of n, the behavior of $y = x^n$ is similar to the graph of $y = x^2$ with variations on the "steepness" of the curve. Figure 3-8 shows that for odd powers, the behavior of $y = x^n$ with $n \geq 3$ is similar to the graph of $y = x^3$. For any power function $y = ax^n$, the coefficient a will impose a vertical shrink or stretch on the graph of $y = x^n$ by a factor of $|a|$. If $a < 0$, then the graph is reflected across the x-axis.

Power functions are helpful to analyze the "end behavior" of a polynomial function with multiple terms. The end behavior is the general direction that the fuction follows as x approaches ∞ or $-\infty$. To describe end behavior, we have the following notation.

Notation for Infinite Behavior of $y = f(x)$	
$x \to \infty$	is read as "x approaches infinity." This means that x becomes infinitely large in the positive direction.
$x \to -\infty$	is read as "x approaches negative infinity." This means that x becomes infinitely "large" in the negative direction.
$f(x) \to \infty$	is read as "$f(x)$ approaches infinity." This means that the y value becomes infinitely large in the positive direction.
$f(x) \to -\infty$	is read as "$f(x)$ approaches negative infinity." This means that the y value becomes infinitely "large" in the negative direction.

Consider the function defined by

$$f(x) = a_n x^n + a_{n-1} x^{n-1} + a_{n-2} x^{n-2} + \cdots + a_1 x + a_0$$

The leading term has the greatest exponent on x.

The leading term has the greatest exponent on x. Therefore, as $|x|$ gets large (that is, as $x \to \infty$ or as $x \to -\infty$), the leading term will be relatively larger in absolute value than all other terms. In fact, x^n will eventually be greater in absolute value than the *sum* of all other terms. Therefore, the end behavior of the function is dictated only by the leading term, and the graph of the function far to the left and far to the right will follow the general behavior of the power function $y = ax^n$.

The Leading Term Test

Consider a polynomial function given by

$$f(x) = a_n x^n + a_{n-1} x^{n-1} + a_{n-2} x^{n-2} + \cdots + a_1 x + a_0.$$

As $x \to \infty$ or as $x \to -\infty$, f eventually becomes forever increasing or forever decreasing and will follow the general behavior of $y = a_n x^n$.

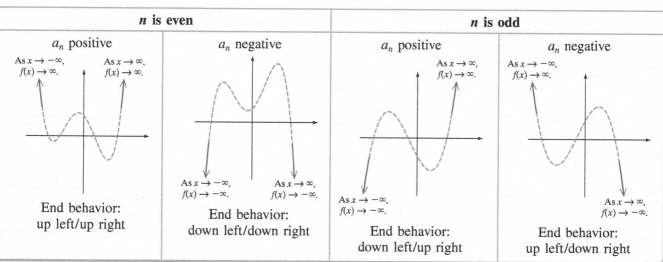

n is even		**n is odd**	
a_n positive	a_n negative	a_n positive	a_n negative
End behavior: up left/up right	End behavior: down left/down right	End behavior: down left/up right	End behavior: up left/down right

> **EXAMPLE 1**   **Determining End Behavior**
>
> Use the leading term to determine the end behavior of the graph of the function.
>
> **a.** $f(x) = -4x^5 + 6x^4 + 2x$ **b.** $g(x) = \frac{1}{4}x(2x - 3)^3(x + 4)^2$
>
> **Solution:**
>
> **a.** $f(x) = -4x^5 + 6x^4 + 2x$
>
>   negative   odd
>
>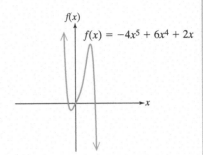
>
> The leading coefficient is negative and the degree is odd. By the leading term test, the end behavior is up to the left and down to the right.
>
> As $x \to -\infty, f(x) \to \infty$.
> As $x \to \infty, f(x) \to -\infty$.
>
> **b.** $g(x) = \frac{1}{4}x(2x - 3)^3(x + 4)^2$
>
>   positive   even
>
> $g(x) = \frac{1}{4}x(2x - 3)^3(x + 4)^2 = 2x^6 + \cdots$
>
> To determine the leading term, multiply the leading terms from each factor. That is,
>
> $\frac{1}{4}x(2x)^3(x)^2 = 2x^6$.
>
> The leading coefficient is positive and the degree is even. By the leading term test, the end behavior is up to the left and up to the right.
>
> As $x \to -\infty, f(x) \to \infty$.
> As $x \to \infty, f(x) \to \infty$.
>
>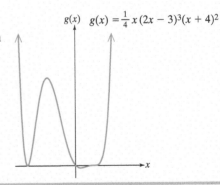
>
> **Skill Practice 1**   Use the leading term to determine the end behavior of the graph of the function.
>
> **a.** $f(x) = -0.3x^4 - 5x^2 - 3x + 4$ **b.** $g(x) = \frac{6}{7}(x - 9)^4(x + 4)^2(3x - 5)$

TIP The graph of $y = f(x)$ from Example 1(a) will exhibit the same behavior as the graph of the power function $y = -4x^5$ for values of x far to the right and far to the left. This is similar to the graph of $y = x^5$ reflected across the x-axis.

2. Identify Zeros and Multiplicities of Zeros

Consider a polynomial function defined by $y = f(x)$. The values of x in the domain of f for which $f(x) = 0$ are called the **zeros** of the function. These are the real solutions (or **roots**) of the equation $f(x) = 0$ and correspond to the x-intercepts of the graph of $y = f(x)$.

Answers
1. a. Down to the left, down to the right.
    As $x \to -\infty, f(x) \to -\infty$.
    As $x \to \infty, f(x) \to -\infty$.

b. Down to the left, up to the right.
    As $x \to -\infty, f(x) \to -\infty$.
    As $x \to \infty, f(x) \to \infty$.

EXAMPLE 2   Determining the Zeros of a Polynomial Function

Find the zeros of the function defined by $f(x) = x^3 + x^2 - 9x - 9$.

Solution:

$$f(x) = x^3 + x^2 - 9x - 9$$
$$0 = x^3 + x^2 - 9x - 9$$

To find the zeros of f, set $f(x) = 0$ and solve for x.

$$0 = x^2(x + 1) - 9(x + 1)$$

Factor by grouping.

$$0 = (x + 1)(x^2 - 9)$$
$$0 = (x + 1)(x - 3)(x + 3)$$

Factor the difference of squares.

$$x = -1, x = 3, x = -3$$

Set each factor equal to zero and solve for x.

The zeros of f are -1, 3, and -3.

The graph of f is shown in Figure 3-9. The zeros of the function are real numbers and correspond to the x-intercepts of the graph. By inspection, we can evaluate $f(0) = -9$, indicating that the y-intercept is $(0, -9)$.

Check:

A table of points can be used to check that $f(-1)$, $f(3)$, and $f(-3)$ all equal 0.

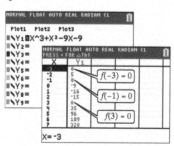

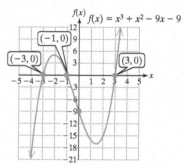

Figure 3-9

Skill Practice 2   Find the zeros of the function defined by
$$f(x) = 4x^3 - 4x^2 - 25x + 25.$$

EXAMPLE 3   Determining the Zeros of a Polynomial Function

Find the zeros of the function defined by $f(x) = -x^3 + 8x^2 - 16x$.

Solution:

$$f(x) = -x^3 + 8x^2 - 16x$$

To find the zeros of f, set $f(x) = 0$ and solve for x.

$$0 = -x(x^2 - 8x + 16)$$

Factor out the GCF.

$$0 = -x(x - 4)^2$$

Factor the perfect square trinomial.

$$x = 0, x = 4$$

Set each factor equal to zero and solve for x.

The zeros of f are 0 and 4.

The graph of f is shown in Figure 3-10. The zeros of the function are real numbers and correspond to the x-intercepts $(0, 0)$ and $(4, 0)$.

The leading term of $f(x)$ is $-x^3$. The coefficient is negative and the exponent is odd. The graph shows the end behavior up to the left and down to the right as expected.

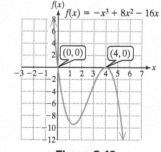

Figure 3-10

Answers

2. $1, \dfrac{5}{2}, -\dfrac{5}{2}$

3. $0, -5$

Skill Practice 3   Find the zeros of the function defined by
$$f(x) = x^3 + 10x^2 + 25x.$$

From Example 3, $f(x) = -x^3 + 8x^2 - 16x$ can be written as a product of linear factors:

$$f(x) = -x(x - 4)^2$$

Notice that the factor $(x - 4)$ appears to the second power. Therefore, we say that the corresponding zero, 4, has a multiplicity of 2. In general, we say that if a polynomial function has a factor $(x - c)$ that appears exactly k times, then c is a **zero of multiplicity k.** For example, consider:

$$g(x) = x^2(x - 2)^3(x + 4)^7 \qquad \text{0 is a zero of multiplicity 2.}$$
$$\text{2 is a zero of multiplicity 3.}$$
$$-4 \text{ is a zero of multiplicity 7.}$$

The graph of a polynomial function behaves in the following manner based on the multiplicity of the zeros.

Touch Points and Cross Points

Let f be a polynomial function and let c be a real zero of f. Then the point $(c, 0)$ is an x-intercept of the graph of f. Furthermore,

- If c is a zero of odd multiplicity, then the graph *crosses* the x-axis at c. The point $(c, 0)$ is called a **cross point.**
- If c is a zero of even multiplicity, then the graph *touches* the x-axis at c and turns back around (does not cross the x-axis). The point $(c, 0)$ is called a **touch point.**

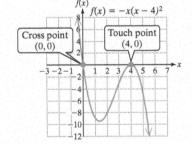

Figure 3-11

To illustrate the behavior of a polynomial function at its real zeros, consider the graph of $f(x) = -x(x - 4)^2$ from Example 3 (Figure 3-11).

- 0 has a multiplicity of 1 (odd multiplicity). The graph *crosses* the x-axis at $(0, 0)$.
- 4 has a multiplicity of 2 (even multiplicity). The graph *touches* the x-axis at $(4, 0)$ and turns back around.

EXAMPLE 4   Determining Zeros and Multiplicities

Determine the zeros and their multiplicities for the given functions.

a. $m(x) = \dfrac{1}{10}(x - 4)^2(2x + 5)^3$ **b.** $n(x) = x^4 - 2x^2$

Solution:

a. $m(x) = \dfrac{1}{10}(x - 4)^{\overset{\text{even}}{2}}(2x + 5)^{\overset{\text{odd}}{3}}$ The function is factored into linear factors. The zeros are 4 and $-\frac{5}{2}$.

The function has a zero of 4 with multiplicity 2 (even). The graph has a touch point at $(4, 0)$.

    The function has a zero of $-\frac{5}{2}$ with multiplicity 3 (odd). The graph has a cross point at $\left(-\frac{5}{2}, 0\right)$.

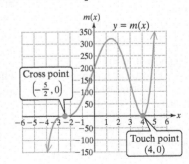

b. $n(x) = x^4 - 2x^2$

$\quad = x^2(x^2 - 2)$

$\quad = x^2(x - \sqrt{2})^1(x + \sqrt{2})^1$

The function has a zero of 0 with multiplicity 2 (even). The graph has a touch point at $(0, 0)$.

The function has a zero of $\sqrt{2}$ with multiplicity 1 (odd). The graph has a cross point at $(\sqrt{2}, 0) \approx (1.41, 0)$.

The function has a zero of $-\sqrt{2}$ with multiplicity 1 (odd). The graph has a cross point at $(-\sqrt{2}, 0) \approx (-1.41, 0)$.

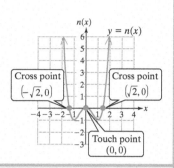

Skill Practice 4   Determine the zeros and their multiplicities for the given functions.

a. $p(x) = -\frac{3}{5}(x + 3)^4(5x - 1)^5$ **b.** $q(x) = 2x^6 - 14x^4$

3. Apply the Intermediate Value Theorem

In Examples 2–4, the zeros of the functions were easily identified by first factoring the polynomial. However, in most cases, the real zeros of a polynomial are difficult or impossible to determine algebraically. For example, the function given by $f(x) = x^4 + 6x^3 - 26x + 15$ has zeros of $-1 \pm \sqrt{6}$ and $-2 \pm \sqrt{7}$. At this point, we do not have the tools to find the zeros of this function analytically. However, we can use the intermediate value theorem to help us search for zeros of a polynomial function and approximate their values.

> **Intermediate Value Theorem**
>
> Let f be a polynomial function. For $a < b$, if $f(a)$ and $f(b)$ have opposite signs, then f has at least one zero on the interval $[a, b]$.

EXAMPLE 5   **Applying the Intermediate Value Theorem**

Show that $f(x) = x^4 + 6x^3 - 26x + 15$ has a zero on the interval $[1, 2]$.

Solution:

$f(x) = x^4 + 6x^3 - 26x + 15$

$f(1) = (1)^4 + 6(1)^3 - 26(1) + 15 = -4$

$f(2) = (2)^4 + 6(2)^3 - 26(2) + 15 = 27$

Since $f(1)$ and $f(2)$ have opposite signs, then by the intermediate value theorem, we know that the function must have at least one zero on the interval $[1, 2]$.

The actual value of the zero on the interval $[1, 2]$ is $-1 + \sqrt{6} \approx 1.45$.

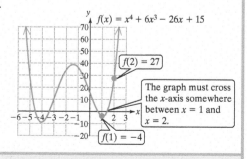

Skill Practice 5   Show that $f(x) = x^4 + 6x^3 - 26x + 15$ has a zero on the interval $[-4, -3]$.

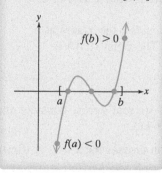
TIP   It is important to note that if the signs of $f(a)$ and $f(b)$ are the same, then the intermediate value theorem is inconclusive.

Answers

4. a. -3 (multiplicity 4) and $\frac{1}{5}$ (multiplicity 5)

b. 0 (multiplicity 4), $\sqrt{7}$ (multiplicity 1), and $-\sqrt{7}$ (multiplicity 1)

5. $f(-4) = -9$ and $f(-3) = 12$. Since $f(-4)$ and $f(-3)$ have opposite signs, then the intermediate value theorem guarantees the existence of at least one zero on the interval $[-4, -3]$.

The intermediate value theorem can be used repeatedly in a technique called the bisection method to approximate the value of a zero. See the online group activity "Investigating the Bisection Method for Finding Zeros."

Point of Interest

The modern definition of a computer is a programmable device designed to carry out a sequence of arithmetic or logical operations. However, the word "computer" originally referred to a person who did such calculations using paper and pencil. "Human computers" were notably used in the eighteenth century to predict the path of Halley's comet and to produce astronomical tables critical to surveying and navigation. Later, during World Wars I and II, human computers developed ballistic firing tables that would describe the trajectory of a shell.

Computing tables of values was very time consuming, and the "computers" would often interpolate to find intermediate values within a table. Interpolation is a method by which intermediate values between two numbers are estimated. Often the interpolated values were based on a polynomial function.

4. Sketch a Polynomial Function

TIP Even with advanced techniques from calculus or the use of a graphing utility, it is often difficult or impossible to find the exact location of the turning points of a polynomial function.

The graph of a polynomial function may also have "turning points." These correspond to relative maxima and minima. For example, consider $f(x) = x(x + 2)(x - 2)^2$. See Figure 3-12.

Multiplying the leading terms within the factors, we have a leading term of $(x)(x)(x)^2 = x^4$. Therefore, the end behavior of the graph is up to the left and up to the right.

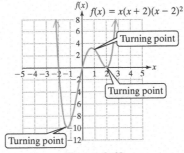

Figure 3-12

Avoiding Mistakes

A polynomial of degree n may have fewer than $n - 1$ turning points. For example, $f(x) = x^3$ is a degree 3 polynomial function (indicating that it could have a maximum of two turning points), yet the graph has no turning points.

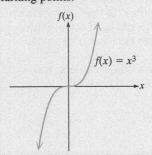

Starting from the far left, the graph of f decreases to the x-intercept of -2. Since -2 is a zero with an odd multiplicity, the graph must cross the x-axis at -2. For the same reason, the graph must cross the x-axis again at the origin. Therefore, somewhere between $x = -2$ and $x = 0$, the graph must "turn around." This point is called a "turning point."

The turning points of a polynomial function are the points where the function changes from increasing to decreasing or vice versa.

Number of Turning Points of a Polynomial Function

Let f represent a polynomial function of degree n. Then the graph of f has at most $n - 1$ turning points.

At this point we are ready to outline a strategy for sketching a polynomial function.

Graphing a Polynomial Function

To graph a polynomial function defined by $y = f(x)$,

1. Use the leading term to determine the end behavior of the graph.
2. Determine the y-intercept by evaluating $f(0)$.
3. Determine the real zeros of f and their multiplicities (these are the x-intercepts of the graph of f).
4. Plot the x- and y-intercepts and sketch the end behavior.
5. Draw a sketch starting from the left-end behavior. Connect the x- and y-intercepts in the order that they appear from left to right using these rules:
   - The curve will cross the x-axis at an x-intercept if the corresponding zero has an odd multiplicity.
   - The curve will touch but not cross the x-axis at an x-intercept if the corresponding zero has an even multiplicity.
6. If a test for symmetry is easy to apply, use symmetry to plot additional points. Recall that
   - f is an even function (symmetric to the y-axis) if $f(-x) = f(x)$.
   - f is an odd function (symmetric to the origin) if $f(-x) = -f(x)$.
7. Plot more points if a greater level of accuracy is desired. In particular, to estimate the location of turning points, find several points between two consecutive x-intercepts.

In Examples 6 and 7, we demonstrate the process of graphing a polynomial function.

EXAMPLE 6   **Graphing a Polynomial Function**

Graph $f(x) = x^3 - 9x$.

Solution:

$f(x) = x^3 - 9x$

1. The leading term is x^3. The end behavior is down to the left and up to the right.

   The exponent on the leading term is odd and the leading coefficient is positive.

2. $f(0) = (0)^3 - 9(0) = 0$
   The y-intercept is $(0, 0)$.

   Determine the y-intercept by evaluating $f(0)$.

3. $0 = x^3 - 9x$
   $0 = x(x^2 - 9)$
   $0 = x(x - 3)(x + 3)$

   Find the real zeros of f by solving for the real solutions to the equation $f(x) = 0$.

   The zeros of the function are 0, 3, and -3, and each has a multiplicity of 1.

   The zeros are real numbers and correspond to x-intercepts on the graph. Since the multiplicity of each zero is an odd number, the graph will cross the x-axis at the zeros.

4.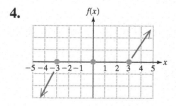

   Plot the x- and y-intercepts and sketch the end behavior.

5. Moving from left to right, the curve increases from the far left and then crosses the x-axis at -3. The graph must have a turning point between $x = -3$ and $x = 0$ so that the curve can pass through the next x-intercept of $(0, 0)$.

The graph crosses the x-axis at $x = 0$. The graph must then have another turning point between $x = 0$ and $x = 3$ so that the curve can pass through the next x-intercept of $(3, 0)$. Finally, the graph crosses the x-axis at $x = 3$ and continues to increase to the far right.

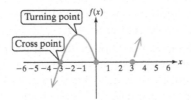

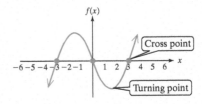

6. $f(x) = x^3 - 9x$

$$f(-x) = (-x)^3 - 9(-x) \qquad -f(x) = -(x^3 - 9x)$$
$$= -x^3 + 9x \xleftarrow{\ \ } \xrightarrow{\ \ } = -x^3 + 9x$$
$$f(-x) = -f(x)$$
(same)

Testing for symmetry, we see that $f(-x) = -f(x)$. Therefore, f is an odd function and is symmetric with respect to the origin.

7. If more accuracy is desired, plot additional points. In this case, since f is symmetric to the origin, if a point (x, y) is on the graph, then so is $(-x, -y)$. The graph of f is shown in Figure 3-13.

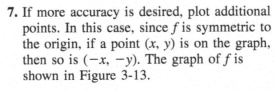

x	$f(x)$
1	-8
2	-10
4	28

Use symmetry.

x	$f(x)$
-1	8
-2	10
-4	-28

Figure 3-13

Skill Practice 6   Graph $g(x) = -x^3 + 4x$.

Answer

6.

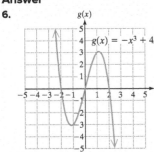

EXAMPLE 7   **Graphing a Polynomial Function**

Graph $g(x) = -0.1(x - 1)(x + 2)(x - 4)^2$.

Solution:

$g(x) = -0.1(x - 1)(x + 2)(x - 4)^2$

1. Multiplying the leading terms within the factors, we have a leading term of $-0.1(x)(x)(x)^2 = -0.1x^4$. The end behavior is down to the left and down to the right.

The exponent on the leading term is even and the leading coefficient is negative.

2. $g(0) = -0.1(0 - 1)(0 + 2)(0 - 4)^2 = 3.2$
The y-intercept is $(0, 3.2)$.

Determine the y-intercept by evaluating $g(0)$.

3. $0 = -0.1(x - 1)(x + 2)(x - 4)^2$
The zeros of the function are 1, -2, and 4.
The multiplicity of 1 is 1.
The multiplicity of -2 is 1.
The multiplicity of 4 is 2.

Find the real zeros of g by solving for the real solutions of the equation $g(x) = 0$.

The zeros are real numbers and correspond to x-intercepts on the graph: $(1, 0)$, $(-2, 0)$, and $(4, 0)$.

4.

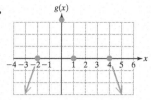

Plot the x- and y-intercepts and sketch the end behavior.

5. Moving from left to right, the curve increases from the far left. It then crosses the x-axis at $x = -2$ and turns back around to pass through the next x-intercept at $x = 1$.

   The curve has another turning point between $x = 1$ and $x = 4$ so that it can touch the x-axis at 4. From there it turns back downward and continues to decrease to the far right.

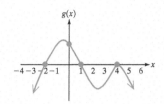

6. From our preliminary sketch in step 5, we see that the function is not symmetric with respect to either the y-axis or origin.

7. If more accuracy is desired, plot additional points. The graph is shown in Figure 3-14.

x	$g(x)$
-3	-19.6
-1	5
2	-1.6
3	-1
5	-2.8

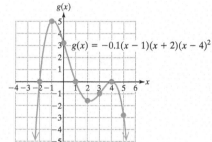

$g(x) = -0.1(x - 1)(x + 2)(x - 4)^2$

Figure 3-14

Skill Practice 7 Graph $h(x) = 0.5x(x - 1)(x + 3)^2$.

TECHNOLOGY CONNECTIONS

Using a Graphing Utility to Graph a Polynomial Function

It is important to have a strong knowledge of algebra to use a graphing utility effectively. For example, consider the graph of $f(x) = 0.005(x - 2)(x + 3)(x - 5)(x + 15)$ on the standard viewing window.

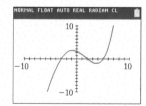

From the leading term, $0.005x^4$, we know that the end behavior should be up to the left and up to the right. Furthermore, the function has four real zeros (2, -3, 5, and -15), and should have four corresponding x-intercepts. Therefore, on the standard viewing window, the calculator does not show the key features of the graph.

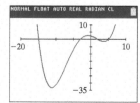

By graphing f on the window $[-20, 10, 2]$ by $[-35, 10, 5]$, we see the end behavior displayed correctly, all four x-intercepts, and the turning points (there should be at most 3).

Answer

7.

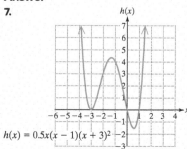

$h(x) = 0.5x(x - 1)(x + 3)^2$

SECTION 3.2 | Practice Exercises

Prerequisite Review

For Exercises R.1–R.2, solve the equation.

R.1. $3x^3 + 21x^2 - 54x = 0$

R.2. $5x^3 + 6x^2 - 20x - 24 = 0$

For Exercises R.3–R.5, use transformations to graph the given function.

R.3. $m(x) = x^3 - 5$

R.4. $f(x) = (x + 2)^2 - 4$

R.5. $g(x) = (3x - 6)^2$

Concept Connections

1. A function defined by $f(x) = a_n x^n + a_{n-1}x^{n-1} + a_{n-2}x^{n-2} + \cdots + a_1 x + a_0$ where $a_n, a_{n-1}, a_{n-2}, \ldots, a_1, a_0$ are real numbers and $a_n \neq 0$ is called a _____ function.

2. The function given by $f(x) = -3x^5 + \sqrt{2}x + \frac{1}{2}x$ (is/is not) a polynomial function.

3. The function given by $f(x) = -3x^5 + 2\sqrt{x} + \frac{2}{x}$ (is/is not) a polynomial function.

4. A quadratic function is a polynomial function of degree _____.

5. A linear function is a polynomial function of degree _____.

6. The values of x in the domain of a polynomial function f for which $f(x) = 0$ are called the _____ of the function.

7. What is the maximum number of turning points of the graph of $f(x) = -3x^6 - 4x^5 - 5x^4 + 2x^2 + 6$?

8. If the graph of a polynomial function has 3 turning points, what is the minimum degree of the function?

9. If c is a real zero of a polynomial function and the multiplicity is 3, does the graph of the function cross the x-axis or touch the x-axis (without crossing) at $(c, 0)$?

10. If c is a real zero of a polynomial function and the multiplicity is 6, does the graph of the function cross the x-axis or touch the x-axis (without crossing) at $(c, 0)$?

11. Suppose that f is a polynomial function and that $a < b$. If $f(a)$ and $f(b)$ have opposite signs, then what conclusion can be drawn from the intermediate value theorem?

12. What is the leading term of $f(x) = -\frac{1}{3}(x - 3)^4(3x + 5)^2$?

Objective 1: Determine the End Behavior of a Polynomial Function

For Exercises 13–20, determine the end behavior of the graph of the function. (See Example 1)

13. $f(x) = -3x^4 - 5x^2 + 2x - 6$

14. $g(x) = -\frac{1}{2}x^6 + 8x^4 - x^3 + 9$

15. $h(x) = 12x^5 + 8x^4 - 4x^3 - 8x + 1$

16. $k(x) = 11x^7 - 4x^2 + 9x + 3$

17. $m(x) = -4(x - 2)(2x + 1)^2(x + 6)^4$

18. $n(x) = -2(x + 4)(3x - 1)^3(x + 5)$

19. $p(x) = -2x^2(3 - x)(2x - 3)^3$

20. $q(x) = -5x^4(2 - x)^3(2x + 5)$

Objective 2: Identify Zeros and Multiplicities of Zeros

21. Given the function defined by $g(x) = -3(x - 1)^3(x + 5)^4$, the value 1 is a zero with multiplicity _____, and the value -5 is a zero with multiplicity _____.

22. Given the function defined by $h(x) = \frac{1}{2}x^5(x + 0.6)^3$, the value 0 is a zero with multiplicity _____, and the value -0.6 is a zero with multiplicity _____.

For Exercises 23–38, find the zeros of the function and state the multiplicities. (See Examples 2–4)

23. $f(x) = x^3 + 2x^2 - 25x - 50$ **24.** $g(x) = x^3 + 5x^2 - x - 5$ **25.** $h(x) = -6x^3 - 9x^2 + 60x$

26. $k(x) = -6x^3 + 26x^2 - 28x$ **27.** $m(x) = x^5 - 10x^4 + 25x^3$ **28.** $n(x) = x^6 + 4x^5 + 4x^4$

29. $p(x) = -3x(x + 2)^3(x + 4)$ **30.** $q(x) = -2x^4(x + 1)^3(x - 2)^2$

31. $t(x) = 5x(3x - 5)(2x + 9)(x - \sqrt{3})(x + \sqrt{3})$ **32.** $z(x) = 4x(5x - 1)(3x + 8)(x - \sqrt{5})(x + \sqrt{5})$

33. $c(x) = [x - (3 - \sqrt{5})][x - (3 + \sqrt{5})]$ **34.** $d(x) = [x - (2 - \sqrt{11})][x - (2 + \sqrt{11})]$

35. $f(x) = 4x^4 - 37x^2 + 9$ **36.** $k(x) = 4x^4 - 65x^2 + 16$

37. $n(x) = x^6 - 7x^4$ **38.** $m(x) = x^5 - 5x^3$

Objective 3: Apply the Intermediate Value Theorem

For Exercises 39–40, determine whether the intermediate value theorem guarantees that the function has a zero on the given interval. (See Example 5)

39. $f(x) = 2x^3 - 7x^2 - 14x + 30$

    a. $[1, 2]$ **b.** $[2, 3]$

    c. $[3, 4]$ **d.** $[4, 5]$

40. $g(x) = 2x^3 - 13x^2 + 18x + 5$

    a. $[1, 2]$ **b.** $[2, 3]$

    c. $[3, 4]$ **d.** $[4, 5]$

For Exercises 41–42, a table of values is given for $Y_1 = f(x)$. Determine whether the intermediate value theorem guarantees that the function has a zero on the given interval.

41. $Y_1 = 21x^4 + 46x^3 - 238x^2 - 506x + 77$

    a. $[-4, -3]$

    b. $[-3, -2]$

    c. $[-2, -1]$

    d. $[-1, 0]$

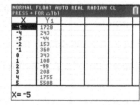

42. $Y_1 = 10x^4 + 21x^3 - 119x^2 - 147x + 343$

    a. $[-4, -3]$

    b. $[-3, -2]$

    c. $[-2, -1]$

    d. $[-1, 0]$

43. Given $f(x) = 4x^3 - 8x^2 - 25x + 50$,

    a. Determine if f has a zero on the interval $[-3, -2]$.

    b. Find a zero of f on the interval $[-3, -2]$.

44. Given $f(x) = 9x^3 - 18x^2 - 100x + 200$,

    a. Determine if f has a zero on the interval $[-4, -3]$.

    b. Find a zero of f on the interval $[-4, -3]$.

Objective 4: Sketch a Polynomial Function

For Exercises 45–52, determine if the graph can represent a polynomial function. If so, assume that the end behavior and all turning points are represented in the graph.

a. Determine the minimum degree of the polynomial.

b. Determine whether the leading coefficient is positive or negative based on the end behavior and whether the degree of the polynomial is odd or even.

c. Approximate the real zeros of the function, and determine if their multiplicities are even or odd.

45.

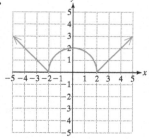

46.

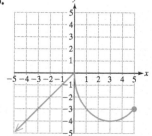

47.

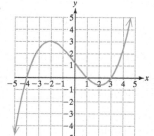

48.

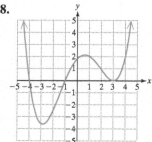

49. 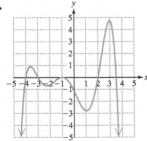   **50.** 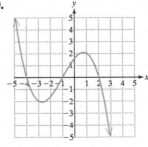   **51.** 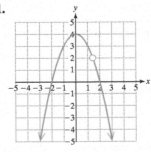   **52.**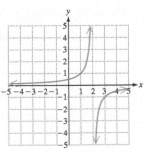

For Exercises 53–58,

a. Identify the power function of the form $y = x^n$ that is the parent function to the given graph.

b. In order, outline the transformations that would be required on the graph of $y = x^n$ to make the graph of the given function. See Section 2.6, page 236.

c. Match the function with the graph of i–vi.

53. $g(x) = -\dfrac{1}{3}x^6 - 2$

54. $f(x) = -\dfrac{1}{2}(x - 3)^4$

55. $k(x) = -(x + 2)^3 + 3$

56. $p(x) = 2(x + 4)^3 - 3$

57. $m(x) = (-x - 3)^5 + 1$

58. $n(x) = (-x + 3)^4 - 1$

i. 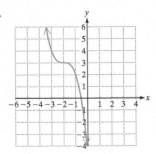   **ii.** 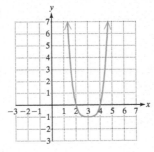   **iii.**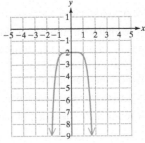

iv. 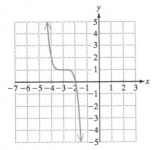   **v.** 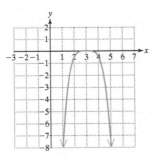   **vi.**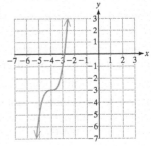

For Exercises 59–76, sketch the function. (See Examples 6–7)

59. $f(x) = x^3 - 5x^2$

60. $g(x) = x^5 - 2x^4$

61. $f(x) = \dfrac{1}{2}(x - 2)(x + 1)(x + 3)$

62. $h(x) = \dfrac{1}{4}(x - 1)(x - 4)(x + 2)$

63. $k(x) = x^4 + 2x^3 - 8x^2$

64. $h(x) = x^4 - x^3 - 6x^2$

65. $k(x) = 0.2(x + 2)^2(x - 4)^3$

66. $m(x) = 0.1(x - 3)^2(x + 1)^3$

67. $p(x) = 9x^5 + 9x^4 - 25x^3 - 25x^2$

68. $q(x) = 9x^5 + 18x^4 - 4x^3 - 8x^2$

69. $t(x) = -x^4 + 11x^2 - 28$

70. $v(x) = -x^4 + 15x^2 - 44$

71. $g(x) = -x^4 + 5x^2 - 4$

72. $h(x) = -x^4 + 10x^2 - 9$

73. $c(x) = 0.1x(x - 2)^4(x + 2)^3$

74. $d(x) = 0.05x(x - 2)^4(x + 3)^2$

75. $m(x) = -\dfrac{1}{10}(x + 3)(x - 3)(x + 1)^3$

76. $f(x) = -\dfrac{1}{10}(x - 1)(x + 3)(x - 4)^2$

Mixed Exercises

For Exercises 77–88, determine if the statement is true or false. If a statement is false, explain why.

77. The function defined by $f(x) = (x + 1)^5(x - 5)^2$ crosses the x-axis at 5.

78. The function defined by $g(x) = -3(x + 4)(2x - 3)^4$ touches but does not cross the x-axis at $\left(\frac{3}{2}, 0\right)$.

79. A third-degree polynomial has three turning points.

80. A third-degree polynomial has two turning points.

81. There is more than one polynomial function with zeros of 1, 2, and 6.

82. There is exactly one polynomial with integer coefficients with zeros of 2, 4, and 6.

83. The graph of a polynomial function with leading term of even degree is up to the far left and up to the far right.

84. If c is a real zero of an even polynomial function, then $-c$ is also a zero of the function.

85. The graph of $f(x) = x^3 - 27$ has three x-intercepts.

86. The graph of $f(x) = 3x^2(x - 4)^4$ has no points in Quadrants III or IV.

87. The graph of $p(x) = -5x^4(x + 1)^2$ has no points in Quadrants I or II.

88. A fourth-degree polynomial has exactly two relative minima and two relative maxima.

89. A rocket will carry a communications satellite into low Earth orbit. Suppose that the thrust during the first 200 sec of flight is provided by solid rocket boosters at different points during liftoff.

The graph shows the acceleration in G-forces (that is, acceleration in 9.8-m/sec² increments) versus time after launch.

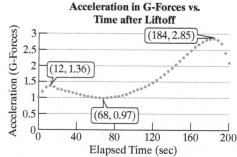

Acceleration in G-Forces vs. Time after Liftoff

a. Approximate the interval(s) over which the acceleration is increasing.

b. Approximate the interval(s) over which the acceleration is decreasing.

c. How many turning points does the graph show?

d. Based on the number of turning points, what is the minimum degree of a polynomial function that could be used to model acceleration versus time? Would the leading coefficient be positive or negative?

e. Approximate the time when the acceleration was the greatest.

f. Approximate the value of the maximum acceleration.

90. Data from a 20-yr study show the number of new AIDS cases diagnosed among 20- to 24-yr-olds in the United States x years after the study began.

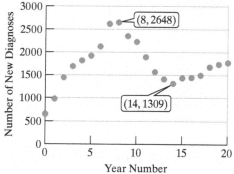

AIDS Diagnoses, 20- to 24-Yr-Olds United States, 1985–2005

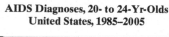

a. Approximate the interval(s) over which the number of new AIDS cases among 20- to 24-yr-olds increased.

b. Approximate the interval(s) over which the number of new AIDS cases among 20- to 24-yr-olds decreased.

c. How many turning points does the graph show?

d. Based on the number of turning points, what is the minimum degree of a polynomial function that could be used to model the data? Would the leading coefficient be positive or negative?

e. How many years after the study began was the number of new AIDS cases among 20- to 24-yr-olds the greatest?

f. What was the maximum number of new cases diagnosed in a single year?

Write About It

91. Given a polynomial function defined by $y = f(x)$, explain how to find the x-intercepts.

92. Given a polynomial function, explain how to determine whether an x-intercept is a touch point or a cross point.

93. Write an informal explanation of what it means for a function to be continuous.

94. Write an informal explanation of the intermediate value theorem.

Expanding Your Skills

The intermediate value theorem given on page 306 is actually a special case of a broader statement of the theorem. Consider the following:

> Let f be a polynomial function. For $a < b$, if $f(a) \neq f(b)$,
> then f takes on every value between $f(a)$ and $f(b)$ on the interval $[a, b]$.

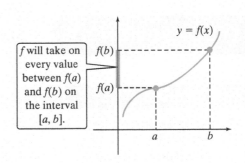

Use this broader statement of the intermediate value theorem for Exercises 95–96.

95. Given $f(x) = x^2 - 3x + 2$,

  a. Evaluate $f(3)$ and $f(4)$.

  b. Use the intermediate value theorem to show that there exists at least one value of x for which $f(x) = 4$ on the interval $[3, 4]$.

  c. Find the value(s) of x for which $f(x) = 4$ on the interval $[3, 4]$.

96. Given $f(x) = -x^2 - 4x + 3$,

  a. Evaluate $f(-4)$ and $f(-3)$.

  b. Use the intermediate value theorem to show that there exists at least one value of x for which $f(x) = 5$ on the interval $[-4, -3]$.

  c. Find the value(s) of x for which $f(x) = 5$ on the interval $[-4, -3]$.

Technology Connections

97. For a certain individual, the volume (in liters) of air in the lungs during a 4.5-sec respiratory cycle is shown in the table for 0.5-sec intervals. Graph the points and then find a third-degree polynomial function to model the volume $V(t)$ for t between 0 sec and 4.5 sec. (*Hint:* Use a CubicReg option or polynomial degree 3 option on a graphing utility.)

Time (sec)	Volume (L)
0.0	0.00
0.5	0.11
1.0	0.29
1.5	0.47
2.0	0.63
2.5	0.76
3.0	0.81
3.5	0.75
4.0	0.56
4.5	0.20

98. The torque (in ft-lb) produced by a certain automobile engine turning at x thousand revolutions per minute is shown in the table. Graph the points and then find a third-degree polynomial function to model the torque $T(x)$ for $1 \leq x \leq 5$.

Engine speed (1000 rpm)	Torque (ft-lb)
1.0	165
1.5	180
2.0	188
2.5	190
3.0	186
3.5	176
4.0	161
4.5	142
5.0	120

99. A solar oven is to be made from an open box with reflective sides. Each box is made from a 30-in. by 24-in. rectangular sheet of aluminum with squares of length x (in inches) removed from each corner. Then the flaps are folded up to form an open box.

  a. Show that the volume of the box is given by
$$V(x) = 4x^3 - 108x^2 + 720x \quad \text{for } 0 < x < 12.$$

  b. Graph the function from part (a) and use a "Maximum" feature on a graphing utility to approximate the length of the sides of the squares that should be removed to maximize the volume. Round to the nearest tenth of an inch.

  c. Approximate the maximum volume. Round to the nearest cubic inch.

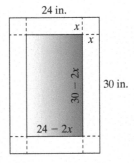

For Exercises 100–101, two viewing windows are given for the graph of $y = f(x)$. Choose the window that best shows the key features of the graph.

100. $f(x) = 2(x - 0.5)(x - 0.1)(x + 0.2)$

  a. $[-10, 10, 1]$ by $[-10, 10, 1]$

  b. $[-1, 1, 0.1]$ by $[-0.05, 0.05, 0.01]$

101. $g(x) = 0.08(x - 16)(x + 2)(x - 3)$

  a. $[-10, 10, 1]$ by $[-10, 10, 1]$

  b. $[-5, 20, 5]$ by $[-50, 30, 10]$

For Exercises 102–103, graph the function defined by $y = f(x)$ on an appropriate viewing window.

102. $k(x) = \dfrac{1}{100}(x - 20)(x + 1)(x + 8)(x - 6)$

103. $p(x) = (x - 0.4)(x + 0.5)(x + 0.1)(x - 0.8)$

Division of Polynomials and the Remainder and Factor Theorems

1. Divide Polynomials Using Long Division

In this section, we use the notation $f(x)$, $g(x)$, and so on to represent polynomials in x. We also present two types of polynomial division: long division and synthetic division. Polynomial division can be used to factor a polynomial, solve a polynomial equation, and find the zeros of a polynomial.

When dividing polynomials, if the divisor has two or more terms we can use a long division process similar to the division of real numbers. This is demonstrated in Examples 1–3.

EXAMPLE 1 **Dividing Polynomials Using Long Division**

Use long division to divide. $(6x^3 - 5x^2 - 3) \div (3x + 2)$

Solution:

First note that the dividend can be written as $6x^3 - 5x^2 + 0x - 3$. The term $0x$ is used as a place holder for the missing power of x. The place holder is helpful to keep the powers of x lined up. We also set up long division with both the dividend and divisor written in descending order.

TIP Take a minute to review long division of whole numbers: $2273 \div 5$

$$
\begin{array}{r}
454 \leftarrow \text{Quotient} \\
5\overline{)2273} \\
-20 \downarrow \\
\overline{27} \\
-25 \downarrow \\
\overline{23} \\
-20 \\
\overline{3} \leftarrow \text{Remainder}
\end{array}
$$

Answer: $454 + \frac{3}{5}$ or $454\frac{3}{5}$

$$3x + 2\overline{)6x^3 - 5x^2 + 0x - 3}$$

Divide the leading term in the dividend by the leading term in the divisor.

$$\frac{6x^3}{3x} = 2x^2$$ This is the first term in the quotient.

$$
\begin{array}{r}
2x^2 \\
3x + 2\overline{)6x^3 - 5x^2 + 0x - 3} \\
-(6x^3 + 4x^2)
\end{array}
$$

Subtract.

Multiply the divisor by $2x^2$: $2x^2(3x + 2) = 6x^3 + 4x^2$, and subtract the result.

$$
\begin{array}{r}
2x^2 \\
3x + 2\overline{)6x^3 - 5x^2 + 0x - 3} \\
-(6x^3 + 4x^2) \\
\overline{-9x^2 + 0x}
\end{array}
$$

Bring down the next term from the dividend and repeat the process.

$$
\begin{array}{r}
2x^2 - 3x \\
3x + 2\overline{)6x^3 - 5x^2 + 0x - 3} \\
-(6x^3 + 4x^2) \\
\overline{-9x^2 + 0x} \\
-(-9x^2 - 6x)
\end{array}
$$

Subtract.

Divide $-9x^2$ by the first term in the divisor.

$$\frac{-9x^2}{3x} = -3x$$

Multiply the divisor by $-3x$: $-3x(3x + 2) = -9x^2 - 6x$, and subtract the result.

$$
\begin{array}{r}
2x^2 - 3x + 2 \\
3x + 2\overline{)6x^3 - 5x^2 + 0x - 3} \\
-(6x^3 + 4x^2) \\
\overline{-9x^2 + 0x} \\
-(-9x^2 - 6x) \\
\overline{6x - 3} \\
-(6x + 4) \\
\overline{-7}
\end{array}
$$

Bring down the next term from the dividend and repeat the process.

Divide $6x$ by the first term in the divisor. $\frac{6x}{3x} = 2$. This is the next term in the quotient.

Multiply the divisor by 2: $2(3x + 2) = 6x + 4$, and subtract the result.

The remainder is -7.

Long division is complete when the remainder is either zero or has degree less than the degree of the divisor.

The quotient is $2x^2 - 3x + 2$.
The remainder is -7.
The divisor is $3x + 2$.
The dividend is $6x^3 - 5x^2 - 3$.

The result of a long division problem is usually written as the quotient plus the remainder divided by the divisor.

$$\underbrace{\boxed{\text{Dividend}} \frac{6x^3 - 5x^2 - 3}{\underset{\boxed{\text{Divisor}}}{3x + 2}}}_{} = \overbrace{2x^2 - 3x + 2}^{\boxed{\text{Quotient}}} + \frac{-7}{3x + 2} \begin{matrix} \boxed{\text{Remainder}} \\ \boxed{\text{Divisor}} \end{matrix}$$

Skill Practice 1 Use long division to divide $(4x^3 - 23x + 3) \div (2x - 5)$.

By clearing fractions, the result of Example 1 can be checked by multiplication.

$$\text{Dividend} = (\text{Divisor})(\text{Quotient}) \quad + \text{Remainder}$$
$$6x^3 - 5x^2 - 3 \overset{?}{=} (3x + 2)(2x^2 - 3x + 2) + (-7)$$
$$\overset{?}{=} 6x^3 - 5x^2 + 4 + (-7)$$
$$\overset{?}{=} 6x^3 - 5x^2 - 3 \checkmark$$

This result illustrates the division algorithm.

Division Algorithm

Suppose that $f(x)$ and $d(x)$ are polynomials where $d(x) \neq 0$ and the degree of $d(x)$ is less than or equal to the degree of $f(x)$. Then there exists unique polynomials $q(x)$ and $r(x)$ such that

$$f(x) = d(x) \cdot q(x) + r(x)$$

where either the degree of $r(x)$ is less than $d(x)$, or $r(x)$ is the zero polynomial.

Note: The polynomial $f(x)$ is the **dividend,** $d(x)$ is the **divisor,** $q(x)$ is the **quotient,** and $r(x)$ is the **remainder.**

EXAMPLE 2   Dividing Polynomials Using Long Division

Use long division to divide $(-5 + x + 4x^2 + 2x^3 + 3x^4) \div (x^2 + 2)$.

Solution:

Write the dividend and divisor in descending order and insert place holders for missing powers of x. $(3x^4 + 2x^3 + 4x^2 + x - 5) \div (x^2 + 0x + 2)$

$$\begin{array}{r} 3x^2 + 2x - 2 \\ x^2 + 0x + 2 \overline{)\, 3x^4 + 2x^3 + 4x^2 + x - 5} \\ \underline{-(3x^4 + 0x^3 + 6x^2)} \\ 2x^3 - 2x^2 + x \\ \underline{-(2x^3 + 0x^2 + 4x)} \\ -2x^2 - 3x - 5 \\ \underline{-(-2x^2 + 0x - 4)} \\ -3x - 1 \end{array}$$

To begin, divide the leading term in the dividend by the leading term in the divisor.
$$\frac{3x^4}{x^2} = 3x^2$$

Multiply the divisor by $3x^2$ and subtract the result.

Bring down the next term from the dividend and repeat the process.

The process is complete when the remainder is either 0 or has degree less than the degree of the divisor.

The result is $3x^2 + 2x - 2 + \dfrac{-3x - 1}{x^2 + 2}$.

Answer

1. $2x^2 + 5x + 1 + \dfrac{8}{2x - 5}$

Check by using the division algorithm.

$$3x^4 + 2x^3 + 4x^2 + x - 5 \stackrel{?}{=} (x^2 + 2)(3x^2 + 2x - 2) + (-3x - 1)$$
$$\stackrel{?}{=} 3x^4 + 2x^3 - 2x^2 + 6x^2 + 4x - 4 + (-3x - 1)$$
$$\stackrel{?}{=} 3x^4 + 2x^3 + 4x^2 + x - 5 \checkmark$$

Skill Practice 2 Use long division to divide.

$(1 - 7x + 5x^2 - 3x^3 + 2x^4) \div (x^2 + 3)$

In Example 3, we discuss the implications of obtaining a remainder of zero when performing division of polynomials.

EXAMPLE 3 Dividing Polynomials Using Long Division

Use long division to divide. $\dfrac{2x^2 + 3x - 14}{x - 2}$

Solution:

$$
\begin{array}{r}
2x + 7 \\
x - 2 \overline{)\, 2x^2 + 3x - 14} \\
-(2x^2 - 4x) \\
\hline
7x - 14 \\
-(7x - 14) \\
\hline
0
\end{array}
$$

To begin, divide the leading term in the dividend by the leading term in the divisor.

$\dfrac{2x^2}{x} = 2x$

Multiply the divisor by $2x$ and subtract the result.

Bring down the next term from the dividend and repeat the process.

The process is complete when the remainder is either 0 or has degree less than the degree of the divisor.

$\dfrac{2x^2 + 3x - 14}{x - 2} = 2x + 7$

The remainder is zero. This implies that the divisor divides evenly into the dividend. Therefore, both the divisor and quotient are factors of the dividend. This is easily verified by the division algorithm.

| Dividend | Divisor | Quotient | Remainder |

$2x^2 + 3x - 14 \stackrel{?}{=} (x - 2)(2x + 7) + 0$
$\stackrel{?}{=} (x - 2)(2x + 7)$

Factored form of $2x^2 + 3x - 14$

Skill Practice 3 Use long division to divide.

$(3x^2 - 14x + 15) \div (x - 3)$

---

2. Divide Polynomials Using Synthetic Division

When dividing polynomials where the divisor is a binomial of the form $(x - c)$ and c is a constant, we can use synthetic division. Synthetic division enables us to find the quotient and remainder more quickly than long division. It uses an algorithm that manipulates the coefficients of the dividend, divisor, and quotient without the accompanying variable factors.

The division of polynomials from Example 3 is shown at the top left of page 319. The equivalent synthetic division is shown on the right. Notice that the same coefficients are used in both cases.

Answers

2. $2x^2 - 3x - 1 + \dfrac{2x + 4}{x^2 + 3}$

3. $3x - 5$

$$\begin{array}{r} 2x + 7 \\ x - 2\overline{)2x^2 + 3x - 14} \\ \underline{-(2x^2 - 4x)} \\ 7x - 14 \\ \underline{-(7x - 14)} \\ 0 \end{array}$$

Coefficients of dividend

$$\begin{array}{r} \underline{2|} \quad 2 \quad 3 \quad -14 \\ \quad 4 \quad 14 \\ \hline 2 \quad 7 \quad \boxed{0} \longleftarrow \text{Remainder} \end{array}$$

Coefficients of quotient

In Example 4, we demonstrate the process to divide polynomials by synthetic division.

EXAMPLE 4   **Dividing Polynomials Using Synthetic Division**

Use synthetic division to divide. $(-10x^2 + 2x^3 - 5) \div (x - 4)$

Solution:

As with long division, the terms of the dividend and divisor must be written in descending order with place holders for missing powers of x.

$$(2x^3 - 10x^2 + 0x - 5) \div (x - 4)$$

To use synthetic division, the divisor must be in the form $x - c$. In this case, $c = 4$.

Step 1: Write the value of c in a box. ⟶ $\underline{4|}$ $\quad 2 \quad -10 \quad 0 \quad -5$ ⟵ **Step 2:** Write the coefficients of the dividend to the right of the box.

$$2$$

Step 3: Skip a line and draw a horizontal line below the list of coefficients.

Step 4: Bring down the leading coefficient from the dividend and write it below the line.

Step 5: Multiply the value of c by the number below the line ($4 \times 2 = 8$). Write the result in the next column above the line.

$$\begin{array}{r} \underline{4|} \quad 2 \quad -10 \quad 0 \quad -5 \\ \quad 8 \\ \hline 2 \quad -2 \end{array}$$

Step 6: Add the numbers in the column above the line ($-10 + 8 = -2$), and write the result below the line.

Repeat steps 5 and 6 until all columns have been completed.

$$\begin{array}{r} \underline{4|} \quad 2 \quad -10 \quad 0 \quad -5 \\ \quad 8 \quad -8 \quad -32 \\ \hline 2 \quad -2 \quad -8 \quad \boxed{-37} \end{array}$$

$$\quad x^2 \quad x \quad \text{constant}$$

A box is often drawn around the remainder.

The rightmost number below the line is the remainder. The other numbers below the line are the coefficients of the quotient in order by the degree of the term.

Since the divisor is linear (first degree), the degree of the quotient is 1 less than the degree of the dividend. In this case, the dividend is of degree 3. Therefore, the quotient will be of degree 2.

The quotient is $2x^2 - 2x - 8$ and the remainder is -37. Therefore,

$$\frac{2x^3 - 10x^2 - 5}{x - 4} = 2x^2 - 2x - 8 + \frac{-37}{x - 4}$$

Skill Practice 4   Use synthetic division to divide.

$$(4x^3 - 28x - 7) \div (x - 3)$$

Avoiding Mistakes

It is important to check that the divisor is in the form $(x - c)$ before applying synthetic division. The variable x in the divisor must be of first degree, and its coefficient must be 1.

Answer

4. $4x^2 + 12x + 8 + \dfrac{17}{x - 3}$

EXAMPLE 5   Dividing Polynomials Using Synthetic Division

Use synthetic division to divide. $(-2x + 4x^3 + 18 + x^4) \div (x + 2)$

Solution:

Write the dividend and divisor in descending order and insert place holders for missing powers of x. $(x^4 + 4x^3 + 0x^2 - 2x + 18) \div (x + 2)$

To use synthetic division, the divisor must be of the form $x - c$. In this case, we have $x + 2 = x - (-2)$. Therefore, $c = -2$.

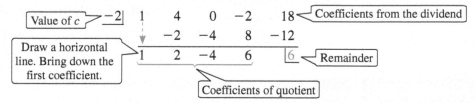

Value of c — -2 | 1   4   0   −2   18 — Coefficients from the dividend

 −2   −4   8   −12

Draw a horizontal line. Bring down the first coefficient. — 1   2   −4   6   ⬚6 — Remainder

Coefficients of quotient

The dividend is a fourth-degree polynomial and the divisor is a first-degree polynomial. Therefore, the quotient is a third-degree polynomial. The coefficients of the quotient are found below the line: 1, 2, −4, 6. The quotient is $x^3 + 2x^2 - 4x + 6$, and the remainder is 6.

$$\frac{x^4 + 4x^3 - 2x + 18}{x + 2} = x^3 + 2x^2 - 4x + 6 + \frac{6}{x + 2}$$

> **Skill Practice 5**   Use synthetic division to divide.
>
> $(-3x + 7x^3 + 5 + 2x^4) \div (x + 1)$

TIP   Given a divisor of the form $(x - c)$, we can determine the value of c by setting the divisor equal to zero and solving for x. In Example 5, we have $x + 2 = 0$, which implies that $x = -2$. The value of c is -2.

3. Apply the Remainder and Factor Theorems

Consider the special case of the division algorithm where $f(x)$ is the dividend and $(x - c)$ is the divisor.

$$f(x) = (x - c) \cdot q(x) + r$$ The remainder r is constant because its degree must be one less than the degree of $x - c$.

Now evaluate $f(c)$: $f(c) = (c - c) \cdot q(c) + r$

$f(c) = 0 \cdot q(c) + r$

$f(c) = r$

This result is stated formally as the remainder theorem.

> ### Remainder Theorem
>
> If a polynomial $f(x)$ is divided by $x - c$, then the remainder is $f(c)$.
>
> *Note:* The remainder theorem tells us that the value of $f(c)$ is the same as the remainder we get from dividing $f(x)$ by $x - c$.

Answer

5. $2x^3 + 5x^2 - 5x + 2 + \dfrac{3}{x + 1}$

| EXAMPLE 6 | Using the Remainder Theorem to Evaluate a Polynomial |

Given $f(x) = x^4 + 6x^3 - 12x^2 - 30x + 35$, use the remainder theorem to evaluate

a. $f(2)$ **b.** $f(-7)$

Solution:

a. If $f(x)$ is divided by $x - 2$, then the remainder is $f(2)$.

$$
\begin{array}{r|rrrrr}
2 & 1 & 6 & -12 & -30 & 35 \\
  &   & 2 & 16 & 8 & -44 \\
\hline
  & 1 & 8 & 4 & -22 & \boxed{-9}
\end{array}
$$

By the remainder theorem, $f(2) = -9$.

b. If $f(x)$ is divided by $x - (-7)$ or equivalently $x + 7$, then the remainder is $f(-7)$.

$$
\begin{array}{r|rrrrr}
-7 & 1 & 6 & -12 & -30 & 35 \\
   &   & -7 & 7 & 35 & -35 \\
\hline
   & 1 & -1 & -5 & 5 & \boxed{0}
\end{array}
$$

By the remainder theorem, $f(-7) = 0$.

The results can be checked by direct substitution.

$$f(2) = (2)^4 + 6(2)^3 - 12(2)^2 - 30(2) + 35 = -9 \checkmark$$
$$f(-7) = (-7)^4 + 6(-7)^3 - 12(-7)^2 - 30(-7) + 35 = 0 \checkmark$$

> **Skill Practice 6**   Given $f(x) = x^4 + x^3 - 6x^2 - 5x - 15$, use the remainder theorem to evaluate
>
> **a.** $f(5)$ **b.** $f(-3)$

TIP From Example 6, the values $f(2) = -9$ and $f(-7) = 0$, imply that $(2, -9)$ and $(-7, 0)$ are on the graph of $y = f(x)$.

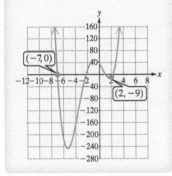

TIP Polynomials with complex coefficients include polynomials with real coefficients and with imaginary coefficients. The following are complex polynomials.

$$f(x) = (2 + 3i)x^2 + 4i$$
$$g(x) = \sqrt{2}x^2 + 3x + 4i$$
$$h(x) = 2x^2 + 3x + 4$$

The division algorithm and remainder theorem can be extended over the set of complex numbers. The definition of a polynomial was given in Section R.4.

$$f(x) = a_n x^n + a_{n-1} x^{n-1} + a_{n-2} x^{n-2} + \cdots + a_1 x + a_0$$

where $a_n \neq 0$ and the coefficients $a_n, a_{n-1}, a_{n-2}, \ldots, a_0$ are real numbers. We now extend our discussion to **complex polynomials.** These are polynomials with complex coefficients.

We will also evaluate polynomials over the set of complex numbers rather than restricting x to the set of real numbers. A complex number $a + bi$ is a zero of a polynomial $f(x)$ if $f(a + bi) = 0$. For example, given $f(x) = x - (5 + 2i)$, we see that the imaginary number $5 + 2i$ is a zero of $f(x)$.

| EXAMPLE 7 | Using the Remainder Theorem to Identify Zeros of a Polynomial |

Use the remainder theorem to determine if the given number c is a zero of the polynomial.

a. $f(x) = 2x^3 - 4x^2 - 13x - 9$; $c = 4$
b. $f(x) = x^3 + x^2 - 3x - 3$; $c = \sqrt{3}$
c. $f(x) = x^3 + x + 10$; $c = 1 + 2i$

Answers
6. a. 560   **b.** 0

Solution:

In each case, divide $f(x)$ by $x - c$ to determine the remainder. If the remainder is 0, then the value c is a zero of the polynomial.

a. Divide $f(x) = 2x^3 - 4x^2 - 13x - 9$ by $x - 4$.

$$
\begin{array}{r|rrrr}
4 & 2 & -4 & -13 & -9 \\
  &   & 8 & 16 & 12 \\
\hline
  & 2 & 4 & 3 & \boxed{3}
\end{array}
$$

By the remainder theorem, $f(4) = 3$. Since $f(4) \neq 0$, 4 is not a zero of $f(x)$.

b. Divide $f(x) = x^3 + x^2 - 3x - 3$ by $x - \sqrt{3}$.

$$
\begin{array}{r|rrrr}
\sqrt{3} & 1 & 1 & -3 & -3 \\
         &   & \sqrt{3} & 3 + \sqrt{3} & 3 \\
\hline
         & 1 & 1 + \sqrt{3} & \sqrt{3} & \boxed{0}
\end{array}
$$

By the remainder theorem, $f(\sqrt{3}) = 0$. Therefore, $\sqrt{3}$ is a zero of $f(x)$.

c. Divide $f(x) = x^3 + x + 10$ by $x - (1 + 2i)$

$$
\begin{array}{r|rrrr}
1 + 2i & 1 & 0 & 1 & 10 \\
       &   & 1 + 2i & -3 + 4i & -10 \\
\hline
       & 1 & 1 + 2i & -2 + 4i & \boxed{0}
\end{array}
$$

Note that $(1 + 2i)(1 + 2i)$
$= 1 + 2i + 2i + 4i^2$
$= 1 + 4i + 4(-1)$
Recall that $i^2 = -1$.
$= -3 + 4i$

Note that $(1 + 2i)(-2 + 4i)$
$= -2 + 4i - 4i + 8i^2$
$= -2 - 8$
$= -10$

By the remainder theorem, $f(1 + 2i) = 0$. Therefore, $1 + 2i$ is a zero of $f(x)$.

Skill Practice 7 Use the remainder theorem to determine if the given number, c, is a zero of the function.

a. $f(x) = 2x^4 - 3x^2 + 5x - 11$; $c = 2$
b. $f(x) = 2x^3 + 5x^2 - 14x - 35$; $c = \sqrt{7}$
c. $f(x) = x^3 - 7x^2 + 16x - 10$; $c = 3 + i$

Suppose that we again apply the division algorithm to a dividend of $f(x)$ and a divisor of $x - c$, where c is a complex number.

$$f(x) = (x - c) \cdot q(x) + r$$

By the remainder theorem, $r = f(c)$.

$$= (x - c) \cdot q(x) + f(c)$$

If $f(c) = 0$, then $f(x) = (x - c) \cdot q(x)$

This tells us that if $f(c)$ is a zero of $f(x)$, then $(x - c)$ is a factor of $f(x)$.

Now suppose that $x - c$ is a factor of $f(x)$. Then for some polynomial $q(x)$,

$$f(x) = (x - c) \cdot q(x)$$
$$f(c) = (c - c) \cdot q(x)$$
$$= 0$$

This tells us that if $(x - c)$ is a factor of $f(x)$, then c is a zero of $f(x)$.

These results can be summarized in the factor theorem.

Factor Theorem

Let $f(x)$ be a polynomial.

1. If $f(c) = 0$, then $(x - c)$ is a factor of $f(x)$.
2. If $(x - c)$ is a factor of $f(x)$, then $f(c) = 0$.

Answers

7. a. No   **b.** Yes   **c.** Yes

EXAMPLE 8 Identifying Factors of a Polynomial

Use the factor theorem to determine if the given polynomials are factors of $f(x) = x^4 - x^3 - 11x^2 + 11x + 12$.

a. $x - 3$ **b.** $x + 2$

Solution:

a. If $f(3) = 0$, then $x - 3$ is a factor of $f(x)$. Using synthetic division we have:

$$
\begin{array}{r|rrrrr}
3 & 1 & -1 & -11 & 11 & 12 \\
  &   & 3 & 6   & -15 & -12 \\
\hline
  & 1 & 2 & -5 & -4 & \boxed{0}
\end{array}
$$

By the factor theorem, since $f(3) = 0$, $x - 3$ is a factor of $f(x)$. $\boxed{f(3) = 0}$

b. If $f(-2) = 0$, then $x + 2$ is a factor of $f(x)$. Using synthetic division we have:

$$
\begin{array}{r|rrrrr}
-2 & 1 & -1 & -11 & 11 & 12 \\
   &   & -2 & 6   & 10 & -42 \\
\hline
   & 1 & -3 & -5 & 21 & \boxed{-30}
\end{array}
$$

By the factor theorem, since $f(-2) \neq 0$, $x + 2$ is not a factor of $f(x)$. $\boxed{f(-2) = -30}$

Skill Practice 8   Use the factor theorem to determine if the given polynomials are factors of $f(x) = 2x^4 - 13x^3 + 10x^2 - 25x + 6$.

a. $x - 6$ **b.** $x + 3$

In Example 9, we illustrate the relationship between the zeros of a polynomial and the solutions (roots) of a polynomial equation.

EXAMPLE 9 Factoring a Polynomial Given a Known Zero

a. Factor $f(x) = 3x^3 + 25x^2 + 42x - 40$, given that -5 is a zero of $f(x)$.
b. Solve the equation. $3x^3 + 25x^2 + 42x - 40 = 0$

Solution:

a. The value -5 is a zero of $f(x)$, which means that $f(-5) = 0$. By the factor theorem, $x - (-5)$ or equivalently $x + 5$ is a factor of $f(x)$. Using synthetic division, we have

$$
\begin{array}{r|rrrr}
-5 & 3 & 25 & 42 & -40 \\
   &   & -15 & -50 & 40 \\
\hline
   & 3 & 10 & -8 & \boxed{0}
\end{array}
$$

This means that $3x^3 + 25x^2 + 42x - 40 = \underset{\text{divisor}}{(x + 5)}\underset{\text{quotient}}{(3x^2 + 10x - 8)} + \underset{\text{remainder}}{0}$

Therefore, $f(x) = (x + 5)(3x - 2)(x + 4)$.

     $\boxed{\text{factors as } (3x - 2)(x + 4)}$

b. $3x^3 + 25x^2 + 42x - 40 = 0$ To solve the equation, set one side equal to zero.

   $(x + 5)(3x - 2)(x + 4) = 0$ Factor the left side.

   $x = -5, \; x = \frac{2}{3}, \; x = -4$ Set each factor equal to zero and solve for x.

The solution set is $\left\{ -5, \frac{2}{3}, -4 \right\}$.

Answers

8. a. Yes   **b.** No

9. a. $f(x) = (x + 4)(x + 2)(2x - 5)$

   b. $\left\{ -4, -2, \frac{5}{2} \right\}$

Skill Practice 9

a. Factor $f(x) = 2x^3 + 7x^2 - 14x - 40$, given that -4 is a zero of f.
b. Solve the equation. $2x^3 + 7x^2 - 14x - 40 = 0$

EXAMPLE 10   Using the Factor Theorem to Build a Polynomial

Write a polynomial $f(x)$ of degree 3 that has the zeros $\frac{1}{2}$, $\sqrt{6}$, and $-\sqrt{6}$.

Solution:

By the factor theorem, if $\frac{1}{2}$, $\sqrt{6}$, and $-\sqrt{6}$ are zeros of a polynomial $f(x)$, then $\left(x - \frac{1}{2}\right)$, $\left(x - \sqrt{6}\right)$, and $\left(x + \sqrt{6}\right)$ are factors of $f(x)$. Therefore, $f(x) = \left(x - \frac{1}{2}\right)\left(x - \sqrt{6}\right)\left(x + \sqrt{6}\right)$ is a third-degree polynomial with the given zeros.

$$f(x) = \left(x - \frac{1}{2}\right)(x^2 - 6) \qquad \text{Multiply conjugates.}$$

$$f(x) = x^3 - \frac{1}{2}x^2 - 6x + 3$$

Skill Practice 10   Write a polynomial $f(x)$ of degree 3 that has the zeros $\frac{1}{3}$, $\sqrt{3}$, and $-\sqrt{3}$.

In Example 10, the polynomial $f(x)$ is not unique. If we multiply $f(x)$ by any nonzero constant a, the polynomial will still have the desired factors and zeros.

$$g(x) = a\left(x - \frac{1}{2}\right)\left(x - \sqrt{6}\right)\left(x + \sqrt{6}\right) \qquad \text{The zeros are still } \frac{1}{2}, \sqrt{6}, \text{ and } -\sqrt{6}.$$

If a is any nonzero multiple of 2, then the polynomial will have integer coefficients. For example:

$$g(x) = 2\left(x - \frac{1}{2}\right)\left(x - \sqrt{6}\right)\left(x + \sqrt{6}\right)$$

$$= 2\left(x^3 - \frac{1}{2}x^2 - 6x + 3\right)$$

$$= 2x^3 - x^2 - 12x + 6$$

The zeros of $f(x)$ and $g(x)$ are real numbers and correspond to the x-intercepts of the graphs of the related functions. The graphs of $y = f(x)$ and $y = g(x)$ are shown in Figure 3-15. Notice that the graphs have the same x-intercepts and differ only by a vertical stretch.

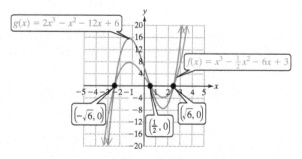

Figure 3-15

Answer

10. $f(x) = x^3 - \frac{1}{3}x^2 - 3x + 1$

SECTION 3.3   Practice Exercises

Prerequisite Review

R.1. Evaluate $(3 + 4i)^2$ and write the answer in standard form, $a + bi$.

R.2. Verify by substitution that the given values of x are solutions to $x^2 - 12x + 39 = 0$.

  a. $6 - i\sqrt{3}$ **b.** $6 + i\sqrt{3}$

R.3. Solve by using the quadratic formula.

  $z^2 + 8z + 19 = 0$

Concept Connections

1. Given the division algorithm, identify the polynomials representing the dividend, divisor, quotient, and remainder.

   $f(x) = d(x) \cdot q(x) + r(x)$

2. Given $\dfrac{2x^3 - 5x^2 - 6x + 1}{x - 3} = 2x^2 + x - 3 + \dfrac{-8}{x - 3}$, use the division algorithm to check the result.

3. The remainder theorem indicates that if a polynomial $f(x)$ is divided by $x - c$, then the remainder is _____.

4. Given a polynomial $f(x)$, the factor theorem indicates that if $f(c) = 0$, then $x - c$ is a _____ of $f(x)$. Furthermore, if $x - c$ is a factor of $f(x)$, then $f(c) =$ _____.

5. Answer true or false. If $\sqrt{5}$ is a zero of a polynomial, then $(x - \sqrt{5})$ is a factor of the polynomial.

6. Answer true or false. If $(x + 3)$ is a factor of a polynomial, then 3 is a zero of the polynomial.

Objective 1: Divide Polynomials Using Long Division

For Exercises 7–8, (See Example 1)

a. Use long division to divide.

b. Identify the dividend, divisor, quotient, and remainder.

c. Check the result from part (a) with the division algorithm.

  7. $(6x^2 + 9x + 5) \div (2x - 5)$

  8. $(12x^2 + 10x + 3) \div (3x + 4)$

For Exercises 9–22, use long division to divide. (See Examples 1–3)

  9. $(3x^3 - 11x^2 - 10) \div (x - 4)$

  10. $(2x^3 - 7x^2 - 65) \div (x - 5)$

  11. $(8 + 30x - 27x^2 - 12x^3 + 4x^4) \div (x + 2)$

  12. $(-48 - 28x + 20x^2 + 17x^3 + 3x^4) \div (x + 3)$

  13. $(-20x^2 + 6x^4 - 16) \div (2x + 4)$

  14. $(-60x^2 + 8x^4 - 108) \div (2x - 6)$

  15. $(x^5 + 4x^4 + 18x^2 - 20x - 10) \div (x^2 + 5)$

  16. $(x^5 - 2x^4 + x^3 - 8x + 18) \div (x^2 - 3)$

  17. $\dfrac{6x^4 + 3x^3 - 7x^2 + 6x - 5}{2x^2 + x - 3}$

  18. $\dfrac{12x^4 - 4x^3 + 13x^2 + 2x + 1}{3x^2 - x + 4}$

  19. $\dfrac{x^3 - 27}{x - 3}$

  20. $\dfrac{x^3 + 64}{x + 4}$

  21. $(5x^3 - 2x^2 + 3) \div (2x - 1)$

  22. $(2x^3 + x^2 + 1) \div (3x + 1)$

Objective 2: Divide Polynomials Using Synthetic Division

For Exercises 23–26, consider the division of two polynomials: $f(x) \div (x - c)$. The result of the synthetic division process is shown here. Write the polynomials representing the

  a. Dividend.   **b.** Divisor.   **c.** Quotient.   **d.** Remainder.

23.
$$
\begin{array}{r|rrrrr}
3 & 2 & -5 & -5 & -4 & 29 \\
  &   & 6 & 3 & -6 & -30 \\
\hline
  & 2 & 1 & -2 & -10 & \boxed{-1}
\end{array}
$$

24.
$$
\begin{array}{r|rrrrr}
2 & 1 & -5 & 2 & -1 & 20 \\
  &   & 2 & -6 & -8 & -18 \\
\hline
  & 1 & -3 & -4 & -9 & \boxed{2}
\end{array}
$$

25.
$$-4 \begin{array}{|rrrr} 1 & -2 & -25 & -4 \\ & -4 & 24 & 4 \\ \hline 1 & -6 & -1 & \underline{|0} \end{array}$$

26.
$$-5 \begin{array}{|rrrr} 3 & 13 & -14 & -20 \\ & -15 & 10 & 20 \\ \hline 3 & -2 & -4 & \underline{|0} \end{array}$$

For Exercises 27–38, use synthetic division to divide the polynomials. (See Examples 4–5)

27. $(4x^2 + 15x + 1) \div (x + 6)$

28. $(6x^2 + 25x - 19) \div (x + 5)$

29. $(5x^2 - 17x - 12) \div (x - 4)$

30. $(2x^2 + x - 21) \div (x - 3)$

31. $(4 - 8x - 3x^2 - 5x^4) \div (x + 2)$

32. $(-5 + 2x + 5x^3 - 2x^4) \div (x + 1)$

33. $\dfrac{4x^5 - 25x^4 - 58x^3 + 232x^2 + 198x - 63}{x - 3}$

34. $\dfrac{2x^5 + 13x^4 - 3x^3 - 58x^2 - 20x + 24}{x - 2}$

35. $\dfrac{x^5 + 32}{x + 2}$

36. $\dfrac{x^4 - 81}{x + 3}$

37. $(2x^4 - 7x^3 - 56x^2 + 37x + 84) \div \left(x - \dfrac{3}{2}\right)$

38. $(-5x^4 - 18x^3 + 63x^2 + 128x - 60) \div \left(x - \dfrac{2}{5}\right)$

Objective 3: Apply the Remainder and Factor Theorems

39. The value $f(-6) = 39$ for a polynomial $f(x)$. What can be concluded about the remainder or quotient of $\dfrac{f(x)}{x + 6}$?

40. Given a polynomial $f(x)$, the quotient $\dfrac{f(x)}{x - 2}$ has a remainder of 12. What is the value of $f(2)$?

41. Given $f(x) = 2x^4 - 5x^3 + x^2 - 7$,

    a. Evaluate $f(4)$.

    b. Determine the remainder when $f(x)$ is divided by $(x - 4)$.

42. Given $g(x) = -3x^5 + 2x^4 + 6x^2 - x + 4$,

    a. Evaluate $g(2)$.

    b. Determine the remainder when $g(x)$ is divided by $(x - 2)$.

For Exercises 43–46, use the remainder theorem to evaluate the polynomial for the given values of x. (See Example 6)

43. $f(x) = 2x^4 + x^3 - 49x^2 + 79x + 15$

    a. $f(-1)$ **b.** $f(3)$ **c.** $f(4)$ **d.** $f\left(\dfrac{5}{2}\right)$

44. $g(x) = 3x^4 - 22x^3 + 51x^2 - 42x + 8$

    a. $g(-1)$ **b.** $g(2)$ **c.** $g(1)$ **d.** $g\left(\dfrac{4}{3}\right)$

45. $h(x) = 5x^3 - 4x^2 - 15x + 12$

    a. $h(1)$ **b.** $h\left(\dfrac{4}{5}\right)$ **c.** $h(\sqrt{3})$ **d.** $h(-1)$

46. $k(x) = 2x^3 - x^2 - 14x + 7$

    a. $k(2)$ **b.** $k\left(\dfrac{1}{2}\right)$ **c.** $k(\sqrt{7})$ **d.** $k(-2)$

For Exercises 47–54, use the remainder theorem to determine if the given number c is a zero of the polynomial. (See Example 7)

47. $f(x) = x^4 + 3x^3 - 7x^2 + 13x - 10$

    a. $c = 2$ **b.** $c = -5$

48. $g(x) = 2x^4 + 13x^3 - 10x^2 - 19x + 14$

    a. $c = -2$ **b.** $c = -7$

49. $p(x) = 2x^3 + 3x^2 - 22x - 33$

    a. $c = -2$ **b.** $c = -\sqrt{11}$

50. $q(x) = 3x^3 + x^2 - 30x - 10$

    a. $c = -3$ **b.** $c = -\sqrt{10}$

51. $m(x) = x^3 - 2x^2 + 25x - 50$

    a. $c = 5i$ **b.** $c = -5i$

52. $n(x) = x^3 + 4x^2 + 9x + 36$

    a. $c = 3i$ **b.** $c = -3i$

53. $g(x) = x^3 - 11x^2 + 25x + 37$

    a. $c = 6 + i$ **b.** $c = 6 - i$

54. $f(x) = 2x^3 - 5x^2 + 54x - 26$

    a. $c = 1 + 5i$ **b.** $c = 1 - 5i$

For Exercises 55–60, use the factor theorem to determine if the given binomial is a factor of $f(x)$. (See Example 8)

55. $f(x) = x^4 + 11x^3 + 41x^2 + 61x + 30$

    a. $x + 5$ **b.** $x - 2$

56. $g(x) = x^4 - 10x^3 + 35x^2 - 50x + 24$

    a. $x - 4$ **b.** $x + 1$

57. $f(x) = x^3 + 64$

    a. $x - 4$ **b.** $x + 4$

58. $f(x) = x^4 - 81$

    a. $x - 3$ **b.** $x + 3$

59. $f(x) = 2x^3 + x^2 - 16x - 8$

    a. $x - 1$ **b.** $x - 2\sqrt{2}$

60. $f(x) = 3x^3 - x^2 - 54x + 18$

    a. $x - 2$ **c.** $x - 3\sqrt{2}$

61. Given $g(x) = x^4 - 14x^2 + 45$,

   a. Evaluate $g(\sqrt{5})$.

   b. Evaluate $g(-\sqrt{5})$.

   c. Solve $g(x) = 0$.

62. Given $h(x) = x^4 - 15x^2 + 44$,

   a. Evaluate $h(\sqrt{11})$.

   b. Evaluate $h(-\sqrt{11})$.

   c. Solve $h(x) = 0$.

63. a. Use synthetic division and the factor theorem to determine if $[x - (2 + 5i)]$ is a factor of $f(x) = x^2 - 4x + 29$.

   b. Use synthetic division and the factor theorem to determine if $[x - (2 - 5i)]$ is a factor of $f(x) = x^2 - 4x + 29$.

   c. Use the quadratic formula to solve the equation. $x^2 - 4x + 29 = 0$

   d. Find the zeros of the polynomial $f(x) = x^2 - 4x + 29$.

64. a. Use synthetic division and the factor theorem to determine if $[x - (3 + 4i)]$ is a factor of $f(x) = x^2 - 6x + 25$.

   b. Use synthetic division and the factor theorem to determine if $[x - (3 - 4i)]$ is a factor of $f(x) = x^2 - 6x + 25$.

   c. Use the quadratic formula to solve the equation. $x^2 - 6x + 25 = 0$

   d. Find the zeros of the polynomial $f(x) = x^2 - 6x + 25$.

65. a. Factor $f(x) = 2x^3 + x^2 - 37x - 36$, given that -1 is a zero. (**See Example 9**)

   b. Solve. $2x^3 + x^2 - 37x - 36 = 0$

66. a. Factor $f(x) = 3x^3 + 16x^2 - 5x - 50$, given that -2 is a zero.

   b. Solve. $3x^3 + 16x^2 - 5x - 50 = 0$

67. a. Factor $f(x) = 20x^3 + 39x^2 - 3x - 2$, given that $\frac{1}{4}$ is a zero.

   b. Solve. $20x^3 + 39x^2 - 3x - 2 = 0$

68. a. Factor $f(x) = 8x^3 - 18x^2 - 11x + 15$, given that $\frac{3}{4}$ is a zero.

   b. Solve. $8x^3 - 18x^2 - 11x + 15 = 0$

69. a. Factor $f(x) = 9x^3 - 33x^2 + 19x - 3$, given that 3 is a zero.

   b. Solve. $9x^3 - 33x^2 + 19x - 3 = 0$

70. a. Factor $f(x) = 4x^3 - 20x^2 + 33x - 18$, given that 2 is a zero.

   b. Solve. $4x^3 - 20x^2 + 33x - 18 = 0$

For Exercises 71–82, write a polynomial $f(x)$ that meets the given conditions. Answers may vary. (See Example 10)

71. Degree 3 polynomial with zeros 2, 3, and -4.

72. Degree 3 polynomial with zeros 1, -6, and 3.

73. Degree 4 polynomial with zeros 1, $\frac{3}{2}$ (each with multiplicity 1), and 0 (with multiplicity 2).

74. Degree 5 polynomial with zeros 2, $\frac{5}{2}$ (each with multiplicity 1), and 0 (with multiplicity 3).

75. Degree 2 polynomial with zeros $2\sqrt{11}$ and $-2\sqrt{11}$.

76. Degree 2 polynomial with zeros $5\sqrt{2}$ and $-5\sqrt{2}$.

77. Degree 3 polynomial with zeros -2, $3i$, and $-3i$.

78. Degree 3 polynomial with zeros 4, $2i$, and $-2i$.

79. Degree 3 polynomial with integer coefficients and zeros of $-\frac{2}{3}$, $\frac{1}{2}$, and 4.

80. Degree 3 polynomial with integer coefficients and zeros of $-\frac{2}{5}$, $\frac{3}{2}$, and 6.

81. Degree 2 polynomial with zeros of $7 + 8i$ and $7 - 8i$.

82. Degree 2 polynomial with zeros of $5 + 6i$ and $5 - 6i$.

Mixed Exercises

83. Given $p(x) = 2x^{452} - 4x^{92}$, is it easier to evaluate $p(1)$ by using synthetic division or by direct substitution? Find the value of $p(1)$.

84. Given $q(x) = 5x^{721} - 2x^{450}$, is it easier to evaluate $q(-1)$ by using synthetic division or by direct substitution? Find the value of $q(-1)$.

85. a. Is $(x - 1)$ a factor of $x^{100} - 1$?

   c. Is $(x - 1)$ a factor of $x^{99} - 1$?

   e. If n is a positive even integer, is $(x - 1)$ a factor of $x^n - 1$?

   b. Is $(x + 1)$ a factor of $x^{100} - 1$?

   d. Is $(x + 1)$ a factor of $x^{99} - 1$?

   f. If n is a positive odd integer, is $(x + 1)$ a factor of $x^n - 1$?

86. If a fifth-degree polynomial is divided by a second-degree polynomial, the quotient is a _____ -degree polynomial.

87. Determine if the statement is true or false: Zero is a zero of the polynomial $3x^5 - 7x^4 - 2x^3 - 14$.

88. Determine if the statement is true or false: Zero is a zero of the polynomial $-2x^4 + 5x^3 + 6x$.

89. Find m so that $x + 4$ is a factor of $4x^3 + 13x^2 - 5x + m$.

90. Find m so that $x + 5$ is a factor of $-3x^4 - 10x^3 + 20x^2 - 22x + m$.

91. Find m so that $x + 2$ is a factor of $4x^3 + 5x^2 + mx + 2$.

92. Find m so that $x - 3$ is a factor of $2x^3 - 7x^2 + mx + 6$.

93. For what value of r is the statement an identity?
$\dfrac{x^2 - x - 12}{x - 4} = x + 3 + \dfrac{r}{x - 4}$ provided that $x \neq 4$

94. For what value of r is the statement an identity?
$\dfrac{x^2 - 5x - 8}{x - 2} = x - 3 + \dfrac{r}{x - 2}$ provided that $x \neq 2$

95. A metal block is formed from a rectangular solid with a rectangular piece cut out.

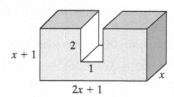

a. Write a polynomial $V(x)$ that represents the volume of the block. All distances in the figure are in centimeters.

b. Use synthetic division to evaluate the volume if x is 6 cm.

96. A wedge is cut from a rectangular solid.

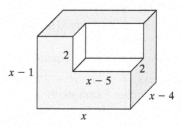

a. Write a polynomial $V(x)$ that represents the volume of the remaining part of the solid. All distances in the figure are in feet.

b. Use synthetic division to evaluate the volume if x is 10 ft.

Write About It

97. Under what circumstances can synthetic division be used to divide polynomials?

98. How can the division algorithm be used to check the result of polynomial division?

99. Given a polynomial $f(x)$ and a constant c, state two methods by which the value $f(c)$ can be computed.

100. Write an informal explanation of the factor theorem.

Expanding Your Skills

101. a. Factor $f(x) = x^3 - 5x^2 + x - 5$ into factors of the form $(x - c)$, given that 5 is a zero.
   b. Solve. $x^3 - 5x^2 + x - 5 = 0$

102. a. Factor $f(x) = x^3 - 3x^2 + 100x - 300$ into factors of the form $(x - c)$, given that 3 is a zero.
   b. Solve. $x^3 - 3x^2 + 100x - 300 = 0$

103. a. Factor $f(x) = x^4 + 2x^3 - 2x^2 - 6x - 3$ into factors of the form $(x - c)$, given that -1 is a zero.
   b. Solve. $x^4 + 2x^3 - 2x^2 - 6x - 3 = 0$

104. a. Factor $f(x) = x^4 + 4x^3 - x^2 - 20x - 20$ into factors of the form $(x - c)$, given that -2 is a zero.
   b. Solve. $x^4 + 4x^3 - x^2 - 20x - 20 = 0$

Technology Connections

For Exercises 105–106,

a. Use the graph to determine a solution to the given equation.

b. Verify your answer from part (a) using the remainder theorem.

c. Find the remaining solutions to the equation.

105. $5x^3 + 7x^2 - 58x - 24 = 0$

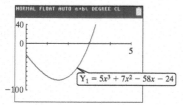

106. $2x^3 - x^2 - 41x + 70 = 0$

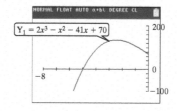

SECTION 3.4 **Zeros of Polynomials**

1. Apply the Rational Zero Theorem

The **zeros of a polynomial** $f(x)$ are the solutions (roots) to the corresponding polynomial equation $f(x) = 0$. For a polynomial function defined by $y = f(x)$, the real zeros of $f(x)$ are the x-intercepts of the graph of the function. Applications of polynomials and polynomial functions arise throughout mathematics. For this reason, it is important to learn techniques to find or approximate the zeros of a polynomial.

The zeros of a polynomial may be real numbers or imaginary numbers. The real zeros can be further categorized as rational or irrational numbers. For example, consider

$$f(x) = 2x^6 - 3x^5 - 7x^4 + 102x^3 - 88x^2 - 279x + 273$$

In factored form this is:

$$f(x) = (x - 1)(2x + 7)(x - \sqrt{3})(x + \sqrt{3})[x - (2 + 3i)][x - (2 - 3i)]$$

rational zeros   irrational zeros

The zeros are: $1, -\dfrac{7}{2}, \overbrace{\sqrt{3}, -\sqrt{3}}, \overbrace{2 + 3i, 2 - 3i}$

real zeros   nonreal zeros

In this section, we develop tools to search for the zeros of polynomials. First we will consider polynomials with integer coefficients. For example, the factored form of $f(x) = 6x^2 + 13x - 5$ is $f(x) = (2x + 5)(3x - 1)$, which leads to zeros of $-\dfrac{5}{2}$ and $\dfrac{1}{3}$. The polynomial $f(x) = 6x^2 + 13x - 5$ is in the form $f(x) = ax^2 + bx + c$, where $a = 6$, $b = 13$, and $c = -5$. Notice that the numerator of each zero is a factor of c, and the denominator of each zero is a factor of a. This observation is consistent with the following theorem to search for zeros that are rational numbers.

> **Rational Zero Theorem**
>
> If $f(x) = a_nx^n + a_{n-1}x^{n-1} + a_{n-2}x^{n-2} + \cdots + a_1x + a_0$ has integer coefficients and $a_n \neq 0$, and if $\frac{p}{q}$ (written in lowest terms) is a rational zero of f, then
>
> - p is a factor of the constant term a_0.
> - q is a factor of the leading coefficient a_n.

TIP Recall that a rational number is a number that can be expressed as a ratio of two integers.

The rational zero theorem does not guarantee the existence of rational zeros. Rather, it indicates that *if* a rational zero exists for a polynomial, then it must be of the form

$$\frac{p}{q} = \frac{\text{Factors of } a_0 \,(\text{constant term})}{\text{Factors of } a_n \,(\text{leading coefficient})}$$

The rational zero theorem is important because it limits our search to find rational zeros (if they exist) to a finite number of choices.

EXAMPLE 1 Listing All Possible Rational Zeros

List all possible rational zeros. $f(x) = -2x^5 + 3x^2 - 2x + 10$

Solution:

First note that the polynomial has integer coefficients.

$$f(x) = -2x^5 + 3x^2 - 2x + 10$$

The constant term is 10. $\pm 1, \pm 2, \pm 5, \pm 10 \longleftarrow$ ⬚Factors of 10

The leading coefficient is -2. $\pm 1, \pm 2 \longleftarrow$ ⬚Factors of -2

$$\frac{\text{Factors of } 10}{\text{Factors of } -2} = \frac{\pm 1, \pm 2, \pm 5, \pm 10}{\pm 1, \pm 2} = \pm\frac{1}{1}, \pm\frac{2}{1}, \pm\frac{5}{1}, \pm\frac{10}{1}, \pm\frac{1}{2}, \pm\frac{2}{2}, \pm\frac{5}{2}, \pm\frac{10}{2}$$

The values $\pm\frac{2}{2}$ and $\pm\frac{10}{2}$ are redundant. They equal ± 1 and ± 5, respectively. The possible rational zeros are ± 1, ± 2, ± 5, ± 10, $\pm\frac{1}{2}$, and $\pm\frac{5}{2}$.

> **Skill Practice 1**   List all possible rational zeros.
> $f(x) = -4x^4 + 5x^3 - 7x^2 + 8$

EXAMPLE 2 Finding the Zeros of a Polynomial

Find the zeros. $f(x) = x^3 - 4x^2 + 3x + 2$

Solution:

We begin by first looking for rational zeros. We can apply the rational zero theorem because the polynomial has integer coefficients.

$$f(x) = 1x^3 - 4x^2 + 3x + 2$$

Possible rational zeros: $\dfrac{\text{Factors of } 2}{\text{Factors of } 1} = \dfrac{\pm 1, \pm 2}{\pm 1} = \pm 1, \pm 2$

Next, use synthetic division and the remainder theorem to determine if any of the numbers in the list is a zero of f.

> **TIP** Since $f(1) = 2$ and $f(-1) = -6$, we know from the intermediate value theorem that $f(x)$ must have a zero between -1 and 1.

Test $x = 1$:

$$\begin{array}{r|rrrr} 1 & 1 & -4 & 3 & 2 \\ & & 1 & -3 & 0 \\ \hline & 1 & -3 & 0 & \boxed{2} \end{array}$$

The remainder is not zero. Therefore, 1 is not a zero of $f(x)$.

Test $x = -1$:

$$\begin{array}{r|rrrr} -1 & 1 & -4 & 3 & 2 \\ & & -1 & 5 & -8 \\ \hline & 1 & -5 & 8 & \boxed{-6} \end{array}$$

The remainder is not zero. Therefore, -1 is not a zero of $f(x)$.

Test $x = 2$:

$$\begin{array}{r|rrrr} 2 & 1 & -4 & 3 & 2 \\ & & 2 & -4 & -2 \\ \hline & 1 & -2 & -1 & \boxed{0} \end{array}$$

The remainder *is* zero. Therefore, 2 is a zero of $f(x)$.

By the factor theorem, since $f(2) = 0$, then $(x - 2)$ is a factor of $f(x)$. The quotient $(x^2 - 2x - 1)$ is also a factor of $f(x)$. We have

$$f(x) = x^3 - 4x^2 + 3x + 2 = (x - 2)(x^2 - 2x - 1)$$

We now have a third-degree polynomial written as the product of a first-degree polynomial and a quadratic polynomial. The quotient $x^2 - 2x - 1$ is called a **reduced polynomial** (or a **depressed polynomial**) of $f(x)$. It has degree 1 less than the degree of $f(x)$, and the remaining zeros of $f(x)$ are the zeros of the reduced polynomial.

Answer

1. $\pm 1, \pm 2, \pm 4, \pm 8, \pm\dfrac{1}{2}$, and $\pm\dfrac{1}{4}$

TIP The graph of $f(x) = x^3 - 4x^2 + 3x + 2$ is shown here. The zeros are $1 + \sqrt{2} \approx 2.41$, $1 - \sqrt{2} \approx -0.41$, and 2.

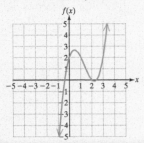

At this point, we no longer need to test for more rational zeros. The reason is that any remaining zeros (whether they be rational, irrational, or imaginary) are the solutions to the quadratic equation $x^2 - 2x - 1 = 0$. There is no guess work because the equation can be solved by using the quadratic formula.

$$x^2 - 2x - 1 = 0$$

$$x = \frac{-(-2) \pm \sqrt{(-2)^2 - 4(1)(-1)}}{2(1)} = \frac{2 \pm \sqrt{8}}{2} = \frac{2 \pm 2\sqrt{2}}{2}$$

$$= 1 \pm \sqrt{2}$$

The zeros of $f(x)$ are 2, $1 + \sqrt{2}$, and $1 - \sqrt{2}$.

Skill Practice 2   Find the zeros. $f(x) = x^3 - x^2 - 4x - 2$

The polynomial in Example 2 has one rational zero. If we had continued testing for more rational zeros, we would not have found others because the remaining two zeros are irrational numbers. In Example 3, we illustrate the case where a function has multiple rational zeros.

TIP It does not matter in which order you test the potential rational zeros.

EXAMPLE 3   Finding the Zeros of a Polynomial

Find the zeros and their multiplicities. $f(x) = 2x^4 + 5x^3 - 2x^2 - 11x - 6$

Solution:

Begin by searching for rational zeros.

$$f(x) = 2x^4 + 5x^3 - 2x^2 - 11x - 6$$

Possible rational zeros:

$$\frac{\text{Factors of } -6}{\text{Factors of } 2} = \frac{\pm 1, \pm 2, \pm 3, \pm 6}{\pm 1, \pm 2} = \pm 1, \pm 2, \pm 3, \pm 6, \pm \frac{1}{2}, \pm \frac{3}{2}$$

We can work methodically through the list of possible rational zeros to determine which if any are actual zeros of $f(x)$. After trying several possibilities, we find that -1 is a zero of $f(x)$.

$$
\begin{array}{r|rrrr}
-1 & 2 & 5 & -2 & -11 & -6 \\
   &   & -2 & -3 & 5 & 6 \\
\hline
   & 2 & 3 & -5 & -6 & \underline{0}
\end{array}
$$

The quotient is $2x^3 + 3x^2 - 5x - 6$.

Since $f(-1) = 0$, then $x + 1$ is a factor of $f(x)$.

$$f(x) = 2x^4 + 5x^3 - 2x^2 - 11x - 6 = (x + 1)\underbrace{(2x^3 + 3x^2 - 5x - 6)}$$

Now find the zeros of the quotient.

TIP The zeros of the polynomial in Example 3 are rational numbers (and therefore real numbers). They correspond to x-intercepts on the graph of $y = f(x)$. Also notice that the graph has a touch point at $(-1, 0)$ because -1 has an even multiplicity.

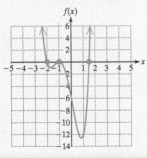

The zeros of $f(x)$ are -1 along with the roots of the equation $2x^3 + 3x^2 - 5x - 6 = 0$. Therefore, we need to find the zeros of $g(x) = 2x^3 + 3x^2 - 5x - 6$.

The possible rational zeros are $\pm 1, \pm 2, \pm 3, \pm 6, \pm \frac{1}{2}, \pm \frac{3}{2}$.

We will test -1 again because it may have multiplicity greater than 1.

$$
\begin{array}{r|rrrr}
-1 & 2 & 3 & -5 & -6 \\
   &   & -2 & -1 & 6 \\
\hline
   & 2 & 1 & -6 & \underline{0}
\end{array}
$$

The quotient is $2x^2 + x - 6$.

Answer

2. $-1, 1 - \sqrt{3}$, and $1 + \sqrt{3}$

The value -1 is a repeated zero. We have:

$$f(x) = 2x^4 + 5x^3 - 2x^2 - 11x - 6 = (x + 1)(x + 1)(2x^2 + x - 6)$$
$$= (x + 1)^2(2x - 3)(x + 2)$$

> $2x^2 + x - 6$ factors as $(2x - 3)(x + 2)$.

The zeros of $f(x)$ are

-1 (multiplicity 2), $\frac{3}{2}$ (multiplicity 1), and -2 (multiplicity 1).

> **Skill Practice 3** Find the zeros and their multiplicities.
> $$f(x) = 2x^4 + 3x^3 - 15x^2 - 32x - 12$$

In Example 4, we have a polynomial with no rational zeros.

EXAMPLE 4 **Finding the Zeros of a Polynomial**

Find the zeros. $\quad f(x) = x^4 - 2x^2 - 3$

Solution:

$f(x) = 1x^4 - 2x^2 - 3$ The possible rational zeros are $\dfrac{\pm 1, \pm 3}{\pm 1} = \pm 1, \pm 3$.

If we apply the rational zero theorem, we see that $f(x)$ has no *rational* zeros.

$$
\begin{array}{r|rrrrr}
1 & 1 & 0 & -2 & 0 & -3 \\
  &   & 1 & 1 & -1 & -1 \\
\hline
  & 1 & 1 & -1 & -1 & \underline{-4}
\end{array}
\qquad
\begin{array}{r|rrrrr}
-1 & 1 & 0 & -2 & 0 & -3 \\
   &   & -1 & 1 & 1 & -1 \\
\hline
   & 1 & -1 & -1 & 1 & \underline{-4}
\end{array}
$$

$$
\begin{array}{r|rrrrr}
3 & 1 & 0 & -2 & 0 & -3 \\
  &   & 3 & 9 & 21 & 63 \\
\hline
  & 1 & 3 & 7 & 21 & \underline{60}
\end{array}
\qquad
\begin{array}{r|rrrrr}
-3 & 1 & 0 & -2 & 0 & -3 \\
   &   & -3 & 9 & -21 & 63 \\
\hline
   & 1 & -3 & 7 & -21 & \underline{60}
\end{array}
$$

However, finding the zeros of $f(x)$ is equivalent to finding the roots of the equation $x^4 - 2x^2 - 3 = 0$.

$$x^4 - 2x^2 - 3 = 0$$
$$(x^2 - 3)(x^2 + 1) = 0 \qquad \text{Factor the trinomial.}$$
$$x^2 - 3 = 0 \quad \text{or} \quad x^2 + 1 = 0 \qquad \text{Set each factor equal to zero.}$$
$$x = \pm\sqrt{3} \quad \text{or} \quad x = \pm i \qquad \text{Apply the square root property.}$$

> **TIP** From the factor theorem,
> $f(x) = (x - \sqrt{3})(x + \sqrt{3})(x - i)(x + i)$.

The zeros are $\sqrt{3}, -\sqrt{3}, i,$ and $-i$.

> **Skill Practice 4** Find the zeros. $\quad f(x) = x^4 - x^2 - 20$

TIP The graph of the function defined by $f(x) = x^4 - 2x^2 - 3$ shows the real zeros of $f(x)$ as x-intercepts.

---

2. Apply the Fundamental Theorem of Algebra

From Examples 2–4, we see that the zeros of a polynomial may be real numbers (either rational or irrational) or nonreal numbers such as $2 + 3i$ or $5i$. In any case, the zeros are all complex numbers.

To find the zeros of a polynomial, it is important to know how many zeros to expect. This is answered by the following three theorems. The first is called the fundamental theorem of algebra because it is so basic to the foundation of algebra.

Answers

3. -2 (multiplicity 2);
   $-\dfrac{1}{2}$ (multiplicity 1); and
   3 (multiplicity 1)

4. $\sqrt{5}, -\sqrt{5}, 2i,$ and $-2i$

> ### Fundamental Theorem of Algebra
>
> If $f(x)$ is a polynomial of degree $n \geq 1$ with complex coefficients, then $f(x)$ has at least one complex zero.

The fundamental theorem of algebra, first proved by German mathematician Carl Friedrich Gauss (1777–1855), guarantees that every polynomial of degree $n \geq 1$ has at least one zero.

Now suppose that $f(x)$ is a polynomial of degree $n \geq 1$ with complex coefficients. The fundamental theorem of algebra guarantees the existence of at least one complex zero, call this c_1. By the factor theorem, we have

$$f(x) = (x - c_1) \cdot q_1(x) \qquad \text{where } q_1(x) \text{ is a polynomial of degree } n - 1.$$

If $q_1(x)$ is of degree 1 or more, then the fundamental theorem of algebra guarantees that $q_1(x)$ must have at least one complex zero, call this c_2. Then,

$$f(x) = (x - c_1) \cdot (x - c_2) \cdot q_2(x) \quad \text{where } q_2(x) \text{ is a polynomial of degree } n - 2.$$

We can continue with this reasoning until the quotient polynomial, $q_n(x)$, is a constant equal to the leading coefficient of $f(x)$.

TIP The set of complex numbers includes the set of real numbers. Therefore, theorems relating to polynomials with complex coefficients also apply to polynomials with real coefficients.

> ### Linear Factorization Theorem
>
> If $f(x) = a_n x^n + a_{n-1} x^{n-1} + a_{n-2} x^{n-2} + \cdots + a_1 x + a_0$, where $n \geq 1$ and $a_n \neq 0$, then
>
> $f(x) = a_n(x - c_1)(x - c_2) \dots (x - c_n)$, where $c_1, c_2, \dots, c_n$ are complex numbers.
>
> *Note*: The complex numbers $c_1, c_2, \dots, c_n$ are not necessarily unique.

The linear factorization theorem tells us that a polynomial of degree $n \geq 1$ with complex coefficients has exactly n linear factors of the form $(x - c)$, where some of the factors may be repeated. The value of c in each factor is a zero of the function, so the function must also have n zeros provided that the zeros are counted according to their multiplicities.

TIP Refer to Examples 2–4. In each case, the number of zeros (including multiplicities) is the same as the degree of the polynomial.

> ### Number of Zeros of a Polynomial
>
> If $f(x)$ is a polynomial of degree $n \geq 1$ with complex coefficients, then $f(x)$ has exactly n complex zeros provided that each zero is counted by its multiplicity.

Now consider the polynomial from Example 4.

$$f(x) = x^4 - 2x^2 - 3 \quad \text{Zeros: } \sqrt{3}, -\sqrt{3}, i, -i$$

Notice that the polynomial has real coefficients. Furthermore, the zeros i and $-i$ appear as a pair. This is not a coincidence. For a polynomial with real coefficients, if $a + bi$ is a zero, then $a - bi$ is a zero.

> ### Conjugate Zeros Theorem
>
> If $f(x)$ is a polynomial with real coefficients and if $a + bi$ ($b \neq 0$) is a zero of $f(x)$, then its conjugate $a - bi$ is also a zero of $f(x)$.

EXAMPLE 5 Finding Zeros and Factoring a Polynomial

Given $f(x) = x^4 - 6x^3 + 28x^2 - 18x + 75$, and that $3 - 4i$ is a zero of $f(x)$,

a. Find the remaining zeros.

b. Factor $f(x)$ as a product of linear factors.

c. Solve the equation. $x^4 - 6x^3 + 28x^2 - 18x + 75 = 0$

Solution:

$f(x)$ is a fourth-degree polynomial, so we expect to find four zeros (including multiplicities). Further note that because $f(x)$ has real coefficients and because $3 - 4i$ is a zero, then the conjugate $3 + 4i$ must also be a zero. This leaves only two remaining zeros to find.

$$
\begin{array}{r|rrrrr}
3-4i & 1 & -6 & 28 & -18 & 75 \\
 & & 3-4i & -25 & 9-12i & -75 \\
\hline
 & 1 & -3-4i & 3 & -9-12i & \boxed{0}
\end{array}
$$

Divide by $[x - (3 - 4i)]$.

coefficients of the quotient

One strategy is to use synthetic division twice using the two known zeros.

Note: $(3 - 4i)(-3 - 4i)$
$= -9 - 12i + 12i + 16i^2$
$= -25$

Note: $(3 - 4i)(-9 - 12i)$
$= -27 - 36i + 36i + 48i^2$
$= -75$

Since $3 + 4i$ is a zero of $f(x)$ it must also be a zero of the quotient.

$$
\begin{array}{r|rrrr}
3+4i & 1 & -3-4i & 3 & -9-12i \\
 & & 3+4i & 0 & 9+12i \\
\hline
 & 1 & 0 & 3 & \boxed{0}
\end{array}
$$

Divide by $[x - (3 + 4i)]$.

Divide the quotient by $[x - (3 + 4i)]$.

The resulting quotient is quadratic: $x^2 + 3$.

Now we have $f(x) = [x - (3 - 4i)][x - (3 + 4i)](x^2 + 3)$.

The remaining two zeros are found by solving $x^2 + 3 = 0$.

$$x^2 + 3 = 0$$
$$x^2 = -3$$
$$x = \pm i\sqrt{3}$$

a. The zeros of $f(x)$ are: $3 - 4i$, $3 + 4i$, $i\sqrt{3}$, and $-i\sqrt{3}$.

b. $f(x)$ factors as four linear factors.

$$f(x) = [x - (3 - 4i)][x - (3 + 4i)]\left(x - i\sqrt{3}\right)\left(x + i\sqrt{3}\right)$$

c. The solution set for $x^4 - 6x^3 + 28x^2 - 18x + 75 = 0$ is $\left\{3 \pm 4i, \pm i\sqrt{3}\right\}$.

Skill Practice 5 Given $f(x) = x^4 - 2x^3 + 28x^2 - 4x + 52$, and that $1 + 5i$ is a zero of $f(x)$,

a. Find the zeros.

b. Factor $f(x)$ as a product of linear factors.

c. Solve the equation. $x^4 - 2x^3 + 28x^2 - 4x + 52 = 0$

Answers

5. a. Zeros:
$1 + 5i, 1 - 5i, i\sqrt{2}, -i\sqrt{2}$

b. $f(x) = [x - (1 + 5i)][x - (1 - 5i)]$
$\left(x - i\sqrt{2}\right)\left(x + i\sqrt{2}\right)$

c. $\left\{1 \pm 5i, \pm i\sqrt{2}\right\}$

> ### EXAMPLE 6   Building a Polynomial with Specified Conditions
>
> **a.** Find a third-degree polynomial $f(x)$ with integer coefficients and with zeros of $\frac{2}{3}$ and $4 + 2i$.
>
> **b.** Find a polynomial $g(x)$ of lowest degree with zeros of -2 (multiplicity 1) and 4 (multiplicity 3), and satisfying the condition that $g(0) = 256$.
>
> **Solution:**
>
> **a.** $f(x)$ is to be a polynomial with integer coefficients (and therefore real coefficients). If $4 + 2i$ is a zero, then $4 - 2i$ must also be a zero. By the linear factorization theorem we have:
>
> $f(x) = a\left(x - \dfrac{2}{3}\right)[x - (4 + 2i)][x - (4 - 2i)]$ a is a nonzero number.
>
> $\qquad = a\left(x - \dfrac{2}{3}\right)[x^2 - x(4 - 2i) - x(4 + 2i) + (4 + 2i)(4 - 2i)]$
>
> $\qquad = a\left(x - \dfrac{2}{3}\right)(x^2 - 4x + 2xi - 4x - 2xi + 16 - 4i^2)$
>
> $\qquad = 3\left(x - \dfrac{2}{3}\right)(x^2 - 8x + 20)$ To give $f(x)$ integer coefficients, choose a to be any multiple of 3. We have chosen $a = 3$.
>
> $\qquad = (3x - 2)(x^2 - 8x + 20)$
>
> $\qquad = 3x^3 - 24x^2 + 60x - 2x^2 + 16x - 40$
>
> $\qquad = 3x^3 - 26x^2 + 76x - 40$
>
> **b.** $g(x) = a(x + 2)^1(x - 4)^3$ -2 is a zero of multiplicity 1.
>
> $\qquad = a(x + 2)(x^3 - 12x^2 + 48x - 64)$ 4 is a zero of multiplicity 3.
>
> $\qquad = a(x^4 - 10x^3 + 24x^2 + 32x - 128)$ *Note:*
> $(x - 4)^3 = (x - 4)^2(x - 4)$
> $\qquad\qquad\quad = (x^2 - 8x + 16)(x - 4)$
> $\qquad\qquad\quad = x^3 - 12x^2 + 48x - 64$
>
> We also have the condition that $g(0) = 256$.
>
> $g(0) = a[(0)^4 - 10(0)^3 + 24(0)^2 + 32(0) - 128] = 256$
>
> $\qquad\qquad\qquad\qquad\qquad\qquad\qquad -128a = 256$
>
> $\qquad\qquad\qquad\qquad\qquad\qquad\qquad\qquad a = -2$
>
> Therefore, $g(x) = -2(x^4 - 10x^3 + 24x^2 + 32x - 128)$
> $\qquad\qquad g(x) = -2x^4 + 20x^3 - 48x^2 - 64x + 256$

> **Skill Practice 6**
>
> **a.** Find a third-degree polynomial $f(x)$ with integer coefficients and with zeros of $2 + i$ and $\frac{4}{3}$.
>
> **b.** Find a polynomial $g(x)$ of lowest degree with zeros of -3 (multiplicity 2) and 5 (multiplicity 2), and satisfying the condition that $g(0) = 450$.

3. Apply Descartes' Rule of Signs

Finding the zeros of a polynomial analytically can be a difficult (or impossible) task. For example, consider

$$f(x) = x^5 - 18x^4 + 128x^3 - 450x^2 + 783x - 540$$

Answers

6. a. $f(x) = 3x^3 - 16x^2 + 31x - 20$

   b. $g(x) = 2x^4 - 8x^3 - 52x^2 + 120x + 450$

Applying the rational zero theorem gives us the following possible rational zeros.

$$\pm1, \pm2, \pm3, \pm4, \pm5, \pm6, \pm9, \pm10, \pm12, \pm15, \pm18, \pm20,$$
$$\pm27, \pm30, \pm36, \pm45, \pm54, \pm60, \pm90, \pm108, \pm135, \pm180, \pm270, \pm540$$

However, we can use the upper and lower bound theorem to show that the real zeros of $f(x)$ are between -1 and 18. Furthermore, we can use a tool called Descartes' rule of signs to show that none of the zeros is negative. This eliminates all possible rational zeros except for 1, 2, 3, 4, 5, 6, 9, 10, 12, and 15.

To study Descartes' rule of signs, we need to establish what is meant by "sign changes" between consecutive terms in a polynomial. For example, the following polynomial is written in descending order and has three changes in sign between consecutive coefficients.

$$2x^6 - 3x^4 - x^3 + 5x^2 - 6x - 4 \qquad \text{(3 sign changes)}$$

positive to negative	negative to positive	positive to negative

TIP For a polynomial with real coefficients, the reason we reduce the number of possible real zeros in increments of 2 is because the alternative, nonreal zeros, occur in conjugate pairs.

Descartes' Rule of Signs

Let $f(x)$ be a polynomial with real coefficients and a nonzero constant term. Then,

1. The number of *positive* real zeros is either
   - the same as the number of sign changes in $f(x)$ or
   - less than the number of sign changes in $f(x)$ by a positive even integer.
2. The number of *negative* real zeros is either
   - the same as the number of sign changes in $f(-x)$ or
   - less than the number of sign changes in $f(-x)$ by a positive even integer.

Descartes' rule of signs is demonstrated in Examples 7 and 8.

EXAMPLE 7 Applying Descartes' Rule of Signs

Determine the number of possible positive and negative real zeros.
$$f(x) = x^5 - 6x^4 + 12x^3 - 12x^2 + 11x - 6$$

Solution:

$f(x)$ has real coefficients and the constant term is nonzero.

To determine the number of possible positive real zeros, determine the number of sign changes in $f(x)$.

$$f(x) = x^5 - 6x^4 + 12x^3 - 12x^2 + 11x - 6 \qquad \text{(5 sign changes)}$$

The number of possible positive real zeros is either 5, 3, or 1.

To determine the number of possible negative real zeros, determine the number of sign changes in $f(-x)$.

$$f(-x) = (-x)^5 - 6(-x)^4 + 12(-x)^3 - 12(-x)^2 + 11(-x) - 6$$
$$= -x^5 - 6x^4 - 12x^3 - 12x^2 - 11x - 6 \qquad \text{(0 sign changes)}$$

TIP Nonreal zeros are numbers that contain the imaginary number i such as $-3i$, $3i$, $4 + 7i$, $4 - 7i$ and so on.

There are no sign changes in $f(-x)$. Therefore, $f(x)$ has no negative real zeros.

Number of possible positive real zeros	5	3	1
Number of possible negative real zeros	0	0	0
Number of nonreal zeros	0	2	4
Total (including multiplicities)	5	5	5

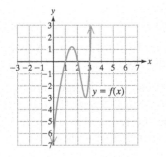

Figure 3-16

The graph of $f(x) = x^5 - 6x^4 + 12x^3 - 12x^2 + 11x - 6$ is shown in Figure 3-16. Notice that there are three positive x-intercepts and therefore, three positive real zeros. There are no negative x-intercepts as expected. The remaining zeros of the polynomial $x^5 - 6x^4 + 12x^3 - 12x^2 + 11x - 6$ are not real numbers.

Skill Practice 7   Determine the number of possible positive and negative real zeros. $f(x) = 4x^5 + 6x^3 + 2x^2 + 6$

EXAMPLE 8   **Applying Descartes' Rule of Signs**

Determine the number of possible positive and negative real zeros.

$$g(x) = 2x^6 - 5x^4 - 3x^3 + 7x^2 + 2x + 5$$

Solution:

$g(x)$ has real coefficients and the constant term is nonzero.

$$g(x) = 2x^6 - 5x^4 - 3x^3 + 7x^2 + 2x + 5 \qquad \text{2 sign changes in } g(x)$$

The number of possible positive real zeros is either 2 or 0.

$$g(-x) = 2(-x)^6 - 5(-x)^4 - 3(-x)^3 + 7(-x)^2 + 2(-x) + 5$$
$$= 2x^6 - 5x^4 + 3x^3 + 7x^2 - 2x + 5 \qquad \text{4 sign changes in } g(-x)$$

The number of possible negative real zeros is either 4, 2, or 0.

Number of possible positive real zeros	2	2	2	0	0	0
Number of possible negative real zeros	4	2	0	4	2	0
Number of nonreal zeros	0	2	4	2	4	6
Total (including multiplicities)	6	6	6	6	6	6

Skill Practice 8   Determine the number of possible positive and negative real zeros. $g(x) = 8x^6 - 5x^7 + 3x^5 - x^2 - 3x + 1$

Descartes' rule of signs stipulates that the constant term of the polynomial $f(x)$ is nonzero. If the constant term is 0, we can factor out the lowest power of x and apply Descartes' rule of signs to the resulting factor.

$$f(x) = x^7 - 8x^6 + 15x^5$$
$$f(x) = x^5(x^2 - 8x + 15)$$

Descartes' rule of signs can be applied to $x^2 - 8x + 15$ to show that there may be 2 or 0 remaining positive real zeros.

The value 0 is a zero of $f(x)$ of multiplicity 5.

4. Find Upper and Lower Bounds

The next theorem helps us limit our search for the real zeros of a polynomial. First we define two key terms.

- A real number b is called an **upper bound** of the real zeros of a polynomial if all real zeros are less than or equal to b.
- A real number a is called a **lower bound** of the real zeros of a polynomial if all real zeros are greater than or equal to a.

Answers
7. Positive: 0; Negative: 1
8. Positive: 4, 2, or 0;
    Negative: 2 or 0

Avoiding Mistakes

It is important to note that upper and lower bounds are not unique. Any number greater than b is also an upper bound for the zeros of the polynomial. Likewise any number less than a is also a lower bound.

Upper and Lower Bound Theorem for the Real Zeros of a Polynomial

Let $f(x)$ be a polynomial of degree $n \geq 1$ with real coefficients and a positive leading coefficient. Further suppose that $f(x)$ is divided by $(x - c)$.

1. If $c > 0$ and if both the remainder and the coefficients of the quotient are nonnegative, then c is an upper bound for the real zeros of $f(x)$.
2. If $c < 0$ and the coefficients of the quotient and the remainder alternate in sign (with 0 being considered either positive or negative as needed), then c is a lower bound for the real zeros of $f(x)$.

The rules for finding upper and lower bounds are stated for polynomial functions having a positive leading coefficient. However, $f(x) = 0$ and $-f(x) = 0$ are equivalent equations. Therefore, if $f(x)$ has a negative leading coefficient, we can factor out -1 from $f(x)$ and apply the rule for upper and lower bounds accordingly.

EXAMPLE 9 Applying the Upper and Lower Bound Theorem

Given $f(x) = 2x^5 + x^4 + 9x^2 - 32x + 20$,

a. Determine if the upper bound theorem identifies 2 as an upper bound for the real zeros of $f(x)$.

b. Determine if the lower bound theorem identifies -2 as a lower bound for the real zeros of $f(x)$.

Solution:

a. Divide $f(x)$ by $(x - 2)$.

First note that the leading coefficient of the polynomial is positive.

$$
\begin{array}{r|rrrrrr}
2 & 2 & 1 & 0 & 9 & -32 & 20 \\
  &   & 4 & 10 & 20 & 58 & 52 \\
\hline
  & 2 & 5 & 10 & 29 & 26 & \underline{|72}
\end{array}
$$

This row nonnegative

The remainder and all coefficients of the quotient are nonnegative.

2 is an upper bound for the real zeros of $f(x)$.

b. Divide $f(x)$ by $(x + 2)$.

$$
\begin{array}{r|rrrrrr}
-2 & 2 & 1 & 0 & 9 & -32 & 20 \\
   &   & -4 & 6 & -12 & 6 & 52 \\
\hline
   & 2 & -3 & 6 & -3 & -26 & \underline{|72} \\
   & + & - & + & - & - & +
\end{array}
$$

No sign change

The signs of the quotient do not alternate. Therefore, we cannot conclude that -2 is a lower bound for the real zeros of $f(x)$.

Skill Practice 9 Given $f(x) = x^4 - 2x^3 - 13x^2 - 4x - 30$,

a. Determine if the upper bound theorem identifies 4 as an upper bound for the real zeros of $f(x)$.

b. Determine if the lower bound theorem identifies -4 as a lower bound for the real zeros of $f(x)$.

Answers

9. a. No   **b.** Yes

TIP From Example 9, although we cannot conclude that -2 is a lower bound for the real zeros of $f(x)$, we can try other negative real numbers. For example, -3 is a lower bound for the real zeros of $f(x)$.

$$
\begin{array}{r|rrrrr}
-3 & 2 & 1 & 0 & 9 & -32 & 20 \\
  &   & -6 & 15 & -45 & 108 & -228 \\
\hline
  & 2 & -5 & 15 & -36 & 76 & \underline{|-208} \\
\end{array}
$$

Therefore, -3 is a lower bound.

signs all alternate

In Example 10, we will use the tools presented in Sections 3.2–3.4 to find all zeros of a polynomial.

EXAMPLE 10   **Finding the Zeros of a Polynomial**

Find the zeros and their multiplicities. $f(x) = 2x^5 + x^4 + 9x^2 - 32x + 20$

Solution:

$f(x)$ is a fifth-degree polynomial and must have five zeros (including multiplicities). We begin by finding the rational zeros (if any exist). By the rational zero theorem, the possible rational zeros are

$$\pm 1, \pm 2, \pm 4, \pm 5, \pm 10, \pm 20, \pm \frac{1}{2}, \pm \frac{5}{2}$$

However, we also know from Example 9 that 2 is not a zero of $f(x)$, but is an upper bound for the real zeros. From the Tip following Example 9, we know that -3 is not a zero of $f(x)$, but is a lower bound for the real zeros. Therefore, we can restrict the list of possible rational zeros to those on the interval $(-3, 2)$.

$$-\frac{5}{2}, -2, \pm\frac{1}{2}, \pm 1$$

After testing several possible rational zeros, we find that 1 is a zero.

$$
\begin{array}{r|rrrrr}
1 & 2 & 1 & 0 & 9 & -32 & 20 \\
  &   & 2 & 3 & 3 & 12 & -20 \\
\hline
  & 2 & 3 & 3 & 12 & -20 & \underline{|0} \\
\end{array}
$$

We have $f(x) = (x - 1)(2x^4 + 3x^3 + 3x^2 + 12x - 20)$. Now look for the zeros of the reduced polynomial. We will try 1 again because it may be a repeated zero.

$$
\begin{array}{r|rrrr}
1 & 2 & 3 & 3 & 12 & -20 \\
  &   & 2 & 5 & 8 & 20 \\
\hline
  & 2 & 5 & 8 & 20 & \underline{|0} \\
\end{array}
$$

The value 1 is a repeated zero.

We have $f(x) = (x - 1)^2(2x^3 + 5x^2 + 8x + 20)$.

Because the polynomial $2x^3 + 5x^2 + 8x + 20$ has no sign changes, Descartes' rule of signs indicates that there are no other positive real zeros. Now the list of possible rational zeros is restricted to $-\frac{5}{2}, -2, -\frac{1}{2}$, and -1. We find that $-\frac{5}{2}$ is a zero of $f(x)$.

$$
\begin{array}{r|rrr}
-\frac{5}{2} & 2 & 5 & 8 & 20 \\
  &   & -5 & 0 & -20 \\
\hline
  & 2 & 0 & 8 & \underline{|0} \\
\end{array}
$$

Thus, $f(x) = (x - 1)^2(x + \frac{5}{2})(2x^2 + 8)$.

$$2x^2 + 8 = 0$$
$$x^2 = -4$$
$$x = \pm 2i$$

The remaining two zeros are found by solving the equation $2x^2 + 8 = 0$.

The zeros are: 1 (multiplicity of 2), $-\frac{5}{2}$, $2i$, and $-2i$ (each with multiplicity of 1).

> **Skill Practice 10**   Find the zeros and their multiplicities.
>
> $$f(x) = x^5 + 6x^3 - 2x^2 - 27x - 18$$

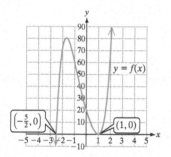

Figure 3-17

The graph of $f(x) = 2x^5 + x^4 + 9x^2 - 32x + 20$ from Example 10 is shown in Figure 3-17. The end behavior is down to the left and up to the right as expected.

The real zeros of $f(x)$ correspond to the x-intercepts $\left(-\frac{5}{2}, 0\right)$ and $(1, 0)$. The point $(1, 0)$ is a touch point because 1 is a zero with an even multiplicity. The graph crosses the x-axis at $-\frac{5}{2}$ because $-\frac{5}{2}$ is a zero with an odd multiplicity.

TECHNOLOGY CONNECTIONS

Applications of Graphing Utilities to Polynomials

A graphing utility can help us analyze a polynomial. For example, given $f(x) = 2x^3 - 11x^2 - 5x + 50$, the possible rational zeros are: $\pm 1, \pm 2, \pm 5, \pm 10,$ $\pm 25, \pm 50, \pm \frac{1}{2}, \pm \frac{5}{2},$ and $\pm \frac{25}{2}$. By graphing the function f on a graphing utility, it appears that the function may cross the x-axis at -2, 2.5, and 5. So we might consider testing these values first.

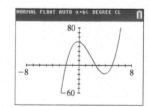

Our knowledge of algebra can also help us use a graphing device effectively. Consider $f(x) = 2x^5 + x^4 + 9x^2 - 32x + 20$ from Examples 9 and 10. We know that -3 is a lower bound for the real zeros of $f(x)$ and that 2 is an upper bound. Therefore, we can set a viewing window showing x between -3 and 2 and be guaranteed to see all the real zeros of $f(x)$.

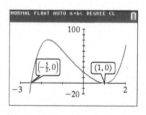

Finally, when analytical methods fail, we can use the Zero feature on a graphing utility to approximate the real zeros of a polynomial. Given $f(x) = x^3 - 7.14x^2 + 25.6x - 40.8$, the calculator approximates a zero of 3.1258745. The real zeros of a function can also be approximated by repeated use of the intermediate value theorem. (See the online group activity "Investigating the Bisection Method for Finding Zeros.")

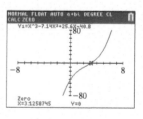

Answer

10. -1 (multiplicity 2), 2, $3i$, $-3i$ (each with multiplicity 1)

SECTION 3.4   Practice Exercises

Prerequisite Review

For Exercises R.1–R.3, perform the indicated operations. Write the answers in standard form $a + bi$.

R.1. $(9i)(-9i)$ **R.2.** $\left(6 + \sqrt{13}i\right)\left(6 - \sqrt{13}i\right)$ **R.3.** $(5 - 6i)^2 + (5 + 6i)^2$

Concept Connections

1. The _____ of a polynomial $f(x)$ are the solutions (or roots) of the equation $f(x) = 0$.

2. If $f(x)$ is a polynomial of degree $n \geq 1$ with complex coefficients, then $f(x)$ has exactly _____ complex zeros, provided that each zero is counted by its multiplicity.

3. The conjugate zeros theorem states that if $f(x)$ is a polynomial with real coefficients, and if $a + bi$ is a zero of $f(x)$, then _____ is also a zero of $f(x)$.

4. A real number b is called an _____ bound of the real zeros of a polynomial $f(x)$ if all real zeros of $f(x)$ are less than or equal to b.

5. A real number a is called a lower bound of the real zeros of a polynomial $f(x)$ if all real zeros of $f(x)$ are _____ or equal to a.

6. Explain why the number 7 cannot be a rational zero of the polynomial $f(x) = 2x^3 + 5x^2 - x + 6$.

Objective 1: Apply the Rational Zero Theorem

For Exercises 7–12, list the possible rational zeros. (See Example 1)

7. $f(x) = x^5 - 2x^3 + 7x^2 + 4$

8. $g(x) = x^3 - 5x^2 + 2x - 9$

9. $h(x) = 4x^4 + 9x^3 + 2x - 6$

10. $k(x) = 25x^7 + 22x^4 - 3x^2 + 10$

11. $m(x) = -12x^6 + 4x^3 - 3x^2 + 8$

12. $n(x) = -16x^4 - 7x^3 + 2x + 6$

13. Which of the following is *not* a possible zero of $f(x) = 2x^3 - 5x^2 + 12$?

$$1, 7, \frac{5}{3}, \frac{3}{2}$$

14. Which of the following is *not* a possible zero of $f(x) = 4x^5 - 2x^3 + 10$?

$$3, 5, \frac{5}{2}, \frac{3}{2}$$

For Exercises 15–16, find all the rational zeros.

15. $p(x) = 2x^4 - x^3 - 5x^2 + 2x + 2$

16. $q(x) = x^4 + x^3 - 7x^2 - 5x + 10$

For Exercises 17–28, find all the zeros. (See Examples 2–4)

17. $c(x) = 2x^4 - 7x^3 - 17x^2 + 58x - 24$

18. $d(x) = 3x^4 - 2x^3 - 21x^2 - 4x + 12$

19. $f(x) = x^3 - 7x^2 + 6x + 20$

20. $g(x) = x^3 - 7x^2 + 14x - 6$

21. $h(x) = 5x^3 - x^2 - 35x + 7$

22. $k(x) = 7x^3 - x^2 - 21x + 3$

23. $m(x) = 3x^4 - x^3 - 36x^2 + 60x - 16$

24. $n(x) = 2x^4 + 9x^3 - 5x^2 - 57x - 45$

25. $q(x) = x^3 - 4x^2 - 2x + 20$

26. $p(x) = x^3 - 8x^2 + 29x - 52$

27. $t(x) = x^4 - x^2 - 90$

28. $v(x) = x^4 - 12x^2 - 13$

Objective 2: Apply the Fundamental Theorem of Algebra

29. Given a polynomial $f(x)$ of degree $n \geq 1$, the fundamental theorem of algebra guarantees at least _____ complex zero.

30. The number of zeros of $f(x) = 4x^3 - 5x^2 + 6x - 3$ is _____, provided that each zero is counted according to its multiplicity.

31. If $f(x)$ is a polynomial with real coefficients and zeros of 5 (multiplicity 2), -1 (multiplicity 1), $2i$, and $3 + 4i$, what is the minimum degree of $f(x)$?

32. If $g(x)$ is a polynomial with real coefficients and zeros of -4 (multiplicity 3), 6 (multiplicity 2), $1 + i$, and $2 - 7i$, what is the minimum degree of $g(x)$?

For Exercises 33–38, a polynomial $f(x)$ and one or more of its zeros is given.

a. Find all the zeros.

b. Factor $f(x)$ as a product of linear factors.

c. Solve the equation $f(x) = 0$. (See Example 5)

33. $f(x) = x^4 - 4x^3 + 22x^2 + 28x - 203$; $2 - 5i$ is a zero

34. $f(x) = x^4 - 6x^3 + 5x^2 + 30x - 50$; $3 - i$ is a zero

35. $f(x) = 3x^3 - 28x^2 + 83x - 68$; $4 + i$ is a zero

36. $f(x) = 5x^3 - 54x^2 + 170x - 104$; $5 + i$ is a zero

37. $f(x) = 4x^5 + 37x^4 + 117x^3 + 87x^2 - 193x - 52$;

$-3 + 2i$ and $-\dfrac{1}{4}$ are zeros

38. $f(x) = 2x^5 - 5x^4 - 4x^3 - 22x^2 + 50x + 75$;

$-1 - 2i$ and $\dfrac{5}{2}$ are zeros

For Exercises 39–48, write a polynomial $f(x)$ that satisfies the given conditions. (See Example 6)

39. Degree 3 polynomial with integer coefficients with zeros $6i$ and $\frac{4}{5}$

40. Degree 3 polynomial with integer coefficients with zeros $-4i$ and $\frac{3}{2}$

41. Polynomial of lowest degree with zeros of -4 (multiplicity 1), 2 (multiplicity 3) and with $f(0) = 160$

42. Polynomial of lowest degree with zeros of 5 (multiplicity 2) and -3 (multiplicity 2) and with $f(0) = -450$

43. Polynomial of lowest degree with zeros of $-\frac{4}{3}$ (multiplicity 2) and $\frac{1}{2}$ (multiplicity 1) and with $f(0) = -16$

44. Polynomial of lowest degree with zeros of $-\frac{5}{6}$ (multiplicity 2) and $\frac{1}{3}$ (multiplicity 1) and with $f(0) = -25$

45. Polynomial of lowest degree with real coefficients and with zeros $7 - 4i$ and 0 (multiplicity 4)

46. Polynomial of lowest degree with real coefficients and with zeros $5 - 10i$ and 0 (multiplicity 3)

47. Polynomial of lowest degree with real coefficients and zeros of $5i$ and $6 - i$.

48. Polynomial of lowest degree with real coefficients and zeros of $-3i$ and $5 + 2i$.

Objective 3: Apply Descartes' Rule of Signs

For Exercises 49–56, determine the number of possible positive and negative real zeros for the given function. (See Examples 7–8)

49. $f(x) = x^6 - 2x^4 + 4x^3 - 2x^2 - 5x - 6$

50. $g(x) = 3x^7 + 4x^4 - 6x^3 + 5x^2 - 6x + 1$

51. $k(x) = -8x^7 + 5x^6 - 3x^4 + 2x^3 - 11x^2 + 4x - 3$

52. $h(x) = -4x^9 + 6x^8 - 5x^5 - 2x^4 + 3x^2 - x + 8$

53. $p(x) = 0.11x^4 + 0.04x^3 + 0.31x^2 + 0.27x + 1.1$

54. $q(x) = -0.6x^4 + 0.8x^3 - 0.6x^2 + 0.1x - 0.4$

55. $v(x) = \frac{1}{8}x^6 + \frac{1}{6}x^4 + \frac{1}{3}x^2 + \frac{1}{10}$

56. $t(x) = \frac{1}{1000}x^6 + \frac{1}{100}x^4 + \frac{1}{10}x^2 + 1$

For Exercises 57–58, use Descartes' rule of signs to determine the total number of real zeros and the number of positive and negative real zeros. (*Hint*: First factor out x to its lowest power.)

57. $f(x) = x^8 + 5x^6 + 6x^4 - x^3$

58. $f(x) = -5x^8 - 3x^6 - 4x^2 + x$

Objective 4: Find Upper and Lower Bounds

For Exercises 59–64, (See Example 9)

a. Determine if the upper bound theorem identifies the given number as an upper bound for the real zeros of $f(x)$.

b. Determine if the lower bound theorem identifies the given number as a lower bound for the real zeros of $f(x)$.

59. $f(x) = x^5 + 6x^4 + 5x^2 + x - 3$
  a. 2   **b.** -5

60. $f(x) = x^4 + 8x^3 - 4x^2 + 7x - 3$
  a. 3   **b.** -4

61. $f(x) = 8x^3 - 42x^2 + 33x + 28$
  a. 6   **b.** -1

62. $f(x) = 6x^3 - x^2 - 57x + 70$
  a. 4   **b.** -4

63. $f(x) = 2x^5 + 11x^4 - 63x^2 - 50x + 40$
  a. 3   **b.** -6

64. $f(x) = 3x^5 - 16x^4 + 5x^3 + 90x^2 - 138x + 36$
  a. 6   **b.** -3

For Exercises 65–68, determine if the statement is true or false. If a statement is false, explain why.

65. If 5 is an upper bound for the real zeros of $f(x)$, then 6 is also an upper bound.

66. If 5 is an upper bound for the real zeros of $f(x)$, then 4 is also an upper bound.

67. If -3 is a lower bound for the real zeros of $f(x)$, then -2 is also a lower bound.

68. If -3 is a lower bound for the real zeros of $f(x)$, then -4 is also a lower bound.

For Exercises 69–84, find the zeros and their multiplicities. Consider using Descartes' rule of signs and the upper and lower bound theorem to limit your search for rational zeros. (See Example 10)

69. $f(x) = 8x^3 - 42x^2 + 33x + 28$
  (*Hint*: See Exercise 61.)

70. $f(x) = 6x^3 - x^2 - 57x + 70$
  (*Hint*: See Exercise 62.)

71. $f(x) = 2x^5 + 11x^4 - 63x^2 - 50x + 40$
  (*Hint*: See Exercise 63.)

72. $f(x) = 3x^5 - 16x^4 + 5x^3 + 90x^2 - 138x + 36$
  (*Hint*: See Exercise 64.)

73. $f(x) = 4x^4 + 20x^3 + 13x^2 - 30x + 9$

74. $f(x) = 9x^4 + 30x^3 + 13x^2 - 20x + 4$

75. $f(x) = 2x^4 - 11x^3 + 27x^2 - 41x + 15$

76. $g(x) = 3x^4 - 20x^3 + 51x^2 - 56x + 20$

77. $h(x) = 4x^4 - 28x^3 + 73x^2 - 90x + 50$

78. $k(x) = 9x^4 - 42x^3 + 70x^2 - 34x + 5$

79. $f(x) = x^6 + 2x^5 + 11x^4 + 20x^3 + 10x^2$

80. $f(x) = x^6 + 6x^5 + 12x^4 + 18x^3 + 27x^2$

81. $f(x) = x^5 - 10x^4 + 34x^3$

82. $f(x) = x^6 - 12x^5 + 40x^4$

83. $f(x) = -x^3 + 3x^2 - 9x - 13$

84. $f(x) = -x^3 + 5x^2 - 11x + 15$

Mixed Exercises

For Exercises 85–90, determine if the statement is true or false. If a statement is false, explain why.

85. A polynomial with real coefficients of degree 4 must have at least one real zero.

86. Given $f(x) = 2ix^4 - (3 + 6i)x^3 + 5x^2 + 7$, if $a + bi$ is a zero of $f(x)$, then $a - bi$ must also be a zero.

87. The graph of a 10th-degree polynomial must cross the x-axis exactly once.

88. Suppose that $f(x)$ is a polynomial, and that a and b are real numbers where $a < b$. If $f(a) < 0$ and $f(b) < 0$, then $f(x)$ has no real zeros on the interval $[a, b]$.

89. If c is a zero of a polynomial $f(x)$, with degree $n \geq 2$ then all other zeros of $f(x)$ are zeros of $\dfrac{f(x)}{x - c}$.

90. If b is an upper bound for the real zeros of a polynomial, then $-b$ is a lower bound for the real zeros of the polynomial.

91. Given that $x - c$ divides evenly into a polynomial $f(x)$, which statements are true?

   a. $x - c$ is a factor of $f(x)$.

   b. c is a zero of $f(x)$.

   c. The remainder of $f(x) \div (x - c)$ is 0.

   d. c is a solution (root) of the equation $f(x) = 0$.

92. a. Use the quadratic formula to solve $x^2 - 7x + 5 = 0$.

   b. Write $x^2 - 7x + 5$ as a product of linear factors.

93. a. Use the intermediate value theorem to show that $f(x) = 2x^2 - 7x + 4$ has a real zero on the interval $[2, 3]$.

   b. Find the zeros.

94. Show that $x - a$ is a factor of $x^n - a^n$ for any positive integer n and constant a.

Write About It

95. Explain why a polynomial with real coefficients of degree 3 must have at least one real zero.

96. Why is it not necessary to apply the rational zero theorem, Descartes' rule of signs, or the upper and lower bound theorem to find the zeros of a second-degree polynomial?

97. Explain why $f(x) = 5x^6 + 7x^4 + x^2 + 9$ has no real zeros.

98. Explain why the fundamental theorem of algebra does not apply to $f(x) = \sqrt{x} + 3$. That is, no complex number c exists such that $f(c) = 0$.

Expanding Your Skills

99. Let n be a positive even integer. Determine the greatest number of possible nonreal zeros of $f(x) = x^n - 1$.

100. Let n be a positive odd integer. Determine the greatest number of possible nonreal zeros of $f(x) = x^n - 1$.

101. The front face of a tent is triangular and the height of the triangle is two-thirds of the base. The length of the tent is 3 ft more than the base of the triangular face. If the tent holds a volume of 108 ft³, determine its dimensions.

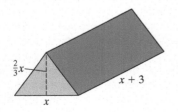

102. An underground storage tank for gasoline is in the shape of a right circular cylinder with hemispheres on each end. If the total volume of the tank is $\dfrac{104\pi}{3}$ ft³, find the radius of the tank.

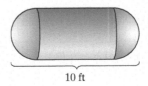

10 ft

103. A food company originally sells cereal in boxes with dimensions 10 in. by 7 in. by 2.5 in. To make more profit, the company decreases each dimension of the box by x inches but keeps the price the same. If the new volume is 81 in.^{3} by how much was each dimension decreased?

104. A truck rental company rents a 12-ft by 8-ft by 6-ft truck for $69.95 per day plus mileage. A customer prefers to rent a less expensive smaller truck whose dimensions are x ft smaller on each side. If the volume of the smaller truck is 240 ft^3, determine the dimensions of the smaller truck.

105. A rectangle is bounded by the x-axis and a parabola defined by $y = 4 - x^2$. What are the dimensions of the rectangle if the area is 6 cm^2? Assume that all units of length are in centimeters.

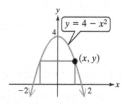

106. A rectangle is bounded by the parabola defined by $y = x^2$, the x-axis, and the line $x = 5$ as shown in the figure. If the area of the rectangle is 12 in.^{2} determine the dimensions of the rectangle.

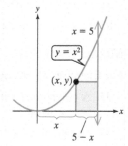

The linear factorization theorem tells us that a polynomial of degree $n \geq 1$ factors into n linear factors over the complex numbers. If we do not factor over the set of complex numbers, then a polynomial with real coefficients can be factored into linear factors and irreducible quadratic factors. An *irreducible quadratic factor* is a quadratic polynomial that does not factor further over the set of real numbers.

For example, consider the polynomial $f(x) = x^4 - 5x^3 + 5x^2 + 25x - 26$.

Factoring over the real numbers, we have two linear factors and one irreducible quadratic factor:

$$x^4 - 5x^3 + 5x^2 + 25x - 26 = \underbrace{(x + 2)(x - 1)}_{\substack{\text{2 linear} \\ \text{factors}}}\underbrace{(x^2 - 6x + 13)}_{\substack{\text{irreducible} \\ \text{quadratic factor}}}$$

Factoring over the complex numbers, we have four linear factors as guaranteed by the linear factorization theorem.

$$x^4 - 5x^3 + 5x^2 + 25x - 26 = (x + 2)(x - 1)[x - (3 + 2i)][x - (3 - 2i)]$$

For Exercises 107–110,

a. Factor the polynomial over the set of real numbers.

b. Factor the polynomial over the set of complex numbers.

107. $f(x) = x^4 + 2x^3 + x^2 + 8x - 12$

108. $f(x) = x^4 - 6x^3 + 9x^2 - 6x + 8$

109. $f(x) = x^4 + 2x^2 - 35$

110. $f(x) = x^4 + 8x^2 - 33$

111. Find all fourth roots of 1, by solving the equation $x^4 = 1$. (*Hint*: Find the zeros of the polynomial $f(x) = x^4 - 1$.)

112. Find all sixth roots of 1, by solving the equation $x^6 = 1$. [*Hint*: Find the zeros of the polynomial $f(x) = x^6 - 1$. Begin by factoring $x^6 - 1$ as $(x^3 - 1)(x^3 + 1)$.]

113. Use the rational zero theorem to show that $\sqrt{5}$ is an irrational number. (*Hint*: Show that $f(x) = x^2 - 5$ has no rational zeros.)

114. a. Given a linear equation $ax + b = 0$ ($a \neq 0$), the solution is given by $x =$ _____.

b. Given a quadratic equation $ax^2 + bx + c = 0$ ($a \neq 0$), the solutions are given by $x =$ _____.

From Exercise 114, we see that linear and quadratic equations have generic formulas that can be used to find the solution sets. But what about a cubic polynomial equation? Mathematicians struggled for centuries to find such a formula. Finally, Italian mathematician Niccolo Tartaglia (1500–1557) developed a method to solve a cubic equation of the form

$$x^3 + mx = n$$

The result was later published in *Ars Magna*, by Gerolamo Cardano (1501–1576).

Foe Exercises 115–116, use the formula

$$x = \sqrt[3]{\sqrt{\left(\frac{n}{2}\right)^2 + \left(\frac{m}{3}\right)^3} + \frac{n}{2}} - \sqrt[3]{\sqrt{\left(\frac{n}{2}\right)^2 + \left(\frac{m}{3}\right)^3} - \frac{n}{2}}$$

to find a solution to the equation $x^3 + mx = n$.

115. $x^3 - 3x = -2$

116. $x^3 + 9x = 26$

Point of Interest

Early in the sixteenth century, Italian mathematicians Niccolo Tartaglia and Gerolamo Cardano solved a general cubic equation in terms of the constants appearing in the equation. Cardano's pupil, Ludovico Ferrari, then solved a general equation of fourth degree. Despite decades of work, no general solution to a fifth-degree equation was found. Finally, Norwegian mathematician Niels Abel and French mathematician Evariste Galois proved that no such solution exists.

SECTION 3.5 | Rational Functions

1. Apply Notation Describing Infinite Behavior of a Function

In this chapter, we have studied polynomials and polynomial functions. Now we look at functions that are defined as the ratio of two polynomials. These are called rational functions.

Definition of a Rational Function

Let $p(x)$ and $q(x)$ be polynomials where $q(x) \neq 0$. A function f defined by

$$f(x) = \frac{p(x)}{q(x)} \text{ is called a \textbf{rational function.}}$$

Note: The domain of a rational function is all real numbers excluding the real zeros of $q(x)$.

Function	Factored Form	Domain
$f(x) = \dfrac{1}{x}$	$f(x) = \dfrac{1}{x}$	$\{x \mid x \neq 0\}$ $(-\infty, 0) \cup (0, \infty)$
$g(x) = \dfrac{5x^2}{2x^2 + 5x - 12}$	$g(x) = \dfrac{5x^2}{(2x - 3)(x + 4)}$	$\{x \mid x \neq \frac{3}{2}, x \neq -4\}$ $(-\infty, -4) \cup \left(-4, \frac{3}{2}\right) \cup \left(\frac{3}{2}, \infty\right)$
$k(x) = \dfrac{x + 3}{x^2 + 4}$	$k(x) = \dfrac{x + 3}{x^2 + 4}$	$\mathbb{R}$ $(-\infty, \infty)$

In this section, we will analyze rational functions. To do so, we want to determine the behavior of the function as x approaches ∞ or $-\infty$ and as x approaches values for which the function is undefined. We will use the following notation (Table 3-1).

Table 3-1

Notation	Meaning
$x \to c^+$	x approaches c from the right (but will not equal c).
$x \to c^-$	x approaches c from the left (but will not equal c).
$x \to \infty$	x approaches infinity (x increases without bound).
$x \to -\infty$	x approaches negative infinity (x decreases without bound).

For example, consider the reciprocal function $f(x) = \frac{1}{x}$ first introduced in Section 2.6 (Figure 3-18).

x	$f(x) = \frac{1}{x}$
-1	-1
-10	-0.1
-100	-0.01
-1000	-0.001

x	$f(x) = \frac{1}{x}$
-1	-1
-0.1	-10
-0.01	-100
-0.001	-1000

x	$f(x) = \frac{1}{x}$
1	1
0.1	10
0.01	100
0.001	1000

x	$f(x) = \frac{1}{x}$
1	1
10	0.1
100	0.01
1000	0.001

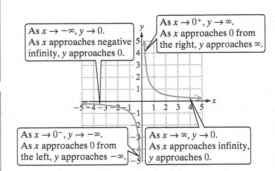

As $x \to -\infty, y \to 0$.
As x approaches negative infinity, y approaches 0.

As $x \to 0^+, y \to \infty$.
As x approaches 0 from the right, y approaches ∞.

As $x \to 0^-, y \to -\infty$.
As x approaches 0 from the left, y approaches $-\infty$.

As $x \to \infty, y \to 0$.
As x approaches infinity, y approaches 0.

Figure 3-18

In Example 1, we study the graph of another basic rational function, $f(x) = \dfrac{1}{x^2}$. From the definition of the function, we make the following observations.

- The domain of $f(x) = \dfrac{1}{x^2}$ is all real numbers excluding zero.

- f is an even function and the graph is symmetric to the y-axis.
- The values of $f(x)$ are positive over the domain of f.

EXAMPLE 1   Investigating the Behavior of a Rational Function

The graph of $f(x) = \dfrac{1}{x^2}$ is given.

Complete the statements.

a. As $x \to -\infty,\, f(x) \to$ _____.
b. As $x \to 0^-,\ f(x) \to$ _____.
c. As $x \to 0^+,\ f(x) \to$ _____.
d. As $x \to \infty,\ f(x) \to$ _____.

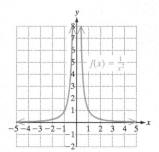

Solution:

a. As $x \to -\infty,\, f(x) \to 0$.

b. As $x \to 0^-,\ f(x) \to \infty$.

c. As $x \to 0^+,\ f(x) \to \infty$.

d. As $x \to \infty,\ f(x) \to 0$.

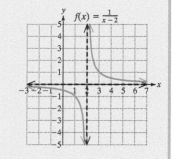

Skill Practice 1   The graph of $f(x) = \dfrac{1}{x - 2}$ is given.

Complete the statements.

a. As $x \to -\infty,\, f(x) \to$ _____.
b. As $x \to 2^-,\ f(x) \to$ _____.
c. As $x \to 2^+,\ f(x) \to$ _____.
d. As $x \to \infty,\ f(x) \to$ _____.

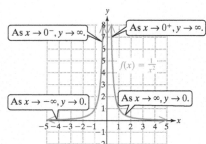

---

2. Identify Vertical Asymptotes

The graphs of $f(x) = \dfrac{1}{x}$ and $f(x) = \dfrac{1}{x^2}$ both approach the y-axis, but do not touch the y-axis. The y-axis is called a *vertical asymptote* of the graphs of the functions.

Answers
1. a. 0   **b.** $-\infty$ **c.** ∞ **d.** 0

Definition of a Vertical Asymptote

The line $x = c$ is a **vertical asymptote** of the graph of a function f if $f(x)$ approaches infinity or negative infinity as x approaches c from either side.

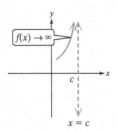 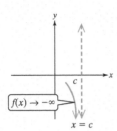 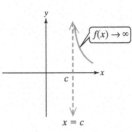 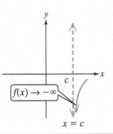

A function may have no vertical asymptotes, one vertical asymptote, or many vertical asymptotes. To locate the vertical asymptotes of a function, determine the real numbers x where the denominator is zero, but the numerator is nonzero.

> **TIP** The case where $p(x)$ and $q(x)$ share a common factor is addressed in Exercises 111–114 on page 367.

Identifying Vertical Asymptotes of a Rational Function

Consider a rational function f defined by $f(x) = \dfrac{p(x)}{q(x)}$, where $p(x)$ and $q(x)$ have no common factors other than 1. If c is a real zero of $q(x)$, then $x = c$ is a vertical asymptote of the graph of f.

EXAMPLE 2   Identifying Vertical Asymptotes

Identify the vertical asymptotes.

a. $f(x) = \dfrac{2}{x - 3}$ **b.** $g(x) = \dfrac{x - 4}{3x^2 + 5x - 2}$ **c.** $k(x) = \dfrac{4x^2}{x^2 + 4}$

Solution:

a. $f(x) = \dfrac{2}{x - 3}$

The expression $\frac{2}{x-3}$ is written in lowest terms. The denominator is zero for $x = 3$.

f has a vertical asymptote of $x = 3$.

> ### Avoiding Mistakes
> A vertical asymptote is a line and should be identified by an equation of the form $x = c$, where c is a constant.

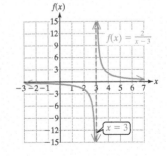

b. $g(x) = \dfrac{x - 4}{3x^2 + 5x - 2}$

Factor the numerator and denominator.

$= \dfrac{x - 4}{(3x - 1)(x + 2)}$

The numerator and denominator share no common factors other than 1. The zeros of the denominator are $\frac{1}{3}$ and -2.

The vertical asymptotes are

$x = \dfrac{1}{3}$ and $x = -2$.

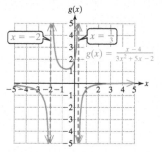

c. $k(x) = \dfrac{4x^2}{x^2 + 4}$

The numerator and denominator are already factored over the real numbers and the rational expression is in lowest terms.

The denominator has no real zeros, so the graph of k has no vertical asymptotes.

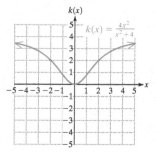

Skill Practice 2 Identify the vertical asymptotes.

a. $f(x) = \dfrac{3}{x + 1}$ **b.** $h(x) = \dfrac{x + 7}{2x^2 - x - 10}$ **c.** $m(x) = \dfrac{5x}{x^4 + 1}$

Avoiding Mistakes

The procedure to find the vertical asymptotes of a rational function is given under the condition that the numerator and denominator share no common factors. This important observation can be illustrated by the graph of

$f(x) = \dfrac{2x^2 + 5x + 3}{x + 1}$ (Figure 3-19). The numerator and denominator share a common factor of $(x + 1)$.

$$f(x) = \dfrac{2x^2 + 5x + 3}{x + 1} = \dfrac{(2x + 3)(x + 1)}{x + 1}$$

The value $x = -1$ is not in the domain of f, but the graph of f has a "hole" at $x = -1$ rather than a vertical asymptote.

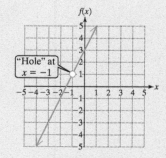

Figure 3-19

Answers

2. a. $x = -1$

  b. $x = \dfrac{5}{2}$ and $x = -2$

  c. No vertical asymptotes

3. Identify Horizontal Asymptotes

Refer to the graph of $f(x) = \frac{1}{x}$ (Figure 3-20). Toward the far left and far right of the graph, $f(x)$ approaches the line $y = 0$ (the x-axis). The x-axis is called a horizontal asymptote of the graph of f.

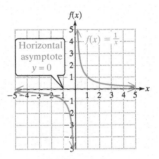

Figure 3-20

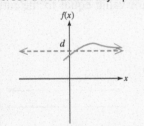

TIP While the graph of a function may not cross a vertical asymptote, it may cross a horizontal asymptote.

Definition of a Horizontal Asymptote

The line $y = d$ is a **horizontal asymptote** of the graph of a function f if $f(x)$ approaches d as x approaches infinity or negative infinity.

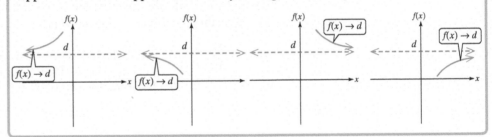

Recall that the leading term determines the far left and far right behavior of the graph of a polynomial function. Since a rational function is the ratio of two polynomials, it seems reasonable that the leading terms of the numerator and denominator determine the end behavior of a rational function.

TIP A rational function may have many vertical asymptotes, but at most one horizontal asymptote.

Identifying Horizontal Asymptotes of a Rational Function

Let f be a rational function defined by

$$f(x) = \frac{a_n x^n + a_{n-1} x^{n-1} + a_{n-2} x^{n-2} + \cdots + a_1 x + a_0}{b_m x^m + b_{m-1} x^{m-1} + b_{m-2} x^{m-2} + \cdots + b_1 x + b_0}$$

The definition of $f(x)$ indicates that n is the degree of the numerator and m is the degree of the denominator.

1. If $n > m$, then f has no horizontal asymptote.
2. If $n < m$, then the line $y = 0$ (the x-axis) is the horizontal asymptote of f.
3. If $n = m$, then the line $y = \dfrac{a_n}{b_m}$ is the horizontal asymptote of f.

1. If the degree of the numerator is greater than the degree of the denominator ($n > m$), then the numerator will "dominate" the quotient. For example:

$$f(x) = \frac{x^4 + \cdots}{x^2 + \cdots}$$ will behave like $y = x^2$ as $|x|$ becomes large. Therefore, f has no horizontal asymptote.

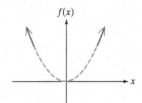

2. If the degree of the numerator is less than the degree of the denominator ($n < m$), then the denominator will "dominate" the quotient. For example:

$$f(x) = \frac{x^2 + \cdots}{x^4 + \cdots}$$ will behave like $y = \frac{1}{x^2}$. The ratio $\frac{1}{x^2}$ tends toward 0 as $|x|$ becomes large. Therefore, f has a horizontal asymptote of $y = 0$.

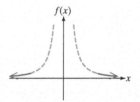

3. If the degree of the numerator is equal to the degree of the denominator ($n = m$), then the magnitude of the numerator and denominator somewhat "offset" each other. As a result, the function tends toward a constant value equal to the ratio of the leading coefficients. For example:

$$f(x) = \frac{4x^2 + \cdots}{3x^2 + \cdots}$$ will behave like $y = \frac{4}{3}$ as $|x|$ becomes large. Therefore, f has a horizontal asymptote of $y = \frac{4}{3}$.

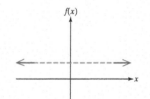

EXAMPLE 3   Identifying Horizontal Asymptotes

Find the horizontal asymptotes (if any) for the given functions.

a. $f(x) = \dfrac{8x^2 + 1}{x^4 + 1}$ **b.** $g(x) = \dfrac{2x^3 - 6x}{x^2 + 4}$ **c.** $h(x) = \dfrac{8x^2 + 9x - 5}{2x^2 + 1}$

Solution:

a. $f(x) = \dfrac{8x^2 + 1}{x^4 + 1}$ The degree of the numerator is 2 ($n = 2$).
The degree of the denominator is 4 ($m = 4$).

Since $n < m$, then the line $y = 0$ (the x-axis) is a horizontal asymptote of f.

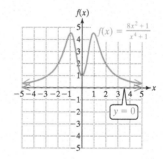

b. $g(x) = \dfrac{2x^3 - 6x}{x^2 + 4}$

The degree of the numerator is 3 ($n = 3$).
The degree of the denominator is 2 ($m = 2$).

Since $n > m$, then the function has no horizontal asymptotes.

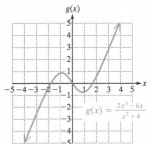

c. $h(x) = \dfrac{8x^2 + 9x - 5}{2x^2 + 1}$

The degree of the numerator is 2 ($n = 2$).
The degree of the denominator is 2 ($m = 2$).

Since $n = m$, then the line $y = \frac{8}{2}$ or equivalently $y = 4$ is a horizontal asymptote of the graph of f.

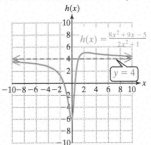

Skill Practice 3   Find the horizontal asymptotes (if any) for the given functions.

a. $f(x) = \dfrac{7x^2 + 2x}{4x^2 - 3}$ **b.** $m(x) = \dfrac{4x^3 + 2}{2x - 1}$ **c.** $n(x) = \dfrac{5}{4x^2 + 9}$

The graph of a rational function may not cross a vertical asymptote. However, as demonstrated in Example 3(c), the graph may cross a horizontal asymptote. For the purpose of graphing a rational function, it is helpful to determine where a graph crosses a horizontal asymptote.

Suppose that the line $y = d$ is a horizontal asymptote of a rational function $y = f(x)$. The solutions to the equation $f(x) = d$ are the values of x where the graph of f crosses its horizontal asymptote. If the equation has no real solution, then the graph does not cross its horizontal asymptote.

EXAMPLE 4   **Determining Where a Graph Crosses a Horizontal Asymptote**

Given $h(x) = \dfrac{8x^2 + 9x - 5}{2x^2 + 1}$, determine the point where the graph of h crosses its horizontal asymptote.

Answers

3. a. $y = \dfrac{7}{4}$
   b. No horizontal asymptotes
   c. $y = 0$

Solution:

$h(x) = \dfrac{8x^2 + 9x - 5}{2x^2 + 1}$

From Example 3(c), the horizontal asymptote is $y = 4$.

$$\dfrac{8x^2 + 9x - 5}{2x^2 + 1} = 4$$ Set $h(x)$ equal to 4.

$$\dfrac{8x^2 + 9x - 5}{2x^2 + 1} \cdot (2x^2 + 1) = 4 \cdot (2x^2 + 1)$$ Clear fractions by multiplying by the LCD.

$$8x^2 + 9x - 5 = 8x^2 + 4$$

$$9x - 5 = 4$$

$$x = 1$$

The function crosses its horizontal asymptote at $(1, 4)$.

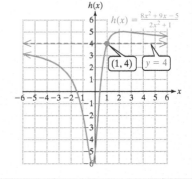

Answer

4. Horizontal asymptote $y = 3$;
   Crosses at $(3, 3)$

Skill Practice 4   Given $g(x) = \dfrac{3x^2 + 4x - 3}{x^2 + 3}$, determine the horizontal asymptote and the point where the graph crosses the horizontal asymptote.

4. Identify Slant Asymptotes

Consider the function defined by $f(x) = \dfrac{x^2 + 1}{x - 1}$. The graph has a vertical asymptote of $x = 1$, but no horizontal asymptote (the degree of the numerator is greater than the degree of the denominator). However, as x approaches infinity and negative infinity, the graph approaches the graph of $y = x + 1$ (shown in red in Figure 3-21). This line is called a **slant asymptote** because it is neither horizontal nor vertical.

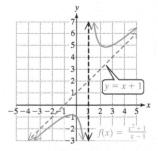

Figure 3-21

Identifying a Slant Asymptote of a Rational Function

- A rational function will have a slant asymptote if the degree of the numerator is exactly one greater than the degree of the denominator.
- To find an equation of a slant asymptote, divide the numerator of the function by the denominator. The quotient will be linear and the slant asymptote will be of the form $y = $ quotient.

TIP Because the divisor is of the form $x - c$, we can also use synthetic division to determine the slant asymptote.

```
1| 1   0   1
       1   1
   1   1   |2
```

The quotient is $x + 1$. The slant asymptote is $y = x + 1$.

For $f(x) = \dfrac{x^2 + 1}{x - 1}$, divide $(x^2 + 1)$ by $(x - 1)$ using long division or synthetic division.

$$
\begin{array}{r}
x + 1 \\
x - 1 \overline{)\, x^2 + 0x + 1} \\
\underline{-(x^2 - x)} \\
x + 1 \\
\underline{-(x - 1)} \\
2
\end{array}
$$

The slant asymptote is $y = x + 1$.

Using the division algorithm, $f(x) = \dfrac{x^2 + 1}{x - 1} = x + 1 + \dfrac{2}{x - 1}$. The expression $\dfrac{2}{x - 1}$ will approach 0 as $|x|$ approaches infinity. Therefore, $f(x)$ will approach the line $y = x + 1$ as $|x|$ approaches infinity.

EXAMPLE 5   **Identifying the Asymptotes of a Rational Function**

Determine the asymptotes. $f(x) = \dfrac{2x^2 - 5x - 3}{x - 2}$

Solution:

$f(x) = \dfrac{2x^2 - 5x - 3}{x - 2}$

The expression $\dfrac{2x^2 - 5x - 3}{x - 2} = \dfrac{(2x + 1)(x - 3)}{x - 2}$ is in lowest terms, and the denominator is zero at $x = 2$.

f has a vertical asymptote of $x = 2$.

f has no horizontal asymptote.

The degree of the numerator is exactly one greater than the degree of the denominator. Therefore, f has no horizontal asymptote, but does have a slant asymptote.

To find the slant asymptote divide $(2x^2 - 5x - 3)$ by $(x - 2)$.

$$
\begin{array}{r}
2x - 1 \\
x - 2 \overline{)\,2x^2 - 5x - 3} \\
-(2x^2 - 4x) \\
\hline
-x - 3 \\
-(-x + 2) \\
\hline
-5
\end{array}
$$

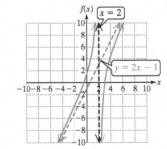

The quotient is $2x - 1$.
The slant asymptote is given by $y = 2x - 1$.

Skill Practice 5   Determine the asymptotes. $g(x) = \dfrac{2x^2 - 9}{x + 1}$

---

5. Graph Rational Functions

We now turn our attention to graphing rational functions. The transformations used in Section 2.6 can be applied to the basic rational functions $y = \dfrac{1}{x}$ and $y = \dfrac{1}{x^2}$.

EXAMPLE 6   **Using Transformations to Graph a Rational Function**

Use transformations to graph $f(x) = \dfrac{1}{(x + 2)^2} + 3$.

Answer

5. Vertical asymptote: $x = -1$;
No horizontal asymptote; Slant asymptote: $y = 2x - 2$

Solution:

$$f(x) = \frac{1}{(x + 2)^2} + 3$$

The graph of f is the graph of $y = \frac{1}{x^2}$ with a shift to the left 2 units and a shift upward 3 units.

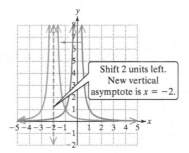

Shift 2 units left. New vertical asymptote is $x = -2$.

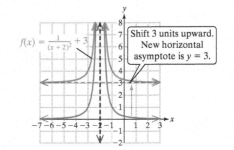

$f(x) = \frac{1}{(x + 2)^2} + 3$

Shift 3 units upward. New horizontal asymptote is $y = 3$.

Skill Practice 6   Use transformations to graph $g(x) = \dfrac{1}{x - 3} - 2$.

To graph a rational function that is not a simple transformation of $y = \dfrac{1}{x}$ or $y = \dfrac{1}{x^2}$, more steps must be employed. Our strategy is to find all asymptotes and key points (intercepts and points where the function crosses a horizontal asymptote). Then determine the behavior of the function on the intervals defined by these key points and the vertical asymptotes.

Graphing a Rational Function

Consider a rational function f defined by $f(x) = \dfrac{p(x)}{q(x)}$, where $p(x)$ and $q(x)$ are polynomials with no common factors.

1. Determine the y-intercept by evaluating $f(0)$.
2. Determine the x-intercept(s) by finding the real solutions of $f(x) = 0$. The value $f(x)$ equals zero when the numerator $p(x) = 0$.
3. Identify any vertical asymptotes and graph them as dashed lines.
4. Determine whether the function has a horizontal asymptote or a slant asymptote (or neither), and graph the asymptote as a dashed line.
5. Determine where the function crosses the horizontal or slant asymptote (if applicable).
6. If a test for symmetry is easy to apply, use symmetry to plot additional points. Recall:
   - f is an even function (symmetric to the y-axis) if $f(-x) = f(x)$.
   - f is an odd function (symmetric to the origin) if $f(-x) = -f(x)$.
7. Plot at least one point on the intervals defined by the x-intercepts, vertical asymptotes, and points where the function crosses a horizontal or slant asymptote.
8. Sketch the function based on the information found in steps 1–7.

Answer

6.

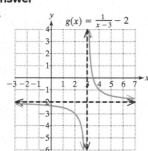

$g(x) = \frac{1}{x - 3} - 2$

EXAMPLE 7   Graphing a Rational Function

Graph $f(x) = \dfrac{x + 3}{x - 2}$.

Solution:

1. Determine the y-intercept.

$$f(0) = \frac{(0) + 3}{(0) - 2} = -\frac{3}{2}$$

The y-intercept is $\left(0, -\frac{3}{2}\right)$.

2. Determine the x-intercept(s).

$\dfrac{x + 3}{x - 2} = 0$ when $x + 3 = 0$ or $x = -3$.

The x-intercept is $(-3, 0)$.

3. Identify the vertical asymptotes.

The polynomial $x - 2$ has a zero at $x = 2$, and the numerator $x + 3$ is nonzero for $x = 2$.

The vertical asymptotes occur at the values of x for which the denominator is zero and the numerator is nonzero.

The graph has one vertical asymptote, $x = 2$.

4. Determine whether f has a horizontal or slant asymptote.

The degree of the numerator is equal to the degree of the denominator. Therefore, the graph has a horizontal asymptote given by the ratio of leading coefficients of the numerator and denominator.

$y = \dfrac{1}{1} = 1$ is the horizontal asymptote.

5. Determine where f crosses its horizontal asymptote (if at all).

Solve the equation $f(x) = 1$.

$\dfrac{x + 3}{x - 2} = 1 \Rightarrow x + 3 = x - 2 \Rightarrow 3 = -2$ (contradiction)

The graph of f does not cross its horizontal asymptote.

6. Test for symmetry.

$f(-x) = \dfrac{-x + 3}{-x - 2}$ does not equal $f(x)$ or $-f(x)$. The function is neither even nor odd and is not symmetric with respect to the y-axis or origin.

7. Determine the behavior of f on each interval.

Determine the sign of the function on the intervals (shown in red) defined by the x-intercept at $x = -3$ and the vertical asymptote $x = 2$.

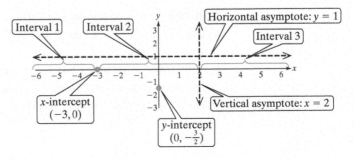

Interval	Test Point	Comments
$(-\infty, -3)$	$\left(-4, \dfrac{1}{6}\right)$	• Since $f(x)$ is positive on this interval, $f(x)$ must approach the horizontal asymptote $y = 1$ from below as $x \to -\infty$.
$(-3, 2)$	$\left(0, -\dfrac{3}{2}\right)$	• Since $f(x)$ is negative on this interval, the graph crosses the x-axis at the intercept $(-3, 0)$ and continues downward (through the y-intercept). As x approaches the vertical asymptote $x = 2$ from the left, $f(x) \to -\infty$.
$(2, \infty)$	$(3, 6)$	• Since $f(x)$ is positive on this interval, as x approaches the vertical asymptote $x = 2$ from the right, $f(x) \to \infty$. • Since $f(x)$ is positive on this interval, $f(x)$ must approach the horizontal asymptote from above as $x \to \infty$.

8. Sketch the function (Figure 3-22).

Plot the x- and y-intercepts $(-3, 0)$ and $(0, -\frac{3}{2})$, and the additional points $(-4, \frac{1}{6})$ and $(3, 6)$.

Graph the horizontal asymptote ($y = 1$) and vertical asymptote $x = 2$ as dashed lines.

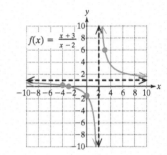

Figure 3-22

> **TIP** Using long division, we can rewrite $f(x) = \frac{x+3}{x-2}$ as $f(x) = 1 + \frac{5}{x-2}$. Then we can graph f by shifting the graph of $y = \frac{1}{x}$ two units to the right, stretching by a factor of 5, and shifting the graph upward 1 unit.

Skill Practice 7   Graph $f(x) = \dfrac{x-1}{x+4}$.

EXAMPLE 8   Graphing a Rational Function

Graph $f(x) = \dfrac{4x}{x^2 - 4}$.

Solution:

1. Determine the y-intercept.

$$f(0) = \frac{4(0)}{(0)^2 - 4} = 0 \qquad \text{The } y\text{-intercept is } (0, 0).$$

2. Determine the x-intercept(s).

$$\frac{4x}{x^2 - 4} = 0 \quad \text{for } x = 0. \qquad \text{The } x\text{-intercept is } (0, 0).$$

3. Identify the vertical asymptotes.

The zeros of $x^2 - 4$ are 2 and -2. Vertical asymptotes: $x = 2$ and $x = -2$.

4. Determine whether f has a horizontal or slant asymptote.

The degree of the numerator is less than the degree of the denominator. The horizontal asymptote is $y = 0$.

Answer

7.

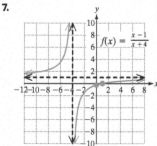

5. Determine where f crosses its horizontal asymptote.

Set $f(x) = 0$. We have $\dfrac{4x}{x^2 - 4} = 0$ for $x = 0$.

Therefore, f crosses its horizontal asymptote at $(0, 0)$.

6. Test for symmetry.

f is an odd function because $f(-x) = \dfrac{4(-x)}{(-x)^2 - 4} = -\dfrac{4x}{x^2 - 4} = -f(x)$.

7. Determine the behavior of f on each interval.

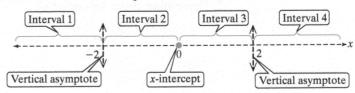

TIP The graph of f is symmetric to the origin because $f(-x) = -f(x)$. Therefore, if $\left(-3, -\frac{12}{5}\right)$ and $\left(-1, \frac{4}{3}\right)$ are points on the graph of f, then $\left(3, \frac{12}{5}\right)$ and $\left(1, -\frac{4}{3}\right)$ are also points on the graph.

Interval	Test Point	Comments
$(-\infty, -2)$	$\left(-3, -\dfrac{12}{5}\right)$	• Since $f(x)$ is negative on this interval, $f(x)$ must approach the horizontal asymptote $y = 0$ from below as $x \to -\infty$. • Since $f(x)$ is negative on this interval, as x approaches the vertical asymptote $x = -2$ from the left, $f(x) \to -\infty$.
$(-2, 0)$	$\left(-1, \dfrac{4}{3}\right)$	• Since $f(x)$ is positive on this interval, as x approaches the vertical asymptote $x = -2$ from the right, $f(x) \to \infty$.
$(0, 2)$	$\left(1, -\dfrac{4}{3}\right)$	• Since $f(x)$ is negative on this interval, as x approaches the vertical asymptote $x = 2$ from the left, $f(x) \to -\infty$.
$(2, \infty)$	$\left(3, \dfrac{12}{5}\right)$	• Since $f(x)$ is positive on this interval, $f(x)$ must approach the horizontal asymptote from above as $x \to \infty$. • Since $f(x)$ is positive on this interval, as x approaches the vertical asymptote $x = 2$ from the right, $f(x) \to \infty$.

8. Sketch the function.

Plot the x- and y-intercept $(0, 0)$.

Graph the asymptotes as dashed lines.

Plot the points.

$\left(-3, -\dfrac{12}{5}\right), \left(-1, \dfrac{4}{3}\right), \left(1, -\dfrac{4}{3}\right),$ and $\left(3, \dfrac{12}{5}\right)$

Sketch the curve.

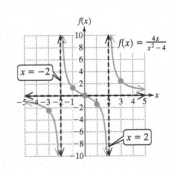

Skill Practice 8 Graph $g(x) = \dfrac{-5x}{x^2 - 9}$.

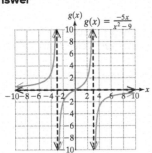

EXAMPLE 9   Graphing a Rational Function

Graph $g(x) = \dfrac{2x^2 - 3x - 5}{x^2 + 1}$.

Solution:

1. Determine the y-intercept.

$$g(0) = \frac{2(0)^2 - 3(0) - 5}{(0)^2 + 1} = -5 \qquad \text{The y-intercept is } (0, -5).$$

2. Determine the x-intercept(s).

$$\frac{2x^2 - 3x - 5}{x^2 + 1} = 0$$
$$2x^2 - 3x - 5 = 0$$
$$(2x - 5)(x + 1) = 0$$
$$x = \frac{5}{2}, \ x = -1 \qquad \text{The x-intercepts are } \left(\tfrac{5}{2}, 0\right) \text{ and } (-1, 0).$$

3. Identify the vertical asymptotes.

$x^2 + 1$ is nonzero for all real numbers. The graph of g has no vertical asymptotes.

4. Determine whether g has a horizontal or slant asymptote.

The degree of the numerator is equal to the degree of the denominator. The horizontal asymptote is $y = \frac{2}{1}$ or simply $y = 2$.

5. Determine where g crosses its horizontal asymptote.

Find the real solutions to the equation $\dfrac{2x^2 - 3x - 5}{x^2 + 1} = 2$.

$$2x^2 - 3x - 5 = 2(x^2 + 1)$$
$$2x^2 - 3x - 5 = 2x^2 + 2$$
$$-3x - 5 = 2$$
$$x = -\frac{7}{3} \qquad g \text{ crosses its horizontal asymptote at } \left(-\frac{7}{3}, 2\right).$$

6. Test for symmetry.

g is neither even nor odd because $g(-x) \neq g(x)$ and $g(-x) \neq -g(x)$.

7. Determine the behavior of g on each interval.

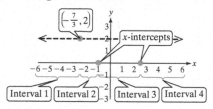

Interval	Test Point	Comments
$\left(-\infty, -\dfrac{7}{3}\right)$	$\left(-3, \dfrac{11}{5}\right)$	• Since $g(-3) = \frac{11}{5}$ is above the horizontal asymptote $y = 2$, $g(x)$ must approach the horizontal asymptote from above as $x \to -\infty$.
$\left(-\dfrac{7}{3}, -1\right)$	$\left(-2, \dfrac{9}{5}\right)$	• Plot the point $\left(-2, \frac{9}{5}\right)$ between the horizontal asymptote and the x-intercept of $(-1, 0)$.
$\left(-1, \dfrac{5}{2}\right)$	$(0, -5)$	• The point $(0, -5)$ is the y-intercept.
$\left(\dfrac{5}{2}, \infty\right)$	$\left(3, \dfrac{2}{5}\right)$	• Since $g(3) = \frac{2}{5}$ is below the horizontal asymptote $y = 2$, $g(x)$ must approach the horizontal asymptote from below as $x \to \infty$.

8. Sketch the function.

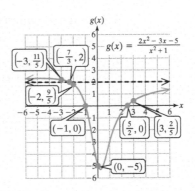

$$\left(-3, \tfrac{11}{5}\right) \quad \left(-\tfrac{7}{3}, 2\right)$$

$$g(x) = \frac{2x^2 - 3x - 5}{x^2 + 1}$$

$$\left(-2, \tfrac{9}{5}\right)$$

$$(-1, 0) \qquad \left(\tfrac{5}{2}, 0\right) \quad \left(3, \tfrac{2}{5}\right)$$

$$(0, -5)$$

Skill Practice 9   Graph $g(x) = \dfrac{4x^2 + 7x - 2}{x^2 + 4}$.

EXAMPLE 10   **Graphing a Rational Function**

Graph. $h(x) = \dfrac{2x^2 + 9x + 4}{x + 3}$

Solution:

1. Determine the y-intercept.

$$h(0) = \frac{2(0)^2 + 9(0) + 4}{(0) + 3} = \frac{4}{3}$$

The y-intercept is $\left(0, \tfrac{4}{3}\right)$.

2. Determine the x-intercept(s).

$$\frac{2x^2 + 9x + 4}{x + 3} = 0$$

$$2x^2 + 9x + 4 = 0$$

$$(2x + 1)(x + 4) = 0$$

$$x = -\frac{1}{2}, \; x = -4$$

The x-intercepts are $\left(-\tfrac{1}{2}, 0\right)$ and $(-4, 0)$.

3. Identify the vertical asymptotes.

$\dfrac{2x^2 + 9x + 4}{x + 3}$ is in lowest terms, and $x + 3$ is zero for $x = -3$.

4. Determine whether h has a horizontal or slant asymptote.

The degree of the numerator is one greater than the degree of the denominator. To find the slant asymptote, divide $(2x^2 + 9x + 4)$ by $(x + 3)$.

$$
\begin{array}{r|rrr}
-3 & 2 & 9 & 4 \\
   &   & -6 & -9 \\
\hline
   & 2 & 3 & \underline{-5} \\
\end{array}
$$

The quotient is $2x + 3$.
The slant asymptote is $y = 2x + 3$.

5. Determine where h will cross the slant asymptote.

Set $h(x) = 2x + 3$. We have $\dfrac{2x^2 + 9x + 4}{x + 3} = 2x + 3$.

$$2x^2 + 9x + 4 = (2x + 3)(x + 3)$$

$$2x^2 + 9x + 4 = 2x^2 + 9x + 9$$

The equation has no solution. Therefore, the graph does not cross the slant asymptote.

$$4 = 9 \quad \text{(No solution)}$$

Answer

9.

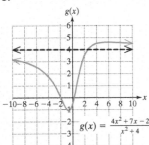

$$g(x) = \frac{4x^2 + 7x - 2}{x^2 + 4}$$

6. Test for symmetry.

h is neither even nor odd because $h(-x) \neq h(x)$ and $h(-x) \neq -h(x)$.

7. Plot test points. Pick values of x on the intervals defined by the x-intercepts and vertical asymptote.

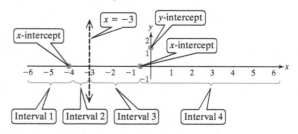

Select test points from each interval.

The graph of $y = h(x)$ passes through $(-5, -4.5)$, $(-3.5, 6)$, $(-2, -6)$, and $(2, 6)$.

8. Sketch the graph.

Plot the x-intercepts: $(-4, 0)$ and $\left(-\frac{1}{2}, 0\right)$.

Plot the y-intercept: $\left(0, \frac{4}{3}\right)$.

Graph the asymptotes as dashed lines.
Plot the points:
$(-5, -4.5)$, $(-3.5, 6)$, $(-2, -6)$, and $(2, 6)$.

Sketch the graph.

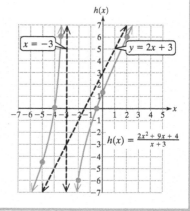

Skill Practice 10 Graph $k(x) = \dfrac{2x^2 - 7x + 3}{x - 2}$.

6. Use Rational Functions in Applications

> **TIP** Recall that variable costs include items such as materials and labor. Fixed costs include overhead costs such as rent and utilities.

In Section 2.5, we presented a linear model for the cost for a business to manufacture x items. The model is $C(x) = mx + b$, where m is the variable cost to produce an individual item, and b is the fixed cost.

The average cost $\overline{C}(x)$ per item manufactured is the sum of all costs (variable and fixed) divided by the total number of items produced x. This is given by

$$\overline{C}(x) = \frac{C(x)}{x}$$

The average cost per item will decrease as more items are produced because the fixed cost will be distributed over a greater number of items. This is demonstrated in Example 11.

Answer

10.

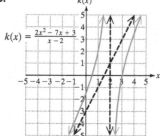

> **EXAMPLE 11**   **Investigating Average Cost**
>
> A cleaning service cleans homes. For each house call, the cost to the company is approximately $40 for cleaning supplies, gasoline, and labor. The business also has fixed monthly costs of $300 from phone service, advertising, and depreciation on the vehicles.
>
> **a.** Write a cost function to represent the cost $C(x)$ (in dollars) for x house calls per month.
> **b.** Write the average cost function that represents the average cost $\overline{C}(x)$ (in $) for x house calls per month.
> **c.** Evaluate $\overline{C}(5)$, $\overline{C}(20)$, $\overline{C}(30)$, and $\overline{C}(100)$.
> **d.** The cleaning service can realistically make a maximum of 160 calls per month. However, if the number of calls were unlimited, what value would the average cost approach? What does this mean in the context of the problem?
>
> **Solution:**
>
> **a.** $C(x) = 40x + 300$ The variable cost is $40 per call ($m = 40$), and the fixed cost is $300 ($b = 300$). $C(x) = mx + b$.
>
> **b.** $\overline{C}(x) = \dfrac{40x + 300}{x}$ The average cost per item is the total cost divided by the total number of items produced.
>
> **c.** $\overline{C}(5) = 100$ The average cost per house call is $100 if 5 calls are made.
>
> $\overline{C}(20) = 55$ The average cost per house call is $55 if 20 calls are made.
>
> $\overline{C}(30) = 50$ The average cost per house call is $50 if 30 calls are made.
>
> $\overline{C}(100) = 43$ The average cost per house call is $43 if 100 calls are made.
>
>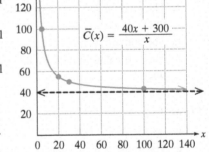
>
> **d.** As x approaches infinity, $\overline{C}(x)$ will approach its horizontal asymptote $y = 40$. This is the cost per house call in the absence of other fixed costs.

Answers

11. a. $C(x) = 50x + 200$
  b. $\overline{C}(x) = \dfrac{50x + 200}{x}$
  c. $\overline{C}(5) = 90$; $\overline{C}(20) = 60$;
     $\overline{C}(30) = 56.67$;
     $\overline{C}(100) = 52$
  d. The average cost $\overline{C}(x)$ will approach $50 per house call.

> **Skill Practice 11**   Repeat Example 11 under the assumption that the company cuts its fixed costs to $200 per month and pays its employees more, leading to a variable cost per house call of $50.

SECTION 3.5   Practice Exercises

Prerequisite Review

For Exercises R.1–R.3, simplify the expression and state the restrictions on the variable.

R.1. $\dfrac{q^2 - 36}{q^2 - 4q - 12}$ **R.2.** $\dfrac{9 - u^2}{2u^2 - 6u}$ **R.3.** $\dfrac{3p - 3m}{xm - xp + 3m - 3p}$

Concept Connections

1. The domain of a rational function defined by $f(x) = \dfrac{p(x)}{q(x)}$ is all real numbers excluding the zeros of _____.

2. The notation $x \to \infty$ is read as _____.

3. The notation $x \to 5^-$ is read as _____.

4. The line $x = c$ is a _____ asymptote of the graph of a function f if $f(x)$ approaches infinity or negative infinity as x approaches _____ from either the left or right.

5. To locate the vertical asymptotes of a function, determine the real numbers x where the denominator is zero, but the numerator is _____.

6. Consider a rational function in which the degree of the numerator is n and the degree of the denominator is m. If n _____ m, then the x-axis is the horizontal asymptote. If n _____ m, then the function has no horizontal asymptote.

Objective 1: Apply Notation Describing Infinite Behavior of a Function

For Exercises 7–12, write the domain of the function in interval notation.

7. $f(x) = \dfrac{x^2 - 25}{x - 5}$

8. $g(x) = \dfrac{x^2 - 9}{x - 3}$

9. $r(x) = \dfrac{2x - 3}{4x^2 + 3x - 1}$

10. $p(x) = \dfrac{3x - 5}{2x^2 + 5x - 7}$

11. $h(x) = \dfrac{18x}{x^2 + 100}$

12. $k(x) = \dfrac{14}{x^2 + 49}$

For Exercises 13–16, refer to the graph of the function and complete the statement. (See Example 1)

13. **a.** As $x \to -\infty$, $f(x) \to$ _____.
    b. As $x \to 4^-$, $f(x) \to$ _____.
    c. As $x \to 4^+$, $f(x) \to$ _____.
    d. As $x \to \infty$, $f(x) \to$ _____.
    e. The graph is increasing over the interval(s) _____.
    f. The graph is decreasing over the interval(s) _____.
    g. The domain is _____.
    h. The range is _____.
    i. The vertical asymptote is the line _____.
    j. The horizontal asymptote is the line _____.

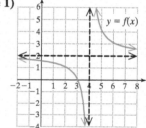

14. **a.** As $x \to -\infty$, $f(x) \to$ _____.
    b. As $x \to -3^-$, $f(x) \to$ _____.
    c. As $x \to -3^+$, $f(x) \to$ _____.
    d. As $x \to \infty$, $f(x) \to$ _____.
    e. The graph is increasing over the interval(s) _____.
    f. The graph is decreasing over the interval(s) _____.
    g. The domain is _____.
    h. The range is _____.
    i. The vertical asymptote is the line _____.
    j. The horizontal asymptote is the line _____.

15. **a.** As $x \to -\infty$, $f(x) \to$ _____.
    b. As $x \to -3^-$, $f(x) \to$ _____.
    c. As $x \to -3^+$, $f(x) \to$ _____.
    d. As $x \to \infty$, $f(x) \to$ _____.
    e. The graph is increasing over the interval(s) _____.
    f. The graph is decreasing over the interval(s) _____.
    g. The domain is _____.
    h. The range is _____.
    i. The vertical asymptote is the line _____.
    j. The horizontal asymptote is the line _____.

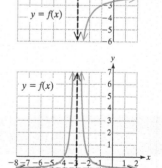

16. a. As $x \to -\infty, f(x) \to$ _____.

   b. As $x \to 1^-, f(x) \to$ _____.

   c. As $x \to 1^+, f(x) \to$ _____.

   d. As $x \to \infty, f(x) \to$ _____.

   e. The graph is increasing over the interval(s) _____.

   f. The graph is decreasing over the interval(s) _____.

   g. The domain is _____.

   h. The range is _____.

   i. The vertical asymptote is the line _____.

   j. The horizontal asymptote is the line _____.

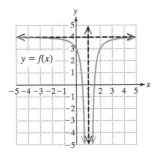

Objective 2: Identify Vertical Asymptotes

For Exercises 17–24, determine the vertical asymptotes of the graph of the function. (See Example 2)

17. $f(x) = \dfrac{8}{x-4}$

18. $g(x) = \dfrac{2}{x+7}$

19. $h(x) = \dfrac{x-3}{2x^2 - 9x - 5}$

20. $k(x) = \dfrac{x+2}{3x^2 + 8x - 3}$

21. $m(x) = \dfrac{x}{x^2 + 5}$

22. $n(x) = \dfrac{6}{x^4 + 1}$

23. $f(t) = \dfrac{t^2 + 2}{2t^2 + 4t - 3}$

24. $k(a) = \dfrac{5 + a^4}{3a^2 + 4a - 1}$

Objective 3: Identify Horizontal Asymptotes

25. The graph of $f(x) = \dfrac{-x^2 + 8}{2x^2 - 3}$ will behave like which function for large values of $|x|$?

   a. $y = -\dfrac{1}{2}$

   b. $y = -\dfrac{x}{2}$

   c. $y = -\dfrac{8}{3}$

   d. $y = -\dfrac{1}{2}x - \dfrac{8}{3}$

26. The graph of $f(x) = \dfrac{2x^3 + 7}{5x^3}$ will behave like which function for large values of $|x|$?

   a. $y = \dfrac{2}{5x}$

   b. $y = \dfrac{2x}{5}$

   c. $y = \dfrac{2}{5}$

   d. $y = \dfrac{2}{5}x^3$

27. The graph of $f(x) = \dfrac{-3x^4 - 2x + 5}{x^5 + x^2 - 2}$ will behave like which function for large values of $|x|$?

   a. $y = -3$

   b. $y = -3x$

   c. $y = -\dfrac{5}{2}$

   d. $y = 0$

28. The graph of $f(x) = \dfrac{x^2 + 7x - 3}{6x^4 + 2}$ will behave like which function for large values of $|x|$?

   a. $y = \dfrac{x^2}{6}$

   b. $y = 0$

   c. $y = -\dfrac{3}{2}$

   d. $y = \dfrac{1}{6}x - \dfrac{3}{2}$

For Exercises 29–36,

a. Identify the horizontal asymptote (if one exists). (**See Example 3**)

b. If the graph of the function has a horizontal asymptote, determine the point (if any) where the graph crosses the horizontal asymptote. (**See Example 4**)

29. $p(x) = \dfrac{5}{x^2 + 2x + 1}$

30. $q(x) = \dfrac{8}{x^2 + 4x + 4}$

31. $h(x) = \dfrac{3x^2 + 8x - 5}{x^2 + 3}$

32. $r(x) = \dfrac{-4x^2 + 5x - 1}{x^2 + 2}$

33. $m(x) = \dfrac{x^4 + 2x + 1}{5x + 2}$

34. $n(x) = \dfrac{x^3 - x^2 + 1}{2x - 3}$

35. $t(x) = \dfrac{2x + 4}{x^2 + 7x - 4}$

36. $s(x) = \dfrac{x + 3}{2x^2 - 3x - 5}$

37. Consider the expression $\dfrac{x^2 + 3x + 1}{2x^2 + 5}$.

    a. Divide the numerator and denominator by the greatest power of x that appears in the denominator. That is, divide numerator and denominator by x^2.

    b. As $|x| \to \infty$ what value will $\dfrac{3}{x}, \dfrac{1}{x^2}$, and $\dfrac{5}{x^2}$ approach?

    (*Hint*: Substitute large values of x such as 100, 1000, 10,000, and so on to help you understand the behavior of each expression.)

    c. Use the results from parts (a) and (b) to identify the horizontal asymptote for the graph of

$$f(x) = \dfrac{x^2 + 3x + 1}{2x^2 + 5}.$$

38. Consider the expression $\dfrac{3x^3 - 2x^2 + 7x}{5x^3 + 1}$.

    a. Divide the numerator and denominator by the greatest power of x that appears in the denominator.

    b. As $|x| \to \infty$ what value will $-\dfrac{2}{x}, \dfrac{7}{x^2}$, and $\dfrac{1}{x^3}$ approach?

    c. Use the results from parts (a) and (b) to identify the horizontal asymptote for the graph of

$$f(x) = \dfrac{3x^3 - 2x^2 + 7x}{5x^3 + 1}.$$

Objective 4: Identify Slant Asymptotes

For Exercises 39–48, identify the asymptotes. (See Example 5)

39. $f(x) = \dfrac{2x^2 + 3}{x}$

40. $g(x) = \dfrac{3x^2 + 2}{x}$

41. $h(x) = \dfrac{-3x^2 + 4x - 5}{x + 6}$

42. $k(x) = \dfrac{-2x^2 - 3x + 7}{x + 3}$

43. $p(x) = \dfrac{x^3 + 5x^2 - 4x + 1}{x^2 - 5}$

44. $q(x) = \dfrac{x^3 + 3x^2 - 2x - 4}{x^2 - 7}$

45. $r(x) = \dfrac{2x + 1}{x^3 + x^2 - 4x - 4}$

46. $t(x) = \dfrac{3x - 4}{x^3 + 2x^2 - 9x - 18}$

47. $f(x) = \dfrac{4x^3 - 2x^2 + 7x - 3}{2x^2 + 4x + 3}$

48. $a(x) = \dfrac{9x^3 - 5x + 4}{3x^2 + 2x + 1}$

Objective 5: Graph Rational Functions

For Exercises 49–56, graph the functions by using transformations of the graphs of $y = \dfrac{1}{x}$ and $y = \dfrac{1}{x^2}$. (See Example 6)

49. $f(x) = \dfrac{1}{x - 3}$

50. $g(x) = \dfrac{1}{x + 4}$

51. $h(x) = \dfrac{1}{x^2} + 2$

52. $k(x) = \dfrac{1}{x^2} - 3$

53. $m(x) = \dfrac{1}{(x + 4)^2} - 3$

54. $n(x) = \dfrac{1}{(x - 1)^2} + 2$

55. $p(x) = -\dfrac{1}{x}$

56. $q(x) = -\dfrac{1}{x^2}$

For Exercises 57–62, for the graph of $y = f(x)$,

    a. Identify the x-intercepts.

    c. Identify the horizontal asymptote or slant asymptote if applicable.

    b. Identify any vertical asymptotes.

    d. Identify the y-intercept.

57. $f(x) = \dfrac{(x + 3)(2x - 7)}{(x + 2)(4x + 1)}$

58. $f(x) = \dfrac{(3x - 4)(x - 6)}{(2x - 3)(x + 5)}$

59. $f(x) = \dfrac{4x - 9}{x^2 - 9}$

60. $f(x) = \dfrac{5x - 8}{x^2 - 4}$

61. $f(x) = \dfrac{(5x - 1)(x + 3)}{x + 2}$

62. $f(x) = \dfrac{(4x + 3)(x + 2)}{x + 3}$

For Exercises 63–66, sketch a rational function subject to the given conditions. Answers may vary.

63. Horizontal asymptote: $y = 2$

    Vertical asymptote: $x = 3$

    y-intercept: $\left(0, \frac{8}{3}\right)$

    x-intercept: $(4, 0)$

64. Horizontal asymptote: $y = 0$

    Vertical asymptote: $x = -1$

    y-intercept: $(0, 1)$

    No x-intercepts

    Range: $(0, \infty)$

65. Horizontal asymptote: $y = 0$

Vertical asymptotes: $x = -2$ and $x = 2$

y-intercept: $(0, 1)$

No x-intercepts

Symmetric to the y-axis

Passes through the point $\left(3, -\frac{4}{5}\right)$

66. Horizontal asymptote: $y = 3$

Vertical asymptotes: $x = -1$ and $x = 1$

y-intercept: $(0, 0)$

x-intercept: $(0, 0)$

Symmetric to the y-axis

Passes through the point $(2, 4)$

For Exercises 67–90, graph the function. (See Examples 7–10)

67. $n(x) = \dfrac{-3}{2x + 7}$

68. $m(x) = \dfrac{-4}{2x - 5}$

69. $f(x) = \dfrac{x - 4}{x - 2}$

70. $g(x) = \dfrac{x - 3}{x - 1}$

71. $h(x) = \dfrac{2x - 4}{x + 3}$

72. $k(x) = \dfrac{3x - 9}{x + 2}$

73. $p(x) = \dfrac{6}{x^2 - 9}$

74. $q(x) = \dfrac{4}{x^2 - 16}$

75. $r(x) = \dfrac{5x}{x^2 - x - 6}$

76. $t(x) = \dfrac{4x}{x^2 - 2x - 3}$

77. $k(x) = \dfrac{5x - 3}{2x - 7}$

78. $h(x) = \dfrac{4x + 3}{3x - 5}$

79. $g(x) = \dfrac{3x^2 - 5x - 2}{x^2 + 1}$

80. $c(x) = \dfrac{2x^2 - 5x - 3}{x^2 + 1}$

81. $n(x) = \dfrac{x^2 + 2x + 1}{x}$

82. $m(x) = \dfrac{x^2 - 4x + 4}{x}$

83. $f(x) = \dfrac{x^2 + 7x + 10}{x + 3}$

84. $d(x) = \dfrac{x^2 - x - 12}{x - 2}$

85. $w(x) = \dfrac{-4x^2}{x^2 + 4}$

86. $u(x) = \dfrac{-3x^2}{x^2 + 1}$

87. $f(x) = \dfrac{x^3 + x^2 - 4x - 4}{x^2 + 3x}$

88. $g(x) = \dfrac{x^3 + 3x^2 - x - 3}{x^2 - 2x}$

89. $v(x) = \dfrac{2x^4}{x^2 + 9}$

90. $g(x) = \dfrac{4x^4}{x^2 + 8}$

Objective 6: Use Rational Functions in Applications

91. A sports trainer has monthly costs of $69.95 for phone service and $39.99 for his website and advertising. In addition he pays a $20 fee to the gym for each session in which he trains a client. **(See Example 11)**

  a. Write a cost function to represent the cost $C(x)$ for x training sessions for a given month.

  b. Write a function representing the average cost $\overline{C}(x)$ for x sessions.

  c. Evaluate $\overline{C}(5)$, $\overline{C}(30)$, and $\overline{C}(120)$.

  d. The trainer can realistically have 120 sessions per month. However, if the number of sessions were unlimited, what value would the average cost approach? What does this mean in the context of the problem?

92. An on-demand printing company has monthly overhead costs of $1200 in rent, $420 in electricity, $100 for phone service, and $200 for advertising and marketing. The printing cost is $40 per thousand pages for paper and ink.

  a. Write a cost function to represent the cost $C(x)$ for printing x thousand pages for a given month.

  b. Write a function representing the average cost $\overline{C}(x)$ for printing x thousand pages for a given month.

  c. Evaluate $\overline{C}(20)$, $\overline{C}(50)$, $\overline{C}(100)$, and $\overline{C}(200)$.

  d. Interpret the meaning of $\overline{C}(200)$.

  e. For a given month, if the printing company could print an unlimited number of pages, what value would the average cost per thousand pages approach? What does this mean in the context of the problem?

93. A parallel circuit is one with several paths through which electricity can travel. The total resistance in a parallel circuit is always less than the resistance in any single branch. The total resistance R for the circuit shown can be computed from the formula $\dfrac{1}{R} = \dfrac{1}{R_1} + \dfrac{1}{R_2}$, where R_1 and R_2 are the resistances in the individual branches. Suppose that a resistor with a fixed resistance of 6 Ω (ohms) is placed in parallel with a variable resistor of resistance x.

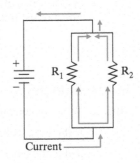

  a. Write R as a function of x.

  b. Complete the table.

x	6	12	18	30
$R(x)$				

  c. What value does $R(x)$ approach as $x \to \infty$? Discuss the significance of this result.

94. The total resistance R of three resistors in parallel is given by

$$R = \frac{R_1 R_2 R_3}{R_1 R_2 + R_1 R_3 + R_2 R_3}$$

Suppose that an 8-Ω and a 12-Ω resistor are placed in parallel with a variable resistor of resistance x.

a. Write R as a function of x.

b. What value does $R(x)$ approach as $x \to \infty$? Write the value in decimal form.

95. The concentration $C(t)$ (in milligrams per liter, mg/L) of a drug in the bloodstream t hours after the drug is administered is modeled by

$$C(t) = \frac{10t}{2t^2 + 1}$$

a. Use a graphing utility to graph the function.

b. What are the domain restrictions on the function?

c. Use the graph to approximate the maximum concentration. Round to the nearest mg/L.

d. What is the limiting concentration?

96. A certain diet pill is designed to delay the administration of the active ingredient for several hours. The concentration $C(t)$ (in mg/L) of the active ingredient in the bloodstream t hours after taking the pill is modeled by

$$C(t) = \frac{3t}{2t^2 - 20t + 51}$$

a. Use a graphing utility to graph the function.

b. What are the domain restrictions on the function?

c. Use the graph to approximate the maximum concentration. Round to the nearest mg/L.

d. What is the limiting concentration?

97. A power company burns coal to generate electricity. The cost $C(x)$ (in $1000) to remove $x\%$ of the air pollutants is given by

$$C(x) = \frac{600x}{100 - x}$$

a. Compute the cost to remove 25% of the air pollutants. (*Hint:* $x = 25$.)

b. Determine the cost to remove 50%, 75%, and 90% of the air pollutants.

c. If the power company budgets $1.4 million for pollution control, what percentage of the air pollutants can be removed?

98. The cost $C(x)$ (in $1000) for a city to remove $x\%$ of the waste from a polluted river is given by

$$C(x) = \frac{80x}{100 - x}$$

a. Determine the cost to remove 20%, 40%, and 90% of the waste. Round to the nearest thousand dollars.

b. If the city has $320,000 budgeted for river cleanup, what percentage of the waste can be removed?

The Doppler effect is a change in the observed frequency of a wave (such as a sound wave or light wave) when the source of the wave and observer are in motion relative to each other. The Doppler effect explains why an observer hears a change in pitch of an ambulance siren as the ambulance passes by the observer. The frequency $F(v)$ of a sound relative to an observer is given by $F(v) = f_a\left(\dfrac{s_0}{s_0 - v}\right)$, where f_a is the actual frequency of the sound at the source, s_0 is the speed of sound in air (772.4 mph), and v is the speed at which the source of sound is moving toward the observer. Use this relationship for Exercises 99–100.

99. Suppose that an ambulance moves toward an observer.

a. Write F as a function of v if the actual frequency of sound emitted by the ambulance is 560 Hz.

b. Use a graphing utility to graph the function from part (a) on the window [0, 1000, 100] by [0, 5000, 1000].

c. As the speed of the ambulance increases, what is the effect of the frequency of sound?

100. Suppose the frequency of sound emitted by a police car siren is 600 Hz.

a. Write F as a function of v if the police car is moving toward an observer.

b. Suppose that the frequency of the siren as heard by an observer is 664 Hz. Determine the velocity of the police car. Round to the nearest tenth of a mph.

c. Although a police car cannot travel close to the speed of sound, interpret the meaning of the vertical asymptote.

Mixed Exercises

101. a. Write an equation for a rational function f whose graph is the same as the graph of $y = \dfrac{1}{x^2}$ shifted up 3 units and to the left 1 unit.

b. Write the domain and range of the function in interval notation.

102. a. Write an equation for a rational function f whose graph is the same as the graph of $y = \frac{1}{x}$ shifted to the right 4 units and down 3 units.

b. Write the domain and range of the function in interval notation.

For Exercises 103–104, given $y = f(x)$,

a. Divide the numerator by the denominator to write $f(x)$ in the form $f(x) = \text{quotient} + \dfrac{\text{remainder}}{\text{divisor}}$.

b. Use transformations of $y = \dfrac{1}{x}$ to graph the function.

103. $f(x) = \dfrac{2x + 7}{x + 3}$

104. $f(x) = \dfrac{5x + 11}{x + 2}$

Write About It

105. Explain why $x = -2$ is not a vertical asymptote of the graph of $f(x) = \dfrac{x^2 + 7x + 10}{x + 2}$.

106. Write an informal definition of a horizontal asymptote of a rational function.

Expanding Your Skills

For Exercises 107–110, write an equation of a function that meets the given conditions. Answers may vary.

107. x-intercepts: $(-3, 0)$ and $(-1, 0)$
vertical asymptote: $x = 2$
horizontal asymptote: $y = 1$
y-intercept: $\left(0, \frac{3}{4}\right)$

108. x-intercepts: $(4, 0)$ and $(2, 0)$
vertical asymptote: $x = 1$
horizontal asymptote: $y = 1$
y-intercept: $(0, 8)$

109. x-intercept: $\left(\frac{3}{2}, 0\right)$
vertical asymptotes: $x = -2$ and $x = 5$
horizontal asymptote: $y = 0$
y-intercept: $(0, 3)$

110. x-intercept: $\left(\frac{4}{3}, 0\right)$
vertical asymptotes: $x = -3$ and $x = -4$
horizontal asymptote: $y = 0$
y-intercept: $(0, -1)$

Graphs with "Holes"

The rational functions studied in this section all have the characteristic that the numerator and denominator do not share a common variable factor. We now investigate rational functions for which this is not the case. For Exercises 111–114,

a. Write the domain of f in interval notation.

b. Simplify the rational expression defining the function.

c. Identify any vertical asymptotes.

d. Identify any other values of x (other than those corresponding to vertical asymptotes) for which the function is discontinuous.

e. Identify the graph of the function.

111. $f(x) = \dfrac{x^2 + x - 6}{x - 2}$

112. $f(x) = \dfrac{-x^2 + 2x + 3}{x + 1}$

113. $f(x) = \dfrac{2x + 10}{x^2 + 9x + 20}$

114. $f(x) = \dfrac{2x - 2}{x^2 + 2x - 3}$

i.

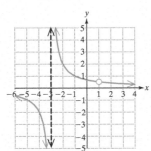

ii.

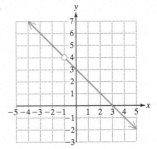

iii.

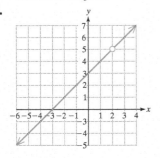

iv.

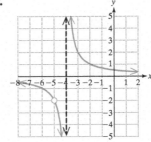

PROBLEM RECOGNITION EXERCISES

Polynomial and Rational Functions

For Exercises 1–8, refer to $p(x) = x^3 + 3x^2 - 6x - 8$ and $q(x) = x^3 - 2x^2 - 5x + 6$.

1. Find the zeros of $p(x)$.

2. Find the zeros of $q(x)$.

3. Find the x-intercept(s) of the graph of $y = q(x)$.

4. Find the x-intercept(s) of the graph of $y = p(x)$.

5. Find the x-intercepts of the graph of
$$f(x) = \frac{p(x)}{q(x)} = \frac{x^3 + 3x^2 - 6x - 8}{x^3 - 2x^2 - 5x + 6}.$$

6. Find the vertical asymptotes of the graph of
$$f(x) = \frac{p(x)}{q(x)} = \frac{x^3 + 3x^2 - 6x - 8}{x^3 - 2x^2 - 5x + 6}.$$

7. Find the horizontal asymptote or slant asymptote of the graph of $f(x) = \dfrac{p(x)}{q(x)} = \dfrac{x^3 + 3x^2 - 6x - 8}{x^3 - 2x^2 - 5x + 6}.$

8. Determine where the graph of $f(x) = \dfrac{x^3 + 3x^2 - 6x - 8}{x^3 - 2x^2 - 5x + 6}$ crosses its horizontal or slant asymptote.

For Exercises 9–16, refer to $c(x) = x^3 - 4x^2 - 2x + 8$ and $d(x) = x^3 + 3x^2 - 4$.

9. Find the zeros of $c(x)$.

10. Find the zeros of $d(x)$.

11. Find the x-intercept(s) of the graph of $y = d(x)$.

12. Find the x-intercept(s) of the graph of $y = c(x)$.

13. Find the x-intercepts of the graph of
$$g(x) = \frac{c(x)}{d(x)} = \frac{x^3 - 4x^2 - 2x + 8}{x^3 + 3x^2 - 4}.$$

14. Find the vertical asymptotes of the graph of
$$g(x) = \frac{c(x)}{d(x)} = \frac{x^3 - 4x^2 - 2x + 8}{x^3 + 3x^2 - 4}.$$

15. Find the horizontal asymptote or slant asymptote of the graph of $g(x) = \dfrac{c(x)}{d(x)} = \dfrac{x^3 - 4x^2 - 2x + 8}{x^3 + 3x^2 - 4}.$

16. Determine where the graph of $g(x) = \dfrac{x^3 - 4x^2 - 2x + 8}{x^3 + 3x^2 - 4}$ crosses its horizontal or slant asymptote.

For Exercises 17–18, use the results from Exercises 5–8 and 13–16 to match the function with its graph.

17. $f(x) = \dfrac{x^3 + 3x^2 - 6x - 8}{x^3 - 2x^2 - 5x + 6}$

18. $g(x) = \dfrac{x^3 - 4x^2 - 2x + 8}{x^3 + 3x^2 - 4}$

a.

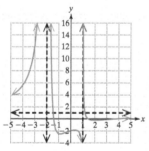

b.

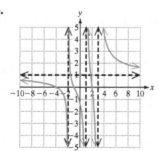

19. Divide $(2x^3 - 4x^2 - 10x + 12) \div (x^2 - 11)$ by using an appropriate method.

   a. Identify the quotient $q(x)$.   **b.** Identify the remainder $r(x)$.

20. Identify the slant asymptote of $f(x) = \dfrac{2x^3 - 4x^2 - 10x + 12}{x^2 - 11}.$

21. Identify the point where the graph of $f(x) = \dfrac{2x^3 - 4x^2 - 10x + 12}{x^2 - 11}$ crosses its slant asymptote.

22. Refer to Exercise 19. Solve the equation $r(x) = 0$. How does the solution to the equation $r(x) = 0$ relate to the point where the graph of f crosses its slant asymptote?

| SECTION 3.6 | Polynomial and Rational Inequalities |

OBJECTIVES

1. Solve Polynomial Inequalities
2. Solve Rational Inequalities
3. Solve Applications Involving Polynomial and Rational Inequalities

1. Solve Polynomial Inequalities

An engineer for a food manufacturer must design an aluminum container for a hot drink mix. The container is to be a right circular cylinder 5.5 in. in height. The surface area represents the amount of aluminum used and is given by

$$S(r) = 2\pi r^2 + 11\pi r \qquad \text{where } r \text{ is the radius of the can.}$$

The engineer wants to limit the surface area so that at most 90 in.^{2} of aluminum is used. To determine the restrictions on the radius, the engineer must solve the inequality $2\pi r^2 + 11\pi r \le 90$ (see Exercise 123). This inequality is a quadratic inequality in the variable r. It is also categorized as a polynomial inequality of degree 2.

TIP If $f(x)$ is a quadratic polynomial, then the inequalities $f(x) < 0$, $f(x) > 0$, $f(x) \le 0$, and $f(x) \ge 0$ are called quadratic inequalities.

Definition of a Polynomial Inequality

Let $f(x)$ be a polynomial. Then an inequality of the form

$f(x) < 0$, $f(x) > 0$, $f(x) \le 0$, or $f(x) \ge 0$ is called a **polynomial inequality.**

Note: A polynomial inequality is nonlinear if $f(x)$ is a polynomial of degree greater than 1.

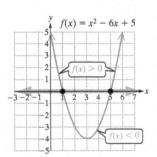

Figure 3-23

Consider the polynomial inequalities $f(x) < 0$ and $f(x) > 0$. We need to determine the intervals over which $f(x)$ is negative or positive. For example, consider the graph of $f(x) = x^2 - 6x + 5$ (Figure 3-23).

The graph shows the solution sets for the following equation and inequalities.

- $f(x) = 0$ for $\{1, 5\}$.
- $f(x) < 0$ on the interval $(1, 5)$. (shown in blue)
- $f(x) > 0$ on the interval $(-\infty, 1) \cup (5, \infty)$. (shown in red)

Notice that the x-intercepts define the endpoints (or "boundary" points) for the solution sets of the inequalities. We can solve a polynomial inequality if we can identify the *sign* of the polynomial for each interval defined by the boundary points. This is the basis on which we solve any nonlinear inequality.

Procedure to Solve a Nonlinear Inequality

1. Express the inequality as $f(x) < 0$, $f(x) > 0$, $f(x) \le 0$, or $f(x) \ge 0$. That is, rearrange the terms of the inequality so that one side is set to zero.
2. Find the real solutions of the related equation $f(x) = 0$ and any values of x that make $f(x)$ undefined. These are the "boundary" points for the solution set to the inequality.
3. Determine the sign of $f(x)$ on the intervals defined by the boundary points.
   - If $f(x)$ is positive, then the values of x on the interval are solutions to $f(x) > 0$.
   - If $f(x)$ is negative, then the values of x on the interval are solutions to $f(x) < 0$.
4. Determine whether the boundary points are included in the solution set.
5. Write the solution set in interval notation or set-builder notation.

EXAMPLE 1   Solving a Quadratic Inequality

Solve the inequality. $3x(x - 1) > 10 - 2x$

Solution:

$3x(x - 1) > 10 - 2x$ **Step 1:** Write the inequality in the form $f(x) > 0$.
$3x^2 - 3x > 10 - 2x$

$$\overbrace{3x^2 - x - 10}^{f(x)} > 0$$
$$3x^2 - x - 10 = 0$$ **Step 2:** Find the real solutions to the related equation $f(x) = 0$.

$(3x + 5)(x - 2) = 0$

$x = -\dfrac{5}{3}$ and $x = 2$ The boundary points are $-\frac{5}{3}$ and 2.

Step 3: Divide the x-axis into intervals defined by the boundary points.

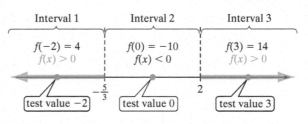

Determine the sign of $f(x) = 3x^2 - x - 10$ on each interval. One method is to evaluate $f(x)$ for a test value x on each interval.

Step 4: The solution set does not include the boundary points because the inequality is strict.

The solution set is $\left(-\infty, -\frac{5}{3}\right) \cup (2, \infty)$ or equivalently in set-builder notation $\left\{x \mid x < -\frac{5}{3} \text{ or } x > 2\right\}$.

Step 5: Write the solution set.

> **TIP** To evaluate the polynomial $f(x) = 3x^2 - x - 10$ at the test points, we can perform direct substitution such as:
>
> $f(3) = 3(3)^2 - (3) - 10$
> $ = 14$
>
> Or use synthetic division and the remainder theorem.
>
> $\begin{array}{r|rrr} 3 & 3 & -1 & -10 \\ & & 9 & 24 \\ \hline & 3 & 8 & \underline{|14} \end{array}$

Skill Practice 1   Solve the inequality. $2x(x - 1) < 21 - x$

From Example 1, the key step is to determine the sign of $f(x)$ on the intervals $\left(-\infty, -\frac{5}{3}\right)$, $\left(-\frac{5}{3}, 2\right)$, and $(2, \infty)$. We can avoid the arithmetic from evaluating $f(x)$ at the test points by creating a sign chart. The inequality $3x^2 - x - 10 > 0$ is equivalent to $(3x + 5)(x - 2) > 0$. We have

> **TIP** The sign of the product in the bottom row of the sign chart is determined by the signs of the individual factors from the rows above.

Sign of $(3x + 5)$:	$-$	$+$	$+$
Sign of $(x - 2)$:	$-$	$-$	$+$
Sign of $(3x + 5)(x - 2)$:	$+$	$-$	$+$

$$\qquad\qquad -\tfrac{5}{3} \qquad 2$$

The sign chart organizes the signs of each factor on the given intervals. Then the sign of the product of factors is given in the bottom row. We see that $f(x) = (3x + 5)(x - 2) > 0$ for $\left(-\infty, -\frac{5}{3}\right) \cup (2, \infty)$.

The result of Example 1 can also be viewed graphically. From Section 3.1, the graph of $f(x) = 3x^2 - x - 10$ is a parabola opening upward (Figure 3-24).

From the factored form $f(x) = (3x + 5)(x - 2)$, the x-intercepts $\left(-\frac{5}{3}, 0\right)$ and $(2, 0)$ mark the points of transition between the intervals where $f(x)$ potentially changes sign.

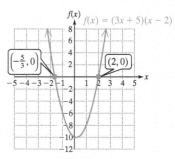

Figure 3-24

Answer

1. Interval notation: $\left(-3, \frac{7}{2}\right)$;
   Set-builder notation:
   $\left\{x \mid -3 < x < \frac{7}{2}\right\}$

> **EXAMPLE 2** **Solving a Polynomial Inequality**
>
> Solve the inequality. $x^4 - 12x \geq 8x^2 - x^3$
>
> **Solution:**
>
> $x^4 - 12x \geq 8x^2 - x^3$ **Step 1:** Write the inequality in the form $f(x) \geq 0$.
>
> $\overbrace{x^4 + x^3 - 8x^2 - 12x}^{f(x)} \geq 0$
>
> $x^4 + x^3 - 8x^2 - 12x = 0$ **Step 2:** Find the real solutions to the related equation $f(x) = 0$.
>
> $x(x^3 + x^2 - 8x - 12) = 0$ Factor the left side of the equation.
>
> The possible rational zeros of $x^3 + x^2 - 8x - 12$ are $\pm 1, \pm 2, \pm 3, \pm 4, \pm 6, \pm 12$.
>
> $\underline{3|}\ \ 1\ \ \ \ 1\ \ \ -8\ \ \ -12$
> $\ \ \ \ \ \ \ \ \ \ \ 3\ \ \ \ 12\ \ \ \ \ 12$
> $\ \ \ \ \overline{1\ \ \ 4\ \ \ \ 4\ \ \ \ \underline{|0}}$
>
> After testing several potential rational zeros, we find that 3 is a zero of $f(x)$.
>
> $x(x - 3)(x^2 + 4x + 4) = 0$ Now factor the quadratic polynomial.
>
> $x(x - 3)(x + 2)^2 = 0$
>
> $x = 0, x = 3, x = -2$ The boundary points are 0, 3, and -2.
>
> **Step 3:** The inequality $x^4 + x^3 - 8x^2 - 12x \geq 0$ is equivalent to $x(x - 3)(x + 2)^2 \geq 0$. Divide the x-axis into intervals defined by the boundary points and determine the sign of $f(x)$ on each interval.
>
> | | Interval 1 | Interval 2 | Interval 3 | Interval 4 |
> |---|---|---|---|---|
> | **Evaluate:** $f(x) = x(x - 3)(x + 2)^2$ | $f(-3) = 18$ $f(x) > 0$ | $f(-1) = 4$ $f(x) > 0$ | $f(1) = -18$ $f(x) < 0$ | $f(4) = 144$ $f(x) > 0$ |
> | Sign of x | $-$ | $-$ | $+$ | $+$ |
> | Sign of $(x - 3)$ | $-$ | $-$ | $-$ | $+$ |
> | Sign of $(x + 2)^2$ | $+$ | $+$ | $+$ | $+$ |
> | Sign of $x(x - 3)(x + 2)^2$ | $+$ | $+$ | $-$ | $+$ |
>
> $\qquad\qquad\qquad -2 \qquad 0 \qquad\quad 3$
>
> The solution set is $(-\infty, 0] \cup [3, \infty)$. In set-builder notation this is $\{x \mid x \leq 0 \text{ or } x \geq 3\}$.
>
> **Step 4:** The solution set includes the boundary points because the inequality sign includes equality. Therefore, the union of intervals 1 and 2 becomes $(-\infty, 0]$.
>
> **Step 5:** Write the solution set.

> **Skill Practice 2**   Solve the inequality. $x^4 - 18x \geq 3x^2 - 4x^3$

Answer

2. Interval notation: $(-\infty, 0] \cup [2, \infty)$;
Set-builder notation: $\{x \mid x \leq 0 \text{ or } x \geq 2\}$

The result of Example 2 can also be interpreted graphically. From Section 3.2, the graph of $f(x) = x^4 + x^3 - 8x^2 - 12x$ is up to the far left and up to the far right.

In factored form $f(x) = x(x - 3)(x + 2)^2$. The x-intercepts are $(0, 0)$, $(3, 0)$, and $(-2, 0)$. Furthermore, the factors x and $(x - 3)$ have odd exponents. This means that the corresponding zeros have odd multiplicities, and that the graph will cross the x-axis at $(0, 0)$ and $(3, 0)$ and change sign. The factor $(x + 2)$ has an even exponent meaning that the corresponding zero has an even multiplicity. The graph will touch the x-axis at $(-2, 0)$ but will *not* change sign. From the sketch in Figure 3-25, we see that $f(x) \geq 0$ on the intervals $(-\infty, 0]$ and $[3, \infty)$.

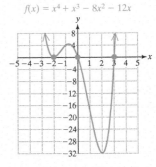

$f(x) = x^4 + x^3 - 8x^2 - 12x$

Figure 3-25

In some situations, the sign of a polynomial may be easily determined by inspection. In such a case, we can abbreviate the procedure to solve a polynomial inequality. This is demonstrated in Example 3.

EXAMPLE 3   Solving Polynomial Inequalities

Solve the inequalities.

a. $4x^2 - 12x + 9 < 0$ **b.** $4x^2 - 12x + 9 \leq 0$

c. $4x^2 - 12x + 9 > 0$ **d.** $4x^2 - 12x + 9 \geq 0$

Solution:

a. $4x^2 - 12x + 9 < 0$

$(2x - 3)^2 < 0$

The solution set is { }.

Factor $4x^2 - 12x + 9$ as $(2x - 3)^2$.
The square of any real number is nonnegative. Therefore, this inequality has no solution.

b. $4x^2 - 12x + 9 \leq 0$

$(2x - 3)^2 \leq 0$

The solution set is $\left\{\frac{3}{2}\right\}$.

The inequality in part (b) is the same as the inequality in part (a) except that equality is included. The expression $(2x - 3)^2 = 0$ for $x = \frac{3}{2}$.

c. $4x^2 - 12x + 9 > 0$

$(2x - 3)^2 > 0$

The solution set is $\left(-\infty, \frac{3}{2}\right) \cup \left(\frac{3}{2}, \infty\right)$.

In set-builder notation: $\left\{x \mid x < \frac{3}{2} \text{ or } x > \frac{3}{2}\right\}$.

The expression $(2x - 3)^2 > 0$ for all real numbers except where $(2x - 3)^2 = 0$. Therefore, the solution set is all real numbers except $\frac{3}{2}$.

d. $4x^2 - 12x + 9 \geq 0$

$(2x - 3)^2 \geq 0$

The solution set is $(-\infty, \infty)$.

The square of any real number is greater than or equal to zero. Therefore, the solution set is all real numbers.

Skill Practice 3   Solve the inequalities.

a. $25x^2 - 10x + 1 < 0$ **b.** $25x^2 - 10x + 1 \leq 0$

c. $25x^2 - 10x + 1 > 0$ **d.** $25x^2 - 10x + 1 \geq 0$

2. Solve Rational Inequalities

We now turn our attention to solving rational inequalities.

Definition of a Rational Inequality

Let $f(x)$ be a rational expression. Then an inequality of the form $f(x) < 0$, $f(x) > 0$, $f(x) \leq 0$, or $f(x) \geq 0$ is called a **rational inequality.**

We solve polynomial and rational inequalities in the same way with one exception. With a rational inequality such as $f(x) < 0$, the list of boundary points must include the real solutions to the related equation $f(x) = 0$ along with the values of x that make $f(x)$ undefined.

Answers

3. a. { }   **b.** $\left\{\frac{1}{5}\right\}$

c. Interval notation:
$\left(-\infty, \frac{1}{5}\right) \cup \left(\frac{1}{5}, \infty\right)$; Set-builder
notation: $\left\{x \mid x < \frac{1}{5} \text{ or } x > \frac{1}{5}\right\}$

d. $(-\infty, \infty)$

Avoiding Mistakes

The graph of a rational function will not always change sign to the left and right of a vertical asymptote or x-intercept. However, since the possibility exists, we must test each interval defined by these values of x.

The graph of $f(x) = \dfrac{2x - 5}{x + 2}$ is shown in Figure 3-26. Notice that the function changes sign to the left and right of the vertical asymptote $x = -2$ and to the left and right of the x-intercept $\left(\frac{5}{2}, 0\right)$.

From the graph of $f(x) = \dfrac{2x - 5}{x + 2}$ we can determine the solution sets for the following inequalities.

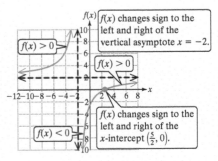

Figure 3-26

$$\frac{2x - 5}{x + 2} < 0 \qquad \text{on the interval } \left(-2, \tfrac{5}{2}\right)$$

$$\frac{2x - 5}{x + 2} > 0 \qquad \text{on the intervals } (-\infty, -2) \cup \left(\tfrac{5}{2}, \infty\right)$$

$$\frac{2x - 5}{x + 2} \leq 0 \qquad \text{on the interval } \left(-2, \tfrac{5}{2}\right]$$

$$\frac{2x - 5}{x + 2} \geq 0 \qquad \text{on the intervals } (-\infty, -2) \cup \left[\tfrac{5}{2}, \infty\right)$$

Avoiding Mistakes

The value -2 is excluded from the solution set because -2 makes the expression $\dfrac{2x - 5}{x + 2}$ undefined.

EXAMPLE 4   **Solving a Rational Inequality**

Solve the inequality. $\dfrac{4x - 5}{x - 2} \leq 3$

Solution:

$$\frac{4x - 5}{x - 2} \leq 3$$

$$\underbrace{\frac{4x - 5}{x - 2} - 3}_{f(x)} \leq 0$$

Step 1: First write the inequality in the form $f(x) \leq 0$. That is, set one side to 0.

$$\frac{4x - 5}{x - 2} - 3 \cdot \frac{x - 2}{x - 2} \leq 0$$

Write each term with a common denominator.

$$\frac{4x - 5 - 3(x - 2)}{x - 2} \leq 0$$

Simplify.

$$\frac{4x - 5 - 3x + 6}{x - 2} \leq 0$$

$$\frac{x + 1}{x - 2} \leq 0$$

The expression $\dfrac{x + 1}{x - 2}$ is undefined for $x = 2$. Therefore, the value $x = 2$ is *not* part of the solution set. However, 2 *is* a boundary point for the solution set.

TIP A rational expression is equal to zero where the numerator is equal to zero.

A rational expression is undefined where the denominator is equal to zero.

$$\frac{x + 1}{x - 2} = 0$$

Step 2: Solve for the real solutions to the equation $f(x) = 0$. The solution is -1, and this is another boundary point.

The boundary points are $x = 2$ and $x = -1$.

$$x = -1$$

Step 3: Divide the x-axis into intervals defined by the boundary points and determine the sign of $f(x)$ on each interval.

	Interval 1	Interval 2	Interval 3
Evaluate: $f(x) = \frac{x+1}{x-2}$	$f(-2) = \frac{1}{4}$ $f(x) > 0$	$f(1) = -2$ $f(x) < 0$	$f(3) = 4$ $f(x) > 0$
Sign of $(x + 1)$	−	+	+
Sign of $(x - 2)$	−	−	+
Sign of $\frac{(x+1)}{(x-2)}$	+	−	+

$$-1 \qquad 2$$

Step 4: Substituting the boundary point $x = -1$ into the inequality $\dfrac{x+1}{x-2} \leq 0$ makes a true statement, $0 \leq 0$. Thus, $x = -1$ is part of the solution set. The boundary point 2 is excluded because $\dfrac{x+1}{x-2}$ is undefined for $x = 2$.

Step 5: Write the solution set.
The solution set is $[-1, 2)$.
In set-builder notation this is $\{x \mid -1 \leq x < 2\}$.

Skill Practice 4 Solve the inequality. $\dfrac{5-x}{x-1} \geq -2$

EXAMPLE 5 Solving a Rational Inequality

Solve the inequality. $\dfrac{1}{x-2} \geq \dfrac{1}{x+3}$

Solution:

$$\frac{1}{x-2} \geq \frac{1}{x+3}$$

Step 1: First write the inequality in the form $f(x) \geq 0$.

$$\frac{1}{x-2} - \frac{1}{x+3} \geq 0$$

Set one side to zero.

$$\frac{1}{x-2} \cdot \frac{x+3}{x+3} - \frac{1}{x+3} \cdot \frac{x-2}{x-2} \geq 0$$

Write each term with a common denominator.

$$\frac{x+3}{(x-2)(x+3)} - \frac{x-2}{(x-2)(x+3)} \geq 0$$

Simplify and write the numerator and denominator in factored form.

$$\frac{x+3-x+2}{(x-2)(x+3)} \geq 0$$

$$\frac{5}{(x-2)(x+3)} \geq 0$$

The expression $\frac{5}{(x-2)(x+3)}$ is undefined for $x = 2$ and $x = -3$. The values 2 and -3 are not included in the solution set, but they are boundary points for the solution set.

$$\frac{5}{(x-2)(x+3)} = 0$$

Step 2: Solve the related equation to determine any additional boundary points.

$$\frac{5}{(x-2)(x+3)} \cdot (x-2)(x+3) = 0 \cdot (x-2)(x+3)$$

$$5 = 0$$
(contradiction)

Clearing fractions results in the contradiction $5 = 0$. There are no solutions to the related equation.

Answer
4. $(-\infty, -3] \cup (1, \infty)$

Step 3: Divide the x-axis into intervals defined by the boundary points and determine the sign of $f(x) = \frac{5}{(x-2)(x+3)}$ on each interval.

	Interval 1	Interval 2	Interval 3
Evaluate: $f(x) = \frac{5}{(x-2)(x+3)}$	$f(-4) = \frac{5}{6}$ $f(x) > 0$	$f(0) = -\frac{5}{6}$ $f(x) < 0$	$f(3) = \frac{5}{6}$ $f(x) > 0$
Sign of 5	+	+	+
Sign of $(x-2)$	−	−	+
Sign of $(x+3)$	−	+	+
Sign of $f(x) = \frac{5}{(x-2)(x+3)}$	+	−	+

$$\xleftarrow{\qquad} \underset{-3}{\qquad} \underset{2}{\qquad} \xrightarrow{\qquad}$$

Now interpret the results to solve $\dfrac{5}{(x-2)(x+3)} \geq 0$.

Step 4: Neither boundary point is included because $\frac{5}{(x-2)(x+3)}$ is undefined for $x = 2$ and $x = -3$.

Step 5: Write the solution set. The expression $\frac{5}{(x-2)(x+3)}$ is greater than zero for $x < -3$ and $x > 2$.

The solution set is $(-\infty, -3) \cup (2, \infty)$.

In set-builder notation, this is $\{x \mid x < -3 \text{ or } x > 2\}$.

Skill Practice 5 Solve the inequality. $\dfrac{1}{x+4} \geq \dfrac{1}{x+2}$

In Example 6 we encounter an inequality in which the signs of the numerator and denominator of the rational expression can be determined by inspection.

EXAMPLE 6 **Solving Rational Inequalities**

Solve the inequalities.

a. $\dfrac{x^2}{x^2+4} \geq 0$ **b.** $\dfrac{x^2}{x^2+4} > 0$

c. $\dfrac{x^2}{x^2+4} \leq 0$ **d.** $\dfrac{x^2}{x^2+4} < 0$

Solution:

The solution to the related equation $\dfrac{x^2}{x^2+4} = 0$ is $x = 0$.

The denominator is nonzero for all real numbers.

Therefore, the only boundary point is $x = 0$.

Sign of x^2	+	+
Sign of $x^2 + 4$	+	+
Sign of $\dfrac{x^2}{x^2+4}$	+	+

$$\underset{0}{\qquad}$$

Therefore, $\dfrac{x^2}{x^2 + 4} = 0$ at $x = 0$ and is positive for all other real numbers.

a. $\dfrac{x^2}{x^2 + 4} \geq 0$ Solution set: $(-\infty, \infty)$

b. $\dfrac{x^2}{x^2 + 4} > 0$ Solution set: $(-\infty, 0) \cup (0, \infty)$

c. $\dfrac{x^2}{x^2 + 4} \leq 0$ Solution set: $\{0\}$

d. $\dfrac{x^2}{x^2 + 4} < 0$ Solution set: $\{\ \}$

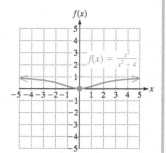

Skill Practice 6   Solve the inequalities.

a. $\dfrac{x^2}{x^4 + 1} \geq 0$ **b.** $\dfrac{x^2}{x^4 + 1} > 0$ **c.** $\dfrac{x^2}{x^4 + 1} \leq 0$ **d.** $\dfrac{x^2}{x^4 + 1} < 0$

3. Solve Applications Involving Polynomial and Rational Inequalities

In Section 1.5, we studied the vertical position $s(t)$ of an object moving upward or downward under the influence of gravity. We use this model to solve the application in Example 7.

$$s(t) = -\frac{1}{2}gt^2 + v_0 t + s_0$$

where

g is the acceleration due to gravity (32 ft/sec^2 or 9.8 m/sec^2).

t is the time of travel.

v_0 is the initial speed.

s_0 is the initial vertical position.

EXAMPLE 7   Solving an Application of a Polynomial Inequality

A toy rocket is shot straight upward from a launch pad 1 ft above ground level with an initial speed of 64 ft/sec.

a. Write a model to express the vertical position $s(t)$ (in ft) of the rocket t seconds after launch.

b. Determine the times at which the rocket is above a height of 50 ft.

TIP Choose $g = 32$ ft/sec^2 because the height is given in feet and speed is given in feet per second.

Solution:

a. $s(t) = -\dfrac{1}{2}gt^2 + v_0 t + s_0$ In this example,

$$s(t) = -\frac{1}{2}(32)t^2 + (64)t + (1)$$

$$s(t) = -16t^2 + 64t + 1$$

$s_0 = 1$ ft
$v_0 = 64$ ft/sec
$g = 32$ ft/sec^2

Answers

6. a. $(-\infty, \infty)$
  b. $(-\infty, 0) \cup (0, \infty)$
  c. $\{0\}$
  d. $\{\ \}$

b. $-16t^2 + 64t + 1 > 50$

$$\overbrace{}^{f(t)}$$

$-16t^2 + 64t - 49 > 0$ Write the inequality in the form $f(t) > 0$.

$-16t^2 + 64t - 49 = 0$ Use the quadratic formula to solve the related equation $f(t) = 0$.

$$t = \frac{-64 \pm \sqrt{(64)^2 - 4(-16)(-49)}}{2(-16)}$$ Evaluate $f(t) = -16t^2 + 64t - 49$ for test points in each interval.

$$= \frac{-64 \pm \sqrt{960}}{-32}$$

$$= \frac{-64 \pm 8\sqrt{15}}{-32}$$

$$= \frac{8 \pm \sqrt{15}}{4} \begin{array}{l} \approx 2.97 \\ \approx 1.03 \end{array}$$

$f(1) = -1$ $f(2) = 15$ $f(3) = -1$
$f(t) < 0$ $f(t) > 0$ $f(t) < 0$

$\frac{8 - \sqrt{15}}{4} \approx 1.03$ $\frac{8 + \sqrt{15}}{4} \approx 2.97$

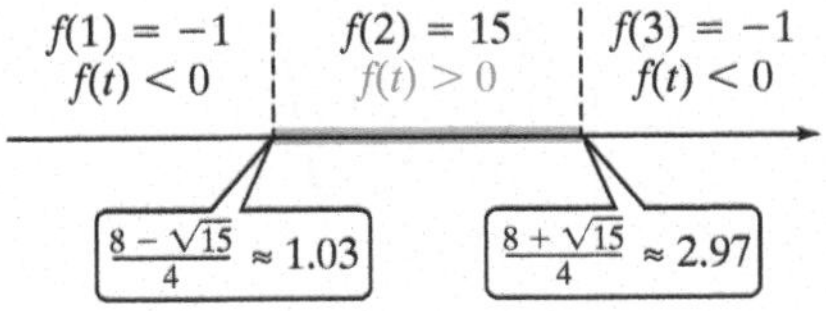

Figure 3-27

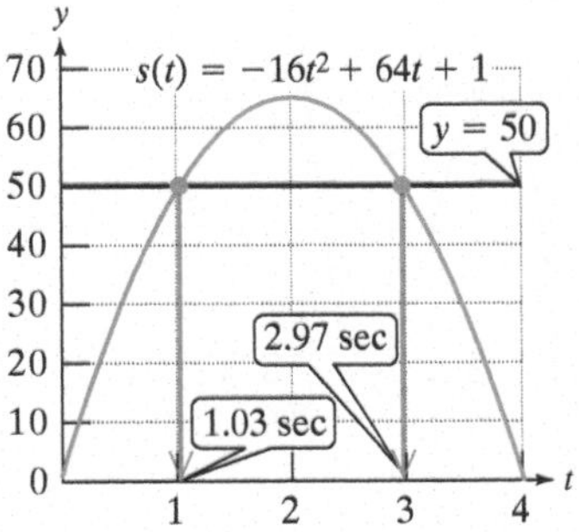

$s(t) = -16t^2 + 64t + 1$

$y = 50$

2.97 sec

1.03 sec

The solution set is $\left(\dfrac{8 - \sqrt{15}}{4}, \dfrac{8 + \sqrt{15}}{4} \right)$ or approximately $(1.03, 2.97)$.

The rocket will be above 50 ft high between 1.03 sec and 2.97 sec after launch.

    The graph of $s(t) = -16t^2 + 64t + 1$ is a parabola opening downward (Figure 3-27). We see that $s(t) > 50$ for t between 1.03 and 2.97 as expected.

Answers

7. a. $s(t) = -16t^2 + 80t + 5$

   b. $\left(\dfrac{10 - \sqrt{55}}{4}, \dfrac{10 + \sqrt{55}}{4} \right)$

Skill Practice 7   Repeat Example 7 under the assumption that the rocket is launched with an initial speed of 80 ft/sec from a height of 5 ft.

SECTION 3.6   Practice Exercises

Prerequisite Review

For Exercises R.1–R.3, solve the inequality and write the solution in interval notation when possible.

R.1. $1 - 3(z - 5) \geq 3 + 3(2z + 7)$ **R.2.** $-4(3x + 2) < 1 - (x - 4) - 9x$ **R.3.** $2w - 3 \geq 9$ or $w < -1$

For Exercises R.4–R.5, add the rational expressions.

R.4. $\dfrac{2}{w + 2} + \dfrac{1}{3w + 2}$ **R.5.** $y + 5 + \dfrac{1}{y - 5}$

Concept Connections

1. Let $f(x)$ be a polynomial. An inequality of the form $f(x) < 0$, $f(x) > 0$, $f(x) \geq 0$, or $f(x) \leq 0$ is called a _____ inequality. If the polynomial is of degree _____, then the inequality is also called a quadratic inequality.

2. Let $f(x)$ be a rational expression. An inequality of the form $f(x) < 0$, $f(x) > 0$, $f(x) \geq 0$, or $f(x) \leq 0$ is called a _____ inequality.

3. The solution set for the inequality $(x + 10)^2 \geq -4$ is _____, whereas the solution set for the inequality $(x + 10)^2 \leq -4$ is _____.

4. The solutions to an inequality $f(x) < 0$ are the values of x on the intervals where $f(x)$ is (positive/negative).

Objective 1: Solve Polynomial Inequalities

For Exercises 5–14, the graph of $y = f(x)$ is given. Solve the inequalities.

a. $f(x) < 0$ **b.** $f(x) \leq 0$ **c.** $f(x) > 0$ **d.** $f(x) \geq 0$

5.
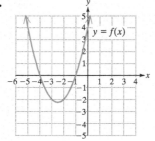

6.
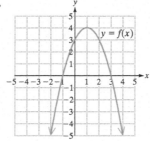

7.
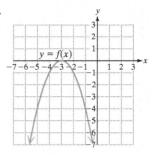

8.
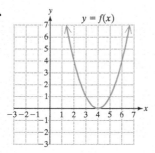

9.
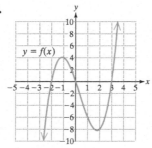

10.
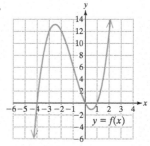

11.
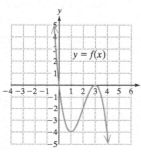

12.
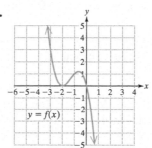

13.
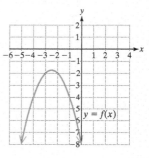

14.
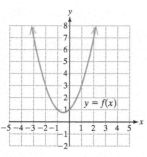

For Exercises 15–20, solve the equations and inequalities. (See Example 3)

15. a. $(5x - 3)(x - 5) = 0$
    b. $(5x - 3)(x - 5) < 0$
    c. $(5x - 3)(x - 5) \leq 0$
    d. $(5x - 3)(x - 5) > 0$
    e. $(5x - 3)(x - 5) \geq 0$

16. a. $(3x + 7)(x - 2) = 0$
    b. $(3x + 7)(x - 2) < 0$
    c. $(3x + 7)(x - 2) \leq 0$
    d. $(3x + 7)(x - 2) > 0$
    e. $(3x + 7)(x - 2) \geq 0$

17. a. $-x^2 + x + 12 = 0$
    b. $-x^2 + x + 12 < 0$
    c. $-x^2 + x + 12 \leq 0$
    d. $-x^2 + x + 12 > 0$
    e. $-x^2 + x + 12 \geq 0$

18. a. $-x^2 - 10x - 9 = 0$
    b. $-x^2 - 10x - 9 < 0$
    c. $-x^2 - 10x - 9 \leq 0$
    d. $-x^2 - 10x - 9 > 0$
    e. $-x^2 - 10x - 9 \geq 0$

19. a. $a^2 + 12a + 36 = 0$
    b. $a^2 + 12a + 36 < 0$
    c. $a^2 + 12a + 36 \leq 0$
    d. $a^2 + 12a + 36 > 0$
    e. $a^2 + 12a + 36 \geq 0$

20. a. $t^2 - 14t + 49 = 0$
    b. $t^2 - 14t + 49 < 0$
    c. $t^2 - 14t + 49 \leq 0$
    d. $t^2 - 14t + 49 > 0$
    e. $t^2 - 14t + 49 \geq 0$

For Exercises 21–54, solve the inequalities. (See Examples 1–2)

21. $4w^2 - 9 \geq 0$

22. $16z^2 - 25 < 0$

23. $3w^2 + w < 2(w + 2)$

24. $5y^2 + 7y < 3(y + 4)$

25. $a^2 \geq 3a$

26. $d^2 \geq 6d$

27. $10 - 6x > 5x^2$

28. $6 - 4x > 3x^2$

29. $m^2 < 49$

30. $y^2 \geq 9$

31. $16p^2 \geq 2$

32. $54q^2 \leq 50$

33. $(x + 4)(x - 1)(x - 3) \geq 0$

34. $(x + 2)(x + 5)(x - 4) \geq 0$

35. $-5c(c + 2)^2(4 - c) > 0$

36. $-6u(u + 1)^2(3 - u) > 0$

37. $t^4 - 10t^2 + 9 \leq 0$

38. $w^4 - 20w^2 + 64 \leq 0$

39. $2x^3 + 5x^2 < 8x + 20$

40. $3x^3 - 3x < 4x^2 - 4$

41. $-2x^4 + 10x^3 - 6x^2 - 18x \geq 0$

42. $-4x^4 + 4x^3 + 64x^2 + 80x \geq 0$

43. $-5u^6 + 28u^5 - 15u^4 \leq 0$

44. $-3w^6 + 8w^5 - 4w^4 \leq 0$

45. $6x(2x - 5)^4(3x + 1)^5(x - 4) < 0$

46. $5x(3x - 2)^2(4x + 1)^3(x - 3)^4 < 0$

47. $(5x - 3)^2 > -2$

48. $(4x + 1)^2 > -6$

49. $-4 \geq (x - 7)^2$

50. $-1 \geq (x + 2)^2$

51. $16y^2 > 24y - 9$

52. $4w^2 > 20w - 25$

53. $(x + 3)(x + 1) \leq -1$

54. $(x + 2)(x + 4) \leq -1$

Objective 2: Solve Rational Inequalities

For Exercises 55–58, the graph of $y = f(x)$ is given. Solve the inequalities.

   a. $f(x) < 0$ **b.** $f(x) \leq 0$ **c.** $f(x) > 0$ **d.** $f(x) \geq 0$

55.

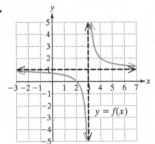

56.

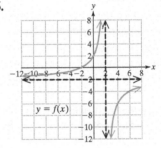

57.

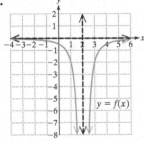

58.

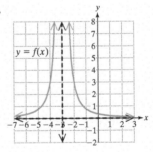

For Exercises 59–62, solve the inequalities. (See Example 6)

59. a. $\dfrac{x + 2}{x - 3} \leq 0$

   b. $\dfrac{x + 2}{x - 3} < 0$

   c. $\dfrac{x + 2}{x - 3} \geq 0$

   d. $\dfrac{x + 2}{x - 3} > 0$

60. a. $\dfrac{x + 4}{x - 1} \leq 0$

   b. $\dfrac{x + 4}{x - 1} < 0$

   c. $\dfrac{x + 4}{x - 1} \geq 0$

   d. $\dfrac{x + 4}{x - 1} > 0$

61. a. $\dfrac{x^4}{x^2 + 9} \leq 0$

   b. $\dfrac{x^4}{x^2 + 9} < 0$

   c. $\dfrac{x^4}{x^2 + 9} \geq 0$

   d. $\dfrac{x^4}{x^2 + 9} > 0$

62. a. $\dfrac{-x^2}{x^4 + 16} \leq 0$

   b. $\dfrac{-x^2}{x^4 + 16} < 0$

   c. $\dfrac{-x^2}{x^4 + 16} \geq 0$

   d. $\dfrac{-x^2}{x^4 + 16} > 0$

For Exercises 63–84, solve the inequalities. (See Examples 4–5)

63. $\dfrac{5 - x}{x + 1} \geq 0$

64. $\dfrac{2 - x}{x + 6} \geq 0$

65. $\dfrac{4 - 2x}{x^2} \leq 0$

66. $\dfrac{9 - 3x}{x^2} \leq 0$

67. $\dfrac{w^2 - w - 2}{w + 3} \geq 0$

68. $\dfrac{p^2 - 2p - 8}{p - 1} \geq 0$

69. $\dfrac{5}{2t - 7} > 1$

70. $\dfrac{4}{3c - 8} > 1$

71. $\dfrac{2x}{x - 2} \leq 2$

72. $\dfrac{3x}{3x - 7} \leq 1$

73. $\dfrac{4 - x}{x + 5} \geq 2$

74. $\dfrac{3 - x}{x + 2} \geq 4$

75. $\dfrac{a - 2}{a^2 + 4} \leq 0$

76. $\dfrac{d - 3}{d^2 + 1} \leq 0$

77. $\dfrac{10}{x + 2} \geq \dfrac{2}{x + 2}$

78. $\dfrac{4}{x - 3} \geq \dfrac{1}{x - 3}$

79. $\dfrac{4}{y + 3} > -\dfrac{2}{y}$

80. $\dfrac{2}{z - 1} > -\dfrac{4}{z}$

81. $\dfrac{3}{4 - x} \leq \dfrac{6}{1 - x}$

82. $\dfrac{5}{2 - x} \leq \dfrac{3}{3 - x}$

83. $\dfrac{(2 - x)(2x + 1)^2}{(x - 4)^4} \leq 0$

84. $\dfrac{(3 - x)(4x - 1)^4}{(x + 2)^2} \leq 0$

Objective 3: Solve Applications Involving Polynomial and Rational Inequalities

85. A professional fireworks team shoots an 8-in. mortar straight upwards from ground level with an initial speed of 216 ft/sec. (**See Example 7**)

  a. Write a function modeling the vertical position $s(t)$ (in ft) of the shell at a time t seconds after launch.

  b. The mortar is designed to explode when the shell is at its maximum height. How long after launch will the shell explode? (*Hint*: Consider the vertex formula from Section 3.1.)

  c. The spectators can see the shell rising once it clears a 200-ft tree line. For what period of time after launch is the shell visible before it explodes?

86. Suppose that a basketball player jumps straight up for a rebound.

  a. If his initial speed leaving the ground is 16 ft/sec, write a function modeling his vertical position $s(t)$ (in ft) at a time t seconds after leaving the ground.

  b. Find the times after leaving the ground when the player will be at a height of more than 3 ft in the air.

87. For a certain stretch of road, the distance d (in ft) required to stop a car that is traveling at speed v (in mph) before the brakes are applied can be approximated by $d(v) = 0.06v^2 + 2v$. Find the speeds for which the car can be stopped within 250 ft.

88. The population $P(t)$ of a bacteria culture is given by $P(t) = -1500t^2 + 60{,}000t + 10{,}000$, where t is the time in hours after the culture is started. Determine the time(s) at which the population will be greater than 460,000 organisms.

89. Suppose that an object that is originally at room temperature of 32°C is placed in a freezer. The temperature $T(x)$ (in °C) of the object can be approximated by the model $T(x) = \dfrac{320}{x^2 + 3x + 10}$, where x is the time in hours after the object is placed in the freezer.

  a. What is the horizontal asymptote of the graph of this function and what does it represent in the context of this problem?

  b. A chemist needs a compound cooled to less than 5°C. Determine the amount of time required for the compound to cool so that its temperature is less than 5°C.

90. The average round trip speed S (in mph) of a vehicle traveling a distance of d miles in each direction is given by

$$S = \dfrac{2d}{\dfrac{d}{r_1} + \dfrac{d}{r_2}}$$

where r_1 and r_2 are the rates of speed for the initial trip and the return trip, respectively.

  a. Suppose that a motorist travels 200 mi from her home to an athletic event and averages 50 mph for the trip to the event. Determine the speeds necessary if the motorist wants the average speed for the round trip to be at least 60 mph.

  b. Would the motorist be traveling within the speed limit of 70 mph?

91. A rectangular quilt is to be made so that the length is 1.2 times the width. The quilt must be between 72 ft² and 96 ft² to cover the bed. Determine the restrictions on the width so that the dimensions of the quilt will meet the required area. Give exact values and the approximated values to the nearest tenth of a foot.

92. A landscaping team plans to build a rectangular garden that is between 480 yd² and 720 yd² in area. For aesthetic reasons, they also want the length to be 1.5 times the width. Determine the restrictions on the width so that the dimensions of the garden will meet the required area. Give exact values and the approximated values to the nearest tenth of a yard.

Mixed Exercises

For Exercises 93–102, write the domain of the function in interval notation.

93. $f(x) = \sqrt{9 - x^2}$

94. $g(t) = \sqrt{1 - t^2}$

95. $h(a) = \sqrt{a^2 - 5}$

96. $f(u) = \sqrt{u^2 - 7}$

97. $p(x) = \sqrt{2x^2 + 9x - 18}$

98. $q(x) = \sqrt{4x^2 + 7x - 2}$

99. $r(x) = \dfrac{1}{\sqrt{2x^2 + 9x - 18}}$

100. $s(x) = \dfrac{1}{\sqrt{4x^2 + 7x - 2}}$

101. $h(x) = \sqrt{\dfrac{3x}{x + 2}}$

102. $k(x) = \sqrt{\dfrac{2x}{x + 1}}$

103. Let a, b, and c represent positive real numbers, where $a < b < c$, and let $f(x) = (x - a)^2(b - x)(x - c)^3$.

   a. Complete the sign chart.

Sign of $(x - a)^2$:			
Sign of $(b - x)$:			
Sign of $(x - c)^3$:			
Sign of $(x - a)^2(b - x)(x - c)^3$:			

   $\qquad\qquad\quad$ $a\qquad b\qquad c$

   b. Solve $f(x) > 0$. $\qquad$ **c.** Solve $f(x) < 0$.

104. Let a, b, and c represent positive real numbers, where $a < b < c$, and let $g(x) = \dfrac{(a - x)(x - b)^2}{(c - x)^5}$.

   a. Complete the sign chart.

Sign of $(a - x)$:			
Sign of $(x - b)^2$:			
Sign of $(c - x)^5$:			
Sign of $\dfrac{(a - x)(x - b)^2}{(c - x)^5}$:			

   $\qquad\qquad\quad$ $a\qquad b\qquad c$

   b. Solve $g(x) > 0$. $\qquad$ **c.** Solve $g(x) < 0$.

Write About It

105. Explain how the solution set to the inequality $f(x) < 0$ is related to the graph of $y = f(x)$.

106. Explain how the solution set to the inequality $f(x) \geq 0$ is related to the graph of $y = f(x)$.

107. Explain why $\dfrac{x^2 + 2}{x^2 + 1} < 0$ has no solution.

108. Given $\dfrac{x - 3}{x - 1} \leq 0$, explain why the solution set includes 3, but does not include 1.

Expanding Your Skills

The procedure to solve a polynomial or rational inequality may be applied to all inequalities of the form $f(x) > 0$, $f(x) < 0$, $f(x) \geq 0$, and $f(x) \leq 0$. That is, find the real solutions to the related equation and determine restricted values of x. Then determine the sign of $f(x)$ on each interval defined by the boundary points. Use this process to solve the inequalities in Exercises 109–120.

109. $\sqrt{2x - 6} - 2 < 0$

110. $\sqrt{3x - 5} - 4 < 0$

111. $\sqrt{4 - x} - 6 \geq 0$

112. $\sqrt{5 - x} - 7 \geq 0$

113. $\dfrac{1}{\sqrt{x - 2} - 4} \leq 0$

114. $\dfrac{1}{\sqrt{x - 3} - 5} \leq 0$

115. $-3 < x^2 - 6x + 5 \leq 5$

116. $8 \leq x^2 + 4x + 3 < 15$

117. $|x^2 - 4| < 5$

118. $|x^2 + 1| < 17$

119. $|x^2 - 18| > 2$

120. $|x^2 - 6| > 3$

Technology Connections

121. Given the inequality,
   $0.552x^3 + 4.13x^2 - 1.84x - 3.5 < 6.7$,

   a. Write the inequality in the form $f(x) < 0$.

   b. Graph $y = f(x)$ on a suitable viewing window.

   c. Use the **Zero** feature to approximate the real zeros of $f(x)$. Round to 1 decimal place.

   d. Use the graph to approximate the solution set for the inequality $f(x) < 0$.

122. Given the inequality,
   $0.24x^4 + 1.8x^3 + 3.3x^2 + 2.84x - 1.8 > 4.5$,

   a. Write the inequality in the form $f(x) > 0$.

   b. Graph $y = f(x)$ on a suitable viewing window.

   c. Use the **Zero** feature to approximate the real zeros of $f(x)$. Round to 1 decimal place.

   d. Use the graph to approximate the solution set for the inequality $f(x) > 0$.

123. An engineer for a food manufacturer designs an aluminum container for a hot drink mix. The container is to be a right circular cylinder 5.5 in. in height. The surface area represents the amount of aluminum used and is given by

$S(r) = 2\pi r^2 + 11\pi r$, where r is the radius of the can.

a. Graph the function $y = S(r)$ and the line $y = 90$ on the viewing window [0, 3, 1] by [0, 150, 10].

b. Use the Intersect feature to approximate the point of intersection of $y = S(r)$ and $y = 90$. Round to 1 decimal place if necessary.

c. Determine the restrictions on r so that the amount of aluminum used is at most 90 in.^{2}. Round to 1 decimal place.

124. The concentration $C(t)$ (in ng/mL) of a drug in the bloodstream t hours after ingestion is modeled by

$$C(t) = \frac{500t}{t^3 + 100}$$

a. Graph the function $y = C(t)$ and the line $y = 4$ on the window [0, 32, 4] by [0, 15, 3].

b. Use the Intersect feature to approximate the point(s) of intersection of $y = C(t)$ and $y = 4$. Round to 1 decimal place if necessary.

c. To avoid toxicity, a physician may give a second dose of the medicine once the concentration falls below 4 ng/mL for increasing values of t. Determine the times at which it is safe to give a second dose. Round to 1 decimal place.

PROBLEM RECOGNITION EXERCISES

Solving Equations and Inequalities

At this point in the text, we have studied several categories of equations and related inequalities. These include

- linear equations and inequalities
- quadratic equations and inequalities
- polynomial equations and inequalities
- rational equations and inequalities
- radical equations and inequalities
- absolute value equations and inequalities
- compound inequalities

For Exercises 1–30, solve the equations and inequalities. Write the solution sets to the inequalities in interval notation if possible.

1. $-\dfrac{1}{2} \le -\dfrac{1}{4}x - 5 < 2$

2. $2x^2 - 6x = 5$

3. $50x^3 - 25x^2 - 2x + 1 = 0$

4. $\dfrac{-5x(x-3)^2}{2+x} \le 0$

5. $\sqrt[4]{m+4} - 5 = -2$

6. $-5 < y$ and $-3y + 4 \ge 7$

7. $|5t - 4| + 2 = 7$

8. $3 - 4\{x - 5[x + 2(3 - 2x)]\} = -2[4 - (x - 1)]$

9. $10x(2x - 14) = -29x^2 - 100$

10. $\dfrac{5}{y-4} = \dfrac{3y}{y+2} - \dfrac{2y^2 - 14y}{y^2 - 2y - 8}$

11. $x(x - 14) \le -40$

12. $\dfrac{1}{x^2 - 14x + 40} \le 0$

13. $|x - 0.15| = |x + 0.05|$

14. $\sqrt{t-1} - 5 \le 1$

15. $n^{1/2} + 7 = 10$

16. $-4x(3 - x)(x + 2)^2(x - 5)^3 \ge 0$

17. $-2x - 5(x + 3) = -4(x + 2) - 3x$

18. $\sqrt{7x + 29} - 3 = x$

19. $(x^2 - 9)^2 - 5(x^2 - 9) - 14 = 0$

20. $2 + 7x^{-1} - 15x^{-2} = 0$

21. $|8x - 3| + 10 \le 7$

22. $2(x - 1)^{3/4} = 16$

23. $x^3 - 3x^2 < 6x - 8$

24. $\dfrac{3 - x}{x + 5} \ge 1$

25. $15 - 3(x - 1) = -2x - (x - 18)$

26. $25x^2 + 70x > -49$

27. $2 < |3 - x| - 9$

28. $-4(x - 3) < 8$ or $-7 > x - 3$

29. $\dfrac{1}{3}x + \dfrac{2}{5} > \dfrac{5}{6}x - 1$

30. $2|2x + 1| - 2 \le 8$

Variation

OBJECTIVES

1. Write Models Involving Direct, Inverse, and Joint Variation
2. Solve Applications Involving Variation

1. Write Models Involving Direct, Inverse, and Joint Variation

The familiar relationship $d = rt$ tells us that distance traveled equals the rate of speed times the time of travel. For a car traveling 60 mph, we have $d = 60t$. From Table 3-2, notice that as the time of travel increases, the distance increases proportionally. We say that d is directly proportional to t, or that d varies directly as t. This is shown graphically in Figure 3-28.

Table 3-2

$d = 60t$

t (hr)	d (mi)
1	60
2	120
3	180
4	240
5	300
6	360

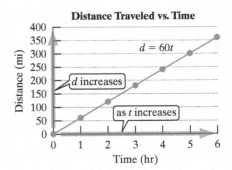

Figure 3-28

Now suppose that a motorist travels a fixed distance of 240 mi. We have

$$d = rt$$

$$240 = rt \longrightarrow t = \frac{240}{r}$$

The time of travel t varies *inversely* as the rate of speed. As the rate r increases, the time of travel will decrease proportionally. Likewise, for slower rates, the time of travel is greater. See Table 3-3 and Figure 3-29.

Table 3-3

$t = \dfrac{240}{r}$

r (mph)	t (hr)
10	24
20	12
30	8
40	6
50	4.8
60	4

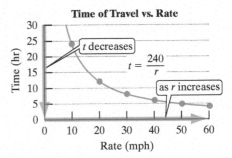

Figure 3-29

Direct and Inverse Variation

Let k be a nonzero constant real number. The statements on the left are equivalent to the equation on the right.

1. y varies **directly** as x.
   y is **directly** proportional to x. $\Big\}\, y = kx$

2. y varies **inversely** as x.
   y is **inversely** proportional to x. $\Big\}\, y = \dfrac{k}{x}$

Note: The value of k is called the **constant of variation.**

The first step in using a variation model is to write an English statement as an equivalent mathematical equation.

EXAMPLE 1 Writing a Variation Model

Write a variation model using k as the constant of variation.

Solution:

a. The amount of medicine A prescribed by a physician varies directly as the weight of the patient w.

$A = kw$

Since the variables are directly related, set up the *product* of k and w.

b. The frequency f in a vibrating string is inversely proportional to the length L of the string.

$f = \dfrac{k}{L}$

Since the variables are inversely related, set up the *quotient* of k and L.

c. The variable y varies directly as the square of x and inversely as the square root of z.

$y = \dfrac{kx^2}{\sqrt{z}}$

Since the square of the variable x is directly related to y, set up the *product* of k and x^2. And since the square root of z is inversely related to y, set up the *quotient* of k and $\sqrt{z}$.

> **TIP** Notice that in each variation model, the constant of variation, k, is always in the numerator.

Skill Practice 1 Write a variation model using k as the constant of variation.

a. The distance d that a spring stretches varies directly as the force F applied to the spring.

b. The force F required to keep a car from skidding on a curved road varies inversely as the radius r of the curve.

c. The variable a varies directly as b and inversely as the cube root of c.

Sometimes a variable varies directly as the product of two or more other variables. In such a case we have joint variation.

Joint Variation

Let k be a nonzero constant real number. The statements on the left are equivalent to the equation on the right.

$\left.\begin{array}{l} y \text{ varies } \textbf{jointly} \text{ as } w \text{ and } x. \\ y \text{ is } \textbf{jointly} \text{ proportional to } w \text{ and } x. \end{array}\right\} y = kwx$

EXAMPLE 2 Writing a Joint Variation Model

Write a variation model using k as the constant of variation.

Solution:

a. y varies jointly as t and the cube root of u.

$y = kt\sqrt[3]{u}$

The variable t and the quantity $\sqrt[3]{u}$ are jointly related to y. Set up the product of k, t, and $\sqrt[3]{u}$.

b. The gravitational force of attraction between two planets varies jointly as the product of their masses and inversely as the square of the distance between them.

$F = \dfrac{km_1m_2}{d^2}$

Let m_1 and m_2 represent the masses of the planets, let d represent the distance between the planets, and let F represent the gravitational force between the planets.

Answers

1. a. $d = kF$ **b.** $F = \dfrac{k}{r}$

c. $a = \dfrac{kb}{\sqrt[3]{c}}$

> **Skill Practice 2** Write a variation model using k as the constant of variation.
>
> **a.** The kinetic energy E of an object varies jointly as the object's mass m and the square of its velocity v.
>
> **b.** z varies jointly as x and y and inversely as the square root of w.

2. Solve Applications Involving Variation

Consider the variation models $y = kx$ and $y = \dfrac{k}{x}$. In either case, if values for x and y are known, we can solve for k. Once k is known, we can write a variation model and use it to find y if x is known, or to find x if y is known. This concept is the basis for solving many applications involving variation.

> **Procedure to Solve an Application Involving Variation**
>
> **Step 1** Write a general variation model that relates the variables given in the problem. Let k represent the constant of variation.
>
> **Step 2** Solve for k by substituting known values of the variables into the model from step 1.
>
> **Step 3** Substitute the value of k into the original variation model from step 1.
>
> **Step 4** Use the variation model from step 3 to solve the application.

EXAMPLE 3   **Solving an Application Involving Direct Variation**

The amount of an allergy medicine that a physician prescribes for a child varies directly as the weight of the child. Clinical research suggests that 13.5 mg of the drug should be given for a 30-lb child.

 a. How much should be prescribed for a 50-lb child?

 b. How much should be prescribed for a 60-lb child?

 c. A nurse wants to double check the dosage on a doctor's order of 18 mg. For a child of what weight is this dosage appropriate?

Solution:

Let A represent the amount of medicine.   Label the variables.

Let w represent the weight of the child.

$A = kw$	**Step 1:** Write a general variation model.
$13.5 = k(30)$	**Step 2:** Substitute known values of A and w into the variation model.
$\dfrac{13.5}{30} = k$	Solve for k by dividing both sides by 30.
$k = 0.45$	
$A = 0.45w$	**Step 3:** Substitute the value of k into the original variation model.
a. $A = 0.45(50)$	**Step 4:** Solve the application by substituting 50 for w.
$A = 22.5$	A 50-lb child would require 22.5 mg of the drug.

Answers

2. a. $E = kmv^2$ **b.** $z = \dfrac{kxy}{\sqrt{w}}$

TIP Notice from Example 3 that more medicine is given in proportion to a patient's weight.

A 60-lb patient weighs twice as much as a 30-lb patient and the amount of medicine given is also twice as much. This is consistent with direct variation.

b. $A = 0.45(60)$ Substitute 60 for w.

    $A = 27$ A 60-lb child would require 27 mg of the drug.

c. $A = 0.45w$

   $18 = 0.45w$ Substitute 18 mg for the amount of medicine.

   $40 = w$ Solve for the weight w. A 40-lb child would receive 18 mg.

Skill Practice 3 The amount of the medicine ampicillin that a physician prescribes for a child varies directly as the weight of the child. A physician prescribes 420 mg for a 35-lb child.

a. How much should be prescribed for a 30-lb child?

b. How much should be prescribed for a 40-lb child?

EXAMPLE 4 Solving an Application Involving Inverse Variation

The loudness of sound measured in decibels (dB) varies inversely as the square of the distance between the listener and the source of the sound. If the loudness of sound is 17.92 dB at a distance of 10 ft from a stereo speaker, what is the decibel level 20 ft from the speaker?

Solution:

Let L represent the loudness of sound in decibels and d represent the distance in feet. The inverse relationship between decibel level and the square of the distance is modeled by:

$$L = \frac{k}{d^2}$$

$$17.92 = \frac{k}{(10)^2} \qquad \text{Substitute } L = 17.92 \text{ dB and } d = 10 \text{ ft.}$$

$$17.92 = \frac{k}{100}$$

$$(17.92)100 = \frac{k}{100} \cdot 100 \qquad \text{Solve for } k \text{ (clear fractions).}$$

$$k = 1792$$

$$L = \frac{1792}{d^2} \qquad \text{Substitute } k = 1792 \text{ into the original model } L = \frac{k}{d^2}.$$

With the value of k known, we can find L for any value of d.

$$L = \frac{1792}{(20)^2} \qquad \text{Find the loudness when } d = 20 \text{ ft.}$$

$$= 4.48 \text{ dB}$$

Notice that the loudness of sound is 17.92 dB at a distance 10 ft from the speaker. When the distance from the speaker is increased to 20 ft, the decibel level decreases to 4.48 dB. This is consistent with an inverse relationship. For $k > 0$, as one variable is increased, the other is decreased. It also seems reasonable that the farther one moves away from the source of a sound, the softer the sound becomes.

Skill Practice 4 The yield on a bond varies inversely as the price. The yield on a particular bond is 4% when the price is $100. Find the yield when the price is $80.

Answers

3. a. 360 mg   **b.** 480 mg

4. 5%

EXAMPLE 5 **Solving an Application Involving Joint Variation**

In the early morning hours of August 29, 2005, Hurricane Katrina plowed into the Gulf Coast of the United States, bringing unprecedented destruction to southern Louisiana, Mississippi, and Alabama. The winds of a hurricane are strong enough to send a piece of plywood through a tree.

The kinetic energy of an object varies jointly as the weight of the object at sea level and as the square of its velocity. During a hurricane, a 0.5-lb stone traveling 60 mph has 81 joules (J) of kinetic energy. Suppose the wind speed doubles to 120 mph. Find the kinetic energy.

Solution:

Let E represent the kinetic energy, let w represent the weight, and let v represent the velocity of the stone. The variation model is

$$E = kwv^2$$
$$81 = k(0.5)(60)^2 \qquad \text{Substitute } E = 81 \text{ J, } w = 0.5 \text{ lb, and } v = 60 \text{ mph.}$$
$$81 = k(0.5)(3600) \qquad \text{Simplify the exponent.}$$
$$81 = k(1800)$$
$$\frac{81}{1800} = \frac{k(\cancel{1800})}{\cancel{1800}} \qquad \text{Divide by 1800.}$$
$$0.045 = k \qquad \text{Solve for } k.$$

With the value of k known, the model $E = kwv^2$ can be written as $E = 0.045wv^2$. We now find the kinetic energy of a 0.5-lb stone traveling 120 mph.

$$E = 0.045(0.5)(120)^2$$
$$= 324$$

The kinetic energy of a 0.5-lb stone traveling 120 mph is 324 J.

Skill Practice 5 The amount of simple interest earned in an account varies jointly as the interest rate and time of the investment. An account earns $200 in 2 yr at 4% interest. How much interest would be earned in 3 yr at a rate of 5%?

Answer

5. $375

SECTION 3.7 Practice Exercises

Prerequisite Review

For Exercises R.1–R.3, solve for the indicated variable.

R.1. $A = Ptr$ for r

R.2. $V = \frac{1}{3}\pi r^2 h$ for h

R.3. $K = \dfrac{IR}{E}$ for E

R.4. Solve for x. $\dfrac{16}{12} = \dfrac{24}{x}$

Concept Connections

1. If k is a nonzero constant real number, then the statement $y = kx$ implies that y varies _____ as x.

2. If k is a nonzero constant real number, then the statement $y = \frac{k}{x}$ implies that y varies _____ as x.

3. The value of k in the variation models $y = kx$ and $y = \frac{k}{x}$ is called the _____ of _____.

4. If y varies directly as two or more other variables such as x and w, then $y = kxw$, and we say that y varies _____ as x and w.

5. **a.** Given $y = 2x$, evaluate y for the given values of x: $x = 1$, $x = 2$, $x = 3$, $x = 4$, and $x = 5$.

   b. How does y change when x is doubled?

   c. How does y change when x is tripled?

   d. Complete the statement. Given $y = 2x$, when x increases, y (increases/decreases) proportionally.

   e. Complete the statement. Given $y = 2x$, when x decreases, y (increases/decreases) proportionally.

6. **a.** Given $y = \frac{24}{x}$, evaluate y for the given values of x: $x = 1$, $x = 2$, $x = 3$, $x = 4$, and $x = 6$.

   b. How does y change when x is doubled?

   c. How does y change when x is tripled?

   d. Complete the statement. Given $y = \frac{24}{x}$, when x increases, y (increases/decreases) proportionally.

   e. Complete the statement. Given $y = \frac{24}{x}$, when x decreases, y (increases/decreases) proportionally.

7. The time required to drive from Atlanta, Georgia, to Nashville, Tennessee, varies _____ as the average speed at which a vehicle travels.

8. The amount of a person's paycheck varies _____ as the number of hours worked.

9. The volume of a right circular cone varies _____ as the square of the radius of the cylinder and as the height of the cylinder.

10. A student's grade on a test varies _____ as the number of hours the student spends studying for the test.

Objective 1: Write Models Involving Direct, Inverse, and Joint Variation

For Exercises 11–20, write a variation model using k as the constant of variation. (See Examples 1–2)

11. The circumference C of a circle varies directly as its radius r.

12. Simple interest I on a loan or investment varies directly as the amount A of the loan.

13. The average cost per minute $\overline{C}$ for a flat rate cell phone plan is inversely proportional to the number of minutes used n.

14. The time of travel t is inversely proportional to the rate of travel r.

15. The volume V of a right circular cylinder varies jointly as the height h of the cylinder and as the square of the radius r of the cylinder.

16. The volume V of a rectangular solid varies jointly as the length l and width w of the solid.

17. The variable E is directly proportional to s and inversely proportional to the square root of n.

18. The variable n is directly proportional to the square of σ and inversely proportional to the square of E.

19. The variable c varies jointly as m and n and inversely as the cube of t.

20. The variable d varies jointly as u and v and inversely as the cube root of T.

Objective 2: Solve Applications Involving Variation

For Exercises 21–26, find the constant of variation k.

21. y varies directly as x. When x is 8, y is 20.

22. m varies directly as x. When x is 10, m is 42.

23. p is inversely proportional to q. When q is 18, p is 54.

24. T is inversely proportional to x. When x is 50, T is 200.

25. y varies jointly as w and v. When w is 40 and v is 0.2, y is 40.

26. N varies jointly as t and p. When t is 2 and p is 2.5, N is 15.

27. The value of y equals 4 when $x = 10$. Find y when $x = 5$ if

   a. y varies directly as x.

   b. y varies inversely as x.

28. The value of y equals 24 when x is $\frac{1}{2}$. Find y when $x = 3$ if

   a. y varies directly as x.

   b. y varies inversely as x.

For Exercises 29–48, use a variation model to solve for the unknown value.

29. The amount of a pain reliever that a physician prescribes for a child varies directly as the weight of the child. A physician prescribes 180 mg of the medicine for a 40-lb child. (**See Example 3**)

   a. How much medicine would be prescribed for a 50-lb child?

   b. How much would be prescribed for a 60-lb child?

   c. How much would be prescribed for a 70-lb child?

   d. If 135 mg of medicine is prescribed, what is the weight of the child?

30. The number of people that a ham can serve varies directly as the weight of the ham. An 8-lb ham feeds 20 people.

   a. How many people will a 10-lb ham serve?

   b. How many people will a 15-lb ham serve?

   c. How many people will an 18-lb ham serve?

   d. If a ham feeds 30 people, what is the weight of the ham?

31. A rental car company charges a fixed amount to rent a car per day. Therefore, the cost per mile to rent a car for a given day is inversely proportional to the number of miles driven. If 100 mi is driven, the average daily cost is $0.80 per mile.

   a. Find the cost per mile if 200 mi is driven.

   b. Find the cost per mile if 300 mi is driven.

   c. Find the cost per mile if 400 mi is driven.

   d. If the cost per mile is $0.16, how many miles were driven?

32. A chef self-publishes a cookbook and finds that the number of books she can sell per month varies inversely as the price of the book. The chef can sell 1500 books per month when the price is set at $8 per book.

   a. How many books would she expect to sell per month if the price were $12?

   b. How many books would she expect to sell per month if the price were $15?

   c. How many books would she expect to sell per month if the price were $6?

   d. If the chef sells 1200 books, what price was set?

33. The distance that a bicycle travels in 1 min varies directly as the number of revolutions per minute (rpm) that the wheels are turning. A bicycle with a 14-in. radius travels approximately 440 ft in 1 min if the wheels turn at 60 rpm. How far will the bicycle travel in 1 min if the wheels turn at 87 rpm?

34. The amount of pollution entering the atmosphere varies directly as the number of people living in an area. If 100,000 people create 71,000 tons of pollutants, how many tons enter the atmosphere in a city with 750,000 people?

35. The stopping distance of a car is directly proportional to the square of the speed of the car.

   a. If a car traveling 50 mph has a stopping distance of 170 ft, find the stopping distance of a car that is traveling 70 mph.

   b. If it takes 244.8 ft for a car to stop, how fast was it traveling before the brakes were applied?

36. The area of a picture projected on a wall varies directly as the square of the distance from the projector to the wall.

   a. If a 15-ft distance produces a 36 ft^2 picture, what is the area of the picture when the projection unit is moved to a distance of 25 ft from the wall?

   b. If the projected image is 144 ft^2, how far is the projector from the wall?

37. The time required to complete a job varies inversely as the number of people working on the job. It takes 8 people 12 days to do a job. (**See Example 4**)

   a. How many days will it take if 15 people work on the job?

   b. If the contractor wants to complete the job in 8 days, how many people should work on the job?

38. The yield on a bond varies inversely as the price. The yield on a particular bond is 5% when the price is $120.

   a. Find the yield when the price is $100.

   b. What price is necessary for a yield of 7.5%?

39. The current in a wire varies directly as the voltage and inversely as the resistance. If the current is 9 amperes (A) when the voltage is 90 volts (V) and the resistance is 10 ohms (Ω), find the current when the voltage is 160 V and the resistance is 5 Ω.

40. The resistance of a wire varies directly as its length and inversely as the square of its diameter. A 50-ft wire with a 0.2-in. diameter has a resistance of 0.0125 Ω. Find the resistance of a 40-ft wire with a diameter of 0.1 in.

41. The amount of simple interest owed on a loan varies jointly as the amount of principal borrowed and the amount of time the money is borrowed. If $4000 in principal results in $480 in interest in 2 yr, determine how much interest will be owed on $6000 in 4 yr.

42. The amount of simple interest earned in an account varies jointly as the amount of principal invested and the amount of time the money is invested. If $5000 in principal earns $750 in 6 yr, determine how much interest will be earned on $8000 in 4 yr.

43. The body mass index (BMI) of an individual varies directly as the weight of the individual and inversely as the square of the height of the individual. The body mass index for a 150-lb person who is 70 in. tall is 21.52. Determine the BMI for an individual who is 68 in. tall and 180 lb. Round to 2 decimal places. (**See Example 5**)

44. The strength of a wooden beam varies jointly as the width of the beam and the square of the thickness of the beam, and inversely as the length of the beam. A beam that is 48 in. long, 6 in. wide, and 2 in. thick can support a load of 417 lb. Find the maximum load that can be safely supported by a board that is 12 in. wide, 72 in. long, and 4 in. thick.

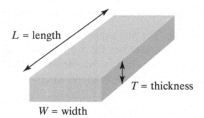

45. The speed of a racing canoe in still water varies directly as the square root of the length of the canoe. A 16-ft canoe can travel 6.2 mph in still water. Find the speed of a 25-ft canoe.

46. The period of a pendulum is the length of time required to complete one swing back and forth. The period varies directly as the square root of the length of the pendulum. If it takes 1.8 sec for a 0.81-m pendulum to complete one period, what is the period of a 1-m pendulum?

47. The cost to carpet a rectangular room varies jointly as the length of the room and the width of the room. A 10-yd by 15-yd room costs $3870 to carpet. What is the cost to carpet a room that is 18 yd by 24 yd?

48. The cost to tile a rectangular kitchen varies jointly as the length of the kitchen and the width of the kitchen. A 10-ft by 12-ft kitchen costs $1104 to tile. How much will it cost to tile a kitchen that is 20 ft by 14 ft?

Mixed Exercises

For Exercises 49–52, use the given data to find a variation model relating y to x.

49.
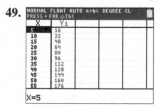

50.

X	Y₁
5	24
15	72
25	120
35	168
45	216
55	264
65	312
75	360
85	408
95	456
105	504

X=5

51.

	A	B	C	D	E
1	**x**	2	4	12	48
2	**y**	6	3	1	0.25

52.

	A	B	C	D	E
1	**x**	4	8	32	100
2	**y**	2	1	0.25	0.08

53. Which formula(s) can represent a variation model?

a. $y = kxyz$

b. $y = kx + yz$

c. $y = \dfrac{kx}{yz}$

d. $y = kx - yz$

54. Which formula(s) can represent a variation model?

a. $y = k\sqrt{x} - z^2$

b. $y = \dfrac{k\sqrt{x}}{z^2}$

c. $y = k\sqrt{xz^2}$

d. $y = k + \sqrt{xz^2}$

Write About It

For Exercises 55–56, write a statement in words that describes the variation model given. Use k as the constant of variation.

55. $P = \dfrac{kv^2}{t}$

56. $E = \dfrac{kc^2}{\sqrt{b}}$

Expanding Your Skills

57. The light from a lightbulb radiates outward in all directions.

a. Consider the interior of an imaginary sphere on which the light shines. The surface area of the sphere is directly proportional to the square of the radius. If the surface area of a sphere with a 10-m radius is 400π m^2, determine the exact surface area of a sphere with a 20-m radius.

b. Explain how the surface area changed when the radius of the sphere increased from 10 m to 20 m.

c. Based on your answer from part (b) how would you expect the intensity of light to change from a point 10 m from the lightbulb to a point 20 m from the lightbulb?

d. The intensity of light from a light source varies inversely as the square of the distance from the source. If the intensity of a lightbulb is 200 lumen/m^2 (lux) at a distance of 10 m, determine the intensity at 20 m.

58. Kepler's third law states that the square of the time T required for a planet to complete one orbit around the Sun is directly proportional to the cube of the average distance d of the planet to the Sun. For the Earth assume that $d = 9.3 \times 10^7$ mi and $T = 365$ days.

    a. Find the period of Mars, given that the distance between Mars and the Sun is 1.5 times the distance from the Earth to the Sun. Round to the nearest day.

    b. Find the average distance of Venus to the Sun, given that Venus revolves around the Sun in 223 days. Round to the nearest million miles.

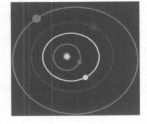

59. The intensity of radiation varies inversely as the square of the distance from the source to the receiver. If the distance is increased to 10 times its original value, what is the effect on the intensity to the receiver?

60. Suppose that y varies inversely as the cube of x. If the value of x is decreased to $\frac{1}{4}$ of its original value, what is the effect on y?

61. Suppose that y varies directly as x^2 and inversely as w^4. If both x and w are doubled, what is the effect on y?

62. Suppose that y varies directly as x^5 and inversely as w^2. If both x and w are doubled, what is the effect on y?

63. Suppose that y varies jointly as x and w^3. If x is replaced by $\frac{1}{3}x$ and w is replaced by $3w$, what is the effect on y?

64. Suppose that y varies jointly as x^4 and w. If x is replaced by $\frac{1}{4}x$ and w is replaced by $4w$, what is the effect on y?

CHAPTER 3   KEY CONCEPTS

SECTION 3.1 Quadratic Functions and Applications	Reference
Quadratic function: The function defined by $f(x) = ax^2 + bx + c$ $(a \neq 0)$ is called a **quadratic function.** A quadratic function can be written in **vertex form:** $f(x) = a(x - h)^2 + k$ by completing the square.	p. 286
• The graph of a quadratic function $f(x) = a(x - h)^2 + k$ is a parabola. • The vertex is (h, k). • If $a > 0$, the parabola opens upward, and the minimum value of the function is k. • If $a < 0$, the parabola opens downward, and the maximum value of the function is k. • The axis of symmetry is the line $x = h$. • The x-intercepts are determined by the real solutions to the equation $f(x) = 0$. • The y-intercept is determined by $f(0)$.	p. 286
Given $f(x) = ax^2 + bx + c$ $(a \neq 0)$, the vertex of the parabola is $\left(\dfrac{-b}{2a},\ f\left(\dfrac{-b}{2a}\right)\right)$.	p. 289

SECTION 3.2 Introduction to Polynomial Functions	Reference
Let n be a whole number and $a_n, a_{n-1}, a_{n-2}, \ldots, a_1, a_0$ be real numbers, where $a_n \neq 0$. Then a function defined by $$f(x) = a_n x^n + a_{n-1} x^{n-1} + a_{n-2} x^{n-2} + \cdots + a_1 x + a_0$$ is called a **polynomial function of degree n.**	p. 300
The far left and far right behavior of the graph of a polynomial function is determined by the leading term of the polynomial, $a_n x^n$. n is even and $a_n > 0$ n is even and $a_n < 0$ n is odd and $a_n > 0$ n is odd and $a_n < 0$	p. 302
The **zeros** of a polynomial function defined by $y = f(x)$ are the values of x in the domain of f for which $f(x) = 0$. These are the real solutions (or **roots**) of the equation $f(x) = 0$.	p. 303

If a polynomial function f has a factor $(x - c)$ that appears exactly k times, then c is a **zero of multiplicity k.**
  • If c is a real zero of odd multiplicity, then the graph of $y = f(x)$ *crosses* the x-axis at c.
  • If c is a real zero of even multiplicity, then the graph of $y = f(x)$ *touches* the x-axis (but does not cross) at c.
	p. 305

Intermediate value theorem: Let f be a polynomial function. For $a < b$, if $f(a)$ and $f(b)$ have opposite signs, then f has at least one zero on the interval $[a, b]$.	p. 306
The graph of a polynomial function of degree n will have at most $n - 1$ turning points.	p. 307

SECTION 3.3 Division of Polynomials and the Remainder and Factor Theorems	Reference
Long division can be used to divide two polynomials.	p. 316
Synthetic division can be used to divide polynomials if the divisor is of the form $x - c$.	p. 319
Remainder theorem: If a polynomial $f(x)$ is divided by $x - c$, then the remainder is $f(c)$.	p. 320
Factor theorem: Let $f(x)$ be a polynomial. 1. If $f(c) = 0$, then $(x - c)$ is a factor of $f(x)$. 2. If $(x - c)$ is a factor of $f(x)$, then $f(c) = 0$.	p. 322

SECTION 3.4 Zeros of Polynomials	Reference
Rational zero theorem: If $f(x) = a_nx^n + a_{n-1}x^{n-1} + \cdots + a_1x + a_0$ has integer coefficients and $a_n \neq 0$, and if $\frac{p}{q}$ (written in lowest terms) is a rational zero of f, then • p is a factor of the constant term, a_0. • q is a factor of the leading coefficient a_n.	p. 329
Fundamental theorem of algebra: If $f(x)$ is a polynomial of degree $n \geq 1$ with complex coefficients, then $f(x)$ has at least one complex zero.	p. 333
Linear factorization theorem: If $f(x) = a_nx^n + a_{n-1}x^{n-1} + \cdots + a_1x + a_0$, where $n \geq 1$ and $a_n \neq 0$, then $f(x) = a_n(x - c_1)(x - c_2) \ldots (x - c_n)$, where $c_1, c_2, \ldots, c_n$ are complex numbers.	p. 333
If $f(x)$ is a polynomial of degree $n \geq 1$ with complex coefficients, then $f(x)$ has exactly n complex zeros provided that each zero is counted by its multiplicity.	p. 333
Conjugate zeros theorem: If $f(x)$ is a polynomial with real coefficients and if $a + bi$ is a zero of $f(x)$, then its conjugate $a - bi$ is also a zero of $f(x)$.	p. 333

Descartes' rule of signs:
Let $f(x)$ be a polynomial with real coefficients and a nonzero constant term. Then,
  1. The number of *positive* real zeros is either the same as the number of sign changes in $f(x)$ or less than the number of sign changes in $f(x)$ by a positive even integer.
  2. The number of *negative* real zeros is either the same as the number of sign changes in $f(-x)$ or less than the number of sign changes in $f(-x)$ by a positive even integer.
	p. 336

Upper and lower bounds:
Let $f(x)$ be a polynomial of degree $n \geq 1$ with real coefficients and a positive leading coefficient. Further suppose that $f(x)$ is divided by $(x - c)$.
  1. If $c > 0$ and if both the remainder and the coefficients of the quotient are nonnegative, then c is an upper bound for the real zeros of $f(x)$.
  2. If $c < 0$ and the coefficients of the quotient and the remainder alternate in sign (with 0 being considered either positive or negative as needed), then c is a lower bound for the real zeros of $f(x)$.
	p. 338

SECTION 3.5 Rational Functions	Reference
Let $p(x)$ and $q(x)$ be polynomials where $q(x) \neq 0$. A function f defined by $f(x) = \dfrac{p(x)}{q(x)}$ is called a **rational function.**	p. 345
The line $x = c$ is a **vertical asymptote** of the graph of $y = f(x)$ if $f(x)$ approaches infinity or negative infinity as x approaches c from either side. To locate the vertical asymptotes of a function, determine the real numbers x where the denominator is zero, but the numerator is nonzero.	p. 347
The line $y = d$ is a **horizontal asymptote** of the graph of $y = f(x)$ if $f(x)$ approaches d as x approaches infinity or negative infinity.	p. 349
Let f be a rational function defined by $$f(x) = \frac{a_n x^n + a_{n-1} x^{n-1} + a_{n-2} x^{n-2} + \cdots + a_1 x + a_0}{b_m x^m + b_{m-1} x^{m-1} + b_{m-2} x^{m-2} + \cdots + b_1 x + b_0}$$ 1. If $n > m$, then f has no horizontal asymptote. 2. If $n < m$, then the line $y = 0$ (the x-axis) is the horizontal asymptote of f. 3. If $n = m$, then the line $y = \dfrac{a_n}{b_m}$ is the horizontal asymptote of f.	p. 349
A rational function will have a slant asymptote if the degree of the numerator is exactly one greater than the degree of the denominator. To find an equation of a slant asymptote of a rational function, divide the numerator by the denominator. The quotient will be linear and the slant asymptote will be of the form $y = $ quotient.	p. 352

SECTION 3.6 Polynomial and Rational Inequalities	Reference
Let $f(x)$ be a polynomial. Then an inequality of the form $f(x) < 0$, $f(x) > 0$, $f(x) \leq 0$, or $f(x) \geq 0$ is called a **polynomial inequality.**	p. 369
Let $f(x)$ be a rational expression. Then an inequality of the form $f(x) < 0$, $f(x) > 0$, $f(x) \leq 0$, or $f(x) \geq 0$ is called a **rational inequality.**	p. 372

Solving nonlinear inequalities:
 1. Express the inequality as $f(x) < 0, f(x) > 0, f(x) \leq 0,$ or $f(x) \geq 0$.
 2. Find the real solutions of the related equation $f(x) = 0$ and the values of x where $f(x)$ is undefined. These are the "boundary" points for the solution set to the inequality.
 3. Determine the sign of $f(x)$ on the intervals defined by the boundary points.
   • If $f(x)$ is positive, then the values of x on the interval are solutions to $f(x) > 0$.
   • If $f(x)$ is negative, then the values of x on the interval are solutions to $f(x) < 0$.
 4. Determine whether the boundary points are included in the solution set.
 5. Write the solution set.
	p. 369

SECTION 3.7 Variation	Reference
Let k be a nonzero constant real number. Then the statements on the left are equivalent to the equations on the right.	p. 383
1. y varies **directly** as x. y is **directly** proportional to x. $\Big\}\, y = kx$	
2. y varies **inversely** as x. y is **inversely** proportional to x. $\Big\}\, y = \dfrac{k}{x}$	
3. y varies **jointly** as w and x. y is **jointly** proportional to w and x. $\Big\}\, y = kwx$	p. 384
The value of k is called the **constant of variation.**	

CHAPTER 3   Review Exercises

SECTION 3.1

1. Given $f(x) = -(x + 5)^2 + 2$, identify the vertex of the graph of the parabola.

For Exercises 2–3,

a. Write the equation in vertex form: $f(x) = a(x - h)^2 + k$.

b. Determine whether the parabola opens upward or downward.

c. Identify the vertex.

d. Identify the x-intercepts.

e. Identify the y-intercept.

f. Sketch the function.

g. Determine the axis of symmetry.

h. Determine the minimum or maximum value of the function.

i. State the domain and range.

2. $f(x) = x^2 - 8x + 15$

3. $f(x) = -2x^2 + 4x + 6$

4. a. Use the vertex formula to determine the vertex of $f(x) = 2x^2 + 12x + 19$.

   b. Based on the location of the vertex and the orientation of the parabola, how many x-intercepts will the graph of $f(x) = 2x^2 + 12x + 19$ have?

5. Suppose that a farmer encloses a corral for cattle adjacent to a river. No fencing is used by the river.

   a. If he has 180 yd of fencing, what dimensions should he use to maximize the area?

   b. What is the maximum area?

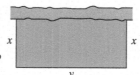

6. Suppose that p is the probability that a randomly selected person is left-handed. The value $(1 - p)$ is the probability that the person is not left-handed. In a sample of 100 people, the function $V(p) = 100p(1 - p)$ represents the variance of the number of left-handed people in a group of 100.

   a. What value of p maximizes the variance?

   b. What is the maximum variance?

7. The annual expenditure for cell phones and cellular service varies in part by the age of an individual. The average annual expenditure $E(a)$ (in $) for individuals of age a (in yr) is given in the table.
(*Source*: U.S. Bureau of Labor Statistics, www.bls.gov)

a	20	30	40	50	60	70
$E(a)$	502	658	649	627	476	213

   a. Use regression to find a quadratic function to model the data.

   b. At what age is the yearly expenditure for cell phones and cellular service the greatest? Round to the nearest year.

   c. What is the maximum yearly expenditure? Round to the nearest dollar.

SECTION 3.2

For Exercises 8–11,

a. Determine the end behavior of the graph of the function.

b. Find all the zeros of the function and state their multiplicities.

c. Determine the x-intercepts.

d. Determine the y-intercept.

e. Is the function even, odd, or neither?

f. Graph the function.

8. $f(x) = -4x^3 + 16x^2 + 25x - 100$

9. $f(x) = x^4 - 10x^2 + 9$

10. $f(x) = x^4 + 3x^3 - 3x^2 - 11x - 6$

11. $f(x) = x^5 - 8x^4 + 13x^3$

12. Determine whether the intermediate value theorem guarantees that the function has a zero on the given interval.

$$f(x) = 2x^3 - 5x^2 - 6x + 2$$

   a. $[-2, -1]$ **b.** $[-1, 0]$ **c.** $[0, 1]$ **d.** $[1, 2]$

For Exercises 13–16, determine if the statement is true or false. If a statement is false, explain why.

13. A fourth-degree polynomial has exactly three turning points.

14. A fourth-degree polynomial has at most three turning points.

15. There is exactly one polynomial with zeros of 2, 3, and 4.

16. If c is a real zero of an odd polynomial function, then $-c$ is also a zero.

SECTION 3.3

For Exercises 17–18,

a. Divide the polynomials.

b. Identify the dividend, divisor, quotient, and remainder.

17. $(-2x^4 + x^3 + 4x - 1) \div (x^2 + x - 3)$

18. $\dfrac{3x^4 - 2x^3 - 15x^2 + 22x - 8}{3x - 2}$

For Exercises 19–20, use synthetic division to divide the polynomials.

19. $(2x^5 + x^2 - 5x + 1) \div (x + 2)$

20. $\dfrac{x^4 + 3x^3 - x^2 + 7x + 2}{x - 3}$

For Exercises 21–22, use the remainder theorem to evaluate the polynomial for the given values of x.

21. $f(x) = 3x^4 + 2x^2 - 4x + 1; f(-2)$

22. $f(x) = x^4 + 2x^3 - 4x^2 - 10x - 5; f(\sqrt{5})$

For Exercises 23–24, use the remainder theorem to determine if the given number c is a zero of the polynomial.

23. $f(x) = 3x^4 + 13x^3 + 2x^2 + 52x - 40$

    a. $c = 2$ **b.** $c = \frac{2}{3}$

24. $f(x) = x^4 + 6x^3 + 9x^2 + 24x + 20$

    a. $c = -5$ **b.** $c = 2i$

For Exercises 25–26, use the factor theorem to determine if the given binomial is a factor of the polynomial.

25. $f(x) = x^3 + 4x^2 + 9x + 36$

    a. $x + 4$ **b.** $x - 3i$

26. $f(x) = x^2 - 4x - 46$

    a. $x + 2$ **b.** $x - \left(2 - 5\sqrt{2}\right)$

27. Factor $f(x) = 15x^3 - 67x^2 + 26x + 8$, given that $\frac{2}{3}$ is a zero of $f(x)$.

28. Write a third-degree polynomial $f(x)$ with zeros -1, $3\sqrt{2}$, and $-3\sqrt{2}$.

29. Write a third-degree polynomial $f(x)$ with integer coefficients and zeros of $\frac{1}{4}$, $-\frac{1}{2}$, and 3.

SECTION 3.4

30. Given $f(x) = 2x^5 - 7x^4 + 9x^3 - 18x^2 + 4x + 40$,

    a. How many zeros does $f(x)$ have (including multiplicities)?

    b. List the possible rational zeros of $f(x)$.

    c. Find all rational zeros of $f(x)$.

    d. Find all the zeros of $f(x)$.

31. Given $f(x) = x^4 + 4x^3 + 2x^2 - 8x - 8$,

    a. How many zeros does $f(x)$ have (including multiplicities)?

    b. List the possible rational zeros of $f(x)$.

    c. Find all rational zeros of $f(x)$.

    d. Find all the zeros of $f(x)$.

32. If $f(x)$ is a polynomial with real coefficients and zeros of 4 (multiplicity 3), -2 (multiplicity 1), and $2 + 7i$ (multiplicity 1), what is the minimum degree of $f(x)$?

33. Given $f(x) = x^4 - 22x^3 + 119x^2 + 66x - 366$ and that $11 - i$ is a zero of $f(x)$,

    a. Find all the zeros of $f(x)$.

    b. Factor $f(x)$ as a product of linear factors.

    c. Solve the equation $f(x) = 0$.

34. Write a polynomial $f(x)$ of lowest degree with real coefficients and with zeros $2 - 3i$ (multiplicity 1) and 0 (multiplicity 2).

35. Write a third-degree polynomial $f(x)$ with integer coefficients and with zeros of $-2i$ and $\frac{5}{3}$.

For Exercises 36–37, determine the number of possible positive and negative real zeros for the given function.

36. $g(x) = -3x^7 + 4x^6 - 2x^2 + 5x - 4$

37. $n(x) = x^6 + \frac{1}{3}x^4 + \frac{2}{7}x^3 + 4x^2 + 3$

For Exercises 38–39,

    a. Determine if the upper bound theorem identifies the given number as an upper bound for the real zeros of $f(x)$.

    b. Determine if the lower bound theorem identifies the given number as a lower bound for the real zeros of $f(x)$.

38. $f(x) = x^4 - 3x^3 + 2x - 3$

    a. 2   **b.** -2

39. $f(x) = x^3 - 4x^2 + 2x + 1$

    a. 5   **b.** -2

SECTION 3.5

40. Refer to the graph of $y = f(x)$ and complete the statements.

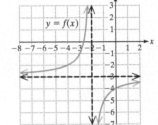

    a. As $x \to -\infty$,

        $f(x) \to$ _____

    b. As $x \to -2^-$,

        $f(x) \to$ _____

    c. As $x \to -2^+$,

        $f(x) \to$ _____

    d. As $x \to \infty$,

        $f(x) \to$ _____

    e. The graph is increasing over the interval(s) _____.

    f. The graph is decreasing over the interval(s) _____.

    g. The domain is _____.

    h. The range is _____.

    i. The vertical asymptote is the line _____.

    j. The horizontal asymptote is the line _____.

For Exercises 41–42, determine the vertical asymptotes of the graph of the function.

41. $f(x) = \dfrac{x + 4}{2x^2 + x - 15}$ **42.** $g(x) = \dfrac{5}{x^2 + 3}$

For Exercises 43–45,

    a. Determine the horizontal asymptotes (if any).

    b. If the graph of the function has a horizontal asymptote, determine the point where the graph crosses the horizontal asymptote.

43. $r(x) = \dfrac{3}{x^2 + 2x + 1}$ **44.** $q(x) = \dfrac{-2x^2 - 3x + 4}{x^2 + 1}$

45. $k(x) = \dfrac{x^3 + 4}{x + 1}$

For Exercises 46–47, identify all asymptotes (vertical, horizontal, and slant).

46. $m(x) = \dfrac{2x^3 - x^2 - 6x + 7}{x^2 - 3}$

47. $n(x) = \dfrac{-4x^2 + 5}{3x^2 - 14x - 5}$

For Exercises 48–51, graph the function.

48. $f(x) = \dfrac{1}{x - 4} + 2$ **49.** $k(x) = \dfrac{x^2}{x^2 - x - 12}$

50. $m(x) = \dfrac{x^2 + 6x + 9}{x}$ **51.** $q(x) = \dfrac{12}{x^2 + 6}$

52. After taking a certain class, the percentage of material retained $P(t)$ decreases with the number of months t after taking the class. $P(t)$ can be approximated by

$$P(t) = \dfrac{t + 90}{0.16t + 1}$$

a. Determine the percentage retained after 1 month, 4 months, and 6 months. Round to the nearest percent.

b. As t becomes infinitely large, what percentage of material will be retained?

SECTION 3.6

53. The graph of $y = f(x)$ is given. Solve the inequalities.

a. $f(x) < 0$

b. $f(x) \le 0$

c. $f(x) > 0$

d. $f(x) \ge 0$

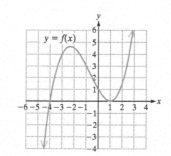

54. The graph of $y = f(x)$ is given. Solve the inequalities.

a. $f(x) < 0$

b. $f(x) \le 0$

c. $f(x) > 0$

d. $f(x) \ge 0$

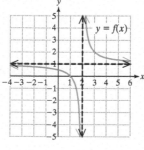

55. Solve the equation and inequalities.

a. $x^2 + 7x + 10 = 0$ **b.** $x^2 + 7x + 10 < 0$

c. $x^2 + 7x + 10 \le 0$ **d.** $x^2 + 7x + 10 > 0$

e. $x^2 + 7x + 10 \ge 0$

56. Solve the inequalities.

a. $\dfrac{x + 1}{x - 5} \le 0$ **b.** $\dfrac{x + 1}{x - 5} < 0$

c. $\dfrac{x + 1}{x - 5} \ge 0$ **d.** $\dfrac{x + 1}{x - 5} > 0$

For Exercises 57–66, solve the inequalities.

57. $t(t - 3) \ge 18$ **58.** $w^3 + w^2 - 9w - 9 > 0$

59. $x^2 - 2x + 4 \le 3$ **60.** $-6x^4(3x - 4)^2(x + 2)^3 \le 0$

61. $z^3 - 3z^2 > 10z - 24$ **62.** $(4x - 5)^4 > 0$

63. $\dfrac{6 - 2x}{x^2} \ge 0$ **64.** $\dfrac{8}{3x - 4} \le 1$

65. $\dfrac{3}{x - 2} < -\dfrac{2}{x}$ **66.** $\dfrac{(1 - x)(3x + 5)^2}{(x - 3)^4} < 0$

67. A sports trainer has monthly costs of $80 for phone service and $40 for his website and advertising. In addition he pays a $15 fee to the gym for each session in which he works with a client.

a. Write a function representing the average cost $\overline{C}(x)$ (in $) for x training sessions.

b. Find the number of sessions the trainer needs if he wants the average cost to drop below $16 per session.

68. A child throws a ball straight upwards to his friend who is sitting in a tree 18 ft above ground level.

a. If the ball leaves the child's hand at a height of 2 ft with an initial speed of 40 ft/sec, write a function representing the vertical position of the ball $s(t)$ (in ft) in terms of the time t after the ball leaves the child's hand. (*Hint*: Use the model $s(t) = -\frac{1}{2}gt^2 + v_0t + s_0$ with $g = 32$ ft/sec^2. See page 375.)

b. Determine the time interval for which the ball will be more than 18 ft high.

SECTION 3.7

For Exercises 69–71, write a variation model using k as the constant of variation.

69. The mass m of an animal varies directly as the weight w of the animal's heart.

70. The value of x varies inversely to the square of p.

71. The variable y is jointly proportional to x and the square root of z, and inversely proportional to the cube of t.

For Exercises 72–73, determine the constant of variation k.

72. The variable Q varies jointly as p and the square root of t. The value of Q is 132 when p is 11 and t is 9.

73. The variable d is directly proportional to c and inversely proportional to the square of x. The value of d is 1.8 when c is 3 and x is 2.

74. The weight of a ball varies directly as the cube of its radius. A weighted exercise ball of radius 3 in. weighs 3.24 lb. How much would a ball weigh if its radius were 5 in.?

75. In karate, the force F required to break a board varies inversely as the length L of the board. If it takes 6.25 lb of force to break a board 1.6 ft long, determine how much force is required to break a 2-ft board.

76. The power in an electric circuit varies jointly as the current and the square of the resistance. If the power is 144 watts (W) when the current is 4 A and the resistance is 6 Ω, find the power when the current is 3 A and the resistance is 10 Ω.

77. Coulomb's law states that the force F of attraction between two oppositely charged particles varies jointly as the magnitude of their electrical charges q_1 and q_2 and inversely as the square of the distance d between the particles. Find the effect on F of doubling q_1 and q_2 and halving the distance between them.

CHAPTER 3 Test

1. Given $f(x) = 2x^2 - 12x + 16$,

   a. Write the equation in vertex form: $f(x) = a(x - h)^2 + k$.

   b. Determine whether the parabola opens upward or downward.

   c. Identify the vertex.

   d. Identify the x-intercepts.

   e. Identify the y-intercept.

   f. Sketch the function.

   g. Determine the axis of symmetry.

   h. Determine the minimum or maximum value of the function.

   i. State the domain and range.

2. Given $f(x) = 2x^4 - 5x^3 - 17x^2 + 41x - 21$,

   a. Determine the end behavior of the graph of the function.

   b. List all possible rational zeros.

   c. Find all the zeros of the function and state their multiplicities.

   d. Determine the x-intercepts.

   e. Determine the y-intercept.

   f. Is the function even, odd, or neither?

   g. Graph the function.

3. Given $f(x) = -0.25x^3(x - 2)^2(x + 1)^4$,

   a. Identify the leading term.

   b. Determine the end behavior of the graph of the function.

   c. Find all the zeros of the function and state their multiplicities.

4. Given $f(x) = x^4 + 5x^2 - 36$,

   a. How many zeros does $f(x)$ have (including multiplicities)?

   b. Find the zeros of $f(x)$.

   c. Identify the x-intercepts of the graph of f.

   d. Is the function even, odd, or neither?

5. Determine whether the intermediate value theorem guarantees that the function has a zero on the given interval.

$$f(x) = x^3 - 5x^2 + 2x + 5$$

   a. $[-2, -1]$ b. $[-1, 0]$ c. $[0, 1]$ d. $[1, 2]$

6. a. Divide the polynomials. $\dfrac{2x^4 - 4x^3 + x - 5}{x^2 - 3x + 1}$

   b. Identify the dividend, divisor, quotient, and remainder.

7. Given $f(x) = 5x^4 + 47x^3 + 80x^2 - 51x - 9$,

   a. Is $\dfrac{3}{5}$ a zero of $f(x)$?

   b. Is -1 a zero of $f(x)$?

   c. Is $(x + 1)$ a factor of $f(x)$?

   d. Is $(x + 3)$ a factor of $f(x)$?

   e. Use the remainder theorem to evaluate $f(-2)$.

8. Given $f(x) = x^4 - 8x^3 + 21x^2 - 32x + 68$ and that $2i$ is a zero of $f(x)$,

   a. Find all zeros of $f(x)$.

   b. Factor $f(x)$ as a product of linear factors.

   c. Solve the equation $f(x) = 0$.

9. Given $f(x) = 3x^4 + 7x^3 - 12x^2 - 14x + 12$,

   a. How many zeros does $f(x)$ have (including multiplicities)?

   b. List the possible rational zeros.

   c. Determine if the upper bound theorem identifies 2 as an upper bound for the real zeros of $f(x)$.

   d. Determine if the lower bound theorem identifies -4 as a lower bound for the real zeros of $f(x)$.

   e. Revise the list of possible rational zeros based on the answers to parts (c) and (d).

   f. Find the rational zeros.

   g. Find all the zeros.

   h. Graph the function.

10. Write a third-degree polynomial $f(x)$ with integer coefficients and zeros of $\frac{1}{5}$, $-\frac{2}{3}$, and 4.

11. Determine the number of possible positive and negative real zeros for $f(x) = -6x^7 - 4x^5 + 2x^4 - 3x^2 + 1$.

For Exercises 12–14, determine the asymptotes (vertical, horizontal, and slant).

12. $r(x) = \dfrac{2x^2 - 3x + 5}{x - 7}$ **13.** $p(x) = \dfrac{-3x + 1}{4x^2 - 1}$

14. $n(x) = \dfrac{5x^2 - 2x + 1}{3x^2 + 4}$

For Exercises 15–17, graph the function.

15. $m(x) = -\dfrac{1}{x^2} + 3$ **16.** $h(x) = \dfrac{-4}{x^2 - 4}$

17. $k(x) = \dfrac{x^2 - 2x + 1}{x}$

For Exercises 18–24, solve the inequality.

18. $c^2 < c + 20$

19. $y^3 > 13y - 12$

20. $-2x(x - 4)^2(x + 1)^3 \le 0$

21. $9x^2 + 42x + 49 > 0$

22. $\dfrac{x + 3}{2 - x} \le 0$

23. $\dfrac{-4}{x^2 - 9} \ge 0$

24. $\dfrac{4}{x - 1} < -\dfrac{3}{x}$

25. Write a variation model using k as the constant of variation: Energy E varies directly as the square of the velocity v of the wind.

26. Solve for the constant of variation k: The variable w varies jointly as y and the square root of x, and inversely as z. The value of w is 7.2 when x is 4, y is 6, and z is 7.

27. The surface area of a cube varies directly as the square of the length of an edge. The surface area is 24 ft^2 when the length of an edge is 2 ft. Find the surface area of a cube with an edge that is 7 ft.

28. The weight of a body varies inversely as the square of its distance from the center of the Earth. The radius of the Earth is approximately 4000 mi. How much would a 180-lb man weigh 20 mi above the surface of the Earth? Round to the nearest pound.

29. The pressure of wind on a wall varies jointly as the area of the wall and the square of the velocity of the wind. If the velocity of the wind is tripled, what is the effect on the pressure on the wall?

30. The population $P(t)$ of rabbits in a wildlife area t years after being introduced to the area is given by

$$P(t) = \dfrac{2000t}{t + 1}$$

   a. Determine the number of rabbits after 1 yr, 5 yr, and 10 yr. Round to the nearest whole unit.

   b. What will the rabbit population approach as t approaches infinity?

31. An agricultural school wants to determine the number of corn plants per acre that will produce the maximum yield. The model $y(n) = -0.103n^2 + 8.32n + 15.1$ represents the yield $y(n)$ (in bushels per acre) based on n thousand plants per acre.

   a. Evaluate $y(20)$, $y(30)$, and $y(60)$ and interpret their meaning in the context of this problem.

   b. Determine the number of plants per acre that will maximize yield. Round to the nearest hundred plants.

   c. What is the maximum yield? Round to the nearest bushel per acre.

32. Suppose that a rocket is shot straight upward from ground level with an initial speed of 98 m/sec.

   a. Write a model that represents the height of the rocket $s(t)$ (in meters) t seconds after launch. (*Hint:* Use the model $s(t) = -\dfrac{1}{2}gt^2 + v_0t + s_0$ with $g = 9.8$ m/sec^2. See page 376.)

   b. When will the rocket reach its maximum height?

   c. What is the maximum height?

   d. Determine the time interval for which the rocket will be more than 200 m high. Round to the nearest tenth of a second.

33. The number of yearly visits to physicians' offices varies in part by the age of the patient. For the data shown in the table, a represents the age of patients (in yr) and $n(a)$ represents the corresponding number of visits to physicians' offices per year. (*Source:* Centers for Disease Control, www.cdc.gov)

a	8	20	35	55	65	85
$n(a)$	2.7	2.0	2.5	3.7	6.7	7.6

   a. Use regression to find a quadratic function to model the data.

   b. At what age is the number of yearly visits to physicians' offices the least? Round to the nearest year of age.

   c. What is the minimum number of yearly visits? Round to 1 decimal place.

CHAPTER 3   Cumulative Review Exercises

1. Given $r(x) = \dfrac{2x^2 - 3}{x^2 - 16}$,

    a. Find the vertical asymptotes.

    b. Find the horizontal asymptote or slant asymptote.

2. Find a polynomial $f(x)$ of lowest degree with real coefficients and with zeros of $3 + 2i$ and 2.

3. Given $f(x) = 2x^3 - x^2 - 8x - 5$,

    a. Determine the end behavior of the graph of f.

    b. Find the zeros and their multiplicities.

    c. Find the x-intercepts.

    d. Find the y-intercept.

    e. Graph the function.

4. Divide and write the answer in the form $a + bi$.

$$\frac{3 + 2i}{4 - i}$$

5. Determine the center and radius of the circle given by

$$x^2 + y^2 + 8x - 14y + 56 = 0$$

6. Write an equation of the line passing through the points $(4, -8)$ and $(2, -3)$. Write the answer in slope-intercept form.

7. Determine the x- and y-intercepts of the graph of $x = y^2 - 9$.

8. Graph $f(x) = \begin{cases} -x - 1 & \text{for } x < 1 \\ \sqrt{x - 1} & \text{for } x \geq 1 \end{cases}$

9. Solve the equation for m. $v_0 t = \sqrt{m - t}$

10. Given $f(x) = 2x^2 - 6x + 1$,

    a. Find the x-intercepts.

    b. Find the y-intercept.

    c. Find the vertex of the parabola.

11. Factor. $125x^6 - y^9$

For Exercises 12–14, simplify the expression.

12. $\left(\dfrac{4x^3 y^{-5}}{z^{-2}}\right)^{-3} \left(\dfrac{4y^{-6}}{x^{-12}}\right)^{1/2}$

13. $\sqrt[3]{250 z^5 x y^{21}}$

14. $\dfrac{\dfrac{1}{3x} - \dfrac{1}{x^2}}{\dfrac{1}{3} - \dfrac{3}{x^2}}$

For Exercises 15–20, solve the equations and inequalities. Write the solutions to the inequalities in interval notation if possible.

15. $-5 \leq -\dfrac{1}{4}x + 3 < \dfrac{1}{2}$

16. $|x - 3| + 4 \leq 10$

17. $|2x + 1| = |x - 4|$

18. $c^2 - 5c + 9 < c(c + 3)$

19. $\dfrac{49x^2 + 14x + 1}{x} > 0$

20. $\sqrt{4x - 3} - \sqrt{x + 12} = 0$

Exponential and Logarithmic Functions

Visible light from the Sun is vitally important for the health of an ocean, lake, or any body of water. In particular, light penetrating through a body of water provides the energy to fuel vast amounts of microscopic plants called phytoplankton that are an essential source of food and oxygen for an aquatic ecosystem. Phytoplankton converts energy from the Sun to usable energy for plant growth. Thus, the amount of light directly affects plant productivity at the base of the food chain and ultimately animal life farther up the food chain.

With increasing depth, the percentage of visible light from the surface of a body of water drops exponentially. This means that light intensity drops quickly at first and then drops more slowly with increasing depth. (More specifically, light intensity decreases at a rate proportional to the intensity at a particular depth.) To study this phenomenon, scientists use exponential functions and their inverses, logarithmic functions. These two important categories of functions have many applications including the study of the decay of radioactive substances, short-term population growth, and the growth of investments subject to compound interest.

Inverse Functions

1. Identify One-to-One Functions

Throughout our study of algebra, we have made use of the fact that the operations of addition and subtraction are inverse operations. For example, adding 5 to a number and then subtracting 5 from the result gives us the original number. Likewise, multiplication and division are inverse operations. We now look at the concept of an inverse function.

Changing currency is an important consideration when traveling abroad. For example, traveling between the United States and several countries in Europe would involve changing American dollars to Euros and then changing back for the return trip. Fortunately, we can use a function to change from one currency to the other, and then use the function's *inverse* to change back again.

Suppose that $1 (American dollar) can be exchanged for 0.8 € (Euro). Then,

$f(x) = 0.8x$ gives the number of Euros $f(x)$ that can be bought from x dollars.

$g(x) = \dfrac{x}{0.8}$ gives the number of dollars $g(x)$ that can be bought from x Euros.

Tables 4-1 and 4-2 show the values of $f(x)$ and $g(x)$ for several values of x.

Table 4-1

x (Dollars)	$f(x) = 0.8x$ (Euros)
100	80
150	120
200	160
250	200

Table 4-2

x (Euros)	$g(x) = \dfrac{x}{0.8}$ (Dollars)
80	100
120	150
160	200
200	250

In this example, functions f and g are inverses of each other, and we observe several interesting characteristics about inverse functions.

- By listing the ordered pairs from Tables 4-1 and 4-2, notice that the x and y values are reversed.

$$f: \quad \{(100, 80), (150, 120), (200, 160), (250, 200)\}$$

$$g: \quad \{(80, 100), (120, 150), (160, 200), (200, 250)\}$$

- For a function and its inverse, the values of x and y are interchanged. This tells us that the domain of a function is the same as the range of its inverse and vice versa.
- From the graphs of f and g (Figure 4-1), we see that the corresponding points on f and g are symmetric with respect to the line $y = x$.
- When we compose functions f and g in both directions, the result is the input value x. In a sense, what function f does to x, function g "undoes" and vice versa.

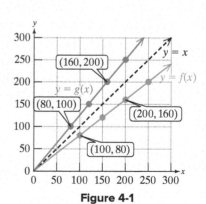

Figure 4-1

$$(f \circ g)(x) = f[g(x)] = 0.8\left(\frac{x}{0.8}\right) = x$$

$$(g \circ f)(x) = g[f(x)] = \frac{(0.8x)}{0.8} = x$$

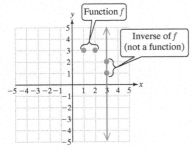

Figure 4-2

The inverse of any relation is found by interchanging the values of x and y in the relation. However, the inverse of a function may itself not be a function. For example, consider $f = \{(1, 3), (2, 3)\}$ shown in blue in Figure 4-2. The inverse is the set of ordered pairs $\{(3, 1), (3, 2)\}$ shown in red. Notice that the relation defining the inverse of f is not a function because it fails the vertical line test.

The function $f = \{(1, 3), (2, 3)\}$ has two points that are aligned horizontally. When the x and y values are reversed to form the inverse, the resulting points will be aligned vertically and will fail the vertical line test. Thus, a function will have an inverse function only if no points in the original function are aligned horizontally. That is, no two distinct points on the function may have the same y value. In such a case, the function is said to be one-to-one.

> **Definition of a One-to-One Function**
>
> A function f is a **one-to-one function,** if for a and b in the domain of f,
>
> if $a \neq b$, then $f(a) \neq f(b)$, or equivalently, if $f(a) = f(b)$, then $a = b$.

The definition of a one-to-one function tells us that each y value in the range is associated with only one x value in the domain. This implies that the graph of a one-to-one function will have no two points aligned horizontally.

> **Horizontal Line Test for a One-to-One Function**
>
> A function defined by $y = f(x)$ is a one-to-one function if no horizontal line intersects the graph in more than one point.

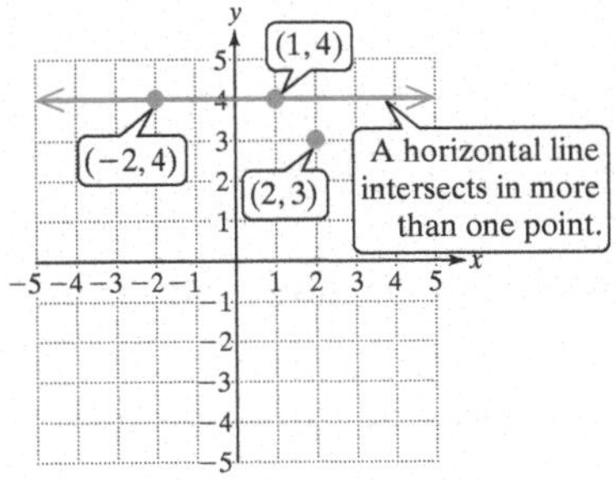

Figure 4-3

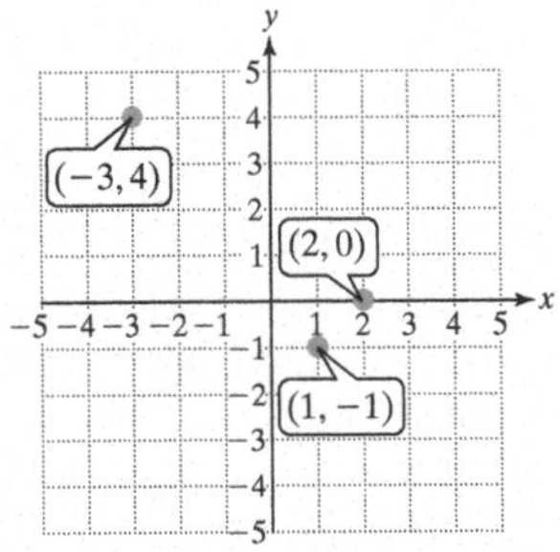

EXAMPLE 1 **Determining Whether a Function is One-to-One**

Determine whether the function is one-to-one.

a. $f = \{(1, 4), (2, 3), (-2, 4)\}$ **b.** $g = \{(-3, 4), (1, -1), (2, 0)\}$

Solution:

same y value

a. $f = \{(1, 4), (2, 3), (-2, 4)\}$

different x value

f is not a one-to-one function.

The ordered pairs $(1, 4)$ and $(-2, 4)$ have the same y value but different x values. That is, $f(1) = f(-2)$, but $1 \neq -2$.

A horizontal line passes through the function in more than one point (Figure 4-3).

All points have different y-values.

b. $g = \{(-3, 4), (1, -1), (2, 0)\}$

g is a one-to-one function.

Each unique ordered pair has a different y value, so the function is one-to-one.

No horizontal line passes through the function in more than one point (Figure 4-4).

Figure 4-4

> **Skill Practice 1** Determine whether the function is one-to-one.
>
> **a.** $h = \{(4, -5), (6, 1), (2, 4), (0, -3)\}$ **b.** $k = \{(1, 0), (3, 0), (4, -5)\}$

Answers
1. a. Yes   **b.** No

EXAMPLE 2 **Using the Horizontal Line Test**

Use the horizontal line test to determine if the graph in blue defines y as a one-to-one function of x.

Solution:

a.
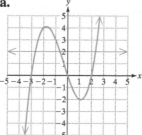

b.
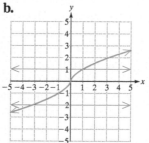

c.
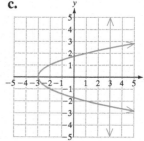

The graph does not define y as a one-to-one function of x because a horizontal line intersects the graph in more than one point.

The graph does define y as a one-to-one function of x because no horizontal line intersects the graph in more than one point.

The relation does not define y as a function of x, because it fails the vertical line test. If the relation is not a function, it is not a one-to-one function.

Skill Practice 2 Use the horizontal line test to determine if the graph defines y as a one-to-one function of x.

a.
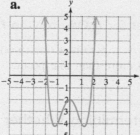

b.
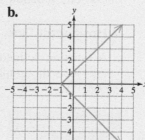

c.
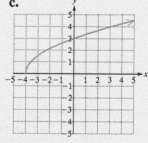

In Example 3, we use algebraic methods to determine whether a function is one-to-one.

EXAMPLE 3 **Determining Whether a Function Is One-to-One**

Use the definition of a one-to-one function to determine whether the function is one-to-one.

a. $f(x) = 2x - 3$ **b.** $f(x) = x^2 + 1$

Solution:

a. We must show that if $f(a) = f(b)$, then $a = b$.

Assume that $f(a) = f(b)$. That is,

$$2a - 3 = 2b - 3$$
$$2a - 3 + 3 = 2b - 3 + 3$$
$$\frac{2a}{2} = \frac{2b}{2}$$
$$a = b$$

The logic of this algebraic proof begins with the assumption that $f(a) = f(b)$, that is, that two y values are equal. For a one-to-one function, this can happen only if the x values (in this case a and b) are the same.

Otherwise, if $a \neq b$, we would have the same y value with two different x values and f would not be one-to-one.

Since $f(a) = f(b)$ implies that $a = b$, then f is one-to-one.

Answers
2. a. No **b.** No **c.** Yes

TIP To show that a function is not one-to-one, we only need one counter-example. That is, any pair of distinct points on the function that have different y values is sufficient to show that the function is not one-to-one.

b. $f(x) = x^2 + 1$

Assume that $\quad f(a) = f(b)$.
$$a^2 + 1 = b^2 + 1$$
$$a^2 = b^2$$
$$a = \pm b$$

For nonzero values of b, $f(a) = f(b)$ does not necessarily imply that $a = b$. Therefore, f is not one-to-one.

From the graph of $f(x) = x^2 + 1$, we see that f is not one-to-one (Figure 4-5). We can also show this algebraically by finding two ordered pairs with the same y value but different x values. From the graph, we have arbitrarily selected $(-2, 5)$ and $(2, 5)$.

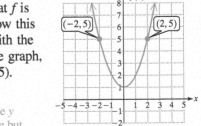

Figure 4-5

If $a = 2$ and $b = -2$, we have:

$$f(a) = f(2) = (2)^2 + 1 = 5 \leftarrow$$
$$f(b) = f(-2) = (-2)^2 + 1 = 5 \leftarrow$$

Same y value but different x values

We have that $f(a) = f(b)$, but $a \neq b$.

Therefore, f fails to be a one-to-one function.

Skill Practice 3   Determine whether the function is one-to-one.

   a. $f(x) = -4x + 1$ **b.** $f(x) = |x| - 3$

---

2. Determine Whether Two Functions Are Inverses

We now have enough background to define an inverse function.

Definition of an Inverse Function

Let f be a one-to-one function. Then g is the **inverse of f** if the following conditions are both true.

   1. $(f \circ g)(x) = x$ for all x in the domain of g.
   2. $(g \circ f)(x) = x$ for all x in the domain of f.

Avoiding Mistakes

Do not confuse inverse notation f^{-1} with exponential notation. The notation f^{-1} does not mean $\frac{1}{f}$.

We should also note that if g is the inverse of f, then f is the inverse of g. Furthermore, given a function f, we often denote its inverse as f^{-1}. So given a function f and its inverse f^{-1}, the definition implies that

$$(f \circ f^{-1})(x) = x \text{ and } (f^{-1} \circ f)(x) = x$$

EXAMPLE 4   **Determining Whether Two Functions Are Inverses**

Determine whether the functions are inverses.

   a. $f(x) = 100 + 12x$ and $g(x) = \dfrac{x - 100}{12}$

   b. $h(x) = \sqrt[3]{x - 1}$ and $k(x) = -1 + x^3$

Answers

3. a. Yes   **b.** No

Solution:

a. $(f \circ g)(x) = f(g(x))$

$$= f\left(\frac{x - 100}{12}\right)$$

$$= 100 + 12\left(\frac{x - 100}{12}\right)$$

$$= 100 + (x - 100)$$

$$= x \checkmark$$

$(g \circ f)(x) = g(f(x))$

$$= g(100 + 12x)$$

$$= \frac{(100 + 12x) - 100}{12}$$

$$= \frac{12x}{12}$$

$$= x \checkmark$$

Since $(f \circ g)(x) = (g \circ f)(x) = x$, f and g are inverses.

b. $(h \circ k)(x) = h(k(x))$

$$= \sqrt[3]{(-1 + x^3)} - 1$$

$$= \sqrt[3]{x^3 - 2} \neq x$$

If either $(h \circ k)(x) \neq x$ or $(k \circ h)(x) \neq x$, then h and k are *not* inverses.

Since $(h \circ k)(x) \neq x$, h and k are not inverses.

Skill Practice 4 Determine whether the functions are inverses.

a. $f(x) = \dfrac{x + 6}{2}$ and $g(x) = 2(x - 6)$ **b.** $m(x) = \dfrac{5}{x - 2}$ and $n(x) = \dfrac{2x + 5}{x}$

---

3. Find the Inverse of a Function

For a one-to-one function defined by $y = f(x)$, the inverse is a function $y = f^{-1}(x)$ that performs the inverse operations in the reverse order. The function given by $f(x) = 100 + 12x$ multiplies x by 12 first, and then adds 100 to the result. Therefore, the inverse function must *subtract* 100 from x first and then *divide* by 12.

$$f^{-1}(x) = \frac{x - 100}{12}$$

To facilitate the process of finding an equation of the inverse of a one-to-one function, we offer the following steps.

Procedure to Find an Equation of an Inverse of a Function

For a one-to-one function defined by $y = f(x)$, the equation of the inverse can be found as follows.

Step 1 Replace $f(x)$ by y.
Step 2 Interchange x and y.
Step 3 Solve for y.
Step 4 Replace y by $f^{-1}(x)$.

EXAMPLE 5 Finding an Equation of an Inverse Function

Write an equation for the inverse function for $f(x) = 3x - 1$.

Answers

4. a. No **b.** Yes

Solution:

Function f is a linear function, and its graph is a nonvertical line. Therefore, f is a one-to-one function.

$$f(x) = 3x - 1$$
$$y = 3x - 1 \qquad \text{\textbf{Step 1:} Replace } f(x) \text{ by } y.$$
$$x = 3y - 1 \qquad \text{\textbf{Step 2:} Interchange } x \text{ and } y.$$
$$x + 1 = 3y \qquad \text{\textbf{Step 3:} Solve for } y. \text{ Add 1 to both sides and divide by 3.}$$
$$\frac{x + 1}{3} = y$$
$$f^{-1}(x) = \frac{x + 1}{3} \qquad \text{\textbf{Step 4:} Replace } y \text{ by } f^{-1}(x).$$

To check the result, verify that $(f \circ f^{-1})(x) = x$ and $(f^{-1} \circ f)(x) = x$.

$$(f \circ f^{-1})(x) = 3\left(\frac{x + 1}{3}\right) - 1 = x \checkmark \quad \text{and} \quad (f^{-1} \circ f)(x) = \frac{(3x - 1) + 1}{3} = x \checkmark$$

Skill Practice 5   Write an equation for the inverse function for $f(x) = 4x + 3$.

TIP We can sometimes find an equation of an inverse function by mentally reversing the operations given in the original function. In Example 5, the function f multiplies x by 3 and then subtracts 1. Therefore, f^{-1} must add 1 to x and then divide by 3.

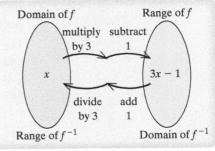

The key step in determining the equation of the inverse of a function is to interchange x and y. By so doing, a point (a, b) on f corresponds to a point (b, a) on f^{-1}. This is why the graphs of f and f^{-1} are symmetric with respect to the line $y = x$. From Example 5, notice that the point $(2, 5)$ on the graph of f corresponds to the point $(5, 2)$ on the graph of f^{-1} (Figure 4-6).

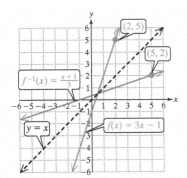

Figure 4-6

EXAMPLE 6   **Finding an Equation of an Inverse Function**

Write an equation for the inverse function for the one-to-one function defined by $f(x) = \dfrac{3 - x}{x + 3}$.

TIP In Example 6, we can show that f is a one-to-one function by graphing the function (see Section 3.5). Or we can show that $f(a) = f(b)$ implies that $a = b$ by solving the equation $\dfrac{3 - a}{a + 3} = \dfrac{3 - b}{b + 3}$ for a or b to show that $a = b$.

Solution:

$$f(x) = \frac{3 - x}{x + 3}$$
$$y = \frac{3 - x}{x + 3} \qquad \text{\textbf{Step 1:} Replace } f(x) \text{ by } y.$$
$$x = \frac{3 - y}{y + 3} \qquad \text{\textbf{Step 2:} Interchange } x \text{ and } y.$$

Answer

5. $f^{-1}(x) = \dfrac{x - 3}{4}$

$$x(y + 3) = 3 - y$$ **Step 3:** Solve for y.
                        Clear fractions (multiply both sides by $y + 3$).

$$xy + 3x = 3 - y$$ Apply the distributive property.

$$xy + y = 3 - 3x$$ Collect the y terms on one side.

$$y(x + 1) = 3 - 3x$$ Factor out y as the greatest common factor.

$$y = \frac{3 - 3x}{x + 1}$$ Divide both sides by $x + 1$.

$$f^{-1}(x) = \frac{3 - 3x}{x + 1}$$ **Step 4:** Replace y by $f^{-1}(x)$.

Skill Practice 6   Write an equation for the inverse function for the one-to-one function defined by $f(x) = \dfrac{x - 2}{x + 2}$.

For a function that is not one-to-one, sometimes we restrict its domain to create a new function that is one-to-one. This is demonstrated in Example 7.

EXAMPLE 7   **Finding an Equation of an Inverse Function**

Given $m(x) = x^2 + 4$ for $x \geq 0$, write an equation of the inverse.

Solution:

The graph of $y = x^2 + 4$ is a parabola with vertex $(0, 4)$. See Figure 4-7. The function is not one-to-one. However, with the restriction on the domain that $x \geq 0$, the graph consists of only the right branch of the parabola (Figure 4-8). This *is* a one-to-one function.

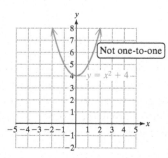

Figure 4-7

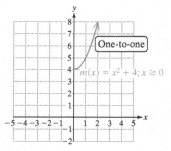

Figure 4-8

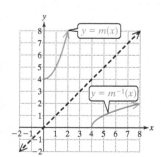

Figure 4-9

To find the inverse, we have

$$m(x) = x^2 + 4; \quad x \geq 0$$

$$y = x^2 + 4; \quad x \geq 0$$ **Step 1:** Replace $m(x)$ by y.

$$x = y^2 + 4 \quad y \geq 0$$ **Step 2:** Interchange x and y. Notice that the restriction $x \geq 0$ becomes $y \geq 0$.

$$x - 4 = y^2$$ **Step 3:** Solve for y by subtracting 4 from both sides.

$$y = \pm\sqrt{x - 4}$$ Apply the square root property.

$$y = +\sqrt{x - 4}$$ Choose the positive square root of $(x - 4)$ because of the restriction $y \geq 0$.

$$m^{-1}(x) = \sqrt{x - 4}$$ **Step 4:** Replace y by $m^{-1}(x)$.

The graphs of m and m^{-1} are symmetric with respect to the line $y = x$ as expected (Figure 4-9).

Answer

6. $f^{-1}(x) = -\dfrac{2x + 2}{x - 1}$

Skill Practice 7 Given $n(x) = x^2 + 1$ for $x \le 0$, write an equation of the inverse.

EXAMPLE 8 **Finding an Equation of an Inverse Function**

Given $f(x) = \sqrt{x - 1}$, find an equation of the inverse.

Solution:

The function f is a one-to-one function and the graph is the same as the graph of $y = \sqrt{x}$ with a shift 1 unit to the right. The domain of f is $\{x \mid x \ge 1\}$ and the range is $\{y \mid y \ge 0\}$. When defining the inverse, we will have the conditions that $x \ge 0$ and $y \ge 1$.

> **TIP** When finding the inverse of a function, the key step of interchanging x and y has the effect of interchanging the domain and range between the function and its inverse.

$$f(x) = \sqrt{x - 1} \qquad \text{Note that } x \ge 1 \text{ and } y \ge 0.$$
$$y = \sqrt{x - 1}$$
$$x = \sqrt{y - 1} \qquad \text{Interchange } x \text{ and } y.$$
$$\qquad\qquad\qquad \text{Note that } y \ge 1 \text{ and } x \ge 0.$$
$$x^2 = y - 1 \qquad \text{Square both sides.}$$
$$y = x^2 + 1$$

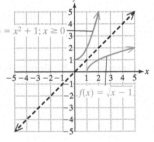

$$f^{-1}(x) = x^2 + 1, \quad x \ge 0 \qquad$$ The restriction $x \ge 0$ on f^{-1} is necessary because f has the restriction that $y \ge 0$. Furthermore, $y = x^2 + 1$ is not a one-to-one function without a restricted domain.

Skill Practice 8 Given $g(x) = \sqrt{x + 2}$, find an equation of the inverse.

Answers
7. $n^{-1}(x) = -\sqrt{x - 1}$
8. $g^{-1}(x) = x^2 - 2; x \ge 0$

SECTION 4.1   Practice Exercises

Prerequisite Review

For Exercises R.1–R.3, find the domain. Write the answer in interval notation.

R.1. $f(x) = \dfrac{x + 4}{x + 1}$ 　　　　**R.2.** $k(p) = \dfrac{p + 1}{p^2 + 4}$ 　　　　**R.3.** $m(x) = \sqrt{2 - 8x}$

For Exercises R.4–R.6, refer to functions n and p to evaluate the function.

$n(x) = x + 1 \quad p(x) = x^2 - 3x$

R.4. $(n \circ p)(x)$ 　　　　**R.5.** $(p \circ n)(x)$ 　　　　**R.6.** $(p \circ p)(x)$

Concept Connections

1. Given the function $f = \{(1, 2), (2, 3), (3, 4)\}$ write the set of ordered pairs representing f^{-1}.

2. The graphs of a function and its inverse are symmetric with respect to the line _____.

3. If no horizontal line intersects the graph of a function f in more than one point, then f is a _____ - _____ - _____ function.

4. Given a one-to-one function f, if $f(a) = f(b)$, then a _____ b.

5. Let f be a one-to-one function and let g be the inverse of f. Then $(f \circ g)(x) =$ _____ and $(g \circ f)(x) =$ _____.

6. If (a, b) is a point on the graph of a one-to-one function f, then the corresponding ordered pair _____ is a point on the graph of f^{-1}.

Objective 1: Identify One-to-One Functions

For Exercises 7–12, a relation in x and y is given. Determine if the relation defines y as a one-to-one function of x. (See Example 1)

7. $\{(6, -5), (4, 2), (3, 1), (8, 4)\}$

8. $\{(-14, 1), (-2, 3), (7, 4), (-9, -2)\}$

9.

x	y
0.6	1.8
1	-1.1
0.5	1.8
2.4	0.7

10.

x	y
12.5	3.21
5.75	-4.5
2.34	7.25
-12.7	3.21

11.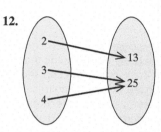

12.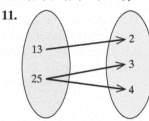

For Exercises 13–22, determine if the relation defines y as a one-to-one function of x. (See Example 2)

13.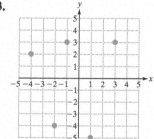

14.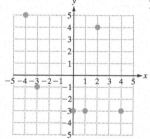

15.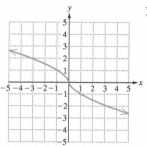

16.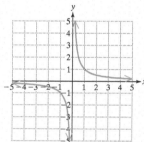

17.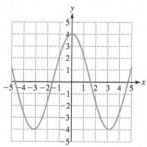

18.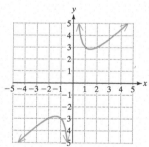

19.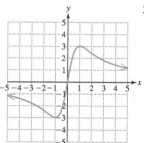

20.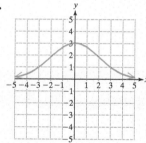

21.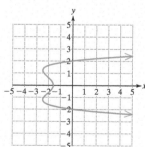

22.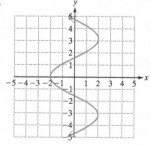

For Exercises 23–30, use the definition of a one-to-one function to determine if the function is one-to-one. (See Example 3)

23. $f(x) = 4x - 7$

24. $h(x) = -3x + 2$

25. $g(x) = x^3 + 8$

26. $k(x) = x^3 - 27$

27. $m(x) = x^2 - 4$

28. $n(x) = x^2 + 1$

29. $p(x) = |x + 1|$

30. $q(x) = |x - 3|$

Objective 2: Determine Whether Two Functions Are Inverses

For Exercises 31–36, determine whether the two functions are inverses. (See Example 4)

31. $f(x) = 5x + 4$ and $g(x) = \dfrac{x - 4}{5}$

32. $h(x) = 7x - 3$ and $k(x) = \dfrac{x + 3}{7}$

33. $m(x) = \dfrac{-2 + x}{6}$ and $n(x) = 6x - 2$

34. $p(x) = \dfrac{-3 + x}{4}$ and $q(x) = 4x - 3$

35. $t(x) = \dfrac{4}{x - 1}$ and $v(x) = \dfrac{x + 4}{x}$

36. $w(x) = \dfrac{6}{x + 2}$ and $z(x) = \dfrac{6 - 2x}{x}$

37. There were 2000 applicants for enrollment to the freshman class at a small college in the year 2010. The number of applications has risen linearly by roughly 150 per year. The number of applications $f(x)$ is given by $f(x) = 2000 + 150x$, where x is the number of years since 2010.

    a. Determine if the function $g(x) = \dfrac{x - 2000}{150}$ is the inverse of f.

    b. Interpret the meaning of function g in the context of this problem.

38. The monthly sales for January for a whole foods market was \$60,000 and has increased linearly by \$2500 per month. The amount in sales $f(x)$ (in \$) is given by $f(x) = 60{,}000 + 2500x$, where x is the number of months since January.

    a. Determine if the function $g(x) = \dfrac{x - 60{,}000}{2500}$ is the inverse of f.

    b. Interpret the meaning of function g in the context of this problem.

Objective 3: Find the Inverse of a Function

39. a. Show that $f(x) = 2x - 3$ defines a one-to-one function.

    b. Write an equation for $f^{-1}(x)$.

    c. Graph $y = f(x)$ and $y = f^{-1}(x)$ on the same coordinate system.

40. a. Show that $f(x) = 4x + 4$ defines a one-to-one function.

    b. Write an equation for $f^{-1}(x)$.

    c. Graph $y = f(x)$ and $y = f^{-1}(x)$ on the same coordinate system.

For Exercises 41–52, a one-to-one function is given. Write an equation for the inverse function. (See Examples 5–6)

41. $f(x) = \dfrac{4 - x}{9}$

42. $g(x) = \dfrac{8 - x}{3}$

43. $h(x) = \sqrt[3]{x - 5}$

44. $k(x) = \sqrt[3]{x + 8}$

45. $m(x) = 4x^3 + 2$

46. $n(x) = 2x^3 - 5$

47. $c(x) = \dfrac{5}{x + 2}$

48. $s(x) = \dfrac{2}{x - 3}$

49. $t(x) = \dfrac{x - 4}{x + 2}$

50. $v(x) = \dfrac{x - 5}{x + 1}$

51. $f(x) = \dfrac{(x - a)^3}{b} - c$

52. $g(x) = b(x + a)^3 + c$

53. a. Graph $f(x) = x^2 - 3$; $x \le 0$. **(See Example 7)**

    b. From the graph of f, is f a one-to-one function?

    c. Write the domain of f in interval notation.

    d. Write the range of f in interval notation.

    e. Write an equation for $f^{-1}(x)$.

    f. Graph $y = f(x)$ and $y = f^{-1}(x)$ on the same coordinate system.

    g. Write the domain of f^{-1} in interval notation.

    h. Write the range of f^{-1} in interval notation.

54. a. Graph $f(x) = x^2 + 1$; $x \le 0$.

    b. From the graph of f, is f a one-to-one function?

    c. Write the domain of f in interval notation.

    d. Write the range of f in interval notation.

    e. Write an equation for $f^{-1}(x)$.

    f. Graph $y = f(x)$ and $y = f^{-1}(x)$ on the same coordinate system.

    g. Write the domain of f^{-1} in interval notation.

    h. Write the range of f^{-1} in interval notation.

55. a. Graph $f(x) = \sqrt{x} + 1$. (**See Example 8**)

    b. From the graph of f, is f a one-to-one function?

    c. Write the domain of f in interval notation.

    d. Write the range of f in interval notation.

    e. Write an equation for $f^{-1}(x)$.

    f. Explain why the restriction $x \geq 0$ is placed on f^{-1}.

    g. Graph $y = f(x)$ and $y = f^{-1}(x)$ on the same coordinate system.

    h. Write the domain of f^{-1} in interval notation.

    i. Write the range of f^{-1} in interval notation.

56. a. Graph $f(x) = \sqrt{x} - 2$.

    b. From the graph of f, is f a one-to-one function?

    c. Write the domain of f in interval notation.

    d. Write the range of f in interval notation.

    e. Write an equation for $f^{-1}(x)$.

    f. Explain why the restriction $x \geq 0$ is placed on f^{-1}.

    g. Graph $y = f(x)$ and $y = f^{-1}(x)$ on the same coordinate system.

    h. Write the domain of f^{-1} in interval notation.

    i. Write the range of f^{-1} in interval notation.

57. Given that the domain of a one-to-one function f is $[0, \infty)$ and the range of f is $[0, 4)$, state the domain and range of f^{-1}.

58. Given that the domain of a one-to-one function f is $[-3, 5)$ and the range of f is $(-2, \infty)$, state the domain and range of f^{-1}.

59. Given $f(x) = |x| + 3$; $x \leq 0$, write an equation for f^{-1}. (*Hint*: Sketch $f(x)$ and note the domain and range.)

60. Given $f(x) = |x| - 3$; $x \geq 0$, write an equation for f^{-1}. (*Hint*: Sketch $f(x)$ and note the domain and range.)

For Exercises 61–66, fill in the blanks and determine an equation for $f^{-1}(x)$ mentally.

61. If function f adds 6 to x, then f^{-1} _____ 6 from x. Function f is defined by $f(x) = x + 6$, and function f^{-1} is defined by $f^{-1}(x) =$ _____.

62. If function f multiplies x by 2, then f^{-1} _____ x by 2. Function f is defined by $f(x) = 2x$, and function f^{-1} is defined by $f^{-1}(x) =$ _____.

63. Suppose that function f multiplies x by 7 and subtracts 4. Write an equation for $f^{-1}(x)$.

64. Suppose that function f divides x by 3 and adds 11. Write an equation for $f^{-1}(x)$.

65. Suppose that function f cubes x and adds 20. Write an equation for $f^{-1}(x)$.

66. Suppose that function f takes the cube root of x and subtracts 10. Write an equation for $f^{-1}(x)$.

For Exercises 67–70, find the inverse mentally.

67. $f(x) = 8x + 1$

68. $p(x) = 2x - 10$

69. $q(x) = \sqrt[5]{x - 4} + 1$

70. $m(x) = \sqrt[3]{4x} + 3$

Mixed Exercises

For Exercises 71–74, the graph of a function is given. Graph the inverse function.

71.

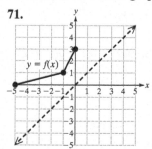

72.

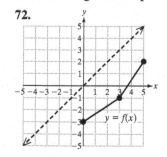

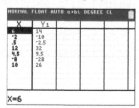

73.

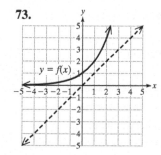

74.

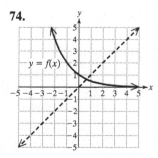

For Exercises 75–76, the table defines $Y_1 = f(x)$ as a one-to-one function of x. Find the values of f^{-1} for the selected values of x.

75. a. $f^{-1}(32)$

    b. $f^{-1}(-2.5)$

    c. $f^{-1}(26)$

NORMAL FLOAT AUTO a+bi DEGREE CL	
X	Y₁
6	14
-2	-10
.5	-2.5
12	32
4.5	9.5
-8	-28
10	26
X=6	

76. a. $f^{-1}(5)$

    b. $f^{-1}(9.45)$

    c. $f^{-1}(8)$

NORMAL FLOAT AUTO a+bi DEGREE CL	
X	Y₁
-3	10.75
4	9
2.2	9.45
8	8
-12	13
20	5
4.6	8.85
X=-3	

For Exercises 77–80, determine if the statement is true or false. If a statement is false, explain why.

77. All linear functions with a nonzero slope have an inverse function.

78. The domain of any one-to-one function is the same as the domain of its inverse function.

79. The range of a one-to-one function is the same as the range of its inverse function.

80. No quadratic function defined by $f(x) = ax^2 + bx + c$ $(a \neq 0)$ is one-to-one.

81. Based on data from Hurricane Katrina, the function defined by $w(x) = -1.17x + 1220$ gives the wind speed $w(x)$ (in mph) based on the barometric pressure x (in millibars, mb).

   a. Approximate the wind speed for a hurricane with a barometric pressure of 1000 mb.

   b. Write a function representing the inverse of w and interpret its meaning in context.

   c. Approximate the barometric pressure for a hurricane with wind speed 100 mph. Round to the nearest mb.

82. The function defined by $F(x) = \dfrac{9}{5}x + 32$ gives the temperature $F(x)$ (in degrees Fahrenheit) based on the temperature x (in Celsius).

   a. Determine the temperature in Fahrenheit if the temperature in Celsius is 25°C.

   b. Write a function representing the inverse of F and interpret its meaning in context.

   c. Determine the temperature in Celsius if the temperature in Fahrenheit is 5°F.

83. Suppose that during normal respiration, the volume of air inhaled per breath (called "tidal volume") by a mammal of any size is 6.33 mL per kilogram of body mass.

   a. Write a function representing the tidal volume $T(x)$ (in mL) of a mammal of mass x (in kg).

   b. Write an equation for $T^{-1}(x)$.

   c. What does the inverse function represent in the context of this problem?

   d. Find $T^{-1}(170)$ and interpret its meaning in context. Round to the nearest whole unit.

84. At a cruising altitude of 35,000 ft, a certain airplane travels 555 mph.

   a. Write a function representing the distance $d(x)$ (in mi) for x hours at cruising altitude.

   b. Write an equation for $d^{-1}(x)$.

   c. What does the inverse function represent in the context of this problem?

   d. Evaluate $d^{-1}(2553)$ and interpret its meaning in context.

85. The millage rate is the amount of property tax per $1000 of the taxable value of a home. For a certain county the millage rate is 24 mil ($24 in tax per $1000 of taxable value of the home). A city within the county also imposes a flat fee of $108 per home.

   a. Write a function representing the total amount of property tax $T(x)$ (in $) for a home with a taxable value of x thousand dollars.

   b. Write an equation for $T^{-1}(x)$.

   c. What does the inverse function represent in the context of this problem?

   d. Evaluate $T^{-1}(2988)$ and interpret its meaning in context.

86. Beginning on January 1, park rangers in Everglades National Park began recording the water level for one particularly dry area of the park. The water level was initially 2.5 ft and decreased by approximately 0.015 ft/day.

   a. Write a function representing the water level $L(x)$ (in ft), x days after January 1.

   b. Write an equation for $L^{-1}(x)$.

   c. What does the inverse function represent in the context of this problem?

   d. Evaluate $L^{-1}(1.9)$ and interpret its meaning in context.

Write About It

87. Explain the relationship between the domain and range of a one-to-one function f and its inverse f^{-1}.

88. Write an informal definition of a one-to-one function.

89. Explain why if a horizontal line intersects the graph of a function in more than one point, then the function is not one-to-one.

90. Explain why the domain of $f(x) = x^2 + k$ must be restricted to find an inverse function.

Expanding Your Skills

91. Consider a function defined as follows: Given x, the value $f(x)$ is the exponent above the base of 2 that produces x. For example, $f(16) = 4$ because $2^4 = 16$. Evaluate

a. $f(8)$ **b.** $f(32)$

c. $f(2)$ **d.** $f\left(\frac{1}{8}\right)$

93. Show that every increasing function is one-to-one.

92. Consider a function defined as follows: Given x, the value $f(x)$ is the exponent above the base of 3 that produces x. For example, $f(9) = 2$ because $3^2 = 9$. Evaluate

a. $f(27)$ **b.** $f(81)$

c. $f(3)$ **d.** $f\left(\frac{1}{9}\right)$

94. A function is said to be periodic if there exists some nonzero real number p, called the period, such that $f(x + p) = f(x)$ for all real numbers x in the domain of f. Explain why no periodic function is one-to-one.

SECTION 4.2   Exponential Functions

OBJECTIVES

1. **Graph Exponential Functions**
2. **Evaluate the Exponential Function Base e**
3. **Use Exponential Functions to Compute Compound Interest**
4. **Use Exponential Functions in Applications**

1. Graph Exponential Functions

The concept of a function was first introduced in Section 2.3. Since then we have learned to recognize several categories of functions. In this section and the next, we will define two new types of functions called exponential functions and logarithmic functions.

To introduce exponential functions, consider two salary plans for a new job. Plan A pays $1 million for 1 month's work. Plan B starts with 2¢ on the first day, and every day thereafter the salary is doubled. At first glance, the million-dollar plan appears to be more favorable. However, Table 4-3 shows otherwise. The daily payments for 30 days are listed for Plan B.

Table 4-3

Day	Payment	Day	Payment	Day	Payment
1	2¢	11	$20.48	21	$20,971.52
2	4¢	12	$40.96	22	$41,943.04
3	8¢	13	$81.92	23	$83,886.08
4	16¢	14	$163.84	24	$167,772.16
5	32¢	15	$327.68	25	$335,544.32
6	64¢	16	$655.36	26	$671,088.64
7	$1.28	17	$1310.72	27	$1,342,177.28
8	$2.56	18	$2621.44	28	$2,684,354.56
9	$5.12	19	$5242.88	29	$5,368,709.12
10	$10.24	20	$10,485.76	30	$10,737,418.24

TIP Consider the pattern involved for the payment for day x = 1, 2, 3, 4, 5, ...

2^1 ¢ = 2¢
2^2 ¢ = 4¢
2^3 ¢ = 8¢
2^4 ¢ = 16¢
2^5 ¢ = 32¢

The salary for the 30th day for Plan B is over $10 million. Taking the sum of the payments, we see that the total salary for the 30-day period is $21,474,836.46.

The daily salary $S(x)$ (in ¢) for Plan B can be represented by the function $S(x) = 2^x$, where x is the number of days on the job. An interesting characteristic of this function is that for every positive 1-unit change in x, the function value doubles. The function $S(x) = 2^x$ is called an exponential function.

Definition of an Exponential Function

Let b be a constant real number such that $b > 0$ and $b \neq 1$. Then for any real number x, a function of the form $f(x) = b^x$ is called an **exponential function of base b.**

An exponential function is recognized as a function with a constant base (positive and not equal to 1) with a variable exponent, x.

Exponential Functions	Not Exponential Functions	
$f(x) = 3^x$	$m(x) = x^2$	base is not constant
$g(x) = \left(\dfrac{1}{3}\right)^x$	$n(x) = \left(-\dfrac{1}{3}\right)^x$	base is negative
$h(x) = \left(\sqrt{2}\right)^x$	$p(x) = 1^x$	base is 1

Avoiding Mistakes

- The base of an exponential function must not be negative to avoid situations where the function values are not real numbers. For example, $f(x) = (-4)^x$ is not defined for $x = \frac{1}{2}$ because $\sqrt{-4}$ is not a real number.
- The base of an exponential function must not equal 1 because $f(x) = 1^x = 1$ for all real numbers x. This is a constant function, not an exponential function.

At this point in the text, we have evaluated exponential expressions with integer exponents and with rational exponents. For example,

$$4^2 = 16 \qquad\qquad 4^{1/2} = \sqrt{4} = 2$$

$$4^{-1} = \frac{1}{4} \qquad\qquad 4^{10/23} = \sqrt[23]{4^{10}} \approx 1.827112184$$

However, how do we evaluate an exponential expression with an *irrational* exponent such as 4^π? In such a case, the exponent is a nonterminating, nonrepeating decimal. We define an exponential expression raised to an irrational exponent as a sequence of approximations using rational exponents. For example:

$$4^{3.14} \approx 77.7084726$$
$$4^{3.141} \approx 77.81627412$$
$$4^{3.1415} \approx 77.87023095$$
$$\cdots$$
$$4^\pi \approx 77.88023365$$

With this definition of a base raised to an irrational exponent, we can define an exponential function over the entire set of real numbers. In Example 1, we graph two exponential functions by plotting points.

EXAMPLE 1 Graphing Exponential Functions

Graph the functions.

a. $f(x) = 2^x$ **b.** $g(x) = \left(\dfrac{1}{2}\right)^x$

Solution:

Table 4-4 shows several function values $f(x)$ and $g(x)$ for both positive and negative values of x.

Table 4-4

x	$f(x) = 2^x$	$g(x) = \left(\frac{1}{2}\right)^x$
-3	$\frac{1}{8}$	8
-2	$\frac{1}{4}$	4
-1	$\frac{1}{2}$	2
0	1	1
1	2	$\frac{1}{2}$
2	4	$\frac{1}{4}$
3	8	$\frac{1}{8}$

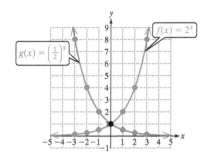

Figure 4-10

> **TIP** The values of $f(x)$ become closer and closer to 0 as $x \to -\infty$. This means that the x-axis is a horizontal asymptote.
> Likewise, the values of $g(x)$ become closer to 0 as $x \to \infty$. The x-axis is a horizontal asymptote.

Notice that $g(x) = \left(\frac{1}{2}\right)^x$ is equivalent to $g(x) = 2^{-x}$. Therefore, the graph of $g(x) = \left(\frac{1}{2}\right)^x = 2^{-x}$ is the same as the graph of $f(x) = 2^x$ with a reflection across the y-axis (Figure 4-10).

Skill Practice 1 Graph the functions.

a. $f(x) = 5^x$ **b.** $g(x) = \left(\frac{1}{5}\right)^x$

The graphs in Figure 4-10 illustrate several important features of exponential functions.

Graphs of $f(x) = b^x$

The graph of an exponential function defined by $f(x) = b^x$ ($b > 0$ and $b \neq 1$) has the following properties.

1. If $b > 1$, f is an *increasing* exponential function, sometimes called an **exponential growth function.**
   If $0 < b < 1$, f is a *decreasing* exponential function, sometimes called an **exponential decay function.**
2. The domain is the set of all real numbers, $(-\infty, \infty)$.
3. The range is $(0, \infty)$.
4. The line $y = 0$ (x-axis) is a horizontal asymptote.
5. The function passes through the point $(0, 1)$ because $f(0) = b^0 = 1$.

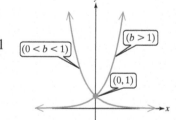

Answer

1.

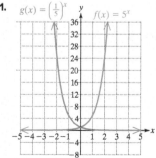

These properties indicate that the graph of an exponential function is an increasing function if the base is greater than 1. Furthermore, the base affects the rate of increase. Consider the graphs of $f(x) = 2^x$ and $k(x) = 5^x$ (Figure 4-11). For every positive 1-unit change in x, $f(x) = 2^x$ is 2 times as great and $k(x) = 5^x$ is 5 times as great (Table 4-5).

Table 4-5

x	$f(x) = 2^x$	$k(x) = 5^x$
-3	$\frac{1}{8}$	$\frac{1}{125}$
-2	$\frac{1}{4}$	$\frac{1}{25}$
-1	$\frac{1}{2}$	$\frac{1}{5}$
0	1	1
1	2	5
2	4	25
3	8	125

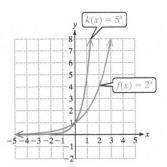

Figure 4-11

In Example 2, we use the transformations of functions learned in Section 2.6 to graph an exponential function.

If $h > 0$, shift to the right.
If $h < 0$, shift to the left.

$$f(x) = ab^{x-h} + k$$

If $a < 0$, reflect across the x-axis.
Shrink vertically if $0 < |a| < 1$.
Stretch vertically if $|a| > 1$.

If $k > 0$, shift upward.
If $k < 0$, shift downward.

EXAMPLE 2   **Graphing an Exponential Function**

Graph. $f(x) = 3^{x-2} + 4$

Solution:

The graph of f is the graph of the parent function $y = 3^x$ shifted 2 units to the right and 4 units upward.

The parent function $y = 3^x$ is an increasing exponential function. We can plot a few points on the graph of $y = 3^x$ and use these points and the horizontal asymptote to form the outline of the transformed graph.

x	$y = 3^x$
-2	$\frac{1}{9}$
-1	$\frac{1}{3}$
0	1
1	3
2	9

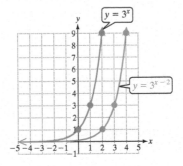

Shift 2 units to the right.
For example, the point
$(0, 1)$ on $y = 3^x$
corresponds to $(2, 1)$
on $y = 3^{x-2}$.

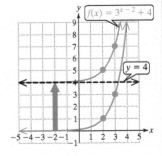

Shift the graph of
$y = 3^{x-2}$ up 4 units.
Notice that with the
vertical shift, the new
horizontal asymptote
is $y = 4$.

Answer

2.

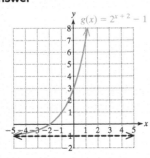

Skill Practice 2   Graph. $g(x) = 2^{x+2} - 1$

2. Evaluate the Exponential Function Base e

We now introduce an important exponential function whose base is an irrational number called e. Consider the expression $\left(1 + \dfrac{1}{x}\right)^x$. The value of the expression for increasingly large values of x approaches a constant (Table 4-6).

As $x \to \infty$, the expression $\left(1 + \dfrac{1}{x}\right)^x$ approaches

a constant value that we call e. From Table 4-6, this value is approximately 2.718281828.

$$e \approx 2.718281828$$

The value of e is an irrational number (a non-terminating, nonrepeating decimal) and like the number π, it is a universal constant. The function defined by $f(x) = e^x$ is called the exponential function base e or the **natural exponential function.**

Table 4-6

x	$\left(1 + \dfrac{1}{x}\right)^x$
100	2.70481382942
1000	2.71692393224
10,000	2.71814592683
100,000	2.71826823717
1,000,000	2.71828046932
1,000,000,000	2.71828182710

EXAMPLE 3 Graphing $f(x) = e^x$

Graph the function. $f(x) = e^x$

Solution:

Because the base e is greater than 1 ($e \approx 2.718281828$), the graph is an increasing exponential function. We can use a calculator to evaluate $f(x) = e^x$ at several values of x. On many calculators, the exponential function, base e, is invoked by selecting 2ND LN or by accessing e^x on the keyboard.

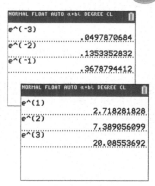

x	$f(x) = e^x$
-3	0.050
-2	0.135
-1	0.368
0	1.000
1	2.718
2	7.389
3	20.086

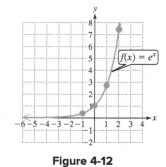

Figure 4-12

TIP In Section 4.3, we will see that the exponential function base e is the inverse of the natural logarithmic function, $y = \ln x$. This is why the exponential function base e is accessed with the 2ND LN keys.

The graph of $f(x) = e^x$ is shown in Figure 4-12.

Skill Practice 3 Explain how the graph of $f(x) = -e^{x-1}$ is related to the graph of $y = e^x$.

---

3. Use Exponential Functions to Compute Compound Interest

Recall that simple interest is interest computed on the principal amount invested (or borrowed). Compound interest is interest computed on both the original principal and the interest already accrued.

Answer

3. The graph of $f(x) = -e^{x-1}$ is the graph of $y = e^x$ with a shift to the right 1 unit and a reflection across the x-axis.

Suppose that interest is compounded annually (one time per year) on an investment of P dollars at an annual interest rate r for t years. Then the amount A (in \$) in the account after 1, 2, and 3 yr is computed as follows.

After 1 yr: $\begin{pmatrix} \text{Total} \\ \text{amount} \end{pmatrix} = \begin{pmatrix} \text{Initial} \\ \text{principal} \end{pmatrix} + (\text{Interest})$

$$A = P + Pr$$

> The interest is given by $I = Prt$, where $t = 1$ yr. So $I = Pr$.

$$= P(1 + r) \qquad \text{Factor out } P.$$

After 2 yr: $\begin{pmatrix} \text{Total} \\ \text{amount} \end{pmatrix} = \begin{pmatrix} \text{Year 1} \\ \text{balance} \end{pmatrix} + \begin{pmatrix} \text{Interest on} \\ \text{Year 1 balance} \end{pmatrix}$

$$A = P(1 + r) + [P(1 + r)]r$$

$$= P(1 + r)(1 + r) \qquad \text{Factor out } P(1 + r).$$

$$= P(1 + r)^2$$

After 3 yr: $\begin{pmatrix} \text{Total} \\ \text{amount} \end{pmatrix} = \begin{pmatrix} \text{Year 2} \\ \text{balance} \end{pmatrix} + \begin{pmatrix} \text{Interest on} \\ \text{Year 2 balance} \end{pmatrix}$

$$A = P(1 + r)^2 + [P(1 + r)^2]r$$

$$= P(1 + r)^2(1 + r) \qquad \text{Factor out } P(1 + r)^2.$$

$$= P(1 + r)^3$$

...

After t years: $\quad A = P(1 + r)^t$

> Amount in an account with interest compounded annually.

Compound interest is often computed more frequently during the course of 1 yr. Let n represent the number of compounding periods per year. For example:

$n = 1$ for interest compounded annually
$n = 4$ for interest compounded quarterly
$n = 12$ for interest compounded monthly
$n = 365$ for interest compounded daily

Each compounding period represents a fraction of a year, and the interest rate is scaled accordingly for each compounding period as $\frac{1}{n} \cdot r$ or $\frac{r}{n}$. The number of compounding periods over the course of the investment is nt. Therefore, to determine the amount in an account where interest is compounded n times per year, we have

replace t by nt

$$A = P(1 + r)^t \qquad\qquad A = P\left(1 + \frac{r}{n}\right)^{nt}$$

> Amount in an account with interest compounded n times per year.

replace r by $\frac{r}{n}$

Now suppose it were possible to compute interest continuously, that is, for $n \to \infty$. If we use the substitution $x = \frac{n}{r}$ (which implies that $n = xr$), the formula for compound interest becomes

$$A = P\left(1 + \frac{r}{n}\right)^{nt} \xrightarrow{\text{Substitute } x = \frac{n}{r}} P\left(1 + \frac{1}{x}\right)^{xrt} = P\left[\left(1 + \frac{1}{x}\right)^x\right]^{rt}$$

For a fixed interest rate r, as n approaches infinity, x also approaches infinity. Since the expression $\left(1 + \frac{1}{x}\right)^x$ approaches e as $x \to \infty$, we have

$$A = Pe^{rt}$$

> Amount in an account with interest compounded continuously.

Summary of Formulas Relating to Simple and Compound Interest

Suppose that P dollars in principal is invested (or borrowed) at an annual interest rate r for t years. Then

- $I = Prt$ Amount of simple interest I (in \$).

- $A = P\left(1 + \dfrac{r}{n}\right)^{nt}$ The future value A (in \$) of the account after t years with n compounding periods per year.

- $A = Pe^{rt}$ The future value A (in \$) of the account after t years under continuous compounding.

In Example 4, we compare the value of an investment after 10 yr under several different compounding options.

EXAMPLE 4   Computing the Balance on an Account

Suppose that \$5000 is invested and pays 6.5% per year under the following compounding options.

  a. Compounded annually   **b.** Compounded quarterly
  c. Compounded monthly   **d.** Compounded daily
  e. Compounded continuously

Determine the total amount in the account after 10 yr with each option.

Solution:

Using $A = P\left(1 + \dfrac{r}{n}\right)^{nt}$ and $A = Pe^{rt}$, we have

Compounding Option	n Value	Formula	Result
Annually	$n = 1$	$A = 5000\left(1 + \dfrac{0.065}{1}\right)^{(1 \cdot 10)}$	\$9385.69
Quarterly	$n = 4$	$A = 5000\left(1 + \dfrac{0.065}{4}\right)^{(4 \cdot 10)}$	\$9527.79
Monthly	$n = 12$	$A = 5000\left(1 + \dfrac{0.065}{12}\right)^{(12 \cdot 10)}$	\$9560.92
Daily	$n = 365$	$A = 5000\left(1 + \dfrac{0.065}{365}\right)^{(365 \cdot 10)}$	\$9577.15
Continuously	Not applicable	$A = 5000e^{(0.065 \cdot 10)}$	\$9577.70

Notice that there is a \$192.01 difference in the account balance between annual compounding and continuous compounding. The table also supports our finding that

$$A = P\left(1 + \frac{r}{n}\right)^{nt} \quad \text{converges to} \quad A = Pe^{rt} \quad \text{as } n \to \infty.$$

Skill Practice 4   Suppose that \$8000 is invested and pays 4.5% per year under the following compounding options.

  a. Compounded annually   **b.** Compounded quarterly
  c. Compounded monthly   **d.** Compounded daily
  e. Compounded continuously

Determine the total amount in the account after 5 yr with each option.

Answers
4. a. \$9969.46   **b.** \$10,006.00
  c. \$10,014.37   **d.** \$10,018.44
  e. \$10,018.58

4. Use Exponential Functions in Applications

Increasing and decreasing exponential functions can be used in a variety of real-world applications. For example:

- Population growth can often be modeled by an exponential function.
- The growth of an investment under compound interest increases exponentially.
- The mass of a radioactive substance decreases exponentially with time.
- The temperature of a cup of coffee decreases exponentially as it approaches room temperature.

A substance that undergoes radioactive decay is said to be radioactive. The **half-life** of a radioactive substance is the amount of time it takes for one-half of the original amount of the substance to change into something else. That is, after each half-life, the amount of the original substance decreases by one-half.

EXAMPLE 5   **Using an Exponential Function in an Application**

The half-life of radium 226 is 1620 yr. In a sample originally having 1 g of radium 226, the amount $A(t)$ (in grams) of radium 226 present after t years is given by $A(t) = \left(\frac{1}{2}\right)^{t/1620}$ where t is the time in years after the start of the experiment. How much radium will be present after

a. 1620 yr?   **b.** 3240 yr?   **c.** 4860 yr?

Solution:

a. $A(t) = \left(\frac{1}{2}\right)^{t/1620}$

$$A(1620) = \left(\frac{1}{2}\right)^{1620/1620}$$
$$= \left(\frac{1}{2}\right)^{1}$$
$$= 0.5$$

b. $A(t) = \left(\frac{1}{2}\right)^{t/1620}$

$$A(3240) = \left(\frac{1}{2}\right)^{3240/1620}$$
$$= \left(\frac{1}{2}\right)^{2}$$
$$= 0.25$$

c. $A(t) = \left(\frac{1}{2}\right)^{t/1620}$

$$A(4860) = \left(\frac{1}{2}\right)^{4860/1620}$$
$$= \left(\frac{1}{2}\right)^{3}$$
$$= 0.125$$

The half-life of radium is 1620 yr. Therefore, we can interpret these results as follows.

After 1620 yr (1 half-life), 0.5 g remains ($\frac{1}{2}$ of the original amount remains).

After 3240 yr (2 half-lives), 0.25 g remains ($\frac{1}{4}$ of the original amount remains).

After 4860 yr (3 half-lives), 0.125 g remains ($\frac{1}{8}$ of the original amount remains).

Skill Practice 5   Cesium-137 is a radioactive metal with a short half-life of 30 yr. In a sample originally having 2 g of cesium-137, the amount $A(t)$ (in grams) of cesium-137 present after t years is given by $A(t) = 2\left(\frac{1}{2}\right)^{t/30}$. How much cesium-137 will be present after

a. 30 yr?   **b.** 60 yr?   **c.** 90 yr?

Point of Interest

In 1898, Marie Curie discovered the highly radioactive element radium. She shared the 1903 Nobel Prize in physics for her research on radioactivity and was awarded the 1911 Nobel Prize in chemistry for her discovery of radium and polonium. Marie Curie died in 1934 from complications of excessive exposure to radiation.

Marie and Pierre Curie

Answers
5. a. 1 g   **b.** 0.5 g   **c.** 0.25 g

TECHNOLOGY CONNECTIONS

Graphing an Exponential Function

A graphing utility can be used to graph and analyze exponential functions. The table shows several values of $A(x) = \left(\frac{1}{2}\right)^{x/1620}$ for selected values of x.

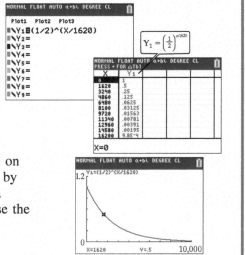

The graph of $A(x) = \left(\frac{1}{2}\right)^{x/1620}$ is shown on the viewing window [0, 10,000, 1000] by [0, 1.2, 0.2]. Notice that the graph is a decreasing exponential function because the base is between 0 and 1.

SECTION 4.2   Practice Exercises

Prerequisite Review

R.1. Determine the domain and range.

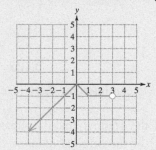

R.2. Refer to the graph of the function and complete the statement.

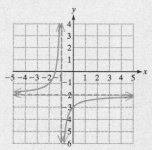

R.3. Use transformations to graph $q(x) = -(x - 3)^2 + 4$.

a. As $x \to -\infty, f(x) \to$ _____

b. As $x \to -1^+, f(x) \to$ _____

c. As $x \to \infty, f(x) \to$ _____

d. As $x \to -1^-, f(x) \to$ _____

e. The graph is decreasing over the interval(s) _____.

f. The graph is increasing over the interval(s) _____.

Concept Connections

1. The function defined by $y = x^3$ (is/is not) an exponential function, whereas the function defined by $y = 3^x$ (is/is not) an exponential function.

2. The graph of $f(x) = \left(\dfrac{5}{3}\right)^x$ is (increasing/decreasing) over its domain.

3. The graph of $f(x) = \left(\dfrac{3}{5}\right)^x$ is (increasing/decreasing) over its domain.

4. The domain of an exponential function $f(x) = b^x$ is _____.

5. The range of an exponential function $f(x) = b^x$ is _____.

6. All exponential functions $f(x) = b^x$ pass through the point _____.

7. The horizontal asymptote of an exponential function $f(x) = b^x$ is the line _____.

8. As $x \to \infty$, the value of $\left(1 + \dfrac{1}{x}\right)^x$ approaches _____.

Objective 1: Graph Exponential Functions

For Exercises 9–12, evaluate the functions at the given values of x. Round to 4 decimal places if necessary.

9. $f(x) = 5^x$
   a. $f(-1)$
   b. $f(4.8)$
   c. $f(\sqrt{2})$
   d. $f(\pi)$

10. $g(x) = 7^x$
    a. $g(-2)$
    b. $g(5.9)$
    c. $g(\sqrt{11})$
    d. $g(e)$

11. $h(x) = \left(\dfrac{1}{4}\right)^x$
    a. $h(-3)$
    b. $h(1.4)$
    c. $h(\sqrt{3})$
    d. $h(0.5e)$

12. $k(x) = \left(\dfrac{1}{6}\right)^x$
    a. $k(-3)$
    b. $k(1.4)$
    c. $k(\sqrt{0.5})$
    d. $k(0.5\pi)$

13. Which functions are exponential functions?
    a. $f(x) = 4.2^x$ b. $g(x) = x^{4.2}$ c. $h(x) = 4.2x$
    d. $k(x) = \left(\sqrt{4.2}\right)^x$ e. $m(x) = (-4.2)^x$

14. Which functions are exponential functions?
    a. $v(x) = (-\pi)^x$ b. $t(x) = \pi^x$ c. $w(x) = \pi x$
    d. $n(x) = \left(\sqrt{\pi}\right)^x$ e. $p(x) = x^\pi$

For Exercises 15–22, graph the functions and write the domain and range in interval notation. (See Example 1)

15. $f(x) = 3^x$

16. $g(x) = 4^x$

17. $h(x) = \left(\dfrac{1}{3}\right)^x$

18. $k(x) = \left(\dfrac{1}{4}\right)^x$

19. $m(x) = \left(\dfrac{3}{2}\right)^x$

20. $n(x) = \left(\dfrac{5}{4}\right)^x$

21. $b(x) = \left(\dfrac{2}{3}\right)^x$

22. $c(x) = \left(\dfrac{4}{5}\right)^x$

For Exercises 23–32,
   a. Use transformations of the graphs of $y = 3^x$ (see Exercise 15) and $y = 4^x$ (see Exercise 16) to graph the given function. (See Example 2)
   b. Write the domain and range in interval notation.
   c. Write an equation of the asymptote.

23. $f(x) = 3^x + 2$

24. $g(x) = 4^x - 3$

25. $m(x) = 3^{x+2}$

26. $n(x) = 4^{x-3}$

27. $p(x) = 3^{x-4} - 1$

28. $q(x) = 4^{x+1} + 2$

29. $k(x) = -3^x$

30. $h(x) = -4^x$

31. $t(x) = 3^{-x}$

32. $v(x) = 4^{-x}$

For Exercises 33–36,
   a. Use transformations of the graphs of $y = \left(\frac{1}{3}\right)^x$ (see Exercise 17) and $y = \left(\frac{1}{4}\right)^x$ (see Exercise 18) to graph the given function. (See Example 2)
   b. Write the domain and range in interval notation.
   c. Write an equation of the asymptote.

33. $f(x) = \left(\dfrac{1}{3}\right)^{x+1} - 3$

34. $g(x) = \left(\dfrac{1}{4}\right)^{x-2} + 1$

35. $k(x) = -\left(\dfrac{1}{3}\right)^x + 2$

36. $h(x) = -\left(\dfrac{1}{4}\right)^x - 2$

Objective 2: Evaluate the Exponential Function Base e

For Exercises 37–38, evaluate the functions for the given values of x. Round to 4 decimal places.

37. $f(x) = e^x$

   a. $f(4)$ **b.** $f(-3.2)$

   c. $f(\sqrt{13})$ **d.** $f(\pi)$

38. $f(x) = e^x$

   a. $f(-3)$ **b.** $f(6.8)$

   c. $f(\sqrt{7})$ **d.** $f(e)$

For Exercises 39–44,

   a. Use transformations of the graph of $y = e^x$ to graph the given function. (**See Example 3**)

   b. Write the domain and range in interval notation.

   c. Write an equation of the asymptote.

39. $f(x) = e^{x-4}$ **40.** $g(x) = e^{x-2}$ **41.** $h(x) = e^x + 2$

42. $k(x) = e^x - 1$ **43.** $m(x) = -e^x - 3$ **44.** $n(x) = -e^x + 4$

Objective 3: Use Exponential Functions to Compute Compound Interest

For Exercises 45–46, complete the table to determine the effect of the number of compounding periods when computing interest. (**See Example 4**)

45. Suppose that $10,000 is invested at 4% interest for 5 yr under the following compounding options. Complete the table.

	Compounding Option	n Value	Result
a.	Annually		
b.	Quarterly		
c.	Monthly		
d.	Daily		
e.	Continuously		

46. Suppose that $8000 is invested at 3.5% interest for 20 yr under the following compounding options. Complete the table.

	Compounding Option	n Value	Result
a.	Annually		
b.	Quarterly		
c.	Monthly		
d.	Daily		
e.	Continuously		

For Exercises 47–48, suppose that P dollars in principal is invested for t years at the given interest rates with continuous compounding. Determine the amount that the investment is worth at the end of the given time period.

47. $P = \$20,000$, $t = 10$ yr

   a. 3% interest

   b. 4% interest

   c. 5.5% interest

48. $P = \$6000$, $t = 12$ yr

   a. 1% interest

   b. 2% interest

   c. 4.5% interest

49. Bethany needs to borrow $10,000. She can borrow the money at 5.5% simple interest for 4 yr or she can borrow at 5% with interest compounded continuously for 4 yr.

   a. How much total interest would Bethany pay at 5.5% simple interest?

   b. How much total interest would Bethany pay at 5% interest compounded continuously?

   c. Which option results in less total interest?

50. Al needs to borrow $15,000 to buy a car. He can borrow the money at 6.7% simple interest for 5 yr or he can borrow at 6.4% interest compounded continuously for 5 yr.

   a. How much total interest would Al pay at 6.7% simple interest?

   b. How much total interest would Al pay at 6.4% interest compounded continuously?

   c. Which option results in less total interest?

51. Jerome wants to invest $25,000 as part of his retirement plan. He can invest the money at 5.2% simple interest for 30 yr, or he can invest at 3.8% interest compounded continuously for 30 yr. Which option results in more total interest?

52. Heather wants to invest $35,000 of her retirement. She can invest at 4.8% simple interest for 20 yr, or she can choose an option with 3.6% interest compounded continuously for 20 yr. Which option results in more total interest?

Objective 4: Use Exponential Functions in Applications

53. Strontium-90 (^{90}Sr) is a by-product of nuclear fission with a half-life of approximately 28.9 yr. After the Chernobyl nuclear reactor accident in 1986, large areas surrounding the site were contaminated with ^{90}Sr. If 10 μg (micrograms) of ^{90}Sr is present in a sample, the function $A(t) = 10\left(\dfrac{1}{2}\right)^{t/28.9}$ gives the amount $A(t)$ (in μg) present after t years. Evaluate the function for the given values of t and interpret the meaning in context. Round to 3 decimal places if necessary. **(See Example 5)**

   a. $A(28.9)$ **b.** $A(57.8)$ **c.** $A(100)$

54. In 2006, the murder of Alexander Litvinenko, a Russian dissident, was thought to be by poisoning from the rare and highly radioactive element polonium-210 (^{210}Po). The half-life of ^{210}Po is 138.4 yr. If 0.1 mg of ^{210}Po is present in a sample then $A(t) = 0.1\left(\dfrac{1}{2}\right)^{t/138.4}$ gives the amount $A(t)$ (in mg) present after t years. Evaluate the function for the given values of t and interpret the meaning in context. Round to 3 decimal places if necessary.

   a. $A(138.4)$ **b.** $A(276.8)$ **c.** $A(500)$

55. According to the CIA's *World Fact Book*, in 2010, the population of the United States was approximately 310 million with a 0.97% annual growth rate. (*Source:* www.cia.gov) At this rate, the population $P(t)$ (in millions) can be approximated by $P(t) = 310(1.0097)^t$, where t is the time in years since 2010.

   a. Is the graph of P an increasing or decreasing exponential function?

   b. Evaluate $P(0)$ and interpret its meaning in the context of this problem.

   c. Evaluate $P(10)$ and interpret its meaning in the context of this problem. Round the population value to the nearest million.

   d. Evaluate $P(20)$ and $P(30)$.

   e. Evaluate $P(200)$ and use this result to determine if it is reasonable to expect this model to continue indefinitely.

56. The population of Canada in 2010 was approximately 34 million with an annual growth rate of 0.804%. At this rate, the population $P(t)$ (in millions) can be approximated by $P(t) = 34(1.00804)^t$, where t is the time in years since 2010. (*Source:* www.cia.gov)

   a. Is the graph of P an increasing or decreasing exponential function?

   b. Evaluate $P(0)$ and interpret its meaning in the context of this problem.

   c. Evaluate $P(5)$ and interpret its meaning in the context of this problem. Round the population value to the nearest million.

   d. Evaluate $P(15)$ and $P(25)$.

57. The atmospheric pressure on an object decreases as altitude increases. If a is the height (in km) above sea level, then the pressure $P(a)$ (in mmHg) is approximated by $P(a) = 760e^{-0.13a}$.

   a. Find the atmospheric pressure at sea level.

   b. Determine the atmospheric pressure at 8.848 km (the altitude of Mt. Everest). Round to the nearest whole unit.

58. The function defined by $A(t) = 100e^{0.0318t}$ approximates the equivalent amount of money needed t years after the year 2010 to equal $100 of buying power in the year 2010. The value 0.0318 is related to the average rate of inflation.

   a. Evaluate $A(15)$ and interpret its meaning in the context of this problem.

   b. Verify that by the year 2032, more than $200 will be needed to have the same buying power as $100 in 2010.

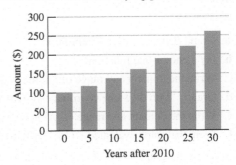

Newton's law of cooling indicates that the temperature of a warm object, such as a cake coming out of the oven, will decrease exponentially with time and will approach the temperature of the surrounding air. The temperature $T(t)$ is modeled by $T(t) = T_a + (T_0 - T_a)e^{-kt}$. In this model, T_a represents the temperature of the surrounding air, T_0 represents the initial temperature of the object, and t is the time after the object starts cooling. The value of k is a constant of proportion relating the temperature of the object to its rate of temperature change. Use this model for Exercises 59–60.

59. A cake comes out of the oven at 350°F and is placed on a cooling rack in a 78°F kitchen. After checking the temperature several minutes later, the value of k is measured as 0.046.

   a. Write a function that models the temperature $T(t)$ (in °F) of the cake t minutes after being removed from the oven.

   b. What is the temperature of the cake 10 min after coming out of the oven? Round to the nearest degree.

   c. It is recommended that the cake should not be frosted until it has cooled to under 100°F. If Jessica waits 1 hr to frost the cake, will the cake be cool enough to frost?

61. A farmer depreciates a $120,000 tractor. He estimates that the resale value $V(t)$ (in $1000) of the tractor t years after purchase is 80% of its value from the previous year. Therefore, the resale value can be approximated by $V(t) = 120(0.8)^t$.

   a. Find the resale value 5 yr after purchase. Round to the nearest $1000.

   b. The farmer estimates that the cost to run the tractor is $18/hr in labor, $36/hr in fuel, and $22/hr in overhead costs (for maintenance and repair). Estimate the farmer's cost to run the tractor for the first year if he runs the tractor for a total of 800 hr. Include hourly costs and depreciation.

60. Water in a water heater is originally 122°F. The water heater is shut off and the water cools to the temperature of the surrounding air, which is 60°F. The water cools slowly because of the insulation inside the heater, and the value of k is measured as 0.00351.

   a. Write a function that models the temperature $T(t)$ (in °F) of the water t hours after the water heater is shut off.

   b. What is the temperature of the water 12 hr after the heater is shut off? Round to the nearest degree.

   c. Dominic does not like to shower with water less than 115°F. If Dominic waits 24 hr, will the water still be warm enough for a shower?

62. A veterinarian depreciates a $10,000 X-ray machine. He estimates that the resale value $V(t)$ (in $) after t years is 90% of its value from the previous year. Therefore, the resale value can be approximated by $V(t) = 10,000(0.9)^t$.

   a. Find the resale value after 4 yr.

   b. If the veterinarian wants to sell his practice 8 yr after the X-ray machine was purchased, how much is the machine worth? Round to the nearest $100.

Mixed Exercises

For Exercises 63–64, solve the equations in parts (a)–(c) by inspection. Then estimate the solutions to parts (d) and (e) between two consecutive integers.

63. a. $2^x = 4$
   b. $2^x = 8$
   c. $2^x = 16$
   d. $2^x = 7$
   e. $2^x = 10$

64. a. $3^x = 3$
   b. $3^x = 9$
   c. $3^x = 27$
   d. $3^x = 7$
   e. $3^x = 10$

65. a. Graph $f(x) = 2^x$. (**See Example 1**)
   b. Is f a one-to-one function?
   c. Write the domain and range of f in interval notation.
   d. Graph f^{-1} on the same coordinate system as f.
   e. Write the domain and range of f^{-1} in interval notation.
   f. From the graph evaluate $f^{-1}(1), f^{-1}(2)$, and $f^{-1}(4)$.

66. a. Graph $g(x) = 3^x$ (see Exercise 15).
   b. Is g a one-to-one function?
   c. Write the domain and range of g in interval notation.
   d. Graph g^{-1} on the same coordinate system as g.
   e. Write the domain and range of g^{-1} in interval notation.
   f. From the graph evaluate $g^{-1}(1), g^{-1}(3)$, and $g^{-1}(\frac{1}{3})$.

67. Refer to the graphs of $f(x) = 2^x$ and the inverse function, $y = f^{-1}(x)$ from Exercise 65. Fill in the blanks.

   a. As $x \to \infty$, $f(x) \to$ _____.
   b. As $x \to -\infty$, $f(x) \to$ _____.
   c. As $x \to \infty$, $f^{-1}(x) \to$ _____.
   d. As $x \to 0^+$, $f^{-1}(x) \to$ _____.

68. Refer to the graphs of $g(x) = 3^x$ and the inverse function, $y = g^{-1}(x)$ from Exercise 66. Fill in the blanks.

   a. As $x \to \infty$, $g(x) \to$ _____.
   b. As $x \to -\infty$, $g(x) \to$ _____.
   c. As $x \to \infty$, $g^{-1}(x) \to$ _____.
   d. As $x \to 0^+$, $g^{-1}(x) \to$ _____.

Write About It

69. Explain why the equation $2^x = -2$ has no solution.

70. Explain why the $f(x) = x^2$ is not an exponential function.

Expanding Your Skills

For Exercises 71–72, find the real solutions to the equation.

71. $3x^2 e^{-x} - 6xe^{-x} = 0$

72. $x^2 e^x - e^x = 0$

73. Use the properties of exponents to simplify.

  a. $e^x e^h$ **b.** $(e^x)^2$ **c.** $\dfrac{e^x}{e^h}$

  d. $e^x \cdot e^{-x}$ **e.** e^{-2x}

74. Factor.

  a. $e^{x+h} - e^x$

  b. $e^{4x} - e^{2x}$

75. Multiply. $(e^x + e^{-x})^2$

76. Multiply. $(e^x - e^{-x})^2$

77. Show that $\left(\dfrac{e^x + e^{-x}}{2}\right)^2 - \left(\dfrac{e^x - e^{-x}}{2}\right)^2 = 1$.

78. Show that $2\left(\dfrac{e^x - e^{-x}}{2}\right)\left(\dfrac{e^x + e^{-x}}{2}\right) = \dfrac{e^{2x} - e^{-2x}}{2}$.

For Exercises 79–80, find the difference quotient $\dfrac{f(x+h) - f(x)}{h}$. **Write the answers in factored form.**

79. $f(x) = e^x$

80. $f(x) = 2^x$

Technology Connections

81. Graph the following functions on the window $[-3, 3, 1]$ by $[-1, 8, 1]$ and comment on the behavior of the graphs near $x = 0$.

  $Y_1 = e^x$

  $Y_2 = 1 + x + \dfrac{x^2}{2}$

  $Y_3 = 1 + x + \dfrac{x^2}{2} + \dfrac{x^3}{6}$

SECTION 4.3   Logarithmic Functions

OBJECTIVES

1. **Convert Between Logarithmic and Exponential Forms**
2. **Evaluate Logarithmic Expressions**
3. **Apply Basic Properties of Logarithms**
4. **Graph Logarithmic Functions**
5. **Use Logarithmic Functions in Applications**

1. Convert Between Logarithmic and Exponential Forms

Consider the following equations in which the variable is located in the exponent of an expression. In some cases, the solution can be found by inspection.

Equation	Solution
$5^x = 5$	$x = 1$
$5^x = 20$	$x = ?$
$5^x = 25$	$x = 2$
$5^x = 60$	$x = ?$
$5^x = 125$	$x = 3$

The equation $5^x = 20$ cannot be solved by inspection. However, we suspect that x is between 1 and 2 because $5^1 = 5$ and $5^2 = 25$. To solve for x explicitly, we must isolate x by performing the inverse operation of 5^x. Fortunately, all exponential functions $y = b^x$ $(b > 0, b \neq 1)$ are one-to-one and therefore have inverse functions. The inverse of an exponential function, base b, is the *logarithmic* function base b which we define here.

Definition of a Logarithmic Function

If x and b are positive real numbers such that $b \neq 1$, then $y = \log_b x$ is called the **logarithmic function base b,** where

$$y = \log_b x \text{ is equivalent to } b^y = x$$

Notes:

- Given $y = \log_b x$, the value y is the exponent to which b must be raised to obtain x.
- The value of y is called the **logarithm,** b is called the **base,** and x is called the **argument.**
- The equations $y = \log_b x$ and $b^y = x$ both define the same relationship between x and y. The expression $y = \log_b x$ is called the **logarithmic form,** and $b^y = x$ is called the **exponential form.**

The logarithmic function base b is defined as the inverse of the exponential function base b.

exponential function	$f(x) = b^x$	First replace $f(x)$ by y.
	$y = b^x$	Next, interchange x and y.
inverse of exponential function	$x = b^y$ ⟵	This equation provides an implicit relationship between x and y. To solve for y explicitly (that is, to isolate y), we
logarithmic function	$y = \log_b x$	must use logarithmic notation.

To be able to solve equations involving logarithms, it is often advantageous to write a logarithmic expression in its exponential form.

EXAMPLE 1 Writing Logarithmic Form and Exponential Form

Write each equation in exponential form.

a. $\log_2 16 = 4$ **b.** $\log_{10}\left(\dfrac{1}{100}\right) = -2$ **c.** $\log_7 1 = 0$

Solution:

Logarithmic form $y = \log_b x$ **Exponential form** $b^y = x$

The logarithm is the exponent to which the base is raised to obtain x.

a. $\log_2 16 = 4$ ⟺ $2^4 = 16$

b. $\log_{10}\left(\dfrac{1}{100}\right) = -2$ ⟺ $10^{-2} = \dfrac{1}{100}$

c. $\log_7 1 = 0$ ⟺ $7^0 = 1$

Skill Practice 1 Write each equation in exponential form.

a. $\log_3 9 = 2$ **b.** $\log_{10}\left(\dfrac{1}{1000}\right) = -3$ **c.** $\log_6 1 = 0$

In Example 2 we reverse this process and write an exponential equation in its logarithmic form.

Answers

1. a. $3^2 = 9$ **b.** $10^{-3} = \dfrac{1}{1000}$

c. $6^0 = 1$

EXAMPLE 2   Writing Exponential Form and Logarithmic Form

Write each equation in logarithmic form.

a. $3^4 = 81$ **b.** $10^6 = 1,000,000$ **c.** $\left(\dfrac{1}{5}\right)^{-1} = 5$

Solution:

Exponential form $b^y = x$ **Logarithmic form** $\log_b x = y$

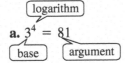

a. $3^4 = 81$ $\Leftrightarrow$ $\log_3 81 = 4$

b. $10^6 = 1,000,000$ $\Leftrightarrow$ $\log_{10} 1,000,000 = 6$

c. $\left(\dfrac{1}{5}\right)^{-1} = 5$ $\Leftrightarrow$ $\log_{1/5} 5 = -1$

Skill Practice 2   Write each equation in logarithmic form.

a. $2^5 = 32$ **b.** $10^4 = 10,000$ **c.** $\left(\dfrac{1}{8}\right)^{-2} = 64$

---

2. Evaluate Logarithmic Expressions

To evaluate a logarithmic expression, we can write the expression in exponential form. Then we make use of the equivalence property of exponential expressions. This states that if two exponential expressions of the same base are equal, then their exponents must be equal.

Equivalence Property of Exponential Expressions

If b, x, and y are real numbers, with $b > 0$ and $b \neq 1$, then

$$b^x = b^y \text{ implies that } x = y.$$

In Example 3, we evaluate several logarithmic expressions.

EXAMPLE 3   Evaluating a Logarithmic Expression

Evaluate each expression.

a. $\log_4 16$ **b.** $\log_2 8$ **c.** $\log_{1/2} 8$

Solution:

Let y represent the value of the logarithm.

a. $\log_4 16$ is the exponent to which 4 must be raised to equal 16. That is, $4^{\square} = 16$.

$\log_4 16 = y$

     $4^y = 16$ or equivalently $4^y = 4^2$ Write the equivalent exponential form.

     $y = 2$

Therefore, $\log_4 16 = 2$.   <u>Check</u>: $4^2 = 16$ ✓

Answers

2. a. $\log_2 32 = 5$

   b. $\log_{10} 10,000 = 4$

   c. $\log_{1/8} 64 = -2$

TIP Once you become comfortable with the concept of a logarithm, you can take fewer steps to evaluate a logarithm.

To evaluate the expression $\log_4 16$ we ask $4^\square = 16$. The exponent is 2, so $\log_4 16 = 2$.

Likewise, to evaluate $\log_2 8$ we ask $2^\square = 8$. So $\log_2 8 = 3$.

b. $\log_2 8$ is the exponent to which 2 must be raised to obtain 8. That is, $2^\square = 8$.

$\log_2 8 = y$

$\quad 2^y = 8$ or equivalently $2^y = 2^3$ Write the equivalent exponential form.

$\quad\quad y = 3$

Therefore, $\log_2 8 = 3$.   Check: $2^3 = 8$ ✓

c. $\log_{1/2} 8 = y$

$\quad \left(\frac{1}{2}\right)^y = 8$ or equivalently $\left(\frac{1}{2}\right)^y = \left(\frac{1}{2}\right)^{-3}$ Write the equivalent exponential form.

$\quad\quad y = -3$

Therefore, $\log_{1/2} 8 = -3$.   Check: $\left(\frac{1}{2}\right)^{-3} = 8$ ✓

Skill Practice 3   Evaluate each expression.

    a. $\log_5 125$ **b.** $\log_3 81$ **c.** $\log_4\left(\dfrac{1}{64}\right)$

The statement $y = \log_b x$ represents a family of logarithmic functions where the base is any positive real number except 1. Two specific logarithmic functions that come up often in applications are the logarithmic functions base 10 and base e.

TIP To help you remember the notation $y = \ln x$, think of "ln" as "log natural."

Definition of Common and Natural Logarithmic Functions

- The logarithmic function base 10 is called the **common logarithmic function.** The common logarithmic function is denoted by $y = \log x$. Notice that the base 10 is not explicitly written; that is, $y = \log_{10} x$ is written simply as $y = \log x$.
- The logarithmic function base e is called the **natural logarithmic function.** The natural logarithmic function is denoted by $y = \ln x$; that is, $y = \log_e x$ is written as $y = \ln x$.

EXAMPLE 4   **Evaluating Common and Natural Logarithms**

Evaluate.

    a. $\log 100{,}000$ **b.** $\log 0.001$ **c.** $\ln e^4$ **d.** $\ln\left(\dfrac{1}{e}\right)$

Solution:

Let y represent the value of the logarithm.

    a. $\log 100{,}000 = y$

$\quad\quad 10^y = 100{,}000$ or equivalently $10^y = 10^5$ Write the exponential form.

$\quad\quad\quad y = 5$

Thus, $\log 100{,}000 = 5$ because $10^5 = 100{,}000$.

    b. $\log 0.001 = y$

$\quad\quad 10^y = 0.001$ or equivalently $10^y = 10^{-3}$ Write the exponential form.

$\quad\quad\quad y = -3$

Thus, $\log 0.001 = -3$ because $10^{-3} = 0.001$.

Answers

3. a. 3   **b.** 4   **c.** -3

c. $\ln e^4 = y$

$\quad e^y = e^4$ Write the equivalent exponential form.

$\quad y = 4$

Therefore, $\ln e^4 = 4$.

d. $\ln\left(\dfrac{1}{e}\right) = y$

$\quad e^y = \left(\dfrac{1}{e}\right)$ or equivalently $e^y = e^{-1}$ Write the equivalent exponential form.

$\quad y = -1$

Therefore, $\ln\left(\dfrac{1}{e}\right) = -1$.

Skill Practice 4   Evaluate.

a. log 10,000,000   **b.** log 0.1   **c.** ln e^5 **d.** ln e

---

Most scientific calculators have a key for the common logarithmic function **LOG** and a key for the natural logarithmic function **LN**. We demonstrate their use in Example 5.

EXAMPLE 5   **Approximating Common and Natural Logarithms**

Approximate the logarithms.

a. log 5809   **b.** $\log(4.6 \times 10^7)$ **c.** log 0.003
d. ln 472   **e.** ln 0.05   **f.** $\ln\sqrt{87}$

Solution:

For parts (a)–(c), use the **LOG** key.   For parts (d)–(f), use the **LN** key.

```
NORMAL FLOAT AUTO a+bi DEGREE CL
log(5809)
                    3.764101376
log(4.6E7)
                    7.662757832
log(.003)
                   -2.522878745
```

```
NORMAL FLOAT AUTO a+bi DEGREE CL
ln(472)
                    6.156978986
ln(0.05)
                   -2.995732274
ln(√(87))
                    2.232954059
```

When using a calculator, there is always potential for user-input error. Therefore, it is good practice to estimate values when possible to confirm the reasonableness of an answer from a calculator. For example,

For part (a), $10^3 < 5809 < 10^4$. Therefore, $3 < \log 5809 < 4$.
For part (b), $10^7 < 4.6 \times 10^7 < 10^8$. Therefore, $7 < \log(4.6 \times 10^7) < 8$.
For part (c), $10^{-3} < 0.003 < 10^{-2}$. Therefore, $-3 < \log 0.003 < -2$.

Skill Practice 5   Approximate the logarithms. Round to 4 decimal places.

a. log 229   **b.** $\log(3.76 \times 10^{12})$ **c.** log 0.0216
d. ln 87   **e.** ln 0.0032   **f.** ln π

Answers

4. a. 7   **b.** −1   **c.** 5   **d.** 1
5. a. 2.3598   **b.** 12.5752
  c. −1.6655   **d.** 4.4659
  e. −5.7446   **f.** 1.1447

3. Apply Basic Properties of Logarithms

From the definition of a logarithmic function, we have the following basic properties.

> ### Basic Properties of Logarithms
>
> | Property | Example |
> |---|---|
> | **1.** $\log_b 1 = 0$ because $b^0 = 1$ | $\log_5 1 = 0$ because $5^0 = 1$ |
> | **2.** $\log_b b = 1$ because $b^1 = b$ | $\log_3 3 = 1$ because $3^1 = 3$ |
> | **3.** $\log_b b^x = x$ because $b^x = b^x$ | $\log_2 2^x = x$ because $2^x = 2^x$ |
> | **4.** $b^{\log_b x} = x$ because $\log_b x = \log_b x$ | $7^{\log_7 x} = x$ because $\log_7 x = \log_7 x$ |

Properties 3 and 4 follow from the fact that a logarithmic function is the inverse of an exponential function of the same base. Given $f(x) = b^x$ and $f^{-1}(x) = \log_b x$,

$$(f \circ f^{-1})(x) = b^{(\log_b x)} = x \qquad \text{(Property 4)}$$

$$(f^{-1} \circ f)(x) = \log_b(b^x) = x \qquad \text{(Property 3)}$$

EXAMPLE 6   Applying the Properties of Logarithms

Simplify.

a. $\log_3 3^{10}$ **b.** $\ln e^2$ **c.** $\log_{11} 11$ **d.** $\log 10$

e. $\log_{\sqrt{7}} 1$ **f.** $\ln 1$ **g.** $5^{\log_5(c^2+4)}$ **h.** $10^{\log(a^2+b^2)}$

Solution:

a. $\log_3 3^{10} = 10$	Property 3	**b.** $\ln e^2 = \log_e e^2 = 2$	Property 3
c. $\log_{11} 11 = 1$	Property 2	**d.** $\log 10 = \log_{10} 10 = 1$	Property 2
e. $\log_{\sqrt{7}} 1 = 0$	Property 1	**f.** $\ln 1 = 0$	Property 1
g. $5^{\log_5(c^2+4)} = c^2 + 4$	Property 4	**h.** $10^{\log(a^2+b^2)} = a^2 + b^2$	Property 4

Skill Practice 6   Simplify.

a. $\log_{13} 13$ **b.** $\ln e$ **c.** $a^{\log_a 3}$ **d.** $e^{\ln 6}$

e. $\log_\pi 1$ **f.** $\log 1$ **g.** $\log_9 9^{\sqrt{2}}$ **h.** $\log 10^e$

---

4. Graph Logarithmic Functions

Since a logarithmic function $y = \log_b x$ is the inverse of the corresponding exponential function $y = b^x$, their graphs must be symmetric with respect to the line $y = x$. See Figures 4-13 and 4-14.

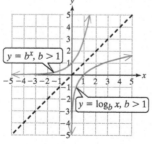

Figure 4-13

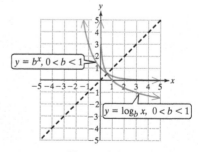

Figure 4-14

From Figures 4-13 and 4-14, the range of $y = b^x$ is the set of positive real numbers. As expected, the domain of its inverse function $y = \log_b x$ is the set of positive real numbers.

EXAMPLE 7   **Graphing Logarithmic Functions**

Graph the functions.

a. $y = \log_2 x$ **b.** $y = \log_{1/4} x$

Solution:

To find points on a logarithmic function, we can interchange the x- and y-coordinates of the ordered pairs on the corresponding exponential function.

a. To graph $y = \log_2 x$, interchange the x- and y-coordinates of the ordered pairs from its inverse function $y = 2^x$. The graph of $y = \log_2 x$ is shown in Figure 4-15.

Exponential Function

x	$y = 2^x$
-3	$\frac{1}{8}$
-2	$\frac{1}{4}$
-1	$\frac{1}{2}$
0	1
1	2
2	4
3	8

Logarithmic Function

x	$y = \log_2 x$
$\frac{1}{8}$	-3
$\frac{1}{4}$	-2
$\frac{1}{2}$	-1
1	0
2	1
4	2
8	3

Switch x and y.

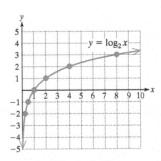

Figure 4-15

b. To graph $y = \log_{1/4} x$, interchange the x- and y-coordinates of the ordered pairs from its inverse function $y = \left(\frac{1}{4}\right)^x$. See Figure 4-16.

Exponential Function

x	$y = \left(\dfrac{1}{4}\right)^x$
-3	64
-2	16
-1	4
0	1
1	$\frac{1}{4}$
2	$\frac{1}{16}$
3	$\frac{1}{64}$

Logarithmic Function

x	$y = \log_{1/4} x$
64	-3
16	-2
4	-1
1	0
$\frac{1}{4}$	1
$\frac{1}{16}$	2
$\frac{1}{64}$	3

Switch x and y.

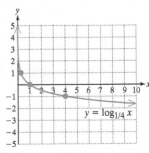

Figure 4-16

Skill Practice 7   Graph the functions.

a. $y = \log_4 x$ **b.** $y = \log_{1/2} x$

Based on the graphs in Example 7 and our knowledge of exponential functions, we offer the following summary of the characteristics of logarithmic and exponential functions.

Answers

7. a.

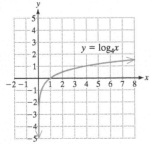

b.

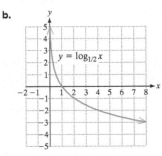

Graphs of Exponential and Logarithmic Functions

Exponential Functions	Logarithmic Functions
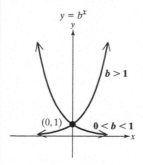	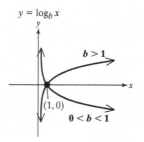

Domain: $(-\infty, \infty)$ Domain: $(0, \infty)$
Range: $(0, \infty)$ Range: $(-\infty, \infty)$
Horizontal asymptote: $y = 0$ Vertical asymptote: $x = 0$
Passes through $(0, 1)$ Passes through $(1, 0)$
If $b > 1$, the function is increasing.   If $b > 1$, the function is increasing.
If $0 < b < 1$, the function is   If $0 < b < 1$, the function is
decreasing.   decreasing.

The roles of x and y are reversed between a function and its inverse. Therefore, it is not surprising that the domain and range are reversed between exponential and logarithmic functions. Furthermore, an exponential function passes through $(0, 1)$, whereas a logarithmic function passes through $(1, 0)$. An exponential function has a horizontal asymptote of $y = 0$, whereas a logarithmic function has a vertical asymptote of $x = 0$.

In Example 8 we use the transformations of functions learned in Section 2.6 to graph a logarithmic function.

If $h > 0$, shift to the right.   If $k > 0$, shift upward.
If $h < 0$, shift to the left.   If $k < 0$, shift downward.

$$f(x) = a \log_b(x - h) + k$$

If $a < 0$, reflect across the x-axis.
Shrink vertically if $0 < |a| < 1$.
Stretch vertically if $|a| > 1$.

EXAMPLE 8   **Using Transformations to Graph Logarithmic Functions**

Graph the function. Identify the vertical asymptote and write the domain in interval notation.

$$f(x) = \log_2(x + 3) - 2$$

Solution:

The graph of the "parent" function $y = \log_2 x$ was presented in Example 7. The graph of $f(x) = \log_2(x + 3) - 2$ is the graph of $y = \log_2 x$ shifted to the left 3 units and down 2 units.

We can plot a few points on the graph of $y = \log_2 x$ and use these points and the vertical asymptote to form an outline of the transformed graph.

x	$y = \log_2 x$
$\frac{1}{4}$	-2
$\frac{1}{2}$	-1
1	0
2	1
4	2

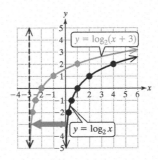

The graph of $f(x) = \log_2(x + 3) - 2$ is shown in blue.
The vertical asymptote is $x = -3$. The domain is $(-3, \infty)$.

Skill Practice 8   Graph the function. Identify the vertical asymptote and write the domain in interval notation. $g(x) = \log_3(x - 4) + 1$

The domain of $f(x) = \log_b x$ is restricted to $x > 0$. In Example 8, this graph was shifted to the left 3 units, restricting the domain of $f(x) = \log_2(x + 3) - 2$ to $x > -3$. The domain of a logarithmic function is the set of real numbers that make the argument positive.

EXAMPLE 9   **Identifying the Domain of a Logarithmic Function**

Write the domain in interval notation.

a. $f(x) = \log_2(2x + 4)$ **b.** $g(x) = \ln(5 - x)$ **c.** $h(x) = \log(x^2 - 9)$

Solution:

a. $f(x) = \log_2(2x + 4)$

$\quad 2x + 4 > 0$ Set the argument greater than zero.

$\quad\quad 2x > -4$ Solve for x.

$\quad\quad\quad x > -2$

The domain is $(-2, \infty)$.
The graph of f is shown in Figure 4-17.
The vertical asymptote is $x = -2$.

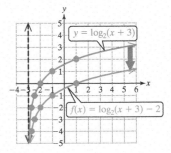

Figure 4-17

b. $g(x) = \ln(5 - x)$

$\quad 5 - x > 0$ Set the argument greater than zero.

$\quad\quad -x > -5$ Subtract 5 and divide by -1
$\quad\quad\quad x < 5$ (reverse the inequality sign).

The domain is $(-\infty, 5)$.
The graph of g is shown in Figure 4-18.
The vertical asymptote is $x = 5$.

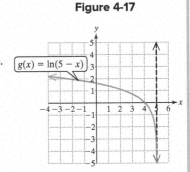

Figure 4-18

Answer

8. Vertical asymptote: $x = 4$
Domain: $(4, \infty)$

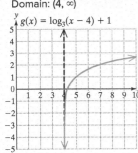

c. $h(x) = \log(x^2 - 9)$

$x^2 - 9 > 0$

Set the argument greater than zero. The result is a polynomial inequality (Section 3.6).

$(x - 3)(x + 3) = 0$

Solve the related equation by setting one side equal to zero and factoring the other side.

Sign of $(x - 3)$:	−	−	+
Sign of $(x + 3)$:	−	+	+
Sign of $(x - 3)(x + 3)$:	+	−	+

The boundary points for the solution set are the solutions to the equation: $x = 3$ and $x = -3$.

$\overset{\displaystyle}{\underset{-3 \qquad 3}{\longleftrightarrow}}$

The domain is $(-\infty, -3) \cup (3, \infty)$.

The graph of h is shown in Figure 4-19.

The vertical asymptotes are $x = -3$ and $x = 3$.

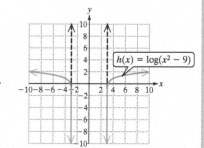

$h(x) = \log(x^2 - 9)$

Figure 4-19

Skill Practice 9   Write the domain in interval notation.

a. $\log_4(1 - 3x)$ **b.** $\log(2 + x)$ **c.** $m(x) = \ln(64 - x^2)$

---

5. Use Logarithmic Functions in Applications

When physical quantities vary over a large range, it is often convenient to take a logarithm of the quantity to have a more manageable set of numbers. For example, suppose a set of data values consists of 10, 100, 1000, and 10,000. The corresponding common logarithms are 1, 2, 3, and 4. The latter list of numbers is easier to manipulate and to visualize on a graph. For this reason, logarithmic scales are used in applications such as

- measuring pH (representing hydrogen ion concentration from 10^{-14} to 1 moles per liter).
- measuring wave energy from an earthquake (often ranging from 10^6 J to 10^{17} J).
- measuring loudness of sound on the decibel scale (representing sound intensity from 10^{-12} to 10^2 Watts per square meter).

Point of Interest

In 1935, American geologist Charles Richter developed the local magnitude (M_L) scale, or Richter scale, for measuring the intensity of moderate-sized earthquakes ($3 < M_L < 7$) in southern California. Today, seismologists no longer follow Richter's methodology because it does not give reliable results for earthquakes of higher magnitude. The magnitudes of modern earthquakes are based on a variety of data types from numerous seismic stations. However, both the Richter scale and modern magnitude scales use a base 10 logarithmic scale to compare amplitudes of waves on a seismogram. This means that an increase of 1 unit in magnitude represents a 10-fold increase in the amplitude of the waves on a seismogram.

Answers

9. a. $\left(-\infty, \dfrac{1}{3}\right)$ **b.** $(-2, \infty)$

c. $(-8, 8)$

EXAMPLE 10   Using a Logarithmic Function in an Application

The intensity I of an earthquake is measured by a seismograph—a device that measures amplitudes of shock waves. I_0 is a minimum reference intensity of a "zero-level" earthquake against which the intensities of other earthquakes may be compared. The magnitude M of an earthquake of intensity I is given by

$$M = \log\left(\frac{I}{I_0}\right).$$

a. Determine the magnitude of the earthquake that devastated Haiti on January 12, 2010, if the intensity was approximately $10^{7.0}$ times I_0.

b. Determine the magnitude of the earthquake that occurred near Washington, D.C., on August 23, 2011, if the intensity was approximately $10^{5.8}$ times I_0.

c. How many times more intense was the earthquake that hit Haiti than the earthquake that hit Washington, D.C.? Round to the nearest whole unit.

Solution:

a. $M = \log\left(\dfrac{I}{I_0}\right)$ **b.** $M = \log\left(\dfrac{I}{I_0}\right)$

$\qquad M = \log\left(\dfrac{10^{7.0} \cdot I_0}{I_0}\right)$ $M = \log\left(\dfrac{10^{5.8} \cdot I_0}{I_0}\right)$

$\qquad\quad = \log 10^{7.0}$ $= \log 10^{5.8}$

$\qquad\quad = 7.0$ $= 5.8$

c. Using the intensities given in parts (a) and (b) we have

$$\frac{10^{7.0}\, I_0}{10^{5.8}\, I_0} = 10^{7.0-5.8} = 10^{1.2} \approx 16$$

The earthquake in Haiti was approximately 16 times more intense.

Skill Practice 10

a. Determine the magnitude of an earthquake that is $10^{5.2}$ times I_0.

b. Determine the magnitude of an earthquake that is $10^{4.2}$ times I_0.

c. How many times more intense is a 5.2-magnitude earthquake than a 4.2-magnitude earthquake?

Answers

10. a. 5.2   **b.** 4.2
     c. 10 times more intense

SECTION 4.3   Practice Exercises

Prerequisite Review

For Exercises R.1–R.4, simplify each expression.

R.1. $9^{-3/2}$ **R.2.** $64^{2/3}$ **R.3.** $(-8)^{2/3}$ **R.4.** $\dfrac{1}{25^{-1/2}}$

For Exercises R.5–R.6, solve the polynomial inequality. Write the answer in interval notation.

R.5. $y^2 + 20y \geq -75$ **R.6.** $x^2 - 8x + 16 > 0$

Concept Connections

1. Given positive real numbers x and b such that $b \neq 1$, $y = \log_b x$ is the _____ function base b and is equivalent to $b^y = x$.

2. Given $y = \log_b x$, the value y is called the _____, b is called the _____, and x is called the _____.

3. The logarithmic function base 10 is called the _____ logarithmic function, and the logarithmic function base e is called the _____ logarithmic function.

4. Given $y = \log x$, the base is understood to be _____. Given $y = \ln x$, the base is understood to be _____.

5. $\log_b 1 =$ _____ because $b^{\square} = 1$.

6. $\log_b b =$ _____ because $b^{\square} = b$.

7. $f(x) = \log_b x$ and $g(x) = b^x$ are inverse functions. Therefore, $\log_b b^x =$ _____ and $b^{\log_b x} =$ _____.

8. The graph of $y = \log_b x$ passes through the point $(1, 0)$ and the line _____ is a (horizontal/vertical) asymptote.

For the exercises in this set, assume that all variable expressions represent positive real numbers.

Objective 1: Convert Between Logarithmic and Exponential Forms

For Exercises 9–16, write the equation in exponential form. (See Example 1)

9. $\log_8 64 = 2$

10. $\log_9 81 = 2$

11. $\log\left(\dfrac{1}{10,000}\right) = -4$

12. $\log\left(\dfrac{1}{1,000,000}\right) = -6$

13. $\ln 1 = 0$

14. $\log_8 1 = 0$

15. $\log_a b = c$

16. $\log_x M = N$

For Exercises 17–24, write the equation in logarithmic form. (See Example 2)

17. $5^3 = 125$

18. $2^5 = 32$

19. $\left(\dfrac{1}{5}\right)^{-3} = 125$

20. $\left(\dfrac{1}{2}\right)^{-5} = 32$

21. $10^9 = 1,000,000,000$

22. $e^1 = e$

23. $a^7 = b$

24. $M^3 = N$

Objective 2: Evaluate Logarithmic Expressions

For Exercises 25–50, simplify the expression without using a calculator. (See Examples 3–4)

25. $\log_3 9$

26. $\log_2 16$

27. $\log_5 5$

28. $\log_6 6$

29. $\log 100,000,000$

30. $\log 10,000,000$

31. $\log_2\left(\dfrac{1}{16}\right)$

32. $\log_3\left(\dfrac{1}{9}\right)$

33. $\log\left(\dfrac{1}{10}\right)$

34. $\log\left(\dfrac{1}{10,000}\right)$

35. $\ln e^6$

36. $\ln e^{10}$

37. $\ln\left(\dfrac{1}{e^3}\right)$

38. $\ln\left(\dfrac{1}{e^8}\right)$

39. $\log_{1/7} 49$

40. $\log_{1/4} 16$

41. $\log_{1/2}\left(\dfrac{1}{32}\right)$

42. $\log_{1/6}\left(\dfrac{1}{36}\right)$

43. $\log 0.00001$

44. $\log 0.0001$

45. $\log_{3/2}\dfrac{4}{9}$

46. $\log_{3/2}\dfrac{9}{4}$

47. $\log_3 \sqrt[5]{3}$

48. $\log_2 \sqrt[3]{2}$

49. $\log_5 \sqrt{\dfrac{1}{5}}$

50. $\log \sqrt{\dfrac{1}{1000}}$

For Exercises 51–52, estimate the value of each logarithm between two consecutive integers. Then use a calculator to approximate the value to 4 decimal places. For example, $\log 8970$ is between 3 and 4 because $10^3 < 8970 < 10^4$. (See Example 5)

51. a. $\log 46,832$
  b. $\log 1,247,310$
  c. $\log 0.24$
  d. $\log 0.0000032$
  e. $\log(5.6 \times 10^5)$
  f. $\log(5.1 \times 10^{-3})$

52. a. $\log 293,416$
  b. $\log 897$
  c. $\log 0.038$
  d. $\log 0.00061$
  e. $\log(9.1 \times 10^8)$
  f. $\log(8.2 \times 10^{-2})$

For Exercises 53–54, approximate $f(x) = \ln x$ for the given values of x. Round to 4 decimal places. (See Example 5)

53. a. $f(94)$

   b. $f(0.182)$

   c. $f(\sqrt{155})$

   d. $f(4\pi)$

   e. $f(3.9 \times 10^9)$

   f. $f(7.1 \times 10^{-4})$

54. a. $f(1860)$

   b. $f(0.0694)$

   c. $f(\sqrt{87})$

   d. $f(2\pi)$

   e. $f(1.3 \times 10^{12})$

   f. $f(8.5 \times 10^{-17})$

Objective 3: Apply Basic Properties of Logarithms

For Exercises 55–64, simplify the expression without using a calculator. (See Example 6)

55. $\log_4 4^{11}$

56. $\log_6 6^7$

57. $\log_c c$

58. $\log_d d$

59. $5^{\log_5(x+y)}$

60. $4^{\log_4(a-c)}$

61. $\ln e^{a+b}$

62. $\ln e^{x^2+1}$

63. $\log_{\sqrt{5}} 1$

64. $\log_\pi 1$

Objective 4: Graph Logarithmic Functions

For Exercises 65–70, graph the function. (See Example 7)

65. $y = \log_3 x$

66. $y = \log_5 x$

67. $y = \log_{1/3} x$

68. $y = \log_{1/5} x$

69. $y = \ln x$

70. $y = \log x$

For Exercises 71–78, (See Example 8)

a. Use transformations of the graphs of $y = \log_2 x$ (see Example 7) and $y = \log_3 x$ (see Exercise 65) to graph the given functions.

b. Write the domain and range in interval notation.

c. Write an equation of the asymptote.

71. $y = \log_3(x + 2)$

72. $y = \log_2(x + 3)$

73. $y = 2 + \log_3 x$

74. $y = 3 + \log_2 x$

75. $y = \log_3(x - 1) - 3$

76. $y = \log_2(x - 2) - 1$

77. $y = -\log_3 x$

78. $y = -\log_2 x$

For Exercises 79–92, write the domain in interval notation. (See Example 9)

79. $f(x) = \log(8 - x)$

80. $g(x) = \log(3 - x)$

81. $h(x) = \log_2(6x + 7)$

82. $k(x) = \log_3(5x + 6)$

83. $m(x) = \ln(x^2 + 14)$

84. $n(x) = \ln(x^2 + 11)$

85. $f(x) = \log_4(x^2 - 16)$

86. $g(x) = \log_7(x^2 - 49)$

87. $m(x) = 3 + \ln\dfrac{1}{\sqrt{11 - x}}$

88. $n(x) = 4 - \log\dfrac{1}{\sqrt{x + 5}}$

89. $p(x) = \log(x^2 - x - 12)$

90. $q(x) = \log(x^2 + 10x + 9)$

91. $r(x) = \log_3(4 - x)^2$

92. $s(x) = \log_5(3 - x)^2$

Objective 5: Use Logarithmic Functions in Applications

93. In 1989, the Loma Prieta earthquake damaged the city of San Francisco with an intensity of approximately $10^{6.9}I_0$. Film footage of the 1989 earthquake was captured on a number of video cameras including a broadcast of Game 3 of the World Series played at Candlestick Park. (**See Example 10**)

   a. Determine the magnitude of the Loma Prieta earthquake.

   b. Smaller earthquakes occur daily in the San Francisco area and most are not detectable without a seismograph. Determine the magnitude of an earthquake with an intensity of $10^{3.2} I_0$.

   c. How many times more intense was the Loma Prieta earthquake than an earthquake with a magnitude of 3.2? Round to the nearest whole unit.

94. The intensities of earthquakes are measured with seismographs all over the world at different distances from the epicenter. Suppose that the intensity of a medium earthquake is originally reported as $10^{5.4}$ times I_0. Later this value is revised as $10^{5.8}$ times I_0.

   a. Determine the magnitude of the earthquake using the original estimate for intensity.

   b. Determine the magnitude using the revised estimate for intensity.

   c. How many times more intense was the earthquake than originally thought? Round to 1 decimal place.

Sounds are produced when vibrating objects create pressure waves in some medium such as air. When these variations in pressure reach the human eardrum, it causes the eardrum to vibrate in a similar manner and the ear detects sound. The intensity of sound is measured as power per unit area. The threshold for hearing (minimum sound detectable by a young, healthy ear) is defined to be $I_0 = 10^{-12}$ W/m^2 (watts per square meter). The sound level L, or "loudness" of sound, is measured in decibels (dB) as $L = 10 \log\left(\dfrac{I}{I_0}\right)$, where I is the intensity of the given sound. Use this formula for Exercises 95–96.

95. a. Find the sound level of a jet plane taking off if its intensity is 10^{15} times the intensity of I_0.

    b. Find the sound level of the noise from city traffic if its intensity is 10^9 times I_0.

    c. How many times more intense is the sound of a jet plane taking off than noise from city traffic?

96. a. Find the sound level of a motorcycle if its intensity is 10^{10} times I_0.

    b. Find the sound level of a vacuum cleaner if its intensity is 10^7 times I_0.

    c. How many times more intense is the sound of a motorcycle than a vacuum cleaner?

Scientists use the pH scale to represent the level of acidity or alkalinity of a liquid. This is based on the molar concentration of hydrogen ions, [H$^+$]. Since the values of [H$^+$] vary over a large range, 1×10^0 mole per liter to 1×10^{-14} mole per liter (mol/L), a logarithmic scale is used to compute pH. The formula

$$\text{pH} = -\log[\text{H}^+]$$

represents the pH of a liquid as a function of its concentration of hydrogen ions, [H$^+$].

The pH scale ranges from 0 to 14. Pure water is taken as neutral having a pH of 7. A pH less than 7 is acidic. A pH greater than 7 is alkaline (or basic). For Exercises 97–98, use the formula for pH. Round pH values to 1 decimal place.

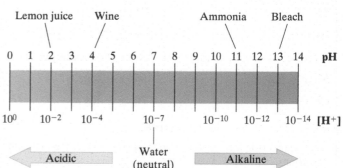

97. Vinegar and lemon juice are both acids. Their [H$^+$] values are 5.0×10^{-3} mol/L and 1×10^{-2} mol/L, respectively.

    a. Find the pH for vinegar.

    b. Find the pH for lemon juice.

    c. Which substance is more acidic?

98. Bleach and milk of magnesia are both bases. Their [H$^+$] values are 2.0×10^{-13} mol/L and 4.1×10^{-10} mol/L, respectively.

    a. Find the pH for bleach.

    b. Find the pH for milk of magnesia.

    c. Which substance is more basic?

Mixed Exercises

For Exercises 99–102,

a. Write the equation in exponential form.

b. Solve the equation from part (a).

c. Verify that the solution checks in the original equation.

99. $\log_3(x + 1) = 4$ **100.** $\log_2(x - 5) = 4$ **101.** $\log_4(7x - 6) = 3$ **102.** $\log_5(9x - 11) = 2$

For Exercises 103–106, evaluate the expressions.

103. $\log_3(\log_4 64)$ **104.** $\log_2\left[\log_{1/2}\left(\dfrac{1}{4}\right)\right]$ **105.** $\log_{16}(\log_{81} 3)$ **106.** $\log_4(\log_{16} 4)$

107. a. Evaluate $\log_2 2 + \log_2 4$

    b. Evaluate $\log_2(2 \cdot 4)$

    c. How do the values of the expressions in parts (a) and (b) compare?

108. a. Evaluate $\log_3 3 + \log_3 27$

    b. Evaluate $\log_3(3 \cdot 27)$

    c. How do the values of the expressions in parts (a) and (b) compare?

109. a. Evaluate $\log_4 64 - \log_4 4$

    b. Evaluate $\log_4\left(\dfrac{64}{4}\right)$

    c. How do the values of the expressions in parts (a) and (b) compare?

110. a. Evaluate $\log 100{,}000 - \log 100$

    b. Evaluate $\log\left(\dfrac{100{,}000}{100}\right)$

    c. How do the values of the expressions in parts (a) and (b) compare?

111. a. Evaluate $\log_2 2^5$

    b. Evaluate $5 \cdot \log_2 2$

    c. How do the values of the expressions in parts (a) and (b) compare?

112. a. Evaluate $\log_7 7^6$

    b. Evaluate $6 \cdot \log_7 7$

    c. How do the values of the expressions in parts (a) and (b) compare?

113. The time t (in years) required for an investment to double with interest compounded continuously depends on the interest rate r according to the function $t(r) = \dfrac{\ln 2}{r}$.

    a. If an interest rate of 3.5% is secured, determine the length of time needed for an initial investment to double. Round to 1 decimal place.

    b. Evaluate $t(0.04)$, $t(0.06)$, and $t(0.08)$.

114. The number n of monthly payments of P dollars each required to pay off a loan of A dollars in its entirety at interest rate r is given by

$$n = -\frac{\log\left(1 - \dfrac{Ar}{12P}\right)}{\log\left(1 + \dfrac{r}{12}\right)}$$

    a. A college student wants to buy a car and realizes that he can only afford payments of $200 per month. If he borrows $3000 and pays it off at 6% interest, how many months will it take him to retire the loan? Round to the nearest month.

    b. Determine the number of monthly payments of $611.09 that would be required to pay off a home loan of $128,000 at 4% interest.

For Exercises 115–116, use a calculator to approximate the given logarithms to 4 decimal places.

115. a. Avogadro's number is 6.022×10^{23}. Approximate $\log(6.022 \times 10^{23})$.

    b. Planck's constant is 6.626×10^{-34} J $\cdot$ sec. Approximate $\log(6.626 \times 10^{-34})$.

    c. Compare the value of the common logarithm to the power of 10 used in scientific notation.

116. a. The speed of light is 2.9979×10^8 m/sec. Approximate $\log(2.9979 \times 10^8)$.

    b. An elementary charge is 1.602×10^{-19} C. Approximate $\log(1.602 \times 10^{-19})$.

    c. Compare the value of the common logarithm to the power of 10 used in scientific notation.

Expanding Your Skills

For Exercises 117–122, write the domain in interval notation.

117. $t(x) = \log_4\left(\dfrac{x-1}{x-3}\right)$

118. $r(x) = \log_5\left(\dfrac{x+2}{x-4}\right)$

119. $s(x) = \ln(\sqrt{x+5} - 1)$

120. $v(x) = \ln(\sqrt{x-8} - 1)$

121. $c(x) = \log\left(\dfrac{1}{\sqrt{x-6}}\right)$

122. $d(x) = \log\left(\dfrac{1}{\sqrt{x+8}}\right)$

Technology Connections

123. a. Graph $f(x) = \ln x$ and
$$g(x) = (x - 1) - \frac{(x-1)^2}{2} + \frac{(x-1)^3}{3} - \frac{(x-1)^4}{4}$$
on the viewing window $[-2, 4, 1]$ by $[-5, 2, 1]$. How do the graphs compare on the interval $(0, 2)$?

    b. Use function g to approximate $\ln 1.5$. Round to 4 decimal places.

124. Compare the graphs of $Y_1 = \dfrac{e^x - e^{-x}}{2}$, $Y_2 = \ln(x + \sqrt{x^2 + 1})$, and $Y_3 = x$ on the viewing window $[-16.1, 16.1, 1]$ by $[-10, 10, 1]$. Based on the graphs, how do you suspect that the functions are related?

125. Compare the graphs of the functions.
$$Y_1 = \ln(2x) \quad \text{and} \quad Y_2 = \ln 2 + \ln x$$

126. Compare the graphs of the functions.
$$Y_1 = \ln\left(\frac{x}{2}\right) \quad \text{and} \quad Y_2 = \ln x - \ln 2$$

PROBLEM RECOGNITION EXERCISES

Analyzing Functions

For Exercises 1–14,

a. Write the domain.   **b.** Write the range.   **c.** Find the x-intercept(s).   **d.** Find the y-intercept.

e. Determine the asymptotes if applicable.   **f.** Determine the intervals over which the function is increasing.

g. Determine the intervals over which the function is decreasing.   **h.** Match the function with its graph.

1. $f(x) = 3$

2. $g(x) = 2x - 3$

3. $d(x) = (x - 3)^2 - 4$

4. $h(x) = \sqrt[3]{x - 2}$

5. $k(x) = \dfrac{2}{x - 1}$

6. $z(x) = \dfrac{3x}{x + 2}$

7. $p(x) = \left(\dfrac{4}{3}\right)^x$

8. $q(x) = -x^2 - 6x - 9$

9. $m(x) = |x - 4| - 1$

10. $n(x) = -|x| + 3$

11. $r(x) = \sqrt{3 - x}$

12. $s(x) = \sqrt{x - 3}$

13. $t(x) = e^x + 2$

14. $v(x) = \ln(x + 2)$

A.

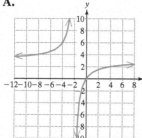

B.

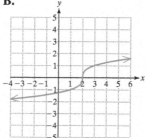

C.

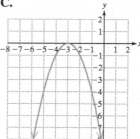

D.

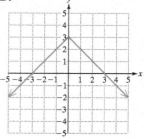

E.

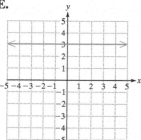

F.

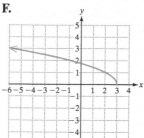

G.

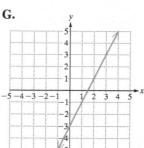

H.

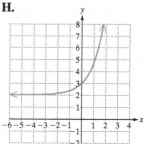

I.

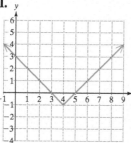

J.

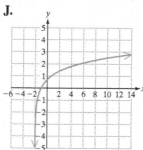

K.

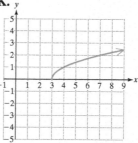

L.

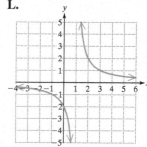

M.

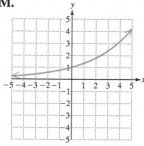

N.

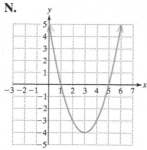

| **SECTION 4.4** | Properties of Logarithms |

OBJECTIVES

1. Apply the Product, Quotient, and Power Properties of Logarithms
2. Write a Logarithmic Expression in Expanded Form
3. Write a Logarithmic Expression as a Single Logarithm
4. Apply the Change-of-Base Formula

1. Apply the Product, Quotient, and Power Properties of Logarithms

By definition, $y = \log_b x$ is equivalent to $b^y = x$. Because a logarithm is an exponent, the properties of exponents can be applied to logarithms. The first is called the product property of logarithms.

> **Product Property of Logarithms**
>
> Let b, x, and y be positive real numbers where $b \neq 1$. Then
>
> $$\log_b(xy) = \log_b x + \log_b y.$$
>
> The logarithm of a product equals the sum of the logarithms of the factors.

TIP When two factors of the same base are multiplied, the base is unchanged and we add the exponents. This is the underlying principle for the product property of logarithms.

Proof:

Let $M = \log_b x$, which implies $b^M = x$.
Let $N = \log_b y$, which implies $b^N = y$.
Then $xy = b^M b^N = b^{M+N}$.

Writing the expression $xy = b^{M+N}$ in logarithmic form, we have,

$$\log_b(xy) = M + N$$
$$\log_b(xy) = \log_b x + \log_b y \checkmark$$

To demonstrate the product property of logarithms, simplify the following expressions by using the order of operations.

$$\log_3(3 \cdot 9) \overset{?}{=} \log_3 3 + \log_3 9$$
$$\log_3 27 \overset{?}{=} 1 + 2$$
$$3 \overset{?}{=} 3 \checkmark \text{ True}$$

| **EXAMPLE 1** | **Applying the Product Property of Logarithms** |

Write the logarithm as a sum and simplify if possible. Assume that x and y represent positive real numbers.

   a. $\log_2(8x)$ **b.** $\ln(5xy)$

Solution:

  a. $\log_2(8x) = \log_2 8 + \log_2 x$ Product property of logarithms
             $= 3 + \log_2 x$ Simplify. $\log_2 8 = \log_2 2^3 = 3$

  b. $\ln(5xy) = \ln 5 + \ln x + \ln y$

> **Skill Practice 1** Write the logarithm as a sum and simplify if possible. Assume that a, c, and d represent positive real numbers.
>
>   **a.** $\log_4(16a)$ **b.** $\log(12cd)$

Answers
1. a. $2 + \log_4 a$
   b. $\log 12 + \log c + \log d$

The quotient rule of exponents tells us that $\dfrac{b^M}{b^N} = b^{M-N}$ for $b \neq 0$. This property can be applied to logarithms.

Quotient Property of Logarithms

Let b, x, and y be positive real numbers where $b \neq 1$. Then

$$\log_b\left(\frac{x}{y}\right) = \log_b x - \log_b y.$$

The logarithm of a quotient equals the difference of the logarithm of the numerator and the logarithm of the denominator.

The proof of the quotient property for logarithms is similar to the proof of the product property (see Exercise 107). To demonstrate the quotient property for logarithms, simplify the following expressions by using the order of operations.

$$\log\left(\frac{1{,}000{,}000}{100}\right) \overset{?}{=} \log 1{,}000{,}000 - \log 100$$

$$\log 10{,}000 \overset{?}{=} 6 - 2$$

$$4 \overset{?}{=} 4 \checkmark \text{ True}$$

EXAMPLE 2　**Applying the Quotient Property of Logarithms**

Write the logarithm as the difference of logarithms and simplify if possible. Assume that the variables represent positive real numbers.

a. $\log_3\left(\dfrac{c}{d}\right)$　　**b.** $\log\left(\dfrac{x}{1000}\right)$

Solution:

a. $\log_3\left(\dfrac{c}{d}\right) = \log_3 c - \log_3 d$ 　　　　Quotient property of logarithms.

b. $\log\left(\dfrac{x}{1000}\right) = \log x - \log 1000$ 　　　Quotient property of logarithms.

$\qquad\qquad\quad = \log x - 3$ 　　　　　　　　Simplify.　$\log 1000 = \log 10^3 = 3$

Skill Practice 2　Write the logarithm as the difference of logarithms and simplify if possible. Assume that t represents a positive real number.

a. $\log_6\left(\dfrac{8}{t}\right)$　　**b.** $\ln\left(\dfrac{e}{12}\right)$

The last property we present here is the power property of logarithms. The power property of exponents tells us that $(b^M)^N = b^{MN}$. The same principle can be applied to logarithms.

Power Property of Logarithms

Let b and x be positive real numbers where $b \neq 1$. Let p be any real number. Then

$$\log_b x^p = p \log_b x.$$

The power property of logarithms is proved in Exercise 108.

Answers
2. a. $\log_6 8 - \log_6 t$ 　**b.** $1 - \ln 12$

EXAMPLE 3   **Applying the Power Property of Logarithms**

Apply the power property of logarithms.   **a.** $\ln \sqrt[5]{x^2}$ **b.** $\log x^2$

Solution:

a. $\ln \sqrt[5]{x^2} = \ln x^{2/5}$ Write $\sqrt[5]{x^2}$ using rational exponents.

  $= \dfrac{2}{5} \ln x$ provided that $x > 0$ Apply the power rule.

b. $\log x^2 = 2 \log x$ provided that $x > 0$ Apply the power rule.

In both parts (a) and (b), the condition that $x > 0$ is mandatory. The properties of logarithms hold true only for values of the variable for which the logarithms are defined. That is, the arguments must be positive.

From the graphs of $y = \log x^2$ and $y = 2 \log x$, we see that the domains are different. Therefore, the statement $\log x^2 = 2 \log x$ is true only for $x > 0$.

$y = \log x^2$ Domain: $(-\infty, 0) \cup (0, \infty)$ $y = 2 \log x$ Domain: $(0, \infty)$

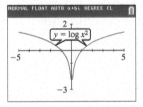   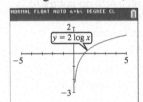

Skill Practice 3   Apply the power property of logarithms.

  a. $\log_5 \sqrt[5]{x^4}$ **b.** $\ln x^4$

At this point, we have learned seven properties of logarithms. The properties hold true for values of the variable for which the logarithms are defined. **Therefore, in the examples and exercises, we will assume that the variable expressions within the logarithms represent positive real numbers.**

Properties of Logarithms

Let b, x, and y be positive real numbers where $b \neq 1$, and let p be a real number. Then the following properties of logarithms are true.

1. $\log_b 1 = 0$ **5.** $\log_b(xy) = \log_b x + \log_b y$ **Product property**

2. $\log_b b = 1$ **6.** $\log_b\left(\dfrac{x}{y}\right) = \log_b x - \log_b y$ **Quotient property**

3. $\log_b b^p = p$ **7.** $\log_b x^p = p \log_b x$ **Power property**

4. $b^{\log_b x} = x$

2. Write a Logarithmic Expression in Expanded Form

Properties 5, 6, and 7 can be used in either direction. For example,

$$\log\left(\frac{ab}{c}\right) = \log a + \log b - \log c \quad \text{or} \quad \log a + \log b - \log c = \log\left(\frac{ab}{c}\right).$$

In some applications of algebra and calculus, the "condensed" form of the logarithm is preferred. In other applications, the "expanded" form is preferred. In Examples 4–6, we practice manipulating logarithmic expressions in both forms.

Answers

3. a. $\dfrac{4}{5} \log_5 x$ provided that $x > 0$

  b. $4 \ln x$ provided that $x > 0$

EXAMPLE 4   Writing a Logarithmic Expression in Expanded Form

Write the expression as the sum or difference of logarithms.

a. $\log_2\left(\dfrac{z^3}{xy^5}\right)$ **b.** $\log\sqrt[3]{\dfrac{(x+y)^2}{10}}$

Solution:

a. $\log_2\left(\dfrac{z^3}{xy^5}\right) = \log_2 z^3 - \log_2(xy^5)$ Apply the quotient property.

$= \log_2 z^3 - (\log_2 x + \log_2 y^5)$ Apply the product property.

$= \log_2 z^3 - \log_2 x - \log_2 y^5$ Apply the distributive property.

$= 3\log_2 z - \log_2 x - 5\log_2 y$ Apply the power property.

Avoiding Mistakes

In Example 4(b) do not try to simplify $\log(x+y)$. The argument contains a sum, not a product.

$\log(x+y)$ cannot be
  $\underbrace{\qquad}_{\text{sum}}$ simplified.

Compare to the logarithm of a product which can be simplified.

$\log(xy) = \log x + \log y$
  $\underbrace{\quad}_{\text{product}}$

---

b. $\log\sqrt[3]{\dfrac{(x+y)^2}{10}} = \log\left[\dfrac{(x+y)^2}{10}\right]^{1/3}$ Write the radical expression with rational exponents.

$= \dfrac{1}{3}\log\left[\dfrac{(x+y)^2}{10}\right]$ Apply the power property.

$= \dfrac{1}{3}[\log(x+y)^2 - \log 10]$ Apply the quotient property.

$= \dfrac{1}{3}[2\log(x+y) - 1]$ Apply the power property and simplify: $\log 10 = 1$.

$= \dfrac{2}{3}\log(x+y) - \dfrac{1}{3}$ Apply the distributive property.

Skill Practice 4   Write the expression as the sum or difference of logarithms.

a. $\ln\left(\dfrac{a^4 b}{c^9}\right)$ **b.** $\log_5\sqrt[3]{\dfrac{25}{(a^2+b)^2}}$

3. Write a Logarithmic Expression as a Single Logarithm

In Examples 5 and 6, we demonstrate how to write a sum or difference of logarithms as a single logarithm. We apply Properties 5, 6, and 7 of logarithms in reverse.

EXAMPLE 5   Writing the Sum or Difference of Logarithms as a Single Logarithm

Write the expression as a single logarithm and simplify the result if possible.

$$\log_2 560 - \log_2 7 - \log_2 5$$

Solution:

$\log_2 560 - \log_2 7 - \log_2 5$

$= \log_2 560 - (\log_2 7 + \log_2 5)$ Factor out -1 from the last two terms.

$= \log_2 560 - \log_2(7 \cdot 5)$ Apply the product property.

$= \log_2\left(\dfrac{560}{7 \cdot 5}\right)$ Apply the quotient property.

$= \log_2 16$ Simplify within the argument.

$= 4$ Simplify. $\log_2 16 = \log_2 2^4 = 4$

Answers

4. a. $4\ln a + \ln b - 9\ln c$

b. $\dfrac{2}{3} - \dfrac{2}{3}\log_5(a^2 + b)$

Skill Practice 5   Write the expression as a single logarithm and simplify the result if possible. $\log_3 54 + \log_3 10 - \log_3 20$

EXAMPLE 6   **Writing the Sum or Difference of Logarithms as a Single Logarithm**

Write the expression as a single logarithm and simplify the result if possible.

a. $3 \log a - \dfrac{1}{2}\log b - \dfrac{1}{2}\log c$ **b.** $\dfrac{1}{2}\ln x + \ln(x^2 - 1) - \ln(x + 1)$

Solution:

a. $3 \log a - \dfrac{1}{2}\log b - \dfrac{1}{2}\log c$

$= 3 \log a - \dfrac{1}{2}(\log b + \log c)$ Factor out $-\frac{1}{2}$ from the last two terms.

$= 3 \log a - \dfrac{1}{2}\log(bc)$ Apply the product property.

$= \log a^3 - \log(bc)^{1/2}$

$= \log a^3 - \log\sqrt{bc}$ Apply the power property.

$= \log\left(\dfrac{a^3}{\sqrt{bc}}\right)$ Apply the quotient property.

Avoiding Mistakes

In all examples and exercises in which we manipulate logarithmic expressions, it is important to note that the equivalences are true only for the values of the variables that make the expressions defined. In Example 6(b) we have the restriction that $x > 1$.

b. $\dfrac{1}{2}\ln x + \ln(x^2 - 1) - \ln(x + 1)$

$= \ln x^{1/2} + \ln(x^2 - 1) - \ln(x + 1)$ Apply the power property.

$= \ln[x^{1/2}(x^2 - 1)] - \ln(x + 1)$ Apply the product property.

$= \ln\left[\dfrac{\sqrt{x}(x^2 - 1)}{x + 1}\right]$ Apply the quotient property.

$= \ln\left[\dfrac{\sqrt{x}(x + 1)(x - 1)}{x + 1}\right]$ Factor the numerator of the argument.

$= \ln\left[\sqrt{x}(x - 1)\right]$ Simplify the argument.

Skill Practice 6   Write the expression as a single logarithm and simplify the result if possible.

a. $3 \log x - \dfrac{1}{3}\log y - \dfrac{2}{3}\log z$ **b.** $\dfrac{1}{3}\ln t + \ln(t^2 - 9) - \ln(t - 3)$

EXAMPLE 7   **Applying Properties of Logarithms**

Given that $\log_b 2 \approx 0.356$ and $\log_b 3 \approx 0.565$, approximate the value of $\log_b 36$.

Solution:

$\log_b 36 = \log_b(2 \cdot 3)^2$ Write the argument as a product of the factors 2 and 3.

$= 2 \log_b(2 \cdot 3)$ Apply the power property of logarithms.

$= 2(\log_b 2 + \log_b 3)$ Apply the product property of logarithms.

$\approx 2(0.356 + 0.565)$ Simplify.

≈ 1.842

Answers

5. $\log_3 27 = 3$

6. a. $\log\left(\dfrac{x^3}{\sqrt[3]{yz^2}}\right)$

b. $\ln\left[\sqrt[3]{t}\,(t + 3)\right]$

4. Apply the Change-of-Base Formula

A calculator can be used to approximate the value of a logarithm base 10 or base e by using the **LOG** key or the **LN** key, respectively. However, to use a calculator to evaluate a logarithmic expression with a different base, we must use the change-of-base formula.

Change-of-Base Formula

Let a and b be positive real numbers such that $a \neq 1$ and $b \neq 1$. Then for any positive real number x,

$$\log_b x = \frac{\log_a x}{\log_a b}$$

Note: The change-of-base formula converts a logarithm of one base to a ratio of logarithms of a different base. For the purpose of using a calculator, we often apply the change-of-base formula with base 10 or base e.

$$\log_b x = \frac{\log x}{\log b}$$
Original base is b.   Ratio of base 10 logarithms

$$\log_b x = \frac{\ln x}{\ln b}$$
Original base is b.   Ratio of base e logarithms

To derive the change-of-base formula, assume that a and b are positive real numbers with $a \neq 1$ and $b \neq 1$. Begin by letting $y = \log_b x$. If $y = \log_b x$, then

$$b^y = x$$ Write the original logarithm in exponential form.
$$\log_a b^y = \log_a x$$ Take the logarithm base a on both sides.
$$y \cdot \log_a b = \log_a x$$ Apply the power property of logarithms.
$$y = \frac{\log_a x}{\log_a b}$$ Solve for y.
$$\log_b x = \frac{\log_a x}{\log_a b}$$ Replace y by $\log_b x$.
This is the change-of-base formula.

EXAMPLE 8   Applying the Change-of-Base Formula

a. Estimate $\log_4 153$ between two consecutive integers.

b. Use the change-of-base formula to approximate $\log_4 153$ by using base 10. Round to 4 decimal places.

c. Use the change-of-base formula to approximate $\log_4 153$ by using base e.

d. Check the result by using the related exponential form.

Solution:

a. $64 < 153 < 256$
$4^3 < 153 < 4^4$
$3 < \log_4 153 < 4$ $\log_4 153$ is between 3 and 4.

b. $\log_4 153 = \dfrac{\log 153}{\log 4} \approx \dfrac{2.184691431}{0.6020599913} \approx 3.6287$

c. $\log_4 153 = \dfrac{\ln 153}{\ln 4} \approx \dfrac{5.030437921}{1.386294361} \approx 3.6287$

d. Check: $4^{3.6287} \approx 153$ ✓

> **Skill Practice 8**
>
> **a.** Estimate $\log_6 23$ between two consecutive integers.
> **b.** Use the change-of-base formula to evaluate $\log_6 23$ by using base 10. Round to 4 decimal places.
> **c.** Use the change-of-base formula to evaluate $\log_6 23$ by using base e. Round to 4 decimal places.
> **d.** Check the result by using the related exponential form.

TIP Although the numerators and denominators in parts (b) and (c) are different, their ratios are the same.

TECHNOLOGY CONNECTIONS

Using the Change-of-Base Formula to Graph a Logarithmic Function

The change-of-base formula can be used to graph logarithmic functions using a graphing utility. For example, to graph $Y_1 = \log_2 x$, enter the function as

$$Y = \log(x)/\log(2) \quad \text{or} \quad Y = \ln(x)/\ln(2)$$

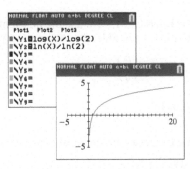

Point of Interest

The slide rule, first built in England in the early 17th century, is a mechanical computing device that uses logarithmic scales to perform operations involving multiplication, division, roots, logarithms, exponentials, and trigonometry. Amazingly, slide rules were used into the space age by engineers in the 1960's to help send astronauts to the moon. It was only with the invention of the pocket calculator that slide rules were replaced by modern computing devices.

Answers
8. a. Between 1 and 2
   b. 1.7500
   c. 1.7500
   d. $6^{1.7500} \approx 23$

SECTION 4.4   Practice Exercises

Prerequisite Review

For Exercises R.1–R.4, use the properties of exponents to simplify the expression.

R.1. $x^{-3} \cdot x^5 \cdot x^7$ **R.2.** $\dfrac{y^{-2}\,y^{10}}{y^3}$ **R.3.** $(4w^{-3}\,z^4)^2$ **R.4.** $\left(\dfrac{7k^4}{n}\right)^{-3}$

Concept Connections

1. The product property of logarithms states that $\log_b(xy) =$ _____ for positive real numbers b, x, and y, where $b \neq 1$.

2. The _____ property of logarithms states that $\log_b\left(\dfrac{x}{y}\right) =$ _____ for positive real numbers b, x, and y, where $b \neq 1$.

3. The power property of logarithms states that for any real number p, $\log_b x^p =$ _____ for positive real numbers b, x, and y, where $b \neq 1$.

4. The change-of-base formula states that $\log_b x$ can be written as a ratio of logarithms with base a as

$$\log_b x = \frac{\boxed{}}{\boxed{}}$$

5. The change-of-base formula is often used to convert a logarithm to a ratio of logarithms with base _____ or base _____ so that a calculator can be used to approximate the logarithm.

6. To use a graphing utility to graph the function defined by $y = \log_5 x$, use the change-of-base formula to write the function as $y =$ _____ or $y =$ _____.

For the exercises in this set, assume that all variable expressions represent positive real numbers.

Objective 1: Apply the Product, Quotient, and Power Properties of Logarithms

For Exercises 7–12, use the product property of logarithms to write the logarithm as a sum of logarithms. Then simplify if possible. (See Example 1)

7. $\log_5(125z)$

8. $\log_7(49k)$

9. $\log(8cd)$

10. $\log(24vw)$

11. $\log_2[(x + y) \cdot z]$

12. $\log_3[(a + b) \cdot c]$

For Exercises 13–18, use the quotient property of logarithms to write the logarithm as a difference of logarithms. Then simplify if possible. (See Example 2)

13. $\log_{12}\left(\dfrac{p}{q}\right)$

14. $\log_9\left(\dfrac{m}{n}\right)$

15. $\ln\left(\dfrac{e}{5}\right)$

16. $\ln\left(\dfrac{x}{e}\right)$

17. $\log\left(\dfrac{m^2 + n}{100}\right)$

18. $\log\left(\dfrac{1000}{c^2 + 1}\right)$

For Exercises 19–24, apply the power property of logarithms. (See Example 3)

19. $\log(2x - 3)^4$

20. $\log(8t - 3)^2$

21. $\log_6 \sqrt[7]{x^3}$

22. $\log_8 \sqrt[4]{x^3}$

23. $\ln 2^{kt}$

24. $\ln(0.5)^{rt}$

Objective 2: Write a Logarithmic Expression in Expanded Form

For Exercises 25–44, write the logarithm as a sum or difference of logarithms. Simplify each term as much as possible. (See Example 4)

25. $\log_4(7yz)$

26. $\log_2(5ab)$

27. $\log_7\left(\dfrac{1}{7}mn^2\right)$

28. $\log_4\left(\dfrac{1}{16}t^3v\right)$

29. $\log_2\left(\dfrac{x^{10}}{yz}\right)$

30. $\log_5\left(\dfrac{p^5}{mn}\right)$

31. $\log_6\left(\dfrac{p^5}{qt^3}\right)$

32. $\log_8\left(\dfrac{a^4}{b^9c}\right)$

33. $\log\left(\dfrac{10}{\sqrt{a^2 + b^2}}\right)$

34. $\log\left(\dfrac{\sqrt{d^2 + 1}}{10{,}000}\right)$

35. $\ln\left(\dfrac{\sqrt[3]{xy}}{wz^2}\right)$

36. $\ln\left(\dfrac{\sqrt[4]{pq}}{t^3m}\right)$

37. $\ln\sqrt[4]{\dfrac{a^2 + 4}{e^3}}$

38. $\ln\sqrt[5]{\dfrac{e^2}{c^2 + 5}}$

39. $\log\left[\dfrac{2x(x^2 + 3)^8}{\sqrt{4 - 3x}}\right]$

40. $\log\left[\dfrac{5y(4x + 1)^7}{\sqrt[3]{2 - 7x}}\right]$

41. $\log_5 \sqrt[3]{x\sqrt{5}}$

42. $\log_2 \sqrt[4]{y\sqrt{2}}$

43. $\log_2\left[\dfrac{4a^2\sqrt{3 - b}}{c(b + 4)^2}\right]$

44. $\log_3\left[\dfrac{27x^3\sqrt{y^2 - 1}}{y(x - 1)^2}\right]$

Objective 3: Write a Logarithmic Expression as a Single Logarithm

For Exercises 45–68, write the logarithmic expression as a single logarithm with coefficient 1, and simplify as much as possible. (See Examples 5–6)

45. $\ln y + \ln 4$

46. $\log 5 + \log p$

47. $\log_{15} 3 + \log_{15} 5$

48. $\log_{12} 8 + \log_{12} 18$

49. $\log_7 98 - \log_7 2$

50. $\log_6 144 - \log_6 4$

51. $\log 150 - \log 3 - \log 5$

52. $\log_3 693 - \log_3 33 - \log_3 7$

53. $2 \log_2 x + \log_2 t$

54. $5 \log_4 y + \log_4 w$

55. $4 \log_8 m - 3 \log_8 n - 2 \log_8 p$

56. $8 \log_3 x - 2 \log_3 z - 7 \log_3 y$

57. $3[\ln x - \ln(x + 3) - \ln (x - 3)]$

58. $2[\log(p - 4) - \log(p - 1) - \log(p + 4)]$

59. $\dfrac{1}{2}\ln(x + 1) - \dfrac{1}{2}\ln(x - 1)$

60. $\dfrac{1}{3}\ln(x^2 + 1) - \dfrac{1}{3}\ln(x + 1)$

61. $6 \log x - \dfrac{1}{3}\log y - \dfrac{2}{3}\log z$

62. $15 \log c - \dfrac{1}{4}\log d - \dfrac{3}{4}\log k$

63. $\dfrac{1}{3}\log_4 p + \log_4(q^2 - 16) - \log_4(q - 4)$

64. $\dfrac{1}{4}\log_2 w + \log_2(w^2 - 100) - \log_2(w + 10)$

65. $\dfrac{1}{2}[6 \ln(x + 2) + \ln x - \ln x^2]$

66. $\dfrac{1}{3}[12 \ln(x - 5) + \ln x - \ln x^3]$

67. $\log(8y^2 - 7y) + \log y^{-1}$

68. $\log(9t^3 - 5t) + \log t^{-1}$

For Exercises 69–78, use $\log_b 2 \approx 0.356$, $\log_b 3 \approx 0.565$, and $\log_b 5 \approx 0.827$ to approximate the value of the given logarithms. (See Example 7)

69. $\log_b 15$

70. $\log_b 10$

71. $\log_b 81$

72. $\log_b 125$

73. $\log_b 50$

74. $\log_b 12$

75. $\log_b\left(\dfrac{15}{2}\right)$

76. $\log_b\left(\dfrac{6}{5}\right)$

77. $\log_b 100$

78. $\log_b 225$

Objective 4: Apply the Change-of-Base Formula

For Exercises 79–84, (See Example 8)

a. Estimate the value of the logarithm between two consecutive integers. For example, $\log_2 7$ is between 2 and 3 because $2^2 < 7 < 2^3$.

b. Use the change-of-base formula and a calculator to approximate the logarithm to 4 decimal places.

c. Check the result by using the related exponential form.

79. $\log_2 15$

80. $\log_3 15$

81. $\log_5 3$

82. $\log_8 5$

83. $\log_2 0.3$

84. $\log_2 0.2$

For Exercises 85–88, use the change-of-base formula and a calculator to approximate the given logarithms. Round to 4 decimal places. Then check the answer by using the related exponential form. (See Example 8)

85. $\log_2(4.68 \times 10^7)$

86. $\log_2(2.54 \times 10^{10})$

87. $\log_4(5.68 \times 10^{-6})$

88. $\log_4(9.84 \times 10^{-5})$

Mixed Exercises

For Exercises 89–98, determine if the statement is true or false. For each false statement, provide a counterexample. For example, $\log(x + y) \neq \log x + \log y$ because $\log(2 + 8) \neq \log 2 + \log 8$ (the left side is 1 and the right side is approximately 1.204).

89. $\log e = \dfrac{1}{\ln 10}$

90. $\ln 10 = \dfrac{1}{\log e}$

91. $\log_5\left(\dfrac{1}{x}\right) = \dfrac{1}{\log_5 x}$

92. $\log_6\left(\dfrac{1}{t}\right) = \dfrac{1}{\log_6 t}$

93. $\log_4\left(\dfrac{1}{p}\right) = -\log_4 p$

94. $\log_8\left(\dfrac{1}{w}\right) = -\log_8 w$

95. $\log(xy) = (\log x)(\log y)$

96. $\log\left(\dfrac{x}{y}\right) = \dfrac{\log x}{\log y}$

97. $\log_2(7y) + \log_2 1 = \log_2(7y)$

98. $\log_4(3d) + \log_4 1 = \log_4(3d)$

Write About It

99. Explain why the product property of logarithms does not apply to the following statement.

$$\log_5(-5) + \log_5(-25)$$
$$= \log_5[(-5)(-25)]$$
$$= \log_5 125 = 3$$

100. Explain how to use the change-of-base formula and explain why it is important.

Expanding Your Skills

101. a. Write the difference quotient for $f(x) = \ln x$.

b. Show that the difference quotient from part (a) can be written as $\ln\left(\dfrac{x+h}{x}\right)^{1/h}$.

102. Show that
$$-\ln\left(x - \sqrt{x^2 - 1}\right) = \ln\left(x + \sqrt{x^2 - 1}\right)$$

103. Show that
$$\log\left(\frac{-b + \sqrt{b^2 - 4ac}}{2a}\right) + \log\left(\frac{-b - \sqrt{b^2 - 4ac}}{2a}\right)$$
$$= \log c - \log a$$

104. Show that
$$\ln\left(\frac{c + \sqrt{c^2 - x^2}}{c - \sqrt{c^2 - x^2}}\right) = 2\ln\left(c + \sqrt{c^2 - x^2}\right) - 2\ln x$$

105. Use the change-of-base formula to write $(\log_2 5)(\log_5 9)$ as a single logarithm.

106. Use the change-of-base formula to write $(\log_3 11)(\log_{11} 4)$ as a single logarithm.

107. Prove the quotient property of logarithms:

$$\log_b\left(\frac{x}{y}\right) = \log_b x - \log_b y.$$

(*Hint*: Modify the proof of the product property given on page 443.)

108. Prove the power property of logarithms:
$$\log_b x^p = p \log_b x.$$

Technology Connections

For Exercises 109–112, graph the function.

109. $f(x) = \log_5(x + 4)$

110. $g(x) = \log_7(x - 3)$

111. $k(x) = -3 + \log_{1/2} x$

112. $h(x) = 4 + \log_{1/3} x$

113. a. Graph $Y_1 = \log|x|$ and $Y_2 = \dfrac{1}{2}\log x^2$. How are the graphs related?

b. Show algebraically that $\dfrac{1}{2}\log x^2 = \log|x|$.

114. Graph $Y_1 = \ln(0.1x)$, $Y_2 = \ln(0.5x)$, $Y_3 = \ln x$, and $Y_4 = \ln(2x)$. How are the graphs related? Support your answer algebraically.

| **SECTION 4.5** | **Exponential and Logarithmic Equations and Applications** |

OBJECTIVES

1. **Solve Exponential Equations**
2. **Solve Logarithmic Equations**
3. **Use Exponential and Logarithmic Equations in Applications**

1. Solve Exponential Equations

A couple invests $8000 in a bond fund. The expected yield is 4.5% and the earnings are reinvested monthly. The growth of the investment is modeled by

$$A = 8000\left(1 + \frac{0.045}{12}\right)^{12t}$$ where A is the amount in the account after t years.

If the couple wants to know how long it will take for the investment to double, they would solve the equation:

$$16{,}000 = 8000\left(1 + \frac{0.045}{12}\right)^{12t}$$ (See Example 11.)

This equation is called an **exponential equation** because the equation contains a variable in the exponent. To solve an exponential equation first note that all exponential functions are one-to-one. Therefore, $b^x = b^y$ implies that $x = y$. This is called the equivalence property of exponential expressions.

TIP The equivalence property tells us that if two exponential expressions with the same base are equal, then their exponents must be equal.

> **Equivalence Property of Exponential Expressions**
>
> If b, x, and y are real numbers with $b > 0$ and $b \neq 1$, then
> $$b^x = b^y \quad \text{implies that } x = y.$$

EXAMPLE 1 **Solving Exponential Equations Using the Equivalence Property**

Solve. **a.** $3^{2x-6} = 81$ **b.** $25^{4-t} = \left(\dfrac{1}{5}\right)^{3t+1}$

Solution:

a.
$$3^{2x-6} = 81$$
$$3^{2x-6} = 3^4 \qquad \text{Write 81 as an exponential expression with a base of 3.}$$
$$2x - 6 = 4 \qquad \text{Equate the exponents.}$$
$$x = 5$$

The solution set is $\{5\}$.

Check: $3^{2x-6} = 81$
$$3^{2(5)-6} \stackrel{?}{=} 81$$
$$3^4 \stackrel{?}{=} 81 \checkmark$$

Avoiding Mistakes

When writing the expression $(5^2)^{4-t}$ as $5^{2(4-t)}$, it is important to use parentheses around the quantity $(4 - t)$. The exponent of 2 must be multiplied by the entire quantity $(4 - t)$. Likewise, parentheses are used around $(3t + 1)$ in the expression $5^{-1(3t+1)}$.

b.
$$25^{4-t} = \left(\dfrac{1}{5}\right)^{3t+1}$$
$$(5^2)^{4-t} = (5^{-1})^{3t+1} \qquad \text{Express both 25 and } \tfrac{1}{5} \text{ as integer powers of 5.}$$
$$5^{2(4-t)} = 5^{-1(3t+1)} \qquad \text{Apply the power property of exponents: } (b^m)^n = b^{m \cdot n}.$$
$$5^{8-2t} = 5^{-3t-1} \qquad \text{Apply the distributive property within the exponents.}$$
$$8 - 2t = -3t - 1 \qquad \text{Equate the exponents.}$$
$$t = -9 \qquad \text{The solution checks in the original equation.}$$

The solution set is $\{-9\}$.

Skill Practice 1 Solve. **a.** $4^{2x-3} = 64$ **b.** $27^{2w+5} = \left(\dfrac{1}{3}\right)^{2-5w}$

In Example 1, we were able to write the left and right sides of the equation with a common base. However, most exponential equations cannot be written in this form by inspection. For example:

$$7^x = 60$$
$$7^x = 7^?$$

60 is not a recognizable power of 7.

To solve such an equation, we can take a logarithm of the same base on each side of the equation, and then apply the power property of logarithms. This is demonstrated in Examples 2–4.

Answers
1. a. $\{3\}$ **b.** $\{-17\}$

> **Steps to Solve Exponential Equations by Using Logarithms**
>
> 1. Isolate the exponential expression on one side of the equation.
> 2. Take a logarithm of the same base on both sides of the equation.
> 3. Use the power property of logarithms to "bring down" the exponent.
> 4. Solve the resulting equation.

EXAMPLE 2 Solving an Exponential Equation Using Logarithms

Solve. $7^x = 60$

Solution:

$$7^x = 60$$ The exponential expression 7^x is isolated.

$$\log 7^x = \log 60$$ Take a logarithm of the same base on both sides of the equation. In this case, we have chosen base 10.

$$x \log 7 = \log 60$$ Apply the power property of logarithms.

This equation is now linear.

$$x = \frac{\log 60}{\log 7} \approx 2.1041$$ Divide both sides by $\log 7$.

Avoiding Mistakes

While 2.1041 is only an approximation, it is useful to check the result.

$$7^{2.1041} \approx 60$$

It is important to note that the exact solution to this equation is $\dfrac{\log 60}{\log 7}$ or equivalently by the change-of-base formula, $\log_7 60$. The value 2.1041 is merely an approximation.

The solution set is $\left\{ \dfrac{\log 60}{\log 7} \right\}$ or $\{\log_7 60\}$.

Skill Practice 2 Solve. $5^x = 83$

To solve the equation from Example 2, we can take a logarithm of any base. For example:

$$7^x = 60$$
$$\log_7 7^x = \log_7 60$$
$$x = \log_7 60 \text{ (solution)}$$

Take the logarithm base 7 on both sides.

$$7^x = 60$$
$$\ln 7^x = \ln 60$$
$$x \ln 7 = \ln 60$$
$$x = \frac{\ln 60}{\ln 7} \text{ (solution)}$$

Take the natural logarithm on both sides.

The values $\log_7 60$, $\dfrac{\log 60}{\log 7}$, and $\dfrac{\ln 60}{\ln 7}$ are all equivalent. However, common logarithms and natural logarithms are often used to express the solution to an exponential equation so that the solution can be approximated on a calculator.

Answer

2. $\left\{ \dfrac{\log 83}{\log 5} \right\}$ or $\{\log_5 83\}$

EXAMPLE 3   **Solving Exponential Equations Using Logarithms**

Solve.   **a.** $10^{5+2x} + 820 = 49{,}600$ **b.** $2000 = 18{,}000e^{-0.4t}$

Solution:

a. $10^{5+2x} + 820 = 49{,}600$ Isolate the exponential expression
$\qquad\quad 10^{5+2x} = 48{,}780$ on the left by subtracting 820 on
                              both sides.

$\log 10^{5+2x} = \log 48{,}780$ Since the exponential expression
                              on the left has a base of 10, take
                              the log base 10 on both sides.

$\quad 5 + 2x = \log 48{,}780$ On the left, $\log 10^{5+2x} = 5 + 2x$.

$\qquad\quad 2x = \log 48{,}780 - 5$ Solve the linear equation by
                              subtracting 5 and dividing by 2.

$\qquad x = \dfrac{\log 48{,}780 - 5}{2} \approx -0.1559$ The solution checks in the
                              original equation.

The solution set is $\left\{\dfrac{\log 48{,}780 - 5}{2}\right\}$.

b. $2000 = 18{,}000e^{-0.4t}$ Isolate the exponential expression on the right by dividing
                              both sides by 18,000.

$\dfrac{1}{9} = e^{-0.4t}$ Since the exponential expression on the left has a base of e,
                              take the log base e on both sides.

$\ln\left(\dfrac{1}{9}\right) = \ln e^{-0.4t}$ On the right, $\ln e^{-0.4t} = -0.4t$.

$\ln\left(\dfrac{1}{9}\right) = \underbrace{-0.4t}_{\text{(linear equation)}}$ Solve the linear equation by dividing by -0.4.

$\dfrac{\ln\left(\dfrac{1}{9}\right)}{-0.4} = t$ The exact solution to the equation can be written in a variety
                              of forms by applying the properties of logarithms:

$$\dfrac{\ln\left(\dfrac{1}{9}\right)}{-0.4} = \dfrac{\ln 1 - \ln 9}{-0.4} = \dfrac{0 - \ln 9}{-0.4} = \dfrac{\ln 9}{0.4} \approx 5.4931$$

Alternatively, $\dfrac{\ln 9}{0.4} = \dfrac{\ln 9}{\frac{2}{5}} = \dfrac{5 \ln 9}{2} \approx 5.4931$

The solution set is $\left\{\dfrac{\ln 9}{0.4}\right\}$ or $\left\{\dfrac{5 \ln 9}{2}\right\}$.

Skill Practice 3   Solve.

a. $400 + 10^{4x-1} = 63{,}000$ **b.** $100 = 700e^{-0.2k}$

In Example 4, we have an equation with two exponential expressions involving different bases.

EXAMPLE 4 **Solving an Exponential Equation**

Solve. $4^{2x-7} = 5^{3x+1}$

Solution:

$$4^{2x-7} = 5^{3x+1}$$

$$\ln 4^{2x-7} = \ln 5^{3x+1}$$ Take a logarithm of the same base on both sides.

$$(2x - 7)\ln 4 = (3x + 1)\ln 5$$ Apply the power property of logarithms.

$$2x \ln 4 - 7 \ln 4 = 3x \ln 5 + \ln 5$$ Apply the distributive property.

$$2x \ln 4 - 3x \ln 5 = \ln 5 + 7 \ln 4$$ Collect x terms on one side of the equation.

$$x(2 \ln 4 - 3 \ln 5) = \ln 5 + 7 \ln 4$$ Factor out x on the left.

$$x = \frac{\ln 5 + 7 \ln 4}{2 \ln 4 - 3 \ln 5} \approx -5.5034$$ Divide by $(2 \ln 4 - 3 \ln 5)$.

The solution set is $\left\{ \dfrac{\ln 5 + 7 \ln 4}{2 \ln 4 - 3 \ln 5} \right\}$. The solution checks in the original equation.

Skill Practice 4 Solve. $3^{5x-6} = 2^{4x+1}$

In Example 5, we look at an exponential equation in quadratic form.

EXAMPLE 5 **Solving an Exponential Equation in Quadratic Form**

Solve. $e^{2x} + 5e^x - 36 = 0$

Solution:

$$e^{2x} + 5e^x - 36 = 0$$

$$(e^x)^2 + 5(e^x) - 36 = 0$$ Note that $e^{2x} = (e^x)^2$.

$$u^2 + 5u - 36 = 0$$ The equation is in quadratic form. Let $u = e^x$.

$$(u - 4)(u + 9) = 0$$ Factor.

$$u = 4 \quad \text{or} \quad u = -9$$

$$e^x = 4 \quad \text{or} \quad e^x = -9$$ Back substitute. The second equation $e^x = -9$ has no solution.

$$\ln e^x = \ln 4$$

$$x = \ln 4 \approx 1.3863$$ No solution to this equation because $\ln(-9)$ is undefined.

The solution set is $\{\ln 4\}$. The solution checks in the original equation.

Skill Practice 5 Solve. $e^{2x} - 5e^x - 14 = 0$

Avoiding Mistakes

Recall that the range of $f(x) = e^x$ is the set of positive real numbers. Therefore, $e^x \neq -9$.

Answers

4. $\left\{ \dfrac{\ln 2 + 6 \ln 3}{5 \ln 3 - 4 \ln 2} \right\}$

5. $\{\ln 7\}$

2. Solve Logarithmic Equations

An equation containing a variable within a logarithmic expression is called a **logarithmic equation.** For example:

$$\log_2(3x - 4) = \log_2(x + 2) \quad \text{and} \quad \ln(x + 4) = 7 \quad \text{are logarithmic equations.}$$

Given an equation in which two logarithms of the same base are equated, we can apply the equivalence property of logarithms. Since all logarithmic functions are one-to-one, $\log_b x = \log_b y$ implies that $x = y$.

> **TIP** The equivalence property tells us that if two logarithmic expressions with the same base are equal, then their arguments must be equal.

Equivalence Property of Logarithmic Expressions

If b, x, and y are positive real numbers with $b \neq 1$, then

$$\log_b x = \log_b y \quad \text{implies that } x = y.$$

EXAMPLE 6 **Solving a Logarithmic Equation Using the Equivalence Property**

Solve. $\log_2(3x - 4) = \log_2(x + 2)$

Solution:

$\log_2(3x - 4) = \log_2(x + 2)$	Two logarithms of the same base are equated.
$3x - 4 = x + 2$	Equate the arguments.
$2x = 6$	Solve for x.
$x = 3$	Because the domain of a logarithmic function is restricted, it is mandatory that we check all potential solutions to a logarithmic equation.

Check: $\log_2(3x - 4) = \log_2(x + 2)$
$$\log_2[3(3) - 4] \stackrel{?}{=} \log_2[(3) + 2]$$

The solution set is $\{3\}$. $\log_2 5 \stackrel{?}{=} \log_2 5 \checkmark$

Skill Practice 6 Solve. $\log_2(7x - 4) = \log_2(2x + 1)$

TECHNOLOGY CONNECTIONS

Using a Calculator to View the Potential Solutions to a Logarithmic Equation

The solution to the equation in Example 6 is the x-coordinate of the point of intersection of $Y_1 = \log_2(3x - 4)$ and $Y_2 = \log_2(x + 2)$. The domain of $Y_1 = \log_2(3x - 4)$ is $\{x \mid x > \frac{4}{3}\}$ and the domain of $Y_2 = \log_2(x + 2)$ is $\{x \mid x > -2\}$. The solution to the equation $Y_1 = Y_2$ may not lie outside the domain of either function. This is why it is mandatory to check all potential solutions to a logarithmic equation.

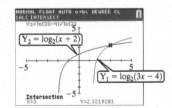

In Example 7, we encounter a logarithmic equation in which one or more solutions does not check.

EXAMPLE 7   **Solving a Logarithmic Equation**

Solve. $\ln(x - 4) = \ln(x + 6) - \ln x$

Solution:

$$\ln(x - 4) = \ln(x + 6) - \ln x$$

$$\ln(x - 4) = \ln\left(\frac{x + 6}{x}\right) \qquad \text{Combine the two logarithmic terms on the right.}$$

$$x - 4 = \frac{x + 6}{x} \qquad \text{Apply the equivalence property of logarithms.}$$

$$x^2 - 4x = x + 6 \qquad \text{Clear fractions by multiplying both sides by } x.$$

$$x^2 - 5x - 6 = 0 \qquad \text{The resulting equation is quadratic.}$$

$$(x - 6)(x + 1) = 0$$

$$x = 6 \quad \text{or} \quad x = -1 \qquad \text{The potential solutions are 6 and } -1.$$

Check:

$\ln(x - 4) = \ln(x + 6) - \ln x$	$\ln(x - 4) = \ln(x + 6) - \ln x$
$\ln(6 - 4) \stackrel{?}{=} \ln(6 + 6) - \ln 6$	$\ln(-1 - 4) \stackrel{?}{=} \ln(-1 + 6) - \ln(-1)$
$\ln 2 \stackrel{?}{=} \ln 12 - \ln 6$	$\ln(-5) \stackrel{?}{=} \ln 5 - \ln(-1)$
$\ln 2 \stackrel{?}{=} \ln\left(\frac{12}{6}\right) \checkmark$	undefined undefined

The only solution that checks is 6.
The solution set is $\{6\}$.

Skill Practice 7   Solve. $\ln x + \ln(x - 8) = \ln(x - 20)$

Many logarithmic equations, such as $4 \log_3 (2t - 7) = 8$ and $\log_2 x = 3 - \log_2 (x - 2)$, involve logarithmic terms and constant terms. In such a case, we can apply the properties of logarithms to write the equation in the form $\log_b x = k$, where k is a constant. At this point, we can solve for x by writing the equation in its equivalent exponential form $x = b^k$.

Solving Logarithmic Equations by Using Exponential Form

Step 1   Given a logarithmic equation, isolate the logarithms on one side of the equation.

Step 2   Use the properties of logarithms to write the equation in the form $\log_b x = k$, where k is a constant.

Step 3   Write the equation in exponential form.

Step 4   Solve the equation from step 3.

Step 5   Check the potential solution(s) in the original equation.

Answer

7. $\{ \ \}$; The values 4 and 5 do
not check.

EXAMPLE 8 **Solving a Logarithmic Equation**

Solve. $4 \log_3(2t - 7) = 8$

Solution:

$4 \log_3(2t - 7) = 8$

$\log_3(2t - 7) = 2$ Isolate the logarithm by dividing both sides by 4.

The equation is in the form $\log_b x = k$, where $x = 2t - 7$.

$2t - 7 = 3^2$ Write the equation in exponential form.

$2t - 7 = 9$ Check: $4 \log_3(2t - 7) = 8$

$t = 8$ $4 \log_3[2(8) - 7] \stackrel{?}{=} 8$

$4 \log_3 9 \stackrel{?}{=} 8$

$4 \cdot 2 \stackrel{?}{=} 8 \checkmark$

The solution set is $\{8\}$.

Skill Practice 8 Solve. $8 \log_4(w + 6) = 24$

EXAMPLE 9 **Solving a Logarithmic Equation**

Solve. $\log(w + 47) = 2.6$

Solution:

$\log(w + 47) = 2.6$ The equation is in the form $\log_b x = k$ where $x = w + 47$ and $b = 10$.

$w + 47 = 10^{2.6}$ Write the equation in exponential form.

$w = 10^{2.6} - 47 \approx 351.1072$ Solve the resulting linear equation.

Check: $\log(w + 47) = 2.6$

$\log[(10^{2.6} - 47) + 47] \stackrel{?}{=} 2.6$

The solution set is $\{10^{2.6} - 47\}$. $\log 10^{2.6} \stackrel{?}{=} 2.6 \checkmark$

Skill Practice 9 Solve. $\log(t - 18) = 1.4$

Example 10 contains multiple logarithmic terms and a constant term. We apply the strategy of collecting the logarithmic terms on one side and the constant term on the other side. Then after combining the logarithmic terms, we write the equation in exponential form.

Answers

8. $\{58\}$

9. $\{10^{1.4} + 18\}$

EXAMPLE 10 **Solving a Logarithmic Equation**

Solve. $\log_2 x = 3 - \log_2(x - 2)$

Solution:

$\log_2 x = 3 - \log_2(x - 2)$

$\log_2 x + \log_2(x - 2) = 3$ Isolate the logarithms on one side of the equation.

$\log_2[x(x - 2)] = 3$ Use the product property of logarithms to write a single logarithm.

$x(x - 2) = 2^3$ Write the equation in exponential form.

$x^2 - 2x = 8$

$x^2 - 2x - 8 = 0$ Set one side equal to zero.

$(x - 4)(x + 2) = 0$

$x = 4$ $x = -2$ Check: $\log_2 x = 3 - \log_2(x - 2)$ $\log_2 x = 3 - \log_2(x - 2)$

$\log_2 4 \overset{?}{=} 3 - \log_2(4 - 2)$ $\log_2(-2) \overset{?}{=} 3 - \log_2(-2 - 2)$

$\log_2 4 \overset{?}{=} 3 - \log_2 2$ $\log_2(-2) \overset{?}{=} 3 - \log_2(-4)$

$2 \overset{?}{=} 3 - 1$ ✓   undefined   undefined

The only solution that checks is $x = 4$.

The solution set is $\{4\}$.

Skill Practice 10   Solve. $2 - \log_7 x = \log_7(x - 48)$

3. Use Exponential and Logarithmic Equations in Applications

In Examples 11 and 12, we solve applications involving exponential and logarithmic equations.

EXAMPLE 11 **Using an Exponential Equation in a Finance Application**

A couple invests $8000 in a bond fund. The expected yield is 4.5% and the earnings are reinvested monthly.

a. Use $A = P\left(1 + \dfrac{r}{n}\right)^{nt}$ to write a model representing the amount A (in $) in the account after t years. The value r is the interest rate and n is the number of times interest is compounded per year.

b. Determine how long it will take the initial investment to double. Round to 1 decimal place.

> **TIP** Recall that monthly compounding indicates that interest is computed $n = 12$ times per year.

Solution:

a. $A = P\left(1 + \dfrac{r}{n}\right)^{nt}$

$A = 8000\left(1 + \dfrac{0.045}{12}\right)^{12t}$ Substitute $P = 8000$, $r = 0.045$, and $n = 12$.

Answer

10. $\{49\}$; The value -1 does not check.

b. $16{,}000 = 8000\left(1 + \dfrac{0.045}{12}\right)^{12t}$

The couple wants to double their money from $8000 to $16,000. Substitute $A = 16{,}000$ and solve for t.

$2 = \left(1 + \dfrac{0.045}{12}\right)^{12t}$

Isolate the exponential expression by dividing both sides by 8000.

$\ln 2 = \ln\left(1 + \dfrac{0.045}{12}\right)^{12t}$

Take a logarithm of the same base on both sides. We have chosen to use the natural logarithm.

$\ln 2 = 12t \ln\left(1 + \dfrac{0.045}{12}\right)$

Apply the power property of logarithms. The equation is now linear in the variable t.

$\dfrac{\ln 2}{12 \ln\left(1 + \dfrac{0.045}{12}\right)} = t$

Divide both sides by $12 \ln\left(1 + \frac{0.045}{12}\right)$.

$t \approx 15.4$

It will take approximately 15.4 yr for the investment to double.

Skill Practice 11 Determine how long it will take $8000 compounded monthly at 6% to double. Round to 1 decimal place.

In Example 12, we use a logarithmic equation in an application.

EXAMPLE 12 Using a Logarithmic Equation in a Medical Application

Suppose that the sound at a rock concert measures 124 dB (decibels).

a. Use the formula $L = 10 \log\left(\frac{I}{I_0}\right)$ to find the intensity of sound I (in W/m^2). The variable L represents the loudness of sound (in dB) and $I_0 = 10^{-12}$ W/m^2.

b. If the threshold at which sounds become painful is 1 W/m^2, will the music at this concert be physically painful? (Ignore the quality of the music.)

Solution:

a. $L = 10 \log\left(\dfrac{I}{I_0}\right)$

$124 = 10 \log\left(\dfrac{I}{10^{-12}}\right)$ Substitute 124 for L and 10^{-12} for I_0.

$12.4 = \log\left(\dfrac{I}{10^{-12}}\right)$ Divide both sides by 10. The logarithm is now isolated.

$10^{12.4} = \dfrac{I}{10^{-12}}$ Write the equation in exponential form.

$10^{12.4} \cdot 10^{-12} = I$ Multiply both sides by 10^{-12}.

$I = 10^{0.4} \approx 2.5$ W/m^2 Simplify.

b. The intensity of sound at the rock concert is approximately 2.5 W/m^2. This is above the threshold for pain.

Skill Practice 12

a. Find the intensity of sound from a leaf blower if the decibel level is 115 dB.

b. Is the intensity of sound from a leaf blower above the threshold for pain?

Answers
11. 11.6 yr
12. a. $10^{-0.5}$ W/m$^2 \approx 0.3$ W/m^2
  b. No

TECHNOLOGY CONNECTIONS

Using a Calculator to Approximate the Solutions to Exponential and Logarithmic Equations

There are many situations in which analytical methods fail to give a solution to a logarithmic or exponential equation. To find solutions graphically,

Enter the left side of the equation as Y_1.

Enter the right side of the equation as Y_2.

Then determine the point(s) of intersection of the graphs.

Example: $4 \ln x - 3x = -8$
$Y_1 = 4 \ln x - 3x$
$Y_2 = -8$ Solutions: $x \approx 0.1516$ and $x \approx 4.7419$

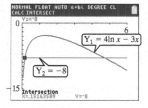 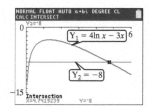

SECTION 4.5   Practice Exercises

Prerequisite Review

For Exercises R.1–R.6, solve the equation.

R.1. $7 = 5 - 2(4q - 1)$

R.2. $15n(n + 3) = 14n - 10$

R.3. $y^2 - 5y - 9 = 0$

R.4. $\dfrac{20}{n^2 - 2n} + 5 = \dfrac{10}{n - 2}$

R.5. $\sqrt{a + 18} + 2 = a$

R.6. $36d^{-2} - 5d^{-1} - 1 = 0$

Concept Connections

1. An equation such as $4^x = 9$ is called an _____ equation because the equation contains a variable in the exponent.

2. The equivalence property of exponential expressions states that if $b^x = b^y$, then _____ = _____.

3. The equivalence property of logarithmic expressions states that if $\log_b x = \log_b y$, then _____ = _____.

4. An equation containing a variable within a logarithmic expression is called a _____ equation.

Objective 1: Solve Exponential Equations

For Exercises 5–16, solve the equation. (See Example 1)

5. $3^x = 81$

6. $2^x = 32$

7. $\sqrt[3]{5} = 5^t$

8. $\sqrt{3} = 3^w$

9. $2^{-3y+1} = 16$

10. $5^{2z+2} = 625$

11. $11^{3c+1} = \left(\dfrac{1}{11}\right)^{c-5}$

12. $7^{2x-3} = \left(\dfrac{1}{49}\right)^{x+1}$

13. $8^{2x-5} = 32^{x-6}$

14. $27^{x-4} = 9^{2x+1}$

15. $100^{3t-5} = 1000^{3-t}$

16. $100{,}000^{2w+1} = 10{,}000^{4-w}$

For Exercises 17–34, solve the equation. Write the solution set with the exact values given in terms of common or natural logarithms. Also give approximate solutions to 4 decimal places. (See Examples 2–5)

17. $6^t = 87$

18. $2^z = 70$

19. $1024 = 19^x + 4$

20. $801 = 23^y + 6$

21. $10^{3+4x} - 8100 = 120{,}000$

22. $10^{5+8x} + 4200 = 84{,}000$

23. $21{,}000 = 63{,}000e^{-0.2t}$

24. $80 = 320e^{-0.5t}$

25. $4e^{2n-5} + 3 = 11$

26. $5e^{4m-3} - 7 = 13$

27. $3^{6x+5} = 5^{2x}$

28. $7^{4x-1} = 3^{5x}$

29. $2^{1-6x} = 7^{3x+4}$

30. $11^{1-8x} = 9^{2x+3}$

31. $e^{2x} - 9e^x - 22 = 0$

32. $e^{2x} - 6e^x - 16 = 0$

33. $e^{2x} = -9e^x$

34. $e^{2x} = -7e^x$

Objective 2: Solve Logarithmic Equations

For Exercises 35–36, determine if the given value of x is a solution to the logarithmic equation.

35. $\log_2(x - 31) = 5 - \log_2 x$

    a. $x = 16$

    b. $x = 32$

    c. $x = -1$

36. $\log_4 x = 3 - \log_4(x - 63)$

    a. $x = 64$

    b. $x = -1$

    c. $x = 32$

For Exercises 37–60, solve the equation. Write the solution set with the exact solutions. Also give approximate solutions to 4 decimal places if necessary. (See Examples 6–10)

37. $\log_4(3w + 11) = \log_4(3 - w)$

38. $\log_7(12 - t) = \log_7(t + 6)$

39. $\log(x^2 + 7x) = \log 18$

40. $\log(p^2 + 6p) = \log 7$

41. $6 \log_5(4p - 3) - 2 = 16$

42. $5 \log_6(7w + 1) + 3 = 13$

43. $2 \log_8(3y - 5) + 20 = 24$

44. $5 \log_3(7 - 5z) + 2 = 17$

45. $\log(p + 17) = 4.1$

46. $\log(q - 6) = 3.5$

47. $2 \ln(4 - 3t) + 1 = 7$

48. $4 \ln(6 - 5t) + 2 = 22$

49. $\log_2 w - 3 = -\log_2(w + 2)$

50. $\log_3 y + \log_3(y + 6) = 3$

51. $\log_6(7x - 2) = 1 + \log_6(x + 5)$

52. $\log_4(5x - 13) = 1 + \log_4(x - 2)$

53. $\log_5 z = 3 - \log_5(z - 20)$

54. $\log_2 x = 4 - \log_2(x - 6)$

55. $\ln x + \ln(x - 4) = \ln(3x - 10)$

56. $\ln x + \ln(x - 3) = \ln(5x - 7)$

57. $\log x + \log(x - 7) = \log(x - 15)$

58. $\log x + \log(x - 10) = \log(x - 18)$ **59.** $\log_8(6 - m) + \log_8(-m - 1) = 1$

60. $\log_3(n - 5) + \log_3(n + 3) = 2$

Objective 3: Use Exponential and Logarithmic Equations in Applications

For Exercises 61–70, use the model $A = Pe^{rt}$ or $A = P\left(1 + \dfrac{r}{n}\right)^{nt}$, where A is the future value of P dollars invested at interest rate r compounded continuously or n times per year for t years. (See Example 11)

61. If $10,000 is invested in an account earning 5.5% interest compounded continuously, determine how long it will take the money to triple. Round to the nearest year.

62. If a couple has $80,000 in a retirement account, how long will it take the money to grow to $1,000,000 if it grows by 6% compounded continuously? Round to the nearest year.

63. A $2500 bond grows to $3729.56 in 10 yr under continuous compounding. Find the interest rate. Round to the nearest whole percent.

64. $5000 grows to $5438.10 in 2 yr under continuous compounding. Find the interest rate. Round to the nearest tenth of a percent.

65. An $8000 investment grows to $9289.50 at 3% interest compounded quarterly. For how long was the money invested? Round to the nearest year.

66. $20,000 is invested at 3.5% interest compounded monthly. How long will it take for the investment to double? Round to the nearest tenth of a year.

67. A $25,000 inheritance is invested for 15 yr compounded quarterly and grows to $52,680. Find the interest rate. Round to the nearest percent.

68. A $10,000 investment grows to $11,273 in 4 yr compounded monthly. Find the interest rate. Round to the nearest percent.

69. If $4000 is put aside in a money market account with interest compounded continuously at 2.2%, find the time required for the account to *earn* $1000. Round to the nearest month.

70. Victor puts aside $10,000 in an account with interest compounded continuously at 2.7%. How long will it take for him to *earn* $2000? Round to the nearest month.

71. Physicians often treat thyroid cancer with a radioactive form of iodine called iodine-131 (^{131}I). The radiological half-life of ^{131}I is approximately 8 days, but the biological half-life for most individuals is 4.2 days. The biological half-life is shorter because in addition to ^{131}I being lost to decay, the iodine is also excreted from the body in urine, sweat, and saliva.

For a patient treated with 100 mCi (millicuries) of ^{131}I, the radioactivity level R (in mCi) after t days is given by $R = 100(2)^{-t/4.2}$.

a. State law mandates that the patient stay in an isolated hospital room for 2 days after treatment with ^{131}I. Determine the radioactivity level at the end of 2 days. Round to the nearest whole unit.

b. After the patient is released from the hospital, the patient is directed to avoid direct human contact until the radioactivity level drops below 30 mCi. For how many days *after* leaving the hospital will the patient need to stay in isolation? Round to the nearest tenth of a day.

72. Caffeine occurs naturally in a variety of food products such as coffee, tea, and chocolate. The kidneys filter the blood and remove caffeine and other drugs through urine. The biological half-life of caffeine is approximately 6 hr. If one cup of coffee has 80 mg of caffeine, then the amount of caffeine C (in mg) remaining after t hours is given by $C = 80(2)^{-t/6}$.

a. How long will it take for the amount of caffeine to drop below 60 mg? Round to 1 decimal place.

b. Laura has trouble sleeping if she has more than 30 mg of caffeine in her bloodstream. How many hours after drinking a cup of coffee would Laura have to wait so that the coffee would not disrupt her sleep? Round to 1 decimal place.

Sunlight is absorbed in water, and as a result the light intensity in oceans, lakes, and ponds decreases exponentially with depth. The percentage of visible light, P (in decimal form), at a depth of x meters is given by $P = e^{-kx}$, where k is a constant related to the clarity and other physical properties of the water. The graph shows models for the open ocean, Lake Tahoe, and Lake Erie for data taken under similar conditions. Use these models for Exercises 73–76.

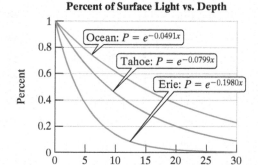

Percent of Surface Light vs. Depth

Ocean: $P = e^{-0.0491x}$

Tahoe: $P = e^{-0.0799x}$

Erie: $P = e^{-0.1980x}$

73. Determine the depth at which the light intensity is half the value from the surface for each body of water given. Round to the nearest tenth of a meter.

74. Determine the depth at which the light intensity is 20% of the value from the surface for each body of water given. Round to the nearest tenth of a meter.

75. The *euphotic* depth is the depth at which light intensity falls to 1% of the value at the surface. This depth is of interest to scientists because no appreciable photosynthesis takes place. Find the euphotic depth for the open ocean. Round to the nearest tenth of a meter.

76. Refer to Exercise 75, and find the euphotic depth for Lake Tahoe and for Lake Erie. Round to the nearest tenth of a meter.

77. Forge welding is a process in which two pieces of steel are joined together by heating the pieces of steel and hammering them together. A welder takes a piece of steel from a forge at 1600°F and places it on an anvil where the outdoor temperature is 50°F. The temperature of the steel T (in °F) can be modeled by $T = 50 + 1550e^{-0.05t}$, where t is the time in minutes after the steel is removed from the forge. How long will it take for the steel to reach a temperature of 100°F so that it can be handled without heat protection? Round to the nearest minute.

78. A pie comes out of the oven at 325°F and is placed to cool in a 70°F kitchen. The temperature of the pie T (in °F) after t minutes is given by $T = 70 + 255e^{-0.017t}$. The pie is cool enough to cut when the temperature reaches 110°F. How long will this take? Round to the nearest minute.

For Exercises 79–80, the formula $L = 10 \log \left(\frac{I}{I_0}\right)$ gives the loudness of sound L (in dB) based on the intensity of sound I (in W/m²). The value $I_0 = 10^{-12}$ W/m² is the minimal threshold for hearing for midfrequency sounds. Hearing impairment is often measured according to the minimal sound level (in dB) detected by an individual for sounds at various frequencies. For one frequency, the table depicts the level of hearing impairment.

Category	Loudness (dB)
Mild	$26 \le L \le 40$
Moderate	$41 \le L \le 55$
Moderately severe	$56 \le L \le 70$
Severe	$71 \le L \le 90$
Profound	$L > 90$

79. a. If the minimum intensity heard by an individual is 3.4×10^{-8} W/m², determine if the individual has a hearing impairment.

b. If the minimum loudness of sound detected by an individual is 30 dB, determine the corresponding intensity of sound. (**See Example 12**)

80. Determine the range that represents the intensity of sound that can be heard by an individual with severe hearing impairment.

For Exercises 81–82, use the formula pH $= -\log[\text{H}^+]$. The variable pH represents the level of acidity or alkalinity of a liquid on the pH scale, and H^+ is the concentration of hydronium ions in the solution. Determine the value of H^+ (in mol/L) for the following liquids, given their pH values.

81. a. Seawater pH $= 8.5$

b. Acid rain pH $= 2.3$

82. a. Milk pH $= 6.2$

b. Sodium bicarbonate pH $= 8.4$

83. A new teaching method to teach vocabulary to sixth-graders involves having students work in groups on an assignment to learn new words. After the lesson was completed, the students were tested at 1-month intervals. The average score for the class $S(t)$ can be modeled by

$$S(t) = 94 - 18 \ln(t + 1)$$

where t is the time in months after completing the assignment. If the average score is 65, how many months had passed since the students completed the assignment? Round to the nearest month.

84. A company spends x hundred dollars on an advertising campaign. The amount of money in sales $S(x)$ (in $1000) for the 4-month period after the advertising campaign can be modeled by

$$S(x) = 5 + 7 \ln(x + 1)$$

If the sales total $19,100, how much was spent on advertising? Round to the nearest dollar.

85. Radiated seismic energy from an earthquake is estimated by $\log E = 4.4 + 1.5M$, where E is the energy in Joules (J) and M is surface wave magnitude.

a. How many times more energy does an 8.2-magnitude earthquake have than a 5.5-magnitude earthquake? Round to the nearest thousand.

b. How many times more energy does a 7-magnitude earthquake have than a 6-magnitude earthquake? Round to the nearest whole number.

86. On August 31, 1854, an epidemic of cholera was discovered in London, England, resulting from a contaminated community water pump. By the end of September, more than 600 citizens who drank water from the pump had died. The cumulative number of deaths $D(t)$ at a time t days after August 31 is given by $D(t) = 91 + 160 \ln(t + 1)$.

a. Determine the cumulative number of deaths by September 15. Round to the nearest whole unit.

b. Approximately how many days after August 31 did the cumulative number of deaths reach 600?

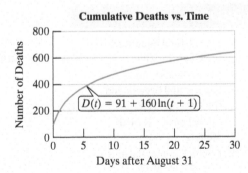

Cumulative Deaths vs. Time

$D(t) = 91 + 160 \ln(t + 1)$

Days after August 31

Mixed Exercises

For Exercises 87–94, find an equation for the inverse function.

87. $f(x) = 2^x - 7$

88. $f(x) = 5^x + 6$

89. $f(x) = \ln(x + 5)$

90. $f(x) = \ln(x - 7)$

91. $f(x) = 10^{x-3} + 1$

92. $f(x) = 10^{x+2} - 4$

93. $f(x) = \log(x + 7) - 9$

94. $f(x) = \log(x - 11) + 8$

For Exercises 95–112, solve the equation. Write the solution set with exact solutions. Also give approximate solutions to 4 decimal places if necessary.

95. $5^{|x|} - 3 = 122$

96. $11^{|x|} + 9 = 130$

97. $\log x - 2 \log 3 = 2$

98. $\log y - 3 \log 5 = 3$

99. $6^{x^2-2} = 36$

100. $8^{y^2-7} = 64$

101. $\log_9 |x + 4| = \log_9 6$

102. $\log_8 |3 - x| = \log_8 5$

103. $x^2 e^x = 9e^x$

104. $x^2 6^x = 6^x$

105. $\log_3(\log_3 x) = 0$

106. $\log_5(\log_5 x) = 1$

107. $3|\ln x| - 12 = 0$

108. $7|\ln x| - 14 = 0$

109. $\log_3 x - \log_3(2x + 6) = \dfrac{1}{2}\log_3 4$

110. $\log_5 x - \log_5(x + 1) = \dfrac{1}{3}\log_5 8$

111. $2e^x(e^x - 3) = 3e^x - 4$

112. $3e^x(e^x - 6) = 4e^x - 7$

Write About It

113. Explain the process to solve the equation $4^x = 11$.

114. Explain the process to solve the equation $\log_b 5 + \log_b(x - 3) = 4$.

Expanding Your Skills

For Exercises 115–126, solve the equation.

115. $\dfrac{10^x - 13 \cdot 10^{-x}}{3} = 4$

116. $\dfrac{e^x - 9e^{-x}}{2} = 4$

117. $(\ln x)^2 - \ln x^5 = -4$

118. $(\ln x)^2 + \ln x^3 = -2$

119. $(\log x)^2 = \log x^2$

120. $(\log x)^2 = \log x^3$

121. $\log w + 4\sqrt{\log w} - 12 = 0$

122. $\ln x + 3\sqrt{\ln x} - 10 = 0$

123. $e^{2x} - 8e^x + 6 = 0$

124. $e^{2x} - 6e^x + 4 = 0$

125. $\log_5 \sqrt{6c + 5} + \log_5 \sqrt{c} = 1$

126. $\log_3 \sqrt{x - 8} + \log_3 \sqrt{x} = 1$

Technology Connections

For Exercises 127–130, an equation is given in the form $Y_1(x) = Y_2(x)$. Graph Y_1 and Y_2 on a graphing utility on the window [10, 10, 1] by [10, 10, 1]. Then approximate the point(s) of intersection to approximate the solution(s) to the equation. Round to 4 decimal places.

127. $4x - e^x + 6 = 0$

128. $x^3 - e^{2x} + 4 = 0$

129. $x^2 + 5 \log x = 6$

130. $x^2 - 0.05 \ln x = 4$

SECTION 4.6   Modeling with Exponential and Logarithmic Functions

OBJECTIVES

1. Solve Literal Equations for a Specified Variable
2. Create Models for Exponential Growth and Decay
3. Apply Logistic Growth Models
4. Create Exponential and Logarithmic Models Using Regression

1. Solve Literal Equations for a Specified Variable

A short-term model to predict the U.S. population P is $P = 310e^{0.00965t}$, where t is the number of years since 2010. If we solve this equation for t, we have

$$t = \frac{\ln\left(\dfrac{P}{310}\right)}{0.00965} \quad \text{or equivalently} \quad t = \frac{\ln P - \ln 310}{0.00965}.$$

This is a model that predicts the time required for the U.S. population to reach a value P. Manipulating an equation for a specified variable was first introduced in Section 1.1. In Example 1, we revisit this skill using exponential and logarithmic equations.

| EXAMPLE 1 | Solving an Equation for a Specified Variable |

a. Given $P = 100e^{kx} - 100$, solve for x. (Used in geology)

b. Given $L = 8.8 + 5.1 \log D$, solve for D. (Used in astronomy)

Solution:

a. $P = 100e^{kx} - 100$

$P + 100 = 100e^{kx}$ Add 100 to both sides to isolate the x term.

$\dfrac{P + 100}{100} = e^{kx}$ Divide by 100.

$\ln\left(\dfrac{P + 100}{100}\right) = \ln e^{kx}$ Take the natural logarithm of both sides.

$\ln\left(\dfrac{P + 100}{100}\right) = kx$ Simplify: $\ln e^{kx} = kx$

$x = \dfrac{\ln\left(\dfrac{P + 100}{100}\right)}{k}$ Divide by k.

$x = \dfrac{\ln\left(\dfrac{P + 100}{100}\right)}{k}$ or equivalently $x = \dfrac{\ln(P + 100) - \ln 100}{k}$

b. $L = 8.8 + 5.1 \log D$

$\dfrac{L - 8.8}{5.1} = \log D$ Subtract 8.8 from both sides and divide by 5.1.

$D = 10^{(L-8.8)/5.1}$ Write the equation in exponential form.

Skill Practice 1

a. Given $T = 78 + 272e^{-kt}$, solve for k.

b. Given $S = 90 - 20 \ln(t + 1)$, solve for t.

---

2. Create Models for Exponential Growth and Decay

In Section 4.2, we defined an exponential function as $y = b^x$, where $b > 0$ and $b \neq 1$. Throughout the chapter, we have used transformations of basic exponential functions to solve a variety of applications. The following variation of the general exponential form is used to solve applications involving exponential growth and decay.

Answers

1. a. $k = -\dfrac{\ln\left(\dfrac{T - 78}{272}\right)}{t}$ or

$k = \dfrac{\ln 272 - \ln(T - 78)}{t}$

b. $t = e^{(90-S)/20} - 1$

Exponential Growth and Decay Models

Let y be a variable changing exponentially with respect to t, and let y_0 represent the initial value of y when $t = 0$. Then for a constant k:

If $k > 0$, then $y = y_0e^{kt}$ is a model for exponential growth.	If $k < 0$, then $y = y_0e^{kt}$ is a model for exponential decay.
Example:	**Example:**
$y = 2000e^{0.06t}$ represents the value of a \$2000 investment after t years with interest compounded continuously.	$y = 100e^{-0.165t}$ represents the radioactivity level t hours after a patient is treated for thyroid cancer with 100 mCi of radioactive iodine.
(*Note*: $k = 0.06 > 0$)	(*Note*: $k = -0.165 < 0$)

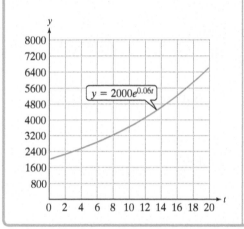

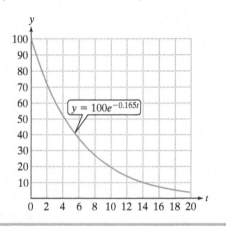

The model $y = y_0e^{kt}$ is often presented with different letters or symbols in place of y, y_0, k, and t to convey their meaning in the context of the application. For example, to compute the value of an investment under continuous compounding, we have

$$A = Pe^{rt}$$

P (for principal) is used in place of y_0.
r (for the annual interest rate) is used in place of k.
A (for the future value of the investment) is used in place of y.

We can also use function notation when expressing a model for exponential growth or decay. For example, consider the model for population growth.

$$P(t) = P_0e^{kt}$$

P_0 (for initial population) is used in place of y_0.
$P(t)$ represents the population as a function of time and is used in place of y.

EXAMPLE 2 Creating a Model for Growth of an Investment

Suppose that \$15,000 is invested and at the end of 3 yr, the value of the account is \$19,356.92. Use the model $A = Pe^{rt}$ to determine the average rate of return r under continuous compounding.

Solution:

$A = Pe^{rt}$	Begin with an appropriate model.
$A = 15{,}000e^{rt}$	P represents the initial value of the account (initial principal). Substitute 15,000 for P.
$19{,}356.92 = 15{,}000e^{r(3)}$	We have a known data point where $A = 19{,}356.92$ when $t = 3$. Substituting these values into the formula enables us to solve for r.
$\dfrac{19{,}356.92}{15{,}000} = e^{3r}$	Divide both sides by 15,000.
$\ln\left(\dfrac{19{,}356.92}{15{,}000}\right) = \ln(e^{3r})$	Take the natural logarithm of both sides.
$\ln\left(\dfrac{19{,}356.92}{15{,}000}\right) = 3r$	Simplify: $\ln e^{3r} = 3r$
$r = \dfrac{\ln\left(\dfrac{19{,}356.92}{15{,}000}\right)}{3}$	Divide by 3 to isolate r.
$r \approx 0.085$	

The average rate of return is approximately 8.5%.

Skill Practice 2   Suppose that $10,000 is invested and at the end of 5 yr, the value of the account is $13,771.28. Use the model $A = Pe^{rt}$ to determine the average rate of return r under continuous compounding.

In Example 3, we build a model to predict short-term population growth.

EXAMPLE 3   **Creating a Model for Population Growth**

On January 1, 2000, the population of California was approximately 34 million. On January 1, 2010, the population was 37.3 million.

a. Write a function of the form $P(t) = P_0e^{kt}$ to represent the population of California $P(t)$ (in millions), t years after January 1, 2000. Round k to 5 decimal places.

b. Use the function in part (a) to predict the population on January 1, 2018. Round to 1 decimal place.

c. Use the function from part (a) to determine the year during which the population of California will be twice the value from the year 2000.

Solution:

TIP The value of k in the model $P(t) = P_0e^{kt}$ is called a parameter and is related to the growth rate of the population being studied. The value of k will be different for different populations.

a.

$P(t) = P_0e^{kt}$	Begin with an appropriate model.
$P(t) = 34e^{kt}$	The initial population is $P_0 = 34$ million.
$37.3 = 34e^{k(10)}$	We have a known data point $P(10) = 37.3$. Substituting these values into the function enables us to solve for k.
$\dfrac{37.3}{34} = e^{k(10)}$	Divide both sides by 34.
$\ln\left(\dfrac{37.3}{34}\right) = 10k$	Take the natural logarithm of both sides.
$k = \dfrac{\ln\left(\dfrac{37.3}{34}\right)}{10} \approx 0.00926$	Divide by 10 to isolate k.
$P(t) = 34e^{0.00926t}$	This model gives the population as a function of time.

Answer

2. 6.4%

b. $P(t) = 34e^{0.00926t}$

$P(18) = 34e^{0.00926(18)}$ Substitute 18 for t.

       $= 40.2$

The population in California on January 1, 2018, will be approximately 40.2 million if this trend continues.

c. $P(t) = 34e^{0.00926t}$

    $68 = 34e^{0.00926t}$ Substitute 68 for $P(t)$.

    $\dfrac{68}{34} = e^{0.00926t}$ Divide both sides by 34.

  $\ln 2 = 0.00926t$ Take the natural logarithm of both sides.

     $t = \dfrac{\ln 2}{0.00926} \approx 74.85$ Divide by 0.00926 to isolate t.

The population of California will reach 68 million toward the end of the year 2074 if this trend continues.

Skill Practice 3   On January 1, 2000, the population of Texas was approximately 21 million. On January 1, 2010, the population was 25.2 million.

  a. Write a function of the form $P(t) = P_0 e^{kt}$ to represent the population $P(t)$ of Texas t years after January 1, 2000. Round k to 5 decimal places.

  b. Use the function in part (a) to predict the population on January 1, 2020. Round to 1 decimal place.

  c. Use the function in part (a) to determine the year during which the population of Texas will reach 40 million if this trend continues.

An exponential model can be presented with a base other than base e. For example, suppose that a culture of bacteria begins with 5000 organisms and the population doubles every 4 hr. Then the population $P(t)$ can be modeled by

     $P(t) = 5000(2)^{t/4}$, where t is the time in hours after the culture was started.

Notice that this function is defined using base 2. It is important to realize that any exponential function of one base can be rewritten in terms of an exponential function of another base. In particular we are interested in expressing the function with base e.

Writing an Exponential Expression Using Base e

Let t and b be real numbers, where $b > 0$ and $b \neq 1$. Then,

$$b^t \text{ is equivalent to } e^{(\ln b)t}.$$

To show that $e^{(\ln b)t} = b^t$, use the power property of exponents; that is,

$$e^{(\ln b)t} = (e^{\ln b})^t = b^t$$

EXAMPLE 4   **Writing an Exponential Function with Base e**

  a. The population $P(t)$ of a culture of bacteria is given by $P(t) = 5000(2)^{t/4}$, where t is the time in hours after the culture was started. Write the rule for this function using base e.

  b. Find the population after 12 hr using both forms of the function from part (a).

Answers

3. a. $P(t) = 21e^{0.01823t}$

  b. 30.2 million

  c. 2035

Solution:

a. $P(t) = 5000(2)^{t/4}$

Note that $2^{t/4} = (2^t)^{1/4}$

$$= \left[e^{(\ln 2)t} \right]^{1/4} \qquad \text{Apply the property that } e^{(\ln b)t} = b^t.$$

$$= e^{[(\ln 2)/4]t} \qquad \text{Apply the power rule of exponents.}$$

Therefore, $P(t) = 5000(2)^{t/4}$

$$= 5000e^{[(\ln 2)/4]t}$$

$$\approx 5000e^{0.17329t}$$

b. $P(t) = 5000(2)^{t/4}$ $P(t) \approx 5000e^{0.17329t}$

$P(12) = 5000(2)^{(12)/4}$ $P(12) \approx 5000e^{0.17329(12)}$

$= 40{,}000$ $\approx 40{,}000$

Skill Practice 4

a. Given $P(t) = 10{,}000(2)^{-0.4t}$, write the rule for this function using base e.

b. Find the function value for $t = 10$ for both forms of the function from part (a).

In Example 5, we apply an exponential decay function to determine the age of a bone through radiocarbon dating. Animals ingest carbon through respiration and through the food they eat. Most of the carbon is carbon-12 (^{12}C), an abundant and stable form of carbon. However, a small percentage of carbon is the radioactive isotope, carbon-14 (^{14}C). The ratio of carbon-12 to carbon-14 is constant for all living things. When an organism dies, it no longer takes in carbon from the environment. Therefore, as the carbon-14 decays, the ratio of carbon-12 to carbon-14 changes. Scientists know that the half-life of ^{14}C is 5730 years and from this, they can build a model to represent the amount of ^{14}C remaining t years after death. This is illustrated in Example 5.

EXAMPLE 5   **Creating a Model for Exponential Decay**

a. Carbon-14 has a half-life of 5730 yr. Write a model of the form $Q(t) = Q_0 e^{-kt}$ to represent the amount $Q(t)$ of carbon-14 remaining after t years if no additional carbon is ingested.

b. An archeologist uncovers human remains at an ancient Roman burial site and finds that 76.6% of the carbon-14 still remains in the bone. How old is the bone? Round to the nearest hundred years.

TIP   Given the half-life of a radioactive substance, we can also write an exponential model using base $\frac{1}{2}$. The format is

$$Q(t) = Q_0 \left(\frac{1}{2} \right)^{t/h}$$

where h is the half-life of the substance.

In Example 5, we have

$$Q(t) = Q_0 \left(\frac{1}{2} \right)^{t/5730}$$

Solution:

a. $Q(t) = Q_0 e^{-kt}$ Begin with a general exponential decay model.

$0.5Q_0 = Q_0 e^{-k(5730)}$ Substitute the known data value. One-half of the original quantity Q_0 is present after 5730 yr.

$0.5 = e^{-k(5730)}$ Divide by Q_0 on both sides.

$\ln 0.5 = -5730k$ Take the natural logarithm of both sides.

$k = \dfrac{\ln 0.5}{-5730}$ Divide by -5730.

≈ 0.000121

$Q(t) = Q_0 e^{-0.000121t}$

Answers

4. a. $P(t) = 10{,}000e^{-0.27726t}$

    b. 625

b. $0.766Q_0 = Q_0e^{-0.000121t}$ The quantity $Q(t)$ of carbon-14 in the bone is 76.6% of Q_0.

$0.766 = e^{-0.000121t}$ Divide by Q_0 on both sides.

$\ln 0.766 = -0.000121t$ Take the natural logarithm of both sides.

$t = \dfrac{\ln 0.766}{-0.000121} \approx 2200$ Divide by -0.000121 to isolate t.

The bone is approximately 2200 years old.

> **Skill Practice 5**   Use the function $Q(t) = Q_0e^{-0.000121t}$ to determine the age of a piece of wood that has 42% of its carbon-14 remaining. Round to the nearest 10 yr.

3. Apply Logistic Growth Models

In Examples 3 and 4, we used a model of the form $P(t) = P_0e^{kt}$ to predict population as an exponential function of time. However, unbounded population growth is not possible due to limited resources. A growth model that addresses this problem is called logistic growth. In particular, a logistic growth model imposes a limiting value on the dependent variable.

> **Logistic Growth Model**
>
> A logistic growth model is a function written in the form
>
> $$y = \frac{c}{1 + ae^{-bt}}$$
>
> where a, b, and c are positive constants.

The general logistic growth equation can be written with a complex fraction.

$$y = \frac{c}{1 + \dfrac{a}{e^{bt}}}$$ This term approaches 0 as t approaches ∞.

In this form, we can see that for large values of t, the term $\dfrac{a}{e^{bt}}$ approaches 0, and the function value y approaches $\frac{c}{1}$.

The line $y = c$ is a horizontal asymptote of the graph, and c represents the limiting value of the function (Figure 4-20).

Notice that the graph of a logistic curve is increasing over its entire domain. However, the *rate* of increase begins to decrease as the function levels off and approaches the horizontal asymptote $y = c$.

In Example 3 we created a function to approximate the population of California assuming unlimited growth. In Example 6, we use a logistic growth model.

> **TIP** The rate of increase of a logistic curve changes from increasing to decreasing to the left and right of a point called the *point of inflection*.

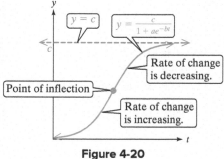

Figure 4-20

EXAMPLE 6   **Using Logistic Growth to Model Population**

The population of California $P(t)$ (in millions) can be approximated by the logistic growth function

$$P(t) = \frac{95.2}{1 + 1.8e^{-0.018t}}, \text{ where } t \text{ is the number of years since the year 2000.}$$

a. Determine the population in the year 2000.

b. Use this function to determine the time required for the population of California to double from its value in 2000. Compare this with the result from Example 3(c).

c. What is the limiting value of the population of California under this model?

Solution:

a. $P(t) = \dfrac{95.2}{1 + 1.8e^{-0.018t}}$

$P(0) = \dfrac{95.2}{1 + 1.8e^{-0.018(0)}} = \dfrac{95.2}{1 + 1.8(1)} = 34$ Substitute 0 for t. Recall that $e^0 = 1$.

The population was approximately 34 million in the year 2000.

b. $68 = \dfrac{95.2}{1 + 1.8e^{-0.018t}}$ Substitute 68 for $P(t)$.

$68(1 + 1.8e^{-0.018t}) = 95.2$ Multiply both sides by $(1 + 1.8e^{-0.018t})$.

$1 + 1.8e^{-0.018t} = 1.4$ Divide by 68 on both sides.

$1.8e^{-0.018t} = 1.4 - 1$ Subtract 1 from both sides.

$e^{-0.018t} = \dfrac{0.4}{1.8}$ Divide by 1.8 on both sides.

$-0.018t = \ln\left(\dfrac{0.4}{1.8}\right)$ Take the natural logarithm of both sides.

$t = \dfrac{\ln\left(\dfrac{0.4}{1.8}\right)}{-0.018} \approx 83.6$ Divide by -0.018 on both sides.

The population will double in approximately 83.6 yr. This is 9 yr later than the predicted value from Example 3(c).

The graphs of $P(t) = \dfrac{95.2}{1 + 1.8e^{-0.018t}}$ and $P(t) = 34e^{0.00926t}$ are shown in Figure 4-21. Notice that the two models agree relatively closely for short-term population growth (out to about 2060). However, in the long term, the unbounded exponential model breaks down. The logistic growth model approaches a limiting population, which is reasonable due to the limited resources to sustain a large human population.

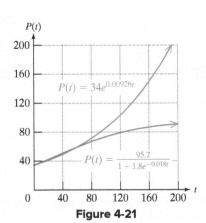

Figure 4-21

c. $P(t) = \dfrac{95.2}{1 + 1.8e^{-0.018t}} = \dfrac{95.2}{1 + \dfrac{1.18}{e^{0.018t}}}$ As $t \to \infty$, the term $\dfrac{1.18}{e^{0.018t}} \to 0$.

As t becomes large, the denominator of $\dfrac{1.18}{e^{0.018t}}$ also becomes large. This causes the quotient to approach zero. Therefore, as t approaches infinity, $P(t)$ approaches 95.2. Under this model, the limiting value for the population of California is 95.2 million.

> **Skill Practice 6**   The score on a test of dexterity is given by
> $$P(t) = \frac{100}{1 + 19e^{-0.354x}},$$ where x is the number of times the test is taken.
>
> **a.** Determine the initial score.
> **b.** Use the function to determine the minimum number of times required for the score to exceed 90.
> **c.** What is the limiting value of the scores?

4. Create Exponential and Logarithmic Models Using Regression

In Examples 7 and 8, we use a graphing utility and regression techniques to find an exponential model or logarithmic model based on observed data.

EXAMPLE 7   **Creating an Exponential Model from Observed Data**

The amount of sunlight y [in langleys (Ly)—a unit used to measure solar energy in calories/cm^2] is measured for six different depths x (in meters) in Lake Lyndon B. Johnson in Texas.

x (m)	1	3	5	7	9	11
y (Ly)	300	161	89	50	27	15

a. Graph the data.
b. From visual inspection of the graph, which model would best represent the data? Choose from $y = mx + b$ (linear), $y = ab^x$ (exponential), or $y = a + b \ln x$ (logarithmic).
c. Use a graphing utility to find a regression equation that fits the data.

Solution:

a. Enter the data in two lists.

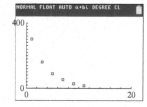

b. Note that for large depths, the amount of sunlight approaches 0. Therefore, the curve is asymptotic to the x-axis. This is consistent with a decreasing exponential model. The exponential model $y = ab^x$ appears to fit.

c. Under the STAT menu, choose CALC, ExpReg, and then Calculate.

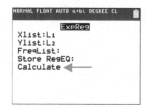

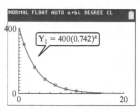

Answers
6. a. 5   **b.** 15   **c.** 100

The equation $y = ab^x$ is $y = 400(0.742)^x$.

475

> **Skill Practice 7**   For the given data,
>
> | x | 1 | 3 | 5 | 7 | 9 | 11 |
> |---|---|---|---|---|---|---|
> | y | 2.9 | 5.6 | 11.1 | 22.4 | 43.0 | 85.0 |
>
> **a.** Graph the data points.
> **b.** Use a graphing utility to find a model of the form $y = ab^x$ to fit the data.

EXAMPLE 8   **Creating a Logarithmic Model from Observed Data**

The diameter x (in mm) of a sugar maple tree, along with the corresponding age y (in yr) of the tree is given for six different trees.

x (mm)	1	50	100	200	300	400
y (yr)	4	60	72	82	89	94

a. Graph the data.
b. From visual inspection of the graph, which model would best represent the data? Choose from $y = mx + b$ (linear), $y = ab^x$ (exponential), or $y = a + b \ln x$ (logarithmic).
c. Use a graphing utility to find a regression equation that fits the data.

Solution:

a. Enter the data into two lists.

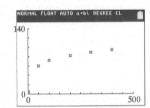

b. By inspection of the graph, the logarithmic model $y = a + b \ln x$ appears to fit.
c. Under the STAT menu, choose CALC, and then LnReg.

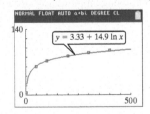

Answers

7. a–b.

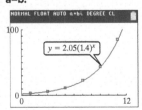

8. a–b.

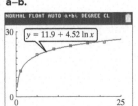

> **Skill Practice 8**   For the given data,
>
> | x | 1 | 5 | 9 | 13 | 17 | 21 |
> |---|---|---|---|---|---|---|
> | y | 11.9 | 19.3 | 21.9 | 23.5 | 24.7 | 25.7 |
>
> **a.** Graph the data points.
> **b.** Use a graphing utility to find a model of the form $y = a + b \ln x$ to fit the data.

Prerequisite Review

For Exercises R.1–R.3, solve for the indicated variable.

R.1. $P = \frac{1}{3}Lm$ for L **R.2.** $Q = \frac{3}{4}m^3$ for m, $m > 0$ **R.3.** $9 + \sqrt{a^2 - b^2} = c$ for b

R.4. Given $f(x) = 5x^2 + 2x$, evaluate

  a. $f(-2)$ **b.** $f(-1)$ **c.** $f(0)$ **d.** $f(1)$

Concept Connections

1. If $k > 0$, the equation $y = y_0 e^{kt}$ is a model for exponential (growth/decay), whereas if $k < 0$, the equation is a model for exponential (growth/decay).

2. A function defined by $y = ab^x$ can be written in terms of an exponential function base e as _____.

3. A function defined by $y = \dfrac{c}{1 + ae^{-bt}}$ is called a _____ growth model and imposes a limiting value on y.

4. Given a logistic growth function $y = \dfrac{c}{1 + ae^{-bt}}$, the limiting value of y is _____.

Objective 1: Solve Literal Equations for a Specified Variable

For Exercises 5–14, solve for the indicated variable. (See Example 1)

5. $Q = Q_0 e^{-kt}$ for k (used in chemistry)

6. $N = N_0 e^{-0.025t}$ for t (used in chemistry)

7. $M = 8.8 + 5.1 \log D$ for D (used in astronomy)

8. $\log E - 12.2 = 1.44M$ for E (used in geology)

9. $\text{pH} = -\log[\text{H}^+]$ for $[\text{H}^+]$ (used in chemistry)

10. $L = 10 \log\left(\dfrac{I}{I_0}\right)$ for I (used in medicine)

11. $A = P(1 + r)^t$ for t (used in finance)

12. $A = Pe^{rt}$ for r (used in finance)

13. $\ln\left(\dfrac{k}{A}\right) = \dfrac{-E}{RT}$ for k (used in chemistry)

14. $-\dfrac{1}{k}\ln\left(\dfrac{P}{14.7}\right) = A$ for P (used in meteorology)

Objective 2: Create Models for Exponential Growth and Decay

15. Suppose that \$12,000 is invested in a bond fund and the account grows to \$14,309.26 in 4 yr. (See Example 2)

  a. Use the model $A = Pe^{rt}$ to determine the average rate of return under continuous compounding. Round to the nearest tenth of a percent.

  b. How long will it take the investment to reach \$20,000 if the rate of return continues? Round to the nearest tenth of a year.

16. Suppose that \$50,000 from a retirement account is invested in a large cap stock fund. After 20 yr, the value is \$194,809.67.

  a. Use the model $A = Pe^{rt}$ to determine the average rate of return under continuous compounding. Round to the nearest tenth of a percent.

  b. How long will it take the investment to reach one-quarter million dollars? Round to the nearest tenth of a year.

17. Suppose that P dollars in principal is invested in an account earning 3.2% interest compounded continuously. At the end of 3 yr, the amount in the account has earned \$806.07 in interest.

  a. Find the original principal. Round to the nearest dollar. (*Hint:* Use the model $A = Pe^{rt}$ and substitute $P + 806.07$ for A.)

  b. Using the original principal from part (a) and the model $A = Pe^{rt}$, determine the time required for the investment to reach \$10,000. Round to the nearest year.

18. Suppose that P dollars in principal is invested in an account earning 2.1% interest compounded continuously. At the end of 2 yr, the amount in the account has earned \$193.03 in interest.

  a. Find the original principal. Round to the nearest dollar. (*Hint:* Use the model $A = Pe^{rt}$ and substitute $P + 193.03$ for A.)

  b. Using the original principal from part (a) and the model $A = Pe^{rt}$, determine the time required for the investment to reach \$6000. Round to the nearest tenth of a year.

19. The populations of two countries are given for January 1, 2000, and for January 1, 2010.

  a. Write a function of the form $P(t) = P_0 e^{kt}$ to model each population $P(t)$ (in millions) t years after January 1, 2000. (**See Example 3**)

Country	Population in 2000 (millions)	Population in 2010 (millions)	$P(t) = P_0 e^{kt}$
Australia	19.0	22.6	
Taiwan	22.9	23.7	

  b. Use the models from part (a) to approximate the population on January 1, 2020, for each country. Round to the nearest hundred thousand.

  c. Australia had fewer people than Taiwan in the year 2000, yet from the result of part (b), Australia would have more people in the year 2020? Why?

  d. Use the models from part (a) to predict the year during which each population would reach 30 million if this trend continues.

21. A function of the form $P(t) = ab^t$ represents the population (in millions) of the given country t years after January 1, 2000. (**See Example 4**)

  a. Write an equivalent function using base e; that is, write a function of the form $P(t) = P_0 e^{kt}$. Also, determine the population of each country for the year 2000.

Country	$P(t) = ab^t$	$P(t) = P_0 e^{kt}$	Population in 2000
Costa Rica	$P(t) = 4.3(1.0135)^t$		
Norway	$P(t) = 4.6(1.0062)^t$		

  b. The population of the two given countries is very close for the year 2000, but their growth rates are different. Use the model to approximate the year during which the population of each country reached 5 million.

  c. Costa Rica had fewer people in the year 2000 than Norway. Why would Costa Rica reach a population of 5 million sooner than Norway?

20. The populations of two countries are given for January 1, 2000, and for January 1, 2010.

  a. Write a function of the form $P(t) = P_0 e^{kt}$ to model each population $P(t)$ (in millions) t years after January 1, 2000.

Country	Population in 2000 (millions)	Population in 2010 (millions)	$P(t) = P_0 e^{kt}$
Switzerland	7.3	7.8	
Israel	6.7	7.7	

  b. Use the models from part (a) to approximate the population on January 1, 2020, for each country. Round to the nearest hundred thousand.

  c. Israel had fewer people than Switzerland in the year 2000, yet from the result of part (b), Israel would have more people in the year 2020? Why?

  d. Use the models from part (a) to predict the year during which each population would reach 10 million if this trend continues.

22. A function of the form $P(t) = ab^t$ represents the population (in millions) of the given country t years after January 1, 2000.

  a. Write an equivalent function using base e; that is, write a function of the form $P(t) = P_0 e^{kt}$. Also, determine the population of each country for the year 2000.

Country	$P(t) = ab^t$	$P(t) = P_0 e^{kt}$	Population in 2000
Haiti	$P(t) = 8.5(1.0158)^t$		
Sweden	$P(t) = 9.0(1.0048)^t$		

  b. The population of the two given countries is very close for the year 2000, but their growth rates are different. Use the model to approximate the year during which the population of each country would reach 10.5 million.

  c. Haiti had fewer people in the year 2000 than Sweden. Why would Haiti reach a population of 10.5 million sooner?

For Exercises 23–24, refer to the model $Q(t) = Q_0 e^{-0.000121t}$ used in Example 5 for radiocarbon dating.

23. A sample from a mummified bull was taken from a pyramid in Dashur, Egypt. The sample shows that 78% of the carbon-14 still remains. How old is the sample? Round to the nearest year. (**See Example 5**)

24. At the "Marmes Man" archeological site in southeastern Washington State, scientists uncovered the oldest human remains yet to be found in Washington State. A sample from a human bone taken from the site showed that 29.4% of the carbon-14 still remained. How old is the sample? Round to the nearest year.

25. The isotope of plutonium ^{238}Pu is used to make thermoelectric power sources for spacecraft. Suppose that a space probe was launched in 2012 with 2.0 kg of ^{238}Pu.

   a. If the half-life of ^{238}Pu is 87.7 yr, write a function of the form $Q(t) = Q_0 e^{-kt}$ to model the quantity $Q(t)$ of ^{238}Pu left after t years.

   b. If 1.6 kg of ^{238}Pu is required to power the spacecraft's data transmitter, for how long after launch would scientists be able to receive data? Round to the nearest year.

26. Technetium-99 (^{99m}Tc) is a radionuclide used widely in nuclear medicine. ^{99m}Tc is combined with another substance that is readily absorbed by a targeted body organ. Then, special cameras sensitive to the gamma rays emitted by the technetium are used to record pictures of the organ. Suppose that a technician prepares a sample of ^{99m}Tc-pyrophosphate to image the heart of a patient suspected of having had a mild heart attack.

   a. At noon, the patient is given 10 mCi (millicuries) of ^{99m}Tc. If the half-life of ^{99m}Tc is 6 hr, write a function of the form $Q(t) = Q_0 e^{-kt}$ to model the radioactivity level $Q(t)$ after t hours.

   b. At what time will the level of radioactivity reach 3 mCi? Round to the nearest tenth of an hour.

27. Fluorodeoxyglucose is a derivative of glucose that contains the radionuclide fluorine-18 (^{18}F). A patient is given a sample of this material containing 300 MBq of ^{18}F (a megabecquerel is a unit of radioactivity). The patient then undergoes a PET scan (positron emission tomography) to detect areas of metabolic activity indicative of cancer. After 174 min, one-third of the original dose remains in the body.

   a. Write a function of the form $Q(t) = Q_0 e^{-kt}$ to model the radioactivity level $Q(t)$ of fluorine-18 at a time t minutes after the initial dose.

   b. What is the half-life of ^{18}F? Round to the nearest minute.

28. Painful bone metastases are common in advanced prostate cancer. Physicians often order treatment with strontium-89 (^{89}Sr), a radionuclide with a strong affinity for bone tissue. A patient is given a sample containing 4 mCi of ^{89}Sr.

   a. If 20% of the ^{89}Sr remains in the body after 90 days, write a function of the form $Q(t) = Q_0 e^{-kt}$ to model the amount $Q(t)$ of radioactivity in the body t days after the initial dose.

   b. What is the biological half-life of ^{89}Sr under this treatment? Round to the nearest tenth of a day.

29. Two million *E. coli* bacteria are present in a laboratory culture. An antibacterial agent is introduced and the population of bacteria $P(t)$ decreases by half every 6 hr. The population can be represented by $P(t) = 2{,}000{,}000\left(\frac{1}{2}\right)^{t/6}$.

   a. Convert this to an exponential function using base e.

   b. Verify that the original function and the result from part (a) yield the same result for $P(0)$, $P(6)$, $P(12)$, and $P(60)$. (*Note*: There may be round-off error.)

30. The half-life of radium-226 is 1620 yr. Given a sample of 1 g of radium-226, the quantity left $Q(t)$ (in g) after t years is given by $Q(t) = \left(\frac{1}{2}\right)^{t/1620}$.

   a. Convert this to an exponential function using base e.

   b. Verify that the original function and the result from part (a) yield the same result for $Q(0)$, $Q(1620)$, and $Q(3240)$. (*Note*: There may be round-off error.)

Objective 3: Apply Logistic Growth Models

31. The population of the United States $P(t)$ (in millions) since January 1, 1900, can be approximated by

$$P(t) = \frac{725}{1 + 8.295e^{-0.0165t}}$$

where t is the number of years since January 1, 1900. (**See Example 6**)

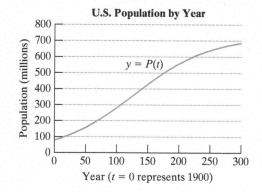

U.S. Population by Year

$y = P(t)$

Year ($t = 0$ represents 1900)

   a. Evaluate $P(0)$ and interpret its meaning in the context of this problem.

   b. Use the function to approximate the U.S. population on January 1, 2020. Round to the nearest million.

   c. Use the function to approximate the U.S. population on January 1, 2050.

   d. From the model, during which year would the U.S. population reach 500 million?

   e. What value will the term $\dfrac{8.295}{e^{0.0165t}}$ approach as $t \to \infty$?

   f. Determine the limiting value of $P(t)$.

32. The population of Canada $P(t)$ (in millions) since January 1, 1900, can be approximated by

$$P(t) = \frac{55.1}{1 + 9.6e^{-0.02515t}}$$

where t is the number of years since January 1, 1900.

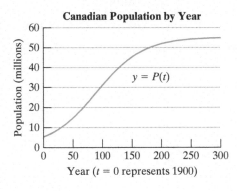

Canadian Population by Year

Year ($t = 0$ represents 1900)

$y = P(t)$

a. Evaluate $P(0)$ and interpret its meaning in the context of this problem.

b. Use the function to approximate the Canadian population on January 1, 2015. Round to the nearest tenth of a million.

c. Use the function to approximate the Canadian population on January 1, 2040.

d. From the model, during which year would the Canadian population reach 45 million?

e. What value will the term $\dfrac{9.6}{e^{0.02515t}}$ approach as $t \to \infty$?

f. Determine the limiting value of $P(t)$.

33. The number of computers $N(t)$ (in millions) infected by a computer virus can be approximated by

$$N(t) = \frac{2.4}{1 + 15e^{-0.72t}}$$

where t is the time in months after the virus was first detected.

a. Determine the number of computers initially infected when the virus was first detected.

b. How many computers were infected after 6 months? Round to the nearest hundred thousand.

c. Determine the amount of time required after initial detection for the virus to affect 1 million computers. Round to the nearest tenth of a month.

d. What is the limiting value of the number of computers infected according to this model?

34. After a new product is launched the cumulative sales $S(t)$ (in $1000) t weeks after launch is given by

$$S(t) = \frac{72}{1 + 9e^{-0.36t}}$$

a. Determine the cumulative amount in sales 3 weeks after launch. Round to the nearest thousand.

b. Determine the amount of time required for the cumulative sales to reach $70,000.

c. What is the limiting value in sales?

Objective 4: Create Exponential and Logarithmic Models Using Regression

For Exercises 35–38, a graph of data is given. From visual inspection, which model would best fit the data? Choose from

$y = mx + b$ (linear) $y = ab^x$ (exponential)

$y = a + b \ln x$ (logarithmic) $y = \dfrac{c}{1 + ae^{-bx}}$ (logistic)

35.

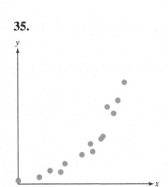

36.

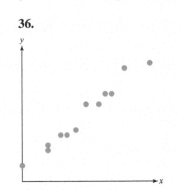

37.

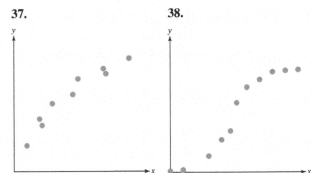

38.

For Exercises 39–46, a table of data is given.

a. Graph the points and from visual inspection, select the model that would best fit the data. Choose from

$$y = mx + b \text{ (linear)} \qquad\qquad y = ab^x \text{ (exponential)}$$

$$y = a + b \ln x \text{ (logarithmic)} \qquad y = \frac{c}{1 + ae^{-bx}} \text{ (logistic)}$$

b. Use a graphing utility to find a function that fits the data. (*Hint:* For a logistic model, go to STAT, CALC, Logistic.)

39.

x	y
0	2.3
4	3.6
8	5.7
12	9.1
16	14
20	22

40.

x	y
0	52
1	67
2	87
3	114
4	147
5	195

41.

x	y
3	2.7
7	12.2
13	25.7
15	30
17	34
21	44.4

42.

x	y
0	640
20	530
40	430
50	360
80	210
100	90

43.

x	y
10	43.3
20	50
30	53
40	56.8
50	58.8
60	60.8

44.

x	y
5	29
10	40
15	45.6
20	50
25	53.3
30	56

45.

x	y
2	0.326
4	2.57
6	10.8
8	16.8
10	17.9
5	6
7	14.8

46.

x	y
0	0.05
2	0.45
4	2.94
5	5.8
6	8.8
7	10.6
8	11.5
10	11.9

47. During a recent outbreak of Ebola in western Africa, the cumulative number of cases y was reported t months after April 1. (**See Example 7**)

Month Number (t)	Cumulative Cases (y)
0	18
1	105
2	230
3	438
4	752
5	1437
6	2502

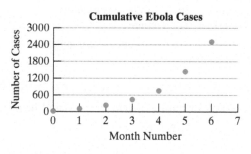

a. Use a graphing utility to find a model of the form $y = ab^t$. Round a to 1 decimal place and b to 3 decimal places.

b. Write the function from part (a) as an exponential function of the form $y = ae^{bt}$.

c. Use either model to predict the number of Ebola cases 8 months after April 1 if this trend continues. Round to the nearest thousand.

d. Would it seem reasonable for this trend to continue indefinitely?

e. Use a graphing utility to find a logistic model $y = \dfrac{c}{1 + ae^{-bt}}$. Round a and c to the nearest whole number and b to 2 decimal places.

f. Use the logistic model from part (e) to predict the number of Ebola cases 8 months after April 1. Round to the nearest thousand.

48. The monthly costs for a small company to do business has been increasing over time due in part to inflation. The table gives the monthly cost y (in $) for the month of January for selected years. The variable t represents the number of years since 2016.

Year (t = 0 is 2016)	Monthly Costs ($) y
0	12,000
1	12,400
2	12,800
3	13,300

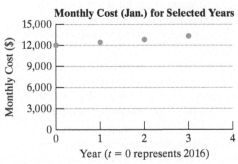

Monthly Cost (Jan.) for Selected Years

a. Use a graphing utility to find a model of the form $y = ab^t$. Round a to the nearest whole unit and b to 3 decimal places.

b. Write the function from part (a) as an exponential function with base e.

c. Use either model to predict the monthly cost for January in the year 2023 if this trend continues. Round to the nearest hundred dollars.

49. The age of a tree t (in yr) and its corresponding height $H(t)$ are given in the table. (**See Example 8**)

Age of Tree (yr) t	Height (ft) H(t)
1	5
2	8.3
3	11.6
4	14.6
5	15.4
6	16.5
7	17.5
8	18.3
9	19
10	19.4
11	19.7
12	20

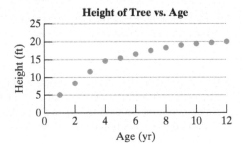

Height of Tree vs. Age

a. Write a model of the form $H(t) = a + b \ln t$. Round a and b to 2 decimal places.

b. Use the model to predict the age of a tree if it is 25 ft high. Round to the nearest year.

c. Is it reasonable to assume that this logarithmic trend will continue indefinitely? Why or why not?

50. The sales of a book tend to increase over the short-term as word-of-mouth makes the book "catch on." The number of books sold $N(t)$ for a new novel t weeks after release at a certain book store is given in the table for the first 6 weeks.

Weeks t	Number Sold N(t)
1	20
2	27
3	31
4	35
5	38
6	39

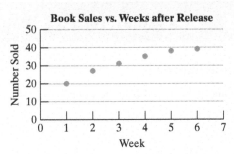

Book Sales vs. Weeks after Release

a. Find a model of the form $N(t) = a + b \ln t$. Round a and b to 1 decimal place.

b. Use the model to predict the sales in week 7. Round to the nearest whole unit.

c. Is it reasonable to assume that this logarithmic trend will continue? Why or why not?

Mixed Exercises

51. A van is purchased new for $29,200.

  a. Write a linear function of the form $y = mt + b$ to represent the value y of the vehicle t years after purchase. Assume that the vehicle is depreciated by $2920 per year.

  b. Suppose that the vehicle is depreciated so that it holds only 80% of its value from the previous year. Write an exponential function of the form $y = V_0b^t$, where V_0 is the initial value and t is the number of years after purchase.

  c. To the nearest dollar, determine the value of the vehicle after 5 yr and after 10 yr using the linear model.

  d. To the nearest dollar, determine the value of the vehicle after 5 yr and after 10 yr using the exponential model.

52. A delivery truck is purchased new for $54,000.

  a. Write a linear function of the form $y = mt + b$ to represent the value y of the vehicle t years after purchase. Assume that the vehicle is depreciated by $6750 per year.

  b. Suppose that the vehicle is depreciated so that it holds 70% of its value from the previous year. Write an exponential function of the form $y = V_0b^t$, where V_0 is the initial value and t is the number of years after purchase.

  c. To the nearest dollar, determine the value of the vehicle after 4 yr and after 8 yr using the linear model.

  d. To the nearest dollar, determine the value of the vehicle after 4 yr and after 8 yr using the exponential model.

Write About It

53. Why is it important to graph a set of data before trying to find an equation or function to model the data.

55. Explain the difference between an exponential growth model and a logistic growth model.

54. How does the average rate of change differ for a linear function versus an increasing exponential function?

56. Explain how to convert an exponential expression b^t to an exponential expression base e.

Expanding Your Skills

57. The monthly payment P (in $) to pay off a loan of amount A (in $) at an interest rate r in t years is given by

$$P = \frac{\dfrac{Ar}{12}}{1 - \left(1 + \dfrac{r}{12}\right)^{-12t}}.$$

  a. Solve for t (note that there are numerous equivalent algebraic forms for the result).

  b. Interpret the meaning of the resulting relationship.

58. Suppose that a population follows a logistic growth pattern, with a limiting population N. If the initial population is denoted by P_0, and t is the amount of time elapsed, then the population P can be represented by

$$P = \frac{P_0 N}{P_0 + (N - P_0)e^{-kt}}.$$

  where k is a constant related to the growth rate.

  a. Solve for t (note that there are numerous equivalent algebraic forms for the result).

  b. Interpret the meaning of the resulting relationship.

CHAPTER 4   KEY CONCEPTS

SECTION 4.1 Inverse Functions	Reference
A function f is **one-to-one** if for a and b in the domain of f, if $a \neq b$, then $f(a) \neq f(b)$, or equivalently, if $f(a) = f(b)$, then $a = b$.	p. 403
Horizontal line test: A function defined by $y = f(x)$ is one-to-one if no horizontal line intersects the graph in more than one point.	p. 403
Function **g is the inverse of f** if $(f \circ g)(x) = x$ for all x in the domain of g and $(g \circ f)(x) = x$ for all x in the domain of f.	p. 405
Procedure to find $f^{-1}(x)$: 1. Replace $f(x)$ by y. 2. Interchange x and y. 3. Solve for y. 4. Replace y by $f^{-1}(x)$.	p. 406

SECTION 4.2 Exponential Functions	Reference
Let b be a real number with $b > 0$ and $b \neq 1$. Then for any real number x, a function of the form $f(x) = b^x$ is an **exponential function of base b.**	p. 415
For the graph of an exponential function $f(x) = b^x$, • If $b > 1$, f is an increasing function. • If $0 < b < 1$, f is a decreasing function. • The domain is $(-\infty, \infty)$. • The range is $(0, \infty)$. • The line $y = 0$ is a horizontal asymptote. • The function passes through $(0, 1)$.	p. 416
The irrational number e is the limiting value of the expression $\left(1 + \frac{1}{x}\right)^x$ as x approaches ∞. $e \approx 2.71828$	p. 418
If P dollars in principal is invested or borrowed at an annual interest rate r for t years, then $I = Prt$ Simple interest $A = P\left(1 + \dfrac{r}{n}\right)^{nt}$ Future value A with interest compounded n times per year. $A = Pe^{rt}$ Future value A with interest compounded continuously.	p. 420

SECTION 4.3 Logarithmic Functions	Reference
If x and b are positive real numbers such that $b \neq 1$, then $y = \log_b x$ is called the **logarithmic function** base b, where $$y = \log_b x \text{ is equivalent to } b^y = x.$$ logarithmic form exponential form	p. 428
The functions $f(x) = \log_b x$ and $g(x) = b^x$ are inverses.	p. 428
Basic properties of logarithms: 1. $\log_b 1 = 0$ 2. $\log_b b = 1$ 3. $\log_b b^x = x$ 4. $b^{\log_b x} = x$	p. 432
$y = \log_{10} x$ is written as $y = \log x$ and is called the **common logarithmic function.**	p. 430
$y = \log_e x$ is written as $y = \ln x$ and is called the **natural logarithmic function.**	
Given $f(x) = \log_b x$, • If $b > 1$, f is an increasing function. • If $0 < b < 1$, f is a decreasing function. • The domain is $(0, \infty)$. • The range is $(-\infty, \infty)$. • The line $x = 0$ is a vertical asymptote. • The function passes through $(1, 0)$.	p. 434
The domain of $f(x) = \log_b x$ is $\{x \mid x > 0\}$.	p. 435

SECTION 4.4 **Properties of Logarithms**	Reference
Let b, x, and y be positive real numbers with $b \neq 1$. Then, $\log_b(xy) = \log_b x + \log_b y$ (Product property) $\log_b\left(\dfrac{x}{y}\right) = \log_b x - \log_b y$ (Quotient property) $\log_b x^p = p \log_b x$ (Power property)	p. 445
Change-of-base formula: For positive real numbers a and b, where $a \neq 1$ and $b \neq 1$, $\log_b x = \dfrac{\log_a x}{\log_a b}$.	p. 448

SECTION 4.5 **Exponential and Logarithmic Equations and Applications**	Reference
Equivalence property of exponential expressions: If b, x, and y are real numbers with $b > 0$ and $b \neq 1$, then $b^x = b^y$ implies that $x = y$.	p. 453
Equivalence property of logarithmic expressions: If b, x, and y are positive real numbers and $b \neq 1$, then $\log_b x = \log_b y$ implies that $x = y$.	p. 457
Steps to solve exponential equations by using logarithms: 1. Isolate the exponential expression on one side of the equation. 2. Take a logarithm of the same base on both sides of the equation. 3. Use the power property of logarithms to "bring down" the exponent. 4. Solve the resulting equation.	p. 454
Guidelines to solve a logarithmic equation: 1. Isolate the logarithms on one side of the equation. 2. Use the properties of logarithms to write the equation in the form $\log_b x = k$, where k is a constant. 3. Write the equation in exponential form. 4. Solve the equation from step 3. 5. Check the potential solution(s) in the original equation.	p. 458

SECTION 4.6 **Modeling with Exponential and Logarithmic Functions**	Reference
The function defined by $y = y_0 e^{kt}$ represents exponential growth if $k > 0$ and exponential decay if $k < 0$.	p. 468
An exponential expression can be rewritten as an expression of a different base. In particular, to convert to base e, we have b^t is equivalent to $e^{(\ln b)t}$.	p. 470
A **logistic growth function** is a function of the form $$y = \dfrac{c}{1 + ae^{-bt}}.$$ A logistic growth function imposes a limiting value on the dependent variable.	p. 472

CHAPTER 4   Review Exercises

SECTION 4.1

For Exercises 1–2, determine if the relation defines y as a one-to-one function of x.

1.

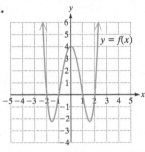

2.

x	y
5	7
-3	1
-4	-2
6	0

For Exercises 3–4, use the definition of a one-to-one function to determine if the function is one-to-one. Recall that f is one-to-one if $a \neq b$ implies that $f(a) \neq f(b)$, or equivalently, if $f(a) = f(b)$, then $a = b$.

3. $f(x) = x^3 - 1$ **4.** $f(x) = x^2 - 1$

For Exercises 5–6, determine if the functions are inverses.

5. $f(x) = 4x - 3$ and $g(x) = \dfrac{x + 3}{4}$

6. $m(x) = \sqrt[3]{x + 1}$ and $n(x) = (x - 1)^3$

For Exercises 7–8, a one-to-one function is given. Write an equation for the inverse function.

7. $f(x) = 2x^3 - 5$ **8.** $f(x) = \dfrac{2}{x + 7}$

9. a. Graph $f(x) = x^2 - 9$, $x \leq 0$.

 b. Is f a one-to-one function?

 c. Write the domain of f in interval notation.

 d. Write the range of f in interval notation.

 e. Find an equation for f^{-1}.

 f. Graph $y = f(x)$ and $y = f^{-1}(x)$ on the same coordinate system.

 g. Write the domain of f^{-1} in interval notation.

 h. Write the range of f^{-1} in interval notation.

10. a. Graph $g(x) = \sqrt{x + 4}$.

 b. Is g a one-to-one function?

 c. Write the domain of g in interval notation.

 d. Write the range of g in interval notation.

 e. Find an equation for g^{-1}.

 f. Graph $y = g(x)$ and $y = g^{-1}(x)$ on the same coordinate system.

 g. Write the domain of g^{-1} in interval notation.

 h. Write the range of g^{-1} in interval notation.

11. The function $f(x) = 5280x$ provides the conversion from x miles to $f(x)$ feet.

 a. Write an equation for f^{-1}.

 b. What does the inverse function represent in the context of this problem?

 c. Determine the number of miles represented by 22,176 ft.

SECTION 4.2

12. Which of the functions is an exponential function?

 a. $f(x) = x^4$ **b.** $h(x) = 4^{-x}$ **c.** $g(x) = \left(\dfrac{4}{3}\right)^x$

 d. $k(x) = \dfrac{4x}{3}$ **e.** $n(x) = \dfrac{4}{3x}$ **f.** $r(x) = \left(-\dfrac{4}{3}\right)^x$

For Exercises 13–16,

 a. Graph the function.

 b. Write the domain in interval notation.

 c. Write the range in interval notation.

 d. Write an equation of the asymptote.

13. $f(x) = \left(\dfrac{5}{2}\right)^x$ **14.** $g(x) = \left(\dfrac{5}{2}\right)^{-x}$

15. $k(x) = -3^x + 1$ **16.** $h(x) = 2^{x-3} - 4$

17. Is the graph of $y = e^x$ an increasing or decreasing exponential function?

For Exercises 18–19, use the formulas on page 420.

18. Suppose that $24,000 is invested at the given interest rates and compounding options. Determine the amount that the investment is worth at the end of t years.

 a. 5% interest compounded monthly for 10 yr

 b. 4.5% interest compounded continuously for 30 yr

19. Jorge needs to borrow $12,000 to buy a car. He can borrow the money at 7.2% simple interest for 4 yr or he can borrow at 6.5% interest compounded continuously for 4 yr.

 a. How much total interest would Jorge pay at 7.2% simple interest?

 b. How much total interest would Jorge pay at 6.5% interest compounded continuously?

 c. Which option results in less total interest?

20. A patient is treated with 128 mCi (millicuries) of iodine-131 (^{131}I). The radioactivity level $R(t)$ (in mCi) after t days is given by $R(t) = 128(2)^{-t/4.2}$. (In this model, the value 4.2 is related to the biological half-life of radioactive iodine in the body.)

 a. Determine the radioactivity level of ^{131}I in the body after 6 days. Round to the nearest whole unit.

b. Evaluate $R(4.2)$ and interpret its meaning in the context of this problem.

c. After how many half-lives will the radioactivity level be 16 mCi?

SECTION 4.3

For Exercises 21–22, write the equation in exponential form.

21. $\log_b(x^2 + y^2) = 4$ **22.** $\ln x = (c + d)$

For Exercises 23–24, write the equation in logarithmic form.

23. $10^6 = 1{,}000{,}000$ **24.** $8^{-1/3} = \dfrac{1}{2}$

For Exercises 25–32, simplify the logarithmic expression without using a calculator.

25. $\log_3 81$ **26.** $\log 100{,}000$

27. $\log_2\left(\dfrac{1}{64}\right)$ **28.** $\log_{1/4}(16)$

29. $\log_{11} 1$ **30.** $\log_5 5$

31. $4^{\log_4 7}$ **32.** $\ln e^{11}$

For Exercises 33–37, write the domain of the function in interval notation.

33. $f(x) = \log(x - 4)$ **34.** $g(x) = \ln(3 - 2x)$

35. $h(x) = \log_2(x^2 + 4)$ **36.** $k(x) = \log_2(x^2 - 4)$

37. $m(x) = \log_2(x - 4)^2$

For Exercises 38–39,

a. Graph the function.

b. Write the domain in interval notation.

c. Write the range in interval notation.

d. Write an equation of the asymptote.

38. $f(x) = \log_2(x - 3)$ **39.** $g(x) = 2 + \ln x$

For Exercises 40–41, use the formula pH $= -\log[H^+]$ to compute the pH of a liquid as a function of its concentration of hydronium ions, $[H^+]$ in mol/L. If the pH is less than 7, then the substance is acidic. If the pH is greater than 7, then the substance is alkaline (or basic).

a. Find the pH. Round to 1 decimal place.

b. Determine whether the substance is acidic or alkaline.

40. Baking soda: $[H^+] = 5.0 \times 10^{-9}$ mol/L

41. Tomatoes: $[H^+] = 3.16 \times 10^{-5}$ mol/L

SECTION 4.4

For Exercises 42–48, fill in the blanks to state the basic properties of logarithms. Assume that x, y, and b are positive real numbers with $b \neq 1$.

42. $\log_b 1 =$ _____   **43.** $\log_b b =$ _____

44. $\log_b b^p =$ _____   **45.** $b^{\log_b x} =$ _____

46. $\log_b(xy) =$ _____   **47.** $\log_b\left(\dfrac{x}{y}\right) =$ _____

48. $\log_b x^p =$ _____

For Exercises 49–52, write the logarithm as a sum or difference of logarithms. Simplify each term as much as possible.

49. $\log\left(\dfrac{100}{\sqrt{c^2 + 10}}\right)$ **50.** $\log_2\left(\dfrac{1}{8}a^2 b\right)$

51. $\ln\left(\dfrac{\sqrt[3]{ab^2}}{cd^5}\right)$ **52.** $\log\left(\dfrac{x^2(2x + 1)^5}{\sqrt{1 - x}}\right)$

For Exercises 53–55, write the logarithmic expression as a single logarithm with coefficient 1, and simplify as much as possible.

53. $4 \log_5 y - 3 \log_5 x + \dfrac{1}{2}\log_5 z$

54. $\log 250 + \log 2 - \log 5$

55. $\dfrac{1}{4}\ln(x^2 - 9) - \dfrac{1}{4}\ln(x - 3)$

For Exercises 56–58, use $\log_b 2 \approx 0.289$, $\log_b 3 \approx 0.458$, and $\log_b 5 \approx 0.671$ to approximate the value of the given logarithms.

56. $\log_b 8$ **57.** $\log_b 45$ **58.** $\log_b\left(\dfrac{1}{9}\right)$

For Exercises 59–60, use the change-of-base formula and a calculator to approximate the given logarithms. Round to 4 decimal places. Then check the answer by using the related exponential form.

59. $\log_7 596$ **60.** $\log_4 0.982$

SECTION 4.5

For Exercises 61–80, solve the equation. Write the solution set with exact values and give approximate solutions to 4 decimal places.

61. $4^{2y-7} = 64$ **62.** $1000^{2x+1} = \left(\dfrac{1}{100}\right)^{x-4}$

63. $7^x = 51$ **64.** $516 = 11^w - 21$

65. $3^{2x+1} = 4^{3x}$ **66.** $2^{c+3} = 7^{2c+5}$

67. $400e^{-2t} = 2.989$ **68.** $2 \cdot 10^{1.2t} = 58$

69. $e^{2x} - 3e^x - 40 = 0$ **70.** $e^{2x} = -10e^x$

71. $\log_5(4p + 7) = \log_5(2 - p)$

72. $\log_2(m^2 + 10m) = \log_2 11$

73. $2 \log_6(4 - 8y) + 6 = 10$

74. $5 = -4 \log_3(2 - 5x) + 1$

75. $3 \ln(n - 8) = 6.3$

76. $-4 + \log_2 x = -\log_2(x + 6)$

77. $\log_6(3x + 2) = \log_6(x + 4) + 1$

78. $\ln x + \ln(x + 2) = \ln(x + 6)$

79. $\log_5(\log_2 x) = 1$

80. $(\log x)^2 - \log x^2 = 35$

For Exercises 81–82, find the inverse of the function.

81. $f(x) = 4^x$

82. $g(x) = \log(x - 5) - 1$

83. The percentage of visible light P (in decimal form) at a depth of x meters for Long Island Sound can be approximated by $P = e^{-0.5x}$.

  a. Determine the depth at which the light intensity is half the value from the surface. Round to the nearest hundredth of a meter. Based on your answer, would you say that Long Island Sound is murky or clear water?

  b. Determine the euphotic depth for Long Island Sound. That is, find the depth at which the light intensity falls below 1%. Round to the nearest tenth of a meter.

SECTION 4.6

For Exercises 84–85, solve for the indicated variable.

84. $\log B - 1.7 = 2.3M$ for B

85. $T = T_f + T_0 e^{-kt}$ for t

86. Suppose that $18,000 is invested in a bond fund and the account grows to $23,344.74 in 5 yr.

  a. Use the model $A = Pe^{rt}$ to determine the average rate of return under continuous compounding. Round to the nearest tenth of a percent.

  b. How long will it take the investment to reach $30,000 if the rate of return continues? Round to the nearest tenth of a year.

87. The population of Germany in 2011 was approximately 85.5 million. The model $P = 85.5e^{-0.00208t}$ represents a short-term model for the population, t years after 2011.

  a. Based on this model, is the population of Germany increasing or decreasing?

  b. Determine the number of years after 2011 at which the population of Germany would decrease to 80 million if this trend continues. Round to the nearest year.

88. The population of Chile was approximately 16.9 million in the year 2011, with an annual growth rate of 0.836%. The population $P(t)$ (in millions) can be modeled by

$P(t) = 16.9(1.00836)^t$, where t is the number of years since 2011.

  a. Write a function of the form $P(t) = P_0 e^{kt}$ to model the population.

  b. Determine the amount of time required for the population to grow to 20 million if this trend continues. Round to the nearest year.

89. A sample from human remains found near Stonehenge in England shows that 71.2% of the carbon-14 still remains. Use the model $Q(t) = Q_0 e^{-0.000121t}$ to determine the age of the sample. In this model, $Q(t)$ represents the amount of carbon-14 remaining t years after death, and Q_0 represents the initial amount of carbon-14 at the time of death. Round to the nearest 100 yr.

90. A lake is stocked with bass by the U.S. Park Service. The population of bass is given by $P(t) = \dfrac{3000}{1 + 2e^{-0.37t}}$, where t is the time in years after the lake was stocked.

  a. Evaluate $P(0)$ and interpret its meaning in the context of this problem.

  b. Use the function to predict the bass population 2 yr after being stocked. Round to the nearest whole unit.

  c. Use the function to predict the bass population 4 yr after being stocked.

  d. Determine the number of years required for the bass population to reach 2800. Round to the nearest year.

  e. What value will the term $\dfrac{2}{e^{0.37t}}$ approach as $t \to \infty$?

  f. Determine the limiting value of $P(t)$.

91. For the given data,

  a. Use a graphing utility to find an exponential function $Y_1 = ab^x$ that fits the data.

  b. Graph the data and the function from part (a) on the same coordinate system.

x	y
0	2.4
1	3.5
2	5.5
3	8.1
4	12.0
5	18.4

CHAPTER 4 Test

1. Given $f(x) = 4x^3 - 1$,

  a. Write an equation for $f^{-1}(x)$.

  b. Verify that $(f \circ f^{-1})(x) = (f^{-1} \circ f)(x) = x$.

2. The graph of f is given.

  a. Is f a one-to-one function?

  b. If f is a one-to-one function, graph f^{-1} on the same coordinate system as f.

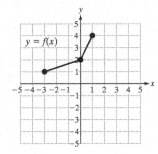

3. Given $f(x) = \dfrac{x + 3}{x - 4}$, write an equation for the inverse function.

For Exercises 4–7,

  a. Write the domain and range of f in interval notation.

  b. Write an equation of the inverse function.

  c. Write the domain and range of f^{-1} in interval notation.

4. $f(x) = -x^2 + 1, x \le 0$

5. $f(x) = \log x$

6. $f(x) = 3^x + 1$

7. $f(x) = \sqrt{x + 5}$

For Exercises 8–11,

a. Graph the function.

b. Write the domain in interval notation.

c. Write the range in interval notation.

d. Write an equation of the asymptote.

8. $f(x) = \left(\dfrac{1}{3}\right)^x + 2$ 　　　　**9.** $g(x) = 2^{x-4}$

10. $h(x) = -\ln x$ 　　　　**11.** $k(x) = \log_2(x + 1) - 3$

12. Write the statement in exponential form.　$\ln(x + y) = a$

For Exercises 13–18, evaluate the logarithmic expression without using a calculator.

13. $\log_9 \dfrac{1}{81}$ 　　　**14.** $\log_6 216$ 　　　**15.** $\ln e^8$

16. $\log 10^{-4}$ 　　　**17.** $10^{\log(a^2 + b^2)}$ 　　　**18.** $\log_{1/2} 1$

For Exercises 19–20, write the domain of the function in interval notation.

19. $f(x) = \log(7 - 2x)$ 　　　　**20.** $g(x) = \log_4(x^2 - 25)$

For Exercises 21–22, write the logarithm as a sum or difference of logarithms. Simplify each term as much as possible.

21. $\ln\left(\dfrac{x^5 y^2}{w\sqrt[3]{z}}\right)$ 　　　　**22.** $\log\left(\dfrac{\sqrt{a^2 + b^2}}{10^4}\right)$

For Exercises 23–24, write the logarithmic expression as a single logarithm with coefficient 1, and simplify as much as possible.

23. $6 \log_2 a - 4 \log_2 b + \dfrac{2}{3}\log_2 c$

24. $\dfrac{1}{2}\ln(x^2 - x - 12) - \dfrac{1}{2}\ln(x - 4)$

For Exercises 25–26, use $\log_b 2 \approx 0.289$, $\log_b 3 \approx 0.458$, and $\log_b 5 \approx 0.671$ to approximate the value of the given logarithms.

25. $\log_b 72$ 　　　　**26.** $\log_b\left(\dfrac{1}{125}\right)$

For Exercises 27–36, solve the equation. Write the solution set with exact values and give approximate solutions to 4 decimal places.

27. $2^{5y+1} = 4^{y-3}$ 　　　　**28.** $5^{x+3} + 3 = 56$

29. $2^{c+7} = 3^{2c+3}$ 　　　　**30.** $7e^{4x} - 2 = 12$

31. $e^{2x} + 7e^x - 8 = 0$

32. $\log_5(3 - x) = \log_5(x + 1)$

33. $5 \ln(x + 2) + 1 = 16$

34. $\log x + \log(x - 1) = \log 12$

35. $-3 + \log_4 x = -\log_4(x + 30)$

36. $\log 3 + \log(x + 3) = \log(4x + 5)$

For Exercises 37–38, solve for the indicated variable.

37. $S = 92 - k \ln(t + 1)$　for t

38. $A = P\left(1 + \dfrac{r}{n}\right)^{nt}$　for t

39. Suppose that $10,000$ is invested and the account grows to $13,566.25$ in 5 yr.

　a. Use the model $A = Pe^{rt}$ to determine the average rate of return under continuous compounding. Round to the nearest tenth of a percent.

　b. Using the interest rate from part (a), how long will it take the investment to reach $50,000$? Round to the nearest tenth of a year.

40. The number of bacteria in a culture begins with approximately 10,000 organisms at the start of an experiment. If the bacteria doubles every 5 hr, the model $P(t) = 10,000(2)^{t/5}$ represents the population $P(t)$ after t hours.

　a. Write a function of the form $P(t) = P_0 e^{kt}$ to model the population.

　b. Determine the amount of time required for the population to grow to 5 million. Round to the nearest hour.

41. The population $P(t)$ of a herd of deer on an island can be modeled by $P(t) = \dfrac{1200}{1 + 2e^{-0.12t}}$, where t represents the number of years since the park service has been tracking the herd.

　a. Evaluate $P(0)$ and interpret its meaning in the context of this problem.

　b. Use the function to predict the deer population after 4 yr. Round to the nearest whole unit.

　c. Use the function to predict the deer population after 8 yr.

　d. Determine the number of years required for the deer population to reach 900. Round to the nearest year.

　e. What value will the term $\dfrac{2}{e^{0.12t}}$ approach as $t \to \infty$?

　f. Determine the limiting value of $P(t)$.

42. The number N of visitors to a new website is given in the table t weeks after the website was launched.

t	0	1	2	3	4
N	24	50	121	270	640

　a. Use a graphing utility to find an equation of the form $N = ab^t$ to model the data. Round a to 1 decimal place and b to 3 decimal places.

　b. Use a graphing utility to graph the data and the model from part (a).

　c. Use the model to predict the number of visitors to the website 10 weeks after launch. Round to the nearest thousand.

For Exercises 1–2, simplify the expression.

1. $\dfrac{3x^{-1} - 6x^{-2}}{2x^{-2} - x^{-1}}$

2. $\dfrac{5}{\sqrt[3]{2x^2}}$

3. Factor. $a^3 - b^3 - a + b$

4. Perform the operations and write the answer in scientific notation. $\dfrac{(3.0 \times 10^7)(8.2 \times 10^{-3})}{1.23 \times 10^{-5}}$

For Exercises 5–13, solve the equations and inequalities. Write the solution sets to the inequalities in interval notation.

5. $5 \le 3 + |2x - 7|$

6. $3x(x - 1) = x + 6$

7. $\sqrt{t + 3} + 4 = t + 1$

8. $9^{2m-3} = 27^{m+1}$

9. $-x^3 - 5x^2 + 4x + 20 < 0$

10. $|5x - 1| = |3 - 4x|$

11. $(x^2 - 9)^2 - 2(x^2 - 9) - 35 = 0$

12. $\log_2(3x - 1) = \log_2(x + 1) + 3$

13. $\dfrac{x - 4}{x + 2} \le 0$

14. Find all the zeros of $f(x) = x^4 + 10x^3 + 10x^2 + 10x + 9$

15. Given $f(x) = x^2 - 16x + 55$,

   a. Does the graph of the parabola open upward or downward?

   b. Find the vertex of the parabola.

   c. Identify the maximum or minimum point.

   d. Identify the maximum or minimum value of the function.

   e. Identify the x-intercept(s).

   f. Identify the y-intercept.

   g. Write an equation for the axis of symmetry.

   h. Write the domain in interval notation.

   i. Write the range in interval notation.

16. Graph. $f(x) = -1.5x^2(x - 2)^3(x + 1)$

17. Given $f(x) = \dfrac{3x + 6}{x - 2}$,

   a. Write an equation of the vertical asymptote(s).

   b. Write an equation of the horizontal or slant asymptote.

   c. Graph the function.

18. Given $f(x) = 2^{x+2} - 3$,

   a. Write an equation of the asymptote.

   b. Write the domain in interval notation.

   c. Write the range in interval notation.

19. Write the expression as a single logarithm and simplify. $\log 40 + \log 50 - \log 2$

20. Given the one-to-one function defined by $f(x) = \sqrt[3]{x - 4} + 1$, write an equation for $f^{-1}(x)$.

Systems of Equations and Inequalities

The economy influences how people spend their money, but it also impacts how people save. When comparing options for investments such as stocks, bonds, and mutual funds, an individual can easily become overwhelmed by the possible scenarios. Collecting data on the rates of return on different investments is helpful in making an informed decision. In some scenarios, we turn to systems of linear equations to analyze such data.

In this chapter, we will solve systems involving both linear and nonlinear equations. In addition, we will use systems of linear inequalities in manufacturing applications. In such applications, the goal is to determine constraints on the production process (such as limits on the amount of material, labor, and equipment) and then maximize profit or minimize cost subject to these constraints.

| SECTION 5.1 | Systems of Linear Equations in Two Variables and Applications |

1. Identify Solutions to Systems of Linear Equations in Two Variables

The point of intersection between two curves is important in the analysis of supply and demand. For example, suppose that a small theater company wants to set an optimal price for tickets to a show. If the theater sells tickets for $1 each, there would be a high demand and many people would buy tickets. However, the revenue brought in would not cover the expense of the show. Therefore, the theater is not willing to offer (supply) tickets at this low price. If the theater sells tickets for $100 each, chances are that few people would buy tickets (demand would be low).

In an open market, the price y of an item is dependent on the number of items x supplied by the producer and demanded by the consumers. Competition between buyers and sellers steers the price toward an equilibrium price. This is the point where the supply curve and demand curve intersect (Figure 5-1).

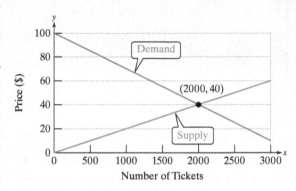

Figure 5-1

Suppose that the following linear equations represent the supply and demand curves for theater tickets. Taken together these equations form a *system* of linear equations.

$$\left. \begin{array}{l} \text{Supply: } y = 0.02x \\ \text{Demand: } y = -0.03x + 100 \end{array} \right\} \quad \begin{array}{l}\text{System of linear}\\\text{equations}\end{array}$$

Two or more linear equations make up a **system of linear equations.** A **solution** to a system of equations in two variables is an ordered pair that is a solution to each individual equation. Graphically, this is a point of intersection of the graphs of the equations. For the system given here, the solution is (2000, 40). This means that 2000 tickets are bought and sold at the equilibrium price of $40. The **solution set** to a system of equations is the set of all solutions to the system. In this case, the solution set is {(2000, 40)}.

TIP A solution to a system of linear equations in two variables is an ordered pair. To express the solution set, we enclose the solution(s) within set brackets.

| EXAMPLE 1 | Determining if an Ordered Pair Is a Solution to a System of Equations |

Determine if the ordered pair is a solution to the system. $6x + y = -2$
$4x - 3y = 17$

a. $\left(\dfrac{1}{2}, -5 \right)$ **b.** $(0, -2)$

Solution:

a. First equation   Second equation   Test the ordered pair $\left(\frac{1}{2}, -5 \right)$ in each equation.

$6x + y = -2$ $4x - 3y = 17$

$6\left(\frac{1}{2}\right) + (-5) \overset{?}{=} -2$ $4\left(\frac{1}{2}\right) - 3(-5) \overset{?}{=} 17$

$\quad\quad 3 + (-5) \overset{?}{=} -2$ ✓ true $\quad\quad 2 + 15 \overset{?}{=} 17$ ✓ true

The ordered pair $\left(\frac{1}{2}, -5 \right)$ is a solution to the system.

The ordered pair is a solution to both equations.

b. <u>First equation</u>

$$6x + y = -2$$

$$6(0) + (-2) \overset{?}{=} -2$$

$$0 + (-2) \overset{?}{=} -2$$

$$-2 \overset{?}{=} -2 \checkmark \text{ true}$$

<u>Second equation</u>

$$4x - 3y = 17$$

$$4(0) - 3(-2) \overset{?}{=} 17$$

$$0 + 6 \overset{?}{=} 17$$

$$6 \overset{?}{=} 17 \text{ false}$$

Test the ordered pair $(0, -2)$ in each equation.

The ordered pair is *not* a solution to the second equation.

The ordered pair $(0, -2)$ is *not* a solution to the system.

Skill Practice 1

Determine if the ordered pair is a solution to the system.

a. $(2, -4)$ **b.** $\left(\dfrac{1}{3}, -9\right)$

$$3x - y = 10$$

$$x + \dfrac{1}{4}y = 1$$

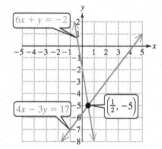

Figure 5-2

The lines from Example 1 are shown in Figure 5-2. From the graph, we can verify that $\left(\frac{1}{2}, -5\right)$ is the only solution to the system of equations.

There are three different possibilities regarding the number of solutions to a system of linear equations.

Solutions to Systems of Linear Equations in Two Variables

One Unique Solution	**No Solution**	**Infinitely Many Solutions**
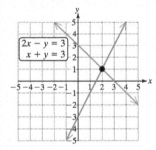	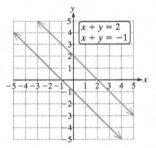	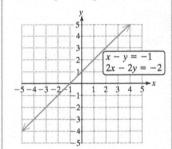
If a system of linear equations represents intersecting lines, then it has exactly one solution.	If a system of linear equations represents parallel lines, then the lines do not intersect, and the system has no solution. In such a case, we say that the system is **inconsistent.**	If a system of linear equations represents the same line, then all points on the common line satisfy each equation. Therefore, the system has infinitely many solutions. In such a case, we say that the equations are **dependent.**

---

2. Solve Systems of Linear Equations in Two Variables

Graphing a system of equations is one method to find the solution(s) to the system. However, sometimes it is difficult to determine the solution(s) using this method because of limitations in the accuracy of the graph. Instead we often use algebraic methods to solve a system of equations. The first method we present is called the **substitution method.**

Answers

1. a. Yes   **b.** No

Solving a System of Equations by Using the Substitution Method

Step 1 Isolate one of the variables from one equation.

Step 2 Substitute the expression found in step 1 into the *other* equation.

Step 3 Solve the resulting equation.

Step 4 Substitute the value found in step 3 back into the equation in step 1 to find the value of the remaining variable.

Step 5 Check the ordered pair in each equation and write the solution as an ordered pair in set notation.

EXAMPLE 2 Solving a System of Equations by the Substitution Method

Solve the system by using the substitution method.
$$-5x - 4y = 2$$
$$4x + y = 5$$

Solution:

$$-5x - 4y = 2$$
$$4x + y = 5 \longrightarrow y = \underline{-4x + 5}$$

Step 1: Isolate one of the variables from one of the equations. A variable with coefficient 1 or -1 is easily isolated.

$$-5x - 4(-4x + 5) = 2$$

Step 2: Substitute the expression from step 1 into the other equation.

$$-5x + 16x - 20 = 2$$
$$11x = 22$$
$$x = 2$$

Step 3: Solve for the remaining variable.

$$y = -4x + 5$$
$$y = -4(2) + 5$$
$$y = -3$$

Step 4: Substitute the known value of x into the equation where y is isolated. From step 1, this is $y = -4x + 5$.

Step 5: Check the ordered pair $(2, -3)$ in each original equation.

Check $(2, -3)$.

$-5x - 4y = 2$	$4x + y = 5$
$-5(2) - 4(-3) \stackrel{?}{=} 2$	$4(2) + (-3) \stackrel{?}{=} 5$
$-10 + 12 \stackrel{?}{=} 2$ ✓ true	$8 - 3 \stackrel{?}{=} 5$ ✓ true

The solution set is $\{(2, -3)\}$.

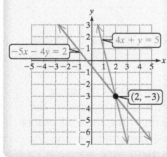

TIP The lines from Example 2 are shown here. The graph shows the point of intersection at $(2, -3)$.

Skill Practice 2 Solve the system by using the substitution method.
$$3x + 4y = 5$$
$$x - 3y = 6$$

Now consider the following system of equations.

$$5x - 4y = 6$$
$$-3x + 7y = 1$$

None of the variable terms has a coefficient of 1 or -1. Therefore, if we isolate x or y from either equation, the resulting equation will have one or more terms with fractional coefficients. To avoid this scenario, we can use another method called the **addition method** (also called the elimination method).

Answer

2. $\{(3, -1)\}$

> **Solving a System of Equations by Using the Addition Method**
>
> **Step 1** Write both equations in standard form: $Ax + By = C$.
> **Step 2** Clear fractions or decimals (optional).
> **Step 3** Multiply one or both equations by nonzero constants to create opposite coefficients for one of the variables.
> **Step 4** Add the equations from step 3 to eliminate one variable.
> **Step 5** Solve for the remaining variable.
> **Step 6** Substitute the known value found in step 5 into one of the original equations to solve for the other variable.
> **Step 7** Check the ordered pair in each equation and write the solution set.

TIP The addition method is sometimes called the *elimination method*. However, since both the substitution method and addition method eliminate a variable, we use the name addition method to emphasize the technique used.

EXAMPLE 3 Solving a System of Equations by the Addition Method

Solve the system by using the addition method. $5x = 4y + 6$
$-3x + 7y = 1$

Solution:

$5x = 4y + 6$ — Subtract $4y$. → $5x - 4y = 6$
$-3x + 7y = 1$ $-3x + 7y = 1$

Step 1: Write each equation in standard form: $Ax + By = C$.
Step 2: There are no decimals or fractions.

$5x - 4y = 6$ — Multiply by 3. → $15x - 12y = 18$
$-3x + 7y = 1$ — Multiply by 5. → $\underline{-15x + 35y = 5}$
                          $23y = 23$

Step 3: Multiply the first equation by 3. Multiply the second equation by 5.
Step 4: Add the equations to eliminate x.

$y = 1$ **Step 5:** Solve for y.

$5x = 4y + 6$
$5x = 4(1) + 6$
$5x = 10$
$x = 2$

Step 6: Substitute $y = 1$ into one of the original equations to solve for x.

Step 7: Check the ordered pair $(2, 1)$ in each original equation.

$5x = 4y + 6$ $-3x + 7y = 1$
$5(2) \overset{?}{=} 4(1) + 6$ $-3(2) + 7(1) \overset{?}{=} 1$
$10 \overset{?}{=} 4 + 6$ ✓ true $-6 + 7 \overset{?}{=} 1$ ✓ true

The solution set is $\{(2, 1)\}$.

TIP In Example 3, the variable x was eliminated.

Alternatively, we could have eliminated y by multiplying the first equation by 7 and the second equation by 4. This would create new equations with coefficients of -28 and 28 on the y terms.

Skill Practice 3 Solve the system by using the addition method.
$2x - 9y = 1$
$3x = 17 - 2y$

Answer
3. $\{(5, 1)\}$

TECHNOLOGY CONNECTIONS

Solving a System of Linear Equations in Two Variables Using Intersect

The solution to a system of linear equations can be checked on a graphing calculator by first writing the equations in slope-intercept form. From Example 3 we have

$$5x = 4y + 6 \longrightarrow y = \frac{5}{4}x - \frac{3}{2}$$

$$-3x + 7y = 1 \longrightarrow y = \frac{3}{7}x + \frac{1}{7}$$

Then graph the equations and use the Intersect feature to approximate the point of intersection. The Intersect feature gives the solution $(2, 1)$.

It is important to write the individual equations in a system of equations in standard form so that the variables line up. Also consider clearing decimals or fractions within an equation to make integer coefficients.

EXAMPLE 4   Solving a System of Equations by the Addition Method

Solve the system by using the addition method.
$$\frac{2}{5}x - y = \frac{19}{10}$$
$$5(x + y) = -7y - 41$$

Solution:

$$\frac{2}{5}x - y = \frac{19}{10} \xrightarrow{\text{Multiply by 10.}} 4x - 10y = 19$$

$$5(x + y) = -7y - 41 \xrightarrow{\text{Simplify.}} 5x + 12y = -41$$

Clear the fractions in the first equation. Write the second equation in standard form.

$$4x - 10y = 19 \xrightarrow{\text{Multiply by } -5.} -20x + 50y = -95$$

$$5x + 12y = -41 \xrightarrow[\text{Multiply by 4.}]{} \underline{20x + 48y = -164}$$

$$98y = -259$$

The LCM of the x coefficients, 4 and 5, is 20. Create opposite coefficients on x of 20 and -20.

$$y = \frac{-259}{98} \qquad \text{Solve for } y.$$

$$y = -\frac{37}{14} \qquad \text{Simplify to lowest terms.}$$

Substituting $y = -\frac{37}{14}$ back into one of the original equations to solve for x would be cumbersome. Alternatively, we can solve for x by repeating the addition method. This time we will eliminate y by creating opposite coefficients on the y terms and then solving for x.

$$4x - 10y = 19 \xrightarrow{\text{Multiply by 6.}} 24x - 60y = 114$$
$$5x + 12y = -41 \xrightarrow[\text{Multiply by 5.}]{} 25x + 60y = -205$$
$$49x \qquad\quad = -91$$

The LCM of 10 and 12 is 60. Create opposite coefficients of -60 and 60 on the y terms.

$$x = \frac{-91}{49} \qquad \text{Solve for } y.$$

$$x = -\frac{13}{7} \qquad \text{Simplify.}$$

The ordered pair $\left(-\frac{13}{7}, -\frac{37}{14}\right)$ checks in both original equations.

The solution set is $\left\{\left(-\frac{13}{7}, -\frac{37}{14}\right)\right\}$.

> **Skill Practice 4**   Solve the system by using the addition method.
>
> $$2(x - 2y) = y + 14$$
> $$\frac{1}{2}x + \frac{7}{6}y = -\frac{13}{3}$$

The systems in Examples 1–4 each have one unique solution. That is, the lines represented by the two equations intersect in exactly one point. In Examples 5 and 6, we investigate systems with no solution or infinitely many solutions.

EXAMPLE 5   **Identifying a System of Equations with No Solution**

Solve the system. $2x + y = 4$
$\qquad\qquad\qquad\quad\; 6x + 3y = 6$

Solution:

$$2x + y = 4 \xrightarrow{\text{Multiply by } -3.} -6x - 3y = -12$$
$$6x + 3y = 6 \xrightarrow{\qquad\qquad} \underline{\;\;6x + 3y = \;\;\;6\;}$$
$$0 = -6$$

We can eliminate either the x terms or y terms by multiplying the first equation by -3.

Both the x and y terms are eliminated, leading to the contradiction $0 = -6$.

The system of equations reduces to a contradiction. This indicates that there is no solution and the system is inconsistent. The equations represent parallel lines and two parallel lines do not intersect.

The solution set is { }.

> **Skill Practice 5**   Solve the system. $3x - y = 2$
> $\qquad\qquad\qquad\qquad\qquad\qquad\quad -9x + 3y = 4$

Answers

4. $\left\{\left(-\dfrac{32}{29}, -\dfrac{94}{29}\right)\right\}$ **5.** { }

Solving a System of Dependent Equations

Solve the system. $y = 2x - 1$
$8x - 4y = 4$

Solution:

$y = \overbrace{2x - 1}$
$8x - 4y = 4$ $8x - 4(\underline{2x - 1}) = 4$
$8x - 8x + 4 = 4$
$4 = 4$

With y already isolated in the first equation, apply the substitution method.

Solve the resulting equation.

Identity.

Notice that both the variable terms and the constant terms were eliminated. The system of equations is reduced to the identity $4 = 4$. Therefore, the two original equations are dependent and represent the same line. The solution set consists of an infinite number of ordered pairs (x, y) that fall on the common line of intersection, $y = 2x - 1$. Therefore, the solutions are ordered pairs of the form $(x, 2x - 1)$.

The solution set is $\{(x, 2x - 1) \mid x \text{ is any real number}\}$.

> **Skill Practice 6**   Solve the system. $x = 5 - 3y$
> $2x + 6y = 10$

> ### Avoiding Mistakes
> To verify the solution to Example 6, write the two equations in slope-intercept form.
>
> $y = 2x - 1 \Rightarrow y = 2x - 1$
> $8x - 4y = 4 \Rightarrow y = 2x - 1$
>
> The equations have the same slope-intercept form and define the same line.

> ### TIP Sometimes the solution
> to a system of dependent equations is written with an arbitrary variable called a **parameter.** For example, letting t represent any real number, then the solution set to Example 6 can be written as $\{(t, 2t - 1) \mid t \text{ is a real number}\}$.

The solution set $\{(x, 2x - 1) \mid x \text{ is any real number})\}$ is called the **general solution** to the system in Example 6. By varying the value of x, we can produce any number of specific solutions to the system of equations. For example:

	$(x, 2x - 1)$	**Solution**
If $x = 1$	$(1, 2(1) - 1)$	$(1, 1)$
If $x = 2$	$(2, 2(2) - 1)$	$(2, 3)$
If $x = 3$	$(3, 2(3) - 1)$	$(3, 5)$
If $x = -1$	$(-1, 2(-1) - 1)$	$(-1, -3)$

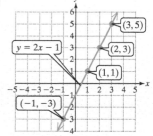

Figure 5-3

Notice that the solutions fall on the common line of intersection as expected (Figure 5-3).

> ### TIP The equations
> $y = 2x - 1$ and $8x - 4y = 4$ represent the same relationship between x and y. Therefore, we have only one unique equation, but two variables. As a result, y depends on the choice of x and vice versa.

We should also note that the general solution to a system of dependent equations can be written in a variety of forms. For instance, the solutions to the system in Example 6 fall on the line $y = 2x - 1$. By solving the equation for x, we have $x = \dfrac{y + 1}{2}$. Thus, the solution set can also be written as $\left\{ \left(\dfrac{y + 1}{2}, y \right) \middle| y \text{ is any real number} \right\}$.

3. Use Systems of Linear Equations in Applications

When solving an application involving two unknowns, sometimes it is convenient to use a system of two independent equations as demonstrated in Examples 7–9.

Solving an Application Involving Mixtures

A hospital uses a 15% bleach solution to disinfect a quarantine area. How much 6% bleach solution must be mixed with an 18% bleach solution to make 50 L of a 15% bleach solution?

Answer

6. $\left\{ \left(x, \dfrac{-x + 5}{3} \right) \middle| x \text{ is any real number} \right\}$
or $\{(5 - 3y, y) \mid y \text{ is any real number}\}$

Solution:

Let x represent the amount of 6% bleach solution.

Let y represent the amount of 18% bleach solution.

The amount of pure bleach in each mixture is found by multiplying the amount of solution by the concentration rate. This information can be organized in a table.

	6% Solution	18% Solution	15% Solution
Amount of mixture	x	y	50
Amount of pure bleach	$0.06x$	$0.18y$	$0.15(50)$

There are two unknown quantities. We will set up a system of two independent equations relating x and y.

$$\left(\begin{array}{c}\text{Amount of}\\ \text{6\% mixture}\end{array}\right) + \left(\begin{array}{c}\text{Amount of}\\ \text{18\% mixture}\end{array}\right) = \left(\begin{array}{c}\text{Amount of}\\ \text{15\% mixture}\end{array}\right)$$

$$\left(\begin{array}{c}\text{Amount of}\\ \text{pure bleach in}\\ \text{6\% mixture}\end{array}\right) + \left(\begin{array}{c}\text{Amount of}\\ \text{pure bleach in}\\ \text{18\% mixture}\end{array}\right) = \left(\begin{array}{c}\text{Amount of}\\ \text{pure bleach in}\\ \text{15\% mixture}\end{array}\right)$$

$$x + y = 50$$
$$0.06x + 0.18y = 0.15(50)$$

$$
\begin{array}{llll}
x + y = 50 & & x + y = 50 & x + y = 50 \\
0.06x + 0.18y = 7.5 \xrightarrow{\text{Multiply by 100.}} & 6x + 18y = 750 \xrightarrow{\text{Divide by } -6.} & \underline{-x - 3y = -125} \\
& & & -2y = -75 \\
& & & y = 37.5
\end{array}
$$

Substitute $y = 37.5$ back into the equation $x + y = 50$.

$$x + y = 50$$
$$x + 37.5 = 50$$
$$x = 12.5$$

Therefore, 12.5 L of the 6% bleach solution should be mixed with 37.5 L of the 18% bleach solution to make 50 L of 15% bleach solution.

Avoiding Mistakes

Check that the answer is reasonable. The total amount of the resulting solution is $12.5\ \text{L} + 37.5\ \text{L}$, which is 50 L.

Amount of pure bleach:
$0.06(12.5\ \text{L}) = 0.75\ \text{L}$
$0.18(37.5\ \text{L}) = 6.75\ \text{L}$
$0.15(50\ \text{L}) = 7.5\ \text{L}$ ✓

Skill Practice 7 How many ounces of 20% and 35% acid solution should be mixed to produce 15 oz of 30% acid solution?

EXAMPLE 8   Solving an Application Involving Uniform Motion

A riverboat traveling upstream against the current on the Mississippi River takes 3 hr to travel 24 mi. The return trip downstream with the current takes only 2 hr. Find the speed of the boat in still water and the speed of the current.

Solution:

Let b represent the speed of the boat in still water.

Let c represent the speed of the current.

The given information can be organized in a table.

Answer

7. Mix 5 oz of 20% acid solution with 10 oz of 35% acid solution to make 15 oz of 30% acid solution.

	Distance (mi)	Rate (mph)	Time (hr)
Upstream	24	$b - c$	3
Downstream	24	$b + c$	2

Use the relationship $d = rt$; that is, distance = (rate)(time).

$$\begin{pmatrix} \text{Distance} \\ \text{upstream} \end{pmatrix} = \begin{pmatrix} \text{Rate} \\ \text{upstream} \end{pmatrix}\begin{pmatrix} \text{Time} \\ \text{upstream} \end{pmatrix} \longrightarrow 24 = (b - c) \cdot 3$$

$$\begin{pmatrix} \text{Distance} \\ \text{downstream} \end{pmatrix} = \begin{pmatrix} \text{Rate} \\ \text{downstream} \end{pmatrix}\begin{pmatrix} \text{Time} \\ \text{downstream} \end{pmatrix} \longrightarrow 24 = (b + c) \cdot 2$$

$$24 = 3b - 3c \xrightarrow{\text{Divide by 3.}} 8 = b - c$$
$$24 = 2b + 2c \xrightarrow{\text{Divide by 2.}} \underline{12 = b + c}$$
$$20 = 2b$$
$$10 = b$$

Substitute $b = 10$ into the equation $12 = b + c$, which gives $c = 2$.
The boat's speed in still water is 10 mph and the speed of the current is 2 mph.

Skill Practice 8   A boat takes 3 hr to go 24 mi upstream against the current. It can go downstream with the current a distance of 48 mi in the same amount of time. Determine the speed of the boat in still water and the speed of the current.

EXAMPLE 9   **Applying a System of Equations in Business**

A lawn service company has fixed monthly costs of $500 and variable costs (labor, gasoline, and depreciation) of $40 per lawn. If the service charges $60 per lawn,

a. Write a cost function representing the cost $C(x)$ to the company to service x lawns per month.

b. Write a revenue function representing the revenue $R(x)$ to service x lawns per month.

c. Determine the number of lawns that must be serviced in a month for the company to break even.

Solution:

Let x represent the number of lawns serviced for a given month.

a. $\begin{pmatrix} \text{Monthly} \\ \text{cost} \end{pmatrix} = \begin{pmatrix} \text{Fixed} \\ \text{cost} \end{pmatrix} + \begin{pmatrix} \text{Variable} \\ \text{cost} \end{pmatrix} \longrightarrow C(x) = 500 + 40x$

b. $\begin{pmatrix} \text{Monthly} \\ \text{Revenue} \end{pmatrix} = \begin{pmatrix} \text{Revenue} \\ \text{per lawn} \end{pmatrix}\begin{pmatrix} \text{Number} \\ \text{of lawns} \end{pmatrix} \longrightarrow R(x) = 60x$

c. To break even, the cost must equal revenue, $C(x) = R(x)$.
$$500 + 40x = 60x$$
$$500 = 20x$$
$$x = 25$$
To break even, the company must service 25 lawns.

Skill Practice 9   A storage company rents its units for $120 per month. The company has fixed monthly costs of $2100 and variable costs (air-conditioning and service) of $50 per unit.

a. Write a cost function representing the monthly cost $C(x)$ to the company for x units.

b. Write a revenue function representing the revenue $R(x)$ when x units per month are rented.

c. Determine the number of units that must be rented in a month for the company to break even.

Answers
8. The boat's speed in still water is 12 mph and the speed of the current is 4 mph.
9. a. $C(x) = 2100 + 50x$
  b. $R(x) = 120x$
  c. 30 units

Prerequisite Review

For Exercises R.1–R.4, solve the equation.

R.1. $4 = 3 + 3(3x + 1)$

R.2. $0.65(x - 6) + 0.35(x + 4) = 0.5$

R.3. $3(7 - 4n) + 2 = -2n - 3 - 10n$

R.4. $-4 + 3x = 3(x - 3) + 5$

R.5. Graph the equation and identify the x- and y-intercepts. $4x + y = 4$

R.6. Graph the equation and identify the slope and y-intercept. $-x + 5y = -5$

Concept Connections

1. Two or more linear equations taken together make up a _____ of linear equations.

2. A _____ to a system of equations in two variables is an ordered pair that is a solution to each individual equation in the system.

3. Two algebraic methods to solve a system of linear equations in two variables are the _____ method and the _____ method.

4. A system of linear equations in two variables may have no solution. In such a case, the equations represent _____ lines.

5. A system of equations that has no solution is called an _____ system.

6. A system of linear equations in two variables may have infinitely many solutions. In such a case, the equations are said to be _____.

Objective 1: Identify Solutions to Systems of Linear Equations in Two Variables

For Exercises 7–10, determine if the ordered pair is a solution to the system of equations. (See Example 1)

7. $3x - 5y = -7$
$x - 4y = -7$

  a. $(1, 2)$

  b. $\left(-\dfrac{2}{3}, 1\right)$

8. $-11x + 6y = -4$
$7x + 3y = 23$

  a. $\left(1, \dfrac{7}{6}\right)$

  b. $(2, 3)$

9. $y = \dfrac{3}{2}x - 5$
$6x - 4y = 20$

  a. $(2, -2)$

  b. $(-4, -11)$

10. $y = -\dfrac{1}{5}x + 2$
$2x + 10y = 10$

  a. $(5, 1)$

  b. $(-10, 4)$

For Exercises 11–14, a system of equations is given in which each equation is written in slope-intercept form. Determine the number of solutions. If the system does not have one unique solution, state whether the system is inconsistent or whether the equations are dependent.

11. $y = \dfrac{2}{5}x - 7$
$y = \dfrac{1}{4}x + 7$

12. $y = 6x - \dfrac{2}{3}$
$y = 6x + 4$

13. $y = 8x - \dfrac{1}{2}$
$y = 8x - \dfrac{1}{2}$

14. $y = \dfrac{1}{2}x + 3$
$y = 2x + \dfrac{1}{3}$

Objective 2: Solve Systems of Linear Equations in Two Variables

For Exercises 15–20, solve the system of equations by using the substitution method. (See Example 2)

15. $x + 3y = 5$
$3x - 2y = -18$

16. $2x + y = 2$
$5x + 3y = 9$

17. $2x + 7y = 1$
$3y - 7 = 2$

18. $3x = 2y - 11$
$6 + 5x = 1$

19. $2(x + y) = 2 - y$
$4x - 1 = 2 - 5y$

20. $5(x + y) = 9 + 2y$
$6y - 2 = 10 - 7x$

For Exercises 21–28, solve the system of equations by using the addition method. (See Examples 3–4)

21. $3x - 7y = 1$
$6x + 5y = -17$

22. $5x - 2y = -2$
$3x + 4y = 30$

23. $11x = -5 - 4y$
$2(x - 2y) = 22 + y$

24. $-3(x - y) = y - 14$
$2x + 2 = 7y$

25. $0.6x + 0.1y = 0.4$
$2x - 0.7y = 0.3$

26. $0.25x - 0.04y = 0.24$
$0.15x - 0.12y = 0.12$

27. $2x + 11y = 4$
$3x - 6y = 5$

28. $3x - 4y = 9$
$2x + 9y = 2$

For Exercises 29–34, solve the system by using any method. If a system does not have one unique solution, state whether the system is inconsistent or whether the equations are dependent. (See Examples 5–6)

29. $3x - 4y = 6$
$9x = 12y + 4$

30. $-4x - 8y = 2$
$2x = 8 - 4y$

31. $3x + y = 6$
$x + \frac{1}{3}y = 2$

32. $2x - y = 8$
$x - \frac{1}{2}y = 4$

33. $2x + 4 = 4 - 5y$
$2 + 4(x + y) = 7y + 2$

34. $3(x - 3y) = 2y$
$2x + 5 = 5 - 7y$

For Exercises 35–36,
a. Write the general solution.
b. Find three individual solutions. Answers will vary.

35. $-5x - y = 6$
$10x = -2(y + 6)$

36. $2y = 6 - 4x$
$8x = 12 - 4y$

Mixed Exercises

For Exercises 37–50, solve the system using any method.

37. $3x - 10y = 1900$
$5y + 800 = x$

38. $2x - 7y = 2400$
$-4x + 1800 = y$

39. $5(2x + y) = y - x - 8$
$x - \frac{3}{2}y = \frac{5}{2}$

40. $3(2x - y) = 2 - x$
$x + \frac{5}{4}y = \frac{3}{2}$

41. $y = \frac{2}{3}x - 1$
$y = \frac{1}{6}x + 2$

42. $y = -\frac{1}{4}x + 7$
$y = -\frac{3}{2}x + 17$

43. $4(x - 2) = 6y + 3$
$\frac{1}{4}x - \frac{3}{8}y = -\frac{1}{2}$

44. $\frac{1}{14}x - \frac{1}{7}y = \frac{1}{2}$
$2(x - 2y) + 3 = 20$

45. $2x = \frac{y}{2} + 1$
$0.04x - 0.01y = 0.02$

46. $0.05x + 0.01y = 0.03$
$x + \frac{y}{5} = \frac{3}{5}$

47. $y = 2.4x - 1.54$
$y = -3.5x + 7.9$

48. $y = -0.18x + 0.129$
$y = -0.15x + 0.1275$

49. $\frac{x - 2}{8} + \frac{y + 1}{2} = -6$
$\frac{x - 2}{2} - \frac{y + 1}{4} = 12$

50. $\frac{x + 1}{2} - \frac{y - 2}{10} = -1$
$\frac{x + 1}{6} + \frac{y - 2}{2} = 21$

Objective 3: Use Systems of Linear Equations in Applications

51. One antifreeze solution is 36% alcohol and another is 20% alcohol. How much of each mixture should be added to make 40 L of a solution that is 30% alcohol? **(See Example 7)**

52. A pharmacist wants to mix a 30% saline solution with a 10% saline solution to get 200 mL of a 12% saline solution. How much of each solution should she use?

53. A radiator has 16 L of a 36% antifreeze solution. How much must be drained and replaced by pure antifreeze to bring the concentration level up to 50%?

54. Jonas performed an experiment for his science fair project. He learned that rinsing lettuce in vinegar kills more bacteria than rinsing with water or with a popular commercial product. As a follow-up to his project, he wants to determine the percentage of bacteria killed by rinsing with a diluted solution of vinegar.

   a. How much water and how much vinegar should be mixed to produce 10 cups of a mixture that is 40% vinegar?

   b. How much pure vinegar and how much 40% vinegar solution should be mixed to produce 10 cups of a mixture that is 60% vinegar?

55. Michelle borrows a total of $5000 in student loans from two lenders. One charges 4.6% simple interest and the other charges 6.2% simple interest. She is not required to pay off the principal or interest for 3 yr. However, at the end of 3 yr, she will owe a total of $762 for the interest from both loans. How much did she borrow from each lender?

56. Juan borrows $100,000 to pay for medical school. He borrows part of the money from the school whereby he will pay 4.5% simple interest. He borrows the rest of the money through a government loan that will charge him 6% interest. In both cases, he is not required to pay off the principal or interest during his 4 yr of medical school. However, at the end of 4 yr, he will owe a total of $19,200 for the interest from both loans. How much did he borrow from each source?

57. Stuart pays back two student loans over a 4-yr period. One loan charges the equivalent of 3% simple interest and the other charges the equivalent of 5.5% simple interest. If the total amount borrowed was $24,000 and the total amount of interest paid after 4 yr is $3280, find the amount borrowed from each loan.

58. A total of $6000 is invested for 5 yr with a total return of $1080. Part of the money is invested in a fund that returns the equivalent of 2% simple interest. The rest of the money is invested at 4% simple interest. Determine the amount invested in each account.

59. Monique and Tara each make an ice cream sundae. Monique gets 2 scoops of Cherry ice cream and 1 scoop of Mint Chocolate Chunk ice cream for a total of 43 g of fat. Tara has 1 scoop of Cherry and 2 scoops of Mint Chocolate Chunk for a total of 47 g of fat. How many grams of fat does 1 scoop of each type of ice cream have?

60. Bryan and Jadyn had barbeque potato chips and soda at a football party. Bryan ate 3 oz of chips and drank 2 cups of soda for a total of 700 mg of sodium. Jadyn ate 1 oz of chips and drank 3 cups of soda for a total of 350 mg of sodium. How much sodium is in 1 oz of chips and how much is in 1 cup of soda?

61. The average weekly salary of two employees is $1350. One makes $300 more than the other. Find their salaries.

62. The average of an electrician's hourly wage and a plumber's hourly wage is $33. One day a contractor hires the electrician for 8 hr of work and the plumber for 5 hr of work and pays a total of $438 in wages. Find the hourly wage for the electrician and for the plumber.

63. A moving sidewalk in an airport moves people between gates. It takes Jason's 9-year-old daughter Josie 40 sec to travel 200 ft walking with the sidewalk. It takes her 30 sec to walk 90 ft against the moving sidewalk (in the opposite direction). Find the speed of the sidewalk and find Josie's speed walking on non-moving ground. (**See Example 8**)

64. A fishing boat travels along the east coast of the United States and encounters the Gulf Stream current. It travels 44 mi north with the current in 2 hr. It travels 56 mi south against the current in 4 hr. Find the speed of the current and the speed of the boat in still water.

65. Two runners begin at the same point on a 390-m circular track and run at different speeds. If they run in opposite directions, they pass each other in 30 sec. If they run in the same direction, they meet each other in 130 sec. Find the speed of each runner.

66. Two particles begin at the same point and move at different speeds along a circular path of circumference 280 ft. Moving in opposite directions, they pass in 10 sec. Moving in the same direction, they pass in 70 sec. Find the speed of each particle.

67. A cleaning company charges $100 for each office it cleans. The fixed monthly cost of $480 for the company includes telephone service and the depreciation on cleaning equipment and a van. The variable cost is $52 per office and includes labor, gasoline, and cleaning supplies. (**See Example 9**)

   a. Write a linear cost function representing the cost $C(x)$ (in $) to the company to clean x offices per month.

   b. Write a linear revenue function representing the revenue $R(x)$ (in $) for cleaning x offices per month.

   c. Determine the number of offices to be cleaned per month for the company to break even.

   d. If 28 offices are cleaned, will the company make money or lose money?

68. A vendor at a carnival sells cotton candy and caramel apples for $2.00 each. The vendor is charged $100 to set up his booth. Furthermore, the vendor's average cost for each product he produces is approximately $0.75.

   a. Write a linear cost function representing the cost $C(x)$ (in $) to the vendor to produce x products.

   b. Write a linear revenue function representing the revenue $R(x)$ (in $) for selling x products.

   c. Determine the number of products to be produced and sold for the vendor to break even.

   d. If 60 products are sold, will the vendor make money or lose money?

For Exercises 69–70, refer to Figure 5-1 and the narrative at the beginning of this section.

69. Suppose that the price p (in $) of theater tickets is influenced by the number of tickets x offered by the theater and demanded by consumers.

   Supply: $p = 0.025x$

   Demand: $p = -0.04x + 104$

   a. Solve the system of equations defined by the supply and demand models.

   b. What is the equilibrium price?

   c. What is the equilibrium quantity?

70. The price p (in $) of a cookbook is determined by the number of cookbooks x demanded by consumers and supplied by the publisher.

   Supply: $p = 0.002x$

   Demand: $p = -0.005x + 70$

   a. Solve the system of equations defined by the supply and demand models.

   b. What is the equilibrium price?

   c. What is the equilibrium quantity?

71. a. Sketch the lines defined by $y = 2x$ and $y = -\frac{1}{2}x + 5$.

   b. Find the area of the triangle bounded by the lines in part (a) and the x-axis.

72. a. Sketch the lines defined by $y = x + 2$ and $y = -\frac{1}{2}x + 2$.

   b. Find the area of the triangle bounded by the lines in part (a) and the x-axis.

73. The **centroid** of a region is the geometric center. For the region shown, the centroid is the point of intersection of the diagonals of the parallelogram.

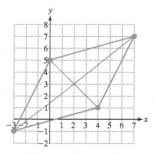

   a. Find an equation of the line through the points $(-3, -1)$ and $(7, 7)$.

   b. Find an equation of the line through the points $(0, 5)$ and $(4, 1)$.

   c. Find the centroid of the region.

74. The centroid of the region shown is the point of intersection of the diagonals of the parallelogram.

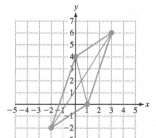

   a. Find an equation of the line through the points $(-2, -2)$ and $(3, 6)$.

   b. Find an equation of the line through the points $(1, 0)$ and $(0, 4)$.

   c. Find the centroid of the region.

75. Two angles are complementary. The measure of one angle is 6° less than twice the measure of the other angle. Find the measure of each angle.

76. Two angles are supplementary. The measure of one angle is 12° more than 5 times the measure of the other angle. Find the measure of each angle.

For Exercises 77–78, find the measure of angles x and y.

77.

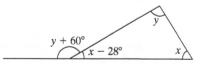

78.

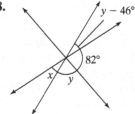

Mixed Exercises

79. Write a system of linear equations with solution set $\{(-3, 5)\}$.

80. Write a system of linear equations with solution set $\{(4, -3)\}$.

81. Find C and D so that the solution set to the system is $\{(4, 1)\}$.

$$Cx + 5y = 13$$
$$-2x + Dy = -5$$

82. Find A and B so that the solution set to the system is $\{(-5, 2)\}$.

$$3x + Ay = -3$$
$$Bx - y = -12$$

83. Given $f(x) = mx + b$, find m and b if $f(3) = -3$ and $f(-12) = -8$.

84. Given $g(x) = mx + b$, find m and b if $g(2) = 1$ and $g(-4) = 10$.

For Exercises 85–86, use the substitution $u = \frac{1}{x}$ and $v = \frac{1}{y}$ to rewrite the equations in the system in terms of the variables u and v. Solve the system in terms of u and v. Then back substitute to determine the solution set to the original system in terms of x and y.

85. $\dfrac{1}{x} + \dfrac{2}{y} = 1$

$-\dfrac{1}{x} + \dfrac{4}{y} = -7$

86. $-\dfrac{3}{x} + \dfrac{4}{y} = 11$

$\dfrac{1}{x} - \dfrac{2}{y} = -5$

87. During a race, Marta bicycled 12 mi and ran 4 mi in a total of 1 hr 20 min $\left(\frac{4}{3}\,\text{hr}\right)$. In another race, she bicycled 21 mi and ran 3 mi in 1 hr 40 min $\left(\frac{5}{3}\,\text{hr}\right)$. Determine the speed at which she bicycles and the speed at which she runs. Assume that her bicycling speed was the same in each race and that her running speed was the same in each race.

88. Shelia swam 1 mi and ran 6 mi in a total of 1 hr 15 min $\left(\frac{5}{4}\,\text{hr}\right)$. In another training session she swam 2 mi and ran 8 mi in a total of 2 hr. Determine the speed at which she swims and the speed at which she runs. Assume that her swimming speed was the same each day and that her running speed was the same each day.

89. A certain pickup truck gets 16 mpg in the city and 22 mpg on the highway. If a driver drives 254 mi on 14 gal of gas, determine the number of city miles and highway miles that the truck was driven.

90. A sedan gets 12 mpg in the city and 18 mpg on the highway. If a driver drives a total of 420 mi on 26 gal of gas, how many miles in the city and how many miles on the highway did he drive?

Write About It

91. A system of linear equations in x and y can represent two intersecting lines, two parallel lines, or a single line. Describe the solution set to the system in each case.

92. When solving a system of linear equations in two variables using the substitution or addition method, explain how you can detect whether the equations are dependent.

93. When solving a system of linear equations in two variables using the substitution or addition method, explain how you can detect whether the system is inconsistent.

94. Consider a system of linear equations in two variables in which the solution set is $\{(x, x + 2) \mid x \text{ is any real number}\}$. Why do we say that the equations in the system are dependent?

Expanding Your Skills

95. A 50-lb weight is supported from two cables and the system is in equilibrium. The magnitudes of the forces on the cables are denoted by $|F_1|$ and $|F_2|$, respectively. An engineering student knows that the horizontal components of the two forces (shown in red) must be equal in magnitude. Furthermore, the sum of the magnitudes of the vertical components of the forces (shown in blue) must be equal to 50 lb to offset the downward force of the weight. Find the values of $|F_1|$ and $|F_2|$. Write the answers in exact form with no radical in the denominator. Also give approximations to 1 decimal place.

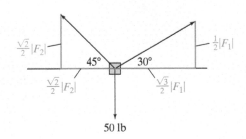

Technology Connections

For Exercises 96–99, use a graphing utility to approximate the solution to the system of equations. Round the x and y values to 3 decimal places.

96. $y = -3.729x + 6.958$

$y = 2.615x - 8.713$

97. $y = -0.041x + 0.068$

$y = 0.019x - 0.053$

98. $-0.25x + 0.04y = -0.42$

$6.775x + 2.5y = -38.1$

99. $0.36x - 0.075y = -0.813$

$0.066x + 0.008y = 0.194$

SECTION 5.2	# Systems of Linear Equations in Three Variables and Applications

1. Identify Solutions to a System of Linear Equations in Three Variables

In Section 5.1 we solved systems of linear equations in two variables. In this section, we expand the discussion to solving systems involving three variables. A **linear equation in three variables** is an equation that can be written in the form

$$Ax + By + Cz = D, \text{ where } A, B, \text{ and } C \text{ are not all zero.}$$

For example, $x + 2y + z = 4$ is a linear equation in three variables. A solution to a linear equation in three variables is an ordered triple (x, y, z) that satisfies the equation. For example, several solutions to $x + 2y + z = 4$ are given here.

Solution	Check: $x + 2y + z = 4$
$(1, 1, 1)$	$(1) + 2(1) + (1) \stackrel{?}{=} 4$ ✓ true
$(4, 0, 0)$	$(4) + 2(0) + (0) \stackrel{?}{=} 4$ ✓ true
$(0, 2, 0)$	$(0) + 2(2) + (0) \stackrel{?}{=} 4$ ✓ true
$(0, 0, 4)$	$(0) + 2(0) + (4) \stackrel{?}{=} 4$ ✓ true

Point of Interest

Beautiful planar surfaces are seen in both art and architecture. For example, the great pyramid of Khufu outside Cairo, Egypt, has four triangular sides. The planes representing the sides form an angle of 52° with the ground.

There are infinitely many solutions to the equation $x + 2y + z = 4$. The set of all solutions to a linear equation in three variables can be represented graphically by a plane in space. Figure 5-4 shows a portion of the plane defined by $x + 2y + z = 4$.

In many applications, we are interested in determining the point or points of intersection of two or more planes. This is given by the solutions to a system of linear equations in three variables. For example:

$$2x + y - 3z = -2$$
$$x - 4y + z = 24$$
$$-3x - y + 4z = 0$$

A solution to a system of linear equations in three variables is an **ordered triple** (x, y, z) that satisfies each equation in the system. Geometrically, a solution is a point of intersection of the planes represented by the equations in the system (Figure 5-5).

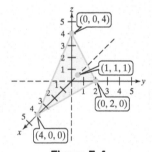

Figure 5-4

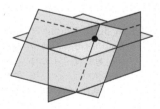

Figure 5-5

EXAMPLE 1	**Determining if an Ordered Triple Is a Solution to a System of Equations**

Determine if the ordered triple is a solution to the system.

$$2x + y - 3z = -2$$
$$x - 4y + z = 24$$
$$-3x - y + 4z = 0$$

a. $(3, -5, 1)$ **b.** $(2, -3, 1)$

Solution:

Test the ordered triple in each equation.

a. First equation

$2x + y - 3z = -2$

$2(3) + (-5) - 3(1) \stackrel{?}{=} -2$

$-2 \stackrel{?}{=} -2$ ✓ true

Second equation

$x - 4y + z = 24$

$(3) - 4(-5) + (1) \stackrel{?}{=} 24$

$24 \stackrel{?}{=} 24$ ✓ true

Third equation

$-3x - y + 4z = 0$

$-3(3) - (-5) + 4(1) \stackrel{?}{=} 0$

$0 \stackrel{?}{=} 0$ ✓ true

The ordered triple $(3, -5, 1)$ is a solution to the system of equations.

b. First equation

$2x + y - 3z = -2$

$2(2) + (-3) - 3(1) \stackrel{?}{=} -2$

$-2 \stackrel{?}{=} -2$ ✓ true

Second equation

$x - 4y + z = 24$

$(2) - 4(-3) + (1) \stackrel{?}{=} 24$

$15 \stackrel{?}{=} 24$ false

Third equation

$-3x - y + 4z = 0$

$-3(2) - (-3) + 4(1) \stackrel{?}{=} 0$

$1 \stackrel{?}{=} 0$ false

If an ordered triple fails to be a solution to any of the equations in the system, then it is not a solution to the system. The ordered triple $(2, -3, 1)$ is *not* a solution to the second or third equation. Therefore, $(2, -3, 1)$ is *not* a solution to the system of equations.

Skill Practice 1   Determine if the ordered triple is a solution to the system.

$$5x - y + 3z = -7$$
$$3x + 4y - z = 5$$
$$9x + 5y + 7z = 1$$

a. $(-2, -6, 3)$ **b.** $(-1, 2, 0)$

---

2. Solve Systems of Linear Equations in Three Variables

To solve a system of three linear equations in three variables, we first eliminate one variable. The system is then reduced to a two-variable system that can be solved by the techniques learned in Section 5.1.

Solving a System of Three Linear Equations in Three Variables

Step 1 Write each equation in standard form $Ax + By + Cz = D$.

Step 2 Choose a pair of equations and eliminate one of the variables by using the addition method.

Step 3 Choose a different pair of equations and eliminate the *same* variable.

Step 4 Once steps 2 and 3 are complete, the resulting system has two equations in two variables. Solve this system by using the substitution or addition method.

Step 5 Substitute the values of the variables found in step 4 into any of the three original equations that contain the third variable. Solve for the third variable.

Step 6 Check the ordered triple in each original equation. Then write the solution as an ordered triple in set notation.

EXAMPLE 2   **Solving a System of Equations in Three Variables**

Solve the system. $3x - 2y + z = 2$

$5x + y - 2z = 1$

$4x - 3y + 3z = 7$

Answers

1. a. No   **b.** Yes

TIP In Example 2, the y terms can also be eliminated easily because the coefficient on y in equation $\boxed{B}$ is 1. Therefore, y can be eliminated from equations $\boxed{A}$ and $\boxed{B}$ by multiplying equation $\boxed{B}$ by 2. Likewise y can be eliminated from equations $\boxed{B}$ and $\boxed{C}$ by multiplying equation $\boxed{B}$ by 3.

Solution:

$\boxed{A}$ $\quad 3x - 2y + z = 2$
$\boxed{B}$ $\quad 5x + y - 2z = 1$
$\boxed{C}$ $\quad 4x - 3y + 3z = 7$

Step 1: The equations are already in standard form.
- It is helpful to label the equations $\boxed{A}$, $\boxed{B}$, and $\boxed{C}$.
- The z variable can easily be eliminated from equations $\boxed{A}$ and $\boxed{B}$ and from equations $\boxed{A}$ and $\boxed{C}$. This is accomplished by creating opposite coefficients for the z terms and then adding the equations.

Step 2: Eliminate z from equations $\boxed{A}$ and $\boxed{B}$.

$\boxed{A}$ $\quad 3x - 2y + z = 2 \xrightarrow{\text{Multiply by 2.}} 6x - 4y + 2z = 4$
$\boxed{B}$ $\quad 5x + y - 2z = 1 \qquad\qquad\qquad\quad \underline{5x + y - 2z = 1}$
$$11x - 3y \qquad = 5 \quad \boxed{D}$$

Step 3: Eliminate z from equations $\boxed{A}$ and $\boxed{C}$.

$\boxed{A}$ $\quad 3x - 2y + z = 2 \xrightarrow{\text{Multiply by }-3.} -9x + 6y - 3z = -6$
$\boxed{C}$ $\quad 4x - 3y + 3z = 7 \qquad\qquad\qquad \underline{4x - 3y + 3z = 7}$
$$-5x + 3y \qquad = 1 \quad \boxed{E}$$

Step 4: $\boxed{D} \quad 11x - 3y = 5 \qquad \boxed{D} \quad 11(1) - 3y = 5$
$\qquad\quad\; \boxed{E} \quad \underline{-5x + 3y = 1} \qquad\qquad\qquad 11 - 3y = 5$
$\qquad\qquad\qquad\quad\; 6x \qquad\; = 6 \qquad\qquad\qquad\quad\; -3y = -6$
$\qquad\qquad\qquad\qquad\quad\; x = 1 \qquad\qquad\qquad\qquad\quad\;\; y = 2$

Solve the system of equations $\boxed{D}$ and $\boxed{E}$.

$\boxed{A}$ $\quad 3x - 2y + z = 2$
$\qquad 3(1) - 2(2) + z = 2$
$\qquad\quad\; 3 - 4 + z = 2$
$\qquad\qquad\; -1 + z = 2$
$\qquad\qquad\qquad\;\; z = 3$

Step 5: Substitute the values of the known variables x and y back into one of the original equations. We have chosen equation $\boxed{A}$.

Step 6: Check the ordered triple $(1, 2, 3)$ in the three original equations.

$\boxed{A}$ $\;3x - 2y + z = 2 \qquad \boxed{B}$ $\;5x + y - 2z = 1 \qquad \boxed{C}$ $\;4x - 3y + 3z = 7$
$\quad 3(1) - 2(2) + (3) \overset{?}{=} 2 \qquad 5(1) + (2) - 2(3) \overset{?}{=} 1 \qquad 4(1) - 3(2) + 3(3) \overset{?}{=} 7$
$\qquad\qquad\quad 2 \overset{?}{=} 2 \;\checkmark \text{ true} \qquad\qquad\quad 1 \overset{?}{=} 1 \;\checkmark \text{ true} \qquad\qquad\qquad 7 \overset{?}{=} 7 \;\checkmark \text{ true}$

The solution set is $\{(1, 2, 3)\}$.

Skill Practice 2 Solve the system.
$$2x - y + 5z = -7$$
$$x + 4y - 2z = 1$$
$$3x + 2y + z = -7$$

In Example 3, we solve a system of linear equations in which one or more equations has a missing term.

EXAMPLE 3 Solving a System of Equations in Three Variables

Solve the system.
$$2x + y = -2$$
$$3y = 5z - 12$$
$$5(x + z) = 2z + 5$$

Solution:

Step 1: Write the equations in standard form.

$\boxed{A}$ $\quad 2x + y = -2 \qquad\longrightarrow\qquad 2x + y \qquad\;\; = -2$
$\boxed{B}$ $\quad 3y = 5z - 12 \qquad\longrightarrow\qquad\quad\; 3y - 5z = -12$
$\boxed{C}$ $\quad 5(x + z) = 2z + 5 \longrightarrow\quad 5x \qquad + 3z = 5$

Notice that the equations already have missing variable terms.

Answer

2. $\{(-3, 1, 0)\}$

Steps 2 and 3: This system of equations has several missing terms. For example, equation $\boxed{C}$ is missing the variable y. If we eliminate y from equations $\boxed{A}$ and $\boxed{B}$, then we will have a second equation with variable y missing.

$\boxed{A}$ $2x + y \quad\quad = -2$ $\xrightarrow{\text{Multiply by } -3.}$ $-6x - 3y \quad\quad = 6$

$\boxed{B}$ $\quad\quad 3y - 5z = -12$ $\quad\quad\quad\quad\quad\quad\quad\quad \underline{\quad\quad 3y - 5z = -12}$

$\quad\quad\quad\quad\quad\quad\quad\quad\quad\quad\quad\quad\quad\quad\quad -6x \quad\quad - 5z = -6 \quad \boxed{D}$

Step 4: Pair up equations $\boxed{C}$ and $\boxed{D}$. These equations form a system of linear equations in two variables. To solve the system with equations $\boxed{C}$ and $\boxed{D}$ we have chosen to eliminate the z variable.

$\boxed{C}$ $5x + 3z = 5$ $\xrightarrow{\text{Multiply by 5.}}$ $25x + 15z = 25$

$\boxed{D}$ $-6x - 5z = -6$ $\xrightarrow[\text{Multiply by 3.}]{}$ $\underline{-18x - 15z = -18}$

$\quad\quad\quad\quad\quad\quad\quad\quad\quad\quad\quad\quad\quad\quad\quad 7x \quad\quad\quad = 7$

$\quad\quad\quad\quad\quad\quad\quad\quad\quad\quad\quad\quad\quad\quad\quad\quad\quad x = 1$

$\boxed{C}$ $5x + 3z = 5$

$\quad\quad 5(1) + 3z = 5$

$\quad\quad\quad\quad\quad z = 0$

$\boxed{B}$ $3y = 5z - 12$ **Step 5:** Substitute the values of the known variables x

$\quad\quad 3y = 5(0) - 12$ and z back into one of the original equations containing

$\quad\quad 3y = -12$ y. We have chosen equation $\boxed{B}$.

$\quad\quad\quad y = -4$

Step 6: The ordered triple $(1, -4, 0)$ checks in each original equation.

The solution set is $\{(1, -4, 0)\}$.

Skill Practice 3   Solve the system. $a \quad\quad + 3c = 4$

$\quad\quad\quad\quad\quad\quad\quad\quad\quad\quad\quad\quad\quad\quad\quad b + 2c = -1$

$\quad\quad\quad\quad\quad\quad\quad\quad\quad\quad\quad\quad 2a - 4b \quad\quad = 14$

A system of linear equations in three variables may have no solution. This occurs if the equations represent planes that do not all intersect (Figure 5-6). In such a case, we say that the system is **inconsistent.**

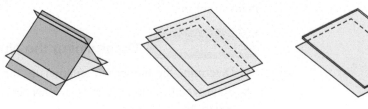

Figure 5-6

A system of linear equations in three variables may also have infinitely many solutions. This occurs if the equations represent planes that intersect in a common line or common plane (Figure 5-7). In such a case, we say that the equations are **dependent.**

Figure 5-7

Answer

3. $\{(1, -3, 1)\}$

EXAMPLE 4 **Determining the Number of Solutions to a System**

a. Determine the number of solutions to the system. $-x + 6y - 3z = -8$
b. State whether the system is inconsistent or the $x - 2y + 2z = 3$
equations are dependent. $3x + 2y + 4z = -6$
c. Write the solution set.

Solution:

Begin by eliminating a variable from two different pairs of equations. We will eliminate x from $\boxed{A}$ and $\boxed{B}$ and from $\boxed{A}$ and $\boxed{C}$.

$\boxed{A}$ $-x + 6y - 3z = -8 \longrightarrow -x + 6y - 3z = -8$ Add equations
$\boxed{B}$ $x - 2y + 2z = 3 \longrightarrow \underline{x - 2y + 2z = 3}$ $\boxed{A}$ and $\boxed{B}$ to
$\boxed{C}$ $3x + 2y + 4z = -6$ $4y - z = -5$ $\boxed{D}$ eliminate x.

$\boxed{A}$ $-x + 6y - 3z = -8 \xrightarrow{\text{Multiply by 3.}} -3x + 18y - 9z = -24$ Multiply
$\boxed{C}$ $3x + 2y + 4z = -6 \longrightarrow \underline{3x + 2y + 4z = -6}$ equation $\boxed{A}$ by
$20y - 5z = -30$ $\boxed{E}$ 3 and add the result to $\boxed{C}$.

$\boxed{D}$ $4y - z = -5 \xrightarrow{\text{Multiply by }-5.} -20y + 5z = 25$ Solving the system of
$\boxed{E}$ $20y - 5z = -30$ $\underline{20y - 5z = -30}$ equations $\boxed{D}$ and $\boxed{E}$
$0 = -5$ results in a contradiction.

The system of equations reduces to a contradiction.
a. There is no solution.
b. The system is inconsistent.
c. The solution set is { }.

Skill Practice 4 Repeat Example 4 with the given system.

$x + y + 4z = -1$
$3x + y - 4z = 3$
$-4x - y + 8z = -2$

In Example 5, we investigate the case in which a system of equations has infinitely many solutions.

EXAMPLE 5 **Determining the Number of Solutions to a System**

a. Determine the number of solutions to the system. $2x + y = -3$
b. State whether the system is inconsistent, or the $2y + 16z = -10$
equations are dependent. $-7x - 3y + 4z = 8$
c. Write the solution set.

Solution:

Eliminate variable z from equations $\boxed{B}$ and $\boxed{C}$.

$\boxed{A}$ $2x + y = -3$
$\boxed{B}$ $2y + 16z = -10 \longrightarrow 2y + 16z = -10$
$\boxed{C}$ $-7x - 3y + 4z = 8 \xrightarrow{\text{Multiply by }-4.} \underline{28x + 12y - 16z = -32}$
$28x + 14y = -42$ $\boxed{D}$

Pair up equations $\boxed{A}$ and $\boxed{D}$ to solve for x and y.

$\boxed{A}$ $\quad 2x + y = -3 \quad\longrightarrow\quad 2x + y = -3$

$\boxed{D}$ $\quad 28x + 14y = -42 \quad\underset{\text{Divide by } -14.}{\longrightarrow}\quad \dfrac{-2x - y = 3}{0 = 0}$

The system reduces to the identity $0 = 0$. This implies that

a. There are infinitely many solutions.

b. The equations are dependent.

c. To find the general solution, we need to express the dependency among the variables as an ordered triple. Note that from equation $\boxed{A}$, we can solve for x in terms of y, and from equation $\boxed{B}$, we can solve for z in terms of y. Thus,

Equation $\boxed{A}$: $2x = -y - 3 \Rightarrow x = \dfrac{-y - 3}{2} \Rightarrow x = -\dfrac{y + 3}{2}$

Equation $\boxed{B}$: $16z = -2y - 10 \Rightarrow z = \dfrac{-2y - 10}{16} \Rightarrow z = -\dfrac{y + 5}{8}$

Therefore, the solution set is $\left\{ \left(-\dfrac{y + 3}{2}, y, -\dfrac{y + 5}{8} \right) \middle| y \text{ is any real number} \right\}$.

Skill Practice 5   Repeat Example 5 with the given system.

$$5y + z = 0$$
$$-x + 4z = 0$$
$$-x + 5y + 5z = 0$$

The general solution to the system in Example 5 can be written in a number of forms.

Solve for x and z in terms of y: $\left\{ \left(-\dfrac{y + 3}{2}, y, -\dfrac{y + 5}{8} \right) \middle| y \text{ is any real number} \right\}$

Solve for y and z in terms of x: $\left\{ \left(x, -2x - 3, \dfrac{x - 1}{4} \right) \middle| x \text{ is any real number} \right\}$

Solve for x and y in terms of z: $\left\{ \left(4z + 1, -8z - 5, z \right) \middle| z \text{ is any real number} \right\}$

TIP Any form of the general solution can be checked by substitution in the original three equations.

<u>Check:</u> $\left\{ \left(-\dfrac{y + 3}{2}, y, -\dfrac{y + 5}{8} \right) \middle| y \text{ is any real number} \right\}$

$\boxed{A} \qquad\qquad 2x + y = -3$

$\quad 2\left(-\dfrac{y + 3}{2} \right) + y \overset{?}{=} -3$

$\qquad\qquad -y - 3 + y = -3 \checkmark$

$\boxed{B} \qquad\qquad 2y + 16z = -10$

$\quad 2y + 16\left(-\dfrac{y + 5}{8} \right) \overset{?}{=} -10$

$\qquad\quad 2y - 2y - 10 = -10 \checkmark$

$\boxed{C} \qquad\qquad\qquad -7x - 3y + 4z = 8$

$\quad -7\left(-\dfrac{y + 3}{2} \right) - 3y + 4\left(-\dfrac{y + 5}{8} \right) \overset{?}{=} 8$

$\qquad\qquad \dfrac{7}{2}y + \dfrac{21}{2} - 3y - \dfrac{1}{2}y - \dfrac{5}{2} \overset{?}{=} 8$

$\qquad\qquad\qquad \dfrac{6}{2}y - 3y + \dfrac{16}{2} = 8 \checkmark$

Answers

5. a. There are infinitely many solutions.

b. The equations are dependent.

c. $\left\{ \left(4z, -\dfrac{1}{5}z, z \right) \middle| z \text{ is any real number} \right\}$ or $\left\{ \left(x, -\dfrac{x}{20}, \dfrac{x}{4} \right) \middle| x \text{ is any real number} \right\}$ or $\{ (-20y, y, -5y) \mid y \text{ is any real number} \}$

The topic of dependent equations will be discussed in more detail when we learn matrix methods to solve systems of linear equations. With additional tools available, we can investigate whether the dependent equations represent planes that intersect in a line or whether the equations all represent the same plane (Figure 5-7).

3. Use Systems of Linear Equations in Applications

When solving an application involving three unknowns, sometimes it is convenient to use a system of three independent equations, as demonstrated in Examples 6 and 7.

EXAMPLE 6   **Solving an Application Involving Finance**

Janette invested a total of $18,000 in three different mutual funds. She invested in a bond fund that returned 4% the first year. An aggressive growth fund lost 8% for the year, and an international fund returned 2%. Janette invested $2000 more in the growth fund than in the other two funds combined. If she had a net loss of −$540 for the year, how much did she invest in each fund?

Solution:

Let x represent the amount invested in the bond fund.   Label the variables.
Let y represent the amount invested in the growth fund.   With three unknowns,
Let z represent the amount invested in the international fund.   we need three independent equations.

$x + y + z = 18{,}000$ $\longrightarrow$ The total amount invested was $18,000.
$y = (x + z) + 2000$ $\longrightarrow$ The amount invested in the growth fund was $2000 more than the combined amount in the other two funds.
$0.04x - 0.08y + 0.02z = -540 \longrightarrow$ The sum of the gain and loss from each fund equals −$540.

$\boxed{A}$ $x + y + z = 18{,}000$
$\boxed{B}$ $y = (x + z) + 2000$ $\xrightarrow[\text{by 100.}]{\text{Standard form}}$ $\boxed{A}$ $x + y + z = 18{,}000$
$\boxed{C}$ $0.04x - 0.08y + 0.02z = -540$ $\xrightarrow{\text{Multiply}}$ $\boxed{B}$ $-x + y - z = 2000$
$\boxed{C}$ $4x - 8y + 2z = -54{,}000$

Eliminate x from equations $\boxed{A}$ and $\boxed{B}$ and equations $\boxed{B}$ and $\boxed{C}$.

$\boxed{A}$ $x + y + z = 18{,}000$ $4 \cdot \boxed{B}$ $-4x + 4y - 4z = 8000$
$\boxed{B}$ $\underline{-x + y - z = 2000}$ $\boxed{C}$ $\underline{4x - 8y + 2z = -54{,}000}$
    $2y = 20{,}000$ $-4y - 2z = -46{,}000$ $\boxed{D}$
    $y = 10{,}000$

Back substitute.

$\boxed{D}$ $-4y - 2z = -46{,}000$ $\boxed{A}$ $x + y + z = 18{,}000$
     $-4(10{,}000) - 2z = -46{,}000$ $x + (10{,}000) + (3000) = 18{,}000$
               $-2z = -6000$ $x + 13{,}000 = 18{,}000$
                  $z = 3000$ $x = 5000$

Janette invested $5000 in the bond fund, $10,000 in the aggressive growth fund, and $3000 in the international fund.

Skill Practice 6   Nicolas mixes three solutions of acid with concentrations of 10%, 15%, and 5%. He wants to make 30 L of a mixture that is 12% acid and he uses four times as much of the 15% solution as the 5% solution. How much of each of the three solutions must he use?

Avoiding Mistakes

For Example 6, verify that the conditions of the problem have been met.

Principal:
$5000 + $10,000 + $3000
   = $18,000 ✓

Return:
  0.04($5000)
−0.08($10,000)
+0.02($3000)
−$540 ✓

More invested in growth:
($5000 + $3000) + $2000
   = $10,000 ✓

Answer

6. Nicolas mixes 10 L of the 10% solution, 16 L of the 15% solution, and 4 L of the 5% solution.

4. Modeling with Linear Equations in Three Variables

In Section 3.1, we used regression to find a quadratic model $y = ax^2 + bx + c$ from observed data points. We will now learn how to find a quadratic model by using a system of linear equations in three variables. The premise is that any three noncollinear points (points that do not all fall on the same line) define a unique parabola given by $y = ax^2 + bx + c\,(a \neq 0)$.

> **EXAMPLE 7**   **Using a System of Linear Equations to Create a Quadratic Model**

Given the noncollinear points $(4, 2)$, $(1, -1)$, and $(-1, 7)$, find an equation of the form $y = ax^2 + bx + c$ that defines the parabola through the points.

Solution:

$$y = ax^2 + bx + c$$

Substitute $(4, 2)$: $2 = a(4)^2 + b(4) + c$ ⟶ $\boxed{A}$ $16a + 4b + c = 2$
Substitute $(1, -1)$: $-1 = a(1)^2 + b(1) + c$ ⟶ $\boxed{B}$ $a + b + c = -1$
Substitute $(-1, 7)$: $7 = a(-1)^2 + b(-1) + c$ ⟶ $\boxed{C}$ $a - b + c = 7$

Eliminate variable b from equations $\boxed{A}$ and $\boxed{C}$ and from equations $\boxed{B}$ and $\boxed{C}$.

$\boxed{A}$ $16a + 4b + c = 2$	$\boxed{B}$ $a + b + c = -1$
$4\cdot\boxed{C}$ $\dfrac{4a - 4b + 4c = 28}{20a \quad\;\; + 5c = 30}$ $\boxed{D}$	$\boxed{C}$ $\dfrac{a - b + c = 7}{2a \quad\; + 2c = 6}$ $\boxed{E}$

$\boxed{D}$ $20a + 5c = 30$ $\xrightarrow{\text{Divide by 5.}}$ $4a + c = 6$
$\boxed{E}$ $2a + 2c = 6$ $\xrightarrow[\text{by }-2.]{\text{Divide}}$ $\dfrac{-a - c = -3}{\dfrac{3a \quad\;\; = 3}{a = 1}}$ $\Big\}$ Back substitute.

$$\begin{cases} \boxed{E}\;\; 2a + 2c = 6 \\ \quad\;\; 2(1) + 2c = 6 \\ \quad\quad\quad\quad c = 2 \\[4pt] \boxed{B}\;\;\;\; a + b + c = -1 \\ \quad\;\; (1) + b + (2) = -1 \\ \quad\quad\quad\quad b = -4 \end{cases}$$

Substituting $a = 1$, $b = -4$, and $c = 2$ into the equation $y = ax^2 + bx + c$ gives

$$y = x^2 - 4x + 2$$

The graph of $y = x^2 - 4x + 2$ passes through the points $(4, 2)$, $(1, -1)$, and $(-1, 7)$ as shown in Figure 5-8.

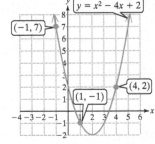

Figure 5-8

The results can also be verified by using the **Table** feature of a graphing utility. Enter $Y_1 = x^2 - 4x + 2$.

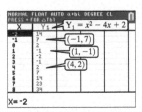

TIP We can verify that the three points given in Example 7 are *not* collinear by showing that the slopes between two pairs of points are different.

For $(4, 2)$ and $(1, -1)$,
$$m = \frac{-1 - 2}{1 - 4} = 1$$
For $(4, 2)$ and $(-1, 7)$,
$$m = \frac{7 - 2}{-1 - 4} = -1$$

> **Skill Practice 7**   Given the noncollinear points $(-3, 2)$, $(-4, 1)$, and $(-6, -7)$, find an equation of the form $y = ax^2 + bx + c$ that defines the parabola through the points.

Answer

7. $y = -x^2 - 6x - 7$

SECTION 5.2 Practice Exercises

Prerequisite Review

R.1. How much interest will Roxanne have to pay if she borrows $2000 for 2 yr at a simple interest rate of 3%?

R.2. Julie needs to have a toilet repaired in her house. The cost of the new plumbing fixtures is $110 and labor is $70/hr. Write a model that represents the cost of the repair C (in $) in terms of the number of hours of labor x.

R.3. Solve the equation. $\dfrac{v-4}{5} + \dfrac{v}{8} = \dfrac{v-2}{5} - 7$

Concept Connections

1. The graph of a linear equation in two variables is a line in a two-dimensional coordinate system. The graph of a linear equation in three variables is a _____ in a three-dimensional coordinate system.

2. A solution to a system of linear equations in three variables is an ordered _____ that satisfies each equation in the system. Graphically, this is a point of _____ of three planes.

Objective 1: Identify Solutions to a System of Linear Equations in Three Variables

For Exercises 3–4, find three ordered triples that are solutions to the linear equation in three variables.

3. $2x + 4y - 6z = 12$

4. $3x - 5y + z = 15$

For Exercises 5–8, determine if the ordered triple is a solution to the system of equations. (See Example 1)

5. $-x + 3y - 7z = 7$
$2x + 4y + z = 16$
$3x - 5y + 6z = -9$

   a. $(2, 3, 0)$
   b. $(-2, 4, 1)$

6. $2x - 3y + z = -12$
$x + y - 2z = 9$
$-3x + 2y - z = 7$

   a. $(2, 5, -1)$
   b. $(1, 4, -2)$

7. $x + y + z = 2$
$x + 2y - z = 2$
$3x + 5y - z = 6$

   a. $(2, 0, 0)$
   b. $(-1, 2, 1)$

8. $-x - y + z = 3$
$3x + 4y - z = 1$
$5x + 7y - z = -1$

   a. $(1, 2, 6)$
   b. $(3, -1, 5)$

Objective 2: Solve Systems of Linear Equations in Three Variables

For Exercises 9–32, solve the system. If a system has one unique solution, write the solution set. Otherwise, determine the number of solutions to the system, and determine whether the system is inconsistent, or the equations are dependent. (See Examples 2–5)

9. $x - 2y + z = -9$
$3x + 4y + 5z = 9$
$-2x + 3y - z = 12$

10. $2x - y + z = 6$
$-x + 5y - z = 10$
$3x + y - 3z = 12$

11. $4x = 3y - 2z - 5$
$2(x + y) = y + z - 6$
$6(x - y) + z = x - 5y - 8$

12. $3x = 5y - z + 13$
$-(x - y) - z = x - 3$
$5(x + y) = 3y - 3z - 4$

13. $2x + 5z = 2$
$3y - 7z = 9$
$-5x + 9y = 22$

14. $3x - 2y = -8$
$5y + 6z = 2$
$7x + 11z = -33$

15. $-4x - 3y = 0$
$3y + z = -1$
$4x - z = 12$

16. $4x - y + 2z = 1$
$3x + 5y - z = -2$
$-9x - 15y + 3z = 0$

17. $2x = 3y - 6z - 1$
$6y = 12z - 10x + 9$
$3z = 6y - 3x - 1$

18. $5x = 2y - 3z - 3$
$4y = -1 - 10x - 5z$
$2z = 5x - 6y$

19. $x + 2y + 4z = 3$
$y + 3z = 5$
$x - 2z = -7$

20. $3x + 2y + 5z = 6$
$3y - z = 4$
$3x + 17y = 26$

21. $0.2x = 0.1y - 0.6z$
$0.004x + 0.005y - 0.001z = 0$
$30x = 50z - 20y$

22. $0.3x = 0.5y - 1.2z$
$0.05x + 0.1y = 0.04z$
$100x = 300y - 700z$

23. $\dfrac{1}{12}x + \dfrac{1}{4}y + \dfrac{1}{3}z = \dfrac{7}{12}$
$-\dfrac{1}{10}x + \dfrac{1}{2}y - \dfrac{1}{5}z = -\dfrac{17}{10}$
$\dfrac{1}{2}x + \dfrac{1}{4}y + z = 3$

24. $x + \frac{7}{2}y + \frac{1}{2}z = 4$

$\frac{3}{4}x + y + \frac{1}{2}z = -1$

$\frac{1}{10}x - \frac{2}{5}y - \frac{3}{10}z = 1$

25. $3x + 2y + 5z = 12$

$3y + 8z = -8$

$10z = 20$

26. $-4x + 6y + z = -4$

$-2y - 4z = -24$

$6z = 48$

27. $\frac{x+2}{3} + \frac{y-4}{2} + \frac{z+1}{6} = 8$

$-\frac{x+2}{3} + \frac{z+1}{2} = 8$

$\frac{y-4}{4} - \frac{z+1}{6} = -1$

28. $\frac{x-1}{7} + \frac{y-2}{3} + \frac{z+2}{4} = 13$

$\frac{y-2}{9} - \frac{z+2}{8} = 3$

$\frac{x-1}{7} + \frac{z+2}{2} = 3$

29. $3(x + y) = 6 - 4z + y$

$4 = 6y + 5z$

$-3x + 4y + z = 0$

30. $4(x - y) = 8 - z - y$

$3 = 3x + 4z$

$-x + 3y + 3z = 1$

31. $-3x + 4y - z = -4$

$x + 2y + z = 4$

$-12x + 16y - 4z = -16$

32. $x + 2y - 3z = 5$

$-2x - 5y + 4z = -6$

$3x + 6y - 9z = 15$

For Exercises 33–36, solve the system from the indicated exercise and write the general solution. (See **Example 5**)

33. Exercise 19   **34.** Exercise 20   **35.** Exercise 31   **36.** Exercise 32

For Exercises 37–38, the general solution is given for a system of linear equations. Find three individual solutions to the system.

37. $-x + 4y + 2z = 4$

$x - 3y - z = -2$

$-x + y - z = -2$

Solution: $\{(-2z + 4, -z + 2, z) \mid z \text{ is any real number}\}$

38. $2x - 3y + z = 1$

$x + 4y - z = 3$

$5x - 2y + z = 5$

Solution: $\{(x, -3x + 4, -11x + 13) \mid x \text{ is any real number}\}$

Objective 3: Use Systems of Linear Equations in Applications

39. Devon invested $8000 in three different mutual funds. A fund containing large cap stocks made 6.2% return in 1 yr. A real estate fund lost 13.5% in 1 yr, and a bond fund made 4.4% in 1 yr. The amount invested in the large cap stock fund was twice the amount invested in the real estate fund. If Devon had a net return of $66 across all investments, how much did he invest in each fund? (See **Example 6**)

40. Pierre inherited $120,000 from his uncle and decided to invest the money. He put part of the money in a money market account that earns 2.2% simple interest. The remaining money was invested in a stock that returned 6% in the first year and a mutual fund that lost 2% in the first year. He invested $10,000 more in the stock than in the mutual fund, and his net gain for 1 yr was $2820. Determine the amount invested in each account.

41. A basketball player scored 26 points in one game. In basketball, some baskets are worth 3 points, some are worth 2 points, and free-throws are worth 1 point. He scored four more 2-point baskets than he did 3-point baskets. The number of free-throws equaled the sum of the number of 2-point and 3-point shots made. How many free-throws, 2-point shots, and 3-point shots did he make?

42. A sawmill cuts boards for a lumber supplier. When saws A, B, and C all work for 6 hr, they cut 7200 linear board-ft of lumber. It would take saws A and B working together 9.6 hr to cut 7200 ft of lumber. Saws B and C can cut 7200 ft of lumber in 9 hr. Find the rate (in ft/hr) that each saw can cut lumber.

43. Plant fertilizers are categorized by the percentage of nitrogen (N), phosphorus (P), and potassium (K) they contain, by weight. For example, a fertilizer that has N-P-K numbers of 8-5-5 has 8% nitrogen, 5% phosphorus, and 5% potassium by weight. Suppose that a fertilizer has twice as much potassium by weight as phosphorus. The percentage of nitrogen equals the sum of the percentages of phosphorus and potassium. If nitrogen, phosphorus, and potassium make up 42% of the fertilizer, determine the proper N-P-K label on the fertilizer.

   a. 14-7-14   **b.** 21-7-14   **c.** 14-7-21   **d.** 14-21-21

44. A theater charges $50 per ticket for seats in Section A, $30 per ticket for seats in Section B, and $20 per ticket for seats in Section C. For one play, 4000 tickets were sold for a total of $120,000 in revenue. If 1000 more tickets in Section B were sold than the other two sections combined, how many tickets in each section were sold?

45. The perimeter of a triangle is 55 in. The shortest side is 7 in. less than the longest side. The middle side is 19 in. less than the combined lengths of the shortest and longest sides. Find the lengths of the three sides.

46. A package in the shape of a rectangular solid is to be mailed. The combination of the girth (perimeter of a cross section defined by w and h) and the length of the package is 48 in. The width is 2 in. greater than the height, and the length is 12 in. greater than the width. Find the dimensions of the package.

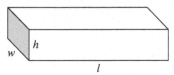

47. The measure of the largest angle in a triangle is $100°$ larger than the sum of the measures of the other two angles. The measure of the smallest angle is two-thirds the measure of the middle angle. Find the measure of each angle.

48. The measure of the largest angle in a triangle is $18°$ more than the sum of the measures of the other two angles. The measure of the smallest angle is one-half the measure of the middle angle. Find the measure of each angle.

Objective 4: Modeling with Linear Equations in Three Variables

49. a. Show that the points $(1, 0)$, $(3, 10)$, and $(-2, 15)$ are not collinear by finding the slope between $(1, 0)$ and $(3, 10)$, and the slope between $(3, 10)$ and $(-2, 15)$. (**See Example 7**)

b. Find an equation of the form $y = ax^2 + bx + c$ that defines the parabola through the points.

c. Use a graphing utility to verify that the graph of the equation in part (b) passes through the given points.

50. a. Show that the points $(2, 9)$, $(-1, -6)$, and $(-4, -3)$ are not collinear by finding the slope between $(2, 9)$ and $(-1, -6)$, and the slope between $(2, 9)$ and $(-4, -3)$.

b. Find an equation of the form $y = ax^2 + bx + c$ that defines the parabola through the points.

c. Use a graphing utility to verify that the graph of the equation in part (b) passes through the given points.

For Exercises 51–52, find an equation of the form $y = ax^2 + bx + c$ that defines the parabola through the three noncollinear points given.

51. $(0, 6)$, $(2, -6)$, $(-1, 9)$

52. $(0, -4)$, $(2, -6)$, $(-3, -31)$

The motion of an object traveling along a straight path is given by $s(t) = \frac{1}{2}at^2 + v_0t + s_0$, where $s(t)$ is the position relative to the origin at time t. For Exercises 53–54, three observed data points are given. Find the values of a, v_0, and s_0.

53. $s(1) = 30$, $s(2) = 54$, $s(3) = 82$

54. $s(1) = -7$, $s(2) = 12$, $s(3) = 37$

Many statistics courses cover a topic called *multiple regression*. This provides a means to predict the value of a dependent variable y based on two or more independent variables $x_1, x_2, …, x_n$. The model $y = ax_1 + bx_2 + c$ is a linear model that predicts y based on two independent variables x_1 and x_2. While statistical techniques may be used to find the values of a, b, and c based on a large number of data points, we can form a crude model given three data values (x_1, x_2, y). Use the information given in Exercises 55–56 to form a system of three equations and three variables to solve for a, b, and c.

55. The selling price of a home y (in $1000) is given based on the living area x_1 (in 100 ft^2) and on the lot size x_2 (in acres).

Living Area (100 ft^2) x_1	Lot Size (acres) x_2	Selling Price ($1000) y
28	0.5	225
25	0.8	207
18	0.4	154

a. Use the data to create a model of the form $y = ax_1 + bx_2 + c$.

b. Use the model from part (a) to predict the selling price of a home that is 2000 ft^2 on a 0.4-acre lot.

56. The gas mileage y (in mpg) for city driving is given based on the weight of the vehicle x_1 (in lb) and on the number of cylinders.

Weight (lb) x_1	Cylinders x_2	Mileage (mpg) y
3500	6	20
3200	4	26
4100	8	18

a. Use the data to create a model of the form $y = ax_1 + bx_2 + c$.

b. Use the model from part (a) to predict the gas mileage of a vehicle that is 3800 lb and has 6 cylinders.

Write About It

57. Give a geometric description of the solution set to a linear equation in three variables.

58. If a system of linear equations in three variables has no solution, then what can be said about the three planes represented by the equations in the system?

59. Explain the procedure presented in this section to solve a system of linear equations in three variables.

60. Explain how to check a solution to a system of linear equations in three variables.

Expanding Your Skills

For Exercises 61–62, find all solutions of the form (a, b, c, d).

61.
$$2a + b - c + d = 7$$
$$3b + 2c - 2d = -11$$
$$a + 3c + 3d = 14$$
$$4a + 2b - 5c = 6$$

62.
$$3a - 4b + 2c + d = 8$$
$$2a + 3b + 2d = 7$$
$$5b - 3c + 4d = -4$$
$$-a + b - 2c = -7$$

For Exercises 63–64, find all solutions of the form (u, v, w).

63.
$$\frac{u-3}{4} + \frac{v+1}{3} + \frac{w-2}{8} = 1$$
$$\frac{u-3}{2} + \frac{v+1}{2} + \frac{w-2}{4} = 0$$
$$\frac{u-3}{4} - \frac{v+1}{2} + \frac{w-2}{2} = -6$$

64.
$$\frac{u+1}{6} + \frac{v-1}{6} + \frac{w+3}{4} = 11$$
$$\frac{u+1}{3} - \frac{v-1}{2} + \frac{w+3}{4} = 7$$
$$\frac{u+1}{2} - \frac{v-1}{6} + \frac{w+3}{2} = 20$$

Recall that an equation of a circle can be written in the form $(x - h)^2 + (y - k)^2 = r^2$, where (h, k) is the center and r is the radius. Expanding terms, the equation can also be written in the form $x^2 + y^2 + Ax + By + C = 0$. For Exercises 65–66,

a. Find an equation of the form $x^2 + y^2 + Ax + By + C = 0$ that represents the circle that passes through the given points.

b. Find the center and radius of the circle.

65. $(2, 2), (6, 0), (7, -3)$

66. $(-1, 12), (5, 10), (9, 2)$

For Exercises 67–68, find the constants A **and** B **so that the two polynomials are equal.** (*Hint:* **Create a system of linear equations by equating the constant terms and by equating the coefficients on the** x **terms and** x^2 **terms.**)

67. $11x^2 + 26x - 5 = 2Ax^2 + 5Ax + 3A + Bx^2 - 2Bx - 8B + 2Cx^2 - 7Cx - 4C$

68. $3x^2 + 37x - 82 = Ax^2 + Ax - 12A + 3Bx^2 - 10Bx + 3B + 3Cx^2 + 11Cx - 4C$

| **SECTION 5.3** | Partial Fraction Decomposition |

1. Set Up a Partial Fraction Decomposition

In Section R.6 we learned how to add and subtract rational expressions. For example:

$$\frac{5}{x+2} + \frac{3}{x-5} = \frac{5(x-5)}{(x+2)(x-5)} + \frac{3(x+2)}{(x-5)(x+2)}$$

$$= \frac{5(x-5) + 3(x+2)}{(x+2)(x-5)}$$

$$= \frac{8x-19}{(x+2)(x-5)}$$

The fraction $\dfrac{8x-19}{(x+2)(x-5)}$ is the result of adding two simpler fractions, $\dfrac{5}{x+2}$ and $\dfrac{3}{x-5}$. The sum $\dfrac{5}{x+2} + \dfrac{3}{x-5}$ is called the **partial fraction decomposition** of $\dfrac{8x-19}{(x+2)(x-5)}$. In some applications in higher mathematics, it is more convenient to work with the partial fraction decomposition than the more complicated single fraction. Therefore, in this section, we will learn the technique of partial fraction decomposition to write a rational expression as a sum of simpler fractions. That is, we will reverse the process of adding two or more fractions. There are two parts to this process.

I. First we set up the "form" or "structure" for the partial fraction decomposition into simpler fractions. For example, the denominator of $\dfrac{8x - 19}{(x + 2)(x - 5)}$ consists of the distinct linear factors $(x + 2)$ and $(x - 5)$. From the preceding discussion, the partial fraction decomposition must be of the form:

$$\frac{8x - 19}{(x + 2)(x - 5)} = \frac{A}{x + 2} + \frac{B}{x - 5}$$

> The expression on the right is the "form" or "structure" for the partial fraction decomposition of $\dfrac{8x - 19}{(x + 2)(x - 5)}$.

II. Next, we solve for the constants A and B. To do so, multiply both sides of the equation by the LCD, and set up a system of linear equations.

$$(x + 2)(x - 5) \cdot \left[\frac{8x - 19}{(x + 2)(x - 5)} \right] = (x + 2)(x - 5) \cdot \left[\frac{A}{x + 2} + \frac{B}{x - 5} \right]$$

Multiply by the LCD to clear fractions.

$$8x - 19 = A(x - 5) + B(x + 2)$$

$$8x - 19 = Ax - 5A + Bx + 2B \qquad \text{Apply the distributive property.}$$

$$8x - 19 = (A + B)x + (-5A + 2B) \qquad \text{Simplify and combine like terms.}$$

x coefficients are equal.

$$8x - 19 = (A + B)x + (-5A + 2B) \qquad \text{Two polynomials are equal if and only if the coefficients on like terms are equal.}$$

Constants are equal.

$$A + B = 8 \qquad \text{Equate the coefficients on } x.$$
$$-5A + 2B = -19 \qquad \text{Equate the constant terms.}$$

Solve the system of linear equations. Then substitute the values of A and B into the partial fraction decomposition.

$$\boxed{1} \quad A + B = 8 \xrightarrow{\text{Multiply by 5.}} 5A + 5B = 40$$
$$\boxed{2} \quad -5A + 2B = -19 \qquad \qquad \underline{-5A + 2B = -19}$$
$$\qquad\qquad\qquad\qquad\qquad\qquad\qquad 7B = 21$$
$$\boxed{1} \quad A + B = 8 \qquad\qquad\qquad\qquad B = 3$$
$$\qquad A + 3 = 8$$
$$\qquad A = 5$$

$A = 5$ and $B = 3$

$$\frac{8x - 19}{(x + 2)(x - 5)} = \frac{A}{x + 2} + \frac{B}{x - 5} = \frac{5}{x + 2} + \frac{3}{x - 5}$$

We begin partial fraction decomposition by factoring the denominator into linear factors $(ax + b)$ and quadratic factors $(ax^2 + bx + c)$ that are irreducible over the integers. A quadratic factor that is irreducible over the integers cannot be factored as a product of binomials with integer coefficients. From the factorization of the denominator, we then determine the proper form for the partial fraction decomposition using the following guidelines.

Decomposition of $\frac{f(x)}{g(x)}$ into Partial Fractions

Consider a rational expression $\frac{f(x)}{g(x)}$, where $f(x)$ and $g(x)$ are polynomials with real coefficients, $g(x) \neq 0$, and the degree of $f(x)$ is less than the degree of $g(x)$.

PART I:

Step 1 Factor the denominator $g(x)$ completely into linear factors of the form $(ax + b)^m$ and quadratic factors of the form $(ax^2 + bx + c)^n$ that are not further factorible over the integers (irreducible over the integers).

Step 2 Set up the form for the decomposition. That is, write the original rational expression $\frac{f(x)}{g(x)}$ as a sum of simpler fractions using these guidelines. Note that $A_1, A_2, ..., A_m, B_1, B_2, ..., B_n,$ and $C_1, C_2, ..., C_n$ are constants.

- **Linear factors of $g(x)$:** For each linear factor of $g(x)$, the partial fraction decomposition must include the sum:

$$\frac{A_1}{(ax + b)^1} + \frac{A_2}{(ax + b)^2} + \cdots + \frac{A_m}{(ax + b)^m}$$

- **Quadratic factors of $g(x)$:** For each quadratic factor of $g(x)$, the partial fraction decomposition must include the sum:

$$\frac{B_1 x + C_1}{(ax^2 + bx + c)^1} + \frac{B_2 x + C_2}{(ax^2 + bx + c)^2} + \cdots + \frac{B_n x + C_n}{(ax^2 + bx + c)^n}$$

PART II:

Step 3 With the form of the partial fraction decomposition set up, multiply both sides of the equation by the LCD to clear fractions.

Step 4 Using the equation from step 3, set up a system of linear equations by equating the constant terms and equating the coefficients of like powers of x.

Step 5 Solve the system of equations from step 4 and substitute the solutions to the system into the partial fraction decomposition.

In Examples 1 and 2, we focus on setting up the proper form for a partial fraction decomposition (Part I). In each example, note that the factors of the denominator will fall into one of the following categories:

$$\begin{aligned}
\text{Linear factors:} \quad & ax + b \\
\text{Repeated linear factors:} \quad & (ax + b)^m \ (m \geq 2, \text{ an integer}) \\
\text{Quadratic factors (irreducible over the integers):} \quad & ax^2 + bx + c \\
\text{Repeated quadratic factors (irreducible over the integers):} \quad & (ax^2 + bx + c)^n \ (n \geq 2, \text{ an integer})
\end{aligned}$$

EXAMPLE 1   Setting Up the Form for a Partial Fraction Decomposition

Set up the form for the partial fraction decomposition for the given rational expressions.

a. $\dfrac{4x - 15}{(2x + 3)(x - 2)}$ **b.** $\dfrac{4x^2 + 10x + 9}{x^3 + 6x^2 + 9x}$

Solution:

a. $\dfrac{4x - 15}{(2x + 3)^1(x - 2)^1} = \dfrac{A}{(2x + 3)^1} + \dfrac{B}{(x - 2)^1}$

The denominator has distinct linear factors of the form $(2x + 3)^1$ and $(x - 2)^1$. Since each linear factor is raised to the first power, only one fraction is needed for each factor.

b. $\dfrac{4x^2 + 10x + 9}{x^3 + 6x^2 + 9x} = \dfrac{4x^2 + 10x + 9}{x(x^2 + 6x + 9)}$

Factor the denominator completely.

The denominator has a linear factor of x^1 and a *repeated* linear factor $(x + 3)^2$.

$= \dfrac{4x^2 + 10x + 9}{x^1(x + 3)^2} = \dfrac{A}{x^1} + \dfrac{B}{(x + 3)^1} + \dfrac{C}{(x + 3)^2}$

For a repeated factor that occurs m times, one fraction must be given for each power less than or equal to m.

> Include one fraction with $(x + 3)$ raised to each positive integer up to and including 2.

Skill Practice 1 Set up the form for the partial fraction decomposition for the given rational expressions.

a. $\dfrac{-x + 18}{(3x + 1)(x + 4)}$ **b.** $\dfrac{-x^2 + 3x + 8}{x^3 + 4x^2 + 4x}$

In Example 2, we practice setting up the form for the partial decomposition of a rational expression that contains irreducible quadratic factors in the denominator.

EXAMPLE 2   **Setting Up the Form for a Partial Fraction Decomposition**

Set up the form for the partial fraction decomposition for the given rational expressions.

a. $\dfrac{2x^2 - 3x + 4}{x^3 + 4x}$ **b.** $\dfrac{3x^2 + 8x + 14}{(x^2 + 2x + 5)^2}$

Solution:

a. $\dfrac{2x^2 - 3x + 4}{x^3 + 4x} = \dfrac{2x^2 - 3x + 4}{x(x^2 + 4)}$

Factor the denominator completely. The denominator has one linear factor x^1 and one irreducible quadratic factor $(x^2 + 4)^1$.

$= \dfrac{2x^2 - 3x + 4}{x^1(x^2 + 4)^1} = \dfrac{A}{x^1} + \dfrac{Bx + C}{(x^2 + 4)^1}$

Since each factor is raised to the first power, only one fraction is needed for each factor.

TIP For a first-degree (linear) denominator, the numerator is constant (degree 0). For a second-degree (quadratic) denominator, the numerator is linear (degree 1).

$= \dfrac{2x^2 - 3x + 4}{x(x^2 + 4)} = \dfrac{A}{x} + \dfrac{Bx + C}{(x^2 + 4)}$

b. $\dfrac{3x^2 + 8x + 14}{(x^2 + 2x + 5)^2}$

The quadratic factor $x^2 + 2x + 5$ does not factor further over the integers.

$= \dfrac{Ax + B}{(x^2 + 2x + 5)^1} + \dfrac{Cx + D}{(x^2 + 2x + 5)^2}$

The factor $(x^2 + 2x + 5)$ appears to the *second* power in the denominator. Therefore, in the partial fraction composition, one fraction must have $(x^2 + 2x + 5)^1$ in the denominator, and one fraction must have $(x^2 + 2x + 5)^2$ in the denominator.

Answers

1. a. $\dfrac{A}{3x + 1} + \dfrac{B}{x + 4}$

   b. $\dfrac{A}{x} + \dfrac{B}{x + 2} + \dfrac{C}{(x + 2)^2}$

Skill Practice 2   Set up the form for the partial fraction decomposition for the given rational expressions.

a. $\dfrac{7x^2 + 2x + 12}{x^3 + 3x}$ **b.** $\dfrac{-3x^2 - 5x - 19}{(x^2 + 3x + 6)^2}$

2. Decompose $\dfrac{f(x)}{g(x)}$, Where $g(x)$ Is a Product of Linear Factors

In Example 3, we find the partial fraction decomposition of a rational expression in which the denominator is a product of distinct linear factors.

EXAMPLE 3   **Decomposing $\dfrac{f(x)}{g(x)}$, Where $g(x)$ Has Distinct Linear Factors**

Find the partial fraction decomposition. $\dfrac{4x - 15}{(2x + 3)(x - 2)}$

Solution:

$\dfrac{4x - 15}{(2x + 3)(x - 2)} = \dfrac{A}{2x + 3} + \dfrac{B}{x - 2}$

From Example 1(a), we have the form for the partial fraction decomposition.

$(2x + 3)(x - 2)\left[\dfrac{4x - 15}{(2x + 3)(x - 2)}\right] = (2x + 3)(x - 2)\left[\dfrac{A}{2x + 3} + \dfrac{B}{x - 2}\right]$

To solve for A and B, first multiply both sides by the LCD to clear fractions.

$4x - 15 = A(x - 2) + B(2x + 3)$ Apply the distributive property.

$4x - 15 = Ax - 2A + 2Bx + 3B$

$4x - 15 = (A + 2B)x + (-2A + 3B)$ Combine like terms.

$\boxed{1}\quad 4 = A + 2B$ Equate the x term coefficients.

$\boxed{2}\; -15 = -2A + 3B$ Equate the constant terms.

Solve the system of linear equations by using the substitution method or addition method.

$\boxed{1}\quad 4 = A + 2B \xrightarrow{\text{Multiply by 2.}} 2A + 4B = 8$

$\boxed{2}\; -15 = -2A + 3B \qquad\qquad\quad \underline{-2A + 3B = -15}$

$\qquad\qquad\qquad\qquad\qquad\qquad\qquad 7B = -7$

$\qquad\qquad\qquad\qquad\qquad\qquad\qquad\; B = -1$

$\boxed{1}\quad 4 = A + 2B$

$\qquad\quad 4 = A + 2(-1)$

$A = 6$ and $B = -1$. $A = 6$

$\dfrac{4x - 15}{(2x + 3)(x - 2)} = \dfrac{A}{2x + 3} + \dfrac{B}{x - 2}$

Substitute $A = 6$ and $B = -1$ into the partial fraction decomposition.

$\dfrac{4x - 15}{(2x + 3)(x - 2)} = \dfrac{6}{2x + 3} + \dfrac{-1}{x - 2}$ or equivalently $\dfrac{6}{2x + 3} - \dfrac{1}{x - 2}$

Skill Practice 3   Find the partial fraction decomposition. $\dfrac{-x + 18}{(3x + 1)(x + 4)}$

Answers

2. a. $\dfrac{A}{x} + \dfrac{Bx + C}{x^2 + 3}$

b. $\dfrac{Ax + B}{x^2 + 3x + 6} + \dfrac{Cx + D}{(x^2 + 3x + 6)^2}$

3. $\dfrac{5}{3x + 1} + \dfrac{-2}{x + 4}$

TIP Always remember that the result of a partial fraction decomposition can be checked by adding the partial fractions and verifying that the sum equals the original rational expression.

To verify the result of Example 3, we can add the rational expressions.

$$\frac{6}{2x+3} + \frac{-1}{x-2} = \frac{6(x-2)}{(2x+3)(x-2)} + \frac{-1(2x+3)}{(x-2)(2x+3)}$$

$$= \frac{6(x-2) - 1(2x+3)}{(2x+3)(x-2)}$$

$$= \frac{4x-15}{(2x+3)(x-2)} \checkmark$$

In Example 4, we perform partial fraction decomposition with a rational expression that has repeated linear factors in the denominator.

EXAMPLE 4 Decomposing $\frac{f(x)}{g(x)}$, Where $g(x)$ Has Repeated Linear Factors

Find the partial fraction decomposition. $\dfrac{4x^2 + 10x + 9}{x^3 + 6x^2 + 9x}$

Solution:

$$\frac{4x^2 + 10x + 9}{x(x+3)^2} = \frac{A}{x} + \frac{B}{(x+3)^1} + \frac{C}{(x+3)^2}$$

From Example 1(b), we have the form for the partial fraction decomposition.

$$x(x+3)^2\left[\frac{4x^2 + 10x + 9}{x(x+3)^2}\right] = x(x+3)^2\left[\frac{A}{x} + \frac{B}{(x+3)^1} + \frac{C}{(x+3)^2}\right]$$

$$4x^2 + 10x + 9 = A(x+3)^2 + Bx(x+3) + Cx$$
$$4x^2 + 10x + 9 = A(x^2 + 6x + 9) + Bx^2 + 3Bx + Cx$$
$$4x^2 + 10x + 9 = Ax^2 + 6Ax + 9A + Bx^2 + 3Bx + Cx$$
$$4x^2 + 10x + 9 = (A+B)x^2 + (6A + 3B + C)x + 9A$$

$$4 = A + B$$ Equate the x^2 term coefficients.
$$10 = 6A + 3B + C$$ Equate the x term coefficients.
$$9 = 9A$$ Equate the constant terms.
$$A = 1, B = 3, \text{ and } C = -5$$ Solve the system of linear equations.

$$\frac{4x^2 + 10x + 9}{x(x+3)^2} = \frac{A}{x} + \frac{B}{(x+3)^1} + \frac{C}{(x+3)^2}$$

Substitute $A = 1$, $B = 3$, and $C = -5$ into the partial fraction decomposition.

$$\frac{4x^2 + 10x + 9}{x(x+3)^2} = \frac{1}{x} + \frac{3}{(x+3)^1} + \frac{-5}{(x+3)^2} = \frac{1}{x} + \frac{3}{x+3} - \frac{5}{(x+3)^2}$$

Skill Practice 4 Find the partial fraction decomposition. $\dfrac{-x^2 + 3x + 8}{x^3 + 4x^2 + 4x}$

3. Decompose $\frac{f(x)}{g(x)}$, Where $g(x)$ Has Irreducible Quadratic Factors

We now turn our attention to performing partial fraction decomposition where the denominator of a rational expression contains quadratic factors irreducible over the integers. In Example 5, we also address the situation in which the given rational expression is an **improper rational expression;** that is, the degree of the numerator

Answer

4. $\dfrac{2}{x} + \dfrac{-3}{x+2} + \dfrac{1}{(x+2)^2}$

is greater than or equal to the degree of the denominator. In such a case, we use long division to write the expression in the form:

$$(\text{polynomial}) + (\text{proper rational expression})$$

where a **proper rational expression** is one in which the degree of the numerator is less than the degree of the denominator.

EXAMPLE 5   Decomposing $\dfrac{f(x)}{g(x)}$, Where $g(x)$ Has an Irreducible Quadratic Factor

Find the partial fraction decomposition. $\dfrac{x^4 + 3x^3 + 6x^2 + 9x + 4}{x^3 + 4x}$

Solution:

First note that the degree of the numerator is not less than the degree of the denominator. Therefore, perform long division first.

$$\frac{x^4 + 3x^3 + 6x^2 + 9x + 4}{x^3 + 4x} \xrightarrow{\text{Long division}}$$

$$\begin{array}{r} x + 3 \\ x^3 + 4x\overline{)x^4 + 3x^3 + 6x^2 + 9x + 4} \\ -(x^4 \qquad + 4x^2) \\ \hline 3x^3 + 2x^2 + 9x \\ -(3x^3 \qquad + 12x) \\ \hline 2x^2 - 3x + 4 \end{array}$$

$$= x + 3 + \frac{2x^2 - 3x + 4}{x^3 + 4x} \xleftarrow{\text{Equivalent form}}$$

$$= x + 3 + \overbrace{}^{\text{polynomial}} \quad \overbrace{\frac{2x^2 - 3x + 4}{x(x^2 + 4)}}^{\substack{\text{proper rational} \\ \text{expression}}} \qquad \text{Factor the denominator.}$$

$$\frac{2x^2 - 3x + 4}{x(x^2 + 4)} = \frac{A}{x} + \frac{Bx + C}{(x^2 + 4)}$$

Perform partial fraction decomposition on the proper fraction. From Example 2(a), we have the form for the partial fraction decomposition.

$$x(x^2 + 4)\left[\frac{2x^2 - 3x + 4}{x(x^2 + 4)}\right] = x(x^2 + 4)\left[\frac{A}{x} + \frac{Bx + C}{(x^2 + 4)}\right]$$

To solve for A, B, and C, multiply both sides by the LCD to clear fractions.

$$2x^2 - 3x + 4 = A(x^2 + 4) + (Bx + C)x \qquad \text{Apply the distributive property.}$$

$$2x^2 - 3x + 4 = Ax^2 + 4A + Bx^2 + Cx$$

$$2x^2 - 3x + 4 = (A + B)x^2 + Cx + 4A \qquad \text{Combine like terms.}$$

$$\left.\begin{array}{r} 2 = A + B \\ -3 = C \\ 4 = 4A \end{array}\right\} \begin{array}{l} A = 1, B = 1, \\ \text{and } C = -3 \end{array}$$

Equate the x^2 term coefficients.

Equate the x term coefficients.

Equate the constant terms.

Solve the system of linear equations.

Substitute $A = 1$, $B = 1$, and $C = -3$.

$$\frac{2x^2 - 3x + 4}{x(x^2 + 4)} = \frac{A}{x} + \frac{Bx + C}{(x^2 + 4)}$$

$$\frac{2x^2 - 3x + 4}{x(x^2 + 4)} = \frac{1}{x} + \frac{1x + (-3)}{x^2 + 4} \quad \text{or} \quad \frac{1}{x} + \frac{x - 3}{x^2 + 4}$$

Therefore, $\dfrac{x^4 + 3x^3 + 6x^2 + 9x + 4}{x^3 + 4x} = x + 3 + \dfrac{1}{x} + \dfrac{x - 3}{x^2 + 4}$.

Skill Practice 5   Find the partial fraction decomposition.

$$\frac{x^4 + 2x^3 + 10x^2 + 8x + 12}{x^3 + 3x}$$

Answer

5. $x + 2 + \dfrac{4}{x} + \dfrac{3x + 2}{x^2 + 3}$

In Example 6, we demonstrate the case in which a rational expression contains a repeated quadratic factor.

EXAMPLE 6 Decomposing $\dfrac{f(x)}{g(x)}$, Where $g(x)$ Has a Repeated Irreducible Quadratic Factor

Find the partial fraction decomposition. $\dfrac{3x^2 + 8x + 14}{(x^2 + 2x + 5)^2}$

Solution:

$\dfrac{3x^2 + 8x + 14}{(x^2 + 2x + 5)^2} = \dfrac{Ax + B}{(x^2 + 2x + 5)^1} + \dfrac{Cx + D}{(x^2 + 2x + 5)^2}$ From Example 2(b), we have the form for the partial fraction decomposition.

To solve for A, B, C, and D, multiply both sides by the LCD to clear fractions.

$(x^2 + 2x + 5)^2\left[\dfrac{3x^2 + 8x + 14}{(x^2 + 2x + 5)^2}\right] = (x^2 + 2x + 5)^2\left[\dfrac{Ax + B}{x^2 + 2x + 5} + \dfrac{Cx + D}{(x^2 + 2x + 5)^2}\right]$

$3x^2 + 8x + 14 = (Ax + B)(x^2 + 2x + 5) + (Cx + D)$

$3x^2 + 8x + 14 = Ax^3 + 2Ax^2 + 5Ax + Bx^2 + 2Bx + 5B + Cx + D$

$3x^2 + 8x + 14 = Ax^3 + (2A + B)x^2 + (5A + 2B + C)x + 5B + D$ Combine like terms.

$\begin{aligned} 0 &= A && \text{Equate the } x^3 \text{ term coefficients.} \\ 3 &= 2A + B && \text{Equate the } x^2 \text{ term coefficients.} \\ 8 &= 5A + 2B + C && \text{Equate the } x \text{ term coefficients.} \\ 14 &= 5B + D && \text{Equate the constant terms.} \end{aligned}$

$A = 0$, $B = 3$, $C = 2$, and $D = -1$ Solve the system of linear equations.

$\dfrac{3x^2 + 8x + 14}{(x^2 + 2x + 5)^2} = \dfrac{Ax + B}{x^2 + 2x + 5} + \dfrac{Cx + D}{(x^2 + 2x + 5)^2}$ Substitute $A = 0$, $B = 3$, $C = 2$, and $D = -1$ into the partial fraction decomposition.

$\dfrac{3x^2 + 8x + 14}{(x^2 + 2x + 5)^2} = \dfrac{(0)x + (3)}{x^2 + 2x + 5} + \dfrac{(2)x + (-1)}{(x^2 + 2x + 5)^2}$ or $\dfrac{3}{x^2 + 2x + 5} + \dfrac{2x - 1}{(x^2 + 2x + 5)^2}$

Skill Practice 6 Find the partial fraction decomposition.
$\dfrac{-3x^2 - 5x - 19}{(x^2 + 3x + 6)^2}$

Answer

6. $\dfrac{-3}{x^2 + 3x + 6} + \dfrac{4x - 1}{(x^2 + 3x + 6)^2}$

SECTION 5.3   Practice Exercises

Prerequisite Review

For Exercises R.1–R.2, factor completely.

R.1. $12y^3 - 16y - 9y^2 + 12$

R.2. $64u^2 + 80u + 25$

R.3. Add. $\dfrac{4}{4a + 3} + \dfrac{4}{a - 3}$

R.4. Use long division to divide.
$\dfrac{25x^4 - 5x^3 + 15x^2 + 4x - 2}{5x^2 - x + 4}$

R.5. Solve. $\dfrac{16}{x} - \dfrac{16}{x - 2} = \dfrac{4}{x}$

Concept Connections

1. The process of decomposing a rational expression into two or more simpler fractions is called partial _____.

2. When setting up a partial fraction decomposition, if a fraction has a linear denominator, then the numerator should be (constant/linear). That is, should the numerator be set up as A or $Ax + B$?

3. When setting up a partial fraction decomposition, if the denominator of a fraction is a quadratic polynomial irreducible over the integers, then the numerator should be (constant/linear). That is, should the numerator be set up as A or $Ax + B$?

4. In what situation should long division be used before attempting to decompose a rational expression into partial fractions?

Objective 1: Set Up a Partial Fraction Decomposition

For Exercises 5–20, set up the form for the partial fraction decomposition. Do not solve for A, B, C, and so on. (See Examples 1–2)

5. $\dfrac{-x - 37}{(x + 4)(2x - 3)}$

6. $\dfrac{20x - 4}{(x - 5)(3x + 1)}$

7. $\dfrac{8x - 10}{x^2 - 2x}$

8. $\dfrac{y - 12}{y^2 + 3y}$

9. $\dfrac{6w - 7}{w^2 + w - 6}$

10. $\dfrac{-10t - 11}{t^2 + 5t - 6}$

11. $\dfrac{x^2 + 26x + 100}{x^3 + 10x^2 + 25x}$

12. $\dfrac{-3x^2 + 2x + 8}{x^3 + 4x^2 + 4x}$

13. $\dfrac{13x^2 + 2x + 45}{2x^3 + 18x}$

14. $\dfrac{17x^2 - 7x + 18}{7x^3 + 42x}$

15. $\dfrac{2x^3 - x^2 + 13x - 5}{x^4 + 10x^2 + 25}$

16. $\dfrac{3x^3 - 4x^2 + 11x - 12}{x^4 + 6x^2 + 9}$

17. $\dfrac{5x^2 - 4x + 8}{(x - 4)(x^2 + x + 4)}$

18. $\dfrac{x^2 + 15x - 6}{(x + 6)(x^2 + 2x + 6)}$

19. $\dfrac{2x^5 + 3x^3 + 4x^2 + 5}{x(x + 2)^3(x^2 + 2x + 7)^2}$

20. $\dfrac{6x^4 - 5x^3 + 2x^2 - 5}{(x - 3)(2x + 9)^2(x^2 + 1)^2}$

Objectives 2 and 3: Decompose $\dfrac{f(x)}{g(x)}$ into Partial Fractions

For Exercises 21–42, find the partial fraction decomposition. (See Examples 3–6)

21. $\dfrac{-x - 37}{(x + 4)(2x - 3)}$

22. $\dfrac{20x - 4}{(x - 5)(3x + 1)}$

23. $\dfrac{8x - 10}{x^2 - 2x}$

24. $\dfrac{y - 12}{y^2 + 3y}$

25. $\dfrac{6w - 7}{w^2 + w - 6}$

26. $\dfrac{-10t - 11}{t^2 + 5t - 6}$

27. $\dfrac{x^2 + 26x + 100}{x^3 + 10x^2 + 25x}$

28. $\dfrac{-3x^2 + 2x + 8}{x^3 + 4x^2 + 4x}$

29. $\dfrac{13x^2 + 2x + 45}{2x^3 + 18x}$

30. $\dfrac{17x^2 - 7x + 18}{7x^3 + 42x}$

31. $\dfrac{x^4 - 3x^3 + 13x^2 - 28x + 28}{x^3 + 7x}$

32. $\dfrac{x^4 - 4x^3 + 11x^2 - 13x + 12}{x^3 + 2x}$

33. $\dfrac{2x^3 - x^2 + 13x - 5}{x^4 + 10x^2 + 25}$

34. $\dfrac{3x^3 - 4x^2 + 11x - 12}{x^4 + 6x^2 + 9}$

35. $\dfrac{5x^2 - 4x + 8}{(x - 4)(x^2 + x + 4)}$

36. $\dfrac{x^2 + 15x - 6}{(x + 6)(x^2 + 2x + 6)}$

37. $\dfrac{4x^3 - 4x^2 + 11x - 7}{x^4 + 5x^2 + 6}$

38. $\dfrac{3x^3 - 4x^2 + 6x - 7}{x^4 + 5x^2 + 4}$

39. $\dfrac{2x^3 - 11x^2 - 4x + 24}{x^2 - 3x - 10}$

40. $\dfrac{3x^3 + 11x^2 + x + 10}{x^2 + 3x - 4}$

41. $\dfrac{3x^3 + 2x^2 - x - 5}{x^2 + 2x + 1}$

42. $\dfrac{2x^3 - 17x^2 + 54x - 68}{x^2 - 6x + 9}$

43. a. Factor. $x^3 - x^2 - 21x + 45$
(*Hint*: Use the rational zero theorem.)

b. Find the partial fraction decomposition for
$$\frac{-3x^2 + 35x - 70}{x^3 - x^2 - 21x + 45}.$$

44. a. Factor. $x^3 + 2x^2 - 7x + 4$

b. Find the partial fraction decomposition for
$$\frac{10x^2 + 17x - 17}{x^3 + 2x^2 - 7x + 4}.$$

45. a. Factor. $x^3 + 6x^2 + 12x + 8$

b. Find the partial fraction decomposition for
$$\frac{3x^2 + 8x + 5}{x^3 + 6x^2 + 12x + 8}.$$

46. a. Factor. $x^3 - 9x^2 + 27x - 27$

b. Find the partial fraction decomposition for
$$\frac{2x^2 - 17x + 37}{x^3 - 9x^2 + 27x - 27}.$$

Write About It

47. Write an informal explanation of partial fraction decomposition.

48. Suppose that a proper rational expression has a single repeated linear factor $(ax + b)^3$ in the denominator. Explain how to set up the partial fraction decomposition.

49. What is meant by a *proper* rational expression?

50. Given an improper rational expression, what must be done first before the technique of partial fraction decomposition may be performed?

Expanding Your Skills

51. a. Determine the partial fraction decomposition for $\dfrac{2}{n(n + 2)}$.

b. Use the partial fraction decomposition for $\dfrac{2}{n(n + 2)}$ to rewrite the infinite sum
$$\frac{2}{1(3)} + \frac{2}{2(4)} + \frac{2}{3(5)} + \frac{2}{4(6)} + \frac{2}{5(7)} \cdots$$

c. Determine the value of $\dfrac{1}{n + 2}$ as $n \to \infty$.

d. Find the value of the sum from part (b).

52. a. Determine the partial fraction decomposition for $\dfrac{3}{n(n + 3)}$.

b. Use the partial fraction decomposition for $\dfrac{3}{n(n + 3)}$ to rewrite the infinite sum
$$\frac{3}{1(4)} + \frac{3}{2(5)} + \frac{3}{3(6)} + \frac{3}{4(7)} + \frac{3}{5(8)} \cdots$$

c. Determine the value of $\dfrac{1}{n + 3}$ as $n \to \infty$.

d. Find the value of the sum from part (b).

For Exercises 53–54, find the partial fraction decomposition. Assume that a and b are nonzero constants.

53. $\dfrac{1}{x(a + bx)}$

54. $\dfrac{1}{a^2 - x^2}$

For Exercises 55–56, find the partial fraction decomposition for the given expression. [*Hint*: Use the substitution $u = e^x$ and recall that $e^{2x} = (e^x)^2$.]

55. $\dfrac{5e^x + 7}{e^{2x} + 3e^x + 2}$

56. $\dfrac{-3e^x - 22}{e^{2x} + 3e^x - 4}$

Systems of Nonlinear Equations in Two Variables

OBJECTIVES

1. Solve Nonlinear Systems of Equations by the Substitution Method
2. Solve Nonlinear Systems of Equations by the Addition Method
3. Use Nonlinear Systems of Equations to Solve Applications

1. Solve Nonlinear Systems of Equations by the Substitution Method

The attending physician in an emergency room treats an unconscious patient suspected of a drug overdose. The physician needs to know the concentration of the drug in the bloodstream at the time the drug was taken to determine the extent of damage to the kidneys. The patient's family does not know the original amount of the drug taken, but believes that he took the drug by injection 3 hr before arriving at the hospital. Blood work at the time of arrival ($t = 3$ hr after the patient had taken the drug) showed that the drug concentration in the bloodstream was 0.69 μg/dL. One hour later ($t = 4$ hr), the level had dropped to 0.655 μg/dL.

The physician can solve the following system of nonlinear equations to determine the concentration of the drug in the bloodstream at the time of injection. The value A_0 represents the initial concentration of the drug, and the value k is related to the rate at which the kidneys can remove the drug.

$$0.69 = A_0 e^{-3k}$$
$$0.655 = A_0 e^{-4k}$$ The solution to this problem is discussed in Exercise 59.

A **nonlinear system of equations** is a system in which one or more equations is nonlinear. For example:

$$\begin{aligned} -x + 7y &= 50 \\ x^2 + y^2 &= 100 \end{aligned}$$ Second equation nonlinear $$\left.\begin{aligned} 2x^2 + y^2 &= 17 \\ x^2 + 2y^2 &= 22 \end{aligned}\right\}$$ Both equations nonlinear

A **solution** to a nonlinear system of equations in two variables is an ordered pair with real-valued coordinates that satisfies each equation in the system. Graphically, these are the points of intersection of the graphs of the equations. A nonlinear system of equations may have no solution or one or more solutions. See Figure 5-9 through Figure 5-11.

TIP A nonlinear system of equations may also have infinitely many solutions. In the graph shown, the "wave" pattern extends infinitely far in both directions.

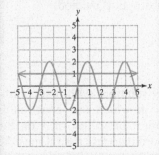

Two solutions

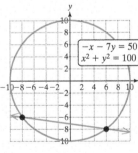

Figure 5-9

Four solutions

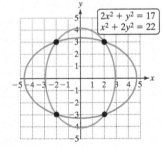

Figure 5-10

No solution

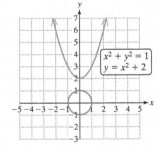

Figure 5-11

We will solve nonlinear systems of equations by the substitution method and by the addition method. In Example 1, we begin with the substitution method.

EXAMPLE 1   Solving a System of Nonlinear Equations by Using the Substitution Method

Solve the system by using the substitution method. $-x - 7y = 50$

$$x^2 + y^2 = 100$$

Solution:

Figure 5-12

$\boxed{A}$ $-x - 7y = 50$ Equation $\boxed{A}$ is a line and can be written in slope-intercept form as
$\boxed{B}$ $x^2 + y^2 = 100$ $y = -\frac{1}{7}x - \frac{50}{7}$.

Equation $\boxed{B}$ represents a circle centered at $(0, 0)$ with radius 10.

A sketch of the two equations suggests that the curves intersect at $(-8, -6)$ and $(6, -8)$. See Figure 5-12.

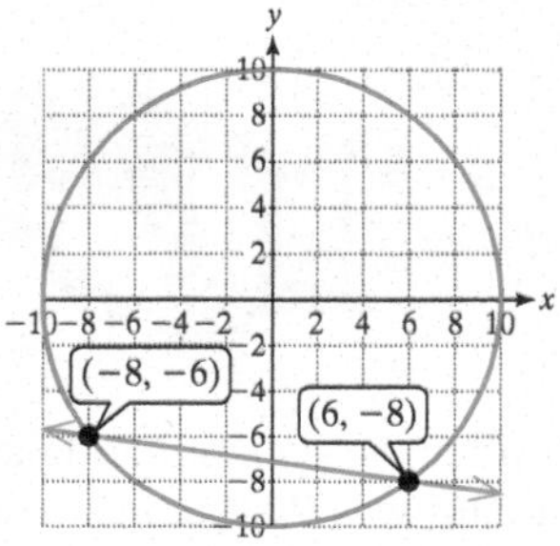

$\boxed{A}$ $-x - 7y = 50 \longrightarrow x = -7y - 50$ To solve the system algebraically by the
$\boxed{B}$ $x^2 + y^2 = 100$ substitution method, first solve for x or y from either equation.

$\boxed{B}$ $(-7y - 50)^2 + y^2 = 100$ Substitute $x = -7y - 50$ from equation $\boxed{A}$ into equation $\boxed{B}$.

$$49y^2 + 700y + 2500 + y^2 = 100$$
$$50y^2 + 700y + 2400 = 0 \qquad \text{Solve the resulting equation for } y.$$
$$50(y^2 + 14y + 48) = 0 \qquad \text{Factor out the GCF of 50.}$$
$$50(y + 6)(y + 8) = 0$$
$$y = -6 \quad \text{or} \quad y = -8$$

For each value of y, find the corresponding x value by substituting y into the equation in which x is isolated: $x = -7y - 50$

$y = -6$: $x = -7(-6) - 50 = -8$ The solution is $(-8, -6)$.
$y = -8$: $x = -7(-8) - 50 = 6$ The solution is $(6, -8)$.

Check: $(-8, -6)$ Check: $(6, -8)$
$\boxed{A}$ $-(-8) - 7(-6) = 50$ ✓ $\boxed{A}$ $-(6) - 7(-8) = 50$ ✓   The solutions both check in each equation.
$\boxed{B}$ $(-8)^2 + (-6)^2 = 100$ ✓ $\boxed{B}$ $(6)^2 + (-8)^2 = 100$ ✓

The solution set is $\{(-8, -6), (6, -8)\}$.

Skill Practice 1   Solve the system by using the substitution method.

$2x + y = 5$
$x^2 + y^2 = 50$

As we solve systems of equations, we will consider only solutions with real coordinates. In Example 2, we have a system of equations in which one equation is $y = 3\sqrt{x - 8}$. The expression $\sqrt{x - 8}$ is a real number for values of x on the interval $[8, \infty)$. Therefore, any ordered pair with an x-coordinate less than 8 must be rejected as a potential solution.

EXAMPLE 2   Solving a System of Nonlinear Equations by Using the Substitution Method

Solve the system by using the substitution method. $(x - 5)^2 + y^2 = 25$
$y = 3\sqrt{x - 8}$

Answer
1. $\{(5, -5), (-1, 7)\}$

Solution:

A̲ $(x - 5)^2 + y^2 = 25$
B̲ $y = 3\sqrt{x - 8}$

Label the equations. The graphs of the equations are shown in Figure 5-13. The graph suggests that there is only one solution: (9, 3).

A̲ $(x - 5)^2 + \left(3\sqrt{x - 8}\right)^2 = 25$

Substitute $3\sqrt{x - 8}$ for y in equation A̲.

$x^2 - 10x + 25 + 9(x - 8) = 25$ Square each term.

$$x^2 - x - 72 = 0$$
$$(x - 9)(x + 8) = 0$$
$$x = 9 \quad \text{or} \quad \cancel{x = -8}$$

Reject $x = -8$ because $3\sqrt{x - 8}$ is not a real number for $x = -8$.

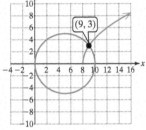

Figure 5-13

Given $x = 9$, solve for y:

B̲ $y = 3\sqrt{x - 8}$
    $y = 3\sqrt{9 - 8} = 3$

The solution is (9, 3) and checks in each original equation.

The solution set is $\{(9, 3)\}$.

Skill Practice 2   Solve the system by using the substitution method.

$$x^2 + y^2 = 90$$
$$y = \sqrt{x}$$

---

2. Solve Nonlinear Systems of Equations by the Addition Method

The substitution method is used most often to solve a system of nonlinear equations. In some situations, however, the addition method is an efficient way to find a solution. Examples 3 and 4 demonstrate that we can eliminate a variable from both equations in a system provided the terms containing the corresponding variables are like terms.

EXAMPLE 3   **Solving a System of Nonlinear Equations by Using the Addition Method**

Solve the system by using the addition method.
$$2x^2 + y^2 = 17$$
$$x^2 + 2y^2 = 22$$

Solution:

Using the addition method, the goal is to create opposite coefficients on either the x^2 terms or the y^2 terms. In this case, we have chosen to eliminate the x^2 terms.

A̲ $2x^2 + y^2 = 17$ $2x^2 + y^2 = 17$
B̲ $x^2 + 2y^2 = 22$ 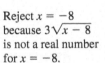 $\underline{-2x^2 - 4y^2 = -44}$
                    Multiply by −2. $-3y^2 = -27$
                                         $y^2 = 9$
                                         $y = \pm 3$

$y = 3$:   B̲ $x^2 + 2(3)^2 = 22$
              $x^2 = 4$
              $x = \pm 2$ The solutions are $(2, 3)$, $(-2, 3)$.

Substitute $y = \pm 3$ into either equation A̲ or B̲ to solve for the corresponding values of x.

$y = -3$: $\boxed{B}$ $x^2 + 2(-3)^2 = 22$

$\qquad\qquad\qquad x^2 = 4$

$\qquad\qquad\qquad x = \pm 2$ The solutions are
$\qquad\qquad\qquad\qquad\qquad (2, -3), (-2, -3).$

The solutions all check in the original equations.

The solution set is $\{(2, 3), (-2, 3), (2, -3), (-2, -3)\}$.

Skill Practice 3   Solve the system by using the addition method.

$x^2 + y^2 = 17$
$x^2 - 2y^2 = -31$

TECHNOLOGY CONNECTIONS

Solving a System of Nonlinear Equations Using Intersect

The equations in Example 3 each represent a curve called an ellipse. We do not yet know how to graph an ellipse; however, we can graph the curves on a graphing calculator. First solve each equation for y. Enter the resulting functions in the calculator and use the Intersect feature to approximate the points of intersection (Figure 5-14).

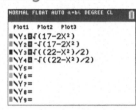

$2x^2 + y^2 = 17 \xrightarrow{\text{Solve for } y.} y = \pm\sqrt{17 - 2x^2}$

$x^2 + 2y^2 = 22 \longrightarrow y = \pm\sqrt{\dfrac{22 - x^2}{2}}$

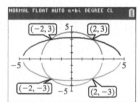

Figure 5-14

> **TIP** The image produced by a graphing calculator in Figure 5-14 does not show the curves touching the x-axis, when indeed they do.

Example 4 illustrates that a nonlinear system of equations may have no solution.

EXAMPLE 4   Solving an Inconsistent System by Using the Addition Method

Solve the system by using the addition method. $x^2 + 4y^2 = 4$
$\qquad\qquad\qquad\qquad\qquad\qquad\qquad\qquad\qquad\qquad\quad x^2 - \;\; y^2 = 9$

Solution:

$x^2 + 4y^2 = 4$

$x^2 - \;\; y^2 = 9 \xrightarrow{\text{Multiply by } -1.}$

$\begin{aligned} x^2 + 4y^2 &= 4 \\ -x^2 + \;\; y^2 &= -9 \\ \hline 5y^2 &= -5 \\ y^2 &= -1 \\ y &= \pm i \end{aligned}$

We use the addition method because like terms are aligned vertically.

The values for y are not real numbers. Therefore, there is no solution to the system of equations over the set of real numbers.

The solution set is $\{\ \}$.

> **TIP** Solutions to a system of equations are limited to ordered pairs with real-valued coordinates because we are interested in the points of intersection of the graphs of the equations.

Skill Practice 4   Solve the system by using the addition method.

$x^2 + \;\; y^2 = 16$
$4x^2 + 9y^2 = 36$

Answers

3. $\{(1, 4), (-1, 4), (1, -4), (-1, -4)\}$

4. $\{\ \}$

TECHNOLOGY CONNECTIONS

Solving a Nonlinear System of Equations

The equations in Example 4 represent two curves that have not yet been studied: an ellipse and a hyperbola. However, we can graph the curves on a graphing calculator by solving for y and entering the functions into the calculator.

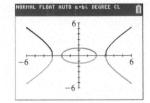

$$x^2 + 4y^2 = 4 \xrightarrow[\text{Solve for } y.]{} y = \pm\sqrt{\dfrac{4 - x^2}{4}}$$

$$x^2 - y^2 = 9 \xrightarrow{} y = \pm\sqrt{x^2 - 9}$$

From the graph, we see that the curves do not intersect.

3. Use Nonlinear Systems of Equations to Solve Applications

In Example 5, we set up a system of nonlinear equations to model an application involving two independent relationships between two variables.

EXAMPLE 5   Solving an Application of a Nonlinear System

The perimeter of a television screen is 140 in. The area is 1200 in.^{2}.

a. Find the length and width of the screen.

b. Find the length of the diagonal.

Solution:

Let x represent the length of the screen.

Let y represent the width of the screen.

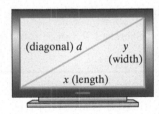

The statement of the problem gives two independent relationships between the length and width of the screen.

The perimeter of a television screen is 140 in. $\longrightarrow 2x + 2y = 140$

The area is 1200 in.^{2}. $\xrightarrow{} xy = 1200$

a. Solve the nonlinear system of equations for x and y.

$\boxed{A}\quad 2x + 2y = 140$	
$\boxed{B}\quad xy = 1200 \xrightarrow[]{\text{Solve for } y.} y = \dfrac{1200}{x}$	Using the substitution method, solve for x or y in either equation.

$$\boxed{A}\quad 2x + 2\left(\frac{1200}{x}\right) = 140$$

Substitute $y = \frac{1200}{x}$ from equation $\boxed{B}$ into equation $\boxed{A}$.

$$2x + \frac{2400}{x} = 140$$

$$x \cdot \left(2x + \frac{2400}{x}\right) = x \cdot (140)$$

Multiply both sides by the LCD to clear fractions.

$$2x^2 + 2400 = 140x$$

The resulting equation is quadratic.

$$2x^2 - 140x + 2400 = 0$$

$$2(x^2 - 70x + 1200) = 0$$

Factor the left side.

$$2(x - 40)(x - 30) = 0$$

$$x = 40 \quad \text{or} \quad x = 30$$

There are two possible values for the length x.

Avoiding Mistakes

Check the solution to Example 5(a) by computing the perimeter and area.

Perimeter:
2(40 in.) + 2(30 in.)
= 140 in. ✓

Area:
(40 in.)(30 in.)
= 1200 in.^{2} ✓

Substitute $x = 40$ and $x = 30$ into the equation $y = \dfrac{1200}{x}$.

$x = 40$: $y = \dfrac{1200}{40} = 30$ If the length x is 40 in., then the width y is 30 in.

$x = 30$: $y = \dfrac{1200}{30} = 40$ If the length x is 30 in., then the width y is 40 in.

Taking the length to be the longer side, we have that the length is 40 in. and the width is 30 in.

b. $x^2 + y^2 = d^2$ Use the Pythagorean theorem to determine the
$(40)^2 + (30)^2 = d^2$ measure of the diagonal of the screen.
$1600 + 900 = d^2$
$2500 = d^2$
$d = \pm 50$

Excluding the negative solution for d, the diagonal is 50 in.

Answer

5. The length of the rug is 12 ft and the width is 8 ft.

Skill Practice 5   The perimeter of a rectangular rug is 40 ft and the area is 96 ft^2. Find the dimensions of the rug.

SECTION 5.4   Practice Exercises

Prerequisite Review

For Exercises R.1–R.3, graph the equation.

R.1. $y = x^2 + 4x + 3$ **R.2.** $(x + 4)^2 + (y - 1)^2 = 4$ **R.3.** $y = \sqrt{x + 2}$

For Exercises R.4–R.5, solve the equation.

R.4. $\ln x = 17$ **R.5.** $3^x = 81$

Concept Connections

1. A _____ system of equations in two variables is a system in which one or more equations in the system is nonlinear.

2. A solution to a nonlinear system of equations in two variables is an _____ pair with real-valued coordinates that satisfies each equation in the system. Graphically, a solution is a point of _____ of the graphs of the equations.

Objective 1: Solve Nonlinear Systems of Equations by the Substitution Method

For Exercises 3–14,

a. Graph the equations in the system.

b. Solve the system by using the substitution method. **(See Examples 1–2)**

3. $y = x^2 - 2$
   $2x - y = 2$

4. $y = -x^2 + 3$
   $y - 2x = 0$

5. $x^2 + y^2 = 25$
   $x + y = 1$

6. $x^2 + y^2 = 25$
   $3y = 4x$

7. $y = \sqrt{x}$
   $x^2 + y^2 = 20$

8. $x^2 + y^2 = 10$
   $y = \sqrt{x - 2}$

9. $(x + 2)^2 + y^2 = 9$
   $y = 2x - 4$

10. $x^2 + (y - 3)^2 = 4$
   $y = -x - 4$

11. $y = x^3$
   $y = x$

12. $y = \sqrt[3]{x}$
   $y = x$

13. $y = -(x - 2)^2 + 5$
   $y = 2x + 1$

14. $y = (x + 3)^2 - 1$
   $y = 2x + 5$

Objective 2: Solve Nonlinear Systems of Equations by the Addition Method

For Exercises 15–22, solve the system by using the addition method. (See Examples 3–4)

15. $2x^2 + 3y^2 = 11$
$\quad\ x^2 + 4y^2 = 8$

16. $3x^2 + \ y^2 = 21$
$\quad\ 4x^2 - 2y^2 = -2$

17. $\quad x^2 - \ xy = 20$
$\quad\ -2x^2 + 3xy = -44$

18. $4xy + 3y^2 = -9$
$\quad\ 2xy + \ y^2 = -5$

19. $5x^2 - 2y^2 = 1$
$\quad\ 2x^2 - 3y^2 = -4$

20. $6x^2 + 5y^2 = 38$
$\quad\ 7x^2 - 3y^2 = 9$

21. $x^2 = 1 - y^2$
$\quad\ 9x^2 - 4y^2 = 36$

22. $4x^2 = 4 - y^2$
$\quad\ 16y^2 = 144 + 9x^2$

Mixed Exercises

For Exercises 23–34, solve the system by using any method.

23. $x^2 - 4xy + 4y^2 = 1$
$\quad\ x + y = 4$

24. $x^2 - 6xy + 9y^2 = 0$
$\quad\ x - y = 2$

25. $y = x^2 + 4x + 5$
$\quad\ y = 4x + 5$

26. $y = x^2 - 6x + 9$
$\quad\ y = -2x + 5$

27. $y = x^2$
$\quad\ y = \dfrac{1}{x}$

28. $y = \dfrac{1}{x}$
$\quad\ y = \sqrt{x}$

29. $x^2 + (y - 4)^2 = 25$
$\quad\ y = -x^2 + 9$

30. $(x - 10)^2 + y^2 = 100$
$\quad\ x = y^2$

31. $y = -x^2 + 6x - 7$
$\quad\ y = x^2 - 10x + 23$

32. $y = -x^2 + 6x - 9$
$\quad\ y = x^2 - 2x - 3$

33. $\dfrac{x^2}{4} + \dfrac{y^2}{16} = 1$
$\quad\ x^2 + y = 4$

34. $\dfrac{x^2}{4} + y^2 = 1$
$\quad\ x = -2y^2 + 2$

For Exercises 35–36, use the substitutions $u = \dfrac{1}{x^2}$ and $v = \dfrac{1}{y^2}$ to solve the system of equations.

35. $\dfrac{4}{x^2} - \dfrac{3}{y^2} = -23$
$\quad\ \dfrac{5}{x^2} + \dfrac{1}{y^2} = 14$

36. $-\dfrac{3}{x^2} + \dfrac{1}{y^2} = 13$
$\quad\ \dfrac{5}{x^2} - \dfrac{1}{y^2} = -5$

Objective 3: Use Nonlinear Systems of Equations to Solve Applications

37. Find two numbers whose sum is 12 and whose product is 35.

38. Find two numbers whose sum is 9 and whose product is -36.

39. The sum of the squares of two positive numbers is 29 and the difference of the squares of the numbers is 21. Find the numbers.

40. The sum of the squares of two negative numbers is 145 and the difference of the squares of the numbers is 17. Find the numbers.

41. The difference of two positive numbers is 2 and the difference of their squares is 44. Find the numbers.

42. The sum of two numbers is 4 and the difference of their squares is 64. Find the numbers.

43. The ratio of two numbers is 3 to 4 and the sum of their squares is 225. Find the numbers.

44. The ratio of two numbers is 5 to 12 and the sum of their squares is 676. Find the numbers.

45. Find the dimensions of a rectangle whose perimeter is 36 m and whose area is 80 m^2.

46. Find the dimensions of a rectangle whose perimeter is 56 cm and whose area is 192 cm^2.

47. The floor of a rectangular bedroom requires 240 ft^2 of carpeting. Molding is placed around the base of the floor except at two 3-ft doorways. If 58 ft of molding is required around the base of the floor, determine the dimensions of the floor. **(See Example 5)**

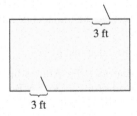
3 ft

3 ft

48. An electronic sign for a grocery store is in the shape of a rectangle. The perimeter of the sign is 72 ft and the area is 320 ft^2. Find the length and width of the sign.

49. A rental truck has a cargo capacity of 288 ft^3. A 10-ft pipe just fits resting diagonally on the floor of the truck. If the cargo space is 6 ft high, find the dimensions of the truck.

6 ft

10 ft

50. A rectangular window has a 15-yd diagonal and an area of 108 yd^2. Find the dimensions of the window.

51. An aquarium is 16 in. high with volume of 4608 in.^{3} (approximately 20 gal). If the amount of glass used for the bottom and four sides is 1440 in.^{2}, determine the dimensions of the aquarium.

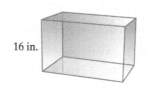

16 in.

52. A closed box is in the shape of a rectangular solid with height 3 m. Its surface area is 268 m^2. If the volume is 240 m^3, find the dimensions of the box.

53. The hypotenuse of a right triangle is $\sqrt{65}$ ft. The sum of the lengths of the legs is 11 ft. Find the lengths of the legs.

54. The hypotenuse of a right triangle is $\sqrt{73}$ in. The sum of the lengths of the legs is 11 in. Find the lengths of the legs.

55. A ball is kicked off the side of a hill at an angle of elevation of 30°. The hill slopes downward 30° from the horizontal. Consider a coordinate system in which the origin is the point on the edge of the hill from which the ball is kicked. The path of the ball and the line of declination of the hill can be approximated by

$$y = -\frac{x^2}{192} + \frac{\sqrt{3}}{3}x \qquad \text{Path of the ball}$$

$$y = -\frac{\sqrt{3}}{3}x \qquad \text{Line of declination of the hill}$$

Solve the system to determine where the ball will hit the ground.

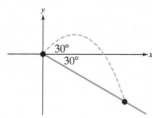

56. A child kicks a rock off the side of a hill at an angle of elevation of 60°. The hill slopes downward 30° from the horizontal. Consider a coordinate system in which the origin is the point on the edge of the hill from which the rock is kicked. The path of the rock and the line of declination of the hill can be approximated by

$$y = -\frac{x^2}{36} + \sqrt{3}x \qquad \text{Path of the rock}$$

$$y = -\frac{\sqrt{3}}{3}x \qquad \text{Line of declination of the hill}$$

Solve the system to determine where the rock will hit the ground.

Write About It

57. What is the difference between a system of linear equations and a system of nonlinear equations?

58. Describe a situation in which the addition method is an efficient technique to solve a system of nonlinear equations.

Expanding Your Skills

59. The attending physician in an emergency room treats an unconscious patient suspected of a drug overdose. The physician does not know the initial concentration A_0 of the drug in the bloodstream at the time of injection. However, the physician knows that after 3 hr, the drug concentration in the blood is 0.69 μg/dL and after 4 hr, the concentration is 0.655 μg/dL. The model $A(t) = A_0 e^{-kt}$ represents the drug concentration $A(t)$ (in μg/dL) in the bloodstream t hours after injection. The value of k is a constant related to the rate at which the drug is removed by the body.

   a. Substitute 0.69 for $A(t)$ and 3 for t in the model and write the resulting equation.

   b. Substitute 0.655 for $A(t)$ and 4 for t in the model and write the resulting equation.

   c. Use the system of equations from parts (a) and (b) to solve for k. Round to 3 decimal places.

   d. Use the system of equations from parts (a) and (b) to approximate the initial concentration A_0 (in μg/dL) at the time of injection. Round to 2 decimal places.

   e. Determine the concentration of the drug after 12 hr. Round to 2 decimal places.

60. A patient undergoing a heart scan is given a sample of fluorine-18 (^{18}F). After 4 hr, the radioactivity level in the patient is 44.1 MBq (megabecquerel). After 5 hr, the radioactivity level drops to 30.2 MBq. The radioactivity level $Q(t)$ can be approximated by $Q(t) = Q_0 e^{-kt}$, where t is the time in hours after the initial dose Q_0 is administered.

   a. Determine the value of k. Round to 4 decimal places.

   b. Determine the initial dose, Q_0. Round to the nearest whole unit.

   c. Determine the radioactivity level after 12 hr. Round to 1 decimal place.

61. The population $P(t)$ of a culture of bacteria grows exponentially for the first 72 hr according to the model $P(t) = P_0 e^{kt}$. The variable t is the time in hours since the culture is started. The population of bacteria is 60,000 after 7 hr. The population grows to 80,000 after 12 hr.

   a. Determine the constant k to 3 decimal places.

   b. Determine the original population P_0. Round to the nearest thousand.

   c. Determine the time required for the population to reach 300,000. Round to the nearest hour.

62. An investment grows exponentially under continuous compounding. After 2 yr, the amount in the account is \$7328.70. After 5 yr, the amount in the account is \$8774.10. Use the model $A(t) = Pe^{rt}$ to

   a. Find the interest rate r. Round to the nearest percent.

   b. Find the original principal P. Round to the nearest dollar.

   c. Determine the amount of time required for the account to reach a value of \$15,000. Round to the nearest year.

For Exercises 63–64, determine the number of solutions to the system of equations.

63.
$$y = 2^{x+1}$$
$$-1 + \log_2 y = x$$

64. $x^2 - y^2 = 0$
$$|x| = |y|$$

For Exercises 65–70, solve the system.

65. $\log x + 2 \log y = 5$
$$2 \log x - \log y = 0$$

66. $\log_2 x + 3 \log_2 y = 6$
$$\log_2 x - \log_2 y = 2$$

67. $2^x + 2^y = 6$
$$4^x - 2^y = 14$$

68. $3^x - 9^y = 18$
$$3^x + 3^y = 30$$

69. $(x - 1)^2 + (y + 1)^2 = 5$
$$x^2 + (y + 4)^2 = 29$$

70. $(x + 3)^2 + (y - 2)^2 = 4$
$$(x - 1)^2 + y^2 = 8$$

For Exercises 71–72, use substitution to solve the system for the set of ordered triples (x, y, λ) that satisfy the system.

71. $2 = 2\lambda x$
$$6 = 2\lambda y$$
$$x^2 + y^2 = 10$$

72. $8 = 4\lambda x$
$$2 = 2\lambda y$$
$$2x^2 + y^2 = 9$$

73. Two circles intersect as shown.

   a. Find the points of intersection.
$$x^2 + y^2 = 25$$
$$(x - 4)^2 + (y + 2)^2 = 25$$

   b. Find an equation of the chord common to both circles (shown in black). (*Hint:* A chord is a line segment on the interior of a circle with both endpoints on the circle.)

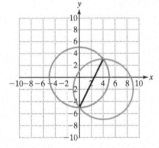

74. The minimum and maximum distances from a point P to a circle are found using the line determined by the given point and the center of the circle. Given the circle defined by $x^2 + y^2 = 9$ and the point $P(4, 5)$,

   a. Find the point on the circle closest to the point $(4, 5)$.

   b. Find the point on the circle farthest from the point $(4, 5)$.

Technology Connections

For Exercises 75–80, use a graphing utility to approximate the solution(s) to the system of equations. Round the coordinates to 3 decimal places.

75. $y = -0.6x + 7$
$$y = e^x - 5$$

76. $y = -0.7x + 4$
$$y = \ln x$$

77. $x^2 + y^2 = 40$
$$y = -x^2 + 8.5$$

78. $x^2 + y^2 = 32$
$$y = 0.8x^2 - 9.2$$

79. $y = x^2 - 8x + 20$
$$y = 4 \log x$$

80. $y = 0.2e^x$
$$y = -0.6x^2 - 2x - 3$$

SECTION 5.5   Inequalities and Systems of Inequalities in Two Variables

1. Solve Linear Inequalities in Two Variables

Adriana estimates that she has 12 hr of available study time before she takes tests in algebra and biology in back-to-back classes. Suppose that x represents the time she spends studying algebra and y represents the time she spends studying biology. Then the inequality $x + y \le 12$, where $x \ge 0$ and $y \ge 0$ represents the distribution of time she can allocate studying for each subject.

An inequality of the form $Ax + By < C$, where A and B are not both zero, is called a **linear inequality in two variables.** (Note that the symbols $>$, $\le$, and $\ge$ can be used in place of $<$ in the definition.) A **solution** to an inequality in two variables is an ordered pair that satisfies the inequality. The set of all such ordered pairs is called the **solution set** to the inequality. Graphically, the solution set is a region in the xy-plane.

To graph the solution set to a linear inequality in two variables, follow these guidelines.

> **TIP** To graph the line in step 1, we can
> • Use the slope-intercept form of the equation of the line.
> • Graph the x- and y-intercepts.
> • Create a table of points.

Graphing a Linear Inequality in Two Variables

Step 1   Graph the related equation. That is, replace the inequality sign with an $=$ sign and graph the line represented by the equation.
  • If the inequality is strict (stated with the symbols $<$ or $>$), then draw the line as a dashed line to indicate that the line *is not* part of the solution set.
  • If the inequality is stated with the symbols $\le$ or $\ge$, then draw the line as a solid line to indicate that the line *is* part of the solution set.

Step 2   Choose a test point from either side of the line (not a point on the line itself) and substitute the ordered pair into the inequality.
  • If a true statement results, then shade the region (half-plane) from which the test point was taken.
  • If a false statement results, then shade the region (half-plane) on the opposite side of the line from which the test point was taken.

EXAMPLE 1   **Graphing a Linear Inequality in Two Variables**

Graph the solution set. $3x - 2y < 6$

Solution:

$3x - 2y < 6 \xrightarrow[\text{equation}]{\text{related}} 3x - 2y = 6$ **Step 1:** Graph the related equation $3x - 2y = 6$ using any technique for graphing. In this case, we have chosen to find the x- and y-intercepts.

x-intercept: _y-intercept_:

$3x - 2(0) = 6$ $3(0) - 2y = 6$

$x = 2$ $y = -3$

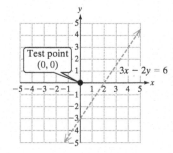

The _x_- and _y_-intercepts are (2, 0) and (0, −3).
Graph the line through the intercepts (Figure 5-15).
Because the inequality is strict, draw the line as a
dashed line.

TIP We can also graph the
inequality from Example 1 by
solving the inequality for _y_.

$3x - 2y < 6$

$-2y < -3x + 6$

$y > \tfrac{3}{2}x - 3$

Then shade the half-plane
above the bounding line
because this region contains
points with _y_-coordinates
greater than those on the
bounding line.

Figure 5-15

Test (0, 0):

$3x - 2y < 6$

$3(0) - 2(0) \overset{?}{<} 6$

$0 \overset{?}{<} 6 \ \checkmark \ \text{true}$

Step 2: Select a test point either above or below the line and
test the ordered pair in the original inequality. In
Figure 5-16, we have chosen (0, 0) as a test point.

The test point (0, 0) is a representative point above
the line. Since (0, 0) satisfies the original inequality,
then it and all other points above the line are
solutions.

The solution set is the set of ordered pairs in the
region (half-plane) above the line (Figure 5-16).

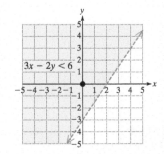

Figure 5-16

Skill Practice 1   Graph the solution set. $4x - y > 3$

TECHNOLOGY CONNECTIONS

Graphing a System of Inequalities in Two Variables

A graphing utility can be used to graph an
inequality in two variables. In most cases,
we solve for _y_ first and enter the related
equation in the graphing editor. Place the
cursor to the left of Y_1 and press ENTER
two times. This will set the graph style to
shade the region above the line ◥. Notice
that the calculator image does not
differentiate between a solid and dashed
bounding line (Figure 5-17).

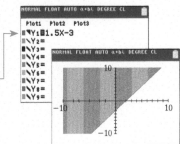

Figure 5-17

Note: With the cursor placed to the left of Y_1,

Answer

1.

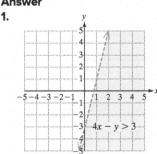

- Select the upper right triangle ◥ for inequalities of the form $Y_1 > f(x)$ and
  $Y_1 \geq f(x)$.
- Select the lower left triangle ◣ for inequalities of the form $Y_1 < f(x)$ and
  $Y_1 \leq f(x)$.

In Example 2, we graph the solutions to an inequality in which the bounding line passes through the origin.

EXAMPLE 2 **Graphing a Linear Inequality**

Graph the solution set. $4y \leq 3x$

Solution:

$4y \leq 3x \xrightarrow[\substack{\text{related} \\ \text{equation}}]{} 4y = 3x$

$$y = \frac{3}{4}x + 0$$

Graph the line having y-intercept $(0, 0)$ and slope $\frac{3}{4}$ (Figure 5-18). Because the inequality symbol $\leq$ allows for equality, draw the line as a solid line.

Step 1: Graph the related equation $4y = 3x$. In this case, we have chosen to find the slope-intercept form of the equation and then graph the line using the slope and y-intercept.

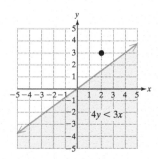

Figure 5-18

TIP As a check, we can select a test point *below* the line, such as (1, −1) and verify that it does indeed satisfy the original inequality.

Test (1, −1):
$4(-1) \overset{?}{\leq} 3(1)$
$-4 \overset{?}{\leq} 3 \checkmark$ true

Test (2, 3):
$4y \leq 3x$
$4(3) \overset{?}{\leq} 3(2)$
$12 \overset{?}{\leq} 6$ false

Step 2: Select a test point. We have chosen (2, 3).

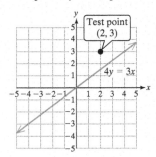

Figure 5-19

The test point (2, 3) is a representative point above the line. Since (2, 3) does *not* satisfy the original inequality, then points on the other side of the line are solutions. Shade below the line.

The solution set is the set of ordered pairs on and below the line (Figure 5-19).

Skill Practice 2 Graph the solution set. $2y \geq 5x$

Recall that for a constant k, the equation $x = k$ represents a vertical line in the xy-plane. The inequalities $x < k$ and $x > k$ represent half-planes to the left or right of the vertical line $x = k$ (Figure 5-20). Likewise, $y = k$ represents a horizontal line in the xy-plane. The inequalities $y < k$ and $y > k$ represent half-planes below or above the line $y = k$ (Figure 5-21).

Answer
2.

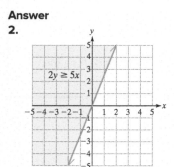

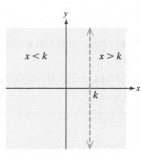

Figure 5-20

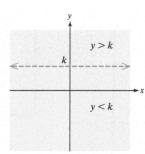

Figure 5-21

EXAMPLE 3   **Graphing Linear Inequalities with a Horizontal or Vertical Bounding Line**

Graph the solution set.

  a. $x \le -1$ **b.** $3y > 5$

Solution:

  a. $x \le -1$

   • The related equation $x = -1$ is a vertical line.

   • The inequality $x \le -1$ represents all points to the *left* of or on the line $x = -1$.

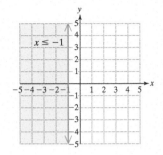

  b. $3y > 5$

   • The inequality is equivalent to $y > \frac{5}{3}$. The related equation $y = \frac{5}{3}$ is a horizontal line.

   • The inequality $y > \frac{5}{3}$ represents all points strictly above the line $y = \frac{5}{3}$.

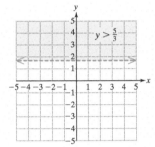

Skill Practice 3   Graph the solution set.

  a. $2y < 6$ **b.** $x \ge \frac{9}{4}$

2. Solve Nonlinear Inequalities in Two Variables

The same approach used to graph a linear inequality in two variables is used to graph a *nonlinear* inequality in two variables. This is demonstrated in Example 4.

EXAMPLE 4   **Graphing a Nonlinear Inequality**

Graph the solution set. $(x - 2)^2 + y^2 > 9$

Solution:

$(x - 2)^2 + y^2 > 9 \xrightarrow[\text{equation}]{\text{related}} (x - 2)^2 + y^2 = 9$

Step 1: Graph the related equation. The equation represents a circle centered at (2, 0) with radius 3. Because the inequality is strict, draw the circle as a dashed curve (Figure 5-22).

Center: (2, 0)
Radius: 3

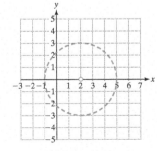

Figure 5-22

Answers

3. a.

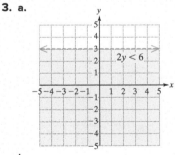

b.

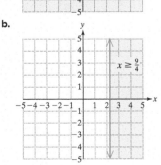

Test (2, 0):

$(x - 2)^2 + y^2 > 9$

$(2 - 2)^2 + (0)^2 \overset{?}{>} 9$

$0 \overset{?}{>} 9$ false

The test point inside the circle does not satisfy the original inequality. Therefore, the solution set consists of the points strictly outside the circle (Figure 5-23).

Step 2: Select a test point. We have chosen the center (2, 0).

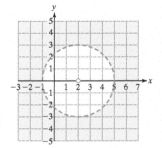

Figure 5-23

Skill Practice 4 Graph the solution set. $x^2 + (y + 1)^2 < 16$

---

3. Solve Systems of Inequalities in Two Variables

Two or more inequalities in two variables make up a system of inequalities in two variables. The solution set is the set of ordered pairs that satisfy each inequality in the system. To graph the solution set to a system of inequalities, graph the solution sets to the individual inequalities first. The solution to the system of inequalities is the *intersection* of the graphs. This is demonstrated in Example 5.

EXAMPLE 5 **Graphing the Solution Set to a System of Linear Inequalities**

Graph the solution set to the system of inequalities.
$$y \leq \tfrac{1}{2}x + 2$$
$$3x - y < 3$$

Solution:

$$y \leq \tfrac{1}{2}x + 2$$
$$3x - y < 3$$

First graph the solutions to the individual inequalities (Figures 5-24 and 5-25). Next, find the intersection (area of overlap) of the solution sets shown in purple in Figure 5-26.

Answer
4.

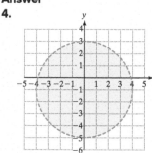

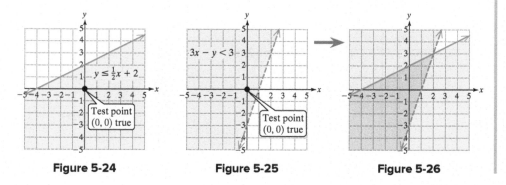

Figure 5-24　　　　**Figure 5-25**　　　　**Figure 5-26**

Notice that the point of intersection between the two bounding lines is graphed as an open dot (Figure 5-27). This indicates that it is *not* part of the solution set. The reason is that it is not a solution to the strict inequality $3x - y < 3$.

The point of intersection can be found by solving the system of related equations. We have used the substitution method.

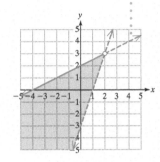

Figure 5-27

$$y = \tfrac{1}{2}x + 2$$
$$3x - y = 3 \qquad 3x - \left(\tfrac{1}{2}x + 2\right) = 3$$
$$3x - \tfrac{1}{2}x - 2 = 3$$
$$\tfrac{5}{2}x = 5$$
$$x = 2 \longrightarrow y = \tfrac{1}{2}(2) + 2$$
$$y = 3$$

The point of intersection is (2, 3) and is excluded from the solution set.

Skill Practice 5   Graph the solution set to the system of inequalities.

$$y < -\tfrac{1}{3}x + 1$$
$$-2x + y \le 1$$

TECHNOLOGY CONNECTIONS

Graphing a System of Inequalities in Two Variables

To graph the system of inequalities from Example 5 on a graphing calculator, solve each inequality for y.

$$y \le \tfrac{1}{2}x + 2$$
$$3x - y < 3 \rightarrow -y < -3x + 3 \rightarrow y > 3x - 3$$

Then enter the related equations in the graphing editor. Choose the appropriate graphing style ◥ or ◣. For $Y_1 \le \tfrac{1}{2}x + 2$, choose ◣. For $Y_2 > 3x - 3$, choose ◥.

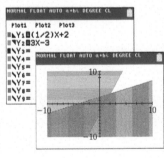

In Example 6, we graph a system of nonlinear inequalities in two variables. This is a system of inequalities in which one or more of the individual inequalities is nonlinear.

Answer

5.

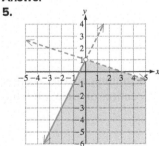

EXAMPLE 6 Solving a System of Nonlinear Inequalities

Graph the solution set to the system of inequalities.

$$y \leq -x^2 + 4$$
$$x - y \geq -2$$
$$y > -5$$

Solution:

To graph the solution set to the given system, first graph each individual inequality.

$y \leq -x^2 + 4$ $\qquad\qquad$ $x - y \geq -2$ $\qquad\qquad$ $y > -5$

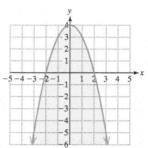 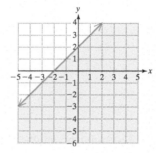 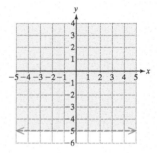

The solution set to the system is the intersection of the three shaded regions (Figure 5-28). The points of intersection are found by pairing up the related equations and solving the system of equations. The inequalities $y \leq -x^2 + 4$ and $x - y \geq -2$ include equality. Therefore, the intersection points between the parabola and slanted line are solutions to the system. These points are plotted as closed dots.

On the other hand, because the inequality $y > -5$ is strict, the intersection points between the parabola and horizontal line are *not* solutions to the system. These points are plotted as open dots.

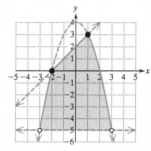

Figure 5-28

To find the points of intersection between the parabola $y = -x^2 + 4$ and the line $x - y = -2$, solve the system:

$$y = -x^2 + 4$$
$$x - y = -2 \qquad \text{The solutions are } (-2, 0) \text{ and } (1, 3).$$

To find the points of intersection between the parabola $y = -x^2 + 4$ and the line $y = -5$, solve the system:

$$y = -x^2 + 4$$
$$y = -5 \qquad \text{The solutions are } (-3, -5) \text{ and } (3, -5).$$

Skill Practice 6 Graph the solution set to the system of inequalities.

$$x^2 + y^2 \leq 25$$
$$-x + y < 1$$
$$y \geq -4$$

Answer

6.

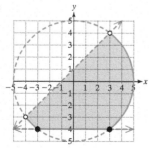

In Example 7, we refer to the problem addressed at the beginning of this section and set up a system of linear inequalities to model the allocation of time studying algebra and biology.

EXAMPLE 7   **Solving a System of Inequalities in an Application**

Adriana has 12 available study hours for algebra and biology.

- Let x represent the number of hours she spends studying algebra.
- Let y represent the number of hours that she studies biology.

a. Set up an inequality that indicates that the number of hours spent studying algebra cannot be negative.

b. Set up an inequality that indicates that the number of hours spent studying biology cannot be negative.

c. Set up an inequality that indicates that the combined number of hours she spends studying for these two classes is at most 12 hr.

d. Graph the solution set to the system of inequalities from parts (a)–(c).

Solution:

a. $x \geq 0$ The number of hours spent studying algebra is 0 or more.

b. $y \geq 0$ The number of hours spent studying biology is 0 or more.

c. $x + y \leq 12$ The sum of time spent studying algebra and the time spent studying biology cannot exceed the maximum number of study hours available. Therefore, the sum is less than or equal to 12 hr.

> **TIP** The points of intersection of the bounding lines are closed dots because they are part of the solution set.

d. $x \geq 0$
$\quad y \geq 0$
$\quad x + y \leq 12$

The inequalities $x \geq 0$ and $y \geq 0$ together represent the set of points in the first quadrant, including the bounding axes.

To graph the inequality $x + y \leq 12$, we graph the related equation $x + y = 12$. A test point $(0, 0)$ taken below the line results in a true statement. Therefore, the inequality $x + y \leq 12$ represents the set of points on and below the line $x + y = 12$.

The solution to the system of inequalities is shown in Figure 5-29.

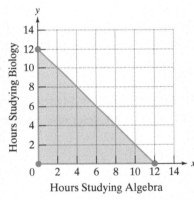

Figure 5-29

Skill Practice 7   A family plans to spend two 8-hr days at Disney World and will split time between the Magic Kingdom and Epcot Center. Let x represent the number of hours spent at the Magic Kingdom and let y represent the number of hours spent at Epcot Center.

a. Set up two inequalities that indicate that the number of hours spent at the Magic Kingdom and the number of hours spent at Epcot cannot be negative.

b. Set up an inequality that indicates that the combined number of hours spent at the two parks is at most 16 hr.

c. Graph the solution set to the system of inequalities.

Answers

7. a. $x \geq 0$; $y \geq 0$
b. $x + y \leq 16$
c.

SECTION 5.5   Practice Exercises

Prerequisite Review

R.1. Use the graph to solve the equation and inequalities. Write the solutions to the inequalities in interval notation.

   a. $x - 1 = 2x - 3$

   b. $x - 1 < 2x - 3$

   c. $x - 1 \geq 2x - 3$

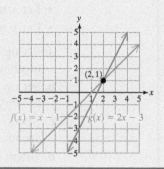

Concept Connections

1. An inequality that can be written in the form $Ax + By < C$ (where A and B are not both zero) is called a _____ inequality in two variables.

2. For a constant real number k, the inequality $x < k$ represents the half-plane to the (left/right) of the (horizontal/vertical) line $x = k$.

3. For a constant real number k, the inequality $y > k$ represents the half-plane (above/below) the (horizontal/vertical) line $y = k$.

4. Given the inequality $y \leq 2x + 1$, the bounding line $y = 2x + 1$ is drawn as a (dashed/solid) line.

5. The solution set to the system of inequalities $x < 0$, $y > 0$ represents the points in quadrant (I, II, III, IV).

6. The equation $x^2 + y^2 = 4$ is a circle centered at _____ with radius _____. The solution set to the inequality $x^2 + y^2 < 4$ represents the set of points (inside/outside) the circle $x^2 + y^2 = 4$.

Objective 1: Solve Linear Inequalities in Two Variables

For Exercises 7–10, determine whether the ordered pair is a solution to the inequality.

7. $3x + 4y < 12$

   a. $(-1, 3)$

   b. $(5, 1)$

   c. $(4, 0)$

8. $2x + 3y > 6$

   a. $(-3, 3)$

   b. $(5, -1)$

   c. $(0, 2)$

9. $y \geq (x - 3)^2$

   a. $(-3, 30)$

   b. $(1, 4)$

   c. $(5, 5)$

10. $y \leq x^3 - 1$

   a. $(-1, -2)$

   b. $(2, 6)$

   c. $(-4, -50)$

11. a. Graph the solution set. $4x - 5y \leq 20$ (**See Example 1**)

   b. Explain how the graph would differ for the inequality $4x - 5y < 20$.

   c. Explain how the graph would differ for the inequality $4x - 5y > 20$.

12. a. Graph the solution set. $2x + 5y > 10$

   b. Explain how the graph would differ for the inequality $2x + 5y \geq 10$.

   c. Explain how the graph would differ for the inequality $2x + 5y < 10$.

For Exercises 13–24, graph the solution set. (See Examples 1–3)

13. $2x + 5y > 5$

14. $-5x + 4y \leq 8$

15. $-30x \geq 20y + 600$

16. $-400x < 100y + 8000$

17. $5x \leq 6y$

18. $3x > 2y$

19. $3 + 2(x + y) > y + 3$

20. $-4 - 3(x - y) < 2y - 4$

21. $x < 6$

22. $y \leq 5$

23. $-\dfrac{1}{2}y + 4 \leq 5$

24. $-\dfrac{1}{3}x + 2 < 4$

Objective 2: Solve Nonlinear Inequalities in Two Variables

25. a. Graph the solution set. $x^2 + y^2 < 4$ (**See Example 4**)

   b. Explain how the graph would differ for the inequality $x^2 + y^2 > 4$.

   c. Explain how the graph would differ for the inequality $x^2 + y^2 \geq 4$.

26. a. Graph the solution set. $y \geq x^2 - 1$

   b. Explain how the graph would differ for the inequality $y \leq x^2 - 1$.

   c. Explain how the graph would differ for the inequality $y > x^2 - 1$.

For Exercises 27–36, graph the solution set. (See Example 4)

27. $y < -x^2$

28. $x^2 + y^2 \geq 16$

29. $y \leq (x - 2)^2 + 1$

30. $y \geq -(x + 1)^2 - 2$

31. $|x| \leq 3$

32. $|y| \leq 2$

33. $2|y| > 2$

34. $|x| + 1 > 3$

35. $y \geq \sqrt{x}$

36. $y < \sqrt{x - 1}$

Objective 3: Solve Systems of Inequalities in Two Variables

37. a. Is the point $(2, 1)$ a solution to the inequality $y < 2x + 3$?

  b. Is the point $(2, 1)$ a solution to the inequality $x + y \leq 1$?

  c. Is the point $(2, 1)$ a solution to the system of inequalities?

$$y < 2x + 3$$
$$x + y \leq 1$$

38. a. Is the point $(3, 2)$ a solution to the inequality $y < -x + 5$?

  b. Is the point $(3, 2)$ a solution to the inequality $3x + y \geq 11$?

  c. Is the point $(3, 2)$ a solution to the system of inequalities?

$$y < -x + 5$$
$$3x + y \geq 11$$

For Exercises 39–40, determine whether the ordered pair is a solution to the system of inequalities.

39. $x + y < 4$
  $y \leq 2x + 1$
  a. $(0, 1)$ **b.** $(3, 1)$ **c.** $(2, 0)$ **d.** $(1, 4)$

40. $y < -x^2 + 3$
  $x + 2y \leq 2$
  a. $(-2, -1)$ **b.** $(0, -2)$ **c.** $(0, 1)$ **d.** $(3, -6)$

For Exercises 41–58, graph the solution set. If there is no solution, indicate that the solution set is the empty set. (See Examples 5–6)

41. $y < \dfrac{1}{2}x - 4$
  $y > -2x + 1$

42. $y \geq \dfrac{1}{3}x - 2$
  $y \leq x - 4$

43. $2x + 5y \leq 5$
  $-3x + 4y \geq 4$

44. $4x - 3y > 3$
  $x + 4y < -4$

45. $x^2 + y^2 \geq 9$
  $x^2 + y^2 \leq 16$

46. $x^2 + y^2 \geq 1$
  $x^2 + y^2 < 25$

47. $y \geq 3x + 3$
  $-3x + y < 1$

48. $y < 2x - 4$
  $-2x + y \geq 2$

49. $|x| < 3$
  $|y| < 3$

50. $|x| \geq 2$
  $|y| \geq 2$

51. $y \geq x^2 - 2$
  $y > x$
  $y \leq 4$

52. $y \leq -x^2 + 7$
  $y \leq -x + 5$
  $y > 1$

53. $x^2 + y^2 \leq 100$
  $y < \dfrac{4}{3}x$
  $x \leq 8$

54. $x^2 + y^2 < 100$
  $y \geq x$
  $y \geq 1$

55. $y < e^x$
  $y > 1$
  $x < 2$

56. $y \leq \dfrac{2}{x}$
  $y > 0$
  $y < x$

57. $(x + 2)^2 + (y - 3)^2 \leq 9$
  $x - y > 2$

58. $(x - 4)^2 + (y + 1)^2 < 25$
  $2x - y < -4$

Mixed Exercises

For Exercises 59–64, write an inequality to represent the statement.

59. x is at most 6.

60. y is no more than 7.

61. y is at least -2.

62. x is no less than $\frac{1}{2}$.

63. The sum of x and y does not exceed 18.

64. The difference of x and y is not less than 4.

65. Let x represent the number of hours that Trenton spends studying algebra, and let y represent the number of hours he spends studying history. For parts (a)–(e), write an inequality to represent the given statement. **(See Example 7)**

  a. Trenton has a total of at most 9 hr to study for both algebra and history combined.

  b. Trenton will spend at least 3 hr studying algebra.

  c. Trenton will spend no more than 4 hr studying history.

  d. The number of hours spent studying algebra cannot be negative.

  e. The number of hours spent studying history cannot be negative.

  f. Graph the solution set to the system of inequalities from parts (a)–(e).

66. Let x represent the number of country songs that Sierra puts on a playlist on her portable media player. Let y represent the number of rock songs that she puts on the playlist. For parts (a)–(e), write an inequality to represent the given statement.

   a. Sierra will put at least 6 country songs on the playlist.

   b. Sierra will put no more than 10 rock songs on the playlist.

   c. Sierra wants to limit the length of the playlist to at most 20 songs.

   d. The number of country songs cannot be negative.

   e. The number of rock songs cannot be negative.

   f. Graph the solution set to the system of inequalities from parts (a)–(e).

67. A couple has $60,000 to invest for retirement. They plan to put x dollars in stocks and y dollars in bonds. For parts (a)–(d), write an inequality to represent the given statement.

   a. The total amount invested is at most $60,000.

   b. The couple considers stocks a riskier investment, so they want to invest at least twice as much in bonds as in stocks.

   c. The amount invested in stocks cannot be negative.

   d. The amount invested in bonds cannot be negative.

   e. Graph the solution set to the system of inequalities from parts (a)–(d).

68. A college theater has a seating capacity of 2000. It reserves x tickets for students and y tickets for general admission. For parts (a)–(d) write an inequality to represent the given statement.

   a. The total number of seats available is at most 2000.

   b. The college wants to reserve at least 3 times as many student tickets as general admission tickets.

   c. The number of student tickets cannot be negative.

   d. The number of general admission tickets cannot be negative.

   e. Graph the solution set to the system of inequalities from parts (a)–(d).

69. Write a system of inequalities that represents the points in the first quadrant less than 3 units from the origin.

70. Write a system of inequalities that represents the points in the second quadrant more than 4 units from the origin.

71. Write a system of inequalities that represents the points inside the triangle with vertices $(-3, -4)$, $(3, 2)$, and $(-5, 4)$.

72. Write a system of inequalities that represents the points inside the triangle with vertices $(-4, -4)$, $(1, 1)$, and $(5, -1)$.

73. A weak earthquake occurred roughly 9 km south and 12 km west of the center of Hawthorne, Nevada. The quake could be felt 16 km away. Suppose that the origin of a map is placed at the center of Hawthorne with the positive x-axis pointing east and the positive y-axis pointing north.

   a. Find an inequality that describes the points on the map for which the earthquake could be felt.

   b. Could the earthquake be felt at the center of Hawthorne?

74. A coordinate system is placed at the center of a town with the positive x-axis pointing east, and the positive y-axis pointing north. A cell tower is located 4 mi west and 5 mi north of the origin.

   a. If the tower has a 8-mi range, write an inequality that represents the points on the map serviced by this tower.

   b. Can a resident 5 mi east of the center of town get a signal from this tower?

Write About It

75. Under what circumstances should a dashed line or curve be used when graphing the solution set to an inequality in two variables?

76. Explain how test points are used to determine the region of the plane that represents the solution to an inequality in two variables.

77. Explain how to find the solution set to a system of inequalities in two variables.

78. Describe the solution set to the system of inequalities. $x \geq 0, y \geq 0, x \leq 1, y \leq 1$

Expanding Your Skills

For Exercises 79–80, graph the solution set.

79. $|x| \geq |y|$

80. $|x| + |y| \leq 1$

Technology Connections

For Exercises 81–82, use a graphing utility to graph the solution set to the system of inequalities.

81. $y \geq 0.4e^x$
    $y \leq 0.25x^3 - 4x$

82. $y < \dfrac{4}{x^2 + 1}$
    $y > \dfrac{-2}{x^2 + 0.5}$

PROBLEM RECOGNITION EXERCISES

Equations and Inequalities in Two Variables

For Exercises 1–2, for parts (a) and (b), graph the equation. For part (c), solve the system of equations. For parts (d) and (e) graph the solution set to the system of inequalities. If there is no solution, indicate that the solution set is the empty set.

1. **a.** $y = -3x + 5$ **b.** $-2x + y = 0$ **c.** $y = -3x + 5$ **d.** $y > -3x + 5$ **e.** $y < -3x + 5$
$\qquad\qquad\qquad\qquad\qquad\qquad\qquad\qquad\qquad -2x + y = 0 \qquad\quad -2x + y < 0 \qquad\quad -2x + y > 0$

2. **a.** $y = 2x - 3$ **b.** $4x - 2y = -2$ **c.** $y = 2x - 3$ **d.** $y \geq 2x - 3$ **e.** $y \leq 2x - 3$
$\qquad\qquad\qquad\qquad\qquad\qquad\qquad\qquad\qquad 4x - 2y = -2 \qquad\quad 4x - 2y \geq -2 \qquad\quad 4x - 2y \leq -2$

For Exercises 3–4, for part (a), graph the equations in the system and determine the solution set. For parts (b) and (c), graph the solution set to the inequality.

3. **a.** $y = x^2$ **b.** $y \leq x^2$ **c.** $y \geq x^2$
$\qquad y = \frac{1}{2}x^2 \qquad\quad y \geq \frac{1}{2}x^2 \qquad\quad y \leq \frac{1}{2}x^2$

4. **a.** $x - y = 1$ **b.** $x - y \geq 1$ **c.** $x - y \leq 1$
$\qquad y = (x - 3)^2 \qquad\quad y \geq (x - 3)^2 \qquad\quad y \leq (x - 3)^2$

SECTION 5.6   Linear Programming

OBJECTIVES

1. Write an Objective Function
2. Solve a Linear Programming Application

TIP The notation $z = f(x, y)$ is read as "z is a function of x and y."

1. Write an Objective Function

When a company manufactures a product, the goal is to obtain maximum profit at minimum cost. However, the production process is often limited by certain constraints such as the amount of labor available, the capacity of machinery, and the amount of money available for the company to invest in the process. In this section, we will study a process called **linear programming** that enables us to maximize or minimize a function under specified constraints.

The function to be optimized (maximized or minimized) in a linear programming application is called the **objective function.** The objective function often has two independent variables. For example, given $z = f(x, y)$, the variable z is dependent on the two independent variables x and y.

EXAMPLE 1   Writing an Objective Function

Suppose that a college wants to rent several buses to transport students to a championship college football game. A large bus costs $1200 to rent and a small bus costs $800 to rent.

Let x represent the number of large buses.

Let y represent the number of small buses.

Write an objective function that represents the total cost z (in $) to rent x large buses and y small buses.

TIP An objective function represents a quantity that is to be maximized or minimized. In Example 1, the school will want to minimize the cost to transport the students.

Solution:

$$\begin{pmatrix} \text{Total} \\ \text{cost} \end{pmatrix} = \begin{pmatrix} \text{Cost to rent} \\ x \text{ large buses} \end{pmatrix} + \begin{pmatrix} \text{Cost to rent} \\ y \text{ small buses} \end{pmatrix}$$

Cost: $z = 1200x + 800y$

The objective function is defined by $z = 1200x + 800y$.

Skill Practice 1 An office manager needs to staff the office. She hires full-time employees at \$36 per hour and part-time employees at \$24 per hour. Write an objective function that represents the total cost (in \$) to staff the office with x full-time employees and y part-time employees for 1 hr.

2. Solve a Linear Programming Application

In Example 2, we identify constraints imposed on the resources that affect the number of buses that the college can rent.

EXAMPLE 2 **Writing a System of Constraints**

Refer to the scenario from Example 1. We now set several constraints that affect the number of buses that can be rented.

Let x represent the number of large buses.
Let y represent the number of small buses.

For parts (a)–(d), write an inequality that represents the given statement.

a. The number of large buses cannot be negative.
b. The number of small buses cannot be negative.
c. Large buses can carry 60 people and small buses can carry 45 people. The college must transport at least 3600 students.
d. The number of available bus drivers is at most 75.

Solution:

a. $x \geq 0$ The number of large buses cannot be negative.
b. $y \geq 0$ The number of small buses cannot be negative.

c. $\begin{pmatrix} \text{Number of students} \\ \text{carried by large buses} \end{pmatrix} + \begin{pmatrix} \text{Number of students} \\ \text{carried by small buses} \end{pmatrix} \underset{\text{is at least}}{\geq} 3600$ Large buses hold 60 people and

$ 60x + 45y \geq 3600$ small buses hold 45 people.

d. $\begin{pmatrix} \text{Number of} \\ \text{large buses} \end{pmatrix} + \begin{pmatrix} \text{Number of} \\ \text{small buses} \end{pmatrix} \underset{\text{is at most}}{\leq} 75$ Because each bus has only one driver, the

$ x + y \leq 75$ total number of buses is the same as the total number of drivers.

Skill Practice 2 Refer to Skill Practice 1. Suppose that the office manager needs at least 20 employees, but not more than 24 full-time employees. Furthermore, to make the office run smoothly, the manager knows that the number of full-time employees must always be greater than or equal to the number of part-time employees. Write a system of inequalities that represents the constraints on the number of full-time employees x and the number of part-time employees y.

Answers
1. $z = 36x + 24y$
2. $x \geq 0, y \geq 0, x \leq 24,$
   $x + y \geq 20, x \geq y$

The constraints in Example 2 make up a system of linear inequalities.

The number of large buses is nonnegative.	$x \geq 0$
The number of small buses is nonnegative.	$y \geq 0$
The school must transport at least 3600 students.	$60x + 45y \geq 3600$
The number of drivers (and therefore buses) is at most 75.	$x + y \leq 75$

The region in the plane that represents the solution set to the system of constraints is called the **feasible region** (Figure 5-30). The points of intersection of the bounding lines in the feasible region are called the **vertices** of the feasible region.

The vertices in Figure 5-30 are (60, 0), (75, 0), and (15, 60).

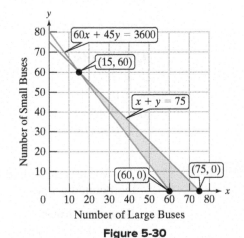

Figure 5-30

Points within the shaded region meet all the constraints in the problem. For example, the ordered pair (65, 5) represents 65 large buses and 5 small buses.

65 large buses and 5 small buses ⟶ 70 buses (≤ 75 drivers) ✓
65 large buses and 5 small buses ⟶ transports 4125 students (≥ 3600) ✓

However, points such as (65, 15) and (40, 20) are *outside* the feasible region and do not satisfy all constraints.

65 large buses and 15 small buses ⟶ 80 buses (exceeds the number of drivers)
40 large buses and 20 small buses ⟶ transports only 3300 students

The goal of a linear programming application is to find the maximum or minimum value of the objective function $z = f(x, y)$ when x and y are restricted to the ordered pairs in the feasible region. Fortunately, it has been proven mathematically that if a maximum or minimum value of a function exists, it occurs at one or more of the vertices of the feasible region. This is the basis for the following procedure to solve a linear programming application.

Solving an Application Involving Linear Programming

Step 1 Write an objective function $z = f(x, y)$.
Step 2 Write a system of inequalities defining the constraints on x and y.
Step 3 Graph the feasible region and identify the vertices.
Step 4 Evaluate the objective function at each vertex of the feasible region. Use the results to identify the values of x and y that optimize the objective function. Identify the optimal value of z.

In Example 3, we optimize the objective function found in Example 1 subject to the constraints defined in Example 2.

> **EXAMPLE 3** Solving a Linear Programming Application
>
> A college wants to rent buses to transport at least 3600 students to a championship college football game. Large buses hold 60 people and small buses hold 45 people. Furthermore, the number of available bus drivers is at most 75. Each large bus costs $1200 to rent and each small bus costs $800 to rent. Find the optimal number of large and small buses that will minimize cost.
>
> **Solution:**
>
> Let x represent the number of large buses. Define the relevant variables.
> Let y represent the number of small buses.
>
> Cost: $z = 1200x + 800y$ **Step 1:** Write an objective function. Since cost is to be minimized, write a cost function. (See Example 1.)
>
> **Constraints:** **Step 2:** Write a system of constraints on the relevant variables. (See Example 2.)
>
> $x \geq 0$
> $y \geq 0$
> $60x + 45y \geq 3600$
> $x + y \leq 75$
>
> **Step 3:** Graph the feasible region and identify the vertices.
>
> The vertices are found by identifying the points of intersection between the bounding lines.
>
>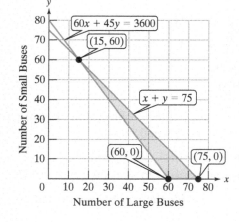
>
> **Bounding lines:**
>
> $60x + 45y = 3600$
> $\quad x + \quad y = 75$ Point of intersection (15, 60)
>
> $60x + 45y = 3600$
> $\qquad\quad y = 0$ Point of intersection (60, 0)
>
> $x + y = 75$
> $\quad\ \ y = 0$ Point of intersection (75, 0)
>
> Cost function: $z = 1200x + 800y$
>
> at (15, 60) $z = 1200(15) + 800(60) = \boxed{\$66{,}000}$ **Step 4:** Evaluate the objective function $z = 1200x + 800y$ at each vertex.
> at (60, 0) $z = 1200(60) + 800(0)\ = \$72{,}000$
> at (75, 0) $z = 1200(75) + 800(0)\ = \$90{,}000$
>
> The cost would be minimized if the college rents 15 large buses and 60 small buses. The minimum cost is $66,000.

> **Skill Practice 3** Refer to Skill Practices 1 and 2. The office manager needs at least 20 employees, but no more than 24 full-time employees. Furthermore, to make the office run smoothly, the manager knows that the number of full-time employees must always be greater than or equal to the number of part-time employees. If she pays full-time employees $36 per hour and part-time employees $24 per hour, determine the number of full-time and part-time employees she should hire to minimize total labor cost per hour.

Answer

3. The cost would be minimized at $600/hr if she hires 10 full-time employees and 10 part-time employees.

In Example 4, we investigate a situation in which we maximize profit.

EXAMPLE 4   Solving a Linear Programming Application

A baker produces whole wheat bread and cheese bread to sell at the farmer's market. The whole wheat bread is denser and requires more baking time, whereas the cheese bread requires more labor. The baking times and average amount of labor per loaf are given in the table along with the profit for each loaf.

	Time to Bake	Labor	Profit
Wheat bread	1.5 hr	$\frac{1}{3}$ hr	$1.20
Cheese bread	1 hr	$\frac{1}{2}$ hr	$1.00

The oven space restricts the baker from baking more than 120 loaves. Furthermore, the amount of oven time for baking is no more than 165 hr and the amount of available labor is at most 55 hr. Determine the number of loaves of each type of bread that the baker should bake to maximize his profit. Assume that all loaves of bread produced are sold.

Solution:

Let x represent the number of loaves of wheat bread.

Let y represent the number of loaves of cheese bread.

$$\text{Profit} = \left(\begin{matrix}\text{Profit from}\\\text{wheat bread}\end{matrix}\right) + \left(\begin{matrix}\text{Profit from}\\\text{cheese bread}\end{matrix}\right)$$

Step 1: Write an objective function. In this example, we need to maximize profit.

$z = 1.20x + 1.00y$

Step 2: Write a system of constraints on the independent variables.

$x \geq 0$ Number of loaves of wheat bread cannot be negative.

$y \geq 0$ Number of loaves of cheese bread cannot be negative.

$x + y \leq 120$ Total number of loaves is no more than 120.

$1.5x + y \leq 165$ The total amount of baking time is no more than 165 hr.

$\frac{1}{3}x + \frac{1}{2}y \leq 55$ The total amount of available labor is at most 55 hr.

Step 3: Graph the feasible region and identify the vertices. Find the points of intersection between pairs of bounding lines:

$x + y = 120$
$\frac{1}{3}x + \frac{1}{2}y = 55$ Point of intersection $(30, 90)$

$x + y = 120$
$1.5x + y = 165$ Point of intersection $(90, 30)$

$1.5x + y = 165$
$y = 0$ Point of intersection $(110, 0)$

$\frac{1}{3}x + \frac{1}{2}y = 55$
$x = 0$ Point of intersection $(0, 110)$

$x = 0$
$y = 0$ Point of intersection $(0, 0)$

Point of Interest

Linear programming was first introduced by Russian mathematician Leonid Kantorovich. The technique was used in World War II to minimize costs to the army. Later in 1975, Kantorovich won the Nobel Prize in economics for his contributions to the theory of optimum allocation of resources.

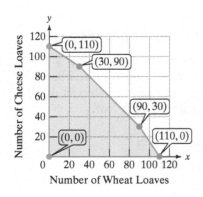

Step 4: Evaluate the objective function at each vertex.

Profit: $z = 1.2x + y$

at $(0, 0)$ $z = 1.2(0) + (0) = \$0$

at $(0, 110)$ $z = 1.2(0) + (110) = \$110$

at $(30, 90)$ $z = 1.2(30) + (90) = \$126$

at $(90, 30)$ $z = 1.2(90) + (30) = \boxed{\$138}$

at $(110, 0)$ $z = 1.2(110) + (0) = \$132$

The profit will be maximized if the baker bakes 90 whole wheat loaves and 30 cheese loaves. The maximum profit is \$138.

Skill Practice 4   A manufacturer produces two sizes of leather handbags. It takes longer to cut and dye the leather for the smaller bag, but it takes more time sewing the larger bag. The production constraints and profit for each type of bag are given in the table.

	Cutting and Dying	Sewing	Profit
Large bag	0.6 hr	2 hr	\$30
Small bag	1 hr	1.5 hr	\$25

The machinery limits the number of bags produced to at most 1000 per week. If the company has 900 hr per week available for cutting and dying and 1800 hr available per week for sewing, determine the number of each type of bag that should be produced weekly to maximize profit. Assume that all bags produced are also sold.

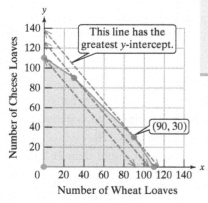

Figure 5-31

Answer

4. The maximum profit of \$28,000 is realized when the company produces 600 large bags and 400 small bags.

To find the maximum or minimum value of an objective function, we evaluate the function at the vertices of the feasible region. It seems reasonable that the profit would be maximized at a point on the upper edge of the feasible region. These are the points in the feasible region where the combined values of x and y are the greatest.

The goal of Example 4 was to find the values of x and y that maximized the profit function $z = 1.2x + y$. To see why the profit z was maximized at a vertex of the feasible region, write the equation in slope-intercept form:

$$y = -1.2x + z$$

In this form, the objective function represents a family of parallel lines with slope -1.2 and y-intercept $(0, z)$. To maximize z, we want the line with the greatest y-intercept that still remains in contact with the feasible region. In Figure 5-31, we see that this occurs for the line passing through the point $(90, 30)$ as expected.

SECTION 5.6   Practice Exercises

Prerequisite Review

R.1. A salesperson makes a base salary of \$500 per week plus 11% commission on sales.

   a. Write a linear function to model the salesperson's weekly salary $S(x)$ for x dollars in sales.

   b. Evaluate $S(8000)$ and interpret the meaning in the context of this problem.

R.2. A luxury car rental company charges \$123 per day in addition to a flat fee of \$75 for insurance.

   a. Write an equation that represents the cost y (in \$) to rent the car for x days.

   b. What is the y-intercept and what does it mean in the context of this problem?

Concept Connections

1. The process that maximizes or minimizes a function subject to linear constraints is called _____ programming.

2. The function to be optimized in a linear programming application is called the _____ function.

3. The region in the plane that represents the solution set to a system of constraints is called the _____ region.

4. The points of intersection of a feasible region are called the _____ of the region.

Objective 1: Write an Objective Function

5. A diner makes a profit of $0.80 for a cup of coffee and $1.10 for a cup of tea. Write an objective function $z = f(x, y)$ that represents the total profit for selling x cups of coffee and y cups of tea. (**See Example 1**)

6. Rita burns 10 calories per minute running and 8 calories per minute lifting weights. Write an objective function $z = f(x, y)$ that represents the total number of calories burned by running for x minutes and lifting weights for y minutes.

7. A courier company makes deliveries with two different trucks. Truck A costs $0.62/mi to operate and truck B costs $0.50/mi to operate. Write an objective function $z = f(x, y)$ that represents the total cost for driving truck A for x miles and driving truck B for y miles.

8. The cost for an animal shelter to spay a female cat is $82 and the cost to neuter a male cat is $55. Write an objective function $z = f(x, y)$ that represents the total cost for spaying x female cats and neutering y male cats.

Objective 2: Solve a Linear Programming Application

For Exercises 9–12,

a. Determine the values of x and y that produce the maximum or minimum value of the objective function on the given feasible region.

b. Determine the maximum or minimum value of the objective function on the given feasible region.

9. Maximize: $z = 3x + 2y$

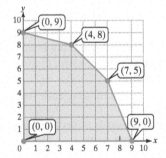

10. Maximize: $z = 1.8x + 2.2y$

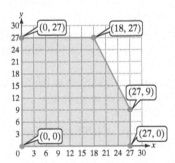

11. Minimize: $z = 1000x + 900y$

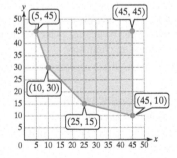

12. Minimize: $z = 6x + 9y$

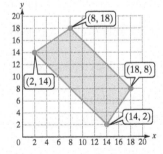

For Exercises 13–18,

a. For the given constraints, graph the feasible region and identify the vertices.

b. Determine the values of x and y that produce the maximum or minimum value of the objective function on the feasible region.

c. Determine the maximum or minimum value of the objective function on the feasible region.

13. $x \geq 0,\ y \geq 0$

$x + y \leq 60$

$y \leq 2x$

Maximize: $z = 250x + 150y$

14. $x \geq 0,\ y \geq 0$

$2x + y \leq 40$

$x + 2y \leq 50$

Maximize: $z = 9.2x + 8.1y$

15. $x \geq 0,\ y \geq 0$

$3x + y \geq 50$

$2x + y \geq 40$

Minimize: $z = 3x + 2y$

16. $x \geq 0,\ y \geq 0$

$4x + 3y \geq 60$

$2x + 3y \geq 36$

Minimize: $z = 4.5x + 6y$

17. $x \geq 0,\ y \geq 0$

$x \leq 36$

$y \leq 40$

$x + y \leq 48$

Maximize: $z = 150x + 90y$

18. $x \geq 0,\ y \geq 0$

$x \leq 10$

$y \leq 8$

$x + y \leq 12$

Maximize: $z = 50x + 70y$

For Exercises 19–20, use the given constraints to find the maximum value of the objective function and the ordered pair (x, y) that produces the maximum value.

19. $x \geq 0,\ y \geq 0$

$3x + 4y \leq 48$

$2x + y \leq 22$

$y \leq 9$

a. Maximize: $z = 100x + 120y$

b. Maximize: $z = 100x + 140y$

20. $x \geq 0,\ y \geq 0$

$x + y \leq 20$

$x + 2y \leq 36$

$x \leq 14$

a. Maximize: $z = 12x + 15y$

b. Maximize: $z = 15x + 12y$

21. A furniture manufacturer builds tables. The cost for materials and labor to build a kitchen table is $240 and the profit is $160. The cost to build a dining room table is $320 and the profit is $240. **(See Examples 2–3)**

Let x represent the number of kitchen tables produced per month. Let y represent the number of dining room tables produced per month.

a. Write an objective function representing the monthly profit for producing and selling x kitchen tables and y dining room tables.

b. The manufacturing process is subject to the following constraints. Write a system of inequalities representing the constraints.

- The number of each type of table cannot be negative.
- Due to labor and equipment restrictions, the company can build at most 120 kitchen tables.
- The company can build at most 90 dining room tables.
- The company does not want to exceed a monthly cost of $48,000.

c. Graph the system of inequalities represented by the constraints.

d. Find the vertices of the feasible region.

e. Test the objective function at each vertex.

f. How many kitchen tables and how many dining room tables should be produced to maximize profit? (Assume that all tables produced will be sold.)

g. What is the maximum profit?

22. Guyton makes $24/hr tutoring chemistry and $20/hr tutoring math.

Let x represent the number of hours per week he spends tutoring chemistry. Let y represent the number of hours per week he spends tutoring math.

a. Write an objective function representing his weekly income for tutoring x hours of chemistry and y hours of math.

b. The time that Guyton devotes to tutoring is limited by the following constraints. Write a system of inequalities representing the constraints.

- The number of hours spent tutoring each subject cannot be negative.
- Due to the academic demands of his own classes he tutors at most 18 hr per week.
- The tutoring center requires that he tutors math at least 4 hr per week.
- The demand for math tutors is greater than the demand for chemistry tutors. Therefore, the number of hours he spends tutoring math must be at least twice the number of hours he spends tutoring chemistry.

c. Graph the system of inequalities represented by the constraints.

d. Find the vertices of the feasible region.

e. Test the objective function at each vertex.

f. How many hours tutoring math and how many hours tutoring chemistry should Guyton work to maximize his income?

g. What is the maximum income?

h. Explain why Guyton's maximum income is found at a point on the line $x + y = 18$.

23. A plant nursery sells two sizes of oak trees to landscapers. Large trees cost the nursery $120 from the grower. Small trees cost the nursery $80. The profit for each large tree sold is $35 and the profit for each small tree sold is $30. The monthly demand is at most 400 oak trees. Furthermore, the nursery does not want to allocate more than $43,200 each month on inventory for oak trees.

   a. Determine the number of large oak trees and the number of small oak trees that the nursery should have in its inventory each month to maximize profit. (Assume that all trees in inventory are sold.)

   b. What is the maximum profit?

   c. If the profit on large trees were $50, and the profit on small trees remained the same, then how many of each should the nursery have to maximize profit?

24. A sporting goods store sells two types of exercise bikes. The deluxe model costs the store $400 from the manufacturer and the standard model costs the store $300 from the manufacturer. The profit that the store makes on the deluxe model is $180 and the profit on the standard model is $120. The monthly demand for exercise bikes is at most 30. Furthermore, the store manager does not want to spend more than $9600 on inventory for exercise bikes.

   a. Determine the number of deluxe models and the number of standard models that the store should have in its inventory each month to maximize profit. (Assume that all exercise bikes in inventory are sold.)

   b. What is the maximum profit?

   c. If the profit on the deluxe bikes were $150 and the profit on the standard bikes remained the same, how many of each should the store have to maximize profit?

25. A paving company delivers gravel for a road construction project. The company has a large truck and a small truck. The large truck has a greater capacity, but costs more for fuel to operate. The load capacity and cost to operate each truck per load are given in the table.

	Load Capacity	Cost per Load
Small truck	18 yd^3	$120
Large truck	24 yd^3	$150

   The company must deliver at least 288 yd^3 of gravel to stay on schedule. Furthermore, the large truck takes longer to load and cannot make as many trips as the small truck. As a result, the number of trips made by the large truck is at most $\frac{3}{4}$ times the number of trips made by the small truck.

   a. Determine the number of trips that should be made by the large truck and the number of trips that should be made by the small truck to minimize cost.

   b. What is the minimum cost to deliver gravel under these constraints?

26. A large department store needs at least 3600 labor hours covered per week. It employs full-time staff 40 hr/wk and part-time staff 25 hr/wk. The cost to employ a full-time staff member is more because the company pays benefits such as health care and life insurance.

	Hours per Week	Cost per Hour
Full time	40 hr	$25
Part time	25 hr	$18

   The store manager also knows that to make the store run efficiently, the number of full-time employees must be at least 1.25 times the number of part-time employees.

   a. Determine the number of full-time employees and the number of part-time employees that should be used to minimize the weekly labor cost.

   b. What is the minimum weekly cost to staff the store under these constraints?

27. A manufacturer produces two models of a gas grill. Grill A requires 1 hr for assembly and 0.4 hr for packaging. Grill B requires 1.2 hr for assembly and 0.6 hr for packaging. The production information and profit for each grill are given in the table. **(See Example 4)**

	Assembly	Packaging	Profit
Grill A	1 hr	0.4 hr	$90
Grill B	1.2 hr	0.6 hr	$120

   The manufacturer has 1200 hr of labor available for assembly and 540 hr of labor available for packaging.

   a. Determine the number of grill A units and the number of grill B units that should be produced to maximize profit assuming that all grills will be sold.

   b. What is the maximum profit under these constraints?

   c. If the profit on grill A units is $110 and the profit on grill B units is unchanged, how many of each type of grill unit should the manufacturer produce to maximize profit?

28. A manufacturer produces two models of patio furniture. Model A requires 2 hr for assembly and 1.2 hr for painting. Model B requires 3 hr for assembly and 1.5 hr for painting. The production information and profit for selling each model are given in the table.

	Assembly	Painting	Profit
Model A	2 hr	1.2 hr	$150
Model B	3 hr	1.5 hr	$200

   The manufacturer has 1200 hr of labor available for assembly and 660 hr of labor available for painting.

   a. Determine the number of model A units and the number of model B units that should be produced to maximize profit assuming that all furniture will be sold.

   b. What is the maximum profit under these constraints?

   c. If the profit on model A units is $180 and the profit on model B units remains the same, how many of each type should the manufacturer produce to maximize profit?

29. A farmer has 1200 acres of land and plans to plant corn and soybeans. The input cost (cost of seed, fertilizer, herbicide, and insecticide) for 1 acre for each crop is given in the table along with the cost of machinery and labor. The profit for 1 acre of each crop is given in the last column.

	Input Cost per Acre	Labor/Machinery Cost per Acre	Profit per Acre
Corn	$180	$80	$120
Soybeans	$120	$100	$100

Suppose the farmer has budgeted a maximum of $198,000 for input costs and a maximum of $110,000 for labor and machinery.

a. Determine the number of acres of each crop that the farmer should plant to maximize profit. (Assume that all crops will be sold.)

b. What is the maximum profit?

c. If the profit per acre were reversed between the two crops (that is, $100 per acre for corn and $120 per acre for soybeans), how many acres of each crop should be planted to maximize profit?

30. To protect soil from erosion, some farmers plant winter cover crops such as winter wheat and rye. In addition to conserving soil, cover crops often increase crop yields in the row crops that follow in spring and summer. Suppose that a farmer has 800 acres of land and plans to plant winter wheat and rye. The input cost for 1 acre for each crop is given in the table along with the cost for machinery and labor. The profit for 1 acre of each crop is given in the last column.

	Input Cost per Acre	Labor/Machinery Cost per Acre	Profit per Acre
Wheat	$90	$50	$42
Rye	$120	$40	$35

Suppose the farmer has budgeted a maximum of $90,000 for input costs and a maximum of $36,000 for labor and machinery.

a. Determine the number of acres of each crop that the farmer should plant to maximize profit. (Assume that all crops will be sold.)

b. What is the maximum profit?

c. If the profit per acre for wheat were $40 and the profit per acre for rye were $45, how many acres of each crop should be planted to maximize profit?

Write About It

31. What is the purpose of linear programming?

33. How is the feasible region determined?

32. What is an objective function?

34. If an optimal value exists for an objective function, it exists at one of the vertices of the feasible region. Explain how to find the vertices.

CHAPTER 5   KEY CONCEPTS

SECTION 5.1 Systems of Linear Equations in Two Variables and Applications	Reference
Two or more linear equations taken together form a **system of linear equations.** A **solution** to a system of equations in two variables is an ordered pair that is a solution to each individual equation. Graphically, this is a point of intersection of the graphs of the equations.	p. 492
The substitution method and the addition method are often used to solve a system of linear equations in two variables.	pp. 493, 494
A system of linear equations in two variables will have no solution if the equations in the system represent parallel lines. In such a case, we say that the system is **inconsistent.**	p. 493
A system of linear equations in two variables will have infinitely many solutions if the equations represent the same line. In such a case, we say that the equations are **dependent.**	p. 493

SECTION 5.2 Systems of Linear Equations in Three Variables and Applications	Reference
A **linear equation in three variables** is an equation that can be written in the form $$Ax + By + Cz = D$$ where A, B, and C are not all zero.	p. 506
A solution to a system of linear equations in three variables is an **ordered triple** (x, y, z) that satisfies each equation in the system. Geometrically, a solution is a point of intersection of the planes represented by the equations in the system.	p. 506

SECTION 5.3 Partial Fraction Decomposition	Reference
Partial fraction decomposition is used to write a rational expression as a sum of simpler fractions.	p. 517
There are two basic parts to find the partial fraction decomposition of a rational expression.	p. 519
I. Factor the denominator of the expression into linear factors and quadratic factors that are not further factorable over the integers. Then set up the "form" or "structure" for the partial fraction decomposition into simpler fractions.	
II. Next, multiply both sides of the equation by the LCD. Then set up a system of linear equations to find the coefficients of the terms in the numerator of each fraction.	
Note: The numerator of the original rational expression must be of lesser degree than the denominator. If this is not the case, first use long division.	

SECTION 5.4 Systems of Nonlinear Equations in Two Variables	Reference
A **nonlinear system of equations** is a system in which one or more equations is nonlinear.	p. 527
The substitution method is often used to solve a nonlinear system of equations.	p. 528
In some cases, the addition method can be used provided that the terms containing the corresponding variables are like terms.	p. 529

SECTION 5.5 Inequalities and Systems of Inequalities in Two Variables	Reference
An inequality of the form $Ax + By < C$, where A and B are not both zero, is called a **linear inequality in two variables.** (The symbols $>$, $\leq$, and $\geq$ can be used in place of $<$ in the definition.)	p. 536
The basic steps to solve a linear inequality in two variables are as follows.	p. 536
1. Graph the related equation. The resulting line is drawn as a dashed line if the inequality is strict, and is otherwise drawn as a solid line.	
2. Select a test point from either side of the line. If the ordered pair makes the original inequality true, then shade the half-plane from which the point was taken. Otherwise, shade the other half-plane.	
A nonlinear inequality in two variables is solved using the same basic procedure.	p. 539
Two or more inequalities in two variables make up a system of inequalities in two variables. The solution set to the system is the region of overlap (intersection) of the solution sets of the individual inequalities.	p. 540

SECTION 5.6 Linear Programming	Reference
A process called **linear programming** enables us to maximize or minimize a function under specified constraints. The function to be maximized or minimized is called the **objective function.**	p. 547
The steps to solve a linear programming application are outlined here.	p. 549
Step 1 Write an objective function, $z = f(x, y)$.	
Step 2 Write a system of inequalities defining the constraints on x and y.	
Step 3 Graph the feasible region and identify the vertices.	
Step 4 Evaluate the objective function at each vertex of the feasible region. Use the results to identify the values of x and y that optimize the objective function and identify the optimal value of z.	

Expanded Chapter Summary available at http://www.mhhe.com/millercollegealgebra

CHAPTER 5   Review Exercises

SECTION 5.1

1. Determine if the ordered pair is a solution to the system.
$$2x - 3y = 0$$
$$-5x + 6y = -1$$

a. $\left(1, \dfrac{2}{3}\right)$ **b.** $(6, 4)$

For Exercises 2–3, based on the slope-intercept form of the equations, determine the number of solutions to the system.

2. $y = -\dfrac{3}{5}x - 4$ **3.** $y = 2x + 6$

$y = -\dfrac{3}{5}x + 1$ $y = \dfrac{1}{2}x - 6$

For Exercises 4–8, solve the system by using any method. If the system does not have one unique solution, state whether the system is inconsistent, or whether the equations are dependent.

4. $4x - y = 7$ **5.** $5(x - y) = 19 - 2y$
   $-2x + 5y = 19$ $0.2x + 0.7y = -1.7$

6. $9x - 2y = 4$ **7.** $\frac{1}{10}x - \frac{1}{2}y = 1$
   $2x + 4y = 7$ $2x = 10y + 6$

8. $y = \frac{3}{4}x$
   $4(y - x) = -x$

9. Shenika wants to monitor her daily calcium intake. One day she had 3 cups of milk and 1 cup of cooked spinach for a total of 1140 mg of calcium. The next day, she had 2 cups of milk and $1\frac{1}{2}$ cups of cooked spinach for a total of 960 mg of calcium. How much calcium is in 1 cup of milk and how much is in 1 cup of cooked spinach?

10. How many liters of a 40% acid mixture and how many liters of a 10% acid mixture should be mixed to obtain 20 L of a 22% acid mixture?

11. A plane can travel 960 mi in 2 hr with a tail wind. The return trip against the wind takes 2 hr and 40 min. Find the speed of the plane in still air and the speed of the wind.

12. A fishing boat captain charges $250 for an excursion. His fixed monthly expenses are $1200 for insurance, rent for the dock, and minor office expenses. He also has variable costs of  $100 per excursion to cover gasoline, bait, and other equipment.

   a. Write a linear cost function representing the cost $C(x)$ (in $) for the fishing boat captain to run x excursions per month.

b. Write a linear revenue function representing the revenue $R(x)$ (in $) for x excursions per month.

c. Determine the number of excursions per month for the captain to break even.

d. If 18 excursions are run in a given month, how much money will the fishing boat captain earn or lose?

SECTION 5.2

For Exercises 13–16, solve the system. If a system has one unique solution, write the solution set. Otherwise, determine the number of solutions to the system, and determine whether the system is inconsistent, or the equations are dependent.

13. $3a - 4b + 2c = -17$ **14.** $6x = 24 - 5y$
   $2a + 3b + c = 1$ $14 = 7z - 3y$
   $4a + b - 3c = 7$ $4x - 3z = 10$

15. $x + 2y + z = 5$ **16.** $u + v + 2w = 1$
   $x + y - z = 1$ $2v - 5w = 2$
   $4x + 7y + 2z = 16$ $3u + 5v + w = 1$

17. Solve the system and write the general solution.
   $5x + 2y + z = 0$
   $-4x + 3y - z = 0$
   $6x + 7y + z = 0$

18. An arena that hosts sporting events and concerts has three sections for three levels of seating. For a basketball game, seats in Section A cost $90, seats in Section B cost $65, and seats in Section C cost $40. The number of seats in Section C equals the number of seats in Sections A and B combined. The arena holds 12,000 seats and the game is sold out. If the total revenue from ticket sales is $655,000, determine the number of seats in each section.

19. Emily receives an inheritance of $20,000 and decides to invest the money. She puts some money in her savings account that earns 1.5% simple interest per year. The remaining money is invested in a bond fund that returns 4.5% and a stock fund that returns 6.2%. She makes a total of $942 at the end of 1 yr. If she invested twice as much in the bond fund as the stock fund, determine the amount that she invested in each fund.

For Exercises 20–21, use a system of linear equations in three variables to find an equation of the form $y = ax^2 + bx + c$ that defines the parabola through the points.

20. $(-1, -4), (1, 6), (3, 8)$ **21.** $(1, -2), (2, 1), (3, 10)$

SECTION 5.3

For Exercises 22–27, set up the form for the partial fraction decomposition. Do not solve for A, B, C, and so on.

22. $\dfrac{5x + 22}{x^2 + 8x + 16}$ **23.** $\dfrac{-x - 11}{(x + 2)(x - 1)}$

24. $\dfrac{2x^2 + x - 10}{x^3 + 5x}$

25. $\dfrac{7x^2 + 19x + 15}{2x^3 + 3x^2}$

26. $\dfrac{4x^4 - 3x^2 + 2x + 5}{x(2x + 5)^3(x^2 + 2)^2}$

27. $\dfrac{2x^3 - x^2 + 8x - 16}{x^4 + 5x^2 + 4}$

For Exercises 28–32, perform the partial fraction decomposition.

28. $\dfrac{-x - 11}{(x + 2)(x - 1)}$

29. $\dfrac{5x + 22}{x^2 + 8x + 16}$

30. $\dfrac{2x^4 + 7x^3 + 13x^2 + 19x + 15}{2x^3 + 3x^2}$

31. $\dfrac{2x^2 + x - 10}{x^3 + 5x}$

32. $\dfrac{2x^3 - x^2 + 8x - 16}{x^4 + 5x^2 + 4}$

SECTION 5.4

For Exercises 33–34,

a. Graph the equations.

b. Solve the system.

33. $y - x^2 = 1$
      $x - y = -3$

34. $y = \sqrt{x - 1}$
      $x^2 + y^2 = 5$

For Exercises 35–37, solve the system.

35. $3x^2 - y^2 = -4$
      $x^2 + 2y^2 = 36$

36. $2x^2 - xy = 24$
      $x^2 + 3xy = -9$

37. $y = \dfrac{8}{x}$
      $y = \sqrt{x}$

38. The sum of the squares of two negative numbers is 97 and the difference of their squares is 65. Find the numbers.

39. The ratio of two numbers is 4 to 3. The sum of the squares of the numbers is 100. Find the numbers.

40. The hypotenuse of a right triangle is $\sqrt{74}$ ft and the sum of the lengths of the legs is 12 ft. Find the lengths of the legs.

41. A rectangular billboard has a perimeter of 72 ft and an area of 288 ft^2. Find the dimensions of the billboard.

SECTION 5.5

42. Graph the solution set to the inequality.

   a. $3x + 4y \le 8$

   b. $3x + 4y > 8$

43. Graph the solution set to the inequality.

   a. $y < (x - 4)^2$

   b. $y \ge (x - 4)^2$

For Exercises 44–48, graph the solution set.

44. $5(x + y) \ge 8x + 15$

45. $x \le 3.5$

46. $-\dfrac{3}{2}y + 1 < 4$

47. $x^2 + (y + 2)^2 < 4$

48. $|y| > 2$

49. Determine if the given ordered pair is a solution to the system of inequalities.
$$x + 2y < 4$$
$$3x - 4y \ge 6$$

   a. $(0, 1)$ **b.** $(1, -4)$

For Exercises 50–53, graph the solution set. If there is no solution, indicate that the solution set is the empty set.

50. $y > \dfrac{1}{2}x + 1$
      $3x + 2y < 4$

51. $x^2 + y^2 \le 9$
      $(x - 1)^2 + y^2 \ge 4$

52. $y \ge x^2 - 3$
      $y > 1$
      $x + y \le 3$

53. $y > e^x$
      $y < -x^2 - 1$

54. Let x represent the number of hours that Gordon spends tutoring math, and let y represent the number of hours that he spends tutoring English. For parts (a)–(d), write an inequality to represent the given statement.

   a. Gordon has at most 12 hr to tutor per week.

   b. The amount of time that Gordon spends tutoring English is at least twice the amount of time he spends tutoring math.

   c. The number of hours spent tutoring math cannot be negative.

   d. The number of hours spent tutoring English cannot be negative.

   e. Graph the solution set to the system of inequalities from parts (a)–(d).

SECTION 5.6

55. At a home store, one sheet of $\frac{3}{8}$-in. sanded pine plywood costs $24. One sheet of $\frac{1}{4}$-in. sanded pine plywood costs $20. Write an objective function $z = f(x, y)$ that represents the total cost for x $\frac{3}{8}$-in. sheets and y $\frac{1}{4}$-in. sheets.

56. For the feasible region given in the figure and the objective function $z = 36x + 50y$,

   a. Determine the values of x and y that produce the maximum value of the objective function.

   b. Determine the maximum value of the objective function.

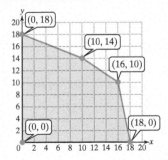

57. For the given constraints and the objective function, $z = 55x + 40y$,

  a. Graph the feasible region and identify the vertices.

$$x \geq 0, \; y \geq 0$$
$$2x + y \geq 18$$
$$5x + 4y \geq 60$$

  b. Determine the values of x and y that produce the minimum value of the objective function on the feasible region.

  c. Determine the minimum value of the objective function on the feasible region.

58. A fitness instructor wants to mix two brands of protein powder to form a blend that limits the amount of fat and carbohydrate but maximizes the amount of fiber.

The nutritional information is given in the table for a single scoop of protein powder.

	Fat	Carbohydrates	Fiber
Brand A	3 g	3 g	10 g
Brand B	2 g	4 g	8 g

Suppose that the fitness instructor wants to make at most 180 scoops of the mixture. She also wants to limit the amount of fat to 480 g and she wants to limit the amount of carbohydrate to 696 g.

  a. Determine the number of scoops of each type of powder that will maximum the amount of fiber.

  b. What is the maximum amount of fiber?

  c. If the fiber content were reversed between the two brands (that is, 8 g for brand A and 10 g for brand B), then how much of each type of protein powder should be used to maximize the amount of fiber?

CHAPTER 5 Test

For Exercises 1–3, determine if the ordered pair or ordered triple is a solution to the system.

1. $x - 5y = -3$
  $y = 2x - 12$
  a. $(7, 2)$
  b. $(-3, 0)$

2. $2x - 3y + z = -5$
  $5x + y - 3z = -18$
  $-x + 2y + 5z = 8$
  a. $(0, 1, -2)$
  b. $(-3, 0, 1)$

3. $2x - 4y < 9$
  $-3x + y \geq 4$
  a. $(-6, 1)$
  b. $(1, 4)$

For Exercises 4–14, solve the system. If the system does not have one unique solution, also state whether the system is inconsistent or whether the equations are dependent.

4. $x = 5 - 4y$
  $-3x + 7y = 4$

5. $0.2x = 0.35y - 2.5$
  $0.16x + 0.5y = 5.8$

6. $x - \dfrac{2}{5}y = \dfrac{3}{10}$
  $5x = 2y + \dfrac{3}{2}$

7. $7(x - y) = 3 - 5y$
  $4(3x - y) = -2x$

8. $a + 6b + 3c = -14$
  $2a + b - 2c = -8$
  $-3a + 2b + c = -8$

9. $x + 4z = 10$
  $3y - 2z = 9$
  $2x + 5y = 21$

10. $2x - y + z = -3$
  $x - 3y = 2$
  $x + 2y + z = -7$

11. $(x - 4)^2 + y^2 = 25$
  $x - y = 3$

12. $5x^2 + y^2 = 14$
  $x^2 - 2y^2 = -17$

13. $2xy - y^2 = -24$
  $-3xy + 2y^2 = 38$

14. $\dfrac{1}{x + 3} - \dfrac{2}{y - 1} = -7$
  $\dfrac{3}{x + 3} + \dfrac{1}{y - 1} = 7$

15. Solve the system and write the general solution.
$$x - 2z = 6$$
$$y + 3z = 2$$
$$x + y + z = 8$$

16. At a candy and nut shop, the manager wants to make a nut mixture that is 56% peanuts. How many pounds of peanuts must be added to an existing mixture of 45% peanuts to make 20 lb of a mixture that is 56% peanuts?

17. Two runners begin at the same point on a 400-m track. If they run in opposite directions they pass each other in 40 sec. If they run in the same direction, they will meet again in 200 sec. Find the speed of each runner.

18. Dylan invests $15,000 in three different stocks. One stock is very risky and after 1 yr loses 8%. The second stock returns 3.2%, and a third stock returns 5.8%. At the end of 1 yr, the total return is $274. If he invested $2000 more in the second stock than in the third stock, determine the amount he invested in each stock.

19. The difference of two positive numbers is 3 and the difference of their squares is 33. Find the numbers.

20. A rectangular television screen has a perimeter of 154 in. and an area of 1452 in.^{2}. Find the dimensions of the screen.

21. Use a system of linear equations in three variables to find an equation of the form $y = ax^2 + bx + c$ that defines the parabola through the points $(1, -1)$, $(2, 1)$, and $(-1, 7)$.

For Exercises 22–23, set up the form for the partial fraction decomposition. Do not solve for A, B, C, and so on.

22. $\dfrac{-15x + 15}{3x^2 + x - 2}$

23. $\dfrac{5x^6 + 3x^5 - 4x^3 + x - 3}{x^3(x - 3)(x^2 + 5x + 1)^2}$

For Exercises 24–28, perform the partial fraction decomposition.

24. $\dfrac{-12x - 29}{2x^2 + 11x + 15}$

25. $\dfrac{6x + 8}{x^2 + 4x + 4}$

26. $\dfrac{x^4 - 6x^3 + 4x^2 + 20x - 32}{x^3 - 4x^2}$

27. $\dfrac{x^2 - 2x - 21}{x^3 + 7x}$

28. $\dfrac{7x^3 + 4x^2 + 63x + 15}{x^4 + 11x^2 + 18}$

For Exercises 29–33, graph the solution set.

29. $2(x + y) > 6 - y$

30. $(x + 3)^2 + y^2 \geq 9$

31. $|x| < 4$

32. $\begin{aligned} x + y &\leq 4 \\ 2x - y &> -2 \end{aligned}$

33. $\begin{aligned} y &\leq -x^2 + 5 \\ y &> 1 \\ x + y &\leq 3 \end{aligned}$

34. A donut shop makes a profit of $2.40 on a dozen donuts and $0.55 per muffin. Write an objective function $z = f(x, y)$ that represents the total profit for selling x dozen donuts and y muffins.

35. For the feasible region given and the objective function $z = 4x + 5y$,

  a. Determine the values of x and y that produce the minimum value of the objective function on the feasible region.

  b. Determine the minimum value of the objective function on the feasible region.

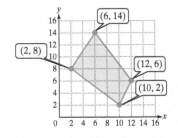

36. For the given constraints and objective function, $z = 600x + 850y$,

  a. Graph the feasible region and identify the vertices.
  $$x \geq 0,\ y \geq 0$$
  $$x + y \leq 48$$
  $$y \leq 3x$$

  b. Determine the values of x and y that produce the maximum value of the objective function on the feasible region.

  c. Determine the maximum value of the objective function on the feasible region.

37. A weight lifter wants to mix two types of protein powder. One is a whey protein and one is a soy protein. The fat, carbohydrate, and protein content (in grams) for 1 scoop of each powder is given in the table.

	Fat	Carbohydrates	Protein
Whey	3 g	3 g	20 g
Soy	2 g	4 g	18 g

Suppose that the weight lifter wants to make at most 60 scoops of a protein powder mixture. Furthermore, he wants to limit the total fat content to at most 150 g and the total carbohydrate content to at most 216 g.

  a. Determine the number of scoops of each type of powder that will maximize the total protein content under these constraints.

  b. What is the maximum total protein content?

  c. If the protein content were reversed between the two brands (that is, 18 g for the whey protein and 20 g for the soy protein), then how much of each type of protein powder should be used to maximize the amount of protein?

CHAPTER 5   Cumulative Review Exercises

For Exercises 1–5, solve the equation.

  1. $2x(x + 2) = 5x + 7$

  2. $\sqrt{2t + 8} - t = 4$

  3. $(x^2 - 4)^2 - 7(x^2 - 4) - 60 = 0$

  4. $\log_2(x - 4) = \log_2(x + 1) + 3$

  5. $50e^{2x+1} = 2000$

For Exercises 6–7, solve the inequality. Write the solution set in interval notation.

  6. $\dfrac{x + 4}{x - 2} \leq 1$

  7. $3|x + 2| - 1 > 8$

  8. Find the partial fraction decomposition. $\dfrac{-5x + 17}{x^2 - 6x + 9}$

9. Given $f(x) = 2x^2 - 3x$ and $g(x) = 5x + 1$,

   a. Find $(f \circ g)(x)$.

   b. Find $(g \circ f)(x)$.

10. Given $f(x) = \sqrt[3]{x - 2}$, write an equation for $f^{-1}(x)$.

11. Use a calculator to approximate the value of $\log_5 256$. Round to 4 decimal places.

12. Given $f(x) = -\dfrac{1}{2}x^3 + 4x^2 + 2$, find the average rate of change on the interval $[1, 3]$.

13. Write an equation of the line perpendicular to the line $x + 3y = 6$ and passing through the point $(2, -1)$.

14. Find all zeros of $f(x) = x^4 - 2x^3 + 10x^2 - 18x + 9$ and state the multiplicity of each zero.

For Exercises 15–16,

a. Write the domain in interval notation.

b. Write the range in interval notation.

  15. $f(x) = 2|x - 1| - 3$

  16. $f(x) = \ln(x - 3)$

For Exercises 17–19, solve the system.

  17. $3x = 5y + 1$

     $y = \dfrac{3}{5}x + 4$

18. $5a - 2b + 3c = 10$

     $-3a + b - 2c = -7$

      $a + 4b - 4c = -3$

19. $-2x^2 + 3y^2 = 10$

    $5x^2 + 2y^2 = 13$

20. Given $f(x) = -x^2 + 5x + 1$, find the difference quotient.

21. Shen invested $8000. After 5 yr with interest compounded continuously, the account is worth $10,907.40.

  a. Write a model of the form $A(t) = Pe^{rt}$, where $A(t)$ represents the amount (in $) in the account if P dollars in principal is invested at interest rate r for t years.

  b. How long will it take for the investment to double? Round to the nearest tenth of a year.

22. The variable y varies jointly as x and the square of z. If y is 36 when x is 10 and z is 3, find the value of y when $x = 12$ and z is 4.

Matrices and Determinants and Applications

Chapter Outline

For matters of security, e-mail messages, websites, and other electronic media often use encryption techniques to encode data into an obscure form. This prevents those who do not know the code from understanding what has been transmitted. A computer program can encrypt words by using a matrix (a two-dimensional array) to encode the message, and then use the inverse of that matrix to decode it.

In this chapter, we will use matrices in a variety of ways. Matrices can be used to model systems of linear equations. Furthermore, a variety of techniques involving the algebra and manipulation of matrices can be used to solve the systems. The first two techniques are called Gaussian elimination and Gauss-Jordan elimination. These methods reduce a system of equations to a simpler system of equations so that the solution can be found by inspection. Next, we use the inverse of a matrix to solve a related matrix equation. Finally, we present Cramer's rule. This method provides convenient formulas to solve a system of linear equations based on the coefficients of the system.

Solving Systems of Linear Equations Using Matrices

OBJECTIVES

1. Write an Augmented Matrix
2. Use Elementary Row Operations
3. Use Gaussian Elimination and Gauss-Jordan Elimination

1. Write an Augmented Matrix

The data in Table 6-1 represent the body mass index (BMI) for people of selected heights and weights. The National Institutes of Health suggests that a BMI should ideally be between 18.5 and 24.9. A table of data values such as Table 6-1 can be represented as a rectangular array of elements called a **matrix.**

Table 6-1   Body Mass Index

Height \ Weight	140 lb	160 lb	180 lb	200 lb
5'7"	21.9	25.1	28.2	31.3
5'10"	20.1	23.0	25.8	28.7
6'1"	18.5	21.1	23.7	26.4

$$\xrightarrow{\text{matrix}} \begin{bmatrix} 21.9 & 25.1 & 28.2 & 31.3 \\ 20.1 & 23.0 & 25.8 & 28.7 \\ 18.5 & 21.1 & 23.7 & 26.4 \end{bmatrix}$$

(*Source*: National Institutes of Health, www.nih.gov)

A matrix can be used to represent a system of linear equations written in standard form. To do so, we extract the coefficients of each term in the equation to form an **augmented matrix.** For example:

System of Equations

$$3x + 2y \quad = 5$$
$$x - y + 3z = 1$$
$$2x + y + z = 4$$

Augmented Matrix

$$\begin{bmatrix} 3 & 2 & 0 & | & 5 \\ 1 & -1 & 3 & | & 1 \\ 2 & 1 & 1 & | & 4 \end{bmatrix}$$

The vertical bar within the augmented matrix separates the coefficients of the variable terms in the equations from the constant terms.

EXAMPLE 1   Writing and Interpreting an Augmented Matrix

a. Write the augmented matrix for the system. $2x = 5y + 5$
$3(x + y) = 17 + y$

b. Write a system of linear equations represented by the augmented matrix. $\begin{bmatrix} 1 & 0 & 0 & | & 6 \\ 0 & 1 & 0 & | & -10 \\ 0 & 0 & 1 & | & 4 \end{bmatrix}$

Solution:

a. Write each equation in standard form.

$$2x = 5y + 5 \quad \longrightarrow \quad 2x - 5y = 5$$
$$3(x + y) = 17 + y \longrightarrow 3x + 2y = 17$$

$\left.\right\}$ augmented matrix $\longrightarrow \begin{bmatrix} 2 & -5 & | & 5 \\ 3 & 2 & | & 17 \end{bmatrix}$

b. $\begin{bmatrix} 1 & 0 & 0 & | & 6 \\ 0 & 1 & 0 & | & -10 \\ 0 & 0 & 1 & | & 4 \end{bmatrix}$ $\begin{array}{l} \longrightarrow x + 0y + 0z = 6 \\ \longrightarrow 0x + y + 0z = -10 \\ \longrightarrow 0x + 0y + z = 4 \end{array}$ or simply $\begin{array}{l} x = 6 \\ y = -10 \\ z = 4 \end{array}$

Skill Practice 1

a. Write the augmented matrix. $7x = 9 + 2y$
$2(x - y) = 4$

b. Write a system of linear equations represented by the augmented matrix. $\begin{bmatrix} 1 & 0 & | & -8 \\ 0 & 1 & | & 3 \end{bmatrix}$

Answers

1. **a.** $\begin{bmatrix} 7 & -2 & | & 9 \\ 2 & -2 & | & 4 \end{bmatrix}$

   b. $x = -8, y = 3$

2. Use Elementary Row Operations

We can use an augmented matrix to solve a system of linear equations in much the same way as we use the addition method. When using the addition method, notice that the following operations produce an equivalent system of equations.

- Interchange two equations.

  Example: $\begin{array}{r} -3x - 5y = -13 \\ x + 2y = 5 \end{array}$ $\xrightarrow[\text{equations.}]{\text{Interchange}}$ $\begin{array}{r} x + 2y = 5 \\ -3x - 5y = -13 \end{array}$

- Multiply an equation by a nonzero constant.

  Example: $\begin{array}{r} x + 2y = 5 \\ -3x - 5y = -13 \end{array}$ $\xrightarrow{\text{Multiply by 3.}}$ $\begin{array}{r} 3x + 6y = 15 \\ -3x - 5y = -13 \end{array}$

- Add a nonzero multiple of one equation to another equation.

  Example: $\begin{array}{r} x + 2y = 5 \\ -3x - 5y = -13 \end{array}$ $\xrightarrow{\text{Multiply by 3.}}$ $\left. \begin{array}{r} 3x + 6y = 15 \\ \underline{-3x - 5y = -13} \\ y = 2 \end{array} \right\} \begin{array}{l} \text{Add the} \\ \text{equations.} \end{array}$

These same operations performed on a matrix are called **elementary row operations.**

Elementary Row Operations

1. Interchange two rows.

Example: $\begin{bmatrix} -3 & -5 & | & -13 \\ 1 & 2 & | & 5 \end{bmatrix}$ $\xrightarrow[\substack{\text{Interchange rows} \\ \text{1 and 2.}}]{R_1 \Leftrightarrow R_2}$ $\begin{bmatrix} 1 & 2 & | & 5 \\ -3 & -5 & | & -13 \end{bmatrix}$

2. Multiply a row by a nonzero constant.

Example: $\begin{bmatrix} 1 & 2 & | & 5 \\ -3 & -5 & | & -13 \end{bmatrix}$ $\xrightarrow[\substack{\text{Multiply} \\ \text{row 1 by 3.}}]{3R_1 \rightarrow R_1}$ $\begin{bmatrix} 3 & 6 & | & 15 \\ -3 & -5 & | & -13 \end{bmatrix}$

3. Add a nonzero multiple of one row to another row.

Example: $\begin{bmatrix} 1 & 2 & | & 5 \\ -3 & -5 & | & -13 \end{bmatrix}$ $\xrightarrow[\substack{\text{Add 3 times} \\ \text{row 1 to row 2.}}]{3R_1 + R_2 \rightarrow R_2}$ $\begin{bmatrix} 1 & 2 & | & 5 \\ 0 & 1 & | & 2 \end{bmatrix}$

$\begin{array}{r} 3R_1 \\ +R_2 \\ \hline \rightarrow R_2 \end{array} \begin{array}{rr|r} 3 & 6 & 15 \\ -3 & -5 & -13 \\ \hline 0 & 1 & 2 \end{array}$

> **Avoiding Mistakes**
>
> When we add a constant multiple of one row to another row, we do not change the row that was multiplied by the constant.

Two matrices are said to be **row equivalent** if one matrix can be transformed into the other matrix through a series of elementary row operations.

EXAMPLE 2 **Performing Elementary Row Operations**

Given $\begin{bmatrix} 4 & 3 & 0 & | & 5 \\ 1 & 4 & -1 & | & 9 \\ 2 & 0 & -3 & | & -8 \end{bmatrix}$, perform the following elementary row operations.

a. $R_1 \Leftrightarrow R_2$ **b.** $\dfrac{1}{4} R_1 \rightarrow R_1$ **c.** $-2R_2 + R_3 \rightarrow R_3$

Solution:

a. $\begin{bmatrix} 4 & 3 & 0 & | & 5 \\ 1 & 4 & -1 & | & 9 \\ 2 & 0 & -3 & | & -8 \end{bmatrix}$ $\xrightarrow[\substack{\text{Interchange rows} \\ \text{1 and 2.}}]{R_1 \Leftrightarrow R_2}$ $\begin{bmatrix} 1 & 4 & -1 & | & 9 \\ 4 & 3 & 0 & | & 5 \\ 2 & 0 & -3 & | & -8 \end{bmatrix}$

TIP The notation $\frac{1}{4}R_1 \to R_1$ means to multiply row 1 by $\frac{1}{4}$ and then *replace* row 1 by the result.

b. $\begin{bmatrix} 4 & 3 & 0 & | & 5 \\ 1 & 4 & -1 & | & 9 \\ 2 & 0 & -3 & | & -8 \end{bmatrix}$ $\xrightarrow[\substack{\text{Multiply} \\ \text{row 1 by } \frac{1}{4}.}]{\frac{1}{4}R_1 \to R_1}$ $\begin{bmatrix} 1 & \frac{3}{4} & 0 & | & \frac{5}{4} \\ 1 & 4 & -1 & | & 9 \\ 2 & 0 & -3 & | & -8 \end{bmatrix}$

c. $\begin{bmatrix} 4 & 3 & 0 & | & 5 \\ 1 & 4 & -1 & | & 9 \\ 2 & 0 & -3 & | & -8 \end{bmatrix}$

$\begin{array}{rrrrr} -2R_2 & -2 & -8 & 2 & | & -18 \\ +\quad R_3 & 2 & 0 & -3 & | & -8 \\ \hline \to R_3 & 0 & -8 & -1 & | & -26 \end{array}$ This now replaces row 3.

$\begin{bmatrix} 4 & 3 & 0 & | & 5 \\ 1 & 4 & -1 & | & 9 \\ 2 & 0 & -3 & | & -8 \end{bmatrix}$ $\xrightarrow[\substack{\text{Add } -2 \text{ times row 2 to row 3.} \\ \text{The result replaces row 3.}}]{-2R_2 + R_3 \to R_3}$ $\begin{bmatrix} 4 & 3 & 0 & | & 5 \\ 1 & 4 & -1 & | & 9 \\ 0 & -8 & -1 & | & -26 \end{bmatrix}$

Skill Practice 2   Use the matrix from Example 2 to perform the given row operations.

a. $R_2 \Leftrightarrow R_3$ **b.** $-\frac{1}{2}R_3$ **c.** $-4R_2 + R_1 \to R_1$

3. Use Gaussian Elimination and Gauss-Jordan Elimination

When an elementary row operation is performed on an augmented matrix, a new row-equivalent augmented matrix is obtained that represents an equivalent system of equations. If we perform repeated row operations, we can form an augmented matrix that represents a system of equations that is easier to solve than the original system. In particular, it is easy to solve a system whose augmented matrix is in *row-echelon form* or *reduced row-echelon form*.

Row-Echelon Form and Reduced Row-Echelon Form

A matrix is in **row-echelon form** if it satisfies the following conditions.

1. Any rows consisting entirely of zeros are at the bottom of the matrix.
2. For all other rows, the first nonzero entry is 1. This is called the leading 1.
3. The leading 1 in each nonzero row is to the right of the leading 1 in the row immediately above.

Note: A matrix is in **reduced row-echelon form** if it is in row-echelon form with the added condition that each row with a leading entry of 1 has zeros above the leading 1.

Answers

2. a. $\begin{bmatrix} 4 & 3 & 0 & | & 5 \\ 2 & 0 & -3 & | & -8 \\ 1 & 4 & -1 & | & 9 \end{bmatrix}$

b. $\begin{bmatrix} 4 & 3 & 0 & | & 5 \\ 1 & 4 & -1 & | & 9 \\ -1 & 0 & \frac{3}{2} & | & 4 \end{bmatrix}$

c. $\begin{bmatrix} 0 & -13 & 4 & | & -31 \\ 1 & 4 & -1 & | & 9 \\ 2 & 0 & -3 & | & -8 \end{bmatrix}$

The following matrices illustrate row-echelon form and reduced row-echelon form.

Row-Echelon Form

$\begin{bmatrix} 1 & 5 & | & 3 \\ 0 & 1 & | & 6 \end{bmatrix}$ $\begin{bmatrix} 1 & -4 & -\frac{1}{2} & | & 6 \\ 0 & 1 & 9 & | & -\frac{1}{3} \\ 0 & 0 & 1 & | & 2 \end{bmatrix}$ $\begin{bmatrix} 1 & 4.1 & 1.2 & | & 3.1 \\ 0 & 1 & 0.6 & | & 4.7 \\ 0 & 0 & 0 & | & 0 \end{bmatrix}$

Reduced Row-Echelon Form

$$\begin{bmatrix} 1 & 0 & | & -27 \\ 0 & 1 & | & 6 \end{bmatrix} \qquad \begin{bmatrix} 1 & 0 & 0 & | & 150 \\ 0 & 1 & 0 & | & -85 \\ 0 & 0 & 1 & | & 12 \end{bmatrix} \qquad \begin{bmatrix} 1 & 0 & 2 & | & 25 \\ 0 & 1 & -3 & | & -4 \\ 0 & 0 & 0 & | & 0 \end{bmatrix}$$

After writing an augmented matrix in row-echelon form, the corresponding system of linear equations can be solved using back substitution. This method is called **Gaussian elimination.**

Solving a System of Linear Equations Using Gaussian Elimination

1. Write the augmented matrix for the system.
2. Use elementary row operations to write the augmented matrix in row-echelon form.
3. Use back substitution to solve the resulting system of equations.

When writing an augmented matrix in row-echelon form, the goal is to make the elements along the main diagonal 1 and the entries below the main diagonal 0. The **main diagonal** refers to the elements on the diagonal from the upper left to the lower right all to the left of the vertical bar.

The main diagonal stretches from the upper left to the lower right.

$$\begin{bmatrix} 1 & \square & \square & | & \square \\ 0 & 1 & \square & | & \square \\ 0 & 0 & 1 & | & \square \end{bmatrix}$$

To write an augmented matrix in row-echelon form, work one column at a time from left to right. This is demonstrated in Example 3.

EXAMPLE 3 Solving a System Using Gaussian Elimination

Solve the system by using Gaussian elimination.

$$\begin{aligned} 3x + 7y - 15z &= -12 \\ x + 2y - 4z &= -3 \\ -4x - 6y + 15z &= 16 \end{aligned}$$

Solution:

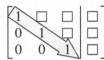
Need a 1 here

$$\begin{aligned} 3x + 7y - 15z &= -12 \\ x + 2y - 4z &= -3 \\ -4x - 6y + 15z &= 16 \end{aligned}$$
$$\begin{bmatrix} ③ & 7 & -15 & | & -12 \\ 1 & 2 & -4 & | & -3 \\ -4 & -6 & 15 & | & 16 \end{bmatrix}$$

Set up the augmented matrix.

Use elementary row operations to write the augmented matrix in row-echelon form.

$$R_1 \Leftrightarrow R_2 \longrightarrow \begin{bmatrix} 1 & 2 & -4 & | & -3 \\ 3 & 7 & -15 & | & -12 \\ -4 & -6 & 15 & | & 16 \end{bmatrix}$$

Begin working with column 1.

To obtain a leading 1 in the first row, interchange rows 1 and 2.

TIP To obtain a leading entry of 1 in the first row, we have the option of multiplying the first row by $\frac{1}{3}$. However, this would present fractions in the first row. Interchanging rows 1 and 2 is "cleaner."

Now we want zeros below the leading 1 in the first row.

- Multiply row 1 by -3 and add the result to row 2.
- Multiply row 1 by 4 and add the result to row 3.

$$\begin{bmatrix} 1 & 2 & -4 & | & -3 \\ ③ & 7 & -15 & | & -12 \\ ④ & -6 & 15 & | & 16 \end{bmatrix}$$

Need 0's here

$$\begin{array}{c|cccc} -3R_1 & -3 & -6 & 12 & 9 \\ +R_2 & 3 & 7 & -15 & -12 \\ \hline \to R_2 & 0 & 1 & -3 & -3 \end{array} \qquad \begin{array}{c|cccc} 4R_1 & 4 & 8 & -16 & -12 \\ +R_3 & -4 & -6 & 15 & 16 \\ \hline \to R_3 & 0 & 2 & -1 & 4 \end{array}$$

$$\begin{aligned} -3R_1 + R_2 \to R_2 \longrightarrow \\ 4R_1 + R_3 \to R_3 \longrightarrow \end{aligned} \begin{bmatrix} 1 & 2 & -4 & | & -3 \\ 0 & 1 & -3 & | & -3 \\ 0 & 2 & -1 & | & 4 \end{bmatrix}$$

Next, work with column 2.
We already have a leading 1 in row 2.
We need 0 below the leading 1 in row 2.

$$\begin{bmatrix} 1 & 2 & -4 & | & -3 \\ 0 & 1 & -3 & | & -3 \\ 0 & ② & -1 & | & 4 \end{bmatrix}$$

Need 0 here

- Multiply row 2 by -2 and add the result to row 3.

$$\begin{array}{c} -2R_2 \\ +R_3 \\ \rightarrow R_3 \end{array} \begin{array}{ccc|c} 0 & -2 & 6 & 6 \\ 0 & 2 & -1 & 4 \\ \hline 0 & 0 & 5 & 10 \end{array}$$

$-2R_2 + R_3 \rightarrow R_3 \quad \longrightarrow \begin{bmatrix} 1 & 2 & -4 & | & -3 \\ 0 & 1 & -3 & | & -3 \\ 0 & 0 & 5 & | & 10 \end{bmatrix}$

$\frac{1}{5}R_3 \rightarrow R_3 \quad \longrightarrow \begin{bmatrix} 1 & 2 & -4 & | & -3 \\ 0 & 1 & -3 & | & -3 \\ 0 & 0 & 1 & | & 2 \end{bmatrix}$

Now work with column 3. We need a leading 1 in row 3.
Multiply row 3 by $\frac{1}{5}$.

$\begin{bmatrix} 1 & 2 & -4 & | & -3 \\ 0 & 1 & -3 & | & -3 \\ 0 & 0 & 1 & | & 2 \end{bmatrix} \begin{array}{l} \longrightarrow x + 2y - 4z = -3 \\ \longrightarrow y - 3z = -3 \\ \longrightarrow z = 2 \end{array}$

We now have row-echelon form.
The corresponding system of equations has z isolated.

Use back substitution to find x and y.

$\begin{aligned} y - 3z &= -3 \\ y - 3(2) &= -3 \\ y &= 3 \end{aligned}$
Substitute $z = 2$ in the second equation.

$\begin{aligned} x + 2y - 4z &= -3 \\ x + 2(3) - 4(2) &= -3 \\ x + 6 - 8 &= -3 \\ x &= -1 \end{aligned}$
Substitute $z = 2$ and $y = 3$ into the first equation.

The solution set is $\{(-1, 3, 2)\}$. 　The solution $(-1, 3, 2)$ checks in each original equation.

Skill Practice 3 Solve the system by using Gaussian elimination.

$\begin{aligned} 2x + 7y + z &= 14 \\ x + 3y - z &= 2 \\ x + 7y + 12z &= 45 \end{aligned}$

Example 3 illustrates that a system of equations represented by an augmented matrix in row-echelon form is easily solved by using back substitution. If we write an augmented matrix in *reduced* row-echelon form, we can solve the corresponding system of equations by inspection. This is called the Gauss-Jordan elimination method [named after Carl Friedrich Gauss (1777–1855) and German scientist Wilhelm Jordan (1842–1899)].

When writing an augmented matrix in *reduced* row-echelon form, the goal is to make the elements along the main diagonal 1 and the entries above and below the main diagonal 0. This is demonstrated in Examples 4 and 5.

$$\begin{bmatrix} 1 & 0 & | & \square \\ 0 & 1 & | & \square \end{bmatrix} \qquad \begin{bmatrix} 1 & 0 & 0 & | & \square \\ 0 & 1 & 0 & | & \square \\ 0 & 0 & 1 & | & \square \end{bmatrix}$$

Answer

3. $\{(2, 1, 3)\}$

Order of Row Operations to Obtain Reduced Row-Echelon Form

To write an augmented matrix with n rows in reduced row-echelon form, transform the entries in the matrix in the following order.

Column 1:   Obtain a leading element of 1 in row 1. Then obtain 0's below this element.

Column 2:   Obtain a leading element of 1 in row 2. Then obtain 0's above and below this element.

Column 3:   Obtain a leading element of 1 in row 3. Then obtain 0's above and below this element.

⋮

Column n: Obtain a leading element of 1 in row n. Then obtain 0's above this element.

EXAMPLE 4   **Solving a System Using Gauss-Jordan Elimination**

Solve the system by using Gauss-Jordan elimination.
$$x = 17 - 2y$$
$$3(x + 2y) = 47 - y$$

Solution:

$x = 17 - 2y \longrightarrow x + 2y = 17$

$3(x + 2y) = 47 - y \longrightarrow 3x + 7y = 47$

To use Gaussian elimination or Gauss-Jordan elimination, always begin by writing each equation in standard form.

$$\begin{aligned} x + 2y &= 17 \\ 3x + 7y &= 47 \end{aligned} \qquad \begin{bmatrix} 1 & 2 & | & 17 \\ ③ & 7 & | & 47 \end{bmatrix}$$

Need 0 here

Set up the augmented matrix.
The leading element in row 1 is already 1.
Now we need a zero below the leading 1 in row 1.

Multiply row 1 by -3 and add the result to row 2.

$$\begin{array}{rrr|r} -3R_1 & -3 & -6 & -51 \\ +R_2 & 3 & 7 & 47 \\ \hline \to R_2 & 0 & 1 & -4 \end{array}$$

$-3R_1 + R_2 \to R_2 \longrightarrow \begin{bmatrix} 1 & 2 & | & 17 \\ 0 & 1 & | & -4 \end{bmatrix}$

Need 0 here

$$\begin{bmatrix} 1 & ② & | & 17 \\ 0 & 1 & | & -4 \end{bmatrix}$$

Now work with column 2. The leading element in row 2 is already 1.
Now we need a zero above the leading 1 in row 2.

Multiply row 2 by -2 and add the result to row 1.

$$\begin{array}{rrr|r} R_1 & 1 & 2 & 17 \\ + -2R_2 & 0 & -2 & 8 \\ \hline \to R_1 & 1 & 0 & 25 \end{array}$$

$-2R_2 + R_1 \to R_1 \longrightarrow \begin{bmatrix} 1 & 0 & | & 25 \\ 0 & 1 & | & -4 \end{bmatrix}$

$\begin{bmatrix} 1 & 0 & | & 25 \\ 0 & 1 & | & -4 \end{bmatrix} \longrightarrow \begin{array}{l} x = 25 \\ y = -4 \end{array}$

From the corresponding system of equations, we can determine the solution $(25, -4)$ by inspection.

The solution set is $\{(25, -4)\}$.

The solution checks in each original equation.

Skill Practice 4 Solve the system by using Gauss-Jordan elimination.

$x - 2y = -1$
$4x - 7y = 1$

In Example 5, we use the Gauss-Jordan elimination method to solve a three-variable system of linear equations.

EXAMPLE 5   Solving a System Using Gauss-Jordan Elimination

Solve the system by using Gauss-Jordan elimination.

$2x - 5y - 21z = 39$
$x - 3y - 10z = 22$
$x + 3y + 2z = -8$

Solution:

$2x - 5y - 21z = 39$
$x - 3y - 10z = 22$
$x + 3y + 2z = -8$

$$\left[\begin{array}{ccc|c} 2 & -5 & -21 & 39 \\ 1 & -3 & -10 & 22 \\ 1 & 3 & 2 & -8 \end{array}\right]$$

Set up the augmented matrix.

$R_1 \Leftrightarrow R_2 \longrightarrow$
$$\left[\begin{array}{ccc|c} 1 & -3 & -10 & 22 \\ 2 & -5 & -21 & 39 \\ 1 & 3 & 2 & -8 \end{array}\right]$$

Use elementary row operations to write the augmented matrix in reduced row-echelon form.

To obtain a leading 1 in the first row, interchange rows 1 and 2.

$-2R_1 + R_2 \rightarrow R_2 \longrightarrow$
$-R_1 + R_3 \rightarrow R_3 \longrightarrow$
$$\left[\begin{array}{ccc|c} 1 & -3 & -10 & 22 \\ 0 & 1 & -1 & -5 \\ 0 & 6 & 12 & -30 \end{array}\right]$$

Multiply row 1 by -2 and add the result to row 2. Multiply row 1 by -1 and add the result to row 3.

This results in zeros below the leading 1 in the first row.

$3R_2 + R_1 \rightarrow R_1 \longrightarrow$
$-6R_2 + R_3 \rightarrow R_3 \longrightarrow$
$$\left[\begin{array}{ccc|c} 1 & 0 & -13 & 7 \\ 0 & 1 & -1 & -5 \\ 0 & 0 & 18 & 0 \end{array}\right]$$

Row 2 already has a leading 1. Multiply row 2 by 3 and add the result to row 1. Multiply row 2 by -6 and add the result to row 3.

$\frac{1}{18}R_3 \rightarrow R_3 \longrightarrow$
$$\left[\begin{array}{ccc|c} 1 & 0 & -13 & 7 \\ 0 & 1 & -1 & -5 \\ 0 & 0 & 1 & 0 \end{array}\right]$$

Multiply row 3 by $\frac{1}{18}$ to obtain a leading 1 in row 3.

$13R_3 + R_1 \rightarrow R_1 \longrightarrow$
$R_3 + R_2 \rightarrow R_2 \longrightarrow$
$$\left[\begin{array}{ccc|c} 1 & 0 & 0 & 7 \\ 0 & 1 & 0 & -5 \\ 0 & 0 & 1 & 0 \end{array}\right]$$

Multiply row 3 by 13 and add the result to row 1.

Multiply row 3 by 1 and add the result to row 2.

$$\left[\begin{array}{ccc|c} 1 & 0 & 0 & 7 \\ 0 & 1 & 0 & -5 \\ 0 & 0 & 1 & 0 \end{array}\right] \begin{array}{l} \longrightarrow x = 7 \\ \longrightarrow y = -5 \\ \longrightarrow z = 0 \end{array}$$

The augmented matrix is in reduced row-echelon form.

From the corresponding system of equations, we can determine the solution $(7, -5, 0)$ by inspection.

The solution set is $\{(7, -5, 0)\}$.

The solution checks in each original equation.

Skill Practice 5 Solve the system by using Gauss-Jordan elimination.

$2x + 7y + 11z = 11$
$x + 2y + 8z = 14$
$x + 3y + 6z = 8$

Answers
4. $\{(9, 5)\}$
5. $\{(14, -4, 1)\}$

TECHNOLOGY CONNECTIONS

Finding the Row-Echelon Form and Reduced Row-Echelon Form of a Matrix

Many graphing utilities have the capability to enter and manipulate matrices. To enter the augmented matrix from Example 5, select the MATRIX menu and then the EDIT menu. Select a name for the matrix such as A. Then enter the dimensions of the matrix (in this case 3 × 4) followed by the individual elements.

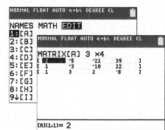

To access the commands for row-echelon form and reduced row-echelon form, select the (MATH) menu from within the MATRIX menu. Select either *ref* for row-echelon form or *rref* for reduced row-echelon form.

In the home screen, insert the matrix name within the parentheses for the *ref* and *rref* functions.

Alternatively, we can find the row-echelon form or reduced row-echelon form of a matrix if the calculator is in "MATHPRINT" mode. Select (MODE), and using the arrow keys to navigate through the menu, highlight "MATHPRINT."

In the home screen, select the MATRIX menu, followed by MATH. Then select either *ref* or *rref*. Next, press the (ALPHA) key followed by F3. Enter the dimensions of the matrix (in this case, 3 × 4).

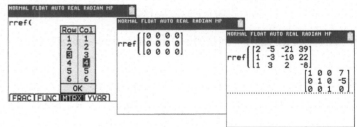

Then enter the elements of the matrix using the arrow keys to navigate through the matrix. Hit (ENTER) to complete the calculation.

SECTION 6.1   Practice Exercises

Prerequisite Review

R.1. Solve the equation and check the solution.

$$3n + 25 = 73$$

R.2. How much interest will Roxanne have to pay if she borrows $1000 for 2 yr at a simple interest rate of 5%?

For Exercises R.3–R.4, set up the form for the partial fraction decomposition. Do not solve for A, B, C, and so on.

R.3. $\dfrac{-12t - 11}{4t^2 - 16t + 15}$

R.4. $\dfrac{-5x^2 - 10x - 13}{7x^3 + 7x}$

Concept Connections

1. A rectangular array of elements is called a _____.

2. Identify the elements on the main diagonal. $\begin{bmatrix} -4 & -1 & 1 & | & 8 \\ 2 & 0 & 5 & | & 11 \\ 0 & 1 & -7 & | & -6 \end{bmatrix}$

3. Explain the meaning of the notation $R_2 \Leftrightarrow R_3$.

4. Explain the meaning of the notation $-R_2 \to R_2$.

5. Explain the meaning of the notation $3R_1 \to R_1$.

6. Explain the meaning of $R_1 \Leftrightarrow R_3$.

7. Explain the meaning of the notation $3R_1 + R_2 \to R_2$.

8. Explain the meaning of the notation $4R_2 + R_3 \to R_3$.

Objective 1: Write an Augmented Matrix

For Exercises 9–14, write the augmented matrix for the given system. (See Example 1)

9. $\begin{aligned} -3x + 2y - z &= 4 \\ 8x + 4z &= 12 \\ 2y - 5z &= 1 \end{aligned}$

10. $\begin{aligned} -4x - y + z &= 8 \\ 2x + 5z &= 11 \\ y - 7z &= -6 \end{aligned}$

11. $\begin{aligned} 4(x - 2y) &= 6y + 2 \\ 3x &= 5y + 7 \end{aligned}$

12. $\begin{aligned} 2(y - x) &= 4 - 8x \\ 5y &= 6 - x \end{aligned}$

13. $\begin{aligned} x &= 2 \\ y &= \tfrac{6}{7} \\ z &= 12 \end{aligned}$

14. $\begin{aligned} x &= 4 \\ y &= 2 \\ z &= -\tfrac{1}{2} \end{aligned}$

For Exercises 15–20, write a system of linear equations represented by the augmented matrix. (See Example 1)

15. $\begin{bmatrix} -4 & 6 & | & 11 \\ -3 & 9 & | & 1 \end{bmatrix}$

16. $\begin{bmatrix} -3 & 7 & | & -2 \\ -1 & 1 & | & 4 \end{bmatrix}$

17. $\begin{bmatrix} 1 & 4 & 3 & | & 8 \\ 0 & 1 & 2 & | & 12 \\ 0 & 0 & 1 & | & 6 \end{bmatrix}$

18. $\begin{bmatrix} 1 & 1 & -2 & | & -4 \\ 0 & 1 & 3 & | & 8 \\ 0 & 0 & 1 & | & 2 \end{bmatrix}$

19. $\begin{bmatrix} 1 & 0 & 0 & | & 8 \\ 0 & 1 & 0 & | & -9 \\ 0 & 0 & 1 & | & \tfrac{3}{2} \end{bmatrix}$

20. $\begin{bmatrix} 1 & 0 & 0 & | & 2 \\ 0 & 1 & 0 & | & 6 \\ 0 & 0 & 1 & | & -\tfrac{1}{2} \end{bmatrix}$

Objective 2: Use Elementary Row Operations

For Exercises 21–26, perform the elementary row operations on $\begin{bmatrix} 1 & 4 & | & 2 \\ -3 & 6 & | & 6 \end{bmatrix}$. (See Example 2)

21. $R_1 \Leftrightarrow R_2$

22. $-\tfrac{1}{3}R_2 \to R_2$

23. $3R_1 \to R_1$

24. $-3R_1 \to R_1$

25. $\tfrac{1}{3}R_2 + R_1 \to R_1$

26. $3R_1 + R_2 \to R_2$

For Exercises 27–32, perform the elementary row operations on $\begin{bmatrix} 1 & 5 & 6 & | & 2 \\ 2 & 1 & 5 & | & 1 \\ 4 & -2 & -3 & | & 10 \end{bmatrix}$. (See Example 2)

27. $R_2 \Leftrightarrow R_3$

28. $R_1 \Leftrightarrow R_2$

29. $\tfrac{1}{4}R_3 \to R_3$

30. $\tfrac{1}{2}R_2 \to R_2$

31. $-2R_1 + R_2 \to R_2$

32. $-4R_1 + R_3 \to R_3$

Objective 3: Use Gaussian Elimination and Gauss-Jordan Elimination

For Exercises 33–36, determine if the matrix is in row-echelon form. If not, explain why.

33. $\begin{bmatrix} 1 & 5 & | & 4 \\ 0 & 2 & | & 6 \end{bmatrix}$

34. $\begin{bmatrix} 1 & 6 & 4 & | & 2 \\ 0 & 1 & 0 & | & -1 \\ 0 & 3 & 1 & | & 3 \end{bmatrix}$

35. $\begin{bmatrix} 1 & 3 & 2 & | & 6 \\ 0 & 1 & 5 & | & 9 \\ 0 & 0 & 0 & | & 0 \end{bmatrix}$

36. $\begin{bmatrix} 1 & 4 & 2 & | & -6 \\ 0 & 1 & -3 & | & 2 \\ 0 & 0 & 1 & | & 0 \end{bmatrix}$

For Exercises 37–40, determine if the matrix is in reduced row-echelon form. If not, explain why.

37. $\begin{bmatrix} 1 & 0 & 0 & | & 3 \\ 1 & 0 & 0 & | & 4 \\ 1 & 0 & 0 & | & 5 \end{bmatrix}$

38. $\begin{bmatrix} 1 & 0 & 2 & | & 3 \\ 0 & 1 & 0 & | & 4 \\ 0 & 0 & 1 & | & 5 \end{bmatrix}$

39. $\begin{bmatrix} 1 & 0 & 0 & 0 & | & 1 \\ 0 & 1 & 0 & 0 & | & 2 \\ 0 & 0 & 1 & 0 & | & -7 \\ 0 & 0 & 0 & 1 & | & 4 \end{bmatrix}$

40. $\begin{bmatrix} 1 & 0 & 0 & -1 & | & 5 \\ 0 & 1 & 0 & 0 & | & 20 \\ 0 & 0 & 1 & 4 & | & -1 \\ 0 & 0 & 0 & 0 & | & 0 \end{bmatrix}$

For Exercises 41–60, solve the system by using Gaussian elimination or Gauss-Jordan elimination. (See Examples 3–5)

41. $2x + 3y = -13$
    $x + 4y = -14$

42. $-3x + 11y = 58$
    $x - 3y = -16$

43. $2x - 7y = -41$
    $3x - 9y = -51$

44. $-2x + 15y = 6$
    $3x - 12y = -9$

45. $-3(x - 6y) = -167 - y$
    $14y = 2x - 122$

46. $2(x - y) = 4x + y - 40$
    $9y = 105 - 3x$

47. $3x + 7y + 22z = 83$
    $x + 3y + 10z = 37$
    $-2x - 5y - 18z = -66$

48. $3x + 5y + 9z = 3$
    $x + 3y + 7z = 5$
    $-2x - 8y - 15z = -11$

49. $-2x + 4y + z = 7$
    $4x - 13y + 10z = 17$
    $3x - 9y + 6z = 9$

50. $2x - 8y + 54z = -4$
    $x - 2y + 14z = -1$
    $x - 3y + 19z = -3$

51. $-3x + 4y - 15z = -44$
    $x - y + 4z = 13$
    $x - 3y + 14z = 27$

52. $-2x + 5y - 4z = -4$
    $x - 2y + z = 3$
    $x - 5y + 9z = -5$

53. $2x + 8z = 7y - 46$
    $x = 3y - 3z - 18$
    $6z = 5y - x - 34$

54. $2x = 7 - y - 3z$
    $x + y = z + 5$
    $16z = 2y - x - 4$

55. $11y + 65 = 3x + 13z$
    $x + 3z = 3y + 15$
    $-2x + 4y - 7z = -25$

56. $2(x - 6z) = 3y + x + 17$
    $2x - 19 = 3y + 18z$
    $-3x + 7y + 36z = -41$

57. $w + 3x - 3z = -5$
    $x - 2z = -6$
    $-2w - 4x + y + 2z = 1$
    $x + y = 5$

58. $w - 2x + 5y = 20$
    $-x + 2y = 9$
    $x - y = -5$
    $2w - x + 7y + z = 25$

59. $x_1 + x_2 + 5x_4 = -4$
    $x_2 + 2x_4 = 3$
    $-2x_2 + x_3 - 3x_4 = -5$
    $3x_1 + 3x_2 + 17x_4 = -10$

60. $x_1 + x_3 - 5x_4 = 1$
    $-2x_1 + x_2 - 2x_3 + 16x_4 = -3$
    $x_1 + 2x_3 - 10x_4 = 5$
    $x_1 - x_3 + 7x_4 = -7$

Mixed Exercises

For Exercises 61–64, set up a system of linear equations to represent the scenario. Solve the system by using Gaussian elimination or Gauss-Jordan elimination.

61. Andre borrowed $20,000 to buy a truck for his business. He borrowed from his parents who charge him 2% simple interest. He borrowed from a credit union that charges 4% simple interest, and he borrowed from a bank that charges 5% simple interest. He borrowed five times as much from his parents as from the bank, and the amount of interest he paid at the end of 1 yr was $620. How much did he borrow from each source?

62. Sylvia invested a total of $40,000. She invested part of the money in a certificate of deposit (CD) that earns 2% simple interest per year. She invested in a stock that returns the equivalent of 8% simple interest, and she invested in a bond fund that returns 5%. She invested twice as much in the stock as she did in the CD, and earned a total of $2300 at the end of 1 yr. How much principal did she put in each investment?

63. Danielle stayed in three different cities (Washington, D.C., Atlanta, Georgia, and Dallas, Texas) for a total of 14 nights. She spent twice as many nights in Dallas as she did in Washington. The total cost for 14 nights (excluding tax) was $2200. Determine the number of nights that she spent in each city.

City	Cost per Night
Washington	$200
Atlanta	$100
Dallas	$150

64. Three pumps (A, B, and C) work to drain water from a retention pond. Working together the pumps can pump 1500 gal/hr of water. Pump C works at a rate of 100 gal/hr faster than pump B. In 3 hr, pump C can pump as much water as pumps A and B working together in 2 hr. Find the rate at which each pump works.

For Exercises 65–66, find the partial fraction decomposition for the given rational expression. Use the technique of Gaussian elimination to find A, B, and C.

65. $\dfrac{5x^2 - 6x - 13}{(x + 3)(x - 2)^2} = \dfrac{A}{x + 3} + \dfrac{B}{x - 2} + \dfrac{C}{(x - 2)^2}$

66. $\dfrac{2x^2 + 17x + 3}{(x + 5)(x + 1)^2} = \dfrac{A}{x + 5} + \dfrac{B}{x + 1} + \dfrac{C}{(x + 1)^2}$

Write About It

67. Explain why interchanging two rows of an augmented matrix results in an augmented matrix that represents an equivalent system of equations.

68. Explain why multiplying a row of an augmented matrix by a nonzero constant results in an augmented matrix that represents an equivalent system of equations.

69. Explain the difference between a matrix in row-echelon form and reduced row-echelon form.

70. Consider the matrix $\begin{bmatrix} 5 & -9 & | & -57 \\ 1 & -2 & | & -12 \end{bmatrix}$. Identify two row operations that could be used to obtain a leading entry of 1 in the first row. Also indicate which operation would be less cumbersome as a first step toward writing the matrix in reduced row-echelon form.

Technology Connections

For Exercises 71–72, use a calculator to approximate the reduced row-echelon form of the augmented matrix representing the given system. Give the solution set where x, y, and z are rounded to 2 decimal places.

71. $0.52x - 3.71y - 4.68z = 9.18$
    $0.02x + 0.06y + 0.11z = 0.56$
    $0.972x + 0.816y + 0.417z = 0.184$

72. $-3.61x + 8.17y - 5.62z = 30.2$
    $8.04x - 3.16y + 9.18z = 28.4$
    $-0.16x + 0.09y + 0.55z = 4.6$

73. A small grocer finds that the monthly sales y (in \$) can be approximated as a function of the amount spent advertising on the radio x_1 (in \$) and the amount spent advertising in the newspaper x_2 (in \$) according to $y = ax_1 + bx_2 + c$.

    The table gives the amounts spent in advertising and the corresponding monthly sales for 3 months.

Radio Advertising, x_1	Newspaper Advertising, x_2	Monthly sales, y
\$2400	\$800	\$36,000
\$2000	\$500	\$30,000
\$3000	\$1000	\$44,000

   a. Use the data to write a system of linear equations to solve for a, b, and c.

   b. Use a graphing utility to find the reduced row-echelon form of the augmented matrix.

   c. Write the model $y = ax_1 + bx_2 + c$.

   d. Predict the monthly sales if the grocer spends \$2500 advertising on the radio and \$500 advertising in the newspaper for a given month.

74. The purchase price of a home y (in \$1000) can be approximated based on the annual income of the buyer x_1 (in \$1000) and on the square footage of the home x_2 (in 100 ft^2) according to $y = ax_1 + bx_2 + c$.

    The table gives the incomes of three buyers, the square footages of the home purchased, and the corresponding purchase prices of the home.

Income (\$1000) x_1	Square Footage (100 ft^2) x_2	Price (\$1000) y
80	21	180
150	28	250
75	18	160

   a. Use the data to write a system of linear equations to solve for a, b, and c.

   b. Use a graphing utility to find the reduced row-echelon form of the augmented matrix.

   c. Write the model $y = ax_1 + bx_2 + c$.

   d. Predict the purchase price for a buyer who makes \$100,000 per year and wants a 2500 ft^2 home.

For Exercises 75–76, the given function values satisfy a function defined by $f(x) = ax^2 + bx + c$.

a. Set up a system of equations to solve for a, b, and c.

b. Use a graphing utility to find the reduced row-echelon form of the augmented matrix.

c. Write a function of the form $f(x) = ax^2 + bx + c$ that fits the data.

75. $f(-3) = -7.28$
    $f(-1) = 3.68$
    $f(10) = 18.2$

76. $f(3) = 6.95$
    $f(-2) = 20.2$
    $f(12) = 39.8$

Inconsistent Systems and Dependent Equations

OBJECTIVES

1. Identify Inconsistent Systems
2. Solve Systems with Dependent Equations
3. Solve Applications of Systems of Equations

1. Identify Inconsistent Systems

When we studied systems of linear equations in two and three variables in Chapter 5, we learned that a system may have no solution. Such a system is said to be **inconsistent,** and we recognize an inconsistent system if the system reduces to a contradiction.

EXAMPLE 1   **Identifying an Inconsistent System**

Solve the system.
$$\begin{aligned} x - 3y - 17z &= -59 \\ x - 2y - 12z &= -41 \\ -2y - 10z &= 20 \end{aligned}$$

Solution:

$$\begin{aligned} x - 3y - 17z &= -59 \\ x - 2y - 12z &= -41 \\ -2y - 10z &= 20 \end{aligned} \qquad \left[\begin{array}{ccc|c} 1 & -3 & -17 & -59 \\ 1 & -2 & -12 & -41 \\ 0 & -2 & -10 & 20 \end{array}\right]$$ Write the augmented matrix. The leading entry in row 1 is already 1.

$$-R_1 + R_2 \rightarrow R_2 \longrightarrow \left[\begin{array}{ccc|c} 1 & -3 & -17 & -59 \\ 0 & 1 & 5 & 18 \\ 0 & -2 & -10 & 20 \end{array}\right]$$ Multiply row 1 by -1 and add the result to row 2.

$$\begin{aligned} 3R_2 + R_1 &\rightarrow R_1 \\ 2R_2 + R_3 &\rightarrow R_3 \end{aligned} \longrightarrow \left[\begin{array}{ccc|c} 1 & 0 & -2 & -5 \\ 0 & 1 & 5 & 18 \\ 0 & 0 & 0 & 56 \end{array}\right]$$ Multiply row 2 by 3 and add the result to row 1. Multiply row 2 by 2 and add the result to row 3.

The last row of the matrix cannot be written with a leading 1 along the main diagonal. The last row represents a contradiction, $0 = 56$.

$$\left[\begin{array}{ccc|c} 1 & 0 & -2 & -5 \\ 0 & 1 & 5 & 18 \\ 0 & 0 & 0 & 56 \end{array}\right] \xrightarrow[\text{system}]{\text{equivalent}} \begin{cases} x - 2z = -5 \\ y + 5z = 18 \\ \qquad\quad 0 = 56 \end{cases}$$

The contradiction indicates that the system is inconsistent and that there is no solution. The solution set is { }.

Skill Practice 1   Solve the system.
$$\begin{aligned} 5x - 9y - 33z &= 3 \\ x - 2y - 7z &= 0 \\ -2x + y + 8z &= -12 \end{aligned}$$

Answer

1. { }

TECHNOLOGY CONNECTIONS

Recognizing an Inconsistent System on a Calculator

We can verify that the system of equations in Example 1 has no solution by using a calculator to find the reduced row-echelon form of the augmented matrix. Notice that the calculator also displays a contradiction in the third row, $0 = 1$.

In Example 1, once we reached the contradiction $0 = 56$, we stopped manipulating the augmented matrix. However, we have the option of simplifying the augmented matrix to reduced row-echelon form. To match the result given in the calculator, we would multiply row 3 by $\frac{1}{56}$. Then add -18 times row 3 to row 2, and add 5 times row 3 to row 1.

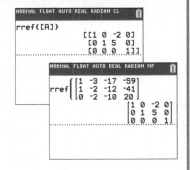

2. Solve Systems with Dependent Equations

Recall that a linear equation in two variables defines a line in a plane. A solution to a system of two linear equations in two variables is a point of intersection of the lines. If the two lines are parallel, the system is inconsistent and has no solution. If the equations in the system represent the same line, we say that the equations are **dependent** and the solution set is the set of all points on the line.

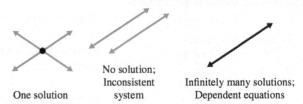

One solution

No solution;
Inconsistent
system

Infinitely many solutions;
Dependent equations

EXAMPLE 2   **Solving a System of Dependent Equations with Two Variables**

Solve the system. $0.25x - 0.75y = 1$
$3y = x - 4$

Solution:

$0.25x - 0.75y = 1$ $\xrightarrow{\text{Multiply by 100.}}$ $25x - 75y = 100$ Write the equations in standard form. As an option, consider clearing decimals in the first equation.

$3y = x - 4$ $\xrightarrow[\text{Standard form}]{}$ $-x + 3y = -4$

$\begin{bmatrix} 25 & -75 & | & 100 \\ -1 & 3 & | & -4 \end{bmatrix}$ $\xrightarrow{\frac{1}{25}R_1 \to R_1}$ $\begin{bmatrix} 1 & -3 & | & 4 \\ -1 & 3 & | & -4 \end{bmatrix}$ To obtain a 1 in the first row, first column, multiply row 1 by $\frac{1}{25}$.

$\xrightarrow{R_1 + R_2 \to R_2}$ $\begin{bmatrix} 1 & -3 & | & 4 \\ 0 & 0 & | & 0 \end{bmatrix}$ To obtain a 0 in the second row, first column, add row 1 to row 2.

The second row of the augmented matrix represents the equation $0 = 0$, which is true regardless of the values of x and y. This means that the original system reduces to the single equation $x - 3y = 4$. That is, the two original equations each represent the same line and all points on the line are solutions to the system. Solving the equation $x - 3y = 4$ for x or y illustrates the dependency between x and y.

Solving the equation $x - 3y = 4$ for x gives: $x = 3y + 4$ ⟵ x depends on the choice of y.

Solving the equation $x - 3y = 4$ for y gives: $y = \dfrac{x - 4}{3}$ ⟵ y depends on the choice of x.

The solution set can be written as

$$\{(3y + 4, y) \mid y \text{ is any real number}\} \quad \text{or} \quad \left\{ \left(x, \frac{x - 4}{3} \right) \,\middle|\, x \text{ is any real number} \right\}.$$

Skill Practice 2   Solve the system. $0.3x - 0.1y = -2$
$y - 20 = 3x$

TIP   In Example 2, we wrote the solution set in two ways: one with an arbitrary value of y and one with an arbitrary value of x. However, when using reduced row-echelon form to find the solution set to a system of dependent equations, we generally let the *last* variable in the ordered pair (ordered triple, etc.) be arbitrary. In Example 2, this is $\{(3y + 4, y) \mid y \text{ is any real number}\}$.

Answer

2. $\left\{ \left(\dfrac{y - 20}{3}, y \right) \,\middle|\, y \text{ is any real number} \right\}$

or $\{(x, 3x + 20) \mid x \text{ is any real number}\}$

A linear equation in three variables represents a plane in space. A solution to a system of equations in three variables is a common point of intersection among all the planes in the system. From Example 1, we have a system with no solution. This means that the planes do not all intersect (Figure 6-1).

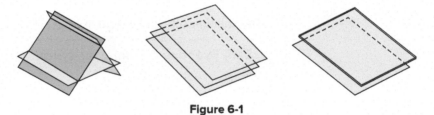

Figure 6-1

A system of linear equations may have infinitely many solutions. In such a case, the equations are dependent. For a system of three equations in three variables, this means that the planes intersect in a common line in space (Figure 6-2 and Figure 6-3), or the three planes all coincide (Figure 6-4).

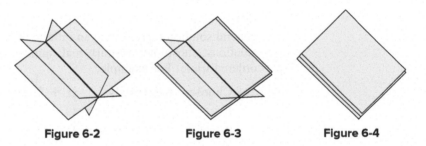

Figure 6-2   **Figure 6-3**   **Figure 6-4**

EXAMPLE 3   **Solving a System of Dependent Equations with Three Variables**

Solve the system.
$$x - 4y + 7z = 14$$
$$-2x + 9y - 16z = -31$$
$$x - 7y + 13z = 23$$

Solution:

$$\begin{aligned} x - 4y + 7z &= 14 \\ -2x + 9y - 16z &= -31 \\ x - 7y + 13z &= 23 \end{aligned}$$
$$\left[\begin{array}{ccc|c} 1 & -4 & 7 & 14 \\ -2 & 9 & -16 & -31 \\ 1 & -7 & 13 & 23 \end{array}\right]$$
Set up the augmented matrix.

$$\xrightarrow[\;-R_1 + R_3 \to R_3\;]{\;2R_1 + R_2 \to R_2\;} \left[\begin{array}{ccc|c} 1 & -4 & 7 & 14 \\ 0 & 1 & -2 & -3 \\ 0 & -3 & 6 & 9 \end{array}\right] \xrightarrow[\;3R_2 + R_3 \to R_3\;]{\;4R_2 + R_1 \to R_1\;} \left[\begin{array}{ccc|c} 1 & 0 & -1 & 2 \\ 0 & 1 & -2 & -3 \\ 0 & 0 & 0 & 0 \end{array}\right]$$

The last row of the matrix cannot be written with a leading 1 along the main diagonal. The last row represents an identity, $0 = 0$.
$$\left[\begin{array}{ccc|c} 1 & 0 & -1 & 2 \\ 0 & 1 & -2 & -3 \\ 0 & 0 & 0 & 0 \end{array}\right] \xrightarrow[\text{system}]{\text{equivalent}} \begin{cases} x - z = 2 \\ y - 2z = -3 \\ 0 = 0 \end{cases}$$

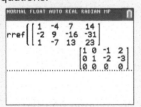

The system of three equations and three variables reduces to a system of two equations and three variables. The third equation, $0 = 0$, is true regardless of the values of x, y, and z. The top two equations each represent a plane in space. Furthermore, two nonparallel planes intersect in a line. All points on the line are solutions to the system, indicating that there are infinitely many solutions.

TIP In Example 3, we can also write the solution set by using a parameter. By choosing any real number t for z, we have:

$$\{(t + 2, 2t - 3, t)\}$$

Since the top two equations both contain the variable z, we can express x and y in terms of z.

$$x - z = 2 \xrightarrow{\text{Express } x \text{ in terms of } z.} x = z + 2$$

$$y - 2z = -3 \xrightarrow[\text{Express } y \text{ in terms of } z.]{} y = 2z - 3$$

The solution set can be written as $\{(z + 2, 2z - 3, z)\,|\,z \text{ is any real number}\}$.

Skill Practice 3   Solve the system.

$$\begin{aligned} -4x - 11y + 3z &= -24 \\ x + 3y - z &= 7 \\ 3x + 11y - 5z &= 29 \end{aligned}$$

We can verify the solution to Example 3, by substituting $(z + 2, 2z - 3, z)$ back into each original equation.

$$x - 4y + 7z = 14 \longrightarrow (z + 2) - 4(2z - 3) + 7z \overset{?}{=} 14 \longrightarrow 14 = 14 \checkmark$$

$$-2x + 9y - 16z = -31 \longrightarrow -2(z + 2) + 9(2z - 3) - 16z \overset{?}{=} -31 \longrightarrow -31 = -31 \checkmark$$

$$x - 7y + 13z = 23 \longrightarrow (z + 2) - 7(2z - 3) + 13z \overset{?}{=} 23 \longrightarrow 23 = 23 \checkmark$$

In Example 3, the solution set $\{(z + 2, 2z - 3, z)\,|\,z \text{ is any real number}\}$ is the general solution representing an infinite number of ordered triples. To find individual solutions, substitute arbitrary real numbers for z, and construct the corresponding ordered triple. For example:

Choose z Arbitrarily	$(z + 2, 2z - 3, z)$	**Solution**
If $z = 1$ $\longrightarrow$	$(1 + 2, 2(1) - 3, 1)$	$(3, -1, 1)$
If $z = 2$ $\longrightarrow$	$(2 + 2, 2(2) - 3, 2)$	$(4, 1, 2)$
If $z = 3$ $\longrightarrow$	$(3 + 2, 2(3) - 3, 3)$	$(5, 3, 3)$
If $z = -1$ $\longrightarrow$	$(-1 + 2, 2(-1) - 3, -1)$	$(1, -5, -1)$

A system of linear equations that has the same number of equations as variables is called a **square system.** In Example 3, we were presented with three equations and three variables in the original system. However, after writing the augmented matrix in row-echelon form, we see that the system reduces to a system of two equations and three variables. A system of linear equations cannot have a unique solution unless there are at least the same number of equations as variables.

In Example 4, we investigate a nonsquare system that has fewer equations than variables.

EXAMPLE 4   **Solving a System with Fewer Equations than Variables**

Solve the system.

$$\begin{aligned} 3x - 8y + 18z &= 15 \\ x - 3y + 4z &= 6 \end{aligned}$$

Solution:

The two given equations each represent a plane in space. By finding the row-echelon form of the augmented matrix, we can determine the geometrical relationship between the planes.

- If we encounter a contradiction, then the system has no solution and the planes are parallel.
- If we encounter an identity, such as $0 = 0$, then the two equations must represent the same plane.
- Otherwise, the planes must meet in a line.

Answer

3. $\{(-2z - 5, z + 4, z)\,|\,z$ is any real number$\}$

$$3x - 8y + 18z = 15 \longrightarrow \begin{bmatrix} 3 & -8 & 18 & | & 15 \\ 1 & -3 & 4 & | & 6 \end{bmatrix}$$
$$x - 3y + 4z = 6 \longrightarrow$$

$$R_1 \Leftrightarrow R_2 \begin{bmatrix} 1 & -3 & 4 & | & 6 \\ 3 & -8 & 18 & | & 15 \end{bmatrix}$$

$$\xrightarrow{-3R_1 + R_2 \rightarrow R_2} \begin{bmatrix} 1 & -3 & 4 & | & 6 \\ 0 & 1 & 6 & | & -3 \end{bmatrix} \xrightarrow{3R_2 + R_1 \rightarrow R_1} \begin{bmatrix} 1 & 0 & 22 & | & -3 \\ 0 & 1 & 6 & | & -3 \end{bmatrix}$$

The augmented matrix is in reduced row-echelon form.

$$\begin{bmatrix} 1 & 0 & 22 & | & -3 \\ 0 & 1 & 6 & | & -3 \end{bmatrix} \longrightarrow \begin{cases} x + 22z = -3 \\ y + 6z = -3 \end{cases}$$

The system did not reduce to a contradiction or an identity. Therefore, the two planes must intersect in a line and there must be infinitely many solutions.

To write the general solution, write x and y in terms of z.

$$x + 22z = -3 \xrightarrow{\text{Express } x \text{ in terms of } z.} x = -22z - 3$$
$$y + 6z = -3 \xrightarrow{\hspace{3cm}} y = -6z - 3$$
$$\text{Express } y \text{ in terms of } z.$$

The solution set is $\{(-22z - 3, -6z - 3, z) \mid z \text{ is any real number}\}$.

> **TIP** In parametric form, the solution to Example 4 is given by:
> $$\{(-22t - 3, -6t - 3, t)\}$$

Skill Practice 4   Solve the system. $x - 3y - 17z = -17$
$$-2x + 7y + 38z = 40$$

In Example 5, we investigate a situation in which the solution set to a system of three variables and three equations consists of all points on a common plane in space.

EXAMPLE 5   **Solving a System of Dependent Equations Representing the Same Plane**

Solve the system. $x + 2y + 3z = 6$
$$-x - 2y - 3z = -6$$
$$2x + 4y + 6z = 12$$

Solution:

$\boxed{A}$ $x + 2y + 3z = 6$
$\boxed{B}$ $-x - 2y - 3z = -6$
$\boxed{C}$ $2x + 4y + 6z = 12$

Upon inspection, you might notice that each equation is a constant multiple of the others. That is, equation $\boxed{B}$ is -1 times equation $\boxed{A}$. Equation $\boxed{C}$ is 2 times equation $\boxed{A}$. This means that equations $\boxed{A}$, $\boxed{B}$, and $\boxed{C}$ represent the same plane in space.

> **TIP** The solution to Example 5 tells us that $x = 6 - 2y - 3z$. That is, x depends on both y and z. If we write the solution set in parametric form, we would need *two* parameters.
> $$\{(6 - 2s - 3t, s, t)\}$$

$$\begin{bmatrix} 1 & 2 & 3 & | & 6 \\ -1 & -2 & -3 & | & -6 \\ 2 & 4 & 6 & | & 12 \end{bmatrix} \xrightarrow[{-2R_1 + R_3 \rightarrow R_3}]{R_1 + R_2 \rightarrow R_2} \begin{bmatrix} 1 & 2 & 3 & | & 6 \\ 0 & 0 & 0 & | & 0 \\ 0 & 0 & 0 & | & 0 \end{bmatrix}$$

The reduced row-echelon form tells us that the system reduces to an equivalent system of one equation with three variables.

The system is equivalent to the single equation $x + 2y + 3z = 6$ and the solution set is the set of all ordered triples that satisfy this equation. Since we have one unique relationship among three variables, one variable is dependent on the other two variables. Letting two variables be arbitrary real numbers, the common equation defines the value of the third variable. For example, solving the equation $x + 2y + 3z = 6$ for x yields: $x = 6 - 2y - 3z$.

The solution set is $\{(6 - 2y - 3z, y, z) \mid y \text{ and } z \text{ are any real numbers}\}$.

Alternatively, the solution set can be expressed as $\{(x, y, z) \mid x + 2y + 3z = 6\}$.

Answer
4. $\{(5z + 1, -4z + 6, z) \mid z \text{ is any real number}\}$

Skill Practice 5   Solve the system.

$$2x - 3y + 5z = 3$$
$$4x - 6y + 10z = 6$$
$$20x - 30y + 50z = 30$$

3. Solve Applications of Systems of Equations

A city planner can study the flow of traffic through a network of streets by measuring the flow rates (number of vehicles per unit time) at various points. For traffic to flow freely, we use the principle that the flow rate into an intersection is equal to the flow rate out of the intersection. This must be true for all intersections in the network.

EXAMPLE 6   Solving an Application Involving Dependent Equations

Consider the network of four one-way streets shown in Figure 6-5. In the figure, x_1, x_2, x_3, and x_4 indicate flow rates (in vehicles per hour) along the stretches of roads AB, BC, CD, and DA, respectively. The other numbers in the figure indicate other flow rates moving into and out of intersections A, B, C, and D.

a. Set up a system of equations that represents traffic flowing freely.

b. Write the augmented matrix for the system of equations in reduced row-echelon form.

c. If the traffic between intersections D and A is 260 vehicles per hour, determine the flow rates x_1, x_2, and x_3.

d. If the traffic between intersections D and A is between 250 and 300 vehicles per hour, inclusive, determine the flow rates x_1, x_2, and x_3.

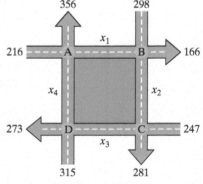

Figure 6-5

Solution:

a.

Intersection	Flow in	=	Flow out
A	$216 + x_4$	=	$356 + x_1$
B	$298 + x_1$	=	$166 + x_2$
C	$x_2 + 247$	=	$281 + x_3$
D	$x_3 + 315$	=	$273 + x_4$

The number of cars flowing into each intersection must equal the number of cars leaving the intersection for traffic to flow freely.

b.
$$\begin{aligned} -x_1 \qquad\qquad + x_4 &= 140 \\ x_1 - x_2 \qquad\qquad &= -132 \\ x_2 - x_3 \qquad &= 34 \\ x_3 - x_4 &= -42 \end{aligned}\Bigg\} \rightarrow \left[\begin{array}{cccc|c} -1 & 0 & 0 & 1 & 140 \\ 1 & -1 & 0 & 0 & -132 \\ 0 & 1 & -1 & 0 & 34 \\ 0 & 0 & 1 & -1 & -42 \end{array}\right]$$

Write the equations in standard form. Set up the augmented matrix.

Answer

5. $\left\{ \left(\dfrac{3 + 3y - 5z}{2}, y, z \right) \middle|\, y \text{ and } z \right.$

are any real numbers $\Big\}$ or

$\{(x, y, z) \,|\, 2x - 3y + 5z = 3\}$

$$\xrightarrow{-1R_1 \to R_1} \begin{bmatrix} 1 & 0 & 0 & -1 & | & -140 \\ 1 & -1 & 0 & 0 & | & -132 \\ 0 & 1 & -1 & 0 & | & 34 \\ 0 & 0 & 1 & -1 & | & -42 \end{bmatrix} \xrightarrow{-1R_1 + R_2 \to R_2} \begin{bmatrix} 1 & 0 & 0 & -1 & | & -140 \\ 0 & -1 & 0 & 1 & | & 8 \\ 0 & 1 & -1 & 0 & | & 34 \\ 0 & 0 & 1 & -1 & | & -42 \end{bmatrix}$$

$$\xrightarrow{-1R_2 \to R_2} \begin{bmatrix} 1 & 0 & 0 & -1 & | & -140 \\ 0 & 1 & 0 & -1 & | & -8 \\ 0 & 1 & -1 & 0 & | & 34 \\ 0 & 0 & 1 & -1 & | & -42 \end{bmatrix} \xrightarrow{-1R_2 + R_3 \to R_3} \begin{bmatrix} 1 & 0 & 0 & -1 & | & -140 \\ 0 & 1 & 0 & -1 & | & -8 \\ 0 & 0 & -1 & 1 & | & 42 \\ 0 & 0 & 1 & -1 & | & -42 \end{bmatrix}$$

$$\xrightarrow{-1R_3 \to R_3} \begin{bmatrix} 1 & 0 & 0 & -1 & | & -140 \\ 0 & 1 & 0 & -1 & | & -8 \\ 0 & 0 & 1 & -1 & | & -42 \\ 0 & 0 & 1 & -1 & | & -42 \end{bmatrix} \xrightarrow{-1R_3 + R_4 \to R_4} \begin{bmatrix} 1 & 0 & 0 & -1 & | & -140 \\ 0 & 1 & 0 & -1 & | & -8 \\ 0 & 0 & 1 & -1 & | & -42 \\ 0 & 0 & 0 & 0 & | & 0 \end{bmatrix}$$

c. From the reduced row-echelon form, we see that the flow rates x_1, x_2, and x_3 can all be expressed in terms of the flow rate x_4.

$$\begin{bmatrix} 1 & 0 & 0 & -1 & | & -140 \\ 0 & 1 & 0 & -1 & | & -8 \\ 0 & 0 & 1 & -1 & | & -42 \\ 0 & 0 & 0 & 0 & | & 0 \end{bmatrix} \longrightarrow \begin{matrix} x_1 - x_4 = -140 \\ x_2 - x_4 = -8 \\ x_3 - x_4 = -42 \end{matrix} \longrightarrow \begin{matrix} x_1 = x_4 - 140 \\ x_2 = x_4 - 8 \\ x_3 = x_4 - 42 \end{matrix}$$

If x_4 is 260 vehicles per hour, we have the following values for x_1, x_2, and x_3.

Flow rate between A and B: $x_1 = 260 - 140 \longrightarrow x_1 = 120$ vehicles per hour

Flow rate between B and C: $x_2 = 260 - 8 \longrightarrow x_2 = 252$ vehicles per hour

Flow rate between C and D: $x_3 = 260 - 42 \longrightarrow x_3 = 218$ vehicles per hour

d. If the traffic flow between intersections D and A is given by $250 \le x_4 \le 300$, we can solve the following inequalities to determine the flow rates x_1, x_2, and x_3.

$$250 \le x_1 + 140 \le 300 \longrightarrow 110 \le x_1 \le 160$$
$$250 \le x_2 + 8 \le 300 \longrightarrow 242 \le x_2 \le 292$$
$$250 \le x_3 + 42 \le 300 \longrightarrow 208 \le x_3 \le 258$$

Skill Practice 6 Refer to the figure. Assume that traffic flows freely with flow rates given in vehicles per hour.

a. If the traffic between intersections D and A is 400 vehicles per hour, determine the flow rates x_1, x_2, and x_3.

b. If the traffic between intersections D and A is between 380 and 420 vehicles per hour, determine the flow rates x_1, x_2, and x_3.

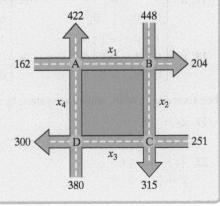

Answers

6. a. $x_1 = 140$, $x_2 = 384$, and $x_3 = 320$

b. $120 \le x_1 \le 160$, $364 \le x_2 \le 404$, $300 \le x_3 \le 340$

SECTION 6.2   Practice Exercises

Prerequisite Review

For Exercises R.1–R.2, solve for the indicated variable.

R.1. $5x + y = -18$ for y **R.2.** $3x - y = -8$ for x

For Exercises R.3–R.4, determine whether the graph of the equation is symmetric with respect to the x-axis or y-axis.

R.3. $y = -|x| - 6$ **R.4.** $x = y^2 + 7$

Concept Connections

1. True or false? A system of linear equations in three variables may have no solution.

2. True or false? A system of linear equations in three variables may have exactly one solution.

3. True or false? A system of linear equations in three variables may have exactly two solutions.

4. True or false? A system of linear equations in three variables may have infinitely many solutions.

5. If a system of linear equations has no solution, then the system is said to be _____.

6. If a system of linear equations has infinitely many solutions, then the equations are said to be _____.

Objectives 1–2: Identify Inconsistent Systems and Solve Systems with Dependent Equations

For Exercises 7–14, an augmented matrix is given. Determine the number of solutions to the corresponding system of equations.

7. $\begin{bmatrix} 1 & 2 & | & 4 \\ 0 & 0 & | & 5 \end{bmatrix}$

8. $\begin{bmatrix} 1 & 0 & 2 & | & 5 \\ 0 & 1 & 4 & | & -2 \\ 0 & 0 & 0 & | & -1 \end{bmatrix}$

9. $\begin{bmatrix} 1 & 0 & 4 & | & 3 \\ 0 & 1 & -1 & | & 6 \\ 0 & 0 & 0 & | & 0 \end{bmatrix}$

10. $\begin{bmatrix} 1 & 3 & | & 5 \\ 0 & 0 & | & 0 \end{bmatrix}$

11. $\begin{bmatrix} 1 & 0 & 0 & | & -3 \\ 0 & 1 & 0 & | & 4 \\ 0 & 0 & 1 & | & 0 \end{bmatrix}$

12. $\begin{bmatrix} 1 & 0 & | & 3 \\ 0 & 1 & | & 0 \end{bmatrix}$

13. $\begin{bmatrix} 1 & 2 & 5 & | & -1 \\ 0 & 0 & 0 & | & 0 \\ 0 & 0 & 0 & | & 0 \end{bmatrix}$

14. $\begin{bmatrix} 1 & 0 & 6 & | & 7 \\ 0 & 0 & 0 & | & 0 \\ 0 & 0 & 0 & | & 0 \end{bmatrix}$

For Exercises 15–18, determine the solution set for the system represented by each augmented matrix.

15. a. $\begin{bmatrix} 1 & 2 & | & 5 \\ 0 & 1 & | & 0 \end{bmatrix}$ **b.** $\begin{bmatrix} 1 & 2 & | & 5 \\ 0 & 0 & | & 0 \end{bmatrix}$ **c.** $\begin{bmatrix} 1 & 2 & | & 5 \\ 0 & 0 & | & 1 \end{bmatrix}$

16. a. $\begin{bmatrix} 1 & 3 & | & -4 \\ 0 & 1 & | & 1 \end{bmatrix}$ **b.** $\begin{bmatrix} 1 & 3 & | & -4 \\ 0 & 0 & | & 1 \end{bmatrix}$ **c.** $\begin{bmatrix} 1 & 3 & | & -4 \\ 0 & 0 & | & 0 \end{bmatrix}$

17. a. $\begin{bmatrix} 1 & 0 & 6 & | & 3 \\ 0 & 1 & 4 & | & 5 \\ 0 & 0 & 1 & | & 0 \end{bmatrix}$ **b.** $\begin{bmatrix} 1 & 0 & 6 & | & 3 \\ 0 & 1 & 4 & | & 5 \\ 0 & 0 & 0 & | & 1 \end{bmatrix}$ **c.** $\begin{bmatrix} 1 & 0 & 6 & | & 3 \\ 0 & 1 & 4 & | & 5 \\ 0 & 0 & 0 & | & 0 \end{bmatrix}$

18. a. $\begin{bmatrix} 1 & 0 & -2 & | & 3 \\ 0 & 1 & 3 & | & 5 \\ 0 & 0 & 1 & | & 1 \end{bmatrix}$ **b.** $\begin{bmatrix} 1 & 0 & -2 & | & 3 \\ 0 & 1 & 3 & | & 5 \\ 0 & 0 & 0 & | & 0 \end{bmatrix}$ **c.** $\begin{bmatrix} 1 & 0 & -2 & | & 3 \\ 0 & 1 & 3 & | & 5 \\ 0 & 0 & 0 & | & 1 \end{bmatrix}$

For Exercises 19–38, solve the system by using Gaussian elimination or Gauss-Jordan elimination. (See Exercises 1–5)

19. $2x + 4y = 5$
     $x + 2y = 4$

20. $4x + 16y = 21$
     $x + 4y = -1$

21. $2x + 7y = 10$
     $\frac{1}{5}x = 1 - \frac{7}{10}y$

22. $4x - 3y = 6$
     $y = \frac{4}{3}x - 2$

23. $x - 3y + 14z = -9$
     $-2x + 7y - 31z = 21$
     $x - 5y + 20z = -14$

24. $x - 3y + 17z = 1$
     $x - y + 7z = 2$
     $2x - 5y + 29z = 5$

25. $5x + 7y - 11z = 45$
$3x + 5y - 9z = 23$
$x + y - z = 11$

26. $x + 3y + 9z = 12$
$2x + 7y + 22z = 26$
$-5x - 17y - 53z = -64$

27. $2x = 5y - 16z + 40$
$2(x + y) = 4z$
$x - 2y + 7z = 18$

28. $x = 2y + 4z + 5$
$y - 4 = -3z$
$5(x - y) - 9z = 4x - 7$

29. $2x - 5y - 20z = -24$
$x - 3y - 11z = -15$

30. $2x - y - 5z = -3$
$x - 2y - 7z = -12$

31. $2x + 3y + 4z = 12$
$-4x - 6y - 8z = -24$
$x + 1.5y + 2z = 6$

32. $-x + 2y + 7z = 14$
$10x - 20y - 70z = -140$
$-\frac{1}{7}x + \frac{2}{7}y + z = 2$

33. $2x - 3y + 9z = -2$
$x = 5y - 8z - 15$
$3(x - y) + 6z = 2x - 7$

34. $3x + 11y - 3z = -13$
$x + y = z - 15$
$3(x + y) = z + 2x - 7$

35. $-5x + 12y - 20z = -11$
$x + 4z = 3y + 1$

36. $x = 4y + 20z - 1$
$-2x + 5y + 25z = -1$

37. $x_1 - 3x_2 + 9x_3 - 14x_4 = 32$
$x_2 - 3x_3 + 6x_4 = -10$
$x_2 - x_3 + 2x_4 = -4$
$x_1 - 2x_2 + 8x_3 - 12x_4 = 24$

38. $x_1 - 3x_3 - 12x_4 = -15$
$x_2 + x_3 + 6x_4 = 8$
$x_2 - 2x_3 - 6x_4 = -7$
$-2x_1 + 4x_3 + 16x_4 = 22$

For Exercises 39–40, the solution set to a system of dependent equations is given. Write the specific solutions corresponding to the given values of z.

39. $\{(2z + 1, z - 4, z) \mid z \text{ is any real number}\}$
   a. $z = 1$ **b.** $z = 4$ **c.** $z = -2$

40. $\{(z + 5, 3z - 2, z) \mid z \text{ is any real number}\}$
   a. $z = -3$ **b.** $z = 1$ **c.** $z = 0$

For Exercises 41–44, the solution set to a system of dependent equations is given. Write three ordered triples that are solutions to the system. Answers may vary.

41. $\{(4z, 6 - z, z) \mid z \text{ is any real number}\}$

42. $\{(2z, z - 3, z) \mid z \text{ is any real number}\}$

43. $\left\{ \left(\dfrac{6 - 3y - 6z}{2}, y, z \right) \,\middle|\, y \text{ and } z \text{ are any real numbers} \right\}$

44. $\{(4y + 2z - 20, y, z) \mid y \text{ and } z \text{ are any real numbers}\}$

Objective 3: Solve Applications of Systems of Equations

For Exercises 45–48, assume that traffic flows freely through the intersections A, B, C, and D. The values x_1, x_2, x_3, and x_4 and the other numbers in the figures represent flow rates in vehicles per hour. (See Example 6)

45.

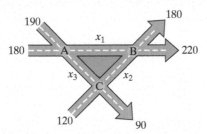

  a. Write an equation representing equal flow into and out of intersection A.

  b. Write an equation representing equal flow into and out of intersection B.

  c. Write an equation representing equal flow into and out of intersection C.

  d. Write the system of equations from parts (a)–(c) in standard form.

  e. Write the reduced row-echelon form of the augmented matrix representing the system of equations from part (d).

  f. If the flow rate between intersections A and C is 120 vehicles per hour, determine the flow rates x_1 and x_2.

  g. If the flow rate between intersections A and C is between 100 and 150 vehicles per hour, inclusive, determine the flow rates x_1 and x_2.

46.

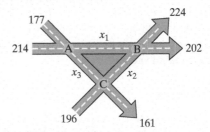

  a. Write an equation representing equal flow into and out of intersection A.

  b. Write an equation representing equal flow into and out of intersection B.

  c. Write an equation representing equal flow into and out of intersection C.

  d. Write the system of equations from parts (a)–(c) in standard form.

  e. Write the reduced row-echelon form of the augmented matrix representing the system of equations in part (d).

  f. If the flow rate between intersections A and C is 156 vehicles per hour, determine the flow rates x_1 and x_2.

  g. If the flow rate between intersections A and C is between 100 and 200 vehicles per hour, inclusive, determine the flow rates x_1 and x_2.

47.

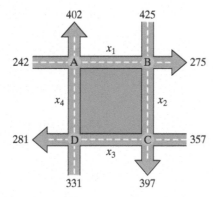

48.

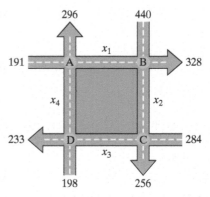

a. Assume that traffic flows at a rate of 220 vehicles per hour on the stretch of road between intersections D and A. Find the flow rates x_1, x_2, and x_3.

b. If traffic flows at a rate of between 200 and 250 vehicles per hour inclusive between intersections D and A, find the flow rates x_1, x_2, and x_3.

a. Assume that traffic flows at a rate of 180 vehicles per hour on the stretch of road between intersections D and A. Find the flow rates x_1, x_2, and x_3.

b. If traffic flows at a rate of between 150 and 200 vehicles per hour inclusive between intersections D and A, find the flow rates x_1, x_2, and x_3.

49. An accountant checks the reported earnings for a theater for three nightly performances against the number of tickets sold.

Night	Children Tickets	Student Tickets	General Admission	Total Revenue
1	80	400	480	$9280
2	50	350	400	$7800
3	75	525	600	$10,500

a. Let x, y, and z represent the cost for children tickets, student tickets, and general admission tickets, respectively. Set up a system of equations to solve for x, y, and z.

b. Set up the augmented matrix for the system and solve the system. (*Hint*: To make the augmented matrix simpler to work with, consider dividing each linear equation by an appropriate constant.)

c. Explain why the auditor knows that there was an error in the record keeping.

50. A concession stand at a city park sells hamburgers, hot dogs, and drinks. Three patrons buy the following food and drink combinations for the following prices.

Patron	Hamburgers	Hot Dogs	Drinks	Total Revenue
1	1	1	5	$11
2	0	1	2	$5
3	3	1	11	$22

a. Let x, y, and z represent the cost for a hamburger, a hot dog, and a drink, respectively. Set up a system of equations to solve for x, y, and z.

b. Set up the augmented matrix for the system and solve the system.

c. Explain why the concession stand manager knows that there was an error in the record keeping.

Mixed Exercises

51. The solution set to Exercise 25 is $\{(-2z + 16, 3z - 5, z) \mid z$ is any real number$\}$. Verify the solution by substituting the ordered triple into each individual equation.

52. The solution set to Exercise 26 is $\{(3z + 6, -4z + 2, z) \mid z$ is any real number$\}$. Verify the solution by substituting the ordered triple into each individual equation.

The systems in Exercises 53–56 are called homogeneous systems because each has $(0, 0, 0)$ as a solution. However, if a system is made up of dependent equations, it will have infinitely many more solutions. For each system, determine whether $(0, 0, 0)$ is the only solution or if the system has infinitely many solutions. If the system has infinitely many solutions, give the solution set.

53.
$$x + 2y - 8z = 0$$
$$-5x - 11y + 43z = 0$$
$$x + 5y - 12z = 0$$

54.
$$x - 2y - 7z = 0$$
$$-3x + 8y + 31z = 0$$
$$-2x + 5y + 22z = 0$$

55.
$$x - 5y + 13z = 0$$
$$-3x + 17y - 45z = 0$$
$$x - 4y + 10z = 0$$

56.
$$-2x + 15y - 83z = 0$$
$$x - 7y + 39z = 0$$
$$x - 5y + 29z = 0$$

Write About It

57. Explain how you can determine from the reduced row-echelon form of a matrix whether the corresponding system of equations is inconsistent.

58. What can you conclude about a system of equations if the corresponding reduced row-echelon form consists of a row entirely of zeros?

59. Consider the following system. By inspection describe the geometrical relationship among the planes represented by the three equations.

$$x + y + z = 1$$
$$2x + 2y + 2z = 2$$
$$3x + 3y + 3z = 3$$

60. Explain why a system of two equations with three variables cannot have exactly one ordered triple as its solution.

Technology Connections

For Exercises 61–66, use a graphing utility to find the reduced row-echelon form of the augmented matrix for the system in the given exercise. Use the result to verify the answer to the given exercise.

61. Exercise 23

62. Exercise 24

63. Exercise 25

64. Exercise 26

65. Exercise 31

66. Exercise 32

SECTION 6.3   Operations on Matrices

OBJECTIVES

1. Determine the Order of a Matrix
2. Add and Subtract Matrices
3. Multiply a Matrix by a Scalar
4. Multiply Matrices
5. Apply Operations on Matrices

1. Determine the Order of a Matrix

We have seen how the manipulation of an augmented matrix can be used to solve a system of linear equations. Matrices have many other useful mathematical applications, particularly when manipulating tables or databases. In particular, matrices play a useful role in digital photography, film, and computer animation.

The **order of a matrix** is determined by the number of rows and number of columns. A matrix with m rows and n columns is an $m \times n$ matrix (read as "m by n" matrix).

EXAMPLE 1   Determining the Order of a Matrix

Determine the order of each matrix.

a. $\begin{bmatrix} 3 & \pi & 1.7 \\ -1 & 6 & 10 \end{bmatrix}$
b. $\begin{bmatrix} -7 \\ 2 \\ 4 \\ \sqrt{2} \end{bmatrix}$
c. $\begin{bmatrix} 1 & 0 & 0 \\ 0 & 1 & 0 \\ 0 & 0 & 1 \end{bmatrix}$
d. $\begin{bmatrix} x & y & z \end{bmatrix}$

Solution:

a. The matrix has 2 rows and 3 columns. Therefore, it is a 2 × 3 matrix.

b. The matrix has 4 rows and 1 column. Therefore, it is a 4 × 1 matrix. A matrix with only 1 column is called a **column matrix.**

c. The matrix has 3 rows and 3 columns. Therefore, it is a 3 × 3 matrix. A matrix with the same number of rows and columns is called a **square matrix.**

d. The matrix has 1 row and 3 columns. Therefore, it is a 1 × 3 matrix. A matrix with only 1 row is called a **row matrix.**

Skill Practice 1 Determine the order of each matrix.

a. $\begin{bmatrix} -9 & 2 \\ 4.1 & -3 \\ \sqrt{3} & 4 \end{bmatrix}$
b. $\begin{bmatrix} -4 \\ 11 \end{bmatrix}$
c. $\begin{bmatrix} -0.1 & 0.4 \\ 0.5 & 0.2 \end{bmatrix}$
d. $[105 \quad 311]$

A matrix can be represented generically as follows:

TIP The notation $[a_{ij}]$ is a generic way to denote a matrix with elements of the form a_{ij}, where i and j represent generic row and column numbers.

$$[a_{ij}] = \begin{bmatrix} a_{11} & a_{12} & \cdots & a_{1n} \\ a_{21} & a_{22} & \cdots & a_{2n} \\ \vdots & & & \\ a_{m1} & a_{m2} & \cdots & a_{mn} \end{bmatrix}$$

Using the double subscript notation, a_{43} represents the element in the 4th row, 3rd column. The notation a_{ij} represents the element in the ith row, jth column.

EXAMPLE 2 Using Matrix Notation

Consider $A = \begin{bmatrix} -4 & 5 & \sqrt{2} & -\pi \\ 10 & \frac{1}{2} & 0 & 6 \\ -\frac{7}{6} & 8 & 12 & -9 \end{bmatrix}$, where $A = [a_{ij}]$. Determine the value of each element.

a. a_{32}
b. a_{23}
c. a_{14}

Solution:

a. a_{32} represents the element in the 3rd row, 2nd column: 8

b. a_{23} represents the element in the 2nd row, 3rd column: 0

c. a_{14} represents the element in the 1st row, 4th column: $-\pi$

Skill Practice 2 Given matrix A from Example 2, determine the value of each element.

a. a_{13}
b. a_{31}
c. a_{34}

Answers

1. a. 3 × 2   b. 2 × 1

   c. 2 × 2   d. 1 × 2

2. a. $\sqrt{2}$ b. $-\dfrac{7}{6}$ c. -9

2. Add and Subtract Matrices

We now consider the conditions under which two matrices are equal.

Equality of Matrices

Two matrices are equal if and only if they have the same order and if their corresponding elements are equal.

Example:

- The statement $\begin{bmatrix} 3 & -4 \\ x & z \end{bmatrix} = \begin{bmatrix} 3 & y \\ 9 & 7 \end{bmatrix}$ implies that $x = 9$, $y = -4$, and $z = 7$.

- Conversely, if $x = 9$, $y = -4$, and $z = 7$, then $\begin{bmatrix} 3 & -4 \\ x & z \end{bmatrix} = \begin{bmatrix} 3 & y \\ 9 & 7 \end{bmatrix}$.

Next, we define addition and subtraction of matrices.

Addition and Subtraction of Matrices

Let $A = [a_{ij}]$ and $B = [b_{ij}]$ be m by n matrices. Then,

$$A + B = [a_{ij} + b_{ij}] \text{ for } i = 1, 2, \ldots, m \text{ and } j = 1, 2, \ldots, n$$
$$A - B = [a_{ij} - b_{ij}] \text{ for } i = 1, 2, \ldots, m \text{ and } j = 1, 2, \ldots, n$$

That is, to add or subtract two matrices, the matrices must have the same order, and the sum or difference is found by adding or subtracting the corresponding elements.

EXAMPLE 3   Adding and Subtracting Matrices

Given $A = \begin{bmatrix} 3 & -7 & 0 \\ 1 & \frac{4}{3} & -6 \end{bmatrix}$, $B = \begin{bmatrix} 8 & -3 & 4 \\ 0 & 1 & -9 \end{bmatrix}$, and $C = \begin{bmatrix} 3 & -5 \\ 4 & 18 \end{bmatrix}$,

find the sum or difference if possible.

a. $A + B$ **b.** $B - A$ **c.** $A + C$

Solution:

a. $A + B = \begin{bmatrix} 3 & -7 & 0 \\ 1 & \frac{4}{3} & -6 \end{bmatrix} + \begin{bmatrix} 8 & -3 & 4 \\ 0 & 1 & -9 \end{bmatrix}$ Matrix A and matrix B both have the same order (2×3), so the sum is defined.

$= \begin{bmatrix} 3 + 8 & -7 + (-3) & 0 + 4 \\ 1 + 0 & \frac{4}{3} + 1 & -6 + (-9) \end{bmatrix}$ Add the elements in the corresponding positions.

$= \begin{bmatrix} 11 & -10 & 4 \\ 1 & \frac{7}{3} & -15 \end{bmatrix}$ Simplify.

b. $B - A = \begin{bmatrix} 8 & -3 & 4 \\ 0 & 1 & -9 \end{bmatrix} - \begin{bmatrix} 3 & -7 & 0 \\ 1 & \frac{4}{3} & -6 \end{bmatrix}$ Matrix A and matrix B both have the same order (2×3), so the difference is defined.

$= \begin{bmatrix} 8 - 3 & -3 - (-7) & 4 - 0 \\ 0 - 1 & 1 - \frac{4}{3} & -9 - (-6) \end{bmatrix}$ Subtract the elements in the corresponding positions.

$= \begin{bmatrix} 5 & 4 & 4 \\ -1 & -\frac{1}{3} & -3 \end{bmatrix}$ Simplify.

c. Matrix A is a 2×3 matrix, whereas matrix C is a 2×2 matrix. The matrices are of different orders and therefore cannot be added or subtracted.

Skill Practice 3 Given $A = \begin{bmatrix} 3 & 4 & -7 \\ 9 & 2 & 0 \end{bmatrix}$, $B = \begin{bmatrix} -8 & 0 \\ 1 & 5 \\ -4 & \frac{1}{2} \end{bmatrix}$, and $C = \begin{bmatrix} 4 & -5 \\ 12 & 3 \\ -2 & 2 \end{bmatrix}$, find the sum or difference if possible.

a. $A + B$ **b.** $B + C$ **c.** $C - B$

The opposite of a real number a is $-a$. The value $-a$ is also called the additive inverse of a. A matrix A also has an additive inverse, denoted by $-A$.

Additive Inverse of a Matrix

Given $A = [a_{ij}]$, the **additive inverse of A**, denoted by $-A$, is $-A = [-a_{ij}]$.

Verbal Interpretation	Example
The additive inverse of a matrix is found by taking the opposite of each element in the matrix.	Given $A = \begin{bmatrix} -4 & 5 & -1 \\ 0 & -3 & 2 \end{bmatrix}$, $-A = \begin{bmatrix} 4 & -5 & 1 \\ 0 & 3 & -2 \end{bmatrix}$.

The number 0 is the identity element under the addition of real numbers because $a + (-a) = 0$. A matrix in which all elements are zero is called a **zero matrix** and is denoted by 0. The sum of a matrix A and its additive inverse $-A$ is the zero matrix of the same order. For example,

$$\underset{A}{\begin{bmatrix} -4 & 5 & -1 \\ 0 & -3 & 2 \end{bmatrix}} + \underset{(-A)}{\begin{bmatrix} 4 & -5 & 1 \\ 0 & 3 & -2 \end{bmatrix}} = \underset{0}{\begin{bmatrix} 0 & 0 & 0 \\ 0 & 0 & 0 \end{bmatrix}} \quad 2 \times 3 \text{ zero matrix}$$

Properties of Matrix Addition

Let A, B, and C be matrices of order $m \times n$, and let 0 be the zero matrix of order $m \times n$. Then,

1. $A + B = B + A$ Commutative property of matrix addition
2. $A + (B + C) = (A + B) + C$ Associative property of matrix addition
3. $A + (-A) = 0$ Inverse property of matrix addition
4. $A + 0 = 0 + A = A$ Identity property of matrix addition

3. Multiply a Matrix by a Scalar

Next we look at the product of a real number k and a matrix. This is called **scalar multiplication.** The real number k is called a **scalar** to distinguish it from a matrix.

Answers

3. a. Not possible

b. $\begin{bmatrix} -4 & -5 \\ 13 & 8 \\ -6 & \frac{5}{2} \end{bmatrix}$

c. $\begin{bmatrix} 12 & -5 \\ 11 & -2 \\ 2 & \frac{3}{2} \end{bmatrix}$

Scalar Multiplication

Let $A = [a_{ij}]$ be an $m \times n$ matrix and let k be a real number. Then,
$kA = [ka_{ij}]$.

Verbal Interpretation	Example
To multiply a matrix A by a scalar k, multiply each element in the matrix by k.	Given $A = \begin{bmatrix} -4 & 5 & -1 \\ 0 & -3 & 2 \end{bmatrix}$, $5A = \begin{bmatrix} -20 & 25 & -5 \\ 0 & -15 & 10 \end{bmatrix}$.

The following properties of scalar multiplication are similar to those of the multiplication of real numbers.

Properties of Scalar Multiplication

Let A and B be $m \times n$ matrices and let c and d be real numbers. Then,

1. $c(A + B) = cA + cB$ Distributive property of scalar multiplication
2. $(c + d)A = cA + dA$ Distributive property of scalar multiplication
3. $c(dA) = (cd)A$ Associative property of scalar multiplication

---

EXAMPLE 4　**Multiplying a Matrix by a Scalar**

Given $A = \begin{bmatrix} 3 & -5 \\ 0 & 1 \\ -4 & 2 \end{bmatrix}$ and $B = \begin{bmatrix} -8 & 9 \\ 4 & -1 \\ 0 & 3 \end{bmatrix}$, find $2A + 4(A + B)$.

Solution:

$2A + 4(A + B) = 2A + 4A + 4B$ Apply the distributive property of scalar
$\qquad\qquad\qquad = 6A + 4B$ multiplication.

$6A + 4B = 6\begin{bmatrix} 3 & -5 \\ 0 & 1 \\ -4 & 2 \end{bmatrix} + 4\begin{bmatrix} -8 & 9 \\ 4 & -1 \\ 0 & 3 \end{bmatrix} = \begin{bmatrix} 18 & -30 \\ 0 & 6 \\ -24 & 12 \end{bmatrix} + \begin{bmatrix} -32 & 36 \\ 16 & -4 \\ 0 & 12 \end{bmatrix}$

$\qquad\qquad = \begin{bmatrix} -14 & 6 \\ 16 & 2 \\ -24 & 24 \end{bmatrix}$

Skill Practice 4　Given $A = [4 \quad -3 \quad 9]$ and $B = [-2 \quad 0 \quad 3]$, find $-5A - 2(A + B)$.

Answer

4. $[-24 \quad 21 \quad -69]$

Adding Matrices and Multiplying a Scalar by a Matrix

Many graphing utilities have the capability to perform operations on matrices. From Example 4, we can enter matrix A and matrix B into the calculator using the EDIT feature under the MATRIX menu. Then on the home screen enter $6[A] + 4[B]$ and press ⟨ENTER⟩.

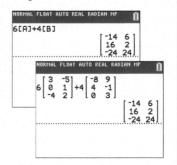

Alternatively, in "MATHPRINT" mode, enter the matrices on the home screen using ⟨ALPHA⟩ F3.

As seen, many of the same properties that are true for addition and multiplication of real numbers are true for the addition of matrices and for scalar multiplication. As a result, we can solve a matrix equation involving these operations in the same way as we solve a linear equation.

EXAMPLE 5   **Solving a Matrix Equation**

Solve the matrix equation for X, given that $A = \begin{bmatrix} 2 & 5 \\ 1 & -4 \end{bmatrix}$ and $B = \begin{bmatrix} 6 & -1 \\ -3 & -8 \end{bmatrix}$.

$4X - A = B$

Solution:

$4X - A = B$

$4X = A + B$ Add matrix A to both sides.

$X = \dfrac{1}{4}(A + B)$ Perform scalar multiplication by $\frac{1}{4}$ on both sides.

$X = \dfrac{1}{4}\left(\begin{bmatrix} 2 & 5 \\ 1 & -4 \end{bmatrix} + \begin{bmatrix} 6 & -1 \\ -3 & -8 \end{bmatrix}\right)$ Add the matrices within parentheses.

$X = \dfrac{1}{4}\begin{bmatrix} 8 & 4 \\ -2 & -12 \end{bmatrix}$ Multiply the matrix by the scalar $\frac{1}{4}$.

$X = \begin{bmatrix} 2 & 1 \\ -\frac{1}{2} & -3 \end{bmatrix}$ Simplify.

> **TIP** We can substitute matrix X back into the equation $4X - A = B$ to verify that it is a solution to the equation.

Skill Practice 5   Given $A = \begin{bmatrix} 4 & -3 \\ 1 & 11 \end{bmatrix}$ and $B = \begin{bmatrix} 2 & -1 \\ 6 & 5 \end{bmatrix}$, solve $2X + A = B$ for X.

---

4. Multiply Matrices

Finding the product of two matrices is more complicated than finding the product of a scalar and a matrix. We will demonstrate the process to multiply two matrices and then offer a formal definition.

Answer

5. $X = \begin{bmatrix} -1 & 1 \\ \frac{5}{2} & -3 \end{bmatrix}$

Consider $A = \begin{bmatrix} 2 & -3 & 1 \\ -4 & 7 & 0 \end{bmatrix}$ and $B = \begin{bmatrix} -1 & -6 \\ 10 & 5 \\ 8 & -2 \end{bmatrix}$.

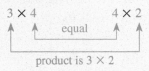

- To multiply AB, we require that the number of columns in A be equal to the number of rows in B.

- The resulting matrix will have dimensions equal to the number of rows of A by the number of columns of B.

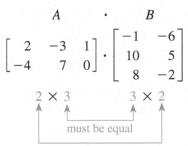

The product AB will be a 2 × 2 matrix.

The product will be a 2 × 2 matrix.

$$\begin{bmatrix} 2 & -3 & 1 \\ -4 & 7 & 0 \end{bmatrix} \cdot \begin{bmatrix} -1 & -6 \\ 10 & 5 \\ 8 & -2 \end{bmatrix} = \begin{bmatrix} \square & \square \\ \square & \square \end{bmatrix}$$

To find the entry in the first row, first column of the product, multiply the corresponding elements in the first row of A by the elements in the first column of B, and take the sum.

$$\begin{bmatrix} 2 & -3 & 1 \\ -4 & 7 & 0 \end{bmatrix} \cdot \begin{bmatrix} -1 & -6 \\ 10 & 5 \\ 8 & -2 \end{bmatrix} = \begin{bmatrix} 2(-1) + (-3)(10) + 1(8) & \square \\ \square & \square \end{bmatrix}$$

To find the entry in the first row, second column of the product, take the sum of the products of the corresponding elements in the first row of A and second column of B.

$$\begin{bmatrix} 2 & -3 & 1 \\ -4 & 7 & 0 \end{bmatrix} \cdot \begin{bmatrix} -1 & -6 \\ 10 & 5 \\ 8 & -2 \end{bmatrix} = \begin{bmatrix} 2(-1) + (-3)(10) + 1(8) & 2(-6) + (-3)(5) + 1(-2) \\ \square & \square \end{bmatrix}$$

To find the entry in the second row, first column of the product, take the sum of the products of the corresponding elements in the second row of A and first column of B.

$$\begin{bmatrix} 2 & -3 & 1 \\ -4 & 7 & 0 \end{bmatrix} \cdot \begin{bmatrix} -1 & -6 \\ 10 & 5 \\ 8 & -2 \end{bmatrix} = \begin{bmatrix} 2(-1) + (-3)(10) + 1(8) & 2(-6) + (-3)(5) + 1(-2) \\ -4(-1) + 7(10) + 0(8) & \square \end{bmatrix}$$

To find the entry in the second row, second column of the product, take the sum of the products of the corresponding elements in the second row of A and second column of B.

$$\begin{bmatrix} 2 & -3 & 1 \\ -4 & 7 & 0 \end{bmatrix} \cdot \begin{bmatrix} -1 & -6 \\ 10 & 5 \\ 8 & -2 \end{bmatrix} = \begin{bmatrix} 2(-1) + (-3)(10) + 1(8) & 2(-6) + (-3)(5) + 1(-2) \\ -4(-1) + 7(10) + 0(8) & -4(-6) + 7(5) + 0(-2) \end{bmatrix}$$

Simplifying, we have

$$AB = \begin{bmatrix} 2 & -3 & 1 \\ -4 & 7 & 0 \end{bmatrix} \cdot \begin{bmatrix} -1 & -6 \\ 10 & 5 \\ 8 & -2 \end{bmatrix} = \begin{bmatrix} -24 & -29 \\ 74 & 59 \end{bmatrix}$$

From this example, note that each element in the product AB is of the form $a_1b_1 + a_2b_2 + \ldots + a_nb_n$, where the elements $a_1, a_2, \ldots a_n$ are elements in a row of A and $b_1, b_2, \ldots b_n$ are elements in a column of B. The expression $a_1b_1 + a_2b_2 + \ldots + a_nb_n$ is called an **inner product.**

> ### Matrix Multiplication
>
> Let A be an $m \times p$ matrix and let B be a $p \times n$ matrix, then the product AB is an $m \times n$ matrix. For the matrix AB, the element in the ith row and jth column is the sum of the products of the corresponding elements in the ith row of A and the jth column of B (the inner product of the ith row and jth column).
>
> Formally, if
>
> $$A = \begin{bmatrix} a_{11} & a_{12} & \cdots & a_{1p} \\ a_{21} & a_{22} & \cdots & a_{2p} \\ \vdots & & & \\ a_{m1} & a_{m2} & \cdots & a_{mp} \end{bmatrix} \text{ and } B = \begin{bmatrix} b_{11} & b_{12} & \cdots & b_{1n} \\ b_{21} & b_{22} & \cdots & b_{2n} \\ \vdots & & & \\ b_{p1} & b_{p2} & \cdots & b_{pn} \end{bmatrix}$$
>
> then the elements in the matrix $C = AB$ are given by
>
> $$c_{ij} = a_{i1} \cdot b_{1j} + a_{i2} \cdot b_{2j} + \cdots + a_{ip} \cdot b_{pj}$$
>
> *Note*: If the number of columns in A does not equal the number of rows in B, then it is not possible to compute the product AB.

EXAMPLE 6 Multiplying Matrices

Given $A = \begin{bmatrix} 2 & 5 & 6 \\ -3 & 0 & 1 \end{bmatrix}$, $B = \begin{bmatrix} 1 \\ 3 \\ -4 \end{bmatrix}$, and $C = \begin{bmatrix} -2 & 10 & 1 \end{bmatrix}$,

find the following products if possible.

a. AB **b.** BC **c.** AC

Solution:

a.

$$\begin{array}{cc} A & B \\ 2 \times 3 & 3 \times 1 \end{array}$$

equal

The product is a 2×1 matrix.

Matrix A is a 2×3 matrix and matrix B is a 3×1 matrix. The number of columns in A is equal to the number of rows in B. The resulting matrix will be a 2×1 matrix.

$$= \begin{bmatrix} 2 & 5 & 6 \\ -3 & 0 & 1 \end{bmatrix} \cdot \begin{bmatrix} 1 \\ 3 \\ -4 \end{bmatrix} = \begin{bmatrix} 2(1) + 5(3) + 6(-4) \\ -3(1) + 0(3) + 1(-4) \end{bmatrix}$$

$$= \begin{bmatrix} -7 \\ -7 \end{bmatrix}$$

Multiply the elements in the first row of A by the elements in the first column of B.

Multiply the elements in the second row of A by the elements in the first column of B.

b.

$$\begin{array}{cc} B & C \\ 3 \times 1 & 1 \times 3 \end{array}$$

equal

The product is a 3×3 matrix.

Matrix B is a 3×1 matrix and matrix C is a 1×3 matrix. The number of columns in B is equal to the number of rows in C. The resulting matrix will be a 3×3 matrix.

$$B \cdot C = \begin{bmatrix} 1 \\ 3 \\ -4 \end{bmatrix} \cdot \begin{bmatrix} -2 & 10 & 1 \end{bmatrix} = \begin{bmatrix} 1(-2) & 1(10) & 1(1) \\ 3(-2) & 3(10) & 3(1) \\ -4(-2) & -4(10) & -4(1) \end{bmatrix}$$

$$= \begin{bmatrix} -2 & 10 & 1 \\ -6 & 30 & 3 \\ 8 & -40 & -4 \end{bmatrix}$$

c. $\quad A \quad \cdot \quad C$

$\qquad 2 \times 3 \qquad 1 \times 3 \qquad$ The number of columns of A does not equal the number of rows of C. Therefore, it is not possible to compute the product AC.

not equal

> **Skill Practice 6** Given $A = \begin{bmatrix} 2 & -5 \\ 1 & 0 \end{bmatrix}$, $B = \begin{bmatrix} 3 & 5 & 4 \\ 6 & 0 & -8 \end{bmatrix}$, and $C = \begin{bmatrix} 1 \\ 0 \\ 6 \end{bmatrix}$,
>
> find the products if possible. **a.** AB **b.** BC **c.** AC

EXAMPLE 7 Multiplying Matrices

Given $A = \begin{bmatrix} 3 & -4 \\ 1 & 5 \end{bmatrix}$ and $B = \begin{bmatrix} 1 & 0 \\ -3 & 6 \end{bmatrix}$, find the following products if possible.

a. AB **b.** BA

Avoiding Mistakes

Example 7 shows that for two matrices A and B, the product AB does not necessarily equal BA. That is, matrix multiplication is *not* commutative.

Solution:

a. $\quad A \quad \cdot \quad B$

$\qquad 2 \times 2 \qquad 2 \times 2$

equal

The product is a 2×2 matrix.

$AB = \begin{bmatrix} 3 & -4 \\ 1 & 5 \end{bmatrix} \cdot \begin{bmatrix} 1 & 0 \\ -3 & 6 \end{bmatrix}$

$= \begin{bmatrix} 3(1) + (-4)(-3) & 3(0) + (-4)(6) \\ 1(1) + 5(-3) & 1(0) + 5(6) \end{bmatrix}$

$= \begin{bmatrix} 15 & -24 \\ -14 & 30 \end{bmatrix}$

b. $\quad B \quad \cdot \quad A$

$\qquad 2 \times 2 \qquad 2 \times 2$

equal

The product is a 2×2 matrix.

$BA = \begin{bmatrix} 1 & 0 \\ -3 & 6 \end{bmatrix} \cdot \begin{bmatrix} 3 & -4 \\ 1 & 5 \end{bmatrix}$

$= \begin{bmatrix} 1(3) + 0(1) & 1(-4) + 0(5) \\ -3(3) + 6(1) & -3(-4) + 6(5) \end{bmatrix}$

$= \begin{bmatrix} 3 & -4 \\ -3 & 42 \end{bmatrix}$

> **Skill Practice 7** Given $A = \begin{bmatrix} 1 & -3 \\ 0 & 5 \end{bmatrix}$ and $B = \begin{bmatrix} 2 & 4 \\ 6 & 1 \end{bmatrix}$, find the
>
> following products if possible. **a.** AB **b.** BA

TECHNOLOGY CONNECTIONS

Multiplying Matrices

To multiply two matrices on a graphing utility, first enter the matrices into the calculator using the EDIT feature under the MATRIX menu. Then on the home screen, enter the product. The products AB and BA from Example 7 are shown here.

Alternatively, in "MATHPRINT" mode, enter the matrices on the home screen using 〈ALPHA〉 F3. The products AB and BA are shown here.

Answers

6. a. $AB = \begin{bmatrix} -24 & 10 & 48 \\ 3 & 5 & 4 \end{bmatrix}$

b. $BC = \begin{bmatrix} 27 \\ -42 \end{bmatrix}$

c. Not possible

7. a. $AB = \begin{bmatrix} -16 & 1 \\ 30 & 5 \end{bmatrix}$

b. $BA = \begin{bmatrix} 2 & 14 \\ 6 & -13 \end{bmatrix}$

5. Apply Operations on Matrices

In applications, a matrix gives mathematicians a systematic format in which to represent data. Multiplication of matrices is a tool to manipulate these data sets. For example, in business, we might multiply the number of items sold by the unit price per item to determine total revenue.

$$\begin{bmatrix} \text{Quantity} \\ \text{matrix} \end{bmatrix} \cdot \begin{bmatrix} \text{Unit price} \\ \text{matrix} \end{bmatrix} = \begin{bmatrix} \text{Total revenue} \\ \text{matrix} \end{bmatrix}$$

$m \times p \qquad\qquad p \times n \qquad\qquad\qquad m \times n$

same

order of product

TIP The number of columns p of the quantity matrix must equal the number of rows p of the unit price matrix for the product to be defined.

EXAMPLE 8 **Multiplying Matrices in a Business Application**

A company owns two coffee shops. The number of donuts, coffee cakes, hot drinks, and cold drinks sold for each shop is given in matrix Q. The price per item is given in matrix P. Find the product QP and interpret the result.

$$Q = \begin{bmatrix} & \text{Coffee} & \text{Hot} & \text{Cold} \\ \text{Donuts} & \text{cakes} & \text{drinks} & \text{drinks} \\ 162 & 34 & 120 & 44 \\ 186 & 50 & 145 & 62 \end{bmatrix} \begin{matrix} \\ \\ \text{Shop 1} \\ \text{Shop 2} \end{matrix} \qquad P = \begin{bmatrix} \$0.40 \\ \$2.50 \\ \$3.50 \\ \$1.50 \end{bmatrix} \begin{matrix} \text{Donuts} \\ \text{Coffee cakes} \\ \text{Hot drinks} \\ \text{Cold drinks} \end{matrix}$$

Solution:

Matrix Q is a 2×4 matrix and matrix P is a 4×1 matrix. Matrix Q has 4 columns, which match the number of rows of P, so the product QP is defined. The product matrix will be a 2×1 matrix. The product QP is

162 donuts times $0.40 per donut (Shop 1).

$$\begin{bmatrix} 162 & 34 & 120 & 44 \\ 186 & 50 & 145 & 62 \end{bmatrix} \cdot \begin{bmatrix} 0.40 \\ 2.50 \\ 3.50 \\ 1.50 \end{bmatrix} = \begin{bmatrix} 162(0.40) + 34(2.50) + 120(3.50) + 44(1.50) \\ 186(0.40) + 50(2.50) + 145(3.50) + 62(1.50) \end{bmatrix}$$

$$= \begin{bmatrix} \$635.80 \\ \$799.90 \end{bmatrix}$$

Row 1: Revenue for Shop 1
Row 2: Revenue for Shop 2

The product QP is a matrix representing the total revenue from these four items for each shop.

Skill Practice 8 A farmer sells organic zucchini, yellow squash, and corn in two different roadside stands. Matrix Q represents the number of pounds of each type of vegetable sold at each stand. Matrix P gives the price per pound of each item. Find the product QP and interpret the result.

$$Q = \begin{bmatrix} \text{Zucchini} & \begin{matrix} \text{Yellow} \\ \text{squash} \end{matrix} & \text{Corn} \\ 42 & 40 & 84 \\ 30 & 36 & 90 \end{bmatrix} \begin{matrix} \\ \\ \text{Stand 1} \\ \text{Stand 2} \end{matrix} \qquad P = \begin{bmatrix} \$4.00 \\ \$3.40 \\ \$4.20 \end{bmatrix} \begin{matrix} \text{Zucchini} \\ \text{Yellow squash} \\ \text{Corn} \end{matrix}$$

Answer

8. $\begin{bmatrix} \$656.80 \\ \$620.40 \end{bmatrix}$; The product QP is a matrix representing the total revenue from these three items for each stand.

In Chapter 2 we learned how to apply transformations to the graphs of functions. In Example 9 we see how matrices can also be used to transform images in a rectangular coordinate system. This is particularly helpful to computer programmers who write software for computer gaming.

EXAMPLE 9   Using Matrix Operations to Transform a Graph

Consider the triangle shown in Figure 6-6. The vertices of the triangle can be represented in a matrix where the first row represents the x-coordinates and the second row represents the corresponding y-coordinates.

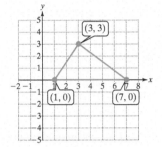

Figure 6-6

$$A = \begin{bmatrix} 1 & 3 & 7 \\ 0 & 3 & 0 \end{bmatrix} \begin{array}{l} \longleftarrow x\text{-coordinates} \\ \longleftarrow y\text{-coordinates} \end{array}$$

a. Use addition of matrices to shift the triangle 3 units to the left and 2 units upward.

b. Find the product $\begin{bmatrix} -1 & 0 \\ 0 & 1 \end{bmatrix} \cdot A$ and determine the effect on the graph.

c. Find the product $\begin{bmatrix} 1 & 0 \\ 0 & -1 \end{bmatrix} \cdot A$ and determine the effect on the graph.

Solution:

a. To shift the graph to the left, subtract 3 from each x-coordinate. To shift the graph upward, add 2 to each y-coordinate.

$$\begin{bmatrix} 1 & 3 & 7 \\ 0 & 3 & 0 \end{bmatrix} + \begin{bmatrix} -3 & -3 & -3 \\ 2 & 2 & 2 \end{bmatrix} \begin{array}{l} \longleftarrow \text{This row subtracts 3 from each } x\text{-coordinate.} \\ \longleftarrow \text{This row adds 2 to each } y\text{-coordinate.} \end{array}$$

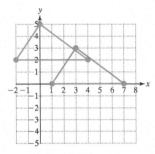

Figure 6-7

$$= \begin{bmatrix} -2 & 0 & 4 \\ 2 & 5 & 2 \end{bmatrix}$$

The vertices of the new triangle are $(-2, 2)$, $(0, 5)$, and $(4, 2)$. See Figure 6-7.

b. $\begin{bmatrix} -1 & 0 \\ 0 & 1 \end{bmatrix} \cdot A = \begin{bmatrix} -1 & 0 \\ 0 & 1 \end{bmatrix}\begin{bmatrix} 1 & 3 & 7 \\ 0 & 3 & 0 \end{bmatrix}$

$$= \begin{bmatrix} -1(1) + 0(0) & -1(3) + 0(3) & -1(7) + 0(0) \\ 0(1) + 1(0) & 0(3) + 1(3) & 0(7) + 1(0) \end{bmatrix}$$

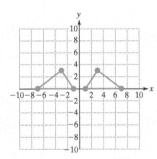

Figure 6-8

$$= \begin{bmatrix} -1 & -3 & -7 \\ 0 & 3 & 0 \end{bmatrix} \quad \begin{array}{l} \text{The corresponding points are } (-1, 0), \\ (-3, 3), \text{ and } (-7, 0). \end{array}$$

The triangle has been reflected across the y-axis. See Figure 6-8.

c. $\begin{bmatrix} 1 & 0 \\ 0 & -1 \end{bmatrix} \cdot A = \begin{bmatrix} 1 & 0 \\ 0 & -1 \end{bmatrix}\begin{bmatrix} 1 & 3 & 7 \\ 0 & 3 & 0 \end{bmatrix}$

$$= \begin{bmatrix} 1(1) + 0(0) & 1(3) + 0(3) & 1(7) + 0(0) \\ 0(1) + -1(0) & 0(3) + -1(3) & 0(7) + -1(0) \end{bmatrix}$$

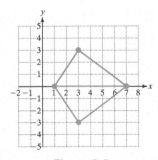

Figure 6-9

$$= \begin{bmatrix} 1 & 3 & 7 \\ 0 & -3 & 0 \end{bmatrix} \quad \begin{array}{l} \text{The corresponding points are} \\ (1, 0), (3, -3), \text{ and } (7, 0). \end{array}$$

The triangle has been reflected across the x-axis. See Figure 6-9.

Skill Practice 9   Use matrix A from Example 9. Find the product $\begin{bmatrix} -1 & 0 \\ 0 & -1 \end{bmatrix} \cdot A$ and determine the effect on the graph of the triangle in Figure 6-6.

Answer

9. $\begin{bmatrix} -1 & -3 & -7 \\ 0 & -3 & 0 \end{bmatrix}$; The triangle is reflected across both the x- and y-axes.

Prerequisite Review

R.1. Identify the additive inverse of 9.

R.2. Apply the commutative property of addition.
$$-9 + x$$

R.3. Solve the equation. $5z - 2 = -22$

Concept Connections

1. If the _____ of a matrix is $p \times q$, then p represents the number of _____ and q represents the number of _____.

2. A matrix with the same number of rows and columns is called a _____ matrix.

3. What are the requirements for two matrices to be equal?

4. An $m \times n$ matrix whose elements are all zero is called a _____ matrix.

5. To multiply two matrices A and B, the number of _____ of A must equal the number of _____ of B.

6. If A is a 5×3 matrix and B is a 3×7 matrix, then the product AB will be a matrix of order _____. The product BA (is/is not) defined.

7. True or false: Matrix multiplication is a commutative operation.

8. True or false: If a row matrix A and a column matrix B have the same number of elements, then the product AB is defined.

Objective 1: Determine the Order of a Matrix

9. What is a row matrix?

10. What is a column matrix?

For Exercises 11–16,

a. Give the order of the matrix.

b. Classify the matrix as a square matrix, row matrix, column matrix, or none of these. (**See Example 1**)

11. $\begin{bmatrix} 3 & 5 & -1 \\ \frac{1}{2} & \sqrt{3} & 1.7 \end{bmatrix}$

12. $\begin{bmatrix} 1 & 5 & 6 & 2 \\ -1 & 3 & -\frac{2}{3} & \pi \\ 0 & 1 & -6.1 & 12 \end{bmatrix}$

13. $\begin{bmatrix} 3 \\ 1 \\ 7 \end{bmatrix}$

14. $\begin{bmatrix} 2.4 & 6.9 \end{bmatrix}$

15. $\begin{bmatrix} 4 & 2 \\ 8 & 4 \end{bmatrix}$

16. $\begin{bmatrix} -4 & 10 & 3 & 0 \\ 0 & 1 & 0 & 0 \\ 7 & 0 & 9 & -1 \\ 0 & 2 & 0 & 1 \end{bmatrix}$

For Exercises 17–22, determine the value of the given element of matrix $A = [a_{ij}]$. (See Example 2)

$$A = \begin{bmatrix} 3 & -6 & \frac{1}{3} \\ 2 & 4 & 0 \\ \sqrt{5} & 11 & 8.6 \\ \frac{1}{2} & 4 & 2 \end{bmatrix}$$

17. a_{31}

18. a_{32}

19. a_{13}

20. a_{23}

21. a_{43}

22. a_{42}

Objective 2: Add and Subtract Matrices

23. Given $A = \begin{bmatrix} 2 & x \\ z & -5 \end{bmatrix}$ and $B = \begin{bmatrix} y & 4 \\ 10 & -5 \end{bmatrix}$, for what values of x, y, and z will $A = B$?

24. Given $A = \begin{bmatrix} 4 & a \\ b & 7 \end{bmatrix}$ and $B = \begin{bmatrix} c & 12 \\ -1 & 6 \end{bmatrix}$, is it possible that $A = B$? Explain.

25. Given $B = \begin{bmatrix} -4 & 6 & 9 \\ \frac{3}{5} & 1 & 7 \end{bmatrix}$, find the additive inverse of B.

26. Given $C = \begin{bmatrix} -1 & 6 \\ \sqrt{3} & 9 \end{bmatrix}$, find the additive inverse of C.

For Exercises 27–32, add or subtract the given matrices if possible. (See Example 3)

$$A = \begin{bmatrix} 6 & -1 \\ 7 & \frac{1}{2} \\ 2 & \sqrt{2} \end{bmatrix} \quad B = \begin{bmatrix} -9 & 2 \\ 6.2 & 2 \\ \frac{1}{3} & \sqrt{8} \end{bmatrix} \quad C = \begin{bmatrix} 11 & 4 \\ 1 & -\frac{1}{3} \\ 1 & 6 \end{bmatrix} \quad D = \begin{bmatrix} 2 & 3 & 8 \\ -1 & 6 & \frac{1}{6} \end{bmatrix}$$

27. $A + B$

28. $A + C$

29. $C - A + B$

30. $B - A - C$

31. $B + D$

32. $C + D$

Objective 3: Multiply a Matrix by a Scalar

33. Explain how to multiply a matrix by a scalar.

34. Given matrix A, explain how to find its additive inverse $-A$.

For Exercises 35–42, use $A = \begin{bmatrix} 2 & 4 & -9 \\ 1 & \sqrt{3} & \frac{1}{2} \end{bmatrix}$ **and** $B = \begin{bmatrix} -1 & 0 & 4 \\ 2 & 9 & \frac{2}{3} \end{bmatrix}$. **(See Example 4)**

35. $3A$

36. $-6B$

37. $-2A - 7B$

38. $4A - 3B$

39. $-4(A + B)$

40. $2(A - B)$

41. $-3A + 5(A - B)$

42. $-8A - 2(A + B)$

For Exercises 43–48, use $A = \begin{bmatrix} 1 & 6 \\ 4 & -2 \end{bmatrix}$ **and** $B = \begin{bmatrix} 2 & -4 \\ 6 & 9 \end{bmatrix}$ **and solve for X. (See Example 5)**

43. $2X - B = A$

44. $3X + A = B$

45. $A + 5X = B$

46. $B - 4X = A$

47. $2A - B = 10X$

48. $3B - A = -2X$

Objective 4: Multiply Matrices

49. Given that A is a 4×2 matrix and B is a 2×1 matrix,
  a. Is AB defined? If so, what is the order of AB?
  b. Is BA defined? If so, what is the order of BA?

50. Given that C is a 3×7 matrix and D is a 7×2 matrix,
  a. Is CD defined? If so, what is the order of CD?
  b. Is DC defined? If so, what is the order of DC?

51. Given that E is a 5×1 matrix and F is a 1×5 matrix,
  a. Is EF defined? If so, what is the order of EF?
  b. Is FE defined? If so, what is the order of FE?

52. Given that G is a 1×6 matrix and H is a 6×1 matrix,
  a. Is GH defined? If so, what is the order of GH?
  b. Is HG defined? If so, what is the order of HG?

For Exercises 53–64, (See Examples 6–7)
  a. Find AB if possible.
  b. Find BA if possible.
  c. Find A^2 if possible. (*Hint*: $A^2 = A \cdot A$.)

53. $A = \begin{bmatrix} 2 & 3 \\ 5 & 7 \end{bmatrix}$ and $B = \begin{bmatrix} 1 & 4 \\ -1 & 3 \end{bmatrix}$

54. $A = \begin{bmatrix} 1 & -6 \\ 5 & 10 \end{bmatrix}$ and $B = \begin{bmatrix} -2 & 3 \\ 7 & -1 \end{bmatrix}$

55. $A = \begin{bmatrix} 2 & 4 \\ -6 & 3 \\ 1 & 7 \end{bmatrix}$ and $B = \begin{bmatrix} 1 & 4 & -1 \\ -2 & 0 & 10 \end{bmatrix}$

56. $A = \begin{bmatrix} 1 & 3 & 4 \\ -2 & 5 & 6 \end{bmatrix}$ and $B = \begin{bmatrix} 1 & 3 \\ 9 & 2 \\ 0 & 4 \end{bmatrix}$

57. $A = \begin{bmatrix} -9 & 2 & 3 \\ -1 & 5 & 4 \\ 0 & 1 & 7 \end{bmatrix}$ and $B = \begin{bmatrix} -1 \\ 5 \\ 0 \end{bmatrix}$

58. $A = \begin{bmatrix} 2 \\ -1 \\ 3 \end{bmatrix}$ and $B = \begin{bmatrix} 2 & 7 & -1 \\ 0 & 4 & 1 \\ 0 & 3 & -6 \end{bmatrix}$

59. $A = \begin{bmatrix} 1 & \frac{1}{2} \end{bmatrix}$ and $B = \begin{bmatrix} -\frac{1}{3} \\ 2 \end{bmatrix}$

60. $A = \begin{bmatrix} 4 \\ \frac{3}{4} \\ 1 \end{bmatrix}$ and $B = \begin{bmatrix} -5 & \frac{1}{2} & \frac{1}{3} \end{bmatrix}$

61. $A = \begin{bmatrix} 4 \\ -6 \end{bmatrix}$ and $B = \begin{bmatrix} 1 & 2 & 5 & 6 \end{bmatrix}$

62. $A = [4]$ and $B = \begin{bmatrix} -3 & 4 & 1 \end{bmatrix}$

63. $A = [-5]$ and $B = [5]$

64. $A = [6]$ and $B = [-2]$

For Exercises 65–68, find AB and BA.

65. $A = \begin{bmatrix} 3.1 & -2.3 \\ 1.1 & 6.5 \end{bmatrix}$ and $B = \begin{bmatrix} 1 & 0 \\ 0 & 1 \end{bmatrix}$

66. $A = \begin{bmatrix} 1 & 0 \\ 0 & 1 \end{bmatrix}$ and $B = \begin{bmatrix} 0.05 & -0.07 \\ 0.16 & 0.09 \end{bmatrix}$

67. $A = \begin{bmatrix} 1 & 0 & 0 \\ 0 & 1 & 0 \\ 0 & 0 & 1 \end{bmatrix}$ and $B = \begin{bmatrix} \frac{9}{5} & -3 & \sqrt{6} \\ 5 & \frac{1}{2} & 2 \\ 3 & 0 & 1 \end{bmatrix}$

68. $A = \begin{bmatrix} \frac{2}{3} & 5 & \sqrt{2} \\ 1 & 0 & 3 \\ -\frac{1}{4} & 0 & 8 \end{bmatrix}$ and $B = \begin{bmatrix} 1 & 0 & 0 \\ 0 & 1 & 0 \\ 0 & 0 & 1 \end{bmatrix}$

Objective 5: Apply Operations on Matrices

69. Matrix D gives the dealer invoice prices for sedan and hatchback models of a car with manual transmission or automatic transmission. Matrix M gives the MSRP (manufacturer's suggested retail price) for the cars.

$$D = \begin{array}{c} \\ \begin{bmatrix} \$29{,}000 & \$27{,}500 \\ \$28{,}500 & \$26{,}900 \end{bmatrix} \end{array} \begin{array}{l} \text{Manual} \\ \text{Automatic} \end{array}$$

with columns **Sedan   Hatchback**

$$M = \begin{bmatrix} \$32{,}600 & \$29{,}900 \\ \$31{,}900 & \$28{,}900 \end{bmatrix} \begin{array}{l} \text{Manual} \\ \text{Automatic} \end{array}$$

with columns **Sedan   Hatchback**

a. Compute $M - D$ and interpret the result.

b. A buyer thinks that a fair price is 6% above dealer invoice. Use scalar multiplication to determine a matrix F that gives the fair price for these cars for each type of transmission.

70. In matrix C, a coffee shop records the cost to produce a cup of standard Columbian coffee and the cost to produce a cup of hot chocolate. Matrix P contains the selling prices to the customer.

with columns **Coffee   Chocolate**

$$C = \begin{bmatrix} \$0.90 & \$0.84 \\ \$1.26 & \$1.15 \\ \$1.64 & \$1.50 \end{bmatrix} \begin{array}{l} \text{Small} \\ \text{Medium} \\ \text{Large} \end{array}$$

with columns **Coffee   Chocolate**

$$P = \begin{bmatrix} \$3.05 & \$2.25 \\ \$3.65 & \$3.05 \\ \$4.15 & \$3.65 \end{bmatrix} \begin{array}{l} \text{Small} \\ \text{Medium} \\ \text{Large} \end{array}$$

a. Compute $P - C$ and interpret its meaning.

b. If the tax rate in a certain city is 7%, use scalar multiplication to find a matrix F that gives the final price to the customer (including sales tax) for both beverages for each size. Round each entry to the nearest cent.

71. A street vendor at a parade sells fresh lemonade, soda, bottled water, and iced-tea, and the unit price for each item is given in matrix P. The number of units sold of each item is given in matrix N. Compute NP and interpret the result.

$$N = \begin{array}{cccc} \text{Lemonade} & \text{Soda} & \text{Water} & \text{Tea} \\ [150 & 270 & 440 & 80], \end{array}$$

$$P = \begin{bmatrix} \$2.50 \\ \$1.50 \\ \$2.00 \\ \$2.00 \end{bmatrix} \begin{array}{l} \text{Lemonade} \\ \text{Soda} \\ \text{Water} \\ \text{Tea} \end{array}$$

72. A math course has 4 exams weighted 20%, 25%, 25%, and 30%, respectively, toward the final grade. Suppose that a student earns grades of 75, 84, 92, and 86, respectively, on the four exams. The weights are given in matrix W and the test grades are given in matrix G. Compute WG and interpret the result.

$$W = \begin{array}{cccc} \text{Test 1} & \text{Test 2} & \text{Test 3} & \text{Test 4} \\ [0.20 & 0.25 & 0.25 & 0.30], \end{array} \quad G = \begin{bmatrix} 75 \\ 84 \\ 92 \\ 86 \end{bmatrix} \begin{array}{l} \text{Test 1} \\ \text{Test 2} \\ \text{Test 3} \\ \text{Test 4} \end{array}$$

73. An electronics store sells three models of tablets. The number of each model sold during "Black Friday" weekend is given in matrix A. The selling price and profit for each model are given in matrix B. **(See Example 8)**

$$A = \begin{array}{c} \\ \begin{bmatrix} 84 & 70 & 32 \\ 62 & 48 & 16 \\ 70 & 40 & 12 \end{bmatrix} \end{array} \begin{array}{l} \text{Friday} \\ \text{Saturday} \\ \text{Sunday} \end{array}$$

with columns **Model: A B C**

$$B = \begin{bmatrix} \$499 & \$200 \\ \$599 & \$240 \\ \$629 & \$280 \end{bmatrix} \begin{array}{l} \text{A} \\ \text{B Model} \\ \text{C} \end{array}$$

with columns **Selling Price   Profit**

a. Compute AB and interpret the result.

b. Determine the total revenue for Sunday.

c. Determine the total profit for the 3-day period for these three models.

74. A gas station manager records the number of gallons of Regular, Plus, and Premium gasoline sold during the week (Monday–Friday) and on the weekends (Saturday–Sunday) in matrix A. The selling price and profit for 1 gal of each type of gasoline are given in matrix B.

with columns **Regular Plus Premium**

$$A = \begin{bmatrix} 4600 & 1850 & 720 \\ 2300 & 620 & 480 \end{bmatrix} \begin{array}{l} \text{Weekdays} \\ \text{Weekend} \end{array}$$

with columns **Selling Price   Profit**

$$B = \begin{bmatrix} \$3.59 & \$0.21 \\ \$3.79 & \$0.24 \\ \$4.19 & \$0.18 \end{bmatrix} \begin{array}{l} \text{Regular} \\ \text{Plus} \\ \text{Premium} \end{array}$$

a. Compute AB and interpret the result.

b. Determine the profit for the weekend.

c. Determine the revenue for the entire week.

75. The labor costs per hour for an electrician, plumber, and air-conditioning/heating expert are given in matrix L. The time required from each specialist for three new model homes is given in matrix T.

$$L = \begin{bmatrix} \$45 \\ \$38 \\ \$35 \end{bmatrix} \begin{matrix} \textbf{Electrician} \\ \textbf{Plumber} \\ \textbf{AC/heating} \end{matrix}$$

Time (hr)

	Electrician	Plumber	AC/heating	
	22	16	14	Model 1
$T =$	28	21	18	Model 2
	18	14	9	Model 3

a. Which product LT or TL gives the total cost for these three services for each model?

b. Find a matrix that gives the total cost for these three services for each model.

77. A student researches the cost for three cell phone plans. Matrix C contains the cost per text message and the cost per minute over the maximum number of minutes allowed in each plan. Matrix N_1 contains the number of text messages and the number of minutes over the maximum incurred for 1 month. Matrix N_3 represents the number of text messages and number of minutes over the maximum for 3 months.

	Cost/text	Cost/min	
	\$0.25	\$0.40	Plan A
$C =$	\$0	\$0.40	Plan B
	\$0.10	\$0	Plan C

$$N_1 = \begin{bmatrix} 24 \\ 100 \end{bmatrix} \begin{matrix} \textbf{Number of texts} \\ \textbf{Minutes over} \end{matrix}$$

	Month 1	Month 2	Month 3	
$N_3 =$	24	56	30	Number of texts
	100	24	0	Minutes over

a. Find the product CN_1 and interpret its meaning.

b. Find the product CN_3 and interpret its meaning.

76. The number of calories burned per hour for three activities is given in matrix N for a 140-lb woman training for a triathlon. The time spent on each activity for two different training days is given in matrix T.

$$N = \begin{bmatrix} 540 \\ 400 \\ 360 \end{bmatrix} \begin{matrix} \textbf{Running} \\ \textbf{Bicycling} \\ \textbf{Swimming} \end{matrix}$$

Time

	Running	Bicycling	Swimming	
$T =$	45 min	1 hr	30 min	Day 1
	1 hr	1 hr 30 min	45 min	Day 2

a. Which product NT or TN gives the total number of calories burned from these activities for each day?

b. Find a matrix that gives the total number of calories burned from these activities for each day.

78. Refer to Exercise 77. Suppose that matrix B represents the base cost for each cell phone plan and matrix T represents the tax for each plan.

$$B = \begin{bmatrix} \$39.99 \\ \$49.99 \\ \$59.99 \end{bmatrix} \begin{matrix} \textbf{Plan A} \\ \textbf{Plan B} \\ \textbf{Plan C} \end{matrix}$$

$$T = \begin{bmatrix} \$11.96 \\ \$13.04 \\ \$14.91 \end{bmatrix} \begin{matrix} \textbf{Plan A} \\ \textbf{Plan B} \\ \textbf{Plan C} \end{matrix}$$

a. Compute $B + CN_1 + T$ and interpret its meaning.

b. Which cell phone plan is the least expensive if the student has 60 text messages and talks 20 min more than the maximum?

79. a. Write a matrix A that represents the coordinates of the vertices of the triangle. Place the x-coordinate of each point in the first row of A and the corresponding y-coordinate in the second row of A.
(See Example 9)

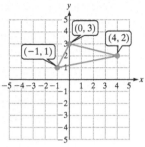

b. Use addition of matrices to shift the triangle 2 units to the right and 4 units downward.

c. Find the product $\begin{bmatrix} -1 & 0 \\ 0 & 1 \end{bmatrix} \cdot A$ and explain the effect on the graph of the triangle.

d. Find the product $\begin{bmatrix} 1 & 0 \\ 0 & -1 \end{bmatrix} \cdot A$ and explain the effect on the graph of the triangle.

e. Find $\begin{bmatrix} 1 & 0 \\ 0 & -1 \end{bmatrix} \cdot A + \begin{bmatrix} -1 & -1 & -1 \\ 2 & 2 & 2 \end{bmatrix}$ and explain the effect on the graph of the triangle.

80. a. Write a matrix A that represents the coordinates of the vertices of the triangle. Place the x-coordinate of each point in the first row of A and the corresponding y-coordinate in the second row of A.

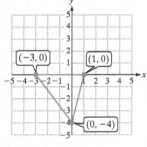

b. Use addition of matrices to shift the triangle 1 unit to the left and 3 units upward.

c. Find the product $\begin{bmatrix} -1 & 0 \\ 0 & 1 \end{bmatrix} \cdot A$ and explain the effect on the graph of the triangle.

d. Find the product $\begin{bmatrix} 1 & 0 \\ 0 & -1 \end{bmatrix} \cdot A$ and explain the effect on the graph of the triangle.

e. Find $\begin{bmatrix} -1 & 0 \\ 0 & 1 \end{bmatrix} \cdot A + \begin{bmatrix} 2 & 2 & 2 \\ -5 & -5 & -5 \end{bmatrix}$ and explain the effect on the graph of the triangle.

81. a. Write a matrix A that represents the coordinates of the vertices of the quadrilateral.

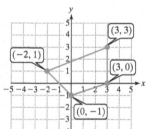

b. What operation on A will shift the graph of the quadrilateral 3 units downward?

c. What operation on A will shift the graph 4 units to the left?

d. Use matrix multiplication to reflect the graph across the x-axis.

e. Use matrix multiplication to reflect the graph across the y-axis.

82. a. Write a matrix A that represents the coordinates of the vertices of the quadrilateral.

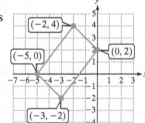

b. What operation on A will shift the graph of the quadrilateral 6 units upward?

c. What operation on A will shift the graph 2 units to the right?

d. Use matrix multiplication to reflect the graph across the x-axis.

e. Use matrix multiplication to reflect the graph across the y-axis.

83. a. Write a matrix A that represents the coordinates of the vertices of the triangle.

b. Multiply $\begin{bmatrix} \frac{\sqrt{3}}{2} & -\frac{1}{2} \\ \frac{1}{2} & \frac{\sqrt{3}}{2} \end{bmatrix} \cdot A$ and round each entry to 1 decimal place.

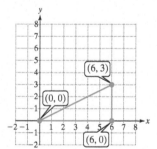

c. Graph the figure represented by the matrix from part (b). What effect does this product have on the graph of the triangle?

84. a. Write a matrix A that represents the coordinates of the vertices of the triangle.

b. Multiply $\begin{bmatrix} \frac{1}{2} & -\frac{\sqrt{3}}{2} \\ \frac{\sqrt{3}}{2} & \frac{1}{2} \end{bmatrix} \cdot A$ and round each entry to 1 decimal place.

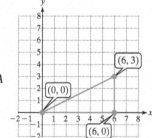

c. Graph the figure represented by the matrix from part (b). What effect does this product have on the graph of the triangle?

For Exercises 85–86, use the following gray scale.

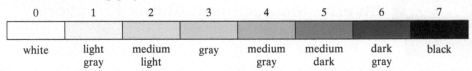

0	1	2	3	4	5	6	7
white	light gray	medium light	gray	medium gray	medium dark	dark gray	black

85. a. Write a 5 × 3 matrix that represents the letter E in dark gray on a white background.

   b. Use matrix addition to change the pixels so that the letter E is medium dark on a light gray background.

86. a. Write a 5 × 3 matrix that represents the letter T in medium gray on a medium light background.

   b. Use matrix addition to change the pixels so that the letter T is dark gray on a light gray background.

Mixed Exercises

For Exercises 87–92, use matrices A, B, and C to prove the given properties. Assume that the elements within A, B, and C are real numbers.

$$A = \begin{bmatrix} a_1 & a_2 \\ a_3 & a_4 \end{bmatrix} \quad B = \begin{bmatrix} b_1 & b_2 \\ b_3 & b_4 \end{bmatrix} \quad C = \begin{bmatrix} c_1 & c_2 \\ c_3 & c_4 \end{bmatrix}$$

87. Commutative property of matrix addition
$$A + B = B + A$$

88. Associative property of matrix addition
$$A + (B + C) = (A + B) + C$$

89. Inverse property of matrix addition
$$A + (-A) = 0$$

90. Identity property of matrix addition
$$A + 0 = A$$

91. Associative property of scalar multiplication
$$s(tA) = (st)A$$

92. Distributive property of scalar multiplication
$$t(A + B) = tA + tB$$

93. Given $A = \begin{bmatrix} i & 0 \\ 0 & i \end{bmatrix}$, find A^2, A^3, and A^4.

(*Hint*: Recall that $i = \sqrt{-1}$.) Discuss the similarities between A^n and i^n, where n is a positive integer.

94. Given $B = \begin{bmatrix} 0 & -i \\ i & 0 \end{bmatrix}$, find B^2.

95. a. For real numbers a, b, c, and d, find the product.
$$\begin{bmatrix} a & 0 \\ 0 & b \end{bmatrix}\begin{bmatrix} c & 0 \\ 0 & d \end{bmatrix}$$

   b. Based on the form of the product of part (a), compute the following product mentally.
$$\begin{bmatrix} 3 & 0 \\ 0 & 7 \end{bmatrix}\begin{bmatrix} 1 & 0 \\ 0 & 2 \end{bmatrix}$$

96. Find the product.
$$\begin{bmatrix} a & a \\ b & b \end{bmatrix}\begin{bmatrix} 1 & -1 \\ -1 & 1 \end{bmatrix}$$

Write About It

97. Explain why the product of a 3 × 2 matrix by a 3 × 2 matrix is undefined.

98. Explain how to add or subtract matrices.

99. Given a matrix, $A = [a_{ij}]$, explain how to find the additive inverse of A.

100. Given two matrices A and B, how can you determine the order of AB, assuming that the product is defined?

Technology Connections

For Exercises 101–104, refer to matrices A, B, and C and perform the indicated operations on a calculator.

$$A = \begin{bmatrix} 1.05 & 3.9 \\ 4.12 & -9.4 \\ -2.4 & 1.5 \end{bmatrix} \quad B = \begin{bmatrix} -10 & 30 \\ 24 & -36 \\ 18 & -8 \end{bmatrix} \quad C = \begin{bmatrix} 6.2 \\ 4.9 \end{bmatrix}$$

101. $2.5A - 3.6B$ **102.** $-6.4(A + B)$ **103.** $-3AC$ **104.** $7.5BC$

SECTION 6.4   Inverse Matrices and Matrix Equations

1. Identify Identity and Inverse Matrices

The identity element under the multiplication of real numbers is 1 because $a \cdot 1 = a$ and $1 \cdot a = a$. We now investigate a similar property for the product of square matrices. The **identity matrix** I_n for matrix multiplication is the $n \times n$ square matrix with 1's along the main diagonal and 0's for all other elements. For example:

The identity matrix of order 2

The identity matrix of order 3

$$I_2 = \begin{bmatrix} 1 & 0 \\ 0 & 1 \end{bmatrix} \quad \text{and} \quad I_3 = \begin{bmatrix} 1 & 0 & 0 \\ 0 & 1 & 0 \\ 0 & 0 & 1 \end{bmatrix}$$

For an $n \times n$ square matrix A, we have that

$$AI_n = A \quad \text{and} \quad I_nA = A \quad \text{(Identity property of matrix multiplication)}$$

In Example 1, we illustrate the identity property of matrix multiplication using a 2×2 matrix.

EXAMPLE 1   Illustrating the Identity Property of Matrix Multiplication

Given $A = \begin{bmatrix} a & b \\ c & d \end{bmatrix}$ show that

a. $AI_2 = A$ **b.** $I_2A = A$

Solution:

a. $AI_2 = \begin{bmatrix} a & b \\ c & d \end{bmatrix}\begin{bmatrix} 1 & 0 \\ 0 & 1 \end{bmatrix} = \begin{bmatrix} a(1) + b(0) & a(0) + b(1) \\ c(1) + d(0) & c(0) + d(1) \end{bmatrix} = \begin{bmatrix} a & b \\ c & d \end{bmatrix}$ ✓

b. $I_2A = \begin{bmatrix} 1 & 0 \\ 0 & 1 \end{bmatrix}\begin{bmatrix} a & b \\ c & d \end{bmatrix} = \begin{bmatrix} 1(a) + 0(c) & 1(b) + 0(d) \\ 0(a) + 1(c) & 0(b) + 1(d) \end{bmatrix} = \begin{bmatrix} a & b \\ c & d \end{bmatrix}$ ✓

Skill Practice 1 Given $A = \begin{bmatrix} 3 & -4 \\ -2 & 10 \end{bmatrix}$ show that

a. $AI_2 = A$ **b.** $I_2A = A$

For a nonzero real number a, the multiplicative inverse of a is $\frac{1}{a}$ because $a \cdot \frac{1}{a} = 1$ and $\frac{1}{a} \cdot a = 1$. We define the multiplicative inverse of a square matrix in a similar fashion.

Answers

1. a. $\begin{bmatrix} 3 & -4 \\ -2 & 10 \end{bmatrix}\begin{bmatrix} 1 & 0 \\ 0 & 1 \end{bmatrix} = \begin{bmatrix} 3 & -4 \\ -2 & 10 \end{bmatrix}$

b. $\begin{bmatrix} 1 & 0 \\ 0 & 1 \end{bmatrix}\begin{bmatrix} 3 & -4 \\ -2 & 10 \end{bmatrix} = \begin{bmatrix} 3 & -4 \\ -2 & 10 \end{bmatrix}$

Avoiding Mistakes

Be sure to interpret the notation A^{-1} as the *inverse* of A, rather than the reciprocal of A. That is, $A^{-1} \neq \frac{1}{A}$.

Multiplicative Inverse of a Square Matrix

Let A be an $n \times n$ matrix and let I_n be the identity matrix of order n. If there exists an $n \times n$ matrix A^{-1} such that

$$AA^{-1} = I_n \quad \text{and} \quad A^{-1}A = I_n$$

then A^{-1} (read as "A inverse") is the **multiplicative inverse** of A.

EXAMPLE 2   **Determining Whether Two Matrices Are Inverses**

Determine whether $A = \begin{bmatrix} 3 & 2 \\ 7 & 5 \end{bmatrix}$ and $B = \begin{bmatrix} 5 & -2 \\ -7 & 3 \end{bmatrix}$ are inverses.

Solution:

We must show that $AB = I_2$ and $BA = I_2$.

$$AB = \begin{bmatrix} 3 & 2 \\ 7 & 5 \end{bmatrix}\begin{bmatrix} 5 & -2 \\ -7 & 3 \end{bmatrix} = \begin{bmatrix} 3(5) + 2(-7) & 3(-2) + 2(3) \\ 7(5) + 5(-7) & 7(-2) + 5(3) \end{bmatrix} = \begin{bmatrix} 1 & 0 \\ 0 & 1 \end{bmatrix} \checkmark$$

$$BA = \begin{bmatrix} 5 & -2 \\ -7 & 3 \end{bmatrix}\begin{bmatrix} 3 & 2 \\ 7 & 5 \end{bmatrix} = \begin{bmatrix} 5(3) + (-2)(7) & 5(2) + (-2)(5) \\ -7(3) + 3(7) & -7(2) + 3(5) \end{bmatrix} = \begin{bmatrix} 1 & 0 \\ 0 & 1 \end{bmatrix} \checkmark$$

Skill Practice 2

> Determine whether $A = \begin{bmatrix} 4 & 3 \\ 13 & 10 \end{bmatrix}$ and $B = \begin{bmatrix} 10 & -3 \\ -13 & 4 \end{bmatrix}$ are inverses.

---

2. Determine the Inverse of a Matrix

In Example 2, we showed that two given matrices are inverses. Now we look at the task of finding the inverse of a matrix. In Example 3, we will find the inverse of a 2×2 matrix directly from the definition. Then, we develop a general procedure to find the inverse of an $n \times n$ matrix, if the inverse exists.

EXAMPLE 3   **Finding the Inverse of a Matrix**

Given $A = \begin{bmatrix} 2 & 9 \\ 1 & 5 \end{bmatrix}$, find A^{-1}.

Solution:

We need to find a matrix $A^{-1} = \begin{bmatrix} x_1 & x_2 \\ x_3 & x_4 \end{bmatrix}$ such that $AA^{-1} = I_2$. That is,

$$\begin{bmatrix} 2 & 9 \\ 1 & 5 \end{bmatrix}\begin{bmatrix} x_1 & x_2 \\ x_3 & x_4 \end{bmatrix} = \begin{bmatrix} 1 & 0 \\ 0 & 1 \end{bmatrix} \longrightarrow \begin{bmatrix} 2x_1 + 9x_3 & 2x_2 + 9x_4 \\ x_1 + 5x_3 & x_2 + 5x_4 \end{bmatrix} = \begin{bmatrix} 1 & 0 \\ 0 & 1 \end{bmatrix}$$

Answer

2. Yes; $AB = I_2$ and $BA = I_2$.

For the matrices to be equal, their corresponding elements must be equal. This results in two systems of equations.

Equating the elements from the 1st column
$$\begin{cases} 2x_1 + 9x_3 = 1 \\ x_1 + 5x_3 = 0 \end{cases} \quad \text{and} \quad \begin{cases} 2x_2 + 9x_4 = 0 \\ x_2 + 5x_4 = 1 \end{cases}$$
Equating the elements from the 2nd column

The corresponding augmented matrices are

$$\begin{bmatrix} 2 & 9 & | & 1 \\ 1 & 5 & | & 0 \end{bmatrix} \quad \text{and} \quad \begin{bmatrix} 2 & 9 & | & 0 \\ 1 & 5 & | & 1 \end{bmatrix}$$

We can solve both systems simultaneously by placing the identity matrix to the right of the array of coefficients.

$$\begin{bmatrix} 2 & 9 & | & 1 & 0 \\ 1 & 5 & | & 0 & 1 \end{bmatrix}$$

Apply Gauss-Jordan elimination.

$$\begin{bmatrix} 2 & 9 & | & 1 & 0 \\ 1 & 5 & | & 0 & 1 \end{bmatrix} \xrightarrow{R_1 \Leftrightarrow R_2} \begin{bmatrix} 1 & 5 & | & 0 & 1 \\ 2 & 9 & | & 1 & 0 \end{bmatrix} \xrightarrow{-2R_1 + R_2 \to R_2} \begin{bmatrix} 1 & 5 & | & 0 & 1 \\ 0 & -1 & | & 1 & -2 \end{bmatrix}$$

$$\xrightarrow{-1R_2 \to R_2} \begin{bmatrix} 1 & 5 & | & 0 & 1 \\ 0 & 1 & | & -1 & 2 \end{bmatrix} \xrightarrow{-5R_2 + R_1 \to R_1} \begin{bmatrix} 1 & 0 & | & 5 & -9 \\ 0 & 1 & | & -1 & 2 \end{bmatrix}$$

This result represents the following two matrices and their corresponding systems of equations.

$$\begin{matrix} x_1 & x_3 \\ \downarrow & \downarrow \end{matrix} \qquad\qquad \begin{matrix} x_2 & x_4 \\ \downarrow & \downarrow \end{matrix}$$

$$\begin{bmatrix} 1 & 0 & | & 5 \\ 0 & 1 & | & -1 \end{bmatrix} \quad \text{and} \quad \begin{bmatrix} 1 & 0 & | & -9 \\ 0 & 1 & | & 2 \end{bmatrix}$$

$$x_1 = 5, x_3 = -1 \qquad\qquad x_2 = -9, x_4 = 2$$

Therefore, we have: $A^{-1} = \begin{bmatrix} x_1 & x_2 \\ x_3 & x_4 \end{bmatrix} = \begin{bmatrix} 5 & -9 \\ -1 & 2 \end{bmatrix}$

Avoiding Mistakes

The solution to Example 3 can be checked by verifying that $AA^{-1} = I_2$ and $A^{-1}A = I_2$.

Skill Practice 3 Given $B = \begin{bmatrix} 9 & 7 \\ 5 & 4 \end{bmatrix}$, find B^{-1}.

In Example 3, we transformed the left side of the augmented matrix into the identity matrix. That is, we performed row operations to make the array on the left of the vertical bar have 1's along the main diagonal, and 0's elsewhere. In the process, we make the following important observation.

- The array of elements to the right of the vertical bar is the inverse of A.

This observation leads to a general procedure to find the multiplicative inverse of a matrix, provided that the inverse exists.

Finding the Multiplicative Inverse of a Square Matrix

Let A be an $n \times n$ matrix for which A^{-1} exists, and let I_n be the $n \times n$ identity matrix. To find A^{-1},

Step 1 Write a matrix of the form $[A \mid I_n]$.

Step 2 Perform row operations to write the matrix in the form $[I_n \mid B]$.

Step 3 The matrix B is A^{-1}.

Answer

3. $B^{-1} = \begin{bmatrix} 4 & -7 \\ -5 & 9 \end{bmatrix}$

It is important to note that not all matrices have a multiplicative inverse. If a matrix A is reducible to a row-equivalent matrix with one or more rows of zeros, then the matrix cannot be written in the form $[I_n \mid B]$. Therefore, the matrix does not have an inverse, and we say that the matrix is **singular.** A matrix that *does* have a multiplicative inverse is said to be **invertible** or **nonsingular.**

In Example 4, we use the general procedure to find the inverse of a 3×3 invertible matrix.

EXAMPLE 4   Finding the Inverse of a Matrix

Given $A = \begin{bmatrix} 1 & 3 & 0 \\ 1 & 1 & -2 \\ -3 & -3 & 5 \end{bmatrix}$, find A^{-1} if possible.

Solution:

$\begin{bmatrix} 1 & 3 & 0 & | & 1 & 0 & 0 \\ 1 & 1 & -2 & | & 0 & 1 & 0 \\ -3 & -3 & 5 & | & 0 & 0 & 1 \end{bmatrix}$ Set up the matrix $[A \mid I_3]$.
Transform the left side of the matrix to the 3×3 identity matrix.

$\xrightarrow[3R_1 + R_3 \to R_3]{-1R_1 + R_2 \to R_2} \begin{bmatrix} 1 & 3 & 0 & | & 1 & 0 & 0 \\ 0 & -2 & -2 & | & -1 & 1 & 0 \\ 0 & 6 & 5 & | & 3 & 0 & 1 \end{bmatrix}$ $\xrightarrow{-\frac{1}{2}R_2 \to R_2} \begin{bmatrix} 1 & 3 & 0 & | & 1 & 0 & 0 \\ 0 & 1 & 1 & | & \frac{1}{2} & -\frac{1}{2} & 0 \\ 0 & 6 & 5 & | & 3 & 0 & 1 \end{bmatrix}$

$\xrightarrow[-6R_2 + R_3 \to R_3]{-3R_2 + R_1 \to R_1} \begin{bmatrix} 1 & 0 & -3 & | & -\frac{1}{2} & \frac{3}{2} & 0 \\ 0 & 1 & 1 & | & \frac{1}{2} & -\frac{1}{2} & 0 \\ 0 & 0 & -1 & | & 0 & 3 & 1 \end{bmatrix}$ $\xrightarrow{-1R_3 \to R_3} \begin{bmatrix} 1 & 0 & -3 & | & -\frac{1}{2} & \frac{3}{2} & 0 \\ 0 & 1 & 1 & | & \frac{1}{2} & -\frac{1}{2} & 0 \\ 0 & 0 & 1 & | & 0 & -3 & -1 \end{bmatrix}$

$\xrightarrow[-1R_3 + R_2 \to R_2]{3R_3 + R_1 \to R_1} \begin{bmatrix} 1 & 0 & 0 & | & -\frac{1}{2} & -\frac{15}{2} & -3 \\ 0 & 1 & 0 & | & \frac{1}{2} & \frac{5}{2} & 1 \\ 0 & 0 & 1 & | & 0 & -3 & -1 \end{bmatrix}$ The left side of the matrix is the 3×3 identity matrix. The right side of the matrix represents the inverse of A.

$A^{-1} = \begin{bmatrix} -\frac{1}{2} & -\frac{15}{2} & -3 \\ \frac{1}{2} & \frac{5}{2} & 1 \\ 0 & -3 & -1 \end{bmatrix}$ ....

Avoiding Mistakes

As a check, verify that $A \cdot A^{-1} = I_3$ and $A^{-1} \cdot A = I_3$.

Skill Practice 4   Given $A = \begin{bmatrix} -2 & 4 & 1 \\ 4 & -13 & 10 \\ 3 & -9 & 6 \end{bmatrix}$ find A^{-1} if possible.

TECHNOLOGY CONNECTIONS

Finding the Inverse of a Nonsingular Square Matrix

To find the inverse of a nonsingular matrix [A] using a calculator, first enter the matrix into the calculator using the MATRIX menu. Then place [A] on the home screen and use the $\boxed{x^{-1}}$ key to take the inverse of [A]. It is very important to realize that in this context, the calculator uses the $\boxed{x^{-1}}$ key for the inverse of a matrix, not the reciprocal of a real number.

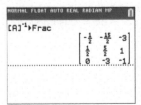

Answer

4. $A^{-1} = \begin{bmatrix} 4 & -11 & \frac{53}{3} \\ 2 & -5 & 8 \\ 1 & -2 & \frac{10}{3} \end{bmatrix}$

In Example 5, we identify a singular matrix—that is, a matrix that has no inverse.

EXAMPLE 5   Identifying a Singular Matrix

Show that matrix A is singular. $A = \begin{bmatrix} 1 & 2 & 0 \\ 4 & 5 & 1 \\ 2 & 1 & 1 \end{bmatrix}$

Solution:

$\begin{bmatrix} 1 & 2 & 0 & | & 1 & 0 & 0 \\ 4 & 5 & 1 & | & 0 & 1 & 0 \\ 2 & 1 & 1 & | & 0 & 0 & 1 \end{bmatrix}$ Set up the matrix $[A\,|\,I_3]$.

Transform the left side of the matrix to the 3×3 identity matrix.

$\xrightarrow[\;-2R_1 + R_3 \to R_3\;]{-4R_1 + R_2 \to R_2} \begin{bmatrix} 1 & 2 & 0 & | & 1 & 0 & 0 \\ 0 & -3 & 1 & | & -4 & 1 & 0 \\ 0 & -3 & 1 & | & -2 & 0 & 1 \end{bmatrix}$ On the left of the vertical bar, the elements in rows 2 and 3 are multiples of each other (in fact they are identical). Therefore, $-1R_2 + R_3 \to R_3$ results in zeros in row 3 to the left of the bar.

$\xrightarrow{-1R_2 + R_3 \to R_3} \begin{bmatrix} 1 & 2 & 0 & | & 1 & 0 & 0 \\ 0 & -3 & 1 & | & -4 & 1 & 0 \\ 0 & 0 & 0 & | & 2 & -1 & 1 \end{bmatrix}$ The left side of the matrix cannot be transformed into the 3×3 identity matrix I_3. Therefore, matrix A does not have an inverse.

A is a singular matrix (does not have an inverse).

Skill Practice 5   Show that matrix A is singular. $A = \begin{bmatrix} 1 & -3 & -7 \\ 3 & -7 & -17 \\ 1 & 0 & -1 \end{bmatrix}$

TECHNOLOGY CONNECTIONS

Identifying a Singular Matrix

A calculator will return an error message if we attempt to find the inverse of a singular matrix. Consider matrix A from Example 5. The calculator returns the error message shown.

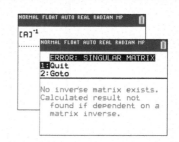

To find the inverse of an invertible 2×2 matrix, we can take a direct approach and apply the following formula.

TIP If A is an invertible matrix then the restriction $ad - bc \neq 0$ is implied.

Formula for the Inverse of a 2 × 2 Invertible Matrix

Let $A = \begin{bmatrix} a & b \\ c & d \end{bmatrix}$ be an invertible matrix. Then the inverse A^{-1} is given by:

$$A^{-1} = \frac{1}{ad - bc} \cdot \begin{bmatrix} d & -b \\ -c & a \end{bmatrix}$$

Answer

5. The matrix $[A\,|\,I_3]$ cannot be written in the form $[I_3\,|\,B]$ by using row operations. Therefore, A is a singular matrix.

This formula can be verified by applying the general procedure to find the inverse of A. See Exercise 67.

$$\begin{bmatrix} a & b & | & 1 & 0 \\ c & d & | & 0 & 1 \end{bmatrix} \xrightarrow[\text{matrix}]{\text{row-equivalent}} \begin{bmatrix} 1 & 0 & | & \frac{d}{ad-bc} & \frac{-b}{ad-bc} \\ 0 & 1 & | & \frac{-c}{ad-bc} & \frac{a}{ad-bc} \end{bmatrix}$$

$$\xrightarrow{\text{inverse of } A} \frac{1}{ad-bc} \cdot \begin{bmatrix} d & -b \\ -c & a \end{bmatrix}$$

EXAMPLE 6   Finding the Inverse of a 2 × 2 Matrix

Given $A = \begin{bmatrix} 2 & 9 \\ 1 & 5 \end{bmatrix}$, find A^{-1} and compare to the result of Example 3.

Solution:

$$\begin{bmatrix} 2 & 9 \\ 1 & 5 \end{bmatrix} \quad a = 2, b = 9, c = 1, d = 5$$

$$A^{-1} = \frac{1}{ad-bc} \cdot \begin{bmatrix} d & -b \\ -c & a \end{bmatrix} = \frac{1}{(2)(5) - (9)(1)} \cdot \begin{bmatrix} 5 & -9 \\ -1 & 2 \end{bmatrix}$$

$$= \frac{1}{1} \cdot \begin{bmatrix} 5 & -9 \\ -1 & 2 \end{bmatrix} = \begin{bmatrix} 5 & -9 \\ -1 & 2 \end{bmatrix} \quad \begin{array}{l}\text{This is the same result as was found in}\\\text{Example 3.}\end{array}$$

Skill Practice 6   Given $B = \begin{bmatrix} 9 & 7 \\ 5 & 3 \end{bmatrix}$, find B^{-1}.

3. Solve Systems of Linear Equations Using the Inverse of a Matrix

A system of linear equations written in standard form can be represented by using matrix multiplication. For example:

$$\begin{array}{l} 3x - 2y + z = 4 \\ x + 4y - z = 2 \\ 2x + 5y - 2z = 1 \end{array} \xrightarrow[\text{matrix equation}]{\text{corresponding}} \begin{bmatrix} 3 & -2 & 1 \\ 1 & 4 & -1 \\ 2 & 5 & -2 \end{bmatrix} \begin{bmatrix} x \\ y \\ z \end{bmatrix} = \begin{bmatrix} 4 \\ 2 \\ 1 \end{bmatrix}$$

The matrix on the left is called the **coefficient matrix**. It consists of the coefficients of the variable terms.

coefficient matrix   column matrix of variables   column matrix of constants

If we denote the coefficient matrix by A, the column matrix of variables as X, and the column matrix of constant terms as B, we can write the equation as $AX = B$. If A^{-1} exists, then we can solve the equation for X by multiplying both sides by A^{-1}.

$$\begin{array}{ll} AX = B & \text{To solve this equation, the goal is to isolate } X. \\ A^{-1}AX = A^{-1}B & \text{Multiply both sides by } A^{-1} \text{ (provided that } A^{-1} \text{ exists).} \\ (A^{-1}A)X = A^{-1}B & \\ I_n X = A^{-1}B & \text{The product of a matrix and its inverse is the identity matrix.} \\ X = A^{-1}B & \end{array}$$

Answer

6. $B^{-1} = \begin{bmatrix} -\frac{3}{8} & \frac{7}{8} \\ \frac{5}{8} & -\frac{9}{8} \end{bmatrix}$

This process is summarized as follows.

Solving Systems of Linear Equations Using Inverse Matrices

Suppose that $AX = B$ represents a system of n linear equations in n variables with a unique solution. Then,

$$X = A^{-1}B$$

where A is the coefficient matrix, B is the matrix of constants, and X is the matrix of variables.

EXAMPLE 7   Solving a System of Equations by Using an Inverse Matrix

Use the inverse of the coefficient matrix to solve the system.

$$\begin{aligned} x + 3y \phantom{{}+ 2z} &= -10 \\ x + y - 2z &= -4 \\ -3x - 3y + 5z &= 11 \end{aligned}$$

Solution:

$$\begin{aligned} x + 3y \phantom{{}+ 2z} &= -10 \\ x + y - 2z &= -4 \\ -3x - 3y + 5z &= 11 \end{aligned} \longrightarrow \underbrace{\begin{bmatrix} 1 & 3 & 0 \\ 1 & 1 & -2 \\ -3 & -3 & 5 \end{bmatrix}}_{A} \cdot \underbrace{\begin{bmatrix} x \\ y \\ z \end{bmatrix}}_{X} = \underbrace{\begin{bmatrix} -10 \\ -4 \\ 11 \end{bmatrix}}_{B}$$

From Example 4, we have that $A^{-1} = \begin{bmatrix} -\frac{1}{2} & -\frac{15}{2} & -3 \\ \frac{1}{2} & \frac{5}{2} & 1 \\ 0 & -3 & -1 \end{bmatrix}$.

$X = A^{-1}B$

$$\begin{bmatrix} x \\ y \\ z \end{bmatrix} = \begin{bmatrix} -\frac{1}{2} & -\frac{15}{2} & -3 \\ \frac{1}{2} & \frac{5}{2} & 1 \\ 0 & -3 & -1 \end{bmatrix} \begin{bmatrix} -10 \\ -4 \\ 11 \end{bmatrix}$$

$$\begin{bmatrix} x \\ y \\ z \end{bmatrix} = \begin{bmatrix} (-\frac{1}{2})(-10) + (-\frac{15}{2})(-4) + (-3)(11) \\ (\frac{1}{2})(-10) + (\frac{5}{2})(-4) + (1)(11) \\ (0)(-10) + (-3)(-4) + (-1)(11) \end{bmatrix} = \begin{bmatrix} 2 \\ -4 \\ 1 \end{bmatrix}$$

$$\begin{bmatrix} x \\ y \\ z \end{bmatrix} = \begin{bmatrix} 2 \\ -4 \\ 1 \end{bmatrix}$$ Therefore, $x = 2$, $y = -4$, and $z = 1$.
The solution set is $\{(2, -4, 1)\}$.

Skill Practice 7   Use the inverse of the coefficient matrix found in Skill Practice 4 to solve the system.

$$\begin{aligned} -2x + 4y + z &= 0 \\ 4x - 13y + 10z &= 24 \\ 3x - 9y + 6z &= 15 \end{aligned}$$

Answer

7. $\{(1, 0, 2)\}$

SECTION 6.4   Practice Exercises

Concept Connections

1. The symbol I_n represents the _____ matrix of order n.

2. I_n is an $n \times n$ matrix with _____'s along the main diagonal and _____'s elsewhere.

3. Given an $n \times n$ matrix A, if there exists a matrix A^{-1} such that $A \cdot A^{-1} = I_n$ and $A^{-1} \cdot A = I_n$, then A^{-1} is called the _____ of A.

4. A matrix that does not have an inverse is called a _____ matrix. A matrix that does have an inverse is said to be invertible or _____.

5. Let $A = \begin{bmatrix} a & b \\ c & d \end{bmatrix}$ be an invertible matrix. Then a formula for the inverse A^{-1} is given by _____.

6. Suppose that the matrix equation $AX = B$ represents a system of n linear equations in n variables with a unique solution. Then $X =$ _____.

Objective 1: Identify Identity and Inverse Matrices

7. Write the matrix I_2.

8. Write the matrix I_3.

For Exercises 9–12, verify that

   a. $AI_n = A$

   b. $I_nA = A$ (See Example 1)

9. $A = \begin{bmatrix} -\frac{7}{8} & \sqrt{5} \\ 5.1 & 8 \end{bmatrix}$ **10.** $A = \begin{bmatrix} \sqrt{3} & 1 \\ \pi & 4 \end{bmatrix}$ **11.** $A = \begin{bmatrix} 1 & -3 & 4 \\ 9 & 5 & 3 \\ 11 & -6 & -4 \end{bmatrix}$ **12.** $A = \begin{bmatrix} -3 & 9 & 1 \\ 0 & 4 & -1 \\ 5 & 0 & 3 \end{bmatrix}$

For Exercises 13–18, determine whether A and B are inverses. (See Example 2)

13. $A = \begin{bmatrix} 10 & -3 \\ 4 & -2 \end{bmatrix}$ and $B = \begin{bmatrix} \frac{1}{4} & -\frac{3}{8} \\ \frac{1}{2} & -\frac{5}{4} \end{bmatrix}$

14. $A = \begin{bmatrix} 4 & -1 \\ -8 & 6 \end{bmatrix}$ and $B = \begin{bmatrix} \frac{3}{8} & \frac{1}{16} \\ \frac{1}{2} & \frac{1}{4} \end{bmatrix}$

15. $A = \begin{bmatrix} -2 & -3 & 1 \\ -3 & -3 & 1 \\ -2 & -4 & 1 \end{bmatrix}$ and $B = \begin{bmatrix} 1 & -1 & 0 \\ 1 & 0 & -1 \\ 6 & -2 & -3 \end{bmatrix}$

16. $A = \begin{bmatrix} -1 & 2 & 1 \\ -5 & 8 & 2 \\ 7 & -11 & -3 \end{bmatrix}$ and $B = \begin{bmatrix} 2 & 5 & 4 \\ 1 & 4 & 3 \\ 1 & -3 & -2 \end{bmatrix}$

17. $A = \begin{bmatrix} 2 & 1 \\ 3 & 4 \end{bmatrix}$ and $B = \begin{bmatrix} 4 & -3 \\ -7 & 6 \end{bmatrix}$

18. $A = \begin{bmatrix} 3 & 2 \\ 10 & 7 \end{bmatrix}$ and $B = \begin{bmatrix} -7 & 10 \\ 11 & -15 \end{bmatrix}$

Objective 2: Determine the Inverse of a Matrix

For Exercises 19–34, determine the inverse of the given matrix if possible. Otherwise, state that the matrix is singular. (See Examples 3–6)

19. $A = \begin{bmatrix} -4 & -3 \\ 6 & 5 \end{bmatrix}$ **20.** $A = \begin{bmatrix} 5 & -3 \\ 10 & -7 \end{bmatrix}$ **21.** $A = \begin{bmatrix} -8 & -2 \\ 10 & 5 \end{bmatrix}$

22. $A = \begin{bmatrix} 0 & 1 \\ 6 & -2 \end{bmatrix}$ **23.** $A = \begin{bmatrix} 3 & 7 \\ 6 & 14 \end{bmatrix}$ **24.** $A = \begin{bmatrix} 2 & -5 \\ -6 & 15 \end{bmatrix}$

25. $A = \begin{bmatrix} 2 & -7 & 8 \\ 1 & -3 & 3 \\ 1 & -5 & 6 \end{bmatrix}$

26. $A = \begin{bmatrix} 1 & -1 & 1 \\ 1 & 1 & -2 \\ -1 & 0 & 1 \end{bmatrix}$

27. $A = \begin{bmatrix} 1 & -2 & 2 \\ 2 & -3 & 1 \\ 0 & 1 & -1 \end{bmatrix}$

28. $A = \begin{bmatrix} -2 & 5 & -4 \\ 1 & -2 & 1 \\ 1 & -5 & 9 \end{bmatrix}$

29. $A = \begin{bmatrix} 5 & 7 & -11 \\ 3 & 5 & -9 \\ 1 & 1 & -1 \end{bmatrix}$

30. $A = \begin{bmatrix} 1 & 3 & 9 \\ 2 & 7 & 22 \\ -5 & -17 & -53 \end{bmatrix}$

31. $A = \begin{bmatrix} 0 & 1 & 1 \\ \frac{5}{3} & -1 & -3 \\ -1 & 1 & 2 \end{bmatrix}$

32. $A = \begin{bmatrix} 3 & -1 & -\frac{1}{2} \\ 2 & -1 & \frac{1}{2} \\ -5 & 2 & \frac{1}{2} \end{bmatrix}$

33. $A = \begin{bmatrix} 1 & -2 & 5 & 0 \\ 0 & -1 & 2 & 0 \\ 0 & 1 & -1 & 0 \\ 2 & -1 & 7 & 1 \end{bmatrix}$

34. $A = \begin{bmatrix} 1 & 3 & 0 & -3 \\ 0 & 1 & 0 & -2 \\ -2 & -4 & 1 & 2 \\ 0 & 1 & 1 & 0 \end{bmatrix}$

Objective 3: Solve Systems of Linear Equations Using the Inverse of a Matrix

For Exercises 35–38, write the system of equations as a matrix equation of the form $AX = B$, where A is the coefficient matrix, X is the column matrix of variables, and B is the column matrix of constants.

35. $3x - 4y = -1$
$\quad\ \ 2x + \ y = 14$

36. $-6x - \ y = 1$
$\quad\ \ \ 2x + 3y = 13$

37. $9x - 6y + 4z = 27$
$\quad\ \ \ 4x \quad\quad - \ z = 1$
$\quad\quad\ \ \ 3y + \ z = 0$

38. $-3x \quad\quad + 8z = 4$
$\quad\quad\ \ \ 6y - \ z = 7$
$\quad\quad 2x - y + 6z = -15$

For Exercises 39–50, solve the system by using the inverse of the coefficient matrix. (See Example 7)

39. $-4x - 3y = -4$
$\quad\ \ 6x + 5y = 8$
(See Exercise 19 for A^{-1}.)

40. $5x - 3y = -20$
$\quad 10x - 7y = -50$
(See Exercise 20 for A^{-1}.)

41. $3x + 7y = -5$
$\quad 4x + 9y = -7$

42. $\quad 3x - \ 8y = -1$
$\quad -4x + 11y = 2$

43. $2x - 7y + 8z = 1$
$\quad\ x - 3y + 3z = 0$
$\quad\ x - 5y + 6z = 2$
(See Exercise 25 for A^{-1}.)

44. $\quad x - y + \ z = -4$
$\quad\ x + y - 2z = 12$
$\quad -x \quad\ \ + z = -5$
(See Exercise 26 for A^{-1}.)

45. $\quad x - 2y + 2z = -12$
$\quad 2x - 3y + \ z = -10$
$\quad\quad\ \ \ y - \ z = 6$
(See Exercise 27 for A^{-1}.)

46. $-2x + 5y - 4z = -20$
$\quad\ x - 2y + \ z = 8$
$\quad\ x - 5y + 9z = 24$
(See Exercise 28 for A^{-1}.)

47. $\quad r - 2s + t = 2$
$\quad -r + 4s + t = 3$
$\quad 2r - 2s - t = -1$

48. $\quad a - \ b + 3c = 2$
$\quad 2a + \ b + 2c = 2$
$\quad -2a - 2b + \ c = 3$

49. $w - 2x + 5y \quad\quad = 3$
$\quad -x + 2y \quad\quad = 1$
$\quad\quad x - \ y \quad\quad = -1$
$\quad 2w - x + 7y + z = 5$
(See Exercise 33 for A^{-1}.)

50. $\quad w + 3x \quad\quad - 3z = 8$
$\quad\quad\quad x \quad\quad - 2z = 4$
$\quad -2w - 4x + y + 2z = -6$
$\quad\quad\quad x + y \quad\quad = 0$
(See Exercise 34 for A^{-1}.)

Mixed Exercises

For Exercises 51–58, determine whether the statement is true or false. If a statement is false, explain why.

51. A 3×2 matrix has a multiplicative inverse.

52. A 2×3 matrix has a multiplicative inverse.

53. Every square matrix has an inverse.

54. Every singular matrix has an inverse.

55. Every nonsingular matrix has an inverse.

56. The matrix $\begin{bmatrix} -a & -b \\ a & b \end{bmatrix}$ is invertible.

57. The matrix $\begin{bmatrix} x & y \\ 2x & 2y \end{bmatrix}$ is invertible.

58. The inverse of the matrix I_4 is itself.

59. Find a 2×2 matrix that is its own inverse. Answers will vary.

60. Given an invertible 2×2 matrix A and the nonzero real number k, find the inverse of kA in terms of A^{-1}.

61. Given $A = \begin{bmatrix} 3 & 2 \\ 5 & 6 \end{bmatrix}$,

    a. Find A^{-1}.   **b.** Find $(A^{-1})^{-1}$.

62. Given $B = \begin{bmatrix} -3 & 2 \\ -5 & 4 \end{bmatrix}$,

    a. Find B^{-1}.   **b.** Find $(B^{-1})^{-1}$.

63. Given $A = \begin{bmatrix} a & 0 \\ 0 & b \end{bmatrix}$, where a and b are nonzero real numbers, find A^{-1}.

64. Given $A = \begin{bmatrix} a & 0 & 0 \\ 0 & b & 0 \\ 0 & 0 & c \end{bmatrix}$, where a, b, and c are nonzero real numbers, find A^{-1}.

Write About It

65. Explain how you can determine if two $n \times n$ matrices, A and B, are inverses.

66. Given a matrix $A = \begin{bmatrix} a & b \\ c & d \end{bmatrix}$, for what conditions on a, b, c, and d will the matrix not have an inverse?

Expanding Your Skills

67. Given $A = \begin{bmatrix} a & b \\ c & d \end{bmatrix}$, perform row operations on the matrix $[A \mid I_n]$ to find A^{-1}. This confirms the formula for the inverse of an invertible 2×2 matrix. (See page 606.)

68. Show by counterexample that $(AB)^{-1} \neq A^{-1}B^{-1}$. That is, find two matrices A and B for which $(AB)^{-1} \neq A^{-1}B^{-1}$.

Physicists know that if each edge of a thin conducting plate is kept at a constant temperature, then the temperature at the interior points is the mean (average) of the four surrounding points equidistant from the interior point. Use this principle in Exercise 69 to find the temperature at points x_1, x_2, x_3, and x_4.

69. *Hint:* Set up four linear equations to represent the temperature at points x_1, x_2, x_3, and x_4. Then solve the system. For example, one equation would be:

$$x_1 = \frac{1}{4}(36 + 32 + x_2 + x_3)$$

Technology Connections

For Exercises 70–71, use a graphing utility to find the inverse of the given matrix. Round the elements in the inverse to 2 decimal places.

70. $A = \begin{bmatrix} 0.04 & 0.13 & -0.08 & -0.43 \\ 0.19 & 0.33 & 0.06 & -0.84 \\ 0.01 & 0.08 & -0.11 & 0.46 \\ 0.37 & -1.42 & 0.03 & 0.52 \end{bmatrix}$

71. $A = \begin{bmatrix} 3.5 & 2.1 & 1.6 & 2.4 \\ -4.5 & 3.9 & -9.1 & 3.8 \\ 5.7 & 6.8 & -1.4 & -7.7 \\ 3.3 & 2.8 & 4.6 & -1.0 \end{bmatrix}$

For Exercises 72–73, use a graphing calculator and the inverse of the coefficient matrix to find the solution to the given system. Round to 2 decimal places.

72. $(\log 2)x + (\sqrt{7})y + z = 4.1$
$(e^2)x - (\sqrt{3})y - \pi z = -3.7$
$(\ln 10)\, x + y - 2.2z = 7.2$

73. $(\sqrt{11})x + y - (\ln 5)z = 52.3$
$(\sqrt{7})x - \pi y + (e^3)z = -27.5$
$-x + (\log 81)\, y - z = 69.8$

SECTION 6.5   Determinants and Cramer's Rule

1. Evaluate the Determinant of a 2 × 2 Matrix

In this section, we present yet another method to solve a linear system with one unique solution. This method, called Cramer's rule, enables us to write a formula to solve a system based on the coefficients of the equations in the system.

Before studying Cramer's rule, we need to become familiar with the determinant of a square matrix. Associated with every square matrix is a real number called the *determinant* of the matrix. A determinant of a square matrix A, denoted $|A|$, is written by enclosing the elements of the matrix within two vertical bars. For example:

$$\text{Given } A = \begin{bmatrix} 3 & -1 \\ 4 & 0 \end{bmatrix}, \text{ then } |A| = \begin{vmatrix} 3 & -1 \\ 4 & 0 \end{vmatrix}$$

Avoiding Mistakes

Notice that a determinant is denoted using vertical bars, whereas a matrix is denoted using brackets.

Determinants have many applications in mathematics, including solving systems of linear equations, finding the area of a triangle, determining whether three points are collinear, and finding an equation of a line through two points (see Exercises 51–58). The determinant of a 2 × 2 matrix is defined as follows.

Determinant of a 2 × 2 Matrix

The **determinant** of the matrix $A = \begin{bmatrix} a & b \\ c & d \end{bmatrix}$ is $ad - bc$. That is,

$$|A| = \begin{vmatrix} a & b \\ c & d \end{vmatrix} = ad - bc.$$

EXAMPLE 1   Evaluating a 2 × 2 Determinant

Evaluate the determinant of each matrix.

a. $A = \begin{bmatrix} 6 & -2 \\ 5 & \frac{1}{3} \end{bmatrix}$ **b.** $B = \begin{bmatrix} 2 & -11 \\ 0 & 0 \end{bmatrix}$

Solution:

TIP Example 1(b) illustrates that the determinant of a matrix having a row of all zeros is 0. The same is true for a matrix having a column of all zeros.

a. $|A| = \begin{vmatrix} 6 & -2 \\ 5 & \frac{1}{3} \end{vmatrix} = \overset{a \cdot d}{(6)(\tfrac{1}{3})} - \overset{b \cdot c}{(-2)(5)} = 12$ $a = 6, b = -2, c = 5,$ and $d = \frac{1}{3}$

b. $|B| = \begin{vmatrix} 2 & -11 \\ 0 & 0 \end{vmatrix} = \overset{a \cdot d}{(2)(0)} - \overset{b \cdot c}{(-11)(0)} = 0$ $a = 2, b = -11, c = 0,$ and $d = 0$

Skill Practice 1   Evaluate the determinant.

a. $\begin{vmatrix} 3 & -\frac{1}{2} \\ 10 & 2 \end{vmatrix}$ **b.** $\begin{vmatrix} 9 & 0 \\ -4 & 0 \end{vmatrix}$

---

2. Evaluate the Determinant of an *n* × *n* Matrix

We now develop the tools to compute the determinant of a higher-order square matrix. To find the determinant of an $n \times n$ matrix $A = [a_{ij}]$ we first need to define the minor of an element of the matrix. For any element a_{ij}, the **minor** M_{ij} of that

Answers

1. a. 11   **b.** 0

element is the determinant of the resulting matrix obtained by deleting the ith row and jth column. For example, consider the matrix:

$$A = [a_{ij}] = \begin{bmatrix} 5 & -1 & 6 \\ 0 & -7 & 1 \\ 4 & 2 & 6 \end{bmatrix}$$

The element a_{11} is 5. The minor of this element is found by deleting the first row and first column and then evaluating the determinant of the remaining 2×2 matrix:

$$\begin{bmatrix} 5 & -1 & 6 \\ 0 & -7 & 1 \\ 4 & 2 & 6 \end{bmatrix} \qquad M_{11} = \begin{vmatrix} -7 & 1 \\ 2 & 6 \end{vmatrix} = (-7)(6) - (1)(2) = -44$$

The element a_{32} is 2. The minor of this element is found by deleting the third row and second column and then evaluating the determinant of the remaining 2×2 matrix:

$$\begin{bmatrix} 5 & -1 & 6 \\ 0 & -7 & 1 \\ 4 & 2 & 6 \end{bmatrix} \qquad M_{32} = \begin{vmatrix} 5 & 6 \\ 0 & 1 \end{vmatrix} = (5)(1) - (6)(0) = 5$$

EXAMPLE 2   **Determining the Minor for Elements in a Matrix**

Find the minor for each element in the first column of the matrix.

$$\begin{bmatrix} 3 & 4 & -1 \\ 2 & -4 & 5 \\ 0 & 1 & -6 \end{bmatrix}$$

Solution:

For the element a_{11}:

$$\begin{bmatrix} 3 & 4 & -1 \\ 2 & -4 & 5 \\ 0 & 1 & -6 \end{bmatrix} \qquad M_{11} = \begin{vmatrix} -4 & 5 \\ 1 & -6 \end{vmatrix} = (-4)(-6) - (5)(1) = 19$$

For the element a_{21}:

$$\begin{bmatrix} 3 & 4 & -1 \\ 2 & -4 & 5 \\ 0 & 1 & -6 \end{bmatrix} \qquad M_{21} = \begin{vmatrix} 4 & -1 \\ 1 & -6 \end{vmatrix} = (4)(-6) - (-1)(1) = -23$$

For the element a_{31}:

$$\begin{bmatrix} 3 & 4 & -1 \\ 2 & -4 & 5 \\ 0 & 1 & -6 \end{bmatrix} \qquad M_{31} = \begin{vmatrix} 4 & -1 \\ -4 & 5 \end{vmatrix} = (4)(5) - (-1)(-4) = 16$$

Skill Practice 2   Find the minor for each element in the second row of the matrix from Example 2.

Next, we define the determinant for a 3×3 matrix.

Answer

2. $M_{21} = -23$; $M_{22} = -18$; $M_{23} = 3$

Determinant of a 3 × 3 Matrix

$$\begin{vmatrix} a_1 & b_1 & c_1 \\ a_2 & b_2 & c_2 \\ a_3 & b_3 & c_3 \end{vmatrix} = a_1 \begin{vmatrix} b_2 & c_2 \\ b_3 & c_3 \end{vmatrix} - a_2 \begin{vmatrix} b_1 & c_1 \\ b_3 & c_3 \end{vmatrix} + a_3 \begin{vmatrix} b_1 & c_1 \\ b_2 & c_2 \end{vmatrix} \text{ or equivalently,}$$

$$= a_1(b_2c_3 - c_2b_3) - a_2(b_1c_3 - c_1b_3) + a_3(b_1c_2 - c_1b_2) \quad \text{or}$$

$$= a_1b_2c_3 + b_1c_2a_3 + c_1a_2b_3 - a_3b_2c_1 - b_3c_2a_1 - c_3a_2b_1$$

From this definition, we see that the determinant of the given 3 × 3 matrix can be written as

$$a_1 \cdot (\text{minor of } a_1) - a_2 \cdot (\text{minor of } a_2) + a_3 \cdot (\text{minor of } a_3)$$

EXAMPLE 3   **Evaluating the Determinant of a 3 × 3 Matrix**

Evaluate the determinant of the matrix. $A = \begin{bmatrix} 2 & 4 & 2 \\ 1 & -3 & 0 \\ -5 & 5 & -1 \end{bmatrix}$

Solution:

$$|A| = \begin{vmatrix} 2 & 4 & 2 \\ 1 & -3 & 0 \\ -5 & 5 & -1 \end{vmatrix} = 2 \cdot \overbrace{\begin{vmatrix} -3 & 0 \\ 5 & -1 \end{vmatrix}}^{\text{minor of 2}} - (1)\overbrace{\begin{vmatrix} 4 & 2 \\ 5 & -1 \end{vmatrix}}^{\text{minor of 1}} + (-5) \cdot \overbrace{\begin{vmatrix} 4 & 2 \\ -3 & 0 \end{vmatrix}}^{\text{minor of } -5}$$

$$= 2\big[(-3)(-1) - (0)(5)\big] - 1\big[(4)(-1) - (2)(5)\big] + (-5)\big[(4)(0) - (2)(-3)\big]$$

$$= 2(3) - 1(-14) - 5(6)$$

$$= -10$$

Skill Practice 3   Evaluate the determinant. $\begin{vmatrix} -2 & 4 & 9 \\ 5 & -1 & 2 \\ 1 & 1 & 6 \end{vmatrix}$

Although we defined the determinant of a 3 × 3 matrix by expanding the minors of the elements in the first column, *any row or column can be used*. However, we must choose the correct sign to apply to the product of factors of each term. The following array of signs is helpful.

row 1, column 1   row 1, column 2
1 + 1 = 2 (even)   1 + 2 = 3 (odd)

$$\begin{bmatrix} + & - & + \\ - & + & - \\ + & - & + \end{bmatrix}$$

Notice that the sign is positive if the sum of the row and column numbers is even. The sign is negative if the sum of the row and column numbers is odd.

To evaluate the determinant of an $n \times n$ matrix, first choose any row or column. For each element a_{ij} in the selected row or column, multiply the minor by 1 or -1 depending on whether the sum of the row and column numbers is even or odd. That is, for the element a_{ij}, we have $(-1)^{i+j}M_{ij}$. This product is called the *cofactor* of the element a_{ij}.

Cofactor of an Element of a Matrix

Given a square matrix $A = [a_{ij}]$, the **cofactor** of a_{ij} is $(-1)^{i+j}M_{ij}$, where M_{ij} is the minor of a_{ij}.

Using the definition of the cofactor of an element of an $n \times n$ matrix, we can now present a generic method to find the determinant of the matrix.

Evaluating the Determinant of an $n \times n$ Matrix by Expanding Cofactors

Step 1 Choose any row or column.
Step 2 Multiply each element in the selected row or column by its cofactor.
Step 3 The value of the determinant is the sum of the products from step 2.

It is important to note that this three-step process to evaluate a determinant works for any $n \times n$ matrix, including a 2 × 2 matrix. For example, given $A = \begin{bmatrix} a & b \\ c & d \end{bmatrix}$, we can evaluate the determinant of A by expanding cofactors about the first row. The cofactor of a is d, and the cofactor of b is $-c$. Therefore,

$$|A| = a(d) + b(-c)$$

$$= ad - bc \text{ as expected.}$$

EXAMPLE 4   **Evaluating the Determinant of a 3 × 3 Matrix**

Evaluate the determinant of the matrix from Example 3 by expanding cofactors about the elements in the third row.

$$A = \begin{bmatrix} 2 & 4 & 2 \\ 1 & -3 & 0 \\ -5 & 5 & -1 \end{bmatrix}$$

Solution:

$$\begin{vmatrix} 2 & 4 & 2 \\ 1 & -3 & 0 \\ -5 & 5 & -1 \end{vmatrix} = \overbrace{(-5)(-1)^{3+1}\begin{vmatrix} 4 & 2 \\ -3 & 0 \end{vmatrix}}^{\text{cofactor of } -5} + \overbrace{(5)(-1)^{3+2}\begin{vmatrix} 2 & 2 \\ 1 & 0 \end{vmatrix}}^{\text{cofactor of } 5} + \overbrace{(-1)(-1)^{3+3}\begin{vmatrix} 2 & 4 \\ 1 & -3 \end{vmatrix}}^{\text{cofactor of } -1}$$

$$= (-5)(1)\big[(4)(0) - (2)(-3)\big] + (5)(-1)\big[(2)(0) - (2)(1)\big] + (-1)(1)\big[(2)(-3) - (4)(1)\big]$$

$$= -5(6) - 5(-2) - 1(-10)$$

$$= -30 + 10 + 10$$

$$= -10$$

This is the same result as in Example 3.

Skill Practice 4   Evaluate the determinant by expanding cofactors about the elements in the third column.

$$\begin{vmatrix} -2 & 4 & 9 \\ 5 & -1 & 2 \\ 1 & 1 & 6 \end{vmatrix}$$

TECHNOLOGY CONNECTIONS

Evaluating a Determinant

To find the determinant of a matrix on a graphing calculator, first enter the matrix in the calculator by using the MATRIX menu and EDIT menu. Select a name for the matrix such as [A]. Then access the determinant function det(under the MATRIX and **MATH** menus. Enter det([A]) in the home screen and press **ENTER**.

Alternatively, if the calculator is in "MATHPRINT" mode, enter the matrix by selecting **ALPHA** F3.

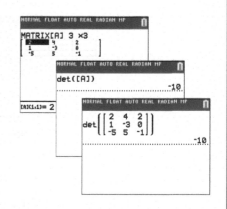

TIP The determinant of a 3 × 3 matrix can also be evaluated by using the "method of diagonals."

Step 1: Recopy columns 1 and 2 to the right of the matrix.

Step 2: Multiply the elements on the diagonals labeled d_1 through d_6 (each diagonal has three elements).

Step 3: The value of the determinant is $(d_1 + d_2 + d_3) - (d_4 + d_5 + d_6)$.

The determinant from Example 4 is evaluated as follows:

$$\begin{bmatrix} 2 & 4 & 2 \\ 1 & -3 & 0 \\ -5 & 5 & -1 \end{bmatrix} \begin{matrix} 2 & 4 \\ 1 & -3 \\ -5 & 5 \end{matrix}$$

$d_4 \quad d_5 \quad d_6 \quad d_1 \quad d_2 \quad d_3$

$$= [(2)(-3)(-1) + (4)(0)(-5) + (2)(1)(5)]$$
$$\quad - [(2)(-3)(-5) + (2)(0)(5) + (4)(1)(-1)]$$
$$= -10$$

Some students find the method of diagonals to be a faster technique to find the determinant of a 3 × 3 matrix. However, it is critical to note that the method of diagonals only works for the determinant of a 3 × 3 matrix.

Perhaps one of the most important applications of the determinant of a matrix is to determine whether the matrix has an inverse. In Section 6.4, we found the inverse of a 2 × 2 matrix $A = \begin{bmatrix} a & b \\ c & d \end{bmatrix}$ by applying the formula $A^{-1} = \dfrac{1}{ad - bc}\begin{bmatrix} d & -b \\ -c & a \end{bmatrix}$. This is equivalent to $A^{-1} = \dfrac{1}{|A|}\begin{bmatrix} d & -b \\ -c & a \end{bmatrix}$. Furthermore, the inverse exists if $ad - bc \neq 0$ or equivalently, if $|A| \neq 0$. This is true in general for an $n \times n$ matrix.

Answer
4. -42

Using Determinants to Determine if a Matrix Is Invertible

Let A be an $n \times n$ matrix. Then A is invertible if and only if $|A| \neq 0$.

In Example 5, we will use the determinant of A to determine if A has an inverse.

EXAMPLE 5   Using a Determinant to Determine if a Matrix Is Invertible

Use $|A|$ to determine if A is invertible. $A = \begin{vmatrix} -2 & 0 & 1 & 2 \\ 6 & 2 & -2 & 5 \\ 5 & 3 & -1 & 1 \\ 0 & 4 & 2 & 1 \end{vmatrix}$

Solution:

To evaluate a determinant, we can simplify the arithmetic by expanding around the row or column that contains the greatest number of 0 elements. In this case, we have chosen row 1.

$$\begin{vmatrix} -2 & 0 & 1 & 2 \\ 6 & 2 & -2 & 5 \\ 5 & 3 & -1 & 1 \\ 0 & 4 & 2 & 1 \end{vmatrix} = -2(-1)^{1+1}\begin{vmatrix} 2 & -2 & 5 \\ 3 & -1 & 1 \\ 4 & 2 & 1 \end{vmatrix} + 0(-1)^{1+2}\begin{vmatrix} 6 & -2 & 5 \\ 5 & -1 & 1 \\ 0 & 2 & 1 \end{vmatrix}$$

$$+ 1(-1)^{1+3}\begin{vmatrix} 6 & 2 & 5 \\ 5 & 3 & 1 \\ 0 & 4 & 1 \end{vmatrix} + 2(-1)^{1+4}\begin{vmatrix} 6 & 2 & -2 \\ 5 & 3 & -1 \\ 0 & 4 & 2 \end{vmatrix}$$

The second term is zero. Evaluating the determinants in the first, third, and fourth terms (shown in red), we have:

$$\begin{vmatrix} -2 & 0 & 1 & 2 \\ 6 & 2 & -2 & 5 \\ 5 & 3 & -1 & 1 \\ 0 & 4 & 2 & 1 \end{vmatrix} = -2(1)(42) + 0 + 1(1)(84) + 2(-1)(0) = 0$$

The determinant of A is zero. Therefore, A is not invertible.

Skill Practice 5   Use $|A|$ to determine if A is invertible. $A = \begin{vmatrix} 3 & 2 & 1 & 0 \\ 4 & 0 & 3 & 1 \\ 2 & 0 & 0 & 5 \\ 3 & -1 & 0 & 9 \end{vmatrix}$

---

3. Apply Cramer's Rule

We have learned several methods to solve a system of linear equations: the substitution method, the addition method, Gaussian elimination, Gauss-Jordan elimination, and the application of matrix inverses. We now present another method called Cramer's rule.

Cramer's rule involves finding the ratio of several determinants derived from the coefficients of the equations within the system. For example, consider the following system of equations.

$$a_1x + b_1y = c_1$$
$$a_2x + b_2y = c_2$$

Answer

5. $|A| = -9$; Since $|A| \neq 0$, A is invertible.

Using the addition method to solve for x, we have

$$a_1x + b_1y = c_1 \xrightarrow{\text{Multiply by } b_2.} \quad a_1b_2x + b_1b_2y = c_1b_2$$

$$a_2x + b_2y = c_2 \xrightarrow{\text{Multiply by } -b_1.} \quad \underline{-a_2b_1x - b_1b_2y = -c_2b_1}$$

$$(a_1b_2 - a_2b_1)x \qquad = c_1b_2 - c_2b_1$$

$$x = \frac{c_1b_2 - c_2b_1}{a_1b_2 - a_2b_1} = \frac{\begin{vmatrix} c_1 & b_1 \\ c_2 & b_2 \end{vmatrix}}{\begin{vmatrix} a_1 & b_1 \\ a_2 & b_2 \end{vmatrix}}$$

Using similar logic, we can show that
$$y = \frac{a_1c_2 - a_2c_1}{a_1b_2 - a_2b_1} = \frac{\begin{vmatrix} a_1 & c_1 \\ a_2 & c_2 \end{vmatrix}}{\begin{vmatrix} a_1 & b_1 \\ a_2 & b_2 \end{vmatrix}}$$

These results are summarized as Cramer's rule for a system of linear equations in two variables.

Cramer's Rule for a System of Two Linear Equations in Two Variables

Given the system
$$a_1x + b_1y = c_1$$
$$a_2x + b_2y = c_2$$

let $D = \begin{vmatrix} a_1 & b_1 \\ a_2 & b_2 \end{vmatrix}$, $D_x = \begin{vmatrix} c_1 & b_1 \\ c_2 & b_2 \end{vmatrix}$, and $D_y = \begin{vmatrix} a_1 & c_1 \\ a_2 & c_2 \end{vmatrix}$.

Then if $D \neq 0$, the system has the unique solution: $x = \dfrac{D_x}{D}$ and $y = \dfrac{D_y}{D}$

TIP Here are some memory tips to remember the patterns presented in Cramer's rule.

$$a_1x + b_1y = c_1$$
$$a_2x + b_2y = c_2$$

Coefficients of
x terms   y terms

1. The determinant D is the determinant of the coefficients of x and y. $\quad D = \begin{vmatrix} a_1 & b_1 \\ a_2 & b_2 \end{vmatrix}$

x-coefficients replaced by c_1 and c_2

2. The determinant D_x has the column of x term coefficients replaced by c_1 and c_2. $\quad D_x = \begin{vmatrix} c_1 & b_1 \\ c_2 & b_2 \end{vmatrix}$

y-coefficients replaced by c_1 and c_2

3. The determinant D_y has the column of y term coefficients replaced by c_1 and c_2. $\quad D_y = \begin{vmatrix} a_1 & c_1 \\ a_2 & c_2 \end{vmatrix}$

EXAMPLE 6   **Solving a 2 × 2 System by Using Cramer's Rule**

Solve the system by using Cramer's rule. $\begin{aligned}-11x + 6y &= 4\\ 2x - 5y &= -3\end{aligned}$

Solution:

For this system, $a_1 = -11$, $b_1 = 6$, $c_1 = 4$, $a_2 = 2$, $b_2 = -5$, and $c_2 = -3$.

$$D = \begin{vmatrix} a_1 & b_1 \\ a_2 & b_2 \end{vmatrix} \longrightarrow D = \begin{vmatrix} -11 & 6 \\ 2 & -5 \end{vmatrix} = (-11)(-5) - (6)(2) = 43$$

$$D_x = \begin{vmatrix} c_1 & b_1 \\ c_2 & b_2 \end{vmatrix} \longrightarrow D_x = \begin{vmatrix} 4 & 6 \\ -3 & -5 \end{vmatrix} = (4)(-5) - (6)(-3) = -2$$

$$D_y = \begin{vmatrix} a_1 & c_1 \\ a_2 & c_2 \end{vmatrix} \longrightarrow D_y = \begin{vmatrix} -11 & 4 \\ 2 & -3 \end{vmatrix} = (-11)(-3) - (4)(2) = 25$$

Therefore, $x = \dfrac{D_x}{D} = \dfrac{-2}{43}$ and $y = \dfrac{D_y}{D} = \dfrac{25}{43}$.

The solution set is $\left\{\left(-\dfrac{2}{43}, \dfrac{25}{43}\right)\right\}$.

Skill Practice 6   Solve the system by using Cramer's rule. $\begin{aligned}3x - 4y &= 9\\ -5x + 6y &= 2\end{aligned}$

The patterns associated with Cramer's rule can be generalized to solve a system of n linear equations in n variables.

Cramer's Rule for a System of n Linear Equations in n Variables

Consider the following system of n linear equations in n variables.

$$\begin{aligned}a_{11}x_1 + a_{12}x_2 + \cdots + a_{1n}x_n &= b_1\\ a_{21}x_1 + a_{22}x_2 + \cdots + a_{2n}x_n &= b_2\\ &\;\;\vdots\\ a_{n1}x_1 + a_{n2}x_2 + \cdots + a_{nn}x_n &= b_n\end{aligned}$$

If the system has a unique solution, then the solution is $(x_1, x_2, \ldots, x_i, \ldots, x_n)$, where

$$x_1 = \frac{D_1}{D}, x_2 = \frac{D_2}{D}, \ldots, x_i = \frac{D_i}{D}, \ldots, x_n = \frac{D_n}{D}.$$

$D \neq 0$ is the determinant of the coefficient matrix, and D_i is the determinant formed by replacing the ith column of the coefficient matrix by the column of constants $b_1, b_2, \ldots, b_n$.

In Example 7, we apply Cramer's rule to a system of equations with three variables.

Answer

6. $\left\{\left(-31, -\dfrac{51}{2}\right)\right\}$

EXAMPLE 7 Solving a 3 × 3 System by Using Cramer's Rule

Solve the system by using Cramer's rule.
$$\begin{aligned} 2x - 3y + 5z &= 11 \\ -5x + 7y - 2z &= -6 \\ 9x - 2y + 3z &= 4 \end{aligned}$$

Solution:

Evaluate the determinants D, D_x, D_y, and D_z.

$$D = \begin{vmatrix} 2 & -3 & 5 \\ -5 & 7 & -2 \\ 9 & -2 & 3 \end{vmatrix} = -222 \qquad D_x = \begin{vmatrix} 11 & -3 & 5 \\ -6 & 7 & -2 \\ 4 & -2 & 3 \end{vmatrix} = 77$$

$$D_y = \begin{vmatrix} 2 & 11 & 5 \\ -5 & -6 & -2 \\ 9 & 4 & 3 \end{vmatrix} = 117 \qquad D_z = \begin{vmatrix} 2 & -3 & 11 \\ -5 & 7 & -6 \\ 9 & -2 & 4 \end{vmatrix} = -449$$

Therefore, $x = \dfrac{D_x}{D} = -\dfrac{77}{222}$, $y = \dfrac{D_y}{D} = -\dfrac{117}{222}$ or $-\dfrac{39}{74}$, and $z = \dfrac{D_z}{D} = \dfrac{449}{222}$.

The solution set is $\left\{ \left(-\dfrac{77}{222}, -\dfrac{39}{74}, \dfrac{449}{222} \right) \right\}$.

Skill Practice 7 Solve the system by using Cramer's rule.
$$\begin{aligned} 5x + 3y - 3z &= -14 \\ 3x - 4y + z &= 2 \\ x + 7y + z &= 6 \end{aligned}$$

Although Cramer's rule may seem cumbersome for solving a 3 × 3 system of linear equations, it provides convenient formulas that can be programmed into a computer or calculator to solve the system. However, it is important to remember that Cramer's rule does not apply if $D = 0$. In such a case, the system of equations is either inconsistent (has no solution) or the equations are dependent (the system has infinitely many solutions).

EXAMPLE 8 Identifying Whether Cramer's Rule Applies

Solve the system by using Cramer's rule if possible. Otherwise, use a different method.

$$\begin{aligned} x + 3y &= 6 \\ -2x - 6y &= -12 \end{aligned}$$

Solution:

Evaluate D. $\quad D = \begin{vmatrix} 1 & 3 \\ -2 & -6 \end{vmatrix} = 1(-6) - 3(-2) = -6 + 6 = 0$

Since $D = 0$, Cramer's rule does not apply. Using Gauss-Jordan elimination, we have

$$\begin{bmatrix} 1 & 3 & | & 6 \\ -2 & -6 & | & -12 \end{bmatrix} \xrightarrow{\ 2R_1 + R_2 \rightarrow R_2\ } \begin{bmatrix} 1 & 3 & | & 6 \\ 0 & 0 & | & 0 \end{bmatrix}$$

The last row of the augmented matrix represents the equation $0 = 0$, which is true for all values of x and y. The system reduces to the single equation $x + 3y = 6$ and there are infinitely many solutions.

The solution set is $\{(6 - 3y, y) \mid y \text{ is any real number}\}$.

Answer

7. $\left\{ \left(-\dfrac{13}{34}, \dfrac{5}{17}, \dfrac{147}{34} \right) \right\}$

Skill Practice 8   Solve the system by using Cramer's rule if possible. Otherwise, use a different method.

$$x + 4y = 2$$
$$3x + 12y = 4$$

Answer

8. { }

| SECTION 6.5 | Practice Exercises |

Prerequisite Review

For Exercises R.1–R.2, simplify the exponential expression.

R.1. $(-1)^4$ **R.2.** $(-1)^7$

R.3. Write an equation of the line that passes through the points $(-3, 5)$ and $(-2, 4)$. Write the equation in slope-intercept form.

Concept Connections

1. Associated with every square matrix A is a real number denoted by $|A|$ called the _____ of A.

2. For a 2×2 matrix $A = \begin{bmatrix} a & b \\ c & d \end{bmatrix}$, $|A| = \begin{vmatrix} a & b \\ c & d \end{vmatrix} = $ _____.

3. Given $A = [a_{ij}]$, the _____ of the element a_{ij} is the determinant obtained by deleting the ith row and jth column.

4. Given $A = [a_{ij}]$, then the value $(-1)^{i+j} M_{ij}$ is called the _____ of the element a_{ij}.

5. The determinant of a 3×3 matrix

$$A = \begin{bmatrix} a_1 & b_1 & c_1 \\ a_2 & b_2 & c_2 \\ a_3 & b_3 & c_3 \end{bmatrix} \text{ is given by}$$

$$|A| = a_1 \begin{vmatrix} \square & \square \\ \square & \square \end{vmatrix} - a_2 \begin{vmatrix} \square & \square \\ \square & \square \end{vmatrix} + a_3 \begin{vmatrix} \square & \square \\ \square & \square \end{vmatrix}$$

6. Suppose that the given system has one solution.

$$a_1x + b_1y = c_1$$
$$a_2x + b_2y = c_2$$

Cramer's rule gives the solution as $x = \dfrac{\square}{\square}$ and $y = \dfrac{\square}{\square}$,

where $D = \begin{vmatrix} \square & \square \\ \square & \square \end{vmatrix}$, $D_x = \begin{vmatrix} \square & \square \\ \square & \square \end{vmatrix}$, and

$D_y = \begin{vmatrix} \square & \square \\ \square & \square \end{vmatrix}$.

Objective 1: Evaluate the Determinant of a 2 × 2 Matrix

For Exercises 7–16, evaluate the determinant of the matrix. (See Example 1)

7. $A = \begin{bmatrix} 3 & -2 \\ 6 & 5 \end{bmatrix}$

8. $B = \begin{bmatrix} 7 & 12 \\ -1 & 4 \end{bmatrix}$

9. $C = \begin{bmatrix} \frac{2}{3} & \frac{1}{5} \\ 10 & 12 \end{bmatrix}$

10. $D = \begin{bmatrix} \frac{8}{9} & 4 \\ \frac{5}{2} & 18 \end{bmatrix}$

11. $E = \begin{bmatrix} -3 & 0 \\ 4 & 0 \end{bmatrix}$

12. $F = \begin{bmatrix} 0 & 9 \\ 0 & 4 \end{bmatrix}$

13. $G = \begin{bmatrix} x & 4 \\ 9 & x \end{bmatrix}$

14. $H = \begin{bmatrix} y & 16 \\ 4 & y \end{bmatrix}$

15. $T = \begin{bmatrix} e^x & e^{2x} \\ 4 & -e^x \end{bmatrix}$

16. $V = \begin{bmatrix} \log x & \log x \\ 2 & 5 \end{bmatrix}$

Objective 2: Evaluate the Determinant of an $n \times n$ Matrix

For Exercises 17–22, refer to the matrix $A = [a_{ij}] = \begin{bmatrix} -6 & 11 & 8 \\ 4 & -2 & -5 \\ -3 & 7 & 10 \end{bmatrix}$.

a. Find the minor of the given element. (**See Example 2**)
b. Find the cofactor of the given element.

17. a_{12}

18. a_{23}

19. a_{31}

20. a_{13}

21. a_{22}

22. a_{33}

For Exercises 23–32, evaluate the determinant of the matrix and state whether the matrix is invertible. (**See Examples 3–5**)

23. $A = \begin{bmatrix} 4 & 1 & 3 \\ 0 & -1 & 2 \\ 5 & 8 & 0 \end{bmatrix}$

24. $B = \begin{bmatrix} 9 & 5 & -1 \\ 2 & 0 & 4 \\ 7 & -2 & 0 \end{bmatrix}$

25. $C = \begin{bmatrix} 5 & 1 & 6 \\ 2 & 3 & 4 \\ 8 & -1 & 7 \end{bmatrix}$

26. $D = \begin{bmatrix} -3 & 1 & -2 \\ 10 & 5 & 8 \\ 6 & 7 & -4 \end{bmatrix}$

27. $E = \begin{bmatrix} 2 & 0 & 1 \\ 1 & -1 & 2 \\ 3 & 1 & 0 \end{bmatrix}$

28. $F = \begin{bmatrix} 1 & -3 & 17 \\ 1 & -1 & 7 \\ 2 & -5 & 29 \end{bmatrix}$

29. $G = \begin{bmatrix} 5 & 6 & 4 & 1 \\ 2 & 0 & 3 & 0 \\ 0 & 1 & 4 & 0 \\ -1 & 2 & 0 & 0 \end{bmatrix}$

30. $H = \begin{bmatrix} 8 & 0 & 5 & 1 \\ 0 & 3 & 4 & -2 \\ 2 & 6 & 3 & 0 \\ -1 & 0 & 0 & 0 \end{bmatrix}$

31. $T = \begin{bmatrix} 3 & 8 & 1 & 4 \\ -2 & 4 & 0 & 5 \\ -1 & 1 & 0 & -1 \\ 0 & 5 & 2 & 3 \end{bmatrix}$

32. $W = \begin{bmatrix} 2 & 5 & 2 & 4 \\ 0 & 0 & -3 & 1 \\ 4 & 8 & 0 & 1 \\ -1 & 2 & 0 & 5 \end{bmatrix}$

Objective 3: Apply Cramer's Rule

For Exercises 33–48, solve the system if possible by using Cramer's rule. If Cramer's rule does not apply, solve the system by using another method. (**See Examples 6–8**)

33. $2x + 10y = 11$
$3x - 5y = 6$

34. $-5x - 8y = 3$
$4x + 7y = 13$

35. $-10x + 4y = 7$
$6x = 7y + 2$

36. $11x + 6y = 8$
$2x = 9y + 5$

37. $3(x - y) = y + 8$
$y = \frac{3}{4}x - 2$

38. $5(x + y) = 7x + 4$
$4x = 10y - 8$

39. $y = -3x + 7$
$\frac{1}{2}x + \frac{1}{6}y = 1$

40. $x = 4y + 5$
$3(x - 4) = 12y$

41. $11x \qquad - 3z = 1$
$2y + 9z = 6$
$4x + 5y \qquad = -9$

42. $-2x + 6y \qquad = 9$
$5y + 7z = 1$
$4x \qquad - 3z = -8$

43. $2x - 5y + z = 11$
$3x + 7y - 4z = 8$
$x - 9y + 2z = 4$

44. $-5x - 6y + 8z = 1$
$2x + y - 4z = 5$
$3x - 4y - z = -2$

45. $2x - 3y + z = 6$
$-4x + 6y - 2z = -12$
$6x - 9y + 3z = 18$

46. $-x + y - 3z = 4$
$3x - 3y + 9z = -12$
$-2x + 2y - 6z = 8$

47. $x - 2y + 3z = -1$
$5x - 7y + 3z = 1$
$x \qquad - 5z = 2$

48. $x + 3y - 5z = 10$
$-2x - 4y + 8z = -14$
$x + y - 3z = 5$

For Exercises 49–50, use Cramer's rule to solve for the indicated variable.

49. $x_1 + 2x_2 + 3x_3 - 4x_4 = 3$
$\qquad 5x_2 \qquad + x_4 = 9$
$\quad x_1 \qquad + 4x_3 \qquad = -1$
$\qquad\qquad 5x_3 - 2x_4 = 8$ Solve for x_2.

50. $-2x_1 - x_2 + x_3 + 3x_4 = 10$
$\quad x_1 \qquad + 5x_3 \qquad = 4$
$\quad 2x_2 \qquad + x_4 = -1$
$\qquad\qquad 4x_3 + 2x_4 = 7$ Solve for x_3.

Mixed Exercises

Determinants can be used to determine whether three points are collinear (lie on the same line). Given the ordered pairs (x_1, y_1), (x_2, y_2), and (x_3, y_3), the points are collinear if the determinant to the right equals zero. For Exercises 51–54, determine if the points are collinear.

$$\begin{vmatrix} x_1 & y_1 & 1 \\ x_2 & y_2 & 1 \\ x_3 & y_3 & 1 \end{vmatrix}$$

51. $(3, 6)$, $(6, 10)$, $(-3, -2)$

52. $(-2, 1)$, $(-4, -4)$, $(4, 16)$

53. $(4, -3)$, $(5, -7)$, $(8, -14)$

54. $(0, 6)$, $(1, 4)$, $(4, -6)$

The equation at the right represents an equation of the line passing through the distinct points (x_1, y_1) and (x_2, y_2). For Exercises 55–56,

a. Use the determinant equation to write an equation of the line passing through the given points.
b. Write the equation of the line in slope-intercept form.

$$\begin{vmatrix} x & y & 1 \\ x_1 & y_1 & 1 \\ x_2 & y_2 & 1 \end{vmatrix} = 0$$

55. $(-3, 2)$ and $(-4, 6)$

56. $(-4, 1)$ and $(-5, 4)$

For Exercises 57–58, use the formula at the right to find the area of a triangle with vertices (x_1, y_1), (x_2, y_2), and (x_3, y_3). Choose the $+$ or $-$ sign so that the value of the area is positive.

$$\text{Area} = \pm \frac{1}{2} \begin{vmatrix} x_1 & y_1 & 1 \\ x_2 & y_2 & 1 \\ x_3 & y_3 & 1 \end{vmatrix}$$

57. $(1, 0)$, $(7, -2)$, $(4, -5)$

58. $(-2, 1)$, $(-1, -6)$, $(-8, -5)$

Given a square matrix A, elementary row operations (or column operations) performed on A affect the value of $|A|$ in the following ways:

- Interchanging any two rows (or columns) of A will change the sign of $|A|$.
- Multiplying a row (or column) of A by a constant real number k multiplies $|A|$ by k.
- Adding a multiple of a row (or column) of A to another row (or column) of A does not change the value of $|A|$.

For Exercises 59–64, demonstrate these three properties.

59. Given $A = \begin{bmatrix} 5 & 2 \\ -3 & 6 \end{bmatrix}$ and $B = \begin{bmatrix} -3 & 6 \\ 5 & 2 \end{bmatrix}$,

a. Evaluate $|A|$.
b. Evaluate $|B|$.
c. How are A and B related and how are $|A|$ and $|B|$ related?

60. Given $A = \begin{bmatrix} 2 & -7 \\ 4 & 10 \end{bmatrix}$ and $B = \begin{bmatrix} 4 & 10 \\ 2 & -7 \end{bmatrix}$,

a. Evaluate $|A|$.
b. Evaluate $|B|$.
c. How are A and B related and how are $|A|$ and $|B|$ related?

61. Given $A = \begin{bmatrix} 1 & -3 \\ 4 & 1 \end{bmatrix}$ and $B = \begin{bmatrix} 2 & -6 \\ 4 & 1 \end{bmatrix}$,

a. Evaluate $|A|$.
b. Evaluate $|B|$.
c. How are A and B related and how are $|A|$ and $|B|$ related?

62. Given $A = \begin{bmatrix} 2 & 1 \\ 5 & 7 \end{bmatrix}$ and $B = \begin{bmatrix} -6 & -3 \\ 5 & 7 \end{bmatrix}$,

a. Evaluate $|A|$.
b. Evaluate $|B|$.
c. How are A and B related and how are $|A|$ and $|B|$ related?

63. Given $A = \begin{bmatrix} 1 & 2 \\ 3 & 4 \end{bmatrix}$ and $B = \begin{bmatrix} 1 & 2 \\ 6 & 10 \end{bmatrix}$,

a. Evaluate $|A|$.
b. Evaluate $|B|$.
c. How are A and B related and how are $|A|$ and $|B|$ related?

64. Given $A = \begin{bmatrix} 1 & 1 \\ 5 & 6 \end{bmatrix}$ and $B = \begin{bmatrix} 1 & 1 \\ 3 & 4 \end{bmatrix}$,

a. Evaluate $|A|$.
b. Evaluate $|B|$.
c. How are A and B related and how are $|A|$ and $|B|$ related?

Given a square matrix A, if either of the following conditions are true, then $|A| = 0$.

- A row (or column) of A consists entirely of zeros.
- One row (or column) is a constant multiple of another row (or column).

For Exercises 65–68, demonstrate these two properties.

65. Given $A = \begin{bmatrix} 3 & 5 \\ 6 & 10 \end{bmatrix}$, find $|A|$.

66. Given $A = \begin{bmatrix} -2 & 7 \\ -6 & 21 \end{bmatrix}$, find $|A|$.

67. Given $A = \begin{bmatrix} 4 & -5 & 0 \\ 3 & -1 & 0 \\ 0 & 1 & 0 \end{bmatrix}$, find $|A|$.

68. Given $A = \begin{bmatrix} 5 & 3 & 1 \\ -1 & 0 & 3 \\ 0 & 0 & 0 \end{bmatrix}$, find $|A|$.

69. Evaluate $|I_2|$.

70. Evaluate $|I_3|$.

71. Evaluate $\begin{vmatrix} a & 0 & 0 \\ 0 & b & 0 \\ 0 & 0 & c \end{vmatrix}$.

72. Evaluate $\begin{vmatrix} x & 0 \\ 0 & x \end{vmatrix}$.

73. If A and B are square matrices, then the product property of determinants indicates that $|AB| = |A| \cdot |B|$. Use matrix A and matrix B to demonstrate this property.

$$A = \begin{bmatrix} 4 & -2 \\ 3 & 1 \end{bmatrix} \text{ and } B = \begin{bmatrix} -5 & 1 \\ 3 & 2 \end{bmatrix}$$

74. The **transpose** of a square matrix A, denoted as A^T, is a square matrix that results by writing the rows of A as the columns of A^T.

   a. Given $A = \begin{bmatrix} 1 & 2 & 5 \\ 0 & 8 & 4 \\ 3 & 7 & 6 \end{bmatrix}$, find A^T.

   b. Show that $|A| = |A^T|$.

Write About It

75. What is the difference between the minor of an element a_{ij} and the cofactor of the element?

76. Explain the difference between the notation $\begin{bmatrix} a & b \\ c & d \end{bmatrix}$ and $\begin{vmatrix} a & b \\ c & d \end{vmatrix}$.

77. The determinant of a square matrix can be computed by expanding the cofactors of the elements in any row or column. How would you choose which row or column?

78. Consider the system shown here. Describe the pattern associated with constructing the determinants used with Cramer's rule.

$$a_1x + b_1y = c_1$$
$$a_2x + b_2y = c_2$$

Technology Connections

For Exercises 79–82, use a graphing utility to evaluate the determinant of the matrix. Round to the nearest whole unit.

79. $\begin{bmatrix} \sqrt{3} & e & 1.6 \\ \log 5 & -2\pi & \ln 3 \\ -4 & 8.4 & -\sqrt{6} \end{bmatrix}$

80. $\begin{bmatrix} 8.9 & -2.3 & 3.8 \\ -1.7 & 0.9 & 4.6 \\ 2.7 & 10.1 & 14.9 \end{bmatrix}$

81. $A = \begin{bmatrix} -0.4 & 1.5 & 9 & 11.3 \\ -3.5 & 0.2 & -1.1 & 3 \\ 8 & 9.4 & -5.4 & 2 \\ -1 & 4.6 & 10.8 & -9.7 \end{bmatrix}$

82. $B = \begin{bmatrix} -2\pi & e^2 & 9.1 & \log 2 \\ \log 50 & -\sqrt{11} & 4.3 & \pi \\ -4.9 & 0 & e^2 & 8.1 \\ \sqrt{7} & \ln 7 & -9.7 & 0 \end{bmatrix}$

PROBLEM RECOGNITION EXERCISES

Using Multiple Methods to Solve Systems of Linear Equations

For Exercises 1–4, solve the system of equations using

    a. The substitution method or the addition method (see Sections 5.1 and 5.2).

    b. Gaussian elimination (see Section 6.1).

    c. Gauss-Jordan elimination (see Section 6.1).

    d. The inverse of the coefficient matrix (see Section 6.4).

    e. Cramer's rule (see Section 6.5).

1. $x = -3y - 10$
    $-3x - 7y = 22$

2. $2x = 2 - 8y$
    $3x + 10y = 5$

3. $x + 2y - z = 0$
    $2x \quad\quad + z = 4$
    $2x - y + 2z = 5$

4. $x + 4y + 2z = 10$
    $2y + z = 4$
    $x + y \quad\quad = 2$

For Exercises 5–8,

    a. Evaluate the determinant of the coefficient matrix.

    b. Based on the value of the determinant from part (a), can an inverse matrix or Cramer's rule be used to solve the system?

    c. Solve the system using an appropriate method.

5. $1.5x - 2y = 3$
    $-3x + 4y = 12$

6. $5x - 2y = 1$
    $x - 0.4y = 4$

7. $x - 3y + 7z = 1$
    $-2x + 5y - 11z = -3$
    $x - 5y + 13z = -1$

8. $x - 2y + 3z = -7$
    $-2x + y \quad\quad = -1$
    $x \quad\quad - z = 3$

CHAPTER 6   KEY CONCEPTS

SECTION 6.1 Solving Systems of Linear Equations Using Matrices	Reference
A system of linear equations can be represented by an **augmented matrix.**	p. 564
Elementary row operations can be used to write a matrix in row-echelon form or reduced row-echelon form. 1. Interchange two rows. 2. Multiply a row by a nonzero constant. 3. Add a nonzero multiple of one row to another row.	p. 565
The method of **Gaussian elimination** uses elementary row operations to write an augmented matrix in row-echelon form so that the system can be solved by back substitution.	p. 567
With the method of **Gauss-Jordan elimination** the augmented matrix is written in *reduced* row-echelon form so that the solution can be found by inspection.	p. 568
SECTION 6.2 Inconsistent Systems and Dependent Equations	Reference
A system of equations that has no solution is called an **inconsistent system.** An inconsistent system is detected algebraically if a contradiction is reached when solving the system.	p. 575
A system of linear equations may have infinitely many solutions. In such a case, the equations are said to be **dependent.** Dependent equations are detected algebraically if an identity is reached when solving the system.	p. 576

SECTION 6.3 Operations on Matrices	Reference
An $m \times n$ matrix has m rows and n columns.	p. 585
Adding and subtracting matrices: If A and B represent two matrices of the same order, then $A + B$ and $A - B$ are found by adding or subtracting the corresponding elements.	p. 587
Scalar multiplication: Let $A = [a_{ij}]$ be an $m \times n$ matrix and let k be a real number. Then, $kA = [ka_{ij}]$.	p. 589
Matrix multiplication: Let A be an $m \times p$ matrix and let B be a $p \times n$ matrix, then the product AB is an $m \times n$ matrix. For the matrix AB, the element in the ith row and jth column is the sum of the products of the corresponding elements in the ith row of A and the jth column of B (the inner product of the ith row and jth column). *Note*: If the number of columns in A does not equal the number of rows in B, then it is not possible to compute the product AB.	p. 591

SECTION 6.4 Inverse Matrices and Matrix Equations	Reference
The **identity matrix** I_n for the multiplication of matrices is the $n \times n$ square matrix with 1's along the main diagonal and 0's for all other elements.	p. 602
Inverse of a square matrix: Let A be an $n \times n$ matrix. If there exists an $n \times n$ matrix A^{-1} such that $$AA^{-1} = I_n \quad \text{and} \quad A^{-1}A = I_n$$ then A^{-1} is the **multiplicative inverse** of A.	p. 603
A matrix that does not have an inverse is called a **singular matrix.** A matrix that does have an inverse is said to be **invertible** or **nonsingular.**	p. 605
Finding the inverse of a nonsingular matrix: Let A be an $n \times n$ matrix for which A^{-1} exists. To find A^{-1}: **Step 1** Write a matrix of the form $[A \mid I_n]$. **Step 2** Perform row operations to write the matrix in the form $[I_n \mid B]$. **Step 3** The matrix B is A^{-1}.	p. 604
The inverse of a nonsingular 2×2 matrix A can also be found as follows. If $A = \begin{bmatrix} a & b \\ c & d \end{bmatrix}$, then $A^{-1} = \dfrac{1}{ad - bc} \cdot \begin{bmatrix} d & -b \\ -c & a \end{bmatrix}$, or equivalently $A^{-1} = \dfrac{1}{\lvert A \rvert} \begin{bmatrix} d & -b \\ -c & a \end{bmatrix}$.	p. 606
Solving a system using inverse matrices: Suppose that $AX = B$ represents a system of n linear equations in n variables with a unique solution. Then, $$X = A^{-1}B$$ where A is the coefficient matrix, B is the matrix of constants, and X is the matrix of variables.	p. 607

SECTION 6.5 Determinants and Cramer's Rule	Reference
Associated with every square matrix is a real number called the determinant of the matrix.	p. 612
Determinant of a 2×2 matrix: Given $A = \begin{bmatrix} a & b \\ c & d \end{bmatrix}$, the determinant of A is denoted by $\lvert A \rvert$ or $\begin{vmatrix} a & b \\ c & d \end{vmatrix}$ and the value is $ad - bc$.	p. 612

The minor and cofactor of an element of a matrix: Given an $n \times n$ matrix $A = [a_{ij}]$, the minor M_{ij} of an element a_{ij} is the determinant of the resulting matrix obtained by deleting the ith row and jth column.	p. 612
The **cofactor** of a_{ij} is $(-1)^{i+j}M_{ij}$ where M_{ij} is the minor of a_{ij}.	p. 615
Determinant of a 3 × 3 matrix: $$\begin{vmatrix} a_1 & b_1 & c_1 \\ a_2 & b_2 & c_2 \\ a_3 & b_3 & c_3 \end{vmatrix} = a_1 \begin{vmatrix} b_2 & c_2 \\ b_3 & c_3 \end{vmatrix} - a_2 \begin{vmatrix} b_1 & c_1 \\ b_3 & c_3 \end{vmatrix} + a_3 \begin{vmatrix} b_1 & c_1 \\ b_2 & c_2 \end{vmatrix}$$	p. 614
Find the determinant of an $n \times n$ matrix: **Step 1** Choose any row or column. **Step 2** Multiply each element in the selected row or column by its cofactor. **Step 3** The value of the determinant is the sum of the products from step 2.	p. 615
Cramer's rule for a system of two equations and two variables: Given $\quad \begin{aligned} a_1x + b_1y &= c_1 \\ a_2x + b_2y &= c_2 \end{aligned}$ if $D \neq 0$, then $x = \dfrac{D_x}{D}$ and $y = \dfrac{D_y}{D}$, where $D = \begin{vmatrix} a_1 & b_1 \\ a_2 & b_2 \end{vmatrix}$, $D_x = \begin{vmatrix} c_1 & b_1 \\ c_2 & b_2 \end{vmatrix}$, and $D_y = \begin{vmatrix} a_1 & c_1 \\ a_2 & c_2 \end{vmatrix}$.	p. 618
The patterns associated with Cramer's rule can be generalized to solve a system of n linear equations in n variables.	p. 619
When applying Cramer's rule, if $D = 0$, then the system has either no solution or infinitely many solutions.	p. 620

Expanded Chapter Summary available at www.mhhe.com/millerca.

CHAPTER 6   Review Exercises

SECTION 6.1

1. Write a system of linear equations represented by the augmented matrix. Then write the solution set.

$$\begin{bmatrix} 1 & -2 & 3 & | & -1 \\ 0 & 1 & 4 & | & -11 \\ 0 & 0 & 1 & | & -2 \end{bmatrix}$$

For Exercises 2–4, perform the elementary row operations on the matrix. $\begin{bmatrix} 2 & -3 & | & 1 \\ 5 & 6 & | & -4 \end{bmatrix}$

2. $R_1 \Leftrightarrow R_2$

3. $\dfrac{1}{2}R_1 \to R_1$

4. $-2R_1 + R_2 \to R_2$

For Exercises 5–8, solve the system by using Gaussian elimination or Gauss-Jordan elimination.

5. $\begin{aligned} -2x + y &= -16 \\ x - 2y &= 17 \end{aligned}$

6. $\begin{aligned} 2(x - 6y) &= 36 \\ 47y &= 7x - 141 \end{aligned}$

7. $\begin{aligned} 2x - 5y + 18z &= 44 \\ x - 3y + 11z &= 27 \\ x - 2y + 11z &= 29 \end{aligned}$

8. $\begin{aligned} w + x \quad - 2z &= 3 \\ 2x \quad - 3z &= 3 \\ 2w \quad + y + z &= 3 \\ 4y - z &= 9 \end{aligned}$

9. Lily borrowed a total of $10,000. She borrowed part of the money from her friend Sly who did not charge her interest. She borrowed part of the money from a credit union at 5% simple interest, and she borrowed the rest of the money from a bank at 7.5% interest. At the end of 1 yr, she owed $500 in interest. If she borrowed $1000 less from her friend than she did from the bank, determine how much she borrowed from each source.

SECTION 6.2

For Exercises 10–13, determine the solution set for the system represented by each augmented matrix.

10. $\begin{bmatrix} 1 & 0 & | & 4 \\ 0 & 1 & | & 0 \end{bmatrix}$

11. $\begin{bmatrix} 1 & -2 & | & 6 \\ 0 & 0 & | & 1 \end{bmatrix}$

12. $\begin{bmatrix} 1 & 4 & | & 0 \\ 0 & 0 & | & 0 \end{bmatrix}$ **13.** $\begin{bmatrix} 1 & 0 & -3 & | & 0 \\ 0 & 1 & 2 & | & 1 \\ 0 & 0 & 0 & | & 0 \end{bmatrix}$

For Exercises 14–19, solve the system by using Gaussian elimination or Gauss-Jordan elimination.

14. $3x + 6y = -9$
    $x + 2y = -3$

15. $-(2x - y) = 8 - y$
    $y = x - 6$

16. $x - 2y = 3z - 10$
    $x - y = z - 7$
    $3x - 7y - 11z = -320$

17. $x \quad\quad - 3z = 5$
    $-2x + y + 10z = -7$
    $x + y + \quad z = 8$

18. $2x = 3y - z - 4$
    $x - 2y = z + 2y - 2$
    $x + y = 2z - 2$

19. $5y = x + 2z + 1$
    $2(x - 5y) + 4z = -2$
    $3(x + 2z) = 15y - 3$

20. The solution set to a system of dependent equations is given. Write three ordered triples that are solutions to the system. Answers may vary.

$\{(2z - 3, z + 2, z) \mid z \text{ is any real number}\}$

21. a. Assume that traffic flows freely around the traffic circle. The flow rates given are measured in vehicles per hour. If the flow rate x_3 is 130 vehicles per hour, determine the flow rates x_1 and x_2.

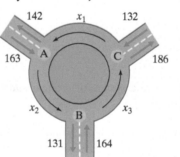

    b. If traffic between intersections B and C flows at a rate of between 100 and 150 vehicles per hour, inclusive, find the range of values for x_1 and x_2.

22. a. Assume that traffic flows freely through intersections A, B, C, and D and that all flow rates are measured in vehicles per hour. If the flow rate x_4 is 220 vehicles per hour, find the flow rates x_1, x_2, and x_3.

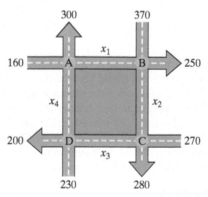

    b. If the flow rate x_4 is between 200 and 250 vehicles per hour, inclusive, find the range of values x_1, x_2, and x_3.

SECTION 6.3

For Exercises 23–26,

    a. Give the order of the matrix.

    b. Classify the matrix as a square matrix, row matrix, column matrix, or none of these.

23. $\begin{bmatrix} 1 & 2 \\ -3 & \pi \\ 4.1 & \sqrt{2} \end{bmatrix}$ **24.** $\begin{bmatrix} 8 & 4 & 1 & -6 \end{bmatrix}$

25. $\begin{bmatrix} -3.1 \\ 8.7 \end{bmatrix}$ **26.** $\begin{bmatrix} -3 & 8 \\ 0 & 0 \end{bmatrix}$

For Exercises 27–28, determine the value of the given element of the matrix $A = [a_{ij}]$.

$$A = \begin{bmatrix} -1 & 8 & -3 \\ 4 & 6 & 9 \end{bmatrix}$$

27. a_{21} **28.** a_{12}

29. For what value of x, y, and z will $A = B$?

$$A = \begin{bmatrix} 3 & -4 \\ x & z \end{bmatrix} \quad \text{and} \quad B = \begin{bmatrix} y & -4 \\ 6 & 8 \end{bmatrix}$$

30. Solve the equation $-3X + A = B$ for X, given that

$$A = \begin{bmatrix} 2 & -7 \\ 2 & -5 \end{bmatrix} \text{ and } B = \begin{bmatrix} 5 & 2 \\ -1 & 7 \end{bmatrix}.$$

For Exercises 31–40, perform the indicated operations if possible.

$$A = \begin{bmatrix} -4 & 1 \\ 6 & -2 \\ 1 & 3 \end{bmatrix} \quad B = \begin{bmatrix} 2 & 3 & -7 \\ 1 & 5 & -6 \end{bmatrix} \quad C = \begin{bmatrix} \pi & 4 \\ -3 & 1 \\ 0 & 5 \end{bmatrix}$$

31. $3A$ **32.** $-2B$ **33.** $A + B$

34. $B + C$ **35.** $2A - C$ **36.** $4A + 3B$

37. AB **38.** BC **39.** AC

40. CA

For Exercises 41–44, perform the indicated operations if possible.

$$A = \begin{bmatrix} 2 & 6 \\ -1 & 4 \end{bmatrix} \quad B = \begin{bmatrix} 1 \\ -3 \end{bmatrix} \quad C = \begin{bmatrix} 2 & 7 \end{bmatrix}$$

41. A^2 **42.** AB **43.** BC **44.** CB

45. A company owns two movie theaters in town. The number of popcorns and drinks sold for each theater is given in matrix Q. The price per item is given in matrix P. Find the product QP and interpret the result.

	Popcorn (small)	Popcorn (large)	Drinks (small)	Drinks (large)	
$Q =$	386	244	418	216	Theater 1
	450	382	476	262	Theater 2

$$P = \begin{bmatrix} \$8.50 \\ \$6.50 \\ \$5.50 \\ \$3.50 \end{bmatrix} \begin{matrix} \textbf{Popcorn (small)} \\ \textbf{Popcorn (large)} \\ \textbf{Drinks (small)} \\ \textbf{Drinks (large)} \end{matrix}$$

46. Matrix M gives the manufacturer price for four models of dining room tables. Matrix P gives the retail price to the customer.

$$M = \begin{array}{c} \quad\;\; \textbf{Wood} \quad \textbf{Metal} \\ \begin{bmatrix} \$1050 & \$940 \\ \$890 & \$800 \end{bmatrix} \begin{array}{l} \textbf{Large} \\ \textbf{Small} \end{array} \end{array}$$

$$P = \begin{array}{c} \quad\;\; \textbf{Wood} \quad \textbf{Metal} \\ \begin{bmatrix} \$1365 & \$1222 \\ \$1157 & \$1040 \end{bmatrix} \begin{array}{l} \textbf{Large} \\ \textbf{Small} \end{array} \end{array}$$

a. Compute $P - M$ and interpret its meaning.

b. If the tax rate in a certain city is 6%, use scalar multiplication to find a matrix F that gives the final price (including sales tax) to the customer for each model.

47. a. Write a matrix A that represents the coordinates of the vertices of the triangle. Place the x-coordinate of each point in the first row of A and the corresponding y-coordinate in the second row of A.

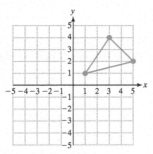

b. Use addition of matrices to shift the triangle 3 units to the left and 1 unit downward.

c. Find the product $\begin{bmatrix} -1 & 0 \\ 0 & 1 \end{bmatrix} \cdot A$ and explain the effect on the graph of the triangle.

d. Find the product $\begin{bmatrix} 1 & 0 \\ 0 & -1 \end{bmatrix} \cdot A$ and explain the effect on the graph of the triangle.

SECTION 6.4

48. Given two $n \times n$ matrices A and B, what are the criteria for the matrices to be inverses?

49. Determine whether A and B are inverses.

$$A = \begin{bmatrix} 4 & 1 \\ 3 & 2 \end{bmatrix} \quad B = \begin{bmatrix} 2 & -3 \\ -1 & 4 \end{bmatrix}$$

50. Determine whether A and B are inverses.

$$A = \begin{bmatrix} 1 & 3 \\ 2 & 1 \end{bmatrix} \quad B = \begin{bmatrix} -\frac{1}{5} & \frac{3}{5} \\ \frac{2}{5} & -\frac{1}{5} \end{bmatrix}$$

For Exercises 51–56, determine the inverse of the given matrix if possible. Otherwise state that the matrix is singular.

51. $A = \begin{bmatrix} 5 & -2 \\ 1 & 2 \end{bmatrix}$

52. $A = \begin{bmatrix} \frac{1}{4} & \frac{3}{8} \\ \frac{1}{8} & -\frac{1}{16} \end{bmatrix}$

53. $A = \begin{bmatrix} 2 & 3 \\ 16 & 24 \end{bmatrix}$

54. $A = \begin{bmatrix} 1 & 0 & 1 \\ -1 & 5 & -3 \\ 1 & -3 & 2 \end{bmatrix}$

55. $A = \begin{bmatrix} -5 & 4 & 1 \\ 15 & -12 & -4 \\ 4 & -3 & -1 \end{bmatrix}$

56. $A = \begin{bmatrix} 1 & -3 & -17 \\ 1 & -2 & -12 \\ 0 & -2 & -10 \end{bmatrix}$

57. Write the system of equations as a matrix equation of the form $AX = B$, where A is the coefficient matrix, X is the column matrix of variables, and B is the column matrix of constants.

$$\begin{aligned} -3x + 7y \quad\;\;\; &= 6 \\ 4x \quad\;\;\; + 2z &= -3 \\ 2x - y + 5z &= -13 \end{aligned}$$

For Exercises 58–61, solve the system using the inverse of the coefficient matrix.

58. $\dfrac{1}{4}x + \dfrac{3}{8}y = 4$ \quad (See Exercise 52 for A^{-1}.)

$\dfrac{1}{8}x - \dfrac{1}{16}y = -2$

59. $5x - 2y = 26$ \quad (See Exercise 51 for A^{-1}.)

$x + 2y = -2$

60. $x \quad\;\; + z = 2$ \quad (See Exercise 54 for A^{-1}.)

$-x + 5y - 3z = -6$

$x - 3y + 2z = 3$

61. $-5x + 4y + z = 6$ \quad (See Exercise 55 for A^{-1}.)

$15x - 12y - 4z = -21$

$4x - 3y - z = -5$

SECTION 6.5

For Exercises 62–65, refer to the matrix

$$A = [a_{ij}] = \begin{bmatrix} -5 & 4 & -2 \\ 1 & 0 & 6 \\ 8 & -9 & 0 \end{bmatrix}.$$

a. Find the minor of the given element.

b. Find the cofactor of the given element.

62. a_{13} \qquad **63.** a_{31} \qquad **64.** a_{32} \qquad **65.** a_{23}

For Exercises 66–71, evaluate the determinant of the given matrix.

66. $A = \begin{bmatrix} 9 & -4 \\ 2 & -3 \end{bmatrix}$

67. $B = \begin{bmatrix} 3 & x \\ x & 27 \end{bmatrix}$

68. $C = \begin{bmatrix} 9 & -15 \\ -3 & 5 \end{bmatrix}$

69. $D = \begin{bmatrix} 4 & -1 & 0 \\ 6 & 8 & -2 \\ 1 & 5 & 3 \end{bmatrix}$

70. $E = \begin{bmatrix} 4 & -9 & 0 \\ -3 & 8 & 0 \\ 6 & 1 & 0 \end{bmatrix}$

71. $F = \begin{bmatrix} -2 & 0 & 3 & 1 \\ 1 & 1 & 0 & 5 \\ 4 & 0 & 0 & -2 \\ 0 & -3 & 0 & 6 \end{bmatrix}$

For Exercises 72–76, solve the system by using Cramer's rule if possible. If Cramer's rule does not apply, use another method to solve the system.

72. $3x - 7y = 11$

$4x + 2y = 3$

73. $9x = 3y + 5$

$-2(x + 3y) = 4$

74. $2x + 5y = 10$
$10y = -4(x - 5)$

75. $3x - 2y + z = 4$
$5x + 3y + 6z = 1$
$-2x + 5z = 7$

76. $2x + y - z = 5$
$7x + 7y - 6z = 5$
$3x + 5y - 4z = -1$

For Exercises 77–78, use Cramer's rule to solve for the indicated variable.

77. $-6x + 7y = 8$ Solve for y.
$2x + 5y + z = -3$
$3x + 2z = 11$

78. $3x_1 + 4x_3 = 6$ Solve for x_4.
$4x_1 + 2x_3 + x_4 = -7$
$x_2 - 3x_4 = 2$
$5x_3 + x_4 = 1$

CHAPTER 6 · Test

For Exercises 1–3, perform the elementary row operations on the matrix $A = \begin{bmatrix} 3 & 1 & 4 & -2 \\ 1 & 5 & -3 & 1 \\ 0 & 4 & 2 & 6 \end{bmatrix}$.

1. $R_1 \leftrightarrow R_2$ **2.** $-3R_2 + R_1 \to R_1$ **3.** $\frac{1}{4}R_3 \to R_3$

4. Explain why the matrix is not in reduced row-echelon form.

$$\begin{bmatrix} 1 & 0 & 6 & 4 \\ 0 & 1 & 0 & 2 \\ 0 & 0 & 1 & -3 \end{bmatrix}$$

For Exercises 5–7, write the solution set for the system represented by each augmented matrix.

5. $\begin{bmatrix} 1 & 4 & 2 \\ 0 & 1 & 3 \end{bmatrix}$

6. $\begin{bmatrix} 1 & 0 & 0 & 4 \\ 0 & 1 & 0 & 2 \\ 0 & 0 & 0 & 1 \end{bmatrix}$

7. $\begin{bmatrix} 1 & 0 & -3 & 0 \\ 0 & 1 & 2 & 5 \\ 0 & 0 & 0 & 0 \end{bmatrix}$

For Exercises 8–9, refer to the matrix $A = \begin{bmatrix} -1 & 3 & 0 \\ 2 & 5 & -4 \\ 6 & 9 & -8 \end{bmatrix}$.

a. Find the minor of the given element.

b. Find the cofactor of the given element.

8. a_{12} **9.** a_{31}

For Exercises 10–11, evaluate the determinant of the matrix.

10. $A = \begin{bmatrix} 3 & 7 \\ 4 & -1 \end{bmatrix}$

11. $B = \begin{bmatrix} -3 & 4 & 7 \\ 1 & -2 & 3 \\ 6 & 5 & 0 \end{bmatrix}$

12. Use the determinant of A to determine whether A has an inverse.

$$A = \begin{bmatrix} 1 & 2 & 3 \\ 1 & 4 & 11 \\ 2 & 5 & 10 \end{bmatrix}$$

For Exercises 13–16, solve the system by using Gaussian elimination or Gauss-Jordan elimination.

13. $-3(x + y) = 3y - 12$
$-3x = 4y - 6$

14. $-3x = 11 - 18y$
$x - 6y = 2$

15. $x - 2y = 5z + 4$
$6y + 18z = 2x - 8$
$-3x + 8y + 20z = -18$

16. $2x + 6y + 30z = 2$
$x + 2y + 11z = 0$
$-3x - 6y - 33z = 0$

17. Solve the system by using Cramer's rule.

$$3x - 5y = 7$$
$$11x + 2y = 8$$

18. Use Cramer's rule to solve for x.

$$2x = 3y - 4z + 11$$
$$9x + y = z - 1$$
$$3x - 4y = 7$$

19. Solve the equation $A - 4X = B$ for X, given that

$$A = \begin{bmatrix} 2 & 5 \\ -2 & -3 \end{bmatrix} \text{ and } B = \begin{bmatrix} 6 & -3 \\ 14 & 5 \end{bmatrix}.$$

For Exercises 20–23, perform the indicated operations if possible.

$$A = \begin{bmatrix} 4 & 1 & -3 \\ 2 & 4 & 6 \end{bmatrix} \quad B = \begin{bmatrix} 1 & 9 \\ 0 & -1 \\ 3 & 5 \end{bmatrix} \quad C = \begin{bmatrix} 0 & 1 & -4 \\ 2 & -1 & 8 \end{bmatrix}$$

20. $2A - 3C$ **21.** $A + B$ **22.** AB **23.** BA

24. Determine whether A and B are inverses.

$$A = \begin{bmatrix} -1 & \frac{2}{3} & -\frac{2}{3} \\ 1 & -\frac{1}{3} & \frac{1}{3} \\ 1 & -\frac{2}{3} & \frac{5}{3} \end{bmatrix} \text{ and } B = \begin{bmatrix} 1 & 2 & 0 \\ 4 & 3 & 1 \\ 1 & 0 & 1 \end{bmatrix}$$

For Exercises 25–27, determine the inverse of the matrix if possible. Otherwise, state that the matrix is singular.

25. $A = \begin{bmatrix} 3 & 2 \\ 5 & 4 \end{bmatrix}$

26. $A = \begin{bmatrix} 2 & 5 & 10 \\ 1 & 3 & 7 \\ 1 & 4 & 11 \end{bmatrix}$

27. $A = \begin{bmatrix} 3 & -1 & -1 \\ 2 & -1 & 1 \\ -5 & 2 & 1 \end{bmatrix}$

For Exercises 28–29, solve the system by using the inverse of the coefficient matrix.

28. $3x + 2y = 13$ (See Exercise 25 for A^{-1}.)
$5x + 4y = 25$

29. $3x - y - z = 8$ (See Exercise 27 for A^{-1}.)
$2x + z = y$
$-5x + 2y + z = -11$

30. **a.** Assume that traffic flows freely around the traffic circle. The flow rates given are measured in vehicles per hour. If the flow rate x_3 is 210 vehicles per hour, determine the flow rates x_1 and x_2.

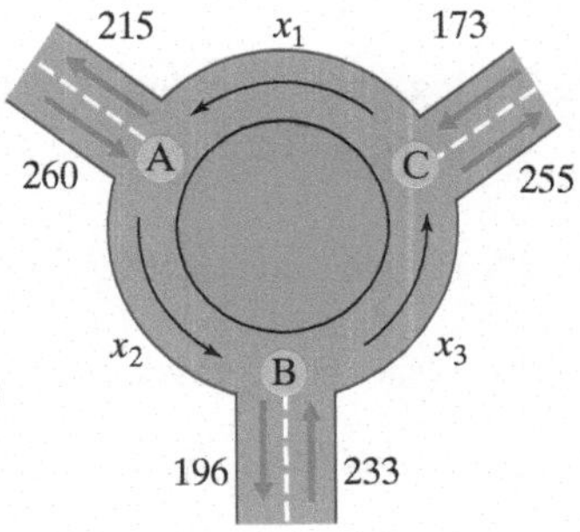

b. If traffic between intersections B and C flows at a rate of between 200 and 250 vehicles per hour, inclusive, find the range of values for x_1 and x_2.

31. Matrix C represents the number of calories burned per hour of exercise riding a bicycle, running, and walking for individuals of two different weights. Matrix N represents the number of hours spent working out with each type of activity for a given week. Find the product CN and interpret its meaning.

$$C = \begin{bmatrix} 400 & 500 & 320 \\ 550 & 780 & 480 \end{bmatrix} \begin{matrix} \textbf{120-lb person} \\ \textbf{180-lb person} \end{matrix}$$
(columns: **Bike**, **Run**, **Walk**)

$$N = \begin{bmatrix} 6 \\ 3 \\ 5 \end{bmatrix} \begin{matrix} \textbf{Bike} \\ \textbf{Run} \\ \textbf{Walk} \end{matrix}$$

32. Given three points (x_1, y_1), (x_2, y_2), and (x_3, y_3), the points are collinear if the following determinant is equal to zero.

$$\begin{vmatrix} x_1 & y_1 & 1 \\ x_2 & y_2 & 1 \\ x_3 & y_3 & 1 \end{vmatrix}$$

Determine if the points $(4, -11)$, $(-1, -1)$, and $(-5, 7)$ are collinear.

CHAPTER 6 Cumulative Review Exercises

1. Given $f(x) = x^3$, find the difference quotient $\dfrac{f(x + h) - f(x)}{h}$.

2. Given $f(x) = -2(x - 4)^2 - 5$,
   a. Determine the vertex of the graph of the parabola.
   b. Determine the intervals over which the graph is increasing.
   c. Determine the intervals over which the graph is decreasing.
   d. Write the domain in interval notation.
   e. Write the range in interval notation.

3. Given $f(x) = \sqrt{2x - 3}$, find the average rate of change on the interval $\left[2, \frac{7}{2}\right]$.

4. Simplify each expression.
   a. i^{57} **b.** i^{22}

5. Divide and write the answer in the form $a + bi$. $\dfrac{3 + 4i}{2 - 3i}$

6. Find the distance between the points $(1, -7)$ and $(4, 2)$.

7. Determine the center and radius of the circle defined by $\left(x + \frac{5}{3}\right)^2 + y^2 = 11$.

8. Technetium-99m (abbreviated ^{99m}Tc) is a short-lived gamma-ray emitter that is used in nuclear medicine. Suppose that a patient is given a small amount of ^{99m}Tc for a diagnostic test. Further suppose that the amount of ^{99m}Tc decays exponentially according to the model $Q(t) = Q_0e^{-kt}$. The value Q_0 is the initial amount administered, and $Q(t)$ represents the amount of ^{99m}Tc remaining after t hours. The value of k is the decay constant.
   a. If 94% of the technetium has decayed after 24 hr (that is, 6% remains), determine the value of k. Round to 3 decimal places.
   b. If 30 mCi is initially given to a patient for blood pool imaging of the heart, determine the amount remaining after 10 hr. Round to 1 decimal place.
   c. Determine the amount of time required for the amount of ^{99m}Tc to fall below 1% of the original amount given. Round to 1 decimal place.

For Exercises 9–12,

a. Graph the function.
b. Write the domain in interval notation.
c. Write the range in interval notation.

9. $f(x) = e^{x-2} + 1$

10. $g(x) = \ln(x + 4)$

11. $h(x) = \sqrt{-x}$

12. $k(x) = \begin{cases} (x - 1)^2 - 2 & \text{for } x \le 1 \\ x + 1 & \text{for } x > 1 \end{cases}$

For Exercises 13–14, solve the system of equations.

13. $3(x - y) = y - 3$

$-2x + \dfrac{1}{2}y = -11$

14. $2x - y + 5z = -2$

$x \qquad + 4z = 0$

$x + y = 2 - 7z$

15. Determine if the graph of the equation $|x| + y^2 = 9$ is symmetric with respect to the x-axis, y-axis, origin, or none of these.

16. Determine if the function defined by $f(x) = 6x^3 - 4x$ is even, odd, or neither.

For Exercises 17–20, refer to the given matrices and perform the indicated operations.

$$A = \begin{bmatrix} 2 & -3 \\ 1 & 6 \end{bmatrix} \quad B = \begin{bmatrix} 1 & 4 \\ 0 & 7 \end{bmatrix} \quad C = \begin{bmatrix} -1 & 4 & 0 \\ 4 & 2 & 6 \\ 3 & -2 & 0 \end{bmatrix}$$

17. AB **18.** $|C|$ **19.** $4A - B$ **20.** A^{-1}

For Exercises 21–23, find the union or intersection of the given sets. Write the answers in set-builder notation.

$A = \{x \mid 3 < x\}, B = \{x \mid -2 < x \le 4\}, C = \{x \mid x \ge 5\}$

21. $A \cup B$ **22.** $A \cap B$ **23.** $A \cap C$

For Exercises 24–27, solve the equation or inequality. Write the solution set to the inequalities in interval notation if possible.

24. $2x - 3 - 5\sqrt{2x - 3} + 4 = 0$

25. $|-x + 4| = |x - 4|$

26. $\dfrac{x + 4}{2x - 1} \ge 1$

27. $\log_2(3x - 4) = \log_2(x + 1) + 1$

28. Find all zeros of $f(x) = 2x^3 - x^2 + 18x - 9$.

29. Find the asymptotes. $r(x) = \dfrac{2x - 1}{x^2 - 9}$

30. Write the expression as a single logarithm.

$-3 \log x - \dfrac{1}{2} \log y + 5 \log z$

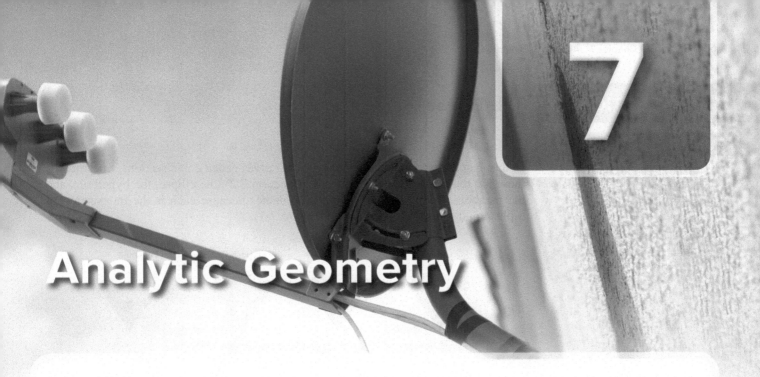

Analytic Geometry

Chapter Outline

Many television viewers today rely on a satellite dish to provide a wide variety of programming options. A satellite dish is a parabolic device that receives microwave signals from broadcast centers via communications satellites. The data are then compressed and rendered into a digital form that can be viewed on an electronic device such as a television, computer, or smartphone.

In this chapter, we study three types of curves called conic sections: the ellipse, the hyperbola, and the parabola. As we work through the chapter, we will learn how the reflective properties of these curves are used in numerous applications. For example, microwaves received by a satellite dish bounce off the parabolic surface to a common point called the focus. It is at this point that the receiver is placed to gather and interpret the signal that ultimately delivers hundreds of channels to our television set.

The Ellipse

1. Graph an Ellipse Centered at the Origin

In this chapter, we investigate four types of curves called conic sections. Specifically, these are the circle (covered in detail in Section 2.2), the ellipse, the hyperbola, and the parabola. Conic sections derive their names because each is the intersection of a plane and a double-napped cone (Figure 7-1).

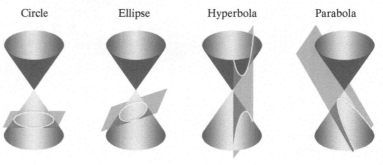

Circle   Ellipse   Hyperbola   Parabola

Figure 7-1

Recall from Section 2.2 that a circle is the set of all points in a plane, r units from a fixed point (the center of the circle), where r is the radius of the circle. We derived the standard form of an equation of a circle from the distance formula by noting that the distance between the center of the circle (h, k) and a point on the circle (x, y) is r units.

Standard Form of an Equation of a Circle

Given a circle centered at (h, k) with radius r, the standard form of an equation of the circle is given by $(x - h)^2 + (y - k)^2 = r^2$ for $r > 0$.

Example
$(x - 3)^2 + (y + 1)^2 = 4$

Standard form
$(x - 3)^2 + [y - (-1)]^2 = 2^2$

Center: $(3, -1)$

Radius: 2

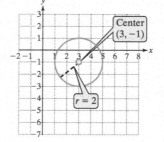

We now study a second type of conic section called an ellipse. An ellipse is an oval-shaped curve that appears in applications in many fields including architecture, acoustics, medicine, and astronomy. To determine an equation of an ellipse, we first need to formalize the definition.

Definition of an Ellipse

An **ellipse** is the set of all points (x, y) in a plane such that the sum of the distances between (x, y) and two fixed points is a constant. The fixed points are called the **foci** (plural of **focus**) of the ellipse.

To visualize the definition of an ellipse, consider the following application. Suppose Sonya wants to cut an elliptical rug from a rectangular rug to avoid a stain made by the family dog. She places two tacks along the center horizontal line. Then she ties the ends of a slack piece of rope to each tack. With the rope pulled tight, she traces out a curve. This curve is an ellipse, and the tacks are located at the foci of the ellipse (Figure 7-2).

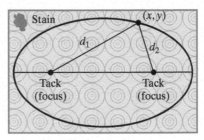

Figure 7-2

An ellipse may be elongated in any direction; however, we will study only those that are elongated horizontally and vertically. Figure 7-3 and Figure 7-4 show ellipses centered at the origin with foci on the x-axis and y-axis, respectively. The line through the foci intersects the ellipse at two points called **vertices.** The line segment with endpoints at the vertices is called the **major axis.** The **center** of the ellipse is the midpoint of the major axis. The line segment perpendicular to the major axis and passing through the center of the ellipse with endpoints on the ellipse is called the **minor axis.**

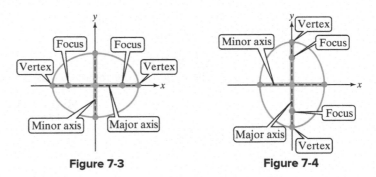

Figure 7-3   **Figure 7-4**

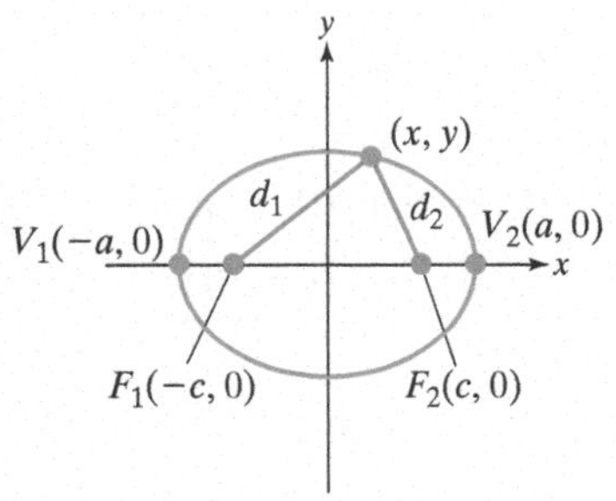

We will derive the standard form of an equation of an ellipse centered at the origin with foci on the x-axis. In Figure 7-5, the foci are labeled as $F_1(-c, 0)$ and $F_2(c, 0)$, and the vertices are labeled as $V_1(-a, 0)$ and $V_2(a, 0)$. Let (x, y) be an arbitrary point on the ellipse and let the distances between (x, y) and the two foci be d_1 and d_2.

From Figure 7-5, the distance from F_1 to V_2 is $(a + c)$. The distance from F_2 to V_2 is $(a - c)$. By the definition of an ellipse, the sum of the distances between a point on the ellipse (such as V_2) and the two foci must be constant. Adding $(a + c)$ and $(a - c)$, we see that this constant is $2a$.

Figure 7-5

$$(a + c) + (a - c) = 2a$$

The sum of the distances between a point on an ellipse and the two foci is equal to the length of the major axis, $2a$.

The sum of the distances between (x, y) and the two foci is equal to $2a$.

$$d_1 \quad + \quad d_2 \quad = 2a$$

$$\sqrt{[(x - (-c)]^2 + (y - 0)^2} + \sqrt{(x - c)^2 + (y - 0)^2} = 2a \quad \text{Apply the distance formula.}$$

$$\sqrt{(x + c)^2 + y^2} + \sqrt{(x - c)^2 + y^2} = 2a \quad \text{Simplify within parentheses.}$$

In Exercise 99, we guide you through a series of algebraic steps to eliminate the radicals. Then collecting the variable terms on one side and the constant terms on the other side, the equation becomes

$$(a^2 - c^2)x^2 + a^2y^2 = a^2(a^2 - c^2) \quad (1)$$

From Figure 7-5, we see that the distance between F_1 and F_2 is $2c$. Furthermore, since $d_1 + d_2 = 2a$, we have that $2c < 2a$, which implies that $c < a$ and that $a^2 - c^2 > 0$. Letting $b^2 = a^2 - c^2$, Equation (1) becomes

$$b^2x^2 + a^2y^2 = a^2b^2$$

$$\frac{b^2x^2}{a^2b^2} + \frac{a^2y^2}{a^2b^2} = \frac{a^2b^2}{a^2b^2} \quad \text{Divide both sides by } a^2b^2.$$

$$\frac{x^2}{a^2} + \frac{y^2}{b^2} = 1 \quad \text{Standard form of an equation of an ellipse}$$

This equation represents an ellipse centered at the origin with foci on the x-axis. The length of the major axis is $2a$ and the length of the minor axis is $2b$. The value a is also referred to as the length of the **semimajor axis** (the distance from the center of an ellipse to a vertex). The value b is referred to as the length of the **semiminor axis** (half the length of the minor axis).

In a similar manner, we can develop the standard form of an equation of an ellipse centered at the origin with foci on the y-axis. These two equations are summarized as follows.

Standard Forms of an Equation of an Ellipse Centered at the Origin

The standard forms of an equation of an ellipse centered at the origin are as follows. Assume that

- $a > b > 0$.
- The length of the major axis is $2a$, and the length of the minor axis is $2b$.

	Major Axis: x-axis	Major Axis: y-axis
Equation:	$\frac{x^2}{a^2} + \frac{y^2}{b^2} = 1$	$\frac{x^2}{b^2} + \frac{y^2}{a^2} = 1$
Center:	$(0, 0)$	$(0, 0)$
Foci: (*Note*: $c^2 = a^2 - b^2$)	$(c, 0)$ and $(-c, 0)$	$(0, c)$ and $(0, -c)$
Vertices: Endpoints: major axis	$(a, 0)$ and $(-a, 0)$	$(0, a)$ and $(0, -a)$
Endpoints: minor axis	$(0, b)$ and $(0, -b)$	$(b, 0)$ and $(-b, 0)$
Graph:		

Avoiding Mistakes

It is important to note that the foci are not actually on the graph of an elliptical curve. However, the foci are used to define the curve.

For an ellipse centered at the origin, the x- and y-intercepts are the endpoints of the major and minor axes. For example, for an ellipse centered at $(0, 0)$ with foci on the x-axis, we have:

Standard form:

x-intercept: substitute 0 for y.

y-intercept: substitute 0 for x.

$$\frac{x^2}{a^2} + \frac{y^2}{b^2} = 1 \longrightarrow \frac{x^2}{a^2} + \frac{(0)^2}{b^2} = 1 \qquad \frac{(0)^2}{a^2} + \frac{y^2}{b^2} = 1$$

$$\frac{x^2}{a^2} = 1 \qquad\qquad \frac{y^2}{b^2} = 1$$

$$x^2 = a^2 \qquad\qquad y^2 = b^2$$

$$x = \pm a \qquad\qquad y = \pm b$$

In Example 1, we graph an ellipse and identify the key features.

EXAMPLE 1   Graphing an Ellipse Centered at the Origin

Given $\dfrac{x^2}{16} + \dfrac{y^2}{9} = 1$, graph the ellipse, and identify the center, vertices, and foci.

Solution:

$$\frac{x^2}{16} + \frac{y^2}{9} = 1$$

The equation is in the standard form of an ellipse. Since $16 > 9$, $a^2 = 16$ and $b^2 = 9$. Since the greater number in the denominator is found in the x^2 term, the ellipse is elongated horizontally and the foci are on the x-axis.

$$\frac{x^2}{(4)^2} + \frac{y^2}{(3)^2} = 1$$

$a^2 = 16$, and because $a > 0$, we have that $a = 4$.
$b^2 = 9$, and because $b > 0$, we have that $b = 3$.

$\boxed{a = 4}$ $\boxed{b = 3}$

> **Point of Interest**
>
>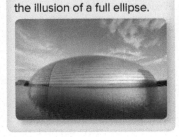
>
> The National Center for the Performing Arts in Beijing, People's Republic of China, is a semielliptical dome made of titanium and glass and surrounded by an artificial lake. On a calm day, the reflection from the lake gives the illusion of a full ellipse.

To graph the ellipse, plot the vertices (endpoints of the major axis) and the endpoints of the minor axis, and sketch the ellipse through the points.

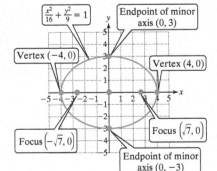

Endpoint of minor axis $(0, 3)$

Vertex $(-4, 0)$

Vertex $(4, 0)$

Focus $\left(-\sqrt{7}, 0\right)$

Focus $\left(\sqrt{7}, 0\right)$

Endpoint of minor axis $(0, -3)$

Center: $(0, 0)$

Vertices: $(a, 0)$ and $(-a, 0)$
$(4, 0)$ and $(-4, 0)$

Minor axis endpoints: $(0, b)$ and $(0, -b)$
$(0, 3)$ and $(0, -3)$

To find the foci, use the relationship
$c^2 = a^2 - b^2$
$c^2 = (4)^2 - (3)^2 = 16 - 9 = 7$
$c^2 = 7$ (Recall that $c > 0$.)
$c = \sqrt{7} \approx 2.65$

Foci: $(c, 0)$ and $(-c, 0)$
$\left(\sqrt{7}, 0\right)$ and $\left(-\sqrt{7}, 0\right)$

It is important to note that the center and foci of the ellipse are not actually part of the curve. For this reason, we graphed the curve in blue and plotted the foci and center in red.

Skill Practice 1   Graph the ellipse, and identify the center, vertices, minor axis endpoints, and foci. $\dfrac{x^2}{9} + \dfrac{y^2}{4} = 1$

The graph of an ellipse does not define y as a function of x (the graph fails the vertical line test). However, an ellipse can be defined by the union of two functions:

Answer

1.

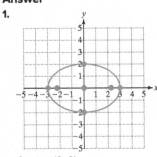

Center: $(0, 0)$
Vertices: $(3, 0)$ and $(-3, 0)$
Minor axis endpoints: $(0, 2)$ and $(0, -2)$
Foci: $\left(\sqrt{5}, 0\right)$ and $\left(-\sqrt{5}, 0\right)$

an upper semiellipse and a lower semiellipse. This is helpful to graph an ellipse on a graphing utility.

From Example 1, we have

$$\frac{x^2}{16} + \frac{y^2}{9} = 1 \longrightarrow y = \pm\sqrt{9 - \frac{9x^2}{16}}$$

The functions represent the top and bottom semiellipses, respectively.

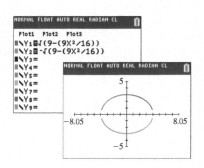

$$y = \sqrt{9 - \frac{9x^2}{16}} \quad \text{top semiellipse}$$

$$y = -\sqrt{9 - \frac{9x^2}{16}} \quad \text{bottom semiellipse}$$

EXAMPLE 2   Graphing an Ellipse Centered at the Origin

Given $25x^2 + 9y^2 = 225$, graph the ellipse and identify the center, vertices, minor axis endpoints, and foci.

Solution:

> **TIP** For an equation of an ellipse written in standard form, the term containing the larger denominator indicates the direction of elongation of the graph. In Example 2, the larger denominator of 25 is located in the y term. Notice that the ellipse is elongated vertically.

$25x^2 + 9y^2 = 225$	For the equation to be in the standard form for an ellipse, we require the constant term to be 1.
$\dfrac{25x^2}{225} + \dfrac{9y^2}{225} = \dfrac{225}{225}$	Divide both sides by 225.
$\dfrac{x^2}{9} + \dfrac{y^2}{25} = 1$	Since $25 > 9$, $a^2 = 25$ and $b^2 = 9$. Since the greater number in the denominator is found in the y^2 term, the ellipse is elongated vertically and the foci are on the y-axis.
$\dfrac{x^2}{(3)^2} + \dfrac{y^2}{(5)^2} = 1$	$a^2 = 25$, and because $a > 0$, we have that $a = 5$. $b^2 = 9$, and because $b > 0$, we have that $b = 3$.

$b = 3$ $a = 5$

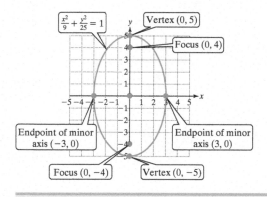

Center: $(0, 0)$

Vertices: $(0, a)$ and $(0, -a)$
$(0, 5)$ and $(0, -5)$

Minor axis endpoints: $(b, 0)$ and $(-b, 0)$
$(3, 0)$ and $(-3, 0)$

To find the foci, use the relationship
$$c^2 = a^2 - b^2$$
$$c^2 = (5)^2 - (3)^2 = 25 - 9 = 16$$
$$c^2 = 16$$
$$c = 4 \qquad \text{(Recall that } c > 0.)$$

Foci: $(0, c)$ and $(0, -c)$
$(0, 4)$ and $(0, -4)$

Skill Practice 2   Graph the ellipse and identify the center, vertices, minor axis endpoints, and foci. $25x^2 + 9y^2 = 900$

Answer

2.

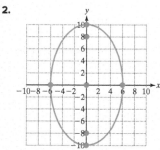

Center: $(0, 0)$
Vertices: $(0, 10), (0, -10)$
Minor axis endpoints: $(6, 0), (-6, 0)$
Foci: $(0, 8), (0, -8)$

2. Graph an Ellipse Centered at (h, k)

In Section 2.2, we learned that the standard form of an equation of a circle centered at the origin with radius r is given by $x^2 + y^2 = r^2$. For a circle centered at (h, k), we have the equation

$$(x - h)^2 + (y - k)^2 = r^2 \qquad \text{Circle centered at } (h, k) \text{ with radius } r$$

Replacing x by $x - h$ and y by $y - k$ shifts the graph of $x^2 + y^2 = r^2$ a total of h units horizontally and k units vertically. The same is true for an equation of an ellipse.

Standard Forms of an Equation of an Ellipse Centered at (h, k)

The standard forms of an equation of an ellipse centered at (h, k) are as follows. Assume that

- $a > b > 0$.
- The length of the major axis is $2a$, and the length of the minor axis is $2b$.

	Major Axis Horizontal	Major Axis Vertical
Equation:	$\dfrac{(x - h)^2}{a^2} + \dfrac{(y - k)^2}{b^2} = 1$	$\dfrac{(x - h)^2}{b^2} + \dfrac{(y - k)^2}{a^2} = 1$
Center:	(h, k)	(h, k)
Foci: (*Note*: $c^2 = a^2 - b^2$)	$(h + c, k)$ and $(h - c, k)$	$(h, k + c)$ and $(h, k - c)$
Vertices: Endpoints: major axis	$(h + a, k)$ and $(h - a, k)$	$(h, k + a)$ and $(h, k - a)$
Endpoints: minor axis	$(h, k + b)$ and $(h, k - b)$	$(h + b, k)$ and $(h - b, k)$
Graph:		

To graph an ellipse centered at (h, k), it is helpful to locate the center first as well as to identify the orientation of the ellipse (elongated horizontally or elongated vertically). Then use the values of a and b to locate the endpoints of the major and minor axes. This is demonstrated in Example 3.

EXAMPLE 3   Graphing an Ellipse Centered at (h, k)

Graph the ellipse and identify the center, vertices, endpoints of the minor axis, and foci. $\dfrac{(x - 3)^2}{16} + \dfrac{(y + 2)^2}{25} = 1$

Solution:

$\dfrac{(x - 3)^2}{16} + \dfrac{(y + 2)^2}{25} = 1$ The standard form of an equation of an ellipse is given. The quantity $(y + 2)^2$ is equivalent to $[y - (-2)]^2$. Therefore, the center of the ellipse is $(3, -2)$.

$\dfrac{(x - 3)^2}{\underbrace{(4)^2}_{b = 4}} + \dfrac{[y - (-2)]^2}{\underbrace{(5)^2}_{a = 5}} = 1$ Since $25 > 16$, $a^2 = 25$ and $b^2 = 16$. $a = 5$ and $b = 4$.

Since the greater number in the denominator is found in the y^2 term, the ellipse is elongated vertically.

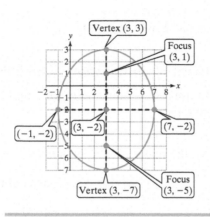

Center: $(3, -2)$

The vertices are a units above and below the center.

Vertices: $(3, -2 + 5)$ and $(3, -2 - 5)$
$(3, 3)$ and $(3, -7)$

The minor axis endpoints are b units to the right and left of the center.

Minor axis endpoints: $(3 + 4, -2)$ and $(3 - 4, -2)$
$(7, -2)$ and $(-1, -2)$

To find the foci, use the relationship
$c^2 = a^2 - b^2$
$c^2 = (5)^2 - (4)^2 = 25 - 16 = 9$
$c = 3$ (Recall that $c > 0$.)

The foci are c units above and below the center. The foci are $(3, 1)$ and $(3, -5)$.

Skill Practice 3 Graph the ellipse and identify the center, vertices, foci, and endpoints of the minor axis. $\dfrac{(x + 1)^2}{25} + \dfrac{(y - 2)^2}{9} = 1$

The equation of the ellipse from Example 3 can be expanded by expanding the binomials and combining like terms.

$$\frac{(x - 3)^2}{16} + \frac{(y + 2)^2}{25} = 1 \qquad \text{Standard form}$$

$$400\left[\frac{(x - 3)^2}{16} + \frac{(y + 2)^2}{25}\right] = 400(1) \qquad \text{Clear fractions.}$$

$$25(x - 3)^2 + 16(y + 2)^2 = 400$$

$$25(x^2 - 6x + 9) + 16(y^2 + 4y + 4) = 400 \qquad \text{Expand the binomials.}$$

$$25x^2 + 16y^2 - 150x + 64y - 111 = 0 \qquad \text{Expanded form}$$

Given an equation of an ellipse in expanded form, we can reverse this process by completing the square to write the equation in standard form. The benefit of writing an equation in standard form is that we can easily identify the center of the ellipse and the values of a and b.

EXAMPLE 4 **Writing an Equation of an Ellipse in Standard Form**

Given $3x^2 + 7y^2 + 6x - 28y + 10 = 0$,

a. Write the equation of the ellipse in standard form.

b. Identify the center, vertices, foci, and endpoints of the minor axis.

Solution:

a. $3x^2 + 7y^2 + 6x - 28y + 10 = 0$

Group the x terms and group the y terms.

$3x^2 + 6x + 7y^2 - 28y = -10$

Move the constant term to the right side of the equation.

$3(x^2 + 2x + \underline{}) + 7(y^2 - 4y + \underline{}) = -10$

Factor out the leading coefficient from the x and y terms. Leave room to complete the square within parentheses.

Answer

3.

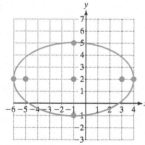

Center: $(-1, 2)$
Vertices: $(4, 2)$, $(-6, 2)$
Minor axis endpoints: $(-1, 5)$, $(-1, -1)$
Foci: $(3, 2)$, $(-5, 2)$

3 · 1 is added to the left side.   7 · 4 is added to the left side.

$$3(x^2 + 2x + 1) + 7(y^2 - 4y + 4) = -10 + \overset{3 \cdot 1}{3} + \overset{7 \cdot 4}{28}$$

To balance the equation add 3 · 1 and 7 · 4 to the right side.

Complete the square.
Note that $\left[\frac{1}{2}(2)\right]^2 = 1$ and $\left[\frac{1}{2}(-4)\right]^2 = 4$.

The values 1 and 4 are added inside parentheses. However, as a result of the factors of 3 and 7 in front of the parentheses, 3 · 1 and 7 · 4 are actually added to the left side of the equation.

$$3(x + 1)^2 + 7(y - 2)^2 = 21$$

To write the equation in standard form, we require that the constant on the right side of the equation be 1.

$$\frac{3(x + 1)^2}{21} + \frac{7(y - 2)^2}{21} = \frac{21}{21}$$

Divide both sides by 21.

$$\frac{(x + 1)^2}{7} + \frac{(y - 2)^2}{3} = 1$$

Standard form

b. $\dfrac{[x - (-1)]^2}{7} + \dfrac{(y - 2)^2}{3} = 1$

The center is $(-1, 2)$.
Since $7 > 3$, $a^2 = 7$ and $b^2 = 3$. The greater number in the denominator is in the x^2 term. Therefore, the ellipse is elongated horizontally.

$$\dfrac{[x - (-1)]^2}{(\sqrt{7})^2} + \dfrac{(y - 2)^2}{(\sqrt{3})^2} = 1$$

$\boxed{a = \sqrt{7}}$ $\boxed{b = \sqrt{3}}$

$a^2 = 7$. Because $a > 0$, $a = \sqrt{7}$.
$b^2 = 3$. Because $b > 0$, $b = \sqrt{3}$.
$c^2 = a^2 - b^2 = 7 - 3 = 4$
$c^2 = 4$, and because $c > 0$, we have $c = 2$.

The center is $(-1, 2)$.
The vertices are $(-1 + \sqrt{7}, 2)$ and $(-1 - \sqrt{7}, 2)$.
The foci are $(1, 2)$ and $(-3, 2)$.
The endpoints of the minor axis are $(-1, 2 + \sqrt{3})$ and $(-1, 2 - \sqrt{3})$.

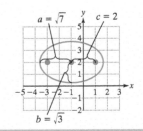

Skill Practice 4   Repeat Example 4 with $2x^2 + 11y^2 - 12x + 44y + 40 = 0$.

Now that we know the standard form of an equation of an ellipse, we can build an equation based on known information about the ellipse. This is demonstrated in Example 5.

EXAMPLE 5   **Finding an Equation of an Ellipse from Given Information**

Determine the standard form of an equation of an ellipse with vertices $(4, 2)$ and $(-4, 2)$ and foci $(\sqrt{15}, 2)$ and $(-\sqrt{15}, 2)$.

Solution:

Plotting the given points tells us that the ellipse is elongated horizontally. Therefore, the standard form is

$$\frac{(x - h)^2}{a^2} + \frac{(y - k)^2}{b^2} = 1$$

The center is midway between the vertices. Therefore, the center is $(0, 2)$, indicating that $h = 0$ and $k = 2$.

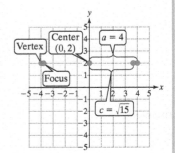

Answers

4. a. $\dfrac{(x - 3)^2}{11} + \dfrac{(y + 2)^2}{2} = 1$

   b. Center: $(3, -2)$; Vertices:
   $(3 + \sqrt{11}, -2)$ and
   $(3 - \sqrt{11}, -2)$
   Foci: $(0, -2)$ and $(6, -2)$; Minor
   axis endpoints: $(3, -2 + \sqrt{2})$
   and $(3, -2 - \sqrt{2})$

The vertices are located 4 units away from the center. This means that $a = 4$. The foci are located $\sqrt{15}$ units from the center, indicating that $c = \sqrt{15}$. The only missing piece of information is the value of b. We can find b from the relationship $c^2 = a^2 - b^2$ or equivalently $b^2 = a^2 - c^2$.

$$b^2 = (4)^2 - \left(\sqrt{15}\right)^2$$
$$b^2 = 1$$
$$b = 1 \qquad \text{(Recall that } b > 0.)$$

$$\frac{(x - 0)^2}{(4)^2} + \frac{(y - 2)^2}{(1)^2} = 1 \qquad \begin{array}{l}\text{Substitute } h = 0, k = 2, a = 4, \text{ and } b = 1 \text{ into the}\\ \text{standard form.}\end{array}$$

$$\frac{x^2}{16} + \frac{(y - 2)^2}{1} = 1 \quad \text{or} \quad \frac{x^2}{16} + (y - 2)^2 = 1 \qquad \text{Standard form}$$

> **Skill Practice 5** Determine the standard form of an equation of an ellipse with vertices $(3, 3)$ and $(-3, 3)$ and foci $\left(2\sqrt{2}, 3\right)$ and $\left(-2\sqrt{2}, 3\right)$.

3. Use Ellipses in Applications

The ellipse and its three-dimensional counterpart, the ellipsoid (an object in the shape of a blimp), have many useful applications that arise from the reflective property of the ellipse. For example, consider an elliptical-shaped dome. Sound or light waves emitted from one focus are reflected off the surface to the other focus. This principle is applied in Example 6.

> ### EXAMPLE 6　Applying the Reflective Property of an Ellipse
>
> A room in a museum has a semielliptical dome with a major axis of 50 ft and a semiminor axis of 12 ft. A light is placed at one focus of the ellipse and the light that reflects off the ceiling will be directed to the other focus. The museum curator wants to place a sculpture so that the sculpture is at the other focus where it will be illuminated by the light.
>
> **a.** Choose a coordinate system so that the ellipse is oriented with horizontal major axis with the center placed at $(0, 0)$. Write an equation of the semiellipse.
>
> **b.** Approximately where should the sculpture be placed?

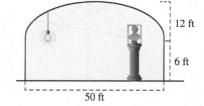

50 ft · 12 ft · 6 ft

Solution:

a. A graph of the semiellipse is shown in Figure 7-6. The length of the semimajor axis is 25 ft ($a = 25$) and the length of the semiminor axis is 12 ft ($b = 12$). An equation of the full ellipse is

$$\frac{x^2}{25^2} + \frac{y^2}{12^2} = 1 \quad \text{or} \quad \frac{x^2}{625} + \frac{y^2}{144} = 1.$$

Figure 7-6

We want the top half only, which is given by $\dfrac{x^2}{625} + \dfrac{y^2}{144} = 1$ for $y \geq 0$,

or equivalently $y = 12\sqrt{1 - \dfrac{x^2}{625}}$.

Answer

5. $\dfrac{x^2}{9} + (y - 3)^2 = 1$

b. To find the location of the sculpture, we must find the location of the other focus.

$$c^2 = a^2 - b^2$$
$$c^2 = (25)^2 - (12)^2 = 625 - 144 = 481$$
$$c = \pm\sqrt{481} \approx \pm 21.9$$

The sculpture should be located approximately 21.9 ft from the center of the room along the major axis, opposite the light source, and 6 ft up.

Skill Practice 6 A tunnel has vertical sides of 7 ft with a semielliptical top. The width of the tunnel is 10 ft, and the height at the top is 10 ft.

a. Write an equation of the semiellipse. For convenience, place the coordinate system with (0, 0) at the center of the ellipse.

b. To construct the tunnel, an engineer needs to find the location of the foci. How far from the center are the foci?

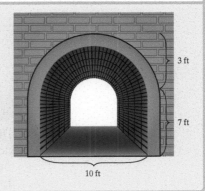

---

4. Determine and Apply the Eccentricity of an Ellipse

The degree of elongation of an ellipse is measured by the eccentricity e of the ellipse.

Eccentricity of an Ellipse

For an ellipse defined by $\dfrac{(x-h)^2}{a^2} + \dfrac{(y-k)^2}{b^2} = 1$ or $\dfrac{(x-h)^2}{b^2} + \dfrac{(y-k)^2}{a^2} = 1$,

the **eccentricity** e is given by $e = \dfrac{c}{a}$, where $a > b > 0$, $c > 0$, and $c^2 = a^2 - b^2$.

Note: The eccentricity of an ellipse is a number between 0 and 1; that is, $0 < e < 1$.

With this definition, we see that the eccentricity of an ellipse is the ratio of the distance between a focus and the center to the distance between a vertex and the center; that is, $e = \frac{c}{a}$.

- If the foci are near the vertices (that is, a is only slightly greater than c), then the ellipse is more elongated (Figure 7-7).
- If the foci are near the center, then the ellipse is more circular (Figure 7-8).

Answers

6. a. $\dfrac{x^2}{25} + \dfrac{y^2}{9} = 1; y \geq 0$ or

equivalently $y = 3\sqrt{1 - \dfrac{x^2}{25}}$

b. The foci are 4 ft to the right and left of the center of the semiellipse. These are at points 7 ft above the ground and 4 ft left and right of the center of the tunnel.

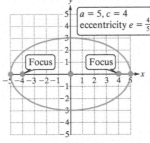

Figure 7-7

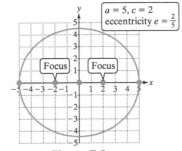

Figure 7-8

The orbits of the planets in our solar system are elliptical with the Sun at one focus. The eccentricities given in Table 7-1 indicate that the orbit of Venus is the most circular and the orbit of Pluto is the most elongated. (*Source*: NASA, www.nasa.gov)

Table 7-1

Planet	Eccentricity
Mercury	0.2056
Venus	0.0067 ←
Earth	0.0167
Mars	0.0935
Jupiter	0.0489
Saturn	0.0565
Uranus	0.0457
Neptune	0.0113
Pluto (dwarf planet)	0.2488 ←

The orbit of Venus is the most circular.

The orbit of Pluto is the most elongated.

EXAMPLE 7 **Using Eccentricity in an Application**

The point in a planet's orbit where it is closest to the Sun is called **perihelion** and the point where it is farthest away is called **aphelion** (Figure 7-9). Suppose that Mars is 2.0662×10^8 km from the Sun at perihelion. Use the eccentricity from Table 7-1 to find the distance between Mars and the Sun at aphelion.

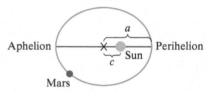

Figure 7-9

Solution:

For convenience, we have labeled the center of the ellipse as point C and the focus at which the Sun resides as point S. Aphelion and perihelion are labeled A and P, respectively (Figure 7-10).

We need to find the distance between Mars and the Sun at aphelion. This is given by $a + c$. Since a and c are two unknown values, we need two independent relationships between a and c.

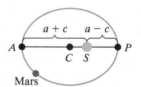

Figure 7-10

The distance at perihelion and the eccentricity give such information.

(1) $a - c = 2.0662 \times 10^8$ (distance at perihelion)   This a nonlinear

(2) $\dfrac{c}{a} = 0.0935 \longrightarrow c = 0.0935a$ (eccentricity)   system of equations.

$$a - 0.0935a = 2.0662 \times 10^8 \quad\quad \text{Use the substitution method.}$$
$$0.9065a = 2.0662 \times 10^8 \quad\quad \text{Combine like terms.}$$
$$a \approx 2.2793 \times 10^8 \quad\quad \text{Divide by 0.9065.}$$
$$c = 0.0935(2.2793 \times 10^8) \quad\quad \text{Back substitute.}$$
$$c = 2.1311 \times 10^7$$

Thus, $a + c = 2.2793 \times 10^8 + 2.1311 \times 10^7$
$$= 2.4924 \times 10^8$$

The distance between Mars and the Sun at aphelion (the farthest point from the Sun) is approximately 2.4924×10^8 km.

Skill Practice 7 Pluto is 7.376×10^9 km at aphelion (farthest point from the Sun). Use the eccentricity of Pluto's orbit of 0.2488 to find the closest distance between Pluto and the Sun.

Answer

7. 4.437×10^9 km

SECTION 7.1   Practice Exercises

Prerequisite Review

For Exercises R.1–R.3, determine the center and radius of the circle.

R.1. $x^2 + (y + 5)^2 = 8$ 　　　　　**R.2.** $x^2 + y^2 = 25$ 　　　　　**R.3.** $(x - 4)^2 + (y + 3)^2 = 81$

For Exercises R.4–R.5, find the value of n so that the expression is a perfect square trinomial. Then factor the trinomial.

R.4. $a^2 - 8a + n$ 　　　　　**R.5.** $y^2 + 7y + n$

R.6. Complete the square and write the equation of the circle in standard form.　$x^2 + y^2 + 10x - 10y + 41 = 0$

Concept Connections

1. The circle, the ellipse, the hyperbola, and the parabola are categories of _____ sections.

2. An _____ is a set of points (x, y) in a plane such that the sum of the distances between (x, y) and two fixed points called _____ is a constant.

3. The line through the foci intersects an ellipse at two points called _____.

4. The line segment with endpoints at the vertices of an ellipse is called the _____ axis.

5. The line segment perpendicular to the major axis, with endpoints on the ellipse, and passing through the center of the ellipse, is called the _____ axis.

6. Given $\dfrac{x^2}{a^2} + \dfrac{y^2}{b^2} = 1$, where $a > b > 0$, the ordered pairs representing the vertices are _____ and _____. The ordered pairs representing the endpoints of the minor axis are _____ and _____.

7. Given $\dfrac{(x - h)^2}{b^2} + \dfrac{(y - k)^2}{a^2} = 1$, where $a > b > 0$, the ordered pairs representing the endpoints of the vertices are _____ and _____. The ordered pairs representing the endpoints of the minor axis are _____ and _____.

8. When referring to the standard form of an equation of an ellipse, the _____ e is defined as $e = \dfrac{\square}{\square}$.

Objective 1: Graph an Ellipse Centered at the Origin

For Exercises 9–10,

a. Use the distance formula to find the distances d_1, d_2, d_3, and d_4. 　　**b.** Find the sum $d_1 + d_2$.

c. Find the sum $d_3 + d_4$. 　　**d.** How do the sums from parts (b) and (c) compare?

e. How do the sums of the distances from parts (b) and (c) relate to the length of the major axis?

9.

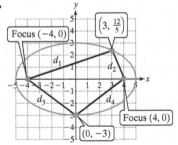

10.

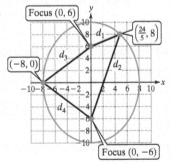

For Exercises 11–12, from the equation of the ellipse, determine if the major axis is horizontal or vertical.

11. a. $\dfrac{x^2}{2} + \dfrac{y^2}{5} = 1$ 　　**b.** $\dfrac{x^2}{5} + \dfrac{y^2}{2} = 1$ 　　　　**12. a.** $\dfrac{x^2}{11} + \dfrac{y^2}{10} = 1$ 　　**b.** $\dfrac{x^2}{10} + \dfrac{y^2}{11} = 1$

For Exercises 13–22,

a. Identify the center of the ellipse.
b. Determine the value of a.
c. Determine the value of b.
d. Identify the vertices.
e. Identify the endpoints of the minor axis.
f. Identify the foci.
g. Determine the length of the major axis.
h. Determine the length of the minor axis.
i. Graph the ellipse. **(See Examples 1–2)**

13. $\dfrac{x^2}{100} + \dfrac{y^2}{25} = 1$ **14.** $\dfrac{x^2}{64} + \dfrac{y^2}{49} = 1$ **15.** $\dfrac{x^2}{25} + \dfrac{y^2}{100} = 1$ **16.** $\dfrac{x^2}{49} + \dfrac{y^2}{64} = 1$

17. $4x^2 + 25y^2 = 100$ **18.** $9x^2 + 64y^2 = 576$ **19.** $-36x^2 - 4y^2 = -36$ **20.** $-64x^2 - 16y^2 = -64$

21. $\dfrac{x^2}{12} + \dfrac{y^2}{5} = 1$ **22.** $\dfrac{x^2}{18} + \dfrac{y^2}{7} = 1$

Objective 2: Graph an Ellipse Centered at (h, k)

For Exercises 23–32,

a. Identify the center of the ellipse.
b. Identify the vertices.
c. Identify the endpoints of the minor axis.
d. Identify the foci.
e. Graph the ellipse. **(See Example 3)**

23. $\dfrac{(x-1)^2}{25} + \dfrac{(y+6)^2}{16} = 1$ **24.** $\dfrac{(x-3)^2}{25} + \dfrac{(y+4)^2}{9} = 1$ **25.** $\dfrac{(x+4)^2}{49} + \dfrac{(y-2)^2}{64} = 1$

26. $\dfrac{(x+1)^2}{36} + \dfrac{(y-5)^2}{81} = 1$ **27.** $(x-6)^2 + \dfrac{y^2}{9} = 1$ **28.** $(x-2)^2 + \dfrac{y^2}{4} = 1$

29. $x^2 + 9(y+1)^2 = 81$ **30.** $x^2 + 4(y+4)^2 = 100$ **31.** $\dfrac{4(x-3)^2}{25} + \dfrac{16(y-2)^2}{49} = 1$

32. $\dfrac{4(x-4)^2}{81} + \dfrac{16(y-3)^2}{225} = 1$

For Exercises 33–42,

a. Write the equation of the ellipse in standard form. **(See Example 4)**
b. Identify the center, vertices, endpoints of the minor axis, and foci.

33. $3x^2 + 5y^2 + 12x - 60y + 177 = 0$ **34.** $7x^2 + 11y^2 + 70x - 66y + 197 = 0$

35. $3x^2 + 2y^2 - 30x - 4y + 59 = 0$ **36.** $5x^2 + 2y^2 - 40x - 12y + 78 = 0$

37. $4x^2 + y^2 + 14y + 45 = 0$ **38.** $x^2 + 49y^2 - 6x - 40 = 0$

39. $36x^2 + 100y^2 - 180x + 800y + 925 = 0$ **40.** $4x^2 + 9y^2 - 12x + 18y - 18 = 0$

41. $4x^2 + 9y^2 - 8x + 3 = 0$ **42.** $25x^2 + 16y^2 + 64y + 63 = 0$

For Exercises 43–56, write the standard form of an equation of an ellipse subject to the given conditions. (See Example 5)

43. Vertices: (4, 0) and (−4, 0);
Foci: (3, 0) and (−3, 0)

44. Vertices: (6, 0) and (−6, 0);
Foci: (5, 0) and (−5, 0)

45. Endpoints of minor axis: $\left(\sqrt{17}, 0\right)$ and $\left(-\sqrt{17}, 0\right)$;
Foci: (0, 9) and (0, −9)

46. Endpoints of minor axis: $\left(\sqrt{21}, 0\right)$ and $\left(-\sqrt{21}, 0\right)$;
Foci: (0, 5) and (0, −5)

47. Major axis parallel to the x-axis; Center: (2, 3);
Length of major axis: 14 units;
Length of minor axis: 10 units

48. Major axis parallel to the y-axis; Center: (1, 5);
Length of major axis: 22 units;
Length of minor axis: 14 units

49. Foci: (0, 1) and (8, 1);
Length of minor axis: 6 units

50. Foci: (−3, 3) and (7, 3);
Length of minor axis: 8 units

51. Vertices: (0, 5) and (0, −5);
Passes through $\left(\frac{16}{5}, 3\right)$

52. Vertices: (0, 13) and (0, −13);
Passes through $\left(\frac{25}{13}, 12\right)$

53. Vertices: (3, 4) and (3, −4);
Foci: $(3, \sqrt{11})$ and $(3, -\sqrt{11})$

54. Vertices: (4, 5) and (−4, 5);
Foci: $(\sqrt{6}, 5)$ and $(-\sqrt{6}, 5)$

55. Vertices: (2, 1) and (−12, 1);
Foci: $(-5 - \sqrt{33}, 1)$ and $(-5 + \sqrt{33}, 1)$

56. Vertices: (−8, 4) and (2, 4);
Foci: $(-3 - \sqrt{21}, 4)$ and $(-3 + \sqrt{21}, 4)$

Objective 3: Use Ellipses in Applications

57. The reflective property
of an ellipse is used in
lithotripsy. Lithotripsy is
a technique for treating
kidney stones without
surgery. Instead, high-
energy shock waves are
emitted from one focus
of an elliptical shell and
reflected painlessly to a
patient's kidney stone
located at the other

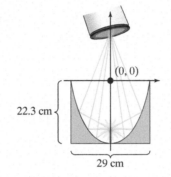

focus. The vibration from the shock waves shatter the
stone into pieces small enough to pass through the
patient's urine.

A vertical cross section of a lithotripter is in the shape
of a semiellipse with the dimensions shown.
Approximate the distance from the center along the
major axis where the patient's kidney stone should be
located so that the shock waves will target the stone.
Round to 2 decimal places. **(See Example 6)**

59. A homeowner wants to make an
elliptical rug from a 12-ft by 10-ft
rectangular piece of carpeting.

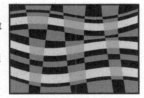

  a. What lengths of the major and
  minor axes would maximize
  the area of the new rug?

  b. Write an equation of the ellipse with maximum area.
  Use a coordinate system with the origin at the center
  of the rug and horizontal major axis.

  c. To cut the rectangular piece of carpeting,
  the homeowner needs to know the location of the
  foci. Then she will insert tacks at the foci, take a
  piece of string the length of the major axis, and
  fasten the ends to the tacks. Drawing the string tight,
  she'll use a piece of chalk to trace the ellipse. At
  what coordinates should the tacks be located?
  Describe the location.

61. Charles and Bernice ("Ray") Eames
were American designers who
made major contributions to
modern architecture and
furniture design. Suppose that a

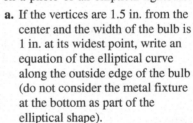

manufacturer wants to make an Eames
elliptical coffee table 90 in. long and 30 in. wide out of
an 8 ft by 4 ft piece of birch plywood. If the center of a
piece of plywood is positioned at (0, 0), determine the
distance from the center at which the foci should be
located to draw the ellipse.

58. The reflective property of an ellipse is the principle
behind "whispering galleries." These are rooms with
elliptically shaped ceilings such that a person standing
at one focus can hear even the slightest whisper spoken
by another person standing at the other focus.

Suppose that a dome has a semielliptical ceiling, 96 ft
long and 23 ft high. Choose a coordinate system so
that the center of the semiellipse is (0, 0) with vertices
(−48, 0) and (48, 0) and with the top of the ceiling
at (0, 23).

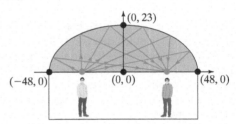

Approximately how far from the center along the major
axis should each person be standing to hear the
"whispering" effect? Round to 1 decimal place.

60. Coordinate axes are superimposed
on a photo of an elliptical lightbulb.

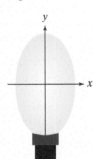

  a. If the vertices are 1.5 in. from the
  center and the width of the bulb is
  1 in. at its widest point, write an
  equation of the elliptical curve
  along the outside edge of the bulb
  (do not consider the metal fixture
  at the bottom as part of the
  elliptical shape).

  b. If the company logo is to be placed at
  one of the foci, at what coordinates could
  the logo be placed? Describe the location.

62. A window above a door is
to be made in the shape of
a semiellipse. If the
window is 10 ft at the base
and 3 ft high at the center,
determine the distance
from the center at which
the foci are located.

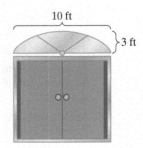

Objective 4: Determine and Apply the Eccentricity of an Ellipse

63. Choose one: The eccentricity of an ellipse is a number
  a. less than 0.
  b. between 0 and 1.
  c. greater than 1.

64. Choose one: An ellipse with eccentricity close to 0 will appear
  a. more elongated.
  b. more circular.

65. Choose one: An ellipse with eccentricity close to 1 will appear
  a. more elongated.
  b. more circular.

66. Choose one: If the foci of an ellipse are close to the center of the ellipse, then the graph will appear
  a. more elongated.
  b. more circular.

For Exercises 67–72, determine the eccentricity of the ellipse.

67. $\dfrac{x^2}{169} + \dfrac{y^2}{25} = 1$

68. $\dfrac{x^2}{100} + \dfrac{y^2}{64} = 1$

69. $\dfrac{\left(x + \frac{4}{5}\right)^2}{144} + \dfrac{\left(y - \frac{5}{3}\right)^2}{225} = 1$

70. $\dfrac{\left(x - \frac{1}{3}\right)^2}{9} + \dfrac{\left(y + \frac{7}{9}\right)^2}{25} = 1$

71. $\dfrac{x^2}{12} + \dfrac{(y - 3)^2}{6} = 1$

72. $\dfrac{(x + 7)^2}{18} + \dfrac{y^2}{12} = 1$

73. Halley's Comet and the Earth both orbit the Sun in elliptical paths with the Sun at one focus. The eccentricity of the comet's orbit is 0.967 and the eccentricity of the Earth's orbit is 0.0167. The eccentricity for the Earth is close to zero, whereas the eccentricity for Halley's Comet is close to 1. Based on this information, how do the orbits compare?

74. Halley's Comet and the comet Hale-Bopp both orbit the Sun in elliptical paths with the Sun at one focus. The eccentricity of Halley's Comet's orbit is 0.967 and the eccentricity of comet Hale-Bopp's orbit is 0.995. Which comet has a more elongated orbit?

75. The moon's orbit around the Earth is elliptical with the Earth at one focus and with eccentricity 0.0549. If the distance between the moon and Earth at the closest point is 363,300 km, determine the distance at the farthest point. Round to the nearest 100 km.
(See Example 7)

76. The planet Saturn orbits the Sun in an elliptical path with the Sun at one focus. The eccentricity of the orbit is 0.0565 and the distance between the Sun and Saturn at perihelion (the closest point) is 1.353×10^9 km. Determine the distance at aphelion (the farthest point). Round to the nearest million kilometers.

77. The Roman Coliseum is an elliptical stone and concrete amphitheater in the center of Rome, built between 70 A.D. and 80 A.D. The Coliseum seated approximately 50,000 spectators and was used among other things for gladiatorial contests.

  a. Using a vertical major axis, write an equation of the ellipse representing the center arena if the maximum length is 287 ft and the maximum width is 180 ft. Place the origin at the center of the arena.

  b. Approximate the eccentricity of the center arena. Round to 2 decimal places.

  c. Find an equation of the outer ellipse if the maximum length is 615 ft and the maximum width is 510 ft.

  d. Approximate the eccentricity of the outer ellipse. Round to 2 decimal places.

  e. Explain how you know that the outer ellipse is more circular than the inner ellipse.

78. *The Ellipse*, also called *President's Park South,* is a park in Washington, D.C. The lawn area is elliptical with a major axis of 1058 ft and minor axis of 903 ft.

  a. Find an equation of the elliptical boundary. Take the horizontal axes to be the major axis and locate the origin of the coordinate system at the center of the ellipse.

  b. Approximate the eccentricity of the ellipse. Round to 2 decimal places.

For Exercises 79–82, write the standard form of an equation of the ellipse subject to the following conditions.

79. Center: $(0, 0)$; Eccentricity: $\frac{15}{17}$;
Major axis vertical of length 34 units

80. Center: $(0, 0)$; Eccentricity: $\frac{40}{41}$;
Major axis vertical of length 82 units

81. Foci: $(0, -1)$ and $(8, -1)$;
Eccentricity: $\frac{4}{5}$

82. Foci: $(0, -1)$ and $(-6, -1)$;
Eccentricity: $\frac{3}{5}$

Mixed Exercises

83. A circular vent pipe is placed on
a flat roof.

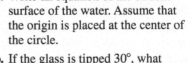

a. Write an equation of the
circular cross section that the
pipe makes with the roof.
Assume that the origin is placed
at the center of the circle.

b. Now suppose that the pipe is
placed on a roof with a slope of $\frac{3}{5}$.
What shape will the cross section
of the pipe form with the plane of
the roof?

c. Determine the length of the major and minor axes.
Find the exact value and approximate to 1 decimal
place if necessary.

84. A cylindrical glass of water with
diameter 3.5 in. sits on a horizontal
counter top.

a. Write an equation of the circular
surface of the water. Assume that
the origin is placed at the center of
the circle.

b. If the glass is tipped 30°, what
shape will the surface of the
water have?

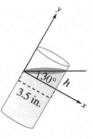

c. With the glass tipped 30°, the
waterline makes a slope of $\frac{1}{2}$ with
the coordinate system shown.
Determine the length of the major
and minor axes. Round to 1
decimal place.

85. The graph of $\dfrac{x^2}{9} + \dfrac{y^2}{4} = 1$ represents an ellipse.

Determine the part of the ellipse represented by the
given equation.

a. $y = 2\sqrt{1 - \dfrac{x^2}{9}}$

b. $y = -2\sqrt{1 - \dfrac{x^2}{9}}$

c. $x = 3\sqrt{1 - \dfrac{y^2}{4}}$

d. $x = -3\sqrt{1 - \dfrac{y^2}{4}}$

86. The graph of $\dfrac{x^2}{16} + \dfrac{y^2}{81} = 1$ represents an ellipse.

Determine the part of the ellipse represented by the
given equation.

a. $x = -4\sqrt{1 - \dfrac{y^2}{81}}$

b. $x = 4\sqrt{1 - \dfrac{y^2}{81}}$

c. $y = 9\sqrt{1 - \dfrac{x^2}{16}}$

d. $y = -9\sqrt{1 - \dfrac{x^2}{16}}$

For Exercises 87–90, solve the system of equations.

87. $\dfrac{x^2}{25} + \dfrac{y^2}{9} = 1$

$3x + 5y = 15$

88. $13y = 12x - 156$

$\dfrac{x^2}{169} + \dfrac{y^2}{144} = 1$

89. $\dfrac{x^2}{4} + \dfrac{y^2}{16} = 1$

$y = -x^2 + 4$

90. $\dfrac{x^2}{64} + \dfrac{y^2}{289} = 1$

$y = \dfrac{13}{64}x^2 - 13$

Given an ellipse with major axis of length 2a and minor axis of length 2b, the area is given by $A = \pi ab$. The perimeter is
approximated **by $P \approx \pi\sqrt{2(a^2 + b^2)}$. For Exercises 91–92,**

a. Determine the area of the ellipse.

b. Approximate the perimeter.

91. $\dfrac{x^2}{8} + \dfrac{(y + 3)^2}{4} = 1$

92. $\dfrac{(x - 1)^2}{9} + \dfrac{y^2}{11} = 1$

93. A line segment with endpoints on an
ellipse, perpendicular to the major axis,
and passing through a focus, is called
a *latus rectum* of the ellipse.
Show that the length of a
latus rectum is $\dfrac{2b^2}{a}$ for the
ellipse.

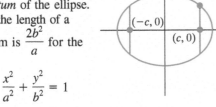

$$\dfrac{x^2}{a^2} + \dfrac{y^2}{b^2} = 1$$

[*Hint*: Substitute (c, y) into the equation and solve
for y. Recall that $c^2 = a^2 - b^2$.]

94. A line segment with endpoints on an ellipse and
passing through a focus of the ellipse is called a focal
chord. Given the ellipse

$$\dfrac{x^2}{25} + \dfrac{y^2}{16} = 1$$

a. Show that one focus of the ellipse lies on the line
$y = \frac{4}{3}x + 4$.

b. Determine the points of intersection between the
ellipse and the line.

c. Approximate the length of the focal chord that lies on
the line $y = \frac{4}{3}x + 4$. Round to 2 decimal places.

Write About It

95. An elliptical pool table is in the shape of an ellipse with one pocket located at one focus of the ellipse. If a ball is located at the other focus, explain why a player can strike the ball in any direction to have the ball land in the pocket.

96. Given an equation of an ellipse in standard form, how do you determine whether the major axis is horizontal or vertical?

97. Explain the difference between the graphs of the two equations.

$$4x^2 + 9y^2 = 36 \qquad 4x + 9y = 36$$

98. Discuss the solution set of the equation.

$$\frac{(x-4)^2}{9} + \frac{(x+2)^2}{16} = -1$$

Expanding Your Skills

99. This exercise guides you through the steps to find the standard form of an equation of an ellipse centered at the origin with foci on the x-axis.

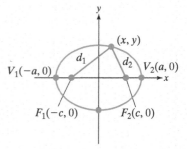

a. Refer to the figure to verify that the distance from F_1 to V_2 is $(a + c)$ and the distance from F_2 to V_2 is $(a - c)$. Verify that the sum of these distances is $2a$.

b. Write an expression that represents the sum of the distances from F_1 to (x, y) and from F_2 to (x, y). Then set this expression equal to $2a$.

c. Given the equation $\sqrt{(x + c)^2 + y^2} + \sqrt{(x - c)^2 + y^2} = 2a$, isolate the leftmost radical and square both sides of the equation. Show that the equation can be written as $a\sqrt{(x - c)^2 + y^2} = a^2 - xc$.

d. Square both sides of the equation $a\sqrt{(x - c)^2 + y^2} = a^2 - xc$ and show that the equation can be written as $(a^2 - c^2)x^2 + a^2y^2 = a^2(a^2 - c^2)$. (*Hint:* Collect variable terms on the left side of the equation and constant terms on the right side.)

e. Replace $a^2 - c^2$ by b^2. Then divide both sides of the equation by a^2b^2. Verify that the resulting equation is $\dfrac{x^2}{a^2} + \dfrac{y^2}{b^2} = 1$.

100. Find the points on the ellipse that are twice the distance from one focus to the other.

$$\frac{x^2}{25} + \frac{y^2}{9} = 1$$

101. An ellipsoid is a three-dimensional surface that resembles the shape of the blimps we see at sporting events. Mathematically, an equation of an ellipsoid centered at the origin of a three-dimensional coordinate system is given by

$$\frac{x^2}{a^2} + \frac{y^2}{b^2} + \frac{z^2}{c^2} = 1.$$

a. Explain how the formula for an ellipsoid is similar to a two-dimensional formula for an ellipse centered at the origin.

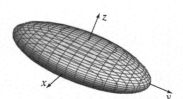

b. The graph of $\dfrac{x^2}{9} + \dfrac{y^2}{36} + \dfrac{z^2}{4} = 1$ can be generated using computer software (see figure). Write an equation that results if $z = 0$. What does this equation represent?

c. Write the equation that results if $x = 0$. What does this equation represent?

d. Write the equation that results if $y = 0$. What does this equation represent?

Technology Connections

For Exercises 102–105, graph the ellipse from the given exercise on a square viewing window.

102. Exercise 16   **103.** Exercise 15   **104.** Exercise 36   **105.** Exercise 35

The Hyperbola

TIP The definition of a hyperbola differs from the definition of an ellipse in that the curve is the set of points for which the difference (rather than the sum) of d_1 and d_2 is a constant.

1. Graph a Hyperbola Centered at the Origin

In this section, we will study another type of conic section called the hyperbola (Figure 7-11).

> **Definition of a Hyperbola**
>
> A **hyperbola** is the set of all points (x, y) in a plane such that the difference in distances between (x, y) and two fixed points (foci) is a positive constant.

In Figure 7-12, we present a hyperbola with foci on the x-axis. By definition, the curve is the set of points such that $d_1 - d_2$ is a positive constant.

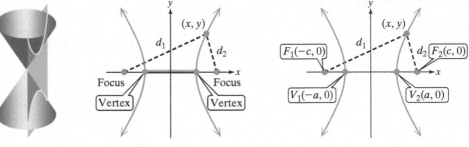

Figure 7-11   **Figure 7-12**   **Figure 7-13**

Notice that the hyperbola consists of two parts called **branches.** The points where the hyperbola intersects the line through the foci are called the **vertices.** The line segment between the vertices (highlighted in red) is called the **transverse axis.** The midpoint of the transverse axis is the **center** of the hyperbola.

To derive the standard form of an equation of a hyperbola, consider a hyperbola centered at the origin with foci on the x-axis (Figure 7-13). The points $F_1(-c, 0)$ and $F_2(c, 0)$ are the foci and the vertices are $V_1(-a, 0)$ and $V_2(a, 0)$.

The distance between F_1 and V_2 is $c + a$. The distance between F_2 and V_2 is $c - a$. By the definition of a hyperbola, the positive difference between a point on the hyperbola (such as V_2) and the two foci is a positive constant. Subtracting $(c + a)$ and $(c - a)$, we have

$$(c + a) - (c - a) = 2a$$

Therefore, the difference of the distances between (x, y) and the two foci is equal to $2a$.

$$d_1 \qquad - \qquad d_2 \qquad = 2a$$

$$\sqrt{[x - (-c)]^2 + (y - 0)^2} - \sqrt{(x - c)^2 + (y - 0)^2} = 2a \qquad \text{Apply the distance formula.}$$

$$\sqrt{(x + c)^2 + y^2} - \sqrt{(x - c)^2 + y^2} = 2a \qquad \text{Simplify within parentheses.}$$

In Exercise 82, we guide you through a series of algebraic steps to eliminate the radicals. Then collecting the variable terms on the left side and the constant terms on the right side, the equation becomes

$$(c^2 - a^2)x^2 - a^2 y^2 = a^2(c^2 - a^2) \qquad (1)$$

From Figure 7-13, we see that $d_1 - d_2 = 2a < 2c$. Therefore, $a < c$ and $c^2 - a^2 > 0$. Substituting $b^2 = c^2 - a^2$ in Equation (1) we have

$$b^2 x^2 - a^2 y^2 = a^2 b^2$$

$$\frac{b^2 x^2}{a^2 b^2} - \frac{a^2 y^2}{a^2 b^2} = \frac{a^2 b^2}{a^2 b^2} \qquad \text{Divide both sides by } a^2 b^2.$$

$$\frac{x^2}{a^2} - \frac{y^2}{b^2} = 1 \qquad \text{Standard form}$$

This equation represents a hyperbola centered at the origin with foci on the x-axis. In a similar manner, we can develop the standard equation of a hyperbola centered at the origin with foci on the y-axis. These two equations are summarized as follows.

Avoiding Mistakes

Notice that we do not have the restriction that $a > b$ as with the ellipse.

Standard Forms of an Equation of a Hyperbola Centered at the Origin

The standard forms of an equation of a hyperbola centered at the origin are as follows. Assume that $a > 0$ and $b > 0$.

	Transverse Axis: x-axis	**Transverse Axis: y-axis**
Equation:	$\dfrac{x^2}{a^2} - \dfrac{y^2}{b^2} = 1$	$\dfrac{y^2}{a^2} - \dfrac{x^2}{b^2} = 1$
Center:	$(0, 0)$	$(0, 0)$
Foci: ($c^2 = a^2 + b^2$)	$(c, 0)$ and $(-c, 0)$	$(0, c)$ and $(0, -c)$
Vertices:	$(a, 0)$ and $(-a, 0)$	$(0, a)$ and $(0, -a)$
Asymptotes:	$y = \dfrac{b}{a}x$ and $y = -\dfrac{b}{a}x$	$y = \dfrac{a}{b}x$ and $y = -\dfrac{a}{b}x$
Graph:		
	Figure 7-14	**Figure 7-15**

TIP Notice that an equation of a hyperbola shows a *difference* between the square terms, whereas an equation of an ellipse shows a *sum* of the square terms.

From the standard forms of an equation of a hyperbola, notice that the variable term with the positive coefficient always contains a^2 in the denominator. This term also indicates the orientation of the hyperbola.

- If the variable term with the positive coefficient is the x^2 term, then the transverse axis is horizontal and the branches of the hyperbola open horizontally.
- If the variable term with the positive coefficient is the y^2 term, then the transverse axis is vertical and the branches of the hyperbola open vertically.

As x or y increase, the branches of a hyperbola approach a pair of asymptotes. For example, given the hyperbola $\dfrac{x^2}{a^2} - \dfrac{y^2}{b^2} = 1$ (horizontal transverse axis),

$$y = \pm b\sqrt{\dfrac{x^2}{a^2} - 1}$$

$$= \pm b\sqrt{\dfrac{x^2 - a^2}{a^2}}$$

$$= \pm \dfrac{b}{a}\sqrt{x^2 - a^2}$$

As $x \to \infty$, the term a^2 is insignificant (very small relative to x^2), so the radical approaches $\sqrt{x^2} = x$. Thus, for large values of x, the graph of the hyperbola approaches $y = \pm\dfrac{b}{a}x$. Likewise, as $x \to -\infty$, the graph of the hyperbola approaches $y = \pm\dfrac{b}{a}x$.

As with the graph of a rational function, the asymptotes are lines of reference that help graph the curve. To identify the asymptotes, first sketch a **reference rectangle** with dimensions $2a$ by $2b$, centered at the center of the hyperbola (Figures 7-14 and 7-15). The asymptotes are the lines through the opposite corners of the rectangle.

The length of the transverse axis is $2a$. The line segment passing through the center of the hyperbola, perpendicular to the transverse axis, and with endpoints on the reference rectangle is called the **conjugate axis.** The length of the conjugate axis is $2b$.

To graph a hyperbola, we suggest the following guidelines.

Graphing a Hyperbola

Step 1 Identify the center and vertices.

Step 2 Draw the reference rectangle centered at the center of the hyperbola, with dimensions $2a$ and $2b$ as shown in Figures 7-14 and 7-15.

Step 3 Draw the asymptotes through the opposite corners of the rectangle.

Step 4 Sketch each branch of the hyperbola starting at the vertices and approaching the asymptotes.

EXAMPLE 1   Graphing a Hyperbola Centered at the Origin

Given $\dfrac{x^2}{4} - \dfrac{y^2}{9} = 1$,

a. Graph the hyperbola.

b. Identify the center, vertices, foci, and equations of the asymptotes.

Point of Interest

The James S. McDonnell Planetarium in St. Louis, Missouri, has vertical cross sections in the shape of a hyperbola.

Solution:

a. $\dfrac{x^2}{4} - \dfrac{y^2}{9} = 1$

Step 1: The equation is in the standard form of a hyperbola centered at $(0, 0)$.

The x^2 term has the positive coefficient, indicating that the transverse axis is horizontal. The denominator of the x^2 term is a^2.

$$\dfrac{x^2}{(2)^2} - \dfrac{y^2}{(3)^2} = 1$$

$\boxed{a = 2}$ $\boxed{b = 3}$

$a^2 = 4$, and because $a > 0$, we have that $a = 2$.
$b^2 = 9$, and because $b > 0$, we have that $b = 3$.

Vertices: $(a, 0)$ and $(-a, 0)$
$\qquad\qquad (2, 0)$ and $(-2, 0)$
Note that $c^2 = a^2 + b^2 = 4 + 9 = 13$.
Since $c > 0$, $c = \sqrt{13}$.
Foci: $(c, 0)$ and $(-c, 0)$
$\qquad (\sqrt{13}, 0)$ and $(-\sqrt{13}, 0)$

Steps 2–3: Draw the reference rectangle and asymptotes.

The reference rectangle is centered at the center of the hyperbola, $(0, 0)$.
Draw the rectangle with width $2a$ in the direction of the transverse axis. The height of the rectangle is the length of the conjugate axis $2b$. Draw the asymptotes through the opposite corners of the reference rectangle.

Step 4: Sketch each branch of the hyperbola starting at the vertices and approaching the asymptotes.

b. The center is $(0, 0)$.
The vertices are $(-2, 0)$ and $(2, 0)$.
The foci are $\left(-\sqrt{13}, 0\right)$ and $\left(\sqrt{13}, 0\right)$.

The equations of the asymptotes are $y = \frac{b}{a}x$ and $y = -\frac{b}{a}x$.
Therefore, the asymptotes are $y = \frac{3}{2}x$ and $y = -\frac{3}{2}x$.

TIP If you find it difficult to memorize the equations of the asymptotes, the equations can be found by using the point-slope formula and the points defining the opposite corners of the reference rectangle.

> **Skill Practice 1**   Repeat Example 1 with the equation: $\dfrac{x^2}{9} - \dfrac{y^2}{4} = 1$

In Example 2, we graph a hyperbola in which the transverse axis is vertical and the conjugate axis is horizontal.

EXAMPLE 2   Graphing a Hyperbola Centered at the Origin

Given $16y^2 - 9x^2 = 144$,

a. Graph the hyperbola.

b. Identify the center, vertices, foci, and equations of the asymptotes.

Solution:

a. $16y^2 - 9x^2 = 144$ For the equation to be in the standard form for a hyperbola, we require the constant term to be 1.

$\dfrac{16y^2}{144} - \dfrac{9x^2}{144} = \dfrac{144}{144}$ Divide both sides by 144.

$\dfrac{y^2}{9} - \dfrac{x^2}{16} = 1$ The y^2 term has the positive coefficient, indicating that the transverse axis is vertical. The denominator of the y^2 term is a^2.

$\dfrac{y^2}{(3)^2} - \dfrac{x^2}{(4)^2} = 1$

$\boxed{a = 3}$ $\boxed{b = 4}$

Step 1: The hyperbola is centered at $(0, 0)$.

$a^2 = 9$, and because $a > 0$, we have that $a = 3$.
$b^2 = 16$, and because $b > 0$, we have that $b = 4$.

Vertices: $(0, a)$ and $(0, -a)$
$(0, 3)$ and $(0, -3)$

Note that $c^2 = a^2 + b^2 = 9 + 16 = 25$ and $c = 5$.

Foci: $(0, c)$ and $(0, -c)$
$(0, 5)$ and $(0, -5)$

Answers

1. a.

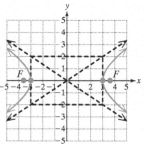

b. Center: $(0, 0)$
Vertices: $(-3, 0)$, $(3, 0)$
Foci: $\left(-\sqrt{13}, 0\right)$, $\left(\sqrt{13}, 0\right)$
Asymptotes: $y = \frac{2}{3}x$ and $y = -\frac{2}{3}x$.

Steps 2–3: Draw the reference rectangle and asymptotes.

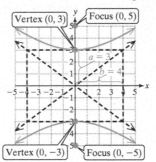

Draw the rectangle centered at the origin and with height $2a$ and width $2b$.

Draw the asymptotes through the opposite corners of the reference rectangle.

Step 4: Sketch each branch of the hyperbola starting at the vertices and approaching the asymptotes.

b. The center is $(0, 0)$.

The vertices are $(0, 3)$ and $(0, -3)$.

The foci are $(0, -5)$ and $(0, 5)$.

The equations of the asymptotes are $y = \frac{a}{b}x$ and $y = -\frac{a}{b}x$.

Therefore, the asymptotes are $y = \frac{3}{4}x$ and $y = -\frac{3}{4}x$.

Skill Practice 2   Repeat Example 2 with the equation: $4y^2 - x^2 = 16$

---

2. Graph a Hyperbola Centered at (h, k)

In the standard form of a hyperbola, if x and y are replaced by $x - h$ and $y - k$, then the graph of the hyperbola is shifted h units horizontally and k units vertically. The hyperbola is then centered at (h, k).

Standard Forms of an Equation of a Hyperbola Centered at (h, k)

The standard forms of an equation of a hyperbola centered at (h, k) are as follows. Assume that $a > 0$ and $b > 0$.

	Transverse Axis Horizontal	**Transverse Axis Vertical**
Equation:	$\dfrac{(x - h)^2}{a^2} - \dfrac{(y - k)^2}{b^2} = 1$	$\dfrac{(y - k)^2}{a^2} - \dfrac{(x - h)^2}{b^2} = 1$
Center:	(h, k)	(h, k)
Foci: (*Note*: $c^2 = a^2 + b^2$)	$(h + c, k)$ and $(h - c, k)$	$(h, k + c)$ and $(h, k - c)$
Vertices:	$(h + a, k)$ and $(h - a, k)$	$(h, k + a)$ and $(h, k - a)$
Asymptotes:	$y - k = \pm \frac{b}{a}(x - h)$	$y - k = \pm \frac{a}{b}(x - h)$
Graph:		

Answers

2. a.

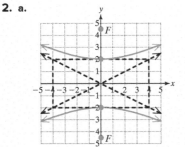

b. Center: $(0, 0)$
Vertices: $(0, 2)$, $(0, -2)$
Foci: $(0, 2\sqrt{5})$, $(0, -2\sqrt{5})$
Asymptotes: $y = \frac{1}{2}x$ and $y = -\frac{1}{2}x$.

To graph a hyperbola centered at (h, k), first locate the center and identify the orientation of the hyperbola (transverse axis horizontal or vertical). Then use the values of a and b to draw the reference rectangle. Draw the asymptotes through the corners of the reference rectangle. Finally, sketch the hyperbola starting at the vertices and approaching the asymptotes.

EXAMPLE 3 Graphing a Hyperbola Centered at (h, k)

Given $-\dfrac{(x - 2)^2}{9} + (y + 3)^2 = 1$,

a. Graph the hyperbola.

b. Identify the center, vertices, foci, and equations of the asymptotes.

Solution:

a.

$$-\frac{(x - 2)^2}{9} + \frac{(y + 3)^2}{1} = 1$$

$$\underbrace{\frac{[y - (-3)]^2}{(1)^2}}_{\boxed{a = 1}} - \underbrace{\frac{(x - 2)^2}{(3)^2}}_{\boxed{b = 3}} = 1$$

Step 1: Center: $(2, -3)$

The y^2 term has a positive coefficient. Therefore, the transverse axis is vertical. We have $a^2 = 1$ and $b^2 = 9$. Thus, $a = 1$ and $b = 3$.

Steps 2–3: Draw the reference rectangle and asymptotes.

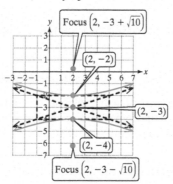

Draw the rectangle centered at $(2, -3)$ with vertical sides of length $2a$ and horizontal sides of length $2b$.

Draw the asymptotes through the opposite corners of the reference rectangle.

The vertices are a units above and below the center.

Vertices: $(2, -3 + 1)$ and $(2, -3 - 1)$
$(2, -2)$ and $(2, -4)$

Step 4: Sketch the hyperbola.

To find the foci, use the relationship:
$$c^2 = a^2 + b^2$$
$$c^2 = 1 + 9 = 10$$
$$c = \sqrt{10} \quad (\text{Recall that } c > 0.)$$

The foci are c units above and below the center.

The foci are $\left(2, -3 + \sqrt{10}\right)$ and $\left(2, -3 - \sqrt{10}\right)$.

b. The center is $(2, -3)$.

The vertices are $(2, -2)$ and $(2, -4)$.

The foci are $\left(2, -3 + \sqrt{10}\right)$ and $\left(2, -3 - \sqrt{10}\right)$.

The equations of the asymptotes can be found by using the point-slope formula and the points through the opposite vertices of the reference rectangle.

Alternatively, the asymptotes can be found by using $y - k = \pm\frac{a}{b}(x - h)$, where h and k are the coordinates of the center.

$$y - k = \frac{a}{b}(x - h) \qquad\qquad y - k = -\frac{a}{b}(x - h)$$

$$y - (-3) = \frac{1}{3}(x - 2) \qquad\qquad y - (-3) = -\frac{1}{3}(x - 2)$$

$$y = \frac{1}{3}x - \frac{11}{3} \qquad\qquad y = -\frac{1}{3}x - \frac{7}{3}$$

Answers

3. a.

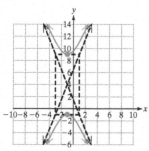

b. Center: $(-1, 4)$
Vertices: $(-1, 9)$, $(-1, -1)$
Foci: $\left(-1, 4 + \sqrt{29}\right)$,
$\left(-1, 4 - \sqrt{29}\right)$
Asymptotes: $y = -\frac{5}{2}x + \frac{3}{2}$,
$y = \frac{5}{2}x + \frac{13}{2}$

Skill Practice 3 Repeat Example 3 with the equation:

$$-\frac{(x + 1)^2}{4} + \frac{(y - 4)^2}{25} = 1$$

TECHNOLOGY CONNECTIONS

Graphing a Hyperbola

To graph the hyperbola from Example 3 on a graphing utility, first solve the equation for y. We have

$$-\frac{(x-2)^2}{9} + (y+3)^2 = 1$$

$$y = -3 \pm \sqrt{1 + \frac{(x-2)^2}{9}}$$

The functions represent the top and bottom branches of the hyperbola, respectively.

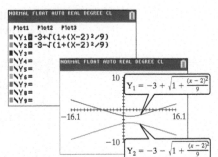

The equation of the hyperbola from Example 3 can be expanded by expanding the binomials and combining like terms.

$$-\frac{(x-2)^2}{9} + (y+3)^2 = 1 \qquad \text{Standard form}$$

$$9 \cdot \left[-\frac{(x-2)^2}{9} + (y+3)^2 \right] = 9 \cdot (1) \qquad \text{Clear fractions.}$$

$$-(x-2)^2 + 9(y+3)^2 = 9$$

$$-x^2 + 4x - 4 + 9y^2 + 54y + 81 = 9 \qquad \text{Expand the binomials.}$$

$$-x^2 + 9y^2 + 4x + 54y + 68 = 0 \qquad \text{Expanded form}$$

Given an equation of a hyperbola in expanded form, we can reverse this process by completing the square to write the equation in standard form. This is shown in Example 4.

EXAMPLE 4   Writing an Equation of a Hyperbola in Standard Form

Write the equation of the hyperbola in standard form. $11x^2 - 2y^2 + 66x - 4y + 75 = 0$

Solution:

$11x^2 - 2y^2 + 66x - 4y + 75 = 0$ Group the x variable terms and group the y variable terms.

$\qquad 11x^2 + 66x - 2y^2 - 4y = -75$ Move the constant term to the right side of the equation.

$11(x^2 + 6x \quad) - 2(y^2 + 2y \quad) = -75$ Factor out the leading coefficient from the x and y terms. Leave room to complete the square within parentheses.

$11 \cdot 9$ is added to the left side. $-2 \cdot 1$ is added to the left side.

$\quad 11 \cdot 9 \quad -2 \cdot 1$ Complete the square.
Note that $\left[\frac{1}{2}(6)\right]^2 = 9$ and $\left[\frac{1}{2}(2)\right]^2 = 1$. The values 9 and 1 were added inside parentheses. However, $11 \cdot 9$ and $-2 \cdot 1$ were actually added to the left side as a result of the factors in front of the parentheses.

$11(x^2 + 6x + 9) - 2(y^2 + 2y + 1) = -75 + 99 - 2$

$11(x + 3)^2 - 2(y + 1)^2 = 22$ To write the equation in standard form, we require a 1 on the right side of the equation.

$$\frac{11(x+3)^2}{22} - \frac{2(y+1)^2}{22} = \frac{22}{22} \qquad \text{Divide both sides by } 22.$$

$$\frac{(x+3)^2}{2} - \frac{(y+1)^2}{11} = 1 \qquad \text{Standard form}$$

Skill Practice 4   Write the equation of the hyperbola in standard form.
$$3x^2 - 7y^2 + 30x + 56y - 58 = 0$$

In Example 5, we determine the standard form of an equation of a hyperbola based on known information about the hyperbola.

EXAMPLE 5   **Finding an Equation of a Hyperbola**

Determine the standard form of an equation of a hyperbola with vertices $(-4, 0)$ and $(8, 0)$ and foci $(-8, 0)$ and $(12, 0)$.

Solution:

Plotting the given points tells us that the foci are aligned horizontally. Therefore, the transverse axis of the hyperbola is horizontal. The standard form is

$$\frac{(x - h)^2}{a^2} - \frac{(y - k)^2}{b^2} = 1$$

The center is midway between the vertices. Therefore, the center is $(2, 0)$, indicating that $h = 2$ and $k = 0$.

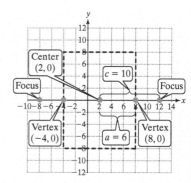

The vertices are located 6 units away from the center. This means that $a = 6$.

The foci are located 10 units from the center, indicating that $c = 10$.

The only missing piece of information is the value of b. We can find b from the relationship $c^2 = a^2 + b^2$ or equivalently $b^2 = c^2 - a^2$.

$$b^2 = (10)^2 - (6)^2$$
$$b^2 = 64$$
$$b = 8 \qquad \text{(Recall that } b > 0.)$$

$$\frac{(x - 2)^2}{(6)^2} - \frac{(y - 0)^2}{(8)^2} = 1 \qquad \text{Substitute } h = 2, k = 0, a = 6, \text{ and } b = 8.$$

$$\frac{(x - 2)^2}{36} - \frac{y^2}{64} = 1 \qquad \text{Standard form}$$

Skill Practice 5   Determine the standard form of an equation of a hyperbola with vertices $(-9, 0)$ and $(1, 0)$ and foci $(-17, 0)$ and $(9, 0)$.

3. Determine the Eccentricity of a Hyperbola

The branches of a hyperbola may be wider or narrower depending on the eccentricity of the hyperbola. Like the ellipse, the eccentricity of a hyperbola is given by $e = \frac{c}{a}$.

Answers

4. $\dfrac{(x + 5)^2}{7} - \dfrac{(y - 4)^2}{3} = 1$

5. $\dfrac{(x + 4)^2}{25} - \dfrac{y^2}{144} = 1$

> **Eccentricity of a Hyperbola**
>
> For a hyperbola defined by $\dfrac{(x-h)^2}{a^2} - \dfrac{(y-k)^2}{b^2} = 1$ or
>
> $\dfrac{(y-k)^2}{a^2} - \dfrac{(x-h)^2}{b^2} = 1$, the **eccentricity** e is given by $e = \dfrac{c}{a}$, where $a > 0$,
>
> $c > 0$, and $c^2 = a^2 + b^2$.
>
> *Note*: The eccentricity of a hyperbola is a number greater than 1; that is, $e > 1$.

In Example 6, we compare the graphs of two hyperbolas with different eccentricities.

EXAMPLE 6   Finding the Eccentricity of a Hyperbola

Determine the eccentricity of the hyperbola.

a. $\dfrac{x^2}{16} - \dfrac{y^2}{9} = 1$ b. $\dfrac{x^2}{9} - \dfrac{y^2}{16} = 1$

Solution:

a. $\dfrac{x^2}{16} - \dfrac{y^2}{9} = 1$

The x^2 term has a positive coefficient. Therefore, $a^2 = 16$.

$\dfrac{x^2}{(4)^2} - \dfrac{y^2}{(3)^2} = 1$

$a = 4$ $b = 3$

$a^2 = 16$, and because $a > 0$, we have that $a = 4$.
$b^2 = 9$, and because $b > 0$, we have that $b = 3$.
$c^2 = a^2 + b^2 = (4)^2 + (3)^2 = 25$
$c = 5$ (Recall that $c > 0$.)

Therefore, $e = \dfrac{c}{a} = \dfrac{5}{4}$.

b. $\dfrac{x^2}{9} - \dfrac{y^2}{16} = 1$

The x^2 term has a positive coefficient. Therefore, $a^2 = 9$.

$\dfrac{x^2}{(3)^2} - \dfrac{y^2}{(4)^2} = 1$

$a = 3$ $b = 4$

$a^2 = 9$, and because $a > 0$, we have that $a = 3$.
$b^2 = 16$, and because $b > 0$, we have that $b = 4$.
$c^2 = a^2 + b^2 = (3)^2 + (4)^2 = 25$
$c = 5$ (Recall that $c > 0$.)

Therefore, $e = \dfrac{c}{a} = \dfrac{5}{3}$.

The graphs of the hyperbolas are shown in Figures 7-16 and 7-17. Compare the graphs, and notice that the hyperbola in Figure 7-17 has a greater eccentricity and that the branches open wider.

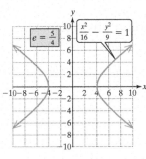

Figure 7-16   **Figure 7-17**

Skill Practice 6   Determine the eccentricity of the hyperbola.

a. $\dfrac{y^2}{25} - \dfrac{x^2}{144} = 1$ b. $\dfrac{y^2}{144} - \dfrac{x^2}{25} = 1$

Answers

6. a. $e = \dfrac{13}{5}$ **b.** $e = \dfrac{13}{12}$

4. Use Hyperbolas in Applications

Applications of hyperbolas arise in many fields. For example, astronomers know that some comets, such as Halley's comet, follow an elliptical path around the Sun. Other comets have a hyperbolic orbit because they are not captured by the Sun's gravity (Figure 7-18).

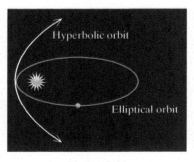

Figure 7-18

Hyperbolas also provide the mathematical basis for LORAN (LOng RAnge Navigation), a technique used for navigation before global positioning systems. A ship using LORAN measures the difference in time between synchronized radio signals from two different transmitters. Using the relationship $d = rt$, the difference in distances between the ship and the two transmitters $(d_1 - d_2)$ can be computed. Thus, the position of the ship is located on the hyperbolic path with the two transmitters at the foci (Figure 7-19).

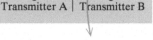

Figure 7-19

The same principle used in LORAN can also be used with sound waves to locate the source of sound heard from two different receivers. This is demonstrated in Example 7.

EXAMPLE 7   Using a Hyperbola in an Application

Suppose that two people located 3 mi (15,840 ft) apart at points A and B hear a clap of thunder (Figure 7-20). The time difference between the sound heard at point A and the sound heard at point B is 10 sec. If sound travels 1100 ft/sec, find an equation of the hyperbola (with foci at A and B) on which the point of origination of the sound must lie.

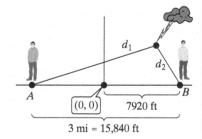

Figure 7-20

Solution:

First, we choose a coordinate system with the origin midway between the two observers. The two observers are at the points (7920, 0) and (−7920, 0). Let d_1 and d_2 represent the distances between each observer and the source of the sound (Figure 7-20).

The difference in distance $d_1 - d_2$ can be computed from the relationship $d = rt$.

$$d_1 - d_2 = (1100 \text{ ft/s})(10 \text{ s}) = 11{,}000 \text{ ft}$$

By the definition of a hyperbola, the source of the sound must be at a point on the hyperbola for which $d_1 - d_2 = 11{,}000$. The center of the hyperbola is (0, 0), and the transverse axis is horizontal with foci (7920, 0) and (−7920, 0). The standard form of the equation of the hyperbola is

$$\frac{x^2}{a^2} - \frac{y^2}{b^2} = 1$$

The value $d_1 - d_2 = 2a$
$$11{,}000 = 2a$$
$$a = 5500 \text{ ft}$$
From the location of the foci, $c = 7920$ ft.
$c^2 = a^2 + b^2$. Taking $b > 0$, we have $b = \sqrt{c^2 - a^2}$.
$$b = \sqrt{(7920)^2 - (5500)^2} \approx 5699 \text{ ft}$$

The source of the sound must be at a point on the curve $\dfrac{x^2}{(5500)^2} - \dfrac{y^2}{(5699)^2} = 1$.

Answer

7. $\dfrac{x^2}{(8800)^2} - \dfrac{y^2}{(5837)^2} = 1$

Skill Practice 7   Repeat Example 7 with two observers 4 mi apart, who hear a clap of thunder 16 sec apart. (*Hint*: 5280 ft = 1 mi.)

The hyperbola, like the ellipse, has an interesting property of reflection. Light directed toward one focus of a hyperbolic mirror is reflected toward the other focus of the hyperbola (Figure 7-21). This property is used in the design of Cassegrain reflecting telescopes. The telescope is a combination of a primary parabolic mirror and a secondary hyperbolic mirror (Figure 7-22). Light from distant stars is reflected off the parabolic mirror toward the focus of the hyperbolic mirror. The light is then reflected again toward an eyepiece located at the other focus of the hyperbola.

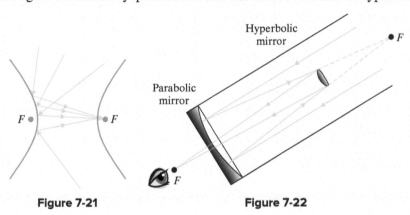

Figure 7-21   **Figure 7-22**

SECTION 7.2   Practice Exercises

Prerequisite Review

R.1. Use the distance formula to find the distance between the points (6, 12) and (−1, 12).

For Exercises R.2–R.3, write an equation of the line satisfying the given conditions. Write the answer in slope-intercept form.

R.2. The line passes through the point (−8, 8) and has a slope of −8.

R.3. The line passes through (−8, 8) and (1, −2).

Concept Connections

1. A _____ is the set of points (x, y) in a plane such that the difference in distances between (x, y) and two fixed points (called _____) is a positive constant.

2. The points where a hyperbola intersects the line through the foci are called the _____.

3. The line segment between the vertices of a hyperbola is called the _____ axis.

4. The line segment perpendicular to the transverse axis passing through the center of a hyperbola, and with endpoints on the reference rectangle is called the _____ axis.

5. The equation $\dfrac{x^2}{a^2} - \dfrac{y^2}{b^2} = 1$ represents a hyperbola with a (horizontal/vertical) transverse axis. The vertices are given by the ordered pairs _____ and _____. The asymptotes are given by the equations _____ and _____.

6. The equation $\dfrac{y^2}{a^2} - \dfrac{x^2}{b^2} = 1$ represents a hyperbola with a (horizontal/vertical) transverse axis. The vertices are given by the ordered pairs _____ and _____. The asymptotes are given by the equations _____ and _____.

Objective 1: Graph a Hyperbola Centered at the Origin

For Exercises 7–8,

a. Use the distance formula to find the distances d_1, d_2, d_3, and d_4.

b. Find the difference of the distances: $d_1 - d_2$.

c. Find the difference of the distances: $d_3 - d_4$.

d. How do the results from parts (b) and (c) compare?

e. How do the results of part (b) and part (c) compare to the length of the transverse axis?

7.

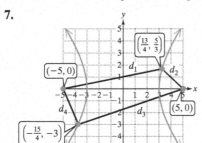

8.

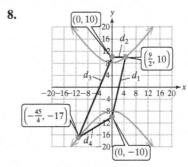

For Exercises 9–12, determine whether the transverse axis and foci of the hyperbola are on the x-axis or the y-axis.

9. $\dfrac{x^2}{15} - \dfrac{y^2}{20} = 1$

10. $\dfrac{y^2}{12} - \dfrac{x^2}{18} = 1$

11. $-x^2 + \dfrac{y^2}{3} = 1$

12. $-\dfrac{y^2}{16} + x^2 = 1$

For Exercises 13–22,

a. Identify the center.   **b.** Identify the vertices.

c. Identify the foci.   **d.** Write equations for the asymptotes.

e. Graph the hyperbola. **(See Examples 1–2)**

13. $\dfrac{x^2}{16} - \dfrac{y^2}{25} = 1$

14. $\dfrac{x^2}{25} - \dfrac{y^2}{36} = 1$

15. $\dfrac{y^2}{4} - \dfrac{x^2}{36} = 1$

16. $\dfrac{y^2}{9} - \dfrac{x^2}{49} = 1$

17. $25y^2 - 81x^2 = 2025$

18. $49y^2 - 16x^2 = 784$

19. $-5x^2 + 7y^2 = -35$

20. $-7x^2 + 11y^2 = -77$

21. $\dfrac{4x^2}{25} - \dfrac{16y^2}{49} = 1$

22. $\dfrac{4x^2}{81} - \dfrac{16y^2}{225} = 1$

Objective 2: Graph a Hyperbola Centered at (h, k)

For Exercises 23–32,

a. Identify the center.   **b.** Identify the vertices.

c. Identify the foci.   **d.** Write equations for the asymptotes.

e. Graph the hyperbola. **(See Example 3)**

23. $\dfrac{(x-4)^2}{9} - \dfrac{(y+2)^2}{16} = 1$

24. $\dfrac{(x-3)^2}{36} - \dfrac{(y+1)^2}{64} = 1$

25. $\dfrac{(y-5)^2}{49} - \dfrac{(x+3)^2}{25} = 1$

26. $\dfrac{(y-4)^2}{36} - \dfrac{(x+5)^2}{16} = 1$

27. $100(y-7)^2 - 81(x+4)^2 = -8100$

28. $49(y-3)^2 - 100(x+6)^2 = -4900$

29. $y^2 - \dfrac{(x-3)^2}{12} = 1$

30. $y^2 - \dfrac{(x-2)^2}{18} = 1$

31. $x^2 - \dfrac{(y-4)^2}{8} = 1$

32. $x^2 - \dfrac{(y-6)^2}{24} = 1$

For Exercises 33–40,

a. Write the equation of the hyperbola in standard form. **(See Example 4)**

b. Identify the center, vertices, and foci.

33. $7x^2 - 5y^2 + 42x + 10y + 23 = 0$

34. $5x^2 - 3y^2 + 10x + 24y - 73 = 0$

35. $-5x^2 + 9y^2 + 20x - 72y + 79 = 0$

36. $-7x^2 + 16y^2 - 70x + 96y - 143 = 0$

37. $4x^2 - y^2 - 10y - 29 = 0$

38. $9x^2 - y^2 - 14y - 58 = 0$

39. $-36x^2 + 64y^2 + 108x + 256y - 401 = 0$

40. $-144x^2 + 25y^2 + 720x - 50y - 4475 = 0$

For Exercises 41–50, write the standard form of the equation of the hyperbola subject to the given conditions. (See Example 5)

41. Vertices: $(12, 0)$, $(-12, 0)$; Foci: $(13, 0)$, $(-13, 0)$

42. Vertices: $(40, 0)$, $(-40, 0)$; Foci: $(41, 0)$, $(-41, 0)$

43. Vertices: $(0, 12)$, $(0, -12)$; Asymptotes: $y = \pm\frac{4}{3}x$

44. Vertices: $(0, 15)$, $(0, -15)$; Asymptotes: $y = \pm\frac{15}{8}x$

45. Vertices: $(-3, -3)$, $(7, -3)$;
Slope of the asymptotes: $\pm\frac{7}{5}$

46. Vertices: $(2, 1)$, $(-10, 1)$;
Slope of the asymptotes: $\pm\frac{5}{6}$

47. Vertices: $(-3, 0)$, $(-3, 8)$;
Foci: $\left(-3, 4 + \sqrt{21}\right), \left(-3, 4 - \sqrt{21}\right)$

48. Vertices: $(2, -1)$, $(2, -11)$;
Foci: $\left(2, -6 + \sqrt{31}\right), \left(2, -6 - \sqrt{31}\right)$

49. Corners of the reference rectangle: $(8, 7)$, $(-6, 7)$, $(8, -3)$, $(-6, -3)$; Horizontal transverse axis

50. Corners of the reference rectangle: $(7, 6)$, $(-1, 6)$, $(7, 0)$, $(-1, 0)$; Horizontal transverse axis

Objective 3: Determine the Eccentricity of a Hyperbola

For Exercises 51–52,

a. Determine the eccentricity of each hyperbola. **(See Example 6)**

b. Based on the eccentricity, match the equation with its graph. The scaling is the same for both graphs.

51. Equation 1: $\dfrac{x^2}{144} - \dfrac{y^2}{81} = 1$

Equation 2: $\dfrac{x^2}{81} - \dfrac{y^2}{144} = 1$

A.

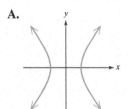

B.

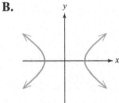

52. Equation 1: $\dfrac{x^2}{225} - \dfrac{y^2}{64} = 1$

Equation 2: $\dfrac{x^2}{64} - \dfrac{y^2}{225} = 1$

A.

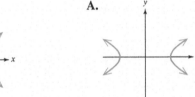

B.

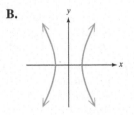

For Exercises 53–54, determine the eccentricity.

53. $\dfrac{\left(y - \frac{2}{3}\right)^2}{1600} - \dfrac{\left(x + \frac{7}{4}\right)^2}{81} = 1$

54. $\dfrac{(y - 3.8)^2}{49} - \dfrac{(x - 2.7)^2}{576} = 1$

55. Determine the eccentricity of a hyperbola with a horizontal transverse axis of length 24 units and asymptotes $y = \pm\frac{3}{4}x$.

56. Determine the eccentricity of a hyperbola with a vertical transverse axis of length 48 units and asymptotes $y = \pm\frac{12}{5}x$.

57. Determine the standard form of an equation of a hyperbola with eccentricity $\frac{5}{4}$ and vertices $(-1, -1)$ and $(7, -1)$.

58. Determine the standard form of an equation of a hyperbola with eccentricity $\frac{13}{12}$ and vertices $(-2, 8)$ and $(-2, -16)$.

Objective 4: Use Hyperbolas in Applications

59. Two radio transmitters are 1000 mi apart at points A and B along a coastline. Using LORAN on the ship, the time difference between the radio signals is 4 milliseconds (0.004 sec). If radio signals travel 186 mi/millisecond, find an equation of the hyperbola with foci at A and B, on which the ship is located. **(See Example 7)**

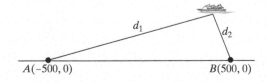

60. Suppose that two microphones 1500 m apart at points A and B detect the sound of a rifle shot. The time difference between the sound detected at point A and the sound detected at point B is 4 sec. If sound travels at approximately 330 m/sec, find an equation of the hyperbola with foci at A and B defining the points where the shooter may be located.

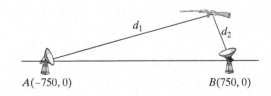

61. In some designs of eyeglasses, the surface is "aspheric," meaning that the contour varies slightly from spherical. An aspheric lens is often used to correct for spherical aberration—a distortion due to increased refraction of light rays when they strike the lens near its edge. Aspheric lenses are often designed with hyperbolic cross sections.

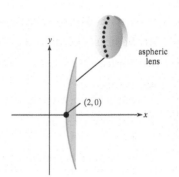

Write an equation of the cross section of the hyperbolic lens shown if the center is (0, 0), one vertex is (2, 0), and the focal length (distance between center and foci) is $\sqrt{85}$. Assume that all units are in millimeters.

62. In 1911, Ernest Rutherford discovered the nucleus of the atom. Experiments leading to this discovery involved the scattering of alpha particles by the heavy nuclei in gold foil. When alpha particles are thrust toward the gold nuclei, the particles are deflected and follow a hyperbolic path.

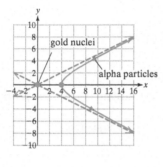

Suppose that the minimum distance that the alpha particles get to the gold nuclei is 4 microns (1 micron is one-millionth of a meter) and that the hyperbolic path has asymptotes of $y = \pm\frac{1}{2}x$. Determine an equation of the path of the particles shown.

63. Atomic particles with like charges tend to repel one another. Suppose that two beams of like-charged particles are hurled toward each other from two parallel atomic accelerators. The path defined by the particles is $x^2 - 4y^2 = 36$, where x and y are measured in microns. What is the minimum distance between the particles?

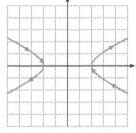

64. A returning boomerang is a V-shaped throwing device made from two wings that are set at a slight tilt and that have an airfoil design. One side is rounded and the other side is flat, similar to an airplane propeller. When thrown properly, a boomerang follows a circular flight path and should theoretically return close to the point of release.

The boomerang pictured is approximately in the shape of one branch of a hyperbola (although the two wings are in slightly different planes). To construct the hyperbola, an engineer needs to know the location of the foci. Determine the location of the focus to the right of the center if the vertex is 7.5 in. from the center and the equations of the asymptotes are $y = \pm\frac{4}{5}x$. Round the coordinates to the nearest tenth of an inch.

65. In September 2009, Australian astronomer Robert H. McNaught discovered comet C/2009 R$_1$ (McNaught). The orbit of this comet is hyperbolic with the Sun at one focus. Because the orbit is not elliptical, the comet will not be captured by the Sun's gravitational pull and instead will pass by the Sun only once. The comet reached perihelion on July 2, 2010. (*Source*: Minor Planet Center, http://minorplanetcenter.net/)

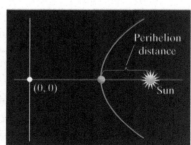

The path of the comet can be modeled by the equation

$$\frac{x^2}{(1191.2)^2} - \frac{y^2}{(30.9)^2} = 1$$

where x and y are measured in AU (astronomical units).

a. Determine the distance (in AU) at perihelion. Round to 1 decimal place.

b. Using the rounded value from part (a), if 1 AU $\approx$ 93,000,000 mi, find the distance in miles.

66. The cross section of a cooling tower of a nuclear power plant is in the shape of a hyperbola, and can be modeled by the equation

$$\frac{x^2}{625} - \frac{(y - 80)^2}{2500} = 1$$

where x and y are measured in meters. The top of the tower is 120 m above the base.

a. Determine the diameter of the tower at the base. Round to the nearest meter.

b. Determine the diameter of the tower at the top. Round to the nearest meter.

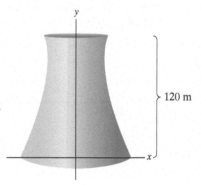

Mixed Exercises

For Exercises 67–70, identify the equation as representing an ellipse or a hyperbola, and match the equation with the graph.

67. $\dfrac{(x-5)^2}{49} + \dfrac{(y+2)^2}{36} = 1$

68. $\dfrac{(x-5)^2}{36} + \dfrac{(y+2)^2}{49} = 1$

69. $\dfrac{(x-5)^2}{49} - \dfrac{(y+2)^2}{36} = 1$

70. $-\dfrac{(x-5)^2}{49} + \dfrac{(y+2)^2}{36} = 1$

A.

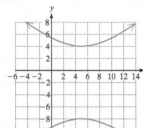

B.

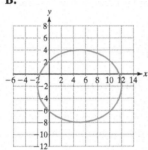

C.

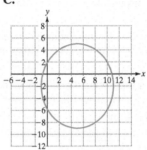

D.

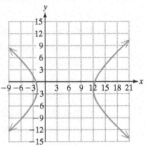

For Exercises 71–74, find the standard form of the equation of the ellipse or hyperbola shown.

71.

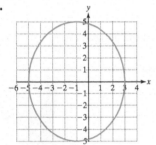

72.

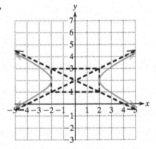

73.

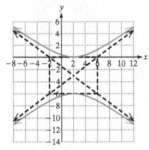

74.

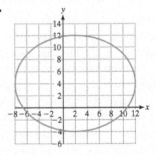

For Exercises 75–76, solve the system of equations.

75. $9x^2 - 4y^2 = 36$
        $-13x^2 + 8y^2 = 8$

76. $x^2 + 4y^2 = 36$
        $4x^2 - y^2 = 8$

Write About It

77. Given an equation of a hyperbola in standard form, how do you determine whether the transverse axis is horizontal or vertical?

78. Discuss the solution set of the equation.
$$\frac{x^2}{4} - \frac{y^2}{9} = 0$$

Expanding Your Skills

79. What is the eccentricity of a hyperbola if the asymptotes are perpendicular?

80. a. Describe the graph of $y = 4\sqrt{1 + \dfrac{x^2}{9}}$.

b. Does the equation $y = 4\sqrt{1 + \dfrac{x^2}{9}}$ define y as a function of x?

c. Sketch $y = 4\sqrt{1 + \dfrac{x^2}{9}}$.

d. Determine the interval(s) over which the function is increasing.

e. Determine the interval(s) over which the function is decreasing.

81. A line segment with endpoints on a hyperbola, perpendicular to the transverse axis, and passing through a focus is called a *latus rectum* of the hyperbola (shown in red). Show that the length of a latus rectum is $\dfrac{2b^2}{a}$ for the hyperbola

$$\frac{x^2}{a^2} - \frac{y^2}{b^2} = 1$$

[*Hint*: Substitute (c, y) into the equation and solve for y. Recall that $c^2 = a^2 + b^2$.]

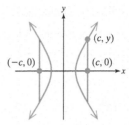

82. This exercise guides you through the steps to find the standard form of an equation of a hyperbola centered at the origin with foci on the x-axis.

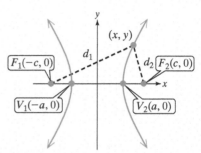

a. Refer to the figure to verify that the distance between F_1 and V_2 is $(c + a)$ and the distance between F_2 and V_2 is $(c - a)$. Verify that the *difference* between these distances is $2a$.

b. Write an expression that represents the difference of the distances from F_1 to (x, y) and from F_2 to (x, y). Then set this expression equal to $2a$.

c. Given the equation $\sqrt{(x + c)^2 + y^2} - \sqrt{(x - c)^2 + y^2} = 2a$, isolate the leftmost radical and square both sides of the equation. Show that the equation can be written as $cx - a^2 = a\sqrt{(x - c)^2 + y^2}$.

d. Square both sides of the equation $cx - a^2 = a\sqrt{(x - c)^2 + y^2}$ and show that the equation can be written as $(c^2 - a^2)x^2 - a^2 y^2 = a^2(c^2 - a^2)$. (*Hint*: Collect variable terms on the left side of the equation and constant terms on the right side.)

e. Replace $c^2 - a^2$ by b^2. Then divide both sides of the equation by $a^2 b^2$. Verify that the resulting equation is $\dfrac{x^2}{a^2} - \dfrac{y^2}{b^2} = 1$.

83. A hyperboloid of one sheet is a three-dimensional surface generated by an equation of the form $\dfrac{x^2}{a^2} + \dfrac{y^2}{b^2} - \dfrac{z^2}{c^2} = 1$. The surface has hyperbolic cross sections and either circular cross sections or elliptical cross sections.

a. Write the equation with $z = 0$. What type of curve is represented by this equation?

b. Write the equation with $x = 0$. What type of curve is represented by this equation?

c. Write the equation with $y = 0$. What type of curve is represented by this equation?

84. a. Radio signals emitted from points $(8, 0)$ and $(-8, 0)$ indicate that a plane is 8 mi closer to $(8, 0)$ than to $(-8, 0)$. Find an equation of the hyperbola that passes through the plane's location and with foci $(8, 0)$ and $(-8, 0)$. All units are in miles.

b. At the same time, radio signals emitted from points $(0, 8)$ and $(0, -8)$ indicate that the plane is 4 mi farther from $(0, 8)$ than from $(0, -8)$. Find an equation of the hyperbola that passes through the plane's location and with foci $(0, 8)$ and $(0, -8)$.

c. From the figure, the plane is located in the fourth quadrant of the coordinate system. Solve the system of equations defining the two hyperbolas for the point of intersection in the fourth quadrant. This is the location of the plane. Then round the coordinates to the nearest tenth of a mile.

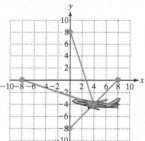

Technology Connections

For Exercises 85–88, graph the hyperbola from the given exercise.

85. Exercise 15   **86.** Exercise 16   **87.** Exercise 37   **88.** Exercise 38

SECTION 7.3   The Parabola

Figure 7-23

1. Identify the Focus and Directrix of a Parabola

A parabola is another type of conic section generated by the cross section of a cone intersected by a plane (Figure 7-23). An equation of the form $y = ax^2 + bx + c$ ($a \neq 0$) is a parabola opening upward if $a > 0$ and opening downward if $a < 0$. We now extend our study of parabolas to include those that open to the left and right. To do so, we will define a parabola geometrically.

> **Definition of a Parabola**
>
> A **parabola** is the set of all points in a plane that are equidistant from a fixed line (called the **directrix**) and a fixed point (called the **focus**).

Figure 7-24 shows a parabola opening upward. The line perpendicular to the directrix and passing through the focus is called the **axis of symmetry.** The **vertex** of the parabola is the point of intersection of the parabola and the axis of symmetry.

A point $P(x, y)$ is on the parabola if $d_1 = d_2$ where d_1 is the distance between the focus and P, and d_2 is the perpendicular distance between P and the directrix.

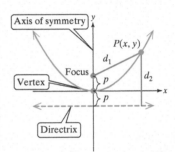

Figure 7-24

To develop an algebraic equation that represents a parabola, we will consider a parabola with vertex at the origin and with a vertical line of symmetry (that is, the parabola opens upward or downward). See Figure 7-25. In the discussion that follows, the same logic can be applied to a parabola with a horizontal line of symmetry (the parabola opens to the left or to the right).

The distance between the vertex and the focus of a parabola is called the **focal length** and is often represented by $|p|$. Suppose that we choose a coordinate system with the vertex at the origin and focus $F(0, p)$. Because the distance between any point on the parabola and the focus must equal the distance between the point and the directrix, the directrix is a horizontal line $|p|$ units from the vertex. The equation of the directrix is $y = -p$.

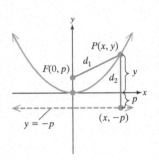

Figure 7-25

Let $P(x, y)$ represent any point on the parabola. Then the distance between $F(0, p)$ and $P(x, y)$ can be found by using the distance formula.

$$d_1 = \sqrt{(x - 0)^2 + (y - p)^2}$$

The distance from $P(x, y)$ to the directrix is the vertical distance between $P(x, y)$ and the point $(x, -p)$ on the directrix.

$$d_2 = y + p$$

Equating d_1 and d_2, we have

$$\sqrt{x^2 + (y - p)^2} = y + p$$
$$x^2 + (y - p)^2 = (y + p)^2 \qquad \text{Square both sides.}$$
$$x^2 + y^2 - 2py + p^2 = y^2 + 2py + p^2 \qquad \text{Expand the binomials.}$$
$$x^2 = 4py$$

> **Standard form**
> If $p > 0$, the parabola opens upward.
> If $p < 0$, the parabola opens downward.

> **TIP** The shortest distance between P and the directrix is the *perpendicular distance*. This is the length of the line segment perpendicular to the directrix with one endpoint at P and the other endpoint on the directrix.

The equation $x^2 = 4py$ represents a parabola with vertex at the origin and focus on the y-axis. Using similar logic, we can derive the standard form of an equation of a parabola with focus on the x-axis.

Standard Forms of an Equation of a Parabola with Vertex at the Origin		
	Axis of Symmetry: y-axis	**Axis of Symmetry: x-axis**
Equation:	$x^2 = 4py$	$y^2 = 4px$
Vertex:	$(0, 0)$	$(0, 0)$
Focus:	$(0, p)$	$(p, 0)$
Directrix:	$y = -p$	$x = -p$
Axis of symmetry:	$x = 0$	$y = 0$
Graph: ($p > 0$)		
Graph: ($p < 0$)		

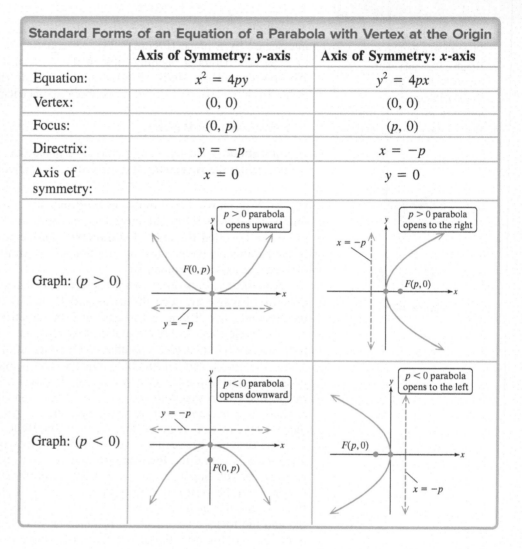

From Section 3.1, we became accustomed to writing an equation of a quadratic function with vertex at the origin in the form $y = ax^2$. The benefit of writing the equation as $x^2 = 4py$ is that we can identify the value of p, which gives us the distance $|p|$ between the vertex and the focus.

Finding the location of the focus is particularly important in applications. For example, a flashlight has a mirror with cross section in the shape of a parabola (Figure 7-26). The bulb is located at the focus. The light emitted from the bulb will reflect off the mirror in lines parallel to the axis of symmetry to form a beam of light. This is a result of the reflective property of a parabola.

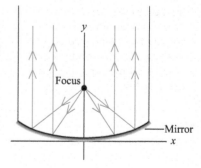

Figure 7-26

| EXAMPLE 1 | **Determining the Focus and Directrix of a Parabola** |

An engineer designs a flashlight with a parabolic mirror. A coordinate system is chosen so that the vertex of a cross section through the center of the mirror is located at $(0, 0)$. The equation of the parabola is modeled by $x^2 = 8y$, where x and y are measured in centimeters.

a. Where should the bulb be placed to make use of the reflective property of the parabola? That is, find the location of the focus.

b. Write an equation of the directrix.

Solution:

a. $x^2 = 8y$ is in the form $x^2 = 4py$ with $4p = 8$.

$4p = 8$

$p = 2$ Since the equation is in the form $x^2 = 4py$, the focus is on the y-axis. Furthermore, since $p > 0$, the parabola opens upward, and the focus is $|p|$ units above the vertex.

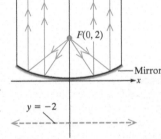

The focus is $(0, 2)$.

The bulb should be placed 2 cm above the vertex.

b. The directrix is the horizontal line $|p|$ units below the vertex.

The equation of the directrix is $y = -2$.

Skill Practice 1 Given a parabola defined by $x^2 = 4y$, find the focus and an equation of the directrix.

2. Graph a Parabola with Vertex at the Origin

In Examples 2 and 3, we practice graphing parabolas with vertex at the origin. To sketch the graph, we will identify the location of the vertex and the orientation of the parabola (up or down versus left or right). To determine how "wide" to graph the parabola, we will determine the length of the line segment passing through the focus, perpendicular to the axis of symmetry with endpoints on the parabola. This line segment is called the **latus rectum.**

For the parabola defined by $x^2 = 4py$, the endpoints of the latus rectum will have the same y-coordinate as the focus. Therefore, to find the endpoints of the latus rectum, we can substitute p for y in the equation (Figure 7-27).

$$x^2 = 4py$$
$$x^2 = 4p(p) \quad \text{Substitute } p \text{ for } y.$$
$$x^2 = 4p^2$$
$$x = \pm 2|p|$$

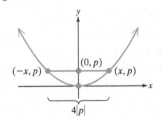

Figure 7-27

Thus, the endpoints of the latus rectum are located $2|p|$ units to the left and right of the focus at $(-2|p|, p)$ and $(2|p|, p)$. The length of the latus rectum is called the **focal diameter** and is equal to $4|p|$.

In Example 2, we graph a parabola by using the vertex, orientation of the parabola, and focal diameter.

Answer

1. Focus: $(0, 1)$; Directrix: $y = -1$

EXAMPLE 2 **Graphing a Parabola with Vertex at the Origin**

Given $x^2 = -12y$,

    a. Identify the vertex, focus, and focal diameter.

    b. Identify the endpoints of the latus rectum and graph the parabola.

    c. Write equations for the directrix and axis of symmetry.

Solution:

  a. $x^2 = -12y$ The equation is in the standard form $x^2 = 4py$. From the standard form, we see that the focus is on the y-axis and that the parabola will open upward or downward. The vertex is $(0, 0)$, with $4p = -12$.

     $4p = -12$

      $p = -3$ Because $p < 0$, the parabola will open downward.

  The vertex is $(0, 0)$.

  The focus is $(0, p)$, which is $(0, -3)$.

  The focal diameter is $4|p|$, which is $4|(-3)| = 12$.

  b. Plot the vertex and locate the focus. Note that the focus is shown in red because it is not actually part of the graph. The focal diameter is 12 units. Therefore, plot points 6 units to the left and right of the focus. These are the endpoints of the latus rectum: $(6, -3)$ and $(-6, -3)$.

> **TIP** The focal diameter gives the length of the latus rectum. This is helpful for graphing a parabola because it tells us the "width" of the parabola at the focus.

  c. Equation of the directrix: $y = 3$
      Equation of the axis of symmetry: $x = 0$

Skill Practice 2   Repeat Example 2 with the equation: $x^2 = -8y$

In Example 3, we graph a parabola opening horizontally.

EXAMPLE 3   **Graphing a Parabola with Vertex at the Origin**

Given $2y^2 = 32x$,

    a. Identify the vertex, focus, and focal diameter.

    b. Identify the endpoints of the latus rectum and graph the parabola.

    c. Write equations for the directrix and axis of symmetry.

Solution:

  a. $2y^2 = 32x$ Divide both sides by 2 to obtain a coefficient of 1 on the square term.

     $y^2 = 16x$ The equation is in the standard form $y^2 = 4px$. From the standard form, we see that the focus is on the x-axis, and that the parabola will open to the left or right. The vertex is $(0, 0)$ and $4p = 16$.

     $4p = 16$

      $p = 4$ Because $p > 0$, the parabola will open to the right.

  The vertex is $(0, 0)$.

  The focus is $(p, 0)$, which is $(4, 0)$.

  The focal diameter is $4|p|$, which is $4|(4)| = 16$.

Answers

2. a. Vertex: $(0, 0)$; Focus: $(0, -2)$; Focal diameter: 8
  b. $(-4, -2)$, $(4, -2)$
  c. Directrix: $y = 2$; Axis of symmetry: $x = 0$

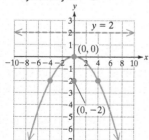

b. Plot the vertex and locate the focus. The focal diameter is 16 units. Therefore, plot points 8 units above and 8 units below the focus. These are the endpoints of the latus rectum, (4, 8) and (4, −8).

c. Equation of the directrix: $x = -4$
Equation of the axis of symmetry: $y = 0$

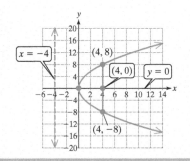

Skill Practice 3   Repeat Example 3 with the equation: $y^2 = -12x$

3. Graph a Parabola with Vertex (*h*, *k*)

In the standard form of a parabola, if x and y are replaced by $x - h$ and $y - k$, then the graph of the parabola is shifted h units horizontally and k units vertically. The vertex is (h, k).

Standard Forms of an Equation of a Parabola with Vertex (*h*, *k*)		
	Vertical Axis of Symmetry	**Horizontal Axis of Symmetry**
Equation:	$(x - h)^2 = 4p(y - k)$	$(y - k)^2 = 4p(x - h)$
Vertex:	(h, k)	(h, k)
Focus:	$(h, k + p)$	$(h + p, k)$
Directrix:	$y = k - p$	$x = h - p$
Axis of symmetry:	$x = h$	$y = k$
Graph:		
Note:	• If $p > 0$, the parabola opens upward as shown. • If $p < 0$, the parabola opens downward.	• If $p > 0$, the parabola opens to the right as shown. • If $p < 0$, the parabola opens to the left.

To graph a parabola with vertex (h, k), first locate the vertex and identify the orientation of the parabola (vertical axis of symmetry or horizontal axis of symmetry). Then identify the focus, directrix, and endpoints of the latus rectum. Finally, sketch the parabola through the vertex and endpoints of the latus rectum.

Answers

3. a. Vertex: (0, 0); Focus: (−3, 0); Focal diameter: 12
b. (−3, 6), (−3, −6)
c. Directrix: $x = 3$; Axis of symmetry: $y = 0$

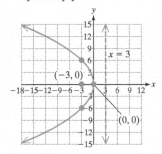

EXAMPLE 4 **Graphing a Parabola with Vertex (h, k)**

Given $(y + 3)^2 = -4(x - 2)$,

a. Identify the vertex, focus, and focal diameter.

b. Identify the endpoints of the latus rectum and graph the parabola.

c. Write equations for the directrix and axis of symmetry.

Solution:

a. $(y + 3)^2 = -4(x - 2)$ The equation is in the form $(y - k)^2 = 4p(x - h)$, where $h = 2$ and $k = -3$. The vertex is $(2, -3)$, and $4p = -4$.

$4p = -4$ This implies that $p = -1$.

$p = -1$ The x term is linear. Therefore, the parabola will open in the x direction (left or right). Furthermore, since $p < 0$, the parabola will open to the left.

The vertex is $(2, -3)$.

The focus is $(1, -3)$.   Since $p = -1$, the focus is 1 unit to the left of the vertex.

The focal diameter is 4.   The focal diameter is $4|p| = 4|(-1)| = 4$.

b. Plot the vertex and locate the focus. The focal diameter is 4 units. Therefore, plot points 2 units above and below the focus. These are the endpoints of the latus rectum: $(1, -1)$ and $(1, -5)$.

Sketch the curve through the vertex and endpoints of the latus rectum.

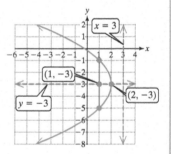

c. Equation of the directrix: $x = 3$

Equation of the axis of symmetry: $y = -3$

Skill Practice 4   Repeat Example 4 with the equation: $(y - 2)^2 = -8(x - 4)$

TECHNOLOGY CONNECTIONS

Graphing a Parabola

To graph the parabola from Example 4 on a graphing utility, first solve the equation for y.

$$(y + 3)^2 = -4(x - 2)$$

$$y = -3 \pm \sqrt{-4(x - 2)} \qquad \text{Solve for } y.$$

The functions represent the top and bottom branches of the parabola, respectively.

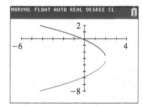

Answers

4. a. Vertex: (4, 2); Focus: (2, 2); Focal diameter: 8

b. (2, 6), (2, −2)

c. Directrix: $x = 6$; Axis of symmetry: $y = 2$

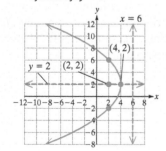

Sometimes an equation of a parabola is given in expanded form. We can complete the square to write the equation in standard form. The process involves completing the square on either the x or y term, depending on which variable is squared in the equation. This is demonstrated in Example 5.

EXAMPLE 5   **Writing an Equation of a Parabola in Standard Form**

Given the equation of the parabola $4x^2 - 20x - 8y + 57 = 0$,

  a. Write the equation in standard form.

  b. Identify the vertex, focus, and focal diameter.

  c. Graph the parabola.

Solution:

a. $4x^2 - 20x - 8y + 57 = 0$ The x term is squared. Therefore, we complete

    $4(x^2 - 5x \qquad) = 8y - 57$ the square on x on the left side. Move the

                                         y term and constant term to the right.

$4 \cdot \frac{25}{4}$ is added   Complete the square.

to the left side. $4 \cdot \frac{25}{4}$ Note that $\left[\frac{1}{2} \cdot (-5)\right]^2 = \left[-\frac{5}{2}\right]^2 = \frac{25}{4}$.

    $4\left(x^2 - 5x + \frac{25}{4}\right) = 8y - 57 + 25$

    $4\left(x - \frac{5}{2}\right)^2 = 8y - 32$ The value $\frac{25}{4}$ was added inside the parentheses.

                                         However, $4 \cdot \frac{25}{4}$ was actually added to the left

                                       side as a result of the factor 4 in front of the

                                     parentheses.

    $\dfrac{4\left(x - \frac{5}{2}\right)^2}{4} = \dfrac{8y - 32}{4}$ To write the equation in standard form, we

                                        require the coefficient of the square term to

                                       be 1. Divide both sides by 4.

    $\left(x - \frac{5}{2}\right)^2 = 2y - 8$

    $\left(x - \frac{5}{2}\right)^2 = 2(y - 4)$ Factor out the leading coefficient on the right

                                         side of the equation. The equation is in the

                                       form $(x - h)^2 = 4p(y - k)$.

b. $4p = 2$ The y term is linear. Therefore, the parabola

    $p = \frac{1}{2}$ will open in the y direction (upward or

                                         downward). Furthermore, since $p > 0$, the

                                         parabola will open upward.

    The vertex is $\left(\frac{5}{2}, 4\right)$.

    The focus is $\left(\frac{5}{2}, 4 + \frac{1}{2}\right) = \left(\frac{5}{2}, \frac{9}{2}\right)$.   Since $p = \frac{1}{2}$, the focus is $\frac{1}{2}$ unit above the vertex.

    The focal diameter is 2.   The focal diameter is $4|p| = 4\left|\frac{1}{2}\right| = 2$.

c. Plot the vertex and locate the focus. The focal diameter is 2 units. Therefore, plot points 1 unit to the left and right of the focus. These are the points $\left(\frac{3}{2}, \frac{9}{2}\right)$ and $\left(\frac{7}{2}, \frac{9}{2}\right)$.

Sketch the curve.

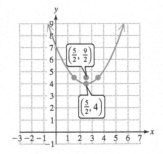

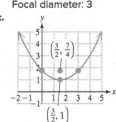
Skill Practice 5   Repeat Example 5 with the equation:

$4x^2 - 12x - 12y + 21 = 0$

In Example 6, we determine the standard form of an equation of a parabola based on known information about the parabola.

EXAMPLE 6   **Writing an Equation of a Parabola**

Determine the standard form of an equation of the parabola with focus (1, 2) and directrix $x = 7$.

Solution:

The directrix is a vertical line $x = 7$ and the focus is 6 units to the left of the directrix (Figure 7-28). The vertex is halfway between the directrix and focus. Therefore, the vertex is 3 units to the right of the focus at (4, 2). This indicates that $h = 4$ and $k = 2$.

The parabola must open away from the directrix and toward the focus which in this case is to the left. The standard equation is

$$(y - k)^2 = 4p(x - h)$$

The distance between the focus and vertex is 3. Therefore, $|p| = 3$. Because the parabola opens to the left, $p < 0$, indicating that $p = -3$.

The equation is: $(y - k)^2 = 4p(x - h)$
$(y - 2)^2 = 4(-3)(x - 4)$
$(y - 2)^2 = -12(x - 4)$

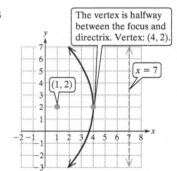

The vertex is halfway between the focus and directrix. Vertex: (4, 2).

$x = 7$

(1, 2)

Figure 7-28

Skill Practice 6   Determine the standard form of an equation of a parabola with focus (−2, 3) and directrix $y = 7$.

---

4. Use Parabolas in Applications

We complete this section by determining an equation of a parabola used to construct a satellite dish. A parabolic satellite dish uses the reflective property of a parabola to gather radio waves and direct the signal to a common focus to make the signal stronger.

EXAMPLE 7   **Applying an Equation of a Parabola**

A satellite dish is in the shape of a paraboloid. Cross sections taken parallel to the direction the dish opens are parabolic. Cross sections taken perpendicular to the direction the dish opens are circular. The diameter of the dish is 30 in. and the depth is 3 in.

a. Find an equation of a parabolic cross section through the vertex of the dish.

b. Where should the receiver be placed?

Solution:

a. Orient the dish with the vertex at the origin of a rectangular coordinate system with the dish opening along the positive x-axis (Figure 7-29). An equation of a parabolic cross section through the origin is of the form $y^2 = 4px$. To solve for the value of p, we can

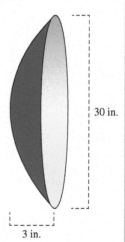

30 in.

3 in.

Answer
6. $(x + 2)^2 = -8(y - 5)$

use a known point on the parabola such as (3, 15) or (3, −15). Using (3, 15), we have

$$y^2 = 4px$$
$$(15)^2 = 4p(3)$$
$$225 = 12p$$
$$p = 18.75$$

Therefore, an equation of a parabolic cross section through the origin is

$$y^2 = 4(18.75)x$$
$$y^2 = 75x$$

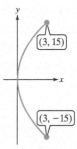

Figure 7-29

b. The receiver should be placed at the focus of the parabola. Since the focal length is $|p| = |18.75| = 18.75$, the receiver should be placed at (18.75, 0).

Skill Practice 7 Repeat Example 7 with a radio telescope of diameter 70 m and depth 14 m.

Answers
7. **a.** $y^2 = 87.5x$
   b. (21.875, 0)

SECTION 7.3 Practice Exercises

Prerequisite Review

R.1. Use the distance formula to find the distance between the points $(-1, -2)$ and $(-2, -9)$.

For Exercises R.2–R.3, write an equation of the line satisfying the given conditions.

R.2. The line is parallel to the line $y = 4$ and passes through $(-4, 3)$.

R.3. The line is perpendicular to the x-axis and passes through (2, 1).

Concept Connections

1. A _____ is the set of all points in a plane that are equidistant from a fixed line (called the _____) and a fixed point (called the _____).

2. The line perpendicular to the directrix and passing through the focus of a parabola is called the axis of _____.

3. The _____ of a parabola is the point of intersection of the parabola and the axis of symmetry.

4. The distance between the vertex and the focus of a parabola is called the _____ length and is often represented by $|p|$.

5. Given $y^2 = 4px$ with $p < 0$, the parabola opens (upward/downward/left/right).

6. Given $x^2 = 4py$ with $p > 0$, the parabola opens (upward/downward/left/right).

7. The line segment perpendicular to the axis of symmetry, passing through the focus and with endpoints on the parabola is called the _____ _____.

8. The length of the latus rectum is called the _____ diameter.

9. Given $(x - h)^2 = 4p(y - k)$, the ordered pairs representing the vertex and focus are _____ and _____, respectively. The directrix is the line defined by the equation _____.

10. Given $(y - k)^2 = 4p(x - h)$, the ordered pairs representing the vertex and focus are _____ and _____, respectively. The directrix is the line defined by the equation _____.

11. If the directrix is horizontal, then the parabola opens (horizontally/vertically).

12. If the line of symmetry is horizontal, then the parabola opens (horizontally/vertically).

For Exercises 13–14, the graph of a parabola is given.

a. Determine the distances d_1, d_2, d_3, and d_4.

b. Compare d_1 and d_2.

c. Compare d_3 and d_4.

13.

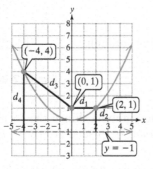

14.

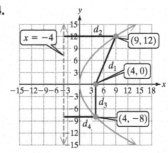

Objective 1: Identify the Focus and Directrix of a Parabola

For Exercises 15–22, a model of the form $x^2 = 4py$ or $y^2 = 4px$ is given.

a. Determine the value of p.

b. Identify the focus of the parabola.

c. Write an equation for the directrix. (**See Example 1**)

15. $x^2 = 24y$

16. $x^2 = 12y$

17. $y^2 = 36x$

18. $y^2 = 48x$

19. $x^2 = -5y$

20. $x^2 = -11y$

21. $-x = y^2$

22. $-3x = y^2$

23. A 20-in. satellite dish for a television has parabolic cross sections. A coordinate system is chosen so that the vertex of a cross section through the center of the dish is located at $(0, 0)$. The equation of the parabola is modeled by $x^2 = 25.2y$, where x and y are measured in inches.

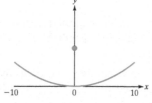

  a. Where should the receiver be placed to maximize signal strength? That is, where is the focus? (**See Example 1**)

  b. Determine the equation of the directrix.

24. Solar cookers provide an alternative form of cooking in regions of the world where consistent sources of fuel are not readily available. Suppose that a 36-in. solar cooker has parabolic cross sections. A coordinate system is chosen with the origin placed at the vertex of a cross section through the center of the mirror. The equation of the parabola is modeled by $x^2 = 82y$, where x and y are measured in inches.

  a. Where should a pot be placed to maximize heat? That is, where is the focus?

  b. Determine the equation of the directrix.

25. If a cross section of the parabolic mirror in a flashlight has an equation $y^2 = 2x$, where should the bulb be placed?

26. A cross section of the parabolic mirror in a car headlight is modeled by $y^2 = 12x$. Where should the bulb be placed?

Objective 2: Graph a Parabola with Vertex at the Origin

For Exercises 27–34, an equation of a parabola $x^2 = 4py$ or $y^2 = 4px$ is given.

a. Identify the vertex, value of p, focus, and focal diameter of the parabola.

b. Identify the endpoints of the latus rectum.

c. Graph the parabola.

d. Write equations for the directrix and axis of symmetry. (**See Examples 2–3**)

27. $x^2 = -4y$ **28.** $x^2 = -20y$ **29.** $10y^2 = 80x$ **30.** $3y^2 = 12x$

31. $4x^2 = 40y$ **32.** $2x^2 = 14y$ **33.** $y^2 = -x$ **34.** $y^2 = -2x$

Objective 3: Graph a Parabola with Vertex (h, k)

For Exercises 35–44, an equation of a parabola $(x - h)^2 = 4p(y - k)$ or $(y - k)^2 = 4p(x - h)$ is given.

a. Identify the vertex, value of p, focus, and focal diameter of the parabola.

b. Identify the endpoints of the latus rectum.

c. Graph the parabola.

d. Write equations for the directrix and axis of symmetry. (**See Example 4**)

35. $(y + 1)^2 = -12(x - 4)$ **36.** $(y + 4)^2 = -16(x - 2)$ **37.** $(x - 1)^2 = -4(y + 5)$

38. $(x - 5)^2 = -8(y + 2)$ **39.** $(x + 3)^2 = 2\left(y - \frac{3}{2}\right)$ **40.** $(x + 2)^2 = \left(y - \frac{7}{4}\right)$

41. $-(y - 3)^2 = 7\left(x - \frac{1}{4}\right)$ **42.** $-(y - 6)^2 = 9\left(x - \frac{3}{4}\right)$ **43.** $2(y - 3) = \frac{1}{10}(x + 6)^2$

44. $8(y - 2) = \frac{1}{2}(x + 3)^2$

For Exercises 45–52, an equation of a parabola is given.

a. Write the equation of the parabola in standard form.

b. Identify the vertex, focus, and focal diameter. (**See Example 5**)

45. $x^2 - 6x - 4y + 5 = 0$ **46.** $x^2 - 4x - 8y - 20 = 0$ **47.** $y^2 + 4y + 8x + 52 = 0$

48. $y^2 + 8y + 4x + 36 = 0$ **49.** $4x^2 - 28x + 24y + 73 = 0$ **50.** $4x^2 + 36x + 40y + 1 = 0$

51. $16y^2 + 24y - 16x + 57 = 0$ **52.** $16y^2 - 56y - 16x + 81 = 0$

For Exercises 53–58, fill in the blanks. Let $|p|$ represent the focal length (distance between the vertex and focus).

53. If the directrix of a parabola is given by $y = 4$ and the focus is $(1, 10)$, then the vertex is given by the ordered pair _____ and the value of p is _____.

54. If the directrix of a parabola is given by $x = 6$ and the focus is $(2, 1)$, then the vertex is given by the ordered pair _____ and the value of p is _____.

55. If the vertex of a parabola is $(-3, 0)$ and the focus is $(-7, 0)$, then the directrix is given by the equation _____, and the value of p is _____.

56. If the vertex of a parabola is $(4, -2)$ and the focus is $(4, -7)$, then the directrix is given by the equation _____, and the value of p is _____.

57. If the focal length of a parabola is 3 units, then the length of the latus rectum is _____ units.

58. If the focal diameter of a parabola is 8 units, then the focal length is _____ units.

59. If the focal length of a parabola is 6 units and the vertex is $(3, 2)$, is it possible to determine whether the parabola opens upward, downward, to the right, or to the left? Explain.

60. If the axis of symmetry of a parabola is given by $x = 4$, is it possible for the directrix to have the equation $x = 3$? Explain.

For Exercises 61–68, determine the standard form of an equation of the parabola subject to the given conditions. (**See Example 6**)

61. Vertex: $(0, 0)$; Directrix: $x = 4$ **62.** Vertex: $(0, 0)$; Directrix: $y = -2$

63. Focus: $(2, 4)$; Vertex: $(2, 1)$ **64.** Focus: $(5, 3)$: Vertex: $(3, 3)$

65. Focus: $(-6, -2)$; Directrix: $y = 0$ **66.** Focus: $(-4, 5)$; Directrix: $x = 0$

67. Vertex: $(2, 3)$; Parabola passes through $(6, 5)$
(*Hint*: There are two possible answers.)

68. Vertex: $(-4, 2)$; Parabola passes through $(8, 14)$
(*Hint*: There are two possible answers.)

Objective 4: Use Parabolas in Applications

69. Suppose that a solar cooker has a parabolic mirror (see figure).

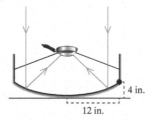

4 in.

12 in.

a. Use a coordinate system with origin at the vertex of the mirror and write an equation of a parabolic cross section of the mirror.

b. Where should a pot be placed so that it receives maximum heat? (**See Example 7**)

70. A solar water heater is made from a long sheet of metal bent so that the cross sections are parabolic. A long tube of water is placed inside the curved surface so that the height of the tube is equal to the focal length of the parabolic cross section. In this way, water in the tube is exposed to maximum heat.

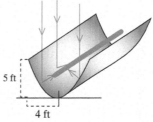

5 ft

4 ft

a. Determine the focal length of the parabolic cross sections so that the engineer knows where to place the tube.

b. Use a coordinate system with origin at the vertex of a parabolic cross section and write an equation of the parabola.

71. The Hubble Space Telescope was launched into space in 1990 and now orbits the Earth at 5 mi/sec at a distance of 353 mi above the Earth. From its location in space, the Hubble is free from the distortion of the Earth's atmosphere, enabling it to return magnificent images from distant stars and galaxies. The quality of the Hubble's images results from a large parabolic mirror, 2.4 m (7.9 ft) in diameter, that collects light from space.

Suppose that a coordinate system is chosen so that the vertex of a cross section through the center of the mirror is located at (0, 0). Furthermore, the focal length is 57.6 m.

a. Assume that x and y are measured in meters. Write an equation of the parabolic cross section of the mirror for $-1.2 \leq x \leq 1.2$. See figure.

b. How thick is the mirror at the edge? That is, what is the y value for $x = 1.2$?

72. The James Webb Space Telescope (JWST) is a new space telescope currently under construction. The JWST will orbit the Sun in an orbit roughly 1.5 million

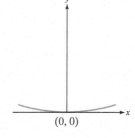

km from the earth, in a position with the Earth between the telescope and the Sun. The primary mirror of the JWST will consist of 18 hexagonally shaped segments that when fitted together will be 6.5 m in diameter (over 2.5 times the diameter of the Hubble). With a larger mirror, the JWST will be able to collect more light to "see" deeper into space.

The primary mirror when pieced together will function as a parabolic mirror with a focal length of 131.4 m. Suppose that a coordinate system is chosen with (0, 0) at the center of a cross section of the primary mirror. Write an equation of the parabolic cross section of the mirror for $-3.25 \leq x \leq 3.25$. Assume that x and y are measured in meters.

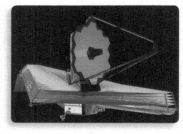

(0, 0)

73. The Subaru telescope is a large optical-infrared telescope at the summit of Mauna Kea, Hawaii. The telescope has a parabolic mirror 8.2 m in diameter with a focal length of 15 m.

   a. Suppose that a cross section of the mirror is taken through the vertex, and that a coordinate system is set up with (0, 0) placed at the vertex. If the focus is (0, 15), find an equation representing the curve.

   b. Determine the vertical displacement of the mirror relative to horizontal at the edge of the mirror. That is, find the y value at a point 4.1 m to the left or right of the vertex.

   c. What is the average slope between the vertex of the parabola and the point on the curve at the right edge?

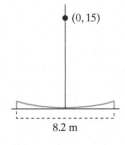

75. A parabolic arch forms a structure that allows equal vertical loading along its length.

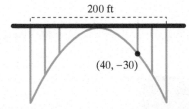

   a. Take the origin at the vertex of the parabolic arch, and write an equation of the parabola.

   b. Determine the focal length.

   c. Determine the length of the vertical support 60 ft from the center.

74. A parabolic mirror on a telescope has a focal length of 16 cm.

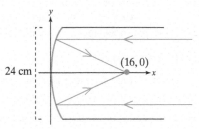

   a. For the coordinate system shown, write an equation of the parabolic cross section of the mirror.

   b. Determine the displacement of the mirror relative to the y-axis at the edge of the mirror. That is, find the x value at a point 12 cm above or below the vertex.

76. A cable hanging freely between two vertical support beams forms a curve called a catenary. The shape of a catenary resembles a parabola but mathematically the two functions are quite different.

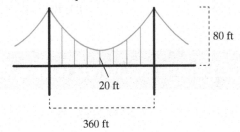

   a. On a graphing utility, graph a catenary defined by $y = \frac{1}{2}(e^x + e^{-x})$ and graph the parabola defined by $y = x^2 + 1$.

   b. A catenary and a parabola are so similar in shape that we can often use a parabolic curve to approximate the shape of a catenary. For example, a bridge has cables suspended from a larger approximately parabolic cable. Take the origin at a point on the road directly below the vertex and write an equation of the parabolic cable.

   c. Determine the focal length of the parabolic cable.

   d. Determine the length of the vertical support cable 100 ft from the vertex. Round to the nearest tenth of a foot.

Mixed Exercises

For Exercises 77–78, solve the system of nonlinear equations.

77. $(y + 3)^2 = 4(x - 4)$

   $2x + y = 9$

78. $(y - 2)^2 = -8(x - 1)$

   $2x - y = -8$

Write About It

79. Given an equation of a parabola $(x - h)^2 = 4p(y - k)$ or $(y - k)^2 = 4p(x - h)$, how can you determine whether the parabola opens vertically or horizontally?

80. Give an example of the reflective property of a parabola.

Expanding Your Skills

81. The surface defined by the equation $z = 4x^2 + y^2$ is called an elliptical paraboloid.

   a. Write the equation with $x = 0$. What type of curve is represented by this equation?

   b. Write the equation with $y = 0$. What type of curve is represented by this equation?

   c. Write the equation with $z = 0$. What type of curve is represented by this equation?

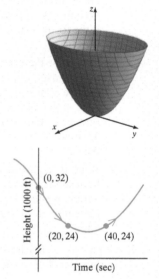

Technology Connections

82. A jet flies in a parabolic arc to simulate partial weightlessness. The curve shown in the figure represents the plane's height y (in 1000 ft) versus the time t (in sec).

   a. For each ordered pair, substitute the t and y values into the model $y = at^2 + bt + c$ to form a linear equation with three unknowns a, b, and c. Together, these form a system of three linear equations with three unknowns.

   b. Use a graphing utility to solve for a, b, and c.

   c. Substitute the known values of a, b, and c into the model $y = at^2 + bt + c$.

   d. Determine the vertex of the parabola.

   e. Determine the focal length of the parabola.

For Exercises 83–86, graph the parabola from the given exercise.

83. Exercise 29   **84.** Exercise 30   **85.** Exercise 41   **86.** Exercise 42

PROBLEM RECOGNITION EXERCISES

Comparing Equations of Conic Sections and the General Equation

For Exercises 1–8, identify each equation as representing a circle, an ellipse, a hyperbola, or a parabola.

- If the equation represents a circle, identify the center and radius.
- If the equation represents an ellipse, identify the center, vertices, endpoints of the minor axis, foci, and eccentricity.
- If the equation represents a hyperbola, identify the center, vertices, foci, equations of the asymptotes, and eccentricity.
- If the equation represents a parabola, identify the vertex, focus, equation of the directrix, and equation of the axis of symmetry.

1. $\dfrac{(x - 2)^2}{16} - \dfrac{(y + 2)^2}{9} = 1$ **2.** $9x^2 + 25(y - 2)^2 = 225$ **3.** $(y - 5)^2 = -(x + 2)$

4. $(x - 3)^2 + (y + 7)^2 = 25$ **5.** $16(x + 1)^2 + y^2 = 16$ **6.** $(x - 1)^2 = 10(y + 3)$

7. $-(x + 1)^2 - (y + 6)^2 = -16$ **8.** $-\dfrac{(x + 4)^2}{25} + \dfrac{(y - 1)^2}{144} = 1$

Suppose that the binomials within an equation of a conic section are expanded and all terms are collected on the left side of the equation. The result will be an equation of the conic section written in **general form:**

$$Ax^2 + Cy^2 + Dx + Ey + F = 0, \text{ where } A \text{ and } C \text{ are not both zero.}$$

Furthermore, given an equation of a conic section in general form, we can complete the square to identify the type of conic that the equation represents. In some cases, however, the graph of an equation in general form is a **degenerate conic section.** Geometrically, a degenerate case occurs when a plane intersects a double-napped cone in a point, a line, or a pair of intersecting lines (Figure 7-30).

Figure 7-30

> ### General Equation of a Conic Section
>
> The graph of the equation $Ax^2 + Cy^2 + Dx + Ey + F = 0$, where A and C are not both zero, is a conic section or a degenerate conic. If the graph represents a conic section, then the graph is
>
> 1. A circle if $A = C$.
> 2. An ellipse if $AC > 0$ (A and C have the same sign).
> 3. A hyperbola, if $AC < 0$ (A and C have opposite signs).
> 4. A parabola if $AC = 0$ (A or C equals zero).

TIP The general form of an equation of a conic section also has a term Bxy, which rotates the curve. Some courses in trigonometry address rotated conics.

For Exercises 9–16, determine whether the equation represents a circle, an ellipse, a hyperbola, or a parabola. Write the equation in standard form.

9. $9x^2 - 16y^2 - 36x - 64y - 172 = 0$

10. $9x^2 + 25y^2 - 100y - 125 = 0$

11. $y^2 - 10y + x + 27 = 0$

12. $x^2 + y^2 - 6x + 14y + 33 = 0$

13. $16x^2 + y^2 + 32x = 0$

14. $x^2 - 2x - 10y - 29 = 0$

15. $x^2 + y^2 + 2x + 12y + 21 = 0$

16. $-144x^2 + 25y^2 - 1152x - 50y - 5879 = 0$

For Exercises 17–22, an equation of a degenerate conic section is given. Complete the square and describe the graph of each equation.

17. $x^2 + y^2 - 4x + 2y + 5 = 0$

18. $9x^2 + 4y^2 - 24y + 36 = 0$

19. $4x^2 - y^2 - 32x - 4y + 60 = 0$

20. $9x^2 - 4y^2 - 90x - 8y + 221 = 0$

21. $9x^2 + 4y^2 - 16y + 52 = 0$

22. $x^2 + y^2 + 12x + 37 = 0$

23. Given $Ax^2 + Cy^2 + Dx + Ey + F = 0$, show that the equation represents a circle if $A = C \neq 0$ and $D^2 + E^2 - 4AF > 0$.

24. Consider $Ax^2 + Cy^2 + Dx + Ey + F = 0$, for $A \neq 0$ and $C \neq 0$.

   a. For A and C of the same sign, show that the equation represents an ellipse if $\dfrac{D^2}{4A} + \dfrac{E^2}{4C} - F$ has the same sign as A.

   b. For A and C of opposite signs, show that the equation represents a hyperbola if $\dfrac{D^2}{4A} + \dfrac{E^2}{4C} - F \neq 0$.

25. Given $Ax^2 + Cy^2 + Dx + Ey + F = 0$, show that the equation represents a parabola if $A = 0$ but $C \neq 0$ and $D \neq 0$.

CHAPTER 7   KEY CONCEPTS

SECTION 7.1 The Ellipse	Reference
An **ellipse** is the set of all points (x, y) in a plane such that the sum of the distances between (x, y) and two fixed points is a constant. The fixed points are called the **foci** of the ellipse.	p. 634

	Reference
Standard forms of an equation of an ellipse: • Major axis horizontal: 　Equation:　$\dfrac{(x-h)^2}{a^2} + \dfrac{(y-k)^2}{b^2} = 1$　　$a > b > 0; c^2 = a^2 - b^2 (c > 0)$ 　Center:　(h, k) 　Vertices:　$(h + a, k), (h - a, k)$ 　Foci:　$(h + c, k), (h - c, k)$ 　Minor axis endpoints:　$(h, k + b), (h, k - b)$ • Major axis vertical: 　Equation:　$\dfrac{(x-h)^2}{b^2} + \dfrac{(y-k)^2}{a^2} = 1$　　$a > b > 0; c^2 = a^2 - b^2 (c > 0)$ 　Center:　(h, k) 　Vertices:　$(h, k + a), (h, k - a)$ 　Foci:　$(h, k + c), (h, k - c)$ 　Minor axis endpoints:　$(h + b, k), (h - b, k)$	p. 639

$2a$ is the length of the major axis. $2b$ is the length of the minor axis.	p. 639

The **eccentricity** $e = \frac{c}{a}$ of an ellipse measures the degree of elongation. Note that for an ellipse, $0 < e < 1$.	p. 643

SECTION 7.2 The Hyperbola	Reference
A **hyperbola** is the set of all points (x, y) in a plane such that the difference in distances between (x, y) and two fixed points (foci) is a positive constant.	p. 651

	Reference
Standard forms of an equation of a hyperbola: • Transverse axis horizontal: 　Equation:　$\dfrac{(x-h)^2}{a^2} - \dfrac{(y-k)^2}{b^2} = 1$　　$a > 0, b > 0; c^2 = a^2 + b^2 (c > 0)$ 　Center:　(h, k) 　Vertices:　$(h + a, k), (h - a, k)$ 　Foci:　$(h + c, k), (h - c, k)$ 　Asymptotes:　$y - k = \pm\frac{b}{a}(x - h)$ • Transverse axis vertical: 　Equation:　$\dfrac{(y-k)^2}{a^2} - \dfrac{(x-h)^2}{b^2} = 1$　　$a > 0, b > 0; c^2 = a^2 + b^2 (c > 0)$ 　Center:　(h, k) 　Vertices:　$(h, k + a), (h, k - a)$ 　Foci:　$(h, k + c), (h, k - c)$ 　Asymptotes:　$y - k = \pm\frac{a}{b}(x - h)$	p. 655

$2a$ is the length of the transverse axis. $2b$ is the length of the conjugate axis.	p. 653

The **eccentricity** of a hyperbola is given by $e = \frac{c}{a}$. Note that for a hyperbola, $e > 1$.	p. 659

SECTION 7.3 The Parabola	Reference
A **parabola** is the set of all points in a plane that are equidistant from a fixed line (called the **directrix**) and a fixed point (called the **focus**).	p. 667

Standard forms of an equation of a parabola:	p. 671

- Vertical axis of symmetry:
  Equation: $(x - h)^2 = 4p(y - k)$
  Vertex: (h, k)
  Focus: $(h, k + p)$
  Directrix: $y = k - p$
  Axis of symmetry: $x = h$

- Horizontal axis of symmetry:
  Equation: $(y - k)^2 = 4p(x - h)$
  Vertex: (h, k)
  Focus: $(h + p, k)$
  Directrix: $x = h - p$
  Axis of symmetry: $y = k$

- If $p > 0$, the parabola opens upward as shown.
- If $p < 0$, the parabola opens downward.

- If $p > 0$, the parabola opens to the right as shown.
- If $p < 0$, the parabola opens to the left.

The value of $|p|$ is the focal length (distance between the vertex and focus).	p. 667
The **latus rectum** is the line segment perpendicular to the axis of symmetry, through the focus, with endpoints on the parabola.	p. 669
The value of $4|p|$ is the focal diameter (length of the latus rectum).	p. 669

Expanded Chapter Summary available at www.mhhe.com/millercollegealgebra.

CHAPTER 7   Review Exercises

SECTION 7.1

1. How are the graphs of these ellipses similar and how are they different?

$$\frac{x^2}{100} + \frac{y^2}{225} = 1 \qquad \frac{(x - 1)^2}{100} + \frac{(y + 7)^2}{225} = 1$$

For Exercises 2–5, an equation of an ellipse is given.

a. Identify the center.

b. Identify the vertices.

c. Identify the endpoints of the minor axis.

d. Identify the foci.

e. Determine the eccentricity.

f. Graph the ellipse.

2. $\dfrac{x^2}{289} + \dfrac{y^2}{64} = 1$

3. $15x^2 + 9y^2 = 135$

4. $\dfrac{(x + 1)^2}{9} + \dfrac{(y - 4)^2}{25} = 1$

5. $\dfrac{(x - 1)^2}{16} + \dfrac{(y - 2)^2}{9} = 1$

For Exercises 6–7,

a. Write the equation of the ellipse in standard form.

b. Identify the center, vertices, endpoints of the minor axis, and foci.

6. $5x^2 + 8y^2 + 40x - 16y + 48 = 0$

7. $100x^2 + 64y^2 - 100x - 1575 = 0$

For Exercises 8–10, write the standard form of an equation of the ellipse subject to the given conditions.

8. Vertices: $(4, 0)$ and $(-4, 0)$;
   Foci: $\left(\sqrt{11}, 0\right)$ and $\left(-\sqrt{11}, 0\right)$

9. Endpoints of minor axis: $(0, 1)$ and $(6, 1)$;
   Foci: $(3, 5)$ and $(3, -3)$

10. Minor axis parallel to x-axis; Center: $(2, -4)$;
    Length of major axis: 20 units;
    Length of minor axis: 12 units

11. Suppose that one ellipse has an eccentricity of $\frac{4}{7}$ and a second ellipse has an eccentricity of $\frac{5}{7}$. Which ellipse is more elongated?

12. Jupiter orbits the Sun in an elliptical path with the Sun at one focus. At perihelion, Jupiter is closest to the Sun at 7.4052×10^8 km. If the eccentricity of the orbit is 0.0489, determine the distance at aphelion (farthest point between Jupiter and the Sun). Round to the nearest million km.

13. A bridge over a gorge is supported by an arch in the shape of a semiellipse. The length of the bridge is 400 ft and the maximum height is 100 ft. Find the height of the arch 50 ft from the center. Round to the nearest foot.

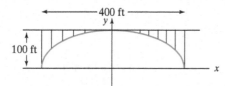

14. Elliptical concrete pipes have a greater capacity for shallow flow than circular pipes. For this reason, elliptical pipes are often used in storm drains, culverts, and sewers.

An engineer wants to design an elliptical concrete pipe with a maximum horizontal opening of 4 ft and a maximum vertical opening of 3 ft.

a. Write an equation for an elliptical cross section of the pipe. For convenience, place the coordinate system with (0, 0) at the center of the pipe.

b. To construct the pipe, the engineer needs to know the location of the foci. How far from the center are the foci?

SECTION 7.2

For Exercises 15–16, an equation of a hyperbola is given. Determine whether the transverse axis is horizontal or vertical.

15. $\dfrac{x^2}{11} - \dfrac{y^2}{16} = 1$ **16.** $-\dfrac{x^2}{11} + \dfrac{y^2}{16} = 1$

For Exercises 17–20, an equation of a hyperbola is given.

a. Identify the center.

b. Identify the vertices.

c. Identify the foci.

d. Write equations for the asymptotes.

e. Graph the hyperbola.

17. $\dfrac{x^2}{9} - \dfrac{y^2}{4} = 1$ **18.** $-3x^2 + 2y^2 = 18$

19. $-\dfrac{(x+3)^2}{16} + \dfrac{(y-2)^2}{9} = 1$

20. $\dfrac{(x-1)^2}{4} - (y+5)^2 = 1$

For Exercises 21–22,

a. Write the equation of the hyperbola in standard form.

b. Identify the center, vertices, and foci.

21. $11x^2 - 7y^2 + 44x - 56y - 145 = 0$

22. $-9x^2 + y^2 + 2y - 8 = 0$

For Exercises 23–25, write the standard form of an equation of a hyperbola subject to the given conditions.

23. Vertices: (4, 0), (−4, 0);
Foci: (5, 0), (−5, 0)

24. Vertices (0, 13), (0, −11);
Slope of asymptotes: $\pm\frac{12}{5}$

25. Corners of reference rectangle:
(19, 15), (19, −1), (−15, −1), (−15, 15);
Horizontal transverse axis

26. Suppose that one hyperbola has an eccentricity of $\frac{41}{40}$ and a second hyperbola has an eccentricity of $\frac{41}{9}$. For which hyperbola do the branches open "wider"?

27. Solve the system of equations.

$$6x^2 - 2y^2 = -12$$
$$2x^2 + y^2 = 11$$

28. Suppose that two people standing 2 mi (10,560 ft) apart at points A and B hear a car backfire. The time difference between the sound heard at point A and the sound heard at point B is 8 sec. If sound travels 1100 ft/sec, find an equation of the hyperbola (with foci at A and B) on which the point of origination of the sound must lie.

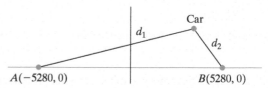

29. A flight control tower in the shape of a hyperboloid has hyperbolic cross sections perpendicular to the ground. Placing the origin at the bottom and center of the tower, the center of a hyperbolic cross section is (0, 30) with one focus at $(15\sqrt{10}, 30)$ and one vertex at (15, 30). All units are in feet.

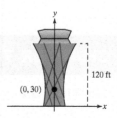

a. Write an equation of a hyperbolic cross section through the origin. Assume that there are no restrictions on x or y.

b. Determine the diameter of the tower at the base. Round to the nearest foot.

c. Determine the diameter of the tower at the top. Round to the nearest foot.

SECTION 7.3

For Exercises 30–33, an equation of a parabola is given in standard form.

a. Determine the value of p.

b. Identify the vertex.

c. Identify the focus.

d. Determine the focal diameter.

e. Determine the endpoints of the latus rectum.

f. Write an equation of the directrix.

g. Write an equation for the axis of symmetry.

h. Graph the parabola.

30. $x^2 = -2y$

31. $\dfrac{1}{10}y^2 = 2x$

32. $(y + 2)^2 = -4(x - 3)$

33. $\left(x - \tfrac{1}{2}\right)^2 = 6(y - 3)$

For Exercises 34–35, an equation of a parabola is given.

a. Write the equation of the parabola in standard form.

b. Identify the vertex, focus, and directrix.

34. $x^2 - 10x - 4y + 17 = 0$

35. $4y^2 - 12y + 56x + 177 = 0$

For Exercises 36–37, fill in the blank. Let $|p|$ represent the focal length (distance between the vertex and focus).

36. If the directrix of a parabola is given by $y = 3$ and the focus is $(4, 1)$, then the vertex is given by the ordered pair _____, and the value of p is _____.

37. If the vertex of a parabola is $(10, -4)$ and the focus is $(2, -4)$, then the directrix is given by the equation _____, and the value of p is _____.

For Exercises 38–40, determine the standard form of the parabola subject to the given conditions.

38. Vertex: $(3, 2)$; Directrix: $x = 5$

39. Vertex: $(-4, 7)$; Focus $(-4, -1)$

40. Focus: $(0, -3)$; Directrix: $y = -6$

41. Solve the system of equations.

$$(y + 2)^2 = 3(x - 5)$$
$$x - y = 7$$

42. The longest span of the Brooklyn Bridge is suspended from cables supported by two towers approximately 1596 ft apart. The tops of the towers are 142 ft above the road. Assuming that the cables are parabolic and are at road level halfway between the towers, find the height of the cables above the roadway 200 ft from either end. Round to the nearest foot.

43. Two large airship hangars were built in Orly, France, in the early 1900s but were destroyed in World War II by American aircraft. The hangars were 175 m in length, 90 m wide, and 60 m high and were formed by a series of parabolic arches.

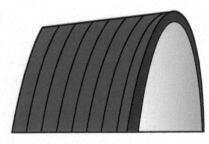

a. Set up a coordinate system with $(0, 0)$ at the vertex of one of the arches and write an equation of the parabola.

b. What is the focal length of an arch?

CHAPTER 7 Test

1. How are the graphs of these ellipses similar, and how are they different?

$$\frac{x^2}{36} + \frac{y^2}{144} = 1 \qquad \frac{x^2}{144} + \frac{y^2}{36} = 1$$

2. Write the standard form of an equation of a circle with center (h, k) and radius r.

3. Write the standard form of an equation of an ellipse with center (h, k) and major axis horizontal.

4. Write the standard form of an equation of an ellipse with center (h, k) and major axis vertical.

5. Write the standard form of an equation of a hyperbola with center (h, k) and transverse axis vertical.

6. Write the standard form of an equation of a hyperbola with center (h, k) and transverse axis horizontal.

7. Write the standard form of an equation of a parabola with vertex (h, k) and vertical axis of symmetry.

8. Write the standard form of an equation of a parabola with vertex (h, k) and horizontal axis of symmetry.

For Exercises 9–16,

a. Identify the equation as representing a circle, an ellipse, a hyperbola, or a parabola.

b. Graph the curve.

c. Identify key features of the graph. That is,

- If the equation represents a circle, identify the center and radius.
- If the equation represents an ellipse, identify the center, vertices, endpoints of the minor axis, foci, and eccentricity.
- If the equation represents a hyperbola, identify the center, vertices, foci, equations of the asymptotes, and eccentricity.
- If the equation represents a parabola, identify the vertex, focus, endpoints of the latus rectum, equation of the directrix, and equation of the axis of symmetry.

9. $\dfrac{(x - 2)^2}{9} + \dfrac{y^2}{16} = 1$

10. $\dfrac{(x - 2)^2}{9} - \dfrac{y^2}{16} = 1$

11. $\dfrac{(x - 2)^2}{16} + y = 0$

12. $x^2 + y^2 - 4x - 6y + 1 = 0$

13. $-9x^2 + 16y^2 + 64y - 512 = 0$

14. $9x^2 + 25y^2 + 72x - 50y - 731 = 0$

15. $\dfrac{(x + 4)^2}{4} + \dfrac{(y - 1)^2}{4} = 1$

16. $y^2 - 8y - 8x + 40 = 0$

17. The entrance to a tunnel is in the shape of a semiellipse over a 24-ft by 8-ft rectangular opening. The height at the center of the opening is 14 ft.

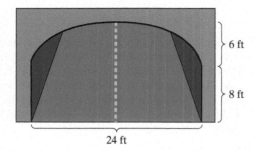

24 ft

a. Determine the height of the opening at a point 6 ft from the edge of the tunnel. Round to 1 decimal place.

b. Can a 10-ft-high truck pass through the opening if the truck's outer wheel passes at a point 3 ft from the edge of the tunnel?

c. Set up a coordinate system with the origin placed on the ground in the center of the roadway. Write a function that represents the height of the opening $h(x)$ (in ft) as a function of the horizontal distance x (in ft) from the center.

18. Neptune orbits the Sun in an elliptical path with the Sun at one focus. At aphelion, Neptune is farthest from the Sun at 4.546×10^9 km. If the eccentricity of the orbit is 0.0113, determine the distance at perihelion (closest point between Neptune and the Sun). Round to the nearest million km.

19. Suppose that two people standing 1.5 mi (7920 ft) apart at points A and B hear a car accident. The time difference between the sound heard at point A and the sound heard at point B is 6 sec. If sound travels 1100 ft/sec, find an equation of the hyperbola (with foci at A and B) at which the accident occurred.

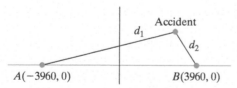

20. A building in the shape of a hyperboloid has hyperbolic cross sections perpendicular to the ground. Placing the origin at the bottom and center of the building, the center of a hyperbolic cross section is $(0, 120)$ with one vertex at $(12, 120)$ and one focus at $\left(12\sqrt{10}, 120\right)$. All units are in feet.

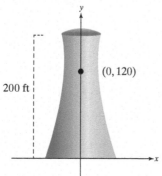

a. Write an equation of a hyperbolic cross section through the origin. Assume that there are no restrictions on x or y.

b. Determine the diameter of the building at the base. Round to the nearest tenth of a foot.

c. Determine the diameter of the building at the top. Round to the nearest tenth of a foot.

21. The middle span of the Golden Gate Bridge in San Francisco is supported by cables from two towers 4200 ft apart. The tops of the towers are 500 ft above the roadway. Assuming that the cables are parabolic and are at road level halfway between the towers, find the height of the cables 1000 ft from the center of the bridge. Round to the nearest foot.

22. Write the standard form of an equation of a parabola with focus $(1, 6)$ and directrix $y = -2$.

23. Write the standard form of an equation of an ellipse with vertices $(2, -3)$ and $(6, -3)$ and foci $(3, -3)$ and $(5, -3)$.

24. Write the standard form of an equation of a hyperbola with vertices $(4, -1)$ and $(4, 7)$ and foci $(4, -3)$ and $(4, 9)$.

25. Write the standard form of an equation of a hyperbola with vertices $(-1, -3)$ and $(5, -3)$ and slope of one asymptote $\frac{4}{3}$.

26. If the major axis of an ellipse is 26 units in length and the minor axis is 10 units in length, determine the eccentricity of the ellipse.

27. If the transverse axis of a hyperbola is 20 units and the conjugate axis is 12 units, determine the eccentricity of the hyperbola.

28. If the directrix of a parabola is $x = 2$, and the focus is $(-4, 0)$, determine the focal length and the focal diameter.

29. If the asymptotes of a hyperbola are $y = \frac{3}{5}x + 1$ and $y = -\frac{3}{5}x + 1$, identify the center of the hyperbola.

30. Solve the system of equations.
$$3x^2 - 4y^2 = -13$$
$$5x^2 + 2y^2 = 13$$

31. Describe the graph of the equation.
$$x = -4\sqrt{1 - \frac{y^2}{9}}$$

CHAPTER 7   Cumulative Review Exercises

1. Use long division to divide.
$(3x^4 - 2x^3 - 5x + 1) \div (x^2 + 3)$

2. Is -4 a zero of the polynomial
$f(x) = 3x^4 - 6x^3 + 5x - 12$?

3. Is $(x - 3)$ a factor of $2x^4 - 7x^3 + 8x^2 - 17x + 6$?

For Exercises 4–5, simplify completely.

4. $\dfrac{\sqrt[3]{54x^4}}{\sqrt[3]{2x^5}}$

5. $\left(\dfrac{-2x^2y^{-1}}{z^3}\right)^{-4}\left(\dfrac{4x^{-5}}{y^7}\right)^2$

6. Find the midpoint of the line segment with endpoints $(-362, 147)$ and $(118, 24)$.

7. Given $f(x) = \sqrt[3]{x} - 7$, write an equation defining $f^{-1}(x)$.

8. Write an equation of the line through the points $(1, 8)$ and $(2, -4)$. Write the answer in slope-intercept form.

9. Is function f even, odd, or neither?
$$f(x) = 3|x| + 5x^3$$

For Exercises 10–15, solve the equation or inequality. Write the solution set to the inequalities in interval notation if possible.

10. $e^{3x+2} = 22$

11. $2 - 3\{4 - [x + 2(x - 1) - 3] + 4\} > 0$

12. $21 \le 9 - 2|x + 1|$ 13. $2x^2 - 8x + 3 \ge 0$

14. $\log_3 x + \log_3(x - 6) = 3$ 15. $e^{2x} - 8e^x + 15 = 0$

16. Graph the solution set.
$$x^2 + (y - 1)^2 \le 9$$

17. Graph the solution set. $\begin{array}{l} 2x - y < 4 \\ 3x + y \le 1 \end{array}$

18. Find the partial fraction decomposition for
$$\dfrac{8x - 27}{x^2 - 7x + 12}$$

19. Given $f(x) = \dfrac{1}{x - 2}$ and $g(x) = \sqrt{x - 1}$,
   a. Find $(f \circ g)(x)$.
   b. Write the domain of $(f \circ g)(x)$ in interval notation.

20. Given $f(x) = \dfrac{x^2 - 4x}{x + 3}$,
   a. Determine the x-intercept(s) of the graph of f.
   b. Determine the y-intercept.
   c. Find the vertical asymptote(s).
   d. Find the horizontal asymptote.
   e. Find the slant asymptote.

For Exercises 21–24, graph the equations and functions.

21. $\dfrac{x^2}{4} + \dfrac{(y - 1)^2}{9} = 1$ 22. $\dfrac{x^2}{4} - \dfrac{(y - 1)^2}{9} = 1$

23. $x^2 = 4(y - 2)$ 24. $f(x) = \left(\dfrac{1}{2}\right)^x + 2$

For Exercises 25–26, solve the system.

25. $\begin{array}{l} 3x - 4y + 2z = 5 \\ 5y - 3z = -12 \\ 7x \qquad + 2z = 1 \end{array}$ 26. $\begin{array}{l} x^2 - 5y^2 = 4 \\ 4x^2 + y^2 = 37 \end{array}$

For Exercises 27–29, refer to the matrices A and B. Simplify each expression.

$$A = \begin{bmatrix} 2 & 4 \\ -1 & 6 \end{bmatrix} \quad B = \begin{bmatrix} -3 & 8 \\ 0 & 4 \end{bmatrix}$$

27. $2AB$ 28. A^{-1}

29. $|B|$

30. Use Cramer's rule to solve for z.
$$\begin{array}{l} 3x - 4y + z = 3 \\ 2x + 3y \qquad = 8 \\ 5y + 8z = 11 \end{array}$$

Sequences, Series, Induction, and Probability

Have you ever wondered why one individual pays more than another for car insurance? Insurance rates are based on a number of variables including age, gender, Zip code, past driving and claims history, and even grades in school. Insurance companies hire actuaries, special statisticians who analyze data to estimate the probability and cost of accidents for people in various demographic groups. In turn, premiums are established to offset the theoretical cost to cover a customer based on the customer's risk factors.

In this chapter, we study sequences and series, the binomial theorem, and a technique of proof called mathematical induction. We complete the chapter by presenting several methods of counting that will ultimately enable us to calculate simple probabilities. The study of probability is the first step toward unraveling the mystery of car insurance premiums.

Sequences and Series

1. Write Terms of a Sequence from the *n*th Term

In day-to-day life, we think of a **sequence** as a progression of elements with some order or pattern. For example, suppose that a new college graduate has to choose between two job offers. Job 1 offers $75,000 the first year with a $4000 raise each year thereafter. Job 2 offers $75,000 with a 5% raise each year. The sequences in Table 8-1 represent the salaries for each job as a function of the year number.

Table 8-1

	Year 1	Year 2	Year 3	Year 4	Year 5	Year 6
Job 1	$75,000	$79,000	$83,000	$87,000	$91,000	$95,000
Job 2	$75,000	$78,750	$82,687.50	$86,821.86	$91,162.97	$95,721.12

By studying the two sequences, we see that by year 5, job 2 has a better salary. If we take the *sum* of the first *n* terms of a sequence, we have a **finite series.** The total salary through the first 5 yr favors job 1 ($415,000 for job 1 versus $414,422.33 for job 2). However, by the end of year 6, job 2 not only pays more each year, but has a greater total salary for the 6-yr period ($510,143.45 for job 2 versus $510,000 for job 1).

From Table 8-1, we can think of a sequence as a function pairing the year number (a positive integer) with the salary. For this reason, it makes sense to define a sequence mathematically as a function whose domain is the set of positive integers.

> ### Finite and Infinite Sequences
>
> An **infinite sequence** is a function whose domain is the set of positive integers. A **finite sequence** is a function whose domain is the set of the first *n* positive integers.

Although a sequence is defined as a function, we typically do not use function notation to denote the terms of a sequence. Instead, the terms of the sequence are denoted by the letter *a* with a subscript representing the term number. The sequence of salaries for job 1 would be denoted by

term 1
$a_1 = 75,000$

term 2
$a_2 = 79,000$

term 3
$a_3 = 83,000$ and so on.

The notation $a_1 = 75,000$ is used in place of function notation $f(1) = 75,000$. Likewise $a_2 = 79,000$ is used instead of $f(2) = 79,000$ and so on. For a positive integer *n*, the value a_n is called the ***n*th term** or **general term** of the sequence. The notation $\{a_n\}$ represents the entire sequence.

$$\{a_n\} = a_1, a_2, a_3, \ldots, a_n, \ldots$$

EXAMPLE 1 **Writing Several Terms of a Sequence**

Write the first four terms of the sequence defined by the nth term.

a. $a_n = 3n^2 - 4$ **b.** $c_n = 9\left(-\frac{1}{3}\right)^n$

Solution:

a. $a_n = 3n^2 - 4$ To find the first four terms, substitute $n = 1, 2, 3,$ and 4.

$a_1 = 3(1)^2 - 4 = -1$

$a_2 = 3(2)^2 - 4 = 8$

$a_3 = 3(3)^2 - 4 = 23$

$a_4 = 3(4)^2 - 4 = 44$ The first four terms are $-1, 8, 23, 44$.

b. $c_n = 9\left(-\frac{1}{3}\right)^n$ To find the first four terms, substitute $n = 1, 2, 3,$ and 4.

$c_1 = 9\left(-\frac{1}{3}\right)^1 = -3$

$c_2 = 9\left(-\frac{1}{3}\right)^2 = 1$

$c_3 = 9\left(-\frac{1}{3}\right)^3 = -\frac{1}{3}$

$c_4 = 9\left(-\frac{1}{3}\right)^4 = \frac{1}{9}$ The first four terms are $-3, 1, -\frac{1}{3}, \frac{1}{9}$.

Skill Practice 1 Write the first four terms of the sequence defined by the nth term.

a. $b_n = 2n + 3$ **b.** $d_n = (-1)^n \cdot n^2$

TECHNOLOGY CONNECTIONS

Displaying Terms of a Sequence

If the nth term of a sequence is known, a seq function on a graphing utility can display a list of terms. Finding the first four terms of the sequence from Example 1(b) is outlined here.

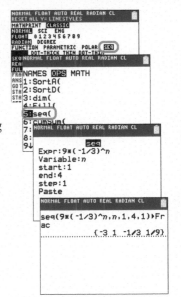

- Select the **MODE** key and highlight SEQ.
- Then access the seq function from the catalog or from the LIST menu (OPS submenu) and enter the parameters as prompted.
- Paste the seq command on the home screen and select **ENTER**.

seq(9*(–1/3)^n, n, 1, 4, 1) > Frac

Notice that the terms of the sequence given in Example 1(b) alternate in signs. Such a sequence is called an **alternating sequence.** The alternation in signs can be represented in the nth term by a factor of -1 raised to a variable power that alternates between an even and odd integer. For example:

$a_n = (-1)^n$ odd-numbered terms, negative; even-numbered terms, positive

$b_n = (-1)^{n+1}$ odd-numbered terms, positive; even-numbered terms, negative

Answers

1. a. $5, 7, 9, 11$ **b.** $-1, 4, -9, 16$

Because a sequence is defined only for positive integers, its graph is a discrete set of points (the points are not connected). Compare the graphs of the sequence defined by $a_n = 8\left(\frac{1}{2}\right)^n$ (see Figure 8-1) versus the function defined by $f(x) = 8\left(\frac{1}{2}\right)^x$ (see Figure 8-2).

> **TIP** In some cases, we may define a sequence with a domain beginning at zero or some other whole number. For example, $a_n = \dfrac{4}{n-1}$ is not defined for $n = 1$. Therefore, we might restrict the domain to the set of integers greater than or equal to 2.

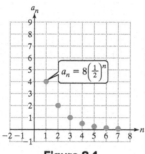

Figure 8-1

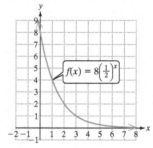

Figure 8-2

The graph of $\{a_n\}$ consists of a discrete set of points that correspond to points on the graph of $f(x) = 8\left(\frac{1}{2}\right)^x$ for positive integers x.

2. Write Terms of a Sequence Defined Recursively

The sequences in Example 1 were defined as a function of the nth term. Now we look at sequences defined *recursively*, using a recursive formula. A **recursive formula** defines the nth term of a sequence as a function of one or more terms preceding it. This is demonstrated in Example 2.

EXAMPLE 2   **Writing Terms of a Sequence Defined Recursively**

Write the first five terms of the sequence defined by $a_1 = 4$ and $a_n = 2a_{n-1} + 1$ for $n > 1$.

Solution:

The first term is given: $a_1 = 4$.

Every term thereafter is defined by $a_n = 2a_{n-1} + 1$

> Value of the nth term is

> two times the value of the preceding term, plus 1.

$a_1 = 4$

$a_2 = 2a_1 + 1 = 2(4) + 1 = 9$ Substitute $a_1 = 4$.

$a_3 = 2a_2 + 1 = 2(9) + 1 = 19$ Substitute $a_2 = 9$.

$a_4 = 2a_3 + 1 = 2(19) + 1 = 39$ Substitute $a_3 = 19$.

$a_5 = 2a_4 + 1 = 2(39) + 1 = 79$ The first five terms are 4, 9, 19, 39, 79.

Skill Practice 2   Write the first five terms of the sequence defined by:

$c_1 = 5,\ c_n = 3c_{n-1} - 4$

Perhaps the most famous sequence is the Fibonacci sequence, named after the twelfth-century Italian mathematician Leonardo of Pisa (known as Fibonacci). The Fibonacci sequence is defined recursively as

$$a_1 = 1, \quad a_2 = 1, \quad a_n = a_{n-2} + a_{n-1}$$

Answer

2. 5, 11, 29, 83, 245

The first two terms of the Fibonacci sequence are 1, and each term thereafter is the sum of its two predecessors. We have:

Fibonacci sequence: 1, 1, 2, 3, 5, 8, 13, 21, 34, 55, 89, …

Point of Interest

The Fibonacci sequence is of particular interest because it is often observed in nature. For example, male honey bees hatch from eggs that have not been fertilized, so each male honey bee has only 1 parent, a female. A female honey bee hatches from a fertilized egg, so she has two parents: one male and one female. Notice that the number of ancestors for n generations for a male honey bee follows the Fibonacci sequence (Figure 8-3).

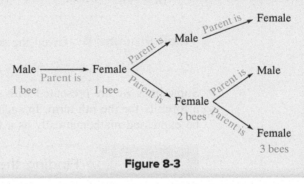

Figure 8-3

3. Use Factorial Notation

We now introduce factorial notation to denote the product of the first n positive integers.

Definition of $n!$

- Let n be a positive integer. The expression $n!$ (read as "n factorial") is defined as $n! = (n)(n - 1)(n - 2) \cdots (2)(1)$
- Zero factorial is defined as $0! = 1$.

Examples:
$0! = 1$ $\qquad\qquad$ $4! = 4 \cdot 3 \cdot 2 \cdot 1 = 24$
$1! = 1$ $\qquad\qquad$ $5! = 5 \cdot 4 \cdot 3 \cdot 2 \cdot 1 = 120$
$2! = 2 \cdot 1 = 2$ $\qquad$ $6! = 6 \cdot 5 \cdot 4 \cdot 3 \cdot 2 \cdot 1 = 720$
$3! = 3 \cdot 2 \cdot 1 = 6$ $\qquad$ $7! = 7 \cdot 6 \cdot 5 \cdot 4 \cdot 3 \cdot 2 \cdot 1 = 5040$

EXAMPLE 3   **Evaluating Expressions with Factorial Notation**

Evaluate.

a. $\dfrac{8!}{3! \cdot 5!}$ $\qquad$ **b.** $\dfrac{n!}{(n + 2)!}$

Solution:

a. $\dfrac{8!}{3! \cdot 5!} = \dfrac{8 \cdot 7 \cdot \overset{1}{\cancel{6}} \cdot \overset{1}{\cancel{5!}}}{(\cancel{3} \cdot \cancel{2} \cdot 1)(5!)} = 56$

b. $\dfrac{n!}{(n + 2)!} = \dfrac{\overset{1}{\cancel{n!}}}{(n + 2)(n + 1)(n!)} = \dfrac{1}{(n + 2)(n + 1)}$

Skill Practice 3   Evaluate.

a. $\dfrac{9!}{2! \cdot 7!}$ $\qquad$ **b.** $\dfrac{(n - 1)!}{n!}$

Answers

3. a. 36 $\quad$ **b.** $\dfrac{1}{n}$

In Example 4, we find a specific term of a sequence in which the expression for the nth term contains factorial notation.

EXAMPLE 4   **Finding a Specific Term of a Sequence**

Given the sequence defined by $b_n = \dfrac{n^2}{(n+1)!}$, find b_6.

Solution:

$$b_n = \frac{n^2}{(n+1)!} \xrightarrow[\text{6 for } n.]{\text{Substitute}} b_6 = \frac{(6)^2}{(6+1)!} = \frac{36}{7!} = \frac{36}{7 \cdot 6 \cdot 5 \cdot 4 \cdot 3 \cdot 2 \cdot 1} = \frac{1}{140}$$

Skill Practice 4   Given the sequence defined by $c_n = \dfrac{2^n}{(n+1)!}$, find c_4.

Sometimes we are presented with several terms of a sequence and are asked to find a formula for the nth term. In such a case, the goal is to look for a pattern that can be expressed mathematically as a function of the term number.

EXAMPLE 5   **Finding the nth Term of a Sequence**

Find the nth term a_n of a sequence whose first four terms are given.

a. $\dfrac{2}{3}, \dfrac{3}{4}, \dfrac{4}{5}, \dfrac{5}{6}, \cdots$ **b.** $\dfrac{1}{5}, -\dfrac{4}{25}, \dfrac{9}{125}, -\dfrac{16}{625}, \cdots$

Solution:

a. Term number: $n = 1\ 2\ 3\ 4$

$\dfrac{2}{3}, \dfrac{3}{4}, \dfrac{4}{5}, \dfrac{5}{6}, \cdots$

$a_n = \dfrac{n+1}{n+2}$

The numerator is 1 more than the term number: $n + 1$
The denominator is 2 more than the term number: $n + 2$

b. Term number: $n = 1\quad 2\quad 3\quad 4$

$\dfrac{1}{5}, -\dfrac{4}{25}, \dfrac{9}{125}, -\dfrac{16}{625}, \cdots$

$a_n = (-1)^{n+1} \cdot \dfrac{n^2}{5^n}$

Terms in the numerator are perfect squares: $1^2, 2^2, 3^2, 4^2, \ldots, n^2$
Terms in the denominator are powers of 5: $5^1, 5^2, 5^3, 5^4, \ldots, 5^n$

The signs are positive for odd-numbered terms. Therefore, we want a factor of -1 raised to an even exponent on odd-numbered terms and an odd exponent for even-numbered terms: $(-1)^{n+1}$

Skill Practice 5   Find the nth term a_n of a sequence whose first four terms are given.

a. $-5, 6, -7, 8, \ldots$ **b.** $\dfrac{1}{3}, \dfrac{1 \cdot 2}{9}, \dfrac{1 \cdot 2 \cdot 3}{27}, \dfrac{1 \cdot 2 \cdot 3 \cdot 4}{81}, \cdots$

Answers

4. $\dfrac{2}{15}$

5. a. $a_n = (-1)^n(n+4)$

 b. $a_n = \dfrac{n!}{3^n}$

4. Use Summation Notation

Consider an infinite sequence $\{a_n\} = a_1, a_2, a_3, \ldots$. The sum S of such a sequence is called an **infinite series** and is represented by $S = a_1 + a_2 + a_3 + \cdots$. The sum of the first n terms of the sequence is called the **nth partial sum** of the sequence and is denoted by S_n. The nth partial sum of a sequence is a **finite series**.

In mathematics, the Greek letter Σ (sigma) is used to represent summations.

Summation Notation

Given a sequence $a_1, a_2, a_3, \ldots$, the *n*th partial sum S_n is a finite series and is represented by

> *n* is the upper limit of summation.

$$S_n = \sum_{i=1}^{n} a_i = a_1 + a_2 + a_3 + \cdots + a_n$$

> *i* is the index of summation.

> 1 is the lower limit of summation.

The sum of *all* terms in the sequence is an infinite series and is given by

$$\sum_{i=1}^{\infty} a_i = a_1 + a_2 + a_3 + \cdots$$

TIP The summation $\sum_{i=1}^{n} a_i$ is read as the sum of a_i for *i* equals 1 to *n*.

Any letter such as i, j, k, and n may be used for the index of summation. However, if i is used for the index of summation, do not confuse it with the imaginary number $i = \sqrt{-1}$.

EXAMPLE 6   Using Summation Notation

Write the terms for each series and evaluate the sum.

a. $\displaystyle\sum_{i=1}^{5} (2i + 3)$ **b.** $\displaystyle\sum_{k=3}^{6} (-1)^k \left(\frac{1}{k}\right)$ **c.** $\displaystyle\sum_{n=1}^{6} 3$

Solution:

$$\qquad\qquad\quad i = 1 \qquad\quad i = 2 \qquad\quad i = 3 \qquad\quad i = 4 \qquad\quad i = 5$$

a. $\displaystyle\sum_{i=1}^{5} (2i + 3) = [2(1) + 3] + [2(2) + 3] + [2(3) + 3] + [2(4) + 3] + [2(5) + 3]$

$$= 5 + 7 + 9 + 11 + 13$$

$$= 45$$

b. $\displaystyle\sum_{k=3}^{6} (-1)^k \left(\frac{1}{k}\right) = \left[(-1)^3\left(\frac{1}{3}\right)\right] + \left[(-1)^4\left(\frac{1}{4}\right)\right] + \left[(-1)^5\left(\frac{1}{5}\right)\right] + \left[(-1)^6\left(\frac{1}{6}\right)\right]$

$$= -\frac{1}{3} + \frac{1}{4} - \frac{1}{5} + \frac{1}{6}$$

$$= -\frac{7}{60}$$

c. $\displaystyle\sum_{n=1}^{6} 3 = \underbrace{3 + 3 + 3 + 3 + 3 + 3}_{6 \text{ terms}} = 18$

Skill Practice 6   Evaluate the sum.

a. $\displaystyle\sum_{i=1}^{5} (3i + 1)$ **b.** $\displaystyle\sum_{n=2}^{5} (-1)^{n+1} \frac{1}{2^n}$ **c.** $\displaystyle\sum_{k=1}^{5} 4$

Answers

6. a. 50   **b.** $-\dfrac{5}{32}$ **c.** 20

TECHNOLOGY CONNECTIONS

Evaluating a Finite Series

A graphing utility can be used to evaluate a finite series if the nth term of the corresponding sequence is known. The Σ symbol can be accessed by selecting (ALPHA) F2.

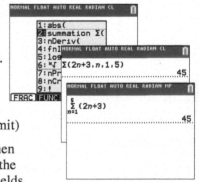

- If the calculator is in "Classic" mode, the sum can be evaluated by entering

  Σ (nth term, variable, lower limit, upper limit)

- If the calculator is in "Mathprint" mode, then the user is prompted to enter the nth term, the variable, and limits of summation in text fields.

It is important to note that the lower limit of summation need not be 1. In Example 6(b) for instance, the index of summation ranges from 3 to 6. In fact, sometimes we adjust the limits of summation and the expression being summed to write the summation in a more convenient form. For example, the following three series are all equivalent.

$$\sum_{i=1}^{5} 3i = 3(1) + 3(2) + 3(3) + 3(4) + 3(5)$$
$$= 3 + 6 + 9 + 12 + 15 = 45$$

$$\sum_{j=0}^{4} 3(j + 1) = 3(0 + 1) + 3(1 + 1) + 3(2 + 1) + 3(3 + 1) + 3(4 + 1)$$
$$= 3 + 6 + 9 + 12 + 15 = 45$$

$$\sum_{k=2}^{6} 3(k - 1) = 3(2 - 1) + 3(3 - 1) + 3(4 - 1) + 3(5 - 1) + 3(6 - 1)$$
$$= 3 + 6 + 9 + 12 + 15 = 45$$

EXAMPLE 7 Write a Series Using Summation Notation

Write each sum using summation notation.

a. $\dfrac{\sqrt{3}}{1!} + \dfrac{\sqrt{4}}{2!} + \dfrac{\sqrt{5}}{3!} + \cdots + \dfrac{\sqrt{n + 2}}{n!}$ **b.** $\dfrac{2}{1} - \dfrac{3}{4} + \dfrac{4}{9} - \dfrac{5}{16} + \dfrac{6}{25}$

Solution:

a. $\dfrac{\sqrt{3}}{1!} + \dfrac{\sqrt{4}}{2!} + \dfrac{\sqrt{5}}{3!} + \cdots + \dfrac{\sqrt{n + 2}}{n!}$ The series consists of n terms where a formula for the nth term is given.

$$= \sum_{i=1}^{n} \dfrac{\sqrt{i + 2}}{i!}$$

We have used n as the upper limit of summation. Therefore, use a different variable for the index of summation. We have chosen i.

b.

Look for a relationship between the term number and the values in the numerator and denominator. Choose a variable such as i for the index of summation.

Term: 1   2   3   4   5

$$\dfrac{2}{1} - \dfrac{3}{4} + \dfrac{4}{9} - \dfrac{5}{16} + \dfrac{6}{25} \longleftarrow$$ The numerator is 1 more than the term number: $(i + 1)$
$\longleftarrow$ The denominator is the square of the term number: i^2

$$\sum_{i=1}^{5} (-1)^{i+1} \cdot \dfrac{i + 1}{i^2}$$

The odd-numbered terms are positive. This can be generated by the factor $(-1)^{i+1}$.

> **Skill Practice 7** Write each sum using summation notation.
>
> **a.** $\dfrac{2!}{7} + \dfrac{3!}{14} + \dfrac{4!}{21} + \cdots + \dfrac{(n+1)!}{7n}$ **b.** $-\dfrac{1}{5} + \dfrac{8}{25} - \dfrac{27}{125} + \dfrac{64}{625}$

Several important properties of summation are given in Table 8-2.

> **Table 8-2   Properties of Summation**
>
> If $\{a_n\}$ and $\{b_n\}$ are sequences, and c is a real number, then:
>
> 1. $\displaystyle\sum_{i=1}^{n} c = cn$ Adding a constant c a total of n times equals cn.
>
> 2. $\displaystyle\sum_{i=1}^{n} ca_i = c\sum_{i=1}^{n} a_i$ A constant factor can be factored out of a summation.
>
> 3. $\displaystyle\sum_{i=1}^{n} (a_i \pm b_i) = \sum_{i=1}^{n} a_i \pm \sum_{i=1}^{n} b_i$ A single sum or difference can be regrouped as two sums or differences.

The proof of Property 2 follows from the distributive property of real numbers.

$$\sum_{i=1}^{n} ca_i = ca_1 + ca_2 + ca_3 + \cdots + ca_n \qquad \text{Expand the terms in the series.}$$

$$= c(a_1 + a_2 + a_3 + \cdots + a_n) \qquad \text{Apply the distributive property.}$$

$$= c\sum_{i=1}^{n} a_i \qquad \text{Write the sum using summation notation.}$$

The proofs of properties 1 and 3 are examined in Exercises 89 and 90.

Answers

7. a. $\displaystyle\sum_{i=1}^{n} \dfrac{(i+1)!}{7i}$ **b.** $\displaystyle\sum_{i=1}^{4} (-1)^i \dfrac{i^3}{5^i}$

SECTION 8.1   Practice Exercises

Prerequisite Review

For Exercises R.1–R.3, evaluate the function for the given value.

R.1. $g(x) = \dfrac{5}{x}$; $g(15)$ **R.2.** $f(x) = 5$; $f(4)$ **R.3.** $h(x) = x^2 + 3x$; $h(3)$

For Exercises R.4–R.5, simplify the rational expression.

R.4. $\dfrac{(r-2)^2(2r-1)^4}{(r-2)^4(2r-1)}$ **R.5.** $\dfrac{p^2(p-6)^5}{p^5(p-6)^2}$

Concept Connections

1. An infinite _____ is a function whose domain is the set of positive integers. A _____ sequence is a function whose domain is the set of the first n positive integers.

2. An _____ sequence is a sequence in which consecutive terms alternate in sign.

3. A _____ formula defines the nth term of a sequence as a function of one or more terms preceding it.

4. For $n \geq 1$, the expression _____ represents the product of the first n positive integers $n(n-1)(n-2) \cdots (2)(1)$. Furthermore, by definition, $0! = $ _____.

5. Given $\sum\limits_{i=1}^{n} a_i$, the variable i is called the _____ of _____. The value 1 is called the _____ limit of summation. The value n is called the upper _____ of summation.

6. One property of summation indicates that $\sum\limits_{i=1}^{n} c =$ _____.

Objective 1: Write Terms of a Sequence from the *n*th Term

For Exercises 7–14, the *n*th term of a sequence is given. Write the first four terms of the sequence. (See Example 1)

7. $a_n = 2n^2 + 3$

8. $b_n = -n^3 + 5$

9. $c_n = 12\left(-\frac{1}{2}\right)^n$

10. $d_n = 64\left(-\frac{1}{4}\right)^n$

11. $a_n = \sqrt{n+3}$

12. $b_n = \sqrt{n-1}$

13. $a_n = e^{3\ln n}$

14. $c_n = \ln e^{2n}$

15. If the *n*th term of a sequence is $(-1)^n n^2$, which terms are positive and which are negative?

16. If the *n*th term of a sequence is $(-1)^{n+1}\dfrac{1}{n}$, which terms are positive and which are negative?

For Exercises 17–20, the *n*th term of a sequence is given. Find the indicated term.

17. $a_n = 2^n - 1$; find a_{10}

18. $b_n = \dfrac{5}{n} - 1$; find b_{20}

19. $c_n = 5n - 4$; find c_{157}

20. $d_n = 6n + 7$; find d_{204}

For Exercises 21–24, match the sequence or function with its graph.

21. $a_n = \sqrt{n}$

22. $f(x) = \sqrt{x}$

23. $f(x) = \dfrac{3}{x}$

24. $a_n = \dfrac{3}{n}$

a.
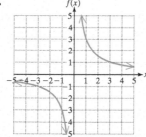

b.
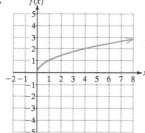

c.
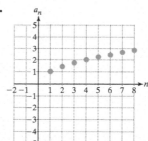

d.
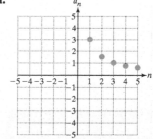

Objective 2: Write Terms of a Sequence Defined Recursively

For Exercises 25–30, write the first five terms of the sequence defined recursively. (See Example 2)

25. $b_1 = -3$; $b_n = 3b_{n-1} + 4$

26. $a_1 = 11$; $a_n = 4a_{n-1} + 3$

27. $c_1 = 4$; $c_n = \frac{1}{2}c_{n-1} + 4$

28. $d_1 = 30$; $d_n = \frac{1}{3}d_{n-1} - 1$

29. $a_1 = 5$; $a_n = \dfrac{1}{a_{n-1}}$

30. $b_1 = 10$; $b_n = -\dfrac{1}{b_{n-1}}$

31. a. Write the first 10 terms of the Fibonacci sequence.

b. The formula $F_n = \dfrac{1}{\sqrt{5}}\left(\dfrac{1+\sqrt{5}}{2}\right)^n - \dfrac{1}{\sqrt{5}}\left(\dfrac{1-\sqrt{5}}{2}\right)^n$ gives the *n*th term of the Fibonacci sequence. Use a calculator to verify this statement for $n = 1$, $n = 2$, and $n = 3$.

32. The numbers in the sequence defined by $a_1 = 1$, $a_2 = 3$, and $a_n = a_{n-1} + a_{n-2}$ for $n \geq 3$ are referred to as Lucas numbers in honor of French mathematician Edouard Lucas (1842–1891).

a. Find the first eight Lucas numbers.

b. The formula $L_n = \left(\dfrac{1+\sqrt{5}}{2}\right)^n + \left(\dfrac{1-\sqrt{5}}{2}\right)^n$ gives the *n*th Lucas number. Use a calculator to verify this statement for $n = 1$, $n = 2$, and $n = 3$.

Objective 3: Use Factorial Notation

For Exercises 33–44, evaluate the expression. (See Example 3)

33. $7!$

34. $8!$

35. $0!$

36. $1!$

37. $\dfrac{8!}{6!}$

38. $\dfrac{12!}{9!}$

39. $\dfrac{9!}{5! \cdot 4!}$

40. $\dfrac{10!}{6! \cdot 4!}$

41. $\dfrac{(n-1)!}{(n+1)!}$

42. $\dfrac{(n-2)!}{n!}$

43. $\dfrac{(2n)!}{(2n+1)!}$

44. $\dfrac{(2n-1)!}{(2n)!}$

For Exercises 45–48, the nth term of a sequence is given. Find the indicated term. (See Example 4)

45. $a_n = \dfrac{2^n}{(n+2)!}$; find a_5

46. $b_n = \dfrac{4n}{(n+1)!}$; find b_3

47. $c_n = \dfrac{(3n)!}{5n}$; find c_3

48. $d_n = \dfrac{(n+3)!}{3^n}$; find d_3

For Exercises 49–56, find the nth term a_n of a sequence whose first four terms are given. (See Example 5)

49. $2, 4, 8, 16, \ldots$

50. $3, 9, 27, 81, \ldots$

51. $-\dfrac{2}{3}, -\dfrac{3}{6}, -\dfrac{4}{9}, -\dfrac{5}{12}, \ldots$

52. $-\dfrac{8}{11}, -\dfrac{9}{22}, -\dfrac{10}{33}, -\dfrac{11}{44}, \ldots$

53. $-1, 4, -9, 16, \ldots$

54. $1, -8, 27, -64, \ldots$

55. $\dfrac{1\cdot2}{2}, \dfrac{1\cdot2\cdot3}{4}, \dfrac{1\cdot2\cdot3\cdot4}{6}, \dfrac{1\cdot2\cdot3\cdot4\cdot5}{8}, \ldots$

56. $\dfrac{5}{4}, \dfrac{10}{9}, \dfrac{15}{16}, \dfrac{20}{25}, \ldots$

Objective 4: Use Summation Notation

For Exercises 57–70, find the sum. (See Example 6)

57. $\displaystyle\sum_{i=1}^{6} (3i - 4)$

58. $\displaystyle\sum_{i=1}^{5} (2i + 7)$

59. $\displaystyle\sum_{j=1}^{5} (-2j^2)$

60. $\displaystyle\sum_{k=1}^{3} (-3k^3)$

61. $\displaystyle\sum_{j=2}^{5} \left(\dfrac{1}{2}\right)^j$

62. $\displaystyle\sum_{n=2}^{4} \left(\dfrac{1}{3}\right)^n$

63. $\displaystyle\sum_{j=1}^{20} 6$

64. $\displaystyle\sum_{i=1}^{50} 4$

65. $\displaystyle\sum_{k=3}^{7} (-1)^k (6k)$

66. $\displaystyle\sum_{n=3}^{8} (-1)^{n-1} (2n)$

67. $\displaystyle\sum_{m=1}^{3} \dfrac{m+1}{m}$

68. $\displaystyle\sum_{n=1}^{4} \dfrac{n-1}{n}$

69. $\displaystyle\sum_{k=1}^{3} (k+2)(k+3)$

70. $\displaystyle\sum_{j=1}^{4} (j+1)(j-1)$

71. a. Evaluate $\displaystyle\sum_{i=1}^{n} (-1)^i$ if n is even.

  b. Evaluate $\displaystyle\sum_{i=1}^{n} (-1)^i$ if n is odd.

72. a. Evaluate $\displaystyle\sum_{i=1}^{n} (-1)^{i+1}$ if n is even.

  b. Evaluate $\displaystyle\sum_{i=1}^{n} (-1)^{i+1}$ if n is odd.

73. Given the sequence defined by $a_n = n^2 - 2n$, find the fifth partial sum.

74. Given the sequence defined by $b_n = n^3 - 3n^2$, find the fourth partial sum.

For Exercises 75–86, write the sum using summation notation. There may be multiple representations. Use i as the index of summation. (See Example 7)

75. $\dfrac{1}{2} + \dfrac{4}{3} + \dfrac{9}{4} + \dfrac{16}{5} + \cdots + \dfrac{n^2}{n+1}$

76. $3 + \dfrac{1}{2} + \dfrac{5}{27} + \dfrac{3}{32} + \cdots + \dfrac{n+2}{n^3}$

77. $1 + 2 + 3 + 4 + 5$

78. $2 + 4 + 6 + 8 + 10 + 12$

79. $8 + 8 + 8 + 8$

80. $11 + 11 + 11 + 11 + 11$

81. $\dfrac{1}{3} - \dfrac{1}{9} + \dfrac{1}{27} - \dfrac{1}{81}$

82. $-\dfrac{1}{2} + \dfrac{1}{4} - \dfrac{1}{8} + \dfrac{1}{16} - \dfrac{1}{32}$

83. $c^3 + c^4 + c^5 + \cdots + c^{20}$

84. $a + ar + ar^2 + \cdots + ar^{12}$

85. $\dfrac{x}{1} + \dfrac{x^2}{2} + \dfrac{x^3}{6} + \dfrac{x^4}{24}$ (*Hint*: Note the pattern produced by $n!$ for $n = 1$, $n = 2$, and so on.)

86. $\dfrac{1}{x + 1} + \dfrac{2}{x + 2} + \dfrac{6}{x + 3} + \dfrac{24}{x + 4} + \dfrac{120}{x + 5}$

For Exercises 87–88, rewrite each series as an equivalent series with the new index of summation.

87. $\displaystyle\sum_{i=1}^{8} i^2 = \sum_{j=0}^{\square} \square = \sum_{k=2}^{\square} \square$

88. $\displaystyle\sum_{i=1}^{5} (4i) = \sum_{j=0}^{\square} \square = \sum_{k=2}^{\square} \square$

Mixed Exercises

89. Prove that $\displaystyle\sum_{i=1}^{n} c = cn$.

90. Prove that $\displaystyle\sum_{i=1}^{n} (a_i + b_i) = \sum_{i=1}^{n} a_i + \sum_{i=1}^{n} b_i$.

For Exercises 91–94, use the sums $\displaystyle\sum_{i=1}^{50} i^2 = 42{,}925$ and $\displaystyle\sum_{i=1}^{50} i = 1275$ and the properties of summation given on page 697 to evaluate the given expression.

91. $\displaystyle\sum_{i=1}^{50} (i^2 + 3i)$

92. $\displaystyle\sum_{i=1}^{50} (2i^2 - i)$

93. $\displaystyle\sum_{i=1}^{50} (5i + 4)$

94. $\displaystyle\sum_{i=1}^{50} (6i - 7)$

For Exercises 95–98, determine whether the statement is true or false. If a statement is false, explain why.

95. $\displaystyle\sum_{i=1}^{n} (3i + 7) = 3 \sum_{i=1}^{n} i + 7n$

96. $\displaystyle\sum_{i=1}^{n} (i^2 - 4i + 5) = \sum_{i=1}^{n} i^2 - 4 \sum_{i=1}^{n} i + 5n$

97. $\displaystyle\sum_{i=1}^{n} a_i b_i = \sum_{i=1}^{n} a_i \sum_{i=1}^{n} b_i$

98. $\displaystyle\sum_{i=1}^{n} \dfrac{a_i}{b_i} = \dfrac{\displaystyle\sum_{i=1}^{n} a_i}{\displaystyle\sum_{i=1}^{n} b_i}$

99. Expenses for a company for year 1 are $24,000. Every year thereafter, expenses increase by $1000 plus 3% of the cost of the prior year. Let a_1 represent the original cost for year 1; that is, $a_1 = 24{,}000$. Use a recursive formula to find the cost a_n in terms of a_{n-1} for each subsequent year, $n \geq 2$.

100. A retirement account initially has $500,000 and grows by 5% per year. Furthermore, the account owner adds $12,000 to the account each year after the first. Let a_1 represent the original amount in the account; that is, $a_1 = 500{,}000$. Use a recursive formula to find the amount in the account a_n in terms of a_{n-1} for each subsequent year, $n \geq 2$.

101. In a business meeting, every person at the meeting shakes every other person's hand exactly one time. The total number of handshakes for n people at the meeting is given by $a_n = \frac{1}{2}n(n - 1)$. Evaluate a_{12} and interpret its meaning in the context of this problem.

102. Given a polygon of $n \geq 3$ sides, the sum of the interior angles within the polygon is given by $s_n = 180(n - 2)$. Evaluate s_{10} and interpret its meaning in the context of this problem.

Given a sequence $a_1, a_2, a_3, \ldots, a_n$, the arithmetic mean $\bar{a}$ is given by $\bar{a} = \dfrac{1}{n} \displaystyle\sum_{i=1}^{n} a_i$. Use the arithmetic mean for Exercises 103–104.

103. Consider the sequence defined by $\{a_n\} = 18, 32, 44, 20,$ $36, 28, 32, 38$. Evaluate $\displaystyle\sum_{i=1}^{8} (a_i - \bar{a})^2$.

104. Show that $\displaystyle\sum_{i=1}^{n} (a_i - \bar{a}) = 0$.

Write About It

105. Explain the difference between the graph of $a_n = n^2$ and $f(x) = x^2$.

106. What is the difference between a sequence and a series?

107. Given the sequence defined by $a_n = \frac{n}{n-1}$, explain why the domain must be restricted to positive integers $n \geq 2$.

108. The value $1206!$ is too large to evaluate on most calculators. Explain how you would evaluate $\frac{1206!}{1204!}$ on a calculator.

Expanding Your Skills

109. For $i = \sqrt{-1}$, find the first eight terms of the sequence defined by $a_n = i^n$.

110. For $i = \sqrt{-1}$, find the first eight terms of the sequence defined by $a_n = 1 - i^n$.

111. The terms of the sequence defined by $a_1 = x$ and $a_n = \dfrac{1}{2}\left(a_{n-1} + \dfrac{x}{a_{n-1}}\right)$ for $n > 1$ give successively better

approximations of $\sqrt{x}$ for $x > 1$. Approximate $\sqrt{2}$ by substituting 2 for x and finding the first four terms of the sequence. Round to 4 decimal places if necessary.

112. Find a formula for the nth term of the sequence. $\sqrt{3}, \sqrt{\sqrt{3}}, \sqrt{\sqrt{\sqrt{3}}}, \ldots$

Technology Connections

For Exercises 113–114, use a graphing utility to find the first four terms of the sequence.

113. $a_n = 12\left(-\frac{1}{2}\right)^n$ (Exercise 9)

114. $a_n = 64\left(-\frac{1}{4}\right)^n$ (Exercise 10)

For Exercises 115–116, use a graphing utility to find the sum.

115. $\displaystyle\sum_{k=3}^{7} (-1)^k (6k)$ (Exercise 65)

116. $\displaystyle\sum_{n=3}^{8} (-1)^{n-1}(2n)$ (Exercise 66)

117. Using calculus, we can show that the series $\displaystyle\sum_{k=0}^{n} \dfrac{1}{k!}$

approaches e as n approaches infinity. Investigate this statement by evaluating the sum for $n = 10$ and $n = 50$.

118. Using calculus, we can show that the series

$\displaystyle\sum_{k=1}^{n} (-1)^{k-1}\dfrac{(0.5)^k}{k}$ approaches $\ln 1.5$ as n approaches

infinity. Investigate this statement by evaluating the sum for $n = 10$ and $n = 50$.

SECTION 8.2   Arithmetic Sequences and Series

OBJECTIVES

1. Identify Specific and General Terms of an Arithmetic Sequence
2. Evaluate a Finite Arithmetic Series
3. Apply Arithmetic Sequences and Series

1. Identify Specific and General Terms of an Arithmetic Sequence

In this section and Section 8.3, we study two special types of sequences. The first is called an arithmetic sequence. For example, consider the salary plan for a job that pays $75,000 the first year with a $4000 raise each year thereafter. The sequence of salaries for the first 5 yr is

Year 1	Year 2	Year 3	Year 4	Year 5
$75,000	$79,000	$83,000	$87,000	$91,000

Notice that each term after the first results from adding a fixed constant ($4000) to its predecessor. This is the characteristic that makes this sequence arithmetic.

Arithmetic Sequence

An **arithmetic sequence** $\{a_n\}$ is a sequence of the form

$$a_1, a_1 + d, a_1 + 2d, a_1 + 3d, a_1 + 4d, \ldots$$

- The value a_1 is the first term, and d is called the **common difference** of the sequence.
- The value of d is the difference of any term after the first and its predecessor. $d = a_{n+1} - a_n$
- The nth term of the sequence is given by $a_n = a_1 + (n-1)d$.
- The sequence can be defined recursively as $a_1, a_n = a_{n-1} + d$ for $n \geq 2$.

EXAMPLE 1   Identifying an Arithmetic Sequence and the Common Difference

Determine whether the sequence is arithmetic. If so, identify the common difference.

   a. 35, 25, 15, 5, −5, ...   **b.** 1, 5, 10, 16, 23, ...

Solution:

   a. 35, 25, 15, 5, −5, ...

$a_2 - a_1 = 25 - 35 = -10$	The sequence is arithmetic because the difference
$a_3 - a_2 = 15 - 25 = -10$	between each term and its predecessor is the same constant.
$a_4 - a_3 = 5 - 15 = -10$	The common difference is $d = -10$.
$a_5 - a_4 = -5 - 5 = -10$	

   b. 1, 5, 10, 16, 23, ...

$a_2 - a_1 = 5 - 1 = 4$	The sequence is *not* arithmetic because the
$a_3 - a_2 = 10 - 5 = 5$	difference between a_2 and a_1 is different than the difference between a_3 and a_2. That is, the difference between consecutive terms is not the same.

Skill Practice 1   Determine whether the sequence is arithmetic. If so, identify the common difference.

   a. 12, 5, −2, −9, −16, ...   **b.** 1, 4, 9, 16, 25, ...

EXAMPLE 2   Writing the Terms of a Sequence

   a. Write the first five terms of an arithmetic sequence with first term −5 and common difference 4.
   b. Write a recursive formula to define the sequence.

Solution:

   a. −5, −1, 3, 7, 11   Because the common difference is 4, each term after the first
           +4   +4 +4 +4   must be 4 *more than* its predecessor.

   b. The recursive formula for an arithmetic sequence is a_1, $a_n = a_{n-1} + d$ for $n \geq 2$.

   The sequence is defined by $a_1 = -5$ and $a_n = a_{n-1} + 4$ for $n \geq 2$.

Skill Practice 2

   a. Write the first four terms of an arithmetic sequence with first term 8 and common difference −3.
   b. Write a recursive formula to define the sequence.

EXAMPLE 3   Applying an Arithmetic Sequence

A park ranger in Everglades National Park measures the water level in one region of the park for a 5-day period during a drought.

Day number	1	2	3	4	5
Water level (in.)	54.0	53.2	52.4	51.6	50.8

Answers
1. a. Arithmetic; $d = -7$
   b. Not arithmetic

2. a. 8, 5, 2, −1
   b. $a_1 = 8$ and
       $a_n = a_{n-1} - 3$ for $n \geq 2$.

a. Based on the given data, does the water level follow an arithmetic progression?

b. Write an expression for the nth term of the sequence where n represents the day number.

c. Predict the water level on day 30 if this trend continues.

Solution:

a. $a_2 - a_1 = 53.2 - 54.0 = -0.8$ The sequence is arithmetic because the
    $a_3 - a_2 = 52.4 - 53.2 = -0.8$ difference between each term and its
    $a_4 - a_3 = 51.6 - 52.4 = -0.8$ predecessor is the same constant.
    $a_5 - a_4 = 50.8 - 51.6 = -0.8$ The common difference is $d = -0.8$ in.

b. $a_n = a_1 + (n - 1)d$ Formula for the nth term of an arithmetic
                                    sequence

    $a_n = 54 + (n - 1)(-0.8)$ Substitute 54 for a_1 and -0.8 for d.

c. $a_{30} = 54 + (30 - 1)(-0.8)$ To find a_{30}, substitute 30 for n.
       $= 30.8$

The water level on day 30 will be 30.8 in. if this trend continues.

Skill Practice 3   A homeowner has kept records of the average monthly electric bill for 4 yr.

Year	1	2	3	4
Amount ($)	102.60	108.00	113.40	118.80

a. Do the average monthly electric bills follow an arithmetic progression?

b. Write an expression for the nth term of the sequence.

c. Predict the average monthly bill for year 6 if this trend continues.

The nth term of the sequence from Example 3 can be written in several equivalent algebraic forms.

$$a_n = 54 + (n - 1)(-0.8)$$
$$a_n = 54 - 0.8n + 0.8 \qquad \text{Apply the distributive property.}$$
$$a_n = -0.8n + 54.8$$

The expression $a_n = -0.8n + 54.8$ resembles the slope-intercept form of a linear function $f(x) = mx + b$. In fact, an arithmetic sequence is a linear function whose domain is the set of positive integers. The graph of the sequence from Example 3 is shown in Figure 8-4. The value of the common difference is negative ($d = -0.8$), indicating that the progression of points slopes downward.

In Example 4, we find specific terms of an arithmetic sequence given information about the sequence.

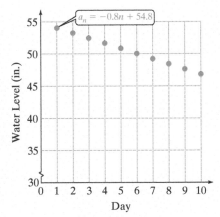

Figure 8-4

EXAMPLE 4   Finding a Specific Term of an Arithmetic Sequence

Find the ninth term of the arithmetic sequence in which $a_1 = -4$ and $a_{22} = 164$.

Solution:

To find the value of a_9, we need to determine the common difference d. To find d, substitute $a_1 = -4$, $n = 22$, and $a_{22} = 164$ into the formula for the nth term.

$$a_n = a_1 + (n - 1)d$$
$$164 = -4 + (22 - 1)d$$
$$164 = -4 + 21d$$
$$d = 8$$

Therefore, $a_n = -4 + (n - 1)(8)$.   With the values of a_1 and d known, the nth term is represented by $a_n = -4 + (n - 1)(8)$.

$$a_n = -4 + 8n - 8$$ Simplify.
$$a_n = 8n - 12$$
$$a_9 = 8(9) - 12$$ To find a_9 substitute $n = 9$.
$$= 60$$

Skill Practice 4   Find the tenth term of the arithmetic sequence in which $a_1 = 12$ and $a_{30} = 128$.

In Example 5, we use the general formula for the nth term of an arithmetic sequence to find the number of terms in an arithmetic sequence.

EXAMPLE 5   Finding the Number of Terms in an Arithmetic Sequence

Find the number of terms of the finite arithmetic sequence 7, 3, -1, -5, ... , -113.

Solution:

$$a_n = a_1 + (n - 1)d$$ The first term of the sequence is 7. The nth term is -113. The common difference is $d = -4$.

$$-113 = 7 + (n - 1)(-4)$$ To find the number of terms n, substitute $a_1 = 7$, $d = -4$, and $a_n = -113$ into the formula for the nth term.
$$-113 = 7 - 4n + 4$$
$$-113 = 11 - 4n$$
$$4n = 124$$
$$n = 31$$ There are 31 terms.

Skill Practice 5   Find the number of terms of the finite arithmetic sequence 16, 11, 6, 1, ... , -239.

Answers

4. $a_{10} = 48$
5. 52 terms

EXAMPLE 6	Finding a Specific Term of an Arithmetic Sequence

For an arithmetic sequence, $a_{15} = 49$ and $a_{27} = 85$. Find the 500th term.

Solution:

Substituting $a_{15} = 49$ and $a_{27} = 85$ into the formula $a_n = a_1 + (n - 1)d$, we can set up a system of linear equations to solve for a_1 and d.

TIP In Example 6, we can also find the value of d by dividing the difference of a_{27} and a_{15} by the number of terms between a_{27} and a_{15}. That is,

$$d = \frac{85 - 49}{27 - 15} = 3$$

$$a_n = a_1 + (n - 1)d \qquad a_n = a_1 + (n - 1)d \qquad \text{Substitute } a_{15} = 49 \text{ and}$$
$$49 = a_1 + (15 - 1)d \qquad 85 = a_1 + (27 - 1)d \qquad a_{27} = 85 \text{ into the formula}$$
$$49 = a_1 + 14d \qquad 85 = a_1 + 26d \qquad \text{for the } n\text{th term.}$$

The two equations form a system of linear equations in two variables.

$$49 = a_1 + 14d \xrightarrow{\text{Multiply by } -1} -49 = -a_1 - 14d \qquad 49 = a_1 + 14d$$
$$85 = a_1 + 26d \qquad\qquad \underline{85 = a_1 + 26d} \qquad 49 = a_1 + 14(3)$$
$$36 = 12d \qquad 7 = a_1$$
$$3 = d$$

With $d = 3$ and $a_1 = 7$, we have an nth term of $a_n = 7 + (n - 1)3$. Thus, the 500th term is given by $a_{500} = 7 + (500 - 1)3 = 1504$.

Skill Practice 6 For an arithmetic sequence, $a_{11} = 65$ and $a_{25} = 149$. Find the 400th term.

2. Evaluate a Finite Arithmetic Series

Consider the finite arithmetic sequence 1, 4, 7, 10, 13. The sum of the first n terms of the sequence is called the **nth partial sum** and is denoted by S_n. In this case, we have

$$S_1 = 1 \qquad\qquad\qquad = 1$$
$$S_2 = 1 + 4 \qquad\qquad\quad = 5$$
$$S_3 = 1 + 4 + 7 \qquad\quad = 12$$
$$S_4 = 1 + 4 + 7 + 10 \quad = 22$$
$$S_5 = 1 + 4 + 7 + 10 + 13 = 35$$

TIP The sum of the first n terms of an arithmetic sequence is called a **finite arithmetic series.**

For a large number of terms, adding the terms individually would be a cumbersome process, so instead, we observe the following pattern. We write the sum of the first five terms of the sequence in both ascending and descending order.

Ascending order: $\quad S_5 = 1 + 4 + 7 + 10 + 13$
Descending order: $\quad \underline{S_5 = 13 + 10 + 7 + 4 + 1}$
$$2S_5 = 14 + 14 + 14 + 14 + 14 \qquad \text{Now add to get } 2S_5.$$

$$2S_5 = 5(14)$$
$$\frac{2S_5}{2} = \frac{5(14)}{2}$$
$$S_5 = 35 \checkmark$$

Answer

6. $a_{400} = 2399$

By adding the terms in ascending and descending order, we double the sum but create a pattern that is easily added. We can use a similar process to find the sum S_n of the first n terms of an arithmetic sequence: $a_1 + a_2 + a_3 + \cdots + a_{n-1} + a_n$.

$$S_n = a_1 \qquad\quad + (a_1 + d) + (a_1 + 2d) + \cdots + a_n \qquad \text{Ascending order}$$

$$\underline{S_n = a_n \qquad\quad + (a_n - d) + (a_n - 2d) + \cdots + a_1} \qquad \text{Descending order}$$

$$2S_n = (a_1 + a_n) + (a_1 + a_n) + (a_1 + a_n) + \cdots + (a_1 + a_n)$$

$$2S_n = n(a_1 + a_n)$$

$$S_n = \frac{n}{2}(a_1 + a_n)$$

---

nth Partial Sum of an Arithmetic Sequence

The sum S_n of the first n terms of an arithmetic sequence is given by

$$S_n = \frac{n}{2}(a_1 + a_n)$$

where a_1 is the first term of the sequence, and a_n is the nth term of the sequence.

In Example 7, we use the formula for the nth partial sum of an arithmetic sequence to find the sum of the first 50 positive even integers.

EXAMPLE 7 **Finding an nth Partial Sum of an Arithmetic Sequence**

Find the sum of the first 50 terms of the sequence. 2, 4, 6, 8, 10, ...

Solution:

The sequence is arithmetic because each term is 2 more than its predecessor. The common difference is 2. To find the sum of the first 50 terms, we need to know the values of a_1 and a_{50}. To find a_{50}, we write a formula for the nth term and then evaluate the expression for $n = 50$.

$$a_n = a_1 + (n - 1)d \qquad \text{General expression for the } n\text{th term}$$

$$a_n = 2 + (n - 1)(2) \qquad \text{Substitute } a_1 = 2 \text{ and } d = 2.$$

$$a_n = 2n$$

$$a_{50} = 2(50) \qquad \text{To find } a_{50}, \text{ substitute } 50 \text{ for } n.$$

$$a_{50} = 100$$

$$S_n = \frac{n}{2}(a_1 + a_n) \qquad \text{Now find the sum of the first 50 terms.}$$

$$S_{50} = \frac{50}{2}(2 + 100) \qquad \text{Substitute } n = 50, a_1 = 2, \text{ and } a_{50} = 100.$$

$$S_{50} = 2550 \qquad \text{The sum of the first 50 positive even integers is 2550.}$$

Skill Practice 7 Find the sum of the first 50 terms of the sequence. 1, 3, 5, 7, 9, ...

Answer

7. 2500

In Example 8, we evaluate an arithmetic series written in summation notation.

EXAMPLE 8   **Evaluating a Finite Arithmetic Series**

Find the sum. $\displaystyle\sum_{i=1}^{60} (3i + 5)$

Solution:

The expression $3i + 5$ is linear in the variable i. This tells us that the terms of the series form an arithmetic progression. To verify, we can write several terms of the sum.

$$\sum_{i=1}^{60} (3i + 5) = [3(1) + 5] + [3(2) + 5] + [3(3) + 5] + \cdots + [3(60) + 5]$$

$$= 8 + 11 + 14 + \cdots + 185$$

The individual terms in the series: 8, 11, 14, ... , 185 form an arithmetic sequence $\{a_n\}$ with a common difference of 3. The first term is $a_1 = 8$, and the 60th term is $a_{60} = 185$. The value of the series is equal to the 60th partial sum of the sequence of terms.

$$S_n = \frac{n}{2}(a_1 + a_n)$$

$$S_{60} = \frac{60}{2}(8 + 185) \qquad \text{Substitute } n = 60,\ a_1 = 8,\ \text{and } a_{60} = 185.$$

$$S_{60} = 5790$$

Avoiding Mistakes

When we apply the formula $S_n = \dfrac{n}{2}(a_1 + a_n)$ to find the sum $\displaystyle\sum_{i=1}^{n} a_n$, the index of summation, i, must begin at 1.

Skill Practice 8   Find the sum. $\displaystyle\sum_{i=1}^{80} (4i + 3)$

---

3. Apply Arithmetic Sequences and Series

In Example 9, we use an arithmetic sequence and series in an application.

EXAMPLE 9   **Applying an Arithmetic Sequence and Series**

Suppose that a job offers a starting salary of $75,000 with a raise of $4000 every year thereafter.

  a. Write an expression for the nth term of an arithmetic sequence that represents the salary as a function of the number of years of employment, n.

  b. Find the total income for an employee who works at the job for 20 yr.

Answer

8. 13,200

Solution:

a. $a_n = a_1 + (n - 1)d$

$a_n = 75,000 + (n - 1)(4000)$ Substitute $a_1 = 75,000$ and $d = 4000$.

$a_n = 75,000 + 4000n - 4000$ Simplify.

$a_n = 4000n + 71,000$

b. $a_{20} = 75,000 + (20 - 1)(4000)$ To find the total income over 20 yr, we need to

$a_{20} = 151,000$ know a_1 and a_{20}. Use the nth term to find a_{20}.

$S_n = \dfrac{n}{2}(a_1 + a_n)$

$S_{20} = \dfrac{20}{2}(75,000 + 151,000)$ Substitute $n = 20$, $a_1 = 75,000$, and

$a_{20} = 151,000$.

$S_{20} = 2,260,000$

Skill Practice 9   A teaching position has a starting salary of \$60,000 with a raise of \$3000 every year thereafter.

a. Write an expression for the nth term of an arithmetic sequence that represents the salary as a function of the number of years of employment, n.

b. Find the total income for an employee who works at the job for 30 yr.

Answers

9 a. $a_n = 3000n + 57,000$

  b. \$3,105,000

SECTION 8.2   Practice Exercises

Prerequisite Review

R.1. Given $f(x) = 7x - 9$, evaluate $f(2)$.

R.2. Determine the slope and y-intercept of the line $y = -\dfrac{1}{4}x + 8$.

For Exercises R.3–R.4, solve the system.

R.3. $6x - 2y = 10$

     $2x - 10y = 22$

R.4. $0.3x - 0.4y = -1.6$

     $0.9x + 0.1y = -3.5$

Concept Connections

1. A(n) _____ sequence is a sequence in which each term after the first is found by adding a fixed constant to its predecessor.

2. The difference between any term after the first and its predecessor in an arithmetic sequence is called the _____ _____ and is denoted by d.

3. Given an arithmetic sequence with first term a_1 and common difference d, the nth term is represented by the formula $a_n =$ _____ or by the recursive formula $a_n =$ _____ for $n \geq 2$.

4. An arithmetic sequence is a linear function whose domain is the set of _____ integers.

5. The sum of the first n terms of a sequence is called the nth _____ sum and is denoted by S_n.

6. Given an arithmetic sequence with first term a_1 and nth term a_n, the nth partial sum is given by the formula $S_n =$ _____.

Objective 1: Identify Specific and General Terms of an Arithmetic Sequence

For Exercises 7–14, determine whether the sequence is arithmetic. If so, find the common difference. (See Example 1)

7. 15, 19, 23, 27, ...   **8.** 256, 268, 280, 292, ...   **9.** 9, −2, −13, −24, ...   **10.** 8, 0, −8, −16, ...

11. 18, 22, 27, 33, ...   **12.** 2, 4, 8, 16, ...   **13.** 4, $\dfrac{14}{3}$, $\dfrac{16}{3}$, 6, ...   **14.** 3, $\dfrac{15}{4}$, $\dfrac{9}{2}$, $\dfrac{21}{4}$, ...

For Exercises 15–18,

a. Write the first five terms of an arithmetic sequence with the given first term and common difference.

b. Write a recursive formula to define the sequence. (**See Example 2**)

15. $a_1 = 3$, $d = 10$ **16.** $a_1 = 6$, $d = 5$ **17.** $a_1 = 4$, $d = -2$ **18.** $a_1 = 5$, $d = -3$

For Exercises 19–24,

a. Write a nonrecursive formula for the nth term of the arithmetic sequence $\{a_n\}$ based on the given information.

b. Find the indicated term.

19. a. $a_1 = 7$, $d = 10$ **20. a.** $a_1 = 102$, $d = 4$ **21. a.** $a_1 = -12$, $d = 5$
  b. Find a_{22}.   **b.** Find a_{43}.   **b.** Find a_{20}.

22. a. $a_1 = -4$, $d = 6$ **23. a.** $a_1 = \dfrac{1}{2}$, $d = \dfrac{1}{3}$ **24. a.** $a_1 = \dfrac{2}{3}$, $d = \dfrac{1}{2}$

  b. Find a_{18}.   **b.** Find a_{10}.   **b.** Find a_8.

25. Jim has 8 unread emails in his inbox before going on vacation. While on vacation, Jim does not read email. If he receives an average of 22 emails each day, write the nth term of a sequence defining the number of unread emails in his box at the end of day n of his vacation.

26. Sandy has a personal trainer who encourages her to get plenty of cardiovascular exercise. In her first week of training, Sandy walks for 10 min on a treadmill every day. Each week thereafter, she increases the time on the treadmill by 5 min. Write the nth term of a sequence defining the number of minutes that Sandy spends on the treadmill per day for her nth week at the gym.

27. A new drug and alcohol rehabilitation program performs outreach for members of the community. The number of participants for a 4-week period is given in the table. (**See Example 3**)

Week number	1	2	3	4
Number of participants	34	50	66	82

  a. Based on the data given, does the number of participants follow an arithmetic progression?

  b. Write an expression for the nth term of the sequence representing the number of participants, where n represents the week number.

  c. Predict the number of participants in week 10 if this trend continues.

28. A student studying to be a veterinarian's assistant keeps track of a kitten's weight each week for a 5-week period after birth.

Week number	1	2	3	4	5
Weight (lb)	0.6	0.88	1.16	1.44	1.72

  a. Based on the data given, does the weight of the kitten follow an arithmetic progression?

  b. Write an expression for the nth term of the sequence representing the kitten's weight, n weeks after birth.

  c. If the weight of the kitten continues to increase linearly for 3 months, predit the kitten's weight 12 weeks after birth.

29. Suppose that an object starts with an initial velocity of v_0 (in ft/sec) and moves under a constant acceleration a (in ft/sec^2). Then the velocity v_n (in ft/sec) after n seconds is given by $v_n = v_0 + an$. Show that this sequence is arithmetic. (*Hint*: Show that $v_{n+1} - v_n$ is constant.)

30. Suppose that an object is dropped from rest from an airplane. Assuming negligible air resistance, the vertical acceleration is 32 ft/sec^2. Using the formula for v_n given in Exercise 29,

  a. Write the nth term v_n of the sequence representing the velocity in the downward direction after n seconds.

  b. Determine the vertical velocity 1 sec after release and 5 sec after release.

31. Find the 8th term of an arithmetic sequence with $a_1 = -2$ and $a_{15} = 68$. (**See Example 4**)

32. Find the 19th term of an arithmetic sequence with $a_1 = -11$ and $a_{30} = 163$.

33. Find the 35th term of an arithmetic sequence with $a_1 = 50$ and $a_{22} = -265$.

34. Find the 46th term of an arithmetic sequence with $a_1 = 210$ and $a_{60} = -262$.

For Exercises 35–38, find the number of terms of the finite arithmetic sequence. (See Example 5)

35. 8, 14, 20, 26, … , 320

36. 7, 16, 25, 34, … , 574

37. 11, 10.7, 10.4, 10.1, … , -3.4

38. 9, 8.4, 7.8, 7.2, … , -39

39. Given an arithmetic sequence with $a_{14} = 148$ and $a_{35} = 316$, find a_1 and d.

40. Given an arithmetic sequence with $a_{12} = -76$ and $a_{51} = -193$, find a_1 and d.

For Exercises 41–46, two terms of an arithmetic sequence are given. Find the indicated term. (See Example 6)

41. $a_{15} = 86$, $a_{34} = 200$; Find a_{150}.

42. $a_{12} = 52$, $a_{51} = 208$; Find a_{172}.

43. $b_{32} = -303$, $b_{54} = -567$; Find b_{214}.

44. $b_{64} = -456$, $b_{81} = -575$; Find b_{105}.

45. $c_{14} = 7.5$, $c_{101} = 29.25$; Find c_{400}.

46. $c_{21} = -11.16$, $c_{116} = -7.36$; Find c_{505}.

47. If the third and fourth terms of an arithmetic sequence are -6 and -9, what are the first and second terms?

48. If the third and fourth terms of an arithmetic sequence are 12 and 16, what are the first and second terms?

Objective 2: Evaluate a Finite Arithmetic Series

For Exercises 49–50, find the sum.

49. $4 + 7 + 10 + 13 + 16 + 19 + 22 + 25 + 28 + 31 + 34 + 37$

50. $8 + 12 + 16 + 20 + 24 + 28 + 32 + 36 + 40 + 44$

51. Find the sum of the first 40 terms of the sequence. $\{1, 6, 11, 16, …\}$ **(See Example 7)**

52. Find the sum of the first 60 terms of the sequence. $\{2, 10, 18, 26, …\}$

53. Find the sum. $5 + 4.5 + 4 + 3.5 + \cdots + (-30.5)$

54. Find the sum. $8 + 7.2 + 6.4 + 5.6 + \cdots + (-45.6)$

For Exercises 55–64, find the sum. (See Example 8)

55. $\displaystyle\sum_{j=1}^{18} (j + 6)$

56. $\displaystyle\sum_{j=1}^{15} (j - 3)$

57. $\displaystyle\sum_{i=1}^{50} (2i + 6)$

58. $\displaystyle\sum_{i=1}^{40} (3i - 7)$

59. $\displaystyle\sum_{k=1}^{162} \left(3 - \tfrac{1}{2}k\right)$

60. $\displaystyle\sum_{k=1}^{141} \left(4 - \tfrac{1}{4}k\right)$

61. $-1 + 4 + 9 + \cdots + 49$

62. $12 + 16 + 20 + \cdots + 84$

63. $-7 + (-11) + (-15) + \cdots + (-39)$

64. $-18 + (-23) + (-28) + \cdots + (-183)$

Objective 3: Apply Arithmetic Sequences and Series

65. Suzanne must choose between two job offers. The first job offers $64,000 for the first year. Each year thereafter, she would receive a $3200 raise. The second job offers $60,000 in the first year. Each year thereafter, she would receive a $5000 raise. **(See Example 9)**

   a. If she anticipates working for the company for 5 yr, find the total amount she would earn from each job.

   b. If she anticipates working for the company for 10 yr, find the total amount she would earn from each job.

66. José must choose between two job offers. The first job pays $50,000 the first year. Each year thereafter, he would receive a raise of $2400. A second job offers $54,000 the first year with a raise of $2000 each year thereafter. However, with the second job, José would have to pay $100 per month out of his paycheck for health insurance.

   a. If José anticipates working for the company for 6 yr, find the total amount he would earn from each job.

   b. If he anticipates working for the company for 12 yr, find the total amount he would earn from each job.

67. An object in free fall is dropped from a tall cliff. It falls 16 ft in the first second, 48 ft in the second second, 80 ft in the third second, and so on.

   a. Write a formula for the nth term of an arithmetic sequence that represents the distance d_n (in ft) that the object will fall in the nth second.

   b. How far will the object fall in the 8th second?

   c. What is the total distance that the object will fall in 8 sec?

68. A ball rolling down an inclined plane rolls 4 in. in the first second, 8 in. in the second second, 12 inches in the third second, and so on.

   a. Write a formula for the nth term of an arithmetic sequence that represents the distance d_n (in inches) that the ball will roll in the nth second.

   b. How far will the ball roll in the 10th second?

   c. What is the total distance that the ball will travel in 10 sec?

69. The students at Prairiewood Elementary plan to make a pyramid out of plastic cups. The bottom row has 15 cups. Moving up the pyramid, the number of cups in each row decreases by 1.

    a. If the students build the pyramid so that there are 12 rows, how many cups will be at the top?

    b. How many total cups will be required?

70. A theater has 32 rows. The first row has 18 seats, and each row that follows has three more seats than the row in front.

    a. Determine the number of seats in row 32.

    b. Determine the total number of seats in the theater.

Mixed Exercises

71. Refer to the graph of the sequence $\{b_n\}$.

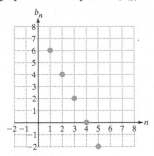

    a. Estimate the first four terms of the sequence.

    b. Write a formula for the nth term of the sequence.

    c. Find the 30th term.

    d. Find the sum of the first 30 terms.

    e. If $b_n = -180$, what is n?

    f. Find the difference between b_{88} and b_{20}.

72. Refer to the graph of the sequence $\{a_n\}$.

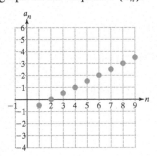

    a. Estimate the first four terms of the sequence.

    b. Write a formula for the nth term of the sequence.

    c. Find the 20th term.

    d. Find the sum of the first 20 terms.

    e. If $a_n = 256$, what is n?

    f. Find the difference between a_{50} and a_{32}.

73. Find the sum of the integers from -20 to 256.

74. Find the sum of the integers from -102 to 57.

75. Compute the sum of the first 50 positive integers that are exactly divisible by 5.

76. Compute the sum of the first 60 positive integers that are exactly divisible by 4.

77. Compute the sum of all integers between 40 and 100 that are exactly divisible by 3.

78. Compute the sum of all integers between 60 and 150 that are exactly divisible by 8.

79. a. Use the formula $S_n = \frac{n}{2}(a_1 + a_n)$ to show that the sum of the first n positive integers is $S_n = \frac{n}{2}(1 + n)$.

    b. Find the sum of the first 100 positive integers.

    c. Find the sum of the first 1000 positive integers.

80. a. Use the formula $S_n = \frac{n}{2}(a_1 + a_n)$ to show that the sum $1, 3, 5, \ldots (2n - 1) = n^2$.

    b. Find the sum of the first 100 positive odd integers.

The arithmetic mean (average) of two numbers c and d is given by $\bar{x} = \dfrac{c + d}{2}$. The value $\bar{x}$ is equidistant between c and d, so the sequence $c, \bar{x}, d$ is an arithmetic sequence. Inserting k equally spaced values between c and d, yields the arithmetic sequence $c, \bar{x}_1, \bar{x}_2, \bar{x}_3, \bar{x}_4, \ldots, \bar{x}_k, d$. Use this information for Exercises 81–82.

81. Insert three arithmetic means between 4 and 28. (*Hint*: There will be a total of five terms. Write the nth term of an arithmetic sequence with $a_1 = 4$ and $a_5 = 28$, and then find a_2, a_3, and a_4.)

82. Insert four arithmetic means between 19 and 64.

Write About It

83. Suppose that $\{a_1, a_2, a_3, \ldots\}$ is an arithmetic sequence with common difference d. Explain why $\{a_1, a_3, a_5, \ldots\}$ is also an arithmetic sequence.

84. Suppose you are helping a friend with the homework for this section. Explain how to construct an arithmetic sequence.

Expanding Your Skills

85. Determine the nth term of the arithmetic sequence whose nth partial sum is $n^2 + 2n$. [*Hint:* The nth term of the sequence is the difference between the sum of the first n terms and the first $(n - 1)$ terms.]

86. Determine the nth term of the arithmetic sequence whose nth partial sum is $2n^2$.

SECTION 8.3	Geometric Sequences and Series

OBJECTIVES

1. Identify Specific and General Terms of a Geometric Sequence
2. Evaluate Finite Geometric Series
3. Evaluate Infinite Geometric Series
4. Find the Value of an Annuity

1. Identify Specific and General Terms of a Geometric Sequence

The sequence 2, 4, 8, 16, 32, … is not an arithmetic sequence because the difference between consecutive terms is not the same constant. However, a different pattern exists. Notice that each term after the first is 2 times the preceding term. This sequence is called a geometric sequence.

> ### Geometric Sequence
>
> A **geometric sequence** $\{a_n\}$ is a sequence of the form
> $$a_1, a_1 r, a_1 r^2, a_1 r^3, a_1 r^4, \ldots$$
>
> - The value a_1 is the first term, and r is called the **common ratio** of the sequence.
> - The value of r is the quotient of any term after the first and its predecessor.
> $$r = \frac{a_{n+1}}{a_n}$$
> - The nth term of the sequence is given by $a_n = a_1 r^{n-1}$.
> - The sequence can be defined recursively as $a_1, a_n = a_{n-1} r$ for $n \geq 2$.

A geometric sequence is recognized by dividing any term after the first by its predecessor. In all cases the quotient is the same ratio r.

EXAMPLE 1	Identifying a Geometric Sequence and the Common Ratio

Determine whether the sequence is geometric. If so, identify the common ratio.

a. $18, 6, 2, \dfrac{2}{3}, \ldots$　　　　**b.** $3, 12, 48, 240, \ldots$

Solution:

a. $18, 6, 2, \dfrac{2}{3}, \ldots$

$\dfrac{a_2}{a_1} = \dfrac{6}{18} = \dfrac{1}{3}, \quad \dfrac{a_3}{a_2} = \dfrac{2}{6} = \dfrac{1}{3},$

$\dfrac{a_4}{a_3} = \dfrac{\frac{2}{3}}{2} = \dfrac{2}{3} \cdot \dfrac{1}{2} = \dfrac{1}{3}$

The sequence is geometric because the ratio between each term and its predecessor is the same constant.

The common ratio is $r = \frac{1}{3}$.

b. $3, 12, 48, 240, \ldots$

$\dfrac{a_2}{a_1} = \dfrac{12}{3} = 4, \quad \dfrac{a_3}{a_2} = \dfrac{48}{12} = 4,$

$\dfrac{a_4}{a_3} = \dfrac{240}{48} = 5$

The sequence is *not* geometric because the ratio between a_4 and a_3 is different than the ratio of other pairs of consecutive terms.

Example 2 illustrates the fundamental characteristic of a geometric sequence; that is, each term of a geometric sequence is a constant multiple of the preceding term.

EXAMPLE 2 **Writing Several Terms of a Geometric Sequence**

Write the first five terms of a geometric sequence with $a_1 = 5$ and $r = -2$.

Solution:

By definition a geometric sequence follows the pattern $a_1, a_1r, a_1r^2, a_1r^3, \ldots$. The first five terms are

$a_1 = 5$
$a_2 = 5(-2) = -10$
$a_3 = 5(-2)^2 = 20$
$a_4 = 5(-2)^3 = -40$
$a_5 = 5(-2)^4 = 80$

TIP Alternatively, we can use a recursive formula:
$a_1 = 5$ and $a_n = a_{n-1} \cdot r$ for $n \geq 2$.

$a_2 = a_1 \cdot (-2) = 5(-2) = -10$
$a_3 = a_2 \cdot (-2) = -10(-2) = 20$
$a_4 = a_3 \cdot (-2) = 20(-2) = -40$
$a_5 = a_4 \cdot (-2) = -40(-2) = 80$

In Example 3, we write an expression for the nth term of a geometric sequence given the first four terms.

EXAMPLE 3 **Writing the nth Term of a Geometric Sequence**

Write a formula for the nth term of the geometric sequence. $2, 3, \dfrac{9}{2}, \dfrac{27}{4}, \ldots$

Solution:

$r = \dfrac{a_2}{a_1} = \dfrac{3}{2}$ — Dividing any term by its predecessor, we have a common ratio of $\frac{3}{2}$.

$a_n = a_1 r^{n-1}$ — Begin with the formula for the nth term of a geometric sequence.

$a_n = 2\left(\dfrac{3}{2}\right)^{n-1}$ — Substitute $a_1 = 2$ and $r = \frac{3}{2}$.

A graph of several terms of the geometric sequence from Example 3 is shown in Figure 8-5. The points representing the sequence coincide with points on the graph of the exponential function $f(x) = 2\left(\frac{3}{2}\right)^{x-1}$ for positive integer values of x (Figure 8-6).

In fact, a geometric sequence with $r > 0$ and $r \neq 1$ is an exponential function whose domain is restricted to the set of positive integers.

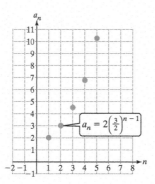

Figure 8-5   Figure 8-6

EXAMPLE 4   **Finding a Specified Term of a Geometric Sequence**

Find the fifth term of a geometric sequence $\{a_n\}$ given that $a_1 = 15$ and $a_2 = -9$.

Solution:

$r = \dfrac{a_2}{a_1} = \dfrac{-9}{15} = -\dfrac{3}{5}$ Divide a_2 by a_1 to obtain the common ratio $-\frac{3}{5}$.

$a_n = a_1 r^{n-1}$ Use the formula for the nth term of a geometric sequence.

$a_n = 15\left(-\dfrac{3}{5}\right)^{n-1}$ Substitute $a_1 = 15$ and $r = -\frac{3}{5}$ to get the nth term for this sequence.

$a_5 = 15\left(-\dfrac{3}{5}\right)^{5-1}$ To find a_5, substitute 5 for n.

$a_5 = \dfrac{243}{125}$

Skill Practice 4   Find the sixth term of a geometric sequence $\{a_n\}$ given that $a_1 = 64$ and $a_2 = -16$.

Example 4 illustrates that a geometric sequence with a negative common ratio is an alternating sequence. The graph of the first 10 terms of the sequence is shown in Figure 8-7.

EXAMPLE 5   **Writing the nth Term of a Geometric Sequence**

Given the terms $a_2 = 80$ and $a_5 = 40.96$ of a geometric sequence, find r, a_1, and a_n.

Solution:

In this example, we need to find two unknown quantities. One strategy is to begin with the formula for the nth term of the geometric sequence and substitute the known values for terms 2 and 5 of the sequence. Then solve the resulting system of nonlinear equations.

$a_n = a_1 r^{n-1}$

$80 = a_1 r^{(2-1)} \longrightarrow 80 = a_1 r$ Substitute $a_2 = 80$.

$40.96 = a_1 r^{(5-1)} \longrightarrow 40.96 = a_1 r^4$ Substitute $a_5 = 40.96$.

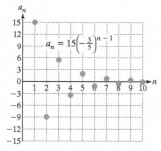

Figure 8-7

Answer

4. $a_6 = -\dfrac{1}{16}$

Substitute $a_1 = \dfrac{80}{r}$ from the first equation into the second equation: $40.96 = a_1 r^4$.

$$40.96 = \dfrac{80}{r} r^4$$
$$40.96 = 80r^3$$
$$0.512 = r^3 \qquad \text{Divide by 80.}$$
$$r = 0.8 \qquad \text{Take the cube root of both sides.}$$
$$a_1 = \dfrac{80}{r} = \dfrac{80}{0.8} = 100 \qquad \text{To solve for } a_1, \text{ substitute } r = 0.8 \text{ into the equation } a_1 = \dfrac{80}{r}.$$

The value of r is 0.8, the value of a_1 is 100, and the nth term is $a_n = 100(0.8)^{n-1}$.

> **Skill Practice 5**   Given the terms $a_2 = 54$ and $a_5 = 182.25$ of a geometric sequence, find r, a_1, and a_n.

2. Evaluate Finite Geometric Series

The nth partial sum S_n of the first n terms of a geometric sequence is a **finite geometric series.** Consider the geometric series:

$$S_n = a_1 + a_1 r + a_1 r^2 + a_1 r^3 + \cdots + a_1 r^{n-1}$$

Suppose that we subtract rS_n from S_n.

$$S_n = a_1 \qquad + a_1 r \qquad + a_1 r^2 \qquad + \cdots + a_1 r^{n-1}$$
$$rS_n = a_1 r \qquad + a_1 r^2 \qquad + a_1 r^3 \qquad + \cdots + a_1 r^{n}$$

$$S_n - rS_n = (a_1 - a_1 r) + (a_1 r - a_1 r^2) + (a_1 r^2 - a_1 r^3) + \cdots + (a_1 r^{n-1} - a_1 r^n)$$
$$S_n - rS_n = a_1 - a_1 r^n \qquad \text{The terms in red form a sum of zero.}$$
$$S_n(1 - r) = a_1(1 - r^n) \qquad \text{Factor each side of the equation.}$$
$$S_n = \dfrac{a_1(1 - r^n)}{1 - r} \qquad \text{Divide by } (1 - r).$$

> **nth Partial Sum of a Geometric Sequence**
>
> The sum S_n of the first n terms of a geometric sequence is given by
> $$S_n = \dfrac{a_1(1 - r^n)}{1 - r}$$
> where a_1 is the first term of the sequence and r is the common ratio, $r \neq 1$.

> **EXAMPLE 6**   **Evaluating a Finite Geometric Series**
>
> Find the sum. $\displaystyle\sum_{i=1}^{6} 4\left(\dfrac{1}{2}\right)^{i-1}$
>
> **Solution:**
>
> $$\sum_{i=1}^{6} 4\left(\dfrac{1}{2}\right)^{i-1} = 4 + 2 + 1 + \dfrac{1}{2} + \dfrac{1}{4} + \dfrac{1}{8}$$
>
> The individual terms in the series form a geometric sequence with $a_1 = 4$ and $r = \dfrac{1}{2}$. The value of the series is the sixth partial sum of the sequence of terms.
>
> $$S_n = \dfrac{a_1(1 - r^n)}{1 - r} = \dfrac{4\left[1 - \left(\frac{1}{2}\right)^6\right]}{1 - \frac{1}{2}} = \dfrac{4\left(1 - \frac{1}{64}\right)}{\frac{1}{2}} = \dfrac{63}{8} \qquad \begin{array}{l}\text{Apply the formula for the}\\ n\text{th partial sum with } n = 6.\end{array}$$

Answers

5. $r = 1.5$; $a_1 = 36$; $a_n = 36(1.5)^{n-1}$

6. $\dfrac{1093}{81}$

> **Skill Practice 6**   Find the sum. $\displaystyle\sum_{i=1}^{7} 9\left(\dfrac{1}{3}\right)^{i-1}$

> **TIP** The equivalence property of exponential expressions indicates that if $b^n = b^m$, then $n = m$. That is, if two exponential expressions with the same base are equal, then their exponents are equal.

EXAMPLE 7   Evaluating a Finite Geometric Series

Find the sum of the finite geometric series. $5 + 10 + 20 + \cdots + 5120$

Solution:

The common ratio is 2 and $a_1 = 5$. The nth term of the sequence of terms can be written as $a_n = 5(2)^{n-1}$. To find the number of terms n, substitute 5120 for a_n.

$$5120 = 5(2)^{n-1}$$
$$1024 = 2^{n-1} \qquad \text{Divide both sides by 5.}$$
$$2^{10} = 2^{n-1} \qquad \text{Apply the equivalence property of exponential expressions.}$$
$$10 = n - 1$$
$$n = 11 \qquad \text{The sequence has 11 terms.}$$

With $a_1 = 5$, $r = 2$, and $n = 11$, we have

$$S_n = \frac{a_1(1 - r^n)}{1 - r} = \frac{5(1 - 2^{11})}{1 - 2} = 10{,}235$$

> **Skill Practice 7**   Find the sum of the finite geometric series.
> $$3 + 6 + 12 + \cdots + 768$$

3. Evaluate Infinite Geometric Series

For a positive integer n, the series $\displaystyle\sum_{i=1}^{n} a_i = a_1 + a_2 + a_3 + \cdots + a_n$ is called a **finite series** because there are a finite number of terms. The series $\displaystyle\sum_{i=1}^{\infty} a_i = a_1 + a_2 + a_3 + \cdots$ is called an **infinite series** because there are an infinite number of terms. Although it is impossible to add an infinite number of terms on a term-by-term basis, we might ask if the sum approaches a limiting value. If the nth partial sum S_n of an infinite series approaches a number L as $n \to \infty$, we say that the series **converges** and we call L the sum of the series. If a series does not converge, we say that the series **diverges** (the sum does not exist).

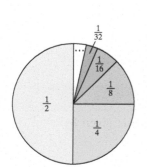

Figure 8-8

Consider the geometric sequence $\frac{1}{2}, \frac{1}{4}, \frac{1}{8}, \frac{1}{16}, \ldots, \left(\frac{1}{2}\right)^n, \ldots$ and the corresponding infinite geometric series $\frac{1}{2} + \frac{1}{4} + \frac{1}{8} + \frac{1}{16} + \cdots + \left(\frac{1}{2}\right)^n + \cdots$. To visualize the infinite series, suppose that we add $\frac{1}{2}$ of a pie plus $\frac{1}{4}$ of a pie plus $\frac{1}{8}$ of a pie and so on. Will we approach some finite amount of pie? From Figure 8-8, it appears that the sum will be 1 whole unit.

For an infinite geometric sequence with $-1 < r < 1$ (equivalently $|r| < 1$), the value of r^n will become smaller as n gets larger. For example:

$$\approx 0.0009766 \qquad \approx 0.0000305$$

$$\frac{1}{2}, \frac{1}{4}, \frac{1}{8}, \frac{1}{16}, \ldots, \left(\frac{1}{2}\right)^{10}, \ldots, \left(\frac{1}{2}\right)^{15}, \ldots$$

In fact, for $|r| < 1$, as $n \to \infty$, $r^n \to 0$. Therefore, as n approaches infinity

$$S_n = \frac{a_1(1 - r^n)}{1 - r} \quad \text{approaches} \quad \frac{a_1(1 - 0)}{1 - r} = \frac{a_1}{1 - r}.$$

Answer

7. 1533

> ## Sum of an Infinite Geometric Series
>
> Given an infinite geometric series $a_1 + a_1 r + a_1 r^2 + a_1 r^3 + \cdots$, with $|r| < 1$, the sum S of all terms in the series is given by
>
> $$S = \frac{a_1}{1 - r}$$
>
> *Note:* If $|r| \geq 1$, then the sum does not exist.

EXAMPLE 8   Evaluating an Infinite Geometric Series

Find the sum if possible.   **a.** $\displaystyle\sum_{i=1}^{\infty} 5\left(-\frac{1}{3}\right)^{i-1}$ **b.** $2 + \dfrac{8}{3} + \dfrac{32}{9} + \dfrac{128}{27} + \cdots$

Solution:

a. $\displaystyle\sum_{i=1}^{\infty} 5\left(-\frac{1}{3}\right)^{i-1} = 5 - \frac{5}{3} + \frac{5}{9} - \frac{5}{27} + \frac{5}{81} + \cdots$

The sum is an infinite geometric series with $a_1 = 5$ and $r = -\frac{1}{3}$.

Because $|r| = \left|-\frac{1}{3}\right| < 1$, we have $S = \dfrac{a_1}{1 - r} = \dfrac{5}{1 - \left(-\frac{1}{3}\right)} = \dfrac{15}{4}$.

The sum is $\dfrac{15}{4}$.

b. $2 + \dfrac{8}{3} + \dfrac{32}{9} + \dfrac{128}{27} + \cdots$ is a geometric series with $a_1 = 2$ and $r = \frac{4}{3}$.

Because $|r| = \left|\frac{4}{3}\right| \geq 1$, the sum does not exist.

> **Skill Practice 8**   Find the sum if possible.
>
> **a.** $\displaystyle\sum_{i=1}^{\infty} 4\left(\frac{3}{4}\right)^{i-1}$ **b.** $2 + \dfrac{5}{2} + \dfrac{25}{8} + \dfrac{125}{32} + \cdots$

In Example 9, we use the formula for the sum of an infinite geometric series to write a repeating decimal as a fraction.

EXAMPLE 9   Writing a Repeating Decimal as a Fraction

a. Write $0.\overline{5}$ as a fraction.   **b.** Write $1.9\overline{75}$ as a fraction.

Solution:

a. $0.\overline{5} = 0.5555\ldots$

$= \underbrace{\dfrac{5}{10} + \dfrac{5}{100} + \dfrac{5}{1000} + \dfrac{5}{10,000} + \cdots}_{\substack{\text{Infinite geometric series} \\ \text{with } a_1 = \frac{5}{10} \text{ and } r = \frac{1}{10}}} \longrightarrow S = \dfrac{a_1}{1 - r} = \dfrac{\frac{5}{10}}{1 - \frac{1}{10}} = \dfrac{\frac{5}{10}}{\frac{9}{10}} = \dfrac{5}{9}$

$0.\overline{5} = \dfrac{5}{9}$

Answers

8. a. 16
   b. $r = \frac{5}{4} \geq 1$, so the sum
   does not exist.

b. $1.9\overline{75} = 1.9757575\ldots$

$$= \frac{19}{10} + \left[\frac{75}{1000} + \frac{75}{100,000} + \cdots \right]$$

Infinite geometric series with $a_1 = \frac{75}{1000}$ and $r = \frac{1}{100}$ $\longrightarrow$ $S = \dfrac{a_1}{1-r} = \dfrac{\frac{75}{1000}}{1 - \frac{1}{100}} = \dfrac{\frac{75}{1000}}{\frac{99}{100}} = \dfrac{75}{990} = \dfrac{5}{66}$

$$1.9\overline{75} = \frac{19}{10} + \frac{5}{66} = \frac{326}{165}$$

> **Skill Practice 9**   **a.** Write $0.\overline{7}$ as a fraction.   **b.** Write $0.3\overline{4}$ as a fraction.

When money is infused into the economy, a percentage of the money received by individuals or businesses is often respent over and over again. Economists call this the **multiplier effect.**

> **EXAMPLE 10**   **Investigating an Application of an Infinite Geometric Series—The Multiplier Effect**
>
> Suppose that $200 million is spent annually by tourists in a certain state. Further suppose that 75% of the money is respent in the state and then respent over and over again, each time at a rate of 75%. Determine the theoretical total amount spent from the initial $200 million if the money can be respent an infinite number of times.
>
> **Solution:**
>
> The total amount spent can be represented by the infinite geometric series where all values are in $ millions.
>
> $\boxed{\text{75\% of 200}}$ $\boxed{\text{75\% of 150}}$
>
> $200 + 150 + 112.5 + \cdots$
>
> We have $a_1 = 200$ and $r = 0.75$. Because $|r| = |0.75| < 1$, we have
>
> $$S = \frac{a_1}{1-r} = \frac{200}{1 - 0.75} = 800.$$
>
> Theoretically, the total amount spent is $800 million.

> **Skill Practice 10**   Suppose that after a tax rebate an individual spends $210. The money is then respent over and over again, each time at a rate of 70%. Determine the total amount spent. Assume that the money can be respent an infinite number of times.

4. Find the Value of an Annuity

In Section 4.2, we studied applications of exponential functions involving compound interest. We learned that if P dollars is invested in an account at interest rate r, compounded annually for t years, then the amount A in the account is given by

$$A = P(1 + r)^t$$

However, rather than making a one-time lump sum payment of P dollars, many individuals will invest smaller amounts at regular and more frequent intervals. Such a sequence of fixed payments made (or received) by an individual over a fixed period of time is called an **annuity.**

Answers

9. a. $\dfrac{7}{9}$ **b.** $\dfrac{31}{90}$

10. $700

Suppose that an individual invests P dollars at the *end* of each year for 4 yr at interest rate r. The money deposited at the end of the first year will have 3 yr in which to earn interest. The money invested at the end of the second year will earn interest for 2 yr, and the money invested at the end of the third year will earn interest for 1 yr. However, the money invested at the end of the fourth year will not earn any interest.

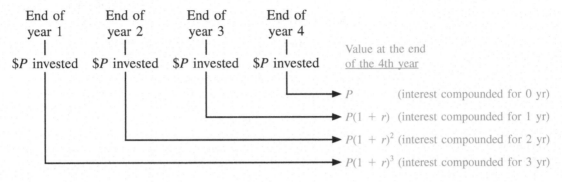

The total amount A, also called the **future value** of the annuity, is given by

$$A = P + P(1 + r) + P(1 + r)^2 + P(1 + r)^3$$

This is a finite geometric series with $a_1 = P$ and common ratio of $(1 + r)$. Using the relationship $S_n = \dfrac{a_1(1 - r^n)}{1 - r}$, the total amount invested for t years is given by

$$A = \frac{P[1 - (1 + r)^t]}{1 - (1 + r)} = \frac{P[1 - (1 + r)^t]}{-r} \text{ or simply } A = \frac{P[(1 + r)^t - 1]}{r}$$

If deposits are made n times per year, then the interest rate per compounding period is $\frac{r}{n}$, and the total number of times the money is compounded is nt.

> **TIP** In this section we study **ordinary annuities**. These are annuities in which money is invested at the *end* of the compounding periods. If the money were invested at the beginning of the compounding periods, such an investment is called an **annuity due** and the future value is computed with a different formula.

Future Value of an Ordinary Annuity

Suppose that P dollars is invested at the end of each compounding period n times per year at interest rate r. Then the value A (in $) of the annuity after t years is given by

$$A = \frac{P\left[\left(1 + \frac{r}{n}\right)^{nt} - 1\right]}{\frac{r}{n}}$$

EXAMPLE 11   **Finding the Value of an Annuity**

Suppose that an employee working for a state college puts aside $150 at the end of each month in a tax-sheltered annuity. The annuity pays 6% annual interest compounded monthly. Suppose that the employee begins contributing at age 28.

a. How much will the annuity be worth by the time the employee reaches age 62?

b. How much interest will have been earned?

Point of Interest

The contributions made to the investment in Example 11 are **tax deferred.** This means that the principal invested each month has not yet been taxed. Because the principal is invested pretax, the individual potentially has more money available to invest. Furthermore, no taxes are paid on either the principal or interest until the money is withdrawn. This type of plan is meant as a long-term investment, and withdrawals are typically taken after age 59½ without penalty.

Solution:

a. $A = \dfrac{P\left[\left(1 + \frac{r}{n}\right)^{nt} - 1\right]}{\frac{r}{n}}$

$A = \dfrac{150\left[\left(1 + \frac{0.06}{12}\right)^{(12)(34)} - 1\right]}{\frac{0.06}{12}}$
$\qquad P = \$150$
$\qquad r = 0.06$
$\qquad n = 12$ (the money is invested monthly)

$A = 199{,}548.50$
$\qquad t = 62 - 28 = 34$ yr

The annuity will be worth \$199,548.50 when the employee reaches age 62.

b. \$150 is invested 12 times per year for 34 yr.
Principal invested: ($\$150)(12)(34) = \$61{,}200$
The amount of interest is $\$199{,}548.50 - \$61{,}200 = \$138{,}348.50$.

Skill Practice 11   Suppose that an employee contributes \$100 to an annuity at the end of each month for 25 yr. If the annuity earns 7%,

a. Determine the value of the annuity at the end of the 25-yr period.

b. How much interest will be earned?

Answers

11. **a.** \$81,007.17   **b.** \$51,007.17

SECTION 8.3   **Practice Exercises**

Prerequisite Review

R.1. Simplify. $-\dfrac{5}{6} \div \dfrac{2}{3}$

R.2. Solve for x. $3^x = 81$

R.3. For $g(x) = \left(\dfrac{1}{4}\right)^x$ find $g(0)$, $g(1)$, $g(2)$, $g(-1)$, and $g(-2)$.

R.4. For $h(x) = 16^x$ find $h(0)$, $h(1)$, and $h(-1)$.

R.5. Suppose that an investor deposits \$14,000 in a savings account for 10 yr at 2% interest. Use the model $A(t) = P\left(1 + \dfrac{r}{n}\right)^{nt}$ for P dollars in principal invested at an interest rate r compounded n times per year for t years for the following compounding options. Round to the nearest dollar.

a. Interest compounded annually

b. Interest compounded quarterly

c. Interest compounded monthly

Concept Connections

1. A _____ sequence is a sequence in which each term after the first is the product of the preceding term and a fixed nonzero real number, called the common _____, r.

2. The nth term of a geometric sequence with first term a_1 and common ratio r is given by $a_n = $ _____.

3. The nth partial sum S_n of the first n terms of a geometric sequence is a (finite/infinite) geometric series.

4. The sum S_n of the first n terms of a geometric sequence with first term a_1 and common ratio r is given by the formula $S_n = $ _____.

5. Given a geometric sequence with $|r| < 1$, the value of $r^n \rightarrow$ _____ as $n \rightarrow \infty$.

6. Given an infinite geometric series with first term a_1 and common ratio r, if $|r| < 1$, then the sum S is given by the formula $S = $ _____. If $|r| \geq 1$, then the sum (does/does not) exist.

7. A sequence of payments made at equal intervals over a fixed period of time is called an _____.

8. Suppose that an infinite series $a_1 + a_2 + a_3 + \cdots + a_n$ approaches a value L as $n \rightarrow \infty$. Then the series _____. Otherwise, the series _____.

Objective 1: Identify Specific and General Terms of a Geometric Sequence

For Exercises 9–18, determine whether the sequence is geometric. If so, find the value of r. (See Example 1)

9. 6, 18, 54, 162, ...

10. 4, 20, 100, 500, ...

11. $-7, \frac{7}{2}, -\frac{7}{4}, \frac{7}{8}, \ldots$

12. $5, -\frac{5}{3}, \frac{5}{9}, -\frac{5}{27}, \ldots$

13. 3, 12, 60, 360, ...

14. 7, 14, 42, 88, ...

15. $\sqrt{5}, 5, 5\sqrt{5}, 25, \ldots$

16. $\sqrt[3]{7}, \sqrt[3]{49}, 7, 7\sqrt[3]{7}, \ldots$

17. $2, \frac{4}{t}, \frac{8}{t^2}, \frac{16}{t^3}, \ldots$

18. $\frac{5}{a^2}, \frac{15}{a^4}, \frac{45}{a^6}, \frac{135}{a^8}, \ldots$

For Exercises 19–24, write the first five terms of a geometric sequence $\{a_n\}$ based on the given information about the sequence. (See Example 2)

19. $a_1 = 7$ and $r = 2$

20. $a_1 = 2$ and $r = 3$

21. $a_1 = 24$ and $r = -\frac{2}{3}$

22. $a_1 = 80$ and $r = -\frac{4}{5}$

23. $a_1 = 36$ and $a_n = \frac{1}{2}a_{n-1}$ for $n \geq 2$

24. $a_1 = 27$, $a_n = \frac{1}{3}a_{n-1}$ for $n \geq 2$

For Exercises 25–30, write a formula for the nth term of the geometric sequence. (See Example 3)

25. 5, 10, 20, 40, ...

26. 3, 6, 12, 24, ...

27. $-2, -1, -\frac{1}{2}, -\frac{1}{4}, \ldots$

28. $-8, -2, -\frac{1}{2}, -\frac{1}{8}, \ldots$

29. $\frac{16}{3}, -4, 3, -\frac{9}{4}, \ldots$

30. $\frac{18}{5}, -\frac{6}{5}, \frac{2}{5}, -\frac{2}{15}, \ldots$

31. A farmer depreciates a \$100,000 tractor. He estimates that the resale value of the tractor n years after purchase is 85% of its value from the previous year.

   a. Write a formula for the nth term of a sequence that represents the resale value of the tractor n years after purchase.

   b. What will the resale value be 5 yr after purchase? Round to the nearest \$1000.

32. A Coulter Counter is a device used to count the number of microscopic particles in a fluid, most notably, cells in blood. A hospital depreciates a \$9000 Coulter Counter at a rate of 75% per year after purchase.

   a. Write a formula for the nth term of a sequence that represents the resale value of the device n years after purchase.

   b. What will the resale value be 4 yr after purchase? Round to the nearest \$100.

33. Doctors in a certain city report 24 confirmed cases of the flu to the health department. At that time, the health department declares a flu epidemic. If the number of reported cases increases by roughly 30% each week thereafter, find the number of cases 10 weeks after the initial report. Round to the nearest whole unit.

34. After a 5-yr slump in the real estate market, housing prices stabilize and even begin to appreciate in value. One homeowner buys a house for \$140,000 and finds that the value of the property increases by 3% per year thereafter. Assuming that the trend continues, find the value of the home 15 yr later. Round to the nearest \$1000.

For Exercises 35–42, find the indicated term of a geometric sequence from the given information. (See Example 4)

35. $a_1 = 12$ and $a_2 = -8$. Find the sixth term.

36. $a_1 = 16$ and $a_2 = -12$. Find the fifth term.

37. $a_1 = 2$ and $a_4 = 16$. Find a_{12}.

38. $a_1 = 4$ and $a_4 = 108$. Find a_{10}.

39. $a_2 = -6$ and $r = \frac{1}{2}$. Find a_7.

40. $a_2 = -15$ and $r = \frac{1}{3}$. Find a_8.

41. $a_5 = -\frac{16}{9}$ and $r = -\frac{2}{3}$. Find a_1.

42. $a_6 = \frac{5}{16}$ and $r = -\frac{1}{2}$. Find a_1.

43. If the second and third terms of a geometric sequence are 15 and 75, what is the first term?

44. If the second and third terms of a geometric sequence are 4 and 1, what is the first term?

For Exercises 45–48, find a_1 and r for a geometric sequence $\{a_n\}$ from the given information. (See Example 5)

45. $a_2 = 18$ and $a_5 = 144$

46. $a_2 = 21$ and $a_7 = 5103$

47. $a_3 = 72$ and $a_6 = -\frac{243}{8}$

48. $a_3 = 45$ and $a_6 = -\frac{243}{25}$

Objectives 2–3: Evaluate Finite and Infinite Geometric Series

For Exercises 49–72, find the sum of the geometric series, if possible. (See Examples 6–8)

49. $\displaystyle\sum_{n=1}^{10} 3(2)^{n-1}$

50. $\displaystyle\sum_{n=1}^{8} 4(3)^{n-1}$

51. $\displaystyle\sum_{k=1}^{7} 6\left(\frac{2}{3}\right)^{k-1}$

52. $\displaystyle\sum_{j=1}^{7} 2\left(\frac{3}{4}\right)^{j-1}$

53. $15 + 5 + \dfrac{5}{3} + \dfrac{5}{9} + \dfrac{5}{27} + \dfrac{5}{81}$

54. $50 + 10 + 2 + \dfrac{2}{5} + \dfrac{2}{25} + \dfrac{2}{125}$

55. $2 + 6 + 18 + \cdots + 13{,}122$

56. $4 + 12 + 36 + \cdots + 78{,}732$

57. $1 + \dfrac{2}{3} + \dfrac{4}{9} + \cdots + \dfrac{32}{243}$

58. $\dfrac{8}{3} + 2 + \dfrac{3}{2} + \cdots + \dfrac{243}{512}$

59. $1 + \dfrac{1}{5} + \dfrac{1}{25} + \dfrac{1}{125} + \cdots$

60. $1 + \dfrac{1}{6} + \dfrac{1}{36} + \dfrac{1}{216} + \cdots$

61. $-2 - \dfrac{1}{2} - \dfrac{1}{8} - \dfrac{1}{32} - \cdots$

62. $-5 - 1 - \dfrac{1}{5} - \dfrac{1}{25} - \cdots$

63. $2 + 8 + 32 + 128 + \cdots$

64. $1 + 6 + 36 + 216 + \cdots$

65. $\displaystyle\sum_{j=1}^{\infty} \left(\frac{2}{3}\right)^{j-1}$

66. $\displaystyle\sum_{i=1}^{\infty} \left(\frac{3}{4}\right)^{i-1}$

67. $\displaystyle\sum_{k=1}^{\infty} \left(\frac{3}{2}\right)^{k-1}$

68. $\displaystyle\sum_{n=1}^{\infty} \left(\frac{4}{3}\right)^{n-1}$

69. $\displaystyle\sum_{i=1}^{12} 4(2)^i$ [*Hint*: Rewrite the expression within the summation so that the base of 2 appears to the $(i-1)$ power.]

70. $\displaystyle\sum_{j=1}^{10} 5(2)^j$

71. $\displaystyle\sum_{n=3}^{\infty} 4\left(\frac{1}{2}\right)^{n-1}$

72. $\displaystyle\sum_{n=4}^{\infty} 18\left(\frac{1}{3}\right)^{n-1}$

(*Hint*: Evaluate the infinite sum from $n = 1$ to infinity. Then subtract the terms corresponding to $n = 1$ and $n = 2$.)

For Exercises 73–80, write the repeating decimal as a fraction. (See Example 9)

73. $0.\overline{8}$

74. $0.\overline{2}$

75. $0.6\overline{4}$

76. $0.7\overline{8}$

77. $0.\overline{81}$

78. $0.\overline{72}$

79. $3.4\overline{25}$

80. $4.1\overline{62}$

81. Bike Week in Daytona Beach brings an estimated 500,000 people to the town. Suppose that each person spends an average of $300.

   a. How much money is infused into the local economy during Bike Week?

   b. If the money is respent in the community over and over again at a rate of 68%, determine the total amount spent. Assume that the money is respent an infinite number of times. (**See Example 10**)

82. An individual with questionable integrity prints and spends $12,000 in counterfeit money. If the "money" is respent over and over again each time at a rate of 76%, determine the total amount spent. Assume that the "money" is respent an infinite number of times without being detected.

83. Rafael received an inheritance of $18,000. He saves $6480 and then spends $11,520 of the money on college tuition, books, and living expenses for school. If the money is respent over and over again in the community an infinite number of times, at a rate of 64%, determine the total amount spent.

84. A tax rebate returns $100 million to individuals in the community. Suppose that $25,000,000 is put into savings, and that $75,000,000 is spent. If the money is spent over and over again an infinite number of times, each time at a rate of 75%, determine the total amount spent.

Objective 4: Find the Value of an Annuity

For Exercises 85–86, find the value of an ordinary annuity in which regular payments of P dollars are made at the end of each compounding period, n times per year, at an interest rate r for t years. (See Example 11)

85. $P = \$200$, $n = 12$, $r = 5\%$, $t = 30$ yr

86. $P = \$100$, $n = 24$, $r = 5.5\%$, $t = 28$ yr

87. a. An employee invests $100 per month in an ordinary annuity. If the interest rate is 6%, find the value of the annuity after 20 yr. (**See Example 11**)

   b. If the employee invests $200 instead of $100 at 6%, find the value of the annuity after 20 yr. Compare the result to part (a).

   c. If the employee invests $100 per month in the annuity at 6% interest, find the value after 40 yr. Compare the result to part (a).

88. a. An employee invests $500 per month in an ordinary annuity. If the interest rate is 5%, find the value of the annuity after 18 yr.

   b. If the employee invests $1000 per month in the annuity instead of $500 at 5% interest, find the value of the annuity after 18 yr. Compare the result to part (a).

   c. If the employee invests $500 per month in the annuity at 5% interest, find the value of the annuity after 36 yr. Compare the result to part (a).

Mixed Exercises

89. a. Given a geometric sequence whose nth term is $a_n = 6(0.4)^n$, are the terms of this sequence increasing or decreasing?

   b. Given a geometric sequence whose nth term is $a_n = 3(1.4)^n$, are the terms of this sequence increasing or decreasing?

90. a. Given a geometric sequence whose nth term is $a_n = -8(0.2)^n$, are the terms of this sequence increasing or decreasing?

   b. Given a geometric sequence whose nth term is $a_n = -5(1.6)^n$, are the terms of this sequence increasing or decreasing?

For Exercises 91–94, match the sequence with its graph.

91. $a_n = 5(0.8)^n$

92. $a_n = 2(1.2)^n$

93. $a_n = 5(-0.8)^n$

94. $a_n = 2(-1.2)^n$

a.

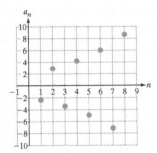

b.

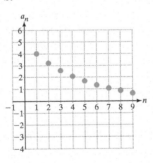

c.

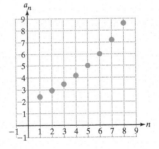

d.

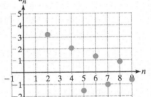

95. The initial swing (one way) of a pendulum makes an arc of 24 in. Each swing (one way) thereafter makes an arc of 98% of the length of the previous swing. What is the total arc length that the pendulum travels?

96. The initial swing (one way) of a pendulum makes an arc of 4 ft. Each swing (one way) thereafter makes an arc of 90% of the length of the previous swing. What is the total arc length that the pendulum travels?

97. A child drops a ball from a height of 4 ft. With each bounce, the ball rebounds to 50% of its original height. After the ball falls from its initial height of 4 ft, the vertical distance traveled for every bounce thereafter is doubled (the ball travels up and down).

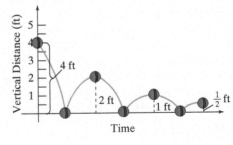

The total vertical distance traveled by the ball is given by

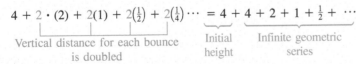

$$4 + 2 \cdot (2) + 2(1) + 2\left(\tfrac{1}{2}\right) + 2\left(\tfrac{1}{4}\right) \cdots = 4 + \underbrace{4 + 2 + 1 + \tfrac{1}{2} + \cdots}$$

Vertical distance for each bounce is doubled   Initial height   Infinite geometric series

Determine the total vertical distance traveled by the ball.

98. A ball is dropped from a height of 12 ft. With each bounce, the ball rebounds to $\frac{3}{4}$ of its height. Determine the total vertical distance traveled by the ball.

99. Suppose that an individual is paid $0.01 on day 1 and every day thereafter, the payment is doubled.

   a. Write a formula for the nth term of a sequence that gives the payment (in $) on day n.

   b. How much will the individual earn on day 10? day 20? and day 30?

   c. What is the total amount earned in 30 days?

100. The vibration of sound is measured in cycles per second, also called hertz (Hz). The frequency for middle C on a piano is 256 Hz. The C above middle C (one octave above) is 512 Hz. The frequencies of musical notes follow a geometric progression.

   a. Find the frequency for C two octaves above middle C.

   b. Find the frequency for C one octave *below* middle C.

101. The yearly salary for job A is $60,000 initially with an annual raise of $3000 every year thereafter. The yearly salary for job B is $56,000 for year 1 with an annual raise of 6%.

a. Consider a sequence representing the salary for job A for year n. Is this an arithmetic or geometric sequence? Find the total earnings for job A over 20 yr.

b. Consider a sequence representing the salary for job B for year n. Is this an arithmetic or geometric sequence? Find the total earnings for job B over 20 yr. Round to the nearest dollar.

c. What is the difference in total salary between the two jobs over 20 yr?

103. If a fair coin is flipped n times, the number of head/tail arrangements follows a geometric sequence. In the figure, if the coin is flipped 1 time, there are two possible outcomes, H or T. If the coin is flipped 2 times, then there are four possible outcomes: HH, HT, TH, and TT.

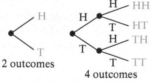

a. Write a formula for the nth term of a sequence representing the number of outcomes if a fair coin is flipped n times.

b. How many outcomes are there if a fair coin is flipped 10 times?

102. a. Jacob has a job that pays $48,000 the first year. He receives a 4% raise each year thereafter. Find the sum of his yearly salaries over a 20-yr period. Round to the nearest dollar.

b. Cherise has a job that pays $48,000 the first year. She receives a 4.5% raise each year thereafter. Find the sum of her yearly salaries over a 20-yr period. Round to the nearest dollar.

c. How much more will Cherise earn than Jacob over the 20-yr period?

104. An ancestor is a person from whom an individual is descended (a parent, a grandparent, a great-grandparent, and so on). Every individual has two biological parents, a mother and a father. The mother and father in turn each have two parents and so on.

1st Generation	2nd Generation	3rd Generation
2 parents	4 grandparents	8 great-grandparents

The terms of the sequence 2, 4, 8, 16, ... give the additional number of ancestors for each generation back in time. Determine the total number of ancestors that an individual has going back 12 generations.

Write About It

105. Explain the difference between an arithmetic sequence and a geometric sequence.

106. Explain why a finite number of terms is not sufficient to determine whether an infinite sequence is arithmetic or geometric. For example, explain why 4, 16, ... can be arithmetic, geometric, or neither.

107. A geometric sequence $a_1, a_2, a_3, \ldots$ has a common ratio r. Explain why the sequence $a_1, a_3, a_5, \ldots$ is also geometric and determine the common ratio.

Expanding Your Skills

108. Show that $x, x + 2, x + 4, x + 6, \ldots$ is *not* a geometric sequence.

110. Suppose that $a_1, a_2, a_3, \ldots$ is an arithmetic sequence with common difference d. Show that $10^{a_1}, 10^{a_2}, 10^{a_3}, 10^{a_4}, \ldots$ is a geometric sequence and find the common ratio r.

112. Given the series $\frac{1}{2} + \frac{1}{4} + \frac{1}{8} + \frac{1}{16} + \cdots$,

a. Find the sum.

b. How many terms must be taken so that the nth partial sum is within $\frac{1}{100}$ of the actual sum?

109. If $a_1, a_2, a_3, a_4, \ldots$ is a geometric sequence with common ratio r, show that $\frac{1}{a_1}, \frac{1}{a_2}, \frac{1}{a_3}, \frac{1}{a_4} \cdots$ is also a geometric sequence and determine the value of r.

111. Suppose that $a_1, a_2, a_3, \ldots$ is a geometric sequence with $r > 0$ and $a_1 > 0$. Show that the sequence $\log a_1, \log a_2, \log a_3, \ldots$ is arithmetic and find the common difference d.

113. Determine whether the sequence $\ln 1, \ln 2, \ln 4, \ln 8, \ldots$ is arithmetic or geometric. If the sequence is arithmetic, find d. If the sequence is geometric, find r.

PROBLEM RECOGNITION EXERCISES

Comparing Arithmetic and Geometric Sequences and Series

For Exercises 1–10, determine if the nth term of the sequence defines an arithmetic sequence, a geometric sequence, or neither. If the sequence is arithmetic, find the common difference d. If the sequence is geometric, find the common ratio r.

1. $a_n = \dfrac{2n}{3} + \dfrac{1}{4}$

2. $a_n = -\dfrac{3}{5}n - \dfrac{1}{3}$

3. $a_n = (-1)^{n-1}$

4. $a_n = (-2)^n$

5. $a_n = \dfrac{n-1}{n+1}$

6. $a_n = \dfrac{3n-4}{2n+1}$

7. $a_n = 5(\sqrt{5})^{n-1}$

8. $a_n = 6(\sqrt{2})^{n+1}$

9. $a_n = 5 + \sqrt{2}n$

10. $a_n = \dfrac{\sqrt{3}}{2}n + 4$

For Exercises 11–28, evaluate the sum if possible.

11. $\displaystyle\sum_{i=1}^{1506} 5$

12. $\displaystyle\sum_{i=1}^{417} (-3)$

13. $\displaystyle\sum_{i=1}^{716} (-1)^{i+1}$

14. $\displaystyle\sum_{j=1}^{2651} (-1)^{j}$

15. $\displaystyle\sum_{n=1}^{3} \dfrac{n-2}{n+1}$

16. $\displaystyle\sum_{j=1}^{4} (j^4 - 2j^2)$

17. $\displaystyle\sum_{n=1}^{4} -3\left(\dfrac{1}{3}\right)^{n-1}$

18. $\displaystyle\sum_{n=1}^{5} -2\left(\dfrac{3}{4}\right)^{n-1}$

19. $\displaystyle\sum_{n=1}^{\infty} 6\left(\dfrac{5}{3}\right)^{n-1}$

20. $\displaystyle\sum_{i=1}^{\infty} 7\left(\dfrac{4}{3}\right)^{i-1}$

21. $\displaystyle\sum_{i=1}^{27} (-6i - 4)$

22. $\displaystyle\sum_{i=1}^{41} (-5i - 3)$

23. $\displaystyle\sum_{n=1}^{\infty} 8\left(\dfrac{1}{2}\right)^{n}$

24. $\displaystyle\sum_{n=1}^{\infty} 12\left(\dfrac{1}{3}\right)^{n}$

25. $36 + 30 + 25 + \dfrac{125}{6} + \cdots$

26. $54 + 36 + 24 + 16 + \cdots$

27. $3 + 11 + 19 + 27 + \cdots + 363$

28. $36 + 29 + 22 + 15 + \cdots + (-419)$

SECTION 8.4   Mathematical Induction

1. Prove a Statement Using Mathematical Induction

In this section, we present a technique of mathematical proof that enables us to prove the validity of a statement that is true over the set of positive integers. This technique is called mathematical induction.

Consider the following sum:

$$\frac{1}{1 \cdot 2} + \frac{1}{2 \cdot 3} + \frac{1}{3 \cdot 4} + \frac{1}{4 \cdot 5} + \cdots + \frac{1}{n(n+1)}$$

The sequence of terms that make up the sum is neither arithmetic nor geometric. Therefore, we have no formula readily available to evaluate the sum of the first n terms.

We might consider evaluating the nth partial sum for several positive integers n to determine if a pattern exists. For each value of n, we have a corresponding statement P_n involving the value of the partial sum.

$n = 1$ $\dfrac{1}{1 \cdot 2} = \dfrac{1}{2}$ Call this statement P_1.

$n = 2$ $\dfrac{1}{1 \cdot 2} + \dfrac{1}{2 \cdot 3} = \dfrac{1}{2} + \dfrac{1}{6} = \dfrac{2}{3}$ Call this statement P_2.

$n = 3$ $\dfrac{1}{1 \cdot 2} + \dfrac{1}{2 \cdot 3} + \dfrac{1}{3 \cdot 4} = \dfrac{1}{2} + \dfrac{1}{6} + \dfrac{1}{12} = \dfrac{3}{4}$ Call this statement P_3.

$n = 4$ $\dfrac{1}{1 \cdot 2} + \dfrac{1}{2 \cdot 3} + \dfrac{1}{3 \cdot 4} + \dfrac{1}{4 \cdot 5} = \dfrac{1}{2} + \dfrac{1}{6} + \dfrac{1}{12} + \dfrac{1}{20} = \dfrac{4}{5}$ Call this statement P_4.

> **TIP** The proof of the statement
> $$\dfrac{1}{1 \cdot 2} + \dfrac{1}{2 \cdot 3} + \cdots + \dfrac{1}{n(n + 1)}$$
> $$= \dfrac{n}{n + 1}$$
> is given in Example 2.

For the first four terms, the numerator is the same as the term number n and the denominator is one more than the term number. From this observation, we might make the following hypothesis, which we call P_n.

$$P_n: = \dfrac{1}{1 \cdot 2} + \dfrac{1}{2 \cdot 3} + \dfrac{1}{3 \cdot 4} + \dfrac{1}{4 \cdot 5} + \cdots + \dfrac{1}{n(n + 1)} = \dfrac{n}{n + 1}$$

In words, the hypothesis P_n suggests that the sum of the first n terms of the sequence $\left\{\dfrac{1}{n(n + 1)}\right\}$ is $\dfrac{n}{n + 1}$. Equivalently, $\displaystyle\sum_{i=1}^{n} \dfrac{1}{i(i + 1)} = \dfrac{n}{n + 1}$.

We have shown that the statement P_n is true for $n = 1, 2, 3,$ and 4, but what about values of n thereafter? Because there are infinitely many positive integers, there are actually infinitely many statements to prove and we cannot approach them on a case-by-case basis. Instead we will use the principle of mathematical induction.

Principle of Mathematical Induction

Let P_n be a statement involving the positive integer n, and let k be an arbitrary positive integer. Then P_n is true for all positive integers n if

1. P_1 is true, and
2. The truth of P_k implies the truth of P_{k+1}.

P_n represents the statements, P_1, P_2, P_3, and so on. Mathematical induction is a two-part process to prove all the statements in the sequence. We first need to show that the statement is true for $n = 1$. That is, prove that P_1 is true. The second part involves proving that if any statement P_k in the sequence is true, then the statement that follows is also true (P_k implies P_{k+1}).

Intuitively, mathematical induction is similar to the sequential effect of falling dominos. Pushing the first domino down is analogous to proving P_1. Then showing that any falling domino in the sequence will cause the next domino to fall is the idea behind part 2 in mathematical induction. That is, if domino 1 falls, and any falling domino makes the next domino in line fall, then all the dominos will fall.

In Examples 1 and 2, we use mathematical induction to prove a statement involving a summation.

EXAMPLE 1 **Using Mathematical Induction**

Use mathematical induction to prove that $1 + 3 + 5 + 7 + \cdots + (2n - 1) = n^2$.

Solution:

Let P_n denote the statement $1 + 3 + 5 + 7 + \cdots + (2n - 1) = n^2$.

1. Show that P_1 is true.
 For $n = 1$, the sum is 1 which equals $(1)^2$. | We need to show that P_1 is true. That is, we need to show that the left and right
 Therefore, P_1 is true. | sides of the statement are equal for $n = 1$.

2. Next, assume that P_k is true and show that P_{k+1} is true.
 Assume that P_k is true.
 Assume that $1 + 3 + 5 + 7 + \cdots + (2k - 1) = k^2$. | The statement that P_k is true
 We must show that P_{k+1} is true. | is called the **inductive hypothesis.**

 We must show that $1 + 3 + 5 + \cdots + (2k - 1) + [2(k + 1) - 1] = (k + 1)^2$. (1)

> **Avoiding Mistakes**
>
> It is important to use parentheses when substituting $(k + 1)$ into a statement for n.

The left side of statement P_{k+1} can be written as

$$\overbrace{[1 + 3 + 5 + 7 + \cdots + (2k - 1)]}^{\text{Sum of first } k \text{ terms}} + \overbrace{[2(k + 1) - 1]}^{(k + 1)\text{th term}}$$

$$= \quad k^2 \qquad\qquad\qquad + [2(k + 1) - 1] \quad \text{By the inductive hypothesis, we can replace the sum of the first } k \text{ terms by } k^2.$$

$$= k^2 + (2k + 2 - 1)$$
$$= k^2 + 2k + 1$$
$$= (k + 1)^2 \text{ as desired. This matches equation (1).}$$

We have shown that P_1 is true, and that if P_k is true, then P_{k+1} is true. By mathematical induction, we conclude that the statement is true for all positive integers.

> **Skill Practice 1** Use mathematical induction to prove
> $P_n: 2 + 4 + 6 + 8 + \cdots + 2n = n(n + 1)$.

In Example 2, we prove the statement introduced at the beginning of this section.

EXAMPLE 2 **Using Mathematical Induction**

Use mathematical induction to prove that

$$\frac{1}{1 \cdot 2} + \frac{1}{2 \cdot 3} + \frac{1}{3 \cdot 4} + \cdots + \frac{1}{n(n + 1)} = \frac{n}{n + 1}.$$

Solution:

Let P_n denote the statement $\dfrac{1}{1 \cdot 2} + \dfrac{1}{2 \cdot 3} + \dfrac{1}{3 \cdot 4} + \cdots + \dfrac{1}{n(n + 1)} = \dfrac{n}{n + 1}$.

1. Show that P_1 is true.

For $n = 1$, the sum equals $\underbrace{\dfrac{1}{1 \cdot 2} = \dfrac{1}{2}}_{\text{left side of statement}}$ which is the same as $\underbrace{\dfrac{n}{n + 1} = \dfrac{1}{1 + 1} = \dfrac{1}{2}}_{\text{right side of statement}}$.

Answer
1. See page SA-46.

2. Next, assume that P_k is true.

Assume that $\dfrac{1}{1 \cdot 2} + \dfrac{1}{2 \cdot 3} + \cdots + \dfrac{1}{k(k+1)} = \dfrac{k}{k+1}$.

Assume that P_k is true. This is the inductive hypothesis.

We must show that P_{k+1} is true.

We must show that

$$\frac{1}{1 \cdot 2} + \frac{1}{2 \cdot 3} + \cdots + \frac{1}{k(k+1)} + \frac{1}{(k+1)[(k+1)+1]} = \frac{k+1}{(k+1)+1} = \frac{k+1}{k+2}. \qquad (1)$$

The left side of statement P_{k+1} can be written as

$$\underbrace{\frac{1}{1 \cdot 2} + \frac{1}{2 \cdot 3} + \cdots + \frac{1}{k(k+1)}}_{\text{Sum of first } k \text{ terms}} + \underbrace{\frac{1}{(k+1)[(k+1)+1]}}_{(k+1)\text{th term}}$$

$$= \frac{k}{k+1} + \frac{1}{(k+1)[(k+1)+1]} \qquad \text{By the inductive hypothesis, replace the sum of the first } k \text{ terms by } \frac{k}{k+1}.$$

$$= \frac{k}{k+1} + \frac{1}{(k+1)(k+2)} \qquad \text{Simplify.}$$

$$= \frac{k}{(k+1)} \cdot \frac{(k+2)}{(k+2)} + \frac{1}{(k+1)(k+2)} \qquad \text{Write terms with a common denominator.}$$

$$= \frac{k^2 + 2k + 1}{(k+1)(k+2)} \qquad \text{Add the fractions.}$$

$$= \frac{(k+1)^2}{(k+1)(k+2)} \qquad \text{Factor the numerator.}$$

$$= \frac{k+1}{k+2} \text{ as desired. This matches equation (1).}$$

We have shown that P_1 is true, and that if P_k is true, then P_{k+1} is true. By mathematical induction, we conclude that the statement is true for all positive integers.

Skill Practice 2   Use mathematical induction to prove that

$$P_n: \frac{1}{1 \cdot 3} + \frac{1}{3 \cdot 5} + \frac{1}{5 \cdot 7} + \cdots + \frac{1}{(2n-1)(2n+1)} = \frac{n}{2n+1}.$$

Mathematical induction can be used to prove the following useful summation formulas involving powers of the first n positive integers (Table 8-3). The proofs of these formulas are addressed in Exercises 17–20.

> **Table 8-3   Sums of Powers**
>
> Let n represent a positive integer. Then,
>
> 1. $\displaystyle\sum_{i=1}^{n} 1 = n$ $\qquad\qquad$ 2. $\displaystyle\sum_{i=1}^{n} i = \frac{n(n+1)}{2}$
>
> 3. $\displaystyle\sum_{i=1}^{n} i^2 = \frac{n(n+1)(2n+1)}{6}$ $\qquad$ 4. $\displaystyle\sum_{i=1}^{n} i^3 = \frac{n^2(n+1)^2}{4}$

Answer

2. See page SA-46.

In Examples 3 and 4, we use mathematical induction to prove statements that do not involve a sum.

EXAMPLE 3   **Using Mathematical Induction**

Use mathematical induction to prove that 4 is a factor of $9^n - 1$.

Solution:

Let P_n be the statement: 4 is a factor of $9^n - 1$.

To show that 4 is a factor of any expression, we need to write the expression as the product of 4 and some positive integer that we call a.

1. Show that P_n is true for $n = 1$.

   P_1 reads: 4 is a factor of $9^1 - 1 = 8$.

   P_1 is true because $8 = 4 \cdot 2$, indicating that 4 is indeed a factor of 8.

2. Next, assume that P_k is true and show that P_{k+1} is true:

   Assume that 4 is a factor of $9^k - 1$.

   This implies that $9^k - 1 = 4a$ or equivalently $9^k = 4a + 1$.   Inductive hypothesis

   We must show that 4 is a factor of $9^{k+1} - 1$.

   $$9^{k+1} - 1 = 9 \cdot 9^k - 1$$

   $\qquad\qquad = 9 \cdot (4a + 1) - 1$ From the inductive hypothesis, replace 9^k by $4a + 1$.

   $\qquad\qquad = 36a + 9 - 1$ Apply the distributive property.

   $\qquad\qquad = 36a + 8$

   $\qquad\qquad = 4(9a + 2)$ Factor. The value 4 is a factor of the expression $4(9a + 2)$. Therefore, 4 is a factor of $9^{k+1} - 1$.

We have shown that P_1 is true, and that if P_k is true, then P_{k+1} is true. By mathematical induction, we conclude that the statement is true for all positive integers.

> **Skill Practice 3**   Use mathematical induction to prove that 2 is a factor of $5^n + 1$.

---

2. Prove a Statement Using the Extended Principle of Mathematical Induction

Mathematical induction can be extended to prove statements that might hold true only for integers greater than or equal to some positive integer j. In such a case, we use the extended principle of mathematical induction.

Extended Principle of Mathematical Induction

Let P_n be a statement involving the positive integer n. Then P_n is true for all positive integers $n \geq j$ if

1. P_j is true, and
2. For an integer k, if $k \geq j$, the truth of P_k implies the truth of P_{k+1}.

Answer

3. See page SA-47.

EXAMPLE 4 **Using the Extended Principle of Mathematical Induction**

Use mathematical induction to prove that $n! > 2^n$ for positive integers $n \geq 4$.

Solution:

Let P_n be the statement: $n! > 2^n$.

1. First show that P_n is true for $n = 4$.

   To prove that P_4 is true, show that $4! > 2^4$.
   $4! = 4 \cdot 3 \cdot 2 \cdot 1 = 24$ and $2^4 = 16$. Therefore, $4! > 2^4$, indicating that P_4 is true.

2. Next, assume that P_k is true for $k \geq 4$, and show that P_{k+1} is true.

   Assume that $k! > 2^k$ and show that $(k + 1)! > 2^{k+1}$ where $k \geq 4$.

   $$\begin{aligned}(k + 1)! &= (k + 1) \cdot k! \\ &> (k + 1)(2^k) \qquad \text{By the inductive hypothesis, } k! > 2^k. \text{ Therefore, we can} \\ & \qquad\qquad\qquad\quad \text{replace } k! \text{ by } 2^k \text{ and replace the } = \text{ sign with } >. \\ &> 2 \cdot 2^k \qquad\quad \text{For } k \geq 4, \text{ the expression } (k + 1) > 2. \text{ Therefore, we can} \\ & \qquad\qquad\qquad\quad \text{replace } (k + 1) \text{ by 2 and maintain the } > \text{ symbol.} \\ &= 2^{k+1}\end{aligned}$$

> **TIP** In part 2 of Example 4, we manipulate the expression $(k + 1)!$ on the left side to show that it is greater than the expression on the right side for $k \geq 4$.

From the string of inequalities we have shown that $(k + 1)! > 2^{k+1}$.

We have shown that P_4 is true, and that if P_k is true, then P_{k+1} is true. By mathematical induction, we conclude that $n! > 2^n$ is true for all integers, $n \geq 4$.

Skill Practice 4 Use mathematical induction to prove that $\left(\dfrac{3}{2}\right)^n > 2n$ for $n \geq 7$.

Point of Interest

Mathematical induction is one type of mathematical proof, but there are many other techniques, including proof by contradiction. This technique uses the premise that a statement's being false would imply a contradiction. This approach was used in the complex proof of Fermat's last theorem.

The equation $x^2 + y^2 = z^2$ has infinitely many positive integer solutions for a, b, and c. Such solutions are called Pythagorean triples, such as $a = 5$, $b = 12$, and $c = 13$. However, French mathematician Pierre de Fermat (1601–1665) posed the statement that the equation $x^n + y^n = z^n$ has *no* such solutions for positive integers $n > 2$. Fermat made this claim, now famously called Fermat's last theorem, while annotating a copy of Diophantus' *Arithmetika*. Fermat included a comment that he had "found a remarkable proof of this fact, but there is not enough space in the margin to write it."

Proof of this theorem eluded mathematicians for over 350 yr until Andrew Wiles of Princeton University announced in June of 1993 that he had solved the problem. Wiles's proof is extremely complex and was revised to correct a slight but critical flaw shortly after publication. The proof involves the use of mathematics that did not exist in Fermat's time, and for this reason mathematicians believe that Fermat did not have a simpler proof.

Answer

4. See page SA-47.

SECTION 8.4 Practice Exercises

Prerequisite Review

For Exercises R.1–R.4, factor completely.

R.1. $x^4 + x^2$

R.2. $-15r^4 - 60r^3$

R.3. $w^2 + 4w + 4$

R.4. $v^2 - 9v + 18$

For Exercises R.5–R.6, simplify.

R.5. $t - 4 + \dfrac{1}{t + 4}$

R.6. $\dfrac{9}{a(a - 3)} + \dfrac{3}{a}$

Concept Connections

1. Let P_n be a statement involving the positive integer n, and let k be an arbitrary positive integer. Proof by mathematical _____ indicates that P_n is true for all positive integers n if (1) _____ is true, and (2) the truth of P_k implies the truth of _____.

2. The statement that P_k is true is called the _____ hypothesis.

Objective 1: Prove a Statement Using Mathematical Induction

For Exercises 3–16, use mathematical induction to prove the given statement for all positive integers n. (See Examples 1–2)

3. $2 + 6 + 10 + \cdots + (4n - 2) = 2n^2$

4. $2 + 8 + 14 + \cdots + (6n - 4) = n(3n - 1)$

5. $5 + 8 + 11 + \cdots + (3n + 2) = \dfrac{n}{2}(3n + 7)$

6. $4 + 9 + 14 + \cdots + (5n - 1) = \dfrac{n}{2}(5n + 3)$

7. $8 + 4 + 0 + \cdots + (-4n + 12) = -2n(n - 5)$

8. $12 + 6 + 0 + \cdots + (-6n + 18) = -3n(n - 5)$

9. $1 + 2 + 2^2 + 2^3 + \cdots + 2^{n-1} = 2^n - 1$

10. $1 + 3 + 3^2 + 3^3 + \cdots + 3^{n-1} = \dfrac{1}{2}(3^n - 1)$

11. $\dfrac{3}{4} + \dfrac{3}{16} + \dfrac{3}{64} + \cdots + \dfrac{3}{4^n} = 1 - \left(\dfrac{1}{4}\right)^n$

12. $\dfrac{1}{2} + \dfrac{1}{4} + \dfrac{1}{8} + \cdots + \dfrac{1}{2^n} = 1 - \left(\dfrac{1}{2}\right)^n$

13. $1 \cdot 2 + 2 \cdot 3 + 3 \cdot 4 + \cdots + n(n + 1)$
$= \dfrac{n(n + 1)(n + 2)}{3}$

14. $1 \cdot 3 + 2 \cdot 5 + 3 \cdot 7 + \cdots + n(2n + 1)$
$= \dfrac{n(n + 1)(4n + 5)}{6}$

15. $\left(1 - \dfrac{1}{2}\right)\left(1 - \dfrac{1}{3}\right)\left(1 - \dfrac{1}{4}\right)\cdots\left(1 - \dfrac{1}{n + 1}\right)$
$= \dfrac{1}{n + 1}$

16. $\left(1 - \dfrac{1}{2^2}\right)\left(1 - \dfrac{1}{3^2}\right)\left(1 - \dfrac{1}{4^2}\right)\cdots\left(1 - \dfrac{1}{(n + 1)^2}\right)$
$= \dfrac{n + 2}{2(n + 1)}$

For Exercises 17–24, use mathematical induction to prove the given statement for all positive integers n. (See Example 3)

17. $\displaystyle\sum_{i=1}^{n} 1 = n$

18. $\displaystyle\sum_{i=1}^{n} i = \dfrac{n(n + 1)}{2}$

19. $\displaystyle\sum_{i=1}^{n} i^2 = \dfrac{n(n + 1)(2n + 1)}{6}$

20. $\displaystyle\sum_{i=1}^{n} i^3 = \dfrac{n^2(n + 1)^2}{4}$

21. 2 is a factor of $5^n - 3$.

22. 2 is a factor of $7^n - 3$.

23. $4^n - 1$ is divisible by 3.

24. $5^n - 1$ is divisible by 4.

Objective 2: Prove a Statement Using the Extended Principle of Mathematical Induction

For Exercises 25–28, use trial-and-error to determine the smallest positive integer n for which the given statement is true.

25. $n! > 3^n$

26. $(n + 1)! > 4^n$

27. $3n < 2^n$

28. $5n < 3^n$

For Exercises 29–32, use mathematical induction to prove the given statement. (See Example 4)

29. $n! > 3^n$ for positive integers $n \geq 7$.

30. $(n + 1)! > 4^n$ for positive integers $n \geq 6$.

31. $3n < 2^n$ for positive integers $n \geq 4$.

32. $5n < 3^n$ for positive integers $n \geq 3$.

Mixed Exercises

For Exercises 33–36, use mathematical induction to prove the given statement for all positive integers n and real numbers x and y.

33. $(xy)^n = x^n y^n$ **34.** $\left(\dfrac{x}{y}\right)^n = \dfrac{x^n}{y^n}$ provided that $y \neq 0$.   **35.** If $x > 1$, then $x^n > x^{n-1}$.   **36.** If $0 < x < 1$, then $x^n < x^{n-1}$.

We have used mathematical induction to prove that a statement is true for all positive integers n. To show that a statement is *not* true, all we need is one case in which the statement is false. This is called a **counterexample**. For Exercises 37–38, find a counterexample to show that the given statement is *not* true.

37. The expression $n^2 - n + 11$ is prime for all positive integers n.

38. The inequality $5^n > n!$ is true for all positive integers n.

Write About It

39. Explain the difference between the principle of mathematical induction and the extended principle of mathematical induction.

40. Suppose that a fellow student showed that the expression $n^2 + n + 1$ is prime for $n = 1, 2,$ and 3. Explain why this is not a sufficient proof that the expression is prime for all positive integers n.

Expanding Your Skills

41. Show that $n^2 - n$ is even for all positive integers n.

42. Show that $n^2 - n + 1$ is odd for all positive integers n.

For Exercises 43–44, use the Fibonacci sequence $\{F_n\} = \{1, 1, 2, 3, 5, 8, 13, \ldots\}$. Recall that the Fibonacci sequence can be defined recursively as $F_1 = 1$, $F_2 = 1$, and $F_n = F_{n-1} + F_{n-2}$ for $n \geq 3$.

43. Prove that $F_1 + F_2 + F_3 + \cdots + F_n = F_{n+2} - 1$ for positive integers $n \geq 3$.

44. Prove that $F_1 + F_3 + F_5 + \cdots + F_{2n-1} = F_{2n}$ for all positive integers n.

SECTION 8.5   The Binomial Theorem

OBJECTIVES

1. Determine Binomial Coefficients
2. Apply the Binomial Theorem
3. Find a Specific Term in a Binomial Expansion

1. Determine Binomial Coefficients

In Section R.4 we learned how to square a binomial.

$$(a + b)^2 = a^2 + 2ab + b^2$$

The expression $a^2 + 2ab + b^2$ is called the **binomial expansion** of $(a + b)^2$. To expand $(a + b)^3$, we can find the product $(a + b)(a + b)^2$.

$$(a + b)(a + b)^2 = (a + b)(a^2 + 2ab + b^2)$$

$$= a^3 + 2a^2b + ab^2 + a^2b + 2ab^2 + b^3$$
$$= a^3 + 3a^2b + 3ab^2 + b^3$$

Similarly, to expand $(a + b)^4$, we can multiply $(a + b)$ by $(a + b)^3$. Using this method, we can expand several powers of $(a + b)$.

$(a + b)^0 = 1$

$(a + b)^1 = 1a + 1b$

$(a + b)^2 = 1a^2 + 2ab + 1b^2$

$(a + b)^3 = 1a^3 + 3a^2b + 3ab^2 + 1b^3$

$(a + b)^4 = 1a^4 + 4a^3b + 6a^2b^2 + 4ab^3 + 1b^4$

$(a + b)^5 = 1a^5 + 5a^4b + 10a^3b^2 + 10a^2b^3 + 5ab^4 + 1b^5$

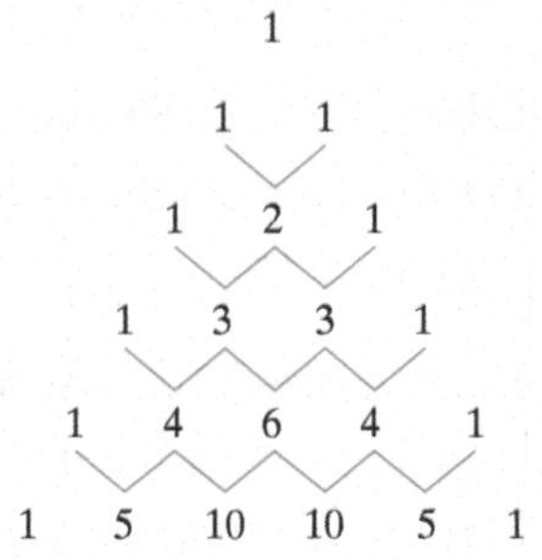

Figure 8-9

From the expansion of $(a + b)^n$, we note the following patterns.

- The exponent on a decreases from n to 0 on sequential terms from left to right.
- The exponent on b increases from 0 to n on sequential terms from left to right.
- The sum of the exponents on each term (that is, the degree of each term) is n.
- The number of terms in the expansion is $(n + 1)$. For example, the expansion of $(a + b)^4$ has five terms.

The coefficients for the expansion of $(a + b)^n$ follow a triangular array of numbers called **Pascal's triangle** (Figure 8-9), named after the French mathematician Blaise Pascal (1623–1662). Each row begins and ends with a 1, and each entry in between is the sum of the two diagonal entries from the row above. For example, in the last row of Figure 8-9, we have

$$1, \quad 1 + 4 = 5, \quad 4 + 6 = 10, \quad 6 + 4 = 10, \quad 4 + 1 = 5, \text{ and } 1$$

EXAMPLE 1   Expanding a Binomial

Expand. $(a + b)^6$

Solution:

The expansion of $(a + b)^6$ will have 7 terms (one more than the exponent of 6). Using the entries of the seventh row of Pascal's triangle as the coefficients, the expansion of $(a + b)^6$ is:

$$1$$
$$1 \quad 1$$
$$1 \quad 2 \quad 1$$
$$1 \quad 3 \quad 3 \quad 1$$
$$1 \quad 4 \quad 6 \quad 4 \quad 1$$
$$1 \quad 5 \quad 10 \quad 10 \quad 5 \quad 1$$
$$1 \quad 6 \quad 15 \quad 20 \quad 15 \quad 6 \quad 1$$

$$(a + b)^6 = 1a^6 + 6a^5b + 15a^4b^2 + 20a^3b^3 + 15a^2b^4 + 6ab^5 + 1b^6$$

Skill Practice 1   Expand. $(a + b)^7$

Determining the coefficients of a binomial expansion using Pascal's triangle would be cumbersome for expansions of higher degree. Instead we can compute the coefficients of a binomial expansion using the following formula.

TIP The expression $\binom{n}{r}$ will also be used in Section 8.6 when we study counting principles. $\binom{n}{r}$ represents the number of ways we can *choose* a group of *r* items in any order from a group of *n* items.

Coefficients of a Binomial Expansion

Let n and r be nonnegative integers with $n \geq r$. The expression $\binom{n}{r}$ (read as "n choose r") is called a **binomial coefficient** and is defined by

$$\binom{n}{r} = \frac{n!}{r!(n - r)!}$$

Note: The notation $_nC_r$ is often used in place of $\binom{n}{r}$.

EXAMPLE 2   Computing Binomial Coefficients

Evaluate.

a. $\binom{6}{0}$ **b.** $\binom{6}{1}$ **c.** $\binom{6}{2}$ **d.** $\binom{6}{6}$

Answer

1. $a^7 + 7a^6b + 21a^5b^2 + 35a^4b^3 + 35a^3b^4 + 21a^2b^5 + 7ab^6 + b^7$

Solution:

For parts (a)–(d), apply $\dbinom{n}{r} = \dfrac{n!}{r!(n-r)!}$

a. $\dbinom{6}{0} = \dfrac{6!}{0! \cdot (6-0)!} = \dfrac{6!}{0! \cdot 6!} = \dfrac{6!}{1 \cdot 6!} = 1$ $n = 6$ and $r = 0$.
Recall that $0! = 1$.

b. $\dbinom{6}{1} = \dfrac{6!}{1! \cdot (6-1)!} = \dfrac{6!}{1! \cdot 5!} = \dfrac{6 \cdot 5!}{1 \cdot 5!} = 6$ $n = 6$ and $r = 1$.

c. $\dbinom{6}{2} = \dfrac{6!}{2! \cdot (6-2)!} = \dfrac{6!}{2! \cdot 4!} = \dfrac{6 \cdot 5 \cdot 4!}{2 \cdot 1 \cdot 4!} = 15$ $n = 6$ and $r = 2$.

d. $\dbinom{6}{6} = \dfrac{6!}{6! \cdot (6-6)!} = \dfrac{6!}{6! \cdot 0!} = \dfrac{6!}{6! \cdot 1} = 1$ $n = 6$ and $r = 6$.

Skill Practice 2   Evaluate.

a. $\dbinom{5}{0}$ **b.** $\dbinom{5}{1}$ **c.** $\dbinom{5}{2}$ **d.** $\dbinom{5}{5}$

TECHNOLOGY CONNECTIONS

Evaluating Binomial Coefficients

To evaluate binomial coefficients on a graphing utility, use the $_nC_r$ function found in the **MATH** menu under PRB.

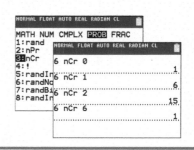

2. Apply the Binomial Theorem

From Example 2, notice that $\dbinom{6}{0} = 1$, $\dbinom{6}{1} = 6$, and $\dbinom{6}{2} = 15$ are the coefficients for the first three terms of the expansion of $(a + b)^6$. The value $\dbinom{6}{6} = 1$ is the coefficient of the last term of the expansion. Using the values of $\dbinom{n}{r}$ for the coefficients, we have a formula for the expansion of the binomial $(a + b)^n$. This is called the **binomial theorem.**

> **TIP** The binomial theorem can be proved by using mathematical induction. See the proof in the online appendix at www.mhhe.com/millercollegealgebra.

The Binomial Theorem

Let n be a positive integer. The expansion of $(a + b)^n$ is given by

$$(a + b)^n = \dbinom{n}{0}a^n + \dbinom{n}{1}a^{n-1}b + \dbinom{n}{2}a^{n-2}b^2 + \cdots + \dbinom{n}{n-1}ab^{n-1} + \dbinom{n}{n}b^n$$

$$= \sum_{r=0}^{n} \dbinom{n}{r}a^{n-r}b^r$$

Answers
2. a. 1   **b.** 5   **c.** 10   **d.** 1

> **EXAMPLE 3**   **Applying the Binomial Theorem**
>
> Expand by using the binomial theorem. $(2x + 3)^4$
>
> **Solution:**
>
> The expression $(2x + 3)^4$ is in the form $(a + b)^4$ with $a = 2x$ and $b = 3$.
>
> $$(a + b)^4 = \binom{4}{0}a^4 + \binom{4}{1}a^3b + \binom{4}{2}a^2b^2 + \binom{4}{3}ab^3 + \binom{4}{4}b^4$$
>
> $$(2x + 3)^4 = \binom{4}{0}(2x)^4 + \binom{4}{1}(2x)^3(3) + \binom{4}{2}(2x)^2(3)^2 + \binom{4}{3}(2x)(3)^3 + \binom{4}{4}(3)^4$$
>
> $$= 1(16x^4) + 4(8x^3)(3) + 6(4x^2)(9) + 4(2x)(27) + 1(81)$$
> $$= 16x^4 + 96x^3 + 216x^2 + 216x + 81$$

> **Skill Practice 3**   Expand by using the binomial theorem. $(4y + 5)^4$

In Example 4, we expand a difference of terms.

> **EXAMPLE 4**   **Applying the Binomial Theorem**
>
> Expand by using the binomial theorem. $(3x^2 - 5y)^3$
>
> **Solution:**
>
> The expression $(3x^2 - 5y)^3 = [3x^2 + (-5y)]^3$ is in the form $(a + b)^3$ with $a = 3x^2$ and $b = -5y$.
>
> $$(a + b)^3 = \binom{3}{0}a^3 + \binom{3}{1}a^2b + \binom{3}{2}ab^2 + \binom{3}{3}b^3$$
>
> $$[3x^2 + (-5y)]^3 = \binom{3}{0}(3x^2)^3 + \binom{3}{1}(3x^2)^2(-5y) + \binom{3}{2}(3x^2)(-5y)^2 + \binom{3}{3}(-5y)^3$$
>
> $$= 1(27x^6) + 3(9x^4)(-5y) + 3(3x^2)(25y^2) + 1(-125y^3)$$
> $$= 27x^6 - 135x^4y + 225x^2y^2 - 125y^3$$

> **Skill Practice 4**   Expand by using the binomial theorem. $(2t^4 - 3v)^3$

3. Find a Specific Term in a Binomial Expansion

Consider the first four terms of the binomial expansion of $(a + b)^n$.

The exponent on b is one less than the term number.

$$\binom{n}{0}a^nb^0 + \binom{n}{1}a^{n-1}b^1 + \binom{n}{2}a^{n-2}b^2 + \binom{n}{3}a^{n-3}b^3$$

These values match the exponent on b.

From the observed pattern, the kth term has a factor of b raised to the $k - 1$ power. That is, the exponent on b is one less than the term number. Furthermore, the sum of the exponents must equal n. Therefore, the factor a is raised to the $n - (k - 1)$ power. That is, the exponent on a is n minus the exponent on b.

Answers

3. $256y^4 + 1280y^3 + 2400y^2$
$+ 2000y + 625$

4. $8t^{12} - 36t^8 v + 54t^4 v^2 - 27v^3$

---

kth Term of a Binomial Expansion

Let n and k be positive integers with $k \leq n + 1$. The kth term of $(a + b)^n$ is

$$\binom{n}{k-1}a^{n-(k-1)}b^{k-1}$$

EXAMPLE 5   Finding a Specific Term of a Binomial Expansion

Find the eighth term of $(2x + y^4)^{10}$.

Solution:

To find the eighth term of $(a + b)^{10}$, we require the exponent on b to be 7 (one less than the term number). Therefore, the exponent on a must be 3 so that the sum of the exponents is 10.

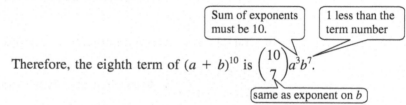

Therefore, the eighth term of $(a + b)^{10}$ is $\binom{10}{7}a^3 b^7$.

The eighth term of $(2x + y^4)^{10}$ is $\binom{10}{7}(2x)^3(y^4)^7 = 120(8x^3)(y^{28}) = 960x^3 y^{28}$.

Alternatively, we can apply the formula for the kth term of a binomial expansion with $n = 10$, $k = 8$, $a = 2x$, and $b = y^4$.

$$\binom{n}{k-1}a^{n-(k-1)}b^{k-1} = \binom{10}{8-1}(2x)^{10-(8-1)}(y^4)^{8-1} = \binom{10}{7}(2x)^3(y^4)^7 = 960x^3 y^{28}$$

The eighth term of the expansion is $960x^3 y^{28}$.

Skill Practice 5   Find the seventh term of $(3c + d^5)^9$.

Answer
5. $2268c^3 d^{30}$

SECTION 8.5   Practice Exercises

Prerequisite Review

For Exercises R.1–R.5, multiply and simplify.

R.1. $(5v + 2)^2$

R.2. $(8a^2 - 9b^3)^2$

R.3. $(4x + \sqrt{15})^2$

R.4. $\left(\frac{1}{5}m + 2\right)^2$

R.5. $(4a - 3b)^3$

Concept Connections

1. The expression $a^3 + 3a^2b + 3ab^2 + b^3$ is called the _____ expansion of $(a + b)^3$.

2. Consider $(a + b)^n$, where n is a whole number. How many terms are in the binomial expansion?

3. Consider $(a + b)^n$, where n is a whole number. What is the degree of each term in the expansion?

4. Consider $(a + b)^n$, where n is a whole number. The coefficients of the terms in the expansion can be found by using _____ triangle or by using $\binom{n}{r}$.

5. For positive integers n and k ($k \leq n + 1$), the kth term of $(a + b)^n$ is given by $\left(\right)a^{\square}b^{\square}$.

6. Given $(a + b)^{17}$, the 12th term is given by $\left(\right)a^{\square}b^{\square}$.

Objective 1: Determine Binomial Coefficients

7. a. Write the first five rows of Pascal's triangle.

   b. Write the expansion of $(x - y)^4$. (**See Example 1**)

8. a. Write the first four rows of Pascal's triangle.

   b. Write the expansion of $(c - d)^3$.

For Exercises 9–10, evaluate the given expressions. Compare the results to the fifth and sixth rows of Pascal's triangle. (See Example 2)

9. a. $\binom{4}{0}$ **b.** $\binom{4}{1}$ **c.** $\binom{4}{2}$

d. $\binom{4}{3}$ **e.** $\binom{4}{4}$

10. a. $\binom{5}{0}$ **b.** $\binom{5}{1}$ **c.** $\binom{5}{2}$

d. $\binom{5}{3}$ **e.** $\binom{5}{4}$ **f.** $\binom{5}{5}$

For Exercises 11–14, evaluate the expression.

11. $\binom{13}{3}$ **12.** $\binom{17}{15}$ **13.** $\binom{11}{5}$ **14.** $\binom{9}{4}$

Objective 2: Apply the Binomial Theorem

For Exercises 15–28, expand the binomial by using the binomial theorem. (See Examples 3–4)

15. $(3x + 1)^5$ **16.** $(5x + 3)^5$ **17.** $(7x + 3)^3$ **18.** $(6x + 5)^3$

19. $(2x - 5)^4$ **20.** $(4x - 1)^4$ **21.** $(2x^3 - y)^5$ **22.** $(3y^2 - z)^5$

23. $(p^2 - w^4)^6$ **24.** $(t^3 - v^5)^6$ **25.** $(0.2 + 0.1k)^4$ **26.** $(0.1 + 0.3m)^4$

27. $\left(\dfrac{c}{2} - d\right)^3$ **28.** $\left(\dfrac{x}{3} - n\right)^3$

Objective 3: Find a Specific Term in a Binomial Expansion

For Exercises 29–40, find the indicated term of the binomial expansion. (See Example 5)

29. $(m + n)^{10}$; seventh term   **30.** $(p + q)^{11}$; ninth term   **31.** $(c - d)^8$; fourth term

32. $(x - y)^9$; sixth term   **33.** $(u^2 + 2v^4)^{15}$; tenth term   **34.** $(y^3 + 2z^2)^{14}$; tenth term

35. $(\sqrt{3}x^2 + y^3)^9$; fourth term   **36.** $(\sqrt{2}u^3 + v^4)^8$; third term   **37.** $(h^4 - 1)^{12}$; middle term

38. $(k^3 - 1)^{10}$; middle term   **39.** $(p^4 + 3q)^8$; term containing p^{12}.   **40.** $(a^3 + 4b)^6$; term containing a^9.

Mixed Exercises

41. Expand $(e^x - e^{-x})^4$.

42. Expand $(e^x + e^{-x})^3$.

43. Use the binomial theorem to expand $(x + y - z)^3$.

44. Use the binomial theorem to expand $(a + b - 2)^3$.

45. Use the binomial theorem to find $(1.01)^4$. [*Hint*: Write the expression as $(1 + 0.01)^4$.]

46. Use the binomial theorem to find $(1.1)^5$.

47. Simplify. $(x + y)^3 - (x - y)^3$

48. Simplify. $(x + 1)^3 - (x - 1)^3$

For Exercises 49–52, simplify the difference quotient: $\dfrac{f(x + h) - f(x)}{h}$

49. $f(x) = 2x^3 + 4$ **50.** $f(x) = 4x^3 + 3$ **51.** $f(x) = x^4 - 5x^2 + 1$ **52.** $f(x) = x^4 - 6x^2 - 4$

For Exercises 53–56, use the binomial theorem to find the value of the complex number raised to the given power. Recall that $i = \sqrt{-1}$.

53. $(2 + i)^3$ **54.** $(3 + i)^3$ **55.** $(5 - 2i)^4$ **56.** $(6 - 5i)^4$

57. Show that $\binom{n}{r}$ and $\binom{n}{n-r}$ are equivalent.

58. Show that $\binom{n}{0} = 1$ and $\binom{n}{n} = 1$.

59. a. Write the sequence corresponding to the sum of the numbers in each row of Pascal's triangle for the first nine rows.

   b. Let n represent the row number in Pascal's triangle. Write a formula for the nth term of the sequence representing the sum of the numbers in row n.

Write About It

60. Explain why $\begin{pmatrix} -13 \\ 3 \end{pmatrix}$ is undefined.

61. Explain why $\begin{pmatrix} 3 \\ 5 \end{pmatrix}$ is undefined.

62. Explain why many graphing utilities give an error message for the expression $\dfrac{120!}{118!}$. How can this expression be evaluated by hand?

Expanding Your Skills

Stirling's formula (named after Scottish mathematician, James Stirling: 1692–1770) is used to approximate large values of $n!$. Stirling's formula is $n! \approx \sqrt{2\pi n}\left(\dfrac{n}{e}\right)^n$. For Exercises 63–64,

a. Use Stirling's formula to approximate the given expression. Round to the nearest whole unit.

b. Compute the actual value of the expression.

c. Determine the percent difference between the approximate value and the actual value. Round to the nearest tenth of a percent.

63. $8!$

64. $12!$

Using techniques from calculus, we can show that $(1 + x)^n = 1 + nx + \dfrac{n(n-1)x^2}{2!} + \dfrac{n(n-1)(n-2)}{3!}x^3 + \cdots$, for $|x| < 1$. This formula can be used to evaluate binomial expressions raised to noninteger exponents. For Exercises 65–66, use the first four terms of this infinite series to approximate the given expression. Round to 3 decimal places if necessary.

65. $(1.4)^{3/2}$ [*Hint*: $(1.4)^{3/2} = (1 + 0.4)^{3/2}$]

66. $\sqrt[4]{(1.3)^3}$

SECTION 8.6   Principles of Counting

OBJECTIVES

1. Apply the Fundamental Principle of Counting
2. Count Permutations
3. Count Combinations

1. Apply the Fundamental Principle of Counting

Many states have lotteries as a means to raise income. The game "Florida Lotto" for example, pays a grand prize to players who choose six distinct numbers from 1 to 53 (in any order) that match the same group of six numbers in the drawing. The number of such six-number combinations is 22,957,480 (see Example 8). If a player chooses one group of six numbers, then the probability (likelihood) of winning the grand prize is $\frac{1}{22,957,480}$. This value (roughly 1 in 23 million) means that it is highly unlikely to win the grand prize. To put this in perspective, the following events are more likely to happen than winning the grand prize in "Florida Lotto."

- A pedestrian being killed by a motor vehicle: Probability $\approx \frac{1}{1,000,000}$
- Flipping a coin 24 times and getting all heads: Probability $= \frac{1}{16,777,216}$
- Dying from a venomous insect or animal bite: Probability $\approx \frac{1}{100,000}$

In this section and in Section 8.7, we present basic principles of counting and how these principles apply to probability. We begin in Example 1, by using a figure called a tree diagram to organize different outcomes of a sequence of events.

EXAMPLE 1 Counting the Outcomes of a
Sequence of Events

Suppose that a frozen custard shop offers 4 flavors of custard, 2 types of syrup, and 2 toppings. If a customer can select 1 item from each group for a sundae, how many different sundaes can be made?

Custard Flavors	Syrups	Toppings
Vanilla (V)	Hot fudge (H)	Nuts (N)
Chocolate (C)	Butterscotch (B)	Granola (G)
Mint chip (M)		
Peanut butter (P)		

Solution:

We can depict the 4 flavors of custard by the left-most branches in Figure 8-10. Then for each branch of custard, there are 2 branches representing the type of syrup. Finally, for each custard-syrup arrangement, there are 2 toppings.

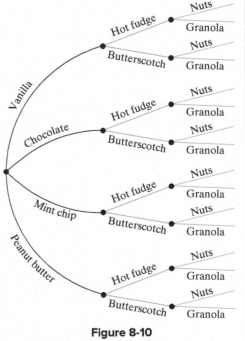

The tree diagram shows 16 different sundaes:

VHN, VHG, VBN, VBG, CHN, CHG, CBN, CBG, MHN, MHG, MBN, MBG, PHN, PHG, PBN, and PBG.

Notice that the number of outcomes is easily found by multiplying the number of choices from each group:

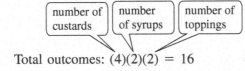

Total outcomes: $(4)(2)(2) = 16$ **Figure 8-10**

Skill Practice 1   A pizza can be made from either thick or thin crust, and a choice of 6 toppings. How many different types of pizza can be made?

The result from Example 1 suggests that the number of outcomes for the sequence of events is the product of the number of outcomes for each individual event. This is generalized as the **fundamental principle of counting**.

Fundamental Principle of Counting

If one event can occur in m different ways and a second event can occur in n different ways, then the number of ways that the two events can occur in sequence is $m \cdot n$.

In Example 2, we demonstrate that the fundamental principle of counting can be applied for a sequence of many events.

Answer

1. 12

EXAMPLE 2 Applying the Fundamental Principle of Counting

A computer password must have three letters, followed by two digits. How many passwords can be made if

a. There are no restrictions on the letters or digits?

b. No letter or digit may be used more than once?

Solution:

a. The computer password is a sequence of five characters. The number of arrangements of letters and digits is given by the product of the number of outcomes for each character. There are 26 letters of the alphabet from which to choose. The digits in the password may be selected from 0, 1, 2, 3, 4, 5, 6, 7, 8, 9. Therefore, the total number of passwords is given by

There are 26 choices for each letter.   There are 10 choices for each digit.

$$26 \cdot 26 \cdot 26 \cdot 10 \cdot 10 = 1{,}757{,}600$$

b. If no letter may be repeated, then the number of letters from which to choose must be decreased by one when the second letter is selected. Likewise, the number of letters must be decreased by one again when the third letter is selected. The same logic is used when selecting the digits. Once the first digit is selected in the password, there are only nine remaining digits available for the second choice.

The number of letters is decreased by one for each letter in the sequence.   There are 10 choices for the first digit, but only 9 remaining for the second.

$$26 \cdot 25 \cdot 24 \cdot 10 \cdot 9 = 1{,}404{,}000$$

Skill Practice 2 A code for an alarm system is made up of two letters, followed by four digits. How many codes can be made if

a. There are no restrictions on the letters or digits?

b. No letter or digit may be used more than once?

EXAMPLE 3 Applying the Fundamental Principle of Counting

A quiz has five true/false questions and five multiple-choice questions. The multiple-choice questions each have four possible answers of which only one is correct. If a student forgets to study and must guess on each question, in how many ways can the student answer the questions on the test?

Solution:

The quiz questions form a 10-stage event. Each true/false question has two possible choices (true or false). Each multiple-choice question has four possible choices (for example: a, b, c, or d). Therefore, the number of ways that the questions can be answered on the test is given by

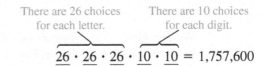

Each true/false question has 2 choices   Each multiple-choice question has 4 choices

$$2 \cdot 2 \cdot 2 \cdot 2 \cdot 2 \cdot 4 \cdot 4 \cdot 4 \cdot 4 \cdot 4 = 32{,}768$$

Answers
2. **a.** $26 \cdot 26 \cdot 10 \cdot 10 \cdot 10 \cdot 10 = 6{,}760{,}000$
   b. $26 \cdot 25 \cdot 10 \cdot 9 \cdot 8 \cdot 7 = 3{,}276{,}000$

> **Skill Practice 3**   A quiz has four true/false questions and six multiple-choice questions. The multiple-choice questions each have five possible answers of which only one is correct. In how many ways can the questions on the test be answered?

2. Count Permutations

We now look at a situation in which n items are to be arranged in order. Such an arrangement is called a **permutation.** For example, consider the different arrangements of the letters in the word FIVE.

FIVE	FIEV	FVIE	FVEI	FEIV	FEVI	
IFVE	IFEV	IVEF	IVFE	IEVF	IEFV	} 24 arrangements
VFIE	VFEI	VEIF	VEFI	VIEF	VIFE	
EFIV	EFVI	EIFV	EIVF	EVIF	EVFI	

There are 24 such arrangements. We can arrive at the same conclusion by applying the fundamental principle of counting. There are four letters available for the first choice in the arrangement. That leaves 3 letters remaining for the second choice, followed by 2 letters for the third choice, and only 1 letter for the last choice. Therefore, the number of permutations of 4 distinct letters is given by

$$4 \cdot 3 \cdot 2 \cdot 1 = 24$$

We generalize this as follows.

> **Number of Permutations of n Distinguishable Elements**
>
> The number of permutations of n distinct elements is $n!$.
>
> *Note*: This means that there are $n!$ ways to arrange n distinguishable items in various orders.

Now suppose that we wanted to arrange the letters in the word NINE. Notice that if we were to switch the first and third letters (both "N's") we would get the same arrangement. This is because the two N's are indistinguishable. This means that when counting the number of permutations we must "divide out" the number of ways that the two N's can be arranged. The two N's can be arranged in 2! different ways. This leads to the following result.

> **Number of Permutations of n Elements, Some Indistinguishable**
>
> Consider a set of n elements with r_1 duplicates of one kind, r_2 duplicates of a second kind, ... , r_k duplicates of a kth kind. Then the number of distinguishable permutations of the n elements of the set is
>
> $$\frac{n!}{r_1! \cdot r_2! \cdot \cdots \cdot r_k!}$$
>
> *Note*: The factors of $r_1!, r_2!, ... , r_k!$ in the denominator remove the repetition of arrangements that arise from the indistinguishable elements.

Answer

3. 250,000

EXAMPLE 4   Counting Permutations

a. Determine the number of ways that 6 people can be arranged in line at a ticket counter.

b. Determine the number of ways that the letters in the word MICROSCOPIC can be arranged.

Solution:

a. No two people can be the same. Therefore, we are arranging 6 different (distinguishable) items in various orders. The number of such permutations is given by

$6! = 720$ There are 720 ways in which 6 people can be arranged in line.

b. The word MICROSCOPIC has 11 letters. However, the letter I appears twice, the letter O appears twice, and the letter C appears 3 times. The number of unique arrangements of these letters is

$$\frac{11!}{2! \cdot 2! \cdot 3!} = 1{,}663{,}200$$

Skill Practice 4

a. In how many ways can 7 different books be arranged on a bookshelf?

b. Determine the number of ways that the letters in the word RIFFRAFF can be arranged.

In Example 5, we determine the number of permutations from a group of n items in which fewer than n items are selected at one time.

EXAMPLE 5   Finding Permutations of n Items Taken r at a Time

Suppose that 5 students (Alberto, Beth, Carol, Dennis, and Erik) submit applications for scholarships. There are two scholarships available. The first is for $1000 and the second is for $500. In how many ways can 2 students from the 5 be selected to receive the scholarships?

Solution:

We begin by labeling the students as A, B, C, D, and E. We can construct a list of the possible outcomes.

AB	AC	AD	AE	BC	BD	BE	CD	CE	DE
BA	CA	DA	EA	CB	DB	EB	DC	EC	ED

Each outcome in the second row involves the same 2 people as the outcome directly above in the first row, but in the reverse order. The order of selection is important. For example:

• AB means that Alberto gets $1000 and Beth gets $500, whereas
• BA means that Beth gets $1000 and Alberto gets $500.

There are 20 possible ways to select 2 people from 5 people in which the order of selection is relevant. That is, there are 20 permutations of 5 people taken 2 at a time.

Answers

4. a. $7! = 5040$ **b.** $\dfrac{8!}{2! \cdot 4!} = 840$

5. 12

Skill Practice 5   Determine the number of ways that 2 students from a group of 4 students can be selected to hold the positions of president and vice president of student government.

The number of permutations of 5 people taken 2 at a time can also be computed by using the fundamental principle of counting. There are 5 candidates for the $1000 scholarship, which leaves 4 students left over for the $500 scholarship. The product is $5 \cdot 4 = 20$.

This result can also be obtained by using factorial notation:

Number of people in the group (5).

$$\frac{5!}{(5-2)!} = \frac{5!}{3!} = \frac{5 \cdot 4 \cdot (3 \cdot 2 \cdot 1)}{(3 \cdot 2 \cdot 1)} = 20$$

Number of people to be selected (2).

Suppose that n represents the number of distinct elements in a group from which r elements will be chosen in a particular order. We call each arrangement a **permutation** of n elements taken r at a time and denote the number of such permutations as $_nP_r$.

Number of Permutations of n Elements Taken r at a Time

The number of permutations of n elements taken r at a time is given by

r factors

$$_nP_r = \frac{n!}{(n-r)!}, \text{ or equivalently, } _nP_r = \overbrace{n(n-1)(n-2)\cdots(n-r+1)}$$

Note: $_nP_r$ counts the number of permutations of n items taken r at a time under the assumption that no item can be selected more than once, and that each of the n items is distinguishable.

EXAMPLE 6   **Counting Permutations in an Application**

If 8 horses enter a race, in how many ways can the horses finish first, second, and third?

Solution:

We must find the number of ways that 3 horses can be selected from 8 horses in a prescribed order (first, second, third). This is given by $_8P_3$.

$$_8P_3 = \frac{8!}{(8-3)!} = \frac{8!}{5!} = \frac{8 \cdot 7 \cdot 6 \cdot 5!}{5!} = 8 \cdot 7 \cdot 6 = 336$$

$n = 8$

Alternatively: $_8P_3 = \underbrace{8 \cdot 7 \cdot 6}_{3 \text{ factors}} = 336$

TIP The alternative formula for $_nP_r$ indicates that we multiply n times the consecutive integers less than n until a total of r factors is reached.

There are 336 possible first-, second-, and third-place arrangements.

Skill Practice 6   A judge at the County Fair must give blue, red, and white ribbons for first-, second-, and third-place entries in a poetry contest. If there are 12 contestants, in how many ways can the judge award the ribbons?

Answer

6. 1320

TECHNOLOGY CONNECTIONS

Evaluating a Number of Permutations, $_nP_r$

Most graphing utilities can evaluate the number of permutations of n elements taken r at a time. For example, use the $_nP_r$ function found in the MATH menu under PRB.

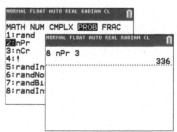

3. Count Combinations

Consider the situation in Example 5 in which 5 students are selected for 2 scholarships. If the scholarships were for *equal* amounts (say for $500 each), then the order in which the 2 students are selected does not matter. This is because the "prizes" are indistinguishable; that is, the outcomes AB and BA are the same because student A and student B would each receive $500.

In this scenario, we want to select a set of 2 people from a set of 5 people without regard to order. In such a case, the outcomes are called *combinations* (rather than permutations). From the list of permutations given in Example 5, we can strike through the redundant cases that arise from the order of the two individuals in the group. As a result, there are 10 combinations of 5 people taken 2 at a time.

AB	AC	AD	AE	BC	BD	BE	CD	CE	DE
~~BA~~	~~CA~~	~~DA~~	~~EA~~	~~CB~~	~~DB~~	~~EB~~	~~DC~~	~~EC~~	~~ED~~

To find the number of combinations, we divide the number of permutations (in this case 20) by 2! because there are 2! ways in which two letters can be arranged in order.

$$\text{Number of combinations} = \frac{_5P_2}{2!} = \frac{20}{2} = 10$$

To generalize, suppose that n represents the number of distinct elements in a group from which r elements will be chosen in *no particular order*. We call each selection a **combination** of n elements taken r at a time and denote the number of such combinations as $_nC_r$.

> **TIP** The number of combinations $_nC_r$ is equal to the number of permutations $_nP_r$ divided by $r!$. The value $r!$ in the denominator "divides out" the redundant cases involving different arrangements of r elements within the same group.

Number of Combinations of n Elements Taken r at a Time

The number of combinations of n elements taken r at a time is given by

$$_nC_r = \frac{n!}{r! \cdot (n-r)!}, \text{ or equivalently, } _nC_r = \frac{_nP_r}{r!}$$

EXAMPLE 7   Comparing Combinations and Permutations

a. In how many ways can 3 students from a group of 15 students be selected to serve on a committee?

b. In how many ways can 3 students from a group of 15 students be selected to serve on a committee, if the students will hold the offices of president, vice president, and treasurer?

Solution:

a. In this situation, the students will serve indistinguishable roles on the committee. Therefore, the order in which the 3 students are selected is not relevant. The number of such committees is the number of *combinations* of $n = 15$ students taken $r = 3$ at a time.

$$_{15}C_3 = \frac{15!}{3! \cdot (15 - 3)!} = \frac{15!}{3! \cdot 12!} = \frac{\overset{5}{\cancel{15}} \cdot \overset{7}{\cancel{14}} \cdot 13 \cdot \overset{1}{\cancel{12!}}}{\cancel{3} \cdot \cancel{2} \cdot 1 \cdot \cancel{12!}} = 455$$

b. Now the students will be selected to a committee and assigned different roles (president, vice president, and treasurer). In this case, the order in which the students are selected matters. We need to count the number of *permutations* of $n = 15$ students taken $r = 3$ at a time.

$$_{15}P_3 = \frac{15!}{(15 - 3)!} = \frac{15!}{12!} = \frac{15 \cdot 14 \cdot 13 \cdot \overset{1}{\cancel{12!}}}{\cancel{12!}} = 2730$$

Skill Practice 7   Suppose that 20 people enter a raffle.

a. In how many ways can four different people among the 20 be selected to receive prizes of $50 each?

b. In how many ways can four different people among the 20 be selected to receive prizes of $50, $25, $10, and $5?

In Example 7, notice that there are $3! = 6$ times as many permutations as combinations. The reason is that for each combination of 3 items, such as A, B, and C, there are $3! = 6$ permutations: ABC, ACB, BAC, BCA, CAB, CBA.

EXAMPLE 8   **Counting Combinations in an Application**

In the game "Florida Lotto," a player must select a group of 6 numbers (without regard to order) from the numbers 1 to 53. If the 6 numbers match the numbers in the drawing, then the player wins the grand prize. How many groups of 6 numbers are possible?

Solution:

In this situation, the order in which the group of 6 numbers is selected does not matter. Therefore, we need to compute the number of combinations of $n = 53$ numbers taken $r = 6$ at a time.

$$_{53}C_6 = \frac{53!}{6! \cdot (53 - 6)!} = \frac{53!}{6! \cdot 47!} = 22{,}957{,}480$$

There are 22,957,480 different groups of 6 numbers taken from 53 numbers.

Skill Practice 8   The California lottery game "Fantasy 5" offers a grand prize to a player who selects the correct group of 5 numbers (in any order) from the numbers 1 to 39. How many groups of 5 numbers are possible?

Point of Interest

As mentioned in the section opener, the probability of winning the grand prize in "Florida Lotto" for a player who plays 1 combination of 6 numbers is $\frac{1}{22,957,480}$.

Answers
7. a. 4845   **b.** 116,280
8. 575,757

TECHNOLOGY CONNECTIONS

Evaluating a Number of Combinations, $_nC_r$

Most graphing utilities can evaluate the number of combinations of n elements taken r at a time. For example, use the $_nC_r$ function found in the **MATH** menu under PRB.

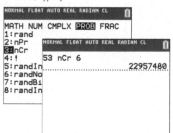

In Example 9 we encounter a situation in which more than one technique of counting must be used.

EXAMPLE 9   Applying Multiple Principles of Counting

Suppose that a committee of 3 women and 2 men must be formed from a group of 8 women and 7 men. In how many ways can such a committee be formed?

Solution:

This situation can be thought of as a sequence of two events in which we apply the fundamental principle of counting.

$$\begin{pmatrix} \text{Total number} \\ \text{of committees} \end{pmatrix} = \begin{pmatrix} \text{Number of ways to select} \\ \text{3 women from 8 women} \\ \text{in any order} \end{pmatrix} \cdot \begin{pmatrix} \text{Number of ways to select} \\ \text{2 men from 7 men} \\ \text{in any order} \end{pmatrix}$$

$$= {_8C_3} \cdot {_7C_2}$$
$$= 56 \cdot 21$$
$$= 1176$$

Avoiding Mistakes

The roles of the men and women on the committee are indistinguishable. Therefore, count the number of combinations rather than permutations.

Skill Practice 9   The coach of a co-ed softball team must select 4 women and 5 men from a group of 7 women and 10 men to play in a game. In how many ways can such a team be formed?

Answer

9. 8820

SECTION 8.6   Practice Exercises

Prerequisite Review

For Exercises R.1–R.4, evaluate the expression.

R.1. $9!$ **R.2.** $0!$ **R.3.** $\dfrac{7!}{5!}$ **R.4.** $\dfrac{10!}{4!\,6!}$

Concept Connections

1. The fundamental _____ of _____ indicates that if one event can occur in m different ways, and a second event can occur in n different ways, then the two events can occur in sequence in _____ different ways.

2. If n items are arranged in order, then each arrangement is called a _____ of n items.

3. The number of ways that n distinguishable items can be arranged in various orders is _____.

4. Consider a set of n elements of which one element is repeated r times. Then the number of permutations of the elements of the set is given by $\frac{\square}{\square}$.

5. Suppose that n represents the number of distinct elements in a group from which r elements will be chosen in a particular order. Then each arrangement is called a _____ of n items taken _____ at a time.

6. Suppose that n represents the number of elements in a group from which r elements will be selected in no particular order. Then each group selected is called a _____ of n elements taken _____ at a time.

7. The number of permutations of n elements taken r at a time is denoted by $_nP_r$ and is computed by
$$_nP_r = \frac{\Box}{\Box} \quad \text{or} \quad _nP_r = n(n-1)(n-2)\cdots(n-r+1).$$

8. The number of combinations of n elements taken r at a time is denoted by $_nC_r$ and is computed by _____.

Objective 1: Apply the Fundamental Principle of Counting

For Exercises 9–14, consider the set of integers from 1 to 20, inclusive. If one number is selected, in how many ways can we obtain

9. an even number?

10. an odd number?

11. a prime number?

12. a number that is a multiple of 5?

13. a number that is a multiple of 10?

14. a number that is divisible by 4?

15. At a hospital, the dinner menu consists of 4 choices of entrée, 3 choices of salad, 6 choices of beverage, and 4 choices of dessert. How many different meals can be formed if a patient chooses one item from each category? (**See Example 1**)

16. Debbie travels several times a year for her job. She takes 2 pairs of slacks, 6 blouses, and 4 scarves, all of different colors. Assuming that the items all match together well, how many different outfits does Debbie have if she selects one item from each category?

17. A license plate has 3 letters followed by 3 digits.
    a. How many license plates can be made if there are no restrictions on the letters or digits? (**See Example 2**)
    b. How many license plates can be made if no digit or letter may be repeated?

18. A security company requires its employees to have a 7-character computer password that must consist of 5 letters and 2 digits.
    a. How many passwords can be made if there are no restrictions on the letters or digits?
    b. How many passwords can be made if no digit or letter may be repeated?

19. The call letters for a radio station must begin with either K or W.
    a. How many 4-letter arrangements are possible assuming that letters may be repeated?
    b. How many 4-letter arrangements are possible assuming that the letters may not be repeated?

20. An employee identification code for a hospital consists of 2 letters from the set {A, B, C, D} followed by 4 digits.
    a. How many identification codes are possible if both letters and digits may be repeated?
    b. How many identification codes are possible if letters and digits may not be repeated?

21. An online survey is used to monitor customer service. The survey consists of 14 questions. Ten questions have 5 possible responses (strongly agree, agree, neutral, disagree, and strongly disagree). The remaining 4 questions are yes/no questions. In how many different ways can the survey be filled in? (**See Example 3**)

22. A test consists of 3 multiple-choice questions, each with four possible responses, and 7 true/false questions. In how many ways can a student answer the questions on the test?

23. On a computer, 1 *bit* is a single binary digit and has two possible outcomes: either 1 or 0. One *byte* is 8 bits. That is, a byte is a sequence of 8 binary digits. How many arrangements of 1's and 0's can be made with one byte?

24. Older models of garage door remote controls have a sequence of 10 switches that are individually placed in an up or down position. The remote control can "talk to" the overhead door unit if the 10 corresponding switches in the unit are in the same up/down sequence. How many up/down sequences are possible in an arrangement of 10 switches?

Objectives 2–3: Count Permutations and Combinations

25. a. In how many ways can the letters in the word FLORIDA be arranged? (**See Example 4**)
    b. In how many ways can the letters in the word MISSISSIPPI be arranged?

26. a. In how many ways can the letters in the word XRAY be arranged?
    b. In how many ways can the letters in the word MAMMOGRAM be arranged?

27. In how many ways can the word WRONG be misspelled?

28. In how many ways can the word EXACTLY be misspelled?

29. A delivery truck must make 4 stops at locations *A*, *B*, *C*, and *D*. Over several weeks, management asks the driver to drive each possible route and record the time required to complete the route. This is to determine the most time-efficient route. How many possible routes are there?

30. In how many ways can 6 people in a family be lined up for a photograph?

For Exercises 31–36, evaluate $_nP_r$.

31. $_6P_4$

32. $_8P_5$

33. $_{12}P_2$

34. $_{11}P_3$

35. $_9P_9$

36. $_5P_5$

37. Evaluate $_{20}P_3$ and interpret its meaning.

38. Evaluate $_{15}P_6$ and interpret its meaning.

For Exercises 39–44, evaluate $_nC_r$.

39. $_6C_4$

40. $_8C_5$

41. $_{12}C_2$

42. $_{11}C_3$

43. $_9C_9$

44. $_5C_5$

45. Evaluate $_{20}C_3$ and interpret its meaning.

46. Evaluate $_{15}C_6$ and interpret its meaning.

47. Given {A, B, C},

    a. List all the permutations of two elements from the set. (**See Example 5**)

    b. List all the combinations of two elements from the set.

48. Given {W, X, Y, Z},

    a. List all the permutations of three elements from the set.

    b. List all the combinations of three elements from the set.

49. Suppose that 9 horses run a race. How many first-, second-, and third-place finishes are possible? (**See Example 6**)

50. In how many ways can a judge award blue, red, and yellow ribbons to 3 films at a film festival if there are 10 films entered?

51. a. In a drama class, 5 students are to be selected from 24 students to perform a synchronized dance. In how many ways can 5 students be selected from the 24? (**See Example 7**)

    b. Determine the number of ways that 5 students can be selected from the class of 24 to play 5 different roles in a short play.

52. a. There are 100 members of the U.S. Senate. Suppose that 4 senators currently serve on a committee. In how many ways can 4 more senators be selected to serve on the committee?

    b. In how many ways can a group of 3 U.S. senators be selected from a group of 7 senators to fill the positions of chair, vice-chair, and secretary for the Ethics Committee?

53. A chess tournament has 16 players. Every player plays every other player exactly once. How many chess matches will be played?

54. Suppose that a tennis tournament has 64 players. In how many ways can a pairing for a first-round match be made between 2 players among the 64 players. Assume that each player can play any other player without regard to seeding.

55. In the Minnesota Lotto game "Gopher 5" a player wins the grand prize by choosing the same group of 5 numbers from 1 through 47 as is chosen by the computer. How many 5-number groups are possible? (**See Example 8**)

56. In the New York state lottery game "Lotto" a player wins the grand prize by choosing the same group of 6 numbers from 1 through 59 as is chosen by the computer. How many 6-number groups are possible?

57. At a ballroom dance lesson, the instructor chooses 3 men and 3 women to demonstrate a new pattern. If there are 9 women and 7 men in the class, in how many ways can the instructor choose 3 men and 3 women? (**See Example 9**)

58. A committee of 4 men and 4 women is to be made from a group of 12 men and 9 women. In how many ways can such a committee be formed?

Mixed Exercises

59. In a "Pick-4" game, a player wins a prize for matching a 4-digit number from 0000 to 9999 with the number randomly selected during the drawing. How many 4-digit numbers can a player choose? Assume that a number can start with a zero or zeros such as 0001.

60. In a "Numbers" game, a player wins a prize for matching a 3-digit number from 000 to 999 with the number randomly selected during the drawing. How many 3-digit numbers can a player choose? Assume that a number can start with a zero or zeros such as 001.

61. Liza is a basketball coach and must select 5 players out of 12 players to start a game. In how many ways can she select the 5 players if each player is equally qualified to play each position?

62. Jean has a list of 8 books that she knows she must read for a class in the upcoming fall semester of school. She wants to get a head start by reading several of the books during the summer. If she has time in the summer to read 5 of the 8 books, in how many ways can she select 5 books from 8 books?

63. In how many ways can a manager assign 5 employees at a coffee shop to 5 different tasks? Assume that each employee is assigned to exactly one task.

64. In how many ways can a class of 12 kindergarten children line up at the cafeteria?

65. Twenty batteries have been sitting in a drawer for 2 yr. There are 4 dead batteries among the 20. If three batteries are selected at random, determine the number of ways in which

   a. 3 dead batteries can be selected.

   b. 3 good batteries can be selected.

   c. 2 good batteries and 1 dead battery can be selected.

66. There are 30 seeds in a package. Five seeds are defective (will not germinate). If four seeds are selected at random, determine the number of ways in which

   a. 4 defective seeds can be selected.

   b. 4 good seeds can be selected.

   c. 2 good seeds and 2 defective seeds can be selected.

67. A "combination" lock is opened by correctly "dialing" 3 numbers from 0 to 39, inclusive. The user who knows the code turns the dial to the right to the first number in the code, then to the left to find the second number in the code, and then back to the right for the third number in the code. If someone does not know the code and tries to guess, how many guesses are possible?

68. A palindrome is an arrangement of letters that reads the same way forward and backward. For example, one five-letter palindrome is: ABCBA.

   a. How many 5-letter palindromes are possible from a 26-letter alphabet?

   b. How many 4-letter palindromes are possible from a 26-letter alphabet?

A line segment connecting any two nonadjacent vertices of a polygon is called a *diagonal* of the polygon. For Exercises 69–72, determine the number of diagonals for the given polygon.

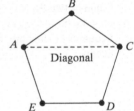

69. quadrilateral (4 sides)

70. pentagon (5 sides)

71. hexagon (6 sides)

72. octagon (8 sides)

For Exercises 73–78, consider the set of numbers {0, 1, 2, 3, 4, 5}. How many 3-digit codes can be formed with the given restrictions?

73. The code has no restrictions.

74. The code may not contain repeated digits.

75. The code must represent a 3-digit number. (*Hint*: This means that the first digit cannot be zero.)

76. The code must represent a 3-digit number that is a multiple of 5.

77. The code must represent an even 3-digit number.

78. All three digits in the code must be the same.

79. In Exercise 103 from Section 8.3, we learned that if a fair coin is flipped n times, the number of head/tail arrangements follows a geometric sequence.

   a. Determine the number of head/tail arrangements if a fair coin is flipped 3 times, 4 times, and 5 times.

   b. If a couple has 3 children, how many boy/girl sequences are possible for the three births? List the outcomes using "B" for boy and "G" for girl.

   c. If a couple has 4 children, how many boy/girl sequences are possible for the four births?

80. Social Security numbers assigned in the United States are comprised of 9 digits of the form _ _ _-_ _-_ _ _ _. Would there be enough Social Security numbers for the population of China if the Chinese used the same system? (*Hint*: The population of China is approximately 1.5 billion.)

81. Airlines often oversell seats on an airplane. This is done so that the seats for the few passengers that are no-shows will still have been sold. Sometimes, however, all passengers show up and there are more ticketed passengers than seats. Suppose that one flight has 160 passengers and only 156 seats. Determine the number of ways that the airline can select 4 people at random to place on a different flight.

82. In how many ways can a platoon leader select 4 soldiers among 15 soldiers to secure a building?

83. A car can comfortably hold a family of five. If two people among the five can drive, how many different seating arrangements are possible?

84. A television station must play twelve 30-sec commercials during a half-hour show. In how many ways can the commercials be aired?

Write About It

85. Consider a horse race with 8 horses. Explain how the fundamental principle of counting or the permutation rule can be used to determine the number of first-, second-, and third-place arrangements.

86. Explain why the number of combinations of n items taken r at a time can be computed by $\dfrac{{}_nP_r}{r!}$.

Expanding Your Skills

87. Three biology books, 4 math books, and 2 physics books are to be placed on a bookshelf where the books in each discipline are grouped together. In how many ways can the books be arranged on the bookshelf?

88. A softball team has 9 players consisting of 3 women and 6 men. In how many ways can the coach arrange the batting order if the men must bat consecutively and the women must bat consecutively?

SECTION 8.7   Introduction to Probability

OBJECTIVES

1. Determine Theoretical Probabilities
2. Determine Empirical Probabilities
3. Find the Probability of the Union of Two Events
4. Find the Probability of Sequential Independent Events

1. Determine Theoretical Probabilities

The Centers for Disease Control publishes National Vital Statistics Reports every year that provide data for birth rates and mortality rates based on gender, race, age, and other factors. The data in Table 8-4 give the 1-yr survival rates for people in the United States for selected ages.

From the table, the probability that a 20-yr-old will live to age 21 is 0.9991. Put another way, this means that approximately 99.91% of 20-yr-olds will live to age 21 (see Example 5).

A **probability** is a value assigned to an event that quantifies the likelihood of the event to occur. To compute the probability of an event, we first need to define several key terms. An **experiment** is a test with an uncertain outcome. The set of all possible outcomes of an experiment is called the **sample space** of the experiment. For example:

Table 8-4

Age (yr)	Probability of 1-yr Survival
10	0.9999
20	0.9991
30	0.9990
40	0.9981
50	0.9959
60	0.9907
70	0.9791
80	0.9434
90	0.8543

Experiment:	**Sample Space:**
Flip a coin	{head, tail}
Roll a single 6-sided die	{1, 2, 3, 4, 5, 6}

An **event** is a subset of the sample space and is often denoted by E or some other capital letter.

Description of Event:	**Set Representing the Event:**
Flip a coin and the outcome is "head"	{head}
Roll a six-sided die and an even number comes up	{2, 4, 6}

The number of elements in an event E is often denoted by $n(E)$. For example, given the event $E = \{2, 4, 6\}$, $n(E) = 3$.

If we assume that all elements in a sample space are equally likely to occur, then we define the theoretical probability of an event as follows.

TIP The word "die" is the singular of the word "dice." That is, we roll one die but we roll two dice.

Theoretical Probability of Event *E*

Let *S* represent a sample space with equally likely outcomes, and let *E* be a subset of *S*. Then the probability of event *E*, denoted by *P(E)*, is given by

$$P(E) = \frac{n(E)}{n(S)}$$ ⟵Number of elements in the event
⟵Number of elements in the sample space

The probability of an event is the relative frequency of the event compared to the sample space. For example, if *E* is the event of rolling an even number on a die, then $E = \{2, 4, 6\}$ and $S = \{1, 2, 3, 4, 5, 6\}$. Then,

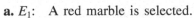 $$P(E) = \frac{n(E)}{n(S)} = \frac{3}{6} = \frac{1}{2}$$

The value of a probability can be written as a fraction, as a decimal, or as a percent. Therefore, $P(E) = \frac{1}{2}$, or 0.5, or 50%.

EXAMPLE 1   Finding the Probability of an Event

Suppose that a box contains two red, three blue, and five green marbles. If one marble is selected at random, find the probability of the event.

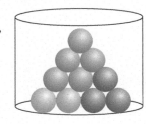

a. E_1:   A red marble is selected.
b. E_2:   A blue marble is selected.
c. E_3:   A white marble is selected.

Solution:

Each of the 10 marbles in the box is equally likely to be selected. Denote the sample space as $S = \{R_1, R_2, B_1, B_2, B_3, G_1, G_2, G_3, G_4, G_5\}$.

a. $E_1 = \{R_1, R_2\}$ | The number of red marbles is 2.

Therefore, $P(E_1) = \dfrac{n(E_1)}{n(S)} = \dfrac{2}{10} = \dfrac{1}{5}$. | The probability of selecting a red marble is $\frac{1}{5}$.

b. $E_2 = \{B_1, B_2, B_3\}$ | The number of blue marbles is 3.

Therefore, $P(E_2) = \dfrac{n(E_2)}{n(S)} = \dfrac{3}{10}$. | The probability of selecting a blue marble is $\frac{3}{10}$.

c. $E_3 = \{\ \}$ | The number of white marbles is 0.

Therefore, $P(E_3) = \dfrac{n(E_3)}{n(S)} = \dfrac{0}{10} = 0$. | It is intuitively obvious that a white marble cannot be selected. Mathematically, a probability of 0 means that the event is impossible.

Skill Practice 1   A sock drawer has 6 blue socks, 2 white socks, and 12 black socks. If one sock is selected at random, find the probability of the event.

a. E_1:   A blue sock is selected.   **b.** E_2:   A black sock is selected.
c. E_3:   A brown sock is selected.

Answers

1. a. $\dfrac{3}{10}$ **b.** $\dfrac{3}{5}$ **c.** 0

The number of outcomes in an event is a nonnegative number less than or equal to the number of outcomes in the sample space. Therefore, the value of a probability is a number between 0 and 1, inclusive. That is, $0 \leq P(E) \leq 1$. Furthermore,

- If $P(E) = 0$, then the event E is called an **impossible event.**
- If $P(E) = 1$, then the event E is called a **certain event.**

In Example 1(c), it is impossible to draw a white marble from the box. However, we would be 100% certain of drawing a red, blue, or green marble from the box and the probability would be 1.

EXAMPLE 2   Finding the Probability of an Event

An American roulette wheel has 38 slots, numbered 1 through 36, 0, and 00. Eighteen slots are red, 18 are black, and 2 are green. The dealer spins the wheel in one direction and rolls a small ball in the opposite direction until both come to rest. The ball is equally likely to fall in any one of the 38 slots. On a given spin of an American roulette wheel, find the probability of the event.

a. E_1: The ball lands on an odd number.
b. E_2: The ball lands on a green slot.
c. E_3: The ball does not land on a green slot.

Solution:

The sample space has 38 outcomes. That is $n(S) = 38$.

a. There are 18 odd numbers. Therefore, $n(E_1) = 18$ and
$$P(E_1) = \frac{n(E_1)}{n(S)} = \frac{18}{38} = \frac{9}{19} \approx 0.4737.$$

b. There are 2 green slots on the wheel. Therefore, $n(E_2) = 2$ and
$$P(E_2) = \frac{n(E_2)}{n(S)} = \frac{2}{38} = \frac{1}{19} \approx 0.0526.$$

c. There are 36 slots that are not green. Therefore, $n(E_3) = 36$ and
$$P(E_3) = \frac{n(E_3)}{n(S)} = \frac{36}{38} = \frac{18}{19} \approx 0.9474.$$

Skill Practice 2   On a given spin of an American roulette wheel, find the probability of the event.

a. E_1: The ball lands on the number 7.
b. E_2: The ball lands on a red slot.
c. E_3: The ball does not land on a red slot.

From Example 2(b) and 2(c), notice that event E_3 consists of all elements in the sample space not in event E_2. Therefore, the union of the two events make up the entire sample space, and $P(E_2) + P(E_3) = \dfrac{1}{19} + \dfrac{18}{19} = 1$. In this case, we say that events E_2 and E_3 are *complementary events.*

Answers

2. a. $\dfrac{1}{38}$ **b.** $\dfrac{9}{19}$ **c.** $\dfrac{10}{19}$

> **Definition of Complementary Events**
>
> Let E be an event relative to sample space S. The **complement of E,** denoted by $\overline{E}$ (or sometimes by $\sim E$ or E') is the set of outcomes in the sample space but not in event E. It follows that
>
> $$P(E) + P(\overline{E}) = 1.$$

EXAMPLE 3   **Finding the Probabilities of Complementary Events**

Suppose that 2 dice are rolled. Determine the probability that

a. The sum of the numbers showing on the dice is 7.

b. The sum of the numbers showing on the dice is *not* 7.

Solution:

Each individual die has 6 equally likely outcomes. Therefore, by the fundamental principle of counting, the number of ways the pair of dice can fall is $6 \cdot 6 = 36$ (Figure 8-11).

Second Die

First Die	1	2	3	4	5	6
1	(1,1)	(1,2)	(1,3)	(1,4)	(1,5)	(1,6)
2	(2,1)	(2,2)	(2,3)	(2,4)	(2,5)	(2,6)
3	(3,1)	(3,2)	(3,3)	(3,4)	(3,5)	(3,6)
4	(4,1)	(4,2)	(4,3)	(4,4)	(4,5)	(4,6)
5	(5,1)	(5,2)	(5,3)	(5,4)	(5,5)	(5,6)
6	(6,1)	(6,2)	(6,3)	(6,4)	(6,5)	(6,6)

Figure 8-11

a. Let E represent the event that the numbers showing on the dice have a sum of 7. A sum of 7 will occur if the dice land on one of the following 6 outcomes.

$E = \{(1, 6), (2, 5), (3, 4), (4, 3), (5, 2), (6, 1)\}$. Therefore, $n(E) = 6$ and

$$P(E) = \frac{n(E)}{n(S)} = \frac{6}{36} = \frac{1}{6}.$$ The probability of rolling a sum of 7 is $\frac{1}{6}$.

b. The event that the numbers on the dice do not present a sum of 7 is the complement of event E. Therefore, we have the option of using the formula $P(E) + P(\overline{E}) = 1$.

$$P(\overline{E}) = 1 - P(E) = 1 - \frac{1}{6} = \frac{5}{6}$$ The probability of rolling a sum other than 7 is $\frac{5}{6}$.

Skill Practice 3   Suppose that two dice are rolled. Determine the probability that

a. The sum of the numbers showing on the dice is 8.

b. The sum of the numbers showing on the dice is *not* 8.

In Example 4, we use the techniques of counting learned in Section 8.6 to determine the probability of an event.

EXAMPLE 4   **Finding the Probability of an Event by Using Counting Techniques**

Suppose that 5 women and 3 men apply for 2 job openings. If the applicants are equally qualified, find the probability that both positions are filled by women.

Answers

3. a. $\frac{5}{36}$ **b.** $\frac{31}{36}$

Solution:

Let E represent the event that two women are chosen. The number of elements in E is the number of possible ways to select 2 women from a group of 5 women without regard to order. That is,

$$n(E) = {}_5C_2 = 10$$

The sample space S consists of all possible ways in which 2 people can be selected from 8 people without regard to order. That is,

$$n(S) = {}_8C_2 = 28$$

The probability that both positions will be filled by women is

$$P(E) = \frac{n(E)}{n(S)} = \frac{{}_5C_2}{{}_8C_2} = \frac{10}{28} = \frac{5}{14} \approx 0.3571$$

Skill Practice 4 Suppose that a committee of three people is to be formed from a group of 8 men and 6 women. Find the probability that the committee will consist of all men.

2. Determine Empirical Probabilities

Suppose that the Centers for Disease Control wants to measure the 1-yr survival rates for Americans for specific ages. There is no way that this can be derived theoretically. Instead, the probability can be approximated through observations. A probability computed in this way is called an **empirical probability.**

Computing Empirical Probability

The empirical probability of an event E is given by

$$P(E) = \frac{\text{Number of times the event } E \text{ occurs}}{\text{Number of times the experiment is performed}}$$

EXAMPLE 5 **Computing an Empirical Probability**

Data from the Centers for Disease Control for a recent year indicate that there were approximately 4,295,000 individuals of age 20 in the United States. One year later 3887 of these individuals had died and 4,291,113 had lived. Determine the probability that a 20-yr-old will survive to age 21. (*Source:* www.cdc.gov)

Solution:

Let E represent the event that a 20-yr-old lives to age 21.

The experiment involves 4,295,000 individuals of age 20. The number of observed outcomes where an individual lives to the age of 21 is 4,291,113. Therefore, the empirical probability is given by

$$P(E) = \frac{4{,}291{,}113}{4{,}295{,}000} \approx 0.9991$$

Skill Practice 5 Suppose that of approximately 4,224,100 individuals of age 40 in the United States, 8038 die before the age of 41 and 4,216,062 survive. Determine the probability that a 40-yr-old will live to age 41.

Answers

4. $\frac{2}{13}$ **5.** 0.9981

3. Find the Probability of the Union of Two Events

We now study the probability of the union of two events. Suppose that one card is drawn from a standard deck of cards. The sample space has the following properties (Figure 8-12).

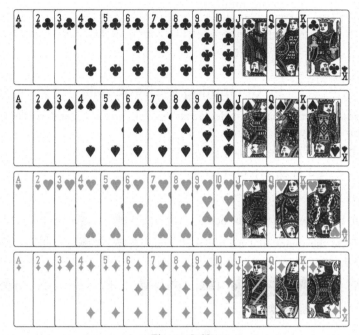

- The deck consists of 52 cards.

- The cards are divided into four suits (or categories) called spades (♠), clubs (♣), hearts (♥), and diamonds (♦).

- Each suit consists of 13 cards labeled: Ace (A), 2, 3, 4, 5, 6, 7, 8, 9, 10, Jack (J), Queen (Q), and King (K).

- There are 26 black cards (spades and clubs) and 26 red cards (hearts and diamonds).

Figure 8-12

Let *A* be the event that an ace is drawn: {A♠, A♣, A♥, A♦}.

Let *K* be the event that a king is drawn: {K♠, K♣, K♥, K♦}.

Events *A* and *K* have no elements in common (a single card cannot be both an ace and a king). Therefore, the intersection of *A* and *K* is the empty set, $A \cap K = \{\}$, and we say that events *A* and *K* are **mutually exclusive** (do not overlap).

Now let event *A* be the event that an ace is drawn, and let *S* be the event that a spade is drawn (Figure 8-13).

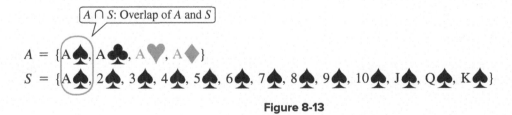

Figure 8-13

Notice that events *A* and *S* share the comment element of the ace of spades, A♠. Therefore, events *A* and *S* are *not* mutually exclusive, meaning that they overlap.

We are now ready to generalize. To find the probability of the event (*A* or *B*), denoted *P*(*A* or *B*), find the probability of the *union* of *A* and *B*.

Probability of (A or B)

Given events A and B in the same sample space, $P(A \text{ or } B)$ is given by

$$P(A \cup B) = P(A) + P(B) - P(A \cap B)$$

Note: If A and B are mutually exclusive, then $P(A \cap B) = 0$ and we have $P(A \cup B) = P(A) + P(B)$.

EXAMPLE 6   **Computing the Probability of A or B**

Suppose that one card is selected at random from a standard deck of cards. Find the probability that

a. The card is an ace or a king.

b. The card is an ace or a spade.

Solution:

Let A be the event that an ace is drawn: {A♠, A♣, A♥, A♦}.

Let K be the event that a king is drawn: {K♠, K♣, K♥, K♦}.

Let S be the event that a spade is drawn:

$$S = \{A\spadesuit, 2\spadesuit, 3\spadesuit, 4\spadesuit, 5\spadesuit, 6\spadesuit, 7\spadesuit, 8\spadesuit, 9\spadesuit, 10\spadesuit, J\spadesuit,$$
$$Q\spadesuit, K\spadesuit\}.$$

> **TIP** In Example 6(a), since A and K are mutually exclusive, then $P(A \cap K) = 0$.

a. Events A and K are mutually exclusive. Therefore, to find $P(A \text{ or } K)$, we have

$$P(A \cup K) = P(A) + P(K)$$

$$= \frac{4}{52} + \frac{4}{52} \qquad \text{There are 4 aces in the deck out of 52 cards. } P(A) = \frac{4}{52}.$$
$$\text{There are 4 kings in the deck out of 52 cards. } P(K) = \frac{4}{52}.$$

$$= \frac{8}{52} \text{ or } \frac{2}{13}$$

b. From Figure 8-13, events A and S are *not* mutually exclusive. Therefore, to find $P(A \text{ or } S)$, we have:

> **TIP** As an alternative to computing $P(A) + P(S) - P(A \cap S)$, count the number of elements in the event $(A \cup S)$, being careful not to count the events common to A and S twice. That is, count 13 spades plus the 3 aces that are not spades. This gives a total of 16 elements in event $(A \cup S)$. Thus,
> $$P(A \cup S) = \frac{16}{52} = \frac{4}{13}$$

$$P(A \cup S) = P(A) + P(S) - P(A \cap S)$$

There are 4 aces in the deck out of 52 cards. $P(A) = \frac{4}{52}$.

$$= \frac{4}{52} + \frac{13}{52} - \frac{1}{52} \boxed{\text{ace of spades}}$$

There are 13 spades in the deck out of 52 cards. $P(S) = \frac{13}{52}$.

$$= \frac{16}{52} \text{ or } \frac{4}{13}$$

There is 1 spade that is an ace in the deck. Therefore, $P(A \cap S) = \frac{1}{52}$.

Skill Practice 6   Suppose that one card is drawn at random from a standard deck. Find the probability that

a. The card is a 2 or a 10.

b. The card is a 2 or a red card.

Answers

6. a. $\dfrac{2}{13}$ **b.** $\dfrac{7}{13}$

EXAMPLE 7   **Computing an Empirical Probability of (A or B)**

The safety and security department at a college asked a sample of 265 students to respond to the following question.

"Do you think that the college has adequate lighting on campus at night?"

The table gives the results of the survey based on gender and response.

	Yes	No	No Opinion	Total
Male	92	7	4	103
Female	36	102	24	162
Total	128	109	28	265

If one student is selected at random from the group, find the probability that

a. The student answered "Yes" or had "No Opinion."

b. The student answered "No" or was female.

Solution:

a. Let Y be the event that the student answered "Yes."

Let O be the event that the student had "No Opinion."

Y and O are mutually exclusive events (do not overlap).

	Yes	No	No Opinion	Total
Male	92	7	4	103
Female	36	102	24	162
Total	128	109	28	265

$P(Y \text{ or } O) = P(Y) + P(O)$

$\qquad = \dfrac{128}{265} + \dfrac{28}{265} \qquad$ $P(Y) = \frac{128}{265}$ and $P(O) = \frac{28}{265}$.

$\qquad = \dfrac{156}{265}$

TIP As an alternative to computing $P(N) + P(F) - P(N \cap F)$, count the number of elements in the event $(N \cup F)$, being careful not to count the elements common to N and F twice; that is, count 162 females plus the 7 males who answered "No" for a total of 169 elements in event $(N \cup F)$. Thus, $P(N \cup F) = \frac{169}{265}$.

b. Let N be the event that the student answered "No."

Let F be the event that the student is female.

N and F are *not* mutually exclusive (the events overlap).

	Yes	No	No Opinion	Total
Male	92	7	4	103
Female	36	102	24	162
Total	128	109	28	265

$P(N \text{ or } F) = P(N) + P(F) - P(N \cap F)$

$\qquad = \dfrac{109}{265} + \dfrac{162}{265} - \dfrac{102}{265} \qquad$ $P(N) = \frac{109}{265}$, $P(F) = \frac{162}{265}$, and $P(N \cap F) = \frac{102}{265}$

$\qquad = \dfrac{169}{265} \qquad$ The 102 females who answered "No" are the elements in the intersection of N and F.

Skill Practice 7   Refer to the data given in Example 7. If one student is selected at random from the group, find the probability that

a. The student answered "Yes" or "No."

b. The student is male or had no opinion.

Answers

7. a. $\dfrac{237}{265}$ **b.** $\dfrac{127}{265}$

4. Find the Probability of Sequential Independent Events

When a coin is tossed, the probability that it lands heads up is $\frac{1}{2}$. Now suppose that the coin is flipped two times in succession. What is the probability that it lands heads up twice in a row? Our intuition tells us that the probability should be less than $\frac{1}{2}$ because it is more unlikely that the desired outcome will happen twice in a row than one time. We also know that the outcome of the first event (heads up) does not affect the outcome of the second event (also heads up). For this reason, the two events are called **independent events.**

Probability of a Sequence of Independent Events

If events A and B are independent events, then the probability that both A and B will occur is

$$P(A \text{ and } B) = P(A) \cdot P(B)$$

For example,

Let A represent the event that a coin lands heads up on the first toss.
Let B represent the event that a coin lands heads up on the second toss.

Then the probability that the coin will land heads up on both tosses is

$$P(A \text{ and } B) = P(A) \cdot P(B) = \frac{1}{2} \cdot \frac{1}{2} = \frac{1}{4}$$

EXAMPLE 8   Finding the Probability of Independent Events

Suppose that a family plans to have three children. Find the probability that all three children will be boys.

Solution:

Each birth is independent of the birth that precedes it.
For each birth, the probability that the child is born a boy is $\frac{1}{2}$.

$P(\text{all boys}) = P(\text{boy on 1st}) \cdot P(\text{boy on 2nd}) \cdot P(\text{boy on 3rd})$
$$= \frac{1}{2} \cdot \frac{1}{2} \cdot \frac{1}{2} = \frac{1}{8}$$

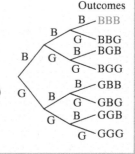

Outcomes

B	BBB
G	BBG
B	BGB
G	BGG
B	GBB
G	GBG
B	GGB
G	GGG

TIP In Example 8 there are 8 elements in the sample space:

BBB, BBG, BGB, BGG,
GBB, GBG, GGB, GGG

The event BBB occurs 1 time. Thus, the probability of all boys is $\frac{1}{8}$.

Skill Practice 8   Suppose that a family plans to have five children. Find the probability that all five children will be girls.

EXAMPLE 9   Finding the Probability of Independent Events

In baseball, a player's batting "average" is the quotient of the number of hits to the number of "at bats." It can also be interpreted as the probability that the player will get a hit on a given time at bat. If Albert Pujols had a batting average of 0.279 for a recent season, determine the probability that he would get a hit on four consecutive times at bat.

Answer

8. $\frac{1}{32}$

Solution:

Let H represent the event that Albert Pujols gets a hit on a given time at bat. Furthermore, each time at bat is independent of the time before.

We have that $P(H) = 0.279$. Therefore,

$$P(4 \text{ consecutive hits}) = (0.279)(0.279)(0.279)(0.279)$$
$$\approx 0.006$$

If Albert Pujols has four times at bat, there is less than a 1% chance that he will get a hit all four times.

Skill Practice 9   Suppose the probability that a person will catch a winter "cold" is 0.16. What is the probability that four unrelated people will all catch winter "colds"?

Answer

9. $(0.16)^4 \approx 0.000655$

SECTION 8.7   Practice Exercises

Prerequisite Review

R.1. Given: $P = \{a, b, c, d, e, f, g, h, i\}$ and $Q = \{c, f, h, o, u\}$
List the elements of the following sets.
   a. $P \cap Q$ **b.** $P \cup Q$

R.2. Given: $M = \{-4, -2, 1, 3, 5\}$ and $N = \{-5, -4, -3, -2, -1\}$
List the elements of the following sets.
   a. $M \cap N$ **b.** $M \cup N$

Concept Connections

1. The _____ _____ of an experiment is the set of all possible outcomes.

2. An _____ is a subset of the sample space of an experiment.

3. If $P(E) = 0$, then E is called an _____ event. If $P(E) = 1$, then E is called a _____ event.

4. The notation $\overline{E}$ represents the _____ of event E. Furthermore, $P(E) + P(\overline{E}) = $ _____.

5. Two events in a sample space are _____ _____ if they do not share any common elements. That is, the two events do not overlap.

6. If two events A and B are not mutually exclusive, then $P(A \cup B)$ can be computed by the formula $P(A \cup B) = $ _____.

7. If two events A and B are mutually exclusive, then $P(A \cap B) = $ _____. As a result, $P(A \cup B)$ can be computed from the formula $P(A \cup B) = $ _____.

8. For two independent events A and B, $P(A \text{ and } B) = $ _____.

Objective 1: Determine Theoretical Probabilities

9. Which of the values can represent the probability of an event?

   a. 1.84   **b.** $-\frac{3}{7}$ **c.** 0.00   **d.** 1.00

   e. 250%   **f.** 6.1   **g.** 0.61   **h.** 6.1%

10. Which of the values can represent the probability of an event?

   a. 2.32   **b.** 0.231   **c.** 2.31%   **d.** -1

   e. $\frac{3}{8}$ **f.** $-\frac{1}{4}$ **g.** $\sqrt{2}$ **h.** 125%

For Exercises 11–16, match the probability with a statement a, b, c, d, e, or f.

a. Event E is certain to happen.

b. Event E cannot happen.

c. Event E is very likely to happen.

d. Event E is very unlikely to happen.

e. Event E is somewhat likely to happen.

f. Event E is as likely to happen as not to happen.

11. $P(E) = 0.994$

12. $P(E) = 1$

13. $P(E) = 0.75$

14. $P(E) = 0.003$

15. $P(E) = 0.5$

16. $P(E) = 0$

For Exercises 17–24, consider an experiment where a single 10-sided die is rolled with the outcomes 1, 2, 3, 4, 5, 6, 7, 8, 9, 10. Determine the probability of each event.

17. A number less than 5 is rolled.

18. A number less than 3 is rolled.

19. A number between 4 and 10, inclusive, is rolled.

20. A number between 2 and 7, inclusive, is rolled.

21. A number greater than 10 is rolled.

22. A number less than 1 is rolled.

23. A number greater than or equal to 1 is rolled.

24. A number less than or equal to 10 is rolled.

25. A course in early civilization has 6 freshmen, 8 sophomores, and 16 juniors. If one student is selected at random, find the probability of the following events. (**See Example 1**)

   a. A junior is selected.

   b. A freshman is selected.

   c. A senior is selected.

26. Suppose that a box contains 4 chocolate chip cookies, 8 molasses cookies, and 12 raisin cookies. If one cookie is selected at random, find the probability of the following events.

   a. A chocolate chip cookie is selected.

   b. A molasses cookie is selected.

   c. A ginger cookie is selected.

For Exercises 27–28, consider an American roulette wheel. (See Example 2) For a given spin of the wheel, find the probability of the following events.

27. a. The ball lands on an even number (do not include 0 and 00).

   b. The ball lands on a number that is a multiple of 5 (do not include 0 and 00).

   c. The ball does not land on the number 8.

28. a. The ball lands on a black slot.

   b. The ball lands on a number that is a multiple of 6 (do not include 0 and 00).

   c. The ball does not land on the number 12.

29. If $P(E) = 0.842$, what is the value of $P(\overline{E})$?

30. If $P(A) = 0.431$, what is the value of $P(\overline{A})$?

31. According to the Centers for Disease Control, the probability that a live birth will be of twins in the United States is 0.016. What is the probability that a live birth will *not* be of twins?

32. A baseball player with a batting average of 0.291 has a probability of 0.291 of getting a hit for a given time at bat. What is the probability that the player will *not* get a hit for a given time at bat?

For Exercises 33–36, consider the sample space when two fair dice are rolled. (See Example 3) Determine the probabilities for the following events.

33. a. The sum of the numbers on the dice is 4.

   b. The sum of the numbers on the dice is *not* 4.

34. a. The sum of the numbers on the dice is 12.

   b. The sum of the numbers on the dice is *not* 12.

35. a. The sum of the numbers on the dice is greater than 9.

   b. The sum of the numbers on the dice is less than 4.

36. a. The sum of the numbers on the dice is greater than or equal to 8.

   b. The sum of the numbers on the dice is less than or equal to 5.

37. After a nationally televised trial, a poll of viewers indicated that 68% thought the defendant was guilty, 22% thought the defendant was not guilty, and the rest were undecided. What is the probability that a person selected from the viewing audience was undecided?

38. If a couple plans to have three children, the probability that all three will be boys is 0.125. What is the probability that the couple will have at least one girl?

39. Suppose that a jury pool consists of 18 women and 16 men.

   a. What is the probability that a jury of 9 people taken at random from the pool will consist only of women? (**See Example 4**)

   b. What is the probability that the jury will consist only of men?

   c. Why do the probabilities from parts (a) and (b) not add up to 1?

40. Suppose that 20 good batteries and 6 defective batteries are in a drawer.

   a. If 4 batteries are drawn at random, what is the probability that all four will be defective?

   b. What is the probability that all four will be good?

   c. Why do the probabilities from parts (a) and (b) not add up to 1?

41. In the Illinois state lottery game "Little Lotto," a player wins the grand prize by choosing the same group of five numbers from 1 through 39 as is chosen by the computer. What is the probability that a player will win the grand prize by playing 1 ticket?

42. In the New York state lottery game "Lotto," a player wins the grand prize by choosing the same group of 6 numbers from 1 through 59 as is chosen by the computer. What is the probability that a player will win the grand prize by playing 5 different tickets?

Scientist Gregor Mendel (1822–1884) is often called the "father of modern genetics" and is famous for his work involving the inheritance of certain traits in pea plants. Suppose that the genes controlling the color of peas are Y for yellow and y for green. Each plant has two genes, one from the female (seed) and one from the male (pollen). The Y gene is dominant, and therefore a plant with genes YY will have yellow peas, a plant with genes Yy or yY will have yellow peas, and a plant with genes yy will have green peas.

	Parent 2	
	y	**y**
Y	Yy	Yy
Y	Yy	Yy

Parent 1

If a plant with two yellow genes (YY) is crossed with a plant with two green genes (yy), the result is four hybrid offspring with genotypes Yy. The offspring will be yellow, but will carry the recessive green gene.

43. Suppose that both parent pea plants are hybrids with genotype Yy.

a. Make a chart showing the possible genotypes of the offspring.

	Parent 2	
	Y	**y**
Y		
y		

Parent 1

b. What is the probability that a given offspring will have green peas?

c. What is the probability that a given offspring will have yellow peas?

44. Suppose that one parent pea plant has genotype YY and the other has genotype Yy.

a. Make a chart showing the possible genotypes of the offspring.

	Parent 2	
	Y	**y**
Y		
Y		

Parent 1

b. What is the probability that a given offspring will have green peas?

c. What is the probability that a given offspring will have yellow peas?

Objective 2: Determine Empirical Probabilities

45. At a hospital specializing in treating heart disease, it was found that 222 out of 4624 patients undergoing open heart mitral valve surgery died during surgery or within 30 days after surgery. Determine the probability that a patient will not survive the surgery or 30 days after the surgery. This is called the mortality rate. Round to 3 decimal places. (**See Example 5**)

46. China has the largest population of any country with approximately 1.5 billion people. In a recent year, census results indicated that 199,500,000 Chinese were over the age of 60. If a person is selected at random from the population of China, what is the probability that the person is over 60 years old? Round to 3 decimal places.

47. For a certain district, a random sample of registered voters results in the distribution by political party given in the graph. Based on these results, what is the probability of selecting a voter at random from the district and getting

a. A Democrat?

b. A voter who is neither Democrat, Republican, nor Independent?

48. The final exam grades for a sample of students in a Freshmen English class at a large university result in the following grade distribution. Based on these results, what is the probability of selecting a student at random taking Freshmen English and getting a student who received

a. An "A."   **b.** A "C."

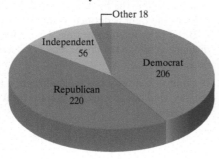

Party Distribution

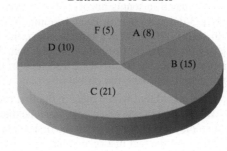

Distribution of Grades

Objective 3: Find the Probability of the Union of Two Events

For Exercises 49–58, consider the sample space for a single card drawn from a standard deck. (**See Example 6 and Figure 8-12**) Find the probability that the card drawn is

49. A jack or a queen.

50. An ace or a 2.

51. A jack or a diamond.

52. A 5 or a heart.

53. A face card (jack, queen, or king).

54. A card numbered between 5 and 10, inclusive.

55. A face card or a red card.

56. A card numbered between 5 and 10, inclusive, or a black card.

57. A heart, club, or spade.

58. An ace, 2, or 3.

For Exercises 59–66, use the data in the table categorizing cholesterol levels by the ages of the individuals in a study. If one person from the study is chosen at random, find the probability of the given event. (See Example 7)

	Normal Cholesterol	Elevated Cholesterol	Total
30 and under	14	4	18
31–60	52	28	80
61 or older	22	80	102
Total	88	112	200

59. The person has elevated cholesterol.

60. The person is 61 or older.

61. The person is 60 or under.

62. The person is 31 or older.

63. The person has normal cholesterol or is 61 or older.

64. The person has elevated cholesterol or is 30 or under.

65. The person is between 31 and 60, inclusive, or has elevated cholesterol.

66. The person is 61 or older or has elevated cholesterol.

Objective 4: Find the Probability of Sequential Independent Events

67. Suppose that a die is rolled followed by the flip of a coin. Find the probability that the outcome is a 5 on the die followed by the coin turning up heads. (**See Example 8**)

68. If a code for an alarm is a 4-digit sequence, determine the probability that someone guesses each digit correctly.

69. The 5-yr survival rate for stage I breast cancer is 88%. (*Source*: American Cancer Society, www.cancer.org) If two women with stage I breast cancer are selected at random, what is the probability that they both survive 5 yr? (**See Example 9**)

70. Basketball player Lebron James makes approximately 76% of free throws. If he plays in a game in which he shoots 6 free throws, what is the probability that he will make all 6?

71. In the 2010 Wimbledon Championships, John Isner from the United States and Nicolas Mahut from France played a first-round tennis match that became the longest match in tennis history. (The match stretched over a 3-day period with Isner winning 70-68 in the fifth set.) In 2011, after a random draw, the two men met again in the first round of Wimbledon. This is highly improbable. If there are 128 men in the tournament, estimate the probability that

  a. Isner and Mahut would meet in the first round at Wimbledon in any given year. Assume that any player can play any other player in the first round (that is, disregard the fact that seeded players do not play one another in the first round).

  b. Isner and Mahut would meet in the first round 2 yr in a row.

72. A traffic light at an intersection has a 120-sec cycle. The light is green for 80 sec, yellow for 5 sec, and red for 35 sec.

  a. When a motorist approaches the intersection, find the probability that the light will be red. (Assume that the color of the light is defined as the color when the car is 100 ft from the intersection. This is the approximate distance at which the driver makes a decision to stop or go.)

  b. If a motorist approaches the intersection twice during the day, find the probability that the light will be red both times.

73. A test has 10 questions. Five questions are true/false and five questions are multiple-choice. Each multiple-choice question has 4 possible responses of which exactly one is correct. Find the probability that a student guesses on each question and gets a perfect score.

74. A quiz has 6 multiple-choice questions and each question has 5 possible responses of which exactly one is correct. Find the probability that a student answers each question *incorrectly*.

Mixed Exercises

The blood type of an individual is classified according to the presence of certain antigens, substances that cause the immune system to produce antibodies. These antigens are denoted by A, B, and Rh. If an individual's blood contains either the A or B antigen, these letters are listed in the blood type. If neither A nor B is present, then the letter O is used. If the Rh antigen is present, the blood is said to be Rh positive (Rh^+); otherwise, the blood is Rh negative (Rh^-). Under this system, a person with AB^+ blood has all three antigens, and group O^- is absent all three antigens.

The distribution of blood types for people living in the United States is given in the table. (*Source*: Stanford School of Medicine, http://bloodcenter.stanford.edu) Refer to the table for Exercises 75–82. Round to 3 decimal places when necessary.

Type	Probability
O^+	0.374
O^-	0.066
A^+	0.357
A^-	0.063
B^+	0.085
B^-	0.015
AB^+	0.034
AB^-	0.006

75. a. If an individual is randomly selected from the population, find the probability that the individual will have the Rh factor.

    b. If three people are selected at random, find the probability that they all have the Rh factor.

77. a. Which blood type is most common ?

    b. Which blood type is most rare?

76. a. If a person is selected at random from the population, find the probability that the individual has the A antigen.

    b. If four people are randomly selected, find the probability that all four have the A antigen.

78. Across all blood types, is a person more likely to have the Rh factor or less likely to have the Rh factor?

Doctors know that certain restrictions apply when considering the administration of blood to a patient. The antigens of the blood donor and recipient must be compatible. If an antigen is absent from the recipient's blood, then the recipient cannot receive blood from a person who has that antigen. For example, a person with B^+ blood cannot receive AB^+ blood because of the presence of the A antigen from the donor. Use this information and the table with Exercises 75–78 for Exercises 79–82.

79. Suppose that a person is randomly selected from the population. What is the probability that this individual's blood can be used for a transfusion for a person with type B^+ blood?

81. a. What is the probability that an individual from the population can donate blood to a person with type O^- blood?

    b. Explain why a person with type O^- blood is called a universal donor.

83. A slot machine in a casino has three wheels that all spin independently. Each wheel has 11 stops, denoted by 0 through 9, and bar. What is the probability that a given outcome is bar-bar-bar?

80. Suppose that a person is randomly selected from the population. What is the probability that this individual's blood can be used for a transfusion for a person with type A^+ blood?

82. a. What is the probability that an individual from the population can donate blood to a person with AB^+ blood?

    b. Explain why a person with AB^+ blood can receive blood from anyone.

84. Airlines often overbook flights because a small percentage of passengers do not show up (perhaps due to missed connections). Past history indicates that for a certain route, the probability that an individual passenger will not show up is 0.04. Suppose that 61 people bought tickets for a flight that has 60 seats. Determine the probability that there will not be enough seats. Round to 3 decimal places.

85. South Florida humorist Dave Barry often wrote about his dog, Zippy. Suppose that the Barry home has 3200 ft^2 of living area with tile floor. Further suppose that an expensive 8 ft by 10 ft oriental rug is placed on the floor. If Zippy had an "accident" in the house, what is the probability that it would happen on the expensive rug?

For Exercises 86–88,

a. Shade the area bounded by the given inequalities on a coordinate grid showing $-5 \le x \le 5$ and $-5 \le y \le 5$.

b. Suppose that an enthusiastic mathematics student makes a square dart board out of the portion of the rectangular coordinate system defined by $-5 \le x \le 5$ and $-5 \le y \le 5$. Find the probability that a dart thrown at the target will land in the shaded region.

86. $y \ge |x|$ and $y \le 4$ **87.** $|y| \le 3$ and $|x| \le 2$ **88.** $x^2 + y^2 \le 9$

Write About It

89. In a carnival game, players win a prize if they can toss a ring around the neck of a bowling pin. How would you go about estimating the probability of winning the game?

90. Give an example of two events that are mutually exclusive.

91. Give an example of two events that are not mutually exclusive.

92. Explain why a probability of $\frac{5}{4}$ is impossible.

Expanding Your Skills

93. Suppose that a box of DVDs contains 10 action movies and 5 comedies.

   a. If two DVDs are selected from the box with replacement, determine the probability that both are comedies.

   b. It probably seems more reasonable that someone would select two *different* DVDs from the box. That is, the first DVD would not be replaced before the second DVD is selected. In such a case, are the events of selecting comedies on the first and second picks independent events?

   c. If two DVDs are selected from the box *without* replacement, determine the probability that both are comedies.

94. Suppose that 12 students (5 freshmen and 7 sophomores) are being considered for two different scholarships. One scholarship is for $500 and the other is for $250.

   a. Two students are selected at random from the group of 12 to receive the scholarships. If a student may receive both scholarships, determine the probability that both students are freshmen.

   b. Now suppose that an individual student may not receive both scholarships. Determine the probability that both students chosen are freshmen.

95. If five cards are dealt from a standard deck of 52 cards, find the probability that

   a. The cards are all hearts.

   b. The cards are all of the same suit.

96. If five cards are dealt from a standard deck of 52 cards, find the probability that

   a. The cards consist of four aces.

   b. The cards are four of a kind (four cards with the same face value).

CHAPTER 8 KEY CONCEPTS

SECTION 8.1 Sequences and Series	Reference
An **infinite sequence** is a function whose domain is the set of positive integers. A **finite sequence** is a function whose domain is the set of the first n positive integers.	p. 690
A sequence in which consecutive terms alternate in sign is an **alternating sequence.**	p. 691
A **recursive formula** defines the nth term of a sequence as a function of one or more terms preceding it.	p. 692
For a positive integer n, the quantity $n!$ ("n factorial") is defined as $n! = (n)(n-1)(n-2) \cdots (2)(1)$ By definition, $0! = 1$.	p. 693
The **nth partial sum** of a sequence $\{a_n\}$ is a finite series and is given by $S_n = \displaystyle\sum_{i=1}^{n} a_i = a_1 + a_2 + a_3 + \cdots + a_n$	p. 694
The sum of *all* terms in an infinite sequence is an **infinite series:** $\displaystyle\sum_{i=1}^{\infty} a_i = a_1 + a_2 + a_3 + \cdots$	p. 695
Properties of summation: If $\{a_n\}$ and $\{b_n\}$ are sequences, and c is a real number, then 1. $\displaystyle\sum_{i=1}^{n} c = cn$ 2. $\displaystyle\sum_{i=1}^{n} ca_i = c\displaystyle\sum_{i=1}^{n} a_i$ 3. $\displaystyle\sum_{i=1}^{n} (a_i \pm b_i) = \displaystyle\sum_{i=1}^{n} a_i \pm \displaystyle\sum_{i=1}^{n} b_i$	p. 697

SECTION 8.2 Arithmetic Sequences and Series	Reference
An **arithmetic sequence** is a sequence in which each term after the first differs from its predecessor by a common difference d.	p. 701
The nth term of an arithmetic sequence: $$a_n = a_1 + (n - 1)d$$ where a_1 is the first term of the sequence and d is the common difference.	p. 701
Sum of the first n terms of an arithmetic sequence (nth partial sum): $$S_n = \frac{n}{2}(a_1 + a_n)$$ where a_1 is the first term of the sequence and a_n is the nth term of the sequence.	p. 706

SECTION 8.3 Geometric Sequences and Series	Reference
A **geometric sequence** is a sequence in which each term after the first is the product of the preceding term and a fixed nonzero real number, called the **common ratio.** If the first term of a geometric sequence is a_1, and the common ratio is r, then the nth term of a geometric sequence is given by $a_n = a_1 r^{n-1}$.	p. 712
nth partial sum of a geometric sequence: The sum S_n of the first n terms of a geometric sequence is given by $S_n = \dfrac{a_1(1 - r^n)}{1 - r}$ where a_1 is the first term of the sequence and r is the common ratio, $r \neq 1$.	p. 715
Sum of an infinite geometric series: Given an infinite geometric series $a_1 + a_1 r + a_1 r^2 + a_1 r^3 + \cdots$ with $\lvert r \rvert < 1$, the sum S of all terms in the series is given by $$S = \frac{a_1}{1 - r}$$ *Note*: If $\lvert r \rvert \geq 1$, then the sum does not exist.	p. 717
Future value of an ordinary annuity: Suppose that P dollars is invested at the end of each compounding period n times per year at interest rate r. Then the value A (in \$) of the annuity after t years is given by $$A = \frac{P\left[\left(1 + \frac{r}{n}\right)^{nt} - 1\right]}{\frac{r}{n}}$$	p. 719

SECTION 8.4 Mathematical Induction	Reference
Mathematical induction is a technique of mathematical proof that is often used to prove that a statement is true for all positive integers.	p. 725
The principle of mathematical induction: Let P_n be a statement involving the positive integer n, and let k be an arbitrary positive integer. Then P_n is true for all positive integers n if 1. P_1 is true, and 2. The truth of P_k implies the truth of P_{k+1}.	p. 726
The assumption that P_k is true is called the **inductive hypothesis.**	p. 727
Extended principle of induction: Mathematical induction can be extended to prove statements that might hold true only for integers greater than or equal to some positive integer j. In such a case, • First prove the statement for $n = j$. • Then show that the truth of the statement for an integer greater than or equal to j implies the truth of the statement for the integer that follows.	p. 729

SECTION 8.5 The Binomial Theorem	Reference

Binomial theorem:

Let n be a positive integer. The expansion of $(a + b)^n$ is given by

$$(a + b)^n = \binom{n}{0}a^n + \binom{n}{1}a^{n-1}b + \binom{n}{2}a^{n-2}b^2 + \cdots + \binom{n}{n-1}ab^{n-1} + \binom{n}{n}b^n$$

$$= \sum_{r=0}^{n} \binom{n}{r}a^{n-r}b^r$$

The values $\binom{n}{0}, \binom{n}{1}, \binom{n}{2}, \ldots, \binom{n}{n}$ are called the **binomial coefficients** of the expansion of $(a + b)^n$.

The binomial coefficients can also be found by using **Pascal's triangle.**

$$1$$
$$1 \quad 1$$
$$1 \quad 2 \quad 1$$
$$1 \quad 3 \quad 3 \quad 1$$
$$1 \quad 4 \quad 6 \quad 4 \quad 1$$
$$1 \ldots$$

pp. 733–734

Finding the *k*th term of a binomial expansion:

Let n and k be positive integers with $k \leq n + 1$. The kth term of $(a + b)^n$ is $\binom{n}{k-1}a^{n-(k-1)}b^{k-1}$

p. 736

SECTION 8.6 Principles of Counting	Reference

Fundamental principle of counting:

If one event can occur in m different ways and a second event can occur in n different ways, then the number of ways that the two events can occur in sequence is $m \cdot n$.

p. 739

A **permutation** is an ordered arrangement of distinct items.

- The number of permutations of n distinct elements is $n!$.

p. 741

- Consider a set of n elements with r_1 duplicates of one kind, r_2 duplicates of a second kind, $\ldots$, r_k duplicates of a kth kind. Then the number of distinguishable permutations of the n elements of the set is:

$$\frac{n!}{r_1! \cdot r_2! \cdot \cdots \cdot r_k!}$$

p. 741

- The number of permutations of n elements taken r at a time is given by

$$_nP_r = \frac{n!}{(n-r)!} \text{ or equivalently, } _nP_r = n(n-1)(n-2)\cdots(n-r+1)$$

p. 743

A **combination** is a collection of distinct items taken without regard to order.

The number of combinations of n elements taken r at a time is given by $_nC_r = \dfrac{n!}{r! \cdot (n-r)!}$.

p. 744

SECTION 8.7 Introduction to Probability	Reference

Theoretical probability of an event:

Let S represent a sample space with equally likely outcomes, and let E be a subset of S. Then the probability of event E, denoted by $P(E)$, is given by

$$P(E) = \frac{n(E)}{n(S)}$$

p. 751

Let E be an event relative to sample space S. The **complement of E,** denoted by $\overline{E}$, is the set of outcomes in the sample space but not in event E. It follows that $P(E) + P(\overline{E}) = 1$.

p. 753

Empirical probability is computed based on observed outcomes of the relative frequency of an event to the number of times an experiment is performed.

p. 754

Two events are **mutually exclusive** if they share no common elements.

The probability of *A* or *B*:

Given events *A* and *B* in a same sample space, $P(A \text{ or } B)$ is given by $P(A \cup B) = P(A) + P(B) - P(A \cap B)$.

If *A* and *B* are mutually exclusive, then $P(A \cup B) = P(A) + P(B)$.

p. 755

Probability of a sequence of independent events:

Two events are **independent** if the probability of one event does not affect the probability of the second event.

If events *A* and *B* are independent events, then the probability that both *A* and *B* will occur is

$$P(A \text{ and } B) = P(A) \cdot P(B).$$

p. 758

Expanded Chapter Summary available at www.mhhe.com/millercollegealgebra.

CHAPTER 8 Review Exercises

SECTION 8.1

For Exercises 1–2, simplify the expression.

1. $\dfrac{8!}{3! \cdot 5!}$

2. $\dfrac{(2n + 1)!}{(2n - 3)!}$

For Exercises 3–4, write the first five terms of the sequence.

3. $a_n = \dfrac{(n + 1)!}{n!}$

4. $c_1 = 5; c_n = -2c_{n-1} + 1$

5. Given the sequence defined by $b_n = (-1)^{n-1} \cdot n$, which terms are positive and which are negative?

6. Given $b_n = \dfrac{n + 4}{3n}$, find b_{45}.

7. a. Write the first five terms of the sequence defined by $a_n = 2n + 3$.

  b. Evaluate $\displaystyle\sum_{i=1}^{5} (2i + 3)$.

  c. Use the formula $S_n = \dfrac{n}{2}(a_1 + a_n)$ to verify the fifth partial sum of the arithmetic sequence from part (a).

8. Write the sum $4 - \dfrac{4}{3} + \dfrac{4}{5} - \dfrac{4}{7} + \cdots$ using summation notation with *n* as the index of summation. (Using techniques from calculus, we can show that this sum converges to π.)

For Exercises 9–11, find the sum.

9. $\displaystyle\sum_{i=3}^{8} 2i$

10. $\displaystyle\sum_{i=1}^{34} 10$

11. $\displaystyle\sum_{i=1}^{60} (-1)^{i+1}$

12. Write an expression for the apparent *n*th term a_n for the sequence. $10, -30, 90, -270, \ldots$

For Exercises 13–14, write the sum using summation notation. Use *i* as the index of summation.

13. $\dfrac{3}{2} + \dfrac{4}{4} + \dfrac{5}{8} + \dfrac{6}{16} + \dfrac{7}{32}$

14. $x^2 + x^2 y + x^2 y^2 + x^2 y^3$

15. Rewrite the series as an equivalent series with the new index of summation.

$$\sum_{i=1}^{10} i^3 = \sum_{j=0}^{\square} \square = \sum_{k=2}^{\square} \square$$

16. Determine if the statement is true or false.

$$\sum_{i=1}^{100} (4i - 2) = -200 + 4 \sum_{i=1}^{100} i$$

17. Suppose that a single cell of bacteria divides every 20 min for 4 hr. Write a formula for the sequence $\{a_n\}$ representing the number of cells after the *n*th cell division.

SECTION 8.2

For Exercises 18–20, determine whether the sequence is arithmetic. If so, identify the common difference *d*.

18. $11, 10.2, 9.4, 8.6, 7.8, \ldots$

19. $4, \dfrac{21}{4}, \dfrac{13}{2}, \dfrac{31}{4}, \ldots$

20. $9, 12, 16, 21, 27, \ldots$

21. Determine the first five terms of the arithmetic sequence $\{a_n\}$ with $a_1 = 4$, and $d = 8$.

22. a. Write an expression for the *n*th term of the arithmetic sequence $\{a_n\}$ with $a_1 = -19$, and $d = 5$.

  b. Find a_{36}.

23. Find the 23rd term of an arithmetic sequence with $a_1 = 15$ and $a_{57} = 239$.

24. Given an arithmetic sequence with $a_{15} = 86$ and $a_{37} = 240$, find the 104th term.

25. Find the number of terms of the arithmetic sequence.
11, 14, 17, 20, 23, ... , 122

26. A sales person working for a heating and air-conditioning company earns an annual base salary of $30,000 plus $500 on every new system he sells. Suppose that $\{a_n\}$ is a sequence representing the sales person's total yearly income based on the number of units sold n.

 a. Write a formula for the nth term of a sequence that represents the sales person's total income for n units sold. In this case, define $\{a_n\}$ with domain $n \geq 0$ to allow for the possibility of 0 units sold.

 b. How much will the sales person earn in a year for selling 42 new units?

27. Find the sum of the first 35 terms of the arithmetic sequence $-1, -9, -17, -25, \ldots$.

For Exercises 28–30, find the sum.

28. $3 + 10 + 17 + 24 + \cdots + 437$

29. $\displaystyle\sum_{n=1}^{36} (2 - 5n)$ **30.** $\displaystyle\sum_{i=1}^{68} \left(5 + \tfrac{1}{2}i\right)$

31. How long will it take to pay off a debt of $3960 if $50 is paid off the first month, $60 is paid off the second month, $70 is paid off the third month, and so on?

SECTION 8.3

For Exercises 32–34, determine whether the sequence is geometric. If so, find the value of r.

32. $\dfrac{3}{10}, \dfrac{3}{1000}, \dfrac{3}{100,000}, \ldots$ **33.** $3, 9, 36, 180, \ldots$

34. $5p^3, -5p^5, 5p^7, -5p^9, \ldots$

35. Write the first five terms of a geometric sequence with $a_1 = 120$, and $r = \tfrac{2}{3}$.

36. Write a formula for the nth term of the geometric sequence.
$$-40, 20, -10, 5, \ldots$$

For Exercises 37–39, find the indicated term of a geometric sequence from the given information.

37. $a_1 = 4$ and $a_2 = 12$. Find a_6.

38. $a_1 = -15$ and $a_4 = -\dfrac{5}{9}$. Find a_7.

39. $a_7 = \dfrac{1}{64}$ and $r = -\dfrac{1}{4}$. Find a_1.

40. Find a_1 and r for a geometric sequence given that $a_3 = 18$ and $a_6 = 486$.

For Exercises 41–46, find the sum of the geometric series, if possible.

41. $\displaystyle\sum_{n=1}^{7} 5(3)^{n-1}$ **42.** $\displaystyle\sum_{i=1}^{6} 12\left(-\dfrac{2}{3}\right)^{i-1}$

43. $\displaystyle\sum_{k=1}^{\infty} 5\left(\dfrac{5}{6}\right)^{k-1}$ **44.** $\displaystyle\sum_{i=1}^{\infty} \dfrac{2}{3}(4)^{i-1}$

45. $\displaystyle\sum_{n=3}^{\infty} 6\left(\dfrac{1}{2}\right)^{n-1}$ **46.** $36 - 12 + 4 - \dfrac{4}{3} + \cdots$

47. Write $0.8\overline{7}$ as a fraction.

48. An estimated 150,000 people attended the Coconut Grove art festival over a 3-day period. Admission to the event is $10 per person. In addition, suppose that each person spends an average of $100 on art, drinks, and food.

 a. How much money is initially infused into the local economy during the festival for admission, art, drinks, and food.

 b. If the money is later respent in the community over and over again at a rate of 70%, determine the total amount spent. Assume that the money is respent an infinite number of times.

For Exercise 49–50, find the value of an ordinary annuity in which regular payments of P dollars are made at the end of each compounding period, n times per year, at an interest rate r for t years.

49. $P = \$150$, $n = 12$, $r = 4\%$, $t = 16$ yr

50. $P = \$300$, $n = 12$, $r = 4\%$, $t = 32$ yr

51. a. At age 28, an employee begins investing $100 each pay period (twice per month) in an ordinary annuity. If the interest rate is 5.5%, find the value of the annuity when the employee retires at age 62.

 b. Determine the value of the annuity if the employee waits to retire at age 65.

52. The initial swing (one way) of a pendulum makes an arc of 2 ft. Each swing (one way) thereafter makes an arc of 85% of the length of the previous swing. What is the total arc length that the pendulum travels?

SECTION 8.4

For Exercises 53–56, use mathematical induction to prove the given statement for all positive integers n.

53. $3 + 7 + 11 + \cdots + (4n - 1) = n(2n + 1)$

54. $-5 + 2 + 9 + \cdots + (7n - 12) = \dfrac{n}{2}(7n - 17)$

55. $1 + 4 + 16 + \cdots + 4^{n-1} = \dfrac{1}{3}(4^n - 1)$

56. $10^n - 1$ is divisible by 3.

57. Use mathematical induction to show that $4^n < (n + 2)!$ for integers $n \geq 2$.

SECTION 8.5

58. a. Write the first four rows of Pascal's triangle.

    b. Expand $(a + b)^3$.

59. Evaluate the given expression. Compare the results to the result of Exercise 58.

  a. $\dbinom{3}{0}$ **b.** $\dbinom{3}{1}$ **c.** $\dbinom{3}{2}$ **d.** $\dbinom{3}{3}$

For Exercises 60–64, expand the binomial by using the binomial theorem.

60. $(4y + 3)^4$ **61.** $(2x - 3)^5$ **62.** $(5c^3 - d^2)^4$

63. $(t^5 + u^3)^6$ **64.** $\left(\dfrac{x}{2} + 2y\right)^3$

For Exercises 65–67, find the indicated term of the binomial expansion.

65. $(5x + 4y)^6$; fifth term

66. $(3x^4 - 2)^8$; middle term

67. $(2c^2 - d^5)^9$; Term containing d^{25}.

68. Given $f(x) = 4x^3 + 2x$, find the difference quotient.

69. Use the binomial theorem to find the value of $(3 + 2i)^4$ where i is the imaginary unit.

SECTION 8.6

70. Gaynelle can travel one of 3 roads from her home to school. From school to work there are 4 different routes. How many different routes are available for Gaynelle to travel from home to school to work?

71. A disc jockey has 7 songs that he must play in a half-hour period. In how many different ways can he arrange the 7 songs?

72. In how many ways can the letters in the word SHUFFLE be arranged?

73. In how many ways can the word SPACE be misspelled?

74. A quiz consists of 6 true/false questions and 4 multiple-choice questions. The multiple-choice questions each have 5 possible responses (a, b, c, d, e) of which only one is correct. In how many different ways can a student fill out the answers to the quiz?

75. A 3-digit code is to be made from the set of digits $\{4, 5, 6, 7, 8\}$.

    a. How many codes can be formed if there are no restrictions?

    b. How many codes can be formed if the corresponding 3-digit number is to be an even number?

    c. How many codes can be formed if the corresponding 3-digit number is to be a multiple of 5 and there can be no repetition of digits?

76. Evaluate $_{21}P_4$ and interpret its meaning.

77. Evaluate $_{21}C_4$ and interpret its meaning.

78. Evaluate $_{10}P_3$ and $_{10}C_3$ and compare the results.

79. In how many ways can a statistician select a sample of 15 people from a population of 90 people?

80. How many triangles can be made if the vertices are from three of the six points on the circle? One such triangle is shown in the figure.

81. The Daytona 500 auto race has 40 cars that initially start the race. How many first-, second-, and third-place finishes can occur?

82. Suppose that a drama class has 22 students.

    a. In how many ways can four students be selected to take part in a survey?

    b. In how many ways can four students be selected to act out a scene from a play involving 4 different parts?

83. To meet the graduation requirements, a student must take 2 English classes out of 10 available, 3 math classes out of 6 available, and 2 history classes out of 7 available. Assuming that the student has met the prerequisites of each course and that there are no scheduling conflicts, determine the number of ways in which the student can select these courses.

SECTION 8.7

84. Which of the following can represent the probability of an event?

    a. 0   **b.** 1   **c.** 1.2   **d.** 0.12

    e. 1.2%   **f.** 120%   **g.** −0.12   **h.** $\frac{4}{5}$

85. If the probability of an event is 0.0042, is the event likely to occur or not likely to occur?

86. If the probability of an event is $\frac{87}{90}$, is the event likely to occur or not likely to occur?

87. If $P(E) = 0.73$, what is the probability of $P(\overline{E})$?

88. Suppose that a box containing music CDs has 4 with country music, 6 with rock music, 3 with jazz music, and 7 with rap music. If one CD is selected at random from the box, determine the probability that

    a. The CD has rap music.

    b. The CD has jazz music.

    c. The CD does not have jazz music.

    d. The CD has classical music.

89. Suppose that two fair dice are rolled. Determine the probability that

  a. The sum of the numbers on the dice is 6.

  b. The sum of the numbers on the dice is greater than 9.

  c. The numbers on the dice form a sum that is a multiple of 5.

90. For a recent season, the batting average for baseball player Jose Iglesias was 0.306. (This means that the probability that Iglesias will get a hit on a given time at bat is 0.306.)

  a. Determine the probability that Iglesias will not get a hit on a given time at bat.

  b. If Iglesias is at bat three times in a game, what is the probability that he will get a hit all three times?

  c. If Iglesias is at bat three times in a game, what is the probability that he will not get a hit on any of the three times at bat?

91. Suppose that 15 lightbulbs are in a cabinet and that 4 are defective. If three bulbs are chosen at random,

  a. What is the probability that all three will be defective?

  b. What is the probability that all three will be good?

  c. Why do the probabilities from parts (a) and (b) not add up to 1?

92. Suppose that a lottery game has the player select 5 distinct numbers from 1 to 30, inclusive. The player wins by choosing numbers that match those randomly selected in a drawing.

  a. What is the probability that a player will win if the numbers do not have to be selected in any particular order?

  b. If the player has to pick the correct 5 numbers in a specific order, what is the probability that the player will win?

93. For a recent year, the Centers for Disease Control reported that the probability that a 50-yr-old will live to age 51 is 0.9959. In a group of ten 50-yr-olds, what is the probability that all ten survive to the age of 51?

For Exercises 94–99, use the data in the table categorizing smokers and nonsmokers according to their blood pressure (BP) levels. If one person is chosen at random, find the probability of the given event.

	Normal BP	Elevated BP	Total
Smokers	42	28	70
Nonsmokers	80	10	90
Total	122	38	160

94. The person has elevated blood pressure.

95. The person is a nonsmoker.

96. The person has elevated blood pressure or is a nonsmoker.

97. The person has normal blood pressure or is a smoker.

98. The person is a smoker or has elevated blood pressure.

99. The person is a nonsmoker or has normal blood pressure.

For Exercises 100–103, refer to the sample space for a card drawn from a standard deck. See page 755.

100. If one card is drawn at random from a standard deck, what is the probability that it is an 8 or a club?

101. If one card is drawn at random from a standard deck, what is the probability that it is a red card or a 5?

102. If *two* cards are drawn at random with replacement from a standard deck, what is the probability that both are hearts?

103. If *two* cards are drawn at random with replacement from a standard deck, what is the probability that both are kings?

CHAPTER 8   Test

1. Write the first six terms of the sequence defined by $a_1 = -2$, $a_2 = 3$, $a_n = a_{n-2} + a_{n-1}$ for $n \geq 3$.

2. Simplify. $\dfrac{(3n + 1)!}{(3n - 1)!}$

For Exercises 3–5,

a. Determine whether the sequence is arithmetic, geometric, or neither.

b. If the sequence is arithmetic, determine d. If the sequence is geometric, determine r.

c. Write an expression a_n for the apparent nth term of the sequence.

  3. 0.139, 0.00139, 0.0000139, …

4. 0.52, 0.68, 0.84, 1.00, …

5. $\dfrac{3}{5}, \dfrac{6}{25}, \dfrac{9}{125}, \dfrac{12}{625}, \dots$

6. Write the first four terms of the geometric sequence with $a_1 = 4$ and $r = \frac{3}{2}$.

7. Write the first five terms of the arithmetic sequence with $a_1 = 10$ and $a_{20} = 67$.

8. Find the 78th term of an arithmetic sequence with $a_1 = 64$ and $d = -11$.

9. Find the 6th term of a geometric sequence with $a_1 = -3$ and $r = 2$.

10. Find the number of terms in the arithmetic sequence.
$$-15, -19, -23, -27, \ldots, -679$$

11. Find the number of terms in the geometric sequence.
$$16, 8, 4, 2, \ldots, \frac{1}{16}$$

12. Given an arithmetic sequence with $a_{12} = -21$ and $a_{50} = -97$, find a_1 and d.

13. Given a geometric sequence with $a_3 = 20$ and $a_8 = 640$, find a_1 and r.

For Exercises 14–18, evaluate the sum if possible.

14. $\dfrac{3}{2} + \dfrac{3}{4} + \dfrac{3}{8} + \dfrac{3}{16} + \cdots$

15. $\displaystyle\sum_{k=1}^{54} (3k + 7)$

16. $\displaystyle\sum_{i=1}^{6} 4\left(\dfrac{3}{2}\right)^{i-1}$

17. $\displaystyle\sum_{k=1}^{4} k!$

18. $\displaystyle\sum_{i=1}^{\infty} \dfrac{1}{2}\left(\dfrac{6}{5}\right)^{i-1}$

19. Suppose that a county fair has an estimated 50,000 people attend over a 2-week period. Admission to the fair is $5.00. In addition, suppose that each person spends an average of $10 on food, drinks, and rides.

 a. How much money is initially infused into the local community for admission, food, drinks, and rides?

 b. If the money is respent in the local community over and over again at a rate of 65%, determine the total amount spent. Assume that the money is respent an infinite number of times. Round to the nearest dollar.

20. An employee invests $400 per month in an ordinary annuity. If the interest rate is 5.2%, find the value of the annuity after 25 yr.

21. Lakeisha wants to put down new tile in her home. The tile costs $3.50/ft² and labor is $4.00/ft². Write the nth term of a sequence representing the cost to tile an n by n square foot area where n is an integer and $n \geq 1$ ft.

22. How many days will it take Johan to read a 920-page book if he reads 20 pages the first day, 25 pages the second day, 30 pages the third day, and so on?

For Exercises 23–25, use mathematical induction to prove the given statement for all positive integers n.

23. $6 + 10 + 14 + \cdots + (4n + 2) = n(2n + 4)$

24. $1 + 5 + 25 + \cdots + 5^{n-1} = \frac{1}{4}(5^n - 1)$

25. 2 is a factor of $7^n - 5$.

For Exercises 26–27, expand the binomial by using the binomial theorem.

26. $\left(\dfrac{x}{2} + 3\right)^4$

27. $(4c^2 - t^4)^5$

For Exercises 28–29, find the indicated term of the binomial expansion.

28. $(-2t + v^2)^{10}$; eighth term

29. $(3x + y^2)^7$; Term containing y^6.

30. Evaluate $_{13}P_5$ and $_{13}C_5$.

31. Explain the difference between a permutation and a combination of n items taken r at a time.

32. A musician plans to perform 9 selections. In how many ways can she arrange the musical selections?

33. How many outfits can be made from 4 pairs of slacks, 5 shirts, and 3 ties, if one selection from each category is made? Assume that all items fashionably match.

34. In how many ways can the word HIPPOPOTAMUS be misspelled?

35. Suppose that a jury pool consists of 30 women and 26 men.

 a. In how many ways can a jury of 12 people be selected?

 b. In how many ways can a jury of 6 women and 6 men be selected?

 c. What is the probability of randomly selecting a jury of 6 women and 6 men?

 d. What is the probability of randomly selecting a jury of all men?

36. A review sheet for a history test has 10 essay questions. Suppose that the professor picks 3 questions from the review sheet to put on the test. In how many ways can the professor choose 3 questions from 10 questions?

37. After a service call by a plumber, the company follows up with a survey to rate the service and professionalism of the technician. The survey has 6 yes/no questions and 4 multiple-choice questions each with 3 possible responses. In how many different ways can a customer fill out the survey?

38. Suppose that 50 people buy raffle tickets.

 a. In how many ways can 4 people who bought tickets be selected if each is to receive a $20 gift certificate to a restaurant?

 b. In how many ways can 4 people who bought tickets be selected if the first person wins a $10 gift certificate, the second person wins a $25 gift certificate, the third person wins a $50 gift certificate, and the fourth person wins a $200 camera?

39. For a recent year, approximately 36,000 people were killed in the United States in motor vehicle accidents. If the population had 300,000,000 people at that time, estimate the probability of being killed in a motor vehicle crash.

40. A cable company advertises short wait times for customer service calls. The graph shows the wait times (in seconds) for a sample of customers. Based on the data given, if a customer is selected at random, find the probability that

Number of Customers vs. Wait Time, x

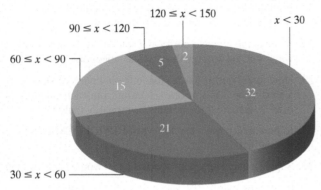

a. The customer will wait less than 30 sec.

b. The customer will wait at least 30 sec.

c. The customer will wait at least 90 sec but less than 120 sec.

d. The customer will wait more than 150 sec.

41. If two fair dice are thrown, find the probability that the sum is between 6 and 8, inclusive.

42. Suppose that two cards are drawn from a standard deck with replacement. What is the probability that an ace is selected, followed by a heart?

43. For a recent year, 31.9% of Americans living below the poverty level were not covered by health insurance. (*Source*: U.S. Census Bureau, www.census.gov) If three people were selected at random from this population, what is the probability that all three would not have coverage?

For Exercises 44–47, use the data in the table categorizing the type of payment used at a grocery store according to the gender of the customer. If one person is chosen at random, find the probability of the given event.

	Cash	Credit Card	Check	Total
Male	19	24	13	56
Female	14	30	20	64
Total	33	54	33	120

44. The customer is female.

45. The customer is male or paid by check.

46. The customer paid by credit card or by cash.

47. The customer paid cash or was female.

CHAPTER 8   Cumulative Review Exercises

For Exercises 1–4, consider sets A and B and determine if the statement is true or false.

$$A = \{x \mid 4 > x\}, B = \{x \mid x \le 1\}$$

1. $4 \in A$

2. $1 \in B$

3. $A \cap B = B$

4. $A \cup B = B$

5. a. Write an expression for the distance between t and 5 on the number line.

  b. Simplify the expression from part (a) for $t < 5$.

6. Simplify without using a calculator.

$$\frac{(6.0 \times 10^{13})(9.0 \times 10^8)}{2.0 \times 10^{-6}}$$

For Exercises 7–10, simplify the expression.

7. $-27^{-4/3}$

8. $\left(\sqrt{x} + 5\sqrt{2}\right)\left(3\sqrt{x} - \sqrt{2}\right)$

9. $\dfrac{4 - x^{-2}}{6 - 3x^{-1}}$

10. i^{127}

For Exercises 11–18, solve the equation or inequality. Write the solution set to the inequalities in interval notation.

11. $\sqrt{2x^2 - x - 14} + 4 = x$

12. $(x^2 + x)^2 - 14(x^2 + x) + 24 = 0$

13. $|3x - 5| = |2x + 1|$

14. $-5x \le 20$ and $3 - [2 - (x - 4)] < 2x + 5$

15. $0 \le |x + 7| - 6$

16. $\dfrac{2x + 1}{x + 4} \ge 1$

17. $\log_4(2x + 7) = 2 + \log_4 x$

18. $5e^{x+1} - 100 = 0$

19. Given $f(x) = 2x^3 - 5x^2 - 28x + 15$,

  a. Find all the zeros of $f(x)$.

  b. Identify the x-intercepts of the graph of f.

  c. Determine the y-intercept of the graph of f.

  d. Graph $y = f(x)$.

  e. Solve the inequality $2x^3 - 5x^2 - 28x + 15 < 0$.

20. An object is launched straight upward with an initial speed of 400 ft/sec from a height of 4 ft. The height of the object t seconds after launch can be modeled by $s(t) = -16t^2 + 400t + 4$, where $s(t)$ is the height measured in feet.

  a. Determine the time required for the object to reach its maximum height.

  b. Determine the maximum height of the object.

  c. Determine the time required for the object to hit the ground. Round to 1 decimal place.

21. Given $x^2 + y^2 - 8x - 2y + 1 = 0$,

   a. Write the equation of the circle in standard form.

   b. Identify the center and radius of the circle.

   c. Graph the circle.

For Exercises 22–25, graph the equation.

22. $\dfrac{y^2}{9} - \dfrac{x^2}{16} = 1$

23. $\dfrac{(x + 1)^2}{4} + \dfrac{(y - 2)^2}{9} = 1$

24. $y = \log_2(x + 3)$

25. $y = \begin{cases} |x| + 1 & \text{for } x \le 1 \\ -x + 1 & \text{for } x > 1 \end{cases}$

26. Explain how the graph of $f(x) = 2\sqrt{x - 1} + 3$ is related to the graph of $y = \sqrt{x}$.

27. Given $f(x) = 4x^3 - 3x$,

   a. Find the difference quotient.

   b. Find the average rate of change on the interval $[1, 3]$.

   c. Determine if the function is even, odd, or neither.

28. Given $g(x) = 5x - 1$,

   a. Is g a one-to-one function?

   b. Write an equation for $g^{-1}(x)$.

29. Given $f(x) = x^3 - x^2 - 7x + 15$,

   a. Is $2 + i$ a zero of $f(x)$?

   b. Is $(x + 3)$ a factor of $f(x)$?

30. Given $r(x) = \dfrac{3x^2 + 2}{x^2 - 5x - 14}$,

   a. Determine the vertical asymptote(s).

   b. Determine the horizontal or slant asymptote if either exist.

31. The population of a city was 320,000 in the year 2000. By 2012, the population reached 360,800.

   a. Write a model of the form $P(t) = P_0 e^{kt}$ to represent the population $P(t)$ for a time t years since 2000.

   b. Approximate the population in the year 2015 assuming that this trend continued. Round to the nearest 100 people.

   c. Determine the amount of time for the population to reach 400,000 if this trend continues. Round to the nearest tenth of a year.

32. Write the expression in terms of $\log x$, $\log y$, and $\log z$.

$$\log\left(\frac{x^3}{y^5\sqrt{z}}\right)$$

33. Simplify. $\quad \log_2 \dfrac{1}{16}$

34. Approximate the value of $\log_5 417$ to 4 decimal places.

For Exercises 35–37, solve the system of equations.

35. $\begin{aligned} 2x - 5y &= 13 \\ -3x + 2y &= -3 \end{aligned}$

36. $\begin{aligned} 4x \quad\quad - 2z &= -4 \\ 6y + 5z &= 8 \\ 7x - 3y \quad\quad &= 13 \end{aligned}$

37. $\begin{aligned} 2x - y &= 6 \\ x^2 + y &= 9 \end{aligned}$

38. Use Gaussian elimination or Gauss-Jordan elimination to solve the system of equations.

$$\begin{aligned} -2x - 9y + 16z &= -15 \\ x + y - z &= 4 \\ x - 2y + 5z &= 1 \end{aligned}$$

39. Use Cramer's rule to solve the system.

$$\begin{aligned} 13x - 2y &= 11 \\ 5x + 3y &= 6 \end{aligned}$$

40. Graph the solution set.

$$\begin{aligned} x^2 + y^2 &\le 16 \\ y &> -x^2 + 4 \end{aligned}$$

For Exercises 41–46, refer to the matrices given and perform the indicated operations, if possible.

$$A = \begin{bmatrix} 4 & -3 \\ 5 & 9 \end{bmatrix} \qquad B = \begin{bmatrix} -1 & 6 \\ 3 & 7 \end{bmatrix}$$

$$C = \begin{bmatrix} -1 & 4 & 0 \\ 3 & 6 & 1 \\ 2 & -5 & 7 \end{bmatrix} \qquad D = \begin{bmatrix} -5 & 1 & 0 \\ 8 & 4 & 6 \end{bmatrix}$$

41. $-5A + 2B$ **42.** $3C - A$

43. $|A|$ **44.** $|C|$

45. CD **46.** AD

47. Given $A = \begin{bmatrix} 1 & 2 & -1 \\ 0 & 1 & 3 \\ 1 & 0 & 2 \end{bmatrix}$, find A^{-1}.

48. Use the inverse matrix from Exercise 47 to solve the system of equations.

$$\begin{aligned} x + 2y - z &= -5 \\ y + 3z &= 13 \\ x \quad\quad + 2z &= 5 \end{aligned}$$

49. Given $(y + 2)^2 = -8(x - 1)$,

   a. Determine the vertex of the graph of the parabola.

   b. Identify the focus.

   c. Write an equation for the directrix.

50. Determine whether the sequence is arithmetic or geometric and find the value of the common difference or the common ratio.

$$-a^2, a^5, -a^8, a^{11}, \dots$$

51. Find the 500th term of an arithmetic sequence with $a_1 = 6.9$ and $d = 0.3$.

For Exercises 52–53, find the sum.

52. $\displaystyle\sum_{k=1}^{74} (5k + 3)$

53. $\displaystyle\sum_{i=1}^{8} 6(2)^{i-1}$

54. Find the sum if possible. $6 + 2 + \dfrac{2}{3} + \dfrac{2}{9} + \cdots$

55. Evaluate the expressions.

  a. $8!$ **b.** $\dbinom{12}{11}$ **c.** $_{15}P_3$

56. Expand the binomial. $(3x - y^2)^5$

57. Simplify. $\dfrac{(3n - 1)!}{2!(3n + 1)!}$

58. In how many ways can 5 children be arranged in a line for a photograph?

59. Suppose that four people are to be randomly selected from a group of 8 women and 5 men. What is the probability of selecting 2 women and 2 men?

60. If one card is selected from a standard deck of cards, what is the probability that the card selected is a diamond or an ace? (*Hint*: Refer to Figure 8-12 from page 755.)

Student Answer Appendix

CHAPTER R

Section R.1 Practice Exercises, pp. 13–18

1. set
3. real, numbers
5. $|a - b|$ or $|b - a|$
7. $(a + b) + c$; $(ab)c$
9. a. 3 is an element of the set of natural numbers.
   b. -3.1 is not an element of the set of whole numbers.
11. a. False   **b.** False   **c.** True   **d.** True
13. a. 6   **b.** 6   **c.** $-12, 6$ **d.** $0.\overline{3}, 0.33, -0.9, -12, \frac{11}{4}, 6$
   e. $\sqrt{5}, \frac{\pi}{6}$ **f.** $\sqrt{5}, 0.\overline{3}, 0.33, -0.9, -12, \frac{11}{4}, 6, \frac{\pi}{6}$
15. $a \geq 5$ **17.** $3c \leq 9$ **19.** $m + 4 > 70$
21. True   **23.** True   **25.** False   **27.** False
29. $(-7, \infty)$; $\{x \mid x > -7\}$ **31.** $(-\infty, 4.1]$; $\{x \mid x \leq 4.1\}$
33. $[-6, 0)$; $\{x \mid -6 \leq x < 0\}$
35. ⟵————┤———→; $(-\infty, 6]$
              6
37. ⟵(————┤——→; $\left(-\frac{7}{6}, \frac{1}{3}\right]$
      $-\frac{7}{6}$ $\frac{1}{3}$
39. ⟵——(————→; $(4, \infty)$
           4
41. ⟵—(————┤—→; $\{x \mid -3 < x \leq 7\}$
        -3 7
43. ⟵————┤——→; $\{x \mid x \leq 6.7\}$
            6.7
45. ⟵———[————→; $\{x \mid x \geq -\frac{3}{5}\}$
      $-\frac{3}{5}$
47. a. $\{0, 3, 4, 6, 8, 9, 12\}$ **b.** $\{0, 12\}$ **c.** $\{-2, 0, 4, 8, 12\}$
   d. $\{4, 8\}$ **e.** $\{-2, 0, 3, 4, 6, 8, 9, 12\}$ **f.** $\{\ \}$
49. a. $\{0, 1, 2, 3, 4, 5, 8, 12\}$ **b.** $\{8\}$ **c.** $\{\ \}$ **d.** $\{3\}$
   e. $\{-2, 4, 6, 7, 8\}$ **f.** $\{0, 4, 6, 7, 8, 12\}$
51. a. $\mathbb{R}$ **b.** $\{x \mid -1 \leq x < 9\}$ **c.** $\{x \mid x < 9\}$ **d.** $\{x \mid x < -8\}$
   e. $\{x \mid x < -8 \text{ or } x \geq -1\}$ **f.** $\{\ \}$
53. a. $(-\infty, 4)$ **b.** $(-2, 1]$ **55. a.** $(-\infty, \infty)$ **b.** $[3, 5)$
57. 6   **59.** 0   **61.** $5 - \sqrt{17}$ **63. a.** $\pi - 3$ **b.** $\pi - 3$
65. a. $x + 2$ **b.** $-x - 2$ **67. a.** 1   **b.** -1
69. $|1 - 6|$ or $|6 - 1|$; 5   **71.** $|3 - (-4)|$ or $|-4 - 3|$; 7
73. $|6 - 2\pi|$ or $|2\pi - 6|$; $2\pi - 6$
75. a. 16   **b.** 16   **c.** -16 **d.** 2   **e.** -2
   f. Not a real number
77. a. 2   **b.** -2 **c.** -2 **d.** 10
   e. Not a real number   **f.** -10
79. -4 **81.** -3 **83.** $-\dfrac{9}{25}$ **85.** 7   **87.** Undefined
89. $x + 7$ **91.** $w + (-3)$ **93.** $t + (3 + 9)$; $t + 12$
95. $\left(\dfrac{1}{5} \cdot 5\right) w$; w **97.** $-16w^3$ **99.** $9x^3y - 6.9xy^3$
101. $-\dfrac{1}{15}c^7 d - \dfrac{3}{2}cd^7$ **103.** $-4x + \pi$ **105.** $-24x^2 - 16x + 8$
107. $-4x^2y - 12yz^2 + \dfrac{4}{3}z^3$ **109.** $22w$ **111.** $-25u + 16v$
113. $88v + 59w - 20$ **115.** $6y^2$
117. a. $J = C - 1$ **b.** $C = J + 1$ **119.** $D \geq 0.25P$
121. $v = \sqrt{2gh}$ **123. a.** $C = 0.12k + 14.89$ **b.** \$158.89
125. a. $C = 640m + 200n + 500$ **b.** \$8780
127. a. $C = 159n + 0.11(159n)$ or $C = 176.49n$ **b.** \$705.96

129. -3 **131.** $\sqrt{34}$ **133.** 308   **135.** 36 gal
137. A parenthesis is used if an endpoint to an interval is *not* included in the set.
139. The commutative property of addition indicates that the order in which two quantities are added does not affect the sum. The associative property of addition indicates that the manner in which quantities are grouped under addition does not affect the sum.
141. 0   **143.** $2n$ **145.** $-n$ **147.** positive   **149.** negative
151. $[1, 2)$ **153.** $[-5, -2)$ **155.** 6744.25   **157.** 0.58

Section R.2 Practice Exercises, pp. 24–27

1. 1   **3.** scientific   **5.** $m - n$
7. a. 1   **b.** -1 **c.** 8   **d.** 1
9. a. 1   **b.** $-\dfrac{1}{3}$ **c.** $-\dfrac{2}{3}$ **d.** 1
11. a. $\dfrac{1}{64}$ **b.** $\dfrac{8}{x^2}$ **c.** $\dfrac{1}{64x^2}$ **d.** $-\dfrac{1}{64}$
13. a. q^2 **b.** $\dfrac{1}{q^2}$ **c.** $\dfrac{5p^3}{q^2}$ **d.** $\dfrac{5q^2}{p^3}$
15. 2^{12} **17.** x^{11} **19.** $-\dfrac{12d^8}{c^3}$ **21.** y **23.** $\dfrac{1}{216}$ **25.** $\dfrac{2p^7}{3k^3}$
27. $\dfrac{n^5}{3m^4}$ **29.** 4^6 **31.** $\dfrac{1}{p^{14}}$ **33.** $-8c^3d^3$ **35.** $\dfrac{49a^2}{b^2}$
37. $\dfrac{16x^4}{y^6}$ **39.** $\dfrac{n^4}{49k^2}$ **41.** -1 **43.** $\dfrac{m^6}{4n^{18}}$ **45.** $\dfrac{27z^{15}}{x^9 y^6}$
47. $-\dfrac{9z^{16}}{2y^7}$ **49.** 27   **51.** 197   **53.** $\dfrac{2y^4}{x^7}$ **55.** $\dfrac{1}{x^{16}}$
57. $-\dfrac{x^{32}}{v^{21}w^{28}}$ **59.** $(3x + 5)^{12}$ **61.** $(6v - 7)^{90}$ **63.** $\dfrac{31}{4}$
65. a. 3.5×10^5 **b.** 3.5×10^{-5} **c.** 3.5×10^0
67. a. 8.6×10^{-1} **b.** 8.6×10^0 **c.** 8.6×10^1
69. 2.998×10^{10} cm/sec   **71.** 1.0×10^{-5} cm   **73.** 4.2×10^0 L
75. a. 0.00000261   **b.** 2,610,000   **c.** 2.61
77. a. 0.6718   **b.** 6.718   **c.** 67.18
79. 1,670,000,000,000,000,000,000 molecules   **81.** 0.000 007 m
83. 8×10^5 **85.** 4×10^{-4} **87.** 1.86×10^{16} **89.** 7.2×10^{-20}
91. 1.24×10^{11} **93. a.** 3.1536×10^7 sec   **b.** 3.1536×10^{11} gal
95. 2×10^4 songs   **97.** 2.5×10^{13} red blood cells
99. In the expression $6x^0$, the exponent 0 applies to x only. In the expression $(6x)^0$, the exponent 0 applies to a base of $(6x)$. The first expression simplifies to 6, and the second expression simplifies to 1.
101. Yes; $(-4)^2 = 16 > 0$ **103.** No
105. 1.55×10^{18} N   **107. a.** $<$ **b.** $>$ **109. a.** $>$ **b.** $=$
111. x^{m+4} **113.** x^{2m+7} **115.** x^{m-8} **117.** x^{m+2}
119. x^{12mn} **121.** $x^{3m+4}y^{2n+5}$

Section R.3 Practice Exercises, pp. 34–37

1. n **3.** $(\sqrt[n]{a})^m$ or $\sqrt[n]{a^m}$ **5.** $|x|$ **7.** $\sqrt[n]{a \cdot b}$ **9.** 3
11. $\dfrac{2}{7}$ **13.** 0.3   **15.** Not a real number   **17.** -3 **19.** $-\dfrac{1}{2}$
21. a. 5   **b.** Undefined   **c.** -5 **23. a.** 3   **b.** -3 **c.** -3
25. a. $\dfrac{11}{13}$ **b.** $\dfrac{13}{11}$
27. a. 8   **b.** $\dfrac{1}{8}$ **c.** -8 **d.** $-\dfrac{1}{8}$
   e. Undefined   **f.** Undefined

29. a. 16 **b.** $\dfrac{1}{16}$ **c.** -16 **d.** $-\dfrac{1}{16}$ **e.** 16 **f.** $\dfrac{1}{16}$

31. a. $\sqrt[11]{y^4}$ or $\left(\sqrt[11]{y}\right)^4$ **b.** $6\sqrt[11]{y^4}$ or $6\left(\sqrt[11]{y}\right)^4$ **c.** $\sqrt[11]{(6y)^4}$ or $\left(\sqrt[11]{6y}\right)^4$

33. $a^{3/5}$ **35.** $(6x)^{1/2}$ **37.** $6x^{1/2}$ **39.** $(a^5 + b^5)^{1/5}$

41. a^2 **43.** $\dfrac{3y^{1/3}}{w^{2/3}}$ **45.** $\dfrac{8y^{3/20}}{x^6}$ **47.** $\dfrac{4m^3}{n^7}$

49. $\dfrac{(m+n)^{1/2}}{m}$ **51. a.** $t \geq 0$ **b.** All real numbers

53. $|y|$ **55.** y **57.** $|2x - 5|$ **59.** w^6

61. a. $c^3\sqrt{c}$ **b.** $c^2\sqrt[3]{c}$ **c.** $c\sqrt[4]{c^3}$ **d.** $\sqrt[9]{c^7}$

63. a. $2\sqrt{6}$ **b.** $2\sqrt[3]{3}$ **65.** $5y^2z^3\sqrt[3]{2x^2z^2}$ **67.** $2p^3q\sqrt[4]{6p^2q^3}$

69. $2(y-2)\sqrt{21(y-2)}$ **71.** $\dfrac{p^3\sqrt{p}}{6}$ **73.** $2wz\sqrt[3]{z^2}$

75. $\dfrac{x}{5y}$ **77.** $2\sqrt{35}$ **79.** xy **81.** $-15a\sqrt{a}$

83. $-\dfrac{4}{3}ac\sqrt[3]{3ab^2}$ **85.** $-5\sqrt[3]{2y^2}$ **87.** $-\sqrt{2}$

89. $-28xy\sqrt[3]{2xy}$ **91.** $(12 + 5y)\sqrt{2y}$ **93.** $(-z + 3)\sqrt{2z}$

95. $2\sqrt{110}$ in.^{2} **97.** $6\sqrt{5}$ m **99.** $16\sqrt{13}$ in. ≈ 58 in.

101. $2\sqrt{29}$ in. ≈ 10.8 in. **103.** 15.9%

105. In each case, add like terms or like radicals by using the distributive property.

107. $\sqrt[20]{x^{17}y^8}$ **109.** $\sqrt[18]{m^5}$ **111.** $\sqrt[8]{x^7}$

113. 2.0×10^4 **115.** $2\sqrt{2}$

117. Yes; The model gives a mean surface temperature of approximately 29.1°C.

Section R.4 Practice Exercises, pp. 44–47

1. polynomial **3.** leading **5.** binomial, trinomial **7.** $(a - b)$

9. a. Yes **b.** Yes **c.** No **d.** No

11. $-18x^7 + 7.2x^3 - 4.1$; Leading coefficient -18; Degree 7

13. $-y^2 + \dfrac{1}{3}y$; Leading coefficient -1; Degree 2 **15.** 11

17. $-6p^7 + 2p^4 + p^2 + 2p - 5$ **19.** $0.08c^3b - 0.06c^2b^2 + 0.01cb^3$

21. $-\dfrac{1}{4}x^2 + \dfrac{11}{8}x + 6\sqrt{2}$ **23.** $-2a^7b^3$

25. $14m^6 - 21m^3 + 28m^2$ **27.** $2x^2 + 3x - 20$

29. $8u^4 + 2u^2v^2 - 15v^4$ **31.** $y^3 - 13y^2 - 42y - 24$

33. $a^2 + 2ab + b^2$ **35.** $a^3 + 3a^2b + 3ab^2 + b^3$ **37.** $16x^2 - 25$

39. $9w^4 - 49z^2$ **41.** $\dfrac{1}{25}c^2 - \dfrac{4}{9}d^6$ **43.** $25m^2 - 30m + 9$

45. $16t^4 + 24t^2p^3 + 9p^6$ **47.** $w^3 + 12w^2 + 48w + 64$

49. $u^2 + 2uv + v^2 - w^2$ **51.** $10x + 22$ **53.** $25y^2 - 4x^2 - 12x - 9$

55. $x^3 + 14x^2 + 64x + 96$ **57.** $ac + bc$ **59.** $x^2 + 6x - 27$

61. a. $x + 1$ **b.** $x + (x + 1); 2x + 1$ **c.** $x(x + 1); x^2 + x$
 d. $x^2 + (x + 1)^2; 2x^2 + 2x + 1$

63. $-y^2 + 26y + 31$ **65.** $x^{2n} - 4x^n - 21$ **67.** $z^{2n} + 2w^mz^n + w^{2m}$

69. $a^{2n} - 25$ **71.** $-60x - 50$ **73.** $20 + 30\sqrt{6} - 20\sqrt{2}$

75. $45\sqrt{2} - 24\sqrt{3} - 18$ **77.** $8y - 2\sqrt{y} - 15$

79. $8\sqrt{15} + 22$ **81.** 5 **83.** $16x^2y - 4xy^2$

85. $36z^2 - 12\sqrt{5}z + 5$ **87.** $25a^4b + 70a^2b^2\sqrt{ab} + 49ab^4$

89. $x - 24$ **91.** $x - 10\sqrt{x + 1} + 26$ **93.** $\sqrt{25 - 4x}$

95. $2x - 2\sqrt{x^2 - y^2}$ **97.** 7 m^2

99. A polynomial consists of a finite number of terms in which the coefficient of each term is a real number, and the variable factor x is raised to an exponent that is a whole number.

101. In each case, multiply by using the distributive property.

103. False **105.** True

Problem Recognition Exercises, p. 47

1. a. 8 **b.** 4 **c.** 16 **d.** $\dfrac{1}{64}$ **e.** -8 **f.** Undefined

 g. 16 **h.** $\dfrac{1}{16}$ **2. a.** $25a^2b^6$ **b.** $25a^2 + 10ab^3 + b^6$

c. $\dfrac{1}{25a^2b^6}$ **d.** $\dfrac{1}{25a^2 + 10ab^3 + b^6}$

3. a. $4x^8y^2$ **b.** $4x^8 - 4x^4y + y^2$ **c.** $\dfrac{1}{4x^8y^2}$

 d. $\dfrac{1}{4x^8 - 4x^4y + y^2}$ **4. a.** x^8 **b.** x^2 **c.** $\dfrac{1}{x^2}$ **d.** $\dfrac{1}{x^{15}}$

5. a. x^4 **b.** $x^2\sqrt[3]{x^2}$ **c.** $x\sqrt[5]{x^3}$ **d.** $\sqrt[9]{x^8}$

6. a. $a + 5b^2$ **b.** $6a^2 + 5ab^2 - 4b^4$

7. a. $2a^2$ **b.** $a^4 - b^4$ **c.** $-4ab$ **d.** $a^4 - 2a^2b^2 + b^4$

8. a. $b - a$ **b.** $a - b$ **9. a.** $x + 2$ **b.** $-x - 2$

10. a. $x + y$ **b.** $\sqrt{x^2 + y^2}$ **c.** $x + 2\sqrt{xy} + y$ **d.** $x - y$

11. a. $\sqrt[3]{4x^2}$ **b.** $2\sqrt[3]{2x}$ **12. a.** $\sqrt[12]{y^7}$ **b.** $\sqrt[4]{y} + \sqrt[3]{y}$

13. a. $\dfrac{1}{4}$ **b.** 9 **c.** 1 **d.** 1 **14. a.** 10 **b.** 14

Section R.5 Practice Exercises, pp. 56–58

1. cubes, $(a + b)(a^2 - ab + b^2)$ **3.** perfect, $(a + b)^2$

5. $5c^3(3c^2 - 6c + 1)$ **7.** $7a^2b(3b^4 - 2ab^3 + 5a^2)$

9. $(x - 6y)(5z + 7)$ **11.** $5k(3k^2 + 7)(2k - 1)$

13. a. $3(-2x^2 + 4x + 3)$ **b.** $-3(2x^2 - 4x - 3)$

15. $-4x^2y(3xy + 2x^2y^2 - 1)$ **17.** $(2a + 5)(4x + 9)$

19. $(3x^2 - 10)(4x - 3)$ **21.** $(c - 2d)(d + 4)$

23. $(p + 9)(p - 7)$ **25.** $2t(t - 4)(t - 10)$ **27.** $(2z + 7)(3z + 2)$

29. $yz(7y + 2z)(y - 6z)$ **31.** $(t - 9)^2$ **33.** $2x(5x + 8y)^2$

35. $(2c^2 - 5d^3)^2$ **37.** $(3w + 8)(3w - 8)$

39. $2(10u^2 + 3v^3)(10u^2 - 3v^3)$ **41.** $(25p^2 + 4)(5p + 2)(5p - 2)$

43. $(y + 4)(y^2 - 4y + 16)$ **45.** $c(c - 3)(c^2 + 3c + 9)$

47. $(2a^2 - 5b^3)(4a^4 + 10a^2b^3 + 25b^6)$

49. $10x(3x + 7)(x + 2)(x - 2)$ **51.** $(a + y - 5)(a - y + 5)$

53. $5xy(3x + 8)(2x + 3)$ **55.** $(x^2 + 2)(x + 3)(x - 3)$

57. $(x^3 + 16)(x + 2)(x^2 - 2x + 4)$ **59.** $(x + y + z)(x + y - z)$

61. $(x + y + z)(x^2 + 2xy + y^2 - xz - yz + z^2)$ **63.** $(3m + 21n + 7)^2$

65. $(3c - 8)(-c + 2)$ or $-(3c - 8)(c - 2)$

67. $(p - 4)(p^2 + 4p + 16)(p^4 + 1)(p^2 + 1)(p + 1)(p - 1)$

69. $(m + 3)(m^2 - 3m + 9)(m - 1)(m^2 + m + 1)$

71. $2z(2x + 3)(4x^2 - 6x + 9)(x - 1)(x^2 + x + 1)$

73. $(x - y)(x + y - 1)$ **75.** $(a - c)(a + 2c + 1)$

77. $\dfrac{x^2 - 7x + 2}{x^4}$ **79.** $\dfrac{(y - 4)(y + 3)}{y^4}$ **81.** $2c^{3/4}(c + 2)$

83. $\dfrac{8x + 5}{3x^{2/3}}$ **85.** $(3x + 1)^{2/3}(8x + 1)$ **87.** $\dfrac{2(2x + 1)}{(3x + 2)^{2/3}}$

89. $A = x^2 - y^2$; $A = (x + y)(x - y)$ **91.** $2\pi r(r + h)$

93. $\dfrac{4}{3}\pi(R - r)(R^2 + Rr + r^2)$ **95.** $(21 + 19)(21 - 19)$; 80

97. Expand the square of any binomial. For example:
 $(2c + 3)^2 = 4c^2 + 12c + 9$

99. In each case, factor out x to the smallest exponent to which it appears in both terms. That is,
 $5x^4 + 4x^3 = x^3(5x + 4)$ and $5x^{-4} + 4x^{-3} = x^{-4}(5 + 4x)$

101. $(x + \sqrt{5})(x - \sqrt{5})$ **103.** $(z^2 + 6)(z + \sqrt{6})(z - \sqrt{6})$

105. $(x - \sqrt{5})^2$ **107. a.** $(x - 1)(x^4 + x^3 + x^2 + x + 1)$
 b. $(x - 1)(x^{n-1} + x^{n-2} + \cdots + 1)$

109. No **111.** Yes **113.** No

Section R.6 Practice Exercises, pp. 68–72

1. rational **3.** $-2, 1$ **5.** complex (or compound)

7. $x \neq -7$ **9.** $a \neq 9, a \neq -9$ **11.** No restricted values

13. $a \neq 0, b \neq 0$ **15.** a and b **17.** $\dfrac{x - 3}{x - 7}$; $x \neq -3, x \neq 7$

19. $-\dfrac{4ac}{b^4}$; $a \neq 0, b \neq 0$ **21.** $\dfrac{2 - \sqrt{6}}{3}$

23. $-\dfrac{2y}{8 + y}$; $y \neq 8, y \neq -8$ **25.** $-\dfrac{4}{x - 2}$; $a \neq b; x \neq 2$

27. $\dfrac{a}{2b^3}$ 29. $\dfrac{2c^3(c-d)}{d(2c-d)}$ 31. a 33. $-\dfrac{x-2}{2x}$

35. $60x^5y^2z^4$ 37. $t(3t+4)^3(t-2)$ 39. $2x(x+10)^2$

41. $m+3$ 43. $\dfrac{10c^2+21}{45c^3}$ 45. $\dfrac{9y-22x}{2x^2y^5}$

47. $\dfrac{5x+21}{x(x+3)}$ 49. $-\dfrac{1}{x(x-y)}$ 51. $\dfrac{7y^2-y-6}{y^2(y+1)}$ 53. 1

55. $\dfrac{1}{(x+1)(x+2)}$ 57. $\dfrac{1}{3}$ 59. $x+2$ 61. $\dfrac{ab}{2b+a}$

63. $-\dfrac{3}{1+h}$ 65. $-\dfrac{7}{x(x+h)}$ 67. $\dfrac{x+2}{3}$ 69. $\dfrac{4\sqrt{y}}{y}$

71. $\dfrac{4\sqrt[3]{y^2}}{y}$ 73. $\dfrac{2\sqrt{3x+3}}{x+1}$ 75. $2(\sqrt{15}+\sqrt{11})$

77. $\sqrt{x}-\sqrt{5}$ 79. $\dfrac{5+4\sqrt{2}}{14}$ 81. $\dfrac{10\sqrt{3x}}{3x}$

83. $-\dfrac{2\sqrt{7}}{7w}$ 85. a. $S=\dfrac{2r_1r_2}{r_1+r_2}$ b. 427.9 mph

87. a. At 1 hr: 4.8 ng/mL; At 12 hr: 3.9 ng/mL; At 24 hr: 1.0 ng/mL; At 48 hr: 0.3 ng/mL b. 0 ng/mL

89. $\dfrac{13x+6}{x(x+1)}$ cm 91. $\dfrac{\sqrt{2x}}{x}$ in.² 93. $\dfrac{x}{5y^4}$ 95. $\dfrac{3t+32}{2t+1}$

97. 1 99. $\dfrac{a+2}{a-1}$ 101. $4\sqrt{5}+2\sqrt{3}$ 103. $\dfrac{4-2\sqrt{3}}{3}$

105. $-\dfrac{x+3}{x-5}$ 107. $-\dfrac{8}{t+2}$ 109. $\dfrac{x^3-x-2}{x^3}$

111. $\dfrac{3(2x+1)}{2\sqrt{3x}}$ 113. $\dfrac{2x^2+1}{(x^2+1)\sqrt{x^2+1}}$ 115. $\dfrac{2(8x^2+9)}{\sqrt{4x^2+9}}$

117. If $x=y$, then the denominator $x-y$ will equal zero. Division by zero is undefined.

119. a. $\dfrac{14}{3}\cdot\dfrac{30}{7}=\dfrac{420}{21}=20$ b. $(5-\sqrt{5})(5+\sqrt{5})$
$=(5)^2-(\sqrt{5})^2$
$=25-5$
$=20$

121. $\dfrac{w^{2n}}{w+z}$ 123. $\dfrac{\sqrt{x^2-y^2}}{x+y}$ 125. $\dfrac{\sqrt[6]{5^3 2^4}}{2}$

127. $\sqrt[3]{a^2}+\sqrt[3]{ab}+\sqrt[3]{b^2}$ 129. $\dfrac{1}{\sqrt{4+h}+2}$

Algebra for Calculus, pp. 72–73

1. $|y-L|<\varepsilon$ or $|L-y|<\varepsilon$
2. $|x-c|<\delta$ or $|c-x|<\delta$
3. a. 1 b. -1 4. a. -1 b. 1
5. a. $4x+2h+3$ b. $4x+3$
6. a. $6x+3h-4$ b. $6x-4$
7. a. $\dfrac{-1}{(x-2)(x+h-2)}$ b. $\dfrac{-1}{(x-2)^2}$
8. a. $\dfrac{-2}{(2x+5)(2x+2h+5)}$ b. $\dfrac{-2}{(2x+5)^2}$
9. a. $3x^2+3xh+h^2$ b. $3x^2$
10. a. $4x^3+6x^2h+4xh^2+h^3$ b. $4x^3$
11. a. $\dfrac{1}{\sqrt{x}+\sqrt{x+h}}$ b. $\dfrac{1}{2\sqrt{x}}$
12. a. $\dfrac{2}{\sqrt{2x+2h}+\sqrt{2x}}$ b. $\dfrac{1}{\sqrt{2x}}$
13. $\dfrac{x^{1/2}(5x+3)}{2}$ 14. $\dfrac{7x-1}{6x^{5/6}}$
15. $6(3x+1)^3(x^2+2)^2(5x^2+x+4)$
16. $-2(-2x+3)^2(4x^2-5)(28x^2-24x-15)$

17. $\dfrac{42(t-1)^5}{(2t+5)^7}$ 18. $\dfrac{24x^5}{(x^2+4)^4}$ 19. $\dfrac{2(x^2+2)}{(x^2+4)^{1/2}}$

20. $-\dfrac{2(x-1)(x+1)}{(2-x^2)^{1/2}}$ 21. $\dfrac{1}{(x^2+1)^{3/2}}$ 22. $\dfrac{6(2x-1)}{(3x-1)^{2/3}}$

23. $\dfrac{x+8}{(x+4)\sqrt{x+4}}$ 24. $-\dfrac{x(x^2-32)}{(16-x^2)\sqrt{16-x^2}}$

25. $\dfrac{4(x-3)}{3(x-4)^{2/3}}$ 26. $\dfrac{5(x+4)}{4(x+5)^{3/4}}$

27. $\dfrac{1}{(2x)^{1/2}(x+1)^{3/2}}$ 28. $-\dfrac{(x-1)(x+1)}{(3x)^{2/3}(x^2+1)^{4/3}}$

Chapter R Review Exercises, pp. 75–78

1. a. $\sqrt{9}$ b. $0,\sqrt{9}$ c. $0,-8,\sqrt{9}$
d. $0,-8,1.\overline{45},\sqrt{9},-\dfrac{2}{3}$ e. $\sqrt{6},3\pi$
f. $\sqrt{6},0,-8,1.\overline{45},\sqrt{9},-\dfrac{2}{3},3\pi$

3. a. $[-3,7)$ $\{x|-3\le x<7\}$
b. $(2.1,\infty)$ $\{x|x>2.1\}$
c. $(-\infty,4]$ $\{x|4\ge x\}$

5. a. $\mathbb{R}$ b. $\{x|-2\le x<7\}$ c. $\{x|x<7\}$
d. $\{x|x<-3\}$ e. $\{x|x<-3 \text{ or } x\ge-2\}$ f. { }
7. a. $|2-\sqrt{5}|$ or $|\sqrt{5}-2|$ b. $\sqrt{5}-2$
9. $\dfrac{5}{2}$ 11. 6 13. a. $J=E+150$ b. $E=J-150$
15. a. $C=3.6s+50p+250n$ b. \$8260
17. $23.9c^2d-16.5cd$ 19. d 21. c 23. e 25. a
27. b 29. a. 1 b. -1 c. 9 d. 1
31. p^3 33. $\dfrac{144b^8}{a^6}$ 35. $\dfrac{2u^{12}}{v^4}$
37. a. 0.98 b. 9.8 c. 98 39. 1.763×10^{12}
41. a. $\sqrt{x^2}$ or $(\sqrt{x})^2$ b. $9\sqrt{x^2}$ or $9(\sqrt{x})^2$
c. $\sqrt{(9x)^2}$ or $(\sqrt{9x})^2$
43. a. -4 b. Not a real number
45. a. 1000 b. $\dfrac{1}{1000}$ c. -1000 d. $-\dfrac{1}{1000}$
e. Undefined f. Undefined
47. $\dfrac{3n^{1/3}}{m^2}$ 49. $3y^4z^4\sqrt[3]{2xz^2}$ 51. $\dfrac{p^6\sqrt{p}}{3}$ 53. $5\sqrt{14}$
55. $\sqrt[12]{c^{11}d^{10}}$ 57. $-51cd\sqrt[3]{2c^2}$ 59. a. Yes b. No c. Yes
61. 8 63. $-8a^2b^3+7.3ab^2-2.9b$ 65. $10w^6+7w^3y^2-6y^4$
67. $81t^2-16$ 69. $25k^2-30k+9$ 71. $4v^2-4v+1-w^2$
73. $30+26\sqrt{15}$ 75. $4c^4d-20c^2d^2\sqrt{cd}+25cd^4$
77. $2x^3+15x^2+37x+30$ 79. $16m^2n(5m^2n^7-3m^3n^2-1)$
81. $(5a+7b)(3c-2)$ 83. $2x(2x-5y)^2$
85. $3k(k-3)(k^2+3k+9)$ 87. $(5n+m+6)(5n-m-6)$
89. $-4(3p-2)(p+3)$ 91. $(x^2+3y)(x^2+3y-1)$
93. $4x^{5/2}(3x-1)$ 95. $-\dfrac{x(3x^2-2)}{(1-x^2)^{1/2}}$ 97. $-\dfrac{3a^2}{2cd^5}; c\ne0, d\ne0$
99. $\dfrac{x^2y^6}{3}$ 101. $-\dfrac{5}{x(x+4)}$ 103. $\dfrac{1}{x(3x+y)}$ 105. 1
107. $\dfrac{mn}{5n+m}$ 109. $\dfrac{5\sqrt[4]{k^3}}{k}$ 111. $\sqrt{x}-2$

Chapter R Test, pp. 78–79

1. a. 8 b. 0, 8 c. 0, 8, -3 d. $0,8,-\dfrac{5}{7},2.1,-0.\overline{4},-3$
e. $\dfrac{\pi}{6}$ f. $0,\dfrac{\pi}{6},8,-\dfrac{5}{7},2.1,-0.\overline{4},-3$

2. a. $M = 2J$ **b.** $J = \dfrac{1}{2}M$

3. a. $\mathbb{R}$ **b.** $\{x \mid 0 \le x < 2\}$ **c.** $\{x \mid x < 2\}$ **d.** $\{x \mid x < -1\}$
e. $\{x \mid x < -1 \text{ or } x \ge 0\}$ **f.** $\{\ \}$
4. $t - 5$ **5. a.** $|\sqrt{2} - 2|$ or $|2 - \sqrt{2}|$ **b.** $2 - \sqrt{2}$

6. a. $y \ne 11, y \ne -2$ **b.** $\dfrac{2}{y+2}$ **7.** $-12x - 49$

8. $-\dfrac{1}{2}a^2 b + \dfrac{7}{2}ab^2$ **9.** $\dfrac{7}{4}$ **10.** $\dfrac{a^8}{b^{15}}$ **11.** 32   **12.** $\dfrac{1}{t^{12}}$

13. $2k^5 n^2 \sqrt[3]{10m^2 n}$ **14.** $\dfrac{3}{2}pq\sqrt[4]{p}$ **15.** $-b\sqrt{5ab}$

16. $6n^4 - 36n^3 + 58n^2 + 12n - 20$ **17.** $9a^2 + 6ab + b^2 - c^2$

18. $\dfrac{1}{16}z - p^4$ **19.** $15x - 7\sqrt{2x} - 4$ **20.** $x - 12z\sqrt{x} + 36z^2$

21. $\dfrac{3x}{2(x+7)}$ **22.** $x - 5$ **23.** $\dfrac{y-3}{y^2}$ **24.** $x + 3$

25. $2(\sqrt{13} - \sqrt{10})$ **26.** $\dfrac{5\sqrt{2}}{2t}$

27. $2x(3x - 1)(5x + 2)$ **28.** $(x + 5a)(y + 2c)$
29. $(x^2 + 9)(x - 3)(x + 3)(x + 2)$ **30.** $(c - 2a - 11)(c + 2a + 11)$

31. $(3u - v^2)(9u^2 + 3uv^2 + v^4)$ **32.** $\dfrac{7w^2 + 2w + 4}{w^6}$

33. $\dfrac{3y - 1}{(2y - 1)^{3/4}}$ **34.** $\dfrac{3x(6 - x^2)}{(9 - x^2)^{1/2}}$ **35.** $x^2 + 11x - 34$

36. 4.5×10^{10}; 1.66×10^6 **37.** $0.000\,000\,8$
38. 1.2×10^{14} **39.** $(2.7, \infty)$ **40.** $\{x \mid -3 \le x < 5\}$

CHAPTER 1

Section 1.1 Practice Exercises, pp. 90–94

R.1. $-13w + 16$ **R.2.** $x^2 + 18x + 81$ **R.3.** $(2p - 5)(p + 3)$
R.4. $42x$ **R.5.** $(5n - 2)(n - 2)(n + 7)$ **R.6.** $(a - 2)(a - 5)$
1. linear   **3.** solution   **5.** identity   **7.** rational
9. a. Linear; $\{-4\}$ **b.** Nonlinear   **c.** Linear; $\{-16\}$
d. Nonlinear   **e.** Linear; $\{10\}$ **11.** $\{-4\}$ **13.** $\{0\}$

15. $\{2\}$ **17.** $\{-6.2\}$ **19.** $\{5000\}$ **21.** $\left\{\dfrac{22}{3}\right\}$ **23.** $\{14\}$

25. $\left\{-\dfrac{33}{2}\right\}$ **27.** $\{-54\}$ **29.** $\{41\}$ **31. a.** $205°\text{F}$
b. 10,400 ft   **33.** 2019   **35. a.** $C = 7x$ **b.** The motorist
will save money beginning on the 16th working day.
37. a. $S_1 = 45{,}000 + 2250x$ **b.** $S_2 = 48{,}000 + 2000x$ **c.** 12 yr
39. Contradiction; $\{\ \}$ **41.** Identity; $\mathbb{R}$ **43.** Conditional equation; $\{-8\}$

45. $x \ne 5, x \ne -4$ **47.** $x \ne \dfrac{3}{2}, x \ne 0, x \ne -3$ **49.** $\{17\}$

51. $\{-23\}$ **53.** $\{\ \}$; The value 3 does not check.   **55.** $\{2\}$
57. $\{-4\}$ **59.** $\{\ \}$; The value -3 does not check.   **61.** $\{16\}$

63. $\{\ \}$; The value -4 does not check.   **65.** $\left\{-\dfrac{5}{13}\right\}$ **67.** $l = \dfrac{A}{w}$

69. $c = P - a - b$ **71.** $s_1 = s_2 - \Delta s$ **73.** $y = -\dfrac{7}{2}x + 4$

75. $y = \dfrac{5}{4}x - \dfrac{1}{2}$ **77.** $y = -\dfrac{3}{2}x + 3$

79. $d = \dfrac{2S}{n} - a$ or $d = \dfrac{2S - an}{n}$ **81.** $h = \dfrac{3V}{\pi r^2}$

83. $x = \dfrac{6}{4 + t}$ **85.** $x = \dfrac{5 - ay}{6 - b}$ or $x = \dfrac{ay - 5}{b - 6}$

87. $P = \dfrac{A}{1 + rt}$ **89.** $\left\{\dfrac{18}{19}\right\}$ **91.** $\{0\}$ **93.** $\{-3\}$ **95.** $\{9\}$

97. $\mathbb{R}$ **99.** $\{1\}$ **101.** $\{\ \}$ **103.** 16 yr   **105.** 33 mi
107. The value 5 is not defined within two of the expressions in the
equation. Substituting 5 into the equation would result in division by 0.

109. The equation cannot be written in the form $ax + b = 0$. The term
$\frac{3}{x} = 3x^{-1}$. Therefore, the term $\frac{3}{x}$ is not first degree and the equation is
not a first-degree equation.   **111.** The equation is a contradiction.
There is no real number x to which we add 1 that will equal the same
real number x to which we add 2.   **113.** $a = 6$ **115.** $a = 3$

Section 1.2 Practice Exercises, pp. 99–104

R.1. $x - 254$ **R.2.** $5x$ **R.3.** $8x - 29$

R.4. $6x + 30$ **R.5.** $\dfrac{\$8.25}{1 \text{ hr}} = \dfrac{\$82.50}{10 \text{ hr}}$

1. \$900   **3.** $\dfrac{d}{r}$ **5.** $P = 2l + 2w$ **7.** Rocco borrowed \$1500 at
3% and \$3500 at 2.5%.   **9.** Fernando invested \$4500 in the 3-yr CD
and \$2500 in the 18-month CD.   **11.** 1250 gal of E5   **13.** 96 ft^3
of sand   **15.** The plane to Los Angeles travels 400 mph and the
plane to New York City travels 460 mph.   **17.** The distance is 24 mi.

19. $\dfrac{220}{7}$ sec or approximately 31.4 sec   **21.** 15 hr   **23.** 62.5 lb of
cement and 225 lb of gravel   **25.** LDL is 144 mg/dL and the total
cholesterol is 204 mg/dL.   **27.** 480 deer   **29.** 300 km
31. \$336   **33. a.** $C = 110 + 60x$ **b.** 4 hr   **35.** 555 ft
37. The pole is 7.2 ft long, and the snow is 4.8 ft deep.
39. 10 L should be drained and replaced by water.   **41.** The easement
is 8 ft.   **43. a.** The kitchen is 14 ft by 10 ft.   **b.** 154 ft^2
c. \$1958.88   **45.** Aliyah invested \$2760 in the stock returning
11% and \$3000 in the stock returning 5%.   **47.** $x = 11.2$ ft and
$y = 7.5$ cm   **49.** No. If x represents the measure of the smallest
angle, then the equation $x + (x + 2) + (x + 4) = 180$ does not result in
an odd integer value for x. Instead the measures of the angles would be
even integers.   **51.** The numbers are 7 and 23.   **53.** The original
number is 68.   **55.** $x_2 = 1.8$ m   **57.** 4 kg

Section 1.3 Practice Exercises, pp. 111–113

R.1. $2\sqrt{7}$ **R.2.** $20\sqrt{6}$ **R.3.** $\dfrac{7\sqrt{3}}{4}$ **R.4.** $p^2 - 3$

R.5. $\dfrac{3\sqrt{2}}{2}$ **R.6.** $3\sqrt{5} + 12$

1. -1 **3.** real, imaginary   **5.** $11i$ **7.** $7i\sqrt{2}$ **9.** $i\sqrt{19}$
11. $-4i$ **13.** -6 **15.** $-5\sqrt{2}$ **17.** $-2\sqrt{21}$ **19.** 7   **21.** $3i$
23. Real part: 3; Imaginary part: -7 **25.** Real part: 0; Imaginary part: 19
27. Real part: $-\frac{1}{4}$; Imaginary part: 0   **29.** $0 + 8i$

31. $2 + 2\sqrt{3}i$ or $2 + 2i\sqrt{3}$ **33.** $\dfrac{4}{7} + \dfrac{3}{14}i$ **35.** $2 + 3i$

37. $-\dfrac{9}{2} + \sqrt{3}i$ or $-\dfrac{9}{2} + i\sqrt{3}$ **39.** $-2 + \sqrt{2}i$ or $-2 + i\sqrt{2}$

41. a. 1   **b.** i **c.** -1 **d.** $-i$ **43. a.** i **b.** $-i$
c. -1 **d.** -1 **45.** $10 - 10i$ **47.** $-3 + 61i$

49. $-\dfrac{1}{3} + \dfrac{7}{12}i$ **51.** $-1.2 + 0i$ **53.** $-2 - 3i$ **55.** $-2 + 10i$

57. $\sqrt{21} + i\sqrt{33}$ **59.** $36 - 57i$ **61.** $-40 - 42i$
63. $17 - i\sqrt{5}$ **65.** $-11 - 7i$ **67.** 6   **69. a.** $3 + 6i$ **b.** 45
71. a. $0 - 8i$ **b.** 64   **73.** 116   **75.** 49   **77.** 5

79. $\dfrac{8}{5} + \dfrac{6}{5}i$ **81.** $\dfrac{94}{173} - \dfrac{81}{173}i$ **83.** $\dfrac{6}{41} - \dfrac{\sqrt{5}}{41}i$ **85.** $0 - \dfrac{5}{13}i$

87. $0 + \dfrac{\sqrt{3}}{3}i$ **89.** $4i\sqrt{2}$ **91.** $2i$

93. a. $(5i)^2 + 25 = 0 \checkmark$ **b.** $(-5i)^2 + 25 = 0 \checkmark$
95. a. $(2 + i\sqrt{3})^2 - 4(2 + i\sqrt{3}) + 7 = 0 \checkmark$
b. $(2 - i\sqrt{3})^2 - 4(2 - i\sqrt{3}) + 7 = 0 \checkmark$
97. $(a + bi)(c + di)$
$\quad = ac + adi + bci + bdi^2$
$\quad = ac + (ad + bc)i + bd(-1)$
$\quad = (ac - bd) + (ad + bc)i$

99. The second step does not follow because the multiplication property
of radicals can be applied only if the individual radicals are real
numbers. Because $\sqrt{-9}$ and $\sqrt{-4}$ are imaginary numbers, the
correct logic for simplification would be
$\sqrt{-9} \cdot \sqrt{-4} = i\sqrt{9} \cdot i\sqrt{4} = i^2\sqrt{36} = -1 \cdot 6 = -6$

101. Any real number. For example: 5. **103.** $a^2 + b^2$

105. a. $(x + 3)(x - 3)$ **b.** $(x + 3i)(x - 3i)$

107. a. $(x + 8)(x - 8)$ **b.** $(x + 8i)(x - 8i)$

109. a. $(x + \sqrt{3})(x - \sqrt{3})$ **b.** $(x + i\sqrt{3})(x - i\sqrt{3})$

111.

```
NORMAL FLOAT AUTO a+bi RADIAN MP
√-16
                              4i
(4-5i)-(2+3i)
                            2-8i
(12-15i)(-2+9i)
                       111+138i
```

113.

```
NORMAL FLOAT AUTO a+bi RADIAN MP
(4-9i)²
                        -65-72i
7/(2i)▶Frac
                           -7/2 i
(14+8i)/(3-i)▶Frac
                       17/5 + 19/5 i
```

Section 1.4 Practice Exercises, pp. 123–125

R.1. a. $x^2 - 8x + 16$ **b.** $(x - 4)^2$ **R.2.** $(5x - 3y)(x + 4y)$

R.3. $4p(p - 6q)^2$ **R.4.** $(4x + 5)^2$ **R.5.** $\dfrac{3 - \sqrt{5}}{2}$

1. quadratic **3.** $\pm\sqrt{k}$ **5.** $x = \dfrac{-b \pm \sqrt{b^2 - 4ac}}{2a}$

7. $\{3, -7\}$ **9.** $\{-8, 3\}$ **11.** $\left\{-\dfrac{5}{2}, -\dfrac{1}{4}\right\}$ **13.** $\left\{\dfrac{3}{2}, -\dfrac{3}{2}\right\}$

15. $\{0, 4\}$ **17.** $\{-3, 4\}$ **19.** $\{9, -9\}$ **21.** $\{\sqrt{7}, -\sqrt{7}\}$

23. $\{4i, -4i\}$ **25.** $\{-2 \pm 2\sqrt{7}\}$ **27.** $\{8, 2\}$

29. $\left\{\dfrac{1}{2} \pm \dfrac{\sqrt{17}}{2}i\right\}$ **31.** $n = 49; (x + 7)^2$ **33.** $n = 169; (p - 13)^2$

35. $n = \dfrac{9}{4}; \left(w - \dfrac{3}{2}\right)^2$ **37.** $n = \dfrac{1}{81}; \left(m + \dfrac{1}{9}\right)^2$

39. $\{-11 \pm 5\sqrt{5}\}$ **41.** $\{4 \pm 2i\sqrt{2}\}$ **43.** $\{-3 \pm i\sqrt{31}\}$

45. $\left\{4, -\dfrac{1}{2}\right\}$ **47.** $\left\{-\dfrac{3}{2} \pm \dfrac{\sqrt{14}}{2}\right\}$ **49.** $\left\{-\dfrac{5}{6} \pm \dfrac{\sqrt{97}}{6}\right\}$

51. $a = 1, b = -7, c = 4$ **53.** $a = 5, b = 3, c = 0$

55. $\left\{\dfrac{3 \pm \sqrt{37}}{2}\right\}$ **57.** $\{-2 \pm i\sqrt{2}\}$ **59.** $\{3 \pm i\}$

61. $\left\{\dfrac{7 \pm i\sqrt{11}}{10}\right\}$ **63.** $\left\{\dfrac{1}{2}, -\dfrac{3}{10}\right\}$ **65.** $\left\{\pm\dfrac{7}{3}i\right\}$

67. $\left\{\dfrac{5 \pm \sqrt{137}}{14}\right\}$ **69.** $\left\{\dfrac{5}{2}\right\}$ **71.** Linear; $\{-2\}$

73. Quadratic; $\{0, -2\}$ **75.** Linear; $\{1\}$ **77.** Neither

79. $\left\{\dfrac{4}{3}\right\}$ **81.** $\{-2 \pm \sqrt{2}\}$ **83.** $\{7 \pm \sqrt{55}\}$ **85.** $\mathbb{R}$

87. $\left\{1, -\dfrac{5}{6}\right\}$ **89.** $\{\ \}$ **91.** $\{-2\}$ **93.** $\left\{\pm\dfrac{\sqrt{35}}{7}i\right\}$

95. $\{\pm\sqrt[4]{5}\}$ **97. a.** -56 **b.** 2 nonreal solutions

99. a. 40 **b.** 2 real solutions **101. a.** 121 **b.** 2 real solutions

103. a. 0 **b.** 1 real solution **105.** $r = \sqrt{\dfrac{A}{\pi}}$ or $r = \dfrac{\sqrt{A\pi}}{\pi}$

107. $t = \sqrt{\dfrac{2s}{g}}$ or $t = \dfrac{\sqrt{2sg}}{g}$ **109.** $a = \sqrt{c^2 - b^2}$

111. $I = \dfrac{1}{c}\sqrt{\dfrac{L}{Rt}}$ or $I = \dfrac{\sqrt{LRt}}{cRt}$ **113.** $w = \dfrac{c \pm \sqrt{c^2 + 4kr}}{2k}$

115. $t = \dfrac{-v_0 \pm \sqrt{v_0^2 + 2as}}{a}$ **117.** $I = \dfrac{-CR \pm \sqrt{C^2R^2 - 4CL}}{2CL}$

119. The right side of the equation is not equal to zero.

121. $x = 2y$ or $x = -y$ **123.** $x^2 - 2x - 8 = 0$

125. $12x^2 - 11x + 2 = 0$ **127.** $x^2 - 5 = 0$ **129.** $x^2 + 4 = 0$

131. $x^2 - 2x + 5 = 0$

133. $x_1 + x_2 = \dfrac{-b + \sqrt{b^2 - 4ac}}{2a} + \dfrac{-b - \sqrt{b^2 - 4ac}}{2a}$

$$= \dfrac{-b + \sqrt{b^2 - 4ac} + (-b) - \sqrt{b^2 - 4ac}}{2a}$$

$$= \dfrac{-2b}{2a} = -\dfrac{b}{a}$$

Problem Recognition Exercises, p. 125

1. a. Expression; $6x^2 - 13x - 5$ **b.** Equation; $\left\{\dfrac{5}{2}, -\dfrac{1}{3}\right\}$

2. a. Expression; $\dfrac{4x + 36}{(x - 3)(x + 7)}$ **b.** Equation; $\{-9\}$

3. a. Equation; $\left\{\dfrac{3 \pm 2\sqrt{2}}{2}\right\}$ **b.** Expression; $4x^2 - 12x + 1$

4. a. Equation; $\{3\}$ **b.** Expression; $15y - 38$

5. a. Equation; $\{7, 4\}$ **b.** Equation; $\left\{\dfrac{11 \pm \sqrt{233}}{2}\right\}$

6. a. Equation; $\left\{\dfrac{2}{3}, -10\right\}$ **b.** Equation; $\left\{-\dfrac{7}{4}\right\}$

7. a. Equation; $\{-7, -5\}$ **b.** Expression; $\dfrac{35 + 12x + x^2}{x}$

8. a. Equation; $\{\ \}$; The value 2 does not check.

  b. Expression; $\dfrac{5}{3}$; for $x \neq 2$

Section 1.5 Practice Exercises, pp. 129–132

R.1. 260 ft^2 **R.2.** 56 ft^3 **R.3.** $\sqrt{55}$ m **R.4.** 121.5 km^2

1. $A = \dfrac{1}{2}bh$ **3.** $V = lwh$

5. a. $x(2x + 3) = 629$ **b.** The width is 17 yd and the length is 37 yd.

7. a. $\dfrac{1}{2}x(x - 2) = 40$ **b.** The base is 10 ft and the height is 8 ft.

9. a. $\dfrac{8}{5}x^2 = 640$ **b.** The length is 20 in., the width is 8 in., and the

height is 4 in. **11. a.** $x^2 + (x + 2)^2 = (2x - 2)^2$ **b.** The legs

are 6 ft and 8 ft, and the hypotenuse is 10 ft. **13.** The width is

approximately 5.4 yd and the length is approximately 7.4 yd.

15. The base is approximately 17.7 ft and the height is approximately

9.7 ft. **17. a.** $x(x + 2) = 120$ **b.** The integers are 10 and 12 or

-10 and -12. **19. a.** $x^2 + (x + 1)^2 = 113$ **b.** The integers are

7 and 8 or -7 and -8. **21.** The dimensions of the cargo space are

6 ft by 7 ft by 12 ft. **23.** The radius is approximately 25 yd.

25. The base is 9 ft and the height is 12 ft. **27.** The distance is $90\sqrt{2}$ ft

or approximately 127.3 ft. **29. a.** The lengths of the sides of the

lower triangle are 6 ft, 8 ft, and 10 ft. **b.** The total area is 44 ft².

31. a. The length is approximately 2.91 in. and the width is

approximately 1.94 in. **b.** Using the rounded values from part (a),

the screen is approximately 949 pixels by 632 pixels.

33. There were 8 players. **35.** There were 600,000 organisms

approximately 9 hr and 39 hr after the culture was started.

37. a. 235 ft **b.** 62 mph **39. a.** $s = -16t^2 + 16t$ **b.** It would

take Michael Jordan 0.5 sec to reach his maximum height of 4 ft.

41. a. $s = -16t^2 + 75t + 4$ **b.** The ball will be at an 80-ft height 1.5 sec

and 3.2 sec after being kicked.

43. a. $L = \dfrac{1 + \sqrt{5}}{2} \approx 1.62$ **b.** 14.6 ft

45. a. $y = \dfrac{160 - 4x}{6}$ or $y = \dfrac{80 - 2x}{3}$ **b.** $A = x\left(\dfrac{80 - 2x}{3}\right)$

c. Each pen can be 25 yd by 10 yd, or it can be 15 yd by $\dfrac{50}{3}$ yd.

Section 1.6 Practice Exercises, pp. 140–144

R.1. $(x - 3)(3x + 4)(3x - 4)$ **R.2.** $a \neq -3$

R.3. There are no restricted values on the variable. **R.4.** $5y(2y + 3)^2$

R.5. $\dfrac{1}{343}$ **R.6.** $\sqrt[4]{c^2}$ or $(\sqrt[4]{c})^2$

1. absolute; $\{k, -k\}$ **3.** quadratic; $m^{1/3}$ **5.** $\left\{0, \dfrac{1}{2}, -6\right\}$

7. $\{\pm\sqrt{7}, \pm 2i\}$ **9.** $\left\{\pm\dfrac{1}{5}, -\dfrac{4}{3}\right\}$ **11.** $\{-7, -2, 2\}$

13. $\{\pm 2i, \pm 2\}$ **15.** $\{0, -4, 2 \pm 2i\sqrt{3}\}$

17. $\left\{\pm\dfrac{2\sqrt{3}}{3}, \pm i\sqrt{5}\right\}$ **19.** $\{2, -1 \pm i\sqrt{2}\}$

21. $\{5\}$; The value -2 does not check.   **23.** $\left\{\dfrac{6 \pm \sqrt{51}}{3}\right\}$

25. $\left\{\dfrac{5}{2}, -1\right\}$ **27.** $\{\ \}$; The value 3 does not check.

29. Jesse travels 6 km/hr in still water.   **31.** Jean runs 8 mph and rides 16 mph.   **33. a.** $\{6, -6\}$ **b.** $\{0\}$ **c.** $\{\ \}$

35. a. $\{7, -1\}$ **b.** $\{3\}$ **c.** $\{\ \}$ **37.** $\left\{\dfrac{5}{3}, 1\right\}$ **39.** $\{11, 3\}$

41. $\{\ \}$ **43.** $\left\{\dfrac{19}{3}, \dfrac{29}{3}\right\}$ **45.** $\left\{-2, -\dfrac{3}{2}\right\}$ **47.** $\{-5, 1\}$

49. $\{0\}$ **51.** $\left\{\dfrac{3}{2}\right\}$ **53.** $\mathbb{R}$ **55.** $\{20\}$ **57.** $\{\ \}$

59. $\{-1, 4\}$ **61.** $\{7\}$; The value -2 does not check.   **63.** $\{2\}$

65. $\left\{\dfrac{4}{3}\right\}$ **67.** $\{-1\}$; The value 4 does not check.   **69.** $\{-2, 1\}$

71. $\{-32, 22\}$ **73.** $\{5\}$ **75.** $\{5\sqrt[3]{5}\}$ or $\{5^{4/3}\}$ **77.** $\left\{\pm\dfrac{1}{32}\right\}$

79. $\{-10\}$ **81.** $\left\{-4, \dfrac{5}{2}\right\}$ **83.** $\{-5, -3, 1, 3\}$ **85.** $\{\pm 3i, \pm 2\}$

87. $\{-2 \pm \sqrt{6}\}$ **89.** $\left\{-3, -\dfrac{8}{5}\right\}$ **91.** $\{-5, -1, 2, 10\}$

93. $\left\{\dfrac{3}{2}, -\dfrac{3}{5}\right\}$ **95.** $\left\{\dfrac{1}{3125}, 32\right\}$ **97.** $\{81\}$ **99.** $\{-3, -1, 1, 3\}$

101. $\left\{0, \dfrac{625}{16}\right\}$ **103.** $\left\{\pm\sqrt{\dfrac{-5 \pm \sqrt{53}}{2}}\right\}$ **105.** $\left\{\dfrac{3}{2}, 10\right\}$

107. $p = \dfrac{fq}{q - f}$ **109.** $T = \sqrt[4]{\dfrac{E}{k}}$ **111.** $m = \dfrac{kF}{a}$

113. $x = \pm\sqrt{(z - 16)^2 + y^2}$ **115.** $T_1 = \dfrac{P_1 V_1 T_2}{P_2 V_2}$ **117.** $g = \dfrac{4\pi^2 L}{T^2}$

119. $\{9\}$ **121.** $\left\{\dfrac{19 - 3\sqrt{5}}{2}\right\}$; The value $\dfrac{19 + 3\sqrt{5}}{2}$ does not check.

123. 288π in.^{3} **125. a.** 55%   **b.** 9.3 hr   **127. a.** 14 m/sec   **b.** 36.6 m   **129. a.** $|x - 4| = 6$ or equivalently $|4 - x| = 6$ **b.** $\{-2, 10\}$ **131.** An equation is in quadratic form if, after a suitable substitution, the equation can be written in the form $au^2 + bu + c = 0$, where u is a variable expression.

133. It would take Joan approximately 5.5 hr working alone, and it would take Henry approximately 6.5 hr.

135. Pam can row to a point $166\dfrac{2}{3}$ ft down the beach or to a point 300 ft down the beach to be home in 5 min.

Section 1.7 Practice Exercises, pp. 152–156

R.1. $\{x \mid x > -1\}$; $(-1, \infty)$ **R.2.** $\{x \mid x \le 5\}$; $(-\infty, 5]$
R.3. $\{x \mid -3 < x \le 4\}$; $(-3, 4]$ **R.4.** $\{12\}$

R.5. $\left\{-\dfrac{2}{13}\right\}$ **R.6.** $\{-7\}$

1. inequality   **3.** $a < x < b$ **5.** $-k; k$ **7.** $\mathbb{R}$
9. $\{x \mid x < -11\}$; $(-\infty, -11)$;

11. $\{w \mid w \le 3\}$; $(-\infty, 3]$;

13. $\{a \mid a \le 8.5\}$; $(-\infty, 8.5]$;

15. $\{c \mid c < 2\}$; $(-\infty, 2)$;

17. $\{x \mid x < -\tfrac{13}{2}\}$; $(-\infty, -\tfrac{13}{2})$;

19. $\{x \mid x \le \tfrac{17}{6}\}$; $(-\infty, \tfrac{17}{6}]$;   **21.** $\{\ \}$

23. $\{x \mid x \ge -\tfrac{5}{6}\}$; $[-\tfrac{5}{6}, \infty)$;

25. $\mathbb{R}$; $(-\infty, \infty)$;
27. a. $[-2, 4)$
   b. $(-\infty, \infty)$
29. a. $(-\infty, 5]$

   b. $(-\infty, -6)$

31. a. $(-\infty, 3.2]$

   b. $(-\infty, 18)$

33. a. $(-\infty, -2] \cup \left(-\dfrac{1}{3}, \infty\right)$ **b.** $\{\ \}$

35. $-2.8 < y$ and $y \le 15$
37. $[-4, 2)$

39. $\left[\dfrac{6}{5}, 2\right)$

41. $\left[-\dfrac{5}{2}, \dfrac{13}{2}\right]$

43. a. $\{7, -7\}$ **b.** $(-7, 7)$ **c.** $(-\infty, -7) \cup (7, \infty)$
45. a. $\{-13, -5\}$ **b.** $[-13, -5]$ **c.** $(-\infty, -13] \cup [-5, \infty)$
47. $(-2, 10)$ **49.** $(-\infty, -8] \cup [2, \infty)$ **51.** $\{\ \}$
53. $\mathbb{R}$; $(-\infty, \infty)$ **55.** $[-10, 6]$ **57.** $\left(-\infty, -\dfrac{12}{5}\right) \cup \left(\dfrac{4}{5}, \infty\right)$
59. $(-15, 9)$ **61. a.** $\{\ \}$ **b.** $\{\ \}$ **c.** $\mathbb{R}$; $(-\infty, \infty)$
63. a. $\{\ \}$ **b.** $\{\ \}$ **c.** $\mathbb{R}$; $(-\infty, \infty)$ **65. a.** $\{0\}$ **b.** $\{\ \}$
   c. $\{0\}$ **d.** $(-\infty, 0) \cup (0, \infty)$ **e.** $\mathbb{R}$; $(-\infty, \infty)$
67. a. $\{-4\}$ **b.** $\{\ \}$ **c.** $\{-4\}$ **d.** $(-\infty, -4) \cup (-4, \infty)$
   e. $\mathbb{R}$; $(-\infty, \infty)$ **69.** $12.0 \le x \le 15.2$ g/dL   **71.** $90 \le d \le 110$ yd
73. Marilee needs to score at least 96 on the final exam.
75. Rita needs to score between 77 and 100, inclusive.
77. It will take more than 1.6 hr or 1 hr 36 min.   **79.** The length must be 300 ft or less.   **81.** An average score in league play between 140 and 220, inclusive, would produce a handicap of 72 or less.
83. a. Donovan would need to sell more than \$250,000 in merchandise.
   b. Job A   **85. a.** $|v - 16| < 0.01$ or equivalently $|16 - v| < 0.01$
   b. $(15.99, 16.01)$ **87. a.** $|x - 4| > 1$ or equivalently $|4 - x| > 1$
   b. $(-\infty, 3) \cup (5, \infty)$ **89. a.** $|t - 36.5| \le 1.5$ or equivalently $|36.5 - t| \le 1.5$ **b.** $[35, 38]$; If the refrigerator is set to 36.5°F, the actual temperature would be between 35°F and 38°F, inclusive.
91. a. $|x - 0.51| \le 0.03$ or equivalently $|0.51 - x| \le 0.03$
   b. $[0.48, 0.54]$; The candidate is expected to receive between 48% of the vote and 54% of the vote, inclusive.
93. a. $\{x \mid x \ge 2\}$ **b.** $\{x \mid x \le 2\}$ **95. a.** $\{x \mid x \ge -4\}$ **b.** $\mathbb{R}$
97. a. $\{x \mid x \ge \tfrac{9}{2}\}$ **b.** $\{x \mid x \ge \tfrac{9}{2}\}$ **99.** False   **101.** True
103. $|x - 2| \le 5$ **105.** $|x - 7| > 3$
107. The steps are the same with the following exception. If both sides of an inequality are multiplied or divided by a negative real number, then the direction of the inequality sign must be reversed.
109. The inequality $|x - 3| \le 0$ will be true only for values of x for which $x - 3 = 0$ (the absolute value will never be less than 0). The solution set is $\{3\}$. The inequality $|x - 3| > 0$ is true for all values of x excluding 3. The solution set is $\{x \mid x < 3 \text{ or } x > 3\}$.

111. $\left(-\infty, \dfrac{11}{2}\right)$ **113.** $(-9, -1) \cup (1, 9)$ **115.** $[-4, -3] \cup [2, 3]$

117. $\hat{p} - z\sqrt{\dfrac{\hat{p}\hat{q}}{n}} < p < \hat{p} + z\sqrt{\dfrac{\hat{p}\hat{q}}{n}}$

Problem Recognition Exercises, p. 156

1. a. Equation in quadratic form and a polynomial equation

b. $\left\{\pm 3, \pm\sqrt{6}\right\}$

2. a. Absolute value inequality   **b.** $\left(-\infty, -\frac{7}{3}\right) \cup [3, \infty)$

3. a. Radical equation   **b.** $\{16\}$

4. a. Absolute value equation   **b.** $\{\ \}$

5. a. Rational equation   **b.** $\left\{\dfrac{9 \pm \sqrt{41}}{2}\right\}$

6. a. Polynomial equation   **b.** $\left\{-\dfrac{5}{3}, \pm\dfrac{1}{4}\right\}$

7. a. Compound inequality   **b.** $(-9, 3]$

8. a. Compound inequality   **b.** $(-\infty, 12)$

9. a. Quadratic equation   **b.** $\left\{\dfrac{5}{2}, -7\right\}$

10. a. Linear equation   **b.** $\{2\}$

11. a. Linear inequality   **b.** $[-18, \infty)$

12. a. Quadratic equation   **b.** $\left\{\pm\dfrac{\sqrt{21}}{3}i\right\}$

13. a. Compound inequality   **b.** $[1, 41]$

14. a. Radical equation   **b.** $\{2\}$; The value 142 does not check.

15. a. Absolute value equation   **b.** $\{1, 3\}$

16. a. Rational equation   **b.** $\{-1\}$; The value $\frac{1}{2}$ does not check.

17. a. Absolute value inequality   **b.** $(-9, 1)$

18. a. Radical equation and an equation in quadratic form   **b.** $\{36\}$

19. a. Radical equation   **b.** $\{\pm 64\}$

20. a. Absolute value inequality   **b.** $(-\infty, \infty)$

Equations and Inequalities for Calculus, p. 157

1. $y' = -\dfrac{9x}{25y}$ **2.** $y' = \dfrac{1 - 2xy^3}{3x^2y^2 - 1}$ **3.** $y' = \dfrac{2y(y - 3x)}{3x^2 - 4xy + 3y^2}$

4. $y' = -\dfrac{y(2x + y)}{x(2y + x)}$ **5.** $\dfrac{x(5x - 6)}{\sqrt{2x - 3}}$ **6.** $\dfrac{x(3x - 14)}{(2x - 7)^{3/2}}$

7. $-\dfrac{9}{(x^2 - 9)^{3/2}}$ **8. a.** $\dfrac{4x(2x - 5)}{(4x - 5)^2}$ **b.** $x = 0, x = \dfrac{5}{2}$

c. $x = \dfrac{5}{4}$ **9. a.** $-\dfrac{6x(3x + 1)}{(6x + 1)^2}$ **b.** $x = 0, x = -\dfrac{1}{3}$

c. $x = -\dfrac{1}{6}$ **10. a.** $\dfrac{2(2 - x^2)}{\sqrt{4 - x^2}}$ **b.** $x = \pm\sqrt{2}$

c. $x = 2$ and $x = -2$ **11.** $(-3, 1)$ **12.** $(-2, 2)$

13. 14.4 ft   **14.** $\dfrac{\sqrt{3}}{4}$ m^3 **15.** 1600 ft^3

Chapter 1 Review Exercises, pp. 160–162

1. $x \neq 2, x \neq -2, x \neq \dfrac{7}{2}$ **3.** $\left\{-\dfrac{40}{3}\right\}$ **5.** $\{\ \}$ **7.** $\left\{\dfrac{1}{2}\right\}$

9. $\{\ \}$; The value 1 does not check.   **11.** $t_2 = 2t_a - t_1$ **13.** 123 mi

15. Cassandra invested $8000 in the Treasury note and $12,000 in the bond.

17. $166\frac{2}{3}$ ft^3 **19.** The northbound boat travels 8 mph and the southbound boat travels 14 mph.   **21. a.** $C = 5x$

b. The dancer will save money on the 17th dance during a 3-month period.   **23.** $\dfrac{55}{3}$ hr $= 18.\overline{3}$ hr   **25.** There are approximately 144 turtles in the pond.   **27.** $2i\sqrt{3}$

29. a. Real part: 3; Imaginary part: -7 **b.** Real part: 0; Imaginary part: 2

31. $\dfrac{1}{2} + \dfrac{1}{5}i$ **33.** $-\sqrt{15} + i\sqrt{55}$ **35.** $-20 - 48i$ **37.** 73

39. $\dfrac{6}{41} + \dfrac{\sqrt{5}}{41}i$ **41.** $\left\{\dfrac{4}{3}, -2\right\}$ **43.** $\{\pm 11i\}$ **45.** $\left\{-\dfrac{3}{2}\right\}$

47. 81; $(x + 9)^2$ **49.** $\{1, 9\}$ **51.** False   **53. a.** 0

b. 1 real solution   **55. a.** -219 **b.** 2 nonreal solutions

57. $y = k \pm \sqrt{r^2 - (x - h)^2}$ **59.** The base is 13 yd and the height is 8 yd.   **61.** The width is 26.5 in. and the length is 42.4 in.

63. a. 230 ft   **b.** 70 mph   **65.** $\left\{\pm\sqrt{5}, \dfrac{3}{2}\right\}$

67. $\{2\}$; The value -6 does not check.   **69.** $\left\{-\dfrac{1}{2}, -11\right\}$

71. $\{\ \}$ **73.** $\{-16, 38\}$ **75.** $\left\{\dfrac{5}{3}, 5\right\}$

77. $\left\{-1, \dfrac{7}{3}\right\}$ **79.** $\left\{-\dfrac{1}{27}, \dfrac{27}{8}\right\}$ **81.** $\{1, -4\}$

83. $a = \sqrt{2m^2 + c^2 - b^2}$ **85.** $v_2 = \dfrac{a_2 t_2 v_1}{a_1 t_1}$

87. $\{x\,|\,x > 2\}$; $(2, \infty)$;

89. $\{t\,|\,t \geq -3\}$; $[-3, \infty)$;

91. a. $\left\{x\,|\,x \geq -\frac{7}{5}\right\}$ **b.** $\mathbb{R}$

93. a. $\{x\,|\,x \leq 6 \text{ or } x > 1\}$; $(-\infty, -6] \cup (1, \infty)$ **b.** $\{\ \}$

95. $\{x\,|\,3 < x < 11\}$; $(3, 11)$

97. More than 8.36 in. is needed.

99. a. $\{-7, 3\}$ **b.** $(-7, 3)$ **c.** $(-\infty, -7] \cup [3, \infty)$

101. a. $\{-5\}$ **b.** $\{\ \}$ **c.** $\{-5\}$ **d.** $(-\infty, -5) \cup (-5, \infty)$

e. $(-\infty, \infty)$ **103.** $(-\infty, -3] \cup [-1, \infty)$

105. $[-6, 7]$ **107. a.** $|t + 2| > 0.01$ or $|-2 - t| > 0.01$

b. $(-\infty, -2.01) \cup (-1.99, \infty)$

Chapter 1 Test, pp. 163–164

1. -10 **2. a.** i **b.** -1 **c.** $-i$ **d.** 1   **e.** i

3. $38 - 34i$ **4.** $-16 - 30i$ **5.** $-\dfrac{7}{29} + \dfrac{26}{29}i$

6. a. -40 **b.** 2 nonreal solutions   **7. a.** 0   **b.** 1 real solution

8. a. 76   **b.** 2 real solutions   **9.** $\left\{\dfrac{83}{2}\right\}$ **10.** $\{25\}$ **11.** $\mathbb{R}$

12. $\{\ \}$; The value -3 does not check.   **13.** $\left\{\dfrac{4 \pm \sqrt{13}}{3}\right\}$

14. $\{-5 \pm \sqrt{29}\}$ **15.** $\left\{\dfrac{1}{3}, -\dfrac{5}{4}\right\}$ **16.** $\left\{\dfrac{2 \pm i\sqrt{2}}{3}\right\}$

17. $\left\{\pm\dfrac{1}{2}, -2\right\}$ **18.** $\{-4\}$ **19.** $\{\ \}$; The values 2 and 18 do not check.   **20.** $\{4\}$; The value -6 does not check.   **21.** $\{\pm 11^{5/4}\}$

22. $\left\{\dfrac{1}{4}, -\dfrac{1}{2}\right\}$ **23.** $\{-1, 7\}$ **24.** $\{-1\}$

25. $P = \dfrac{6}{a - t}$ or $P = -\dfrac{6}{t - a}$ **26.** $b = \sqrt{a^2 - c^2}$

27. $t = \dfrac{-v_0 \pm \sqrt{v_0^2 + 128}}{-32}$ or $t = \dfrac{v_0 \pm \sqrt{v_0^2 + 128}}{32}$

28. $c = \sqrt{49 - a^2 - b^2}$ **29.** $[-14, 10]$ **30.** $[10, \infty)$ **31.** $(6, \infty)$

32. $(-\infty, 0) \cup (\frac{3}{2}, \infty)$ **33.** $[2, 14]$ **34. a.** $\{\ \}$ **b.** $\{\ \}$ **c.** $\mathbb{R}$

35. a. $\{13\}$ **b.** $\{\ \}$ **c.** $\{13\}$ **d.** $(-\infty, 13) \cup (13, \infty)$

e. $(-\infty, \infty)$ **36.** 1 gal of 80% antifreeze should be used.

37. The plane flying to Seattle flies 440 mph, and the plane flying to New York flies 500 mph.   **38.** The second hose can fill the pool in 2 hr.   **39.** The LDL level is 196 mg/dL and the total cholesterol is 266 mg/dL.   **40.** The base of the triangular portions is 5 ft and the height is 12 ft.   **41. a.** $s = -16t^2 + 60t + 2$ **b.** The ball will be at a height of 52 ft at times 1.25 sec and 2.5 sec after being kicked.

42. The golfer would need to score less than 84.   **43.** 2 in.

Chapter 1 Cumulative Review Exercises, p. 164

1. $3600x^2$ **2.** 40   **3.** $\dfrac{3(x+3)}{2}$ **4.** $\dfrac{2x-22}{(x+2)(x-2)}$

5. $\dfrac{1-3x}{10+x}$ **6.** $\dfrac{\sqrt{7}-\sqrt{3}}{2}$ **7.** $3yw^4\sqrt[3]{3y^2z^2}$

8. a. $|4\pi-11|$ or $|11-4\pi|$ **b.** $4\pi-11$

9. $4(x-2y^2)(x^2+2xy^2+4y^4)$ **10.** $-1-i$

11. Stephan borrowed \$6000 at 5% and \$2000 at 4%.

12. $\left\{\dfrac{1\pm\sqrt{3}}{4}\right\}$ **13.** $\left\{\dfrac{5\pm\sqrt{35}}{2}\right\}$ **14.** $\left\{\dfrac{3}{2},-15\right\}$

15. $\{0\}$; The value -3 does not check.   **16.** $\{3,7\}$ **17.** $\{0,17\}$

18. a. $A\cup B=\mathbb{R}$ **b.** $A\cap B=\{x\mid 4\le x<11\}$

   c. $A\cup C=\{x\mid x<11\}$ **d.** $A\cap C=\{x\mid x<2\}$

   e. $B\cup C=\{x\mid x<2 \text{ or } x\ge 4\}$ **f.** $B\cap C=\{\ \}$

19. $[0,11]$ **20.** $(-25,\infty)$

CHAPTER 2

Section 2.1 Practice Exercises, pp. 173–177

R.1. $4\sqrt{3}$ **R.2.** 24 km   **R.3.** $y=\dfrac{c-ax}{b}$ or $y=\dfrac{c}{b}-\dfrac{ax}{b}$ **R.4.** 10

1. origin   **3.** $d=\sqrt{(x_2-x_1)^2+(y_2-y_1)^2}$ **5.** solution   **7.** 0

9.

11. a. $2\sqrt{5}$ **b.** $(-3,9)$ **13. a.** $9\sqrt{2}$ **b.** $\left(-\dfrac{5}{2},\dfrac{1}{2}\right)$

15. a. 5   **b.** $(3.7,-4.4)$ **17. a.** $\sqrt{117}$ **b.** $\left(\dfrac{5\sqrt{5}}{2},-4\sqrt{2}\right)$

19. Yes   **21.** No   **23. a.** Yes   **b.** No   **c.** Yes

25. $\{x\mid x\neq 3\}$ **27.** $\{x\mid x\ge 10\}$ **29.** $\{x\mid x\le 1.5\}$

31.

33.

35.

37.

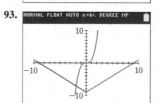

39.

41.

43.

45. x-intercepts: $(-1,0),(9,0)$; y-intercepts: $(0,-3),(0,3)$

47. x-intercept: $(-2,0)$; y-intercept: None

49. x-intercept: $(0,0)$; y-intercept $(0,0)$

51. x-intercept: $(-6,0)$; y-intercept: $(0,3)$

53. x-intercepts: $(-3,0),(3,0)$; y-intercept: $(0,9)$

55. x-intercepts: $(3,0),(7,0)$; y-intercept: $(0,3)$

57. x-intercept: $(-1,0)$; y-intercepts: $(0,-1),(0,1)$

59. x-intercept: $(0,0)$; y-intercept: $(0,0)$

61. x-intercept: None; y-intercept: None

63. $d(A,C)=2\sqrt{26}$ mi and $d(B,C)=2\sqrt{17}$ mi

65. a. 457 pixels   **b.** $(223,184)$ **c.** $(317,119)$

67. From the distance formula, $d(A,B)=|x|$, $d(A,C)=|x|$, and $d(B,C)=|x|$.

69. a. Length: $3\sqrt{2}$ ft; Width: $2\sqrt{2}$ ft

   b. Perimeter: $10\sqrt{2}$ ft; Area: 12 ft^2

71. Center: $(1,2)$; Radius: $\sqrt{10}$ **73.** Area: 25 m^2

75. Collinear   **77.** Not collinear

79. The points (x_1,y_1) and (x_2,y_2) define the endpoints of the hypotenuse d of a right triangle. The lengths of the legs of the triangle are $|x_2-x_1|$ and $|y_2-y_1|$. Applying the Pythagorean theorem produces $d^2=|x_2-x_1|^2+|y_2-y_1|^2$, or equivalently $d=\sqrt{(x_2-x_1)^2+(y_2-y_1)^2}$ for $d\ge 0$.

81. To find the x-intercept(s), substitute 0 for y and solve for x. To find the y-intercept(s), substitute 0 for x and solve for y.

83. $\sqrt{91}$ **85.** $9\sqrt{2}$

87. The viewing window is part of the Cartesian plane shown in the display screen of a calculator. The boundaries of the window are often denoted by [Xmin, Xmax, Xscl] by [Ymin, Ymax, Yscl].

89.

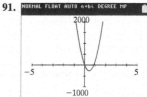

91.

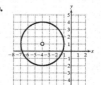

93.

Section 2.2 Practice Exercises, pp. 181–183

R.1. $n=16$; $(c-4)^2$ **R.2.** $n=\dfrac{1}{49}$; $\left(x+\dfrac{1}{7}\right)^2$

R.3. $5\sqrt{2}$ **R.4.** x^2-4x+4

1. circle; center   **3.** $(x-h)^2+(y-k)^2=r^2$ **5.** No   **7.** Yes

9. Center: $(4,-2)$; Radius: 9   **11.** Center: $(0,2.5)$; Radius: 2.5

13. Center: $(0,0)$; Radius: $2\sqrt{5}$ **15.** Center: $\left(\dfrac{3}{2},-\dfrac{3}{4}\right)$; Radius: $\dfrac{9}{7}$

17. a. $(x+2)^2+(y-5)^2=1$ **19. a.** $(x+4)^2+(y-1)^2=9$

   b.

   b.

21. a. $(x+4)^2+(y+3)^2=11$ **23. a.** $x^2+y^2=6.76$

   b.

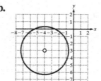

   b.

25. a. $(x - 2)^2 + (y - 1)^2 = 25$ **27. a.** $(x + 2)^2 + (y + 1)^2 = 100$

b.

b.

29. a. $(x - 4)^2 + (y - 6)^2 = 16$ **31. a.** $(x - 5)^2 + (y + 5)^2 = 25$

b.

b.

33. $(x - 8)^2 + (y + 11)^2 = 25$ **35.** $\left(x - \sqrt{7}\right)^2 + \left(y - \sqrt{7}\right) = 7$

37. $\{(-1, 5)\}$ **39.** $\{ \ \}$

41. $(x + 3)^2 + (y - 1)^2 = 4$; Center: $(-3, 1)$; Radius: 2

43. $(x - 11)^2 + (y + 3)^2 = 1$; Center: $(11, -3)$; Radius: 1

45. $x^2 + (y - 10)^2 = 104$; Center: $(0, 10)$; Radius: $2\sqrt{26}$

47. $(x - 4)^2 + (y + 10)^2 = 24$; Center: $(4, -10)$; Radius: $2\sqrt{6}$

49. $(x - 2)^2 + (y - 9)^2 = -4$; Degenerate case: $\{ \ \}$

51. $x^2 + \left(y - \dfrac{5}{2}\right)^2 = 0$; Degenerate case (single point): $\left\{\left(0, \dfrac{5}{2}\right)\right\}$

53. $\left(x - \dfrac{1}{2}\right)^2 + \left(y - \dfrac{3}{4}\right)^2 = \dfrac{25}{16}$; Center: $\left(\dfrac{1}{2}, \dfrac{3}{4}\right)$; Radius: $\dfrac{5}{4}$

55. $(x - 4)^2 + (y - 6)^2 = 2.25$ **57.**

59. A circle is the set of all points in a plane that are equidistant from a fixed point called the center.   **61.** $y = -2$ and $y = 14$

63. $\left(3 + \sqrt{17}, 3 + \sqrt{17}\right)$ and $\left(3 - \sqrt{17}, 3 - \sqrt{17}\right)$

65. a.

b.

c.

d.

67. a.

b.

c.

d.

69. $\sqrt{49 - 12\sqrt{5}}$ or $3\sqrt{5} - 2$

71.
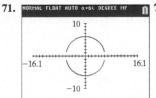

73.
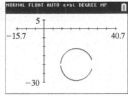

Section 2.3 Practice Exercises, pp. 191–197

R.1. $\left\{-\sqrt{5}, \sqrt{5}\right\}$ **R.2.** $\left\{-\dfrac{5}{4}, 3\right\}$ **R.3.** $[-8, \infty)$

R.4. $\left\{-\dfrac{7}{2}, -\dfrac{3}{2}\right\}$ **R.5. a.** $(10, 0)$ **b.** $(0, -4)$

1. relation; domain; y **3.** y **5.** -5

7. Yes. For a given time after the tree is planted, there cannot be two or more different heights. That is, the height is unique at any given time.

9. a. $\{(\text{Tom Hanks}, 5), (\text{Jack Nicholson}, 12), (\text{Sean Penn}, 5),$ $(\text{Dustin Hoffman}, 7)\}$ **b.** $\{\text{Tom Hanks, Jack Nicholson,}$ Sean Penn, Dustin Hoffman$\}$ **c.** $\{5, 12, 7\}$ **d.** Yes

11. a. $\{(-4, 3), (-2, -3), (1, 4), (3, -2), (3, 1)\}$ **b.** $\{-4, -2, 1, 3\}$

c. $\{3, -3, 4, -2, 1\}$ **d.** No

13. False   **15.** Yes   **17.** No   **19.** Yes   **21.** Yes

23. No   **25.** Yes   **27.** No   **29.** Yes   **31. a.** Yes   **b.** No

33. $(4, 1)$ **35. a.** -2 **b.** -2 **c.** 0   **d.** 4   **e.** 10

37. a. 5   **b.** 5   **c.** 5   **d.** 5   **e.** 5   **39.** $\dfrac{1}{3}$ **41.** 3

43. Undefined   **45.** 3   **47.** $\dfrac{1}{t}$ **49.** $\sqrt{x + h + 1}$

51. $a^2 + 11a + 28$ **53.** Undefined   **55.** $x^2 + 2xh + h^2 + 3x + 3h$

57. $-4x^2 - 8xh - 4h^2 - 5x - 5h + 2$ **59.** $-3x^2 - 6xh - 3h^2 + 7$

61. $x^3 + 3x^2h + 3xh^2 + h^3 + 2x + 2h - 5$ **63.** 7   **65.** 4

67. -1 **69.** 2   **71. a.** $d(2) = 36$; Joe rides 36 mi in 2 hr.

b. 12 mi   **73.** $C(225) = 279$; If the cost of the food is \$225, then the total bill including tax and tip is \$279.

75. x-intercept: $(2, 0)$; y-intercept: $(0, -4)$

77. x-intercepts: $(8, 0)$, $(-8, 0)$; y-intercept: $(0, -8)$

79. x-intercepts: $\left(2\sqrt{3}, 0\right)$, $\left(-2\sqrt{3}, 0\right)$; y-intercept: $(0, 12)$

81. x-intercept: $(8, 0)$; y-intercept: $(0, 8)$

83. x-intercept: $(4, 0)$; y-intercept: $(0, -2)$

85. The y-intercept is $(0, 14{,}820)$ and means that the amount owed after the initial down payment is \$14,820. The t-intercept is $(60, 0)$ and means that after 60 months, the amount owed is \$0.

87. Domain: $\{-3, -2, -1, 2, 3\}$; Range: $\{-4, -3, 3, 4, 5\}$

89. Domain: $(-3, \infty)$; Range: $[1, \infty)$

91. Domain: $(-\infty, \infty)$; Range: $(-\infty, \infty)$

93. Domain: $(-\infty, \infty)$; Range: $[-3, \infty)$

95. Domain: $(-5, 1]$; Range: $\{-1, 1, 3\}$

97. a. $(-\infty, 4) \cup (4, \infty)$ **b.** $(-\infty, -2) \cup (-2, 2) \cup (2, \infty)$ **c.** $(-\infty, \infty)$

99. a. $[-9, \infty)$ **b.** $(-\infty, 9]$ **c.** $(-9, \infty)$

101. a. $(-\infty, \infty)$ **b.** $(-\infty, \infty)$ **c.** $(-\infty, 5) \cup (5, \infty)$

103. a. $(-\infty, \infty)$ **b.** $(-\infty, -4) \cup (-4, 7) \cup (7, \infty)$

c. $(-\infty, -2) \cup (-2, \infty)$

105. a. $(-\infty, \infty)$ **b.** $(-\infty, \infty)$ **c.** $(-\infty, -5) \cup (-5, 3) \cup (3, \infty)$

107. a. $[-15, \infty)$ **b.** $[-15, \infty)$ **c.** $[-15, -11) \cup (-11, \infty)$

109. a. $(-\infty, \infty)$ **b.** $[0, \infty)$ **c.** $[0, 7)$

111. a. -4 **b.** 2   **c.** $x = -3, x = -1, x = 1$

d. $x = -2, x = 2$ **e.** $(0, 0)$ and $\left(-\dfrac{10}{3}, 0\right)$ **f.** $(0, 0)$

g. $(-\infty, \infty)$ **h.** $[-4, \infty)$

113. a. 0   **b.** 5   **c.** $x = -3, x = -1, x = 1$

d. $x = -4, x = 0$ **e.** $(-2, 0)$ and $\left(\dfrac{4}{3}, 0\right)$

f. $(0, -4)$ **g.** $[-4, \infty)$ **h.** $[-4, 5]$

115. $r(x) = 400 - x$ **117.** $P(x) = 3x$ **119.** $C(x) = 90 - x$

121. $f(x) = 3x^2 - 2$

123. If two points in a set of ordered pairs are aligned vertically in a graph, then they have the same x-coordinate but different y-coordinates. This contradicts the definition of a function. Therefore, the points do not define y as a function of x.

125. a. $P(s) = 4s$ **b.** $A(s) = s^2$

    c. $A(P) = \left(\dfrac{P}{4}\right)^2$ or $A(P) = \dfrac{P^2}{16}$ **d.** $P(A) = 4\sqrt{A}$

    e. $d(s) = \sqrt{2}s$ **f.** $s(d) = \dfrac{d}{\sqrt{2}}$ or $s(d) = \dfrac{d\sqrt{2}}{2}$

    g. $P(d) = 2\sqrt{2}d$ **h.** $A(d) = \dfrac{d^2}{2}$

Section 2.4 Practice Exercises, pp. 207–212

R.1. x-intercept $(7, 0)$; y-intercept $(0, -42)$ **R.2.** $y = -\dfrac{7}{8}x - \dfrac{1}{8}$

R.3. $(-2, \infty)$ **R.4.** $[10, \infty)$ **R.5.** -2

1. linear   **3.** horizontal   **5.** zero; undefined

7. $m = \dfrac{f(x_2) - f(x_1)}{x_2 - x_1}$

9. x-intercept: $(-4, 0)$; y-intercept: $(0, 3)$

11. x-intercept: $\left(\frac{2}{5}, 0\right)$; y-intercept: $(0, 1)$

13. x-intercept: $(-6, 0)$; y-intercept: None

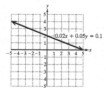

15. x-intercept: None; y-intercept: $(0, 2)$

17. x-intercept: $(5, 0)$; y-intercept: $(0, 2)$

19. x-intercept: $(0, 0)$; y-intercept: $(0, 0)$

21. $m = \dfrac{3}{10}$ **23.** $m = \dfrac{1}{40}$ **25.** $m = -3$ **27.** $m = -\dfrac{3}{5}$

29. $m = -\dfrac{1}{4}$ **31.** $m = -\dfrac{26}{23}$ **33.** $m = -\dfrac{20}{7}$ **35.** $m = \dfrac{\sqrt{30}}{12}$

37. $m = 3$ **39.** $m = -\dfrac{1}{3}$ **41.** $m = 0$ **43.** Undefined

45. 0   **47.** 41.6 ft   **49.** Change in population over change in time

51. a. $y = \frac{1}{2}x - 2$; $m = \frac{1}{2}$; y-intercept: $(0, -2)$ **b.**

53. a. $y = \frac{3}{2}x + 2$; $m = \frac{3}{2}$; y-intercept: $(0, 2)$ **b.**

55. a. $y = \frac{3}{4}x$; $m = \frac{3}{4}$; y-intercept: $(0, 0)$ **b.**

57. a. $y = 7$; $m = 0$; y-intercept: $(0, 7)$ **b.**

59. a. $y = -\frac{1}{3}x + 1$; $m = -\frac{1}{3}$; y-intercept: $(0, 1)$ **b.**

61. a. $y = -\frac{7}{4}x + 7$; $m = -\frac{7}{4}$; y-intercept: $(0, 7)$ **b.**

63. a. Linear   **b.** Linear   **c.** Neither   **d.** Constant

65. a. $y = \dfrac{1}{2}x + 9$ **b.** $f(x) = \dfrac{1}{2}x + 9$

67. a. $y = -3x - 3$ **b.** $f(x) = -3x - 3$

69. a. $y = \dfrac{2}{3}x + \dfrac{1}{3}$ **b.** $f(x) = \dfrac{2}{3}x + \dfrac{1}{3}$

71. a. $y = 5$ **b.** $f(x) = 5$

73. a. $y = 1.2x + 0.78$ **b.** $f(x) = 1.2x + 0.78$

75. a. $y = 2x - 6$ **b.** $f(x) = 2x - 6$

77. a. $y = -\dfrac{4}{3}x + \dfrac{19}{3}$ **b.** $f(x) = -\dfrac{4}{3}x + \dfrac{19}{3}$

79. $m = \dfrac{3}{2}$ **81. a.** \$364.80/yr   **b.** \$772.20/yr   **c.** Increasing

83. a. -262; The number of new flu cases dropped by 262 per month during this time interval.

    b. Between months 4 and 6: -683 cases/month; Between months 10 and 12: -110/month

    c.

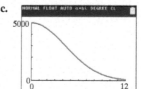

The number of new flu cases dropped slowly during the first two months. Then the rate of new cases dropped more rapidly between months 4 and 6 (perhaps as health department officials managed the outbreak). Finally, the rate of new cases dropped more slowly toward the end of the outbreak.

85. a. 1   **b.** 4   **c.** -2 **87. a.** 1   **b.** 1   **c.** 7

89. a. 1   **b.** $\dfrac{1}{3}$ **c.** $\dfrac{1}{5}$

91. a. $\{-1\}$ **b.** $(-\infty, -1)$ **c.** $[-1, \infty)$

93. a. $\{2\}$ **b.** $(-\infty, 2)$ **c.** $[2, \infty)$

95. a. $\{-5\}$ **b.** $[-5, \infty)$ **c.** $(-\infty, -5]$

97. a. $\{14\}$ **b.** $(14, \infty)$ **c.** $(-\infty, 14)$

99. The line will be slanted if both A and B are nonzero. If A is zero and B is not zero, then the equation can be written in the form $y = k$ and the graph is a horizontal line. If B is zero and A is not zero, then the equation can be written in the form $x = k$, and the graph is a vertical line.   **101.** The slope and y-intercept are easily determined by inspection of the equation.

103. 4 units^2 **105.** 10 units^2

107. a. $y = -\dfrac{A}{B}x + \dfrac{C}{B}$ **b.** $m = -\dfrac{A}{B}$ **c.** $\left(0, \dfrac{C}{B}\right)$

109. a. $\{-1.5\}$

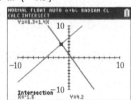

111. a. $\{-4, 7.8\}$

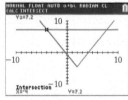   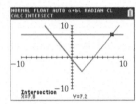

b. $(-\infty, -4] \cup [7.8, \infty)$ **c.** $[-4, 7.8]$

113. The lines are not exactly the same. The slopes are different.

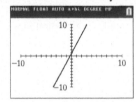

b. $(-\infty, -1.5)$
c. $(-1.5, \infty)$

Section 2.5 Practice Exercises, pp. 222–228

R.1. $y = -5x + 8$ **R.2.** $y = \dfrac{3}{5}x + 3$; Slope $= \dfrac{3}{5}$, y-intercept is $(0, 3)$

R.3. Undefined   **R.4.** 0   **R.5.** $C = 450 + 850m + 250n$

1. $y - y_1 = m(x - x_1)$ **3.** -1 **5.** $y = -2x - 1$

7. $y = \dfrac{2}{3}x + \dfrac{2}{3}$ **9.** $y = 1.2x - 1.48$ **11.** $y = \dfrac{1}{9}x + \dfrac{4}{3}$

13. $y = -\dfrac{8}{5}x + 8$ **15.** $y = 3.5x - 2.95$ **17.** $y = -4$

19. $x = \dfrac{2}{3}$ **21.** Undefined   **23. a.** $m = \dfrac{3}{11}$ **b.** $m = -\dfrac{11}{3}$

25. a. $m = -6$ **b.** $m = \dfrac{1}{6}$ **27. a.** $m = 1$ **b.** $m = -1$

29. Perpendicular   **31.** Parallel   **33.** Perpendicular   **35.** Neither

37. $y = -2x + 9$; $2x + y = 9$ **39.** $y = -5x + 26$; $5x + y = 26$

41. $y = \dfrac{3}{7}x + \dfrac{38}{7}$; $3x - 7y = -38$

43. $y = 0.5x + 5.3$; $5x - 10y = -53$ **45.** $y = 6$

47. $y = -\dfrac{3}{4}$ **49.** $x = -61.5$

51. a. $S(x) = 0.12x + 400$ for $x \geq 0$
b. $S(8000) = 1360$ means that the salesperson will make \$1360 if \$8000 in merchandise is sold for the week.

53. a. $T(x) = 0.019x + 172$ for $x > 0$
b. $T(80{,}000) = 1692$ means that the property tax is \$1692 for a home with a taxable value of \$80,000.

55. a. $C(x) = 34.5x + 2275$ **b.** $R(x) = 80x$
c. $P(x) = 45.5x - 2275$ **d.** 50 items

57. a. $\{730\}$ **b.** $[0, 730)$ **c.** $(730, \infty)$

59. a. $C(x) = 2.88x + 790$ **b.** $R(x) = 6x$
c. $P(x) = 3.12x - 790$
d. The business will make a profit if it produces and sells 254 dozen or more cookies.
e. The business will lose \$322.

61. a. $y = -1.2x + 1250$
b. $m = -1.2$ mph/mb means that for an increase of 1 mb in pressure, the wind speed decreases by 1.2 mph.   **c.** 170 mph
d. No. There is no guarantee that the linear trend continues outside the interval of the observed data points.

63. a. $y = 2.75x + 29.5$
b. $m = 2.75$ means that the average height of girls increased by 2.75 in. per year during this time period.
c. 59.75 in.   **d.** 66.4 in.

65. a.

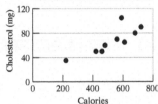

b. $c(x) = 0.125x$ **c.** $m = 0.125$ means that the amount of cholesterol increases at an average rate of 0.125 mg per calorie of hamburger.   **d.** 81.25 mg   **67.** Yes   **69.** No

71. a. $y = -1.22x + 1273$ **b.**
c. 175 mph   **d.** 5 mph

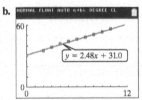

73. a. $y = 2.48x + 31.0$ **b.**
c. 58.28 in.
d. 64.8 in.   **e.** 1.6 in.

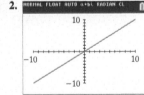

75. a. $y = 0.118x + 4.97$ **b.**
c. Approximately 82 mg

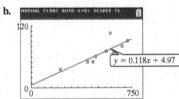

77. $\left(\dfrac{8}{5}, 0\right)$ **79.** $f(x) = \dfrac{7}{3}x + 4$ **81.** $h(x) = x + 5$

83. If the slopes of the two lines are the same and the y-intercepts are different, then the lines are parallel. If the slope of one line is the opposite of the reciprocal of the slope of the other line, then the lines are perpendicular.

85. Profit is equal to revenue minus cost.   **87.** $y = -x + 4$

89. $y = 12x + 17$ **91.** $y = -\dfrac{5}{2}x + \dfrac{21}{2}$ for $1 \leq x \leq 5$

Problem Recognition Exercises, p. 228

1.
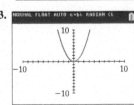

2.
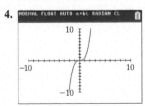

3.
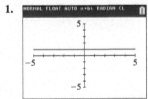

4.

5.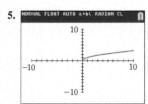

6.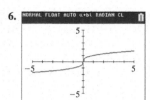

7.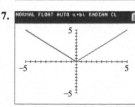

8.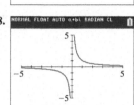

17. The graph of $g(x) = \sqrt{-x}$ has the shape of the graph of $y = \sqrt{x}$ but is reflected across the y-axis.

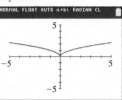

18. The graph of $g(x) = \sqrt[3]{-x}$ has the shape of the graph of $y = \sqrt[3]{x}$ but is reflected across the y-axis.

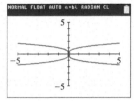

9. The graphs have the shape of $y = x^2$ with a vertical shift.

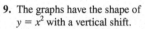

10. The graphs have the shape of $y = |x|$ with a vertical shift.

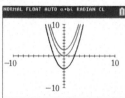

Section 2.6 Practice Exercises, pp. 239–243

R.1.

R.2.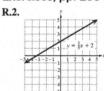

R.3.

1. left   **3.** down   **5.** horizontal shrink   **7.** vertical shrink

9. e   **11.** b   **13.** a

11. The graphs have the shape of $y = \sqrt{x}$ with a horizontal shift.

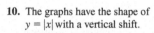

12. The graphs have the shape of $y = x^2$ with a horizontal shift.

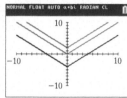

15.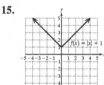

17.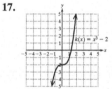

13. The graph of $g(x) = -|x|$ has the shape of the graph of $y = |x|$ but is reflected across the x-axis.

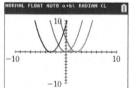

14. The graph of $g(x) = -\sqrt{x}$ has the shape of the graph of $y = \sqrt{x}$ but is reflected across the x-axis.

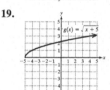

19.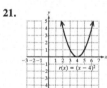

21.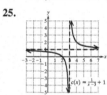

23.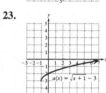

25.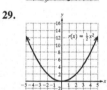

15. The graphs have the shape of $y = x^2$ but show a vertical shrink or stretch.

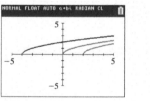

16. The graphs have the shape of $y = |x|$ but show a vertical shrink or stretch.

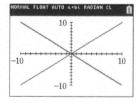

27.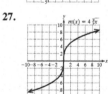

29.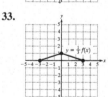

31.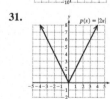

33.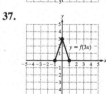

35.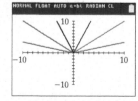

37.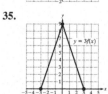

39.
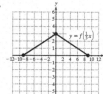

41.

75.

77.

43.

45.

79.

81.

47.

49.

83.

85.

87. $y = (-x + 4.5)^3 + 2.1$ **89.** $y = -\dfrac{2}{x} - 3$

51.

53.

91. As written, $g(x) = |2x|$ is in the form $g(x) = f(ax)$ with $a > 1$. This indicates a horizontal shrink. However, $g(x)$ can also be written as $g(x) = |2| \cdot |x| = 2|x|$. This is written in the form $g(x) = af(x)$ with $a > 1$. This represents a vertical stretch.

93. The graph of f is the same as the graph of $y = |x|$ with a horizontal shift to the right 2 units and a vertical shift downward 3 units. By contrast, the graph of g is the graph of $y = |x|$ with a horizontal shift to the right 3 units and a vertical shift downward 2 units.

55. Parent function: $f(x) = \dfrac{1}{x}$; Shift the graph of f to the left 1 unit, stretch the graph vertically by a factor of 3, and shift the graph downward by 2 units.

57. Parent function: $f(x) = x^2$. Shift the graph of f to the right 2.1 units, shrink the graph vertically by a factor of $\frac{1}{3}$, and shift the graph upward 7.9 units.

59. Parent function: $f(x) = \sqrt{x}$. Shift the graph of f to the left 5 units, shrink the graph horizontally by a factor of $\frac{1}{2}$, stretch the graph vertically by a factor of 2, and reflect the graph across the y-axis.

61. Parent function: $f(x) = \sqrt{x}$. Stretch the graph of f horizontally by a factor of 3, reflect across the x-axis, and shift the graph downward 6 units.

95. $f(x) = (x - 2)^2 - 3$ **97.** $f(x) = \dfrac{1}{x + 3}$

99. $f(x) = -x^3 + 1$ **101.** $y = 2t^2 + 1$

103. a.
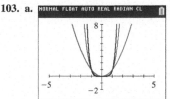
b.

c. The general shape of $y = x^n$ is similar to the graph of $y = x^2$ for even values of n greater than 1.

d. The general shape of $y = x^n$ is similar to the graph of $y = x^3$ for odd values of n greater than 1.

63.

65.

Section 2.7 Practice Exercises, pp. 255–261

R.1. $f(-a) = -7a - 2$

R.2. Interval notation: $(-\infty, 8)$

R.3. Interval notation: $[-2.4, 5.8)$ ——— -2.4   5.8

R.4. Interval notation: $\left[-\dfrac{9}{2}, \infty\right)$ ——— $-\dfrac{9}{2}$

67.

69.
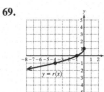

1. y **3.** origin   **5.** origin   **7.** y-axis   **9.** x-axis

11. x-axis, y-axis, and origin   **13.** None of these

15. x-axis, y-axis, and origin   **17.** None of these

19. y-axis symmetry   **21.** Odd   **23.** Even

25. Neither even nor odd   **27. a.** $f(-x) = 4x^2 - 3|x|$ **b.** Yes

c. Even   **29. a.** $h(-x) = -4x^3 + 2x$ **b.** $-h(x) = -4x^3 + 2x$

c. Yes   **d.** Odd   **31. a.** $m(-x) = 4x^2 - 2x - 3$

b. $-m(x) = -4x^2 - 2x + 3$ **c.** No   **d.** No   **e.** Neither

33. Even   **35.** Odd   **37.** Neither   **39.** Even   **41.** Odd

43. Neither   **45.** Odd   **47. a.** 12   **b.** 13   **c.** 4

d. 5   **e.** 5   **49. a.** 1   **b.** 2   **c.** -1 **d.** 1   **e.** -1

51. 15 ft/sec; 30 ft/sec; 2.7 ft/sec; and 1.4 ft/sec   **53.** c   **55.** d

71.

73.

57. a.   **b.** 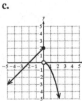   **c.**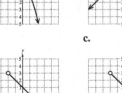

59. a.   **b.**   **c.**

61.   **63.**

65.   **67.**

69.   **71. a.**   **b.** $y = |x|$

73. -5 **75.** -1 **77.** 0 **79.** -9

81. 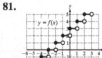   **83.**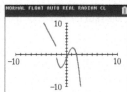

85. $C(x) = \begin{cases} 0.44 & \text{for } 0 < x \le 1 \\ 0.61 & \text{for } 1 < x \le 2 \\ 0.78 & \text{for } 2 < x \le 3 \\ 0.95 & \text{for } 3 < x \le 3.5 \end{cases}$

87. $S(x) = \begin{cases} 2000 & \text{for } 0 \le x < 40{,}000 \\ 2000 + 0.05(x - 40{,}000) & \text{for } x \ge 40{,}000 \end{cases}$

89. a. $(2, \infty)$ **b.** $(-3, -2)$ **c.** $(-2, 2)$

91. a. $(-\infty, \infty)$ **b.** Never decreasing   **c.** Never constant

93. a. $(1, \infty)$ **b.** $(-\infty, 1)$ **c.** Never constant

95. a. $(-\infty, -2) \cup (2, \infty)$ **b.** Never decreasing   **c.** $(-2, 2)$

97. At $x = 1$, the function has a relative minimum of -3.

99. At $x = -2$, the function has a relative minimum of 0. At $x = 0$, the function has a relative maximum of 2. At $x = 2$, the function has a relative minimum of 0.

101. At $x = -2$, the function has a relative minimum of -4. At $x = 0$, the function has a relative maximum of 0. At $x = 2$, the function has a relative minimum of -4.

103. a. $(8, 12)$ and $(18, 20)$ **b.** $(0, 8)$ and $(12, 18)$

c. The function has relative minima of 3 ft and 3.5 ft at approximately 8 days and 18 days after recording began. The function has a relative maximum of 4.5 ft at a time 12 days after recording began.   **d.** The weather was dry on the intervals of decreasing depth and water from the pond evaporated. The weather was rainy during intervals of increasing depth.

105. $f(x) = \begin{cases} -2 & \text{for } x < 1 \\ 3 & \text{for } x \ge 1 \end{cases}$ **107.** $f(x) = \begin{cases} -|x| & \text{for } x < 2 \\ -2 & \text{for } x \ge 2 \end{cases}$

109. $f(x) = \begin{cases} \dfrac{1}{x} & \text{for } x < 0 \\ x & \text{for } x > 0 \end{cases}$

111. a.   **b.** $(-\infty, \infty)$ **c.** $(-\infty, 1] \cup (2, \infty)$
d. $f(-1) = 0, f(1) = 0,$ and $f(2) = 4$
e. $x = 3$ **f.** $x = -2$
g. Increasing: $(-\infty, 0) \cup (1, \infty)$;
Decreasing: $(0, 1)$;
Never constant

113. a. 2 **b.** -4 **c.** 4 **d.** 3 **e.** -3 **f.** 4

115. If replacing y by $-y$ in the equation results in an equivalent equation, then the graph is symmetric to the x-axis. If replacing x by $-x$ in the equation results in an equivalent equation, then the graph is symmetric to the y-axis. If replacing both x by $-x$ and y by $-y$ results in an equivalent equation, then the graph is symmetric to the origin.

117. At $x = 1$, there are two different y values. The relation contains the ordered pairs $(1, 2)$ and $(1, 3)$.

119. A relative maximum of a function is the greatest function value relative to other points on the function nearby.

121. Relative minimum   **123. a.** Concave down   **b.** Decreasing

125. a. Concave up   **b.** Decreasing

127. $f(x) = \begin{cases} 0.1x & \text{if } 0 < x \le 8925 \\ 892.50 + 0.15(x - 8925) & \text{if } 8925 < x \le 36{,}250 \\ 4991.25 + 0.25(x - 36{,}250) & \text{if } 36{,}250 < x \le 87{,}850 \end{cases}$

or

$f(x) = \begin{cases} 0.1x & \text{if } 0 < x \le 8925 \\ 0.15x - 446.25 & \text{if } 8925 < x \le 36{,}250 \\ 0.25x - 4071.25 & \text{if } 36{,}250 < x \le 87{,}850 \end{cases}$

129.   **131.**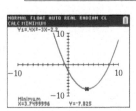

133. a. Relative minimum of -7.825 at $x = 3.750$
b. Increasing on $(3.750, \infty)$; Decreasing on $(-\infty, 3.750)$
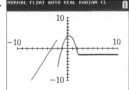

135. a. Relative maximum of 3.726 at $x = 0.667$;
Relative minimum of -2.625 at $x = -2.500$
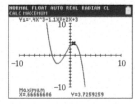
b. Increasing on $(-2.500, 0.667)$;
Decreasing on $(-\infty, -2.500) \cup (0.667, \infty)$

Section 2.8 Practice Exercises, pp. 270–275

R.1. $(-\infty, -2) \cup (-2, \infty)$ **R.2.** $[-3, \infty)$ **R.3.** $(-\infty, 3)$

R.4. $(-\infty, \infty)$ **R.5.** $k(x + 3) = x^2 + 4x + 6$

1. $f(x); g(x)$ **3.** $\dfrac{f(x + h) - f(x)}{h}$

5. $(f + g)(x) = |x| + 3$; Graph d

7. $(f + g)(x) = x^2 - 4$; Graph a

9. -13 **11.** 6   **13.** $\dfrac{11}{3}$ **15.** $-\dfrac{4}{3}$ **17.** Undefined

19. $(r - p)(x) = -x^2 - 6x; (-\infty, \infty)$

21. $(p \cdot q)(x) = (x^2 + 3x)\sqrt{1 - x}; (-\infty, 1]$

23. $\left(\dfrac{q}{p}\right)(x) = \dfrac{\sqrt{1 - x}}{x^2 + 3x}; (-\infty, -3) \cup (-3, 0) \cup (0, 1]$

25. $\left(\dfrac{p}{q}\right)(x) = \dfrac{x^2 + 3x}{\sqrt{1 - x}}; (-\infty, 1)$

27. $(s \cdot t)(x) = \dfrac{-1}{x + 3};$

Domain: $(-\infty, -3) \cup (-3, 2) \cup (2, 3) \cup (3, \infty)$

29. $(s + t)(x) = -\dfrac{x^3 - 4x^2 - 5x + 23}{(x + 3)(x - 3)(x - 2)}$

Domain: $(-\infty, -3) \cup (-3, 2) \cup (2, 3) \cup (3, \infty)$

31. $(s \cdot v)(x) = \dfrac{(x - 2)\sqrt{x + 3}}{(x - 3)(x + 3)};$ Domain: $(-3, 3) \cup (3, \infty)$

33. a. $5x + 5h + 9$ **b.** 5

35. a. $x^2 + 2xh + h^2 + 4x + 4h$ **b.** $2x + h + 4$ **37.** -2

39. $-10x - 5h - 4$ **41.** $3x^2 + 3xh + h^2$ **43.** $-\dfrac{1}{x(x + h)}$

45. a. $\dfrac{4\sqrt{x + h} - 4\sqrt{x}}{h}$ **b.** 1.6569; 1.9524; 1.9950; 1.9995   **c.** 2

47. 48   **49.** -3 **51.** 192   **53.** $\sqrt{210}$ **55.** -27

57. Undefined (not a real number)   **59.** 3315   **61.** 315

63. a. $(f \circ g)(x) = 2x^2 + 4$ **b.** $(g \circ f)(x) = 4x^2 + 16x + 16$

c. No   **65.** $(n \circ p)(x) = x^2 - 9x - 5; (-\infty, \infty)$

67. $(m \circ n)(x) = \sqrt{x + 3}; [-3, \infty)$

69. $(q \circ n)(x) = \dfrac{1}{x - 15}; (-\infty, 15) \cup (15, \infty)$

71. $(q \circ r)(x) = \dfrac{1}{|2x + 3| - 10}; \left(-\infty, -\dfrac{13}{2}\right) \cup \left(-\dfrac{13}{2}, \dfrac{7}{2}\right) \cup \left(\dfrac{7}{2}, \infty\right)$

73. $(n \circ r)(x) = |2x + 3| - 5; (-\infty, \infty)$

75. $(q \circ q)(x) = -\dfrac{x - 10}{10x - 101}; (-\infty, 10) \cup \left(10, \dfrac{101}{10}\right) \cup \left(\dfrac{101}{10}, \infty\right)$

77. $(f \circ g)(x) = -\dfrac{3}{x + 14}; (-\infty, -14) \cup (-14, 2]$

79. $(f \circ g)(x) = \dfrac{9}{25 - x^2}; (-\infty, -5) \cup (-5, -4) \cup (-4, 4) \cup (4, 5) \cup (5, \infty)$

81. $(f \circ f)(x) = \dfrac{x - 2}{-2x + 5}; (-\infty, 2) \cup \left(2, \dfrac{5}{2}\right) \cup \left(\dfrac{5}{2}, \infty\right)$

83. $(f \circ g \circ h)(x) = 2(\sqrt[3]{x})^2 + 1$ **85.** $(h \circ g \circ f)(x) = \sqrt[3]{(2x + 1)^2}$

87. a. $C(x) = 21.95x$ **b.** $T(a) = 1.06a + 10.99$

c. $(T \circ C)(x) = 23.267x + 10.99$ **d.** $(T \circ C)(4) = 104.058$; The total cost to purchase 4 boxes of stationery is $104.06.

89. a. $r(t) = 80t$ **b.** $d(r) = 7.2r$ **c.** $(d \circ r)(t) = 576t$ represents the distance traveled (in ft) in t minutes.   **d.** $(d \circ r)(30) = 17{,}280$ means that the bicycle will travel 17,280 ft (approximately 3.27 mi) in 30 min.   **91.** $f(x) = x^2$ and $g(x) = x + 7$ **93.** $f(x) = \sqrt[3]{x}$ and $g(x) = 2x + 1$ **95.** $f(x) = |x|$ and $g(x) = 2x^2 - 3$

97. $f(x) = \dfrac{5}{x}$ and $g(x) = x + 4$ **99. a.** 1   **b.** -1 **c.** -6

d. $-\dfrac{1}{2}$ **e.** 1   **f.** 1   **g.** -2 **101. a.** -1 **b.** 0

c. Undefined   **d.** -3 **e.** -3 **f.** 0   **g.** 2

103. 1   **105.** 3   **107.** 6   **109.** Undefined

111. a. $d(r) = 2r$ **b.** $r(d) = \dfrac{d}{2}$ **c.** $(V \circ r)(d) = \dfrac{1}{6}\pi d^3$ is the volume of the sphere as a function of its diameter.

113. $(A \circ A)(x) = (1.045)^2 x$ represents the amount of money in the account after 2 yr compounded annually.

115. $\left(\dfrac{H + L}{2}\right)(x)$ represents the average of the high and low temperatures for day x.

117. a. $S_1(x) = x^2 + 4x$ **b.** $S_2(x) = \dfrac{1}{8}\pi x^2$

c. $(S_1 - S_2)(x) = x^2 + 4x - \dfrac{1}{8}\pi x^2$ and represents the area of the region outside the semicircle, but inside the rectangle.

119. The domain of $(f \circ g)(x)$ is the set of real numbers x in the domain of g such that $g(x)$ is in the domain of f.

121. a. $\dfrac{\sqrt{x + h + 3} - \sqrt{x + 3}}{h}$ **b.** $\dfrac{1}{\sqrt{x + h + 3} + \sqrt{x + 3}}$

c. $\dfrac{1}{2\sqrt{x + 3}}$

123. a. $-9.68t - 4.84h + 88$ **b.** 78.32 ft/sec   **c.** 58.96 ft/sec
d. 39.6 ft/sec   **e.** 20.24 ft/sec

125. $c(b) = (b + 8)^2$ **127.** $z(y) = 2y - 4$

129. $m(x) = \sqrt[3]{x}, n(x) = x + 1, h(x) = 4x, k(x) = x^2$

Chapter 2 Review Exercises, pp. 277–281

1. a. $5\sqrt{5}$ **b.** $\left(\dfrac{3}{2}, 3\right)$ **3. a.** Yes   **b.** No

5. x-intercept: (4, 0); y-intercepts: (0, -4), (0, -10)

7.   **9.** Center: (4, -3); Radius: 2

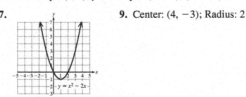

11. a. $(x + 3)^2 + (y - 1)^2 = 11$ **13. a.** $(x - 4)^2 + (y - 1)^2 = 25$
b.   **b.**

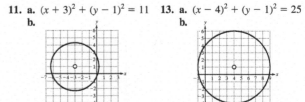

15. a. $(x + 5)^2 + (y - 1)^2 = 9$ **b.** Center: (-5, 1); Radius: 3

17. $\{(-3, 5)\}$ **19. a.** {(Dara Torres, 12), (Carl Lewis, 10), (Bonnie Blair, 6), (Michael Phelps, 16)}   **b.** {Dara Torres, Carl Lewis, Bonnie Blair, Michael Phelps}   **c.** {12, 10, 6, 16}   **d.** Yes

21. No   **23.** Yes   **25. a.** 5   **b.** 4   **c.** $x = 3$

27. x-intercepts: (4, 0), (2, 0); y-intercept: (0, 2)

29. Domain: $\{-4, -2, 0, 2, 3, 5\}$; Range: $\{-3, 0, 2, 1\}$

31. $(-\infty, 5) \cup (5, \infty)$ **33.** $(-\infty, \infty)$

35. a. -2 **b.** -1 **c.** $x = -1, x = 3$ **d.** $x = -4$
e. (0, 0), (2, 0)   **f.** (0, 0)   **g.** $(-\infty, \infty)$ **h.** $(-\infty, 1]$

37. x-intercept: (-4, 0);   **39.** x-intercept: None;
y-intercept: (0, 2) y-intercept: (0, 2)

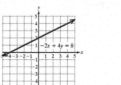

41. $m = \dfrac{1}{8}$ **43.** $m = \dfrac{f(b) - f(a)}{b - a}$ **45.** Undefined

47. $\dfrac{\Delta C}{\Delta t}$ represents the change in cost per change in time.

49. $y = -\dfrac{2}{3}x - \dfrac{13}{3}$; $f(x) = -\dfrac{2}{3}x - \dfrac{13}{3}$ **51.** $m = \dfrac{3}{7}$

53. a. -4 **b.** -28 **55. a.** $\dfrac{2}{3}$ **b.** $-\dfrac{3}{2}$

57. $y = 3x - 1$ **59.** $y = -0.9x + 6.29$ **61.** $y = 2x - 10$

63. $y = 7$ **65. a.** $C(x) = 1500 + 35x$ **b.** $R(x) = 60x$
c. $P(x) = 25x - 1500$ **d.** The studio needs more than 60 private
lessons per month to make a profit.   **e.** \$550

67. a. $y = 11.9x + 169$ **b.**
c. 1002 m

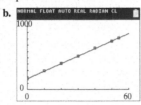

69.

71.

73.

75.

77.

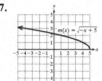

79.

81.

83.

85. y-axis   **87.** None of these   **89.** Odd   **91.** Even
93. Neither   **95. a.** 18   **b.** 1   **c.** 5   **d.** 4
97.

99. a. -3 **b.** -3 **c.** -1 **d.** 5
101. a. $(2, \infty)$ **b.** $(-\infty, 2)$ **c.** Never constant
103. At $x = -2$, the function has a relative maximum of 4.

105. $-\dfrac{19}{3}$ **107.** -28 **109.** 17

111. $(n - m)(x) = x^2$; Domain: $(-\infty, \infty)$

113. $\left(\dfrac{n}{p}\right)(x) = \dfrac{x^2 - 4x}{\sqrt{x - 2}}$; Domain: $(2, \infty)$

115. $(q \circ n)(x) = \dfrac{1}{x^2 - 4x - 5}$; Domain: $(-\infty, -1) \cup (-1, 5) \cup (5, \infty)$

117. -6 **119.** $f(x) = x^2$ and $g(x) = x - 4$

121. a. $d(t) = 60t$ **b.** $n(d) = \dfrac{d}{28}$ **c.** $(n \circ d)(t) = \dfrac{15t}{7}$ represents
the number of gallons of gasoline used in t hours.
d. $(n \circ d)(7) = 15$ means that 15 gal of gasoline is used in 7 hr.

Chapter 2 Test, pp. 282–283

1. a. $(3, -1)$ **b.** $\sqrt{41}$ **c.** $(x - 3)^2 + (y + 1)^2 = 41$
2. a. x-intercept: $(-4, 0)$; y-intercepts: $(0, 4)$, $(0, -4)$ **b.** No
3. a. $(x + 7)^2 + (y - 5)^2 = 4$ **b.** Center: $(-7, 5)$; Radius: 2
4. Yes   **5.** No   **6. a.** -12
b. $-2x^2 - 4xh - 2h^2 + 7x + 7h - 3$ **c.** $-4x - 2h + 7$
d. $\left(\frac{1}{2}, 0\right)$ and $(3, 0)$ **e.** $(0, -3)$ **f.** -1
7. a. -2 **b.** 0   **c.** $x = -2$ and $x = 2$
d. $(-\infty, -2) \cup (0, 2)$ **e.** $(-2, 0) \cup (2, \infty)$
f. At $x = 0$, the function has a relative minimum of -2.
g. At $x = -2$, the function has a relative maximum of 2. At $x = 2$,
the function has a relative maximum of 2.
h. $(-\infty, \infty)$ **i.** $(-\infty, 2]$ **j.** Even

8. $\left(-\infty, -\dfrac{7}{3}\right) \cup \left(-\dfrac{7}{3}, \infty\right)$ **9.** $(-\infty, 4]$

10. a. $m = -\dfrac{3}{4}$ **b.** $(0, 2)$ **c.**
d. $\dfrac{4}{3}$ **e.** $-\dfrac{3}{4}$

11. $y = 3x + 12$ **12. a.** $\{-2\}$ **b.** $(-\infty, -2)$ **c.** $[-2, \infty)$
13.   **14.**

15.   **16.**

17. Symmetric to the y-axis, x-axis, and origin.
18. Odd   **19.** Neither   **20. a.** 4   **b.** -5 **21.** 1   **22.** 0
23. Undefined (not a real number)

24. $(f \cdot g)(x) = \dfrac{x - 4}{x - 3}$; Domain: $(-\infty, 3) \cup (3, \infty)$

25. $\left(\dfrac{g}{f}\right)(x) = \dfrac{1}{(x - 3)(x - 4)}$; Domain: $(-\infty, 3) \cup (3, 4) \cup (4, \infty)$

26. $(g \circ h)(x) = \dfrac{1}{\sqrt{x - 5} - 3}$; Domain: $[5, 14) \cup (14, \infty)$

27. $f(x) = \sqrt[3]{x}$ and $g(x) = x - 7$ **28. a.** -1 **b.** -1
c. -2 **d.** Undefined   **e.** $(-3, 0)$ **f.** $(-\infty, 2)$

29. a.

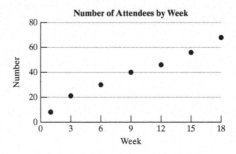
Number of Attendees by Week

b. $y = 4x + 4$ **c.** The slope is 4 and means that the number of attendees has increased by approximately 4 people per week.
d. 100 people

30. a. $y = 3.3x + 8.5$
b.

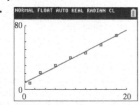

c. 88 people

Chapter 2 Cumulative Review Exercises, pp. 283–284

1. a. -1 **b.** $x = 0$ and $x = 3$ **c.** $(-4, \infty)$ **d.** $[-2, \infty)$
e. $(1, \infty)$ **f.** $(-1, 1)$ **g.** $(-4, -1)$ **h.** -1
2. a. $(x + 6)^2 + (y - 2)^2 = 9$ **b.** Center: $(-6, 2)$; Radius: 3

3. $(g \circ f)(x) = \dfrac{1}{-x^2 + 3x}$; Domain: $(-\infty, 0) \cup (0, 3) \cup (3, \infty)$

4. $(g \cdot h)(x) = \dfrac{\sqrt{x + 2}}{x}$; Domain: $[-2, 0) \cup (0, \infty)$

5. $-2x - h + 3$ **6.** 0
7. x-intercepts: $(0, 0)$ and $(3, 0)$; y-intercept: $(0, 0)$
8.

9.

10. $y = -\dfrac{2}{5}x + \dfrac{1}{5}$ **11.** $|x - 7|$ or $|7 - x|$

12. $2(x - 4)(x^2 + 4x + 16)$ **13.** $\left\{ \dfrac{1}{6} \pm \dfrac{\sqrt{71}}{6}i \right\}$

14. $\{0, 1\}$ **15.** $\{1, 32\}$ **16.** $\left[-4, \dfrac{10}{3} \right]$ **17.** $(-3, -1]$

18. $\dfrac{3\sqrt{15} - 3\sqrt{11}}{2}$ **19.** $9c^2 d\sqrt{2d}$ **20.** $\dfrac{uw}{2w + u}$

CHAPTER 3

Section 3.1 Practice Exercises, pp. 294–299

R.1. $\{-6, 3\}$ **R.2. a.** $x = -\dfrac{5}{3}$ or $x = 4$ **b.** $f(0) = -20$

R.3. $\{-2, -6\}$ **R.4.** $-\dfrac{21}{4}$ **R.5.** Domain $(-4, 4]$; Range $[-1, 4]$

1. quadratic   **3.** (h, k) **5.** upward; minimum; k
7. a. Downward   **b.** $(4, 1)$ **c.** $(3, 0)$ and $(5, 0)$ **d.** $(0, -15)$
e.

f. $x = 4$ **g.** Maximum: 1
h. Domain: $(-\infty, \infty)$; Range: $(-\infty, 1]$

9. a. Upward   **b.** $(-1, -8)$ **c.** $(-3, 0)$ and $(1, 0)$ **d.** $(0, -6)$
e.

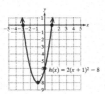

f. $x = -1$ **g.** Minimum: -8
h. Domain: $(-\infty, \infty)$; Range: $[-8, \infty)$

11. a. Upward   **b.** $(1, 0)$ **c.** $(1, 0)$ **d.** $(0, 3)$
e.

f. $x = 1$ **g.** Minimum: 0
h. Domain: $(-\infty, \infty)$; Range: $[0, \infty)$

13. a. Downward   **b.** $(-4, 1)$ **c.** $\left(-4 + \sqrt{5}, 0\right)$ and $\left(-4 - \sqrt{5}, 0\right)$

d. $\left(0, -\dfrac{11}{5}\right)$

e.

f. $x = -4$ **g.** Maximum: 1
h. Domain: $(-\infty, \infty)$; Range: $(-\infty, 1]$

15. a. $f(x) = (x + 3)^2 - 4$ **b.** $(-3, -4)$ **c.** $(-1, 0)$ and $(-5, 0)$
d. $(0, 5)$

e.

f. $x = -3$ **g.** Minimum: -4
h. Domain: $(-\infty, \infty)$; Range: $[-4, \infty)$

17. a. $p(x) = 3(x - 2)^2 - 19$ **b.** $(2, -19)$

c. $\left(\dfrac{6 + \sqrt{57}}{3}, 0\right)$ and $\left(\dfrac{6 - \sqrt{57}}{3}, 0\right)$ **d.** $(0, -7)$

e.

f. $x = 2$ **g.** Minimum: -19
h. Domain: $(-\infty, \infty)$; Range: $[-19, \infty)$

19. a. $c(x) = -2\left(x + \dfrac{5}{2}\right)^2 + \dfrac{33}{2}$ **b.** $\left(-\dfrac{5}{2}, \dfrac{33}{2}\right)$

c. $\left(\dfrac{-5 + \sqrt{33}}{2}, 0\right)$ and $\left(\dfrac{-5 - \sqrt{33}}{2}, 0\right)$ **d.** $(0, 4)$

e.

f. $x = -\dfrac{5}{2}$ **g.** Maximum: $\dfrac{33}{2}$
h. Domain: $(-\infty, \infty)$;

Range: $\left(-\infty, \dfrac{33}{2}\right]$

21. a. $h(x) = -2\left(x - \dfrac{7}{4}\right)^2 + \dfrac{49}{8}$ **b.** $\left(\dfrac{7}{4}, \dfrac{49}{8}\right)$

c. $(0, 0)$ and $\left(\dfrac{7}{2}, 0\right)$ **d.** $(0, 0)$

e.

f. $x = \dfrac{7}{4}$ **g.** Maximum: $\dfrac{49}{8}$
h. Domain: $(-\infty, \infty)$;

Range: $\left(-\infty, \dfrac{49}{8}\right]$

23. a. $p(x) = \left(x + \dfrac{9}{2}\right)^2 - \dfrac{13}{4}$ **b.** $\left(-\dfrac{9}{2}, -\dfrac{13}{4}\right)$

c. $\left(\dfrac{-9 + \sqrt{13}}{2}, 0\right)$ and $\left(\dfrac{-9 - \sqrt{13}}{2}, 0\right)$ **d.** $(0, 17)$

e.

f. $x = -\dfrac{9}{2}$ **g.** Minimum: $-\dfrac{13}{4}$

h. Domain: $(-\infty, \infty)$;

Range: $\left[-\dfrac{13}{4}, \infty\right)$

25. $(7, -238)$ **27.** $(9, 28)$ **29.** $(0, -5)$ **31.** $(-0.75, -4.275)$

33. a. Downward   **b.** $(1, -3)$ **c.** None   **d.** $(0, -4)$

e.

f. $x = 1$ **g.** Maximum: -3

h. Domain: $(-\infty, \infty)$; Range: $(-\infty, -3]$

35. a. Upward   **b.** $\left(\dfrac{3}{2}, -\dfrac{33}{4}\right)$

c. $\left(\dfrac{15 + \sqrt{165}}{10}, 0\right)$ and $\left(\dfrac{15 - \sqrt{165}}{10}, 0\right)$ **d.** $(0, 3)$

e.

f. $x = \dfrac{3}{2}$ **g.** Minimum: $-\dfrac{33}{4}$

h. Domain: $(-\infty, \infty)$;

Range: $\left[-\dfrac{33}{4}, \infty\right)$

37. a. Upward   **b.** $(0, 3)$ **c.** None   **d.** $(0, 3)$

e.

f. $x = 0$ **g.** Minimum: 3

h. Domain: $(-\infty, \infty)$; Range: $[3, \infty)$

39. a. Downward   **b.** $(-5, 0)$ **c.** $(-5, 0)$ **d.** $(0, -50)$

e.

f. $x = -5$ **g.** Maximum: 0

h. Domain: $(-\infty, \infty)$; Range: $(-\infty, 0]$

41. a. Upward   **b.** $\left(\dfrac{1}{2}, \dfrac{11}{4}\right)$ **c.** None   **d.** $(0, 3)$

e.

f. $x = \dfrac{1}{2}$ **g.** Minimum: $\dfrac{11}{4}$

h. Domain: $(-\infty, \infty)$;

Range: $\left[\dfrac{11}{4}, \infty\right)$

43. a. \$17   **b.** \$2450   **c.** \$10 and \$24

45. a. 3.96 m   **b.** 0.72 m   **c.** 7.9 m

47. a. 24 hr   **b.** 988,000   **49.** The numbers are 12 and 12.

51. The numbers are 5 and -5.   **53. a.** 40 ft by 80 ft   **b.** 3200 ft^2

55. a. $V(x) = -40x^2 + 240x$

b. $x = 3$; The sheet of aluminum should be folded 3 in. from each end.

c. 360 in.^{3}

57. a. $y = -0.000838t^2 + 0.0812t + 0.040$ **b.** 48 hr   **c.** 2 g

59. a. 15.3 sec   **b.** 16.7 sec

c. $v(x) = -0.1436x^2 - 0.4413x + 195.7$ **d.** 188.9 cm/sec

61. False. If a relation has two y-intercepts, then it would fail the vertical line test and the relation would not define y as a function of x.

63. True   **65.** Discriminant is 0; one x-intercept

67. Discriminant is 57; two x-intercepts   **69.** Discriminant is -96; no x-intercepts   **71.** Graph g   **73.** Graph h   **75.** Graph f

77. Graph c   **79.** For a parabola opening upward, such as the graph of $f(x) = x^2$, the minimum value is the y-coordinate of the vertex. There is no maximum value because the y values of the function become arbitrarily large for large values of $|x|$.   **81.** No function defined by $y = f(x)$ can have two y-intercepts because the graph would fail the vertical line test.   **83.** Because a parabola is symmetric with respect to the vertical line through the vertex, the x-coordinate of the vertex must be equidistant from the x-intercepts. Therefore, given $y = f(x)$, the x-coordinate of the vertex is 4 because 4 is midway between 2 and 6. The y-coordinate of the vertex is $f(4)$.

85. $f(x) = 2(x - 2)^2 - 3$ **87.** $f(x) = -\dfrac{1}{3}(x - 4)^2 + 6$

89. $c = 9$ **91.** $b = 4$ or $b = -4$

Section 3.2 Practice Exercises, pp. 311–315

R.1. $\{-9, 0, 2\}$ **R.2.** $\left\{-2, -\dfrac{6}{5}, 2\right\}$

R.3.   **R.4.**

R.5.

1. polynomial   **3.** is not   **5.** 1   **7.** 5   **9.** cross

11. f has at least one zero on the interval $[a, b]$.

13. Down left and down right. As $x \to -\infty$, $f(x) \to -\infty$, and as $x \to \infty$, $f(x) \to -\infty$.

15. Down left and up right. As $x \to -\infty$, $f(x) \to -\infty$, and as $x \to \infty$, $f(x) \to \infty$.

17. Up left and down right. As $x \to -\infty$, $f(x) \to \infty$, and as $x \to \infty$, $f(x) \to -\infty$.

19. Up left and up right. As $x \to -\infty$, $f(x) \to \infty$, and as $x \to \infty$, $f(x) \to \infty$.

21. 3; 4   **23.** $-2, 5, -5$; each of multiplicity 1

25. $0, -4, \dfrac{5}{2}$; each of multiplicity 1   **27.** 0 (multiplicity 3), 5 (multiplicity 2)

29. 0 (multiplicity 1), -2 (multiplicity 3), -4 (multiplicity 1)

31. $0, \dfrac{5}{3}, -\dfrac{9}{2}, \pm\sqrt{3}$; each of multiplicity 1

33. $3 \pm \sqrt{5}$; each of multiplicity 1

35. $-3, -\dfrac{1}{2}, \dfrac{1}{2}, 3$; each of multiplicity 1

37. $-\sqrt{7}$ (multiplicity 1), 0 (multiplicity 4), $\sqrt{7}$ (multiplicity 1)

39. a. Yes   **b.** No   **c.** No   **d.** Yes

41. a. Yes   **b.** Yes   **c.** No   **d.** No   **43. a.** Yes   **b.** $-\dfrac{5}{2}$

45. Not a polynomial function. The graph is not smooth.

47. Polynomial function   **a.** Minimum degree 3

b. Leading coefficient positive; degree odd

c. -4 (odd multiplicity), 1 (odd multiplicity), 3 (odd multiplicity)

49. Polynomial function   **a.** Minimum degree 6
  b. Leading coefficient negative; degree even
  c. -4 (odd multiplicity), -3 (odd multiplicity), -1 (even multiplicity), 2 (odd multiplicity), $\frac{7}{2}$ (odd multiplicity)

51. Not a polynomial function. The graph is not continuous.

53. a. $y = x^6$
  b. Shrink $y = x^6$ vertically by a factor of $\frac{1}{3}$.
    Reflect across the x-axis. Shift downward 2 units.
  c. Graph iii

55. a. $y = x^3$
  b. Shift $y = x^3$ to the left 2 units.
    Reflect across the x-axis. Shift upward 3 units.
  c. Graph i

57. a. $y = x^5$
  b. Shift $y = x^5$ to the right 3 units.
    Reflect across the y-axis. Shift upward 1 unit.
  c. Graph iv

59.

61.

63.

65.

67.

69.

71.

73.

75.

77. False. The value 5 is a zero with even multiplicity. Therefore, the graph touches but does not cross the x-axis at 5.

79. False. An nth-degree polynomial has at most $n - 1$ turning points. Therefore, a third-degree polynomial has at most 2 turning points.

81. True   **83.** False. If the leading coefficient is negative, the graph will be down to the far left and down to the far right.

85. False. The only real solution to the equation $x^3 - 27 = 0$ is $x = 3$. Therefore, the graph of $f(x) = x^3 - 27$ has only one x-intercept.

87. True   **89. a.** $(0, 12) \cup (68, 184)$ **b.** $(12, 68) \cup (184, 200)$
  c. 3   **d.** Degree 4; leading coefficient negative
  e. 184 sec after launch   **f.** 2.85 G-forces

91. The x-intercepts are the real solutions to the equation $f(x) = 0$.

93. A function is continuous if its graph can be drawn without lifting the pencil from the paper.

95. a. $f(3) = 2; f(4) = 6$ **b.** By the intermediate value theorem, because $f(3) = 2$ and $f(4) = 6$, then f must take on every value between 2 and 6 on the interval $[3, 4]$.
  c. $x = \dfrac{3 + \sqrt{17}}{2} \approx 3.56$

97. $V(t) = -0.0406t^3 + 0.154t^2 + 0.173t - 0.0024$

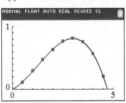

99. a. $V = lwh$
    $= (30 - 2x)(24 - 2x)(x)$
    $= 4x^3 - 108x^2 + 720x$
  The domain is restricted to $0 < x < 12$ because the width of the rectangular sheet is 24 in. The maximum amount that can be removed from each end would be half of 24 in.
  b. 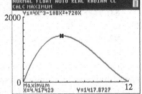   **c.** 1418 in.^{3}

  4.4 in.

101. Window b is better.

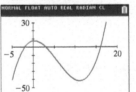

103.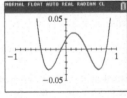

Section 3.3 Practice Exercises, pp. 325–329

R.1. $-7 + 24i$

R.2. a. $(6 - i\sqrt{3})^2 - 12(6 - i\sqrt{3}) + 39 = 0$ ✓
  b. $(6 + i\sqrt{3})^2 - 12(6 + i\sqrt{3}) + 39 = 0$ ✓

R.3. $\{-4 \pm i\sqrt{3}\}$ **1.** Dividend: $f(x)$; Divisor: $d(x)$; Quotient: $q(x)$; Remainder: $r(x)$ **3.** $f(c)$ **5.** True

7. a. $3x + 12 + \dfrac{65}{2x - 5}$ **b.** Dividend: $6x^2 + 9x + 5$; Divisor: $2x - 5$; Quotient: $3x + 12$; Remainder: 65
  c. $(2x - 5)(3x + 12) + 65 = 6x^2 + 9x + 5$ ✓

9. $3x^2 + x + 4 + \dfrac{6}{x - 4}$ **11.** $4x^3 - 20x^2 + 13x + 4$

13. $3x^3 - 6x^2 + 2x - 4$ **15.** $x^3 + 4x^2 - 5x - 2 + \dfrac{5x}{x^2 + 5}$

17. $3x^2 + 1 + \dfrac{5x - 2}{2x^2 + x - 3}$ **19.** $x^2 + 3x + 9$

21. $\dfrac{5}{2}x^2 + \dfrac{1}{4}x + \dfrac{1}{8} + \dfrac{\frac{25}{8}}{2x - 1}$

23. a. $2x^4 - 5x^3 - 5x^2 - 4x + 29$ **b.** $x - 3$
  c. $2x^3 + x^2 - 2x - 10$ **d.** -1

25. a. $x^3 - 2x^2 - 25x - 4$ **b.** $x + 4$
  c. $x^2 - 6x - 1$ **d.** 0

27. $4x - 9 + \dfrac{55}{x + 6}$ **29.** $5x + 3$

31. $-5x^3 + 10x^2 - 23x + 38 + \dfrac{-72}{x + 2}$

33. $4x^4 - 13x^3 - 97x^2 - 59x + 21$

35. $x^4 - 2x^3 + 4x^2 - 8x + 16$ **37.** $2x^3 - 4x^2 - 62x - 56$

39. The remainder is 39.   **41. a.** 201   **b.** 201

43. a. -112 **b.** 0   **c.** 123   **d.** 0

45. a. -2 **b.** 0   **c.** 0   **d.** 18

47. a. No   **b.** Yes   **49. a.** No   **b.** Yes

51. a. Yes   **b.** Yes   **53. a.** Yes   **b.** Yes

55. a. Yes   **b.** No   **57. a.** No   **b.** Yes

59. a. No   **b.** Yes   **61. a.** 0   **b.** 0   **c.** $\{-3, -\sqrt{5}, \sqrt{5}, 3\}$

63. a. Yes   **b.** Yes   **c.** $\{2 \pm 5i\}$ **d.** $2 \pm 5i$

65. a. $f(x) = (x + 1)(2x - 9)(x + 4)$ **b.** $\left\{-1, \dfrac{9}{2}, -4\right\}$

67. a. $f(x) = 4\left(x - \dfrac{1}{4}\right)(5x + 1)(x + 2)$ or $f(x) = (4x - 1)(5x + 1)(x + 2)$

   b. $\left\{\dfrac{1}{4}, -\dfrac{1}{5}, -2\right\}$ **69. a.** $f(x) = (x - 3)(3x - 1)^2$ **b.** $\left\{3, \dfrac{1}{3}\right\}$

71. $f(x) = x^3 - x^2 - 14x + 24$

73. $f(x) = x^4 - \dfrac{5}{2}x^3 + \dfrac{3}{2}x^2$ or $f(x) = 2x^4 - 5x^3 + 3x^2$

75. $f(x) = x^2 - 44$ **77.** $f(x) = x^3 + 2x^2 + 9x + 18$

79. $f(x) = 6x^3 - 23x^2 - 6x + 8$ **81.** $f(x) = x^2 - 14x + 113$

83. Direct substitution; -2 **85. a.** Yes   **b.** Yes   **c.** Yes

   d. No   **e.** Yes   **f.** No   **87.** False   **89.** $m = 28$

91. $m = -5$ **93.** $r = 0$ **95. a.** $V(x) = 2x^3 + 3x^2 - x$ **b.** 534 cm^3

97. The divisor must be of the form $(x - c)$, where c is a constant.

99. Compute $f(c)$ either by direct substitution or by using the remainder theorem. The remainder theorem states that $f(c)$ is equal to the remainder obtained after dividing $f(x)$ by $(x - c)$.

101. a. $f(x) = (x - 5)(x - i)(x + i)$ **b.** $\{5, i, -i\}$

103. a. $f(x) = (x + 1)^2(x - \sqrt{3})(x + \sqrt{3})$ **b.** $\{-1, \sqrt{3}, -\sqrt{3}\}$

105. a. 3   **b.** 3 is a solution.   **c.** $\left\{3, -4, -\dfrac{2}{5}\right\}$

Section 3.4 Practice Exercises, pp. 340–344

R.1. $81 + 0i$ **R.2.** $49 + 0i$ **R.3.** $-22 + 0i$

1. zeros   **3.** $a - bi$ **5.** greater than

7. $\pm 1, \pm 2, \pm 4$ **9.** $\pm 1, \pm 2, \pm 3, \pm 6, \pm\dfrac{1}{2}, \pm\dfrac{3}{2}, \pm\dfrac{1}{4}, \pm\dfrac{3}{4}$

11. $\pm 1, \pm 2, \pm 4, \pm 8, \pm\dfrac{1}{2}, \pm\dfrac{1}{3}, \pm\dfrac{2}{3}, \pm\dfrac{4}{3}, \pm\dfrac{8}{3}, \pm\dfrac{1}{4}, \pm\dfrac{1}{6}, \pm\dfrac{1}{12}$

13. 7 and $\dfrac{5}{3}$ **15.** $-\dfrac{1}{2}, 1$ **17.** $-3, \dfrac{1}{2}, 2, 4$ **19.** $5, 1 \pm \sqrt{5}$

21. $\dfrac{1}{5}, \pm\sqrt{7}$ **23.** 2 (multiplicity 2), $\dfrac{1}{3}, -4$ **25.** $-2, 3 \pm i$

27. $\pm\sqrt{10}, \pm 3i$ **29.** one   **31.** 7

33. a. $2 \pm 5i, \pm\sqrt{7}$

   b. $[x - (2 + 5i)][x - (2 - 5i)](x - \sqrt{7})(x + \sqrt{7})$

   c. $\{2 \pm 5i, \pm\sqrt{7}\}$

35. a. $4 \pm i, \dfrac{4}{3}$ **b.** $[x - (4 + i)][x - (4 - i)](3x - 4)$

   c. $\left\{4 \pm i, \dfrac{4}{3}\right\}$ **37. a.** $-3 \pm 2i, -\dfrac{1}{4}, 1, -4$

   b. $[x - (-3 + 2i)][x - (-3 - 2i)](4x + 1)(x - 1)(x + 4)$

   c. $\left\{-3 \pm 2i, -\dfrac{1}{4}, 1, -4\right\}$

39. $f(x) = 5x^3 - 4x^2 + 180x - 144$

41. $f(x) = -5x^4 + 10x^3 + 60x^2 - 200x + 160$

43. $f(x) = 18x^3 + 39x^2 + 8x - 16$

45. $f(x) = x^6 - 14x^5 + 65x^4$

47. $f(x) = x^4 - 12x^3 + 62x^2 - 300x + 925$

49. Positive: 3 or 1; Negative: 3 or 1   **51.** Positive: 6, 4, 2, or 0; Negative: 1   **53.** Positive: 0; Negative: 4, 2, or 0

55. Positive: 0; Negative: 0   **57.** 4 real zeros; $f(x)$ has 1 positive real zero, no negative real zeros, and the number 0 is a zero of multiplicity 3.

59. a. Yes   **b.** No   **61. a.** Yes   **b.** Yes   **63. a.** Yes

   b. Yes   **65.** True   **67.** False. Only numbers less than -3 are also guaranteed to be lower bounds.

69. $\dfrac{7}{4}, -\dfrac{1}{2}$, and 4 (each with multiplicity 1)

71. $\pm\sqrt{5}, \dfrac{1}{2}, -2$, and -4 (each with multiplicity 1)

73. -3 (multiplicity 2) and $\dfrac{1}{2}$ (multiplicity 2)

75. $\dfrac{1}{2}, 3, 1 \pm 2i$ (each with multiplicity 1)

77. $\dfrac{5}{2}$ (multiplicity 2), $1 \pm i$ (each multiplicity 1)

79. 0 (multiplicity 2), -1 (multiplicity 2), and $\pm i\sqrt{10}$ (each multiplicity 1)

81. 0 (multiplicity 3) and $5 \pm 3i$ (each multiplicity 1)

83. -1 and $2 \pm 3i$ (each multiplicity 1)

85. False. For example, the graph of $f(x) = x^4 + 1$ has no x-intercepts. Thus, $x^4 + 1$ has no real zeros.

87. False. For example, the graph of $f(x) = x^{10} + 1$ has no x-intercepts.

89. True   **91.** All statements are true.

93. a. $f(2) = -2$ and $f(3) = 1$. Since $f(2)$ and $f(3)$ have opposite signs, the intermediate value theorem guarantees that f has at least one real zero between 2 and 3.

   b. $\dfrac{7 \pm \sqrt{17}}{4}$; Furthermore, $\dfrac{7 + \sqrt{17}}{4} \approx 2.78$ is on the interval $[2, 3]$.

95. If a polynomial has real coefficients, then all nonreal zeros must come in conjugate pairs. This means that if the polynomial has nonreal zeros, there would be an even number of them. A third-degree polynomial has 3 zeros (including multiplicities). Therefore, it would have either 2 or 0 nonreal zeros, leaving room for either 1 or 3 real zeros.

97. $f(x)$ has no variation in sign, nor does $f(-x)$. By Descartes' rule of signs, there are no positive or negative real zeros. Furthermore, 0 itself is not a zero of $f(x)$ because x is not a factor of $f(x)$. Therefore, there are no real zeros of $f(x)$.

99. $n - 2$ possible nonreal zeros   **101.** The triangular front has a base of 6 ft and a height of 4 ft. The length is 9 ft.

103. Each dimension was decreased by 1 in.

105. The dimensions are either 2 cm by 3 cm or $-1 + \sqrt{13}$ cm by $\dfrac{1 + \sqrt{13}}{2}$ cm.

107. a. $(x + 3)(x - 1)(x^2 + 4)$ **b.** $(x + 3)(x - 1)(x + 2i)(x - 2i)$

109. a. $(x - \sqrt{5})(x + \sqrt{5})(x^2 + 7)$

   b. $(x - \sqrt{5})(x + \sqrt{5})(x + \sqrt{7}i)(x - \sqrt{7}i)$

111. The fourth roots of 1 are $1, -1, i$, and $-i$.

113. The number $\sqrt{5}$ is a real solution to the equation $x^2 - 5 = 0$ and a zero of the polynomial $f(x) = x^2 - 5$. However, by the rational zeros theorem, the only possible rational zeros of $f(x)$ are ± 1 and ± 5. This means that $\sqrt{5}$ is irrational.   **115.** -2

Section 3.5 Practice Exercises, pp. 361–367

R.1. $\dfrac{q + 6}{q + 2}; q \neq 6, q \neq -2$ **R.2.** $-\dfrac{u + 3}{2u}; u \neq 0, u \neq 3$

R.3. $-\dfrac{3}{x + 3}; m \neq p, x \neq -3$

1. $q(x)$ **3.** x approaches 5 from the left   **5.** nonzero

7. $(-\infty, 5) \cup (5, \infty)$ **9.** $(-\infty, -1) \cup \left(-1, \dfrac{1}{4}\right) \cup \left(\dfrac{1}{4}, \infty\right)$

11. $(-\infty, \infty)$ **13. a.** 2   **b.** $-\infty$ **c.** ∞ **d.** 2

   e. Never increasing   **f.** $(-\infty, 4) \cup (4, \infty)$

   g. $(-\infty, 4) \cup (4, \infty)$ **h.** $(-\infty, 2) \cup (2, \infty)$

   i. $x = 4$ **j.** $y = 2$

15. a. -1 **b.** ∞ **c.** ∞ **d.** -1 **e.** $(-\infty, -3)$

   f. $(-3, \infty)$ **g.** $(-\infty, -3) \cup (-3, \infty)$ **h.** $(-1, \infty)$

   i. $x = -3$ **j.** $y = -1$

17. $x = 4$ **19.** $x = 5$ and $x = -\dfrac{1}{2}$ **21.** None

23. $t = \dfrac{-2 + \sqrt{10}}{2}$ and $t = \dfrac{-2 - \sqrt{10}}{2}$ **25.** a   **27.** d

29. a. $y = 0$ **b.** Graph does not cross $y = 0$.

31. a. $y = 3$ **b.** $\left(\dfrac{7}{4}, 3\right)$ **33. a.** No horizontal asymptote

b. Not applicable   **35. a.** $y = 0$ **b.** $(-2, 0)$

37. a. $\dfrac{1 + \dfrac{3}{x} + \dfrac{1}{x^2}}{2 + \dfrac{5}{x^2}}$ **b.** 0   **c.** $y = \dfrac{1}{2}$

39. Vertical asymptote: $x = 0$; Slant asymptote: $y = 2x$

41. Vertical asymptote: $x = -6$; Slant asymptote: $y = -3x + 22$

43. Vertical asymptotes: $x = \sqrt{5}$ and $x = -\sqrt{5}$; Slant asymptote: $y = x + 5$ **45.** Vertical asymptotes: $x = 2$, $x = -2$, and $x = -1$; Horizontal asymptote: $y = 0$ **47.** Slant asymptote: $y = 2x - 5$

49.
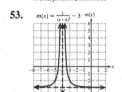

51.

53.
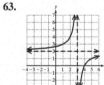

55.

57. a. $(-3, 0)$ and $\left(\dfrac{7}{2}, 0\right)$ **b.** $x = -2$ and $x = -\dfrac{1}{4}$

c. Horizontal asymptote: $y = \dfrac{1}{2}$ **d.** $\left(0, -\dfrac{21}{2}\right)$

59. a. $\left(\dfrac{9}{4}, 0\right)$ **b.** $x = 3$ and $x = -3$

c. Horizontal asymptote: $y = 0$ **d.** $(0, 1)$

61. a. $\left(\dfrac{1}{5}, 0\right)$ and $(-3, 0)$ **b.** $x = -2$

c. Slant asymptote: $y = 5x + 4$ **d.** $\left(0, -\dfrac{3}{2}\right)$

63.
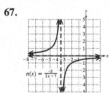

65.
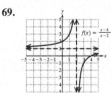

67.
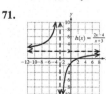

69.
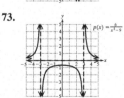

71.
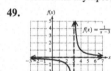

73.
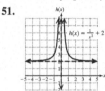

75.
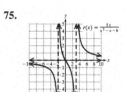

77.
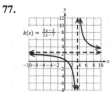

79.
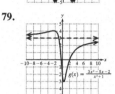

81.
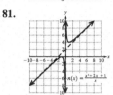

83.
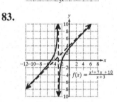

85.
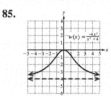

87.
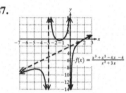

89.
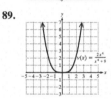

91. a. $C(x) = 109.94 + 20x$ **b.** $\overline{C}(x) = \dfrac{109.94 + 20x}{x}$

c. $\overline{C}(5) = 41.99$; $\overline{C}(30) = 23.67$; $\overline{C}(120) = 20.92$

d. The average cost would approach \$20 per session. This is the same as the fee paid to the gym in the absence of fixed costs.

93. a. $R(x) = \dfrac{6x}{x + 6}$

b.

x	6	12	18	30
$R(x)$	3	4	4.5	5

c. 6 Ω; Even for large values of x, the total resistance will always be less than 6 Ω. This is consistent with the statement that the total resistance is always less than the resistance in any individual branch of the circuit.

95. a.
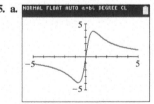

b. $t \geq 0$ **c.** 4 mg/L   **d.** 0 mg/L

97. a. \$200,000   **b.** \$600,000; \$1,800,000; \$5,400,000   **c.** 70%

99. a. $F(v) = 560\left(\dfrac{772.4}{772.4 - v}\right)$

b.
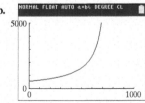

c. The frequency increases, making the pitch of the siren higher to the observer.

101. a. $f(x) = \dfrac{1}{(x+1)^2} + 3$

  b. Domain: $(-\infty, -1) \cup (-1, \infty)$; Range: $(3, \infty)$

103. a. $f(x) = 2 + \dfrac{1}{x+3}$ **b.**

105. The numerator and denominator share a common factor of $x + 2$. The value -2 is not in the domain of f. The graph will have a "hole" at $x = -2$ rather than a vertical asymptote.

107. $f(x) = \dfrac{x^2 + 4x + 3}{x^2 - 4x + 4}$ **109.** $f(x) = \dfrac{20x - 30}{x^2 - 3x - 10}$

111. a. $(-\infty, 2) \cup (2, \infty)$ **b.** $f(x) = x + 3$ where $x \neq 2$ **c.** None
  d. $x = 2$ **e.** Graph iii

113. a. $(-\infty, -5) \cup (-5, -4) \cup (-4, \infty)$

  b. $f(x) = \dfrac{2}{x+4}$ where $x \neq -5$

  c. $x = -4$ **d.** $x = -5$ **e.** Graph iv

Problem Recognition Exercises, p. 368

1. $2, -1,$ and -4 **2.** $-2, 1,$ and 3 **3.** $(-2, 0), (1, 0),$ and $(3, 0)$
4. $(2, 0), (-1, 0),$ and $(-4, 0)$ **5.** $(2, 0), (-1, 0),$ and $(-4, 0)$
6. $x = -2, x = 1,$ and $x = 3$ **7.** Horizontal asymptote: $y = 1$
8. $\left(\dfrac{1 + \sqrt{281}}{10}, 1\right) \approx (1.78, 1)$ and $\left(\dfrac{1 - \sqrt{281}}{10}, 1\right) \approx (-1.58, 1)$
9. $4, \sqrt{2},$ and $-\sqrt{2}$ **10.** $1, -2$ (multiplicity 2)
11. $(1, 0)$ and $(-2, 0)$ **12.** $(4, 0), (\sqrt{2}, 0),$ and $(-\sqrt{2}, 0)$
13. $(4, 0), (\sqrt{2}, 0),$ and $(-\sqrt{2}, 0)$ **14.** $x = 1$ and $x = -2$
15. Horizontal asymptote: $y = 1$
16. $\left(\dfrac{-1 + \sqrt{85}}{7}, 1\right) \approx (1.17, 1)$ and $\left(\dfrac{-1 - \sqrt{85}}{7}, 1\right) \approx (-1.46, 1)$
17. Graph b   **18.** Graph a   **19. a.** $q(x) = 2x - 4$
  b. $r(x) = 12x - 32$ **20.** $y = 2x - 4$ **21.** $\left(\dfrac{8}{3}, \dfrac{4}{3}\right)$

22. $\left\{\dfrac{8}{3}\right\}$; The solution to $r(x) = 0$ gives the x-coordinate of the point where the graph of f crosses its slant asymptote.

Section 3.6 Practice Exercises, pp. 377–382

R.1. $\left(-\infty, -\dfrac{8}{9}\right]$ **R.2.** $\left(-\dfrac{13}{2}, \infty\right)$

R.3. $(-\infty, -1) \cup [6, \infty)$ **R.4.** $\dfrac{7w + 6}{(w + 2)(3w + 2)}$

R.5. $\dfrac{y^2 - 24}{y - 5}$ **1.** polynomial; 2   **3.** $(-\infty, \infty)$; { }

5. a. $(-4, -1)$ **b.** $[-4, -1]$ **c.** $(-\infty, -4) \cup (-1, \infty)$
  d. $(-\infty, -4] \cup [-1, \infty)$ **7. a.** $(-\infty, -3) \cup (-3, \infty)$
  b. $(-\infty, \infty)$ **c.** { }   **d.** $\{-3\}$
9. a. $(-\infty, -2) \cup (0, 3)$ **b.** $(-\infty, -2] \cup [0, 3]$
  c. $(-2, 0) \cup (3, \infty)$ **d.** $[-2, 0] \cup [3, \infty)$
11. a. $(0, 3) \cup (3, \infty)$ **b.** $[0, \infty)$ **c.** $(-\infty, 0)$
  d. $(-\infty, 0] \cup \{3\}$ **13. a.** $(-\infty, \infty)$ **b.** $(-\infty, \infty)$

  c. { }   **d.** { }   **15. a.** $\left\{\dfrac{3}{5}, 5\right\}$ **b.** $\left(\dfrac{3}{5}, 5\right)$ **c.** $\left[\dfrac{3}{5}, 5\right]$

  d. $\left(-\infty, \dfrac{3}{5}\right) \cup (5, \infty)$ **e.** $\left(-\infty, \dfrac{3}{5}\right] \cup [5, \infty)$
17. a. $\{-3, 4\}$ **b.** $(-\infty, -3) \cup (4, \infty)$
  c. $(-\infty, -3] \cup [4, \infty)$ **d.** $(-3, 4)$ **e.** $[-3, 4]$
19. a. $\{-6\}$ **b.** { }   **c.** $\{-6\}$
  d. $(-\infty, -6) \cup (-6, \infty)$ **e.** $(-\infty, \infty)$

21. $\left(-\infty, -\dfrac{3}{2}\right] \cup \left[\dfrac{3}{2}, \infty\right)$ **23.** $\left(-1, \dfrac{4}{3}\right)$ **25.** $(-\infty, 0] \cup [3, \infty)$
27. $\left(\dfrac{-3 - \sqrt{59}}{5}, \dfrac{-3 + \sqrt{59}}{5}\right)$ **29.** $(-7, 7)$
31. $\left(-\infty, -\dfrac{\sqrt{2}}{4}\right] \cup \left[\dfrac{\sqrt{2}}{4}, \infty\right)$ **33.** $[-4, 1] \cup [3, \infty)$
35. $(-\infty, -2) \cup (-2, 0) \cup (4, \infty)$ **37.** $[-3, -1] \cup [1, 3]$
39. $\left(-\infty, -\dfrac{5}{2}\right) \cup (-2, 2)$ **41.** $[-1, 0] \cup \{3\}$
43. $\left(-\infty, \dfrac{3}{5}\right] \cup [5, \infty)$ **45.** $\left(-\infty, -\dfrac{1}{3}\right) \cup \left(0, \dfrac{5}{2}\right) \cup \left(\dfrac{5}{2}, 4\right)$
47. $(-\infty, \infty)$ **49.** { }   **51.** $\left(-\infty, \dfrac{3}{4}\right) \cup \left(\dfrac{3}{4}, \infty\right)$ **53.** $\{-2\}$
55. a. $(2, 3)$ **b.** $[2, 3)$ **c.** $(-\infty, 2) \cup (3, \infty)$
  d. $(-\infty, 2] \cup (3, \infty)$ **57. a.** $(-\infty, 2) \cup (2, \infty)$
  b. $(-\infty, 2) \cup (2, \infty)$ **c.** { }   **d.** { }
59. a. $[-2, 3)$ **b.** $(-2, 3)$ **c.** $(-\infty, -2] \cup (3, \infty)$
  d. $(-\infty, -2) \cup (3, \infty)$ **61. a.** $\{0\}$ **b.** { }   **c.** $(-\infty, \infty)$
  d. $(-\infty, 0) \cup (0, \infty)$ **63.** $(-1, 5]$ **65.** $[2, \infty)$
67. $(-3, -1] \cup [2, \infty)$ **69.** $\left(\dfrac{7}{2}, 6\right)$ **71.** $(-\infty, 2)$ **73.** $(-5, -2]$
75. $(-\infty, 2]$ **77.** $(-2, \infty)$ **79.** $(-3, -1) \cup (0, \infty)$
81. $(-\infty, 1) \cup (4, 7]$ **83.** $[2, 4) \cup (4, \infty)$
85. a. $s(t) = -16t^2 + 216t$ **b.** The shell will explode 6.75 sec after launch.   **c.** The spectators can see the shell between 1 sec and 6.75 sec after launch.   **87.** The car will stop within 250 ft if the car is traveling less than 50 mph.
89. a. The horizontal asymptote is $y = 0$ and means that the temperature will approach 0°C as time increases without bound.
  b. More than 6 hr is required for the temperature to fall below 5°C.
91. The width should be between $2\sqrt{15}$ ft and $4\sqrt{5}$ ft. This is between approximately 7.7 ft and 8.9 ft.
93. $[-3, 3]$ **95.** $(-\infty, -\sqrt{5}] \cup [\sqrt{5}, \infty)$
97. $(-\infty, -6] \cup \left[\dfrac{3}{2}, \infty\right)$ **99.** $(-\infty, -6) \cup \left(\dfrac{3}{2}, \infty\right)$
101. $(-\infty, -2) \cup [0, \infty)$
103. a.

Sign of $(x - a)^2$:	+	+	+	+
Sign of $(b - x)$:	+	+	−	−
Sign of $(x - c)^3$:	−	−	−	+
Sign of $(x - a)^2(b - x)(x - c)^3$:	−	−	+	−

       a b c

  b. (b, c) **c.** $(-\infty, a) \cup (a, b) \cup (c, \infty)$
105. The solution set to the inequality $f(x) < 0$ corresponds to the values of x for which the graph of $y = f(x)$ is below the x-axis.
107. Both the numerator and denominator of the rational expression are positive for all real numbers x. Therefore, the expression cannot be negative for any real number.
109. $[3, 5)$ **111.** $(-\infty, -32]$ **113.** $[2, 18)$ **115.** $[0, 2) \cup (4, 6]$
117. $(-3, 3)$ **119.** $(-\infty, -2\sqrt{5}) \cup (-4, 4) \cup (2\sqrt{5}, \infty)$
121. a. $0.552x^3 + 4.13x^2 - 1.84x - 10.2 < 0$
  b.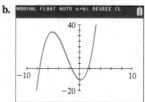

  c. The real zeros are approximately $-7.6, -1.5,$ and 1.6.
  d. $(-\infty, -7.6) \cup (-1.5, 1.6)$

123. a. 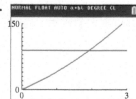   **b.** $(1.9, 90)$

c. The radius should be no more than 1.9 in. to keep the amount of aluminum to at most 90 in.^{2}.

Problem Recognition Exercises, p. 382

1. $(-28, -18]$ **2.** $\left\{\dfrac{3 \pm \sqrt{19}}{2}\right\}$ **3.** $\left\{\pm\dfrac{1}{5}, \dfrac{1}{2}\right\}$

4. $(-\infty, -2) \cup [0, \infty)$ **5.** $\{77\}$ **6.** $(-5, -1]$ **7.** $\left\{-\dfrac{1}{5}, \dfrac{9}{5}\right\}$

8. $\left\{\dfrac{133}{66}\right\}$ **9.** $\left\{\dfrac{10}{7}\right\}$ **10.** $\{5\}$; The value -2 does not check.

11. $[4, 10]$ **12.** $(4, 10)$ **13.** $\{0.05\}$ **14.** $[1, 37]$ **15.** $\{9\}$

16. $\{-2\} \cup [0, 3] \cup [5, \infty)$ **17.** $\{\ \}$

18. $\{5\}$; The value -4 does not check.   **19.** $\{\pm 4, \pm\sqrt{7}\}$

20. $\left\{\dfrac{3}{2}, -5\right\}$ **21.** $\{\ \}$ **22.** $\{17\}$ **23.** $(-\infty, -2) \cup (1, 4)$

24. $(-5, -1]$ **25.** $(-\infty, \infty)$ **26.** $\left(-\infty, -\dfrac{7}{5}\right) \cup \left(-\dfrac{7}{5}, \infty\right)$

27. $(-\infty, -8) \cup (14, \infty)$ **28.** $(-\infty, -4) \cup (1, \infty)$

29. $\left(-\infty, \dfrac{14}{5}\right)$ **30.** $[-3, 2]$

Section 3.7 Practice Exercises, pp. 387–391

R.1. $r = \dfrac{A}{Pt}$ **R.2.** $h = \dfrac{3V}{\pi r^2}$ **R.3.** $E = \dfrac{IR}{K}$ **R.4.** $\{18\}$

1. directly   **3.** constant; variation
5. a. 2; 4; 6; 8; 10   **b.** y is also doubled.   **c.** y is also tripled.
d. increases   **e.** decreases   **7.** inversely   **9.** jointly

11. $C = kr$ **13.** $\bar{C} = \dfrac{k}{n}$ **15.** $V = khr^2$ **17.** $E = \dfrac{ks}{\sqrt{n}}$

19. $c = \dfrac{kmn}{t^3}$ **21.** $k = \dfrac{5}{2}$ **23.** $k = 972$ **25.** $k = 5$

27. a. 2   **b.** 8   **29. a.** 225 mg   **b.** 270 mg   **c.** 315 mg
d. 30 lb   **31. a.** $0.40 per mile   **b.** $0.27 per mile
c. $0.20 per mile   **d.** 500 mi   **33.** 638 ft
35. a. 333.2 ft   **b.** 60 mph   **37. a.** 6.4 days   **b.** 12 people
39. 32 A   **41.** $1440   **43.** 27.37   **45.** 7.75 mph

47. $11,145.60   **49.** $y = 3.2x$ **51.** $y = \dfrac{12}{x}$ **53.** a and c

55. The variable P varies directly as the square of v and inversely as t.
57. a. 1600π m^2 **b.** The surface area is 4 times as great. Doubling the radius results in $(2)^2$ times the surface area of the sphere.
c. The intensity at 20 m should be $\frac{1}{4}$ the intensity at 10 m. This is because the energy from the light is distributed across an area 4 times as great.   **d.** 50 lux
59. The intensity is $\frac{1}{100}$ as great.   **61.** y will be $\frac{1}{4}$ its original value.
63. y will be 9 times its original value.

Chapter 3 Review Exercises, pp. 394–397

1. $(-5, 2)$
3. a. $f(x) = -2(x - 1)^2 + 8$ **b.** Downward   **c.** $(1, 8)$
d. $(-1, 0)$ and $(3, 0)$ **e.** $(0, 6)$
f. 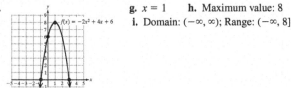   **g.** $x = 1$ **h.** Maximum value: 8
i. Domain: $(-\infty, \infty)$; Range: $(-\infty, 8]$

5. a. 45 yd by 90 yd   **b.** 4050 yd^2
7. a. $E(a) = -0.476a^2 + 37.0a - 44.6$ **b.** 39 yr   **c.** $674
9. a. Up to the left, up to the right; As $x \to -\infty$, $f(x) \to \infty$, and as $x \to \infty$, $f(x) \to \infty$.   **b.** Zeros: 3, -3, 1, -1 (each with multiplicity 1)
c. $(3, 0), (-3, 0), (1, 0), (-1, 0)$ **d.** $(0, 9)$ **e.** Even function
f.

11. a. Down to the left, up to the right; As $x \to -\infty$, $f(x) \to -\infty$, and as $x \to \infty$, $f(x) \to \infty$.
b. Zeros: 0 (with multiplicity 3) and $4 \pm \sqrt{3}$ (each with multiplicity 1)
c. $(0, 0), \left(4 + \sqrt{3}, 0\right), \left(4 - \sqrt{3}, 0\right)$ **d.** $(0, 0)$
e. Neither even nor odd
f.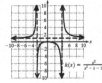

13. False. It may have three or fewer turning points.
15. False. There are infinitely many such polynomials. For example, any polynomial of the form $f(x) = a(x - 2)(x - 3)(x - 4)$ has the required zeros.

17. a. $-2x^2 + 3x - 9 + \dfrac{22x - 28}{x^2 + x - 3}$
b. Dividend: $-2x^4 + x^3 + 4x - 1$; Divisor: $x^2 + x - 3$; Quotient: $-2x^2 + 3x - 9$; Remainder: $22x - 28$

19. $2x^4 - 4x^3 + 8x^2 - 15x + 25 + \dfrac{-49}{x + 2}$ **21.** 65

23. a. No   **b.** Yes   **25. a.** Yes   **b.** Yes
27. $f(x) = (3x - 2)(5x + 1)(x - 4)$ **29.** $f(x) = 8x^3 - 22x^2 - 7x + 3$
31. a. 4   **b.** $\pm 1, \pm 2, \pm 4, \pm 8$ **c.** -2 (multiplicity 2)
d. -2 (multiplicity 2), $\pm\sqrt{2}$
33. a. $11 \pm i, \pm\sqrt{3}$
b. $[x - (11 - i)][x - (11 + i)](x - \sqrt{3})(x + \sqrt{3})$
c. $\{11 \pm i, \pm\sqrt{3}\}$
35. $f(x) = 3x^3 - 5x^2 + 12x - 20$ **37.** Positive: 0; Negative: 2 or 0

39. a. Yes   **b.** Yes   **41.** $x = \dfrac{5}{2}, x = -3$

43. a. $y = 0$ **b.** Graph does not cross $y = 0$.
45. a. No horizontal asymptote   **b.** Not applicable

47. Vertical asymptotes: $x = -\dfrac{1}{3}, x = 5$; Horizontal asymptote: $y = -\dfrac{4}{3}$

49.   **51.**

53. a. $(-\infty, -4)$ **b.** $(-\infty, -4] \cup \{1\}$
c. $(-4, 1) \cup (1, \infty)$ **d.** $[-4, \infty)$
55. a. $\{-5, -2\}$ **b.** $(-5, -2)$ **c.** $[-5, -2]$
d. $(-\infty, -5) \cup (-2, \infty)$ **e.** $(-\infty, -5] \cup [-2, \infty)$
57. $(-\infty, -3] \cup [6, \infty)$ **59.** $\{1\}$ **61.** $(-3, 2) \cup (4, \infty)$

63. $(-\infty, 0) \cup (0, 3]$ **65.** $(-\infty, 0) \cup \left(\dfrac{4}{5}, 2\right)$

67. a. $\bar{C}(x) = \dfrac{120 + 15x}{x}$

b. The trainer must have more than 120 sessions with his clients for his average cost to drop below $16 per session.

69. $m = kw$ **71.** $y = \dfrac{kx\sqrt{z}}{t^3}$

73. $k = 2.4$ **75.** 5 lb   **77.** The force will be 16 times as great.

Chapter 3 Test, pp. 397–398

1. a. $f(x) = 2(x - 3)^2 - 2$ **b.** Upward   **c.** $(3, -2)$
d. $(2, 0), (4, 0)$ **e.** $(0, 16)$
f.

g. $x = 3$
h. Minimum value: -2
i. Domain: $(-\infty, \infty)$; Range: $[-2, \infty)$

$f(x) = 2x^2 - 12x + 16$

2. a. Up to the left and up to the right; As $x \rightarrow -\infty, f(x) \rightarrow \infty$, and as $x \rightarrow \infty, f(x) \rightarrow \infty$.

b. $\pm 1, \pm 3, \pm 7, \pm 21, \pm\dfrac{1}{2}, \pm\dfrac{3}{2}, \pm\dfrac{7}{2}, \pm\dfrac{21}{2}$

c. $\dfrac{7}{2}, -3$ (each multiplicity 1), and 1 (multiplicity 2)

d. $\left(\dfrac{7}{2}, 0\right), (-3, 0), (1, 0)$ **e.** $(0, -21)$ **f.** Neither even nor odd

g.

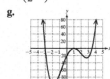

$f(x) = 2x^4 - 5x^3 - 17x^2 + 41x - 21$

3. a. $-0.25x^9$ **b.** Up to the left and down to the right; As $x \rightarrow -\infty$, $f(x) \rightarrow \infty$, and as $x \rightarrow \infty, f(x) \rightarrow -\infty$.   **c.** 0 (multiplicity 3), 2 (multiplicity 2), -1 (multiplicity 4)

4. a. 4   **b.** $2, -2, 3i, -3i$ **c.** $(2, 0)$ and $(-2, 0)$ **d.** Even

5. a. No   **b.** Yes   **c.** No   **d.** Yes

6. a. $2x^2 + 2x + 4 + \dfrac{11x - 9}{x^2 - 3x + 1}$

b. Dividend: $2x^4 - 4x^3 + x - 5$; Divisor: $x^2 - 3x + 1$; Quotient: $2x^2 + 2x + 4$; Remainder: $11x - 9$

7. a. Yes   **b.** No   **c.** No   **d.** Yes   **e.** 117

8. a. $\pm 2i, 4 \pm i$ **b.** $(x - 2i)(x + 2i)[x - (4 + i)][x - (4 - i)]$
c. $\{\pm 2i, 4 \pm i\}$

9. a. 4   **b.** $\pm 1, \pm 2, \pm 3, \pm 4, \pm 6, \pm 12, \pm\dfrac{1}{3}, \pm\dfrac{2}{3}, \pm\dfrac{4}{3}$

c. Yes   **d.** Yes

e. $\pm 1, \pm\dfrac{1}{3}, \pm\dfrac{2}{3}, \pm\dfrac{4}{3}, -2, -3$; From part (c), the value 2 itself is not a zero of $f(x)$. Likewise, from part (d), the value -4 itself is not a zero. Therefore, 2 and -4 are also eliminated from the list of possible rational zeros.

f. $\dfrac{2}{3}$ and -3 **g.** $\dfrac{2}{3}, -3, \sqrt{2}, -\sqrt{2}$

h.

$f(x) = 3x^4 + 7x^3 - 12x^2 - 14x + 12$

10. $f(x) = 15x^3 - 53x^2 - 30x + 8$

11. Positive: 3 or 1; Negative: 2 or 0

12. Vertical asymptote: $x = 7$; Slant asymptote: $y = 2x + 11$

13. Vertical asymptotes: $x = \dfrac{1}{2}, x = -\dfrac{1}{2}$; Horizontal asymptote: $y = 0$

14. Horizontal asymptote: $y = \dfrac{5}{3}$

15.

$m(x) = -\dfrac{1}{x} + 3$

16.

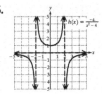

$h(x) = \dfrac{-4}{x^2 - 4}$

17.

$k(x) = \dfrac{x^2 - 2x + 1}{x}$

18. $(-4, 5)$ **19.** $(-4, 1) \cup (3, \infty)$ **20.** $(-\infty, -1] \cup [0, \infty)$

21. $\left(-\infty, -\dfrac{7}{3}\right) \cup \left(-\dfrac{7}{3}, \infty\right)$ **22.** $(-\infty, -3] \cup (2, \infty)$

23. $(-3, 3)$ **24.** $(-\infty, 0) \cup \left(\dfrac{3}{7}, 1\right)$ **25.** $E = kv^2$ **26.** $k = 4.2$

27. 294 ft^2 **28.** 178 lb   **29.** The pressure is 9 times as great.

30. a. 1000 rabbits after 1 yr, 1667 rabbits after 5 yr, and 1818 after 10 yr.
b. The rabbit population will approach 2000 as t increases.

31. a. $y(20) = 140.3$ means that with 20,000 plants per acre, the yield will be 140.3 bushels per acre; $y(30) = 172$ means that with 30,000 plants per acre, the yield will be 172 bushels per acre; $y(60) = 143.5$ means that with 60,000 plants per acre, the yield will be 143.5 bushels per acre.   **b.** 40,400   **c.** 183 bushels

32. a. $s(t) = -4.9t^2 + 98t$ **b.** 10 sec after launch   **c.** 490 m
d. $2.3 < t < 17.7$ sec

33. a. $n(a) = 0.0011a^2 - 0.027a + 2.46$
b. 12 yr   **c.** 2.3 visits per year

Chapter 3 Cumulative Review Exercises, p. 399

1. a. $x = 4, x = -4$ **b.** Horizontal asymptote: $y = 2$

2. $f(x) = x^3 - 8x^2 + 25x - 26$

3. a. Down to the left and up to the right; As $x \rightarrow -\infty, f(x) \rightarrow -\infty$, and as $x \rightarrow \infty, f(x) \rightarrow \infty$.

b. $\dfrac{5}{2}$ (multiplicity 1) and -1 (multiplicity 2)

c. $\left(\dfrac{5}{2}, 0\right)$ and $(-1, 0)$ **d.** $(0, -5)$

e.

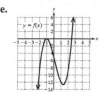

$y = f(x)$

4. $\dfrac{10}{17} + \dfrac{11}{17}i$ **5.** Center: $(-4, 7)$; Radius: 3   **6.** $y = -\dfrac{5}{2}x + 2$

7. x-intercept: $(-9, 0)$; y-intercepts: $(0, 3), (0, -3)$

8.

$y = f(x)$

9. $m = (v_0t)^2 + t$

10. a. $\left(\dfrac{3 + \sqrt{7}}{2}, 0\right), \left(\dfrac{3 - \sqrt{7}}{2}, 0\right)$ **b.** $(0, 1)$ **c.** $\left(\dfrac{3}{2}, -\dfrac{7}{2}\right)$

11. $(5x^2 - y^3)(25x^4 + 5x^2y^3 + y^6)$ **12.** $\dfrac{y^{12}}{32x^3z^6}$ **13.** $5zy^7\sqrt[3]{2z^2x}$

14. $\dfrac{1}{x + 3}$ **15.** $(10, 32]$ **16.** $[-3, 9]$ **17.** $\{-5, 1\}$

18. $\left(\dfrac{9}{8}, \infty\right)$ **19.** $(0, \infty)$ **20.** $\{5\}$

CHAPTER 4

Section 4.1 Practice Exercises, pp. 409–414

R.1. $(-\infty, -1) \cup (-1, \infty)$　　**R.2.** $(-\infty, \infty)$　　**R.3.** $\left(-\infty, \frac{1}{4}\right]$

R.4. $(n \circ p)(x) = x^2 - 3x + 1$　　**R.5.** $(p \circ n)(x) = x^2 - x - 2$
R.6. $(p \circ p)(x) = x^4 - 6x^3 + 6x^2 + 9x$

1. $\{(2, 1), (3, 2), (4, 3)\}$　　**3.** one, to, one　　**5.** x, x

7. Yes　　**9.** No　　**11.** No　　**13.** No　　**15.** Yes　　**17.** No

19. No　　**21.** No

23. Yes; If $f(a) = f(b)$, then $4a - 7 = 4b - 7$, which implies that $a = b$.

25. Yes; If $g(a) = g(b)$, then $a^3 + 8 = b^3 + 8$, which implies that $a = b$.

27. No; For example the points $(1, -3)$ and $(-1, -3)$ have the same y values but different x values. That is, $m(a) = m(b) = -3$, but $a \neq b$.

29. No; For example, the points $(2, 3)$ and $(-4, 3)$ have the same y values but different x values. That is, $p(a) = p(b) = 3$, but $a \neq b$.

31. Yes　　**33.** No　　**35.** Yes

37. a. Yes　　**b.** The value $g(x)$ represents the number of years since the year 2010 based on the number of applicants to the freshman class, x.

39. a. If $f(a) = f(b)$, then $2a - 3 = 2b - 3$, which implies that $a = b$. The function is one-to-one.

b. $f^{-1}(x) = \dfrac{x + 3}{2}$　　**c.**

41. $f^{-1}(x) = 4 - 9x$　　**43.** $h^{-1}(x) = x^3 + 5$

45. $m^{-1}(x) = \sqrt[3]{\dfrac{x - 2}{4}}$　　**47.** $c^{-1}(x) = \dfrac{5 - 2x}{x}$

49. $t^{-1}(x) = -\dfrac{2x + 4}{x - 1}$　　**51.** $f^{-1}(x) = \sqrt[3]{b(x + c)} + a$

53. a.

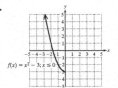

b. Yes　　**c.** $(-\infty, 0]$
d. $[-3, \infty)$
e. $f^{-1}(x) = -\sqrt{x + 3}$
f.

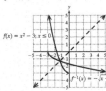

g. $[-3, \infty)$　　**h.** $(-\infty, 0]$

55. a.

b. Yes　　**c.** $[-1, \infty)$
d. $[0, \infty)$
e. $f^{-1}(x) = x^2 - 1; x \geq 0$
f. The range of f is $[0, \infty)$. Therefore, the domain of f^{-1} must be $[0, \infty)$.
g.

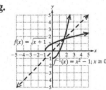

h. $[0, \infty)$　　**i.** $[-1, \infty)$

57. Domain: $[0, 4)$; Range: $[0, \infty)$　　**59.** $f^{-1}(x) = 3 - x; x \geq 3$

61. subtracts, $x - 6$　　**63.** $f^{-1}(x) = \dfrac{x + 4}{7}$　　**65.** $f^{-1}(x) = \sqrt[3]{x - 20}$

67. $f^{-1}(x) = \dfrac{x - 1}{8}$　　**69.** $q^{-1}(x) = (x - 1)^5 + 4$

71.

73.

75. a. 12　　**b.** 0.5　　**c.** 10　　**77.** True

79. False. The range of a one-to-one function is the same as the domain of its inverse.

81. a. 50 mph　　**b.** $w^{-1}(x) = \dfrac{x - 1220}{-1.17}$; The inverse gives the barometric pressure $w^{-1}(x)$ for a given wind speed x.　　**c.** 957 mb

83. a. $T(x) = 6.33x$　　**b.** $T^{-1}(x) = \dfrac{x}{6.33}$　　**c.** $T^{-1}(x)$ represents the mass of a mammal based on the amount of air inhaled per breath, x.
d. $T^{-1}(170) = 27$ means that a mammal that inhales 170 mL of air per breath during normal respiration is approximately 27 kg (this is approximately 60 lb—the size of a Labrador retriever).

85. a. $T(x) = 24x + 108$　　**b.** $T^{-1}(x) = \dfrac{x - 108}{24}$　　**c.** $T^{-1}(x)$ represents the taxable value of a home (in $1000) based on x dollars of property tax paid on the home.　　**d.** $T^{-1}(2988) = 120$ means that if a homeowner is charged $2998 in property taxes, then the taxable value of the home is $120,000.

87. The domain and range of a function and its inverse are reversed.

89. If a horizontal line intersects the graph of a function in more than one point, then the function has at least two ordered pairs with the same y-coordinate but different x-coordinates. This conflicts with the definition of a one-to-one function.

91. a. $f(8) = 3$　　**b.** $f(32) = 5$　　**c.** $f(2) = 1$　　**d.** $f\left(\frac{1}{8}\right) = -3$

93. Let f be an increasing function. Then for every value a and b in the domain of f such that $a < b$ we have $f(a) < f(b)$. Now if $u \neq v$, then either $u < v$ or $v < u$. Then either $f(u) < f(v)$ or $f(v) < f(u)$. In either case, $f(u) \neq f(v)$, and f is one-to-one.

Section 4.2 Practice Exercises, pp. 422–427

R.1. Domain $(-\infty, 3)$, Range $(-\infty, 0]$　　**R.3.**

R.2. a. -2　　**b.** $-\infty$
c. -2　　**d.** ∞
e. Never decreasing
f. $(-\infty, -1) \cup (-1, \infty)$

1. is not; is　　**3.** decreasing　　**5.** $(0, \infty)$　　**7.** $y = 0$

9. a. 0.2　　**b.** 2264.9364　　**c.** 9.7385　　**d.** 156.9925

11. a. 64　　**b.** 0.1436　　**c.** 0.0906　　**d.** 0.1520　　**13.** a, d

15. Domain: $(-\infty, \infty)$;　　**17.** Domain: $(-\infty, \infty)$;
Range: $(0, \infty)$　　　　　　　Range: $(0, \infty)$

19. Domain: $(-\infty, \infty)$;　　**21.** Domain: $(-\infty, \infty)$;
Range: $(0, \infty)$　　　　　　　Range: $(0, \infty)$

23. a.

25. a.

b. Domain: $(-\infty, \infty)$;　　**b.** Domain: $(-\infty, \infty)$;
Range: $(2, \infty)$　　　　　　　Range: $(0, \infty)$
c. $y = 2$　　　　　　　　　　**c.** $y = 0$

27. a.

b. Domain: $(-\infty, \infty)$;
Range: $(-1, \infty)$
c. $y = -1$

29. a.

b. Domain: $(-\infty, \infty)$;
Range: $(-\infty, 0)$
c. $y = 0$

31. a.

b. Domain: $(-\infty, \infty)$;
Range: $(0, \infty)$
c. $y = 0$

33. a.

b. Domain: $(-\infty, \infty)$;
Range: $(-3, \infty)$
c. $y = -3$

35. a.

b. Domain: $(-\infty, \infty)$;
Range: $(-\infty, 2)$
c. $y = 2$

37. a. 54.5982   **b.** 0.0408
c. 36.8020   **d.** 23.1407

39. a.

b. Domain: $(-\infty, \infty)$;
Range: $(0, \infty)$
c. $y = 0$

41. a.

b. Domain: $(-\infty, \infty)$;
Range: $(2, \infty)$
c. $y = 2$

43. a.

b. Domain: $(-\infty, \infty)$;
Range: $(-\infty, -3)$
c. $y = -3$

45. a. 1, \$12,166.53
b. 4, \$12,201.90
c. 12, \$12,209.97
d. 365, \$12,213.89
e. n/a, \$12,214.03

47. a. \$26,997.18   **b.** \$29,836.49   **c.** \$34,665.06
49. a. \$2200   **b.** \$2214.03   **c.** 5.5% simple interest results in less
interest.   **51.** 3.8% compounded continuously for 30 yr results in
more interest.
53. a. $A(28.9) = 5$ means that after 28.9 yr, the amount of ^{90}Sr remaining
is 5 µg. After one half-life, the amount of substance has been
halved.   **b.** $A(57.8) = 2.5$ means that after 57.8 yr, the amount
of ^{90}Sr remaining is 2.5 µg. After two half-lives, the amount of
substance has been halved, twice.   **c.** $A(100) = 0.909$ means that
after 100 yr, the amount of ^{90}Sr remaining is approximately 0.909 µg.
55. a. Increasing   **b.** $P(0) = 310$ means that in the year 2010, the
U.S. population was approximately 310 million. This is the initial
population in 2010.   **c.** $P(10) = 341$ means that in the year 2020,
the U.S. population will be approximately 341 million if this trend
continues.   **d.** $P(20) = 376; P(30) = 414$ **e.** $P(200) = 2137$;
In the year 2210 the U.S population will be approximately 2.137 billion.
The model cannot continue indefinitely because the population will
become too large to be sustained from the available resources.

57. a. 760 mmHg   **b.** 241 mmHg
59. a. $T(t) = 78 + 272e^{-0.046t}$ **b.** 250°F   **c.** Yes; after 60 min, the
cake will be approximately 95.2°F.
61. a. \$39,000   **b.** It costs the farmer \$84,800 to run the tractor for
800 hr during the first year.
63. a. {2}   **b.** {3}   **c.** {4}   **d.** x is between 2 and 3.
e. x is between 3 and 4.
65. a. and **d.**

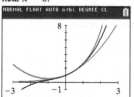

b. Yes   **c.** Domain: $(-\infty, \infty)$;
Range: $(0, \infty)$
e. Domain: $(0, \infty)$; Range: $(-\infty, \infty)$
f. $f^{-1}(1) = 0; f^{-1}(2) = 1$;
$f^{-1}(4) = 2$

67. a. ∞ **b.** 0   **c.** ∞ **d.** $-\infty$
69. The range of an exponential function is the set of positive real
numbers; that is, 2^x is nonnegative for all values of x in the domain.
71. {0, 2}   **73. a.** e^{x+h} **b.** e^{2x} **c.** e^{x-h} **d.** 1   **e.** $\dfrac{1}{e^{2x}}$

75. $e^{2x} + 2 + e^{-2x}$ or $\dfrac{e^{4x} + 2e^{2x} + 1}{e^{2x}}$

77. $\left(\dfrac{e^x + e^{-x}}{2}\right)^2 - \left(\dfrac{e^x - e^{-x}}{2}\right)^2$

$= \dfrac{1}{4}[(e^{2x} + 2 + e^{-2x}) - (e^{2x} - 2 + e^{-2x})]$

$= \dfrac{1}{4}(4) = 1$

79. $\dfrac{e^x(e^h - 1)}{h}$

81. The graphs of Y_2 and Y_3 are close approximations of $Y_1 = e^x$
near $x = 0$.

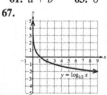

Section 4.3 Practice Exercises, pp. 437–441

R.1. $\dfrac{1}{27}$ **R.2.** 16   **R.3.** 4   **R.4.** 5

R.5. $(-\infty, -15] \cup [-5, \infty)$ **R.6.** $(-\infty, 4) \cup (4, \infty)$
1. logarithmic   **3.** common, natural   **5.** 0; 0   **7.** x, x

9. $8^2 = 64$ **11.** $10^{-4} = \dfrac{1}{10,000}$ **13.** $e^0 = 1$ **15.** $a^c = b$

17. $\log_5 125 = 3$ **19.** $\log_{1/5} 125 = -3$
21. $\log 1,000,000,000 = 9$ **23.** $\log_a b = 7$
25. 2   **27.** 1   **29.** 8   **31.** -4 **33.** -1 **35.** 6
37. -3 **39.** -2 **41.** 5   **43.** -5 **45.** -2
47. $\dfrac{1}{5}$ **49.** $-\dfrac{1}{2}$ **51. a.** Between 4 and 5; 4.6705
b. Between 6 and 7; 6.0960   **c.** Between -1 and 0; -0.6198
d. Between -6 and -5; -5.4949 **e.** Between 5 and 6; 5.7482
f. Between -3 and -2; -2.2924
53. a. 4.5433   **b.** -1.7037 **c.** 2.5217   **d.** 2.5310
e. 22.0842   **f.** -7.2502
55. 11   **57.** 1   **59.** $x + y$ **61.** $a + b$ **63.** 0
65.   **67.**

69.

71. a.

    b. Domain: $(-2, \infty)$;
      Range: $(-\infty, \infty)$
    c. $x = -2$

73. a.

75. a.

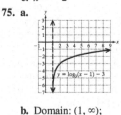

b. Domain: $(0, \infty)$;
   Range: $(-\infty, \infty)$
c. $x = 0$

b. Domain: $(1, \infty)$;
   Range: $(-\infty, \infty)$
c. $x = 1$

77. a.

b. Domain: $(0, \infty)$;
   Range: $(-\infty, \infty)$
c. $x = 0$

79. $(-\infty, 8)$
81. $\left(-\dfrac{7}{6}, \infty\right)$
83. $(-\infty, \infty)$
85. $(-\infty, -4) \cup (4, \infty)$
87. $(-\infty, 11)$
89. $(-\infty, -3) \cup (4, \infty)$
91. $(-\infty, 4) \cup (4, \infty)$

93. a. 6.9   **b.** 3.2   **c.** Approximately 5012 times more intense
95. a. 150 dB   **b.** 90 dB   **c.** 1,000,000 times more intense
97. a. 2.3   **b.** 2   **c.** Lemon juice is more acidic.
99. a. $3^4 = x + 1$ **b.** $\{80\}$ **c.** $\log_3(80 + 1) = \log_3 81 = 4$ ✓
101. a. $4^3 = 7x - 6$ **b.** $\{10\}$ **c.** $\log_4(7 \cdot 10 - 6) = \log_4 64 = 3$ ✓
103. 1   **105.** $-\dfrac{1}{2}$ **107. a.** 3   **b.** 3   **c.** They are the same.
109. a. 2   **b.** 2   **c.** They are the same.
111. a. 5   **b.** 5   **c.** They are the same.
113. a. 19.8 yr   **b.** $t(0.04) = 17.3$; $t(0.06) = 11.6$; $t(0.08) = 8.7$
115. a. 23.7797   **b.** -33.1787 **c.** Given a number $a \times 10^n$, $\log(a \times 10^n)$ is between n and $n + 1$, inclusive.
117. $(-\infty, 1) \cup (3, \infty)$ **119.** $(-4, \infty)$ **121.** $(6, \infty)$
123. a. The graphs match closely on the interval $(0, 2)$.
    b. $\ln 1.5 \approx 0.4010$

125. The graphs are the same.

Problem Recognition Exercises, p. 442

1. a. $(-\infty, \infty)$ **b.** $\{3\}$ **c.** No x-intercept   **d.** $(0, 3)$
    e. No asymptotes   **f.** Never increasing   **g.** Never decreasing
    h. Graph E   **2. a.** $(-\infty, \infty)$ **b.** $(-\infty, \infty)$ **c.** $\left(\dfrac{3}{2}, 0\right)$
    d. $(0, -3)$ **e.** No asymptotes   **f.** $(-\infty, \infty)$
    g. Never decreasing   **h.** Graph G
3. a. $(-\infty, \infty)$ **b.** $[-4, \infty)$ **c.** $(1, 0)$ and $(5, 0)$ **d.** $(0, 5)$
    e. No asymptotes   **f.** $(3, \infty)$ **g.** $(-\infty, 3)$ **h.** Graph N
4. a. $(-\infty, \infty)$ **b.** $(-\infty, \infty)$ **c.** $(2, 0)$ **d.** $\left(0, -\sqrt[3]{2}\right)$
    e. No asymptotes   **f.** $(-\infty, \infty)$ **g.** Never decreasing
    h. Graph B

5. a. $(-\infty, 1) \cup (1, \infty)$ **b.** $(-\infty, 0) \cup (0, \infty)$ **c.** None
    d. $(0, -2)$ **e.** Vertical asymptote: $x = 1$; Horizontal asymptote: $y = 0$ **f.** Never increasing   **g.** $(-\infty, 1) \cup (1, \infty)$
    h. Graph L
6. a. $(-\infty, -2) \cup (-2, \infty)$ **b.** $(-\infty, 3) \cup (3, \infty)$ **c.** $(0, 0)$
    d. $(0, 0)$ **e.** Vertical asymptote: $x = -2$; Horizontal asymptote: $y = 3$ **f.** $(-\infty, -2) \cup (-2, \infty)$ **g.** Never decreasing
    h. Graph A
7. a. $(-\infty, \infty)$ **b.** $(0, \infty)$ **c.** No x-intercept   **d.** $(0, 1)$
    e. Horizontal asymptote: $y = 0$ **f.** $(-\infty, \infty)$
    g. Never decreasing   **h.** Graph M
8. a. $(-\infty, \infty)$ **b.** $(-\infty, 0]$ **c.** $(-3, 0)$ **d.** $(0, -9)$
    e. No asymptotes   **f.** $(-\infty, -3)$ **g.** $(-3, \infty)$ **h.** Graph C
9. a. $(-\infty, \infty)$ **b.** $[-1, \infty)$ **c.** $(3, 0)$ and $(5, 0)$ **d.** $(0, 3)$
    e. No asymptotes   **f.** $(4, \infty)$ **g.** $(-\infty, 4)$ **h.** Graph I
10. a. $(-\infty, \infty)$ **b.** $(-\infty, 3]$ **c.** $(-3, 0)$ and $(3, 0)$ **d.** $(0, 3)$
    e. No asymptotes   **f.** $(-\infty, 0)$ **g.** $(0, \infty)$ **h.** Graph D
11. a. $(-\infty, 3]$ **b.** $[0, \infty)$ **c.** $(3, 0)$ **d.** $\left(0, \sqrt{3}\right)$
    e. No asymptotes **f.** Never increasing **g.** $(-\infty, 3)$ **h.** Graph F
12. a. $[3, \infty)$ **b.** $[0, \infty)$ **c.** $(3, 0)$ **d.** No y-intercept
    e. No asymptotes   **f.** $(3, \infty)$ **g.** Never decreasing   **h.** Graph K
13. a. $(-\infty, \infty)$ **b.** $(2, \infty)$ **c.** No x-intercept   **d.** $(0, 3)$
    e. Horizontal asymptote: $y = 2$ **f.** $(-\infty, \infty)$ **g.** Never decreasing
    h. Graph H
14. a. $(-2, \infty)$ **b.** $(-\infty, \infty)$ **c.** $(-1, 0)$ **d.** $(0, \ln 2)$
    e. Vertical asymptote: $x = -2$ **f.** $(-2, \infty)$ **g.** Never decreasing
    h. Graph J

Section 4.4 Practice Exercises, pp. 449–452

R.1. x^9 **R.2.** y^5 **R.3.** $\dfrac{16z^8}{w^6}$ **R.4.** $\dfrac{n^3}{343k^{12}}$

1. $\log_b x + \log_b y$ **3.** $p \log_b x$ **5.** $10, e$
7. $3 + \log_5 z$ **9.** $\log 8 + \log c + \log d$ **11.** $\log_2(x + y) + \log_2 z$
13. $\log_{12} p - \log_{12} q$ **15.** $1 - \ln 5$ **17.** $\log(m^2 + n) - 2$
19. $4 \log(2x - 3)$ **21.** $\dfrac{3}{7}\log_6 x$ **23.** $kt \ln 2$
25. $\log_4 7 + \log_4 y + \log_4 z$ **27.** $-1 + \log_7 m + 2\log_7 n$
29. $10 \log_2 x - \log_2 y - \log_2 z$ **31.** $5 \log_6 p - \log_6 q - 3 \log_6 t$
33. $1 - \dfrac{1}{2}\log(a^2 + b^2)$ **35.** $\dfrac{1}{3}\ln x + \dfrac{1}{3}\ln y - \ln w - 2 \ln z$
37. $\dfrac{1}{4}\ln(a^2 + 4) - \dfrac{3}{4}$
39. $\log 2 + \log x + 8 \log(x^2 + 3) - \dfrac{1}{2}\log(4 - 3x)$
41. $\dfrac{1}{3}\log_5 x + \dfrac{1}{6}$
43. $2 + 2 \log_2 a + \dfrac{1}{2}\log_2(3 - b) - \log_2 c - 2 \log_2(b + 4)$
45. $\ln 4y$ **47.** 1   **49.** 2   **51.** 1   **53.** $\log_2(x^2 t)$
55. $\log_8\left(\dfrac{m^4}{n^3 p^2}\right)$ **57.** $\ln\left(\dfrac{x}{x^2 - 9}\right)^3$ **59.** $\ln\sqrt{\dfrac{x + 1}{x - 1}}$
61. $\log\left(\dfrac{x^6}{\sqrt[3]{yz^2}}\right)$ **63.** $\log_4[\sqrt[3]{p}(q + 4)]$ **65.** $\ln\left[\dfrac{(x + 2)^3}{\sqrt{x}}\right]$
67. $\log(8y - 7)$ **69.** 1.392   **71.** 2.26   **73.** 2.01   **75.** 1.036
77. 2.366   **79. a.** Between 3 and 4   **b.** 3.9069   **c.** $2^{3.9069} \approx 15$
81. a. Between 0 and 1   **b.** 0.6826   **c.** $5^{0.6826} \approx 3$
83. a. Between -2 and -1 **b.** -1.7370 **c.** $2^{-1.7370} \approx 0.3$
85. 25.4800; $2^{25.4800} \approx 4.68 \times 10^7$
87. -8.7128; $4^{-8.7128} \approx 5.68 \times 10^{-6}$
89. True   **91.** False; $\log_5\left(\dfrac{1}{125}\right) \neq \dfrac{1}{\log_5 125}$ (The left side is -3 and the right side is $\tfrac{1}{3}$.)
93. True   **95.** False; $\log(10 \cdot 10) \neq (\log 10)(\log 10)$ (The left side is 2 and the right side is 1.)   **97.** True

99. The given statement $\log_5(-5) + \log_5(-25)$ is not defined because the arguments to the logarithmic expressions are not positive real numbers.

101. a. $\dfrac{\ln(x + h) - \ln x}{h}$ **b.** $\dfrac{1}{h}[\ln(x + h) - \ln x] = \dfrac{1}{h}\ln\left(\dfrac{x + h}{x}\right)$

$= \ln\left(\dfrac{x + h}{x}\right)^{1/h}$

103. $\log\left(\dfrac{-b + \sqrt{b^2 - 4ac}}{2a}\right) + \log\left(\dfrac{-b - \sqrt{b^2 - 4ac}}{2a}\right)$

$= \log\left(\dfrac{-b + \sqrt{b^2 - 4ac}}{2a} \cdot \dfrac{-b - \sqrt{b^2 - 4ac}}{2a}\right)$

$= \log\left[\dfrac{b^2 - (b^2 - 4ac)}{4a^2}\right]$

$= \log\left(\dfrac{4ac}{4a^2}\right)$

$= \log\left(\dfrac{c}{a}\right) = \log c - \log a$

105. $\log_2 9$

107. Let $M = \log_b x$ and $N = \log_b y$, which implies that $b^M = x$ and $b^N = y$. Then $\dfrac{x}{y} = \dfrac{b^M}{b^N} = b^{M-N}$. Writing the expression $\dfrac{x}{y} = b^{M-N}$ in logarithmic form, we have $\log_b\left(\dfrac{x}{y}\right) = M - N$, or equivalently, $\log_b\left(\dfrac{x}{y}\right) = \log_b x - \log_b y$ as desired.

109. 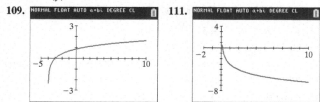 **111.**

113. a. The graphs are the same.

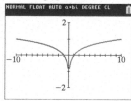

b. $\dfrac{1}{2}\log x^2 = \log(x^2)^{1/2} = \log\sqrt{x^2} = \log|x|$

Section 4.5 Practice Exercises, pp. 462–466

R.1. $\{0\}$ **R.2.** $\left\{-\dfrac{5}{3}, -\dfrac{2}{5}\right\}$ **R.3.** $\left\{\dfrac{5 \pm \sqrt{61}}{2}\right\}$

R.4. $\{\ \}$; The value 2 does not check.   **R.5.** $\{7\}$; The value -2 does not check.   **R.6.** $\{-9, 4\}$

1. exponential   **3.** x; y

5. $\{4\}$ **7.** $\left\{\dfrac{1}{3}\right\}$ **9.** $\{-1\}$ **11.** $\{1\}$ **13.** $\{-15\}$

15. $\left\{\dfrac{19}{9}\right\}$ **17.** $\left\{\dfrac{\ln 87}{\ln 6}\right\}$; $t \approx 2.4925$

19. $\left\{\dfrac{\ln 1020}{\ln 19}\right\}$; $x \approx 2.3528$ **21.** $\left\{\dfrac{\log 128{,}100 - 3}{4}\right\}$; $x \approx 0.5269$

23. $\left\{\dfrac{\ln 3}{0.2}\right\}$ or $\{5\ln 3\}$; $t \approx 5.4931$ **25.** $\left\{\dfrac{5 + \ln 2}{2}\right\}$; $n \approx 2.8466$

27. $\left\{\dfrac{5\ln 3}{2\ln 5 - 6\ln 3}\right\}$; $x \approx -1.6286$

29. $\left\{\dfrac{\ln 2 - 4\ln 7}{3\ln 7 + 6\ln 2}\right\}$; $x \approx -0.7093$ **31.** $\{\ln 11\}$; $x \approx 2.3979$

33. $\{\ \}$

35. a. No   **b.** Yes   **c.** No   **37.** $\{-2\}$ **39.** $\{-9, 2\}$

41. $\{32\}$ **43.** $\{23\}$ **45.** $\{10^{4.1} - 17\}$; $p \approx 12{,}572.2541$

47. $\left\{\dfrac{4 - e^3}{3}\right\}$; $t \approx -5.3618$ **49.** $\{2\}$; The value -4 does not check.

51. $\{32\}$ **53.** $\{25\}$; The value -5 does not check.

55. $\{5\}$; The value 2 does not check.

57. $\{\ \}$; The values 5 and 3 do not check.

59. $\{-2\}$; The value 7 does not check.

61. 20 yr   **63.** 4%   **65.** 5 yr   **67.** 5%   **69.** 10 yr, 2 months

71. a. 72 mCi   **b.** 5.3 days

73. Ocean: 14.1 m; Tahoe: 8.7 m; Erie: 3.5 m   **75.** Ocean: 93.8 m

77. 69 min (1 hr 9 min)

79. a. 3.4×10^{-8} W/m^2 corresponds to 45.3 dB which indicates a moderate hearing impairment.   **b.** 10^{-9} W/m^2

81. a. 3.16×10^{-9} mol/L   **b.** 5.01×10^{-3} mol/L   **83.** 4 months

85. a. 11,000   **b.** 32

87. $f^{-1}(x) = \log_2(x + 7)$ **89.** $f^{-1}(x) = e^x - 5$

91. $f^{-1}(x) = \log(x - 1) + 3$ **93.** $f^{-1}(x) = 10^{x+9} - 7$ **95.** $\{-3, 3\}$

97. $\{900\}$ **99.** $\{2, -2\}$ **101.** $\{-10, 2\}$ **103.** $\{3, -3\}$

105. $\{3\}$ **107.** $\left\{e^4, \dfrac{1}{e^4}\right\}$; $x \approx 54.5982$, $x \approx 0.0183$

109. $\{\ \}$; The value -4 does not check.

111. $\left\{\ln\dfrac{1}{2}, \ln 4\right\}$; $x \approx -0.6931$, $x \approx 1.3863$

113. Take a logarithm of any base b on each side of the equation. Then apply the power property of logarithms to write the product of x and the $\log_b 4$. Finally divide both sides by $\log_b 4$.

115. $\{\log 13\}$; $x \approx 1.1139$ **117.** $\{e, e^4\}$; $x \approx 2.7183$, $x \approx 54.5982$

119. $\{100, 1\}$ **121.** $\{10{,}000\}$

123. $\{\ln(4 \pm \sqrt{10})\}$; $x \approx 1.9688$, $x \approx -0.1771$

125. $\left\{\dfrac{5}{3}\right\}$; The value $-\dfrac{5}{2}$ does not check.

127. $\{-1.4408, 2.8584\}$ **129.** $\{2.0960\}$

Section 4.6 Practice Exercises, pp. 476–482

R.1. $L = \dfrac{3P}{m}$ **R.2.** $m = \sqrt[3]{\dfrac{4Q}{3}}$ **R.3.** $b = \pm\sqrt{a^2 - (c - 9)^2}$

R.4 a. 16   **b.** 3   **c.** 0   **d.** 7

1. growth; decay   **3.** logistic   **5.** $k = -\dfrac{\ln\left(\dfrac{Q}{Q_0}\right)}{t}$ or $\dfrac{\ln Q_0 - \ln Q}{t}$

7. $D = 10^{(M-8.8)/5.1}$ **9.** $[H^+] = 10^{-pH}$

11. $t = \dfrac{\ln\left(\dfrac{A}{P}\right)}{\ln(1 + r)}$ or $\dfrac{\ln A - \ln P}{\ln(1 + r)}$ **13.** $k = Ae^{-E/(RT)}$

15. a. 4.4%   **b.** 11.6 yr   **17. a.** $8000   **b.** 7 yr

19. a. $P(t) = 19e^{0.01735t}$; $P(t) = 22.9e^{0.00343t}$ **b.** Australia: 26.9 million; Taiwan: 24.5 million   **c.** The population growth rate for Australia is greater.   **d.** Australia: 2026; Taiwan: 2078

21. a. $P(t) = 4.3e^{0.01341t}$; $P(t) = 4.6e^{0.00618t}$; 4.3 million; 4.6 million   **b.** Costa Rica: 2011; Norway: 2013   **c.** The population growth rate for Costa Rica is greater.

23. 2053 yr   **25. a.** $Q(t) = 2e^{-0.0079t}$ **b.** 28 yr

27. a. $Q(t) = 300e^{-0.0063t}$ **b.** 110 min

29. a. $P(t) = 2{,}000{,}000e^{-0.1155t}$ **b.** $P(0) = 2{,}000{,}000$; $P(6) = 1{,}000{,}000$; $P(12) = 500{,}000$; $P(60) = 1953$

31. a. $P(0) = 78$ means that on January 1, 1900, the U.S. population was approximately 78 million.   **b.** 338 million   **c.** 427 million   **d.** 2076   **e.** 0   **f.** 725 million

33. a. 150,000   **b.** 2,000,000   **c.** 3.3 months   **d.** 2,400,000

35. exponential   **37.** logarithmic

39. a. exponential

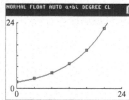

b. $y = 2.3(1.12)^x$

41. a. linear

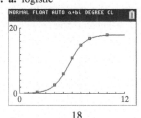

b. $y = 2.28x - 4.08$

43. a. logarithmic

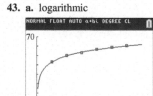

b. $y = 20.7 + 9.72 \ln x$

45. a. logistic

b. $y = \dfrac{18}{1 + 496e^{-1.1x}}$

47. a. $y = 34.9(2.134)^t$ **b.** $y = 34.9e^{0.758t}$ **c.** 15,000 cases
   d. No, eventually the number of cases would exceed the human
   population.   **e.** $y = \dfrac{11{,}731}{1 + 205e^{-0.67t}}$ **f.** 6000 cases

49. a. $H(t) = 4.86 + 6.35 \ln t$ **b.** 24 yr   **c.** No, the tree will
   eventually die.

51. a. $y = -2920t + 29{,}200$ **b.** $y = 29{,}200(0.8)^t$
   c. \$14,600 and \$0   **d.** \$9568 and \$3135

53. A visual representation of the data can be helpful in determining the
   type of equation or function that best models the data.

55. An exponential growth model has unbounded growth, whereas a
   logistic growth model imposes a limiting value on the dependent
   variable. That is, a logistic growth model has an upper bound
   restricting the amount of growth.

57. a. $t = -\dfrac{\ln\left(1 - \dfrac{Ar}{12P}\right)}{12\ln\left(1 + \dfrac{r}{12}\right)}$ **b.** This represents the amount of time
   (in yr) required to completely pay off
   a loan of A dollars at interest rate r,
   by paying P dollars per month.

Chapter 4 Review Exercises, pp. 485–487

1. No

3. Yes; If $f(a) = f(b)$, then $a^3 - 1 = b^3 - 1$, which implies that $a = b$.

5. Yes, because $(f \circ g)(x) = (g \circ f)(x) = x$ **7.** $f^{-1}(x) = \sqrt[3]{\dfrac{x+5}{2}}$

9. a.

b. Yes   **c.** $(-\infty, 0]$ **d.** $[-9, \infty)$
   e. $f^{-1}(x) = -\sqrt{x+9}$

f.

g. $[-9, \infty)$
h. $(-\infty, 0]$

11. a. $f^{-1}(x) = \dfrac{x}{5280}$

b. f^{-1} represents the conversion from x feet to $f^{-1}(x)$ miles.   **c.** 4.2 mi

13. a.

b. $(-\infty, \infty)$ **c.** $(0, \infty)$
d. $y = 0$

15. a.

b. $(-\infty, \infty)$ **c.** $(-\infty, 1)$
d. $y = 1$

17. Increasing   **19. a.** \$3456   **b.** \$3563.16
   c. 7.2% simple interest results in less interest.

21. $b^4 = x^2 + y^2$ **23.** $\log 1{,}000{,}000 = 6$ **25.** 4   **27.** -6

29. 0   **31.** 7   **33.** $(4, \infty)$ **35.** $(-\infty, \infty)$ **37.** $(-\infty, 4) \cup (4, \infty)$

39. a.

b. $(0, \infty)$ **c.** $(-\infty, \infty)$
d. $x = 0$

41. pH ≈ 4.5; acidic   **43.** 1   **45.** x **47.** $\log_b x - \log_b y$

49. $2 - \dfrac{1}{2}\log(c^2 + 10)$ **51.** $\dfrac{1}{3}\ln a + \dfrac{2}{3}\ln b - \ln c - 5\ln d$

53. $\log_5\left(\dfrac{y^4\sqrt{z}}{x^3}\right)$ **55.** $\ln\sqrt[4]{x+3}$ **57.** 1.587

59. 3.2839; $7^{3.2839} \approx 596$ **61.** $\{5\}$ **63.** $\left\{\dfrac{\ln 51}{\ln 7}\right\}$; $x \approx 2.0206$

65. $\left\{\dfrac{\ln 3}{3\ln 4 - 2\ln 3}\right\}$; $x \approx 0.5600$

67. $\left\{\dfrac{\ln\left(\frac{2.989}{400}\right)}{-2}\right\}$ or $\left\{\dfrac{\ln 400 - \ln 2.989}{2}\right\}$; $t \approx 2.4483$

69. $\{\ln 8\}$; $x \approx 2.0794$ **71.** $\{-1\}$ **73.** $\{-4\}$

75. $\{e^{2.1} + 8\}$; $n \approx 16.1662$ **77.** $\{\ \}$; The value $-\frac{22}{3}$ does not check.

79. $\{32\}$ **81.** $f^{-1}(x) = \log_4 x$ **83. a.** 1.39 m; murky   **b.** 9.2 m

85. $t = -\dfrac{1}{k}\ln\left(\dfrac{T - T_f}{T_0}\right)$ or $\dfrac{1}{k}\left[\ln T_0 - \ln(T - T_f)\right]$

87. a. Decreasing   **b.** 32 yr   **89.** 2800 yr

91. a. $Y_1 = 2.38(1.5)^x$ **b.**

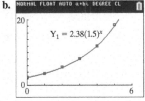

Chapter 4 Test, pp. 487–488

1. a. $f^{-1}(x) = \sqrt[3]{\dfrac{x+1}{4}}$

b. $(f \circ f^{-1})(x) = 4\left(\sqrt[3]{\dfrac{x+1}{4}}\right)^3 - 1 = 4\left(\dfrac{x+1}{4}\right) - 1 = x + 1 - 1 = x$

$(f^{-1} \circ f)(x) = \sqrt[3]{\dfrac{4x^3 - 1 + 1}{4}} = \sqrt[3]{\dfrac{4x^3}{4}} = \sqrt[3]{x^3} = x$

2. a. Yes   **b.**

3. $f^{-1}(x) = \dfrac{4x + 3}{x - 1}$

4. a. Domain: $(-\infty, 0]$; Range: $(-\infty, 1]$ **b.** $f^{-1}(x) = -\sqrt{1 - x}$
    c. Domain: $(-\infty, 1]$; Range: $(-\infty, 0]$
5. a. Domain: $(0, \infty)$; Range: $(-\infty, \infty)$ **b.** $f^{-1}(x) = 10^x$
    c. Domain: $(-\infty, \infty)$; Range: $(0, \infty)$
6. a. Domain: $(-\infty, \infty)$; Range: $(1, \infty)$ **b.** $f^{-1}(x) = \log_3(x - 1)$
    c. Domain: $(1, \infty)$; Range: $(-\infty, \infty)$
7. a. Domain: $[-5, \infty)$; Range $[0, \infty)$ **b.** $f^{-1}(x) = x^2 - 5; x \geq 0$
    c. Domain: $[0, \infty)$; Range: $[-5, \infty)$

8. a.

9. a.

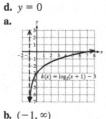

b. $(-\infty, \infty)$ **b.** $(-\infty, \infty)$
c. $(2, \infty)$ **c.** $(0, \infty)$
d. $y = 2$ **d.** $y = 0$

10. a.

11. a.

b. $(0, \infty)$ **b.** $(-1, \infty)$
c. $(-\infty, \infty)$ **c.** $(-\infty, \infty)$
d. $x = 0$ **d.** $x = -1$

12. $e^a = x + y$ **13.** -2 **14.** 3 **15.** 8 **16.** -4
17. $a^2 + b^2$ **18.** 0 **19.** $\left(-\infty, \dfrac{7}{2}\right)$ **20.** $(-\infty, -5) \cup (5, \infty)$

21. $5 \ln x + 2 \ln y - \ln w - \dfrac{1}{3}\ln z$ **22.** $\dfrac{1}{2}\log(a^2 + b^2) - 4$
23. $\log_2\left(\dfrac{a^6 \sqrt[3]{c^2}}{b^4}\right)$ **24.** $\ln\sqrt{x + 3}$ **25.** 1.783 **26.** -2.013
27. $\left\{-\dfrac{7}{3}\right\}$ **28.** $\left\{\dfrac{\ln 53}{\ln 5} - 3\right\}; x \approx -0.5331$
29. $\left\{\dfrac{7 \ln 2 - 3 \ln 3}{2 \ln 3 - \ln 2}\right\}; c \approx 1.0346$ **30.** $\left\{\dfrac{\ln 2}{4}\right\}; x \approx 0.1733$
31. $\{0\}$ **32.** $\{1\}$ **33.** $\{e^3 - 2\}; x \approx 18.0855$
34. $\{4\}$; The value -3 does not check.
35. $\{2\}$; The value -32 does not check.   **36.** $\{4\}$
37. $t = e^{(92 - S)/k} - 1$ **38.** $t = \dfrac{\ln\left(\frac{A}{P}\right)}{n \ln\left(1 + \frac{r}{n}\right)}$ or $\dfrac{\ln A - \ln P}{n \ln\left(l - \frac{r}{n}\right)}$
39. a. 6.1%   **b.** 26.4 yr
40. a. $P(t) = 10{,}000e^{0.1386t}$ **b.** Approximately 45 hr
41. a. 400 deer were present when the park service began tracking the herd.
    b. 536 deer   **c.** 680 deer   **d.** 15 yr   **e.** 0   **f.** 1200 deer
42. a. $N = 23.1(2.283)^t$ **b.**
    c. 89,000

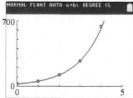

Chapter 4 Cumulative Review Exercises, p. 489

1. -3 **2.** $\dfrac{5\sqrt[3]{4x}}{2x}$ **3.** $(a - b)(a^2 + ab + b^2 - 1)$
4. 2.0×10^{10} **5.** $\left(-\infty, \dfrac{5}{2}\right] \cup \left[\dfrac{9}{2}, \infty\right)$ **6.** $\left\{\dfrac{2 \pm \sqrt{22}}{3}\right\}$
7. $\{6\}$; The value 1 does not check.   **8.** $\{9\}$ **9.** $(-5, -2) \cup (2, \infty)$
10. $\left\{\dfrac{4}{9}, -2\right\}$ **11.** $\{\pm 4, \pm 2\}$
12. $\{ \}$; The value $-\dfrac{9}{5}$ does not check.   **13.** $(-2, 4]$ **14.** $-1, -9, \pm i$
15. a. Upward   **b.** $(8, -9)$ **c.** Minimum point: $(8, -9)$
    d. Minimum value: -9 **e.** $(5, 0)$ and $(11, 0)$ **f.** $(0, 55)$
    g. $x = 8$ **h.** $(-\infty, \infty)$ **i.** $[-9, \infty)$
16.

17. a. $x = 2$ **c.**
    b. Horizontal asymptote: $y = 3$

18. a. $y = -3$ **b.** $(-\infty, \infty)$ **c.** $(-3, \infty)$
19. 3 **20.** $f^{-1}(x) = (x - 1)^3 + 4$

CHAPTER 5

Section 5.1 Practice Exercises, pp. 501–505

R.1. $\left\{-\dfrac{2}{9}\right\}$ **R.2.** $\{3\}$ **R.3.** $\{ \}$ **R.4.** $\mathbb{R}$
R.5.   **R.6.**

x-intercept: $(1, 0)$;   slope: $m = \dfrac{1}{5}$; y-intercept: $(0, -1)$
y-intercept: $(0, 4)$

1. system   **3.** substitution; addition   **5.** inconsistent
7. a. Yes   **b.** No   **9. a.** Yes   **b.** Yes   **11.** One solution
13. Infinitely many solutions; The equations are dependent.
15. $\{(-4, 3)\}$ **17.** $\{(-10, 3)\}$ **19.** $\left\{\left(-\dfrac{1}{2}, 1\right)\right\}$
21. $\{(-2, -1)\}$ **23.** $\{(1, -4)\}$ **25.** $\left\{\left(\dfrac{1}{2}, 1\right)\right\}$ **27.** $\left\{\left(\dfrac{79}{45}, \dfrac{2}{45}\right)\right\}$
29. $\{ \}$; The system is inconsistent.
31. $\{(x, -3x + 6)|x$ is any real number$\}$
    or $\left\{\left(\dfrac{6 - y}{3}, y\right)\Big| y$ is any real number$\right\}$;
    The equations are dependent.
33. $\{(0, 0)\}$
35. a. $\{(x, -5x - 6)|x$ is any real number$\}$
    or $\left\{\left(-\dfrac{y + 6}{5}, y\right)\Big| y$ is any real number$\right\}$
    b. For example: $(0, -6), (1, -11), (-2, 4)$
37. $\{(300, -100)\}$ **39.** $\left\{\left(-\dfrac{4}{41}, -\dfrac{71}{41}\right)\right\}$ **41.** $\{(6, 3)\}$ **43.** $\{ \}$
45. $\{(x, 4x - 2)|x$ is any real number$\}$
    or $\left\{\left(\dfrac{y + 2}{4}, y\right)\Big| y$ is any real number$\right\}$
47. $\{(1.6, 2.3)\}$ **49.** $\{(18, -17)\}$
51. 25 L of 36% solution and 15 L of 20% solution should be mixed.
53. 3.5 L should be replaced.   **55.** She borrowed $3500 at 4.6% and $1500 at 6.2%.   **57.** He borrowed $20,000 at 3% and $4000 at 5.5%.   **59.** Cherry has 13 g of fat and Mint Chocolate Chunk has 17 g of fat.   **61.** One makes $1200 and the other makes $1500.

63. The sidewalk moves at 1 ft/sec and Josie walks 4 ft/sec on nonmoving ground.   **65.** The speeds are 8 m/sec and 5 m/sec.

67. a. $C(x) = 52x + 480$ **b.** $R(x) = 100x$ **c.** 10 offices   **d.** The company will make money.

69. a. $\{(1600, 40)\}$ **b.** \$40   **c.** 1600 tickets

71. a.

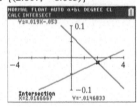

b. 20 square units

73. a. $y = \dfrac{4}{5}x + \dfrac{7}{5}$ **b.** $y = -x + 5$ **c.** $(2, 3)$

75. The angles are $32°$ and $58°$.   **77.** $x = 60°, y = 88°$

79. For example: $x + y = 2$
$$2x + y = -1$$

81. $C = 2$ and $D = 3$ **83.** $m = \dfrac{1}{3}$ and $b = -4$ **85.** $\{(\frac{1}{3}, -1)\}$

87. Marta bicycles 18 mph and runs 6 mph.   **89.** The truck was driven 144 mi in the city and 110 mi on the highway.

91. If the system represents two intersecting lines, then the lines intersect in exactly one point. The solution set consists of the ordered pair representing that point. If the lines in the system are parallel, then the lines do not intersect and the system has no solution. If the equations in a system of linear equations represent the same line, then the solution set is the set of points on the line.

93. If the system of equations reduces to a contradiction such as $0 = 1$, then the system has no solution and is said to be inconsistent.

95. $|F_1| = 50(\sqrt{3} - 1)$ lb ≈ 36.6 lb and
$|F_2| = 25\sqrt{2}(3 - \sqrt{3})$ lb ≈ 44.8 lb

97. $\{(2.017, -0.015)\}$ **99.** $\{(1.028, 15.772)\}$

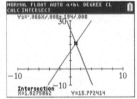

Section 5.2 Practice Exercises, pp. 514–517

R.1. \$120   **R.2.** $C = 110 + 70x$ **R.3.** $\left\{-\dfrac{264}{5}\right\}$

1. plane   **3.** For example: $(0, 0, -2), (0, 3, 0),$ and $(6, 0, 0)$
5. a. Yes   **b.** No   **7. a.** Yes   **b.** Yes
9. $\{(1, 4, -2)\}$ **11.** $\{(-2, 1, 3)\}$ **13.** $\{(1, 3, 0)\}$
15. No solution; The system is inconsistent.   **17.** $\{(\frac{1}{2}, \frac{1}{3}, -\frac{1}{6})\}$
19. Infinitely many solutions; The equations are dependent.
21. $\{(0, 0, 0)\}$ **23.** $\{(1, -2, 3)\}$ **25.** $\{(6, -8, 2)\}$
27. $\{(1, 12, 17)\}$ **29.** No solution; The system is inconsistent.
31. Infinitely many solutions; The equations are dependent.
33. $\left\{\left(x, \dfrac{-3x - 11}{2}, \dfrac{x + 7}{2}\right)\middle| x \text{ is any real number}\right\}$ or
$\left\{\left(\dfrac{-2y - 11}{3}, y, \dfrac{5 - y}{3}\right)\middle| y \text{ is any real number}\right\}$ or
$\{(2z - 7, -3z + 5, z)| z \text{ is any real number}\}$
35. $\left\{\left(x, \dfrac{1}{3}x, \dfrac{12 - 5x}{3}\right)\middle| x \text{ is any real number}\right\}$ or
$\{(3y, y, 4 - 5y)| y \text{ is any real number}\}$ or
$\left\{\left(\dfrac{12 - 3z}{5}, \dfrac{4 - z}{5}, z\right)\middle| z \text{ is any real number}\right\}$
37. For example: $(2, 1, 1), (0, 0, 2), (4, 2, 0)$ **39.** He invested \$4000 in the large cap fund, \$2000 in the real estate fund, and \$2000 in the bond fund.

41. He made eight free-throws, six 2-point shots, and two 3-point shots.   **43.** b   **45.** The sides are 15 in., 18 in., and 22 in.
47. The angles are $16°$, $24°$, and $140°$.
49. a. The slopes are 5 and -1.   **b.** $y = 2x^2 - 3x + 1$
c.

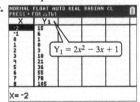

51. $y = -x^2 - 4x + 6$ **53.** $a = 4, v_0 = 18,$ and $s_0 = 10$
55. a. $y = 7x_1 + 10x_2 + 24$ **b.** \$168,000
57. The set of all ordered pairs that are solutions to a linear equation in three variables forms a plane in space.
59. Pair up two equations in the system and eliminate a variable. Choose a different pair of two equations from the system and eliminate the same variable. The result should be a system of two linear equations in two variables. Solve this system using either the substitution or addition method. Then back substitute to find the third variable.
61. $\{(2, -1, 0, 4)\}$ **63.** $\{(-13, 11, 10)\}$
65. a. $x^2 + y^2 - 4x + 6y - 12 = 0$ **b.** Center: $(2, -3)$; Radius: 5
67. $A = 5, B = 3, C = -1$

Section 5.3 Practice Exercises, pp. 524–526

R.1. $(4y - 3)(3y^2 - 4)$ **R.2.** $(8u + 5)^2$
R.3. $\dfrac{20a}{(4a + 3)(a - 3)}$ **R.4.** $5x^2 - 1 + \dfrac{3x + 2}{5x^2 - x + 4}$ **R.5.** $\{-6\}$
1. fraction decomposition   **3.** linear; $Ax + B$
5. $\dfrac{A}{x + 4} + \dfrac{B}{2x - 3}$ **7.** $\dfrac{A}{x} + \dfrac{B}{x - 2}$ **9.** $\dfrac{A}{w - 2} + \dfrac{B}{w + 3}$
11. $\dfrac{A}{x} + \dfrac{B}{x + 5} + \dfrac{C}{(x + 5)^2}$ **13.** $\dfrac{A}{2x} + \dfrac{Bx + C}{x^2 + 9}$
15. $\dfrac{Ax + B}{x^2 + 5} + \dfrac{Cx + D}{(x^2 + 5)^2}$ **17.** $\dfrac{A}{x - 4} + \dfrac{Bx + C}{x^2 + x + 4}$
19. $\dfrac{A}{x} + \dfrac{B}{x + 2} + \dfrac{C}{(x + 2)^2} + \dfrac{D}{(x + 2)^3} + \dfrac{Ex + F}{x^2 + 2x + 7} + \dfrac{Gx + H}{(x^2 + 2x + 7)^2}$
21. $\dfrac{3}{x + 4} + \dfrac{-7}{2x - 3}$ **23.** $\dfrac{5}{x} + \dfrac{3}{x - 2}$ **25.** $\dfrac{1}{w - 2} + \dfrac{5}{w + 3}$
27. $\dfrac{4}{x} + \dfrac{-3}{x + 5} + \dfrac{1}{(x + 5)^2}$ **29.** $\dfrac{5}{2x} + \dfrac{4x + 1}{x^2 + 9}$
31. $x - 3 + \dfrac{4}{x} + \dfrac{2x - 7}{x^2 + 7}$ **33.** $\dfrac{2x - 1}{x^2 + 5} + \dfrac{3x}{(x^2 + 5)^2}$
35. $\dfrac{3}{x - 4} + \dfrac{2x + 1}{x^2 + x + 4}$ **37.** $\dfrac{3x + 1}{x^2 + 2} + \dfrac{x - 5}{x^2 + 3}$
39. $2x - 5 + \dfrac{4}{x + 2} + \dfrac{-3}{x - 5}$ **41.** $3x - 4 + \dfrac{4}{x + 1} + \dfrac{-5}{(x + 1)^2}$
43. a. $(x - 3)^2(x + 5)$ **b.** $\dfrac{2}{x - 3} + \dfrac{1}{(x - 3)^2} + \dfrac{-5}{x + 5}$
45. a. $(x + 2)^3$ **b.** $\dfrac{3}{x + 2} + \dfrac{-4}{(x + 2)^2} + \dfrac{1}{(x + 2)^3}$
47. Partial fraction decomposition is a procedure in which a rational expression is written as a sum of two or more simpler rational expressions.
49. A proper rational expression is a rational expression in which the degree of the numerator is less than the degree of the denominator.
51. a. $\dfrac{1}{n} - \dfrac{1}{n + 2}$
b. $\left(\dfrac{1}{1} - \dfrac{1}{3}\right) + \left(\dfrac{1}{2} - \dfrac{1}{4}\right) + \left(\dfrac{1}{3} - \dfrac{1}{5}\right) + \left(\dfrac{1}{4} - \dfrac{1}{6}\right) + \left(\dfrac{1}{5} - \dfrac{1}{7}\right) + \cdots$
c. 0   **d.** $\dfrac{3}{2}$
53. $\dfrac{1}{ax} - \dfrac{b}{a(a + bx)}$ **55.** $\dfrac{2}{e^x + 1} + \dfrac{3}{e^x + 2}$

Section 5.4 Practice Exercises, pp. 532–535

R.1.   **R.2.**

R.3.   **R.4.** $\{e^{17}\}$ **R.5.** $\{4\}$

1. nonlinear

3. a.   **5. a.**

b. $\{(2, 2), (0, -2)\}$ **b.** $\{(-3, 4), (4, -3)\}$

7. a.   **9. a.**

b. $\{(4, 2)\}$ **b.** $\{\ \}$

11. a.   **13. a.**

b. $\{(-1, -1), (0, 0), (1, 1)\}$ **b.** $\{(0, 1), (2, 5)\}$

15. $\{(2, 1), (2, -1), (-2, 1), (-2, -1)\}$ **17.** $\{(4, -1), (-4, 1)\}$
19. $\{(1, \sqrt{2}), (1, -\sqrt{2}), (-1, \sqrt{2}), (-1, -\sqrt{2})\}$ **21.** $\{\ \}$
23. $\{(3, 1), (\frac{7}{3}, \frac{5}{3})\}$ **25.** $\{(0, 5)\}$ **27.** $\{(1, 1)\}$
29. $\{(0, 9), (-3, 0), (3, 0)\}$ **31.** $\{(3, 2), (5, -2)\}$
33. $\{(-2, 0), (0, 4), (2, 0)\}$ **35.** $\{(1, \frac{1}{3}), (-1, \frac{1}{3}), (1, -\frac{1}{3}), (-1, -\frac{1}{3})\}$
37. The numbers are 5 and 7.   **39.** The numbers are 5 and 2.
41. The numbers are 12 and 10.
43. The numbers are 9 and 12 or -9 and -12.
45. The rectangle is 10 m by 8 m.   **47.** The floor is 20 ft by 12 ft.
49. The truck is 6 ft by 6 ft by 8 ft.
51. The aquarium is 24 in. by 12 in. by 16 in.
53. The legs are 4 ft and 7 ft.   **55.** The ball will hit the ground at the
point $(128\sqrt{3}, -128)$ or approximately $(221.7, -128)$.
57. A system of linear equations contains only linear equations, whereas
a nonlinear system has one or more equations that are nonlinear.
59. a. $0.69 = A_0e^{-3k}$ **b.** $0.655 = A_0e^{-4k}$ **c.** $k \approx 0.052$
   d. $A_0 \approx 0.81\ \mu g/dL$ **e.** $0.43\ \mu g/dL$
61. a. $k \approx 0.058$ **b.** The original population is 40,000.
   c. The population will reach 300,000 approximately 35 hr after the
culture is started.
63. Infinitely many solutions   **65.** $\{(10, 100)\}$ **67.** $\{(2, 1)\}$
69. $\{(2, 1), (\frac{7}{5}, \frac{6}{5})\}$ **71.** $\{(1, 3, 1), (-1, -3, -1)\}$
73. a. $(0, -5)$ and $(4, 3)$ **b.** $y = 2x - 5$ for $0 \le x \le 4$
75. $\{(2.359, 5.584)\}$
77. $\{(1.538, 6.135), (-1.538, 6.135), (3.693, -5.135), (-3.693, -5.135)\}$
79. $\{\ \}$

Section 5.5 Practice Exercises, pp. 544–546

R.1. a. $\{2\}$ **b.** $(2, \infty)$ **c.** $(-\infty, 2]$
1. linear   **3.** above; horizontal   **5.** II
7. a. Yes   **b.** No   **c.** No
9. a. No   **b.** Yes   **c.** Yes

11. a.   **b.** The bounding line would be
drawn as a dashed line.
c. The bounding line would be
dashed and the graph would be
shaded strictly below the line.

13.   **15.**

17.   **19.**

21.   **23.**

25. a.   **b.** The region outside the circle would be
shaded.   **c.** The shaded region would
contain points on the circle (solid curve)
and points outside the circle.

27.   **29.**

31. 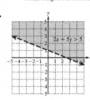   **33.**

35.

37. a. Yes   **b.** No   **c.** No
39. a. Yes   **b.** No   **c.** Yes   **d.** No

41.

43.

45.

47. The solution set is { }.

49.

51.

53.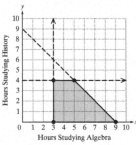

55.

57. The solution set is { }.   **59.** $x \le 6$

61. $y \ge -2$ **63.** $x + y \le 18$

65. **a.** $x + y \le 9$ **b.** $x \ge 3$ **c.** $y \le 4$ **d.** $x \ge 0$ **e.** $y \ge 0$

f.

67. **a.** $x + y \le 60{,}000$ **b.** $y \ge 2x$ **c.** $x \ge 0$ **d.** $y \ge 0$

e.

69. $x^2 + y^2 < 9$, $x > 0$, $y > 0$

71. $y > x - 1$ $y > -4x - 16$ $y < -\frac{1}{4}x + \frac{11}{4}$

73. **a.** $(x + 12)^2 + (y + 9)^2 \le 256$

b. Yes; The center of Hawthorne is 15 km from the earthquake.

75. If the inequality is strict—that is, posed with $<$ or $>$—then the bounding line or curve should be dashed.

77. Find the solution set to each individual inequality in the system. Then to find the solution set for the system of inequalities, take the intersection of the solution sets to the individual inequalities.

79.

81.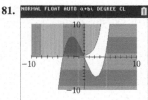

Problem Recognition Exercises, p. 547

1. **a.**   **b.**   **c.** $\{(1, 2)\}$

d.   **e.**

2. **a.**   **b.**   **c.** { }

d. 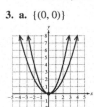   **e.** { }

3. **a.** $\{(0, 0)\}$ **b.**   **c.**

4. **a.** $\{(2, 1), (5, 4)\}$ **b.**   **c.**

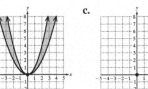

Section 5.6 Practice Exercises, pp. 552–556

R.1. **a.** $S(x) = 0.11x + 500$, for $x \ge 0$ **b.** $S(8000) = 1380$; The salesperson will make $1380 if $8000 in merchandise is sold for the week.

R.2. **a.** $y = 123x + 75$, for $x \ge 0$ **b.** $(0, 75)$; The base cost to rent the car is $75.

1. linear   **3.** feasible

5. $z = 0.80x + 1.10y$ **7.** $z = 0.62x + 0.50y$

9. **a.** $x = 7, y = 5$ **b.** Maximum value: 31

11. **a.** $x = 10, y = 30$ **b.** Minimum value: 37,000

13. a. Vertices:

(0, 0), (20, 40), (60, 0)

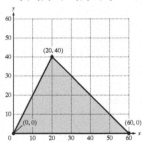

b. $x = 60$, $y = 0$

c. Maximum: 15,000

15. a. Vertices:

(0, 50), (10, 20), (20, 0)

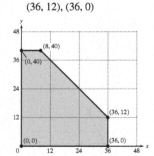

b. $x = 20$, $y = 0$

c. Minimum: 60

17. a. Vertices:

(0, 0), (0, 40), (8, 40),
(36, 12), (36, 0)

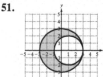

b. $x = 36$, $y = 12$

c. Maximum: 6480

19. a. 1520 at (8, 6)   **b.** 1660 at (4, 9)

21. a. Profit: $z = 160x + 240y$

b. $x \geq 0$ $y \geq 0$ $x \leq 120$ $y \leq 90$ $240x + 320y \leq 48{,}000$

c.

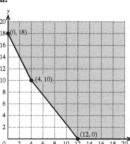

d. (0, 0), (0, 90), (80, 90),
(120, 60), (120, 0)

e. Profit at (0, 0): $z = 0$
Profit at (0, 90): $z = 21{,}600$
Profit at (80, 90): $z = 34{,}400$
Profit at (120, 60): $z = 33{,}600$
Profit at (120, 0): $z = 19{,}200$

f. The greatest profit is realized when 80 kitchen tables and 90 dining room tables are produced.

g. The maximum profit is $34,400.

23. a. 280 large trees and 120 small trees would maximize profit.
b. The maximum profit is $13,400.   **c.** In this case, the nursery should have 360 large trees and no small trees.

25. a. The company should make 8 trips with the small truck and 6 trips with the large truck.   **b.** The minimum cost is $1860.

27. a. The manufacturer should produce 600 grill A units and 500 grill B units to maximize profit.   **b.** The maximum profit is $114,000.
c. In this case, the manufacturer should produce 1200 grill A units and 0 grill B units.

29. a. The farmer should plant 900 acres of corn and 300 acres of soybeans.   **b.** The maximum profit is $138,000.   **c.** In this case, 500 acres of corn and 700 acres of soybeans should be planted.

31. Linear programming is a technique that enables us to maximize or minimize a function under specific constraints.

33. The feasible region for a linear programming application is found by first identifying the constraints on the relevant variables. Then the regions defined by the individual constraints are graphed. The intersection of the constraints defines the feasible region.

Chapter 5 Review Exercises, pp. 558–560

1. a. Yes   **b.** No   **3.** One solution   **5.** $\{(2, -3)\}$

7. { }; The system is inconsistent.

9. Milk has 300 mg per cup and spinach has 240 mg per cup.

11. The speed of the plane in still air is 420 mph and the speed of the wind is 60 mph.   **13.** $\{(-1, 2, -3)\}$

15. Infinitely many solutions; The equations are dependent.

17. $\left\{\left(x, -\dfrac{1}{5}x, -\dfrac{23}{5}x\right)\middle| x \text{ is any real number}\right\}$ or

$\{(-5y, y, 23y)\,|\, y \text{ is any real number}\}$ or

$\left\{\left(-\dfrac{5}{23}z, \dfrac{1}{23}z, z\right)\middle| z \text{ is any real number}\right\}$

19. She put $2000 in savings, and invested $12,000 in the bond fund and $6000 in the stock fund.   **21.** $y = 3x^2 - 6x + 1$

23. $\dfrac{A}{x + 2} + \dfrac{B}{x - 1}$ **25.** $\dfrac{A}{x} + \dfrac{B}{x^2} + \dfrac{C}{2x + 3}$

27. $\dfrac{Ax + B}{x^2 + 1} + \dfrac{Cx + D}{x^2 + 4}$ **29.** $\dfrac{5}{x + 4} + \dfrac{2}{(x + 4)^2}$

31. $\dfrac{-2}{x} + \dfrac{4x + 1}{x^2 + 5}$

33. a.

35. $\{(2, 4), (2, -4), (-2, 4), (-2, -4)\}$

37. $\{(4, 2)\}$

39. The numbers are 8 and 6 or -8 and -6.

41. The billboard is 12 ft by 24 ft.

b. $\{(-1, 2), (2, 5)\}$

43. a.

b.

45.

47.

49. a. No   **b.** Yes

51.

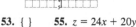

53. { }   **55.** $z = 24x + 20y$

57. a.

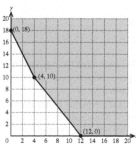

b. $x = 4$, $y = 10$
c. 620

Chapter 5 Test, pp. 560–561

1. a. Yes **b.** No **2. a.** No **b.** Yes **3. a.** Yes **b.** No
4. $\{(1, 1)\}$ **5.** $\{(5, 10)\}$
6. $\left\{\left(x, \dfrac{10x - 3}{4}\right)\middle| x \text{ is any real number}\right\}$

or $\left\{\left(\dfrac{4y + 3}{10}, y\right)\middle| y \text{ is any real number}\right\}$;

The equations are dependent.
7. $\{\ \}$; The system is inconsistent. **8.** $\{(1, -4, 3)\}$ **9.** $\{(-2, 5, 3)\}$
10. $\{\ \}$; The system is inconsistent. **11.** $\{(0, -3), (7, 4)\}$
12. $\{(1, 3), (1, -3), (-1, 3), (-1, -3)\}$
13. $\{(5, -2), (-5, 2)\}$ **14.** $\left\{\left(-2, \dfrac{5}{4}\right)\right\}$
15. $\left\{\left(x, \dfrac{22 - 3x}{2}, \dfrac{x - 6}{2}\right)\middle| x \text{ is any real number}\right\}$ or

$\left\{\left(\dfrac{22 - 2y}{3}, y, \dfrac{2 - y}{3}\right)\middle| y \text{ is any real number}\right\}$ or

$\{(2z + 6, 2 - 3z, z)| z \text{ is any real number}\}$
16. The manager should mix 4 lb of peanuts with 16 lb of the 45% mixture.
17. One runner runs 6 m/sec and the other runs 4 m/sec.
18. Dylan invested \$3000 in the risky stock, \$7000 in the second stock, and \$5000 in the third stock.
19. The numbers are 7 and 4. **20.** The screen is 44 in. by 33 in.
21. $y = 2x^2 - 4x + 1$ **22.** $\dfrac{A}{x + 1} + \dfrac{B}{3x - 2}$
23. $\dfrac{A}{x} + \dfrac{B}{x^2} + \dfrac{C}{x^3} + \dfrac{D}{x - 3} + \dfrac{Ex + F}{x^2 + 5x + 1} + \dfrac{Gx + H}{(x^2 + 5x + 1)^2}$
24. $\dfrac{-7}{x + 3} + \dfrac{2}{2x + 5}$ **25.** $\dfrac{6}{x + 2} + \dfrac{-4}{(x + 2)^2}$
26. $x - 2 - \dfrac{3}{x} + \dfrac{8}{x^2} - \dfrac{1}{x - 4}$ **27.** $\dfrac{-3}{x} + \dfrac{4x - 2}{x^2 + 7}$
28. $\dfrac{7x + 1}{x^2 + 2} + \dfrac{3}{x^2 + 9}$
29.

30.

31.

32.

33.

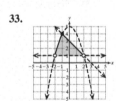

34. $z = 2.4x + 0.55y$ **35. a.** $x = 2, y = 8$
b. Minimum value: 48

36. a.

b. $x = 12, y = 36$
c. Maximum value: 37,800

37. a. 30 scoops of each type of protein powder should be mixed to maximize protein content. **b.** The maximum protein content is 1140 g. **c.** In this case, 24 scoops of whey protein should be mixed with 36 scoops of soy protein.

Chapter 5 Cumulative Review Exercises, pp. 561–562

1. $\left\{\dfrac{1 \pm \sqrt{57}}{4}\right\}$ **2.** $\{-2, -4\}$ **3.** $\{\pm 4, \pm i\}$
4. $\{\ \}$; The value $-\dfrac{12}{7}$ does not check.
5. $\left\{\dfrac{-1 + \ln 40}{2}\right\}$ **6.** $(-\infty, 2)$ **7.** $(-\infty, -5) \cup (1, \infty)$
8. $\dfrac{-5}{x - 3} + \dfrac{2}{(x - 3)^2}$ **9. a.** $(f \circ g)(x) = 50x^2 + 5x - 1$
b. $(g \circ f)(x) = 10x^2 - 15x + 1$ **10.** $f^{-1}(x) = x^3 + 2$
11. 3.4454 **12.** $\dfrac{19}{2}$ **13.** $y = 3x - 7$
14. $3i$ (multiplicity 1); $-3i$ (multiplicity 1); 1 (multiplicity 2)
15. a. $(-\infty, \infty)$ **b.** $[-3, \infty)$ **16. a.** $(3, \infty)$ **b.** $(-\infty, \infty)$
17. $\{\ \}$ **18.** $\{(1, 2, 3)\}$ **19.** $\{(1, 2), (1, -2), (-1, 2), (-1, -2)\}$
20. $-2x - h + 5$ **21. a.** $A(t) = 8000e^{0.062t}$ **b.** 11.2 yr
22. $y = 76.8$

CHAPTER 6

Section 6.1 Practice Exercises, pp. 571–574

R.1. $\{16\}$ **R.2.** \$100
R.3. $\dfrac{A}{2t - 5} + \dfrac{B}{2t - 3}$ **R.4.** $\dfrac{A}{7x} + \dfrac{Bx + C}{x^2 + 1}$
1. matrix **3.** Interchange rows 2 and 3. **5.** Multiply row 1 by 3 and replace the original row 1 with the result. **7.** Add 3 times row 1 to row 2 and replace the original row 2 with the result.
9. $\begin{bmatrix} -3 & 2 & -1 & | & 4 \\ 8 & 0 & 4 & | & 12 \\ 0 & 2 & -5 & | & 1 \end{bmatrix}$ **11.** $\begin{bmatrix} 4 & -14 & | & 2 \\ 3 & -5 & | & 7 \end{bmatrix}$ **13.** $\begin{bmatrix} 1 & 0 & 0 & | & 2 \\ 0 & 1 & 0 & | & \frac{6}{7} \\ 0 & 0 & 1 & | & 12 \end{bmatrix}$
15. $-4x + 6y = 11$ **17.** $x + 4y + 3z = 8$ **19.** $x = 8$
$\quad\ \ -3x + 9y = 1$ $\qquad\quad y + 2z = 12$ $\qquad y = -9$
$\qquad\qquad\qquad\qquad\qquad\qquad z = 6$ $\qquad z = \frac{3}{2}$
21. $\begin{bmatrix} -3 & 6 & | & 6 \\ 1 & 4 & | & 2 \end{bmatrix}$ **23.** $\begin{bmatrix} 3 & 12 & | & 6 \\ -3 & 6 & | & 6 \end{bmatrix}$ **25.** $\begin{bmatrix} 0 & 6 & | & 4 \\ -3 & 6 & | & 6 \end{bmatrix}$
27. $\begin{bmatrix} 1 & 5 & 6 & | & 2 \\ 4 & -2 & -3 & | & 10 \\ 2 & 1 & 5 & | & 1 \end{bmatrix}$ **29.** $\begin{bmatrix} 1 & 5 & 6 & | & 2 \\ 2 & 1 & 5 & | & 1 \\ 1 & -\frac{1}{2} & -\frac{3}{4} & | & \frac{5}{2} \end{bmatrix}$
31. $\begin{bmatrix} 1 & 5 & 6 & | & 2 \\ 0 & -9 & -7 & | & -3 \\ 4 & -2 & -3 & | & 10 \end{bmatrix}$
33. No; The element on the main diagonal in the second row is not 1.
35. Yes **37.** No; The elements on the main diagonal are not 1 with zeros above and below. **39.** Yes
41. $\{(-2, -3)\}$ **43.** $\{(4, 7)\}$ **45.** $\{(5, -8)\}$ **47.** $\{(1, 2, 3)\}$
49. $\{(0, 1, 3)\}$ **51.** $\{(7, -2, 1)\}$ **53.** $\{(0, 2, -4)\}$
55. $\{(-6, -4, 3)\}$ **57.** $\{(1, 2, 3, 4)\}$ **59.** $\{(-10, 1, 0, 1)\}$

61. He borrowed $10,000 from his parents, $8000 from the credit union, and $2000 from the bank.   **63.** She spent 4 nights in Washington, 2 nights in Atlanta, and 8 nights in Dallas.

65. $\dfrac{2}{x+3} + \dfrac{3}{x-2} + \dfrac{-1}{(x-2)^2}$

67. Interchanging two rows in an augmented matrix represents interchanging two equations in a system of equations. This operation does not affect the solution set of the system.   **69.** Reduced row-echelon form is the same format as row-echelon form with the added condition that all elements above the leading 1's must be 0's.

71. $\{(9.32, -17.48, 12.93)\}$

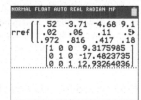

73. a. $2400a + 800b + c = 36{,}000$
$2000a + 500b + c = 30{,}000$
$3000a + 1000b + c = 44{,}000$
c. $y = 12x_1 + 4x_2 + 4000$
d. $36{,}000$

b.

75. a. $9a - 3b + c = -7.28$
$a - b + c = 3.68$
$100a + 10b + c = 18.2$
c. $f(x) = -0.32x^2 + 4.2x + 8.2$

b.

Section 6.2 Practice Exercises, pp. 582–585

R.1. $y = -5x - 18$ **R.2.** $x = \dfrac{y}{3} - \dfrac{8}{3}$ **R.3.** y-axis   **R.4.** x-axis

1. True   **3.** False   **5.** inconsistent   **7.** No solution
9. Infinitely many solutions   **11.** One solution
13. Infinitely many solutions
15. a. $\{(5, 0)\}$ **b.** $\{(5 - 2y, y) \mid y$ is any real number$\}$ **c.** $\{\ \}$
17. a. $\{(3, 5, 0)\}$ **b.** $\{\ \}$ **c.** $\{(-6z + 3, -4z + 5, z) \mid z$ is any real number$\}$
19. $\{\ \}$ **21.** $\left\{\left(\dfrac{10 - 7y}{2}, y\right) \middle| y$ is any real number$\right\}$ **23.** $\{\ \}$
25. $\{(-2z + 16, 3z - 5, z) \mid z$ is any real number$\}$ **27.** $\{(4, 0, 2)\}$
29. $\{(5z + 3, -2z + 6, z) \mid z$ is any real number$\}$
31. $\left\{\left(\dfrac{12 - 3y - 4z}{2}, y, z\right) \middle| y$ and z are any real numbers$\right\}$ or
$\{(x, y, z) \mid 2x + 3y + 4z = 12\}$
33. $\{(-3z + 5, z + 4, z) \mid z$ is any real number$\}$
35. $\{(-4z + 7, 2, z) \mid z$ is any real number$\}$ **37.** $\{\ \}$
39. a. $(-3, -3, 1)$ **b.** $(9, 0, 4)$ **c.** $(-3, -6, -2)$
41. For example: $(0, 6, 0)$, $(4, 5, 1)$, $(8, 4, 2)$
43. For example: $(0, 0, 1)$, $(0, 2, 0)$, $(3, 0, 0)$
45. a. $180 + 190 = x_1 + x_3$ **b.** $x_1 + x_2 = 180 + 220$
c. $x_3 + 120 = x_2 + 90$
d. $x_1 + x_3 = 370$
$x_1 + x_2 = 400$
$ x_2 - x_3 = 30$
e. $\begin{bmatrix} 1 & 0 & 1 & | & 370 \\ 0 & 1 & -1 & | & 30 \\ 0 & 0 & 0 & | & 0 \end{bmatrix}$
f. $x_1 = 250$ vehicles per hour;   **g.** $220 \le x_1 \le 270$ vehicles per hour;
$x_2 = 150$ vehicles per hour $130 \le x_2 \le 180$ vehicles per hour
47. a. $x_1 = 60$ vehicles per hour;   **b.** $40 \le x_1 \le 90$ vehicles per hour;
$x_2 = 210$ vehicles per hour; $190 \le x_2 \le 240$ vehicles per hour;
$x_3 = 170$ vehicles per hour $150 \le x_3 \le 200$ vehicles per hour

49. a. $80x + 400y + 480z = 9280$
$50x + 350y + 400z = 7800$
$75x + 525y + 600z = 10{,}500$
b. $\{\ \}$ **c.** The system of equations reduces to a contradiction. There are no values for x, y, and z that can simultaneously meet the conditions of this problem.
51. $5(-2z + 16) + 7(3z - 5) - 11z = 45$ ✓
$3(-2z + 16) + 5(3z - 5) - 9z = 23$ ✓
$(-2z + 16) + (3z - 5) - z = 11$
53. $(0, 0, 0)$ is the only solution.   **55.** Infinitely many solutions; $\{(2z, 3z, z) \mid z$ is any real number$\}$
57. If a row of the reduced row-echelon form results in a contradiction (that is, zeros to the left of the vertical bar and a nonzero element to the right), then the system is inconsistent.   **59.** The equations are equivalent, meaning that they all have the same solution set. The points in the solution set represent a common plane in space.
61. $\{\ \}$
63. $\{(-2z + 16, 3z - 5, z) \mid z$ is any real number$\}$

65. $\left\{\left(\dfrac{12 - 3y - 4z}{2}, y, z\right) \middle| y$ and z are any real numbers$\right\}$ or
$\{(x, y, z) \mid 2x + 3y + 4z = 12\}$

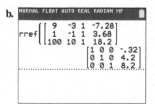

Section 6.3 Practice Exercises, pp. 596–601

R.1. -9 **R.2.** $x + (-9)$ **R.3.** $\{-4\}$
1. order; rows; columns   **3.** The order of the matrices must be the same, and the corresponding elements must be equal.
5. columns; rows   **7.** False   **9.** A row matrix is a matrix with only one row.   **11. a.** 2×3 **b.** None of these   **13. a.** 3×1
b. Column matrix   **15. a.** 2×2 **b.** Square matrix
17. $\sqrt{5}$ **19.** $\frac{1}{3}$ **21.** 2
23. $x = 4$, $y = 2$, $z = 10$ **25.** $\begin{bmatrix} 4 & -6 & -9 \\ -\frac{3}{5} & -1 & -7 \end{bmatrix}$
27. $\begin{bmatrix} -3 & 1 \\ 13.2 & \frac{5}{2} \\ \frac{7}{3} & 3\sqrt{2} \end{bmatrix}$ **29.** $\begin{bmatrix} -4 & 7 \\ 0.2 & \frac{7}{6} \\ -\frac{2}{3} & 6 + \sqrt{2} \end{bmatrix}$ **31.** Not possible
33. Multiply each element in the matrix by the scalar.
35. $\begin{bmatrix} 6 & 12 & -27 \\ 3 & 3\sqrt{3} & \frac{3}{2} \end{bmatrix}$ **37.** $\begin{bmatrix} 3 & -8 & -10 \\ -16 & -2\sqrt{3} - 63 & -\frac{17}{3} \end{bmatrix}$
39. $\begin{bmatrix} -4 & -16 & 20 \\ -12 & -4\sqrt{3} - 36 & -\frac{14}{3} \end{bmatrix}$ **41.** $\begin{bmatrix} 9 & 8 & -38 \\ -8 & 2\sqrt{3} - 45 & -\frac{7}{3} \end{bmatrix}$
43. $X = \begin{bmatrix} \frac{3}{2} & 1 \\ 5 & \frac{7}{2} \end{bmatrix}$ **45.** $X = \begin{bmatrix} \frac{1}{5} & -2 \\ \frac{2}{5} & \frac{11}{5} \end{bmatrix}$ **47.** $X = \begin{bmatrix} 0 & \frac{8}{5} \\ \frac{1}{5} & -\frac{13}{10} \end{bmatrix}$
49. a. Yes; 4×1 **b.** No   **51. a.** Yes; 5×5 **b.** Yes; 1×1
53. a. $\begin{bmatrix} -1 & 17 \\ -2 & 41 \end{bmatrix}$ **b.** $\begin{bmatrix} 22 & 31 \\ 13 & 18 \end{bmatrix}$ **c.** $\begin{bmatrix} 19 & 27 \\ 45 & 64 \end{bmatrix}$
55. a. $\begin{bmatrix} -6 & 8 & 38 \\ -12 & -24 & 36 \\ -13 & 4 & 69 \end{bmatrix}$ **b.** $\begin{bmatrix} -23 & 9 \\ 6 & 62 \end{bmatrix}$ **c.** Not possible

57. a. $\begin{bmatrix} 19 \\ 26 \\ 5 \end{bmatrix}$ **b.** Not possible   **c.** $\begin{bmatrix} 79 & -5 & 2 \\ 4 & 27 & 45 \\ -1 & 12 & 53 \end{bmatrix}$

59. a. $\begin{bmatrix} 2 \\ 3 \end{bmatrix}$ **b.** $\begin{bmatrix} -\frac{1}{3} & -\frac{1}{6} \\ 2 & 1 \end{bmatrix}$ **c.** Not possible

61. a. $\begin{bmatrix} 4 & 8 & 20 & 24 \\ -6 & -12 & -30 & -36 \end{bmatrix}$ **b.** Not possible

   c. Not possible   **63. a.** $[-25]$ **b.** $[-25]$ **c.** $[25]$

65. $\begin{bmatrix} 3.1 & -2.3 \\ 1.1 & 6.5 \end{bmatrix}$ **67.** $\begin{bmatrix} \frac{9}{5} & -3 & \sqrt{6} \\ 5 & \frac{1}{2} & 2 \\ 3 & 0 & 1 \end{bmatrix}$

69. a. $M - D = \begin{bmatrix} \$3600 & \$2400 \\ \$3400 & \$2000 \end{bmatrix}$;

   This represents the profit that the dealer clears for each model.

   b. $F = 1.06D = \begin{bmatrix} \$30,740 & \$29,150 \\ \$30,210 & \$28,514 \end{bmatrix}$

71. [$1820]; The value $1820 represents the total revenue from the sale of these four items.

73. a. $\begin{bmatrix} \$103,974 & \$42,560 \\ \$69,754 & \$28,400 \\ \$66,438 & \$26,960 \end{bmatrix}$; The first column gives the total revenue

   for Friday, Saturday, and Sunday, respectively. The second column gives the profit for Friday, Saturday, and Sunday, respectively.
   b. $66,438   **c.** $97,920

75 a. TL (The product LT is not possible.)   **b.** $TL = \begin{bmatrix} \$2088 \\ \$2688 \\ \$1657 \end{bmatrix}$

77. a. $CN_1 = \begin{bmatrix} \$46 \\ \$40 \\ \$2.40 \end{bmatrix}$; The matrix CN_1 represents the additional cost for

   24 text messages and 100 extra minutes for each of the cell phone plans.

   b. $CN_3 = \begin{bmatrix} \$46 & \$23.60 & \$7.50 \\ \$40 & \$9.60 & \$0 \\ \$2.40 & \$5.60 & \$3 \end{bmatrix}$; The matrix CN_3 represents the

   additional cost per month for each plan. For example, row 1 represents the cost for plan A for months 1, 2, and 3, respectively.

79. a. $A = \begin{bmatrix} -1 & 0 & 4 \\ 1 & 3 & 2 \end{bmatrix}$ **b.** $\begin{bmatrix} -1 & 0 & 4 \\ 1 & 3 & 2 \end{bmatrix} + \begin{bmatrix} 2 & 2 & 2 \\ -4 & -4 & -4 \end{bmatrix}$

   $= \begin{bmatrix} 1 & 2 & 6 \\ -3 & -1 & -2 \end{bmatrix}$

   c. $\begin{bmatrix} 1 & 0 & -4 \\ 1 & 3 & 2 \end{bmatrix}$; This matrix represents the reflection of the

   triangle across the y-axis.   **d.** $\begin{bmatrix} -1 & 0 & 4 \\ -1 & -3 & -2 \end{bmatrix}$; This matrix

   represents the reflection of the triangle across the x-axis.

   e. $\begin{bmatrix} -2 & -1 & 3 \\ 1 & -1 & 0 \end{bmatrix}$; This matrix represents the reflection of the

   triangle across the x-axis, followed by a shift to the left 1 unit and a shift upward 2 units.

81. a. $A = \begin{bmatrix} -2 & 3 & 3 & 0 \\ 1 & 3 & 0 & -1 \end{bmatrix}$ **b.** $A + \begin{bmatrix} 0 & 0 & 0 & 0 \\ -3 & -3 & -3 & -3 \end{bmatrix}$

   c. $A + \begin{bmatrix} -4 & -4 & -4 & -4 \\ 0 & 0 & 0 & 0 \end{bmatrix}$

   d. $\begin{bmatrix} 1 & 0 \\ 0 & -1 \end{bmatrix} \cdot A = \begin{bmatrix} -2 & 3 & 3 & 0 \\ -1 & -3 & 0 & 1 \end{bmatrix}$

   e. $\begin{bmatrix} -1 & 0 \\ 0 & 1 \end{bmatrix} \cdot A = \begin{bmatrix} 2 & -3 & -3 & 0 \\ 1 & 3 & 0 & -1 \end{bmatrix}$

83. a. $\begin{bmatrix} 0 & 6 & 6 \\ 0 & 3 & 0 \end{bmatrix}$ **b.** $\begin{bmatrix} 0 & 3.7 & 5.2 \\ 0 & 5.6 & 3 \end{bmatrix}$

   c. It appears that the triangle was rotated approximately 30° counterclockwise.

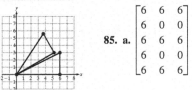

85. a. $\begin{bmatrix} 6 & 6 & 6 \\ 6 & 0 & 0 \\ 6 & 6 & 6 \\ 6 & 0 & 0 \\ 6 & 6 & 6 \end{bmatrix}$

   b. $\begin{bmatrix} 6 & 6 & 6 \\ 6 & 0 & 0 \\ 6 & 6 & 6 \\ 6 & 0 & 0 \\ 6 & 6 & 6 \end{bmatrix} + \begin{bmatrix} -1 & -1 & -1 \\ -1 & 1 & 1 \\ -1 & -1 & -1 \\ -1 & 1 & 1 \\ -1 & -1 & -1 \end{bmatrix} = \begin{bmatrix} 5 & 5 & 5 \\ 5 & 1 & 1 \\ 5 & 5 & 5 \\ 5 & 1 & 1 \\ 5 & 5 & 5 \end{bmatrix}$

87. $A + B = \begin{bmatrix} a_1 & a_2 \\ a_3 & a_4 \end{bmatrix} + \begin{bmatrix} b_1 & b_2 \\ b_3 & b_4 \end{bmatrix}$

   $= \begin{bmatrix} a_1 + b_1 & a_2 + b_2 \\ a_3 + b_3 & a_4 + b_4 \end{bmatrix} = \begin{bmatrix} b_1 + a_1 & b_2 + a_2 \\ b_3 + a_3 & b_4 + a_4 \end{bmatrix}$

   $= B + A$

89. $A + (-A) = \begin{bmatrix} a_1 & a_2 \\ a_3 & a_4 \end{bmatrix} + \begin{bmatrix} -a_1 & -a_2 \\ -a_3 & -a_4 \end{bmatrix}$

   $= \begin{bmatrix} a_1 + (-a_1) & a_2 + (-a_2) \\ a_3 + (-a_3) & a_4 + (-a_4) \end{bmatrix} = \begin{bmatrix} 0 & 0 \\ 0 & 0 \end{bmatrix} = 0$

91. $s(tA) = s \cdot \left(t \begin{bmatrix} a_1 & a_2 \\ a_3 & a_4 \end{bmatrix} \right) = s \cdot \begin{bmatrix} ta_1 & ta_2 \\ ta_3 & ta_4 \end{bmatrix}$

   $= \begin{bmatrix} sta_1 & sta_2 \\ sta_3 & sta_4 \end{bmatrix} = (st) \begin{bmatrix} a_1 & a_2 \\ a_3 & a_4 \end{bmatrix} = (st)A$

93. $A^2 = \begin{bmatrix} -1 & 0 \\ 0 & -1 \end{bmatrix}$, $A^3 = \begin{bmatrix} -i & 0 \\ 0 & -i \end{bmatrix}$, $A^4 = \begin{bmatrix} 1 & 0 \\ 0 & 1 \end{bmatrix}$; The entries

   along the main diagonal in matrix A^n are the same as the value of i^n.

95. a. $\begin{bmatrix} ac & 0 \\ 0 & bd \end{bmatrix}$ **b.** $\begin{bmatrix} 3 & 0 \\ 0 & 14 \end{bmatrix}$

97. The number of columns in the first matrix is not equal to the number of rows in the second matrix.   **99.** To find $-A$, take the additive inverse of each individual element of A. That is, $-A = [-a_{ij}]$.

101. 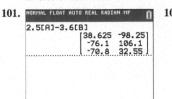   **103.**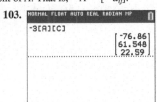

Section 6.4 Practice Exercises, pp. 609–611

R.1. b   **R.2.** c   **R.3.** d   **R.4.** a

1. identity   **3.** inverse   **5.** $\dfrac{1}{ad - bc} \begin{bmatrix} d & -b \\ -c & a \end{bmatrix}$

7. $I_2 = \begin{bmatrix} 1 & 0 \\ 0 & 1 \end{bmatrix}$

9. a. $AI_2 = \begin{bmatrix} -\frac{7}{8} & \sqrt{5} \\ 5.1 & 8 \end{bmatrix} \begin{bmatrix} 1 & 0 \\ 0 & 1 \end{bmatrix} = \begin{bmatrix} -\frac{7}{8}(1) + \sqrt{5}(0) & -\frac{7}{8}(0) + \sqrt{5}(1) \\ 5.1(1) + 8(0) & 5.1(0) + 8(1) \end{bmatrix}$

   $= \begin{bmatrix} -\frac{7}{8} & \sqrt{5} \\ 5.1 & 8 \end{bmatrix} \checkmark$

   b. $I_2A = \begin{bmatrix} 1 & 0 \\ 0 & 1 \end{bmatrix} \begin{bmatrix} -\frac{7}{8} & \sqrt{5} \\ 5.1 & 8 \end{bmatrix} = \begin{bmatrix} 1(-\frac{7}{8}) + 0(5.1) & 1(\sqrt{5}) + 0(8) \\ 0(-\frac{7}{8}) + 1(5.1) & 0(\sqrt{5}) + 1(8) \end{bmatrix}$

   $= \begin{bmatrix} -\frac{7}{8} & \sqrt{5} \\ 5.1 & 8 \end{bmatrix} \checkmark$

11. a. $AI_3 = \begin{bmatrix} 1 & -3 & 4 \\ 9 & 5 & 3 \\ 11 & -6 & -4 \end{bmatrix} \begin{bmatrix} 1 & 0 & 0 \\ 0 & 1 & 0 \\ 0 & 0 & 1 \end{bmatrix}$

$= \begin{bmatrix} 1(1) + -3(0) + 4(0) & 1(0) + -3(1) + 4(0) & 1(0) - 3(0) + 4(1) \\ 9(1) + 5(0) + 3(0) & 9(0) + 5(1) + 3(0) & 9(0) + 5(0) + 3(1) \\ 11(1) - 6(0) - 4(0) & 11(0) - 6(1) - 4(0) & 11(0) - 6(0) - 4(1) \end{bmatrix}$

$= \begin{bmatrix} 1 & -3 & 4 \\ 9 & 5 & 3 \\ 11 & -6 & -4 \end{bmatrix}$ ✓

b. $I_3A = \begin{bmatrix} 1 & 0 & 0 \\ 0 & 1 & 0 \\ 0 & 0 & 1 \end{bmatrix} \begin{bmatrix} 1 & -3 & 4 \\ 9 & 5 & 3 \\ 11 & -6 & -4 \end{bmatrix}$

$= \begin{bmatrix} 1(1) + 0(9) + 0(11) & 1(-3) + 0(5) + 0(-6) & 1(4) + 0(3) + 0(-4) \\ 0(1) + 1(9) + 0(11) & 0(-3) + 1(5) + 0(-6) & 0(4) + 1(3) + 0(-4) \\ 0(1) + 0(9) + 1(11) & 0(-3) + 0(5) + 1(-6) & 0(4) + 0(3) + 1(-4) \end{bmatrix}$

$= \begin{bmatrix} 1 & -3 & 4 \\ 9 & 5 & 3 \\ 11 & -6 & -4 \end{bmatrix}$ ✓

13. Yes   **15.** Yes   **17.** No

19. $A^{-1} = \begin{bmatrix} -\frac{5}{2} & -\frac{3}{2} \\ 3 & 2 \end{bmatrix}$ **21.** $A^{-1} = \begin{bmatrix} -\frac{1}{4} & -\frac{1}{10} \\ \frac{1}{2} & \frac{2}{5} \end{bmatrix}$

23. Singular matrix   **25.** $A^{-1} = \begin{bmatrix} 3 & -2 & -3 \\ 3 & -4 & -2 \\ 2 & -3 & -1 \end{bmatrix}$

27. $A^{-1} = \begin{bmatrix} 1 & 0 & 2 \\ 1 & -\frac{1}{2} & \frac{3}{2} \\ 1 & -\frac{1}{2} & \frac{1}{2} \end{bmatrix}$ **29.** Singular matrix

31. $A^{-1} = \begin{bmatrix} 3 & -3 & -6 \\ -1 & 3 & 5 \\ 2 & -3 & -5 \end{bmatrix}$ **33.** $A^{-1} = \begin{bmatrix} 1 & -3 & -1 & 0 \\ 0 & 1 & 2 & 0 \\ 0 & 1 & 1 & 0 \\ -2 & 0 & -3 & 1 \end{bmatrix}$

35. $\begin{bmatrix} 3 & -4 \\ 2 & 1 \end{bmatrix} \begin{bmatrix} x \\ y \end{bmatrix} = \begin{bmatrix} -1 \\ 14 \end{bmatrix}$ **37.** $\begin{bmatrix} 9 & -6 & 4 \\ 4 & 0 & -1 \\ 0 & 3 & 1 \end{bmatrix} \begin{bmatrix} x \\ y \\ z \end{bmatrix} = \begin{bmatrix} 27 \\ 1 \\ 0 \end{bmatrix}$

39. $\{(-2, 4)\}$ **41.** $\{(-4, 1)\}$ **43.** $\{(-3, -1, 0)\}$

45. $\{(0, 2, -4)\}$ **47.** $\left\{\left(1, \frac{1}{2}, 2\right)\right\}$ **49.** $\{(1, -1, 0, 2)\}$

51. False. If a matrix has an inverse, it must be a square matrix.
53. False. For example, a square matrix with a row or column of all zeros does not have an inverse.   **55.** True   **57.** False. If one row (or column) of a matrix is a multiple of another row (or column), then the matrix does not have an inverse.

59. For example: $\begin{bmatrix} 0 & 1 \\ 1 & 0 \end{bmatrix}$ **61. a.** $A^{-1} = \begin{bmatrix} \frac{3}{4} & -\frac{1}{4} \\ -\frac{5}{8} & \frac{3}{8} \end{bmatrix}$

b. $(A^{-1})^{-1} = A = \begin{bmatrix} 3 & 2 \\ 5 & 6 \end{bmatrix}$ **63.** $A^{-1} = \begin{bmatrix} \frac{1}{a} & 0 \\ 0 & \frac{1}{b} \end{bmatrix}$

65. If $AB = I_n$ and $BA = I_n$, then the matrices are inverses.

67. $\begin{bmatrix} a & b \\ c & d \end{bmatrix} \begin{bmatrix} 1 & 0 \\ 0 & 1 \end{bmatrix} \overset{\frac{1}{a}R_1 \to R_1}{=} \begin{bmatrix} 1 & \frac{b}{a} \\ c & d \end{bmatrix} \begin{bmatrix} \frac{1}{a} & 0 \\ 0 & 1 \end{bmatrix}$

$\overset{-c \cdot R_1 + R_2 \to R_2}{=} \begin{bmatrix} 1 & \frac{b}{a} \\ 0 & d - \frac{cb}{a} \end{bmatrix} \begin{bmatrix} \frac{1}{a} & 0 \\ -\frac{c}{a} & 1 \end{bmatrix} = \begin{bmatrix} 1 & \frac{b}{a} \\ 0 & \frac{ad - bc}{a} \end{bmatrix} \begin{bmatrix} \frac{1}{a} & 0 \\ -\frac{c}{a} & 1 \end{bmatrix}$

$\overset{\frac{a}{ad-bc} \cdot R_2 \to R_2}{=} \begin{bmatrix} 1 & \frac{b}{a} \\ 0 & 1 \end{bmatrix} \begin{bmatrix} \frac{1}{a} & 0 \\ -\frac{c}{ad-bc} & \frac{a}{ad-bc} \end{bmatrix}$

$\overset{-\frac{b}{a} \cdot R_2 + R_1 \to R_1}{=} \begin{bmatrix} 1 & 0 \\ 0 & 1 \end{bmatrix} \begin{bmatrix} \frac{d}{ad-bc} & -\frac{b}{ad-bc} \\ -\frac{c}{ad-bc} & \frac{a}{ad-bc} \end{bmatrix}$

Therefore, $A^{-1} = \dfrac{1}{ad - bc} \begin{bmatrix} d & -b \\ -c & a \end{bmatrix}$, provided $ad - bc \neq 0$.

69. $x_1 = 36.5°F, x_2 = 37.5°F, x_3 = 40.5°F, x_4 = 41.5°F$

71. $A^{-1} = \begin{bmatrix} 0.29 & -0.10 & 0.08 & -0.27 \\ -0.09 & 0.12 & 0.00 & 0.27 \\ -0.12 & 0.00 & -0.07 & 0.25 \\ 0.15 & 0.03 & -0.06 & -0.01 \end{bmatrix}$

73. $\{(5.35, 41.71, 4.45)\}$

Section 6.5 Practice Exercises, pp. 621–624
R.1. 1   **R.2.** -1 **R.3.** $y = -x + 2$

1. determinant   **3.** minor   **5.** $\begin{vmatrix} b_2 & c_2 \\ b_3 & c_3 \end{vmatrix}$; $\begin{vmatrix} b_1 & c_1 \\ b_3 & c_3 \end{vmatrix}$; $\begin{vmatrix} b_1 & c_1 \\ b_2 & c_2 \end{vmatrix}$

7. 27   **9.** 6   **11.** 0   **13.** $x^2 - 36$ **15.** $-5e^{2x}$
17. a. 25   **b.** -25 **19. a.** -39 **b.** -39
21. a. -36 **b.** -36 **23.** -39; Yes   **25.** -13; Yes
27. 0; No   **29.** 13; Yes   **31.** 121; Yes

33. $\left\{\left(\frac{23}{8}, \frac{21}{40}\right)\right\}$ **35.** $\left\{\left(-\frac{57}{46}, -\frac{31}{23}\right)\right\}$

37. $\left\{\left(\frac{4y + 8}{3}, y\right) \middle| y \text{ is any real number}\right\}$ **39.** $\{ \}$

41. $\left\{\left(\frac{63}{157}, -\frac{333}{157}, \frac{536}{471}\right)\right\}$ **43.** $\left\{\left(\frac{13}{2}, \frac{3}{2}, \frac{11}{2}\right)\right\}$

45. $\left\{\left(\frac{3y - z + 6}{2}, y, z\right) \middle| y \text{ and } z \text{ are any real numbers}\right\}$ or

$\{(x, y, z) | 2x - 3y + z = 6\}$ **47.** $\{ \}$ **49.** $x_2 = \dfrac{113}{60}$

51. Yes   **53.** No   **55. a.** $\begin{vmatrix} x & y & 1 \\ -3 & 2 & 1 \\ -4 & 6 & 1 \end{vmatrix} = 0$ **b.** $y = -4x - 10$

57. 12 square units   **59. a.** 36   **b.** -36 **c.** Rows 1 and 2 are interchanged between matrix A and matrix B. The determinants are opposite in sign.   **61. a.** 13   **b.** 26   **c.** Row 1 of matrix B is 2 times row 1 of matrix A. The value $|B| = 2|A|$.
63. a. -2 **b.** -2 **c.** Row 2 of matrix B is the same as the sum of 3 times row 1 of A and row 2 of A. The value of $|A|$ equals $|B|$.
65. 0   **67.** 0   **69.** 1   **71.** abc **73.** $|AB| = -130; |A| = 10$ and $|B| = -13$. So, $|A| \cdot |B| = (10)(-13) = -130$ and, therefore, $|A| \cdot |B| = |AB|$.
75. The minor is the determinant of the matrix obtained by deleting the ith row and jth column of the original matrix. The cofactor is the product of the minor and the factor $(-1)^{i+j}$.
77. Choose the row or column with the greatest number of zero elements.
79. -27 **81.** 10,112

Problem Recognition Exercises, p. 625
1. $\{(2, -4)\}$ **2.** $\{(5, -1)\}$ **3.** $\{(2, -1, 0)\}$ **4.** $\{(2, 0, 4)\}$
5. a. $\begin{vmatrix} 1.5 & -2 \\ -3 & 4 \end{vmatrix} = 0$ **b.** No   **c.** $\{ \}$ **6. a.** $\begin{vmatrix} 5 & -2 \\ 1 & -0.4 \end{vmatrix} = 0$
b. No   **c.** $\{ \}$ **7. a.** $\begin{vmatrix} 1 & -3 & 7 \\ -2 & 5 & -11 \\ 1 & -5 & 13 \end{vmatrix} = 0$
b. No   **c.** $\{(2z + 4, 3z + 1, z) | z \text{ is any real number}\}$
8. a. $\begin{vmatrix} 1 & -2 & 3 \\ -2 & 1 & 0 \\ 1 & 0 & -1 \end{vmatrix} = 0$ **b.** No
c. $\{(z + 3, 2z + 5, z) | z \text{ is any real number}\}$

Chapter 6 Review Exercises, pp. 627–630
1. $x - 2y + 3z = -1$
$\quad\quad y + 4z = -11$
$\quad\quad\quad\quad z = -2;$
Solution set: $\{(-1, -3, -2)\}$

3. $\begin{bmatrix} 1 & -\frac{3}{2} & \frac{1}{2} \\ 5 & 6 & -4 \end{bmatrix}$

5. $\{(5, -6)\}$ **7.** $\{(0, 2, 3)\}$ **9.** Lily borrowed $1000 from her friend, $7000 from the credit union, and $2000 from the bank.

11. $\{\ \}$ **13.** $\{(3z, -2z + 1, z) \mid z \text{ is any real number}\}$

15. $\{\ \}$ **17.** $\{(3z + 5, -4z + 3, z) \mid z \text{ is any real number}\}$

19. $\{(5y - 2z - 1, y, z) \mid y \text{ and } z \text{ are any real numbers}\}$ or $\{(x, y, z) \mid x - 5y + 2z = -1\}$ **21. a.** $x_1 = 76$ vehicles per hour; $x_2 = 97$ vehicles per hour   **b.** $46 \le x_1 \le 96$ vehicles per hour; $67 \le x_2 \le 117$ vehicles per hour   **23. a.** 3×2

b. None of these   **25. a.** 2×1 **b.** Column matrix

27. 4   **29.** $x = 6, y = 3, z = 8$ **31.** $3A = \begin{bmatrix} -12 & 3 \\ 18 & -6 \\ 3 & 9 \end{bmatrix}$

33. Not possible   **35.** $2A - C = \begin{bmatrix} -8 - \pi & -2 \\ 15 & -5 \\ 2 & 1 \end{bmatrix}$

37. $AB = \begin{bmatrix} -7 & -7 & 22 \\ 10 & 8 & -30 \\ 5 & 18 & -25 \end{bmatrix}$ **39.** Not possible

41. $A^2 = \begin{bmatrix} -2 & 36 \\ -6 & 10 \end{bmatrix}$ **43.** $BC = \begin{bmatrix} 2 & 7 \\ -6 & -21 \end{bmatrix}$

45. $QP = \begin{bmatrix} \$7922 \\ \$9843 \end{bmatrix}$; QP is the matrix representing the total revenue from these four items for each theater.

47. a. $A = \begin{bmatrix} 1 & 3 & 5 \\ 1 & 4 & 2 \end{bmatrix}$ **b.** $\begin{bmatrix} 1 & 3 & 5 \\ 1 & 4 & 2 \end{bmatrix} + \begin{bmatrix} -3 & -3 & -3 \\ -1 & -1 & -1 \end{bmatrix}$

$= \begin{bmatrix} -2 & 0 & 2 \\ 0 & 3 & 1 \end{bmatrix}$

c. $\begin{bmatrix} -1 & -3 & -5 \\ 1 & 4 & 2 \end{bmatrix}$; This matrix represents the reflection of the triangle across the y-axis.   **d.** $\begin{bmatrix} 1 & 3 & 5 \\ -1 & -4 & -2 \end{bmatrix}$; This matrix represents the reflection of the triangle across the x-axis.   **49.** No

51. $A^{-1} = \begin{bmatrix} \frac{1}{6} & \frac{1}{6} \\ -\frac{1}{12} & \frac{5}{12} \end{bmatrix}$ **53.** Singular matrix

55. $A^{-1} = \begin{bmatrix} 0 & -1 & 4 \\ 1 & -1 & 5 \\ -3 & -1 & 0 \end{bmatrix}$ **57.** $\begin{bmatrix} -3 & 7 & 0 \\ 4 & 0 & 2 \\ 2 & -1 & 5 \end{bmatrix} \begin{bmatrix} x \\ y \\ z \end{bmatrix} = \begin{bmatrix} 6 \\ -3 \\ -13 \end{bmatrix}$

59. $\{(4, -3)\}$ **61.** $\{(1, 2, 3)\}$ **63. a.** 24   **b.** 24

65. a. 13   **b.** -13 **67.** $81 - x^2$ **69.** 156   **71.** -270

73. $\left\{\left(\frac{3}{10}, -\frac{23}{30}\right)\right\}$ **75.** $\left\{\left(-\frac{7}{25}, -\frac{222}{125}, \frac{161}{125}\right)\right\}$ **77.** $y = -\frac{94}{67}$

Chapter 6 Test, pp. 630–631

1. $\begin{bmatrix} 1 & 5 & -3 & | & 1 \\ 3 & 1 & 4 & | & -2 \\ 0 & 4 & 2 & | & 6 \end{bmatrix}$ **2.** $\begin{bmatrix} 0 & -14 & 13 & | & -5 \\ 1 & 5 & -3 & | & 1 \\ 0 & 4 & 2 & | & 6 \end{bmatrix}$

3. $\begin{bmatrix} 3 & 1 & 4 & | & -2 \\ 1 & 5 & -3 & | & 1 \\ 0 & 1 & \frac{1}{2} & | & \frac{3}{2} \end{bmatrix}$

4. The elements above the leading 1 in the third column are not all zero. Specifically, the element in row 1, column 3 should be 0.

5. $\{(-10, 3)\}$ **6.** $\{\ \}$ **7.** $\{(3z, -2z + 5, z) \mid z \text{ is any real number}\}$

8. a. 8   **b.** -8 **9. a.** -12 **b.** -12 **10.** -31

11. 236   **12.** $|A| = 0$. Therefore, A is singular (does not have an inverse).

13. $\{(-2, 3)\}$ **14.** $\{\ \}$ **15.** $\{(-2, -8, 2)\}$

16. $\{(-3z - 2, -4z + 1, z) \mid z \text{ is any real number}\}$

17. $\left\{\left(\frac{54}{61}, -\frac{53}{61}\right)\right\}$ **18.** $x = \frac{7}{31}$ **19.** $X = \begin{bmatrix} -1 & 2 \\ -4 & -2 \end{bmatrix}$

20. $2A - 3C = \begin{bmatrix} 8 & -1 & 6 \\ -2 & 11 & -12 \end{bmatrix}$

21. Not possible   **22.** $AB = \begin{bmatrix} -5 & 20 \\ 20 & 44 \end{bmatrix}$

23. $BA = \begin{bmatrix} 22 & 37 & 51 \\ -2 & -4 & -6 \\ 22 & 23 & 21 \end{bmatrix}$ **24.** Yes   **25.** $A^{-1} = \begin{bmatrix} 2 & -1 \\ -\frac{5}{2} & \frac{3}{2} \end{bmatrix}$

26. Singular matrix   **27.** $A^{-1} = \begin{bmatrix} 3 & 1 & 2 \\ 7 & 2 & 5 \\ 1 & 1 & 1 \end{bmatrix}$ **28.** $\{(1, 5)\}$

29. $\{(2, 1, -3)\}$ **30. a.** $x_1 = 128$ vehicles per hour; $x_2 = 173$ vehicles per hour   **b.** $118 \le x_1 \le 168$ vehicles per hour; $163 \le x_2 \le 213$ vehicles per hour

31. $CN = \begin{bmatrix} 5500 \\ 8040 \end{bmatrix}$; This represents the total number of calories burned by two individuals with different weights after biking 6 hr, running 3 hr, and walking 5 hr. For example, the element 5500 in the first row tells us that 5500 cal would be burned by a 120-lb individual who biked 6 hr, ran 3 hr, and walked 5 hr in a given week.   **32.** Yes

Chapter 6 Cumulative Review Exercises, pp. 631–632

1. $3x^2 + 3xh + h^2$ **2. a.** $(4, -5)$ **b.** $(-\infty, 4)$ **c.** $(4, \infty)$

d. $(-\infty, \infty)$ **e.** $(-\infty, -5]$ **3.** $\frac{2}{3}$ **4. a.** i **b.** -1

5. $-\frac{6}{13} + \frac{17}{13}i$ **6.** $3\sqrt{10}$ **7.** Center: $\left(-\frac{5}{3}, 0\right)$; Radius: $\sqrt{11}$

8. a. $k = 0.117$ **b.** 9.3 mCi   **c.** More than 39.4 hr

9. a. 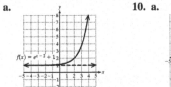   **10. a.**

b. $(-\infty, \infty)$ **c.** $(1, \infty)$ **b.** $(-4, \infty)$ **c.** $(-\infty, \infty)$

11. a. 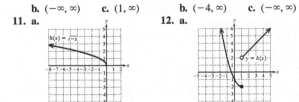   **12. a.**

b. $(-\infty, 0]$ **c.** $[0, \infty)$ **b.** $(-\infty, \infty)$ **c.** $[-2, \infty)$

13. $\{(7, 6)\}$ **14.** $\{(-4z, -3z + 2, z) \mid z \text{ is any real number}\}$

15. x-axis, y-axis, and origin   **16.** odd

17. $AB = \begin{bmatrix} 2 & -13 \\ 1 & 46 \end{bmatrix}$ **18.** 60   **19.** $4A - B = \begin{bmatrix} 7 & -16 \\ 4 & 17 \end{bmatrix}$

20. $A^{-1} = \begin{bmatrix} \frac{2}{5} & \frac{1}{5} \\ -\frac{1}{15} & \frac{2}{15} \end{bmatrix}$ **21.** $\{x \mid x > -2\}$ **22.** $\{x \mid 3 < x \le 4\}$

23. $\{x \mid x \ge 5\}$ **24.** $\left\{\frac{19}{2}, 2\right\}$ **25.** $\mathbb{R}$ **26.** $\left(\frac{1}{2}, 5\right]$

27. $\{6\}$ **28.** The zeros are $\frac{1}{2}, 3i$, and $-3i$.

29. Vertical asymptotes: $x = 3, x = -3$; Horizontal asymptote: $y = 0$

30. $\log\left(\frac{z^5}{x^3\sqrt{y}}\right)$

CHAPTER 7

Section 7.1 Practice Exercises, pp. 645–650

R.1. Center: $(0, -5)$; Radius: $2\sqrt{2}$ **R.2.** Center: $(0, 0)$; Radius: 5

R.3. Center: $(4, -3)$; Radius: 9   **R.4.** $n = 16$; $(a - 4)^2$

R.5. $n = \frac{49}{4}$; $\left(y + \frac{7}{2}\right)^2$ **R.6.** $(x + 5)^2 + (y - 5)^2 = 9$

1. conic   **3.** vertices   **5.** minor   **7.** $(h, k + a)$; $(h, k - a)$; $(h + b, k)$; $(h - b, k)$ **9. a.** $d_1 = \dfrac{37}{5}, d_2 = \dfrac{13}{5}, d_3 = 5, d_4 = 5$ **b.** 10   **c.** 10   **d.** They are the same.   **e.** The sums equal the length of the major axis.   **11. a.** Vertical   **b.** Horizontal

13. a. Center: $(0, 0)$
b. $a = 10$ **c.** $b = 5$
d. Vertices: $(10, 0), (-10, 0)$
e. Endpoints of minor axis: $(0, 5), (0, -5)$
f. Foci: $(5\sqrt{3}, 0), (-5\sqrt{3}, 0)$
g. 20   **h.** 10
i.

15. a. Center: $(0, 0)$
b. $a = 10$ **c.** $b = 5$
d. Vertices: $(0, 10), (0, -10)$
e. Endpoints of minor axis: $(5, 0), (-5, 0)$
f. Foci: $(0, 5\sqrt{3}), (0, -5\sqrt{3})$
g. 20   **h.** 10
i.

17. a. Center: $(0, 0)$
b. $a = 5$ **c.** $b = 2$
d. Vertices: $(5, 0), (-5, 0)$
e. Endpoints of minor axis: $(0, 2), (0, -2)$
f. Foci: $(\sqrt{21}, 0), (-\sqrt{21}, 0)$
g. 10   **h.** 4
i.

19. a. Center: $(0, 0)$
b. $a = 3$ **c.** $b = 1$
d. Vertices: $(0, 3), (0, -3)$
e. Endpoint of minor axis: $(1, 0), (-1, 0)$
f. Foci: $(0, 2\sqrt{2}), (0, -2\sqrt{2})$
g. 6   **h.** 2
i.

21. a. Center: $(0, 0)$
b. $a = 2\sqrt{3}$ **c.** $b = \sqrt{5}$
d. Vertices: $(2\sqrt{3}, 0), (-2\sqrt{3}, 0)$
e. Endpoints of minor axis: $(0, \sqrt{5}), (0, -\sqrt{5})$
f. Foci: $(\sqrt{7}, 0), (-\sqrt{7}, 0)$
g. $4\sqrt{3}$ **h.** $2\sqrt{5}$
i.

23. a. Center: $(1, -6)$
b. Vertices: $(6, -6), (-4, -6)$
c. Endpoints of minor axis: $(1, -2), (1, -10)$
d. Foci: $(4, -6), (-2, -6)$
e.

25. a. Center: $(-4, 2)$
b. Vertices: $(-4, 10), (-4, -6)$
c. Endpoints of minor axis: $(-11, 2), (3, 2)$
d. Foci: $(-4, 2 + \sqrt{15}), (-4, 2 - \sqrt{15})$
e.

27. a. Center: $(6, 0)$
b. Vertices: $(6, 3), (6, -3)$
c. Endpoints of minor axis: $(5, 0), (7, 0)$
d. Foci: $(6, 2\sqrt{2}), (6, -2\sqrt{2})$
e.

29. a. Center: $(0, -1)$
b. Vertices: $(9, -1), (-9, -1)$
c. Endpoints of minor axis: $(0, 2), (0, -4)$
d. Foci: $(6\sqrt{2}, -1), (-6\sqrt{2}, -1)$
e.

31. a. Center: $(3, 2)$
b. Vertices: $(\frac{1}{2}, 2), (\frac{11}{2}, 2)$
c. Endpoints of minor axis: $(3, \frac{1}{4}), (3, \frac{15}{4})$
d. Foci: $(3 + \frac{\sqrt{51}}{4}, 2), (3 - \frac{\sqrt{51}}{4}, 2)$
e.

33. a. $\dfrac{(x + 2)^2}{5} + \dfrac{(y - 6)^2}{3} = 1$
b. Center: $(-2, 6)$;
Vertices: $(-2 - \sqrt{5}, 6); (-2 + \sqrt{5}, 6)$
Endpoints of minor axis: $(-2, 6 + \sqrt{3}), (-2, 6 - \sqrt{3})$
Foci: $(-2 - \sqrt{2}, 6), (-2 + \sqrt{2}, 6)$

35. a. $\dfrac{(x - 5)^2}{6} + \dfrac{(y - 1)^2}{9} = 1$
b. Center: $(5, 1)$
Vertices: $(5, 4), (5, -2)$
Endpoints of minor axis: $(5 + \sqrt{6}, 1), (5 - \sqrt{6}, 1)$
Foci: $(5, 1 + \sqrt{3}), (5, 1 - \sqrt{3})$

37. a. $x^2 + \dfrac{(y + 7)^2}{4} = 1$
b. Center: $(0, -7)$
Vertices: $(0, -5), (0, -9)$
Endpoints of minor axis: $(-1, -7), (1, -7)$
Foci: $(0, -7 + \sqrt{3}), (0, -7 - \sqrt{3})$

39. a. $\dfrac{(x - \frac{5}{2})^2}{25} + \dfrac{(y + 4)^2}{9} = 1$
b. Center: $(\frac{5}{2}, -4)$
Vertices: $(\frac{15}{2}, -4), (-\frac{5}{2}, -4)$
Endpoints of minor axis: $(\frac{5}{2}, -1), (\frac{5}{2}, -7)$
Foci: $(\frac{13}{2}, -4), (-\frac{3}{2}, -4)$

41. a. $\dfrac{(x - 1)^2}{\frac{1}{4}} + \dfrac{y^2}{\frac{1}{9}} = 1$ **b.** Center: $(1, 0)$; Vertices: $(\frac{1}{2}, 0), (\frac{3}{2}, 0)$; Endpoints of minor axis: $(1, \frac{1}{3}), (1, -\frac{1}{3})$; Foci: $(1 - \frac{\sqrt{5}}{6}, 0), (1 + \frac{\sqrt{5}}{6}, 0)$

43. $\dfrac{x^2}{16} + \dfrac{y^2}{7} = 1$ **45.** $\dfrac{x^2}{17} + \dfrac{y^2}{98} = 1$

47. $\dfrac{(x - 2)^2}{49} + \dfrac{(y - 3)^2}{25} = 1$ **49.** $\dfrac{(x - 4)^2}{25} + \dfrac{(y - 1)^2}{9} = 1$

51. $\dfrac{x^2}{16} + \dfrac{y^2}{25} = 1$ **53.** $\dfrac{(x - 3)^2}{5} + \dfrac{y^2}{16} = 1$

55. $\dfrac{(x + 5)^2}{49} + \dfrac{(y - 1)^2}{16} = 1$ **57.** 16.94 cm

59. a. Major axis: 12 ft; Minor axis: 10 ft   **b.** $\dfrac{x^2}{36} + \dfrac{y^2}{25} = 1$
c. Place the tacks at $(\sqrt{11}, 0)$ and $(-\sqrt{11}, 0)$. These are approximately 3.3 ft to the left and right of the center.

61. The foci should be located $\pm 30\sqrt{2}$ in. (approximately 42.4 in.) from the center along the major axis.   **63.** b   **65.** a

67. $e = \dfrac{12}{13}$ **69.** $e = \dfrac{3}{5}$ **71.** $e = \dfrac{\sqrt{2}}{2}$ **73.** The Earth's orbit is more circular, and the orbit for Halley's Comet is very elongated.

75. 405,500 km   **77. a.** $\dfrac{x^2}{90^2} + \dfrac{y^2}{143.5^2} = 1$ **b.** $e \approx 0.78$
c. $\dfrac{x^2}{255^2} + \dfrac{y^2}{307.5^2} = 1$ **d.** $e \approx 0.56$ **e.** The outer ellipse is more circular than the inner ellipse because the eccentricity is closer to zero.

79. $\dfrac{x^2}{64} + \dfrac{y^2}{289} = 1$ **81.** $\dfrac{(x-4)^2}{25} + \dfrac{(y+1)^2}{9} = 1$

83. a. $x^2 + y^2 = 4$ **b.** Elliptical **c.** Major axis: $\dfrac{4\sqrt{34}}{5}$ in. $\approx$ 4.7 in.; Minor axis: 4 in.

85. a. Top semiellipse **b.** Bottom semiellipse **c.** Right semiellipse **d.** Left semiellipse **87.** $\{(0, 3), (5, 0)\}$

89. $\{(0, 4), (-2, 0), (2, 0)\}$ **91. a.** $A = 4\pi\sqrt{2}$ square units
b. $P \approx 2\pi\sqrt{6}$ units

93. $\dfrac{x^2}{a^2} + \dfrac{y^2}{b^2} = 1$ $\dfrac{c^2}{a^2} + \dfrac{y^2}{b^2} = 1$ $y = b\sqrt{1 - \dfrac{c^2}{a^2}}$

$y = b\sqrt{\dfrac{a^2 - c^2}{a^2}}$ $y = \dfrac{b}{a}\sqrt{a^2 - c^2}$

Recall that $c^2 = a^2 - b^2$, or equivalently $b^2 = a^2 - c^2$ and $b > 0$.

Therefore, $y = \dfrac{b}{a}\sqrt{b^2}$ or $y = \dfrac{b^2}{a}$. The length of a latus rectum is

$2y = \dfrac{2b^2}{a}$.

95. By the reflective property of an ellipse, any shot passing through one focus is reflected through the other focus.

97. The first equation represents an ellipse centered at the origin, whereas the second equation represents a line with slope $-\frac{4}{9}$ and y-intercept $(0, 4)$.

99. a. $(a + c) + (a - c) = 2a$
b. $\sqrt{(x + c)^2 + y^2} + \sqrt{(x - c)^2 + y^2} = 2a$
c. $\sqrt{(x + c)^2 + y^2} = 2a - \sqrt{(x - c)^2 + y^2}$
$(x + c)^2 + y^2 = 4a^2 - 4a\sqrt{(x - c)^2 + y^2} + (x - c)^2 + y^2$
$4a\sqrt{(x - c)^2 + y^2} = 4a^2 + (x - c)^2 - (x + c)^2$
$4a\sqrt{(x - c)^2 + y^2} = 4a^2 - 4xc$
$a\sqrt{(x - c)^2 + y^2} = a^2 - xc$
d. $\left(a\sqrt{(x - c)^2 + y^2}\right)^2 = (a^2 - xc)^2$
$a^2[(x - c)^2 + y^2] = a^4 - 2a^2xc + c^2x^2$
$a^2[x^2 - 2xc + c^2 + y^2] = a^4 - 2a^2xc + c^2x^2$
$a^2x^2 - 2a^2xc + a^2c^2 + a^2y^2 = a^4 - 2a^2xc + c^2x^2$
$a^2x^2 - c^2x^2 + a^2y^2 = a^4 - a^2c^2$
$(a^2 - c^2)x^2 + a^2y^2 = a^2(a^2 - c^2)$
e. $b^2x^2 + a^2y^2 = a^2b^2$
$\dfrac{b^2x^2}{a^2b^2} + \dfrac{a^2y^2}{a^2b^2} = \dfrac{a^2b^2}{a^2b^2}$
$\dfrac{x^2}{a^2} + \dfrac{y^2}{b^2} = 1$

101. a. Both equations have a sum of terms with a variable squared in the numerator and a real number squared in the denominator, all equal to 1.
b. $\dfrac{x^2}{9} + \dfrac{y^2}{36} = 1$; This represents the graph of an ellipse in the xy-plane.
c. $\dfrac{y^2}{36} + \dfrac{z^2}{4} = 1$; This represents the graph of an ellipse in the yz-plane.
d. $\dfrac{x^2}{9} + \dfrac{z^2}{4} = 1$; This represents the graph of an ellipse in the xz-plane.

103. 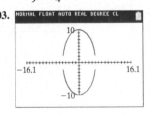 **105.**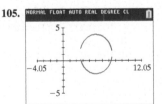

Section 7.2 Practice Exercises, pp. 661–666

R.1. 7 **R.2.** $y = -8x - 56$ **R.3.** $y = -\dfrac{10}{9}x - \dfrac{8}{9}$

1. hyperbola; foci **3.** transverse **5.** horizontal; $(a, 0)$; $(-a, 0)$; $y = \frac{b}{a}x$; $y = -\frac{b}{a}x$

7. a. $d_1 = \dfrac{101}{12}, d_2 = \dfrac{29}{12}, d_3 = \dfrac{37}{4}, d_4 = \dfrac{13}{4}$ **b.** 6 **c.** 6
d. They are the same. **e.** The difference in distances equals the length of the transverse axis. **9.** x-axis **11.** y-axis

13. a. Center: $(0, 0)$ **e.**
b. Vertices: $(4, 0), (-4, 0)$
c. Foci: $(\sqrt{41}, 0), (-\sqrt{41}, 0)$
d. $y = \frac{5}{4}x$ and $y = -\frac{5}{4}x$

15. a. Center: $(0, 0)$ **e.**
b. Vertices: $(0, 2), (0, -2)$
c. Foci: $(0, 2\sqrt{10}), (0, -2\sqrt{10})$
d. $y = \frac{1}{3}x$ and $y = -\frac{1}{3}x$

17. a. Center: $(0, 0)$ **e.**
b. Vertices: $(0, 9), (0, -9)$
c. Foci: $(0, \sqrt{106}), (0, -\sqrt{106})$
d. $y = \frac{9}{5}x$ and $y = -\frac{9}{5}x$

19. a. Center: $(0, 0)$ **e.**
b. Vertices: $(\sqrt{7}, 0), (-\sqrt{7}, 0)$
c. Foci: $(2\sqrt{3}, 0), (-2\sqrt{3}, 0)$
d. $y = \frac{\sqrt{35}}{7}x$ and $y = -\frac{\sqrt{35}}{7}x$

21. a. Center: $(0, 0)$ **e.**
b. Vertices: $(\frac{5}{2}, 0), (-\frac{5}{2}, 0)$
c. Foci: $(\frac{\sqrt{149}}{4}, 0), (-\frac{\sqrt{149}}{4}, 0)$
d. $y = \frac{7}{10}x$ and $y = -\frac{7}{10}x$

23. a. Center: $(4, -2)$ **e.**
b. Vertices: $(1, -2), (7, -2)$
c. Foci: $(-1, -2), (9, -2)$
d. $y = \frac{4}{3}x - \frac{22}{3}$ and $y = -\frac{4}{3}x + \frac{10}{3}$

25. a. Center: $(-3, 5)$ **e.**
b. Vertices: $(-3, -2), (-3, 12)$
c. Foci: $(-3, 5 + \sqrt{74}), (-3, 5 - \sqrt{74})$
d. $y = \frac{7}{5}x + \frac{46}{5}$ and $y = -\frac{7}{5}x + \frac{4}{5}$

27. a. Center: $(-4, 7)$ **e.**
b. Vertices: $(6, 7), (-14, 7)$
c. Foci: $(-4 + \sqrt{181}, 7), (-4 - \sqrt{181}, 7)$
d. $y = \frac{9}{10}x + \frac{53}{5}$ and $y = -\frac{9}{10}x + \frac{17}{5}$

29. a. Center: $(3, 0)$ **e.**
b. Vertices: $(3, 1), (3, -1)$
c. Foci: $(3, \sqrt{13}), (3, -\sqrt{13})$
d. $y = \frac{\sqrt{3}}{6}x - \frac{\sqrt{3}}{2}$ and $y = -\frac{\sqrt{3}}{6}x + \frac{\sqrt{3}}{2}$

31. a. Center: $(0, 4)$

    b. Vertices: $(1, 4), (-1, 4)$

    c. Foci: $(3, 4), (-3, 4)$

    d. $y = 2\sqrt{2}x + 4$ and $y = -2\sqrt{2}x + 4$

e.

33. a. $\dfrac{(x + 3)^2}{5} - \dfrac{(y - 1)^2}{7} = 1$

    b. Center: $(-3, 1)$; Vertices: $(-3 + \sqrt{5}, 1), (-3 - \sqrt{5}, 1)$;

    Foci: $(-3 + 2\sqrt{3}, 1), (-3 - 2\sqrt{3}, 1)$

35. a. $\dfrac{(y - 4)^2}{5} - \dfrac{(x - 2)^2}{9} = 1$

    b. Center: $(2, 4)$; Vertices: $(2, 4 + \sqrt{5}), (2, 4 - \sqrt{5})$;

    Foci: $(2, 4 + \sqrt{14}), (2, 4 - \sqrt{14})$

37. a. $x^2 - \dfrac{(y + 5)^2}{4} = 1$ **b.** Center: $(0, -5)$;

    Vertices: $(1, -5), (-1, -5)$; Foci: $(\sqrt{5}, -5), (-\sqrt{5}, -5)$

39. a. $\dfrac{(y + 2)^2}{9} - \dfrac{(x - \frac{3}{2})^2}{16} = 1$

    b. Center: $(\frac{3}{2}, -2)$; Vertices: $(\frac{3}{2}, 1), (\frac{3}{2}, -5)$; Foci: $(\frac{3}{2}, 3), (\frac{3}{2}, -7)$

41. $\dfrac{x^2}{144} - \dfrac{y^2}{25} = 1$ **43.** $\dfrac{y^2}{144} - \dfrac{x^2}{81} = 1$

45. $\dfrac{(x - 2)^2}{25} - \dfrac{(y + 3)^2}{49} = 1$ **47.** $\dfrac{(y - 4)^2}{16} - \dfrac{(x + 3)^2}{5} = 1$

49. $\dfrac{(x - 1)^2}{49} - \dfrac{(y - 2)^2}{25} = 1$

51. a. Equation 1: $e = \frac{5}{4}$, Equation 2: $e = \frac{5}{3}$ **b.** Graph B represents Equation 1. Graph A represents Equation 2.

53. $e = \dfrac{41}{40}$ **55.** $e = \dfrac{5}{4}$ **57.** $\dfrac{(x - 3)^2}{16} - \dfrac{(y + 1)^2}{9} = 1$

59. $\dfrac{x^2}{(372)^2} - \dfrac{y^2}{(334)^2} = 1$ **61.** $\dfrac{x^2}{4} - \dfrac{y^2}{81} = 1; x \geq 0$ or $x = 2\sqrt{1 + \dfrac{y^2}{81}}$

63. 12 microns   **65. a.** 0.4 AU   **b.** 37,200,000 mi

67. Ellipse; B   **69.** Hyperbola; D

71. $\dfrac{(x + 1)^2}{16} + \dfrac{y^2}{25} = 1$ **73.** $\dfrac{(y + 3)^2}{9} - \dfrac{(x - 2)^2}{16} = 1$

75. $\{(4, 3\sqrt{3}), (4, -3\sqrt{3}), (-4, 3\sqrt{3}), (-4, -3\sqrt{3})\}$

77. The transverse axis is horizontal if the coefficient of the x^2 term is positive. The transverse axis is vertical if the coefficient of the y^2 term is positive.   **79.** $\sqrt{2}$

81. $\dfrac{x^2}{a^2} - \dfrac{y^2}{b^2} = 1$ $\dfrac{c^2}{a^2} - \dfrac{y^2}{b^2} = 1$ $y = b\sqrt{\dfrac{c^2}{a^2} - 1}$

    $y = b\sqrt{\dfrac{c^2 - a^2}{a^2}}$ $y = \dfrac{b}{a}\sqrt{c^2 - a^2}$

    Recall that $c^2 = a^2 + b^2$ or equivalently $b^2 = c^2 - a^2$ and $b > 0$.

    Therefore, $y = \dfrac{b}{a}\sqrt{b^2}$ or $y = \dfrac{b^2}{a}$.

    The length of a latus rectum is $2y = \dfrac{2b^2}{a}$.

83. a. $\dfrac{x^2}{a^2} + \dfrac{y^2}{b^2} = 1$; This is an equation of an ellipse in the xy-plane.

    b. $\dfrac{y^2}{b^2} - \dfrac{z^2}{c^2} = 1$; This is an equation of a hyperbola in the yz-plane with transverse axis on the y-axis.   **c.** $\dfrac{x^2}{a^2} - \dfrac{z^2}{c^2} = 1$; This is an equation of a hyperbola in the xz-plane with transverse axis on the x-axis.

85.
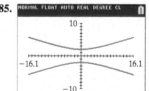
87.

Section 7.3 Practice Exercises, pp. 675–680

R.1. $5\sqrt{2}$ **R.2.** $y = 3$ **R.3.** $x = 2$

1. parabola; directrix; focus   **3.** vertex   **5.** left   **7.** latus; rectum   **9.** (h, k); $(h, k + p)$; $y = k - p$ **11.** vertically

13. a. $d_1 = 2, d_2 = 2, d_3 = 5, d_4 = 5$ **b.** They are the same.   **c.** They are the same.   **15. a.** $p = 6$ **b.** $(0, 6)$ **c.** $y = -6$

17. a. $p = 9$ **b.** $(9, 0)$ **c.** $x = -9$ **19. a.** $p = -\frac{5}{4}$ **b.** $(0, -\frac{5}{4})$ **c.** $y = \frac{5}{4}$ **21. a.** $p = -\frac{1}{4}$ **b.** $(-\frac{1}{4}, 0)$ **c.** $x = \frac{1}{4}$

23. a. Focus: $(0, 6.3)$; Place the receiver 6.3 in. above the center of the dish.   **b.** $y = -6.3$ **25.** $\frac{1}{2}$ unit to the right of the vertex

27. a. Vertex: $(0, 0)$; $p = -1$; Focus: $(0, -1)$; Focal diameter: 4

    b. $(2, -1), (-2, -1)$

    d. Directrix: $y = 1$; Axis of symmetry: $x = 0$

c.

29. a. Vertex: $(0, 0)$; $p = 2$; Focus: $(2, 0)$; Focal diameter: 8

    b. $(2, 4), (2, -4)$

    d. Directrix: $x = -2$; Axis of symmetry: $y = 0$

c.
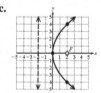

31. a. Vertex: $(0, 0)$; $p = \frac{5}{2}$; Focus: $(0, \frac{5}{2})$; Focal diameter: 10

    b. $(5, \frac{5}{2}), (-5, \frac{5}{2})$

    d. Directrix: $y = -\frac{5}{2}$; Axis of symmetry: $x = 0$

c.
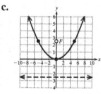

33. a. Vertex: $(0, 0)$; $p = -\frac{1}{4}$; Focus: $(-\frac{1}{4}, 0)$; Focal diameter: 1

    b. $(-\frac{1}{4}, \frac{1}{2}), (-\frac{1}{4}, -\frac{1}{2})$

    d. Directrix: $x = \frac{1}{4}$; Axis of symmetry: $y = 0$

c.

35. a. Vertex: $(4, -1)$; $p = -3$; Focus: $(1, -1)$; Focal diameter: 12

    b. $(1, 5), (1, -7)$

    d. Directrix: $x = 7$; Axis of symmetry: $y = -1$

c.

37. a. Vertex: $(1, -5)$; $p = -1$; Focus: $(1, -6)$; Focal diameter: 4

    b. $(3, -6), (-1, -6)$

    d. Directrix: $y = -4$; Axis of symmetry: $x = 1$

c.
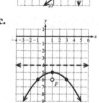

39. a. Vertex: $(-3, \frac{3}{2})$; $p = \frac{1}{2}$; Focus: $(-3, 2)$; Focal diameter: 2

    b. $(-4, 2), (-2, 2)$

    d. Directrix: $y = 1$; Axis of symmetry: $x = -3$

c.
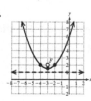

41. a. Vertex: $\left(\frac{1}{4}, 3\right)$; $p = -\frac{7}{4}$;
Focus: $\left(-\frac{3}{2}, 3\right)$;
Focal diameter: 7
b. $\left(-\frac{3}{2}, \frac{13}{2}\right), \left(-\frac{3}{2}, -\frac{1}{2}\right)$
d. Directrix: $x = 2$
Axis of symmetry: $y = 3$

c.

43. a. Vertex: $(-6, 3)$; $p = 5$;
Focus: $(-6, 8)$; Focal diameter: 20
b. $(4, 8), (-16, 8)$
d. Directrix: $y = -2$;
Axis of symmetry: $x = -6$

c.

45. a. $(x - 3)^2 = 4(y + 1)$ **b.** Vertex: $(3, -1)$; Focus: $(3, 0)$;
Focal diameter: 4 **47. a.** $(y + 2)^2 = -8(x + 6)$
b. Vertex: $(-6, -2)$; Focus: $(-8, -2)$; Focal diameter: 8
49. a. $\left(x - \frac{7}{2}\right)^2 = -6(y + 1)$ **b.** Vertex: $\left(\frac{7}{2}, -1\right)$; Focus: $\left(\frac{7}{2}, -\frac{5}{2}\right)$;
Focal diameter: 6 **51. a.** $\left(y + \frac{3}{4}\right)^2 = (x - 3)$
b. Vertex: $\left(3, -\frac{3}{4}\right)$; Focus: $\left(\frac{13}{4}, -\frac{3}{4}\right)$; Focal diameter: 1
53. $(1, 7)$; 3 **55.** $x = 1$; -4 **57.** 12 **59.** No; Focal length is
a distance and gives no information regarding the orientation of a
parabola. **61.** $y^2 = -16x$ **63.** $(x - 2)^2 = 12(y - 1)$
65. $(x + 6)^2 = -4(y + 1)$ **67.** $(x - 2)^2 = 8(y - 3)$ or
$(y - 3)^2 = (x - 2)$ **69. a.** $x^2 = 36y$ for $-12 \le x \le 12$
b. $(0, 9)$; The pot should be placed 9 in. above the vertex.
71. a. $x^2 = 230.4y$ for $-1.2 \le x \le 1.2$ **b.** 0.00625 m or 6.25 mm
73. a. $x^2 = 60y$ for $-4.1 \le x \le 4.1$ **b.** $\frac{1681}{6000}$ m ≈ 0.28 m

c. $m = \frac{41}{600} \approx 0.068$ **75. a.** $x^2 = -\frac{160}{3}y$ **b.** $\frac{40}{3}$ ft

c. 67.5 ft **77.** $\{(5, -1), (8, -7)\}$
79. If the y term is linear, then the parabola opens vertically. If the x term
is linear, then the parabola opens horizontally.

81. a. $z = y^2$; This is an equation of a parabola in the yz-plane.
b. $z = 4x^2$; This is an equation of a parabola in the xz-plane.
c. $0 = 4x^2 + y^2$; This is an equation of a degenerate ellipse.
The solution set is a single point (the origin).

83. 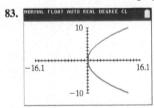 **85.**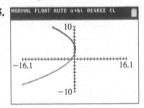

Problem Recognition Exercises, pp. 680–681

1. Hyperbola; Center: $(2, -2)$; Vertices: $(6, -2), (-2, -2)$;
Foci: $(7, -2), (-3, -2)$; Asymptotes: $y = \frac{3}{4}x - \frac{7}{2}$ and $y = -\frac{3}{4}x - \frac{1}{2}$;
Eccentricity: $\frac{5}{4}$ **2.** Ellipse: Center: $(0, 2)$; Vertices: $(5, 2), (-5, 2)$;
Endpoints of minor axis: $(0, 5), (0, -1)$; Foci: $(4, 2), (-4, 2)$;
Eccentricity: $\frac{4}{5}$
3. Parabola; Vertex: $(-2, 5)$; Focus: $\left(-\frac{9}{4}, 5\right)$; Directrix: $x = -\frac{7}{4}$;
Axis of symmetry: $y = 5$
4. Circle; Center: $(3, -7)$; Radius: 5 **5.** Ellipse; Center: $(-1, 0)$;
Vertices: $(-1, 4), (-1, -4)$; Endpoints of minor axis: $(-2, 0), (0, 0)$;
Foci: $\left(-1, \sqrt{15}\right), \left(-1, -\sqrt{15}\right)$; Eccentricity: $\frac{\sqrt{15}}{4}$
6. Parabola; Vertex: $(1, -3)$; Focus: $\left(1, -\frac{1}{2}\right)$;
Directrix: $y = -\frac{11}{2}$; Axis of symmetry: $x = 1$
7. Circle; Center: $(-1, -6)$; Radius: 4
8. Hyperbola; Center: $(-4, 1)$; Vertices: $(-4, 13), (-4, -11)$;
Foci: $(-4, 14), (-4, -12)$; Asymptotes: $y = \frac{12}{5}x + \frac{53}{5}$ and
$y = -\frac{12}{5}x - \frac{43}{5}$; Eccentricity: $\frac{13}{12}$

9. Hyperbola; $\dfrac{(x - 2)^2}{16} - \dfrac{(y + 2)^2}{9} = 1$ **10.** Ellipse;

$\dfrac{x^2}{25} + \dfrac{(y - 2)^2}{9} = 1$ **11.** Parabola; $(y - 5)^2 = -(x + 2)$

12. Circle; $(x - 3)^2 + (y + 7)^2 = 25$ **13.** Ellipse; $(x + 1)^2 + \dfrac{y^2}{16} = 1$

14. Parabola; $(x - 1)^2 = 10(y + 3)$ **15.** Circle; $(x + 1)^2 + (y + 6)^2 = 16$

16. Hyperbola; $\dfrac{(y - 1)^2}{144} - \dfrac{(x + 4)^2}{25} = 1$

17. $(x - 2)^2 + (y + 1)^2 = 0$; The graph is a single point: $(2, -1)$.

18. $\dfrac{x^2}{4} + \dfrac{(y - 3)^2}{9} = 0$; The graph is a single point: $(0, 3)$.

19. $(x - 4)^2 - \dfrac{(y + 2)^2}{4} = 0$; The graph is a pair of intersecting lines:
$y = -2x + 6$ and $y = 2x - 10$.

20. $\dfrac{(x - 5)^2}{4} - \dfrac{(y + 1)^2}{9} = 0$; The graph is a pair of intersecting lines:
$y = \frac{3}{2}x - \frac{17}{2}$ and $y = -\frac{3}{2}x + \frac{13}{2}$.

21. $\dfrac{x^2}{4} + \dfrac{(y - 2)^2}{9} = -1$; No solution. There are no real numbers x and
y that would make the sum of two squares equal to -1.

22. $(x + 6)^2 + y^2 = -1$; No solution. There are no real numbers x and y
that would make the sum of two squares equal to -1.

23. With $C = A$ and completing the square, the equation
$Ax^2 + Cy^2 + Dx + Ey + F = 0$ becomes
$Ax^2 + Ay^2 + Dx + Ey + F = 0$
$x^2 + y^2 + \dfrac{D}{A}x + \dfrac{E}{A}y + \dfrac{F}{A} = 0$
$\left(x + \dfrac{D}{2A}\right)^2 + \left(y + \dfrac{E}{2A}\right)^2 = \dfrac{D^2}{4A^2} + \dfrac{E^2}{4A^2} - \dfrac{F}{A}$
$\left(x + \dfrac{D}{2A}\right)^2 + \left(y + \dfrac{E}{2A}\right)^2 = \dfrac{D^2 + E^2 - 4AF}{4A^2}$
This is the standard form of an equation of circle with center
$\left(-\dfrac{D}{2A}, -\dfrac{E}{2A}\right)$ and radius $\dfrac{\sqrt{D^2 + E^2 - 4AF}}{2A}$ provided
$D^2 + E^2 - 4AF > 0$.

24. Let $k = \dfrac{D^2}{4A} + \dfrac{E^2}{4C} - F$ and assume that $k \ne 0$. Then, by completing
the square,
$Ax^2 + Cy^2 + Dx + Ey + F = 0$ becomes
$A\left(x + \dfrac{D}{2A}\right)^2 + C\left(y + \dfrac{E}{2C}\right)^2 = \dfrac{D^2}{4A} + \dfrac{E^2}{4C} - F$
$A\left(x + \dfrac{D}{2A}\right)^2 + C\left(y + \dfrac{E}{2C}\right)^2 = k$
$\dfrac{\left(x + \dfrac{D}{2A}\right)^2}{\dfrac{k}{A}} + \dfrac{\left(y + \dfrac{E}{2C}\right)^2}{\dfrac{k}{C}} = 1$

a. The numerator of each term on the left side of the equation is
nonnegative. Therefore, the sign of each term is dictated by the
denominator. For A and C of the same sign, if k has the same sign
as A, it also has the same sign as C and is nonzero. Thus, the
denominators $\dfrac{k}{A}$ and $\dfrac{k}{C}$ are both positive. Therefore, the equation
is in the standard form of an ellipse with center $\left(-\dfrac{D}{2A}, -\dfrac{E}{2C}\right)$.

b. For A and C of opposite signs and $k \ne 0$, the denominators $\dfrac{k}{A}$ and $\dfrac{k}{C}$
have opposite signs, indicating that the terms on the left side of the
equation have opposite signs. Therefore, the equation is a
hyperbola with center $\left(-\dfrac{D}{2A}, -\dfrac{E}{2C}\right)$.

25. If $A = 0$, then $Ax^2 + Cy^2 + Dx + Ey + F = 0$ becomes $Cy^2 + Dx + Ey + F = 0$. Completing the square gives

$$C\left(y^2 + \frac{E}{C}y + \frac{E^2}{4C^2}\right) = -Dx - F + \frac{E^2}{4C}$$

$$C\left(y + \frac{E}{2C}\right)^2 = -D\left(x + \frac{F}{D} - \frac{E^2}{4CD}\right)$$

$$\left(y + \frac{E}{2C}\right)^2 = -\frac{D}{C}\left[x - \left(\frac{E^2}{4CD} - \frac{F}{D}\right)\right]$$

Assuming that $C \neq 0$ and $D \neq 0$, this is the standard form of an equation of a parabola opening to the left or right with vertex $\left(\frac{E^2}{4CD} - \frac{F}{D}, -\frac{E}{2C}\right)$.

Chapter 7 Review Exercises, pp. 683–685

1. Each equation represents an ellipse with a vertical major axis of length 30 units and horizontal minor axis of length 20 units. However, the first equation represents an ellipse centered at $(0, 0)$, whereas the second equation represents an ellipse centered at $(1, -7)$.

3. a. $(0, 0)$ **b.** $\left(0, -\sqrt{15}\right), \left(0, \sqrt{15}\right)$ **f.**
c. $(3, 0), (-3, 0)$
d. $\left(0, -\sqrt{6}\right), \left(0, \sqrt{6}\right)$
e. $\dfrac{\sqrt{10}}{5}$

5. a. $(1, 2)$ **b.** $(-3, 2), (5, 2)$ **f.**
c. $(1, -1), (1, 5)$
d. $\left(1 - \sqrt{7}, 2\right), \left(1 + \sqrt{7}, 2\right)$
e. $\dfrac{\sqrt{7}}{4}$

7. a. $\dfrac{\left(x - \frac{1}{2}\right)^2}{16} + \dfrac{y^2}{25} = 1$ **b.** Center: $\left(\frac{1}{2}, 0\right)$; Vertices: $\left(\frac{1}{2}, 5\right), \left(\frac{1}{2}, -5\right)$; Endpoints of minor axis: $\left(-\frac{7}{2}, 0\right), \left(\frac{9}{2}, 0\right)$; Foci: $\left(\frac{1}{2}, 3\right), \left(\frac{1}{2}, -3\right)$

9. $\dfrac{(x - 3)^2}{9} + \dfrac{(y - 1)^2}{25} = 1$ **11.** The second ellipse with eccentricity $\frac{5}{7}$ is more elongated.   **13.** The height 50 ft from the center is approximately 97 ft.   **15.** Horizontal

17. a. $(0, 0)$ **b.** $(3, 0), (-3, 0)$ **e.**
c. $\left(\sqrt{13}, 0\right), \left(-\sqrt{13}, 0\right)$
d. $y = \frac{2}{3}x$ and $y = -\frac{2}{3}x$

19. a. $(-3, 2)$ **b.** $(-3, 5), (-3, -1)$ **e.**
c. $(-3, 7), (-3, -3)$
d. $y = \frac{3}{4}x + \frac{17}{4}$ and $y = -\frac{3}{4}x - \frac{1}{4}$

21. a. $\dfrac{(x + 2)^2}{7} - \dfrac{(y + 4)^2}{11} = 1$
b. Center: $(-2, -4)$;
Vertices: $\left(-2 + \sqrt{7}, -4\right), \left(-2 - \sqrt{7}, -4\right)$
Foci: $\left(-2 + 3\sqrt{2}, -4\right), \left(-2 - 3\sqrt{2}, -4\right)$

23. $\dfrac{x^2}{16} - \dfrac{y^2}{9} = 1$ **25.** $\dfrac{(x - 2)^2}{289} - \dfrac{(y - 7)^2}{64} = 1$

27. $\{(1, 3), (1, -3), (-1, 3), (-1, -3)\}$

29. a. $\dfrac{x^2}{15^2} - \dfrac{(y - 30)^2}{45^2} = 1$ **b.** 36 ft   **c.** 67 ft

31. a. $p = 5$ **b.** $(0, 0)$
c. $(5, 0)$ **d.** 20
e. $(5, 10), (5, -10)$
f. $x = -5$ **g.** $y = 0$
h.

33. a. $p = \frac{3}{2}$ **b.** $\left(\frac{1}{2}, 3\right)$
c. $\left(\frac{1}{2}, \frac{9}{2}\right)$ **d.** 6
e. $\left(-\frac{5}{2}, \frac{9}{2}\right), \left(\frac{7}{2}, \frac{9}{2}\right)$
f. $y = \frac{3}{2}$ **g.** $x = \frac{1}{2}$
h.

35. a. $\left(y - \frac{3}{2}\right)^2 = -14(x + 3)$
b. Vertex: $\left(-3, \frac{3}{2}\right)$; Focus: $\left(-\frac{13}{2}, \frac{3}{2}\right)$; Directrix: $x = \frac{1}{2}$

37. $x = 18; -8$ **39.** $(x + 4)^2 = -32(y - 7)$ **41.** $\{(8, 1), (5, -2)\}$

43. a. $x^2 = -\dfrac{135}{4}y$ or $y = -\dfrac{4}{135}x^2$ for $-45 \leq x \leq 45$
b. $\dfrac{135}{16}$ ft

Chapter 7 Test, pp. 685–686

1. Each equation represents an ellipse centered at the origin with a major axis of length 24 units and minor axis of length 12 units. However, the ellipse represented by the first equation has foci on the y-axis, whereas the second equation represents an ellipse with foci on the x-axis.

2. $(x - h)^2 + (y - k)^2 = r^2$

3. $\dfrac{(x - h)^2}{a^2} + \dfrac{(y - k)^2}{b^2} = 1$ **4.** $\dfrac{(x - h)^2}{b^2} + \dfrac{(y - k)^2}{a^2} = 1$

5. $\dfrac{(y - k)^2}{a^2} - \dfrac{(x - h)^2}{b^2} = 1$ **6.** $\dfrac{(x - h)^2}{a^2} - \dfrac{(y - k)^2}{b^2} = 1$

7. $(x - h)^2 = 4p(y - k)$ **8.** $(y - k)^2 = 4p(x - h)$

9. a. Ellipse   **b.**
c. Center: $(2, 0)$; Vertices: $(2, 4), (2, -4)$;
Endpoints of minor axis: $(-1, 0), (5, 0)$;
Foci: $\left(2, \sqrt{7}\right), \left(2, -\sqrt{7}\right)$;
Eccentricity: $\dfrac{\sqrt{7}}{4}$

10. a. Hyperbola   **b.**
c. Center: $(2, 0)$; Vertices: $(-1, 0), (5, 0)$;
Foci: $(-3, 0), (7, 0)$;
Asymptotes:
$y = \frac{4}{3}x - \frac{8}{3}$ and $y = -\frac{4}{3}x + \frac{8}{3}$;
Eccentricity: $\frac{5}{3}$

11. a. Parabola   **b.**
c. Vertex: $(2, 0)$; Focus: $(2, -4)$;
Endpoints of latus rectum:
$(10, -4), (-6, -4)$
Directrix: $y = 4$; Axis of symmetry: $x = 2$

12. a. Circle   **b.**
c. Center: $(2, 3)$; Radius: $2\sqrt{3}$

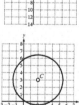

13. a. Hyperbola   **b.**
c. Center: $(0, -2)$; Vertices: $(0, 4), (0, -8)$;
Foci: $(0, 8), (0, -12)$;
Asymptotes:
$y = \frac{3}{4}x - 2$ and $y = -\frac{3}{4}x - 2$;
Eccentricity: $\frac{5}{3}$

14. a. Ellipse
  c. Center: $(-4, 1)$;
   Vertices: $(-14, 1)$, $(6, 1)$;
   Endpoints of minor axis:
   $(-4, 7)$, $(-4, -5)$;
   Foci: $(-12, 1)$, $(4, 1)$; Eccentricity: $\frac{4}{5}$

b.

15. a. Circle
  c. Center: $(-4, 1)$; Radius: 2

b.

16. a. Parabola
  c. Vertex: $(3, 4)$; Focus: $(5, 4)$;
   Endpoints of latus rectum:
   $(5, 0)$, $(5, 8)$
   Directrix: $x = 1$;
   Axis of symmetry: $y = 4$

b.

17. a. 13.2 ft   **b.** Yes; The height of the opening is approximately 12 ft at a point 3 ft from the edge.   **c.** $h(x) = 6\sqrt{1 - \dfrac{x^2}{144}} + 8$

18. 4.444×10^9 km   **19.** $\dfrac{x^2}{(3300)^2} - \dfrac{y^2}{(2189)^2} = 1$

20. a. $\dfrac{x^2}{144} - \dfrac{(y - 120)^2}{1296} = 1$ **b.** 83.5 ft   **c.** 58.5 ft

21. 113 ft   **22.** $(x - 1)^2 = 16(y - 2)$

23. $\dfrac{(x - 4)^2}{4} + \dfrac{(y + 3)^2}{3} = 1$ **24.** $\dfrac{(y - 3)^2}{16} - \dfrac{(x - 4)^2}{20} = 1$

25. $\dfrac{(x - 2)^2}{9} - \dfrac{(y + 3)^2}{16} = 1$ **26.** $\dfrac{12}{13}$ **27.** $\dfrac{\sqrt{34}}{5}$

28. Focal length: 3; Focal diameter: 12   **29.** Center: $(0, 1)$

30. $\{(1, 2), (1, -2), (-1, 2), (-1, -2)\}$ **31.** The graph is a left semiellipse with center at the origin and major axis of length 8 units and minor axis of length 6 units.

Chapter 7 Cumulative Review Exercises, p. 687

1. $3x^2 - 2x - 9 + \dfrac{x + 28}{x^2 + 3}$ **2.** No   **3.** Yes   **4.** $\dfrac{3\sqrt[3]{x^2}}{x}$

5. $\dfrac{z^{12}}{x^{18}y^{10}}$ **6.** $\left(-122, \dfrac{171}{2}\right)$ **7.** $f^{-1}(x) = x^3 + 7$ **8.** $y = -12x + 20$

9. Neither   **10.** $\left\{\dfrac{-2 + \ln 22}{3}\right\}$ **11.** $\left(\dfrac{37}{9}, \infty\right)$ **12.** $\{\ \}$

13. $\left(-\infty, \dfrac{4 - \sqrt{10}}{2}\right] \cup \left[\dfrac{4 + \sqrt{10}}{2}, \infty\right)$

14. $\{9\}$; The value -3 does not check.   **15.** $\{\ln 5, \ln 3\}$

16.   **17.**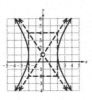

18. $\dfrac{3}{x - 3} + \dfrac{5}{x - 4}$ **19. a.** $(f \circ g)(x) = \dfrac{1}{\sqrt{x - 1} - 2}$
  b. $[1, 5) \cup (5, \infty)$

20. a. $(0, 0)$, $(4, 0)$ **b.** $(0, 0)$ **c.** $x = -3$
  d. None   **e.** $y = x - 7$

21.   **22.**

---

23.   **24.**

25. $\{(-1, 0, 4)\}$ **26.** $\{(3, 1), (3, -1), (-3, 1), (-3, -1)\}$

27. $\begin{bmatrix} -12 & 64 \\ 6 & 32 \end{bmatrix}$ **28.** $\begin{bmatrix} \frac{3}{8} & -\frac{1}{4} \\ \frac{1}{16} & \frac{1}{8} \end{bmatrix}$ **29.** -12 **30.** $z = \dfrac{97}{146}$

CHAPTER 8

Section 8.1 Practice Exercises, pp. 697–701

R.1. $\dfrac{1}{3}$ **R.2.** 5   **R.3.** 18   **R.4.** $\dfrac{(2r - 1)^3}{(r - 2)^2}$ **R.5.** $\dfrac{(p - 6)^3}{p^3}$

1. sequence; finite   **3.** recursive   **5.** index; summation; lower; limit
7. $5, 11, 21, 35$ **9.** $-6, 3, -\frac{3}{2}, \frac{3}{4}$ **11.** $2, \sqrt{5}, \sqrt{6}, \sqrt{7}$
13. $1, 8, 27, 64$ **15.** The odd-numbered terms are negative, and the even-numbered terms are positive.   **17.** 1023   **19.** 781
21. c   **23.** a   **25.** $-3, -5, -11, -29, -83$ **27.** $4, 6, 7, \frac{15}{2}, \frac{31}{4}$
29. $5, \frac{1}{5}, 5, \frac{1}{5}, 5$ **31. a.** $1, 1, 2, 3, 5, 8, 13, 21, 34, 55$
  b. $F_1 = 1, F_2 = 1, F_3 = 2$ **33.** 5040   **35.** 1   **37.** 56
39. 126   **41.** $\dfrac{1}{n(n + 1)}$ **43.** $\dfrac{1}{2n + 1}$ **45.** $a_5 = \dfrac{2}{315}$
47. $c_3 = 24{,}192$ **49.** $a_n = 2^n$ **51.** $a_n = -\dfrac{n + 1}{3n}$
53. $a_n = (-1)^n(n^2)$ **55.** $a_n = \dfrac{(n + 1)!}{2n}$ **57.** 39   **59.** -110
61. $\dfrac{15}{32}$ **63.** 120   **65.** -30 **67.** $\dfrac{29}{6}$ **69.** 62
71. a. 0   **b.** -1 **73.** 25   **75.** $\displaystyle\sum_{i=1}^{n} \dfrac{i^2}{i + 1}$ **77.** $\displaystyle\sum_{i=1}^{5} i$
79. $\displaystyle\sum_{i=1}^{4} 8$ **81.** $\displaystyle\sum_{i=1}^{4} (-1)^{i+1} \dfrac{1}{3^i}$ **83.** $\displaystyle\sum_{i=1}^{18} c^{i+2}$ **85.** $\displaystyle\sum_{i=1}^{4} \dfrac{x^i}{i!}$
87. $\displaystyle\sum_{i=1}^{8} i^2 = \sum_{j=0}^{7} (j + 1)^2 = \sum_{k=2}^{9} (k - 1)^2$

89. Expanding the series, there are n terms and each term is c. Therefore, the sum is cn.
$$\sum_{i=1}^{n} c = c + c + c + \cdots + c = cn$$
91. 46,750   **93.** 6575   **95.** True
97. False; $\displaystyle\sum_{i=1}^{n} a_i b_i = a_1 b_1 + a_2 b_2 + \cdots + a_n b_n$
  $\neq (a_1 + a_2 + \cdots + a_n)(b_1 + b_2 + \cdots + b_n)$
99. $a_n = 1.03a_{n-1} + 1000; n \geq 2$
101. 66; If 12 people are present at the meeting, then there will be 66 handshakes.   **103.** 544   **105.** The graph of the sequence $a_n = n^2$ is a set of discrete points corresponding to $n = 1, 2, 3, \ldots$, whereas the function $f(x) = x^2$ is a continuous curve over the set of real numbers.
107. The sequence defined by $a_n = \dfrac{n}{n - 1}$ is not defined for $n = 1$ because the denominator would be zero.   **109.** $i, -1, -i, 1, i, -1, -i, 1$
111. $2, 1.5, 1.4167, 1.4142$

113.   **115.**

117.

```
NORMAL FLOAT AUTO REAL RADIAN MP
  10
  Σ (1/n!)
 n=0
                    2.718281801
  15
  Σ (1/n!)
 n=0
                    2.718281828
 e¹
                    2.718281828
```

Section 8.2 Practice Exercises, pp. 708–712

R.1. $f(2) = 5$ **R.2.** Slope: $-\dfrac{1}{4}$; y-intercept: $(0, 8)$

R.3. $\{(1, -2)\}$ **R.4.** $\{(-4, 1)\}$

1. arithmetic   **3.** $a_1 + (n-1)d$; $a_{n-1} + d$ **5.** partial

7. Yes; $d = 4$ **9.** Yes; $d = -11$ **11.** No   **13.** Yes; $d = \dfrac{2}{3}$

15. a. 3, 13, 23, 33, 43   **b.** $a_1 = 3$ and $a_n = a_{n-1} + 10$ for $n \geq 2$

17. a. 4, 2, 0, -2, -4 **b.** $a_1 = 4$ and $a_n = a_{n-1} - 2$ for $n \geq 2$

19. a. $a_n = 10n - 3$ **b.** $a_{22} = 217$ **21. a.** $a_n = 5n - 17$

b. $a_{20} = 83$ **23. a.** $a_n = \dfrac{1}{3}n + \dfrac{1}{6}$ **b.** $a_{10} = \dfrac{7}{2}$

25. $a_n = 22n + 8$ **27. a.** Yes   **b.** $a_n = 16n + 18$ **c.** 178

29. $v_{n+1} - v_n = [v_0 + a(n+1)] - (v_0 + an)$
$= v_0 + an + a - v_0 - an = a$
The difference between two consecutive terms is the constant a. Therefore, the sequence is arithmetic.

31. $a_8 = 33$ **33.** $a_{35} = -460$ **35.** $n = 53$ **37.** $n = 49$

39. $a_1 = 44$; $d = 8$ **41.** $a_{150} = 896$ **43.** $b_{214} = -2487$

45. $c_{400} = 104$ **47.** $a_1 = 0$, $a_2 = -3$ **49.** 246   **51.** 3940

53. -918 **55.** 279   **57.** 2850   **59.** -6115.5

61. 264   **63.** -207 **65. a.** Job 1 for 5 yr: \$352,000; Job 2 for 5 yr: \$350,000   **b.** Job 1 for 10 yr: \$784,000; Job 2 for 10 yr: \$825,000   **67. a.** $d_n = 32n - 16$ **b.** 240 ft   **c.** 1024 ft

69. a. 4 cups   **b.** 114 cups   **71. a.** 6, 4, 2, 0   **b.** $b_n = -2n + 8$ **c.** $b_{30} = -52$ **d.** -690 **e.** $n = 94$ **f.** -136

73. 32,686   **75.** 6375   **77.** 1410

79. a. Substitute $a_1 = 1$ and $a_n = n$. $S_n = \dfrac{n}{2}(a_1 + a_n) = \dfrac{n}{2}(1 + n)$ **b.** 5050   **c.** 500,500

81. The three arithmetic means between 4 and 28 are 10, 16, and 22.

83. The terms in the given sequence $a_1, a_2, a_3, \ldots$ differ by d units. Therefore, every other term would differ by $2d$ units. Thus, $a_1, a_3, a_5, \ldots$ is an arithmetic sequence with common difference $2d$.

85. $a_n = 2n + 1$; The sequence is $\{3, 5, 7, \ldots, (2n+1), \ldots\}$.

Section 8.3 Practice Exercises, pp. 720–724

R.1. $-\dfrac{5}{4}$ **R.2.** $\{4\}$

R.3. $g(0) = 1$, $g(1) = \dfrac{1}{4}$, $g(2) = \dfrac{1}{16}$, $g(-1) = 4$, $g(-2) = 16$

R.4. $h(0) = 1$, $h(1) = 16$, $h(-1) = \dfrac{1}{16}$

R.5. a. \$17,066   **b.** \$17,091   **c.** \$17,097

1. geometric; ratio   **3.** finite   **5.** 0   **7.** annuity

9. Yes; $r = 3$ **11.** Yes; $r = -\dfrac{1}{2}$ **13.** No   **15.** Yes; $r = \sqrt{5}$

17. Yes; $r = \dfrac{2}{t}$ **19.** 7, 14, 28, 56, 112   **21.** 24, -16, $\frac{32}{3}$, $-\frac{64}{9}$, $\frac{128}{27}$

23. 36, 18, 9, $\frac{9}{2}$, $\frac{9}{4}$ **25.** $a_n = 5(2)^{n-1}$ **27.** $a_n = -2\left(\dfrac{1}{2}\right)^{n-1}$

29. $a_n = \dfrac{16}{3}\left(-\dfrac{3}{4}\right)^{n-1}$ **31. a.** $a_n = 100,000(0.85)^n$ or $85,000(0.85)^{n-1}$

b. \$44,000   **33.** 331 cases   **35.** $a_6 = -\dfrac{128}{81}$ **37.** $a_{12} = 4096$

39. $a_7 = -\dfrac{3}{16}$ **41.** $a_1 = -9$ **43.** $a_1 = 3$ **45.** $a_1 = 9$; $r = 2$

47. $a_1 = 128$; $r = -\dfrac{3}{4}$ **49.** 3069   **51.** $\dfrac{4118}{243}$ **53.** $\dfrac{1820}{81}$

55. 19,682   **57.** $\dfrac{665}{243}$ **59.** $\dfrac{5}{4}$ **61.** $-\dfrac{8}{3}$

63. Sum does not exist   **65.** 3   **67.** Sum does not exist

69. 32,760   **71.** 2   **73.** $\dfrac{8}{9}$ **75.** $\dfrac{29}{45}$ **77.** $\dfrac{9}{11}$ **79.** $\dfrac{3391}{990}$

81. a. \$150,000,000   **b.** \$468,750,000   **83.** \$32,000

85. \$166,451.73   **87. a.** \$46,204.09   **b.** \$92,408.18; The value of the annuity doubles.   **c.** \$199,149.07; The value of the annuity more than doubles.   **89. a.** Decreasing   **b.** Increasing

91. Graph b   **93.** Graph d   **95.** 1200 in. or 100 ft   **97.** 12 ft

99. a. $a_n = 0.01(2)^{n-1}$ (dollars)   **b.** Day 10: \$5.12; Day 20: \$5242.88; Day 30: \$5,368,709.12   **c.** \$10,737,418.23

101. a. Arithmetic; \$1,770,000   **b.** Geometric; \$2,059,993   **c.** \$289,993   **103. a.** $a_n = 2^n$ **b.** 1024

105. In an arithmetic sequence, the difference between a term and its predecessor is a fixed constant. In a geometric sequence, the ratio between a term and its predecessor is a fixed constant.

107. Given the sequence $a_1, a_2, a_3, \ldots$, each term after the first is obtained by multiplying the preceding term by r. Therefore, every other term is obtained by multiplying by r^2. Therefore, the sequence $a_1, a_3, a_5, \ldots$ is geometric with common ratio r^2.

109. The sequence $\frac{1}{a_1}, \frac{1}{a_2}, \frac{1}{a_3}, \frac{1}{a_4}, \ldots$ can be written as $\frac{1}{a_1}, \frac{1}{a_1 r}, \frac{1}{a_1 r^2}, \frac{1}{a_1 r^3}, \ldots$ or equivalently as $\frac{1}{a_1}, \frac{1}{a_1}\left(\frac{1}{r}\right), \frac{1}{a_1}\left(\frac{1}{r}\right)^2, \frac{1}{a_1}\left(\frac{1}{r}\right)^3, \ldots$. Therefore, the sequence is geometric with common ratio $\frac{1}{r}$.

111. Each term of the geometric sequence $a_1, a_2, a_3, \ldots$ can be written in the form $a_1 r^{n-1}$. Therefore, $\log a_1, \log a_2, \log a_3, \ldots = \log a_1, \log a_1 r, \log a_1 r^2, \ldots, \log a_1 r^{n-1}, \log a_1 r^n, \ldots$.
$a_{n+1} - a_n = \log a_1 r^n - \log a_1 r^{n-1}$
$= (\log a_1 + \log r^n) - (\log a_1 + \log r^{n-1})$
$= (\log a_1 + n \log r) - [\log a_1 + (n-1) \log r]$
$= \log r$
The common difference is $\log r$.

113. Arithmetic; $d = \ln 2$

Problem Recognition Exercises, p. 725

1. Arithmetic; $d = \dfrac{2}{3}$ **2.** Arithmetic; $d = -\dfrac{3}{5}$

3. Geometric; $r = -1$ **4.** Geometric; $r = -2$ **5.** Neither

6. Neither   **7.** Geometric; $r = \sqrt{5}$ **8.** Geometric; $r = \sqrt{2}$

9. Arithmetic; $d = \sqrt{2}$ **10.** Arithmetic; $d = \dfrac{\sqrt{3}}{2}$ **11.** 7530

12. -1251 **13.** 0   **14.** -1 **15.** $-\dfrac{1}{4}$ **16.** 294

17. $-\dfrac{40}{9}$ **18.** $-\dfrac{781}{128}$ **19.** The sum does not exist.

20. The sum does not exist.   **21.** -2376 **22.** -4428 **23.** 8

24. 6   **25.** 216   **26.** 162   **27.** 8418   **28.** $-12,639$

Section 8.4 Skill Practice

Answers 1

Let P_n denote the statement that $2 + 4 + \cdots + 2n = n(n + 1)$.

1. P_1 is true because for $n = 1$, the sum is 2 which equals $1(1 + 1)$.

2. Assume that $2 + 4 + \cdots + 2k = k(k + 1)$ (inductive hypothesis). Show that $2 + 4 + \cdots + 2k + 2(k + 1) = (k + 1)[(k + 1) + 1] = (k + 1)(k + 2)$.
By the inductive hypothesis, $[2 + 4 + \cdots + 2k] + 2(k + 1) = [k(k + 1)] + 2k + 2 = k^2 + 3k + 2 = (k + 1)(k + 2)$ as desired.

Answers 2

Let P_n denote the statement $\dfrac{1}{1 \cdot 3} + \dfrac{1}{3 \cdot 5} + \cdots + \dfrac{1}{(2n - 1)(2n + 1)} = \dfrac{n}{2n + 1}$.

1. P_1 is true because for $n = 1$, the sum is $\dfrac{1}{1 \cdot 3} = \dfrac{1}{3}$ which equals $\dfrac{1}{2(1) + 1} = \dfrac{1}{3}$.

2. Assume that $\dfrac{1}{1 \cdot 3} + \dfrac{1}{3 \cdot 5} + \cdots + \dfrac{1}{(2k - 1)(2k + 1)} = \dfrac{k}{2k + 1}$ (inductive hypothesis). Show that $\dfrac{1}{1 \cdot 3} + \dfrac{1}{3 \cdot 5} + \cdots + \dfrac{1}{(2k - 1)(2k + 1)} + \dfrac{1}{[2(k + 1) - 1][2(k + 1) + 1]} = \dfrac{k + 1}{2(k + 1) + 1} = \dfrac{k + 1}{2k + 3}$.

By the inductive hypothesis,
$$\left[\frac{1}{1\cdot 3} + \frac{1}{3\cdot 5} + \cdots + \frac{1}{(2k-1)(2k+1)}\right] + \frac{1}{[2(k+1)-1][2(k+1)+1]} =$$
$$\frac{k}{2k+1} + \frac{1}{(2k+1)(2k+3)} = \frac{2k^2 + 3k + 1}{(2k+1)(2k+3)} = \frac{(2k+1)(k+1)}{(2k+1)(2k+3)} = \frac{k+1}{2k+3} \text{ as}$$
desired.

Answers 3

Let P_n be the statement that 2 is a factor of $5^n + 1$.

1. P_1 is true because 2 is a factor of $5^1 + 1 = 6$.
2. Assume that 2 is a factor of $5^k + 1$. Then $5^k + 1 = 2a$ for some positive integer a. Equivalently, $5^k = 2a - 1$. Show that 2 is a factor of $5^{k+1} + 1$.
   We have $5^{k+1} + 1 = 5 \cdot 5^k + 1 = 5(2a - 1) + 1 = 10a - 4 = 2(5a - 2)$. The value 2 is a factor of $2(5a - 2)$, and therefore a factor of $5^{k+1} + 1$ as desired.

Answers 4

Let P_n be the statement that $\left(\frac{3}{2}\right)^n > 2n$ for $n \geq 7$.

1. Show that P_7 is true. That is, show that $\left(\frac{3}{2}\right)^7 > 2(7)$. We have $\left(\frac{3}{2}\right)^7 = \frac{2187}{128} = 17\frac{11}{128}$, which is greater than $2(7) = 14$. Therefore, $\left(\frac{3}{2}\right)^7 > 2(7)$.
2. Assume that $\left(\frac{3}{2}\right)^k > 2k$ for $k \geq 7$. Show that $\left(\frac{3}{2}\right)^{k+1} > 2(k+1)$.
   We have $\left(\frac{3}{2}\right)^{k+1} = \frac{3}{2}\left(\frac{3}{2}\right)^k > \frac{3}{2}(2k) = 3k$ for $k \geq 7$. Furthermore, $3k = 2k + k > 2k + 2$ for $k \geq 7$. Therefore, $\left(\frac{3}{2}\right)^{k+1} > 2(k+1)$ as desired.

Section 8.4 Practice Exercises, pp. 731–732

R.1. $x^2(x^2 + 1)$ **R.2.** $-15r^3(r + 4)$ **R.3.** $(w + 2)^2$

R.4. $(v - 3)(v - 6)$ **R.5.** $\dfrac{t^2 - 15}{t + 4}$ **R.6.** $\dfrac{3}{a - 3}$

1. induction; P_1; P_{k+1}
3. Let P_n be the statement $2 + 6 + \cdots + (4n - 2) = 2n^2$.
   1. P_1 is true because $2 = 2(1)^2$.
   2. Assume that $2 + 6 + \cdots + (4k - 2) = 2k^2$ (Inductive hypothesis). Show that $2 + 6 + \cdots + (4k - 2) + [4(k + 1) - 2] = 2(k + 1)^2$.
      By the inductive hypothesis,
      $[2 + 6 + \cdots + (4k - 2)] + [4(k + 1) - 2]$
      $= 2k^2 + (4k + 2) = 2(k^2 + 2k + 1) = 2(k + 1)^2$ as desired.

5. Let P_n be the statement $5 + 8 + \cdots + (3n + 2) = \frac{n}{2}(3n + 7)$.
   1. P_1 is true because $5 = \frac{1}{2}[3(1) + 7]$.
   2. Assume that $5 + 8 + \cdots + (3k + 2) = \frac{k}{2}(3k + 7)$ (Inductive hypothesis).
      Show that $5 + 8 + \cdots + (3k + 2) + [3(k + 1) + 2] = \frac{k+1}{2}[3(k + 1) + 7] = \frac{(k+1)(3k+10)}{2}$.
      By the inductive hypothesis,
      $[5 + 8 + \cdots + (3k + 2)] + [3(k + 1) + 2]$
      $= \frac{k}{2}(3k + 7) + 3k + 5 = \frac{3k^2 + 13k + 10}{2}$
      $= \frac{(k+1)(3k+10)}{2}$ as desired.

7. Let P_n be the statement $8 + 4 + \cdots + (-4n + 12) = -2n(n - 5)$.
   1. P_1 is true because $8 = -2(1)(1 - 5)$.
   2. Assume that $8 + 4 + \cdots + (-4k + 12) = -2k(k - 5)$ (Inductive hypothesis).
      Show that $8 + 4 + \cdots + (-4k + 12) + [-4(k + 1) + 12] = -2(k + 1)[(k + 1) - 5] = -2k^2 + 6k + 8$.
      By the inductive hypothesis,
      $[8 + 4 + \cdots + (-4k + 12)] + [-4(k + 1) + 12]$
      $= -2k(k - 5) + (-4k + 8) = -2k^2 + 6k + 8$ as desired.

9. Let P_n be the statement $1 + 2 + 2^2 + \cdots + 2^{n-1} = 2^n - 1$.
   1. P_1 is true because $1 = 2^1 - 1$.
   2. Assume that $1 + 2 + 2^2 + \cdots + 2^{k-1} = 2^k - 1$ (Inductive hypothesis).

Show that $1 + 2 + 2^2 + \cdots + 2^{k-1} + 2^{(k+1)-1} = 2^{k+1} - 1$.
By the inductive hypothesis,
$[1 + 2 + 2^2 + \cdots + 2^{k-1}] + 2^{(k+1)-1} = (2^k - 1) + 2^k$
$= 2 \cdot 2^k - 1 = 2^{k+1} - 1$ as desired.

11. Let P_n be the statement $\frac{3}{4} + \frac{3}{16} + \cdots + \frac{3}{4^n} = 1 - \left(\frac{1}{4}\right)^n$.
    1. P_1 is true because $\frac{3}{4} = 1 - \left(\frac{1}{4}\right)^1$.
    2. Assume that $\frac{3}{4} + \frac{3}{16} + \cdots + \frac{3}{4^k} = 1 - \left(\frac{1}{4}\right)^k$ (Inductive hypothesis).
       Show that $\frac{3}{4} + \frac{3}{16} + \cdots + \frac{3}{4^k} + \frac{3}{4^{k+1}} = 1 - \left(\frac{1}{4}\right)^{k+1}$.
       By the inductive hypothesis,
       $\left[\frac{3}{4} + \frac{3}{16} + \cdots + \frac{3}{4^k}\right] + \frac{3}{4^{k+1}} = \left[1 - \left(\frac{1}{4}\right)^k\right] + \frac{3}{4}\left(\frac{1}{4}\right)^k$
       $= 1 - \frac{1}{4}\left(\frac{1}{4}\right)^k = 1 - \left(\frac{1}{4}\right)^{k+1}$ as desired.

13. Let P_n be the statement $1 \cdot 2 + 2 \cdot 3 + \cdots + n(n + 1) = \frac{n(n+1)(n+2)}{3}$.
    1. P_1 is true because $1 \cdot 2 = \frac{(1)(1+1)(1+2)}{3}$.
    2. Assume that $1 \cdot 2 + 2 \cdot 3 + \cdots + k(k + 1) = \frac{k(k+1)(k+2)}{3}$ (Inductive hypothesis).
       Show that $[1 \cdot 2 + 2 \cdot 3 + \cdots + k(k + 1)] + (k + 1)[(k + 1) + 1]$
       $= \frac{(k+1)[(k+1)+1][(k+1)+2]}{3} = \frac{k^3 + 6k^2 + 11k + 6}{3}$.
       By the inductive hypothesis,
       $[1 \cdot 2 + 2 \cdot 3 + \cdots + k(k + 1)] + (k + 1)[(k + 1) + 1]$
       $= \frac{k(k+1)(k+2)}{3} + (k + 1)(k + 2)$
       $= \frac{k^3 + 3k^2 + 2k}{3} + k^2 + 3k + 2$
       $= \frac{k^3 + 6k^2 + 11k + 6}{3}$ as desired.

15. Let P_n be the statement $\left(1 - \frac{1}{2}\right)\left(1 - \frac{1}{3}\right)\cdots\left(1 - \frac{1}{n+1}\right) = \frac{1}{n+1}$.
    1. P_1 is true because $\left(1 - \frac{1}{2}\right) = \frac{1}{1+1}$.
    2. Assume that $\left(1 - \frac{1}{2}\right)\left(1 - \frac{1}{3}\right)\cdots\left(1 - \frac{1}{k+1}\right) = \frac{1}{k+1}$ (Inductive hypothesis).
       Show that $\left(1 - \frac{1}{2}\right)\left(1 - \frac{1}{3}\right)\cdots\left(1 - \frac{1}{k+1}\right)\left[1 - \frac{1}{(k+1)+1}\right]$
       $= \frac{1}{(k+1)+1} = \frac{1}{k+2}$.
       By the inductive hypothesis,
       $\left[\left(1 - \frac{1}{2}\right)\left(1 - \frac{1}{3}\right)\cdots\left(1 - \frac{1}{k+1}\right)\right]\left[1 - \frac{1}{(k+1)+1}\right]$
       $= \frac{1}{k+1} \cdot \left(1 - \frac{1}{k+2}\right) = \frac{1}{k+1} \cdot \frac{k+1}{k+2} = \frac{1}{k+2}$ as desired.

17. Let P_n be the statement $\sum_{i=1}^{n} 1 = n$.
    1. P_1 is true because $\sum_{i=1}^{1} 1 = 1$.
    2. Assume that $\sum_{i=1}^{k} 1 = k$ (inductive hypothesis).
       Show that $\sum_{i=1}^{k+1} 1 = (k + 1)$. By the inductive hypothesis,
       $\sum_{i=1}^{k+1} 1 = \left(\sum_{i=1}^{k} 1\right) + 1 = k + 1$ as desired.

19. Let P_n be the statement $\sum_{i=1}^{n} i^2 = \dfrac{n(n+1)(2n+1)}{6}$.

   1. P_1 is true because $\sum_{i=1}^{1} i^2 = (1)^2 = \dfrac{1(1+1)[2(1)+1]}{6}$.

   2. Assume that $\sum_{i=1}^{k} i^2 = \dfrac{k(k+1)(2k+1)}{6}$ (Inductive hypothesis).

   Show that $\sum_{i=1}^{k+1} i^2 = \dfrac{(k+1)[(k+1)+1][2(k+1)+1]}{6} = \dfrac{2k^3 + 9k^2 + 13k + 6}{6}$.

   By the inductive hypothesis,

$$\sum_{i=1}^{k+1} i^2 = \left(\sum_{i=1}^{k} i^2\right) + (k+1)^2 = \dfrac{k(k+1)(2k+1)}{6} + (k+1)^2$$

$$= \dfrac{2k^3 + 9k^2 + 13k + 6}{6} \text{ as desired.}$$

21. Let P_n be the statement 2 is a factor of $5^n - 3$.

   1. P_1 is true because 2 is a factor of $(5)^1 - 3 = 2$.

   2. Assume that P_k is true; that is, assume that 2 is a factor of $5^k - 3$. This implies that $5^k - 3 = 2a$ and that $5^k = 2a + 3$ for some positive integer a.

   Show that P_{k+1} is true; that is, show that 2 is a factor of $5^{k+1} - 3$.

$5^{k+1} - 3 = 5 \cdot 5^k - 3$

       $= 5(2a + 3) - 3$ Replace 5^k by $2a + 3$.

       $= 10a + 12$

       $= 2(5a + 6)$

   Therefore, 2 is a factor of $5^{k+1} - 3$ as desired.

23. Let P_n be the statement $4^n - 1$ is divisible by 3. This is equivalent to saying that 3 is a factor of $4^n - 1$.

   1. P_1 is true because 3 is a factor of $(4)^1 - 1 = 3$.

   2. Assume that P_k is true; that is, assume that 3 is a factor of $4^k - 1$. This implies that $4^k - 1 = 3a$ and that $4^k = 3a + 1$ for some positive integer a.

   Show that P_{k+1} is true; that is, show that 3 is a factor of $4^{k+1} - 1$.

$4^{k+1} - 1 = 4 \cdot 4^k - 1$

       $= 4 \cdot (3a + 1) - 1$ Replace 4^k by $3a + 1$.

       $= 12a + 3$

       $= 3(4a + 1)$

   Therefore, 3 is a factor of $4^{k+1} - 1$ as desired.

25. $n = 7$ **27.** $n = 4$

29. Let P_n be the statement $n! > 3^n$ for $n \geq 7$.

   1. P_7 is true because $7! = 5040$ and $3^7 = 2187$. Therefore, $7! > 3^7$.

   2. Assume that $k! > 3^k$ for a positive integer $k \geq 7$.

   Show that $(k+1)! > 3^{k+1}$.

$(k+1)! = (k+1) \cdot k!$

       $> (k+1)(3^k)$ by the inductive hypothesis.

       $> 3(3^k)$ Since $k \geq 7$, then $(k + 1) > 3$.

       $= 3^{k+1}$

   Therefore, $(k+1)! > 3^{k+1}$ as desired.

31. Let P_n be the statement $3n < 2^n$ for $n \geq 4$.

   1. P_4 is true because $3(4) = 12$ and $2^4 = 16$. Therefore, $3(4) < 2^{(4)}$.

   2. Assume that $3k < 2^k$ for a positive integer $k \geq 4$.

   Show that $3(k+1) < 2^{k+1}$.

$3(k+1) = 3k + 3$.   Furthermore, since $k \geq 4$, $3k > 3$.

       $< 3k + 3k$

       $< 2^k + 2^k$ by the inductive hypothesis.

       $= 2(2^k)$

       $= 2^{k+1}$

   Therefore, $3(k+1) < 2^{k+1}$ as desired.

33. Let P_n be the statement $(xy)^n = x^n y^n$.

   1. P_1 is true because $(xy)^1 = xy = x^1 y^1$.

   2. Assume that $(xy)^k = x^k y^k$ (Inductive hypothesis).

   Show that $(xy)^{k+1} = x^{k+1} y^{k+1}$.

   Multiplying both sides of the equation $(xy)^k = x^k y^k$ by (xy) gives $(xy)^k(xy) = x^k y^k(xy)$. Using the property $a^n a^m = a^{n+m}$ gives $(xy)^{k+1} = x^{k+1} y^{k+1}$ as desired.

35. Let P_n be the statement if $x > 1$, then $x^n > x^{n-1}$.

   1. P_1 is true because $x^1 > 1$ for $x > 1$ and $x^0 = 1$. Therefore, $x^1 > x^0$ for $x > 1$.

   2. Assume that $x^k > x^{k-1}$ for $x > 1$.

   Show that $x^{(k+1)} > x^{(k+1)-1}$ or equivalently, $x^{k+1} > x^k$.

$x^{(k+1)} = x^k \cdot x$

       $> x^{k-1} \cdot x$ by the inductive hypothesis.

       $= x^k x^{-1} \cdot x$

       $= x^k$ as desired.

37. The statement is false for $n = 11$.

39. The principle of mathematical induction has us test the truth of a statement for $n = 1$. The extended principle of mathematical induction has us test the truth of a statement for the first allowable value of n. In each case, the proof is concluded by showing that the truth of a statement for any other positive integer after the first allowable value of n follows directly from its predecessor.

41. Let P_n be the statement that $n^2 - n$ is even.

   1. P_1 is true because $(1)^2 - (1) = 0$, which is even.

   2. Assume that P_k is true; that is, assume that $k^2 - k$ is even. This implies that $k^2 - k = 2a$ for some positive integer a, and that $k = k^2 - 2a$.

   Show that P_{k+1} is true; that is, show that $(k+1)^2 - (k+1)$ is even.

$(k+1)^2 - (k+1)$

       $= k^2 + k$

       $= k^2 + (k^2 - 2a)$ by the inductive hypothesis.

       $= 2k^2 - 2a$

       $= 2(k^2 - a)$, which is an even integer.

43. Let P_n be the statement $F_1 + F_2 + \cdots + F_n = F_{n+2} - 1$ for positive integers $n \geq 3$.

   1. P_3 is true because $F_1 + F_2 + F_3 = 1 + 1 + 2 = 4$ and $F_{3+2} - 1 = F_5 - 1 = 5 - 1 = 4$.

   2. Assume that P_k is true; that is, assume that $F_1 + F_2 + \cdots + F_k = F_{k+2} - 1$.

   Show that P_{k+1} is true; that is, show that $F_1 + F_2 + \cdots + F_k + F_{k+1} = F_{[(k+1)+2]} - 1 = F_{k+3} - 1$.

$F_1 + F_2 + \cdots + F_k + F_{k+1} = (F_1 + F_2 + \cdots + F_k) + F_{k+1}$

       $= (F_{k+2} - 1) + F_{k+1}$ by the inductive hypothesis.

       $= (F_{k+1} + F_{k+2}) - 1$ Replace $F_{k+1} + F_{k+2}$ by F_{k+3}.

       $= F_{k+3} - 1$ as desired.

Section 8.5 Practice Exercises, pp. 736–738

R.1. $25v^2 + 20v + 4$ **R.2.** $64a^4 - 144a^2b^3 + 81b^6$

R.3. $16x^2 + 8\sqrt{15}x + 15$ **R.4.** $\dfrac{1}{25}m^2 + \dfrac{4}{5}m + 4$

R.5. $64a^3 - 144a^2b + 108ab^2 - 27b^3$

   1. binomial   **3.** n **5.** $\dbinom{n}{k-1} a^{n-(k-1)} b^{k-1}$

   7. a. See Figure 8-9 on page 732.   **b.** $x^4 - 4x^3y + 6x^2y^2 - 4xy^3 + y^4$

   9. a. 1   **b.** 4   **c.** 6   **d.** 4   **e.** 1   **11.** 286   **13.** 462

15. $243x^5 + 405x^4 + 270x^3 + 90x^2 + 15x + 1$

17. $343x^3 + 441x^2 + 189x + 27$

19. $16x^4 - 160x^3 + 600x^2 - 1000x + 625$

21. $32x^{15} - 80x^{12}y + 80x^9 y^2 - 40x^6 y^3 + 10x^3 y^4 - y^5$

23. $p^{12} - 6p^{10}w^4 + 15p^8 w^8 - 20p^6 w^{12} + 15p^4 w^{16} - 6p^2 w^{20} + w^{24}$

25. $0.0016 + 0.0032k + 0.0024k^2 + 0.0008k^3 + 0.0001k^4$

27. $\dfrac{1}{8}c^3 - \dfrac{3}{4}c^2 d + \dfrac{3}{2}cd^2 - d^3$ **29.** $210m^4 n^6$ **31.** $-56c^5 d^3$

33. $2,562,560u^{12}v^{36}$ **35.** $2268x^{12}y^9$ **37.** $924h^{24}$
39. $13,608\,p^{12}q^5$ **41.** $e^{4x} - 4e^{2x} + 6 - \dfrac{4}{e^{2x}} + \dfrac{1}{e^{4x}}$
43. $x^3 + 3x^2y + 3xy^2 + y^3 - 3x^2z - 6xyz - 3y^2z + 3xz^2 + 3yz^2 - z^3$
45. 1.04060401 **47.** $6x^2y + 2y^3$ **49.** $6x^2 + 6xh + 2h^2$
51. $4x^3 + 6x^2h + 4xh^2 + h^3 - 10x - 5h$ **53.** $2 + 11i$ **55.** $41 - 840i$
57. $\dbinom{n}{r} = \dfrac{n!}{r! \cdot (n-r)!}$ and
$$\dbinom{n}{n-r} = \dfrac{n!}{(n-r)! \cdot [n-(n-r)]!} = \dfrac{n!}{(n-r)! \cdot r!}.$$
By the commutative property of multiplication,
$$\dfrac{n!}{r! \cdot (n-r)!} = \dfrac{n!}{(n-r)! \cdot r!}.$$
59. a. $1, 2, 4, 8, 16, 32, 64, 128, 256$ **b.** $a_n = 2^{n-1}$
61. In this expression $n = 3$ and $r = 5$. The value $(n-r)!$ is $(3-5)! = (-2)!$, which is undefined.   **63. a.** $39,902$ **b.** $40,320$
c. 1.0% error   **65.** 1.656

Section 8.6 Practice Exercises, pp. 746–750
R.1. $362,880$ **R.2.** 1 **R.3.** 42 **R.4.** 210
1. principle; counting; $m \cdot n$ **3.** $n!$ **5.** permutation; r
7. $\dfrac{n!}{(n-r)!}$ **9.** 10 **11.** 8 **13.** 2
15. $4 \cdot 3 \cdot 6 \cdot 4 = 288$ **17. a.** $26^3 \cdot 10^3 = 17,576,000$
b. $26 \cdot 25 \cdot 24 \cdot 10 \cdot 9 \cdot 8 = 11,232,000$
19. a. $2 \cdot 26 \cdot 26 \cdot 26 = 35,152$ **b.** $2 \cdot 25 \cdot 24 \cdot 23 = 27,600$
21. $5^{10} \cdot 2^4 = 156,250,000$ **23.** $2^8 = 256$ **25. a.** $7! = 5040$
b. $\dfrac{11!}{4! \cdot 4! \cdot 2!} = 34,650$ **27.** $5! - 1 = 119$ **29.** $4! = 24$
31. 360 **33.** 132 **35.** $362,880$ **37.** 6840; There are 6840 ways to select 3 distinct items in a specific order from a group of 20 items.   **39.** 15 **41.** 66 **43.** 1 **45.** 1140; There are 1140 ways to select 3 distinct items in *no specific order* from a group of 20 items.   **47. a.** AB, BA, AC, CA, BC, CB   **b.** AB, AC, BC (*Note*: The order within the individual combinations does not matter. That is, AB or BA represents the same group of two elements.)
49. $_9P_3 = 9 \cdot 8 \cdot 7 = 504$ **51. a.** $_{24}C_5 = 42,504$ **b.** $_{24}P_5 = 5,100,480$
53. $_{16}C_2 = 120$ **55.** $_{47}C_5 = 1,533,939$ **57.** $(_9C_3) \cdot (_7C_3) = 2940$
59. $10^4 = 10,000$ **61.** $_{12}C_5 = 792$ **63.** $_5P_5 = 5! = 120$
65. a. $_4C_3 = 4$ **b.** $_{16}C_3 = 560$ **c.** $(_{16}C_2) \cdot (_4C_1) = 480$
67. $40^3 = 64,000$ **69.** 2 **71.** 9 **73.** $6^3 = 216$
75. $5 \cdot 6 \cdot 6 = 180$ **77.** $5 \cdot 6 \cdot 3 = 90$ **79. a.** 8; 16; 32
b. 8; BBB, BBG, BGB, BGG, GBB, GBG, GGB, GGG   **c.** 16
81. $_{160}C_4 = 26,294,360$ **83.** $2 \cdot 4 \cdot 3 \cdot 2 \cdot 1 = 48$
85. Using the fundamental principle of counting, we have $8 \cdot 7 \cdot 6 = 336$. There are 8 horses that can cross the finish line first. Once the first horse finishes, there are 7 horses remaining that can come in second. Then there are 6 horses that are available for third place. Alternatively, the number of first-, second-, and third-place ordered arrangements can be found by taking the number of permutations of 8 horses taken 3 at a time, $_8P_3$.
87. $[(_3P_3) \cdot (_4P_4) \cdot (_2P_2)] \cdot (_3P_3) = 1728$

Section 8.7 Practice Exercises, pp. 759–764
R.1. a. $\{c, f, h\}$ **b.** $\{a, b, c, d, e, f, g, h, i, o, u\}$
R.2. a. $\{-4, -2\}$ **b.** $\{-5, -4, -3, -2, -1, 1, 3, 5\}$
1. sample; space   **3.** impossible; certain   **5.** mutually; exclusive
7. $0; P(A) + P(B)$ **9.** c, d, g, h   **11.** c   **13.** e   **15.** f
17. $\dfrac{4}{10} = \dfrac{2}{5}$ **19.** $\dfrac{7}{10}$ **21.** 0 **23.** 1 **25. a.** $\dfrac{16}{30} = \dfrac{8}{15}$
b. $\dfrac{6}{30} = \dfrac{1}{5}$ **c.** 0 **27. a.** $\dfrac{18}{38} = \dfrac{9}{19}$ **b.** $\dfrac{7}{38}$ **c.** $\dfrac{37}{38}$
29. 0.158 **31.** 0.984 **33. a.** $\dfrac{3}{36} = \dfrac{1}{12}$ **b.** $\dfrac{33}{36} = \dfrac{11}{12}$
35. a. $\dfrac{6}{36} = \dfrac{1}{6}$ **b.** $\dfrac{3}{36} = \dfrac{1}{12}$ **37.** 0.10

39. a. $\dfrac{_{18}C_9}{_{34}C_9} = \dfrac{48,620}{52,451,256} \approx 0.00093$
b. $\dfrac{_{16}C_9}{_{34}C_9} = \dfrac{11,440}{52,451,256} \approx 0.00022$
c. The events from parts (a) and (b) are not complementary events. There are many other cases to consider regarding the number of male and female jurors: for example, 4 male, 5 female, etc.
41. $\dfrac{1}{_{39}C_5} = \dfrac{1}{575,757}$
43. a.

		Parent 2	
		Y	y
Parent 1	Y	YY	Yy
	y	yY	yy

b. $\dfrac{1}{4}$ **c.** $\dfrac{3}{4}$
45. 0.048 **47. a.** $\dfrac{103}{250}$ or 0.412 **b.** $\dfrac{9}{250}$ or 0.036 **49.** $\dfrac{8}{52} = \dfrac{2}{13}$
51. $\dfrac{16}{52} = \dfrac{4}{13}$ **53.** $\dfrac{12}{52} = \dfrac{3}{13}$ **55.** $\dfrac{32}{52} = \dfrac{8}{13}$ **57.** $\dfrac{39}{52} = \dfrac{3}{4}$
59. $\dfrac{112}{200} = \dfrac{14}{25}$ **61.** $\dfrac{98}{200} = \dfrac{49}{100}$ **63.** $\dfrac{168}{200} = \dfrac{21}{25}$
65. $\dfrac{164}{200} = \dfrac{41}{50}$ **67.** $\dfrac{1}{6} \cdot \dfrac{1}{2} = \dfrac{1}{12}$ **69.** $(0.88)(0.88) = 0.7744$
71. a. $\dfrac{1}{127} \approx 0.007874$ **b.** $\dfrac{1}{127} \cdot \dfrac{1}{127} \approx 0.000062$
73. $\left(\dfrac{1}{2}\right)^5 \cdot \left(\dfrac{1}{4}\right)^5 = \dfrac{1}{32,768} \approx 0.0000305$
75. a. 0.85 **b.** $(0.85)^3 \approx 0.614$ **77. a.** O^+ **b.** AB^-
79. 0.54 **81. a.** 0.066 **b.** O^- blood is absent all three antigens and will not introduce a new antigen to the recipient's blood.
83. $\left(\dfrac{1}{11}\right)^3 = \dfrac{1}{1331}$ **85.** 0.025
87. a.

b. 0.24
89. Observe the game being played by other players. Approximate the probability by dividing the number of times a player wins to the number of games played.   **91.** Answers will vary.
93. a. $\dfrac{5}{15} \cdot \dfrac{5}{15} = \dfrac{1}{9}$ **b.** No   **c.** $\dfrac{5}{15} \cdot \dfrac{4}{14} = \dfrac{2}{21}$
95. a. $\dfrac{_{13}C_5}{_{52}C_5} = \dfrac{1287}{2,598,960} \approx 0.000495$
b. $(4)\dfrac{_{13}C_5}{_{52}C_5} = (4)\dfrac{1287}{2,598,960} \approx 0.001981$

Chapter 8 Review Exercises, pp. 767–770
1. 56 **3.** $2, 3, 4, 5, 6$
5. The odd-numbered terms are positive, and the even-numbered terms are negative.
7. a. $5, 7, 9, 11, 13$ **b.** 45 **c.** $S_5 = 45$ **9.** 66 **11.** 0
13. $\displaystyle\sum_{i=1}^{5} \dfrac{i+2}{2^i}$ **15.** $\displaystyle\sum_{i=1}^{10} i^3 = \sum_{j=0}^{9}(j+1)^3 = \sum_{k=2}^{11}(k-1)^3$
17. $a_n = 2^n; 1 \le n \le 12$ **19.** Yes; $d = \dfrac{5}{4}$
21. $4, 12, 20, 28, 36$ **23.** $a_{23} = 103$ **25.** $n = 38$ **27.** -4795
29. -3258 **31.** 24 months   **33.** No
35. $120, 80, \dfrac{160}{3}, \dfrac{320}{9}, \dfrac{640}{27}$ **37.** $a_6 = 972$ **39.** $a_1 = 64$
41. 5465 **43.** 30 **45.** 3 **47.** $\dfrac{79}{90}$ **49.** $\$40,250.86$

51. a. $238,884.21   **b.** $289,503.57

53. Let P_n be the statement $3 + 7 + \cdots + (4n - 1) = n(2n + 1)$.
  1. P_1 is true because $3 = 1[2(1) + 1]$.
  2. Assume that $3 + 7 + \cdots + (4k - 1) = k(2k + 1)$
    (Inductive hypothesis).
    Show that $3 + 7 + \cdots + (4k - 1) + [4(k + 1) - 1]$
    $= (k + 1)[2(k + 1) + 1] = (k + 1)(2k + 3)$.
    By the inductive hypothesis,
    $[3 + 7 + \cdots + (4k - 1)] + [4(k + 1) - 1] = k(2k + 1) + (4k + 3)$
    $= 2k^2 + 5k + 3$
    $= (k + 1)(2k + 3)$ as desired.

55. Let P_n be the statement $1 + 4 + \cdots + 4^{n-1} = \frac{1}{3}(4^n - 1)$.
  1. P_1 is true because $1 = \frac{1}{3}(4^1 - 1)$.
  2. Assume that $1 + 4 + \cdots + 4^{k-1} = \frac{1}{3}(4^k - 1)$
    (Inductive hypothesis).
    Show that $1 + 4 + \cdots + 4^{k-1} + 4^{(k+1)-1} = \frac{1}{3}(4^{k+1} - 1)$.
    By the inductive hypothesis,
    $[1 + 4 + \cdots + 4^{k-1}] + 4^{(k+1)-1} = \frac{1}{3}(4^k - 1) + 4^k$
    $= \frac{1}{3}4^k - \frac{1}{3} + 4^k$
    $= \frac{4}{3} \cdot 4^k - \frac{1}{3}$
    $= \frac{1}{3}(4 \cdot 4^k - 1)$
    $= \frac{1}{3}(4^{k+1} - 1)$ as desired.

57. Let P_n be the statement $4^n < (n + 2)!$ for $n \geq 2$.
  1. P_2 is true because $4^2 = 16$ and $(2 + 2)! = 24$.
    Therefore, $4^2 < (2 + 2)!$.
  2. Assume that $4^k < (k + 2)!$ for an integer $k \geq 2$. Show that
    $4^{k+1} < [(k + 1) + 2]!$, or equivalently, that $4^{k+1} < (k + 3)!$.
    $4^{k+1} = 4 \cdot 4^k < 4 \cdot (k + 2)!$ by the inductive hypothesis.
    If $k \geq 2$, then $(k + 3) > 4$. Therefore,
    $4^{k+1} < 4(k + 2)! < (k + 3)(k + 2)! = (k + 3)!$ as desired.

59. a. 1   **b.** 3   **c.** 3   **d.** 1

The values of $\binom{3}{0}, \binom{3}{1}, \binom{3}{2},$ and $\binom{3}{3}$ match the entries in the fourth row of Pascal's triangle.

61. $32x^5 - 240x^4 + 720x^3 - 1080x^2 + 810x - 243$
63. $t^{30} + 6t^{25}u^3 + 15t^{20}u^6 + 20t^{15}u^9 + 15t^{10}u^{12} + 6t^5u^{15} + u^{18}$
65. $96,000x^2y^4$ **67.** $-2016c^8d^{25}$ **69.** $-119 + 120i$
71. $7! = 5040$ **73.** $5! - 1 = 119$
75. a. $5^3 = 125$ **b.** $5 \cdot 5 \cdot 3 = 75$ **c.** $4 \cdot 3 \cdot 1 = 12$
77. 5985; There are 5985 ways to select 4 distinct items in no specific order from a group of 21 items.
79. $_{90}C_{15} \approx 4.58 \times 10^{16}$ **81.** $_{40}P_3$ or $40 \cdot 39 \cdot 38 = 59,280$
83. $(_{10}C_2) \cdot (_6C_3) \cdot (_7C_2) = 18,900$ **85.** Not likely   **87.** 0.27
89. a. $\dfrac{5}{36}$ **b.** $\dfrac{6}{36} = \dfrac{1}{6}$ **c.** $\dfrac{7}{36}$
91. a. $\dfrac{_4C_3}{_{15}C_3} = \dfrac{4}{455} \approx 0.00879$ **b.** $\dfrac{_{11}C_3}{_{15}C_3} = \dfrac{165}{455} \approx 0.36264$
  c. The events from parts (a) and (b) are not complementary events. There are many other cases to consider regarding the number of defective and good lightbulbs: for example, 2 good, 1 defective, etc.
93. $(0.9959)^{10} \approx 0.9597$ **95.** $\dfrac{90}{160} = \dfrac{9}{16}$ **97.** $\dfrac{150}{160} = \dfrac{15}{16}$
99. $\dfrac{132}{160} = \dfrac{33}{40}$ **101.** $\dfrac{28}{52} = \dfrac{7}{13}$ **103.** $\dfrac{4}{52} \cdot \dfrac{4}{52} = \dfrac{1}{169}$

Chapter 8 Test, pp. 770–772

1. $-2, 3, 1, 4, 5, 9$ **2.** $(3n + 1)(3n)$ **3. a.** Geometric
  b. $r = 0.01$ **c.** $a_n = 0.139(0.01)^{n-1}$ **4. a.** Arithmetic
  b. $d = 0.16$ **c.** $a_n = 0.16n + 0.36$

5. a. Neither   **b.** Not applicable   **c.** $a_n = \dfrac{3n}{5^n}$ **6.** $4, 6, 9, \dfrac{27}{2}$
7. $10, 13, 16, 19, 22$ **8.** $a_{78} = -783$ **9.** $a_6 = -96$ **10.** 167
11. 9   **12.** $a_1 = 1, d = -2$ **13.** $a_1 = 5, r = 2$ **14.** 3

15. 4833   **16.** $\dfrac{665}{8}$ **17.** 33   **18.** The sum does not exist.
19. a. $750,000   **b.** $2,142,857   **20.** $245,446.68
21. $a_n = 7.50n^2$ **22.** 16 days
23. Let P_n be the statement $6 + 10 + \cdots + (4n + 2) = n(2n + 4)$.
  1. P_1 is true because $6 = 1[2(1) + 4]$.
  2. Assume that $6 + 10 + \cdots + (4k + 2) = k(2k + 4)$
    (Inductive hypothesis).
    Show that $6 + 10 + \cdots + (4k + 2) + [4(k + 1) + 2]$
    $= (k + 1)[2(k + 1) + 4] = (k + 1)(2k + 6)$.
    By the inductive hypothesis,
    $[6 + 10 + \cdots + (4k + 2)] + [4(k + 1) + 2] = k(2k + 4) + (4k + 6)$
    $= 2k^2 + 8k + 6$
    $= (k + 1)(2k + 6)$ as desired.
24. Let P_n be the statement $1 + 5 + \cdots + 5^{n-1} = \frac{1}{4}(5^n - 1)$.
  1. P_1 is true because $1 = \frac{1}{4}(5^1 - 1)$.
  2. Assume that $1 + 5 + \cdots + 5^{k-1} = \frac{1}{4}(5^k - 1)$ (Inductive hypothesis).
    Show that $1 + 5 + \cdots + 5^{k-1} + 5^{(k+1)-1} = \frac{1}{4}(5^{k+1} - 1)$.
    By the inductive hypothesis,
    $[1 + 5 + \cdots + 5^{k-1}] + 5^{(k+1)-1} = \frac{1}{4}(5^k - 1) + 5^k$
    $= \frac{1}{4}5^k - \frac{1}{4} + 5^k$
    $= \frac{5}{4} \cdot 5^k - \frac{1}{4}$
    $= \frac{1}{4}(5 \cdot 5^k - 1)$
    $= \frac{1}{4}(5^{k+1} - 1)$ as desired.
25. Let P_n be the statement that 2 is a factor of $7^n - 5$.
  1. P_1 is true because 2 is a factor of $(7)^1 - 5 = 2$.
  2. Assume that P_k is true; that is, assume that 2 is a factor of $7^k - 5$.
    This implies that $7^k - 5 = 2a$ and that $7^k = 2a + 5$ for some positive integer a.
    Show that P_{k+1} is true; that is, show that 2 is a factor of $7^{k+1} - 5$.
    $7^{k+1} - 5 = 7 \cdot 7^k - 5$
                 $= 7 \cdot (2a + 5) - 5$ Replace 7^k by $2a + 5$.
                 $= 14a + 30$
                 $= 2(7a + 15)$
    Therefore, 2 is a factor of $7^{k+1} - 5$ as desired.
26. $\dfrac{x^4}{16} + \dfrac{3}{2}x^3 + \dfrac{27}{2}x^2 + 54x + 81$
27. $1024c^{10} - 1280c^8t^4 + 640c^6t^8 - 160c^4t^{12} + 20c^2t^{16} - t^{20}$
28. $-960t^3v^{14}$ **29.** $2835x^4y^6$ **30.** $_{13}P_5 = 154,440$ and $_{13}C_5 = 1287$
31. A permutation of n items taken r at a time is an arrangement of r items taken from a group of n items in a specific order. A combination of n items taken r at a time is a group of r items taken from a group of n items in no particular order.
32. $9! = 362,880$ **33.** $4 \cdot 5 \cdot 3 = 60$
34. $\dfrac{12!}{3! \cdot 2!} - 1 = 39,916,799$ **35. a.** $_{56}C_{12} \approx 5.584 \times 10^{11}$
  b. $(_{30}C_6) \cdot (_{26}C_6) \approx 1.367 \times 10^{11}$ **c.** $\dfrac{(_{30}C_6) \cdot (_{26}C_6)}{_{56}C_{12}} \approx 0.245$
  d. $\dfrac{_{26}C_{12}}{_{56}C_{12}} \approx 0.0000173$ **36.** $_{10}C_3 = 120$ **37.** $2^6 \cdot 3^4 = 5184$
38. a. $_{50}C_4 = 230,300$ **b.** $_{50}P_4 = 5,527,200$
39. $\dfrac{36,000}{300,000,000} = 0.00012$ **40. a.** $\dfrac{32}{75}$ **b.** $\dfrac{43}{75}$ **c.** $\dfrac{5}{75} = \dfrac{1}{15}$
  d. 0   **41.** $\dfrac{16}{36} = \dfrac{4}{9}$ **42.** $\dfrac{4}{52} \cdot \dfrac{13}{52} = \dfrac{1}{52}$
43. $(0.319)^3 \approx 0.0325$ **44.** $\dfrac{64}{120} = \dfrac{8}{15}$ **45.** $\dfrac{76}{120} = \dfrac{19}{30}$
46. $\dfrac{87}{120} = \dfrac{29}{40}$ **47.** $\dfrac{83}{120}$

Chapter 8 Cumulative Review Exercises, pp. 772–774

1. False   **2.** True   **3.** True   **4.** False
5. a. $|t - 5|$ or $|5 - t|$ **b.** $5 - t$ **6.** 2.7×10^{28} **7.** $-\dfrac{1}{81}$

8. $3x + 14\sqrt{2x} - 10$ **9.** $\dfrac{2x+1}{3x}$ **10.** $-i$

11. $\{\ \}$; The values 3 and -10 do not check.   **12.** $\{-4, 3, -2, 1\}$

13. $\left\{6, \dfrac{4}{5}\right\}$ **14.** $[-4, \infty)$ **15.** $(-\infty, -13] \cup [-1, \infty)$

16. $(-\infty, -4) \cup [3, \infty)$ **17.** $\left\{\dfrac{1}{2}\right\}$ **18.** $\{-1 + \ln 20\}$

19. a. $-3, 5, \frac{1}{2}$ **b.** $(-3, 0), (5, 0), \left(\frac{1}{2}, 0\right)$ **c.** $(0, 15)$

d.   **e.** $(-\infty, -3) \cup \left(\frac{1}{2}, 5\right)$

20. a. 12.5 sec   **b.** 2504 ft   **c.** 25 sec

21. a. $(x-4)^2 + (y-1)^2 = 16$ **b.** Center: $(4, 1)$; Radius: 4

c.

22.   **23.**

24.   **25.**

26. The graph of f is the graph of $y = \sqrt{x}$ shifted to the right 1 unit, stretched vertically by a factor of 2, and shifted upwards 3 units.

27. a. $12x^2 + 12xh + 4h^2 - 3$ **b.** 49   **c.** Odd

28. a. Yes   **b.** $g^{-1}(x) = \dfrac{x+1}{5}$ **29. a.** Yes   **b.** Yes

30. a. $x = 7, x = -2$ **b.** Horizontal asymptote: $y = 3$

31. a. $P(t) = 320{,}000e^{0.01t}$ **b.** 371,800   **c.** 22.3 yr

32. $3 \log x - 5 \log y - \frac{1}{2} \log z$ **33.** -4 **34.** 3.7486

35. $\{(-1, -3)\}$ **36.** $\{(1, -2, 4)\}$ **37.** $\{(-5, -16), (3, 0)\}$

38. $\{(-z + 3, 2z + 1, z) \mid z \text{ is a real number}\}$ **39.** $\left\{\left(\dfrac{45}{49}, \dfrac{23}{49}\right)\right\}$

40.

41. $\begin{bmatrix} -22 & 27 \\ -19 & -31 \end{bmatrix}$ **42.** Not possible   **43.** 51   **44.** -123

45. Not possible   **46.** $\begin{bmatrix} -44 & -8 & -18 \\ 47 & 41 & 54 \end{bmatrix}$

47. $A^{-1} = \begin{bmatrix} \frac{2}{9} & -\frac{4}{9} & \frac{7}{9} \\ \frac{1}{3} & \frac{1}{3} & -\frac{1}{3} \\ -\frac{1}{9} & \frac{2}{9} & \frac{1}{9} \end{bmatrix}$ **48.** $\{(-3, 1, 4)\}$

49. a. $(1, -2)$ **b.** $(-1, -2)$ **c.** $x = 3$

50. Geometric; $r = -a^3$ **51.** $a_{500} = 156.6$ **52.** 14,097

53. 1530   **54.** 9

55. a. 40,320   **b.** 12   **c.** 2730

56. $243x^5 - 405x^4y^2 + 270x^3y^4 - 90x^2y^6 + 15xy^8 - y^{10}$

57. $\dfrac{1}{2(3n+1)(3n)}$ **58.** $5! = 120$ **59.** $\dfrac{{}_8C_2 \cdot {}_5C_2}{{}_{13}C_4} \approx 0.3916$

60. $\dfrac{16}{52} = \dfrac{4}{13}$

Photo Credits

Subject Index

Tests for Symmetry	Even and Odd Functions
Consider the graph of an equation in x and y. The graph of the equation is • Symmetric to the **y-axis** if substituting $-x$ for x results in an equivalent equation. • Symmetric to the **x-axis** if substituting $-y$ for y results in an equivalent equation. • Symmetric to the **origin** if substituting $-x$ for x and $-y$ for y results in an equivalent equation.	• f is an **even function** if $f(-x) = f(x)$ for all x in the domain of f. • f is an **odd function** if $f(-x) = -f(x)$ for all x in the domain of f.

Properties of Logarithms	Variation
$\log_b 1 = 0 \qquad \log_b (xy) = \log_b x + \log_b y$ $\log_b b = 1 \qquad \log_b\left(\dfrac{x}{y}\right) = \log_b x - \log_b y$ $\log_b b^x = x \qquad \log_b x^p = p \log_b x$ $b^{\log_b x} = x$ Change-of-base formula: $\qquad \log_b x = \dfrac{\log_a x}{\log_a b}$ $b^x = b^y$ implies that $x = y$. $\log_b x = \log_b y$ implies that $x = y$.	$\left.\begin{array}{l} y \text{ varies } \textbf{directly} \text{ as } x. \\ y \text{ is } \textbf{directly} \text{ proportional to } x. \end{array}\right\} y = kx$ $\left.\begin{array}{l} y \text{ varies } \textbf{inversely} \text{ as } x. \\ y \text{ is } \textbf{inversely} \text{ proportional to } x. \end{array}\right\} y = \dfrac{k}{x}$ $\left.\begin{array}{l} y \text{ varies } \textbf{jointly} \text{ as } w \text{ and } x. \\ y \text{ is } \textbf{jointly} \text{ proportional to} \\ w \text{ and } x. \end{array}\right\} y = kwx$

Arithmetic Sequences and Series	Geometric Sequences and Series
nth term: $a_n = a_1 + (n-1)d$ Sum of a finite arithmetic series: $$S_n = \frac{n}{2}(a_1 + a_n)$$	nth term: $a_n = a_1 r^{n-1}$ Finite geometric series: $\displaystyle\sum_{i=1}^{n} a_1 r^{i-1} = \dfrac{a_1(1 - r^n)}{1 - r}$ Infinite geometric series: $$\sum_{i=1}^{\infty} a_1 r^{i-1} = \frac{a_1}{1 - r} \text{ provided that } |r| < 1$$

Binomial Theorem

Factorial notation: $n! = n(n-1)(n-2)\ldots(2)(1)$ and $0! = 1$

Binomial coefficients: $\dbinom{n}{r} = {}_nC_r = \dfrac{n!}{r! \cdot (n-r)!}$

Binomial theorem: $(a + b)^n = \dbinom{n}{0}a^n + \dbinom{n}{1}a^{n-1}b + \dbinom{n}{2}a^{n-2}b^2 + \cdots + \dbinom{n}{n-1}ab^{n-1} + \dbinom{n}{n}b^n = \displaystyle\sum_{r=0}^{n}\dbinom{n}{r}a^{n-r}b^r$

Perimeter and Circumference

Rectangle
$P = 2l + 2w$

Square
$P = 4s$

Triangle
$P = a + b + c$

Circle
Circumference: $C = 2\pi r$

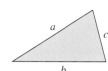

Area

Rectangle
$A = lw$

Square
$A = s^2$

Parallelogram
$A = bh$

Triangle
$A = \frac{1}{2}bh$

Trapezoid
$A = \frac{1}{2}(b_1 + b_2)h$

Circle
$A = \pi r^2$

Volume

Rectangular Solid
$V = lwh$

Right Circular Cylinder
$V = \pi r^2 h$

Right Circular Cone
$V = \frac{1}{3}\pi r^2 h$

Sphere
$V = \frac{4}{3}\pi r^3$

Angles

- Two angles are complementary if the sum of their measures is 90°.

$x + y = 90°$

- Two angles are supplementary if the sum of their measures is 180°.

$x + y = 180°$

Triangles

- The sum of the measures of the angles of a triangle is 180°.

$x + y + z = 180°$

- Given a right triangle with legs of length a and b, and hypotenuse of length c, the Pythagorean theorem indicates that

$a^2 + b^2 = c^2$

ISBN 978-1-259-57511-2
MHID 1-259-57511-X

EAN

9 781259 575112

90000

mheducation.com/highered